# TRANSACTIONS
of the
# FOURTH CIRCUM-PACIFIC ENERGY AND MINERAL RESOURCES CONFERENCE

## AUGUST 17 — 22, 1986, SINGAPORE

**Organized By**

The Circum-Pacific Council
for Energy and Mineral Resources

**Sponsored Jointly By**

The American Association of Petroleum Geologists (AAPG)

The Committee for Coordination of Joint Prospecting
for Mineral Resources in Asian Offshore Areas (CCOP)

The Committee for Coordination of Joint Prospecting
for Mineral Resources in South Pacific Offshore Areas (CCOP/SOPAC)

International Union of Geological Sciences (IUGS)

Pacific Science Association (PSA)

South East Asia Petroleum Exploration Society (SEAPEX)

**Edited By**

Myron K. Horn
Consulting Geologist
Tulsa, Oklahoma

Published December, 1987

**Library of Congress Cataloging-in-Publication Data**

Circum-Pacific Energy and Mineral Resources Conference (4th: 1986: Singapore)
Transactions of the Fourth Circum-Pacific Energy and Mineral Resources Conference: August 17-22, 1986, Singapore / organized by the Circum-Pacific Council for Energy and Mineral Resources; sponsored jointly by the American Association of Petroleum Geologists (AAPG)...[et al.]; edited by Myron K. Horn.
p. cm.
Includes bibliographies.
ISBN 0-933687-08-7: $28.00
1. Power resources—Pacific Area—Congresses. 2. Mines and mineral resources—Pacific Area—Congresses. I. Horn, Myron K. II. Circum-Pacific Council for Energy and Mineral Resources. III. American Association of Petroleum Geologists. IV. Title. TJ163.25.P16C57 1986
553.2'8—dc19 88-16859
CIP

# FOREWORD

In 1974, the Circum-Pacific Council for Energy and Mineral Resources organized a conference in Honolulu, Hawaii, to encourage exploration, assessment, and development of resources in the Pacific region. Similar conferences were held in 1978 and 1982, also in Honolulu. This volume contains the proceedings of the fourth Circum-Pacific Conference, which was held in Singapore, August 17–22, 1986, to focus attention on the resources of the western Pacific and their potential. The conference theme was "Circum-Pacific: Region of Promise."

Co-sponsors of the Fourth Circum-Pacific Conference were the American Association of Petroleum Geologists (AAPG), the Southeast Asia Petroleum Exploration Society (SEAPEX), the Coordinating Committee for Joint Prospecting for Mineral Resources in the Asia Offshore Areas (CCOP) and in South Pacific Offshore Areas (SOPAC), the International Union of Geological Sciences (IUGS), and the United States Agency for International Development (AID), which supported the participation of twenty-two scientists from developing countries of the Pacific region.

Also participating in the conference was the InterUnion Commission on the Lithosphere (ICL), which held its annual meeting in conjunction with the conference and organized a symposium on the work of the International Lithosphere Program. In addition, the Circum-Pacific Council sponsored a meeting of its working group on Volcanic Hazard Reduction, which issued a report that is included as an appendix of this volume.

As part of the conference, the Council's Circum-Pacific Map Committee had an extensive display of the geological, geophysical, and resources maps of the Pacific region being prepared at scales of 1:10 million and 1:17 million. The demand for these maps has increased to such an extent that more than 50 maps are now scheduled under the Map Project, about two-thirds of which have been published or are being printed. A unique adjunct to the conference was a series of displays and poster sessions of new data from recent geophysical surveys of offshore areas of the Pacific. Three such sessions were held in the display rooms adjacent to the conference meetings.

The Fourth Conference was attended by more than six hundred scientists and engineers from forty countries. The conference was regarded as a successful and significant contribution to a better understanding of the Pacific resources potential. The Fifth Conference has been scheduled for 1990 in Honolulu.

Michel T. Halbouty
Chairman and President

# CONTENTS

## GENERAL SESSION



## HYDROCARBONS



## HYDROCARBONS (NEW FRONTIERS)



## SEDIMENTARY BASIN DEVELOPMENT

# GENERAL SESSION

CHAPTER 1

# Opening Address

**Lee Boon Yang**
*Minister of State for Trade and Industry and Home Affairs*
*Singapore*

Distinguished Guests, Ladies, and Gentlemen:

It is indeed a pleasure for me to welcome you to Singapore and to officiate at the opening of the Fourth Conference on Circum-Pacific Energy and Mineral Resources. I have been told that this is the first time that the conference has been held outside the United States. We in Singapore are honored that you should have chosen to come here for such an important conference. Your organizers must have convincing reasons for choosing Singapore, and I can assure you that they are not wrong.

The question might be asked: Why hold an energy and mineral resources conference in Singapore? Singapore is a small and very young nation. It is conspicuously lacking in the energy and mineral resources which form the subject of your conference. These cannot be the reasons which persuaded you to come to Singapore.

Fortunately, we are neither small nor young in world trade. Our total external trade roughly equals that of the People's Republic of China and is more than twice that of India, both of which are much larger and more populous countries. In terms of world trade in services, Singapore is ranked 17th according to the statistics compiled by the U.S. Department of Commerce. These are accomplishments which took us more than a century to build up, and we achieved this level of prominence in world trade by participating in regional growth and development. And few countries in the world can claim greater commitment than Singapore in espousing the policy of free trade, which is fundamental to economic well-being and prosperity of the global community.

Thus, views of the growing importance of the Circum-Pacific Region mean a lot to us. We believe that our strategic location in the region will continue to allow us to play a vital role in this part of the world. Even though lacking in energy and mineral resources, we are in the middle of a region which is well endowed with both. So in deciding to meet in Singapore, you will be able to experience at first hand the nature and importance of strategic location.

You have come to Singapore at a time when our economy is undergoing very dynamic changes. After two decades of relentlessly high growth, it was traumatic for Singapore to experience negative growth for the first time last year. Our economy declined by − 1.8% in 1985. We have to accept that external economic factors had not been favorable. We had control over neither the economies of the OECD countries nor the decline of commodity prices, which seriously plagued the economies of our resource-rich neighbors. There were also internal structural deficiencies and rigidities which reduced our global competitiveness.

This year we have embarked on a major effort to prune the cost of doing business in Singapore in order to arrest the decline. By 1987, we expect that the economy will be able to resume its growth path once again. Part of this optimism may be attributed to our decision to pay more attention to the activities in which we had historically demonstrated strength and resilience. We will actively promote internationally tradeable services.

As some of you may be aware, the creation of an export-oriented manufacturing sector over the last two decades was largely responsible for the uplift of the Singapore economy. By giving the services sector as much attention as we had given to manufacturing, the new economic policy aims ultimately to make Singapore a totally international business center. This re-orientation in our policy is in fact a little overdue. We have for some time been aware that the world trade in services has been growing at twice the merchandise trade, albeit from a smaller base. For a small country with no natural resource, Singapore must hook onto such a new growth trend and adjust our economic base in order to prosper.

To ensure that we will indeed be able to achieve a turnaround, the Singapore Economic Development Board (EDB) has been charged with the task of promoting Singapore as a total business center. The EDB is happy to announce the establishment of a Services Promotion Division. This new Division has been armed with a new set of tax incentives designed to promote both local and international services companies to set up new operations in Singapore or expand existing ones. It will concern itself with many areas of services—computer services, communication and information services, technical and engineering services, leisure and medical services, to name a few.

The Services Promotion Division will also be encouraging companies that operate throughout the region to establish their operational headquarters in Singapore, and through their bases in Singapore to provide a wide range of service activities such as product development, marketing, treasury management, administrative, technical and other management services to their related companies. Many of these service activities are relevant to the exploration and utilization of energy and mineral resources. We hope that we will be able to contribute significantly in your field of resource development.

In summary, ladies and gentlemen, whatever the reason which persuaded you to come to Singapore, I would like to place at your disposal the strategic location of Singapore in this part of the world and our new efforts to encourage a total business concept. It is our hope that you will find Singapore a congenial and fruitful base from which you can support your professional activities in this region. We would like you to visualize Singapore as a gateway to the Asian-Pacific, which has been identified as the region with the highest growth potential.

On this note, it is now my pleasant duty to declare open the Fourth Circum-Pacific Energy and Mineral Resources Conference, and to wish all of you a fruitful conference, and an enjoyable stay in Singapore.

CHAPTER 2

# THE WORLD BANK AND THE PRIVATE SECTOR

**Eric S. Daffern**
*The World Bank*
*Washington, D.C.*

## INTRODUCTION

The activity of the World Bank to encourage private sector investment is not well known. The Bank has long been more associated with the governments of the developing countries, to whom it owes its support. Yet in recent years the Bank has probably done more to encourage international private investment than any other organization. This role is little understood and has been controversial.

## BACKGROUND

Some explanation of the World Bank is necessary so that both countries and investors can understand the range of what it can do and to understand equally what it cannot do. Like every other institution, there are many constraints that limit its actions.

## THE BANK

The Bank is an international public institution. It is owned by 150 countries, excluding mainly the USSR and some of the centrally planned economies, who chose not to join. Overall control of the Bank is through its governors and executive directors, who have a weighted voting system. The majority of the votes are under the control of the industrialized countries. The Bank's clients are the developing countries, those low- and middle-income countries still in need of some kind of international financial assistance, and defined on the basis of average income per capita.

The objective of the Bank is to encourage and facilitate productive investment in developing countries, including making loans where appropriate. Encouraging private sector investment is clearly part of the objective, as is encouraging full use of commercial banks and export credit finance. The Bank also has a clear role to step in and make loans to finance projects where other sources of investment funds are not available on reasonable terms.

The Bank operates through a number of channels to achieve these aims. The main part of the Bank group, the International Bank for Reconstruction and Development, makes long-term loans at market-related interest rates, subject to the constraint that it obtains the guarantee of the government of the country in which the investment is made. The second major part of the Bank group, the International Development Association, is the soft loan arm of the Bank, making 50-year interest-free credits to the poorest countries. Together these two organizations lend about $15 billion a year. There is a better than 90% track record in the success of the projects they support, and the Bank has only a minor problem in getting its loans and credit repaid without rescheduling.

The third main arm of the Bank group is the International Finance Corporation (IFC), an organization whose aim is to lend money to private companies for new projects in LDCs and invest its own equity in developing country projects. IFC lends almost $1 billion annually. In addition, the Bank President is president of the International Centre for the Settlement of Investment Disputes, and is Chairman of a new political risk insurance agency, the Multilateral Investment Guarantee Agency, which should be operational early in 1987.

## THE WORLD BANK AND THE ENERGY SECTOR

Energy has always been an important sector for World Bank support. From the earliest days of the Bank we assisted in financing power projects, coal, pipelines, and the first LNG scheme. In response to the 1973 oil price increase, the Bank stepped up its support of oil and gas projects, a move that brought loud protests from many in the international oil industry. Lending for petroleum projects has ranged from $1/2 billion to $1 billion a year, providing about 30% of project cost. This was primarily focused on areas of little interest to the international industry and is small compared with the industry investment. In total the Bank has lent $6 billion for 100 petroleum projects in 50 countries. These projects have given an important understanding of the petroleum potential and policies of these countries.

Although financing of projects is important and worthwhile in itself, it has been standard policy for many years to seek wider benefits from lending. In some cases this has simply been improvement in technical efficiency, in others institution building, and in others the development and implementation of policies to encourage investment and competition.

It is not well known that more than a third of energy staff time is spent on country policy issues and on energy policy problems. The Energy Department has prepared more than 60 reports on energy issues and strategies of individual countries, and more than 30 papers on specialized aspects of energy, such as how to value natural gas and how to introduce it to a developing country's economy. Most of these reports and papers are available to interested parties.

In addition to the many reports, the Bank has worked closely with many governments in reviewing their public sector investment programs, the efficiency of public enterprise management, the respective roles of the public and private sectors, and the framework for private investment. Much of the background knowledge and experience for this comes from active involvement in lending for projects.

## THE NEEDS OF DEVELOPING COUNTRIES

It is not the purpose of this paper to review what is happening in the international oil industry. It is clear that many exporting countries are suffering from a reduced cash flow and for financial reasons need to encourage international oil companies to step up the rate of exploration and production. Many of the oil importing countries have some oil exploration and production, but they too need to find ways of keeping that exploration going. The international oil industry no longer has the cash to maintain its activity worldwide, and is having to cut back and prioritize even more sharply. Countries need to compete with each other to attract the companies; they need to offer more flexible terms and work programs meeting company circumstances; and they need to find ways of reducing the other problems and risks of investment. It is part of the World Bank's role to help them in this.

## RECENT ACTION BY THE WORLD BANK

In response to the weak oil prices and declining investment trend, in late 1984 the Bank decided to step up and strengthen its role in encouraging investment by the international oil industry. Bank staff have been actively working to ensure that the new political risk insurance agency being established by the Bank is relevant to the oil industry; they have been looking at ways to guarantee contracts and facilitate innovative financing; and they have been looking at simplifying and speeding up internal procedures. At the same time the petroleum staff has visited at the corporate level more than 80 international oil companies and other energy companies to tell them what it can do and what it is doing and to offer to work with them toward investing in developing countries. Many companies now have a better understanding of how the Bank group can help. The rest of this paper explains more fully what it can do.

## WORLD BANK ATTITUDE TO NEW OIL AND GAS INVESTMENT

The World Bank's lending policy for oil and gas investment is aimed at helping countries benefit from the international oil industry's wide experience. Where international companies or local private companies can be encouraged to risk their funds in exploration then often it will be in the host country's interest to cooperate in this. Similarly, when international or local private investors are willing to undertake refining and marketing, then countries can avoid undertaking public investment to do this themselves. The Banks' project lending is therefore focused on (1) helping countries to set the framework, incentives, and regulations that will encourage international investment; (2) supporting local companies in joint ventures for exploration and developing, refining, etc., or, if joint ventures cannot be formed on reasonable terms (e.g., gas exploration), helping the country to undertake the work itself; (3) financing the lower risk/lower reward activities such as enhanced oil recovery; and (4) financing infrastructure, refinery modifications, and marketing. As before, an important element in this is policy reform, institutional strengthening, etc.

## THE BANK AND THE PRIVATE SECTOR

### Encouragement of Exploration

Thirty-three of the Bank's petroleum lending operations have been aimed at helping countries to attract more private investment. These loans have helped countries to update their petroleum legislation, tax legislation, work requirements, incentive packages, and regulations. They have also in many cases brought together the existing petroleum information, identified important gaps and arranged to get it if possible at reasonable cost. The loans have also led to the preparation of habitat studies and the opening of data centers. The resulting new information and incentive packages have then been made available to interested companies. The need now is to help countries get the necessary advice to tailor their terms to today's conditions.

While firmly aimed at helping countries to attract more private investment, these exploration promotion projects are obviously of benefit to the international investor. Within the constraints set by the countries concerned and observing their wishes on the Bank's role, the Bank staff is willing to discuss prospects and possibilities in some detail with interested investors, and to advise investors on contacts to enable them to follow up in the countries. Subject to the same constraints, the Bank staff is also willing to discuss the country's energy sector and market and the general economic history and prospects with a view to reducing the cloud of uncertainty and lack of information that many would-be investors see in trying to make developing country investments.

### Strengthening Local Financial Capability

In its project work the Bank routinely examines the financial capability of its borrowers, e.g., the local energy institutions, and their ability to finance their operations and projects. The Bank has financed the NOC share of joint ventures, either through the cash call system or by earmarking funds for certain items. The Bank has also financed petroleum infrastructure: gas pipelines from the wellhead to the market, storage depots, power plant conversions, and the many advisors and studies needed for action to be taken. There is little difficulty in matching the speed needed by companies provided that the cooperation starts reasonably early. Outside of petroleum the Bank has, of course, financed ports, roads, and other infrastructure used by the oil industry.

While aimed at assisting the local company, the availability of World Bank finance has obvious benefits to international investors: it helps to ensure that a joint venture partner can meet cash calls, it reduces the financial exposure of the international company if some other company provides a pipeline, and it reduces the risks if the Bank as an impartial party has a vested interest in making a project work. The Bank has certain legal remedies in its loan agreements with its borrowers, and it prefers remedial action to make the project work rather than invoking the legal rights. Equally important, but more indirectly, the Bank's concern on a borrower's financial viability does help put the local company into a position where it can meet its energy purchase obligations.

For local private sector companies the IFC has a similar role in financing projects and in helping to ensure the financial viability of its clients.

## RISK REDUCTION

International investors have long regarded developing country investments as subject to greater political risk than those in their home country. In some cases this may be true and well-publicized expropriations or breach of contract in one or two developing countries give the whole group a bad name, and the press rarely publishes the eventual settlement with equal prominence. Equally, the record of some OECD countries in changing laws and tax terms indicates much greater instability than is common or acceptable in LDCs. Nevertheless, fear of political risk in LDCs is clearly seen as a deterrent and cause to seek a higher return and lower commitments.

There is obvious but informal political risk reduction from direct World Bank involvement in a project and the general advisory role of the Bank to governments. In addition, the Bank has operated a slightly more formal system of "letters of cooperation" for some time. Under this system the host government and the international investor have jointly and formally asked the Bank to review the fairness of their proposed concession or contract. Subject to this being satisfactory, the Bank has expressed its views in writing, and has gone on to offer to give active consideration to finance the local partner's part of any subsequent development. Clearly these letters of cooperation do not give formal protection, but they do give the Bank a formal role and an obvious duty to get involved if the agreement is not honored. So far the system has worked well and has obvious merits at the exploration stage.

A new option for political risk reduction is the insurance offered by MIGA, the Multilateral Investment Guarantee Agency. The Agency now has the necessary number of subscribers and is currently drafting the regulations setting out the scope of coverage. MIGA is aimed specifically at the needs of the international oil industry as well as other international business, and includes all the common forms and types of oil industry participation. Coverage will be available, individually or in combination, for expropriation and nationalization, breach of contract, currency convertibility and repatriation, and war and civil commotion. Insurance coverage will be comparable to the best amounts available elsewhere and terms will be competitive. MIGA should be operational early in 1987.

Apart from the above we have been looking to see if there are ways for the Bank to use its power to guarantee loans as part of a solution to contract stability, particularly in ways and amounts that are feasible for the Bank and yet would permit limited recourse financing. This work is still in progress, and we are unsure of the extent it is needed.

## NEW APPROACHES

If another major shock is to be avoided, the level of exploration activity in the developing countries has to be maintained during the present difficult time for the international energy industry. If countries come up with new approaches to attract and maintain a vigorous program from international companies, the Bank will be glad to see what it can do to help. The petroleum staff has come forward with some ideas and is discussing them with a number of countries. If the industry has its own ideas, the Bank will be keen to discuss them.

In addition to maintaining the level of international company activity, Bank policy is to strengthen the local private sector. A small start has been made, but there must be many opportunities in the service industry to form local joint ventures to collaborate with the local companies. Nevertheless, progress is not evident.

As for the state enterprises, several governments have asked the Bank for help in improving the efficiency of the commercial undertakings, including redefinition of roles, restructuring of the institutions and, where the government wishes, privatization. These are fundamental areas for economic well-being, and the solutions adopted in the developed countries have to be examined critically to see if they provide a good model. There is little doubt that many of the state enterprises are not as efficient as they could be, but it may be that privatization is not the short-term solution.

## CONCLUSION

As a development bank the role of the Bank group is to assist the economic development of its member countries. Many of them are over-indebted and have difficulty in getting new money. They need help to make better use of the sources available: the international energy companies, the local private sector, a more efficient state sector, and new forms of financing based on the projects themselves rather than the country creditworthiness.

The World Bank's many contacts with international investors have helped it get a good understanding of the aims and needs of the international investors, and have helped in a mutual understanding of what is possible. At the same time the Bank's long experience of working with its clients, the developing countries, has helped it identify their needs. The Bank can act as an honest broker in introducing "marriage partners" and can help the countries in their approach to setting the terms. It can facilitate the whole process, through encouraging a realistic assessment of what is needed, through information, through finance, through modifying the political risks, and through making possible in developing countries the financing schemes used in North America and Europe. The profits are there to be made in the developing countries; it is now for the investors to take advantage of today's low costs and make those investments.

CHAPTER 3

# WORLD'S OFFSHORE PETROLEUM POTENTIAL

**Michel T. Halbouty**
*Halbouty Energy Company*
*Houston, Texas*

## INTRODUCTION

World population has risen from one billion in 1830 to today's mark of five billion. Projections from the United Nations indicate that within four decades—by the year 2025—in a generation—the world will have a total population of about eight billion people (Figure 1). It took a hundred years to go from 1 billion to 2 billion and only 45 years to go from 2 billion to 4 billion. Thence only 11 years to add another billion—where we are today. In 40 more years, 3 billion more people will be added on this earth. And they all must be supplied with their necessities.

As the global population continues to expand, energy demands will rise accordingly, and until alternate fuel technologies are in place, petroleum—oil and natural gas—will remain the world's mainstay energy fuels.

Our concern now is not the question of world petroleum supply—of which there is a temporary surplus—but rather the issue of most importance is the geologic evaluation of the world's petroleum potential, so that scientists in the future will be prepared to discover the necessary petroleum to supply the needs of the world's people.

Those of us who are looking to the long-term, well into the future, know that probably the only good aspect of the current world oil supply dilemma is that when sanity once again prevails in world markets, the oil and gas prospects that are not now searched will still be there in great numbers, awaiting the drill.

It is my opinion that in the last five years there have been more new conceptual ideas in geoscience than have been promulgated in the preceding fifty years. New concepts will be forthcoming each year—which makes me firmly believe that we will find in the future—worldwide—as much oil and significantly more gas than have been found to date.

Today, approximately 27% of worldwide oil production and 37% of the world's proven petroleum reserves are from offshore basins. Ocean drilling has heretofore been predominantly limited to shallow-water and near-shore areas, yet numerous exploration targets remain farther seaward and in deep-water areas (Petroleum Frontiers, 1985).

It is estimated that a majority of the world's future crude oil and natural gas resources lie offshore, with promising prospects in both moderate and harsh deep-water environments and in the harsh ice-clad regions of the world.

## PRODUCTIVE AND NONPRODUCTIVE REGIONS

A careful examination of the world's sedimentary petroleum basins that are productive and those areas remaining to be explored shows that there are many regions remaining to be tested.

Figure 2 shows both the currently productive and the potential areas for petroleum exploration. It is interesting to note the vast petroleum potential areas of the world—outstanding basins and continental shelves and slopes yet to be explored and tested. Many of these areas are frontiers. Of the world's prospective sedimentary basins, only 27% are currently productive of commercial hydrocarbons. Another 40% of the basins, which are partially or moderately explored, have been tested to some extent but have not yielded commercial quantities of petroleum; and approximately 33% of the world's basins have had little or essentially no exploration activity.

Table 1 shows the approximate area of the sedimentary basins of the world. Of the total prospective area of approximately 77,643,000 $km^2$ (29,978,130 $mi^2$), approximately 26,395,000 $km^2$ (10,191,167 $mi^2$) of this prospective area is in the ocean basins of the world where regions of thick sediments of the continental margins extend far beyond the base of the continental slope into the abyssal oceans.

These basins include some of the world's most productive offshore areas, such as the Gulf of Mexico, North Sea, Caribbean Sea, Persian Gulf, Mediterranean Sea, Lake Maracaibo, and the Indonesian ocean basins. There also remain many essentially unexplored ocean basins, such as the Bering Sea, the Norwegian Sea, and the Scotia Sea, which have high potential for future petroleum discoveries. In many instances, exploration has identified hydrocarbon accumulations, but to date there has been no petroleum production in these regions (Halbouty, 1986).

### Circum-Arctic

Beginning with the provinces at the top of the world, Figure 3 shows a large portion of the Circum-Arctic region, and Figure 4 depicts the productive and nonproductive petroleum provinces in the same area. Provinces shown in red are those that are currently productive or that have production plans in place. Those shown in yellow are areas or provinces that are currently nonproductive. Many of these provinces do contain

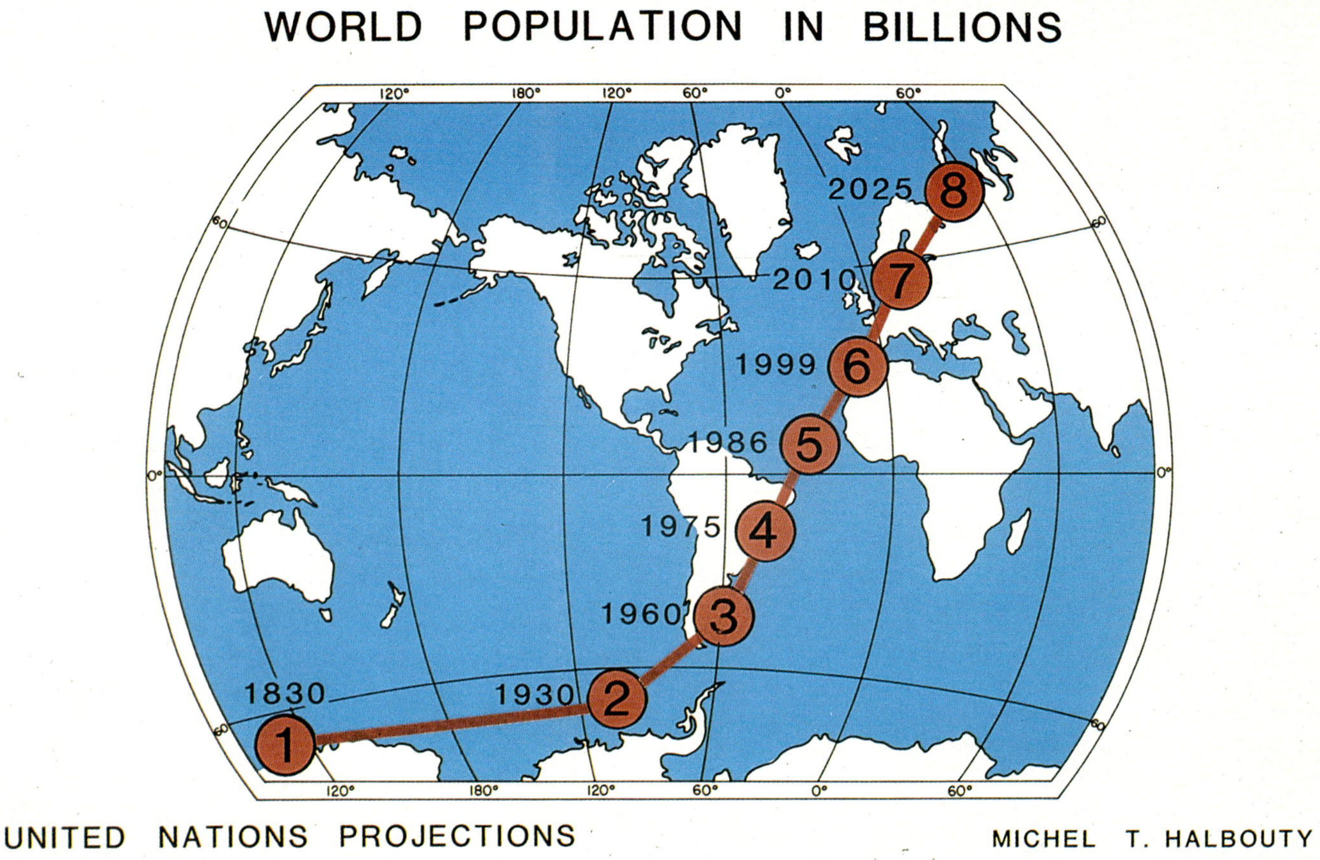


Figure 1. Chart of world population growth over time, with projection to 2025.

commercial quantities of petroleum, but no plans for production are in place.

These provinces contain a wealth of potential structural and stratigraphic traps yet to be explored. Moving counterclockwise from the top of Figure 4, giant and sub-giant oil and gas accumulations have already been discovered in the North Slope of Alaska, the Mackenzie delta, the Sverdrup basin, the Labrador basin, the Norwegian Sea, the Pechora basin and the West Siberian basin (Green, Kaplan, and Vierbuchen, 1986). In the vast offshore reaches of the Soviet Union, whose continental shelf covers an estimated 5.1 million $km^2$ (2.5 million $mi^2$), a recent significant gas discovery has been made in the Pechora basin, and discoveries have been reported at Kolguyev Island in the Barents Sea and another near Beli Island in the Kara Sea (NOROIL, 1986; Sostack, 1986).

The 1984 Seal Island discovery, the first potentially commercial find that may include resources in the U.S. Arctic OCS, has estimated reserves of 300 million barrels of oil (Lynch, Sliton, and Rudolph, 1985).

Many researchers feel that in the combined Arctic basins, including the Arctic areas not shown in Figure 4, there is the potential for accumulations of oil and gas which may equal those of the Middle East (Halbouty, 1986).

## Western Europe, the Mediterranean and Asia

The offshore provinces of Western Europe, parts of Asia, and the Mediterranean Sea are shown in Figure 5. This illustration also provides a different view of some of the Circum-Arctic provinces already discussed.

The search for North Sea oil was prompted by the 1959 discovery of the super-giant gas field near Groningen in north Holland. Following the discovery of several large gas fields in the southern part of the North Sea the exploration effort gradually moved upward into the North Sea. In the early 1970s, a series of large oil fields were discovered in British and Norwegian waters. These finds generated the intensive offshore search which resulted in the gradual shift of interest to the virgin areas west of Britain, and into Irish waters. Offshore Ireland, the recent discovery of what may prove to be a commercial oil field in the Celtic Sea has caused a reappraisal of exploration prospects in the region. Although the North Sea is one of the major petroleum producing regions of the world, it still has potential for discoveries in both producing and untested waters (Naylor, 1984).

To the east, in the Soviet Union, in addition to the previously mentioned Circum-Arctic provinces, promising prospects lie in the shallow waters of the Caspian Sea and off the Black Sea and Azov Sea coasts on the southern Baltic Sea shelf. Another area of great interest to offshore exploration is the circum-Mediterranean region. Significant discoveries offshore Spain, Sicily, Italy, Egypt, and north Africa indicate the tremendous petroleum potential and importance of this entire region to the world's future petroleum supply.

Although there has been little exploration on the eastern edge of the Mediterranean, the geological composition of

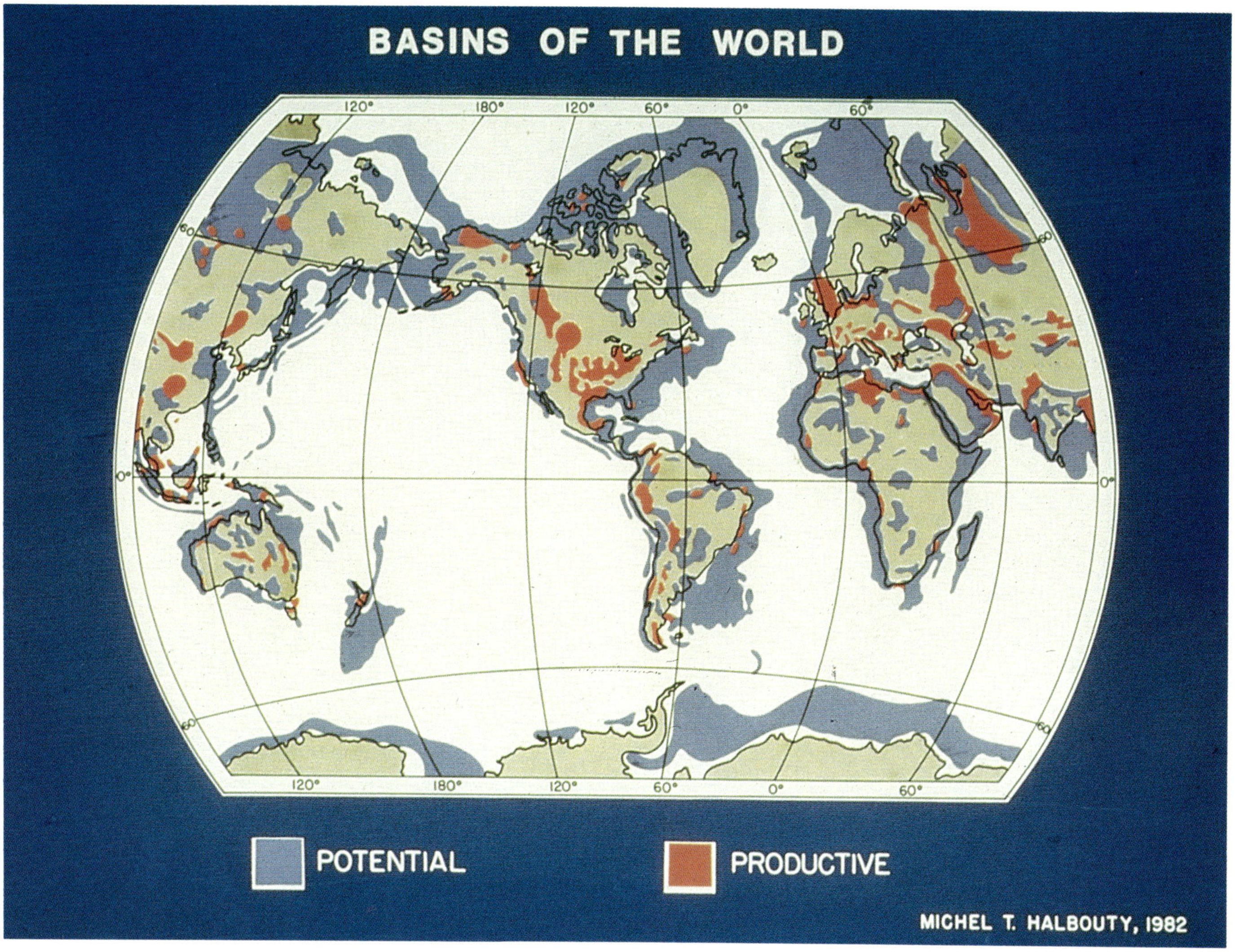


Figure 2. Basins of the world, petroleum exploratory status showing productive and potential areas.

nearshore areas has provided optimism for deep-water reserves. Oil has been tested in the Italian Adriatic in more than 2200 ft of water (Reddin, 1985), and totally unexplored territories remain in the deeper waters along the southern margin of the eastern Mediterranean, especially offshore Libya.

The recently discovered Vega oil field offshore Sicily, with estimated recoverable reserves of 200 to 500 million barrels, indicates that other significant accumulations will be found in Mediterranean Sea basins that have not yet been adequately explored. These regions include the basins offshore north Africa and Turkey, and the central portion basins within the Mediterranean Sea itself, including the unexplored deltas and continental shelves (Halbouty, 1986).

## United States, Mexico and Canada

Figure 6 shows the offshore petroleum provinces of North America (Canada, the United States and Mexico). The U.S. offshore provinces contain more than 5 billion barrels of oil and more than 47 tcf of gas reserves, the majority of which are in the Gulf of Mexico. U.S. Pacific provinces hold another 1.3 billion barrels of oil and 2 tcf of gas. Combined with the Gulf's reserves, these provinces account for 18.9% of the oil reserves and 24.6% of the gas reserves of the United States (Oil Daily, 1986).

Although the Gulf of Mexico is the most prospective area for future offshore discoveries, the basins off central and southern California also hold good promise of substantial oil reserves. The Point Arguello field offshore California, one of the largest discoveries on the U.S. offshore, with recoverable reserves estimated at 400 to 500 million barrels, combined with the Wilmington field and others found off Santa Barbara County, make this offshore area one of the most attractive in the world (Reddin, 1985).

Figure 7 shows a different view of the northernmost region of the U.S. outer continental shelf and the basins offshore Alaska and the Soviet Union. In Alaska, only the Prudhoe Bay and Cook Inlet areas have had any appreciable exploration, yet the Alaskan offshore represents almost 50% of the entire federal offshore reaches of the United States. Offshore western Alaska are four petroleum prospective basins—the St. George, Norton, North Aleutian, and Navarin—where only limited exploration has been conducted.

In the Bering Sea, the Anadyr basin off the Soviet Union and the Navarin basin off Alaska are drawing great attention. Onshore, the Anadyr basin underlies the lowlands that are

## APPROXIMATE PROSPECTIVE AREAS SEDIMENTARY BASINS OF THE WORLD

| | TOTAL (1000 sq km) | ONSHORE (1000 sq km) | OFFSHORE (1000 sq km) |
|---|---|---|---|
| USSR | 14,945 | 10,000 | 4,945 |
| AFRICA - MADAGASCAR | 13,223 | 11,725 | 1,498 |
| S & SE ASIA | 8,916 | 3,705 | 5,211 |
| USA | 8,247 | 6,604 | 1,643 |
| LATIN AMERICA | 7,851 | 4,843 | 3,008 |
| AUSTRALIA - NZ | 6,604 | 4,424 | 2,180 |
| CANADA | 5,167 | 3,084 | 2,083 |
| WESTERN EUROPE | 3,848 | 1,944 | 1,904 |
| MIDDLE EAST | 3,669 | 2,152 | 1,517 |
| PRC | 2,472 | 1,787 | 685 |
| ANTARCTICA | 1,042 | 0 | 1,042 |
| EASTERN EUROPE | 1,015 | 900 | 115 |
| JAPAN | 644 | 80 | 564 |
| TOTAL | 77,643 | 51,248 | 26,395 |

Table 1. Approximate area, sedimentary basins of the world.

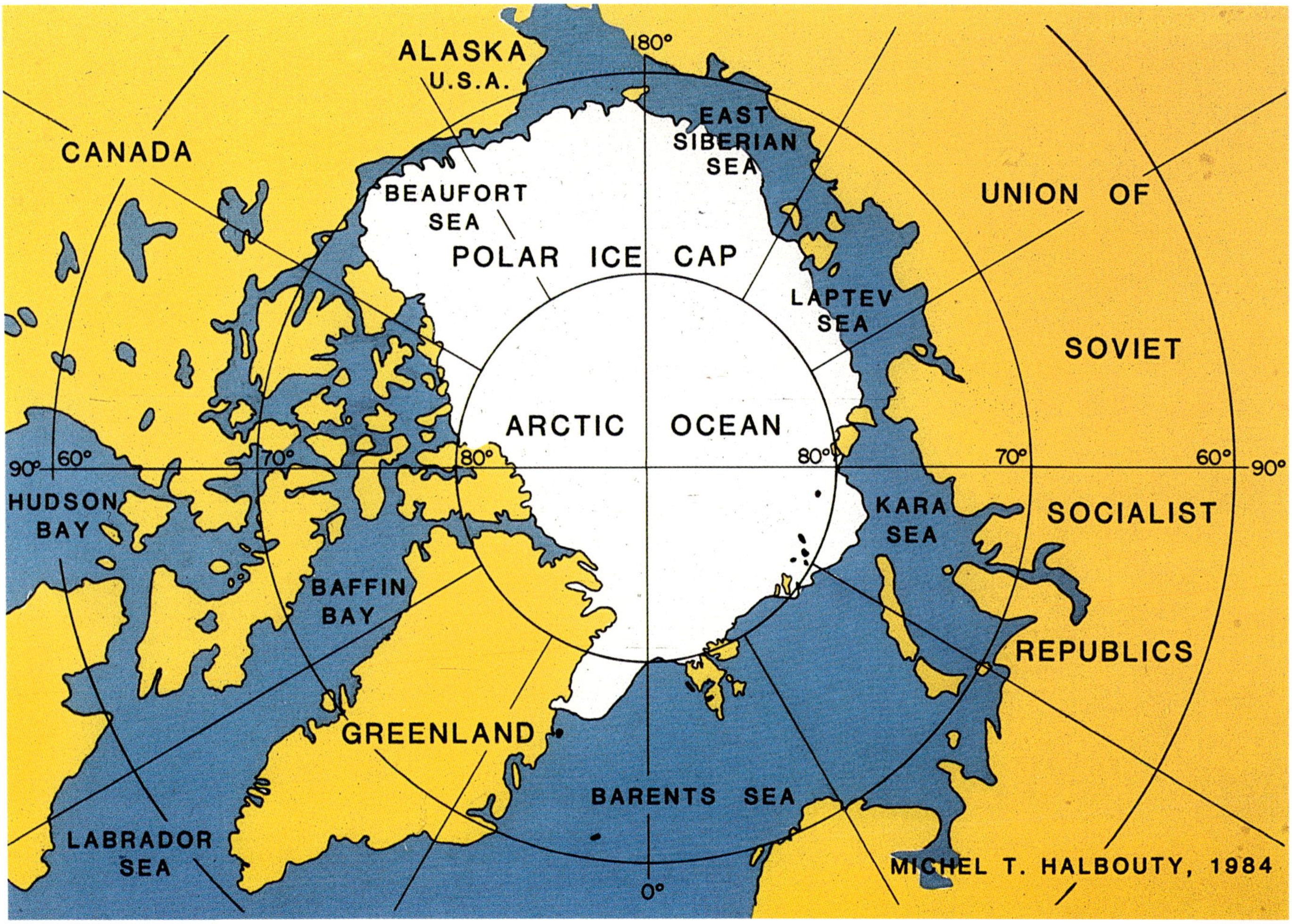


Figure 3. Arctic Ocean and surrounding areas.

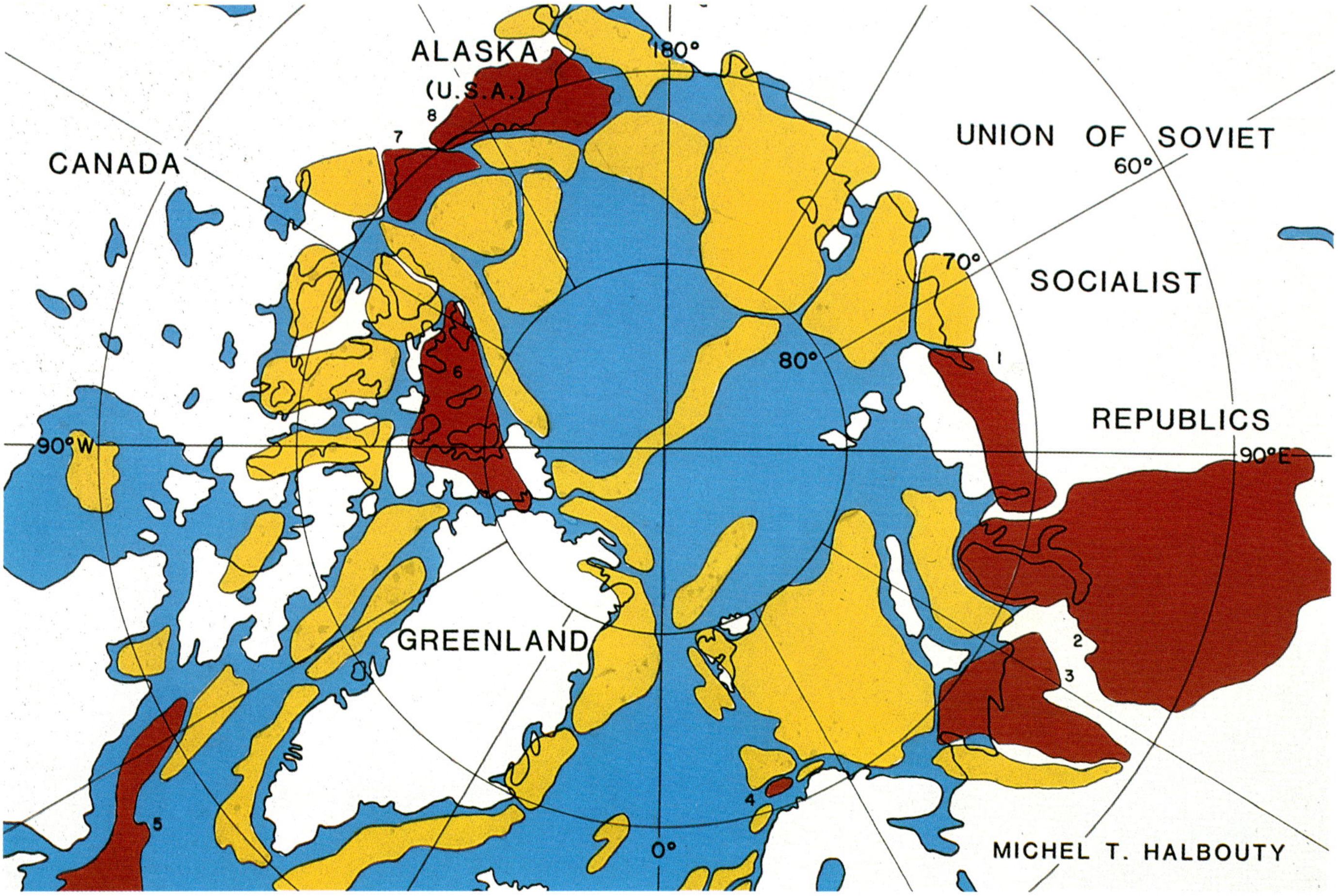


Figure 4. Petroleum productive provinces of the Circum-Arctic region; offshore and offshore/onshore (after St. John, 1984). Productive provinces, shown in red, are numbered as follows: (1) Yenise-Khatarya, (2) West Siberia, (3) Pechora, (4) Tromsø, (5) Labrador, (6) Sverdrup, (7) Mackenzie, and (8) North Slope. Nonproductive provinces are shown in yellow.

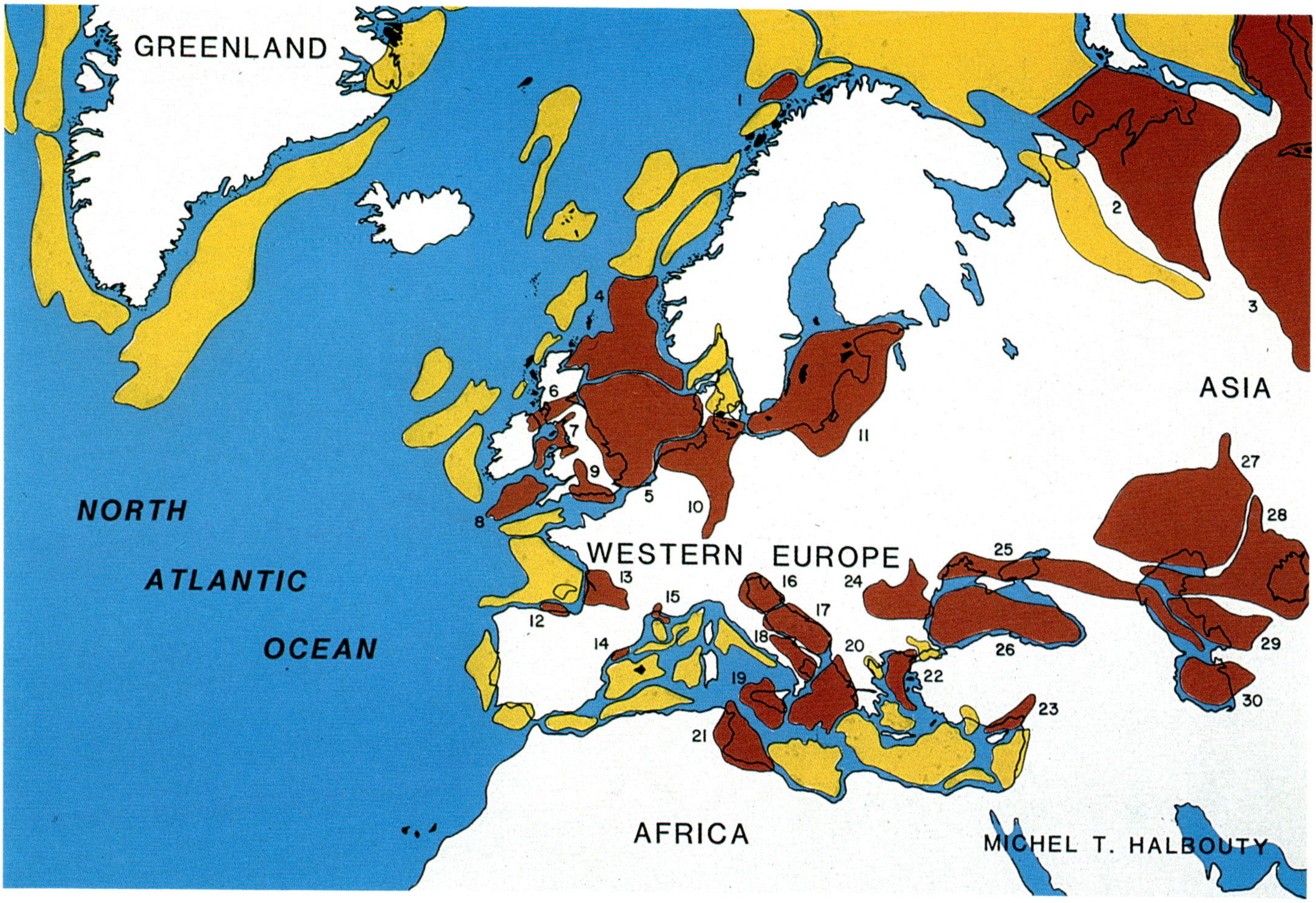


Figure 5. Petroleum productive provinces of Western Europe; offshore and offshore/onshore (after St. John, 1984). Productive provinces, shown in red, are numbered as follows:

1 Tromsø
2 Pechora
3 West Siberia
4 Northern North Sea
5 Southern North Sea
6 Midlands
7 Irish
8 Celtic
9 Wessex
10 Northwest German
11 Baltic
12 Cantabrian
13 Aquitaine
14 Ebro Fan
15 Rhone Delta
16 North Adriatic
17 South Adriatic
18 Taranto
19 Caltanisetta
20 Ionian
21 Pelagian
22 Aegean
23 Adana
24 Moesian
25 Caucasus North
26 Black Sea
27 Caspian North
28 Ust Urt
29 Mangyshlak
30 Caspian South

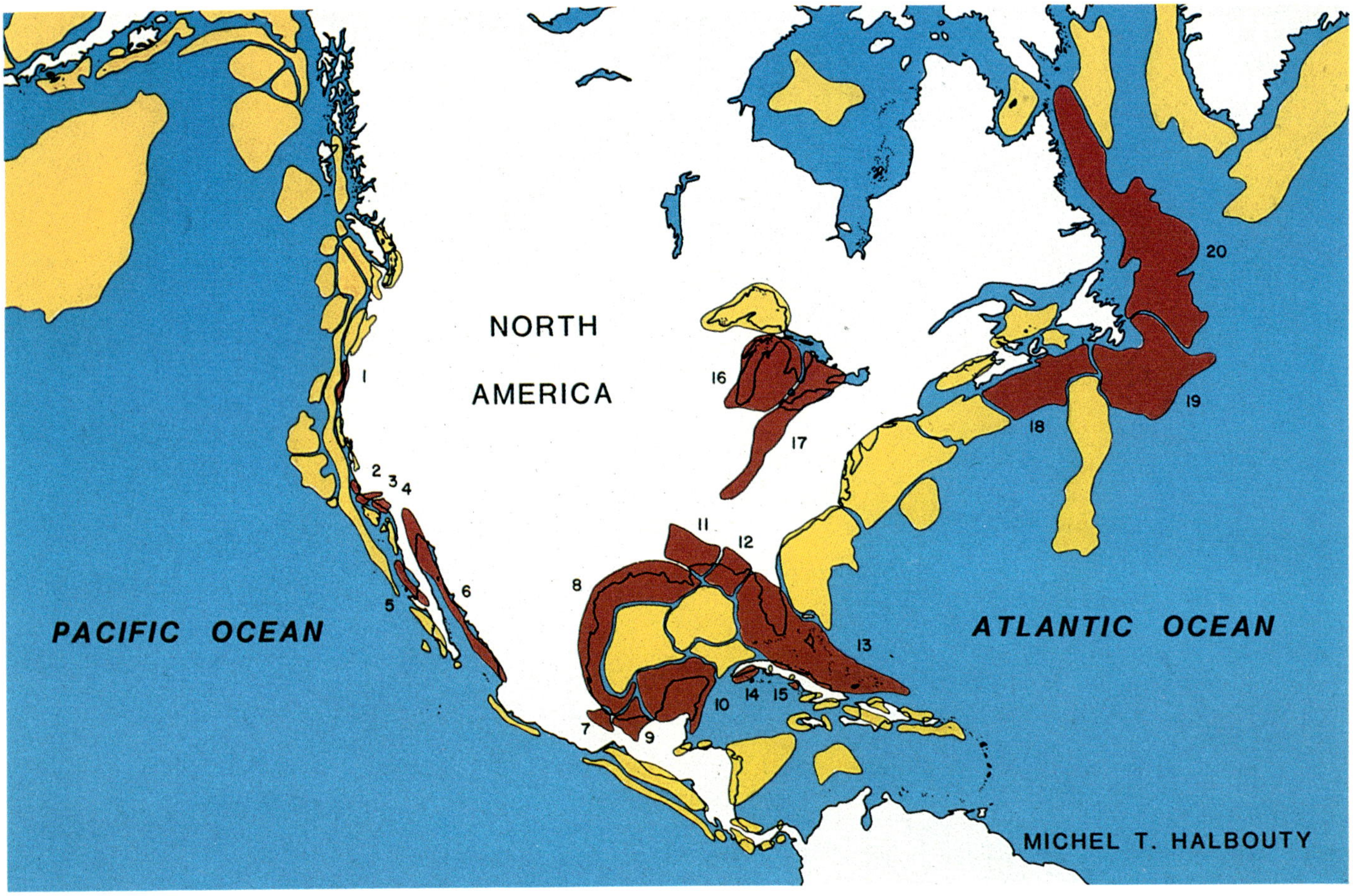


Figure 6. Petroleum productive provinces of North America; offshore and offshore/onshore (after St. John, 1984). Productive provinces, shown in red, are numbered as follows:

1 Eel River
2 Santa Maria
3 Ventura
4 Los Angeles
5 Viscaino
6 Guaymas
7 Vera Cruz
8 Gulf Coast
9 Salinas
10 Campeche
11 Mississippian Salt Dome
12 N.E. Gulf Salt Dome
13 Florida-Bahama
14 Bato Bano
15 Ana Maria
16 Michigan
17 Cincinnati Arch
18 Scotia Shelf
19 Grand Banks
20 Labrador

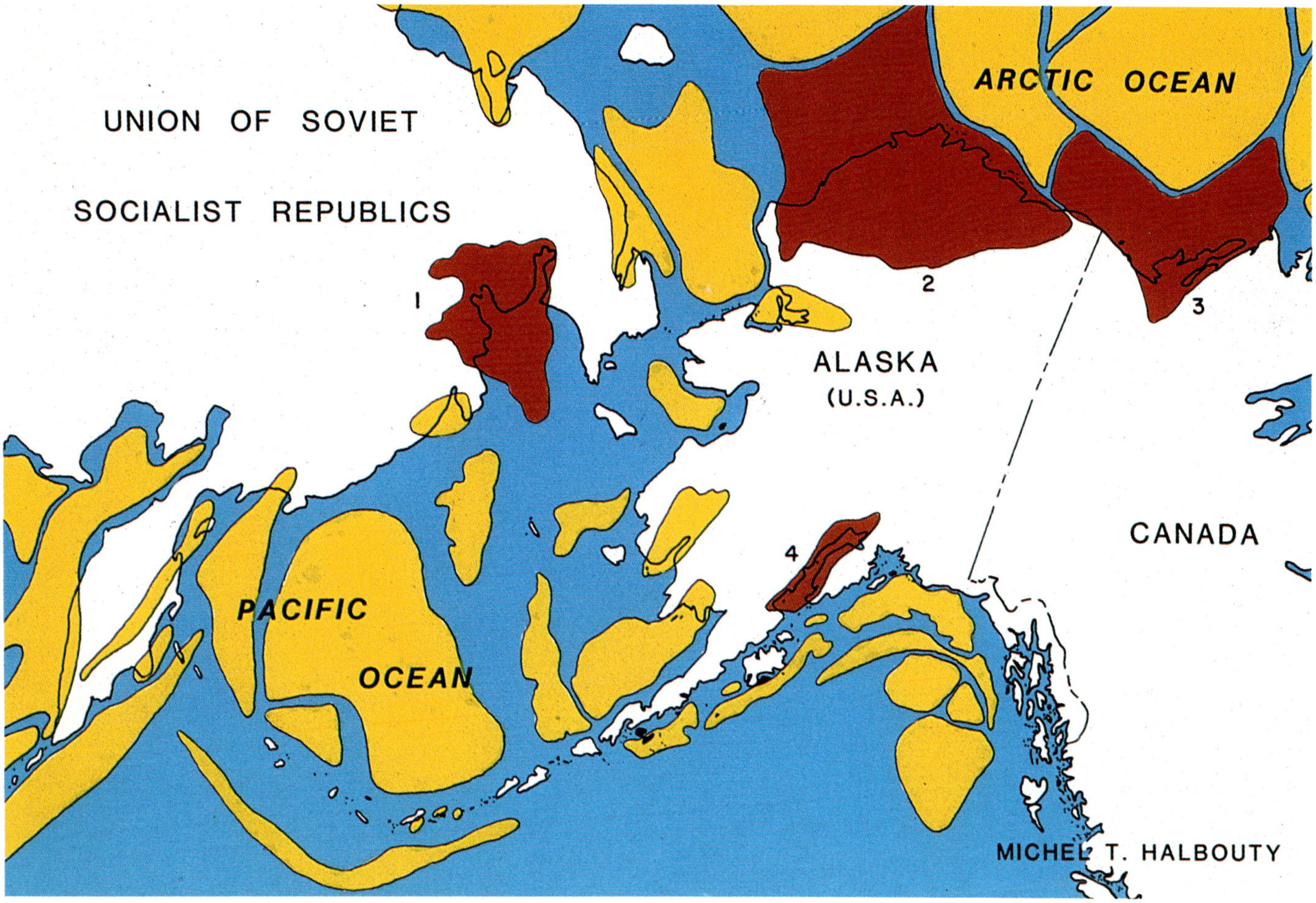


Figure 7. Petroleum productive provinces of portions of the USSR and US northernmost regions; offshore and offshore/onshore (after St. John, 1984). Productive provinces, shown in red, are numbered as follows: (1) Anadyr, (2) North Slope, (3) Mackenzie, and (4) Cook Inlet. Nonproductive provinces are shown in yellow.

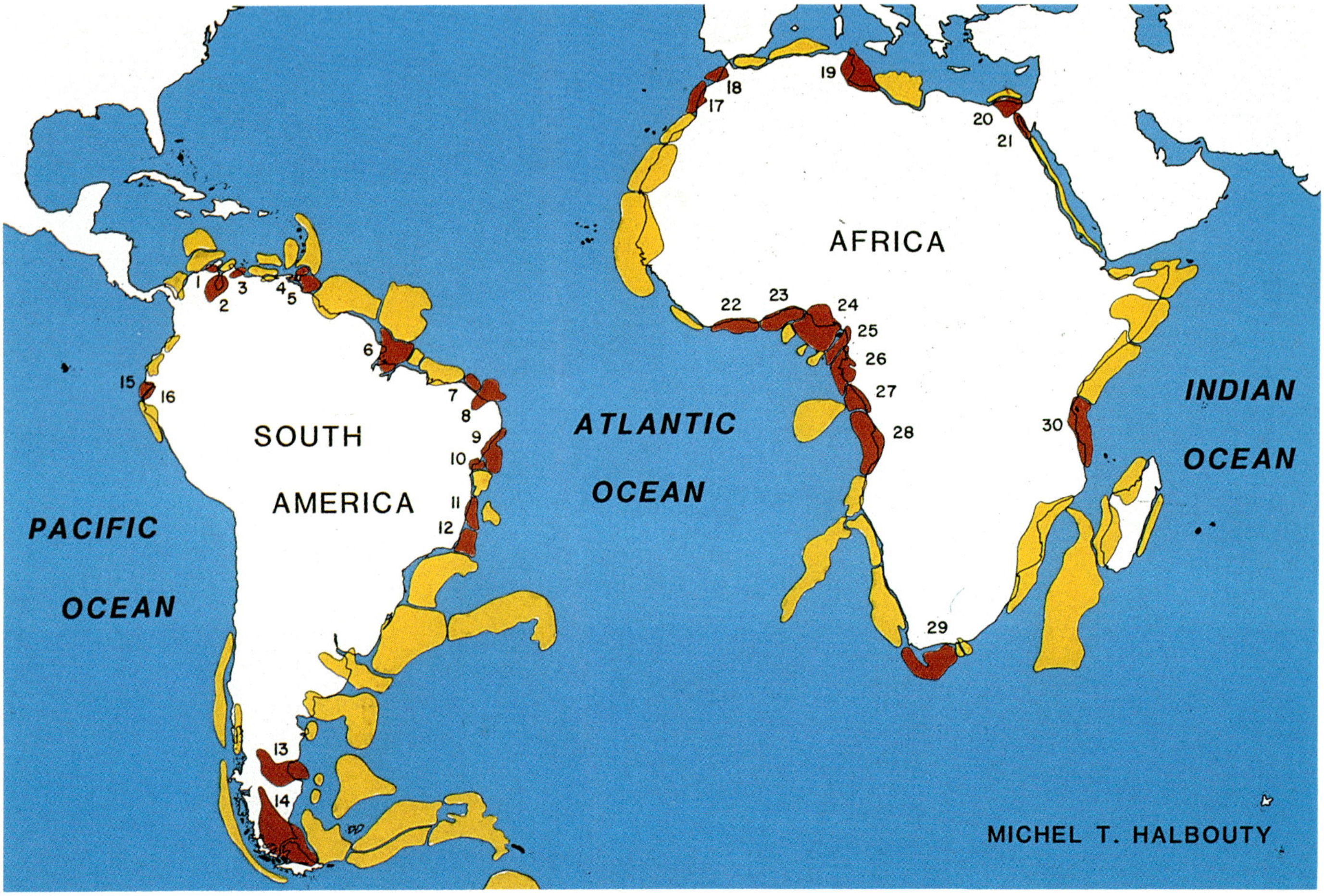


Figure 8. Petroleum productive provinces of South America and Africa; offshore and offshore/onshore (after St. John, 1984). Productive areas, shown in red, are numbered as follows:

1 Gulf of Venezuela
2 Maracaibo
3 Falcon
4 Trinidad-Tobago
5 Orinoco Delta
6 Mouth of the Amazon
7 Ceara
8 Potiguar
9 Sergipe/Alagoas
10 Reconcavo
11 Espirito Santo
12 Campos
13 San Jorge
14 Magallanes
15 Guayaquil
16 Talara Esaouria
17 Esaouria
18 Rharp
19 Pelagian
20 Nile Delta
21 Gulf of Suez
22 Ivory Coast
23 Ghana
24 Niger Delta
25 Cameroon
26 Gabon
27 Cabinda
28 Cuanza
29 South African
30 Tanzania

flanked on the north, west, and south of the Koryak Range. Offshore, the east Anadyr trough and Navarin basin underlie the virtually flat Bering Sea shelf and contain thick sedimentary sequences (Marlowe, Cooper, and Childs, 1983). The Navarin basin, the majority of which lies in U.S. territorial waters, has experienced very little exploration to date. Encouraging oil shows were encountered by the COST wells drilled, but industry's eight wells drilled in 1985 were apparently dry. It is far too soon, however, to be discouraged by this extremely limited exploration level.

The U.S. Atlantic basins (Figure 6) have had no commercial discoveries announced to date. Exploration efforts have been concentrated in five major sedimentary troughs or basins—the Georges Bank Basin, the Baltimore Canyon Trough, the Carolina Trough, the Southeast Georgia Embayment, and the Blake Plateau Basin. The Georges Bank includes the largest and most attractive exploration targets in the North Atlantic, and samples from equivalent strata to the northeast off Canada's Scotian Shelf have shown evidence of hydrocarbons (Rudolph and Havran, 1984). The petroleum geology of the South Atlantic is less well known than other U.S. offshore areas because it has the largest number of basins and the fewest wells.

Since exploration began in Canada's Scotian Basin in 1959, eighteen significant discoveries have been made in the Sable Island area. There have been five significant discoveries on the Labrador Shelf and one on the Southeast Baffin Shelf. In the Grand Banks province of the North Atlantic, 13 significant discoveries have been made in the Jeanne d'Arc Basin. More than 2 tcf of gas resources have been found, most of it associated with oil discoveries at the giant Hibernia field, which has an estimated 1 billion barrels of oil and 2 tcf of gas. Overall, the most important point regarding the petroleum potential of offshore eastern Canada is that exploration of this region is still in a very early stage, and numerous structures remain to be tested in deep water and in subtle stratigraphic traps (Grant, McAlpine, and Wade, 1986).

Turning to Mexico and Central America, all of Mexico's Pacific Ocean basins have been determined to contain potential source and reservoir rocks but there are many petroleum-prospective basins along the Central American Pacific coast that have had little or no exploration. In the southern part of the Gulf of Mexico, a supergiant hydrocarbon province is developing in the Campeche basin, where the potentially productive area is estimated at 8000 $km^2$ (3100 $mi^2$) (Ronfeldt, Nehring, and Gandara, 1980). The entire Gulf region will be a major exploration area for decades to come for both the United States and Mexico.

## South America and Africa

In the productive and potentially productive petroleum provinces offshore South America and Africa (Figure 8), deep-water activity in 1985 in the Campos Basin off Brazil has resulted in discoveries of two giant fields with anticipated reserves of 500 million to 1 billion barrels of oil. It is estimated that these two fields, Albacora and Marlim, will double Brazil's oil reserves (Sostack, 1986; NOROIL, 1986). And despite dropping oil and gas prices, drilling activities are resuming offshore the Santos Basin where previously gas has been found.

In the Tierra del Fuego fields off Argentina, nineteen exploratory wells have found oil and gas with reserves estimated at 65 million barrels of crude and 0.85 tcf of gas. Off Chile geophysical studies have indicated the existence of potential reservoirs at depths exceeding 4000 m (13,124 ft) in the Magellan basin (NOROIL, 1986).

Across the Atlantic Ocean, the offshore African basins show promise for new fields to be found in many areas. It is interesting to note that the areas of separation in South America and Africa are the most prolific petroleum provinces of both continents, which certainly lends significant credence to the separation of continents. Off North Africa, the Pelagian basin of Tunisia, the Gulf of Aden, the Red Sea and the Nile delta have considerable petroleum potential, and the Essaouira basin of Morocco has opened a new play which may yield oil in addition to gas and condensate recently discovered (Clifford, 1986).

Off West Africa, petroleum exploration in the prolific Niger Delta Basin has resulted more than 250 oil and gas field discoveries, with total proven reserves of about 22 billion barrels. The major petroleum reserves are concentrated in a belt extending from east to west as a mirror image of the present coastline (Ejedawe, Abiola, and Young, 1983). Also off West Africa, the significant deep-water discovery of the Ivory Coast's Espoir field has attracted much attention and has raised the possibility that the country may eventually become an oil exporter. Additional drilling off South Africa is estimated to substantially increase the reserves of the Mossel Bay gas field, the country's largest offshore project (NOROIL, 1986).

## Middle East and Southeast Asia

Figure 9 shows the vast expanse of offshore basins extending from the Middle East to India, Southeast Asia and China. The offshore oil and gas fields of the Arabian Basin account for a substantial portion of total Middle East reserves. The world's largest offshore oil field (Saudi Arabia's 70-billion-barrel Safaniya oil field) and the world's largest non-associated gas reservoir (Qatar's North Field, estimated to have recoverable reserves of 150 tcf and 380 tcf probable reserves) are located here. There is good potential for significant future discoveries (NOROIL, 1986, Tiratsoo, 1984).

To the east, many of India's offshore sedimentary basins have yet to be explored for hydrocarbons. Off India's west coast lies the giant Bombay High field, the most significant discovery in Indian waters to date. Off the southeast coast, a discovery in the Cauvery Basin in 1985 has sparked interest in renewed exploration activity.

In the Asia-Pacific region there are numerous producing provinces, yet there are large areas where resource potential remains relatively unknown. Recent exploration has resulted in discoveries of major fields off the coasts of Sumatra, Java and Peninsular Malaysia (Tiratsoo, 1984). Other discoveries have been made off the coasts of Sarawak, Sabah, Brunei and some of the Japanese and Philippine islands. Several large accumulations of offshore gas lie in remote areas, such as the recent find off Papua New Guinea and the Natuna Sea find estimated at 40 tcf (NOROIL, 1986).

The offshore Chinese basins represent some of the world's most interesting petroleum-potential areas. The meandering coastline of China covers more than 6436 km (4000 mi) and the offshore basins offer an estimated 880,000 $km^2$ (339,770 $mi^2$) of prospective area (Halbouty, 1986). Although it is anticipated that the major portion of China's future oil production will come from mature and newly discovered onshore fields whose proven reserves are estimated at more than 18 billion barrels and with an estimated potential of 40 billion barrels,

CHAPTER 4

# ANTARCTIC TREATY

**David Kear**
*Consultant*
*Ohope, New Zealand*

ABSTRACT

International scientific cooperation during IGY (1957-58) included the setting up of Antarctic bases by 12 nations. A strong desire to continue collaboration led to the 1961 Antarctic Treaty, which embraces all land and ice-shelves south of latitude 60 °S.

Treaty provisions include requirements for: only peaceful activities; free and open science; and the setting aside of territorial claims. Two-yearly meetings of the 12 original "Consultative Parties" have led to many wide-ranging recommendations which amplify the Treaty provisions. The original 12 have been joined by other acceding nations, some of whom are now full members following substantial scientific activity in Antarctica.

Considerable international interest was aroused by both the 1980 Convention of Antarctic Marine Resources, and subsequent attempts to establish a parallel "Minerals Regime." The paper ends with a review of the possibility of "economic" minerals in Antarctica.

## BACKGROUND

Antarctica, a continent of nearly 14 million $km^2$ (5.2 million $mi^2$), is one of the Circum-Pacific landmasses, and a sector, therefore, with its own suite of maps of the Circum-Pacific Map Project. However, it differs markedly from other marginal Pacific lands. Climatologically, most of the continent is permanently covered in snow and ice, and much of its coastline is formed of ice cliffs; biologically, it has a very impoverished terrestrial flora and fauna (although a number of marine faunal species make good use of the land); anthropologically, it has effectively no indigenous population; and politically, it has no resident government.

Even before its discovery, Antarctica held a special interest for explorers. Its presence had been postulated widely long before its discovery. It was seen for the first time, however, less than 200 years ago, and was touched by the first human foot less than 100 years ago, in 1895. Its inhospitality rightly conferred the title of "the Heroic Age" on the times of earlier exploration. A number of explorers formally raised their national flags. As a result, some countries laid claim to slices of the continent, and, where those claims were in conflict, there was significant international friction. In contrast, other "non-claimant" nations neither claimed territory nor recognized the claims of those that did. This whole curious political situation was changed completely by a collaborative international scientific effort—"IGY."

This chapter has been compiled principally from personal association with Antarctic activities, and Treaty and Special (earlier, Scientific) Committee on Antarctic Research (SCAR) publications; from a recent discussion paper (Hill, 1983); from the proceedings of a workshop on the Antarctic Treaty System held at Beardmore South Camp in January 1985 (National Research Council, 1986); from an unpublished 2-page paper presented by Dr. J.C. Behrendt to the 1986 SCAR meeting at San Diego; and from personal communications from Drs. Eddie Robertson and Trevor Hatherton, and Bob Thomson and Hugh Logan, who have all had a long association with New Zealand's Antarctic activities.

## IGY AND ESTABLISHING THE TREATY

In the decade after World War II, geophysicists recognized their increasing need for synoptic observations and measurements. This was becoming particularly pressing in those disciplines associated with the growing sophistication of global communications. Thus the International Geophysical Year (IGY, 1957-58) was planned and launched, as a successor in part to the First (1882-83) and Second (1932-33) Polar Years. A full program of synoptic observations and measurements was planned at a large number of permanent and temporary observatories throughout the world. Fields covered included ionospheric and atmospheric sciences, meteorology, oceanography, seismology, gravity measurements, cosmic rays, solar flares, and geodetics.

If IGY were to be a success, it would be essential that Antarctica be fully covered, too. It was well known that polar physical phenomena had profound effects on global communications, and that the ice-cold Antarctic landmass had similarly profound effects on the weather in the Southern Ocean (extending north in some places to 45 °S) and its adjacent lands. Thus a group of 12 nations agreed to establish observatories (55 in all) on the continent and on sub-Antarctic islands. A coordinated series of observations was to be made.

This experiment in worldwide scientific collaboration proved a major success, and showed the way to removing international discord from a significant part of the earth's surface. Scientifically, it led to a strong desire to continue the collaboration and to extend its coverage into appropriate areas of the other earth sciences and biological sciences. Politically, there were significant advantages to many countries of a continuing stable, demilitarized Antarctica. These factors persuaded the United States to call a meeting of all the nations involved in IGY work in Antarctica. At that meeting they agreed to establish an "Antarctic Treaty," and they signed their agreement in December 1959.

The Treaty was ratified by all 12 nations (Argentina, Australia, Belgium, Chile, France, Japan, New Zealand, Norway, South Africa, U.K., U.S.A., and U.S.S.R.), and it came into force in June 1961. Among other things, the nations agreed to consult together regularly. Indeed, they have held "Consultative Meetings" at about 2-yearly intervals since then, and therefore are referred to normally as the "Consultative Parties."

## THE TREATY—ITS PROVISIONS AND MEMBERSHIP

The treaty area comprises all lands and ice shelves south of latitude 60 °S, with the high seas there being covered by International Law.

Other articles of the Treaty include provisions that:

—Although military personnel and equipment may assist scientific work, the Antarctic may be used for peaceful purposes only;

—Scientific investigation and cooperation should continue, and plans, observations, and scientists should be freely exchanged;

—No territorial claims should be asserted, supported, or denied;

—Nuclear explosions and the dumping of radioactive waste materials would be prohibited;

—Free access for the inspection of any or all activities should always be available, subject to prior notice being given; and

—Any member state may request a conference to review the Treaty and its operations after 30 years from the date of its coming into force (i.e., after 1991).

The Treaty may be modified or amended only by unanimity among the Consultative Parties up to June 1991. Thereafter, it would also be necessary to obtain the agreement of the majority of the "Contracting Parties"—i.e., the Consultative Parties, plus all other acceding states. Since 1961, twenty other nations have acceded to the Treaty and its provisions. Membership of the Consultative Parties itself has been likened to that of a club. Certainly the entrance fee is considerable—a country must demonstrate significant and substantial national scientific research activity in Antarctica. This is usually taken to imply that at least a base has been established and occupied regularly. As a result of such work, six additional nations have now joined the group of the original Consultative Parties (Poland, the Federal Republic of Germany, Brazil, India, the Peoples Republic of China, and Uruguay).

A number of non-Treaty countries have undertaken work in Antarctica, or have allowed their nationals to join the scientific teams of Consultative Parties.

## PROCEDURES UNDER THE TREATY, AND RECOMMENDATIONS

The 2-yearly Consultative Meetings of government representatives have traditionally been held in private. However, since 1983, in an attempt to develop and strengthen the operation of the Antarctic Treaty system, nonconsultative members and observers from interested international organizations have been invited to attend. The meeting venue rotates among the Parties, with the host nation providing all facilities, in the absence of a permanent secretariat. The meetings allow for the discussion of topics of interest to the Parties, and those discussions have resulted in many "Recommendations" being formulated. These Recommendations are agreed to by consensus, not by vote, and come into effect when ratified by all governments. They clarify and further define the provisions of the Treaty itself, and serve to control Antarctic activities in ways that give effect to the principles declared in the Treaty. Past Recommendations have embraced a wide range of topics including, for example, the designation of specially protected areas and of sites of special scientific interest; tourism; logistics; the preservation of historic sites; and even postal services.

From time to time, a major topic of interest may cause the Parties to convene a "Special Consultative Meeting." Such meetings have been convened to make recommendations on the admission of new Consultative Parties, and also for the special topics of Antarctic marine living resources and of minerals (see below).

On a more regular procedural basis, each Consultative Party provides other Parties annually with its national program of activities for the coming season. Invitations to other nationals to join a country's programmed activities are commonplace; formal requests for visits and inspections do occur on a bilateral basis; and the 2-yearly Consultative Meetings do allow for the airing and resolution of any potential problems.

## SCIENTIFIC WORK

The International Council of Scientific Unions (ICSU) set up its Special (earlier Scientific) Committee on Antarctic Research (SCAR) as early as 1958, and SCAR "National Committees" were established in many countries. They work typically alongside national scientific academies and funding institutions. SCAR itself has established widely representative working groups which review the progress and needs of work in their specific disciplines. The working groups also hold appropriate symposia. Thus scientific ideas are debated, results are exchanged, and future collaborative work is planned. Antarctic scientific publications are large in number and high in quality. Their authors are probably fitter, and certainly significantly younger on the average, than those in other fields. The number of female authors, although small, is growing. Virtually every scientific discipline has been involved, including meteorology, hydrology, glaciology, oceanography, atmospheric and ionospheric physics, seismology, geology, volcanology, zoology, botany, and psychology. Equipment has ranged from all-sky cameras and magnetometers to scuba-diving equipment and small drilling rigs.

## LOGISTICS

Typically, the strictly logistical effort is much greater in cost and personnel numbers than the purely scientific effort it supports. The sea freezes each winter to a latitude ranging from 65 °S to 45 °S. The circumpolar belt of pack ice varies seasonally from about 20 million km$^2$ in spring to about 4 million km$^2$ in summer (from 7.5 to 1.5 million mi$^2$), effectively doubling the size of the continent in winter (Laws, 1985). This pack-ice has caused a number of support ships to be lost. Air links are therefore as important for transport to the continent as they are within it. The U.S. base at McMurdo Sound in the Ross Sea, for example, uses a sea-ice runway for wheeled aircraft until that is in danger of melting and, later in the season, a bullldozed runway on the less smooth ice shelf for ski-equipped aircraft. Bulk cargo for McMurdo also arrives annually by ship, with ice-breaker support.

Early field expeditions, even in the years immediately following IGY, commonly involved man-hauling or dog-hauling of sledges. This, however, has largely given way now to a wide variety of snow vehicles. They may be landed in the field by air, along with the field party, and may be resupplied also by air. Alternatively, helicopters may be the only "vehicles" used.

Clothing and food for field parties have become specialized products. Bases initially followed northern Arctic designs but have evolved significantly. Those on the Polar Plateau at 3000-3500 m (say 12,000 ft) are either constructed below the level of ice-cover, or slowly sink below it. To ensure that ice distortion does not crush the buildings, the temperature around the buildings has to be kept below -10 °C (say 15 °F), perhaps by refrigeration.

## CONVENTION OF ANTARCTIC MARINE RESOURCES

Antarctic marine living resources comprise a relatively small number of species. Each species, however, is present in relatively large numbers, or would have been if exploitation (e.g., of fur seals and whales) had not taken place. Nutrient-rich upwellings of sea water, coupled with 24-hour summer sunshine, lead to prolific blooms of phytoplankton. These provide food for the equally prolific "krill," pelagic shrimplike crustaceans 5-7 cm (2 in.) long, which themselves become food for whales, seals, penguins, fish, and other marine animals. It has been estimated (Laws, 1985) that penguins alone, directly or indirectly, consume 130 million tonnes of krill annually. Fishing for krill already takes place, and Laws (1985) has estimated that the total catch probably exceeds 600,000 tonnes annually. Of this, 88% would be taken by Russia and 11% by Japan, probably more for animal and poultry feed than for direct human consumption (McElroy, 1982). The annual yield of krill has been estimated as subequal to the world's total annual catch of fish and crustaceans (Hill, 1983).

The Treaty specifies that the Consultative Parties should discuss the preservation and conservation of the living resources of Antarctica. It was therefore appropriate that the Parties negotiated the "Convention for the Conservation of Antarctic Seals," which entered into force in March 1978, and subsequently convened a Special Consultative Meeting which led eventually to the "Convention on the Conservation of Antarctic Marine Living Resources" in 1980. The latter Convention covers all species and effectively requires, for example, that all "harvesting" of marine living resources should recognize the proper principles of conservation; that the permitted harvesting of individual species should be such as not to allow their populations to fall below the level that ensures the greatest net annual increment to those populations; that depleted populations should be raised to those levels; that ecological relationships should be maintained; and that all changes, which might risk consequent changes to the marine ecosystem, that could not be reversed over a period of two or three decades, should be avoided or their effects should be minimized.

A Commission has been established in Hobart (Australia), with functions that include the facilitation of research that is necessary to implement the above conservation principles; the compilation of appropriate data and statistics; and the setting-up of a system of observation and inspection.

The Convention is open to adherence by non-Treaty nations.

## A MINERALS REGIME?

The Treaty does not mention nonliving resources. Nevertheless, interest in Antarctic mineral resources has quickened in recent years, particularly in regard to petroleum.

In response to that interest, the Consultative Parties convened a Special Consultative Meeting, beginning in Wellington (New Zealand) in June 1982, in an attempt to define an acceptable "Minerals Regime," possibly along similar lines to the Convention on marine living resources. They had already foreshadowed such discussions by agreeing in 1977 that there should be a moratorium on mineral exploration and exploitation. By that mechanism, their nationals were to refrain from those activities. In addition, the Parties bought time, both for their discussions and hopefully for agreed solution(s) to the minerals question. The moratorium has been renewed regularly.

The Special Consultative Meeting has held a number of sessions in different locations, a recent session being held in March 1986 in New Zealand. The brief for the Meeting was to negotiate a minerals regime for the exploration and exploitation of minerals within the continent and its adjacent offshore areas (excluding the deep sea bed). The minerals regime should maintain the Antarctic Treaty; protect the Antarctic environment; not prejudice the interest of all humankind in Antarctica; leave intact the basic Treaty principles on territorial claims (and could therefore be agreed to by claimant and nonclaimant states alike); encourage appropriate scientific research, and the assessment of the impact and acceptability/nonacceptability of mineral activities; and allow both adherence by other nations and cooperation with relevant international organizations.

Finally, and perhaps critically, the Special Consultative Meeting would have to consider the relationship of the Consultative Parties to other nations of the world—Would a non-treaty nation, for example, have access to the minerals and/or a share in any royalties or profits?

The question of Antarctic resources was raised at the United Nations General Assembly by the Prime Minister of Malaysia in 1982. He stressed that, in his view, the continent belonged to the international community and that exploitation of its resources should benefit the poor nations as well as the rich. As a result, the Secretary-General agreed to conduct an appropriate background study of Antarctica for the United Nations.

The work of the Special Consultative Meeting continues. Dr. John C. Behrendt (USGS) presented a 2-page unpublished paper to the August 1986 SCAR meeting at San Diego on the "Status of Negotiations of an Antarctic Mineral Resources

Regime." He emphasized that significant progress had been made in the development of what would be a framework agreement, not a detailed mining code. There is already general consensus on the establishment of:

(a) a "Commission" of Consultative Parties which would be the principal decision-making body, and the final authority;
(b) one or more "Regulatory Committees," comprising a smaller number of Consultative Parties, governing particular mineral resources in particular areas, and regulating prospecting, exploration, and development; and
(c) a "Scientific and Technical Advisory Committee" of Consultative Parties and Signatory Parties to the Treaty.

Prospecting would be allowed 6 months after notification by a sponsoring state, which would itself be required to ensure compliance with the guidelines of the Minerals Regime, and with the Agreed Measures of the Antarctic Treaty. "Prospecting" differs from "research" in the permitted retention of proprietary data for a period yet to be determined. No drilling deeper than 25 m (82 ft) would be allowed.

Exploration would be prohibited unless a positive decision authorized it, following a "lengthy legal process requiring the opening of an area for a specific mineral resource by the Commission," and other formal recommendations and actions. Exploration would presume some rights to any discoveries made. Deep test wells would be permitted.

The development stage would follow after the completion of exploration. However, Dr. Behrendt stressed that the general consensus view of knowledgeable geologists and geophysicists was that only prospecting activities are likely for several decades.

## MINERAL RESOURCES

The question of a possible minerals regime raises the question of whether "economic" minerals exist in Antarctica or not.

Antarctica is a remaining fragment of the old continent of Gondwana. Because other Gondwana fragments (e.g., Australia, South Africa) have substantial mineral industries, it seems probable, at first sight, that Antarctica should also be mineral-rich. However, Antarctica differs from the others in that all but 1% (Laws, 1985), or at most 2% (Hill, 1983) of the continent is covered in an ice-sheet averaging 2000 m (6500 ft) in thickness. Most ice-free areas are at high elevations and are commonly far from the sea. Further, the majority of the 1-2% ice-free land exists in West Antarctica (Andean and Ellsworth Geological Provinces) and in the Transantarctic mountains, areas which are composed primarily of Paleozoic and Mesozoic rocks. Only a smaller minority of ice-free land is composed of the much more prospective Precambrian shield rocks (Figure 1). Thus the mineral potential of the small exposed area tends to be more akin to that of eastern Australia, New Zealand, and southern South America—with deposits of petroleum, coal, and other bulk minerals, with some copper, gold, and silver, being most likely to be significant. Thus the mineral potential is not likely to be typical of some of the best known Gondwana mining areas.

In practice many occurrences of a very wide variety of minerals have been noted by the significant number of geologists who have concentrated their surveys on the ice-free areas. Their results have been received at a series of meetings, from Oslo in 1973 (Willett, 1973) and Bellagio (Italy) in 1978, to the Chile Antarctic Station Teniente Marsh in 1982, and have been reviewed by a number of individual authors. The conclusions of those reviews have not changed over the years—that only some coal deposits in the Transantarctic Mountains and some iron ore in the Prince Charles Mountains have the necessary concentration of minerals that would constitute a mineral prospect, even if they were to occur in the most favorable mining environment. In the harsh and remote Antarctic locations, the relatively low-value, high-bulk minerals of coal and iron ore do not constitute economic deposits. Thus it is generally accepted that if economic mineral deposits exist on land in Antarctica, they have remained successfully hidden from the sight, knowledge, and/or recording of field geologists so far.

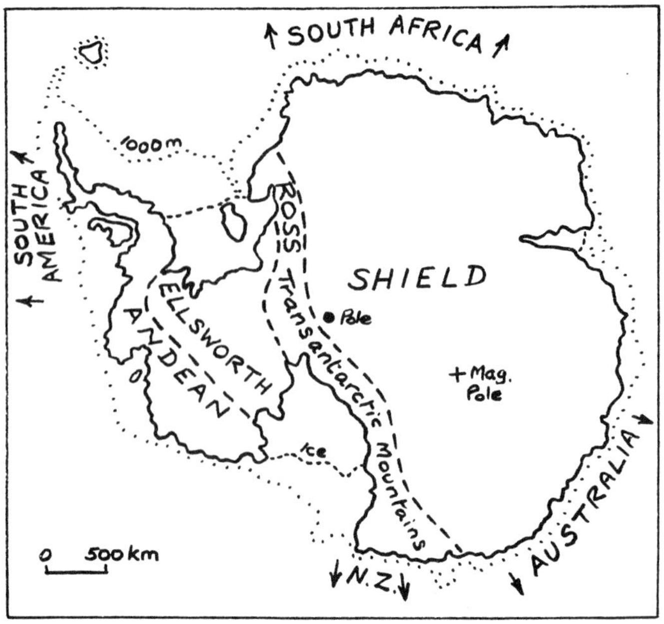


Figure 1. Sketch map of Antarctica showing geological provinces and their relationship to nearest landmasses (Anonymous, 1982).

Offshore petroleum prospects may be different. The story is well-known of "Glomar Challenger's" inability to continue the drilling to planned depths of three out of four DSDP holes in the Ross Sea, because of gas pressure (e.g., Hayes et al., 1973; Willett, 1973). A number of authors (e.g., Lock, 1983) have concluded that several sectors of Antarctic continental margin have geological histories suitable for the development of petroleum basins, perhaps similar to the young basins of southeast Australia and New Zealand. Marine geophysical surveys within the national programs of several Consultative Parties have confirmed the presence of adequate sediment thickness in some such basins (e.g., Davey, Bennett, and Houtz, 1982). However, careful analysts (e.g., Sanderson, 1983) have stressed that it is not yet known what reserves, if any, exist; that accumulations would have to be of oil and to be huge to be economic; and other "less hostile" sources would have to be scarce to attract both the finance and the technical effort for Antarctic petroleum exploration.

There are two additional important factors. First, major gas fields, with only minor oil fields, have been the rule in the comparable basins of southeast Australia and New Zealand. On the other hand, hypothetical marine beds in Antarctic marine sediments, rich in the remains of bountiful Antarctic marine life, might be thought more likely to yield oil than gas. However, Lock (1983), while expecting oil as much as gas, agrees with the probability of a terrestrial plant source. Second, exploration has been abandoned recently in the closest of those basins, the "Great South Basin." The latter basin, to the south of New Zealand, is about the size of Texas. The seas are hostile there, but they lack the ice and access problems of Antarctica.

## CONCLUSIONS

The Antarctica Treaty has proved a workable mechanism by which a group of 18 nations have ensured that only peaceful, free, and open scientific work will be undertaken there. All territorial claims have been set aside, and nations that are antagonistic elsewhere cooperate amicably in Antarctica. The Treaty nations have effectively accepted the key role in the management and stewardship of the continent, particularly in their conclusion of innovative conservation agreements such as the Convention of Marine Living Resources, and in their attempts to develop a comparable minerals regime. Indeed, many people believe that "the best and most realistic hope for the future of the Antarctic lies in the development of the Antarctic Treaty System" (Laws, 1985).

As far as minerals are concerned, it has long been stressed that no economic mineral deposits have yet been found on the land, and there is as yet no good reason to expect that that position will change in the future. Sedimentary basins, possibly suitable for petroleum accumulation, have been located on the continental shelf around Antarctica. However, no evidence exists yet of any economic accumulations, or that the established trend in that part of the globe—of natural gas predominating over oil—would be reversed on the Antarctic continental shelf.

## REFERENCES CITED

Anonymous, 1982, Antarctic achievements 1957-82; New Zealand's role in the Antarctic: Antarctic Division, New Zealand Department of Scientific and Industrial Research, Christchurch, 28 p.

Davey, F.J., D.J. Bennett, and R.E. Houtz, 1982, Sedimentary basins of the Ross Sea, Antarctica: New Zealand Journal of Geology and Geophysics, v. 25, p. 245-255.

Hayes, D.E., L.A. Frakes, P. Barrett, D.A. Burns, P. Chen, A.B. Ford, A.G. Kaneps, E.M. Kemp, D.W. McCollum, D.J.W. Piper, R.E. Wall, and P.N. Webb, 1973, Leg 28 deep-sea drilling in the Southern Ocean: Geotimes, v. 18, p. 19-24.

Hill, J., 1983, New Zealand and Antarctica: New Zealand Commission for the Environment, discussion paper, Government Printer, Wellington, 99 p.

Laws, R.M., 1985, International stewardship of the Antarctic, problems, successes and future options: Marine Pollution Bulletin, v. 16, n. 2, p. 49-55.

Lock, R.G., 1983, Continental margin petroleum potential in the Ross Sea region: New Zealand Antarctic Record, v. 5, n. 1, p. 6-15.

McElroy, J., 1982, Krill, still an enigma: Marine Policy, v. 6, n. 3.

National Research Council, 1986, Antarctic Treaty System, an assessment; proceedings of a workshop held at Beardmore South Field Camp, Antarctica, January 7-13, 1985: National Academy Press, Washington, 435 p.

Sanderson, T.J.O., 1983, Offshore oil development in polar regions: New Zealand Antarctic Record, v. 5, n. 1, p. 30-44.

Willett, R.W., (ed.), 1973, Antarctic resources—Report from the informal meeting of experts 30 May–9 June 1973: The Fridtjof Nansen Foundation at Polhogda, Cyclostyled, 18 p. and appendices.

CHAPTER 5

# GEOTHERMAL DEVELOPMENT: A KEY TO ENERGY SECURITY

**Carel Otte**
*Unocal Corporation, Geothermal Division*
*Los Angeles, California*

You do not have to pay much for energy these days. As a result, energy producers have been hurt severely. Worldwide, the energy industry is in a slump, and geothermal is part of it, but we ought not to be.

Although we are a young industry, we have proven that geothermal is real—we know the resource exists in many places, we know how to find it, we know how to produce electricity from it, and we have proven we can do that at a competitive price. But the low price of crude oil has made geothermal development seemingly less attractive to some nations. The truth, however, is that the low price of crude has created an opportunity for geothermal development.

We are at a point in the worldwide development of geothermal energy where we can do one of two things. We can let short-term market forces dictate a slow-growth pattern for us, or we can use this period of low prices to push for further geothermal development now.

If I can impress you with one thing, it is this—Another energy crisis is on its way, and we must develop all our alternative energy resources before the next crisis hits. We should use the period of grace we have now to ready ourselves—and the world—for the next energy shortage.

We are only experiencing a short-term glut in hydrocarbon fuels. It is important to remember that. It is a highly visible, artificially created, short-term glut that is occurring within a long-term supply shortage.

One of today's popular arguments is that with oil supplies plentiful and prices relatively cheap, it is not economical to develop geothermal energy, or any other alternative source of energy for that matter, but that is a short-sighted argument. The sharp decline in oil and gas production, and the increase in consumption caused by the present low price of crude, are setting the stage for another crisis and higher prices.

To illustrate this point, let us take a closer look at the petroleum output in the United States. During the first year that prices remain in the \$12 to \$13 per barrel range, production is estimated to fall by at least 1 million barrels per day. That fall-off will result from the shutting in of stripper wells and the abandonment of enhanced oil recovery projects that are becoming uneconomic.

If prices remain at levels below about \$15 per barrel, most experts predict that U.S. crude oil and natural gas production will have declined from about 9 million barrels per day to about 6 million barrels per day by the early 1990s. That is a drop of 30%.

It is also estimated that something on the order of 2 million barrels per day of higher-cost production outside the U.S. is in serious jeopardy.

Couple this with a modest increase in consumption of 1 to 1 1/2% per year, and demand would go up by 2 million barrels per day by 1990. Thus, in about 3 to 4 years, the demand supply picture in the free world could change by at least 7 million barrels per day.

With such a swing, we would find ourselves living under the same set of conditions we had before the last energy crisis in 1979. OPEC will be in control of the world markets and prices will rise.

Will we be better prepared to face a crisis this time than last? No, we will not. We no longer have healthy service industries, nor large inventories of exploration prospects to develop. Finally, we will not be able to conserve as much. In the United States, the major conservation efforts have been made—the downsizing of automobiles, the retrofitting of industries for higher energy efficiency, and the construction of more energy-efficient buildings. These have all been accomplished for the most part. We simply will not be able to cut back as much as we did in 1973 and 1979.

In addition, we need to consider another factor. It takes time to switch from economies based on oil and gas to alternative energy projects. We need to plan and build new energy industries now so that they will be operating 5 to 10 years from now. It takes from 12 to 15 years to build a nuclear power plant. It takes 7 to 10 years from the time a geothermal reservoir is discovered to the time of start-up of a geothermal power plant.

And we need to develop the world's energy resources efficiently. Imagine for a moment the world as one huge energy marketplace, not individual countries with political conflicts and varying economies—no oil embargo, no Gulf war, no credit or foreign exchange crises, no anxious waiting for the next OPEC price decision.

What would be the most efficient way for this apolitical world to allocate its various energy resources? Would it distribute Middle Eastern oil to countries to power factories or heat homes? Would it build plants to turn Canadian or American coal into gas? Would it build a nuclear power plant in a country that needed to modernize quickly, or did not require a high base load?

The answer is no. Those are not efficient uses of those forms of energy.

Oil and gas are finite resources that are inefficiently used as boiler fuel to generate electricity. Coal, while plentiful, becomes too expensive when turned into gas or liquid fuels. Nuclear energy is inefficient for areas of low or cyclical demand, or for places that need to develop quickly. Geothermal energy cannot be transported.

So taking into account the characteristics and limitations of each energy source, how would this ideal world put each energy source to its highest and most beneficial use? It would use oil as a liquid transportation fuel. It would use natural gas for domestic needs and chemical raw material. And it would use coal, nuclear, and geothermal as sources for central power generation.

Let us now take a look at the geography of geothermal energy. Many of us who live on the Pacific Rim have a special opportunity—the potential to develop geothermal resources.

The advantages to developing geothermal resources are clear. A productive geothermal resource can help a nation move toward energy self-sufficiency. It can provide the option to sell exportable resources which produce foreign currency. And geothermal is a clean and competitively priced source of electrical generation.

We have certainly come a long way in its development. Since mid-1979, worldwide geothermal plant capacity has grown 269%. That represents an annual growth rate of 15% over the 7-year period.

The United States is the world's largest producer of geothermal energy, having 2006 installed megawatts. The geothermal energy from the geyers in northern California, provides enough electricity to supply cities the size of San Francisco and Oakland.

The Imperial Valley in southern California is also a major resource and in the future will help supply the rapidly growing southern California market. Unocal's 10-megawatt pilot project there has made tremendous progress in controlling the corrosion and scaling that occur with the valley's brine resource. We are confident that we have solved most of the problems, and new projects are being planned. We are also exploring in other areas of the western United States.

New Zealand, too, has long explored for and developed geothermal energy. That country, one of the pioneer developers, now has installed capacity of 167 megawatts. The discovery of natural gas, however, has slowed geothermal development, but the New Zealanders should develop their full complement of energy resources. If they developed their geothermal resources for domestic electrical energy, for example, their natural gas could be sold to foreign markets, earning foreign currency.

A country that develops all of its energy resources opens up options for itself. The more options, the healthier the economy.

Indonesia is a good example of a country with a broader outlook. Indonesia currently produces more than 1.3 million barrels of oil per day. But it is simultaneously moving ahead to develop its significant geothermal potential.

In addition to its 32 megawatts of installed geothermal generating capacity and expansion under way, Indonesia is also exploiting its coal resources for power generation.

The energy policy practiced by Indonesia—that is, developing as many of its energy resources as possible—will allow Indonesia to use less of its exportable energy, its oil and gas, for domestic development. When the price of oil is back up again, say, in the mid-1990s, Indonesia should have in place the geothermal power plants to supply a good portion of Java's electricity. What is most attractive, whether the price of oil is high or low, Indonesia earns valuable foreign currency by exporting more of its oil. That currency can then be used much more wisely for domestic development projects.

The Philippines, with 894 installed megawatts, has developed two primary geothermal fields, at Tiwi and Mak-Ban, each with 330 megawatts of capacity. There is additional potential on the island of Luzon but, for the moment, the local market for electricity is limited pending a decision on the recently completed nuclear power plant and the severe economic crisis. There is a 112-megawatt installation on the island of Leyte and a 118-megawatt installation on southern Negros. Regrettably, both installations are underutilized.

Japan is a country that imports virtually all of its oil, so it should have an overwhelming incentive to develop indigenous energy supplies beyond its present 215 megawatts of installed capacity. Yet Japan is hampered by environmental concerns in developing its geothermal resources.

Mexico is an oil producer, but it has become the world's third largest producer of geothermal energy. While Mexico has severe economic problems, it has 645 megawatts of installed capacity and its geothermal development at Cerro Prieto supplies a substantial amount of electrical power in northern Mexico.

In Central America, El Salvador produces 95 megawatts and Nicaragua, which is expanding its developments at Momotombo, producers 35 megawatts. In both Central and South America, geothermal is a potential energy source for other countries as well—Guatemala and Costa Rica, Chile, Peru, Bolivia, Ecuador, Colombia, and Argentina.

What all of these Pacific Rim countries have in common is the option to expand their sources of energy. That option ought to be enticing at this time of low oil prices.

We must plan now. As prices rise and demand increases, the world will again begin to clamor for alternative energy projects. Governments should be working on developing new energy industries. By establishing a long-term energy policy and by creating a political and investment climate that encourages development, we Pacific Rim nations can take the lead in developing diversified energy resources.

In summary, with an enlightened and farsighted outlook by government and industry, the future for geothermal energy is a bright one. Energy forecasts predict accelerating demand and price increases for hydrocarbon fuels by the early to mid-1990s. Higher prices and greater demand will create a positive economic climate for geothermal energy developments.

And we are ready. Studies show that the resource is plentiful. And it is plentiful in some of the nations that need it most—countries that surround the Pacific Basin. These developing countries can benefit greatly from the increased employment, new skills, and foreign currency savings that developing a geothermal industry offers.

And countries do not have to be at the mercy of wild swings in the energy market. Their governments can take the lead today by establishing a long-term national energy policy and the political and investment climate that encourages the development of diversified energy resources. In so doing, they provide for their energy security. Government and pri vate enterprise can and should do it now, because the lead times for their development are long.

Many of us here at the conference are earth scientists. It is our duty to remind our governments that mineral resources are finite and that we have the obligation to use them wisely and efficiently. Energy minerals should be developed for their highest and best use. That is conservation at its best—*responsible* utilization for the betterment of mankind. Let us take advantage of the breathing room we have to develop the resources we will need for the future. Let us not be lulled into complacency.

Governments, like individuals, should hope for the best but prepare for the worst. In developing our resources now, we not only hope for the best, we help create it.

# GENERAL SESSION ABSTRACTS

GAFFNEY, PETER D., Gaffney, Cline & Associates, Alton, England, United Kingdom

## Asia-Pacific Regional Exploration in an Uncertain Oil Price Market

This discussion provides an overview of exploration successes in the region over the last 15 years and relates them to the relative economics of the time compared to various potential economic circumstances in the future. It addresses the potential availability of indigenous regional exploration funds and presents scenarios for believing excellent exploration prospects can continue to be developed within the region, which would be economically justified for a range of potential oil and gas prices.

Correlations have been made with target depth horizons and play types with potential economics to high-grade basin exploration opportunities for given oil price ranges.

LARSON, R. L., University of Rhode Island, Narragansett, RI, and P. D. RABINOWITZ, Ocean Drilling Program, Texas A&M University, College Station, TX

## Plans for Ocean Drilling in Pacific Ocean

The newly organized Ocean Drilling Program has outfitted the drill ship *Joides Resolution* with state-of-the-art drilling, positioning, laboratory, and geophysical logging equipment that allow coring and logging operations to proceed successfully despite adverse weather conditions and unusual geology. This facility is capable of the largest and most precise suite of geologic and geophysical downhole measurements ever made at sea. The results are expected to define the leading edge of marine geologic research for the next decade.

The *Joides Resolution* will spend the last 4 months of 1986 drilling on the spreading centers and continental/ocean margin regions of the eastern Pacific to address questions relating to the formation of ocean lithosphere, the tectonics of subduction near Peru, and the paleo-oceanographic history of western South America. After an intervening 1½ years in the Indian Ocean and Antarctic areas, *Joides Resolution* will return to the Pacific in late 1988. Plans are tentative at this time; however, the campaign would begin in the Indonesian region, move to other areas of the western Pacific, and finally into the Pacific basin region in the 1989-1991 time frame. General goals for western Pacific drilling will include the dynamics of forearc evolution, the structure and history of island arcs and back-arc basins, and the histories of ocean circulation and sedimentation in this region. These studies will set the framework for subsequent exploration for energy resources in the Pacific.

# HYDROCARBONS

CHAPTER 6

# DEVELOPMENT OF THE SIRIKIT OIL FIELD, THAILAND

**John Brooks**
*Thai Shell Exploration & Production Co., Ltd.*
*Bangkok, Thailand*

ABSTRACT

The Sirikit oil field, Thailand's first significant oil find, was discovered in late 1981 by Thai Shell EP Co. Ltd., with its second exploration well. After the decision to develop the field (named after Thailand's Queen), it took only 1 year to design and install the production station, and organize an unconventional evacuation system (road tanker and railway) before oil came on stream in January 1983. A series of facility upgradings kept pace with the build-up of production to a plateau of about 21,000 b/d.

The crude oil is waxy (pourpoint around 35 ° C) but is light (40 ° API) and has an attractive refinery yield.

Associated gas is sold to the nearby (specially installed) electricity generating station. Gas compression was commissioned in 1985 to increase utilization of gas which was previously flared.

The agricultural environment dictated the need for cluster drilling of deviated wells, as well as highlighting the importance of good relationships with the local population and authorities. Safety and security are two subjects of particular interest.

The field is in a geologically complex, extensively faulted lacustrine environment, and the reservoir quality is extremely stratified and heterogeneous. One of two major reservoirs has a gas cap. After some early surprises in delineating the field, a 3-D seismic survey was carried out which better defined the structure and reserves. Nevertheless, parallel appraisal and development continues on a careful step-by-step approach, using the latest production and pressure data to refine the reservoir-geological model.

In November 1985 the Petroleum Authority of Thailand became a minority partner, with Shell remaining as operator.

## INTRODUCTION

The Sirikit field, located in the Phitsanulok Basin of central Thailand (Figure 1) is estimated to have a STOIIP of 312 MMstb and a recovery expectation of 34 MMstb. Production is presently (May 1986) some 21,000 b/d from about 45 wells, most of them completed on two or more reservoirs with dual string completions. Current field offtake has been at this plateau level since the second quarter of 1985. Produced oil is light (40 ° API) and has a high wax content (15-20 % wt) and a high pourpoint (35 ° C). The crude does not contain any sulphur. The oil is produced from highly stratified sandstones (with thicknesses ranging from centimeters to several meters) deposited in a marginal lacustrine delta environment and vertically separated by shales (Figure 2).

Two main reservoir sequences (K & L) contain virtually all the reserves. The upper reservoir is saturated and has an oilrim of 86 m; the lower reservoir is undersaturated and has an oil column in excess of 200 m. Development concepts for the Sirikit field have matured considerably over the past 4½ years since its discovery. Well data, a 3-D seismic survey, and some 3½ years production history have all assisted in unravelling some of the characteristics of this extremely complex field in terms of structure, stratigraphy, and future reservoir performance.

## HISTORICAL PERSPECTIVE

### 1981—Discovery

After the Phitsanulok Basin in the Central Plains of Thailand was defined by field geological and aeromagnetic surveys, a seismic campaign was carried out during 1980 based on a grid of 1 × 2 km. This led to the proposal of two wells, both of which found producible hydrocarbons (1981). The first well, Pratu Tao-AO1, tested oil and gas in subcommercial quantities. The second well, Lan Krabu-AO1 (LKU-AO1) (Figure 3), proved to be a highly encouraging success, with RFT pressure data suggesting a very long oil column and production test rates of up to 4400 b/d. The extrapolated oil-water contact, together with the structural interpretation (a highly faulted anticline), led to high STOIIP estimates tending toward the 1 billion barrels class.

Recognizing the dual benefits of early production, both in terms of reservoir performance data and improved cash flow, a decision to install early production facilities was made following the results of LKU-AO1. Design and installation of facilities were ready within the 12 months following the decision to go ahead and the field, named after Thailand's Queen Sirikit, went on test production in December 1982.

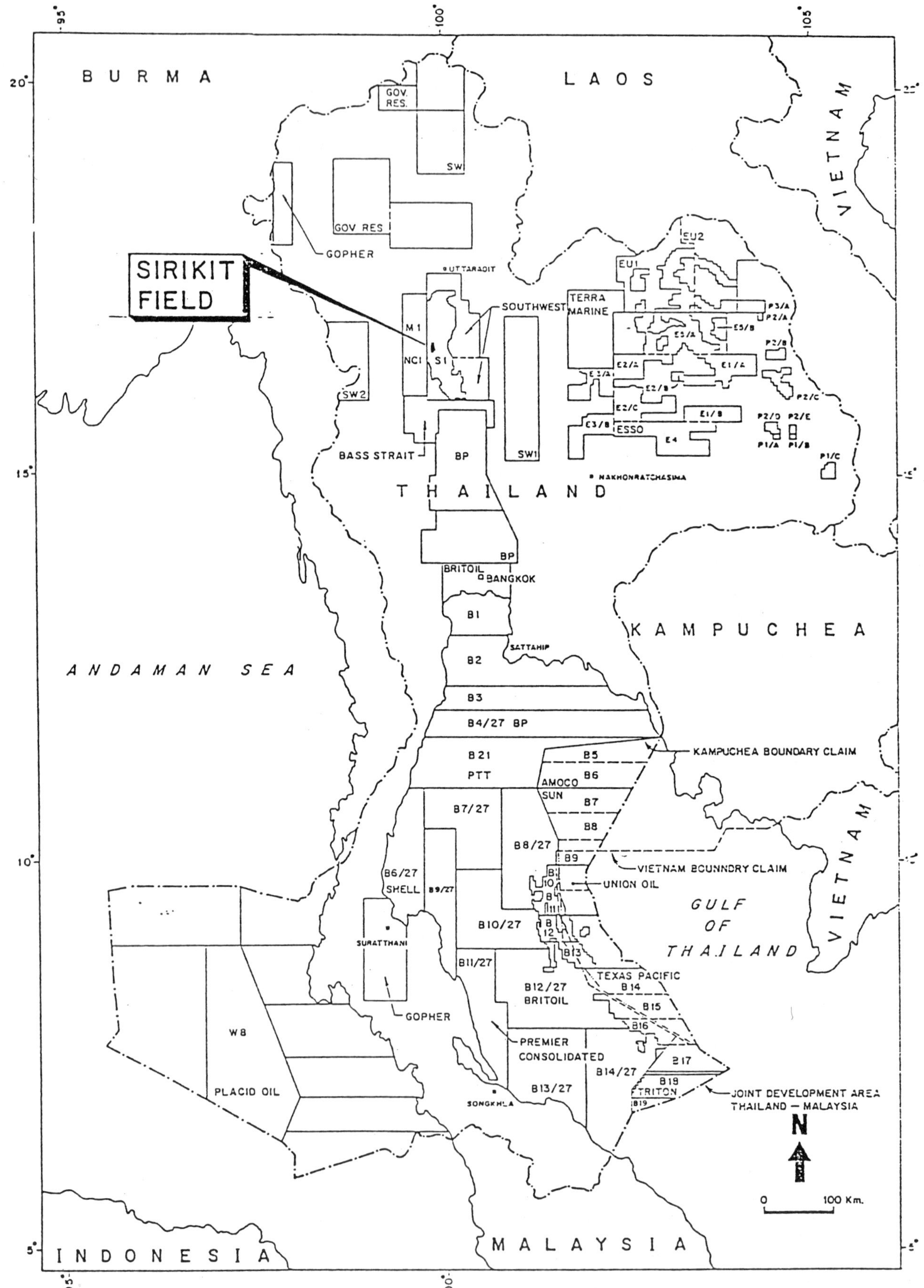


Figure 1. Concession map, Thailand.

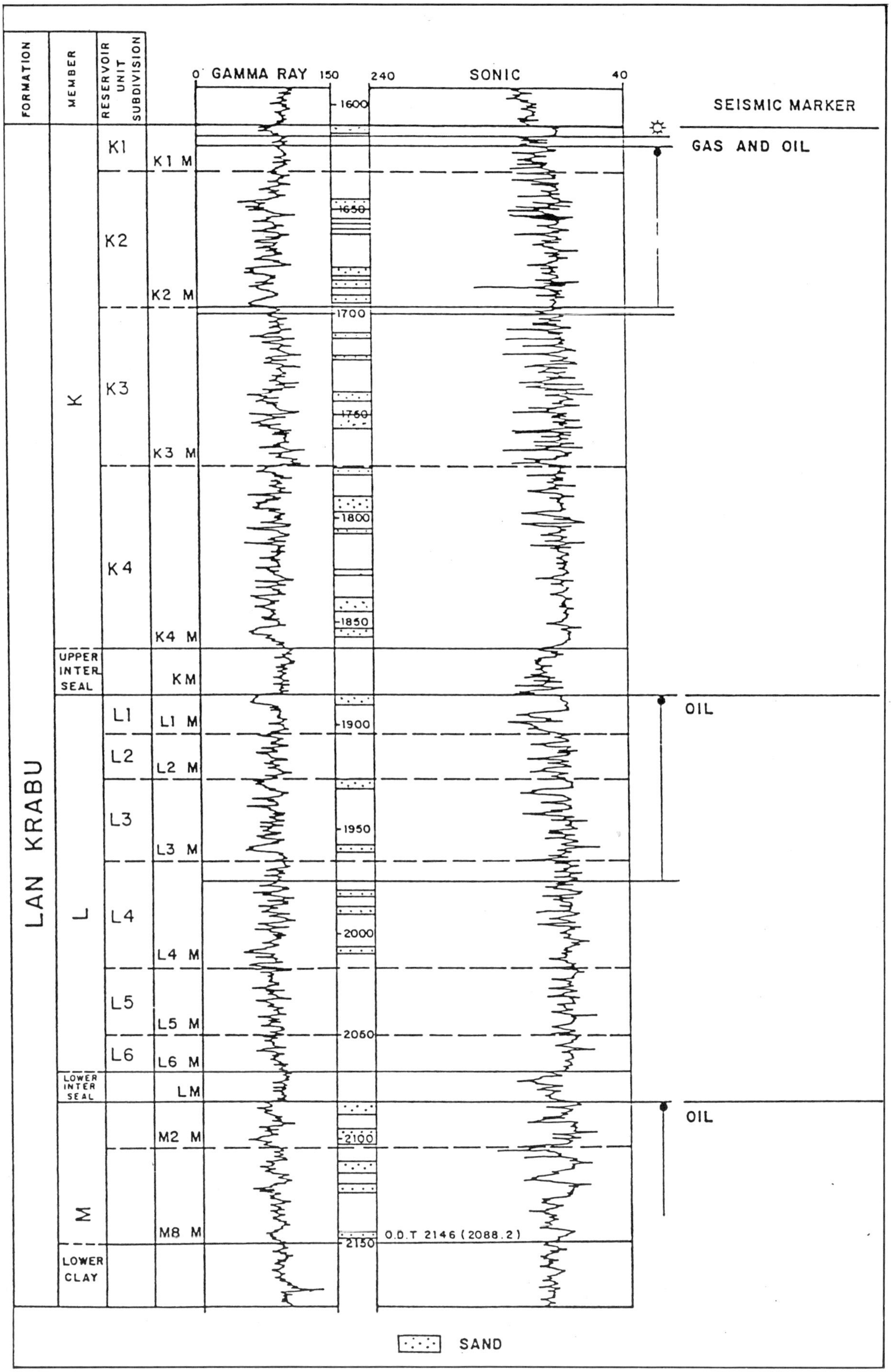


Figure 2. Type log, Sirikit Field.

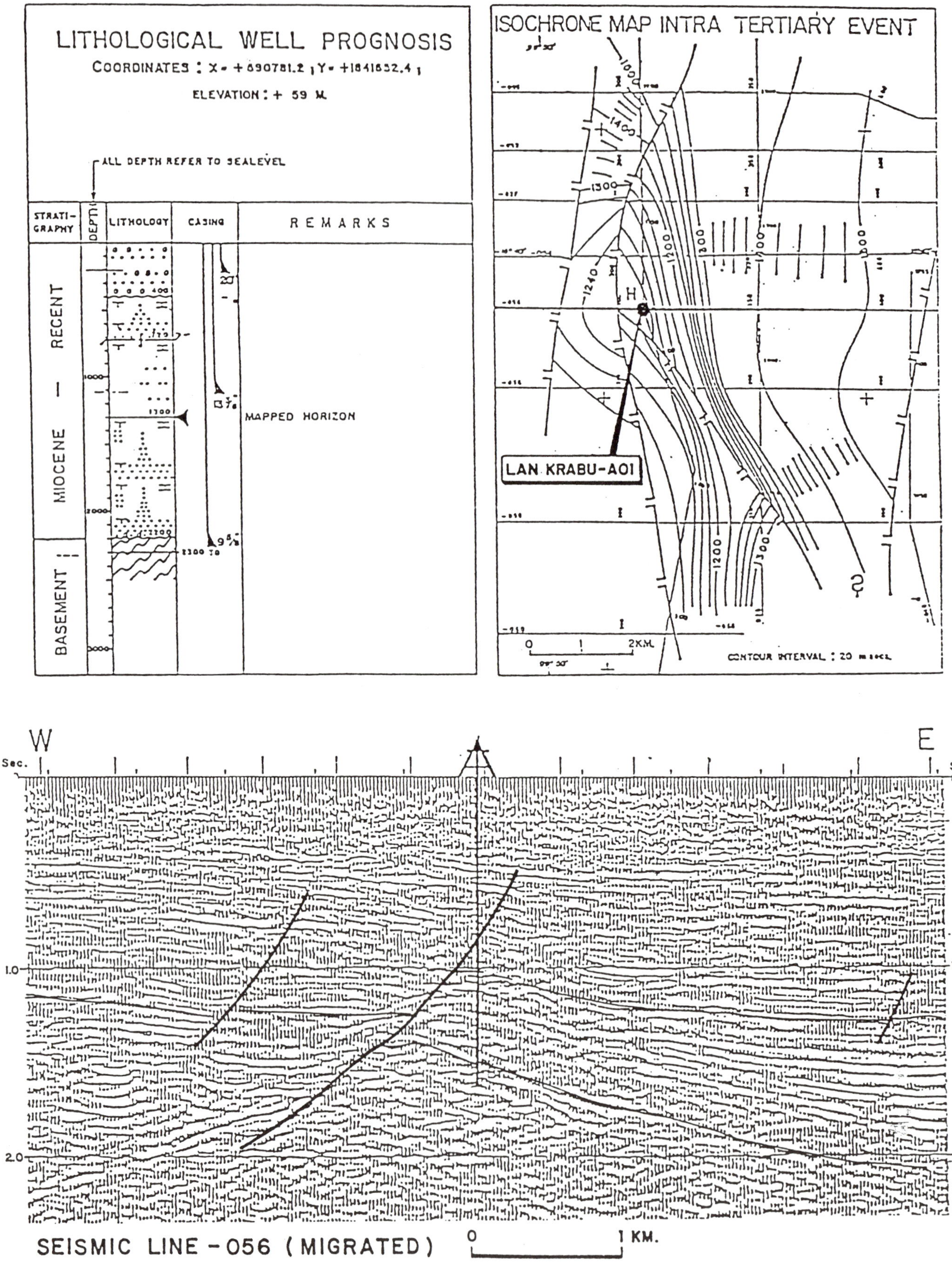


Figure 3. Prospect summary LKU-AO1.

**TABLE 1. SIRIKIT OIL FIELD—HISTORY OF RESERVE ESTIMATION (Million Barrels).**

| END-YEAR | STOIIP | ULTIMATE RECOVERY |
|---|---|---|
| 1981 | 1000? | ? |
| 1982 | 186 | 28 |
| 1983 | 290 | 36 |
| 1984 | 319 | 42 |
| 1985 | 312 | 34 |

## 1982—First Appraisal Campaign

After drilling the first sequence of appraisal wells in the course of 1982, it soon appeared that LKU-AO1 was nontypical for two reasons. First, the observed GOC at 1435 m ss was different from the level of 1555 m ss observed in the rest of the field. Second, the tested productivity was the highest so far observed in the Sirikit field.

This initial appraisal campaign consisted of four successful wells in the northern part of the field (LKU-AO2, -BO1, -CO1 and -DO1) at approximately 2 km spacing plus an outstep well (LKU-JO1), drilled 10 km south of LKU-AO1, which was essentially dry.

These wells put the LKU-AO1 data into their proper perspective by revealing the existence of several separate reservoirs of highly stratified sands. STOIIP and ultimate recovery estimates of 180 MMstb and 30 MMstb, respectively, were based on these new data (Table 1).

## 1983—Further Appraisal/Development

During 1982 eight cluster locations were built at a spacing of 2-3 km covering the Sirikit accumulation as envisaged at that time. However, disappointing results obtained from the first two 1983 wells LKU-FO1 (poor productivity) and LKU-GO1 (wet) significantly downgraded the potential of locations LKU-H and LKU-I. These results clearly illustrated that the existing structural geological model was an inadequate basis for further appraisal and development drilling. A 3-D seismic survey was therefore programmed for the 3rd quarter 1983 to unravel the structural complexity before further appraisal of the southern and western areas.

In order to meet set production targets, a structurally more simple area in the NE (northeast) part of the field had been identified for development since it was of lower risk. Three wells were drilled into this area during 1983 and all were highly successful in terms of productivity and useful information provided.

At year's end, the expectations of STOIIP and ultimate recovery were revised to 290 and 36 MMstb, respectively, mainly as a result of upgrading the NE part of the field and of the lower(L) reservoir being confirmed to be undersaturated (i.e., no gas cap). The results of a general petrophysical review also contributed significantly to the upward revision of STOIIP (Table 1).

## 1984—Continued Appraisal/Development

By early 1984 the 3-D survey giving superior quality data (Figure 4) had been completed and, although the structure was very complex, many of the structural problems in the central and eastern parts of the field had been unravelled. In contrast to the previous year, few wells drilled in 1984 gave major surprises. The only exception was an appraisal well on the western flank of the field, which indicated a densely faulted structure containing gas where oil had been expected. Because additional significant oil accumulations could not yet be ruled out, this part of the field was identified for further appraisal.

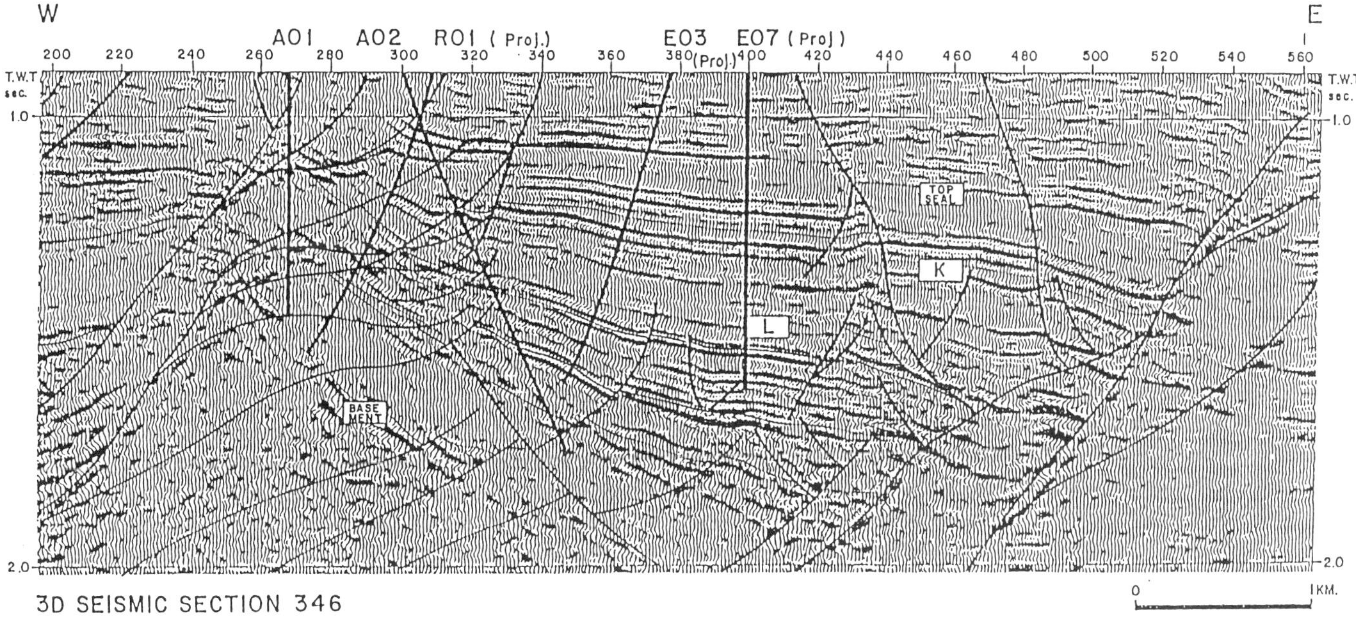


Figure 4. 3-D seismic section.

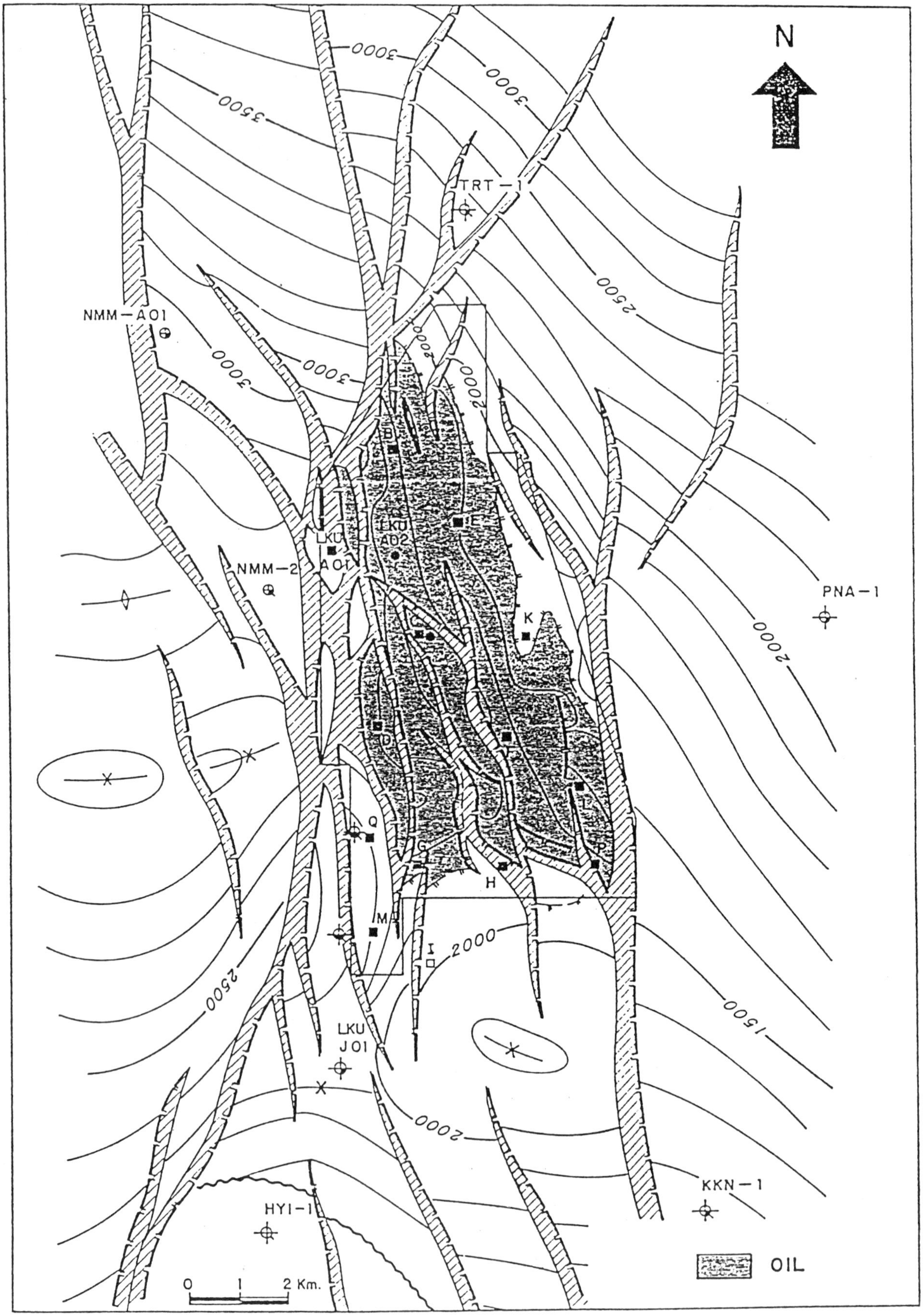


Figure 5. Structural map, Top L Sand

## 1985—19 Wells, Production Plateau Reached

The main target in 1985 was to reach and sustain a plateau level of about 20,000 b/d. This was achieved by an intensive development drilling campaign which, in spite of the 3-D seismic data, resulted in a number of surprises. Some wells were found to have penetrated very small drainage areas because of small faults below the seismic resolution. In other cases, major faults were penetrated significantly shallower than expected, resulting in changes to the planned drainage pattern. Uncharged (i.e., water-bearing) sands were found in places within the oil column and, conversely, oil-bearing sands were found below the general OWC. An appraisal well in the western flank confirmed the presence of a gas-water contact, which significantly downgraded the expectation of oil in this area. Although not part of the western flank, a well at the southern end of the field also indicated a gas-water contact at the same level.

Additional STOIIP was found in deeper reservoirs, although the productivity was poor. A change in the petrophysical saturation exponent following laboratory experiments on core plugs resulted in a 10% reduction in STOIIP and hence ultimate recovery of the main reservoirs. Although the STOIIP decrease was compensated by the additional oil in the deeper sands, the low recovery factor did not allow the same compensation in ultimate recovery. Furthermore, although the shallower-than-expected faults had little effect on the STOIIP, L sand oil with a low recovery factor replaced K sand oil with a much higher recovery factor; this resulted in a reduction in ultimate recovery, although the STOIIP was essentially unchanged (see Table 1).

## GENERAL GEOLOGY

As more results of appraisal and development wells became available, together with the 3-D seismic interpretation, the understanding of the complex geology of the field improved.

The Tertiary clastic sediments comprising the reservoir sequence have been deposited in an intracratonic basin bounded by major basement faults, most of which have a strike-slip component. The Lan Krabu Formation reservoir sequence consists of lacustrine deltaic sands and shales. These form stacked regressive sedimentary cycles which, if complete, contain basal lacustrine clays, delta front silts and sands, mouthbars, channel sands, and fluvial plain deposits at the top. Because of the lacustrine setting, the cycles are relatively small compared with marine equivalents and do not exceed 20 m in thickness. Periods of high water level yielded distinct and laterally extensive shales, which provide both the reservoir seals and the source rocks (Figure 2). Their lateral equivalents are sufficiently buried to reach maturity in a major depression north of the field. Hydrocarbon migration from this area has therefore been assumed.

The reservoir sands of the Sirikit field are, like those in similar intracratonic basins, rich in unstable minerals, and apt to early diagenesis. The porosity decreases therefore rather rapidly with depth, and economic basement is reached relatively shallow, at about 2,500 m.

The hydrocarbon trap is a combined fault/dip closure. Basement faulting occurred during deposition of the Lan Krabu Formation and ceased after the bulk of the fluviatile overburden sediments had been deposited (Late Middle Miocene) (Figures 4, 6). In addition to the main, north-south-

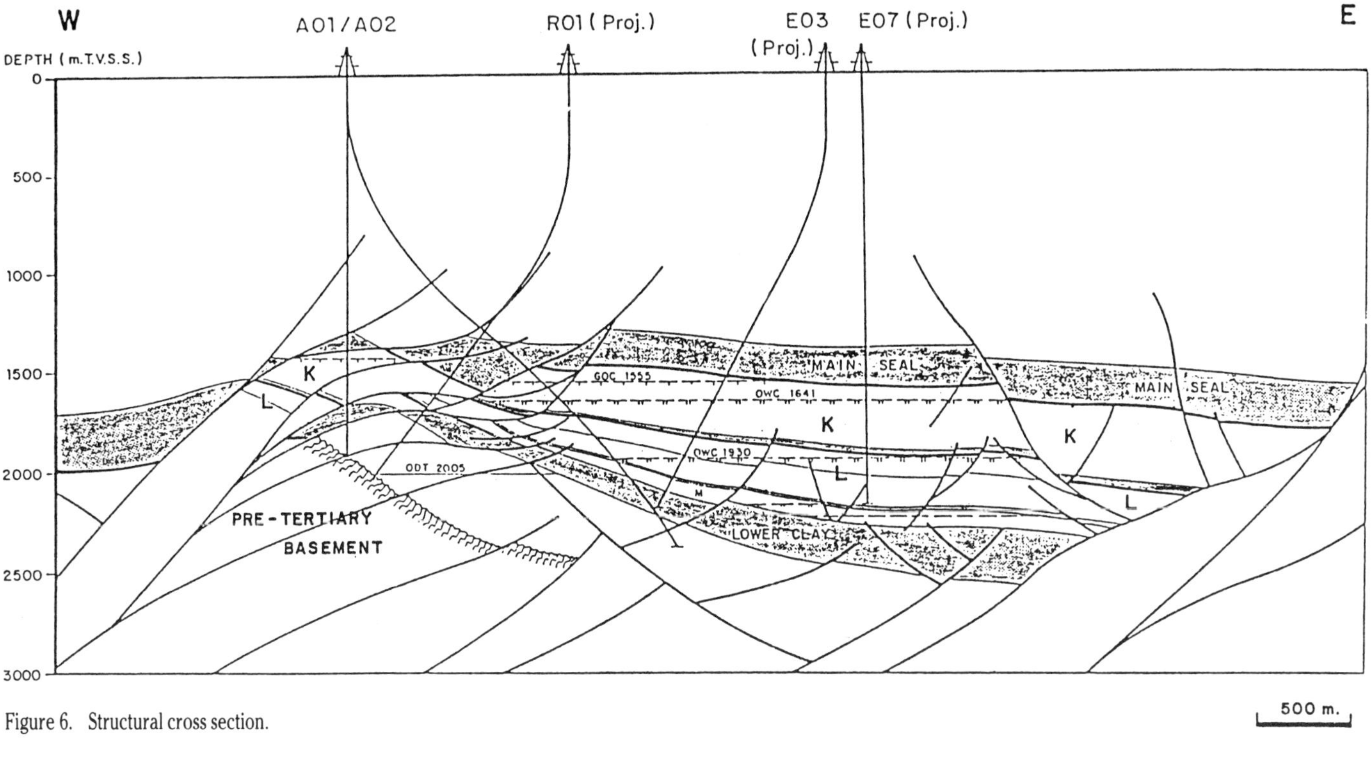


Figure 6. Structural cross section.

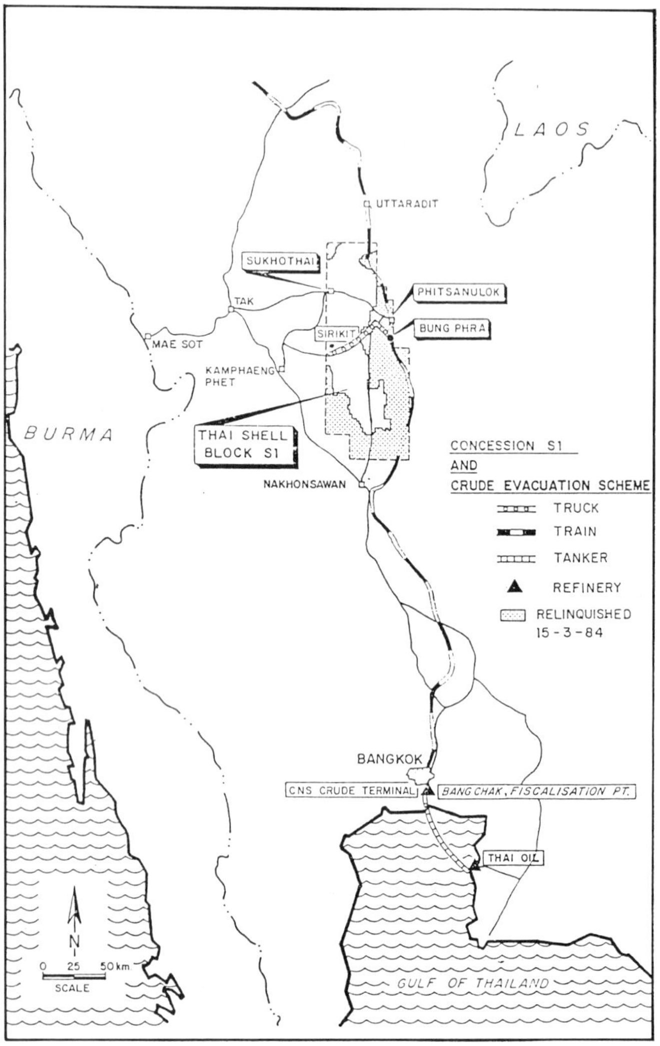


Figure 7. Crude evacuation route.

striking basement faults, Riedel shears and normal faults are present. Their pattern suggests oblique, left-lateral strike-slip movements of the basement faults which control the trap of the Sirikit field (Figure 5).

## 1986—Oil Price Halved: Slowdown in Development Drilling

The reduction in oil price in early 1986 had a significant impact on the development plans for the Sirikit field. One rig was released in April and plans are to release the second rig in September/October and to stop all drilling activity until oil prices recover. This slowdown in drilling will, in principle, not affect the overall development plans, inasmuch as the same number of wells for the primary development is still envisaged (total around 60 producers); i.e., part of the development of the field is merely deferred.

## Production Evacuation

The oil is produced via 3-stage separation into storage tanks. About 20 MMscfd (million standard cubic feet per day) of associated gas is piped to the adjacent, purpose-built generating station of the Electricity Generating Authority of Thailand (EGAT).

Crude oil evacuation involves transport by road to the railway terminal at Bung Phra, a distance of 55 km. A fleet of 30 road tankers of 200 bbl each and a turnaround time of 4 hrs provide a carrying capacity of up to 28,000 b/d, based on 20 working hrs/day operation. For safety reasons, to avoid the dangerous evening peak in road traffic, particularly in local villages, there is a curfew on road evacuation between 16:00 and 20:00 hrs. To date, 18 million barrels have been moved and the total distance covered by the road tankers is more than 11 million km. Some road tankers have already covered up to half a million km. A fleet of 170 rail tank wagons is used for the onward rail transportation from Bung Phra to Bang Chak refinery at Bangkok, 400 km to the south. With a turnaround time of 36 hrs, the maximum carrying capacity (28,000 b/d) requires 4 trains/day. Part of the production is forwarded by coastal tanker to the Thai Oil refinery on the eastern seaboard of the Gulf of Thailand, a journey of some 75 kms. The evacuation system is depicted in Figure 7.

In spite of the evacuation system being potentially hazardous, it is pleasing to note that there have been very few accidents, and the drivers' incentive scheme has helped to achieve an excellent safety performance.

CHAPTER 7

# GEOLOGY AND OIL AND GAS EXPLORATION IN CALIFORNIA'S OFFSHORE BASINS

**William E. Crain**
**Stephen P. Thurston**
*Chevron U.S.A. Inc.*
*San Francisco, CA*

## ABSTRACT

California's continental shelf contains 13 large offshore basins, containing from 5000 to 30,000 ft (1524 to 9144 m) of Neogene sediments, which formed during Neogene transform-margin tectonics. The Neogene sequence generally consists of relatively thin transgressive marine deposits overlain by deep-marine hemipelagic and biogenic sediments, capped by Late Miocene shales and Plio-Pleistocene submarine-fan clastics. The Miocene section contains the basins' primary hydrocarbon source beds, which reached peak levels of maturity for hydrocarbon generation from the Pliocene to present. Structural development within the Neogene basins began in the Early to Middle Miocene with peak structural growth in the Pleistocene.

Only three of California's 13 major offshore basins have been actively explored for hydrocarbons in recent years; each is an extension of a producing onshore basin. California's current offshore production from two of these basins averages 255,000 BOPD and 263,000 MCFD, and cumulative production to date is 2.3 billion bbls oil and 1.7 tcf gas from 29 offshore fields. Estimates of remaining proved reserves from these fields are 803 million bbls oil and 475 bcf gas. Undeveloped reserves of 914 million bbls oil and 2.0 tcf gas are estimated from 20 more offshore fields which await development. In addition, MMS estimates an undiscovered recoverable resource potential as large as 3.0 billion bbls oil and oil equivalent gas remaining in California's offshore basins.

## INTRODUCTION

This paper summarizes the geology of California's offshore basins and their potential for hydrocarbon accumulations by reviewing the tectonic history, stratigraphy, structure, and hydrocarbon maturation history of the basins. In addition, it briefly reviews the exploration and development history, reserves, and resources of the offshore basins.

California's continental shelf is 20 to 50 mi (32 to 80 km) wide along the coastline north of the Santa Barbara Channel (Figure 1). Water depths on this shelf generally exceed 500 ft (152 m), and beyond 10 mi from shore water depths range from 1000 ft (350 m) to 3000 ft (915 m). South of the Santa Barbara Channel the continental shelf is an active borderland that is 150 mi (241 km) wide consisting of fault-bounded basement blocks and islands, separating deep bathymetric troughs as much as 6000 ft (1830 m) deep. This southern borderland area contains five large Neogene basins as well as the offshore extension of the Los Angeles Basin (Figure 1).

California's continental shelf contains 13 sedimentary basins which formed primarily in the early Neogene (Figure 1). Except for Eel River basin, these basins formed along a transform margin which has been the predominant tectonic setting of western California from the Late Oligocene to present (Blake et al., 1978). The transform margin is characterized by right-lateral shear producing northward translation and tectonic disruption of the pre-Neogene convergent margin, and the formation of long, narrow, deep basins along the transform system. The offshore basins contain from 5000 to 30,000 ft (1524 to 9144 m) of Neogene deposits, which in most cases overlie a relatively thin section of Paleogene and/or Cretaceous forearc rocks ranging from less than 1000 ft (350 m) to more than 5000 ft (1524 m) in thickness. The thickness and distribution of the Paleogene and/or Cretaceous forearc deposits are highly variable from basin to basin. The "basement" rocks, where evident, are either Mesozoic continental granitic rocks or accretionary and subduction complex rocks of the Franciscan Complex or equivalent (Blake and Jones, 1981). In most of the offshore basins, the basement rocks and portions of the overlying forearc deposits are parts of allochthonous tectonostratigraphic terranes (Nilsen, 1986; Nilsen, in press). They were accreted to the North American margin in the early Tertiary during a period of northward translation of terranes, some of which were detached and dispersed from the Mesozoic forearc margin of southern North America and Central America (Nilsen, 1986; Nilsen, in press).

The Neogene development of the offshore basins evolved via transform margin tectonic processes, whereby sedimenta-

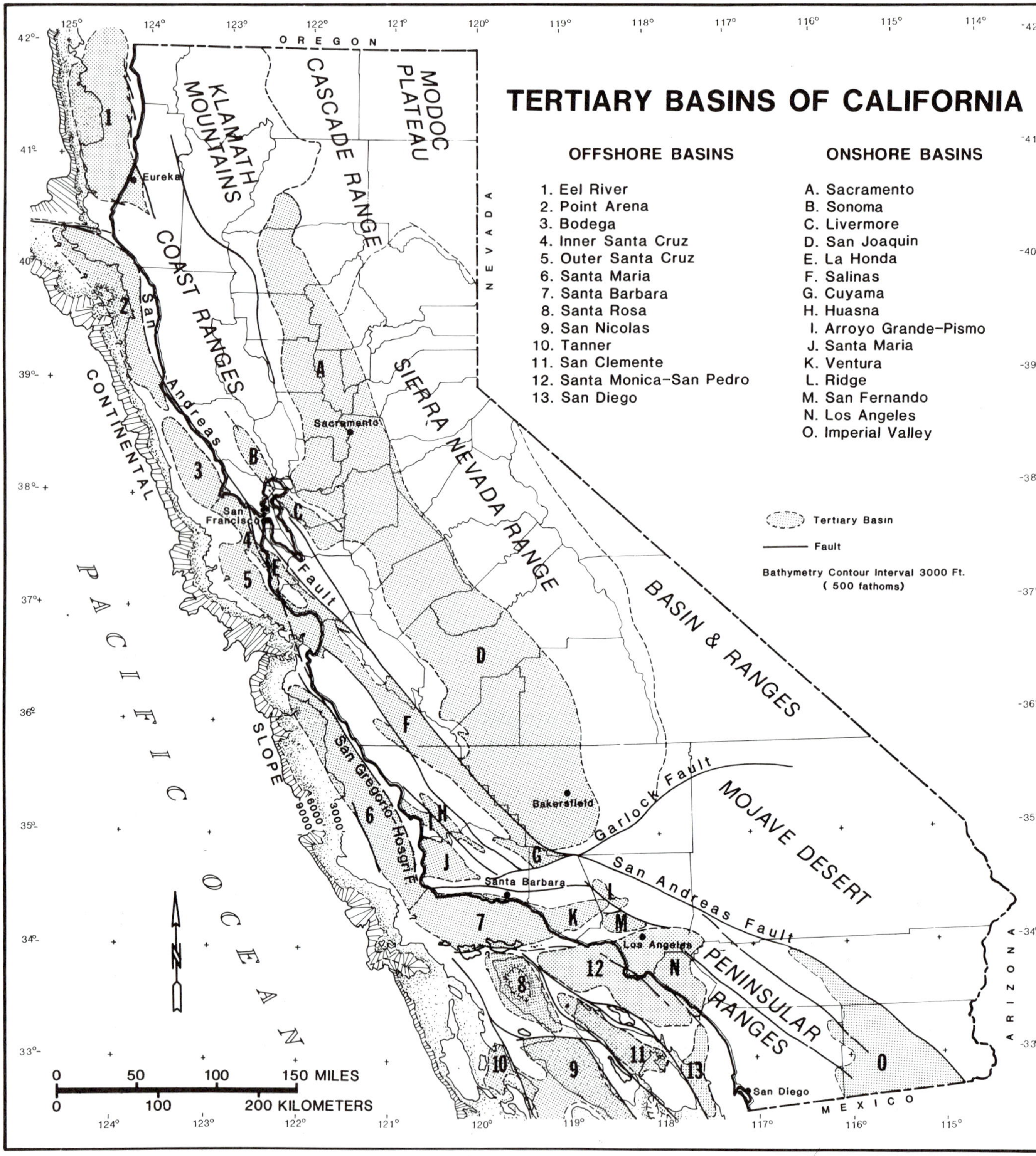


Figure 1. Location of Tertiary basins in California and its continental shelf. Modified from Blake et al. (1978) and Kilkenny (1971).

tion was controlled by tectonic development of narrow basinal troughs and adjacent uplifted blocks within the transform margin. The spatial distribution of the Neogene basins has changed during their evolution because of continual northward translation of the crustal blocks west of the transform boundary, the eastward migration of the transform boundary with time, and in some areas by rotation of crustal blocks and basins, such as in the Transverse Ranges of southern California.

Only three of the 13 major offshore basins have been significantly explored for hydrocarbons, and each of these produce oil. They are the Los Angeles, Santa Barbara (offshore Ventura basin), and the recently explored Santa Maria offshore basin, each an extension of an onshore producing basin (Figure 1). There are five northern and central and two southern California offshore basins which have only been slightly explored. Although results of some of these initial tests have been encouraging, there were no commercial discoveries (Zieglar and Cassell, 1978; Ogle, 1981; Wallis, 1981). Further exploration of the lesser explored basins is at present hampered by public opposition due to environmental concerns, and Federal, State and local regulatory disputes.

## REGIONAL TECTONIC SETTING

### Mesozoic

By the Early Cretaceous, the western margin of North America formed an "Andean-type" convergent margin between the North American continental plate and the subducting Farallon oceanic plate (Schweickert and Cowan, 1975; Dickinson, 1976; Nilsen, 1986). This setting produced the "classic" Mesozoic convergent margin complex here consisting of the Sierran magmatic arc, the Great Valley forearc basin, and the Franciscan subduction complex and accretionary wedge assemblages (Ingersoll, 1979; Dickinson, 1979; Ingersoll and Dickinson, 1981). With time, the arc-trench gap widened (Snyder, Dickinson, and Silberman, 1976), creating a broad convergent margin setting with orthogonal subduction along much of the western edge of the North and Central American continents (Nilsen, 1986). By Late Cretaceous from 85 to 56 myBP, a major shift in the relative plate motions occurred, producing oblique northward subduction of the Farallon plate beneath the North American plate (Engebretson, Cox, and Gordon, 1984). This shift disrupted the "classic" pre-Late Cretaceous convergent margin setting and caused detachment and dispersion of terranes from the convergent margin in the south, northward translation of the terranes, and their eventual accretion to California's forearc margin in the early Paleogene (Vedder, Howell, and McLean, 1983; Dickinson, 1983; Nilsen, 1986; Nilsen, in press).

A major sea level drop and regression in the Late Maestrichtian affected much of the forearc margin, resulting in erosional remnants of Cretaceous sediment and a stratigraphic discontinuity underlying the Tertiary strata in most of the offshore basins (Nilsen, 1986; Nilsen, in press). However, in some bathyal and abyssal portions of the forearc, sedimentation was continuous across the Cretaceous-Paleocene boundary (Ingle, 1980).

### Paleogene

Throughout the Paleocene and Lower Eocene, oblique subduction of the Farallon oceanic plate continued to dominate the tectonic setting of western California (Jurdy, 1984; Page and Engebretson, 1984). During this time, continued northward translation of tectonostratigraphic terranes into California's forearc setting produced regional strike-slip faulting and development of borderland basinal topography within the forearc margin (Nilsen, in press). These terranes, namely the Franciscan, Salinian, Sur-Obispo, Transverse Ranges, and Peninsula Ranges contain Late Cretaceous, and in some places Paleocene, rocks of forearc margin affinity. They are allocthonous to California as evidenced by their paleontologic, stratigraphic, and paleomagnetic characteristics (Vedder, Howell, and McLean, 1983; Nilsen, 1986; Nilsen, in press), and were in most cases accreted onto California's forearc margin in the very Late Paleocene and Early Eocene (Nilsen, in press).

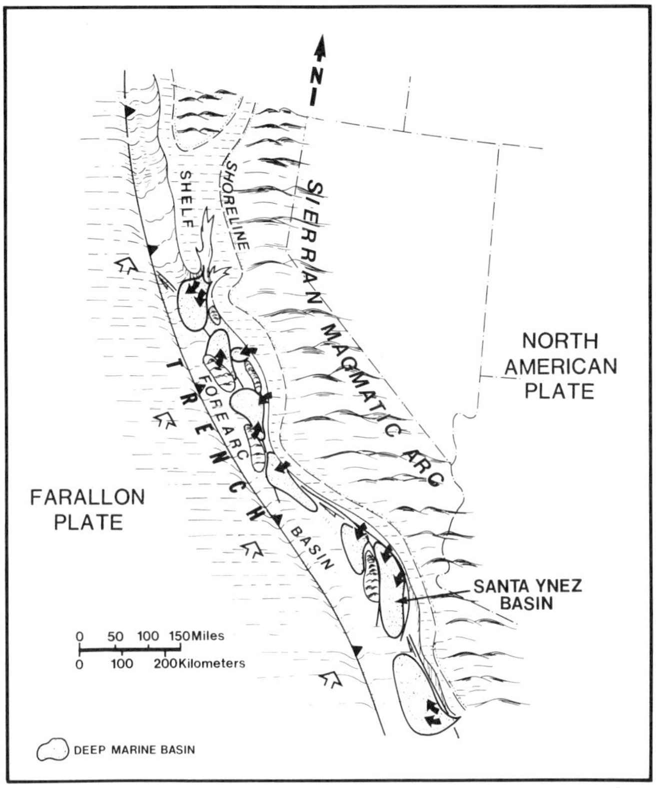


Figure 2. Schematic representation of early Paleogene paleogeography of California shown to indicate formation of borderland topography within the forearc basin during oblique subduction of the Farallon oceanic plate, simplified from Nilsen (in press), Nilsen and McKee (1979), Hornafius (1985), and Nilsen and Clarke (1975). Strike-slip faults within the forearc basin are approximately located (e.g., Suppe, 1970; Nilsen and Clarke, 1975; Hornafius, 1985).

The Salinian terrane was accreted to the California forearc margin in the Early Eocene, as evidenced by the overlap assemblage of northward-directed clastics of the Lower to Middle Eocene Butano Sandstone Formation. The Butano was sourced from within the allocthonous Salinian terrane, and spilled across the suture boundary into the western margin of the autochthonous San Joaquin forearc basin (Nilsen, 1985). The reverse relationship of west-directed sedimentation from the San Joaquin basin overlapping the Salinian terrane in the Lower Eocene through Oligocene is also well documented (Nilsen, 1984a). The Salinian terrane has undergone subsequent northward translation of as much as 320 km since 10 myBP (Nilsen, 1984a).

This episode of early Paleogene transform faulting and accretion of terranes within the forearc margin created several large borderland-like, generally north-south-oriented, Eocene basins within the forearc region, such as the Santa Ynez basin (Figure 2) (Nilsen and Clarke, 1975; Clarke, Howell, and Nilsen, 1975; Hornafius, 1985; Nilsen, in press). Convergent tectonics continued to dominate the margin from the Early Eocene to Oligocene, however, with much less translation of terranes and transform faulting. During this time the Eocene basins were fed with abundant sediment from the Sierran magmatic arc to the east, as well as from uplifted blocks within the forearc area (Figure 2) (Nilsen and Clarke, 1975; Nilsen, in press). This resulted in thick deposits of Eocene deep-water turbidites in the basins nearest the magmatic arcs, including the Santa Ynez basin, and progressively

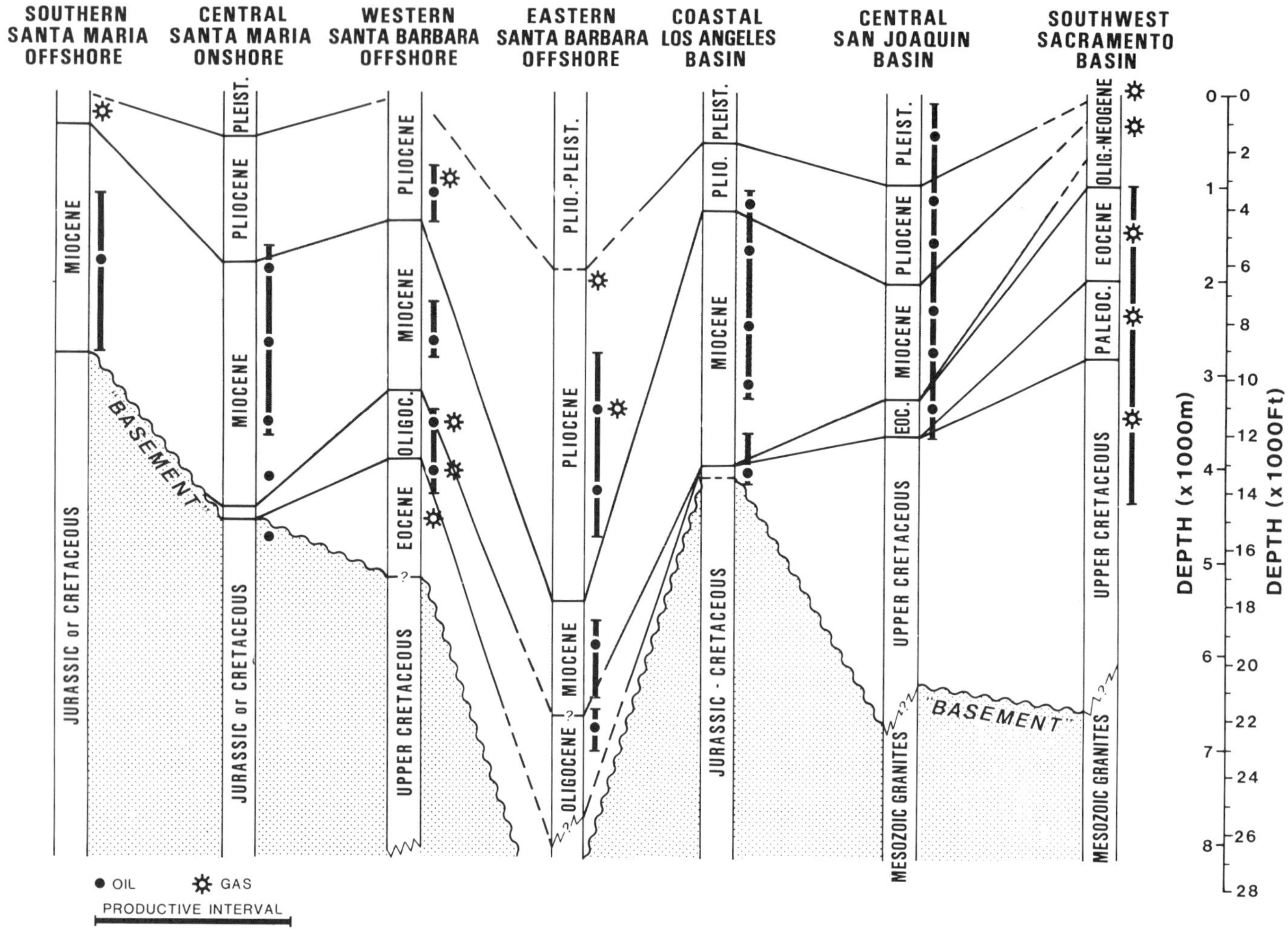


Figure 3. Schematic comparison of relative thicknesses of Tertiary deposits between onshore and offshore basins. See Figure 1 for location of each basin.

thinner Eocene deposits on the outer westernmost margin of the forearc (Figure 2). It is on this outer western margin of the Eocene and Oligocene forearc basin that many of the present-day offshore basins have formed, and they therefore possess only a relatively thin Paleogene section (Figure 3). However, portions of the Neogene Point Arena, Santa Cruz, Santa Barbara, Santa Rosa, and the San Nicolas basins overlie portions of thicker Paleogene basins, which in places contain significant source and reservoir rocks (Hoskins and Griffiths, 1971; Vedder et al., 1974).

The Late Eocene to Early Oligocene marked the second major regression over the forearc, during which the major Paleogene basins filled with progressively shallower facies of marine sediments, and in many cases were capped by extensive shallow-marine or nonmarine deposits (Nilsen, 1984b). It is postulated that this regression was caused by a combination of: (1) eustatic lowering of sea level during worldwide climatic refrigeration (Kennett, 1977; Ingle, Graham, and Dickinson, 1976) and, possibly, (2) regional uplift of the continental margin as the Pacific-Farallon spreading ridge impinged on the continental margin and entered the trench (Nilsen, 1984b; Nilsen, in press). The Oligocene marked the end of the convergent margin tectonic history of western California, and the beginning of transform margin tectonics of the Neogene to present (Blake et al., 1978; Nilsen, 1984b).

## Neogene

The onset of transform margin tectonics in California occurred approximately 29 myBP when the Farallon-Pacific spreading center intersected the trench in southern California (Atwater, 1970; Atwater and Molnar, 1973; Page and Engebretson, 1984). This produced two diverging triple-junctions that migrated north and south with the San Andreas transform system in between (Blake et al., 1978; Dickinson and Snyder, 1979). Throughout the Neogene, the transform system has grown in length as the two triple-junctions spread apart, and the primary zone of right-lateral shear has migrated from the trench to its present position along the San Andreas fault system (Howell et al., 1980; Nilsen, in press).

The transform system has tectonically disrupted the Paleogene and older forearc margin via northward translation, and in places rotations of crustal blocks, along major northwest-trending strike-slip faults. This produced narrow and deep borderland basins which are smaller and have a different orientation than their Paleogene and Late Cretaceous precursors (Silver, 1974; Blake et al., 1978; Nilsen, 1984b; Nilsen, in press). As much as 620 mi (1000 km) of cumulative displacement along the right-lateral strike-slip faults of the transform system has been documented, including 196 mi (315 km) and 71 to 93 mi (115 to 150 km) along the San Andreas and San Gregorio-Hosgri faults, respectively (Figure 1) (Nilsen, in press).

Neogene transform basin development began in the majority of the offshore basins in the Early to Middle Miocene (Blake et al., 1978). The mechanism for basin formation varied throughout the transform system, although extension via transtensional stresses (Stanley, 1986) was the dominant force. These stresses formed either "pull-apart" rift basins between roughly parallel strike-slip faults (Crowell, 1974; Dibblee, 1977; Rodgers, 1980), wrench basins where wrench faults diverge (Howell et al., 1980), or basins adjacent to rotating crustal blocks (Hornafius, 1985; Luyendyk et al., 1985). Most of the offshore basins contain Late Oligocene to Early Miocene alkalic-basaltic volcanics (rift-type) associated with the opening of the transform basins (Hawkins, 1970; Weaver, 1969; Vedder et al., 1974; Snyder, Dickinson, and Silberman, 1976). Similar volcanics have persisted in places within the transform system into the Pleistocene.

The present-day offshore transform basins underwent rapid subsidence in the Early Miocene; in the southern offshore basins the rate of subsidence outpaced sedimentation in the Early and Middle Miocene, averaging 1000 ft (305 m)/my and was as high as 2700 ft (823 m)/my in the coastal Los Angeles area (Ingle, 1980). During the Late Miocene through Pliocene sedimentation exceeded subsidence, with much of the basin-filling deposition occurring after 10 myBP (Blake et al., 1978). Pliocene sedimentation rates were as high as 6000 ft (1829 m)/my during deposition of deep-marine turbidite systems in the Santa Barbara-Ventura basin (Ingle, 1980). During the Middle to Late Miocene structures began to develop within the partially filled basins, because of a shift from predominantly transtensive to transpressive stresses within the transform system (e.g., Sylvester and Smith, 1976). Moderate structural development continued until the Pleistocene, when a significant component of compression within the entire transform produced accelerated structural development and deformation of the basins (e.g., Crouch, Bachman, and Shay, 1984; Hornafius, 1985). This Pleistocene deformation is due to an overall increase in relative eastward movement of the Pacific plate with respect to North America (Dickinson, 1979; Crouch, Bachman, and Shay, 1984; Hornafius, 1985).

## STRATIGRAPHY

The following discussion of stratigraphy is most closely related to the producing offshore basins, although much of it is relevant to the lesser explored northern, central, and southern offshore basins. It reviews the types of sediments and their potential, either as hydrocarbon sources or as reservoirs. A composite stratigraphic section for each of the offshore basins is shown in Figure 4.

### Mesozoic

The Mesozoic rocks of the offshore basins are either forearc basin sedimentary deposits, granitic and/or metamorphic basement rocks, or Jurassic-Cretaceous Franciscan Complex rocks (Figure 4). The Franciscan Complex consist of subduction and accretionary wedge material, including graywacke sandstones and shales, low-grade metasedimentary and metavolcanic rocks, and serpentinites and glaucophane schists (Blake, 1984). The Franciscan is predominantly of Late Jurassic to Lower Cretaceous age throughout central and southern California, and is Mesozoic to at least Miocene age in northern California (Bachman, 1978; McLaughlin et al., 1982) (Figure 4). The plutonic basement rocks of the Salinian terrane have yielded Early Cretaceous emplacement ages and numerous Late Cretaceous cooling ages (Nilsen, in press). The basement rocks of the southern California borderland basins range from island-arc volcanic and ophiolitic rocks, 140 to 160 myBP in age, to glaucophane-bearing schistose rocks of the Franciscan Complex (Howell and Vedder, 1981; Sorensen, 1985).

Most of the offshore basins overlie the Franciscan Complex or granitic basement rocks, and a relatively thin portion of Cretaceous forearc deposits, whereas in most cases the more inboard onshore Tertiary basins overlie thick sections of Cretaceous forearc deposits and continental granitic basement (Figure 3).

In the offshore basins the distribution of the Late Cretaceous forearc deposits is highly variable and in many places unclear; these deposits are in many cases locally discontinuous (Hoskins and Griffiths, 1971; Taylor, 1976) because of nondeposition or Late Mesozoic erosion (Nilsen, 1986; Howell et al., 1980). In the northern and central offshore basins the Pigeon Point and Gualala formations may be locally discontinuous, whereas the Coastal Belt Franciscan likely underlies the entire Eel River basin (Figure 4). The Cretaceous forearc deposits of the Santa Maria offshore basin are also discontinuous; they are absent in wells in the north and present in the south (Figure 4) (Hoskins and Griffiths, 1971; Crain, Mero, and Patterson, 1985). However, in the Santa Barbara offshore, a thick Upper Cretaceous section may exist (Curran, Hall, and Heron, 1971), consisting of deep marine basin plain shales and fan-fringe siltstones (Espada Formation), overlain by submarine fan sandstones (Jalama Formation) (Figure 4) (Dibblee, 1966, Ingle, 1980). The Cretaceous forearc sediments in the offshore basins are not considered to have significant potential as hydrocarbon sources or as reservoirs. The shales tend to be organic-lean and reservoir sands have low porosity and permeability (Crain, Mero, and Patterson, 1985). There are, however, excellent Cretaceous gas reservoirs and potential source rocks in the onshore Sacramento Basin (Figure 3) (Zieglar and Spotts, 1978), and Cretaceous rocks have produced limited quantities of oil in the eastern Ventura basin (Curran, Hall, and Heron, 1971). California's first oil production was in 1865 from fractured Coastal Belt Franciscan within the Mendocino shear zone, onshore near Petrolia in northern California; total production from the area only amounted to a few hundred barrels.

### Paleocene

The Paleocene deposits of the offshore basins, where present, are typically deep marine clastics, locally distributed, which are indicative of deposition in small borderland-type basins that formed within the allocthonous terranes of California's western margin (Nilsen, in press). The distribution of Paleocene sediments is poorly understood because of their relatively limited occurrence in the offshore wells and

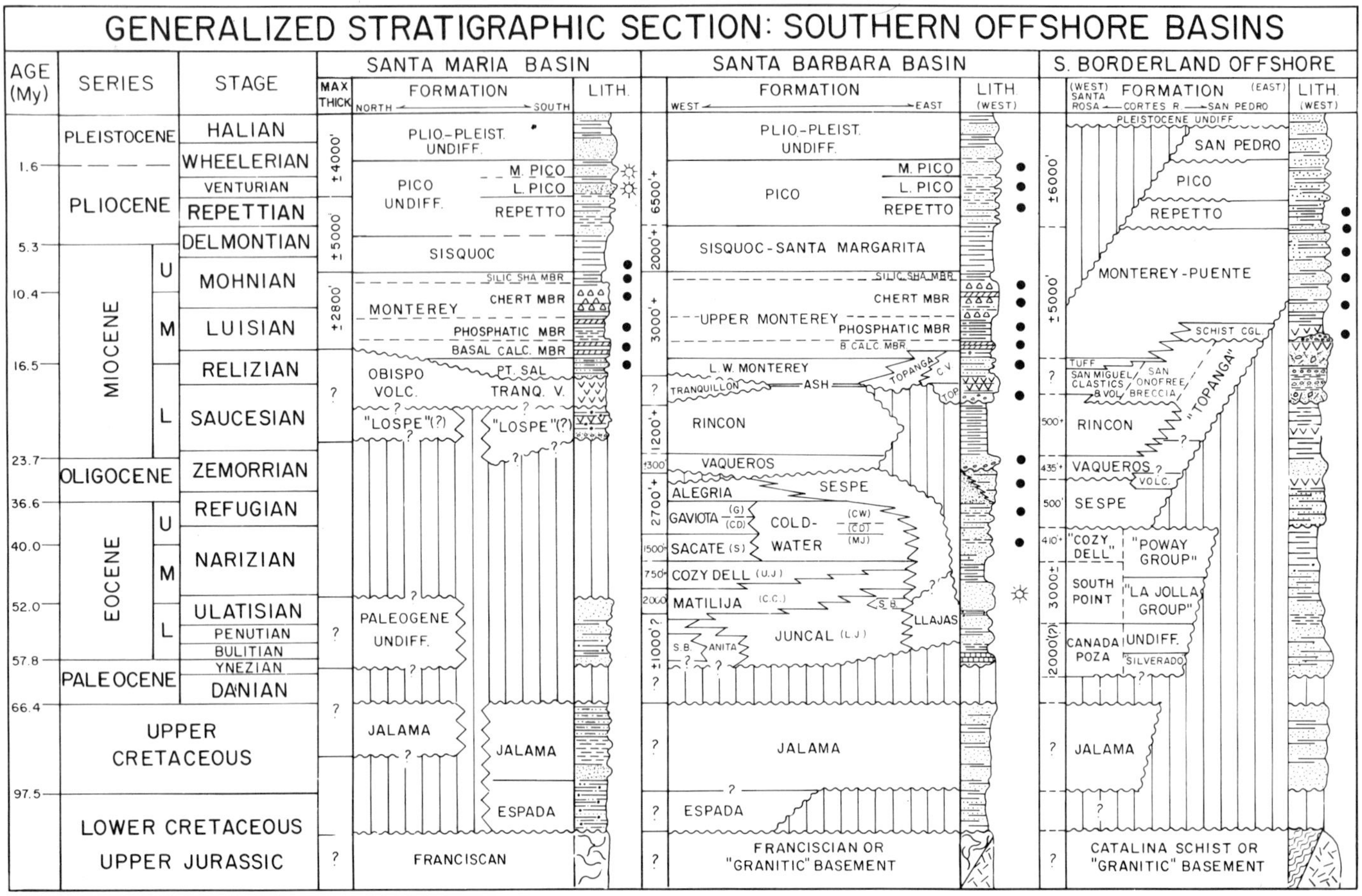

GENERALIZED STRATIGRAPHIC SECTION: SOUTHERN OFFSHORE BASINS
AGE (My)
SERIES
STAGE
SANTA MARIA BASIN
SANTA BARBARA BASIN
S. BORDERLAND OFFSHORE
MAX THICK
FORMATION
NORTH
SOUTH
LITH.
WEST
EAST
LITH. (WEST)
(WEST) SANTA ROSA
CORTES R
SAN PEDRO
(EAST)
PLEISTOCENE
HALIAN
WHEELERIAN
VENTURIAN
PLIOCENE
REPETTIAN
DELMONTIAN
MOHNIAN
LUISIAN
RELIZIAN
SAUCESIAN
MIOCENE
U
M
L
OLIGOCENE
ZEMORRIAN
REFUGIAN
NARIZIAN
ULATISIAN
PENUTIAN
BULITIAN
YNEZIAN
EOCENE
PALEOCENE
DANIAN
UPPER CRETACEOUS
LOWER CRETACEOUS
UPPER JURASSIC
1.6
5.3
10.4
16.5
23.7
36.6
40.0
52.0
57.8
66.4
97.5
±4000'
±5000'
±2800'
PLIO-PLEIST. UNDIFF.
PICO UNDIFF.
M. PICO
L. PICO
REPETTO
SISQUOC
SILIC. SHA. MBR.
CHERT MBR
MONTEREY
PHOSPHATIC MBR
BASAL CALC. MBR
OBISPO VOLC.
PT. SAL
TRANQ. V.
"LOSPE"(?)
PALEOGENE UNDIFF.
JALAMA
ESPADA
FRANCISCAN
6500'+
2000'+
3000'+
1200'+
300
2700'+
1500
750
2000
±1000'?
PLIO-PLEIST. UNDIFF.
PICO
M. PICO
L. PICO
REPETTO
SISQUOC-SANTA MARGARITA
SILIC. SHA. MBR
CHERT MBR
UPPER MONTEREY
PHOSPHATIC MBR
B. CALC. MBR
L. W. MONTEREY
TRANQUILLON
ASH
TOPANGA
C.V.
TOP.
RINCON
VAQUEROS
ALEGRIA
SESPE
GAVIOTA (G) (CD)
COLD-WATER
(CW) (CD) (MJ)
SACATE (S)
COZY DELL (U.J.)
MATILIJA (C.C.)
S.B.
LLAJAS
JUNCAL (L.J.)
S.B.
ANITA
JALAMA
ESPADA
FRANCISCAN OR "GRANITIC" BASEMENT
PLEISTOCENE UNDIFF.
±6000'
±5000'
500+
435+
500'
410+
3000'±
2000'(?)
SAN PEDRO
PICO
REPETTO
MONTEREY-PUENTE
SCHIST CGL.
TUFF
SAN MIGUEL CLASTICS & VOL.
SAN ONOFRE BRECCIA
"TOPANGA"
RINCON
VAQUEROS
VOLC.
SESPE
"COZY DELL"
"POWAY GROUP"
SOUTH POINT
"LA JOLLA GROUP"
CANADA POZA
UNDIFF.
SILVERADO
JALAMA
CATALINA SCHIST OR "GRANITIC" BASEMENT


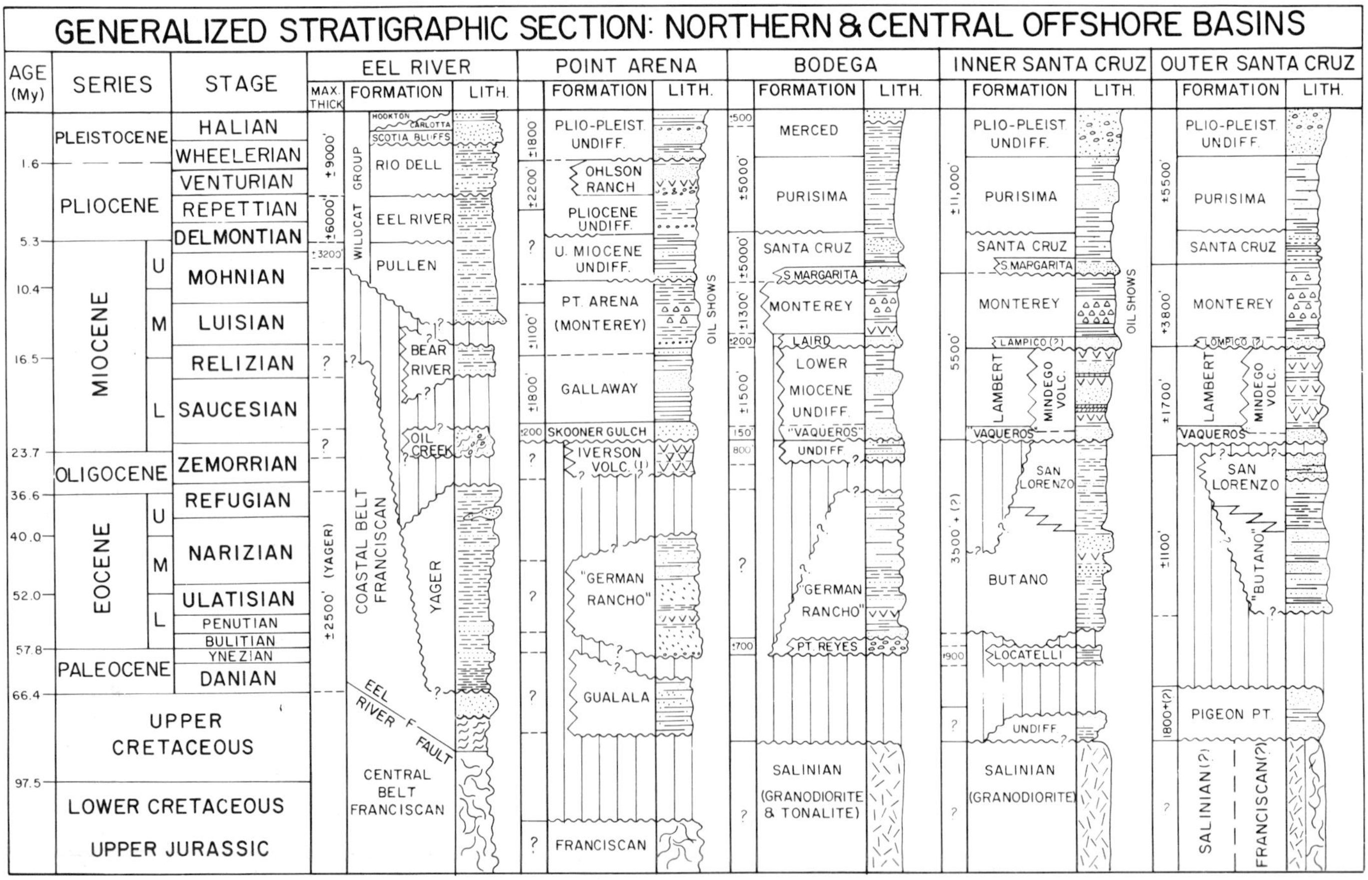

GENERALIZED STRATIGRAPHIC SECTION: NORTHERN & CENTRAL OFFSHORE BASINS
AGE (My)
SERIES
STAGE
EEL RIVER
POINT ARENA
BODEGA
INNER SANTA CRUZ
OUTER SANTA CRUZ
MAX. THICK
FORMATION
LITH.
PLEISTOCENE
HALIAN
WHEELERIAN
VENTURIAN
PLIOCENE
REPETTIAN
DELMONTIAN
MOHNIAN
LUISIAN
RELIZIAN
SAUCESIAN
MIOCENE
U
M
L
OLIGOCENE
ZEMORRIAN
REFUGIAN
NARIZIAN
ULATISIAN
PENUTIAN
BULITIAN
YNEZIAN
EOCENE
PALEOCENE
DANIAN
UPPER CRETACEOUS
LOWER CRETACEOUS
UPPER JURASSIC
1.6
5.3
10.4
16.5
23.7
36.6
40.0
52.0
57.8
66.4
97.5
±9000'
±6000'
3200'
±2500' (YAGER)
HOOKTON
CARLOTTA
SCOTIA BLUFFS
RIO DELL
EEL RIVER
WILDCAT GROUP
PULLEN
BEAR RIVER
OIL CREEK
COASTAL BELT FRANCISCAN
YAGER
EEL RIVER FAULT
CENTRAL BELT FRANCISCAN
±1800
±2200'
±1100'
±1800'
200
PLIO-PLEIST. UNDIFF.
OHLSON RANCH
PLIOCENE UNDIFF.
U. MIOCENE UNDIFF.
PT. ARENA (MONTEREY)
GALLAWAY
SKOONER GULCH
IVERSON VOLC. (J)
"GERMAN RANCHO"
GUALALA
FRANCISCAN
OIL SHOWS
500
±5000'
±5000'
±1300'
200
±1500'
150
800
700
MERCED
PURISIMA
SANTA CRUZ
S. MARGARITA
MONTEREY
LAIRD
LOWER MIOCENE UNDIFF.
"VAQUEROS"
UNDIFF.
"GERMAN RANCHO"
PT. REYES
SALINIAN (GRANODIORITE & TONALITE)
±11,000'
5500'
3500'+(?)
900
PLIO-PLEIST. UNDIFF.
PURISIMA
SANTA CRUZ
S. MARGARITA
MONTEREY
LAMPICO (?)
LAMBERT
MINDEGO VOLC.
VAQUEROS
SAN LORENZO
BUTANO
LOCATELLI
UNDIFF.
SALINIAN (GRANODIORITE)
OIL SHOWS
±5500'
±3800'
±1700'
±1100'
1800+(?)
PLIO-PLEIST. UNDIFF.
PURISIMA
SANTA CRUZ
MONTEREY
LOMPICO (?)
LAMBERT
MINDEGO VOLC.
VAQUEROS
SAN LORENZO
"BUTANO"
PIGEON PT.
SALINIAN (?)
FRANCISCAN (?)

Figure 4. Generalized stratigraphic sections for California's offshore basins. Geochronology from Berggren et al., 1985. Eel River basin: Bachman, Underwood, and Menack, 1984; Field, Clarke, and White, 1980; Hobson, 1971; Hoskins and Griffiths, 1971; Ingle, 1976; Kleinpell, 1938; Ogle, 1953; Underwood, 1983. Point Arena basin: Addicott, 1967; Higgens, 1960; Hobson, 1971; Hoskins and Griffiths, 1971; Miller, 1981; Wentworth, 1968. Iverson Volcanics dated 23.8 myBP (Miller, 1981). Bodega and Santa Cruz basins: Baldwin, 1971; Clark et al., 1984; Clark, 1981; Galloway, 1977; Green and Clark, 1979; Hoskins and Griffiths, 1971. Santa Maria basin: Crain, Mero, and Patterson, 1985; Crawford, 1971; Hoskins and Griffiths, 1971. Santa Barbara basin: Curran, Hall, and Heron, 1971; Dibblee, 1950, 1966; Howell, 1975; Ingle, 1980; Nagle and Parker, 1971; Taylor, 1976; Vedder, Wagner, and Schoellhamer, 1969; Yeats, 1979. For alternate Paleogene nomenclature (shown in parenthesis) see Van de Kamp et al., 1974.
C.V. = Conejo volcanics, Top. = Topanga, S.B. = Sierra Blanca, L.J. = Lower Juncal, C.C. = Camino Ciello, U.J. = Upper Juncal, S = Socate, MJ = Matilija, C.D. = Cozy Dell, G = Gaviota, CW = Coldwater.
Southern Borderland basins and Los Angeles basin offshore: Bishop and Davis, 1984; Howell, 1975; Ingle, 1980; Nardin and Henyey, 1978; Stuart, 1979; Taylor, 1976; Turner, 1970; Vedder et al., 1974; Vedder, Wagner, and Schoellhamer, 1969; Yerkes et al., 1965.
Solid dots = oil productive intervals, star symbols = gas productive intervals.

onshore. Where present, Paleocene strata are relatively thin, and in the onshore are commonly truncated by unconformities (Nilsen and McKee, 1979). Paleocene sedimentation ranges from trench-slope drape mudstones and thin-bedded turbidites (e.g., Yager Formation) in the Eel River Basin (Underwood, 1985) to deep marine shales (e.g., Locatelli and Anita formations) and sands (e.g., Gualala, German Rancho, and Juncal formations) of turbidite fan systems in the Point Arena, Santa Cruz, and Santa Barbara basins (Figure 4) (Miller, 1981; Clark et al., 1984; Ingle, 1980).

Where present, the Paleocene sediments may possess both source and reservoir potential, as in the Eel River Basin (Underwood, 1985). In portions of the Paleogene Santa Ynez basin, shallow marine limestones of the Sierra Blanca Formation (Ingle, 1980) formed on the margin (Nilsen and McKee, 1979). These may have provided algal debris to the coeval basinal shales of the Anita Formation, yielding possibly oil-prone sediments. To date only minor amounts of oil and gas have been produced from Paleocene reservoirs in the offshore or related nearshore basins; however, they are productive onshore in portions of the large Sacramento and San Joaquin basins (Figure 3).

## Eocene

The Lower to Middle Eocene section, where present in the Santa Barbara basin, represents the progressive filling of a deep marine basin, which from bottom to top consists of basinal turbidites, depositional lobe turbidites, and channelized submarine fan deposits, overlain by slope to shallow marine shale and sand deposits. The Upper Eocene to Oligocene section is generally slope to shelf deposits grading upward into coastal and nonmarine depositional environments. In the northern and central offshore basins (Figure 4), the Eocene deposits consist of the Lower and Middle Eocene Butano and German Rancho formations, which are predominantly northerly directed deep-sea turbidite deposits (Wentworth, 1968; Nilsen, 1986) derived from the Salinian terrane (Nilsen, in press). In the Santa Cruz basins these turbidite sequences are overlain by shales and mudstones, with some marine sandstones of the San Lorenzo Formation or equivalent (Figure 4) (Clark et al., 1984; Hoskins and Griffiths, 1971).

The Eocene deposits of the Santa Barbara area consist of deep-water clastic turbidites of the Juncal and Matilija formations, grading upward into shallower water sands of the Gaviota, Alegria, and the nonmarine Sespe formations (Figure 4) (Dibblee, 1966; Ingle, 1980). The Juncal Formation consists of sand-rich prograding outer-fan depositional lobes and shaly interlobe turbidites; it is overlain by outer- and mid-fan turbidite sand and shale sequences of the Middle Eocene Matilija (Ingle, 1980) or Camino Cielo (Van de Kamp et al., 1974) formations. In places the Matilija grades into deltaic deposits (Link and Welton, 1982). Intrachannel and slope shales of the Cozy Dell Formation (Ingle, 1980) or Upper Juncal shale (Van de Kamp et al., 1974) interfinger with thick lobe and channel turbidite sands of the Sacate and Coldwater formations. The turbidite facies are overlain by Upper Eocene slope mudstones and Lower Oligocene littoral shelf sandstones of the Gaviota and Alegria formations, and nonmarine sandstones of the Sespe Formation (Figure 4) (Dibblee, 1966; Ingle, 1980).

The Eocene deposits of the southern borderland offshore are poorly understood because of limited well data. In the western Santa Rosa and San Nicolas basins along the topographic ridge from Santa Rosa Island to the Tanner and Cortes Bank, a thick section of Eocene deep-water turbidite sandstone and shale overlies a thick section of Paleocene and Upper Cretaceous clastic rocks (Figure 4) (Parker, 1971; Wallis, 1981). The Eocene section of the Santa Rosa Ridge and Cortes Bank has been correlated to Poway and La Jolla Group rocks of the San Diego area (Yeats et al., 1974; Bishop and Davis, 1984). Eocene rocks appear to be virtually absent from the inner borderland San Clemente, offshore Los Angeles, and San Diego basins (Howell and Vedder, 1981). This is primarily due to the Neogene age of the opening of these inner basins; either by rifting (Yeats et al., 1974), transcurrent faulting (Howell and Vedder, 1981), or a combination strike-slip faulting and crustal rotation (Kamerling and Luyendyk, 1985). The hydrocarbon potential of the Eocene deposits in some of the offshore basins is excellent because of the interbedding of turbidite reservoir sandstones and basinal and slope shale source rocks.

## Oligocene

During the Oligocene large portions of the western margin of California were uplifted while sea level dropped, producing a regional regression whereby sedimentation in the Paleogene basins progressively shifted to shallow-marine and nonmarine deposits (Nilsen, 1984b). Deep-marine depositional environments persisted in the abyssal portions of the Paleogene basins and on the outer western continental margin, while large portions of the eastern margin, including the Sacramento, San Joaquin, and eastern Santa Ynez basins, underwent significant nonmarine redbed deposition (Nilsen, 1984b). The Eocene to Oligocene regressive section is commonly capped by a nondepositional to erosional unconformity representing the uplift and exposure of the now-filled Paleogene basins on the continental margin. In most offshore basins this unconformity is overlain by the Late Oligocene to Early Miocene transgressive marine deposits which represent the onset of Neogene basin development.

There are fewer Oligocene rocks in the northern and central offshore basins than in the south, likely because of limited deposition or erosion during the Late Oligocene (Hoskins and Griffiths, 1971; Ogle, 1981; Crain, Mero, and Patterson, 1985). Where present, they include deep-marine sandstones and shales of the San Lorenzo or equivalent formations, basaltic volcanic rocks of the Iverson and Mindego volcanic suites, and the basal portions of Vaqueros and equivalent marine sandstones (Figure 4) (Nilsen, 1984a). Oligocene rocks appear to be virtually absent from the offshore Santa Maria basin (Figure 4). Oligocene rocks are more prevalent in the Santa Barbara Channel (Ingle, 1980; Vedder, Wagner, and Schoellhamer, et al., 1969) and in the outer southern California Borderland basins (Vedder et al., 1974; Parker, 1971; Taylor, 1976). In these basins, the Oligocene commonly consists of the upper section of the Sespe (nonmarine) or coeval Alegria (marine) formations, unconformably overlain by the lower portion of the transgressive shallow-marine sandstones of the Vaqueros Formation (Figure 4). Offshore the Vaqueros Formation is in most cases diachronous, containing Oligocene Zemorrian stage foraminifera in the more outboard localities and generally Miocene Saucesian stage foraminifera in the more inboard locations (Ingle, 1980).

The Oligocene sandstones of the marine Alegria and coeval nonmarine Sespe formations, as well as the Vaqueros sandstones, produce oil and gas in the Santa Barbara basin. The hydrocarbon source for these reservoirs, and possibly also some of the basal Miocene sand reservoirs, is thought to be the Eocene Cozy Dell shales and shale-rich facies of the Juncal Formation (Bailey, 1947; Taylor 1976). However, in some cases they may be sourced from Miocene formations.

To date 14 of the 22 developed fields in the offshore Santa Barbara basin have produced more than 100 million bbls of oil from Eocene and Oligocene reservoirs, and six more fields or pools await development. In general, the Eocene and Oligocene reservoirs will continue to be prospective targets in the offshore.

## NEOGENE STRATIGRAPHY

The Neogene stratigraphy of the offshore basins generally follows a three-phase depositional history related directly to the tectonic evolution of the basins. Initially, a marine transgressive depositional episode occurred during the basin opening, with some associated volcanism. This was followed by rapid basinal subsidence and an extended period of restricted deep-marine pelagic deposition in the Middle to Upper Miocene. Late Miocene and Pliocene deposition is characterized by the gradual encroachment of terrigenous fine-grained, progressing to coarse-grained, clastic sediment into the restricted basins via submarine turbidite systems. The following discussion of the Neogene will be most applicable to the Santa Barbara, Santa Maria, and in part the Los Angeles basin areas, because of the abundance of data available. For the northern, central, and southern borderland basins, much of the general characteristics apply; for a more detailed review of the stratigraphy of each offshore basin, the reader is referred to the references listed for each basin in Figure 4.

Neogene reservoirs account for about 97% of all the oil and gas produced from all the basins west of the San Andreas fault, including the prolific Los Angeles Basin (Taylor, 1976), and 98% of the offshore production to date.

### Lower to Middle Miocene

Most of the offshore basins began forming in the Lower to Middle Miocene as evidenced by a basal, transgressive shallow-marine sandstone as much as 300 ft (92 m) thick. It contains Zemorrian to Saucesian foraminifera (Vaqueros Formation or equivalent) and unconformably overlies the pre-Neogene section (Figure 4). The Vaqueros deposition in the Santa Barbara basin was followed by rapid subsidence, with rates as high as 1100 ft/my (Ingle, 1980), and the deposition of slope mudstones and basinal shales of the Rincon and equivalent formations. Volcanics associated with the initial basin opening are Late Oligocene to Lower Miocene intermediate to basaltic shallow intrusive sills, flows, and water-laid volcaniclastics (Figure 4).

Throughout the Late Saucesian to Relezian, deep-marine sediments of the Lower Monterey or equivalent formations consisted of calcareous mudstones, dolostones, and various amounts of terrigenous clastics, including sediment derived from adjacent newly uplifted blocks (e.g., San Onofre Breccia) (Stuart, 1979; Ingle, 1980). By the end of the Relizian continued basin subsidence had left most of the offshore margin with sediment-starved basins 4000 to 8000 ft (1220 to 2440 m) deep. These were protected from terrigenous debris by intervening ridges similar to those found in the modern basins of the southern continental borderland. Furthermore, the eastward transgression of the Miocene shoreline localized clastic sedimentation along the nearshore basins, including the Los Angeles basin, which allowed for the deposition of the thick biogenic and pelagic-rich Upper Monterey Formation in the restricted basins farther offshore (Ingle, 1980).

### Monterey Formation

In the offshore basins the Monterey sediments are predominantly biogenic, indicating the effectiveness of the basin sills; whereas, in the more inboard basins the Monterey section is more clastic rich, as in the onshore Cuyama and Los Angeles basins (Lagoe, 1985; Yerkes et al., 1965). The dominant biogenic detritus is diatomaceous debris and frustules, and benthonic calcareous foraminifera which are characteristics of the low-oxygen conditions created by the worldwide cool glacial climate in the Middle to Late Miocene (Blake, 1981; Keller and Baron, 1981; Ingle, 1981). Thick interbedded deposits of silicic, calcareous, and organic detritus occurred in primarily anaerobic environments. The anaerobic conditions allowed for the formation and preservation of dolomitic and organic-rich phosphatic muds (Pisciotto and Garrison, 1981; Garrison, Kostner, and Zenger, 1984). The diagenetic products of all of these pelagic sediments are interbedded organic-rich phosphatic mudstone, dolostone, porcellanite, chert, siliceous shale, and diatomite, in which the silica has undergone phase transformation from the opal-A to opal-CT or quartz phase, depending on the burial history (Isaacs, 1981; Isaacs and Garrison, 1983; Isaacs, 1984). Most of the productive fractured Monterey reservoirs in the offshore have undergone silica diagenesis into the quartz phase, which has the greatest fracturability. The phosphatic mudstone and other organic-rich layers are the primary Monterey source rock, rich in organic carbon, with TOC values generally from 5 to 20 weight %. The kerogen is oil prone (Type II) and sulfur rich (King and Claypool, 1983; Magoon and Isaacs, 1983). The highly fractured siliceous chert member (Figure 4) is the best reservoir rock in the Monterey, and currently produces 45,000 BOPD and 36,000 MCFPD from 21 wells at the Hondo field in the Santa Barbara basin.

The uppermost siliceous shales of the Monterey Formation grade into the mixed diatomaceous and clay-rich mudstones and shales of the Late Miocene Santa Margarita or equivalent formations. These formations represent the initial dilution of the biogenic cycle of sedimentation with the gradual influx of terrigenous clays, generally devoid of calcareous foraminifera, and eventually coarse clastics into the restricted offshore basins. The Santa Margarita is not noted as a good reservoir or source rock in most cases. The relatively high clay content, and in some cases lack of brittle siliceous material, makes the Santa Margarita less prone to fracturing and thus the cap rock for the underlying fractured Monterey reservoirs.

### Pliocene

The Pliocene represents a major shift in the tectonics and sedimentation of the Neogene basins, as indicated by rapid sedimentation and subsidence, and deposition of thick terrigenous clastics in submarine turbidite systems in most of the offshore basins (Ingle, 1980). During the Late Miocene, uplift of the basin margin areas was concurrent with subsidence in the basins, creating very deep troughs (as much as 20,000 ft deep) adjacent to clastic sources (Dibblee, 1950; Weaver, 1969; Vedder et al., 1974).

Lower Pliocene deposition is characterized by basin plain turbidite deposits containing abyssal and lower bathyal benthonic foraminifera (e.g., Repetto Formation and equivalent; see Figure 4) representing the initial progradation of submarine fan systems into the basins. The Repetto Formation in the Santa Barbara basin commonly consists of basin plain shales and thin-bedded turbidites overlain by laterally continuous sand-rich outer-fan turbidites typical of prograding lobe deposits (for example, see Nilsen, 1980; Mutti, Nilsen, and Lucchi, 1978; Casnedi, 1983; Nilsen, 1985).

The Upper Pliocene in the Santa Barbara basin is characterized by channelized mid-fan or supra-fan thick-bedded turbidites and shaly interchannel deposits. These formed as the submarine fans prograded into the basin, and contain middle to upper bathyal Venturian and Wheelerian Stage benthonic foraminifera. Basin-filling sedimentation rates as high as 6000 ft (1830 m)/my are suggested for the Pico Formation in the eastern Santa Barbara basin (Ingle, 1980). As the basins filled sedimentation rates decreased, and by the Holocene the sedimentation rates in most of the Neogene offshore basins had become far below those characteristic of the Pliocene (Schwalbach and Gorsline, 1985).

## STRUCTURES WITHIN THE OFFSHORE BASINS

Structures relevant to the oil and gas potential of the offshore basins are related to two phases of Neogene transform basin development: (1) an initial phase of rifting and basin opening due to transtension within the transform system, producing extensional faulting and structures in the pre-Neogene section (Blake et al., 1978; Stanley, 1986), and (2) a Middle Miocene to Pleistocene transpressional phase, both syndepositional and postdepositional, which produced large anticlinal structures with a relatively long growth history within the basins. Both phases of deformation in part controlled Neogene sedimentation, produced structural and stratigraphic traps, and provided mechanisms for hydrocarbon migration.

The Early to Middle Miocene basin opening, which was typical of transform basins, was characterized by normal faults in the pre-Neogene section along the basin margins, both subparallel and at high angles to the basin-bounding strike-slip faults (e.g., Crowell, 1974; Sylvester, 1984). This created fault traps within the pre-Neogene section, and stratigraphic traps associated with the onlap of the early Neogene sediments adjacent to the basin-bounding faults. The basin-bounding normal faults in some cases may have been reactivated into reverse faults by the younger Plio-Pleistocene compressional events (e.g., Jackson and Yeats, 1982).

In some cases basin-floor topographic features such as uplifted "basement" blocks formed during the extensional phase, providing intervening sills to sedimentation and/or early formation of anticlinal structures (e.g., Teel and Frost, 1982). In some cases they persisted throughout the late Neogene, as in the basement uplift which cores the giant Wilmington Field of the Los Angeles basin (Yerkes et al., 1965) and the Amberjack High of the southern Santa Maria offshore basin (Crain, Mero, and Patterson, 1985).

The second phase of structural development was compressional and was due to oblique-transpression during dextral shear within the transform system from the Middle Miocene to present (Sylvester and Smith, 1976; Hornafius, 1985; Howell et al., 1980). This resulted in either transform-dominated wrench-faulted en echelon anticlines which formed at low angles to the basin margins (Harding, 1974; Harding and Lowell, 1979; Dibblee, 1977; Wilcox, Harding, and Seely, 1973) and/or more compression-dominated thrust-faulted anticlines and 'conjugate-like' fold systems which formed parallel to the basin margins (Crouch, Bachman, and Shay, 1984; Harding, 1985). The compression-dominated structures are common in basins more distant from the San Andreas transform system, such as the Santa Maria and Santa Barbara basins (Figure 1). In the compression-dominated case, the basin bounding strike-slip faults may also have a significant thrustlike component, as interpreted for the Hosgri fault system (Crouch, Bachman, and Shay, 1984). Yet those basins which lie closer to the San Andreas fault system, such as Los Angeles and Point Arena, possess more typical, strike-slip-dominated, wrench-fault anticlinal structures, and the basins are bounded by high-angle strike-slip faults (e.g., Harding and Lowell, 1979; Harding, 1973; Harding, 1976).

Both the transform and compressional structures form excellent hydrocarbon traps, and in most cases the structures and traps have a long history of development beginning in the Middle to Late Miocene, with maximum development in the Pleistocene. Active periods of structural growth in part correlate with active periods of transform movement (Harding, 1976). Both onlapping and crestward truncation of Middle Miocene and younger sediments onto the growing anticlines, and downflank thickening and changes in the degree of bedding flexure away from the anticlines were common, indicating syndepositional structural growth (Crain, Mero, and Patterson, 1985; Yeats, 1973; Harding, 1973). Although structural development may have begun by Middle to Late Miocene, the peak of the development was within the Pleistocene.

The Pleistocene compressional event was related to the overall increase in relative eastward movement of the Pacific plate against the entire transform margin system (Blake et al., 1978; Dickinson, 1979; Crouch, Bachman, and Shay, 1984; Hornafius, 1985). In most offshore basins this compression was roughly east-west-oriented except for the Santa Barbara basin, where it was north-south. This anomaly is undoubtedly related to the late Neogene rotation of crustal blocks within this portion of the transform system (Luyendyk et al., 1985). The Pleistocene compression developed large, tightly folded anticlinal trends, commonly associated with steep faults with a tectonic transport direction away from the center of the basins. The structures within the Point Arguello field and elsewhere in the Santa Maria offshore are compression-dominated, large, flat-topped anticlines bounded by high-angle reverse faults (Crain, Mero, and Patterson, 1985; Crouch, Bachman, and Shay, 1984), similar to regional conjugate fold systems or positive flower structures (Harding, 1985). Such compression-dominant structures also occur in the Santa Barbara offshore. However, here the compression was in many places more intense, and tighter folds with steep to overturned flanks developed, such as along the giant anticlinal Dos Cuadras-Rincon trend and in the Santa Ynez unit (McCulloh, 1969; Isaacs and Garrison, 1983).

## SOURCE ROCK EVALUATION

### Paleogene

The Paleogene source beds offshore California are less widespread and generally have less total organic carbon (TOC) than those of Neogene age. They do, however, have a longer maturation history and a greater potential to generate significant quantities of light hydrocarbons, such as those produced from Paleogene reservoirs in the Santa Barbara basin. Most of the Paleogene source beds, where present, are predominantly hemipelagic deep-marine shales with adequate oil-prone kerogen for hydrocarbon generation. In most of the offshore basins, the distribution and quality of potential Paleogene source beds is poorly understood.

In the northern and central offshore basins, the Paleogene may in places possess significant hydrocarbon source beds. In the Eel River basin the Yager Formation contains gas-prone, woody, Type III kerogen (Underwood, 1985; Kvenvolden and Field, 1981). In the Santa Cruz basins more than 1000 ft (305 m) of a Paleogene shale and silty shale section was drilled in early test wells, although shows within the interval were poor (Ogle, 1981; Zieglar and Cassell, 1978). The bulk of the known Paleogene section in the Point Arena and Bodega basins may be too sand-rich to contain significant source rocks (Hoskins and Griffiths, 1971).

The primary Paleogene source beds of the Santa Barbara basin are the deep-marine shales within the Anita (Juncal equivalent) and Cozy Dell formations (Figure 4) (Curran, Hall, and Heron, 1971; Ingle, 1980). Analyses from Eocene outcrop and a few offshore wells indicate that in places the Anita Formation and the Cozy Dell Formation have TOC values ranging from 0.5 to 3.6 weight %. Most of the kerogens are oil prone and marine algal related, although various amounts of woody kerogens also are present. The oils recovered from Alegria, Gaviota, and in some cases the Sespe and Vaqueros reservoirs in the Santa Barbara basin (Figure 4) are in most cases chemically distinct from those in the Neogene reservoirs, indicating a Paleogene source. There are no known Paleogene hydrocarbon source rocks in the Los Angeles basin. Marine and nonmarine Paleogene sediments only occur in the southern onshore portion of the Los Angeles basin, and are not petroliferous (Yerkes et al., 1965; Gardett, 1971). In the offshore Santa Rosa basin the available USGS data indicate relatively poor total organic carbon (TOC) for the Paleogene, generally less than 0.4 weight % based on the San Nicolas Island Eocene outcrop (Taylor, 1976). However, significant thicknesses of lower paleogene sediments are present in the western San Nicolas basin and may possess greater source potential (Wallis, 1981).

### Neogene

In general, the Neogene section in all the offshore basins, except the Eel River Basin, includes a Lower Miocene slope and/or deep basinal shale sequence (e.g., Rincon Formation), and Middle to Upper Miocene pelagic and biogenic shales (e.g., Monterey Formation), both of which are excellent hydrocarbon sources (Figure 4). The most significant Neogene source beds are the Monterey Formation (or equivalent), with TOC values from 0.3 to 20.0 weight % and the Rincon Shale (or equivalent), with TOC values from 1.3 to 8.0 weight %; both contain primarily oil-prone, Type II Kerogen (Isaacs and Garrison, 1983). The Monterey kerogen, which is sulfur rich (Magoon and Isaacs, 1983), is capable of generating a low-gravity oil at levels of thermal maturation (Orr, 1984; Heasler and Surdam, 1985) lower than would be considered mature for conventional kerogens (Waples, 1980; King and Claypool, 1983; Kablanow and Surdam, 1983). Monterey oils can contain as much as 7% sulfur by weight and tend to be between 12 °API to 22 °API gravity, although oil gravities up to 36 °API do occur.

## HYDROCARBON MATURATION HISTORY

The thermal maturation history of the Tertiary source rocks varies widely within and between the offshore basins. There are, however, some aspects which are common to most of the offshore basins. The first is that most Paleogene source beds, where present, remained immature because of relatively thin cover until at least Early to Middle Miocene. Secondly, most of the Neogene source beds entered the oil-generative window no earlier than 10 myBP and in most cases closer to 5 myBP (Heasler and Surdam, 1983; Kablanow and Surdam, 1983). Peak oil generation commonly occurred from 2 myBP to present. This period of late Neogene oil and gas generation from multiple source units fortunately coincided with structural development within the basins.

To demonstrate important relationships between thermal maturation and the depositional history of the offshore basins, a geohistory (time-depth) diagram modified from Waples (1980) was constructed for a hypothetical depocenter in the Santa Barbara basin (Figure 5). The stratigraphic model, roughly 24,000 ft (7320 m) thick is a composite of the Tertiary section documented for the western Santa Ynez Range (Dibblee, 1966; Ingle, 1980) and the offshore Santa Ynez unit (Table 1) (Isaacs and Garrison, 1983). The model assumed only minimal erosion during the Oligocene regression. Had more erosion occurred, the pre-Neogene section would have entered the oil-generative window later than depicted at 23 myBP (Figure 5). The model also only depicts 10,000 ft (3048 m) of post-Miocene section; with greater Pliocene deposition, as is common in some depocenters in the offshore basins, the Monterey source beds could enter the oil maturation window several million years earlier than 4 myBP as depicted. For simplicity, a constant geothermal gradient with time of 1.8° F/100 ft + 40° F at the ocean bottom was used to calculate the source rock maturation history. For a detailed evaluation of

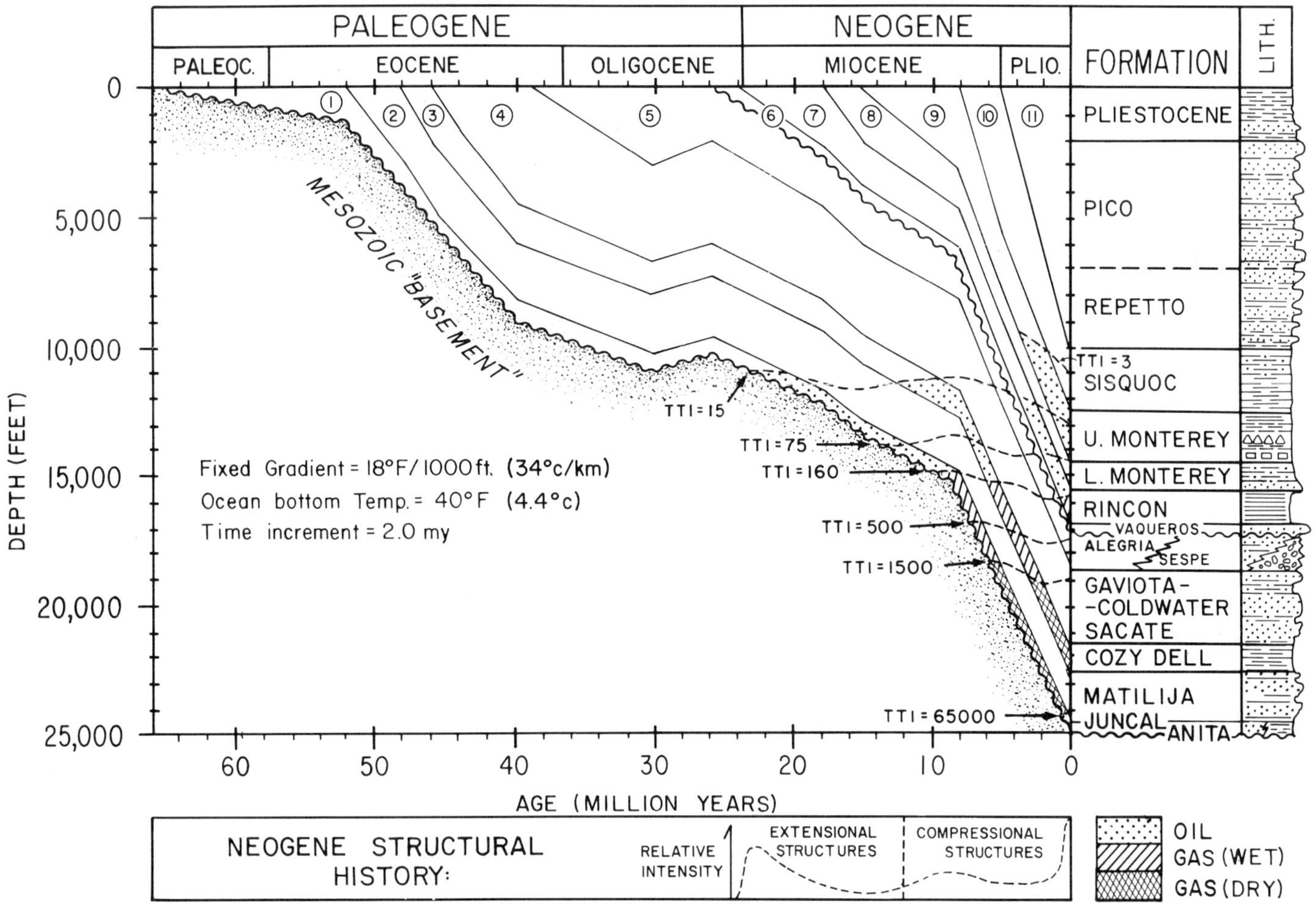


Figure 5. Schematic geohistory diagram for a hypothetical depocenter compiled from stratigraphic sections in the western Transverse Ranges and Santa Barbara Channel (see text for explanation). See Table 1 for stratigraphic model.

the offshore basins, a variable geothermal gradient with time as related to Neogene tectonics should be used as proposed by Heasler and Surdam (1985). The traditional TTI values for oil and gas generation are from Waples (1980), and the low TTI value of 3 for the onset of oil generation from sulfur-rich Monterey kerogens is derived from Orr (1984), Heasler and Surdam (1985), and Waples (1985).

This simplified geohistory diagram (Figure 5) indicates that both the Paleogene and Neogene potential source rocks did not reach the main phase of oil generation (TTI = 75) until the Middle to Late Miocene (8.5 to 5 myBP). Most of the Neogene source rocks are presently in the main phase of oil generation. Secondly, the hydrocarbon generation is either synchronous with or postdates the development of the Neogene structures within the basins. In general, the Monterey source beds enter into the oil-generative window at 10,000 ft (3048 m) of burial assuming a constant geothermal gradient of 18° F/1000 ft (305 m) and a TTI = 3; it would occur as shallow as 7500 ft (2286 m) of burial using a 25° F/1000 ft (46° C/km) gradient (Kablanow and Surdam, 1983). Oil generation within the Monterey Formation may begin at shallower depths, from 6000 to 8000 ft (1830 to 2440 m), when a variable geothermal gradient with time (Heasler and Surdam, 1985) and lower TTI values for oil generation from sulfur-rich type II kerogen are considered (Heasler and Surdam, 1985; Waples, 1985). Geothermal gradients as high as 32° F/1000 ft (55° C/km) are suggested for the early Miocene coinciding with basin extension-related volcanism (Heasler and Surdam, 1985), compared with the present-day gradients in the offshore basins which range from 18° F/1000 ft (31° C/km) to 28° F/1000 ft (51° C/km) (Crain, Mero, and Patterson, 1985; Philippi, 1965; Isaacs and Garrison, 1983).

## EXPLORATION AND PRODUCTION HISTORY

California's offshore exploration began in 1896 at Summerland in the Santa Barbara basin, where possibly the world's first offshore well was drilled from a pier in 15 ft of water (Yerkes, Wagner, and Yenne, 1969). Over the next 60 years, seven more onshore fields were extended into the off-

TABLE 1: STRATIGRAPHIC MODEL FOR "HYPOTHETICAL" DEPOCENTER

| LAYER # (Fig. 5) | THICKNESS (FT) | AGE AT BASE (myBP)[1] | FORMATION(S) | LITHOLOGY |
|---|---|---|---|---|
| 1 | 500 | 65.0 | Anita & Juncal | Shale and Sandstone |
| 2 | 2,000 | 52.0 | Matilija | Sandstone |
| 3 | 1,000 | 48.0 | Cozy Dell | Shale |
| 4 | 3,000 | 46.0 | Sacate & Gaviota and Coldwater | Sandstones and Shales |
| 5 | 1,500[2] | 38.5 | Alegria & Sespe | Sandstone |
| 6 | 300 | 26.0 | Vaqueros | Sandstone |
| 7 | 1,200 | 23.0 | Rincon | Shale |
| 8 | 1,000 | 17.5 | Lw. Monterey | Sandstone and Shale |
| 9 | 2,000 | 15.0 | Up. Monterey | Siliceous Shale, Phosphatic mudstone, dolostones, cherts |
| 10 | 2,500 | 8.0 | Sisquoc | Diatomaceous Shales |
| 11 | 10,000 | 5.0 | Plio-Pleist. (Pico Fm) | Turbidite Sands and Shales |

[1] Derived primarily from Ingle, 1980.

[2] Original thickness = 2100 ft, eroded 600 ft between 30 and 26 myBP.

shore, primarily drilled from onshore sites. To date 13 of the 29 offshore fields are drilled from land or nearshore islands or structures. It was not until 1958, again at Summerland, that California's first true offshore drilling and production platform was placed in 100 ft (30 m) water depth. California's first subsea sea-floor completion was made at the Rincon offshore extension in 1961 (Yerkes, Wagner, and Yenne, 1969), and to date seven other fields in part produce by this method. California's platform count currently stands at 28, and as many as 13 more are planned for the next 5 years (Tables 2, 3).

The first offshore leasing in California occurred in 1921 for state tideland waters within 3 miles of shore. Following this the Rincon, Ellwood, and Capitan fields were extended into the offshore. Leasing of small parcels of state waters adjacent to nearshore production continued until the late 1950s. At that time, large portions of the Santa Barbara area state waters were leased, and 13 offshore fields were discovered shortly thereafter.

Federal leasing for offshore California beyond 3 miles from the coastline began in 1963 (Table 4) for the central and northern offshore basins. Nineteen exploratory wells were subsequently drilled with encouraging hydrocarbon shows but no discoveries. The first discovery in the federal waters off California was in 1968, following the late 1966 drainage leasing adjacent to the Carpenteria field in state waters. Significant leasing of the California OCS began in 1968 with the OCS-P4 sale in the Santa Barbara basin, from which 10 offshore fields were discovered. Currently four of these fields are producing, with the other six awaiting development (Table 4).

By the end of 1968 there were 25 producing offshore fields, of which seven were being produced from platforms. In January 1969 a blowout occurred at the Dos Cuadras offshore field near Santa Barbara. The blowout was controlled at the wellhead, but it charged the shallow oil zone and caused significant seepage of oil from near-surface sands. Multiple seepage rates of up to 500 BOPD were recorded (McCulloh, 1969). After the blowout, all drilling in state waters was suspended until 1974, when drilling from existing structures was allowed. In 1982 new exploratory drilling was again permitted in state waters. Activity in federal waters ceased for only a few months after the blowout, and later in 1969 and the early 1970s eight more fields were discovered in federal waters. At last report (July 1986) there were 51 active state leases and 166 active federal leases, totalling 133,974 state acres and 864,750 federal acres, all in southern California waters.

The most significant exploration effort in recent years occurred just prior to and following the OCS lease sale #53 held in May 1981, during which the Santa Maria offshore was opened to exploration. The sale prompted strong industry response from 27 companies, and 60 tracts were leased for more than $2.0 billion in lease bonuses (2nd largest total bonus in history), including a record bid of $333.6 million for a single tract (Table 4). In the 5 years following this sale, 50 expendable exploratory wells were drilled, resulting in the discovery of nine new fields with potential reserves exceeding 1 billion barrels (Table 3). Oil and gas production from the Santa Maria offshore is expected to begin in late 1986, and in 1987 from the Point Pedernales and Point Arguello fields.

The current cumulative production from the California offshore is 2.3 billion bbls oil and 1.8 tcf gas from 29 fields, of which 19 are still producing. The Division of Oil and Gas estimates reserves from these fields at 803 million bbls of oil and 474 bcf gas (Table 2). The present daily offshore production is averaging 255,000 BOPD and 264,000 MCFGD. The majority of the current offshore oil production is from the nearshore fields of the Los Angeles basin, including the giant Wilmington and Huntington Beach fields. The offshore L.A. basin has contributed 76% of the total cumulative oil production and 59% of the current daily oil production. However, the contribution to the offshore production from the new fields in the Santa Barbara and Santa Maria basins is increasing as each new field goes on production. Federal OCS production began in the Santa Barbara Channel from the Carpenteria field in 1968 and the Dos Cuadras fields in 1969. These fields reached their peak production in 1971 at 85,000 BOPD, and declined to 27,000 BOPD by the early 1980s when the Santa Clara, Hondo, and Hueneme offshore fields came on stream (Table 2). By 1983 the federal OCS production had increased to 82,000 BOPD. With the addition of the Pitas Point field in December 1983 and the gas pipeline to the Hondo field, federal OCS gas production has increased from 27,000 MCFD in 1982 to 142,000 MCFD at last report. Over the next 10 years potential new production of as much as 400,000 BOPD from up to 13 new platforms is possible from the 21 as-yet-undeveloped discoveries in the Santa Maria and Santa Barbara offshore basins (Table 3). The additional reserves from these basins and the potential of the other yet-unexplored areas in the offshore will undoubtedly be California's primary reserves in the future.

Recent Mineral Management Service (MMS) estimates of remaining recoverable reserves from the 24 fields discovered in the federal offshore continental shelf (OCS) are 1.205 billion bbls oil and 2.198 tcf gas. This estimate includes 740 million bbls oil and 1.81 tcf gas from 17 as yet undeveloped field discoveries, and 465 million bbls oil and 388 bcf gas remaining in the seven fields currently producing in the federal OCS. To date these seven fields have produced 340 million bbls oil and 228 bcf gas (Tables 2, 3). Estimates of additional undeveloped reserves for State Tidelands waters are currently 174 million bbls oil and 239 bcf gas from four as-yet-undeveloped field discoveries, and 338 million bbls oil and 87 bcf gas remaining in the State waters' 23 developed fields. Total estimates of remaining offshore reserves from developed and as-yet-undeveloped fields are 1.717 billion bbls oil and 2.524 tcf gas (Tables 2, 3). The timing of development of these additional

TABLE 2: PRODUCTION HISTORY OF EXISTING OFFSHORE FIELDS; SANTA BARBARA CHANNEL

| FIELD (Disc. Date) | DATE OF FIRST PROD. | PRODUCING ZONE | TYPE | OIL GRAVITY (°API) | METHOD OF DEVELOPMENT | # OF WELLS PROD./S.I. or Abd. | STATUS | TOTAL FIELD Avg. Daily Prod. (1/85-9/85) O+COND. (BOPD) | Avg. Daily Prod. GAS (MCFPD) | Cumulative Prod. (Thru 9/85) O+COND. (Mbbls) | Cumulative Prod. GAS (Bcf) | Est. Reserves[1] (9/85) OIL (Mbbls) | Est. Reserves GAS (Bcf) | REMARKS |
|---|---|---|---|---|---|---|---|---|---|---|---|---|---|---|
| Alegria | 3/62 | Vaqueros | DG | -- | sfc | 1/0 | prod. | 19 | 120 | 0.78 | 3.84 | 0.14 | 0.09 | |
| | 3/62 | Alegria | OG | 45 | | | | | | | | | | |
| Caliente (1962) | 3/62 | Vaqueros | DG | -- | sfc | 0/3 | susp. | 0 | 0 | 0 | 32.61 | 0 | 0.19 | |
| | 2/73 | Sespe+Alegria | DG | -- | | | | | | | | | | |
| Capitan (offshore area) | 2/30 | Vaqueros | OG | 16 | dfl | 0/2 | abd. | 0 | 0 | 0.07 | 0.03 | 0 | 0 | Last prod. 1958 |
| | 9/32 | Sespe | OG | 43 | | | | | | | | | | |
| Carpinteria-(state)(1966) | 2/66 | Pico | OG | 19-31 | 2 pl. | 64/17 | prod. | 3,619 | 8,022 | 32.56 | 31.46 | 4.44 | 4.25 | Platforms Hope and Heidi |
| | 2/67 | Subthrust Pico | OG | 33 | | | | | | | | | | |
| Carpinteria-(federal)(6/68) | 1968 | Pico+Repetto | OG | 19-33 | 3 pl. | 67/17 | prod. | 8,059 | 6,850 | 47.23 | 37.81 | 24.70 | 59.09 | Platforms Hogan, Houchin, Henry (P-0166;P-0240) |
| Coal Oil Point (7/61) | 7/61 | Vaqueros+Sespe | OG | 29-30 | sfc | 0/3 | susp. | 0 | 0 | 1.31 | 3.27 | 0 | 0 | Planned for Monterey Devel. |
| Conception (11/59) | 7/61 | Vaqueros | OG | 32-42 | 2 pl.+sfc | 0/48 | susp. | 0 | 0 | 20.93 | 12.33 | 0 | 0 | Platform Herman |
| | 5/61 | Alegria | OG | 32-42 | | | | | | | | | | Platform Harry |
| | 4/61 | Gaviota | OG | 32-42 | | | | | | | | | | removed in 1973 |
| Cuarta (1/59) | 1/61 | Vaqueros | DG | -- | | 0/2 | susp. | 0 | 0 | 0 | 12.00 | 0 | 0 | Tested oil in |
| | 1/61 | Alegria | OG | 28-38 | 1 pl. | | | | | | | | | Matilija and |
| | 3/61 | Gaviota | OG | 32-55 | | 0/9 | abd. | 0 | 0 | 0.61 | 6.77 | 0 | 0 | Anita Fm., but |
| | 11/61 | Sacate+Cozy Dell | OG | 33-36 | | | | | | | | | | not commercial. |
| Dos Cuadras (3/68) | 3/69 | Pico | OG | 21-27 | 4 pl. | 143/13 | prod. | 20,256 | 10,403 | 198.21 | 91.52 | 68.37 | 30.67 | Platforms A,B,C, and Hillhouse (P-0241; P-0240) |
| | 7/72 | Monterey | OG | 34 | | | | | | | | | | |
| Elwood (6/28) | 4/38 | Rincon | OG | 26 | dfl and pier | 9/103 | prod. | 1,330 | 60 | 77.70 | 62.72 | 1.31 | 1.77 | |
| | 11/29 | Vaqueros | OG | 38 | | | | | | | | | | |
| | 5/35 | Upper Sespe | OG | 36 | | | | | | | | | | |
| | 8/34 | Lower Sespe | OG | 42 | | | | | | | | | | |
| | 5/36 | Sespe Gas Zone | DG | -- | | 1/0 | prod. | 0 | 7 | .01 | 4.35 | 0 | .01 | |
| South Elwood (7/65) | 7/69 | Sisquoc | OG | 25-34 | 1 pl. | 28/2 | prod. | 10,733 | 9,121 | 32.95 | 18.85 | 34.07 | 16.72 | Platform Holly |
| | 5/69 | Monterey | | 25-34 | | | | | | | | | | |
| | 1/67 | Rincon | | 32 | | | | | | | | | | |
| | 1/67 | Vaqueros | | 33 | | | | | | | | | | |
| | 1/67 | Sespe | | 33 | | | | | | | | | | |
| Gaviota (11/58) | 7/60 | Vaqueros+Sespe | DG | -- | dfl. | 0/3 | susp. | 0 | 0 | 0 | 69.96 | 0 | n.a. | |
| Hondo (7/69) | 4/81 | Monterey | OG | 14-26 | 1 pl. | 21/7 | prod. | 45,055 | 36,117 | 59.00 | 17.67 | 142.90 | 132.44 | Platform Hondo-production primarily Monterey (OCS-P-0188) |
| | | Vaqueros | | 38 | | | | | | | | | | |
| | | Sespe+Alegria | | 33-52 | | | | | | | | | | |
| | | Alegria+Gaviota | | 29 | | | | | | | | | | |
| Hueneme (7/69) | 1/82 | Sespe | OG | 14-15 | 1 pl. | 6/0 | prod. | 3,249 | 762 | 4.45 | 1.18 | 2.21 | 0.52 | Platform Gina (OCS-P-0202) |
| | | Topanga | OG | | | | | | | | | | | |
| Molino (10/62) | 10/62 | Vaqueros | DG | -- | sfc. | 2/7 | prod. | 334* | 21,612* | 4.49* | 242.69* | 0 | 3.14** | |
| | 10/62 | Sespe | | -- | | | | | | | | | | *Includes Matilija Prod. |
| | 2/83 | Matilija | DG | | sfc. | 3/0 | prod. | -- | -- | -- | -- | -- | -- | **Excludes Matilija |
| Montalvo,West (6/53) | 6/53 | Sespe | OG | 13-32 | dfl. | 2/11 | prod. | 179 | 59 | 5.71 | 3.32 | 0.12 | .01 | |
| Naples (9/60) | 9/60 | Vaqueros | DG | -- | sfc.+ dfl. | 0/5 | abd. | 0 | 0 | 0.22 | 20.82 | 0 | 0 | Abd. 7/68 |
| Pitas Point (1976-Pico) | 12/83 | Pico | DG | -- | 1 pl. | 10/4 | prod. | 126 | 110,256 | 0.05 | 46.21 | 0 | 111.90 | Platform Habitat(P-0234) |
| Point Conception (4/65) | 4/65 | Sacate | OG | 32 | dfl. | 4/2 | prod. | 44 | 95 | 1.11 | 0.60 | .09 | .05 | |
| Rincon(Offshore) (12/27) | 3/28 | Pico + Repetto | OG | 25-30 | 1 isl.+ sfc. | 65/38 | prod. | 813 | 564 | 33.09 | 33.33 | 2.51 | 1.98 | Rincon Island |
| Santa Clara (3/71) | 6/80 | Monterey | OG | 33 | | | | | | | | | | |
| | | Pico | OG | 20-24 | | 45/9 | prod. | 9,487 | 16,227 | 10.97 | 19.55 | 32.70 | 23.00 | Platforms Grace |
| | | Repetto | OG | 20-26 | 2 pl. | | | | | | | | | and Gilda |
| | | Pico gas | DG | -- | | 5/0 | prod. | 0 | 6,227 | .01 | 8.50 | 0 | 13.90 | (P-0217;P-0216) |
| Summerland Offshore (1957) | 11/58 | Vaqueros | OG | 29-60 | 2 pl.+ sfc. | 25/24 | prod. | 1,000 | 1,821 | 26.38 | 95.09 | 2.60 | 10.88 | Platforms Hilda and Hazel |
| Summerland Old offshore (1896) | 1896* | Casitas-Pleist. | OG | 7 | dfl. | 0/0 | abd. | 0 | 0 | 0.31 | n.a. | 0 | 0 | Abd. 1940 |
| | 12/29 | Vaqueros | OG | 16 | | | | | | | | | | |
| | *First offshore oil well in California. | | | | | | | | | | | | | |
| Total for Santa Barbara Channel (22 Fields) | | | | | 20 pl. | 501/329 | | 104,303 | 228,323 | 558.15 | 876.48 | 316.16 | 410.61 | |

All Fields are in state waters except where noted (e.g., OCS-P-0296 or P-0166 are Federal OCS leases).

[1] Reserve estimates from California Division of Oil and Gas.

DG = Dry Gas, OG = Oil and associated gas, sfc. = sea floor completion, dfl. = drilled from land, pl. = platform, abd. = abandoned, susp. = suspended, prod. = producing, S.I. - shut in, isl. = drilling island.

TABLE 2 (continued): PRODUCTION HISTORY OF EXISTING OFFSHORE FIELDS; LOS ANGELES BASIN

| FIELD (Disc. Date) | DATE OF FIRST PROD. | PRODUCING ZONE | TYPE | OIL GRAVITY (°API) | METHOD OF DEVELOPMENT | # OF WELLS PROD./S.I. | STATUS | TOTAL FIELD: Avg. Daily Prod. (1/85-9/85) O+COND. (BOPD) | Avg. Daily Prod. GAS (MCFPD) | Cumulative Prod. (Thru 9/85) O+COND. (Mbbls) | Cumulative Prod. GAS (Bcf) | Est. Reserves[1] (9/85) OIL (Mbbls) | Est. Reserves GAS (Bcf) | REMARKS |
|---|---|---|---|---|---|---|---|---|---|---|---|---|---|---|
| Belmont Offshore (1947) | 7/47<br>11/65<br>9/65 | Ranger<br>Repetto<br>Puenta-Miocene | DG<br><br>OG | --<br>21-26<br>23-29 | dfl.<br><br>2 isl. | 0/0<br><br>29/86 | abd.<br><br>prod. | 0<br><br>2,374 | 0<br><br>244 | 0<br><br>55.61 | 3.65<br><br>34.78 | 0<br><br>12.89 | 0<br><br>3.51 | Islands Monterey and Esther<br>Island Esther S.I. for repair |
| Beta (6/77) (San Pedro Bay) | 4/81 | Puente-U. Miocene | OG | 12-22 | 4 pl. | 42*/6 | prod. | 23,517 | 7,070 | 20.11 | 5.34 | 194.18 | 16.77 | Platforms Ellen Elly (Prod.), Edith, Eureka *79 wells(1/86) (OCS-P-0296; P-0300 P-0301) |
| Huntington Beach Offshore (1930) | 5/30<br>4/64 | Puente<br>Puente-TM | OG | 18-22<br>11-13 | 2 pl. & dfl. | 324/179 | prod. | 21,245 | 4,469 | 527.08 | 301.23 | 55.94 | 7.02 | Platforms Emmy and Eva |
| Newport, West Offshore Area (12/53) | 12/53 | Puente+Mohnian | OG | 19 | dfl. | 16/0 | prod. | 230 | 100 | 4.62 | 1.60 | 1.15 | 0.29 | |
| Torrance Offshore Area (4/56) | 5/56<br>4/56 | Puente-Main<br>Puente-Del Amo | OG | 14-28<br>26-30 | dfl. | 43/5 | prod. | 520 | 1,597 | 7.73 | 10.70 | .70 | 1.80 | |
| Venice Beach (1966) | 8/66 | Puente-Schist sd. | OG | 22 | dfl. | 1/4 | prod. | 102 | 115 | 2.64 | 1.56 | .11 | .17 | |
| Wilmington Offshore Area (1939) | 3/43<br>5/39<br>5/39<br>6/45 | Repetto-Tar<br>Repetto-Puente<br>Puente<br>Schist | OG | 12-15<br>12-25<br>25-32<br>28-32 | 4 isl. | 1,099/408 | prod. | 102,930 | 21,392 | 1,122.47 | 495.52 | 222.29 | 34.77 | |
| Total for Los Angeles Basin Offshore (7 Fields) | | | | | 6 pl. | 1,554/688 | | 150,918 | 34,987 | 1,740.26 | 854.38 | 487.26 | 64.33 | |

[1]Reserve estimates from California Division of Oil and Gas.

TABLE 3: UNDEVELOPED OFFSHORE FIELD OR POOL DISCOVERIES; SANTA BARBARA CHANNEL

| FIELD | DISCOVERY DATE | PRODUCING ZONE | TYPE | OIL GRAVITY (°API) | LEASES | DEVELOPMENT PLAN: PLATFORM (#SLOTS) | DEVELOPMENT PLAN: WATER DEPTH (ft) | PEAK PRODUCTION ANTICIPATED*: OIL (BOPD) | PEAK PRODUCTION ANTICIPATED*: GAS (MCFD) | REMARKS |
|---|---|---|---|---|---|---|---|---|---|---|
| Coal Oil Point | 1982 | Monterey | OG | 29-30 | State 308, 309 | Unnamed<br>Two platforms | 200' | 70,000 | ---- | Est. over 100 MMbbls oil reserves. |
| Cojo (Cojo Bay area) | 1965 | Vaqueros<br>Gaviota | DG(?)<br>OG | -- | State 2879 | Hayley (48) | 285' | 15,000 | 7,000 | Disc. drilled from onshore. |
| Hondo "B" (west half) | 7/69 | Monterey<br>Vaqueros<br>Sespe-Alegria<br>Alegria-Gaviota | OG<br>OG<br>OG<br>OG | 14-26<br>38<br>33-52<br>29 | P-0190 | Harmony (60) | 1,200' | 43,000 | 44,000 | Santa Ynez Unit (1) |
| Hueneme (southwest ext.) | 8/69 | Monterey<br>Topanga | OG<br>OG | *<br>15 | P-0203 | Plans not released | | ---- | ---- | Est. 31.0 MMbbls oil reserves.<br>*Tested gas cap in Monterey |
| Government Point | 8/68 | Vaqueros<br>Alegria-Gaviota<br>Matilija | DG<br>OG<br>OG | --<br>40-46<br>35-49 | P-0197 | Plans not released | | ---- | ---- | Santa Ynez Unit (1) |
| Embarcadero | 3/85 | Monterey | OG | 20-30 | State 208 | Haven (?) | 230' | n.a. | n.a. | West. extension of S. Elwood Field |
| Pescado | 5/70 | Monterey<br>Vaqueros<br>Alegria-Gaviota<br>Matilija | OG<br>OG<br>OG<br>OG | 12-18<br>38<br>33-37<br>50 | P-0182,<br>P-0183 | Heritage (60) | 1,075' | 40,000 | 60,000 | Santa Ynez Unit (1) |
| Sacate | 8/70 | Monterey<br>Vaqueros<br>Alegria-Gaviota<br>Matilija | OG<br>OG<br>OG<br>OG | 16-18<br>33-44<br>32-34<br>36-39 | P-0193<br>P-0194 | Heather (28) | 620' | 15,000 | 35,000 | Santa Ynez Unit (1) |
| Sockeye | 11/70 | Monterey<br>Topanga<br>Sespe | OG<br>OG<br>OG | 16<br>18<br>29 | P-0205 | Gail (36) | 739' | 12,500 | n.a. | Discovered in 1970, confirmed in 1983.<br>Est. reserves vary from 50 to 100 MMbbls oil. |

TABLE 3 (continued): UNDEVELOPED OFFSHORE FIELD OR POOL DISCOVERIES; SANTA BARBARA CHANNEL

| FIELD | DISCOVERY DATE | PRODUCING ZONE | TYPE | OIL GRAVITY (°API) | LEASES | DEVELOPMENT PLAN: PLATFORM (#SLOTS) | DEVELOPMENT PLAN: WATER DEPTH (ft) | PEAK PRODUCTION ANTICIPATED*: OIL (BOPD) | PEAK PRODUCTION ANTICIPATED*: GAS (MCFD) | | REMARKS |
|---|---|---|---|---|---|---|---|---|---|---|---|
| Unnamed (Sword Unit) | 7/78 | Monterey | OG | 12 | P-0232 | Plans not released | | --- | --- | | "Heavy Oil" in deep water (greater than 1,200 ft) |
| Molino | 1983<br>10/83 | Monterey<br>Matilija* | OG<br>DG | (?) | State 2920 and 2933 | Hercules (64) | 237' | 30,000 | 100,000 | | *Est. ± 100 Bcf reserves *Presently producing from 3 sea floor completions |
| Unnamed (Santa Rosa Area) | 9/85 | Monterey | OG | 11-30 | P-0467 | Plans not released | | ---- | ---- | | "Diablo Prospect" |
| Bonito | 3/83 | Monterey | OG | 16-18 | P-0446 | Plans not released | | ---- | ---- | ---- | |
| Electra | 12/83 | Monterey | OG | 14-16 | P-0449 | Plans not released | | ---- | ---- | ---- | |
| Jalama | 5/83 | Monterey | OG | 12-18 | P-0452 | Plans not released | | ---- | ---- | ---- | |
| Point Arguello | 4/81 | Monterey<br>Sisquoc<br>Pico | OG<br>OG<br>DG | 18-21<br><br>11-23<br><br>16-34 | P-0316<br><br>P-0315<br><br>P-0450 | Hermosa (48)<br>Harvest (50)<br>Hildalgo (56) | 602'<br>680'<br>430' | 1986-1987<br>1986-1987<br>1986-1987 | 27,000<br>46,000<br>20,000 | 25,000<br>42,000<br>10,000 | Hermosa installed 10/85<br>Harvest installed 6/85<br>Hidalgo installed late 1986 Total field reserves 300 to 500 MMbbls oil. |
| Point Pedernales | 1/83 | Monterey | OG | 15-16<br><br>15-17 | P-0440<br><br>P-0441 | Irene (72)<br>Independence (60) | 242'<br>277' | April 1986<br>n.a. | 20,000<br>20,000 | 13,300<br>45,000 | Irene installed 8/85 Est. 43 MMbbls + 51 Bcf reserves. Total field res. may be as high as 300 MMbbls oil. |
| Rocky Point | 2/83 | Monterey<br>Sisquoc | OG<br>OG | 25-36 | P-0451<br>P-0452 | Hacienda (48) | 300' | n.a. | n.a. | n.a. | |
| San Miguel | 1/83 | Monterey | OG | 8-12 | P-0409 | Julius (70) | 478' | 1986-1987 | 40,000 | (re-inj) | Total field reserves may be as high as 300 MMbbls oil. |
| Point Sal Unit Area | 9/84 | Monterey | OG | 10-16 | P-0415<br>P-0416 | Plans not released | | ---- | ---- | ---- | |
| Unnamed | 6/82 | Monterey | OG | 11-16 | P-0443 | Plans not released | | ---- | ---- | ---- | |

(1) Undeveloped Santa Ynez Unit reserves estimated at 200 MM bbls oil and 550 Bcf gas based on initial total reserves of 400 MMbbls oil and 700 Bcf gas for the entire Santa Ynez Unit.

* Peak anticipated production rates are preliminary estimates only and subject to change.

TABLE 4: HISTORY OF FEDERAL OCS LEASING OFF CALIFORNIA (As of July, 1986)*

| SALE # (DATE) | AREA OF LEASE SALE | # TRACTS OFFERED (Acres) | # TRACTS LEASED (Acres) | TOTAL BONUS (Million $) | CURRENT STATUS (7/86): # LEASES (Active) | CURRENT STATUS (7/86): # FIELDS DISC'D (# Fields Prod.) |
|---|---|---|---|---|---|---|
| P1 (5/14/63) | Eel River Basin to Morro Bay | 129 (669,777) | 57 (312,945) | $ 12.81 | 0 | 0 |
| P3 (12/15/66) | Santa Barbara Channel | 1 (1,995) | 1 (1,995) | $ 21.19 | 1 | 1 Extension (1) |
| P4 (2/6/68) | Santa Barbara Channel | 110 (540,609) | 71 (363,181) | $ 602.72 | 34 | 10 (4) |
| #35 (12/11/75) | Southern Borderland | 231 (1,257,593) | 56 (310,049) | $ 417.31 | 4 | 1 (1) |
| #48 (6/29/79) | Santa Barbara to Mexico | 148 (792,845) | 54 (288,260) | $ 572.83 | 15 | 2 (1) |
| #53 (5/28/81) | Santa Maria Basin | 111 (603,613) | 60 (320,579) | $ 2,047.80 | 43 | 9 (0) |
| #68 6/11/82) | Santa Barbara to Mexico | 140 (716,866) | 29 (147,066) | $ 117.80 | 29 | 1 (0) |
| #RS-2 (8/5/82) | Santa Maria Basin | 27 (153,716) | 10 (56,932) | $ 4.37 | 9 | 0 |
| #73 (11/30/83) | Santa Maria Basin | 137 (768,341) | 8 (43,801) | $ 16.02 | 8 | 0 |
| #80 (10/17/84) | Santa Barbara to Mexico | 657 (3,147,352) | 23 (114,367) | $ 62.12 | 23 | 0 |
| TOTAL | | | 369 | $ 3,874.97 | 166 | 24 (7) |

*Minerals Management Service Pacific OCS Region - "STATUS OF LEASES" Report.

reserves will rely heavily upon the market price and the federal, state and local permitting process. The MMS estimates for undiscovered recoverable resources remaining in all of California's offshore basins are as large as 3.0 billion bbls of oil and oil equivalent gas (Cooke, 1985).

## ACKNOWLEDGMENTS

We thank D.L. Zieglar, J.R. Carson, E.H. Phillips, W.J. Classen, M.H. Pytte, P.F. Bertucci, and special thanks to T.H. Nilsen for their review of this paper, and B.T. Dao for drafting.

## REFERENCES CITED

Addicott, W.O., 1967, Age of the Skooner Gulch Formation, Mendocino County, California: U.S. Geological Survey Bulletin 1254-C, p. C1-C10.

Atwater, T., 1970, Implications of plate tectonics for the Cenozoic tectonic evolution of western North America: GSA Bulletin, v. 81, p. 3513-3536.

Atwater, T., and P. Molnar, 1973, Relative motion of the Pacific and North American plates deduced from sea-floor spreading in the Atlantic, Indian and South Pacific Oceans, *in* R.L. Kovach, and A. Nur, eds., Proceedings of the Conference on Tectonic Problems of the San Andreas Fault System: Stanford University Publications Geological Science, v. 13, p. 136-148.

Bachman, S.B., 1978, A Cretaceous and early Tertiary subduction complex, Mendocino coast, northern California, *in* D.G. Howell, and K.A. McDougall, eds., Mesozoic paleogeography of the western United States: SEPM, Pacific Section, Pacific Coast Paleogeography Symposium 2, Los Angeles, California, p. 419-430.

Bachman, S.B., M.B. Underwood, and J.S. Menack, 1984, Cenozoic evolution of coastal northern California, *in* J.K. Crouch, and S.B. Bachman, eds., Tectonics and sedimentation along the California Margin: SEPM, Pacific Section, v. 38, p. 55-66.

Bailey, T.L., 1947, Origin and migration of oil into Sespe redbeds, California: AAPG Bulletin, v. 31, p. 1913-1935.

Baldwin, T.A., 1971, Petroleum potential of California Central Coast Ranges, *in* Future petroleum provinces of the United States—Their geology and potential: AAPG Memoir 15, p. 309-315.

Berggren, W.A., D.V. Kent, J.J. Flynn, and J.A. Van Couvering, 1985, Cenozoic geochronology: GSA Bulletin, v. 96, p. 1407-1418.

Bishop, C.C., and J.F. Davis, 1984, Correlation of stratigraphic units of North American (COSUNA) project; southern California province correlation chart: Tulsa, AAPG.

Blake, G.H., 1981, Biostratigraphic relationship of Neogene benthic foraminifera from the southern California outer continental borderland to the Monterey Formation, *in* R.E. Garrison, and R.G. Douglas, eds., The Monterey Formation and related siliceous rocks of California: SEPM, Pacific Section, p. 1-14.

Blake, M.C., Jr. (ed.), 1984, Franciscan geology of northern California: SEPM, Pacific Section, Special Publication 43, Los Angeles, California, 254 p.

Blake, M.C., Jr., and D.L. Jones, 1981, The Franciscan assemblage and related rocks in northern California: A reinterpretation, *in* W.G. Ernst, ed., The geotectonic development of California (Rubey Volume 1): Englewood Cliffs, New Jersey, Prentice-Hall, Inc., p. 307-328.

Blake, M.C., Jr., R.H. Campbell, T.W. Dibblee, Jr., D.G. Howell, T.H. Nilsen, W.R. Normark, J.G. Vedder, and E.A. Silver, 1978, Neogene basin formation in relation to plate-tectonic evolution of San Andreas fault system, California: AAPG Bulletin, v. 62, p. 344-372.

Casnedi, R., 1983, Hydrocarbon-bearing submarine fan system of Cellino Formation, central Italy: AAPG Bulletin, v. 67, p. 359-370.

Clark, J.C., 1981, Stratigraphy, paleontology, and geology of the central Santa Cruz Mountains, California Coast Ranges: U.S. Geological Survey Professional Paper 1168, 51 p.

Clark, J.C., E.E. Brabb, H.G. Greene, and D.C. Ross, 1984, Geology of Point Reyes Peninsula and implications for San Gregorio fault history, *in* J.K. Crouch, and S.B. Bachman, eds., Tectonics and sedimentation along the California margin: Pacific Section, SEPM, v. 38, p. 67-86.

Clarke, S.H., Jr., D.G. Howell, and T.H. Nilsen, 1975, Paleogene geography of California, *in* D.W. Weaver, G.R. Hornaday, and A. Tipton, eds., Paleogene symposium and selected technical papers: Pacific Section AAPG-SEPM-SEG Annual Meeting, Long Beach, California p. 121-145.

Cooke, L.W., 1985, Estimates of undiscovered economically recoverable oil and gas resources for the outer continental shelf as of July 1984: Minerals Management Service OCS Report MMS 85-0012, 45 p.

Crain, W.E., W.E. Mero, and D. Patterson, 1985, Geology of the Point Arguello discovery: AAPG Bulletin, v. 69, p. 537-545.

Crawford, F.D., 1971, Petroleum potential of Santa Maria province, California, *in* Future petroleum provinces of the United States—Their geology and potential: AAPG Memoir 15, p. 316-328.

Crouch, J.K., S.B. Bachman, and J.T. Shay, 1984, Post-Miocene compresssional tectonics along the central California margin, *in* J.K. Crouch, and S.B. Bachman, eds., Tectonics and sedimentation along the California margin: SEPM, Pacific Section, v. 38, p. 37-54.

Crowell, J.C., 1974, Origin of late Cenozoic basins in southern California, *in* Tectonics and sedimentation: SEPM Special Publication 22, p. 190-204.

Curran, J.F., K.B. Hall, and R.F. Heron, 1971, Geology, oil fields, and future petroleum potential of Santa Barbara Channel area, California, *in* Future petroleum provinces of the United States—Their geology and potential: AAPG Memoir 15, p. 192-210.

Dibblee, T.W., Jr., 1950, Geology of southwestern Santa Barbara County, California: California Division of Mines and Geology Bulletin 150, p. 1-95.

Dibblee, T.W., Jr., 1966, Geology of the central Santa Ynez Mountains, Santa Barbara County, California: California Division of Mines and Geology Bulletin 186, p. 1-99.

Dibblee, T.W., Jr., 1977, Strike-slip tectonics of the San Andreas fault and its role in Cenozoic basin evolvement, *in* A.G. Sylvester, ed., Wrench fault tectonics: AAPG Reprint Series 28, p. 159-172.

Dickinson, W.R., 1976, Sedimentary basins developed during evolution of Mesozoic-Cenozoic arc-trench system in western North America: Canadian Journal of Earth Science, v. 13, p. 1268-1287.

Dickinson, W.R., 1979, Cenozoic plate tectonic setting of the Cordilleran region in the United States, *in* J.M. Armentrout, M.R. Cole, and H. TerBest, Jr., eds., Cenozoic paleogeography of the western United States: SEPM, Pacific Section, p. 1-13.

Dickinson, W.R., 1983, Cretaceous sinistral strike slip along Nacimiento fault in coastal California: AAPG Bulletin, v. 67, p. 624-645.

Dickinson, W.R., and W.S. Snyder, 1979, Geometry of triple junctions related to San Andreas transform: Journal of Geophysical Research, v. 84, p. 561-572.

Engebretson, D.C., A. Cox, and R.G. Gordon, 1984, Relative motions between oceanic plates of the Pacific Basin: Journal of Geophysical Research, v. 89, p. 10,291-10,310.

Field, M.E., S.H. Clarke, Jr., and M.E. White, 1980, Geology and geologic hazards of offshore Eel River basin, northern California continental margin: U.S. Geological Survey Open-File Report 80-1080, 80 p.

Galloway, A.J., 1977, Geology of the Point Reyes peninsula, Marin County, California; California Division of Mines and Geology Bulletin 202, 72 p.

Gardett, P.H., 1971, Petroleum potential of Los Angeles basin, California, *in* Future petroleum provinces of the United States—Their geology and potential: AAPG Memoir 15, p. 298-308.

Garrison, R.E., M. Kostner, and D.H. Zenger (eds.), 1984, Dolomites of the Monterey Formation and other organic-rich units: SEPM, Pacific Section, Publication 41, 215 p.

Green, H.G., and J.C. Clark, 1979, Neogene paleogeography of the Monterey Bay area, California, *in* J.M. Armentrout, M.R. Cole, and H. TerBest, Jr., eds., Cenozoic paleogeography of the western United States: SEPM, Pacific Section, p. 277-296.

Harding, T.P., 1973, Newport-Inglewood trend, California—An example of wrenching style of deformation: AAPG Bulletin, v. 57, p. 97-116.

Harding, T.P., 1974, Petroleum traps associated with wrench faults: AAPG Bulletin, v. 58, p. 1290-1304.

Harding, T.P., 1976, Tectonic significance and hydrocarbon trapping consequences of sequential folding synchronous with San Andreas faulting, San Joaquin Valley, California: AAPG Bulletin, v. 60, p. 356-378.

Harding, T.P., 1985, Seismic characteristics and identification of negative flower structures, positive flower structures, and positive structural inversion: AAPG Bulletin, v. 69, p. 582-600.

Harding, T.P., and J.D. Lowell, 1979, Structural styles, their plate-tectonic habitats, and hydrocarbon traps in petroleum provinces: AAPG Bulletin, v. 63, p. 1016-1058.

Hawkins, J.W., 1970, Petrology and possible tectonic significance of late Cenozoic volcanic rocks, southern California and Baja California: GSA Bulletin, v. 81, p. 3323-3338.

Heasler, H.P., and R.C. Surdam, 1983, A thermally subsiding basin model for the maturation of hydrocarbons in the Pismo basin, California, *in* C.M. Isaacs, and R.E. Garrison, eds., Petroleum generation and occurrence in the Miocene Monterey Formation, California: SEPM, Pacific Section, p. 69-74.

Heasler, H.P., and R.C. Surdam, 1985, Thermal evolution of coastal California with application to hydrocarbon maturation: AAPG Bulletin, v. 69, p. 1386-1400.

Higgens, C.G., 1960, Ohlson Ranch Formation, Pliocene, northwestern Sonoma County, California: University of California Publications in Geological Sciences, v. 36, p. 199-232.

Hobson, H.D., 1971, Petroleum potential of northern Coast Ranges, California, *in* Future petroleum provinces of the United States—Their geology and potential: AAPG Memoir 15, p. 339-353.

Hornafius, J.S.., 1985, Neogene tectonic rotation of the Santa Ynez range, western Transverse Ranges, California, suggested by paleomagnetic investigation of the Monterey Formation: Journal of Geophysical Research, v. 90, p. 12,503-12,522.

Hoskins, E.G., and J.R. Griffiths, 1971, Hydrocarbon potential of northern and central California offshore, *in* Future petroleum provinces of the United States—Their geology and potential: AAPG Memoir 15, p . 212-228.

Howell, D.G., 1975, Middle Eocene paleogeography of southern California, *in* D.W. Weaver, G.R. Hornaday, and A. Tipton, eds., Paleogene symposium and selected technical papers: Pacific Section AAPG-SEPM-SEG Annual Meeting, Long Beach, California, p. 272-293.

Howell, D.G., and J.G. Vedder, 1981, Structural implications of stratigraphic discontinuities across the Southern California Borderland, *in* W.G. Ernst, ed., The geotectonic development of California (Rubey Volume I): Englewood, Cliffs, New Jersey, Prentice-Hall, Inc., p. 535-558.

Howell, D.G., J.K. Crouch, H.G. Greene, D.S. McCulloch, and J.G. Vedder, 1980, Basin development along the late Mesozoic and Cainozoic California margin: A plate tectonic margin of subduction, oblique subduction and transform tectonics, *in* P.F. Ballance, and H.G. Reading, eds., Sedimentation in oblique-slip mobile zones: International Association of Sedimentologists, Special Publication, v. 4, p. 43-62.

Ingersoll, R.V., 1979, Evolution of the Late Cretaceous fore-arc basin, northern and central California: GSA Bulletin, v. 90, p. 813-826.

Ingersoll, R.V., and W.R. Dickinson, 1981, Great Valley Group (Sequence) Sacramento Valley, California, *in* V. Frizzel, ed., Upper Mesozoic Franciscan rocks and Great Valley Sequence, central Coast Ranges, California: SEPM, Pacific Section, Field Trips Guide 1 and 4, p. 1-33.

Ingle, J.C., Jr., 1976, Late Neogene paleobathymetry and paleoenvironments of the Humbolt basin, northern California, *in* The Neogene Symposium: SEPM, Pacific Section, Special Publication, p. 53-60.

Ingle, J.C., Jr., 1980, Cenozoic paleobathymetry and depositional history of selected sequences within the southern California continental borderland: Cushman Foundation Special Publication 19, p. 163-195.

Ingle, J.C., Jr., 1981, Origin of Neogene diatomites around the north Pacific rim, *in* R.E. Garrison and R.G. Douglas, eds., The Monterey Formation and related siliceous rocks of California: SEPM, Pacific Section, p. 159-179.

Ingle, J.C., Jr., S.A. Graham, and W.R. Dickinson, 1976, Evidence and implications of worldwide late Paleogene climatic and eustatic events (abs.): GSA Abstracts with Programs, v. 8, p. 934-935.

Isaacs, C.M., 1981, Porosity reduction during diagenesis of the Monterey Formation, Santa Barbara coastal area, California, *in* R.E. Garrison and R.G. Douglas, eds., The Monterey Formation and related siliceous rocks of California: SEPM, Pacific Section, p. 257-271.

Isaacs, C.M., 1984, The Monterey—Key to offshore California boom: Oil and Gas Journal, v. 82, p. 75-81.

Isaacs, C.M., and R.E. Garrison, 1983, Petroleum generation and occurrence in the Miocene Monterey Formation, California: SEPM, Pacific Section, 228 p.

Jackson, P.A., and R.S. Yeats, 1982, Structural evolution of Carpinteria Basin, western Transverse Ranges, California: AAPG Bulletin, v. 66, p. 805-829.

Jurdy, D.M., 1984, The subduction of the Farallon plate beneath North America as derived from relative plate motion: Tectonics, v. 3, p. 107-113.

Kablanow, R.I., and R.C. Surdam, 1983, Diagenesis and hydrocarbon generation in the Monterey Formation, Huasna basin, California, *in* C.M. Isaacs and R.E. Garrison, eds., Petroleum generation and occurrence in the Miocene Monterey Formation, California: SEPM, Pacific Section, p. 53-68.

Kamerling, M.J., and B.P. Luyendyk, 1985, Paleomagnetism and Neogene tectonics of the northern Channel Islands, California: Journal of Geophysical Research, v. 90, p. 12,485-12,502.

Keller, G., and J.A. Barron, 1981, Integrated planktic foraminiferal and diatom biochronology for the northeast Pacific and the Monterey Formation, *in* R.E. Garrison and R.G. Douglas, eds., The Monterey Formation and related siliceous rocks of California: SEPM, Pacific Section, p. 43-54.

Kennett, J.P., 1977, Cenozoic evolution of Antarctic glaciation, the Circum-Antarctic Ocean, and their impact on global paleoceanography: Journal of Geophysical Research, v. 82, p. 3843-3860.

Kilkenny, J.E., 1971, Future petroleum potential of region 2, Pacific coastal states and adjacent continental shelf and slope, *in* Future petroleum provinces of the United States—Their geology and potential: AAPG Memoir 15, p. 170-177.

King, J.D., and G.E. Claypool, 1983, Biological marker compounds and implications for generation and migration of petroleum in rocks of the Point Conception deep-stratigraphic test well, OCS-CAL 78-164 No. 1, offshore California, *in* C.M. Isaacs and R.E. Garrison, eds., Petroleum generation and occurrence in the Miocene Monterey Formation, California: SEPM, Pacific Section, p. 191-200.

Kleinpell, R.M., 1938, Miocene stratigraphy of California: Tulsa, AAPG, 450 p.

Kvenvolden, K.A., and M.E. Field, 1981, Thermogenic hydrocarbons in unconsolidated sediment of Eel River basin, offshore northern California: AAPG Bulletin, v. 65, p. 1642-1646.

Lagoe, M.B., 1985, Depositional environments in the Monterey Formation, Cuyama basin, California: GSA Bulletin, v. 96, p. 1296-1312.

Link, M.H., and J.E. Welton, 1982, Sedimentology and reservoir potential of Matilija sandstone: An Eocene sand-rich deep-sea fan and shallow-marine complex, California: AAPG Bulletin, v. 66, p. 1514-1534.

Luyendyk, B.P., M.J. Kamerling, R.R. Terres, and J.S. Hornafius, 1985, Simple shear of southern California during Neogene time suggested by paleomagnetic declinations: Journal of Geophysical Research, v. 90, p. 12,454-12,466.

Magoon, L.B., and C.M. Isaacs, 1983, Chemical characteristics of some crude oils from the Santa Maria basin, California, *in* C.M. Isaacs and R.E. Garrison, eds., Petroleum generation and occurrence in the Miocene Monterey Formation, California: SEPM, Pacific Section, p. 201-211.

McCulloh, T.H., 1969, Geologic characteristics of the Dos Cuadras offshore oil field: U.S. Geological Survey Professional Paper 679-C, p. 29-46.

McLaughlin, R.J., S.A. Kling, R.Z. Poore, K. McDougall, and E.C. Beutner, 1982, Post-middle Miocene accretion of Franciscan rocks, northwestern California: GSA Bulletin, v. 92, p. 595-605.

Miller, P.L., 1981, Tertiary calcareous nannoplankton and benthic foraminifera biostratigraphy of the Pointe Arena area, California: Micropaleontology, v. 27, p. 419-443.

Mutti, E., T.H. Nilsen, and F. Ricci Lucchi, 1978, Outer fan depositional lobes of the Laga Formation (upper Miocene and lower Pliocene) east-central Italy, *in* D.J. Stanley and G. Kelling, eds., Sedimentation in submarine canyons, fans and trenches: Stroudsburg, Pennsylvania, Dowden, Hutchinson and Ross, p. 210-223.

Nagle, H.E., and E.S. Parker, 1971, Future oil and gas potential of onshore Ventura basin, California, *in* Future petroleum provinces of the United States—Their geology and potential: AAPG Memoir 15, p. 254-297.

Nardin, T.R., and T.L. Henyey, 1978, Pliocene-Pleistocene diastrophism of Santa Monica and San Pedro shelves, California continental borderland: AAPG Bulletin, v. 62, p. 247-272.

Nilsen, T.H., 1980, Modern and ancient submarine fans: Discussion of papers by R.G. Walker and W.R. Normark: AAPG Bulletin, v. 64, p. 1094-1112.

Nilsen, T.H., 1984a, Offset along the San Andreas fault of Eocene strata from the San Juan Bautista area and western San Emigdio Mountains, California: GSA Bulletin, v. 95, p. 599-609.

Nilsen, T.H., 1984b, Oligocene sedimentation and tectonics, California, *in* T.H. Nilsen, ed., Fluvial sedimentation and related tectonic framework, western North America: Sedimentary Geology, v. 38, p. 305-336.

Nilsen, T.H., 1985, Butano turbidite system, *in* A.H. Bouma, W.R. Normark, and N.E. Barnes, eds., Submarine fans and related turbidite systems: New York, Springer-Verlag, p. 173-178.

Nilsen, T.H., 1986, Cretaceous paleogeography of western North America, *in* P.L. Abbott, ed., Cretaceous stratigraphy of western North America: SEPM, Pacific Section, v. 46, p. 1-39.

Nilsen, T.H., in press, Paleogene tectonics and sedimentation of coastal California, *in* W.G. Ernst, and R.V. Ingersoll, eds., Cenozoic basin development of coastal California (Rubey Volume 6): Englewood Cliffs, New Jersey, Prentice-Hall, Inc.

Nilsen, T.H., and S.H. Clarke, Jr., 1975, Sedimentation and tectonics in the early Tertiary continental borderland of central California: U.S. Geological Survey Professional Paper 925, 64 p.

Nilsen, T.H., and E.H. McKee, 1979, Paleogene paleogeography of the western United States, *in* J.M. Armentrout, M.R. Cole, and H. TerBest, Jr., eds., Cenozoic paleogeography of the western United States: SEPM, Pacific Section, p. 257-276.

Ogle, B.A., 1953, Geology of the Eel River valley area, Humboldt County, California: California Division of Mines Bulletin, 164, 128 p.

Ogle, B.A., 1981, Oil and gas exploration offshore central and northern California, *in* M.T. Halbouty, ed., Energy resources of the Pacific region: AAPG Studies in Geology 12, p. 375-381.

Orr, W.L., 1984, Sulfur and sulfur isotope ratios in Monterey oils of the Santa Maria River basin and Santa Barbara channel area (abs.): SEPM Annual Midyear Meeting Abstracts, p. 62.

Page, B.M. and D.C. Engebretson, 1984, Correlation between the geologic record and computed plate motions for central California: Tectonics, v. 3, p. 133-155.

Parker, F.S., 1971, Petroleum potential of southern California offshore, *in* Future petroleum provinces of the United States—Their geology and potential: AAPG Memoir 15, p. 178-191.

Philippi, G.T., 1965, On the depth, time and mechanism of petroleum generation: Geochimica et Cosmochimica Acta, v. 29, p. 1021-1049.

Pisciotto, K.A., and R.E. Garrison, 1981, Lithofacies and depositional environments of the Monterey Formation, California, *in* R.E. Garrison and R.G. Douglas, eds., The Monterey Formation and related siliceous rocks of California: SEPM, Pacific Section, p. 97-122.

Rodgers, D.A., 1980, Analysis of pull-apart basin development produced by en echelon strike-slip faults, *in* A.G. Sylvester, ed., Wrench fault tectonics: AAPG Reprint Series 28, p. 345-359.

Schwalbach, J.R., and D.S. Gorsline, 1985, Holocene sediment budgets for the basins of the California continental borderland: Journal of Sedimentary Petrology, v. 55, p. 829-842.

Schweickert, R.A., and D.S. Cowan, 1975, Early Mesozoic tectonic evolution of the western Sierra Nevada, California: GSA Bulletin, v. 82, p. 1329-1336.

Silver, E.A., 1974, Basin development along translational continental margins, *in* W.R. Dickinson, ed., Geologic interpretations from global tectonics with applications for California geology and petroleum exploration: Bakersfield, California, San Joaquin Geological Society, p. 6-1 to 6-5.

Snyder, W.S., W.R. Dickinson, and M.L. Silberman, 1976, Tectonic implications of space-time patterns of Cenozoic magmatism in the western United States: Earth and Planetary Science Letters, v. 32, p. 91-106.

Sorensen, S., 1985, Petrologic evidence for Jurassic, island-arc-like basement rocks in the southwestern Transverse Ranges and California continental borderland: GSA Bulletin, v. 96, p. 997-1006.

Stanley, R.G., 1986, Evidence for Late Oligocene-Early Miocene episode of transtension along San Andreas fault system in central California (abs.): Pacific Section AAPG-SEPM-SEG 61st Annual Meeting, p. 40.

Stuart, C.J., 1979, Middle Miocene paleogeography of coastal southern California and the California borderland—evidence from schist-bearing sedimentary rocks, *in* J.M. Armentrout, M.R. Cole, and H. TerBest, Jr., eds., Cenozoic paleogeography of the western United States: SEPM, Pacific Section, p. 29-44.

Suppe, J., 1970, Offset of late Mesozoic basement terranes by the San Andreas fault system: GSA Bulletin, v. 81, p. 3253-3258.

Sylvester, A.G., (ed.), 1984, Wrench fault tectonics: AAPG Reprint Series 28, 374 p.

Sylvester, A.G., and R.R. Smith, 1976, Tectonic transpression and basement-controlled deformation in San Andreas fault zone, Salton Trough, California: AAPG Bulletin, v. 60, p. 2081-2102.

Taylor, J.C., 1976, Geologic appraisal of the petroleum potential of offshore southern California: The borderland compared to onshore coastal basins: U.S. Geological Survey Circular 730, 42 p.

Teel, D.B., and E.G. Frost, 1982, Synorogenic evolution of the Copper Basin Formation in the eastern Whipple Mountains, San Bernardino County, California, *in* E.G. Frost and D.L. Martin, eds., Mesozoic-Cenozoic tectonic evolution of the Colorado River region, California, Arizona, and Nevada: GSA Special Publication, Boulder, Colorado, p. 275-285.

Turner, D.L., 1970, Potassium-argo dating of Pacific Coast Miocene foraminiferal stages, *in* O.L. Bandy, ed., Radiometric dating and paleontologic zonation: GSA Special Paper 127, p. 91-129.

Underwood, M.B., 1983, Depositional setting of the Paleogene Yager Formation, northern Coast Ranges of California, *in* D.K. Larue and R.J. Steel, eds., Cenozoic marine sedimentation, Pacific margin, U.S.A.: SEPM, Pacific Section, p. 81-101.

Underwood, M.B., 1985, Sedimentology and hydrocarbon potential of Yager structural complex—Possible Paleogene source rocks in Eel River basin, northern California: AAPG Bulletin, v. 69, p. 1088-1100.

Van de Kamp, P.C., J.D. Harper, J.J. Conniff, and D.A. Morris, 1974, Facies relations in the Eocene-Oligocene in the Santa Ynez Mountains, California: Journal of the Geological Society of London, v. 130, p. 545-565.

Vedder, J.G., H.C. Wagner, and J.E. Schoellhamer, 1969, Geologic framework of the Santa Barbara Channel region: U.S. Geological Survey Professional Paper 679, p. 1-11.

Vedder, J.G., D.G. Howell, and H. McLean, 1983, Stratigraphy, sedimentation, and tectonic accretion of exotic terranes, southern Coast Ranges, California, *in* J.S. Watkins and C.L. Drake, eds., Studies in continental margin geology: AAPG Memoir 34, p. 471-496.

Vedder, J.G., L.A. Beyer, A. Junger, G.W. Moore, A.E. Roberts, J.C. Taylor, and H.C. Wagner, 1974, Preliminary report on the geology of the continental borderland of southern California: U.S. Geological Survey Misc. Field Investigations Map MF-624, 34 p.

Wallis, W.S., 1981, Oil and gas exploration, offshore southern California, *in* M.T. Halbouty, ed., Energy resources of the Pacific region: AAPG Studies in Geology 12, p. 383-389.

Waples, D.W., 1980, Time and temperature in petroleum formation: Application of Lopatin's method to petroleum exploration: AAPG Bulletin, v. 64, p. 916-926.

Waples, D.W., 1985, Geochemistry in petroleum exploration: IHRDC publications, Boston, Massachusetts, 232 p.

Weaver, D.W., 1969, Geology of the northern Channel Islands: Pacific Section AAPG-SEPM Special Publication, 200 p.

Wentworth, C.M., 1986, Upper Cretaceous and lower Tertiary strata near Gualala, California, and inferred large right slip on the San Andreas fault, *in* W.R. Dickinson and A. Grantz, eds., Proceedings of the conference on geologic problems of the San Andreas fault system: Stanford University Publications in Geological Sciences, v. 11, p. 130-143.

Wilcox, R.E., T.P. Harding, and D.R. Seely, 1973, Basic wrench tectonics: AAPG Bulletin, v. 57, p. 74-96.

Yeats, R.S., 1973, Newport-Inglewood fault zone, Los Angeles basin, California: AAPG Bulletin, v. 57, p. 117-135.

Yeats, R.S., 1979, Stratigraphy and paleogeography of the Santa Susana fault zone, Transverse Ranges, California, *in* J.M. Armentrout, M.R. Cole, and H. TerBest, Jr., eds., Cenozoic paleogeography of the western United States: SEPM, Pacific Section, p. 191-204.

Yeats, R.S., M.R. Cole, W.R. Merschat, and R.M. Parsley, 1974, Poway Fan and submarine cone and rifting of the inner southern California borderland: GSA Bulletin, v. 85, p. 293-302.

Yerkes, R.F., H.C. Wagner, and K.A. Yenne, 1969, Petroleum development in the region of the Santa Barbara Channel: U.S. Geological Survey Professional Paper 679-B, p. 13-27.

Yerkes, R.F., T.H. McCulloh, J.E. Schoellhamer, and J.G. Vedder, 1965, Geology of the Los Angeles basin, California—An introduction: U.S. Geological Survey Professional Paper 420-A, p. 1-57.

Zieglar, D.L., and J.K. Cassell, 1978, A synthesis of OCS well information, offshore central and northern California, Oregon, and Washington: Presented at 53rd Annual Meeting Pacific Section AAPG, Sacramento, California, 27 p.

Zieglar, D.L., and J.H. Spotts, 1978, Reservoir and source-bed history of Great Valley, California: AAPG Bulletin, v. 62, p. 813-826.

# CHAPTER 8

# OIL AND GAS POTENTIAL OF PAPUA NEW GUINEA

E.F. Durkee
W.D. Stewart
G. Francis, et al.
*Lundin Group*
*Geneva, Switzerland*

## INTRODUCTION

Papua New Guinea (PNG) including New Britain, New Ireland, Bougainville, Manus, and all other related islands encompasses a total land and water area in excess of 1,833,000 km$^2$ (716,000 mi$^2$). Geographically, PNG has a position nearly equidistant from major population centers in the western Pacific. Geologically, it also has a unique position, being at the east end of one of the world's longest island archipelagos as well as at an apparent point of "collision" between the Australian and Pacific Ocean plates (Figures 1, 2).

To encourage further oil and gas exploration, the Government of Papua New Guinea commissioned a comprehensive technical evaluation of its petroleum potential in 1983. Financed by the World Bank and OPEC, a joint study was undertaken by Robertson Research (Australia) Ltd. and Flower Doery Buchan (Australia) Ltd. Project supervision was provided by the Geological Survey of PNG and Worldwide Exploration Consultants of Denver. All nonconfidential geological and geophysical data available from previous exploration were evaluated and interpreted. New petrographic, biostratigraphic, and geochemical data were acquired.

Oil and gas exploration in PNG dates back to the early 1900s. Overall, the effort has been rather desultory and widespread across the country. Most oil and gas exploration to date has been done in the Papuan Basin, an area of about 212,000 km$^2$ occupying the southern part of mainland PNG and the adjacent offshore area (Figure 2).

To date exploration in PNG has led to nine potential discoveries of gas, condensate, and oil. None of these have been developed for commercial production. However, Juha and Iagifu in the Southern Highlands, northwestern Papuan Basin, are being considered as possible candidates for development. Juha is suggested by various workers and authorities to range between 70 and 100 million barrels recoverable of condensate. In this same general area a recent test, Iagifu 2, flowed 3438 bo/d of 45° API oil from a zone within the earliest Cretaceous Toro Sandstone.

Areas currently being held for exploration in PNG are shown on Figure 3. Papua New Guinea's oil and gas potential is largely confined to four areas—the Papuan, Cape Vogel, New Ireland-Bougainville, and North New Guinea basins. Some of the important, geology-related oil and gas aspects of these four areas are reviewed in this report. The detailed reports and basic geophysical and well data upon which this paper is based are an open file with the PNG Geological Survey in Port Moresby.

## PAPUAN BASIN

### General

The Papuan Basin was the site of littoral, deltaic, open-shelf and continental-slope sedimentation on and around the northern margin of subsiding Australian continental crust during the Mesozoic and Cenozoic. Two major tectonic episodes significantly affected basin evolution during these eras.

First, during the latest Cretaceous to Early Paleocene, the breakup of the northern margin of the Australian continent and subsequent thermal bulging along the axis of the Coral Sea caused uplift and tilting of the basin to the northwest. As a result, much of the Cretaceous section was removed from the southeastern part of the basin. Erosional effects diminish in both northwesterly and northeasterly directions across the basin.

Second, convergence between the Indo-Australian and Pacific plates from the latest Miocene onward led to the deformation of the northern and eastern boundaries of the basin. The deformation front migrated progressively southwestward throughout the Pliocene and Quaternary, causing the formation of a major fold and thrust belt.

### Structure

Three major tectonic provinces have been identified in the Papuan Basin—the Fly Platform, the Papuan Fold Belt, and the Aure Tectonic Belt (Figure 4). Some workers consider the latter to be a zone within the Papuan Fold Belt.

The Fly Platform is an area of flat-lying to gently deformed, slightly to moderately block and wrench-faulted Mesozoic clastics, Tertiary carbonates, and Plio-Pleistocene clastics and volcanics. Sediment thicknesses range from 2 to 3 km onshore and from 3 to 10 km offshore.

Numerous major structural elements within the Fly Platform have influenced sedimentation throughout the history of

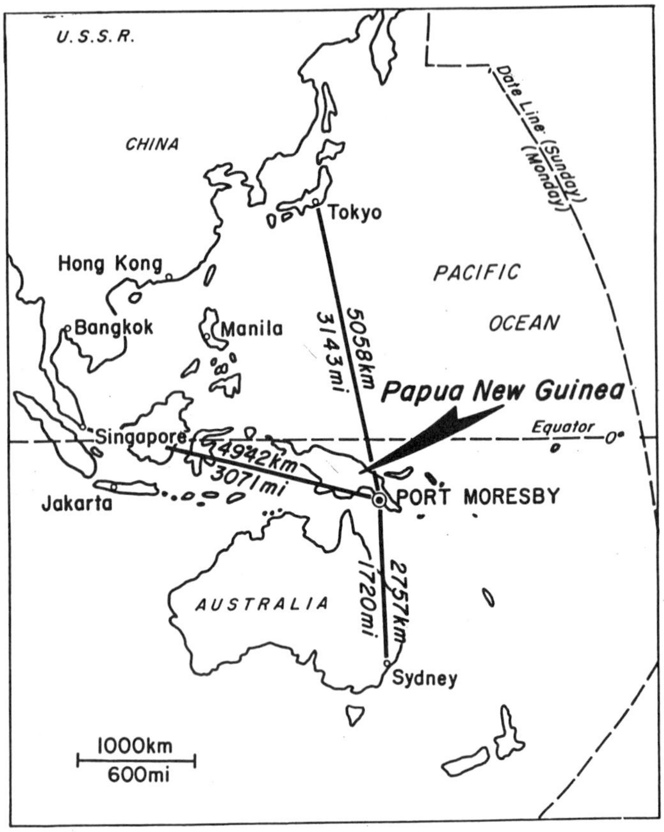


Figure 1. Index map of Papua New Guinea and the western Pacific.

the basin (Figure 4). Prominent basement highs are present, among them the Oriomo and Lake Murray highs, the Pasca Ridge and the Dibiri Spur. The Oriomo High is flanked by the moderately faulted Morehead Subbasin on the west and by the similarly faulted Daru Embayment on the east. To the northeast, the South Turama Low was an important depocenter during the Mesozoic. Dipping toward that feature is the Omati Trough, a linear zone of apparent subsidence at the basement level which extends offshore as a graben at the north end of the Dibiri Spur. The graben is bounded on the north by the Uramu High, which appears to have originated from Early Tertiary upfaulting.

The main areas of normal faulting in the Fly Platform are the Morehead Subbasin, the Daru Embayment, and around the margins of the Pasca Ridge (Figure 4). Although some movement occurred along these faults during the Jurassic, the main phase of normal faulting coincided with the latest Cretaceous to Early Paleocene opening of the Coral Sea. The faults trend dominantly northwest, although orthogonal northeasterly trends also are evident in the Daru Embayment and along the flanks of the Pasca Ridge.

Seismic studies have identified at least two wrench zones in the western Fly Platform—the Komewu and Fly wrench systems (Figure 4). Additional parallel systems are suggested by gravity and aeromagnetic data in the general area. These zones typically appear as "flower structures" as much as 2.5 km wide on seismic profiles. Rapid lateral changes in fault character, from normal to reverse dip-separations, and numerous splay faults are typical. The main phase of move-

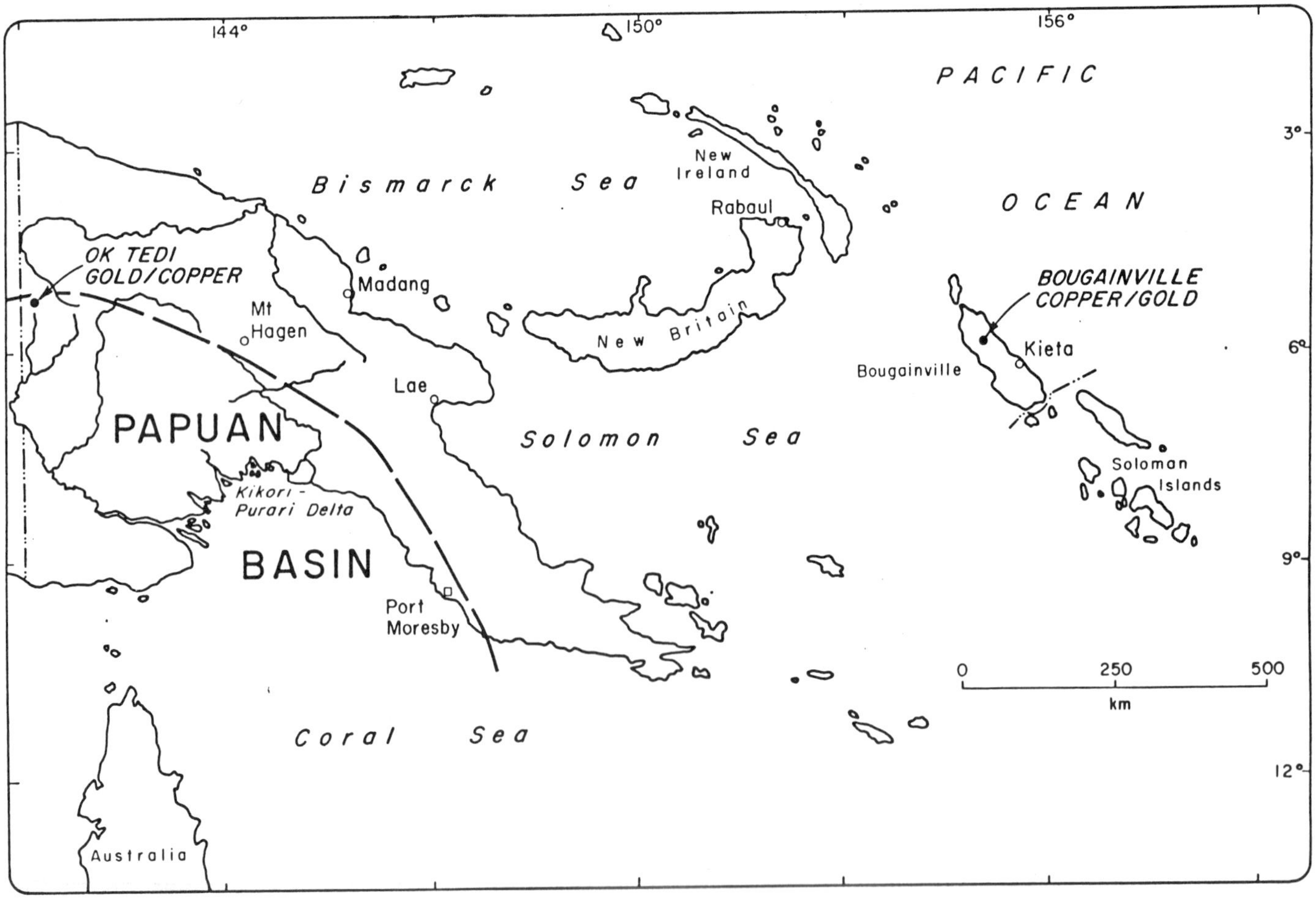


Figure 2. Index map of Papua New Guinea.

ment occurred during the latest Cretaceous to Early Paleocene.

The Papuan Fold Belt is about 100 km wide and extends about 500 km southeasterly from the Irian Jaya border (Figure 4). The sedimentary section in this zone comprises as much as 5000 m of Mesozoic marginal to fully marine clastics, 1500 m of Paleogene to Early Neogene shelf and basinal limestones and basinal clastics, and 600 m of Late Neogene marine to terrestrial clastics.

The structure of the southern Papuan Fold Belt is dominated by a series of subparallel, southward-verging, horizontal to gently plunging, thrust-faulted anticlines traceable along strike for 60-100 km. The associated faults generally dip northeastward and as a group can be traced as far as 200 km along strike. Field evidence suggests basement involvement in thrusting in a few areas. Deformation commenced during the latest Miocene and possibly in places continues to the present.

The Aure Tectonic Belt is about 150 km wide and 400 km long and trends north-south. The known sedimentary succession within it comprises Late Mesozoic fine clastics, Paleogene bathyal limestones and clastics, a thick Miocene turbiditic succession, and Plio-Pleistocene marine to terrestrial clastics. Basement is depressed to a depth of more than 10 km along this zone.

The structural style of the Aure Tectonic Belt bears similarities to that of the Papuan Fold Belt, and the boundary between them is indistinct. However, some significant differences can be noted. In the Aure Tectonic Belt, broad synclines with intervening pinched and faulted anticlines are typical, and associated thrust faults dip both east and west. As the main deformation occurs in the thick Miocene clastic sequence, the presence of overpressured, undercompacted mudstones has led to the development of swells and diapiric anticlines. There were two main phases of thrusting in the latest Miocene and Plio-Pleistocene, with some sporadic and localized movements occurring during the Pliocene.

## Stratigraphy

A generalized stratigraphic section for the Papuan Basin is shown in Figure 5. For the purpose of this paper, the basin has been subdivided into two parts to illustrate differences in the stratigraphic successions in its western and eastern areas.

Sedimentation commenced in the Papuan Basin by Triassic time, but the main phase of clastic littoral, deltaic, open-shelf and continental-slope sedimentation occurred from the mid-Jurassic through the Cretaceous. Throughout this period, sediment supply came largely from the Australian craton to the south and west, and at times locally from basement highs such as the Lake Murray and Oriomo highs.

Predominantly fluviodeltaic sedimentation prevailed over the present area of the western Fly Platform and northwestern Papuan Fold Belt during the Middle to Late Jurassic (Kuabgen Group; Figure 5). The Bol Arkose, Magobu Formation, and Koi-Iange Formation contain both sandy and shaly facies deposited in fluvial, upper and lower delta plain, and prodelta environments. Deposition of these units was interrupted twice by apparently eustatically controlled transgres-

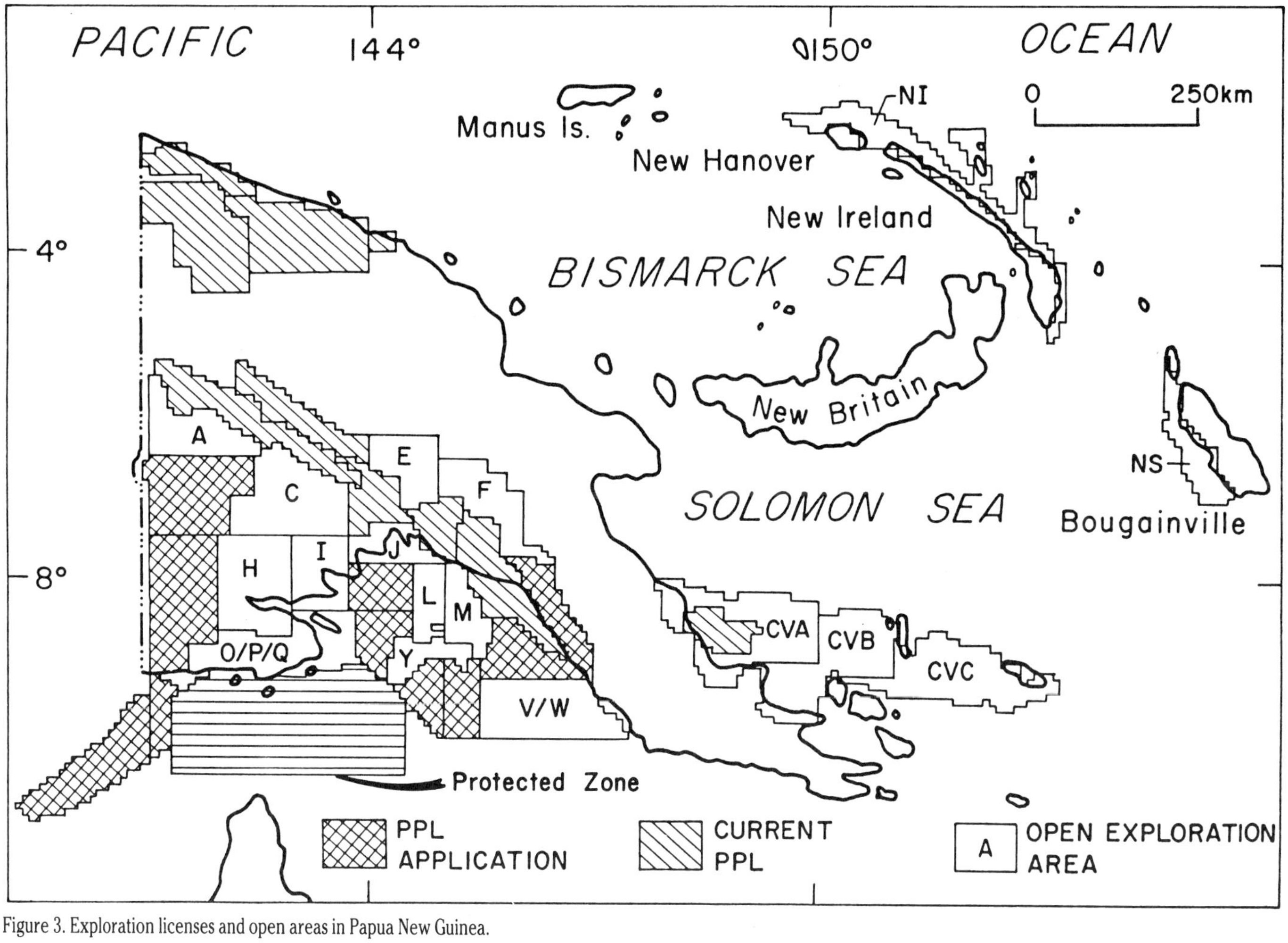


Figure 3. Exploration licenses and open areas in Papua New Guinea.

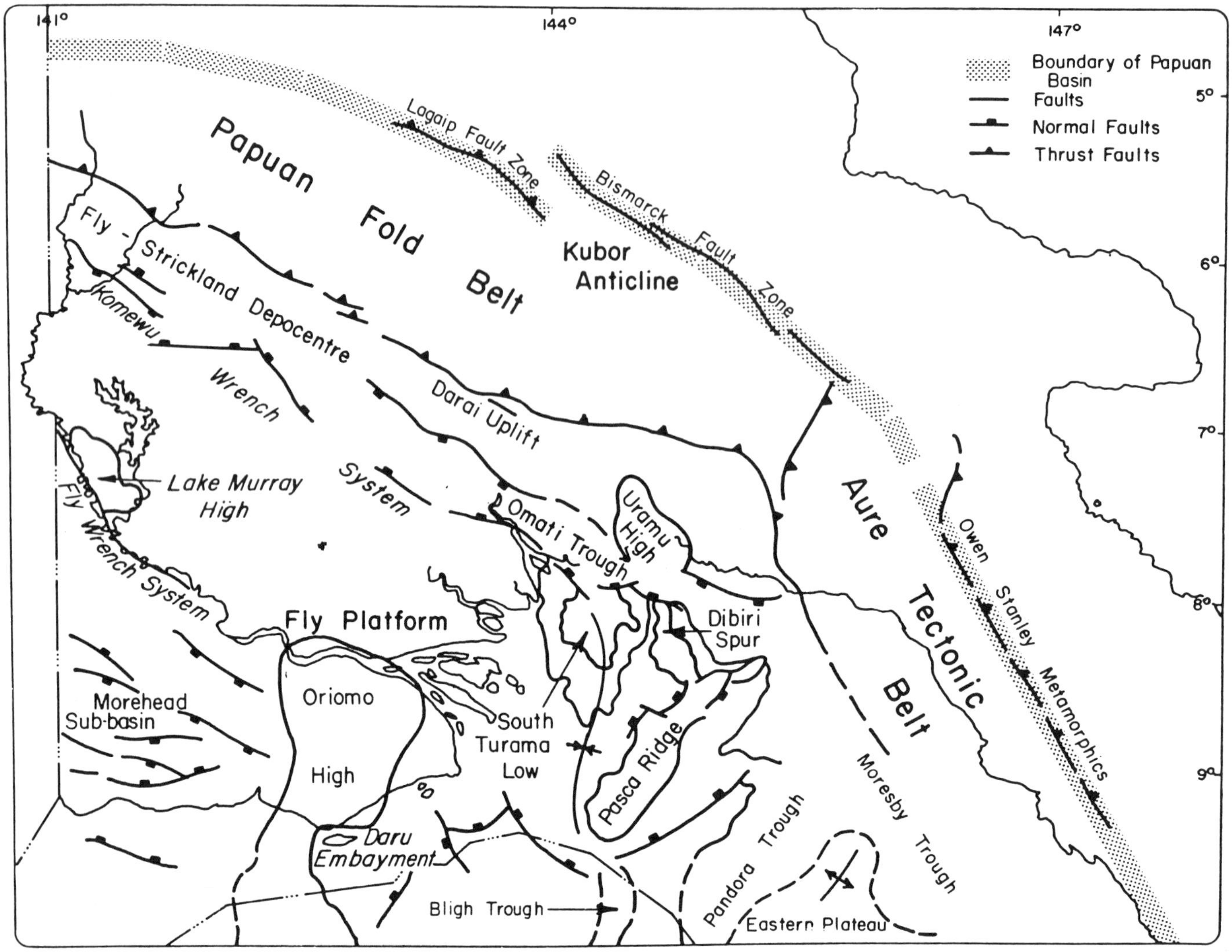


Figure 4. Structural elements of the Papuan Basin.

sions which resulted in widespread mudstone deposition (Barikewa and Imburu Mudstones; Figure 5).

The second transgression at least partially submerged the Oriomo and Lake Murray highs. General regression during the Kimmeridgian led to the deposition of a sandy shoreface to shallow-marine unit, provisionally referred to here as the lower Toro Sandstone, in the Darai Plateau area (Figure 4). This regression culminated in a low sea-level stand, causing a significant depositional hiatus recognizable in most parts of the basin (Omati Break; Figure 5).

Contemporaneously with Kuabgen Group sedimentation, fine clastics were being deposited in more distal parts of the shelf and on the continental slope. These sediments now constitute the Maril Shale, which is widely distributed on the central Fly Platform and in the Papuan Fold Belt.

Widespread sandy deposition resumed during Berriasian time with deposition of the upper Toro Sandstone in shoreface, barrier bar, marine bar and shelf environments. This unit, which is the primary Mesozoic reservoir target in the Papuan Basin, onlaps the Lake Murray High in the west, and shales out basinward. Deposition of the upper Toro Sandstone terminated with a second low sea-level stand from Valanginian to Late Barremian time (Mid-Neocomian Break; Figure 5). The resulting depositional hiatus is widely recognizable in proximal parts of the Papuan Basin.

Shallow-marine shelf shale and local sand deposition became re-established in the central and western parts of the Papuan Basin during latest Barremian time, and persisted until the close of the Cretaceous (Ieru Formation; Figure 5). Fine clastics of the "Kondaku Tuff" and Chim Formation accumulated contemporaneously in more distal areas. With the commencement of uplift and tilting of the Papuan Basin during the latest Cretaceous, substantial thicknesses of Cretaceous sediments were eroded in areas to the southeast.

Early Paleogene sedimentation was restricted, except in the eastern extremity of the basin where bathyal limestones, cherts, and mudstones of the Port Moresby association and Paga beds were accumulating. Widespread carbonate platform sedimentation subsequently became established over the central part of the basin by the Late Eocene (Mendi Group; Figure 5). These sediments are absent because of erosion or nondeposition in the western Fly Platform, except in the Morehead Subbasin, where clastic sedimentation was taking place.

After another period of nondeposition during most of the Oligocene, a broad carbonate platform with a fringing barrier

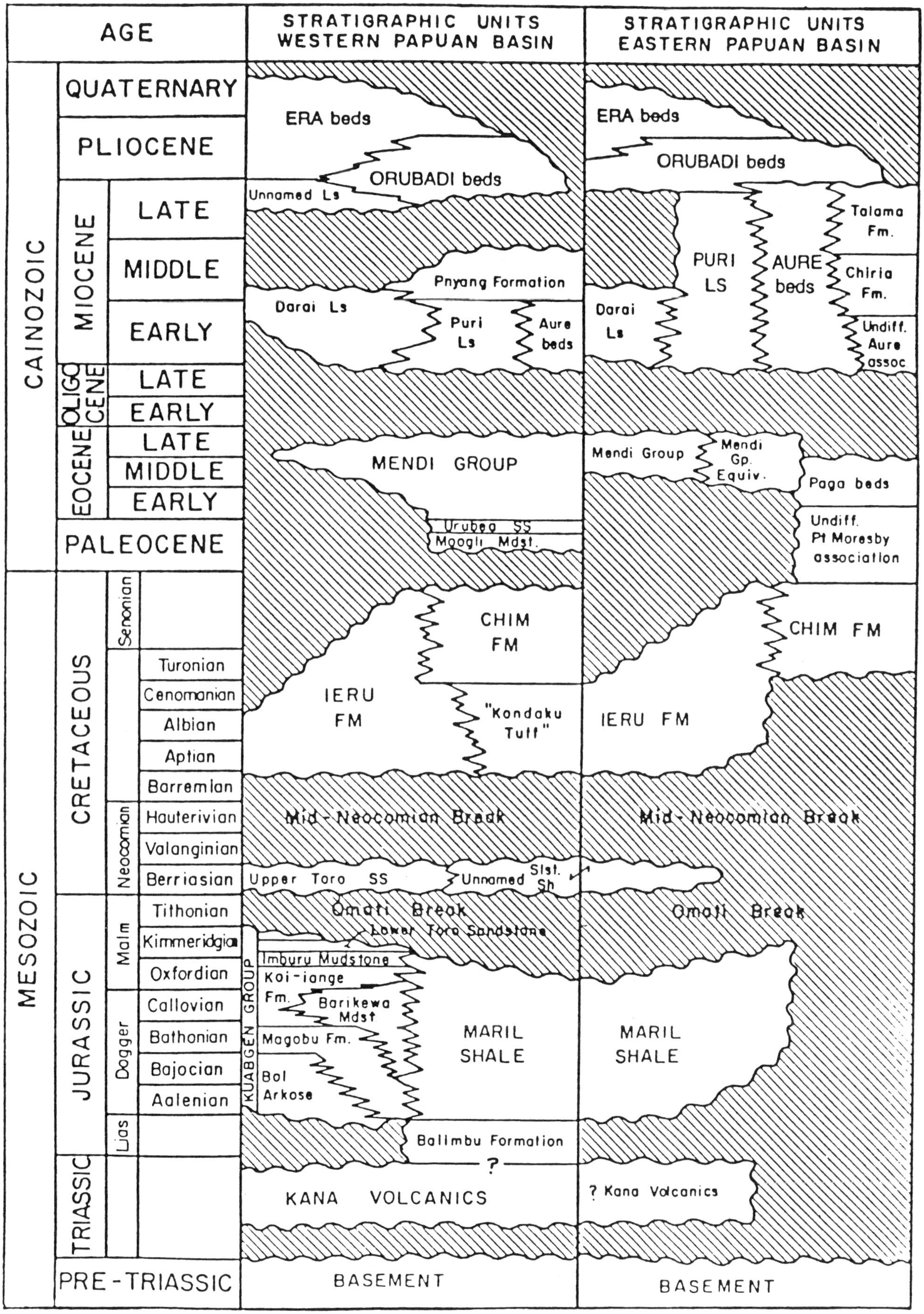


Figure 5. Generalized stratigraphic table for the Papuan Basin.

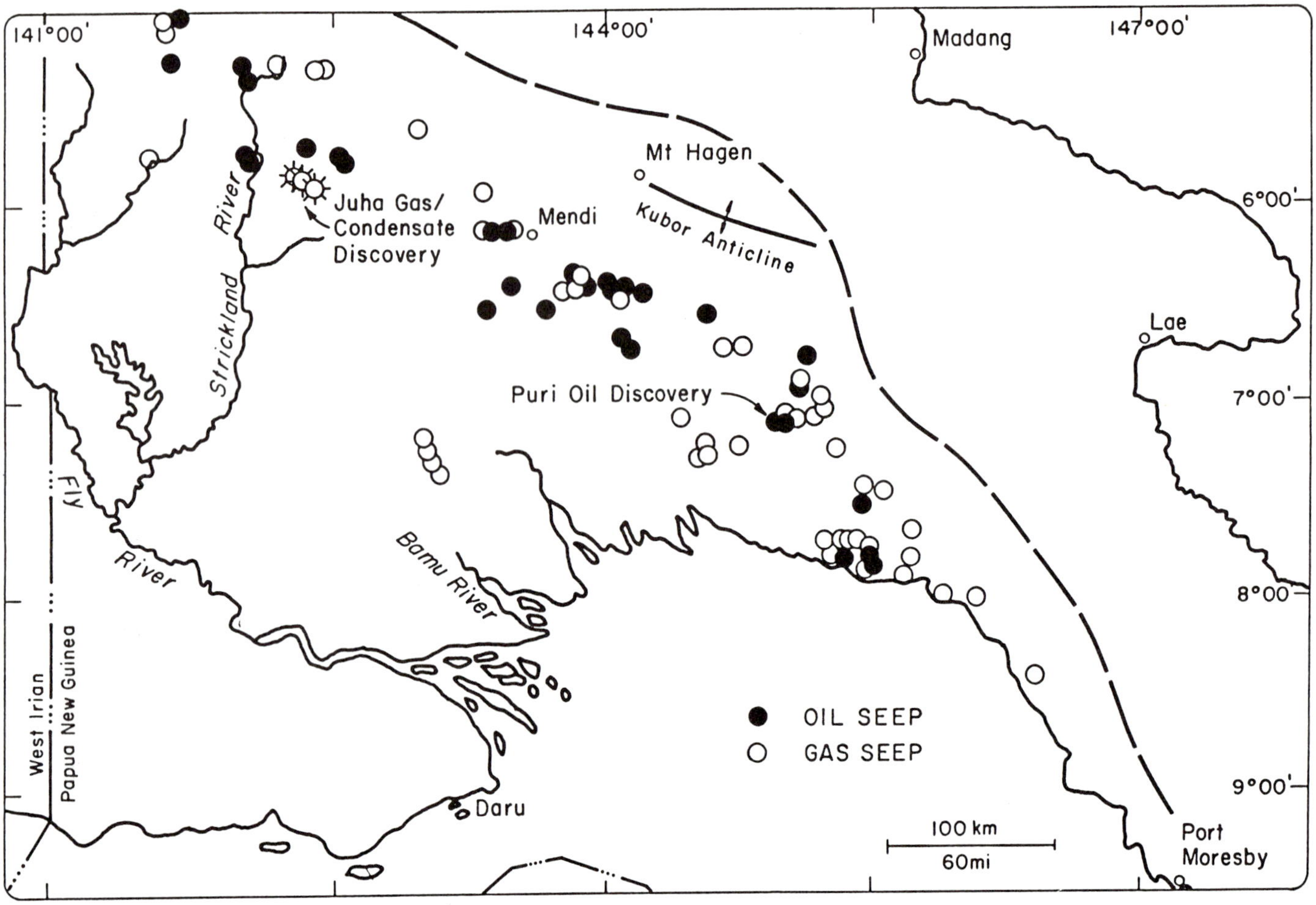


Figure 6. Oil and gas seeps in the Papuan Basin.

reef (Borabi Reef Trend) became established. These platform and reef carbonates now constitute the Darai Limestone (Figure 5), which is widely distributed over the western and central Fly Platform and in the Papuan Fold Belt. East of the Borabi Reef Trend, bathyal Puri limestone and its turbiditic clastic lateral equivalent, the Aure beds, were being deposited. Within this bathyal regime, atoll reefs developed on structural and basement highs such as the Uramu High and Pasca Ridge.

By the late Early to early Middle Miocene, carbonate sedimentation was terminated in the northwestern part of the basin by prograding clastics of the Pnyang Formation, derived from a volcanic arc. Nondeposition and/or erosion affected most of the carbonate platform elsewhere, but Puri limestone and Aure bed deposition persisted in deeper water areas until the Late Miocene. At the close of the Miocene, the emerging Central Highlands supplied vast amounts of clastic detritus which terminated carbonate deposition. Fine marine clastics of the Orubadi beds and marginal marine to terrestrial clastics of the Era beds blanketed the area during the Plio-Pleistocene.

## Oil and Gas

*Known occurrences*—Perhaps the most striking evidence for the presence of hydrocarbons in the Papuan Basin is the occurrence of a large number of oil and gas seeps (Figure 6), some 120 of which are documented in the RRA/FDB (1984) report.

Most seeps occur in the Papuan Fold Belt and Aure Tectonic Belt, and are almost certainly due to the rupture of reservoir seals along these structural trends.

The presence of seeps first stimulated interest in the Papuan Basin, and since World War II some 60 wells have been drilled (Figure 7). Subsequent evaluation of these by RRA/FDB (1984) has indicated that only 21 (33%) could be considered valid tests of their geological objectives. Of the valid tests, nine, or 40%, have resulted in hydrocarbon discoveries (Table 1).

*Reservoirs*—The principal reservoir objective in the western half of the Papuan Basin is the Toro Sandstone.

A total sand thickness of up to 150 m is present in areas of maximum sand development in the northwestern Papuan Fold Belt and Fly River Delta areas. The good initial flow rates from the four discoveries made so far in the Toro Sandstone (Juha, Iagifu, Barikewa, and Iehi, Table 1) attest to its reservoir potential in many areas. This unit continues to be the primary exploration target in PPLs 17, 18, and 27, and it is also widely distributed in areas not currently under license.

Other Meosozic units also contain potential reservoirs. The Magobu, Koi-lange, and Ieru formations locally contain fair to good quality sands which would constitute at least secondary targets over much of the western Fly Platform. In the central Fly Platform, Jurassic shoreface and nearshore sands possibly onlap the Pasca Ridge and Dibiri Spur, providing a further, as yet untested, potential reservoir target.

In the central part of the basin, Miocene atoll reefs form the

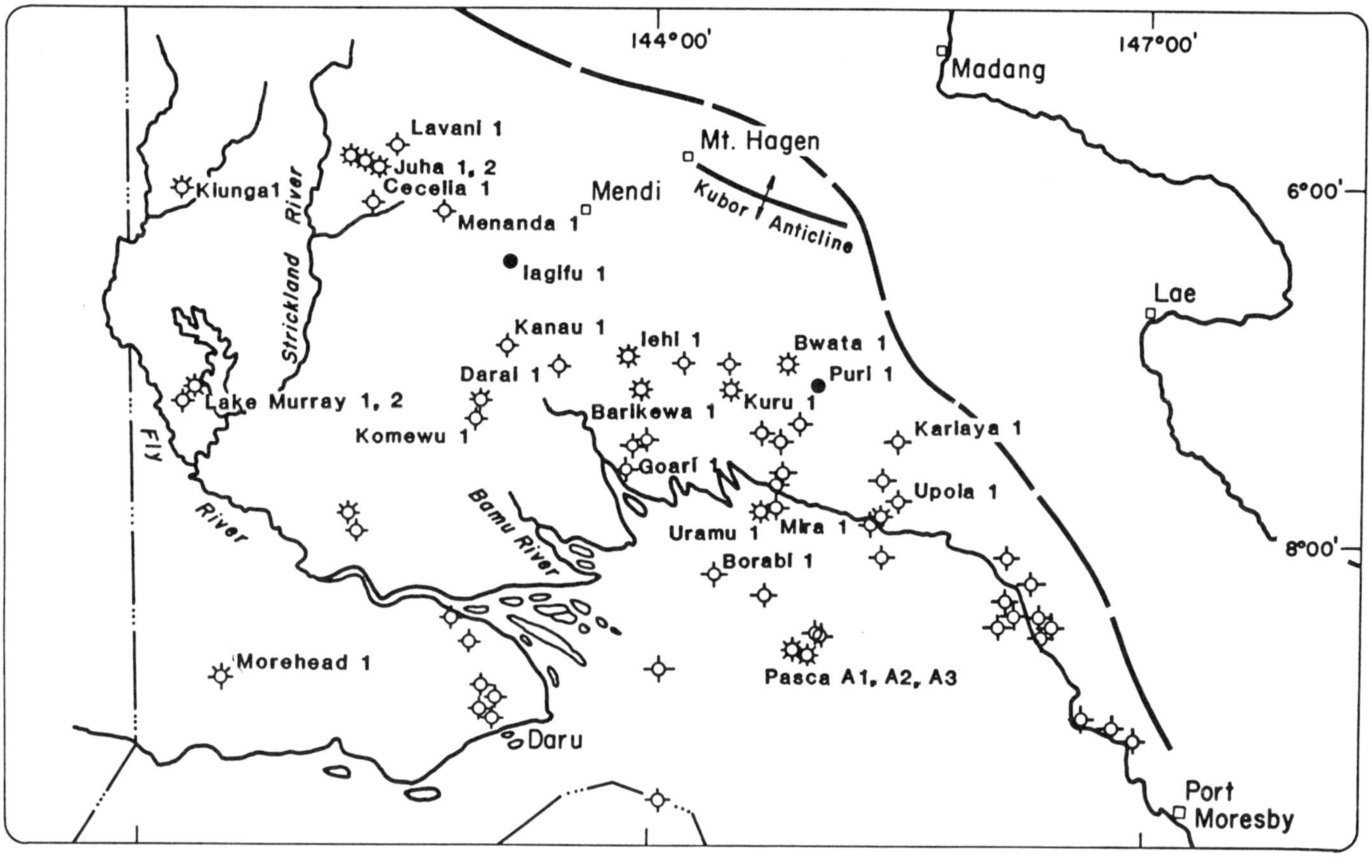


Figure 7. Well location map for Papuan Basin.

reservoirs for two discoveries to date, Pasca and Uramu (Figure 7, Table 1). These features contain as much as 1000 m of reef and reef-related carbonates. The best reservoir characteristics are found in reef-crest and back-reef sediments, which exhibit good mouldic and vugular porosity and which are in part pervasively dolomitized.

In the Papuan Fold Belt, fractured Puri limestone is the reservoir in the Kuru, Bwata, and Puri discoveries (Figure 7, Table 1). The excellent reservoir characteristics of this unit are demonstrated by, for example, the initial flow rate of 1610 bo/d in Puri 1. The Puri limestone continues to be the primary exploration target in PPL 51, and it is also present in currently unlicensed areas to the northwest (Figure 3).

The reservoir potential of Eocene carbonates remains virtually untested. Probable patch reefs have only recently been interpreted from seismic data on the Fly Platform. The reservoir characteristics of such features remain unknown, as they have not been encountered by any well. Eocene platform carbonates in Wana 1 flowed 18,000 barrels of water per day in a drill-stem test. Fractured Eocene limestones in the Papuan Fold Belt and structured areas of the Fly Platform probably possess reservoir characteristics similar to those of fractured Puri limestone.

The Darai Limestone contains a number of potential reservoirs. Coral patch reefs and algal banks are present within the platform carbonates. The platform carbonates are in places overlain by foraminiferal-red algal banks of Late Miocene to Pliocene age. Cores from Borabi 1, which penetrated one of these features, indicate porosities as high as 30% and permeabilities of 16 md, thus demonstrating their reservoir potential. The fringing barrier reef (Borabi Reef Trend) contains sediments with reservoir characteristics similar to those of the Pasca and Uramu reefs. Another untested reservoir possibility is a prominent carbonate detrital wedge developed off the northeast margin of the Borabi Reef Trend. Seismic velocity analysis indicates that velocities are significantly lower for this interval than for laterally equivalent carbonates of the Borabi Reef Trend. This suggests that it is very porous and/or that it contains interbedded slower velocity argillaceous sediments.

TABLE 1

Oil and Gas Discoveries in the Papuan Basin

| Field/Well | Reserve (BCF) | Initial Flow |
|---|---|---|
| Mesozoic | | |
| Iehi | 312 | 32 MMCFG/D |
| Barikewa | 700-900 | 6.1-9.1 MMCFG/D |
| Juha 1 (Press Repts) | 50-100 x $10^6$ Bbls Cond. | 14.8-24.1 MMCFG/D w/991-1680 BC/D |
| Iagifu 1 (Press Repts) | | DST 5-3,438 BC/D w/5.4 MMCFG/D<br>DST 6-3,128 BC/D w/6.56 MMCFG/D |
| Tertiary | | |
| Kuru | 13 | 50 (?) MMCFG/D (Blow-Out) |
| Bwata | 52 | 29 MMCFG/D |
| Puri | | 1610 BO/D |
| Uramu | 213 | 24 MMCFG/D 48 BC/D |
| Pasca | 350 BCF w/ 36 MM Bbls C. | 8.6 MMCFG/D 10008 BC/D |

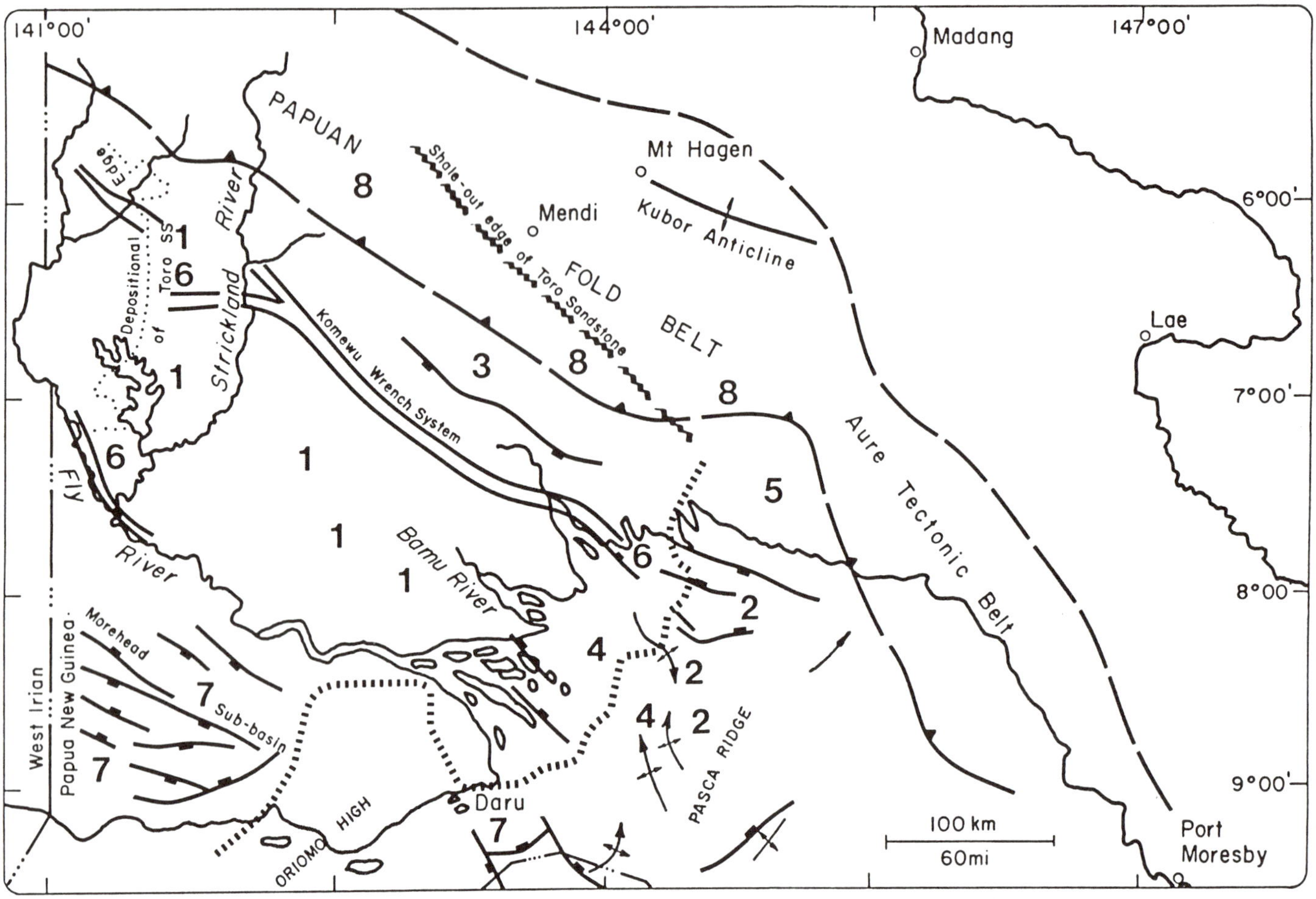


Figure 8. Mesozoic prospects.

In the Aure Tectonic Belt, thick sequences of turbiditic, volcaniclastic sandstone of the Aure beds have some reservoir potential. The reservoir characteristics of this unit differ widely depending upon matrix content and diagenetic history. Detailed data are lacking, but porosities averaging 19-22% were encountered in the Kariava 1 and Upoia 1 wells (Figure 7), although permeabilities were generally less than 100 md.

*Source rocks*—In order to better assess source rock quality and maturity, Robertson Research conducted geochemical studies on samples from 23 wells. The most common kerogen types found were inertinite (coaly kerogen) and vitrinite (woody or humic kerogen) derived from land plant wood tissues. The latter kerogen type has been identified within the Magobu Formation, the Barikewa Mudstone, and the Koi-Iange Formation in parts of the western and central Fly Platform. In areas where vitrinite-dominated kerogen is present, these units are considered to have fair oil- and gas-generating potential.

Data from the westernmost part of the Fly Platform, particularly in the Morehead Subbasin, are virtually lacking and no definite conclusions can be made regarding source rock quality there. Amorphous waxy sappropel was found to be common in samples of the Maril Shale from the central Fly Platform. Mudstones from this unit are described as being of oil shale quality, and are considered to have fair to good oil and gas source potential.

Maturation studies suggest that the Magobu Formation is fully mature over most of its area of occurrence except in the northwestern Papuan Fold Belt, where it is late mature. The Barikewa Mudstone and Koi-Iange Formation are marginally to fully mature over much of the central and western Fly Platform, although data in the latter area are sparse. In the central Papuan Basin, the Maril Shale is fully to late mature and possibly overmature farther east.

The geochemical characteristics and maturation level of the Maril Shale suggest that it may have acted as a major source of hydrocarbons, both in the Papuan Fold Belt and on the Fly Platform. This unit is favorably situated with respect to both Mesozoic and Tertiary reservoirs and could have sourced either depending upon the timing of migration in any one part of the basin. Its oil source potential has probably already been demonstrated by Puri 1, as the geology of the Puri structure strongly suggests that the oil found within it was sourced from the Maril Shale.

## Play Concepts/Prospects

*Mesozoic prospects*—Mesozoic reservoirs are located in a variety of structural settings in the Papuan Basin.

Basement-high-associated traps, simple anticlines, normal faulted anticlines, and fault traps are common on the Fly Platform. In the Papuan Fold Belt, thrust-faulted anticlines are the major targets.

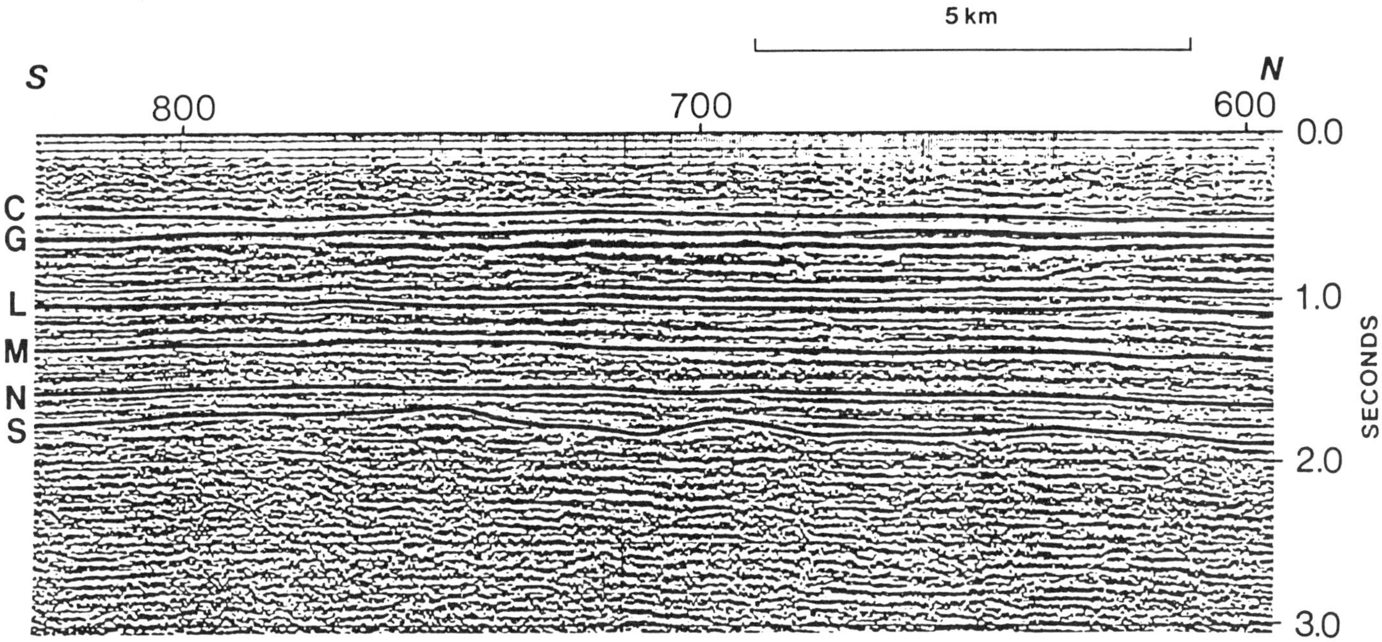


Figure 9. Basement high in Strickland River Area.

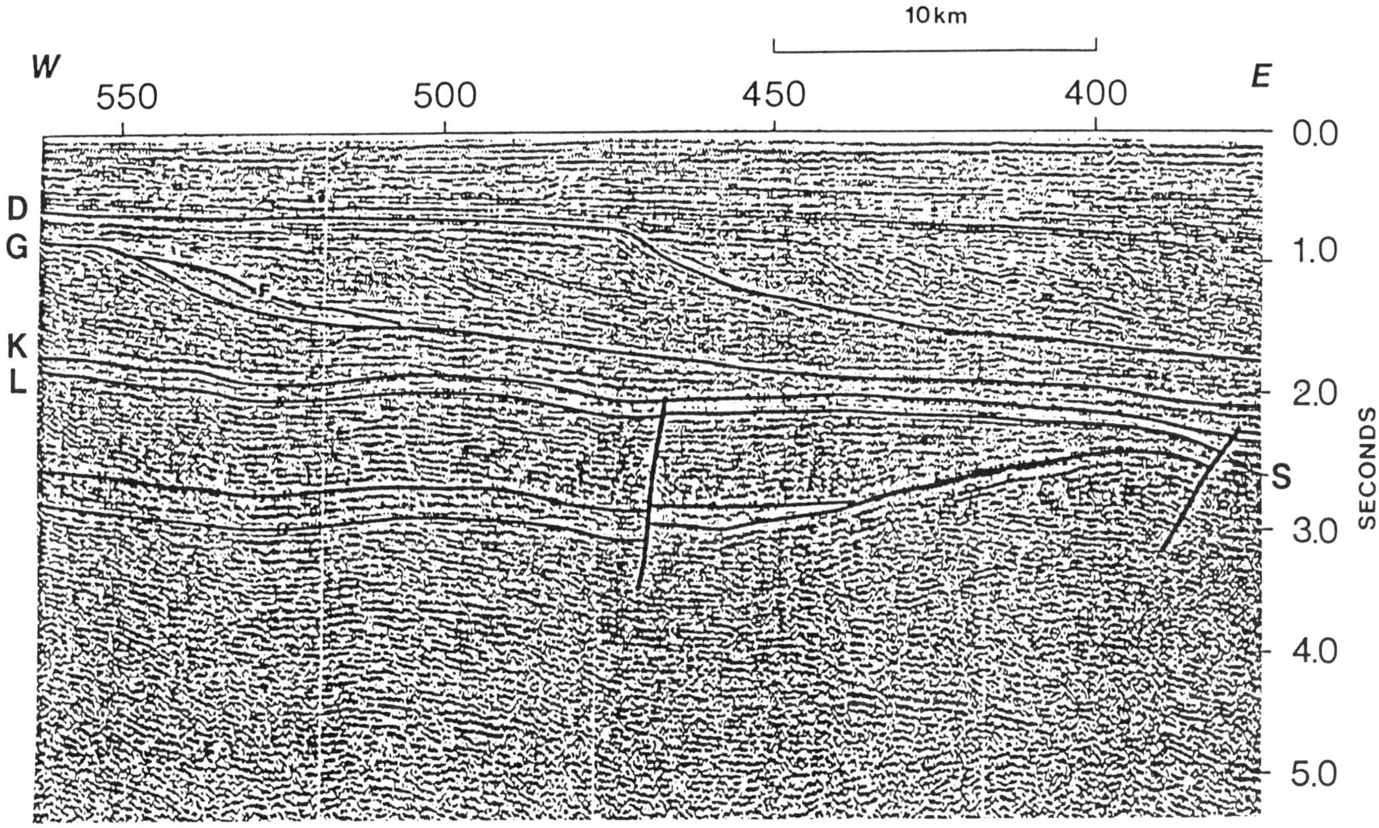


Figure 10. Mesozoic on flank of the Pasca Ridge.

Traps associated with basement highs are common on the western Fly Platform. The two best defined highs are the Lake Murray and Oriomo highs, but seismic and aeromagnetic surveys over the general area suggest the presence of several more unexplored highs of various sizes between these two features and in the northwestern part of the platform. The highs are onlapped by the Toro Sandstone at some localities and are draped by it at others (Prospect Type 1, Figures 8, 9). Some of the drape structures have areal closures of 50 $km^2$, and reservoirs range in depth from 1000 to 2500 m.

Much potential remains for further exploration on and around the margins of these basement highs. Previous tests around the northern margin of the Oriomo High were drilled early in the exploration of the area. These tests were drilled without seismic control. Lateral and vertical closure were therefore not proven. Lateral seals were probably also lacking in the Lake Murray 2 area (Figure 7), where a good quality Early Cretaceous reservoir was encountered just above basement. Use of modern high-resolution seismic in the area will enable more selective positioning of future wells.

In the central Fly Platform, similar plays are possible involving Jurassic sediments onlapping the Pasca Ridge and Dibiri Spur (Prospect Type 2, Figures 8, 10). The potential reservoirs in this case are expected to be clean, shoreface sandstone which probably developed around the margins of the then-emergent basement ridge. These sandstones, if present, would be at depths of 3000 to 4000 m.

Simple anticlines are present along the northern margins of the Fly Platform in the Darai Plateau region (Prospect Type 3, Figure 8). Structural closures have been mapped on these features at the level of the surface Darai Limestone, although their structural configuration at depth is not fully understood because of difficulties in acquiring seismic data over the rugged karst topography. The Toro Sandstone reservoir in this area is expected to occur at depths of about 1000 to 3000 m.

Previous tests on the Darai Plateau include Darai 1 and Kanau 1 (Figure 7), both of which were drilled on the basis of surface geology alone. Provided that hydrocarbon migration took place in post-Late Miocene, these holes may have been dry because they were off-structure at the level of the Mesozoic reservoir. If early migration (post-Cretaceous, pre-Miocene) is favored, the vexing question then arises as to whether there was any closed structure to trap migrating hydrocarbons. The structures appear to be late (post-Miocene).

Broad anticlinal structures have been mapped on the top-Mesozoic unconformity and top-Eocene seismic horizons southwest of the Pasca Ridge (Prospect Type 4, Figures 8, 10). Middle to Late Jurassic sandstone reservoirs may be present in these structures at depths of 3500 to 4000 m. These prospects have never been tested.

Normal faulted anticlines have been identified farther north in the Kikori-Purari Delta region (Prospect Type 5, Figure 8). Trap types include simple anticlinal closures and updip closures against the faults. Middle to Late Jurassic sandstone reservoirs, if present, would occur at depths of 3000 to 4000 m in these as yet untested features.

Numerous trapping possibilities associated with wrench fault movement occur in the Komewu and Fly wrench systems (Prospect Type 6, Figures 8, 11). Trapping along these structural trends has been created by updip closure against faults, by roll-over against faults, and by sandstone deposition against contemporaneous faults. Associated en echelon folds trending obliquely to the fault zone provide a further trapping possibility. Within these fault zones, the Toro Sandstone occurs at a depth of 2000 to 3000 m.

Previous wells drilled along the Komewu Wrench System include Komewu 1 and 2 (Figure 7) drilled in 1957. Subsequent seismic data acquired in 1974 show that neither well drilled a closed structure. Goari 1, situated to the southeast, also may have been drilled outside closure. Kiunga 1, drilled in the northwestern part of the fault zone, encountered an apparently isolated Toro Sandstone reservoir west of the main depositional edge of that unit. More favorable reservoir development and source rock conditions might be expected along trend to the southeast.

Normal-fault-related traps are important in the Morehead Subbasin and in the Daru Embayment (Prospect Type 7, Fig-

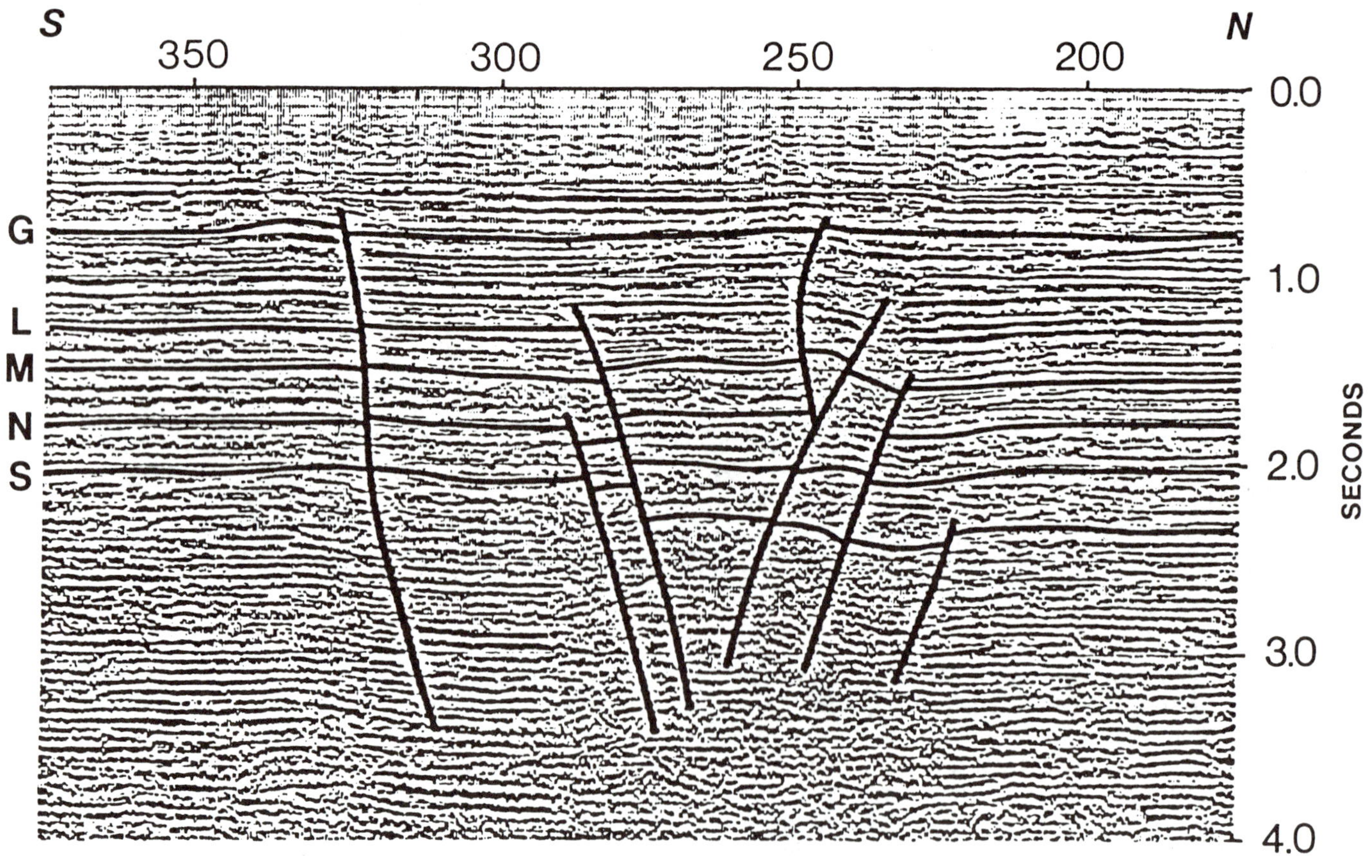


Figure 11. Flower structures, Fly River Platform.

ures 8, 12). The faults are primarily normal, although in the latter area a flower structure and some compressional features have been identified, suggesting the influence of at least some convergent or wrench movement. Updip fault closures are the main trapping possibility in both areas. Reservoir depth would be on the order of 2000 to 3000 m.

The Morehead Subbasin has been tested by only a single well, Morehead 1, drilled in 1957. Only poor refraction seismic control existed in the area at that time. Subsequent reflection seismic data acquired in 1973 demonstrated that the well was not drilled on a closed structure. Both the Morehead Subbasin and the Daru Embayment remain virtually unexplored.

In the Papuan Fold Belt, Mesozoic reservoirs are present within thrust-faulted anticlines (Prospect Type 8, Figures 8, 13). Traps in these structures are formed by roll-over of the overthrust sheet or by closure against either side of the fault. Areal closures of up to 200 km$^2$ and vertical closures of up to 500 m are common, suggesting that these structures are capable of housing very large accumulations of hydrocarbons. The Toro Sandstone reservoir occurs at depths of 1000 to 3000 m deeper along this trend.

The Iehi gas discovery and the Juha gas condensate discovery both occur within this structural setting. Aside from these, very few valid tests of these structures have been made, largely because of an inadequate appreciation for their structural configuration at depth. For example, the Mananda 1 and Cecilia 1 wells (Figure 7) were both dry because they were off structure at depth and passed through the thrust fault before encountering the Toro Sandstone. The Mananda structure and similar structures along trend are being actively explored at present in PPLs 17 and 18 (Figure 3).

*Tertiary prospects*—Tertiary reservoirs are involved in a wide variety of trap types in the central and eastern parts of the Papuan Basin. On the Fly Platform, the main prospects include Eocene and Miocene reef and reef-related traps, basement-high-related traps, and simple structural highs. In

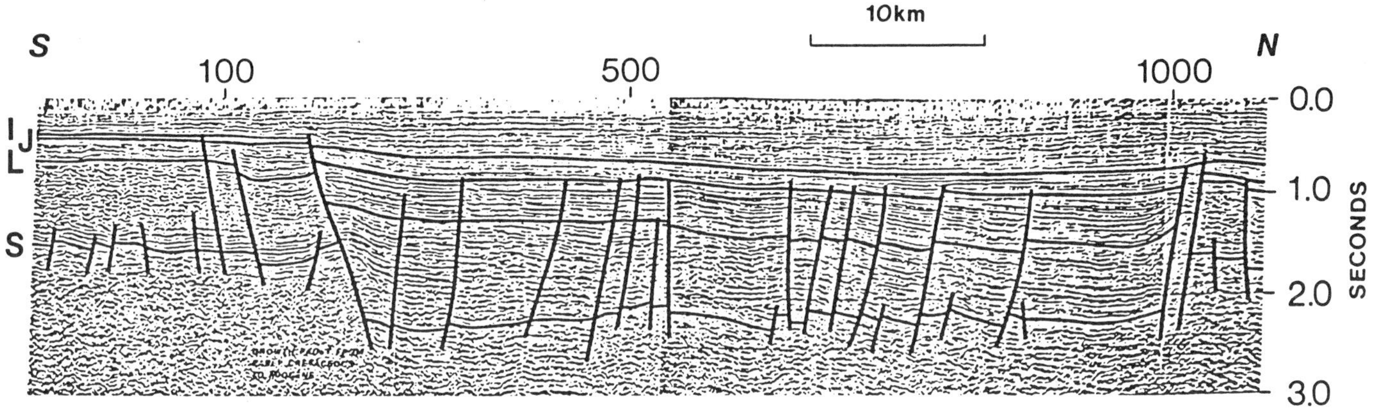


Figure 12. Morehead Basin compressed seismic section.

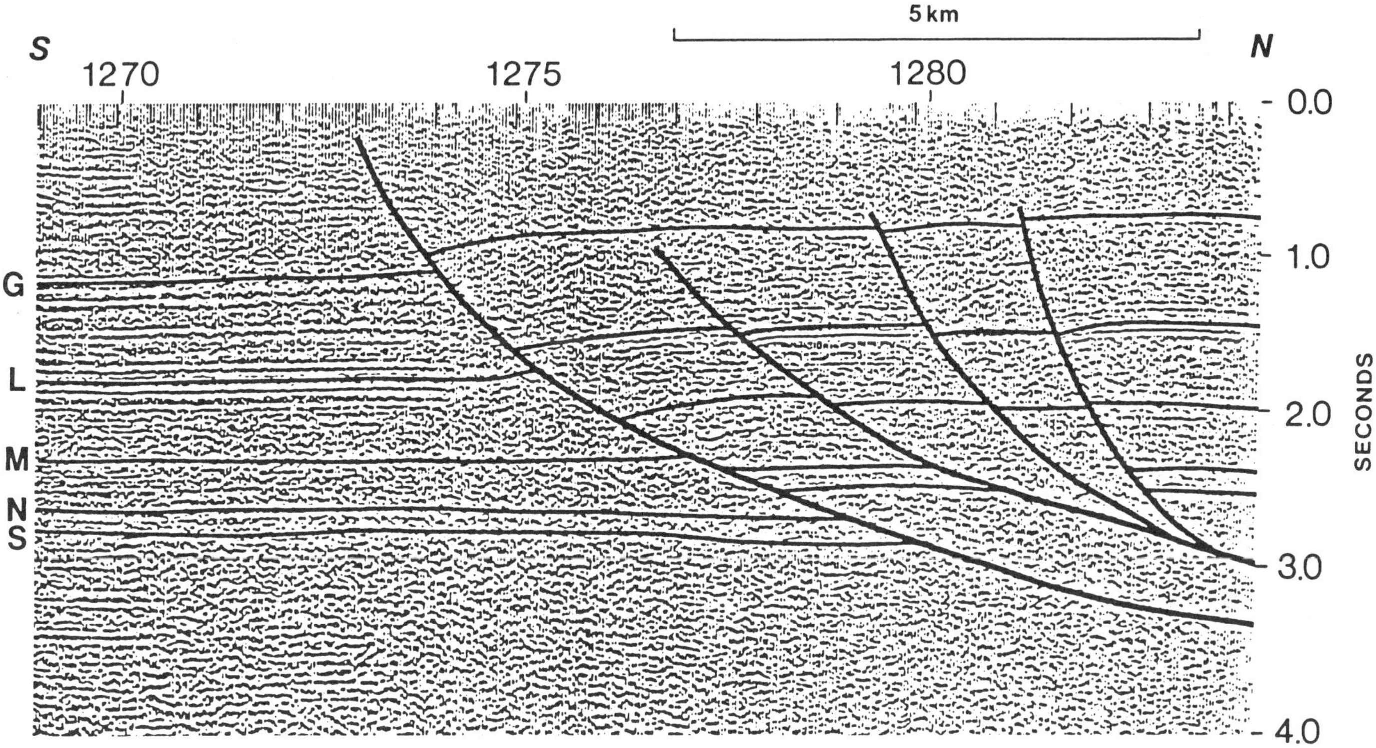


Figure 13. Untested thrust fault trap, western Papua.

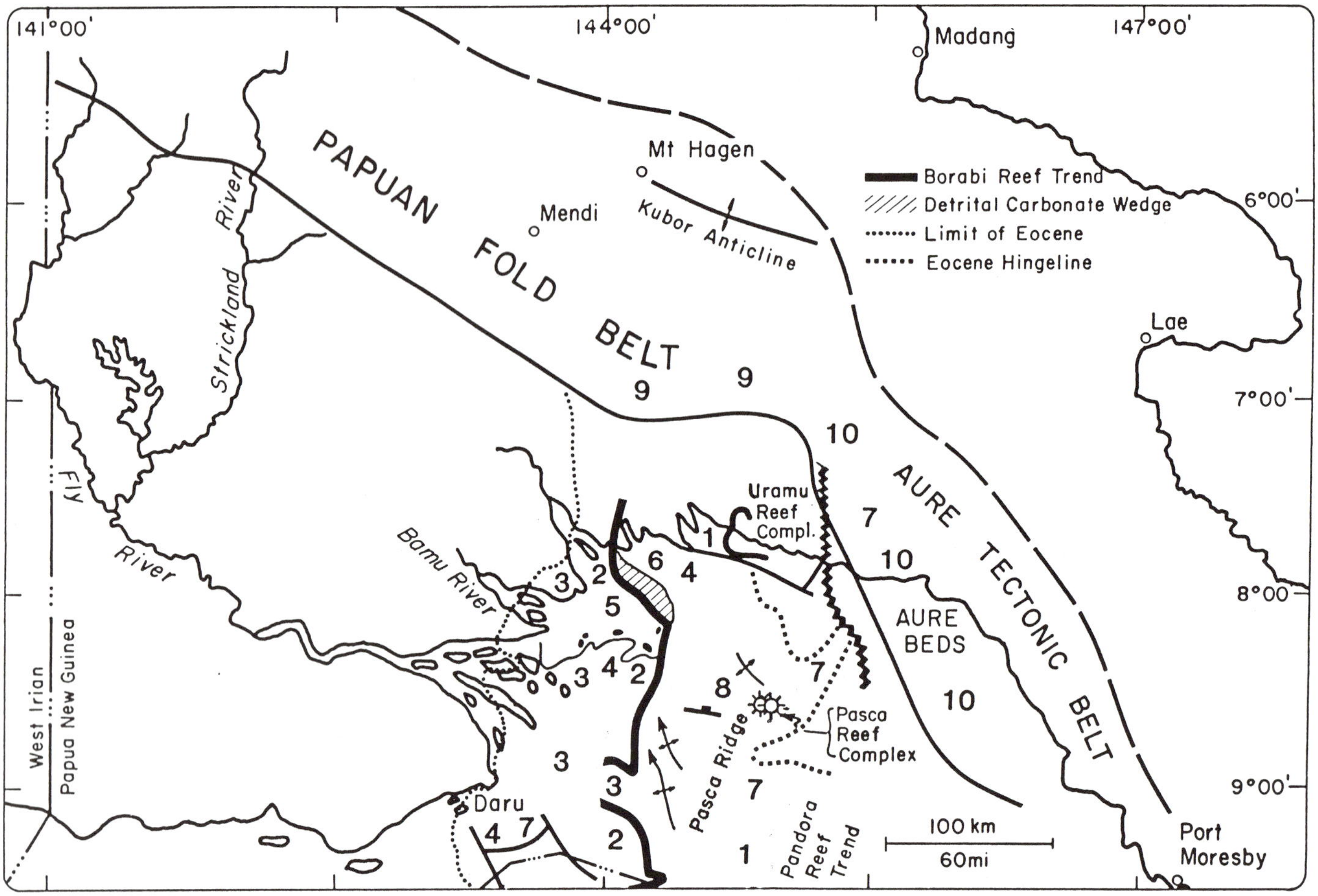


Figure 14. Tertiary prospects in the Papuan Basin.

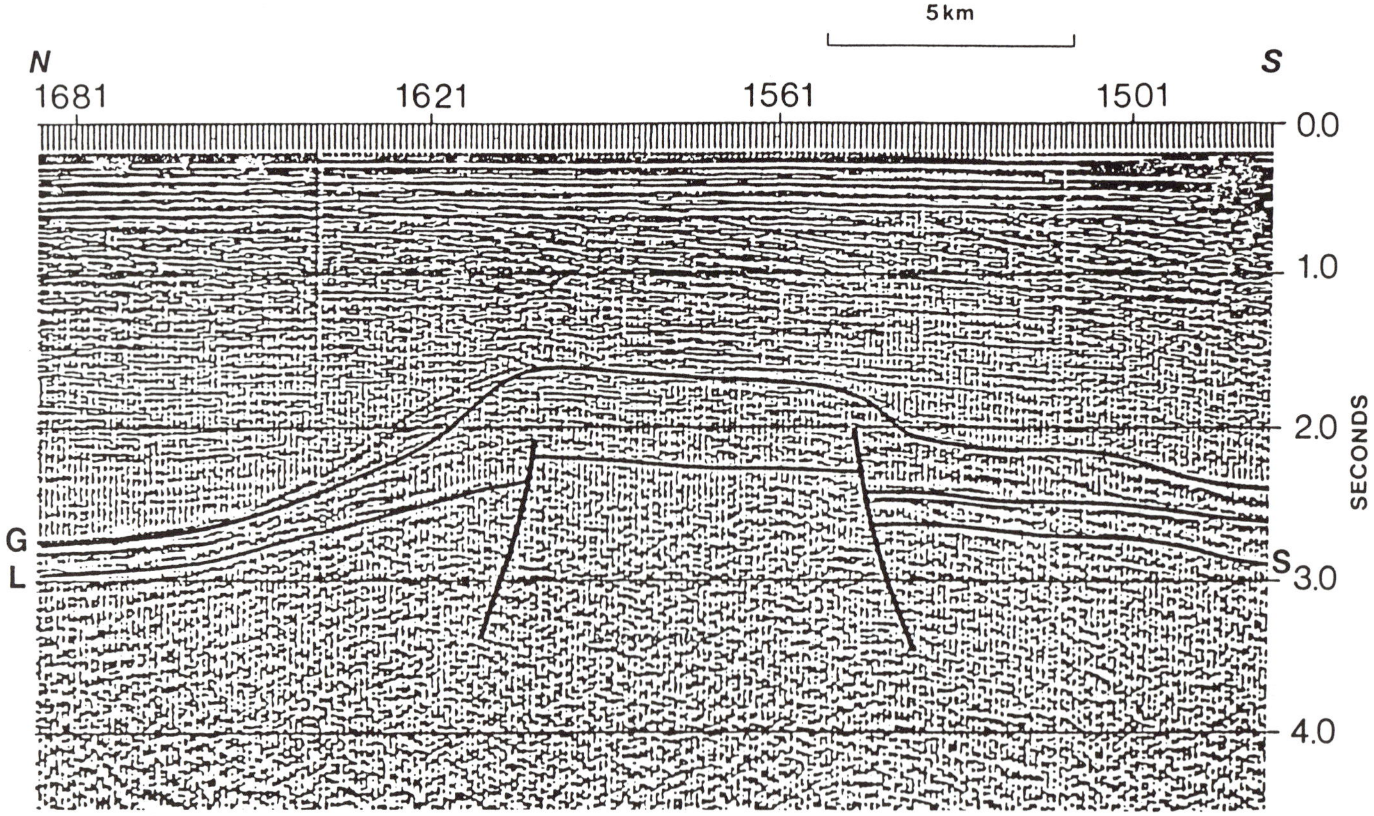


Figure 15. Pandora reef trend.

both the Papuan Fold Belt and the Aure Tectonic Belt, thrust-faulted anticlines are the dominant prospect.

Miocene atoll reefs east of the Borabi Reefs Trend are proven reservoirs, and much potential exists for extending this type of play elsewhere on the Fly Platform (Prospect Type 1, Figure 14). Southeast of and parallel to the Pasca Ridge is the untested Pandora Reef Trend, which consists of atoll reefs situated on an outer continental high (Figure 15). Seismic control over the area is sparse, being limited to only three regional lines run by a research vessel, and the possibilities are good that future detailed seismic over the area will disclose the presence of additional reefs. Reservoir depth along the Pandora Trend is on the order of 2000 m.

To the north, further exploration is warranted in the vicinity of the Uramu High (Figure 14). Untested Eocene and Miocene reef and reef-associated sediments are present at depths of 2500 to 4000 m along the faulted southeastern margin of this feature. In addition, a second major reefal buildup is present immediately east of the Uramu reef (Figure 16). Although targeted by Mira 1, the well missed the reef as shown by subsequent migration of the seismic data, and it therefore remains untested.

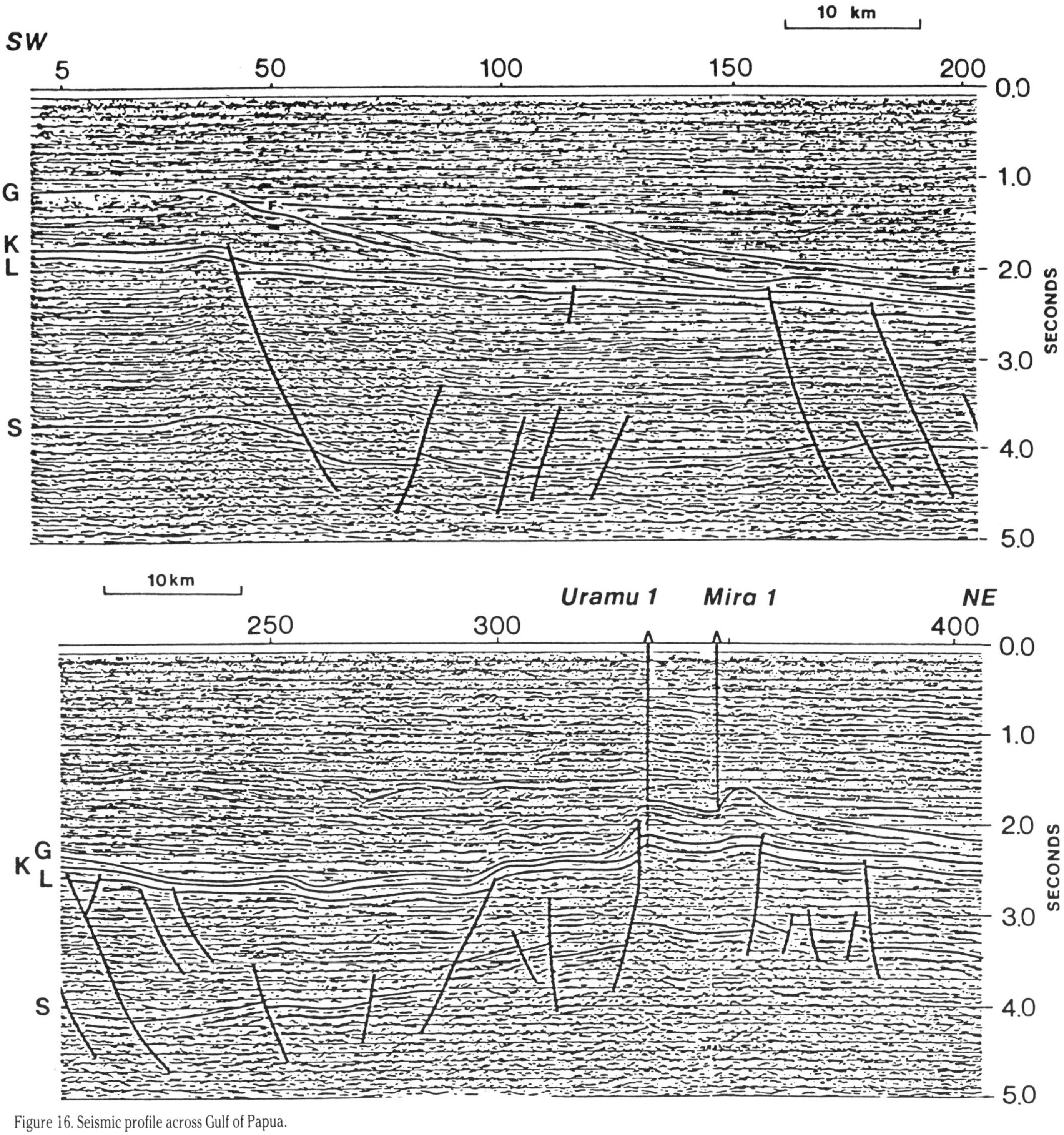


Figure 16. Seismic profile across Gulf of Papua.

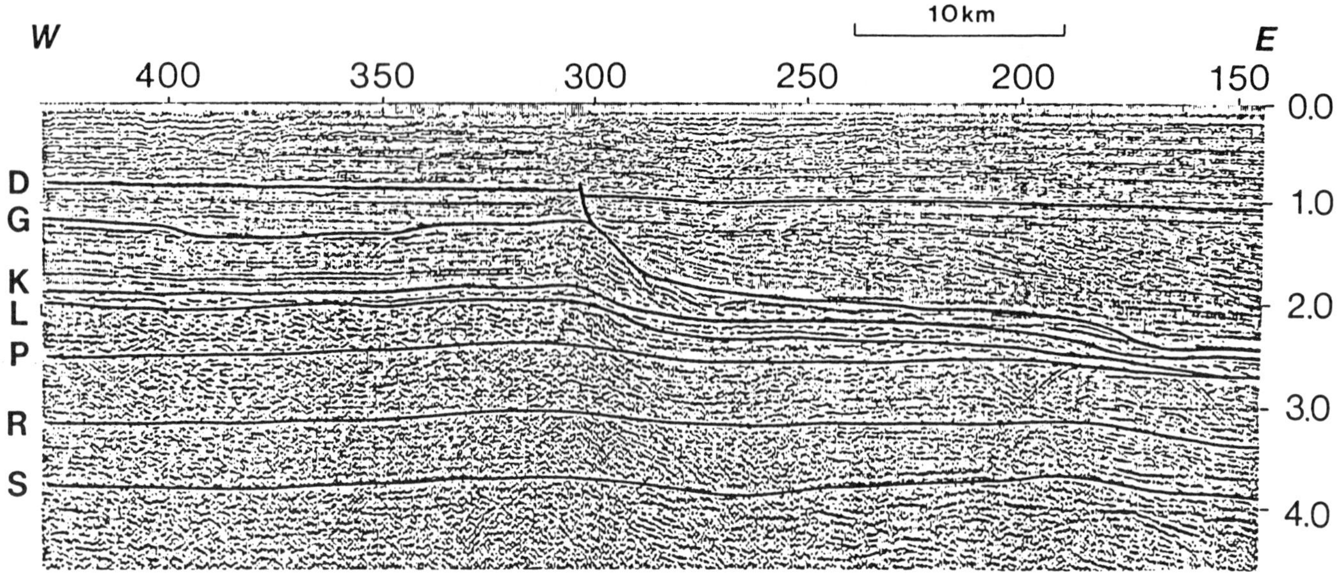


Figure 17. Borabi reef trend.

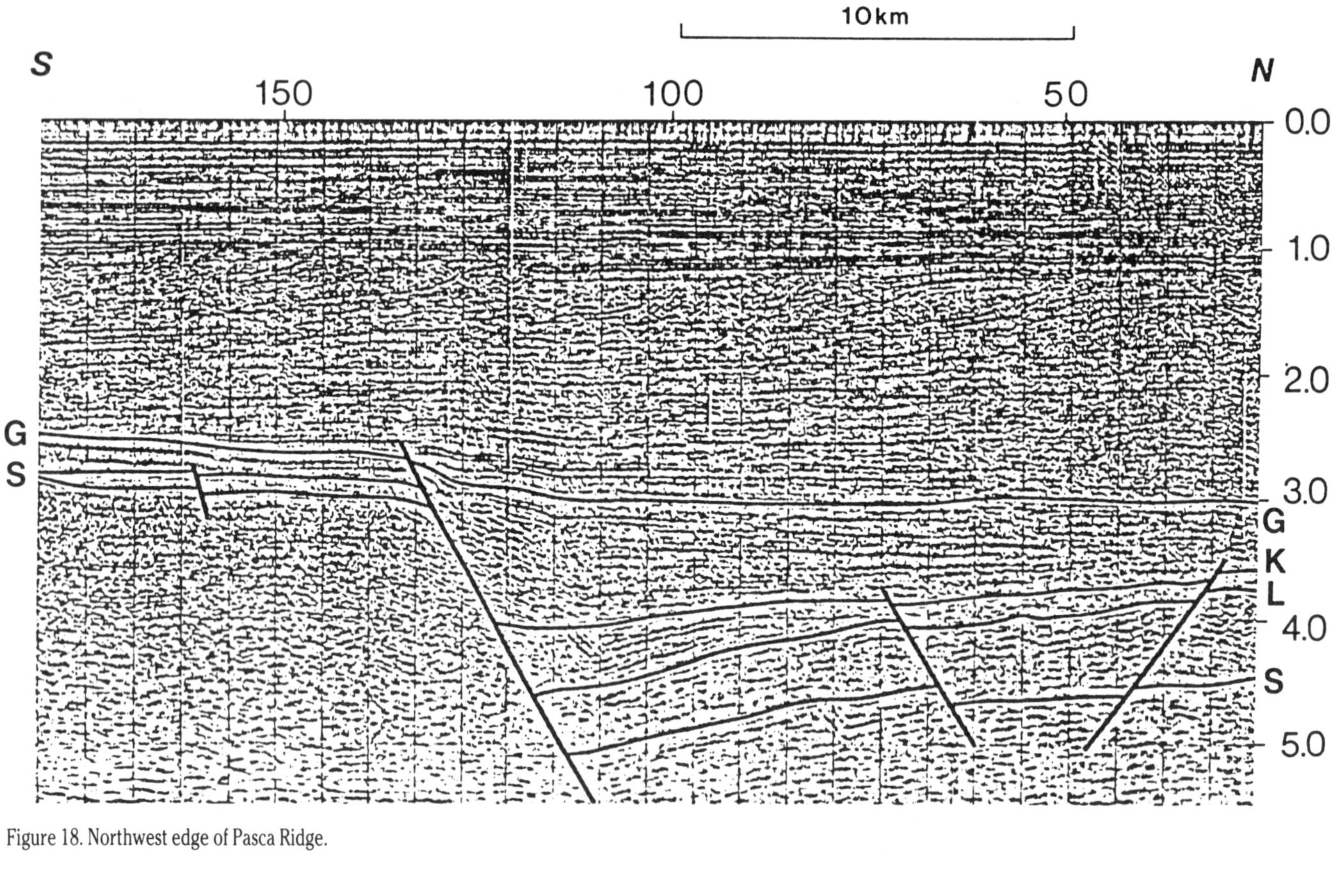


Figure 18. Northwest edge of Pasca Ridge.

Perhaps the largest and potentially most significant Miocene reefs prospect is the virtually unexplored Borabi Reef Trend (Prospect Type 2, Figures 14, 17). More than 1600 m of reef complex has been built up along this feature, which trends north-south for more than 260 km beneath the western Gulf of Papua. The reservoir occurs at depths of 1000 to 1500 m.

Borabi 1 is the only well that has been targeted specifically at this giant feature. Seismic records through the well location are poor, and re-evaluation of available data suggests that it might not have been drilled at an optimal location and/or that lateral seal might have been lacking. A major, regional culmination has been identified by Flower Doery Buchan geophysicists farther south along trend, and this area might be attractive for future exploration.

Potential Miocene intracarbonate platform prospects include coral patch reefs and algal banks, which occur at 750 to 1250 m (Prospect Type 3, Figure 14).

Probably similar, and as yet untested, Eocene patch reefs occur stratigraphically lower at depths of 1500 to 2000 m, par-

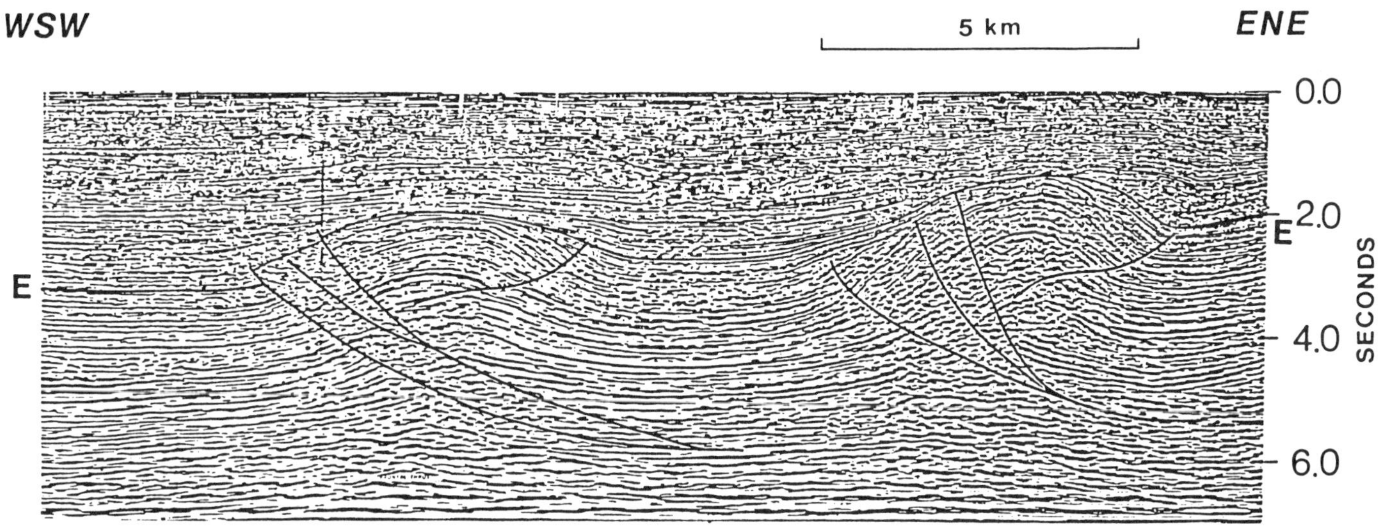


Figure 19. Thrust-faulted anticlines, eastern Gulf of Papua.

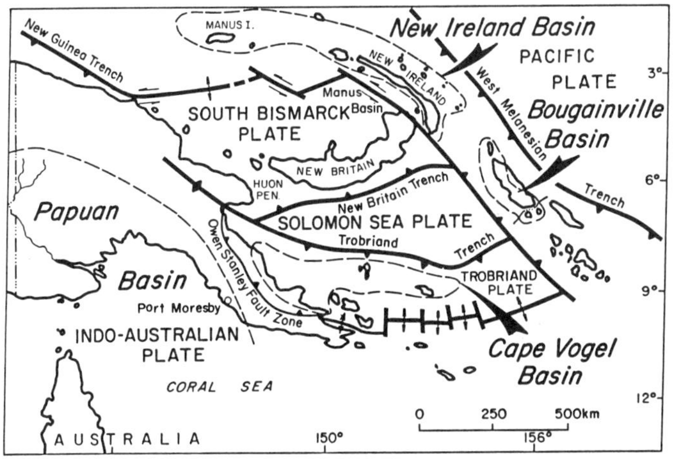


Figure 20. Index map of eastern Papua New Guinea.

ticularly in the area of the Daru Embayment (Prospect Type 4, Figure 14). In addition, immediately west of the southern part of the Borabi Reef Trend, a second intracarbonate barrier reef called the Rui Reef Trend has been identified at a depth of approximately 1500 m. It apparently grew contemporaneously with the lower and middle parts of the Borabi Reef Trend. Finally, foraminiferal-red algal banks of Late Miocene to Pliocene age overlie older Darai Limestone platform carbonates at depths of 600 to 1250 m under the northwestern Gulf of Papua (Prospect Type 5, Figure 14). These have areal extents of 5 to 20 km$^2$ and thickness of up to 100 m. Although a similar feature penetrated by Borabi 1 was found to be waterfilled, the potential still exists for hydrocarbon accumulations in other buildups of this type.

Another potentially large prospect is the major detrital carbonate wedge developed off the northeastern flank of the Borabi Reef Trend (Prospect Type 6, Figures 14, 16). It measures about 15 km wide and 45 km long and probably extends farther south, where seismic control is lacking. Depth to this potential untested reservoir is about 1000 to 2000 m.

Eocene carbonates form potential reservoirs in broad, gentle structural highs commonly associated with normal faulting in the Kikori-Purari Delta region, in the Daru area, and southwest of Pasca Ridge (Prospect Type 7, Figures 10, 14). Reservoir depths range from 1500 to 3500 m depending on locality. This type of prospect would depend upon fractured tight Eocene limestones, which are known to occur in other parts of the basin.

Along the northwestern margin of the Pasca Ridge, tilted half-graben structures contain potential reservoirs of Eocene and Miocene age (Prospect Type 8, Figure 14, 18). As much as 1200 m of sediments are present at a depth of 3500 to 5500 m. The sediments would be composed of carbonate debris derived from the Pasca reef and/or clastics derived from the Pasca Ridge itself. The reservoir units, which have never been tested, either wedge out updip or are truncated by unconformities.

In the Papuan Fold Belt, asymmetrical thrust-faulted anticlines remain the dominant type of structural trap (Prospect Type 9, Figure 14). Fractured Puri limestone is a proven reservoir along this structural trend, and it is from this unit that the only medium-quality oil discovery to date in the Papuan Basin has been made (Puri 1, Figure 7). Fractured Eocene limestones form a secondary target in the same area. Reservoir depths range from 500 to 5000 m. Numerous prospects of this type are present in unlicensed areas to the northwest of PPL 51, where active exploration of the Puri and adjacent structures is currently taking place.

Similar structures involving the Miocene Aure beds and possibly Puri limestone occur in the Aure Tectonic Belt (Prospect Type 10, Figures 14, 19). In the Purari-Vailala River region and offshore along trend to the south, large linear anticlines associated with imbricate thrust sheets have relief of up to 1500 m and wave lengths of 5 to 8 km. Trap depths in this area range from 500 to 3000 m.

## CAPE VOGEL BASIN

### General

During the Oligocene (?) and Miocene, convergence between the Indo-Australian and Pacific plates was partially accommodated by southward subduction along the Trobriand Trench (Figures 20, 21). The Cape Vogel Basin developed in a forearc position between an island arc system and the subduction zone.

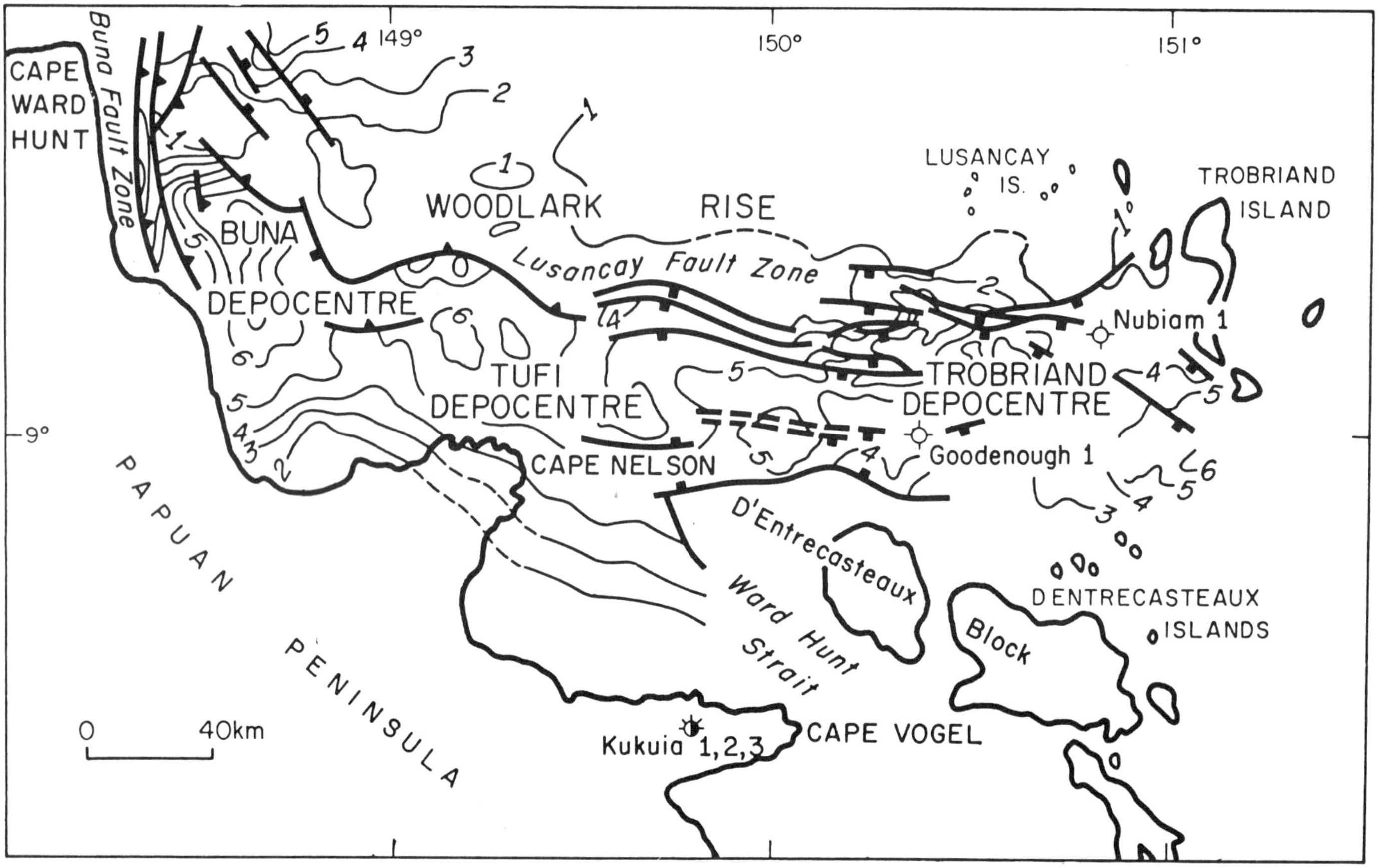


Figure 21. Basement structure and index map of Cape Vogel Basin.

The margins of the Cape Vogel Basin might have been defined as early as the Late Oligocene. Downfaulting produced a series of narrow east-west-trending grabens or proto-depocenters, which persisted throughout the remainder of the history of the basin. Periodic rejuvenation of major fault systems, at times with a wrench component, led to basinal tilting and the formation of widespread unconformities during the Miocene and Pliocene.

## Structure

Major fault systems, including the Lusancay Fault Zone along the northern basin margin, trend dominantly east-west (Figure 21). The majority of the faults intersect basement only, and appear to be normal taphrogenic features formed under a tensional regime at the time of basin formation. Fault density appears to be greatest in the Trobriand part of the basin, although this may in part be a function of denser seismic control in that area compared with the Cape Nelson area to the west. Throughout the basin, the intensity of faulting decreases upward through the sedimentary section.

A secondary west-northwest cross-trend is developed in the northeastern part of the basin near Nubiam 1 (Figure 21). These faults, together with others bordering the D'Entrecasteaux Block along the southern basin margin, also appear to be normal taphrogenic features. Some may have been rejuvenated with a wrench component during the Middle Miocene.

To the west, the style of faulting along the northern basin margin changes to steep reverse and/or oblique slip faults. These perhaps formed in response to latest Miocene-Early Pliocene transcurrent and compressional deformation associated with obduction of the Papuan Ultramafic Belt along the Owen Stanley Fault Zone. Northwest- to north-trending thrust faults were formed at the same time in the Buna Fault Zone to the west.

The principal structural lows in the basin are the Buna, Tufi, and Trobriand depocenters (Figure 21), the northerly trending Buna Depocenter is the deepest, containing as much as 7500 m of sedimentary section. The Tufi Depocenter trends east-west and contains more than 6000 m of sediments. Its margins are formed by the Lusancay Fault Zone in the north, and by gently rising basement in the south. Farther east, the Trobriand Depocenter contains as much as 6300 m of sediments. It is bounded on the north by a series of narrow horsts and grabens, and on the south by the Goodenough Ridge. The latter feature may have formed either as a complex tensional horst block, or as a wrench-related feature.

## Stratigraphy

The Cape Vogel Basin contains a thick sequence of Late Oligocene to Middle Miocene bathyal clastics and minor carbonates, Late Miocene to Pliocene shallow-marine to terrestrial clastics, and Pleistocene to Holocene platform carbonates (Figure 22). The sequence thins significantly toward the basin margins and over basement highs. Numerous unconformities of different magnitudes have been recognized in seismic-reflection profiles.

Basement in the region is formed by the Lokanu association ophiolite sequence of Late Paleocene to Eocene age (Figure 22). The Late Eocene and much of the Oligocene sedimentary record is largely missing because of nondeposition and/or erosion, although areally restricted mid-Oligocene neritic carbonates of the Iarame limestone are known to occur in the Cape Vogel area.

Commencing in the Late Oligocene, widespread volcanism led to the deposition of a thick sequence of pyroclastic and volcaniclastic rocks belonging to the Iauga volcanics (Figure 22). Deposition occurred mainly at bathyal depths, although neritic carbonates of the Armit Caves limestone accumulated locally during the latest Oligocene or Early Miocene in a structurally positive area west of Cape Ward Hunt.

At about the same time, extensive down-to-basin faulting produced protodepocenters, which were infilled by bathyal siltstones, mudstones, and argillaceous sandstones of the "pre-Woruka" and Woruka units (Figure 22). Initial sedimentation was confined to a series of narrow grabens near the northern basin margin. By the late Early Miocene, sedimentation had become basin-wide as the Goodenough Ridge and other basement highs were buried by onlapping sediments. Shallow-water conditions prevailed locally in the northwestern part of the basin, where neritic carbonates of the Robinson Bay limestone accumulated from the latest Early Miocene onward.

After a brief period of faulting and associated folding during the Middle Miocene, bathyal sedimentation continued over a wide area with the deposition of pelagic carbonate and volcaniclastic sediments of the Castle Hill unit. A massive influx of fine-grained clastic sediments belonging to the Nubiam shale followed, marking the beginning of a major regressive phase in the basin.

Another brief period of uplift and erosion ensued during the mid-Late Miocene. Sandstone, shale, conglomerate and rare lignite of the Ruaba unit were subsequently deposited over a wide area in shallow-marine, shoreface and terrestrial environments. Synsedimentary tectonism continued, leading to erosional truncation of this unit during the Early Pliocene.

By the mid-Pliocene, areas proximal to the emergent cordillera in the south and southwest began to receive a thick accumulation of coarse-grained, terrestrial, clastic sediments now constituting the Kwinimage Formation (Figure 22). Shallow-marine sandstone, limestone, and mudstone of the Awaitapu Formation accumulated contemporaneously in more distal areas.

Uplift of the basin margins continued throughout the remainder of the Pleistocene. Shallow-marine conditions returned to most of the basin, and carbonate platform sediments of the Kiriwina Limestone were widely deposited.

## Exploration History

Exploration commenced in the Cape Vogel Basin in 1928, when three shallow wells were drilled on the Cape Vogel Peninsula (Figure 21). The deepest of these, Kukuia 2, reportedly encountered shows of gas and light oil in the Ruaba unit before reaching a total depth of 310 m.

More recent exploration has concentrated on the offshore part of the basin. Amoco Australia Exploration Co. acquired 2500 km of seismic-reflection data in the Trobriand area in 1972-1973, and an additional 1000 km of data were acquired by Texaco Exploration Co. over the Cape Nelson area in 1974. Goodenough 1 and Nubiam 1 (Figure 21) were subsequently drilled in the Trobriand part of the basin in 1973 to test interpreted Miocene reefal anomalies. The wells bottomed at 2835 m and 2366 m, respectively, in basement fault blocks. No further wells were drilled.

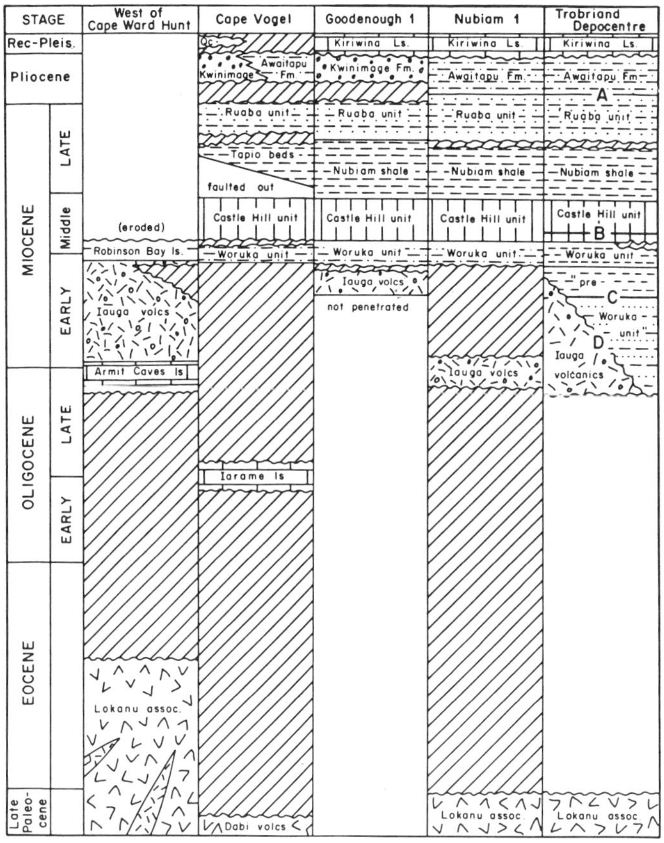


Figure 22. Stratigraphic chart for Cape Vogel Basin.

A single Petroleum Prospecting License (PPL 41; Figure 3) is currently held in the western Cape Vogel Basin by Albion International Resources Inc., Aberford Resources Ltd., and Cordura Exploration. The companies acquired new seismic-reflection data over their license area in November 1985 and were expected to drill an exploratory well in late 1986.

### OIL AND GAS POTENTIAL

*Reservoir potential*—The widespread Ruaba unit probably has the best reservoir potential in the basin, assuming that relatively late stage (post-Miocene) hydrocarbon migration has taken place. Sand bodies within this shallow-marine to terrestrial unit probably have variable thicknesses, and may be stacked vertically. Major sand trends are likely to be oriented east-west, parallel to the paleoshoreline. Log porosities as high as 27% have been encountered in the wells, and permeabilities are reported to be fair.

Tuffaceous clastic turbidite reservoirs are also probably developed in the Woruka and "pre-Woruka" units, particularly on the onlap margins of basement highs. Where encountered by the wells, coarse-grained intervals within these units showed a range of porosities and generally had poor permeabilities. Well-sorted, water-laid tuffs with better reservoir properties may be present elsewhere in the basin.

Potential carbonate reservoirs include the Robinson Bay limestone, although its areal extent offshore is unknown. Local Oligocene carbonate buildups, similar in lithology to the Iarame limestone, may also be associated with upfaulted basement blocks in parts of the basin.

*Source potential*—The source potential of the basin is poorly defined because existing wells have not penetrated the

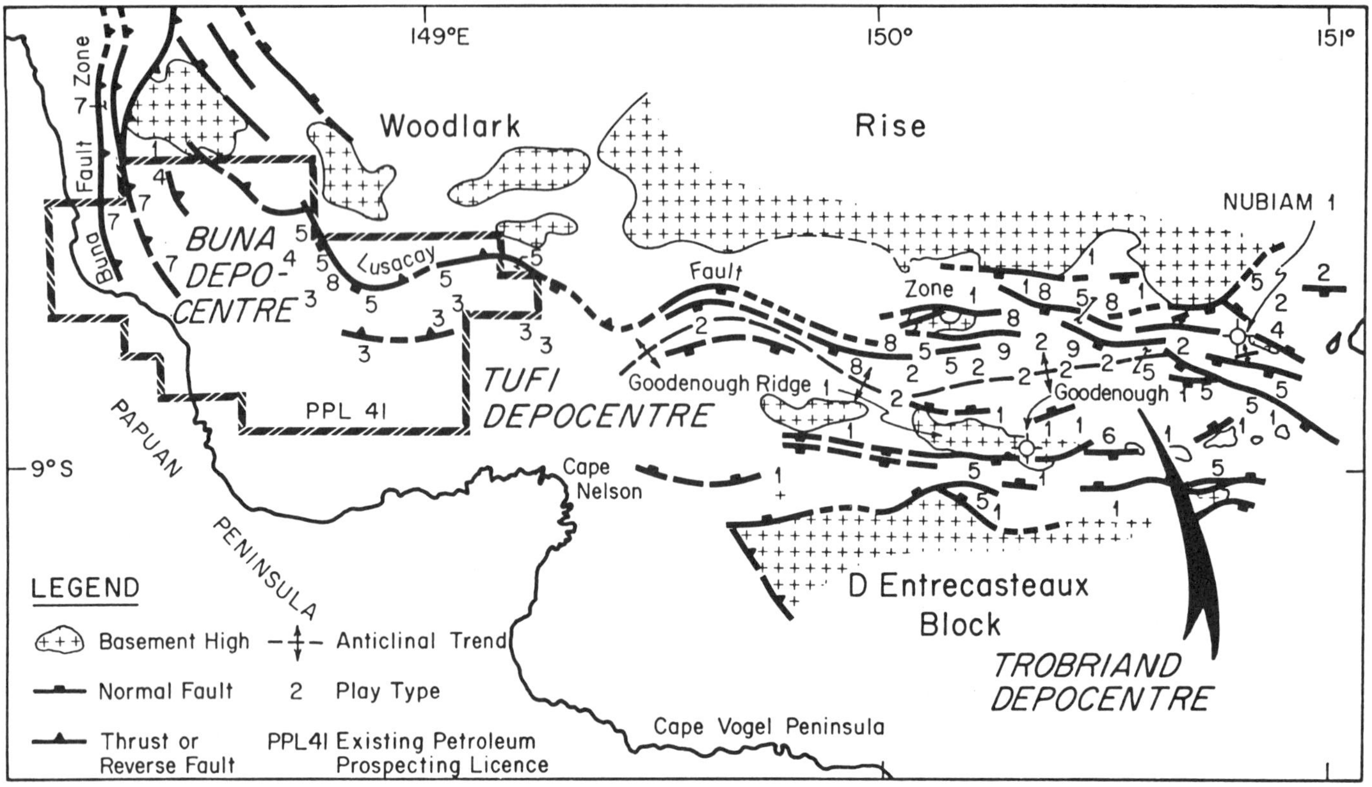


Figure 23. Cape Vogel Basin play map.

entire sedimentary section. This is particularly true in the western part of the basin, where no offshore wells have been drilled. However, an unconfirmed report of an oil seep along the Gira River (west of Cape Ward Hunt) may suggest source potential in that area.

Geochemical analyses of the late Early Miocene to Early Pliocene sequence in Goodenough 1 and Nubiam 1 indicate generally lean to fair total organic carbon (TOC), in the range of 0.2 to 0.9%. However, some samples from the lower Ruaba unit and upper Nubiam shale yielded higher TOC values of 1 to 1.5%.

Humic kerogens predominate, and on pyrolysis these generally yielded low hydrogen and high oxygen indices. However, liptinitic kerogen consisting of cutinite, resinite, and sporinite accounts for as much as 15% of the kerogen composition in parts of the Nubiam shale and Castle Hill unit. If present in sufficient quantities, kerogens of these types have the potential to source gas, condensate, and light oil under suitable maturation conditions.

The present-day geothermal gradient in the Trobriand part of the basin has been measured at 1.6° C/100 m. Vitrinite reflectance and spore color index data from Nubiam 1 and Goodenough 1 suggest that sediments buried deeper than about 3.5 to 4 km should be mature for hydrocarbon generation. Because sedimentary thicknesses exceed 6000 m in each of the three depocenters, a substantial volume of sediment likely lies within the oil generation zone in all three areas.

## Play Concepts/Prospects

Re-evaluation of seismic data in light of well results and onshore geological information has revealed a variety of structural and stratigraphic trapping possibilities. These include onlap/pinchout traps against basement highs; simple anticlines; normal, wrench, and thrust-fault-related traps; unconformity traps, and stratigraphic traps associated with facies changes.

Onlap/pinchout traps are developed against the basin margins in the Trobriand area, and against intrabasinal highs such as Goodenough Ridge (play type 1; Figure 23). Turbidite sandstone reservoirs may be developed in the "pre-Woruka" and Woruka units where they onlap rising basement (Figures 24, 25). Depths to these potential reservoirs would be highly variable depending upon location and stratigraphic position, but would generally range between about 1500 and 4800 m.

Simple anticlines of Early to Middle Miocene age are distributed along an anticlinal trend north of the basin axis (play type 2; Figure 23). Most of the culminations are drape features associated with pre-Miocene or earliest Miocene basement structures (Figures 24, 25). The depth range of potential Early to Middle Miocene clastic turbidite reservoirs in the "pre-Woruka" and Woruka units is about 2000 to 4500 m.

Middle to Late Miocene anticlines are developed both inside and adjacent to PPL 41 in the western Cape Vogel Basin (play type 3; Figure 23). Vertical closures of up to 450 m and areal closures of up to 15 $km^2$ have been mapped. Potential Miocene clastic reservoirs range in depth from about 2500 to 4500 m in the region.

Pliocene anticlines are less common, but have been recognized in the northwestern extremity of the basin and east of Nubiam 1 (play type 4; Figure 23). The culmination mapped in the latter area is developed over a possible Miocene wrench-faulted structure. It has an areal closure of 12 $km^2$, and a vertical closure of about 100 m. Pliocene anticlines in the northwestern part of the basin tend to be broad and gentle and appear to be simple drape features.

The best potential reservoir in these structures would be

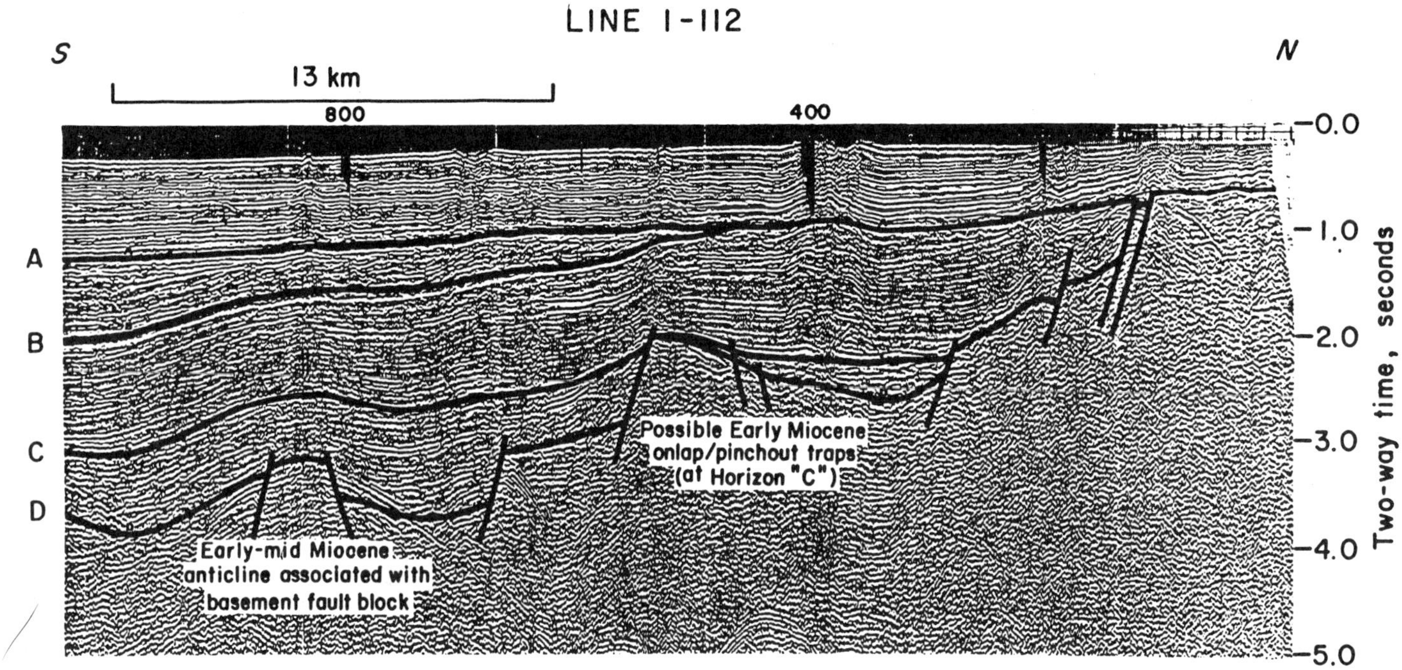


Figure 24. North-South seismic record section across Trobriand Area.

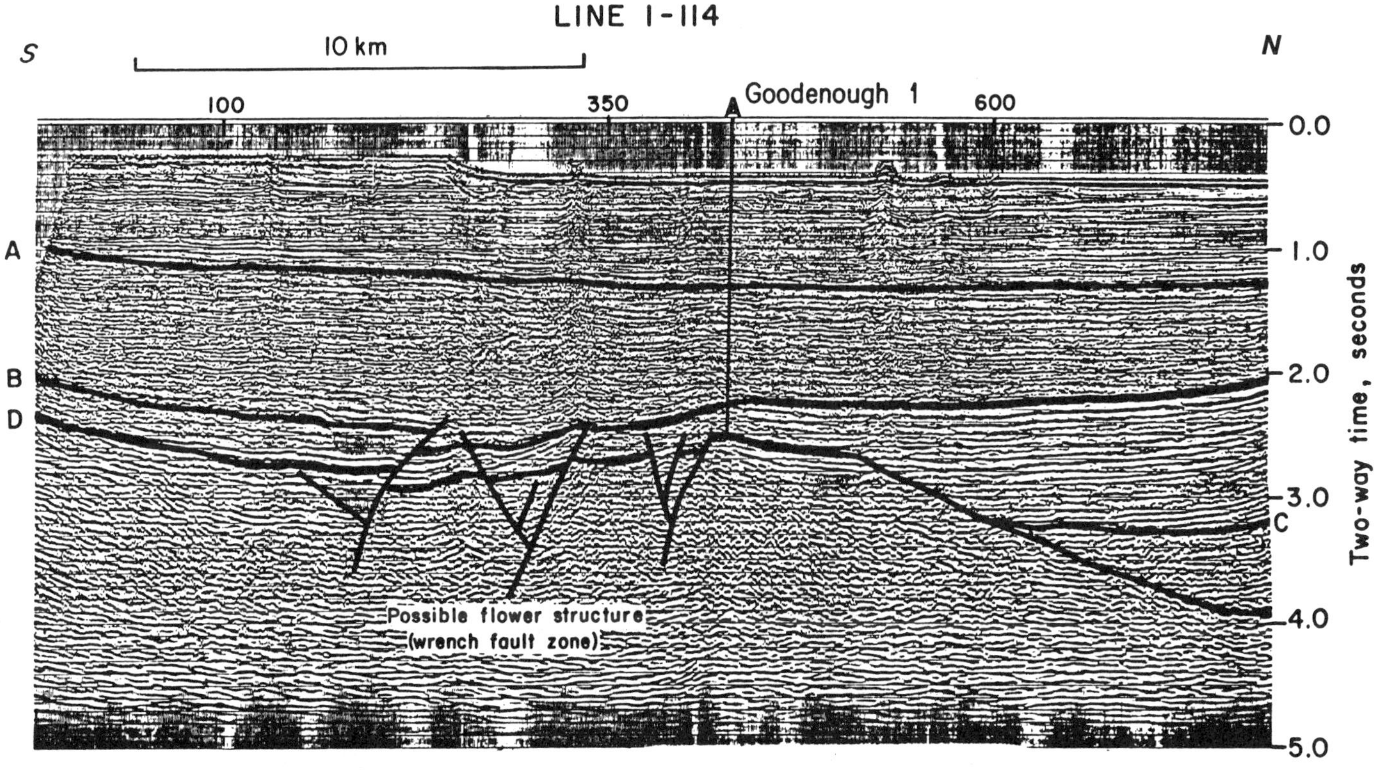


Figure 25. North-South seismic record section near Goodenough 1.

paralic sands in the Ruaba unit, which lie at depths of as little as 700 m in the Nubiam 1 area. Possible clastic turbidite reservoirs in the underlying sequence would extend to a maximum depth of about 2800 m.

Potential fault traps of Early Miocene age are developed along the Lusancay Fault Zone on the northern basin margin, and along the southern margin of the basin south of Goodenough Ridge (play type 5; Figure 23). Updip fault traps along these trends may involve Early to Middle Miocene clastic reservoirs in the "pre-Woruka" and Woruka units, which would lie within a depth range of about 1500 to 4500 m (Figure 25). Vertical closures of up to 450 m and areal closures of as much as 14 km$^2$ have been mapped.

Possible wrench-fault-related plays have been identified in the vicinity of Goodenough Ridge (play type 6; Figure 23). The Middle Miocene and older sequence is complexly faulted in that area, whereas younger sediments are relatively undeformed. Both updip fault traps and rollovers are discernible on individual lines, but detailed, high-resolution seismic-reflection coverage will be necessary to establish individual closures. Turbidite clastic reservoirs in the Woruka and "pre-Woruka" units would lie at depths of 2500 m or more along this trend.

Thrust-faulted anticlines of Late Miocene to Pliocene age are present in the Buna Fault Zone near Cape Ward Hunt (play type 7; Figure 23). Some of the faults terminate below the Early Pliocene (A) reflector (Figure 26), whereas other remained active during the Quaternary and Holocene. Vertical closures on the order of 300 m and areal closures of up to 65 $km^2$ have been mapped, although existing seismic coverage is very sparse. Neritic carbonate reservoirs of the Middle Miocene Robinson Bay Limestone may be involved in the thrusting, provided that this unit extends offshore from Cape Ward Hunt. The top of that stratigraphic interval generally lies at a depth of less than 1000 m on the crests of the structures.

Unconformity traps may be developed in the northeastern part of the basin beneath regional unconformities corresponding to the basal Pliocene (A), Middle Miocene (B), and intra-Early Miocene (C) reflectors (play type 8; Figure 23). Additional unconformities have also been recognized below the C reflector. Traps involving Early-Middle Miocene reservoirs beneath the B and C reflectors are generally subtle, and would require definition by detailed, high-resolution seismic-reflection coverage. Potential traps involving Ruaba unit sands are well defined below the A reflector (Figure 26). Their prospectivity would rely upon relatively late stage (post-Miocene) hydrocarbon migration.

Other stratigraphic traps may be present along the northern flank of the basin in the Middle Miocene and older (B-D) sequence (play type 9; Figure 23). Lateral amplitude changes along individual reflections have been recognized and are thought to represent sand/shale or carbonate/shale facies changes. These features are also visible in reprocessed seismic sections, and are not thought to be a function of processing parameters.

The above play types have been identified from existing seismic control in the Cape Nelson and Trobriand parts of the basin (open exploration areas CVA and CVB, respectively; Figure 3). Little is known about the deeper water areas to the east in open exploration area CVC, where only sparse, regional research seismic-reflection control is available. Interpretation of existing lines suggests that onlap/pinchout traps against basement, large anticlines, and normal fault traps are present. This suggests that the area is worthy of additional exploration.

## BOUGAINVILLE AND SOUTHEASTERN NEW IRELAND BASINS

### General

The Bougainville and New Ireland Basins are situated in a zone of complex interaction between the Pacific and Indo-Australian plates (Figure 20). Oblique convergence between these major features was partially accommodated by subduction along the West Melanesian Trench, resulting in the formation of a volcanic island arc system during the Eocene and Oligocene.

Within this general tectonic setting, the Bougainville Basin developed in a back-arc position along the western margin of the island arc system. At about the same time, the New Ireland Basin formed to the north in a fore-arc position between the island arc and the West Melanesian Trench (Figure 27).

Subduction eventually ceased along the West Melanesian Trench, perhaps because of choking of the subduction zone by anomalously thick oceanic crust of the Ontong Java Plateau. Arc reversal ensued, and a new subduction zone, the New Britain Trench, formed along the western margin of the old island arc. The reversal apparently occurred during the Late Miocene in the Bougainville-New Ireland region, but had commenced as early as the Late Oligocene in parts of the island arc to the northwest. As a result, the two sedimentary basins appear to have been transferred from the Indo-Australian plate to the Pacific plate, and the Bougainville Basin came to occupy a fore-arc position relative to the New Britain Trench (Figure 27).

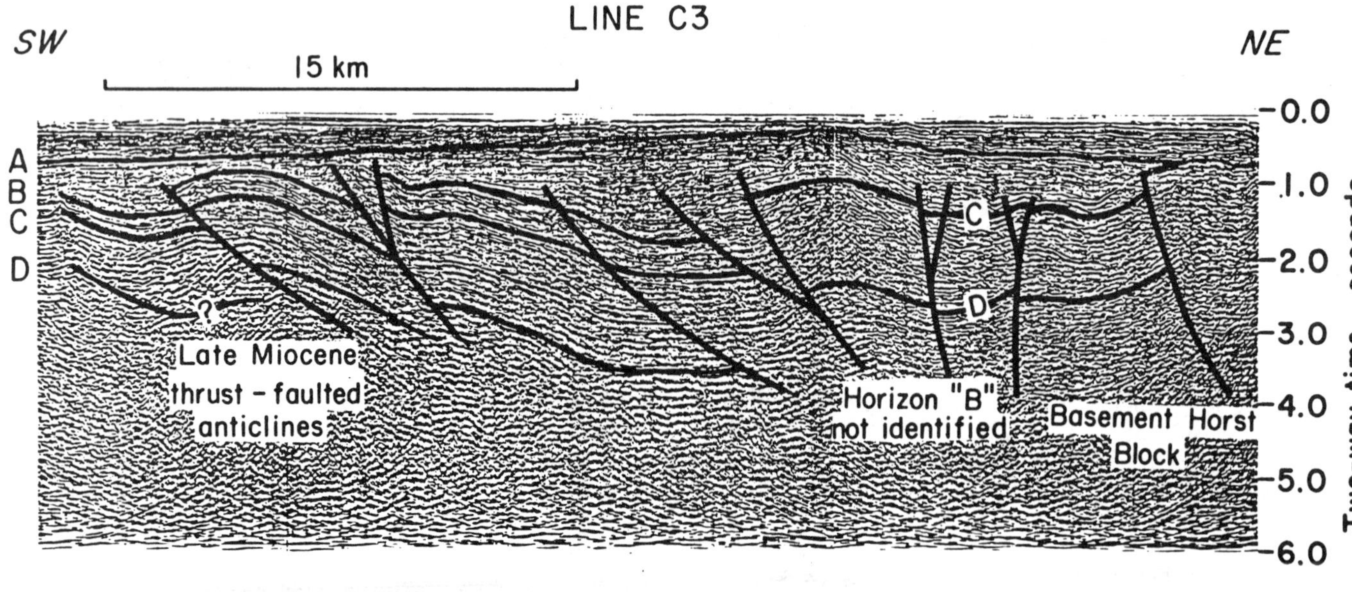


Figure 26. Northeast-Southwest seismic-reflection section Buna Area, Cape Vogel Basin.

Back-arc spreading commenced in the Manus Basin during the mid-Pliocene, causing the New Britain-Huon portion of the arc to rotate anticlockwise and collide with the Papua New Guinea mainland. At the same time, the Bougainville-New Ireland portion of the old island arc system commenced movement to the northwest along a transform fault system. As a result, New Ireland is currently situated well to the northwest of the New Britain Trench (Figure 27).

To date, very little exploration has been undertaken in the potentially prospective Bougainville and New Ireland basins (Figure 27). All nonconfidential geophysical data from previous exploration were re-evaluated and interpreted in light of onshore stratigraphy and well information. In addition, new lithostratigraphic and biostratigraphic data were obtained from field investigations on New Ireland.

Based on the results of these investigations, two open exploration areas have been designated in the Bougainville and New Ireland basins (Figure 3). Area NS, situated mainly offshore to the southwest of Bougainville Island, has an area of approximately 7100 $km^2$. Area NI, which covers the offshore part of the southeastern New Ireland Basin, has an area of about 17,000 $km^2$.

## Bougainville Basin

### Structure

Positive structural elements in the Bougainville Basin include the Kunua and Aitara highs, and the northwest-trending Tarangau Ridge along the western basin margin (Figure 28). To the southeast of the Tarangau Ridge lies a major structural low called the Laruma depocenter, which contains as much as 7600 m of early Miocene to Holocene sediments.

Major normal fault systems in the basin trend dominantly northwest, including the series of faults stepping down into the New Britain Trench along the western margin of the Tarangau Ridge (Figure 28). Displacements of more than 2000 m have been recognized in the Miocene section along some of the faults, a few of which remained active during the Pliocene.

A secondary northeast-southwest normal fault trend is present along the northern margin of the Kunua High (Figure 28). This may be part of an older, pre-Miocene basement trend, which was subsequently reactivated. The intersection of this trend with the primary northwest fault trend has formed triangular-shaped fault blocks.

An important west-northwest-trending reverse fault system lies to the south of the Laruma depocenter. Faults bound the Motupena anticlinorium (Figure 28), which developed during a Late Miocene compressional episode associated with arc reversal. Nearby north-northeast-trending normal faults possibly formed at about the same time.

### Stratigraphy

The Bougainville Basin contains a thick sequence of Early Miocene platform carbonates and laterally equivalent bathyal sediments, Miocene-Pliocene volcaniclastic turbidites, and Pleistocene-Holocene neritic limestones (Figure 29). The pre-

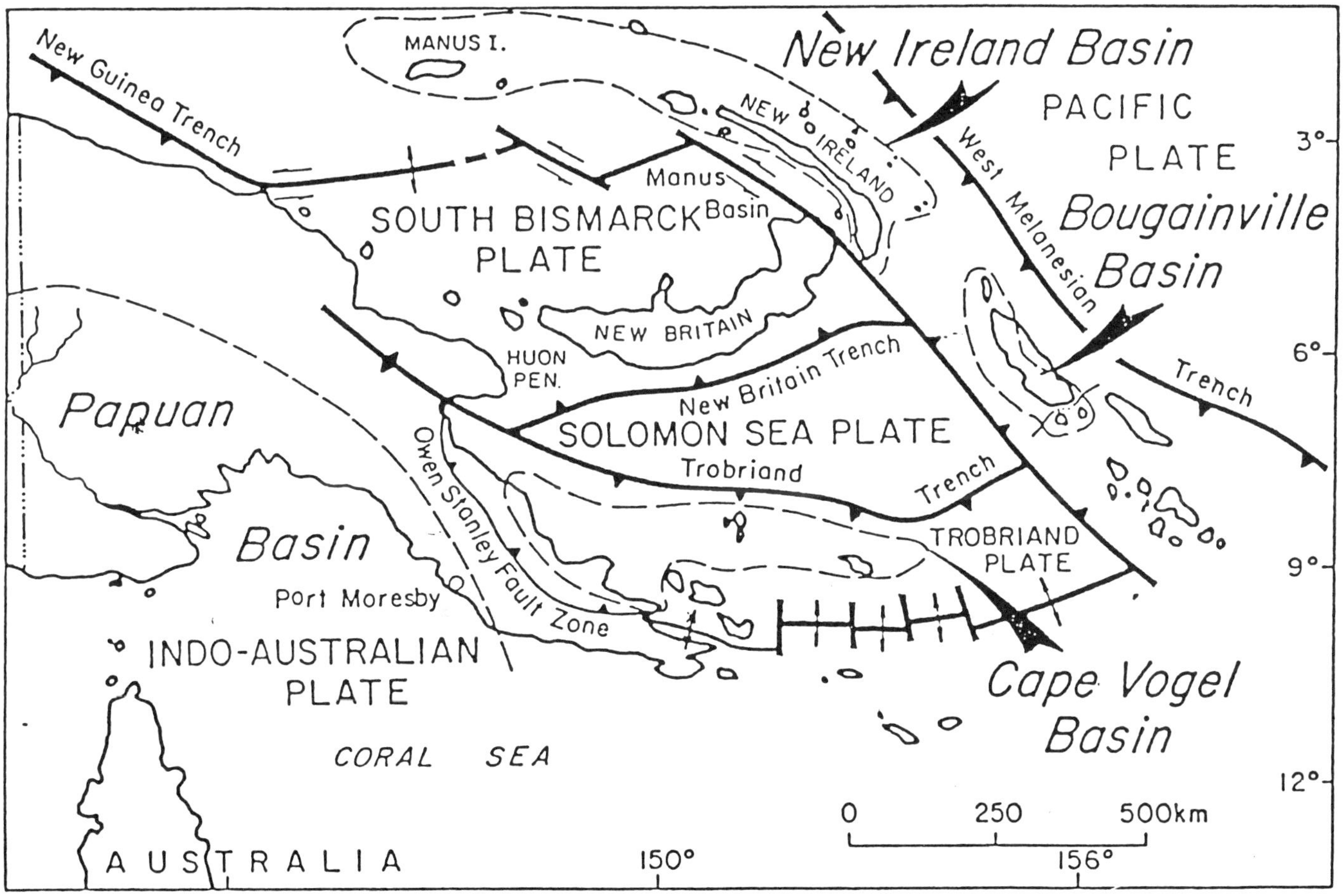


Figure 27. Index map of Bougainville and New Ireland Basins.

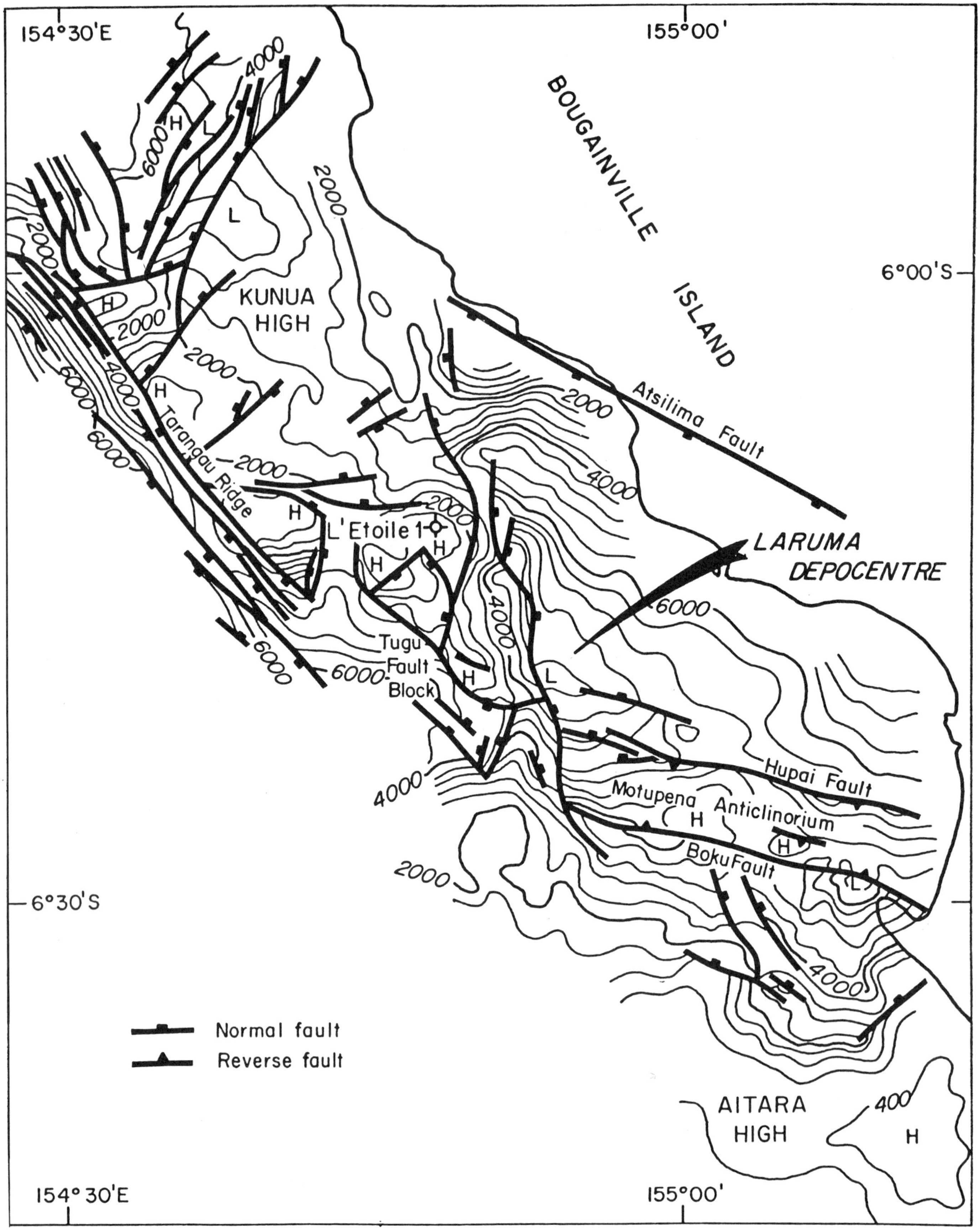


Figure 28. Structural elements of the Bougainville Basin.

Pliocene sequence onlaps positive structural features such as the Kunua High, L'Etoile Structure, and Aitara High. A complete sedimentary sequence appears to be preserved in the adjacent Laruma depocenter. As the entire sedimentary section was not penetrated by the only well drilled in the basin, the composition of the pre-Pliocene section is inferred from a combination of marine seismic data and outcrop data from neighboring islands.

Basement in the Bougainville Basin is formed by Oligocene-Early Miocene andesitic rocks of the Kieta Volcanics, which built upward upon older Paleogene oceanic crust during active island arc volcanism. The Kieta Volcanics are overlain by an additional, unnamed volcanic sequence of earliest Miocene age (Figure 29).

Carbonate platform sedimentation became established in at least the eastern part of the basin by the mid-Early Miocene. Algal-foraminiferal-coralline biocalcarenites and biocalcirudites of the Keriaka LImestone probably extend immediately offshore from Bougainville Island. In more distal areas of the basin, fine-grained volcaniclastic and carbonate sediments probably accumulated contemporaneously in a bathyal environment.

Subsequent sedimentation until the end of the Miocene took place mainly at bathyal depths. The Mole Formation (B-D seismic sequence; Figure 29) is a volcaniclastic turbidite sequence, the upper part of which is dominated by fine-grained sediments. It exceeds 5200 m in thickness in the Laruma depocenter but thins substantially on the margins of basement highs.

Arc reversal during the Late Miocene initiated extensive normal and reverse faulting. By the latest Miocene, large stratovolcanoes developed along the eastern basin margin, where lavas and proximal pyroclastic rocks of the Bougainville Group were laid down. At the same time, a massive influx of fine-grained pyroclastic and volcaniclastic sediments (Mono Siltstone; Figure 29) prograded northwestward across more distal parts of the basin. Carbonate buildups grew locally within this bathyal regime on highs created by synsedimentary tectonism. Pliocene barrier reefs grew along the Tarangau Ridge and on the outer flank of the Aitara High.

Regression became more pronounced during the Late Pliocene and earliest Pleistocene because of a combination of eustatic sea level changes and regional tectonic activity. Neritic carbonates and coarse-grained calcareous clastic sediments were subsequently deposited over the Mono Siltstone.

## Oil and Gas Occurrence

*Exploration History*—Between 1971 and 1974, a total of 1800 km of seismic-reflection data were acquired over the offshore part of the basin by Shell Development (Australian) and Amoco Australian Exploration Co. L'Etoile 1 was subsequently drilled in 1975 to test what the companies believed to be a massive Miocene carbonate buildup in the central part of the basin (Figure 28). The well, however, reached a total depth of 1682 m in the upper part of a basement fault block.

No further exploration has been undertaken despite the identification of numerous other anomalies elsewhere in the basin. Re-evaluation of existing seismic-reflection data and well information by RRA/FDB (1985) indicates that the well penetrated a mainly Pliocene sedimentary section, and that a substantial untested Miocene sequence is present downflank is the Laruma depocenter.

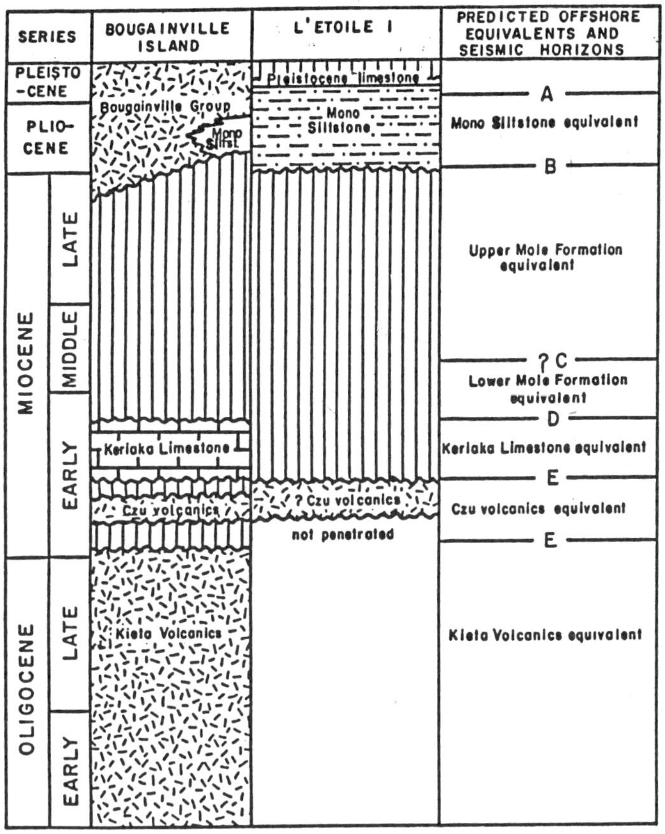


Figure 29. Stratigraphy of the Bougainville Basin.

*Reservoir potential*—Carbonate platform sediments of the Keriaka Limestone probably have good reservoir potential. Correlation between the onshore and offshore areas is complicated by the Atsilima fault (Figure 28), but it is likely that the D-E seismic sequence is composed of this unit in the eastern part of the basin.

Additional reservoirs may be developed in local carbonate buildups associated with synsedimentary highs in both the Miocene Mole Formation and the Pliocene Mono Siltstone. The most important Pliocene reef trends occur along the Tarangau Ridge and on the northwestern margin of the Aitara High.

Tuffaceous clastic turbidite reservoirs are probably present in the Lower Mole Formation, and possibly in bathyal equivalents of the Keriaka Limestone in the western part of the basin. Well-sorted water-laid tuffs, if present, would likely have suitable reservoir properties.

*Source potential*—No quantitative source rock data are available from the pre-Pliocene sedimentary sequence. Fine-grained bathyal clastics in the Keriaka Limestone equivalents and overlying Mole Formation may have source potential. Carbonaceous, fine-grained clastic sediments are known to occur in the latter unit to the south in the Solomon Islands.

The present-day geothermal gradient is anomalously low (1.6 C/100 m). If a similar paleogeothermal gradient prevailed in the past, the top of the oil generation zone would lie at a depth of about 4700 m. However, the paleogeothermal gradient prevailing prior to the Late Miocene arc reversal was probably significantly greater, as the basin was at that time in back-arc position.

The early Miocene (D-E) sequence is overlain by more than 5000 m of sediment in the Laruma depocenter, and could have acted as a significant hydrocarbon source. Younger sediments in the overlying Lower Mole Formation might also have reached a suitable maturation level, assuming that a higher paleogeothermal gradient prevailed prior to arc reversal.

### Play Concepts/Prospects

Re-evaluation of existing seismic-reflection data in light of the L'Etoile 1 well results has led to the development of a number of play concepts, none of which has been tested. They include closures mapped on the Motupena anticlinorium, onlap/pinchout traps around the western margin of the Laruma depocenter, reefal anomalies along the Tarangau Ridge and Aitara High, and normal fault traps along the margins of the Tarangau Ridge and Kunua High.

The largest and most significant potential structural trap in the basin is the Motupena anticlinorium (Figure 30; and play type 1, Figure 31). This Late Miocene compressional feature is bounded by west-northwest-trending reverse faults, and closure has been mapped at all key horizons below the sub-Mono Siltstone unconformity (Figure 32). Vertical relief exceeds 200 m at the level of the C reflector.

Keriaka Limestone reservoirs may be present in the D–E sequence in this structure, although this could not be confirmed from existing seismic data. The D reflector lies at a depth of about 2800 m over the main culmination. Turbidite clastic reservoirs may also be contained within the overlying Lower Mole Formation (C-D sequence). These reservoirs are favorably situated relative to deeply buried, potential source rocks in the adjacent Laruma depocenter.

Potential onlap/pinchout traps have been identified along the western margin of the Laruma depocenter on the east flank of the Tugu fault block (play type 2, Figure 31). This structure is bounded on the west by antithetic normal faults, and has persisted as a basement high since the Early Miocene. The C-D sequence (Lower Mole Formation) contains strong reflectors, possibly indicating the presence of coarse-grained clastic reservoirs onlapping basement (Figure 32). Detailed seismic-reflection coverage will be necessary to establish lateral seal. Depth to the C reflector ranges from 2700 to 4000 m along this trend.

A Neogene barrier reef trend has been recognized along the Tarangau Ridge, which was a persistent positive structural element during the Pliocene (play type 3; Figure 31). Most of the apparent buildups occur within the Middle Pliocene to Pleistocene section. Some of these features are overlain by modern reefs.

One of the buildups, situated at the southern end of the ridge, is about 6 km long and has a vertical relief of approximately 100 milliseconds. As only sparse seismic control is available, additional seismic acquisition will be necessary to define individual prospects which are adequately sealed.

A second apparent barrier reef trend has been tentatively identified on the outer flank of the Aitara High (play type 4; Figure 31). This trend, called the Bubulung lead, is about 23 km long and lies under about 100 m of water. Existing seismic coverage over the area is limited to reconnaissance lines.

Reef development may have been initiated along this trend during the latest Miocene. In the Early Pliocene, carbonate deposition was terminated by rapid subsidence. These potential reservoirs are favorably located updip from the Laruma depocenter, and are sealed by the Mono Siltstone.

Normal-fault traps have been mapped in relatively deep water close to the shelf edge around the northern margin of the Kunua High and on the western flank of the Tarangau Ridge (play type 5; Figure 31). Broad rollovers have been mapped at the level of the C horizon (intra-Mole Formation) on the downthrown sides of the faults. Clastic turbidite units in the lower Mole Formation are the most likely reservoir possibility. The C reflector lies at a depth of about 2400 to 4000 m along the fault trends.

## NEW IRELAND BASIN

### Structure

The structural configuration in the offshore New Ireland Basin is poorly defined because of very sparse seismic-reflection control. Interpretation of individual lines suggests that the basin is extensively faulted. The general style of deformation is probably similar to that seen onshore, where

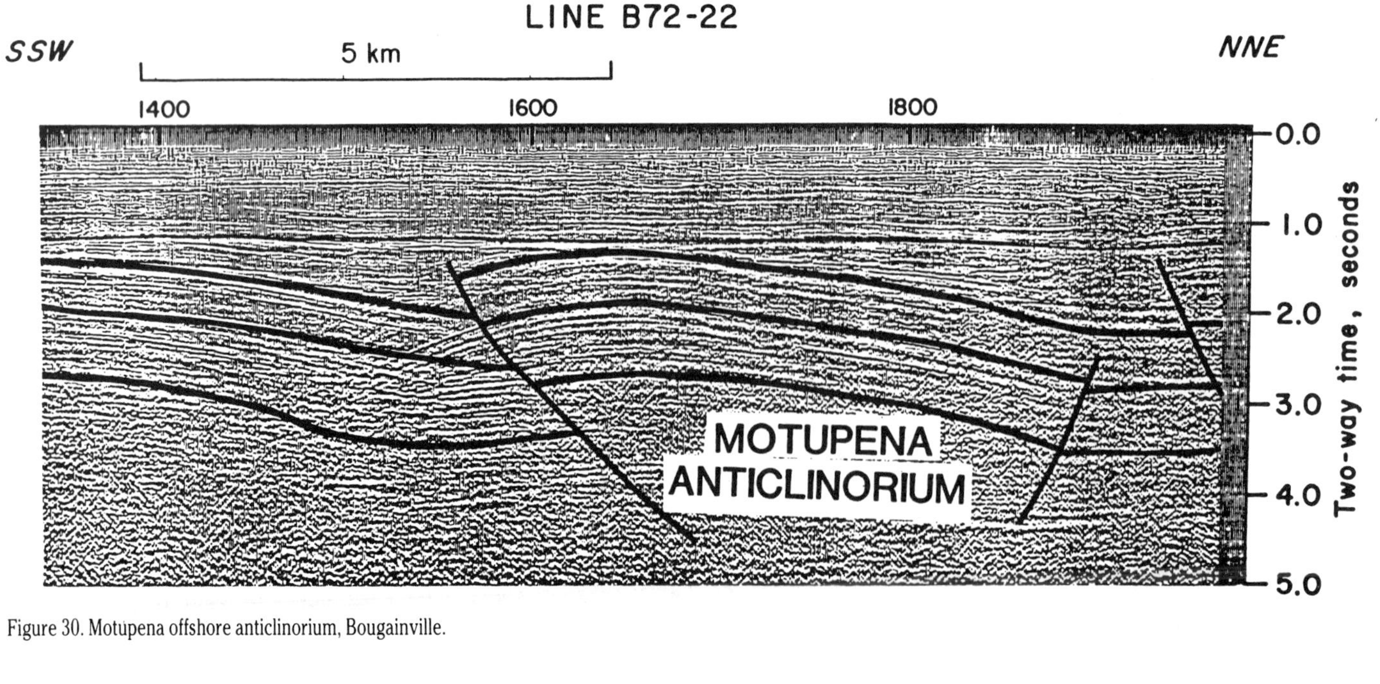


Figure 30. Motupena offshore anticlinorium, Bougainville.

two major northwesterly and northeasterly normal-fault trends have been mapped. Most of the onshore faults were formed during latest Pliocene and Pleistocene uplift along the southwestern basin margin.

A major northwest-trending normal fault, the Tabar Fault System, has been recognized along the basin axis offshore (Figure 34). The trend probably consists of a large number of normal faults with different orientations rather than a single fault. Overall downthrow is to the southwest. These faults were also active during the latest Pliocene and Pleistocene.

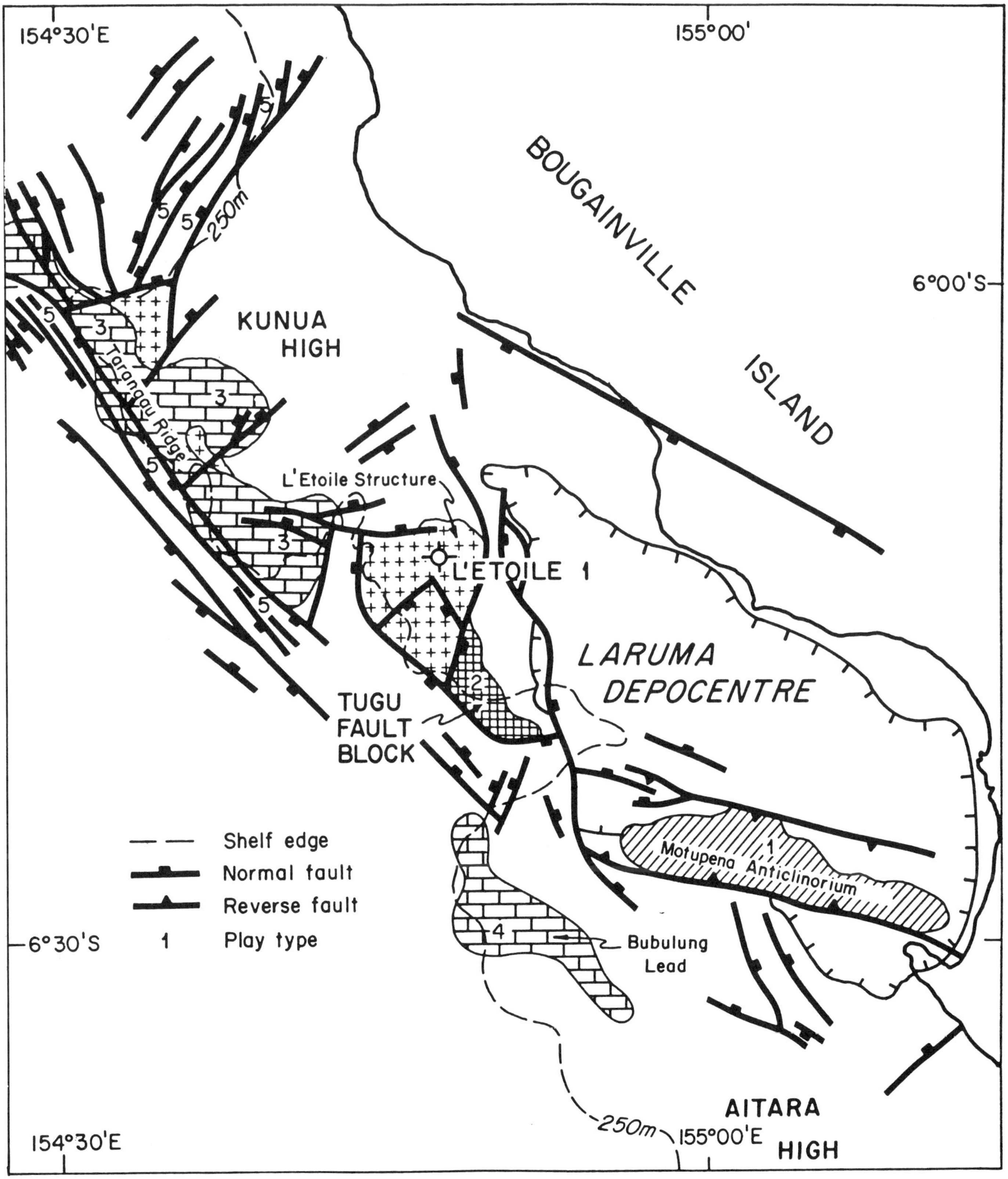


Figure 31. Play map of the Bougainville Basin.

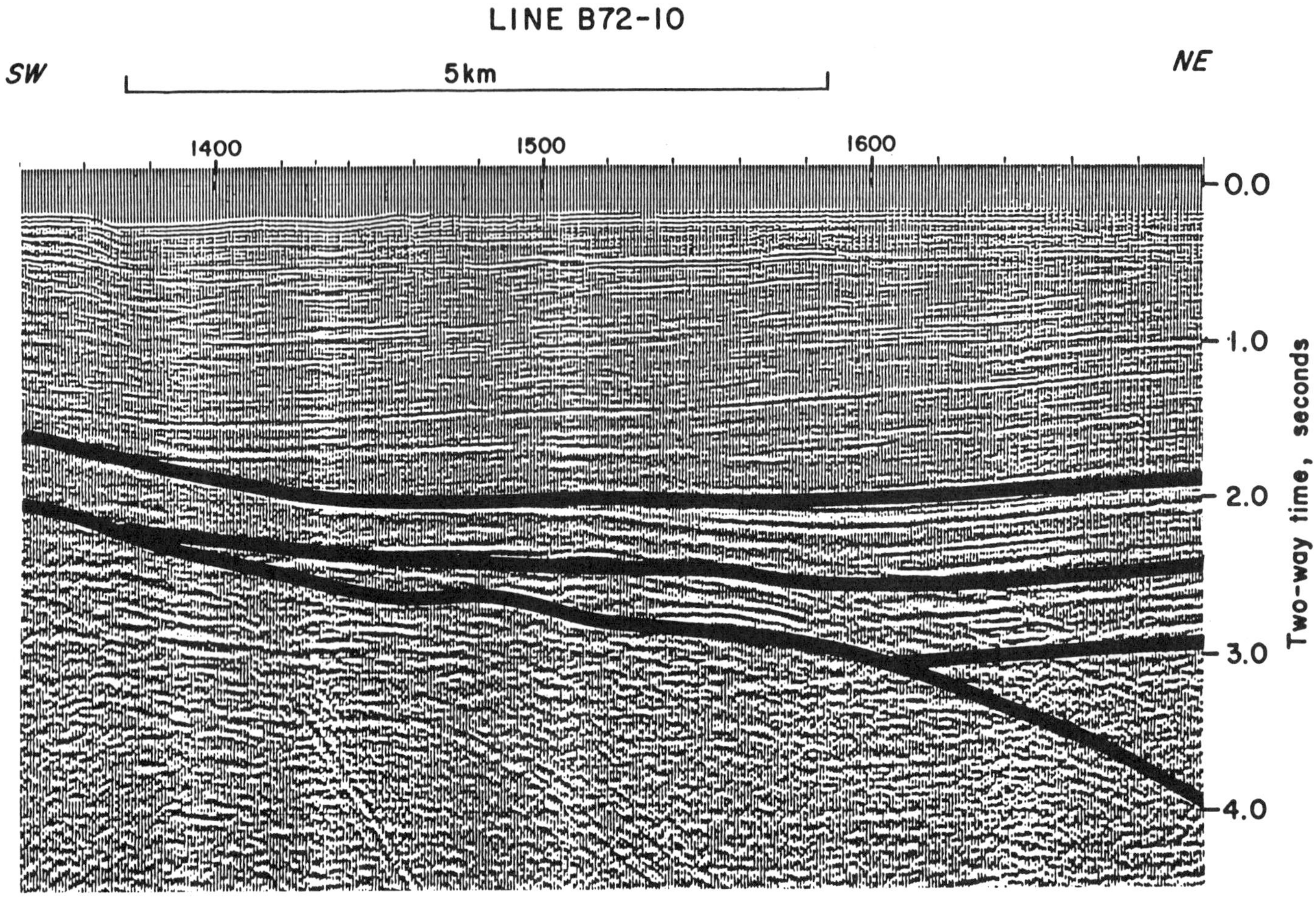


Figure 32. Seismic section, Bougainville Basin.

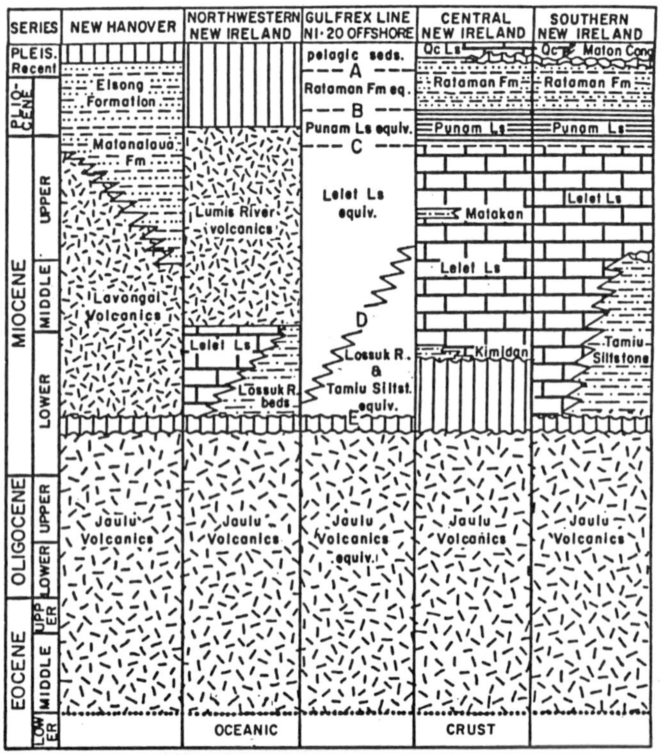


Figure 33. Generalized stratigraphic chart, New Ireland Basin.

Additional, apparently east-west-trending faults have been recognized offshore from southeastern New Ireland (Figure 34). These may be part of an older, basement-related fault trend which remained active during the Early and Middle Miocene.

The principal positive structural elements are the islands of New Ireland and New Hanover along the uplifted basin margin, and the Emirau-Feni Ridge, which marks the northeastern limit of the basin (Figure 34). The latter feature developed as an outer arc high early in the history of the basin. It was uplifted further at the end of the Pliocene.

Three major depocenters have been identified from existing seismic control (Figure 34). The Tatau and New Hanover depocenters in the central and western parts of the basin contain more than 5600 m of Miocene and younger sediments. A comparable thickness of sediment appears to be present in the Tanga depocenter to the southeast, although the basement reflector cannot be accurately mapped in that area because of sparse, poor-quality seismic control.

## Stratigraphy

The sedimentary sequence on central and southern New Ireland consists of Early to Late Miocene carbonate platform sediments, latest Miocene to Early Pliocene bathyal carbonates, Plio-Pleistocene volcaniclastic turbidites, and Pleistocene neritic carbonates and terrestrial sediments (Figure 33). The Middle Miocene and younger sequences on New Hano-

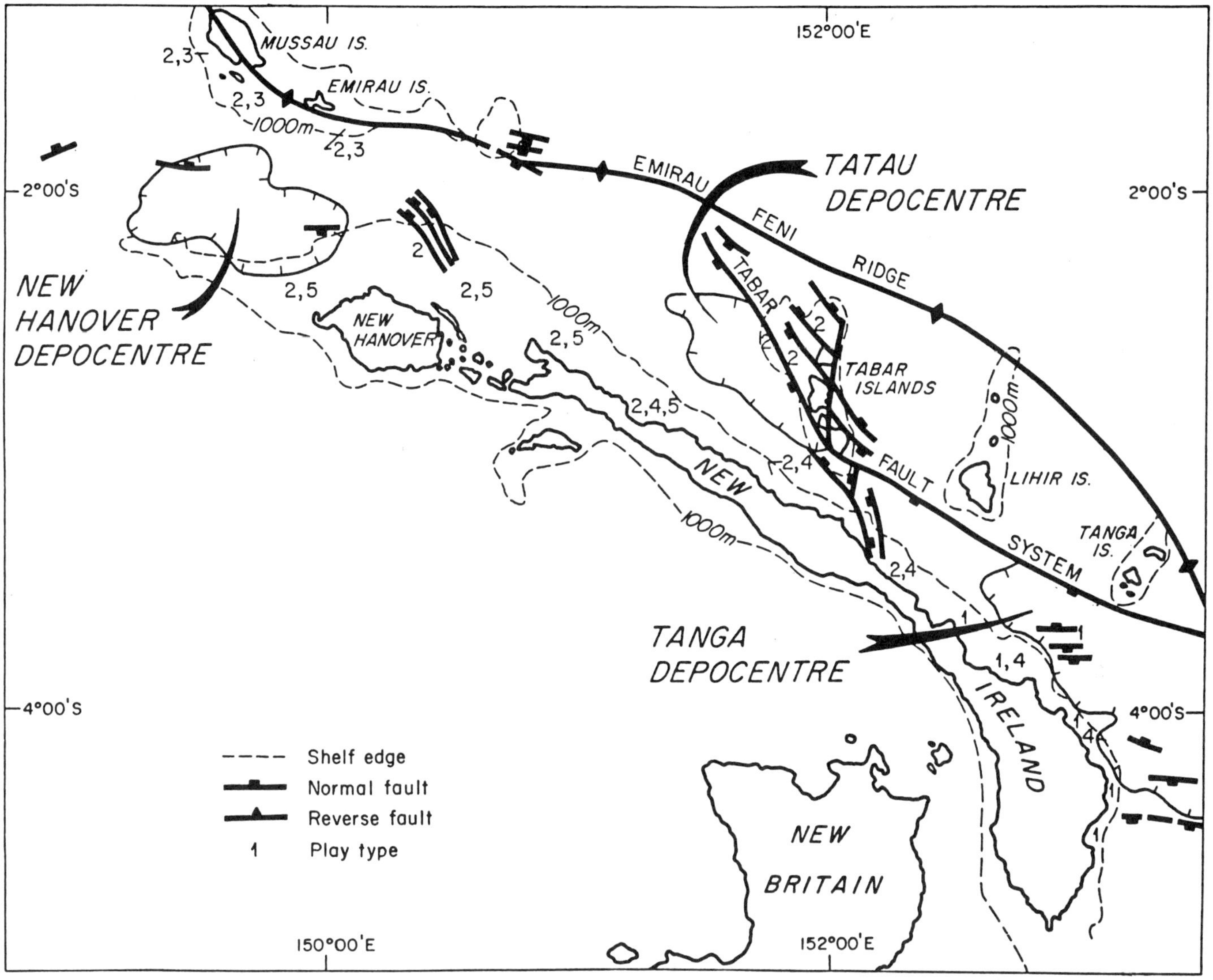


Figure 34. Play map of the New Ireland Basin.

ver and northwestern New Ireland are composed mainly of pyroclastic and volcaniclastic rocks of bathyal origin.

No well data exist to establish the areal extent of these units offshore. However, a good correlation appears to be possible between the sequence exposed on central and southern New Ireland and the seismic sequences identified offshore to the northeast (Figure 33). This suggests that major stratigraphic units such as the Miocene carbonate platform sequence may have a wide distribution. The northeastward extent of these units cannot be resolved from present seismic coverage, but a facies change to a dominantly volcaniclastic sequence probably occurs north of New Hanover.

Basement in the New Ireland Basin is formed by the Jaulu Volcanics (Figure 33), which built upward from older Paleogene sea floor during Middle Eocene to earliest Miocene arc volcanism. Carbonate platform sediments of the Lelet Limestone were subsequently deposited in the western part of the basin, commencing in the Early Miocene. At the same time, fine-grained epiclastic sediments of the Lossuk River beds and Tamiu siltstone accumulated in a bathyal regime seaward of the incipient shelf edge. Carbonate sediment production and accumulation exceeded the subsidence rate, resulting in basinward progradation of the carbonate shelf.

Volcanism persisted to the northwest in the New Hanover area, where pyroclastic and volcaniclastic rocks of the Lavongai Volcanics and Matanalaua Formation were laid down throughout the Miocene (Figure 33). The volcanism spread to northwestern New Ireland by the Middle Miocene. Carbonate deposition was terminated in that area by an influx of pyroclastic and volcaniclastic sediments now constituting the Lumis River volcanics.

Areas to the south were unaffected by the volcanism, and Lelet Limestone deposition persisted until the latest Miocene. By that time, progradation of the carbonate shelf was well advanced, and more than 2000 m of carbonate platform sediments had accumulated in parts of the basin.

Arc reversal during the Late Miocene was followed by marked regional subsidence, and subsequent sedimentation took place at bathyal depths. After an initial phase of pelagic carbonate deposition (Punam Limestone; Figure 33), regional pyroclastic volcanism commenced in association with subduction along the New Britain Trench. A thick sequence of volcaniclastic turbidites belonging to the Rataman Formation was deposited through to the beginning of the Pleistocene.

Regional uplift from the earliest Pleistocene onward elevated the southwestern margin of the basin to neritic depths. Coarse clastic sediments of the Maton Conglomerate were deposited on alluvial fans adjacent to emerging fault scarps, whereas neritic carbonates accumulated nearby. Adjacent areas to the northeast continued to subside, except locally along the basin axis where extensive faulting occurred in association with alkali volcanism.

### Oil and Gas Occurrence

*Exploration history*—Virtually no commercial exploration has been undertaken in the New Ireland Basin, except for brief onshore investigations by Continental Oil Co. and New Guinea Cities Services, Inc. from 1966-1969. Usable seismic control offshore is limited to two reconnaissance seismic-reflection surveys. The first of these was conducted by the vessel "Gulfrex" for the Gulf Research and Development Corporation in 1972. Exon et al. (in press) subsequently correlated the seismic sequences identified in the Gulfrex profiles with the stratigraphic succession exposed on New Ireland. The second survey was carried out by the tripartite (Australia, New Zealand, and U.S.A.) CCOP/SOPAC cruise of the "S.P. Lee" in 1984. Both surveys concentrated mainly on the deep-water areas of the basin, and thus little is known about the relatively shallow-water areas where drilling would be feasible.

*Reservoir potential*—The presence of a thick, laterally extensive Miocene platform carbonate sequence on New Ireland suggests that the laterally equivalent C-D seismic sequence offshore should have good reservoir potential. Onshore studies indicate that red algal-foraminiferal banks and local coral patch reefs were probably the most common types of organic buildups during deposition of this unit.

The northwestward extent of the Lelet Limestone in the offshore area cannot be resolved from existing seismic data. At least part of the Miocene sequence north of New Hanover may be composed of bathyal volcaniclastic sediments similar in composition to the Matanalaua Formation onshore. Coarse-grained volcaniclastic turbidites and submarine fan channel deposits may have reservoir potential in such a sequence.

*Source potential*—Fine-grained, carbonaceous clastic sediments similar to the Lossuk River beds and Tamiu siltstone onshore are the most likely source rock possibility. These units have been correlated with the D-E seismic sequence offshore (Figure 33). Samples form the Lossuk River beds have yielded total organic carbon contents ranging from 1.86-2.74%, and contain organic matter derived from both terrestrial and marine sources.

The geothermal gradient of the basin is unknown, so the depth range of the oil generation zone is conjectural. In the Cape Vogel Basin, which occupied a broadly similar tectonic setting during its early history, vitrinite reflectance and spore color index data suggest that sediments buried deeper than about 3.5 to 4 km should be mature. Assuming that a similar paleotemperature regime prevailed in the New Ireland Basin, it is probable that much of the D-E sequence lies below the top of this zone in all three depocenters. A basement subsidence curve for one of the depocenters suggests that hydrocarbon migration may have been during the earliest Pliocene.

### Play Concepts/Prospects

No specific exploration targets could be identified because existing seismic control covers mainly the deep-water areas of the basin. The present interpretation has, however, identified a series of play concepts which are equally applicable to the relatively shallow-water areas of the basin where drilling would be feasible. These include drape features associated with Miocene fault blocks, Pliocene-Pleistocene fault structures and anticlines, onlap/pinchout traps against the basin margins, Miocene carbonate buildups, and Miocene-Pliocene mound-like features.

A broad anticline associated with a Miocene fault block has been recognized by Marlow and Exon (1985) southwest of the Tanga Islands (play type 1; Figure 34). Faulting occurred contemporaneously with deposition of the C-D seismic sequence (Lelet Limestone equivalent), the upper part of which forms a broad drape over the underlying horst block. Onlap by younger sediments indicates that the structure persisted as a relative high during the Early Pliocene. Similar structures of Miocene age probably occur at drillable depths around the western margin of the Tanga depocenter (Figure 34). Assuming that hydrocarbon migration commenced during the Pliocene, these older structures could prove to be the most prospective in the basin.

Traps associated with Pliocene-Pleistocene normal faults have been identified along the Tabar Fault System and in seismic profiles extending from the New Hanover area to the margin of the Emirau-Feni Ridge (play type 2; Figure 34). Both updip fault traps and broad rollovers in the faults have been recognized (Figure 35). West of the Tabar Islands, possible Miocene carbonate reservoirs in the C–D sequence are favorably situated relative to potential source rocks in the Tatau depocenter. The equivalent sequence north of New Hanover may contain Miocene carbonate and/or volcaniclastic reservoirs. Similar traps are probably developed throughout the relatively shallow water area bordering New Hanover and northwestern New Ireland (Figure 34).

Onlap/pinchout traps involving the B–C and C–D sequences have been recognized along the southwestern margin of the Emirau-Feni Ridge (play type 3; Figures 34, 35). This feature lies under very shallow water near Mussau and Emirau Islands. Traps involving clastic turbidite reservoirs and possibly Miocene carbonates probably occur at drillable depths in that vicinity.

Probable carbonate buildups have been identified in the C–D sequence (Lelet Limestone equivalent) northeast of New Ireland near the Tatau depocenter (play type 4; Figure 34). These may correspond to algal banks of framework coral reefs. Similar buildups can probably be found under relatively shallow water immediately offshore from New Ireland.

Numerous other mound-like features of uncertain origin have been identified at the levels of the A, B and C reflectors offshore from both New Ireland and New Hanover (play type 5; Figure 34). One of these features, situated in the New Hanover depocenter, has an amplitude of about 250 m and a width of about 10 km in the plane of section. Some of the anomalies may represent carbonate buildups, whereas others may be slumps or fault-related features. Detailed, high-resolution seismic coverage will be required to properly evaluate their origins and significance.

## NORTH NEW GUINEA BASIN

### General

Northern New Guinea is here considered as one major sedimentary region, and as such is called the North New Guinea

Basin. It covers some 67,650 km$^2$ (26,120 mi$^2$) (Figure 36). This overall area can be divided into three principal structural subbasins, as follows:

| | km$^2$ | (mi$^2$) |
|---|---|---|
| 1. Aitape | 7,850 | (3,030) |
| 2. Sepik | 35,040 | (13,530) |
| 3. Ramu | 24,760 | (9,560) |
| Total | 67,650 | (26,120) |

These subbasins have their origin due primarily to tectonism associated with major convergence and strike-slip fault movement. These structural activities are related to and are coincident with the collision zone between the South Bismarck plate and the Australian plate. The age of these movements ranges from Oligocene to Recent, with a major pulse of deformation in the Middle to Late Miocene. The general surface geology of the North New Guinea Basin is shown on Figure 37.

Figure 38 shows the well control available in Northern New Guinea. Three of these wells, Boap Creek 1, Pulan 1, and Puwani 1 are within exploration permits that are still active. As such the well results are not available. In the total North New Guinea Basin, excluding the shallow Matapau oil seep bores, there have been seven exploration wells drilled, or one well for every 9664 km$^2$ (3731 mi$^2$). This is a very light test density per unit area. It is likely that the North New Guinea Basin contains significant oil and gas deposits, but further exploration work is required to evaluate this potential.

Initial exploration in Northern New Guinea was done by German scientists and explorers prior to 1914. Subsequently, some reconnaissance geology was done by Anglo Persian geologists about 1925. This led to the drilling in 1928 of the Marienberg 1 by Ormildah Oil Company of Australia.

Although Marienberg 1 was plugged, oil shows were reported and the well still leaks gas-cut saltwater. No further exploration work was done in the North New Guinea Basin until 1965, when Conoco and Cities Service became active in the region.

It was not until 1971, however, 43 years after Marienberg 1, that another well was drilled in the North New Guinea Basin. This test Bongos 1 was drilled on the north margin of the Sepik Subbasin. In 1974 Keram 1 was drilled in the northern Ramu Subbasin. There was no further drilling until 1984, when Coastal States drilled Tsumba 1x in the Ramu Subbasin. In 1985 BHP drilled a series of three wells in the Aitape Subbasin. Perhaps the recent BHP wells were well located geologically (?). However, in retrospect it seems clear that the earlier wells were either not valid structural tests and/or were poorly located stratigraphically in terms of testing a trap.

All told, scattered over nearly 60 years, there have been seven wells drilled over the total area of about 67,650 km$^2$. It is obvious there yet remains a lot of room for new 8½-inch holes in the North New Guinea Basin. Additional exploration is required.

## Stratigraphy

The stratigraphic succession of the North New Guinea Basin is dominantly Miocene and Pliocene in age. Should hydrocar-

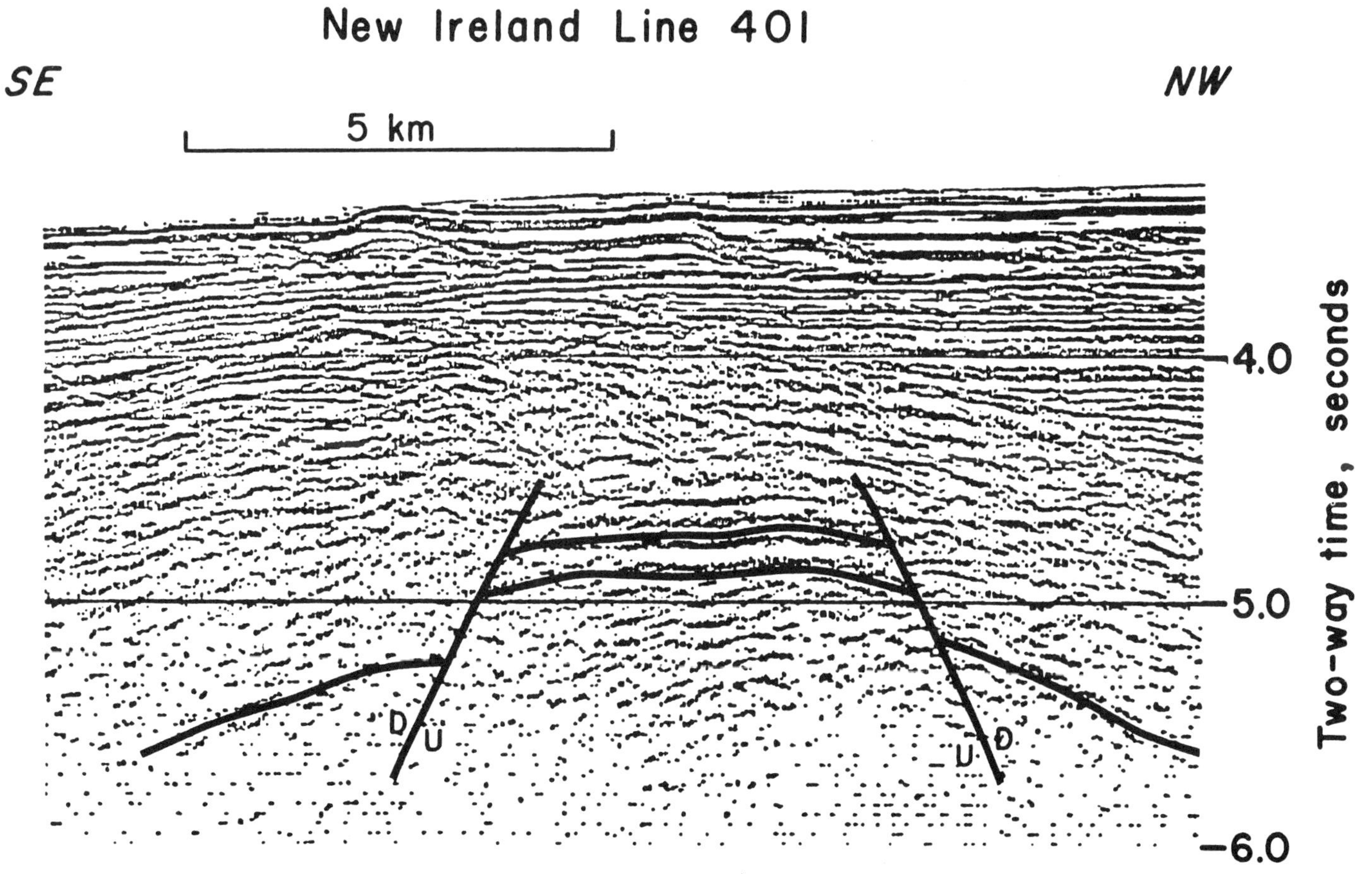


Figure 35. Seismic section, New Ireland Basin.

bon discoveries be made in the subject area, the source, reservoir, seals, and traps are likely to involve Miocene and Pliocene strata. Detailed stratigraphic studies of the basin are in process by the Papua New Guinea Geological Survey. No attempt is made here to further confuse the stratigraphy by introducing yet another nomenclature.

The general time rock sequence is presented in Table 2.

Certain generalities about the stratigraphy are set forth which might provide broad suggestions and clues about possible hydrocarbon occurrence in northern Papua New Guinea.

BMR, PNG, and private industry investigations in the 1960s to early 1980s (see PNG 1:250,000 Geological Sheet Explanatory Notes for Aitape-Vanimo, Ambunti, Wewak, Madang, Ramu, Sepik, and Bogia sheets) point to a basement rock complex ranging in age from perhaps Jurassic to Miocene. This complex includes metamorphics, dolorites, gabbros, diorites, ultramafics and andesitic lavas.

This suite of basement rocks is not unexpected considering the position of the North New Guinea Basin relative to the South Bismarck plate to the north and the Australian plate to the south. Such rocks also imply unstable tectonic conditions and as a consequence the overlying derivative sedimentary cover can be quite different lithologically in short distances. The unstable mineral suites of such basement and source areas can also provide monotonously thick sequences of mudstones in central portions of depositional basins. Sandstone formed in such environments can be quartz deficient and have variable or inconsistent porosities and permeabilities because of unstable clay assemblages.

Limestones probably provide the better possibility for forming important oil and gas reservoir objectives in a basin such as the North New Guinea Basin.

## Structure

Grund (1976) described the North New Guinea Basin as a buffer zone between opposing compressional movements of the Pacific and Australian crustal plates. The details of these movements and resulting trends and structural effects on the basin were more fully studied and elaborated upon by Pigott, Trumbly, and O'Neal (1983).

In general, these workers considered that the evolution of a rift zone with left-lateral movement has been the controlling factor of structural grain, trends, and basement development in the basin. One of the major structural manifestations of this development was thought to be the Ramu-Markham fault zone, which trends northwesterly for some 500 km (300 mi)

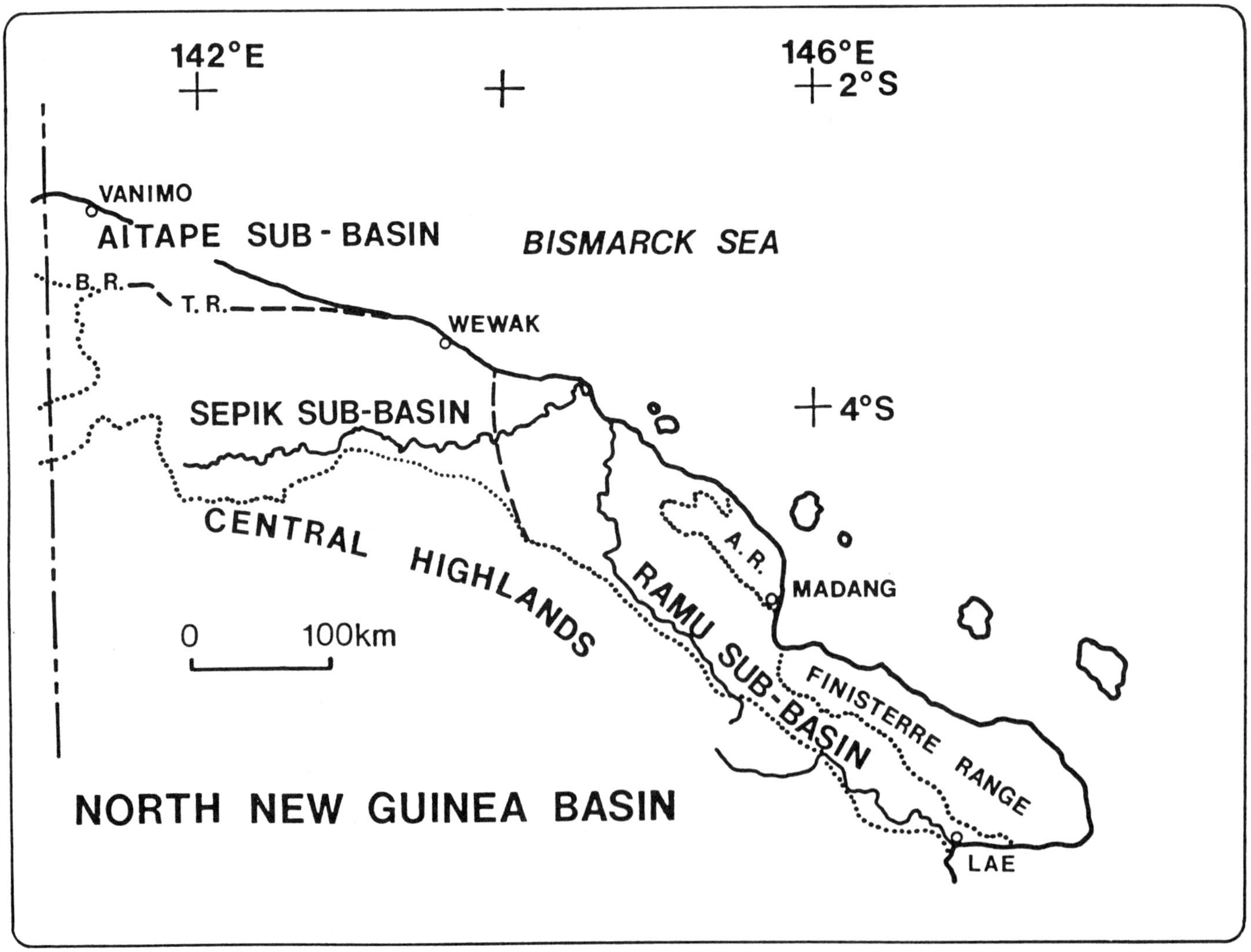


Figure 36. Index map of North Papua New Guinea.

along the south margin of the North New Guinea Basin. It extends from the Gulf of Huon near Lae to the Sepik River area (Figure 39). However, work in progress by the Geological Survey of PNG (R. Rogerson and G. Francis, personal communication) suggests that low-angle thrusting from the north might have played a more prominent role than strike-slip faulting in early basin evolution.

From the general area centering around 5° S and 144° E, the general zone of tectonism trends more northwesterly along the southern quarter of the Ambunti 1:250,000 PNG Geological Sheet.

Although the Sepik lowlands apparently mask the northwest extension of the Ramu-Markham fault zone as such, the major east-west rift zone of "popped-up" basement rocks and structural melange does occur in the Bewani-Torricelli range. This zone then appears to project directly westward into the extension of the major, Soorong left-lateral fault zone of West Irian.

The overall movements, both lateral and vertical, along this major fault system have left their imprint on both structure and stratigraphy in the North New Guinea subbasins. For example, the Ramu Subbasin appears to have a strong northwest-trending fold and fault block control of the Tertiary section in the subsurface and along the flanks of the Adelbert and Finisterre ranges. This appears to be related to deeper basement control.

In the northern Sepik Subbasin there is also a strong northwest trend of surface structures. This is perhaps a manifestation of left-lateral movement along the Bewani-Torricelli fault zone. The northwest-trending folds are exemplified by features such as the Bongos, Nombo, Maimai, and Yukilo anticlines (PNG Aitape-Vanimo and Wewak 1:250,000 Sheets).

The elevation and depression of various segments of basement terrane beneath the general area has affected the nature and thickness of clastic sediments and the distribution of clastic versus carbonate facies across the subbasins.

The presence of higher basement along the coastal regions of the Aitape Subbasin has controlled the distribution of Middle Miocene limestones. These units extend laterally east-west and thin to the south into the sedimentary thick or deep area along the north flank of the Bewani-Torricelli trend. As such, a more detailed understanding of the structural situation and development can play a role in perhaps better predicting location of possible oil-bearing reef or porosity fronts in the carbonate sequence.

Another carbonate-clastic facies change is also noted from

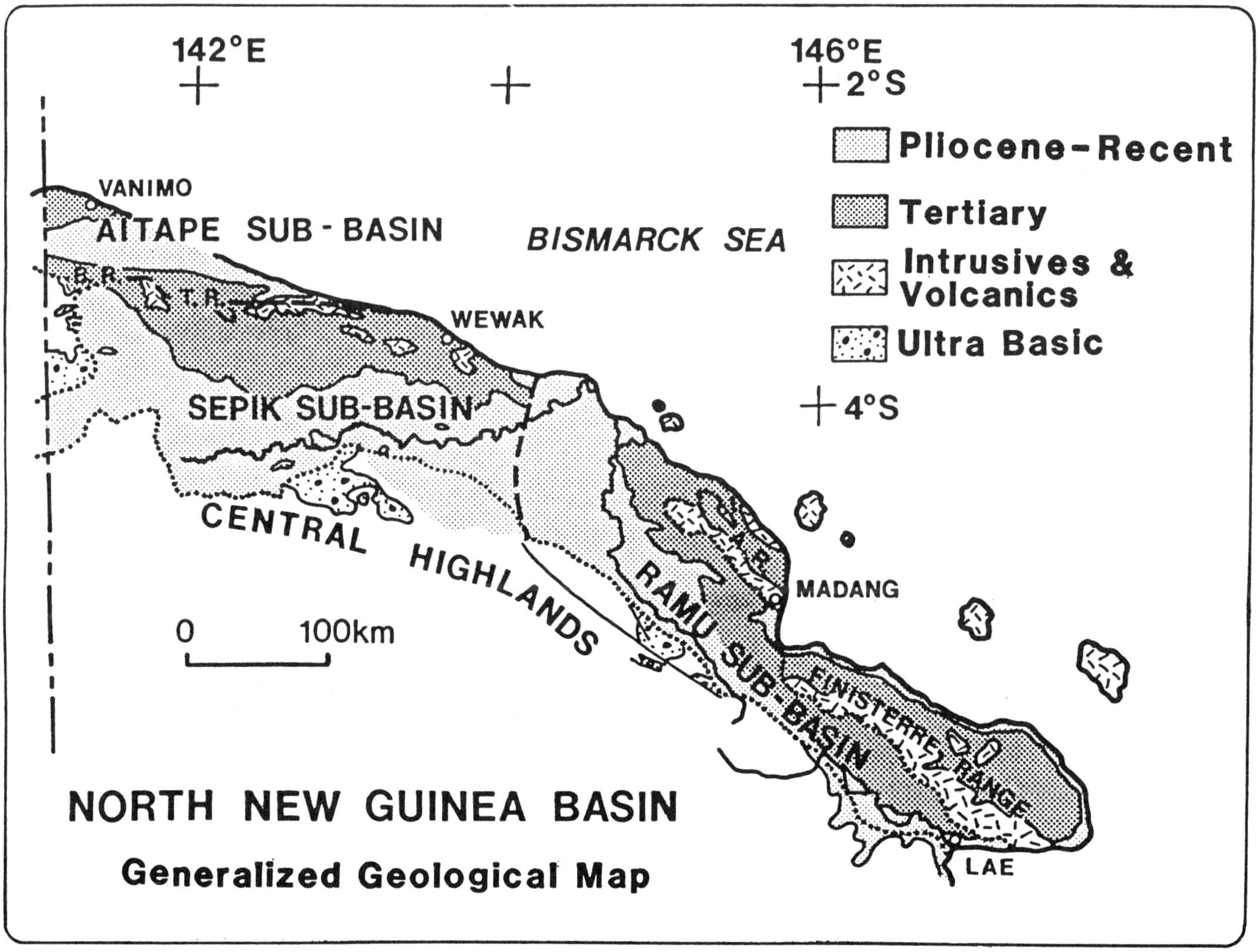


Figure 37. General geologic map of North New Guinea Basin.

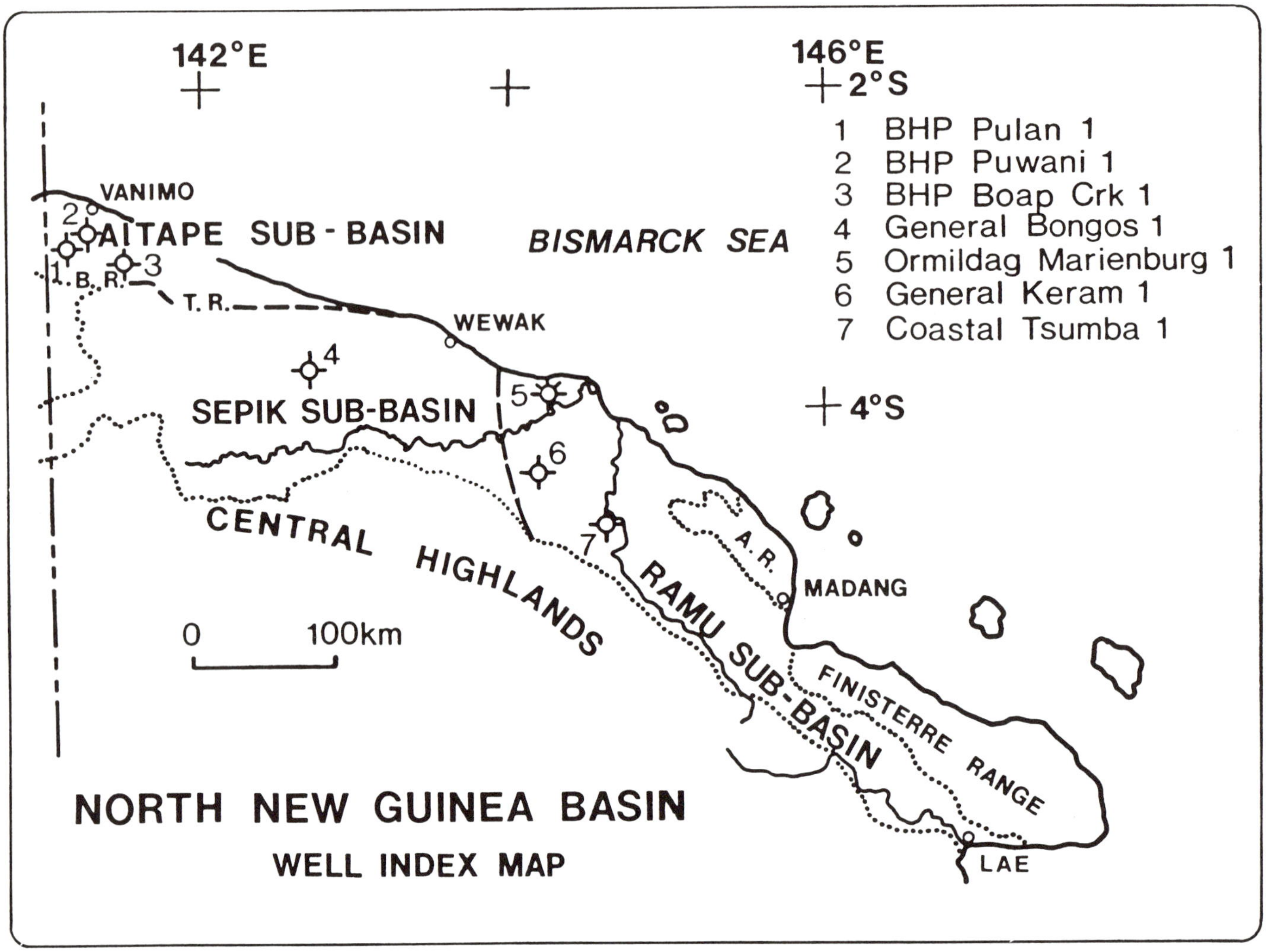


Figure 38. North New Guinea Basin well index map.

**TABLE 2. North New Guinea Basin general time rock sequence.**

| Age | Rock type | Thickness |
|---|---|---|
| Holocene | Sand, gravel, silts, coral/ algal limestone reefs, mudstones | up to 1,000 m |
| Pleistocene | Sand, gravels, silts | 10 m+ |
| Pliocene | Mudstone, siltstones, same carbonaceous streaks, polymictic conglomerates, coral reefs in Aitape sub-basin | up to 2,000 m |
| Miocene | Siltstones, claystones, conglomerates, reef and basinal limestones, volcanics | up to 3,000 m |
| Oligocene | Coral/algal limestones sandstones, volcanics | +300 m |
| Eocene - Middle Cretaceous | Grey carbonaceous shale (often metamorphosed), radiolarian siltstone, lava, pyroclastics, basic igneous and ultrabasic rocks | (est.) 5,000-10,000 m |

north to south in the northern Ramu Subbasin (Figure 40).

*Ramu Subbasin*—About 1117 line km of seismic data were acquired by Coastal States et al. in the northwest Ramu Subbasin between 1981 and 1983. Figure 41 shows a generalized seismic isochrone interpretation after Coastal's final reports in the PNG files. The horizon shown is the base "Upper Miocene."

This interpretation suggests a basement in the north part of the map area, east of the Keram and Tsumba wells, which dips eastward. This picture is complicated by fault blocks bounded by a set of major northwest-trending faults with a possible conjugate set of right-lateral strike-slip or tear faults which strike northeast.

The deepest structural lows are in a general way traversed by the north-flowing Ramu River in the mapped area. A major northwest-trending structurally high trend is present to the southeast, along the northeast margin of the ex-Coastal permit. This feature, about 39 km long, has some 1.0 to 1.5 sec of two-way time vertical relief (3500 to 4000 feet or 1067 to 1220 m).

Other isolated fault-block closures and small four-way dip closures are suggested in the Miocene northeast of Keram 1.

The interpretations suggest to the writer that there has not yet been a valid structural closure tested in the Ramu Subbasin. This is especially true in that area where Miocene car-

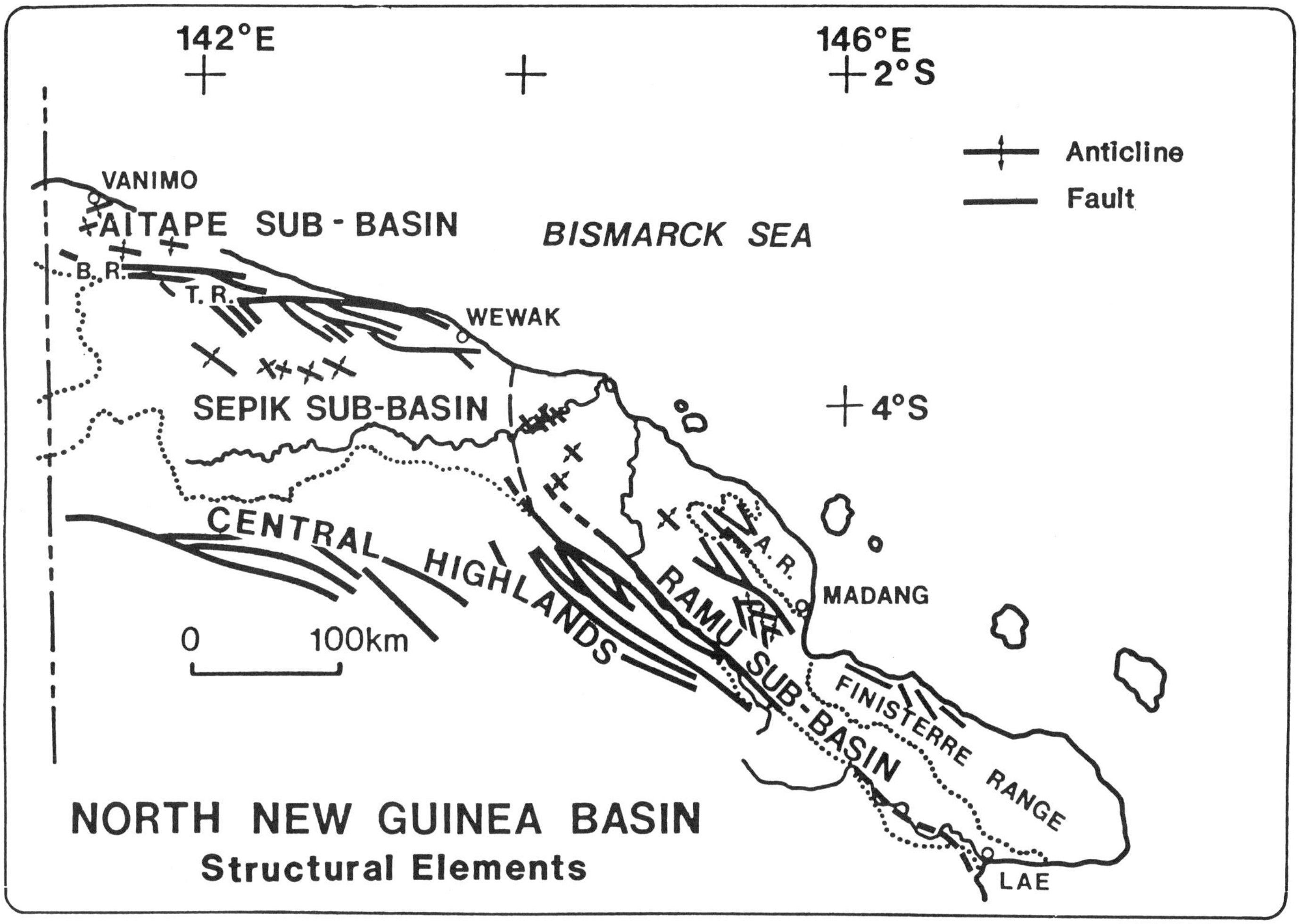


Figure 39. Structural elements of Northern Papua New Guinea.

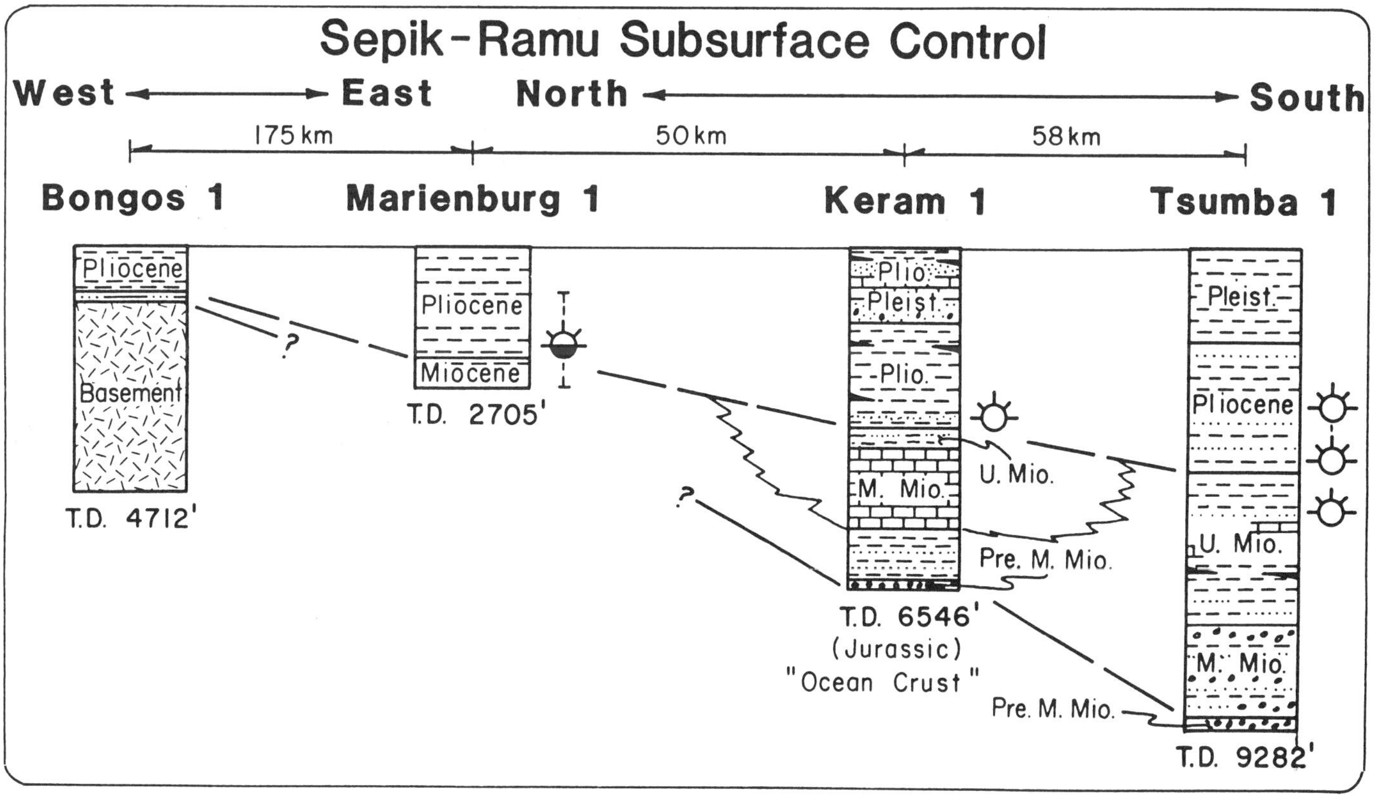


Figure 40. Cross section through Sepik-Ramu subbasins wells.

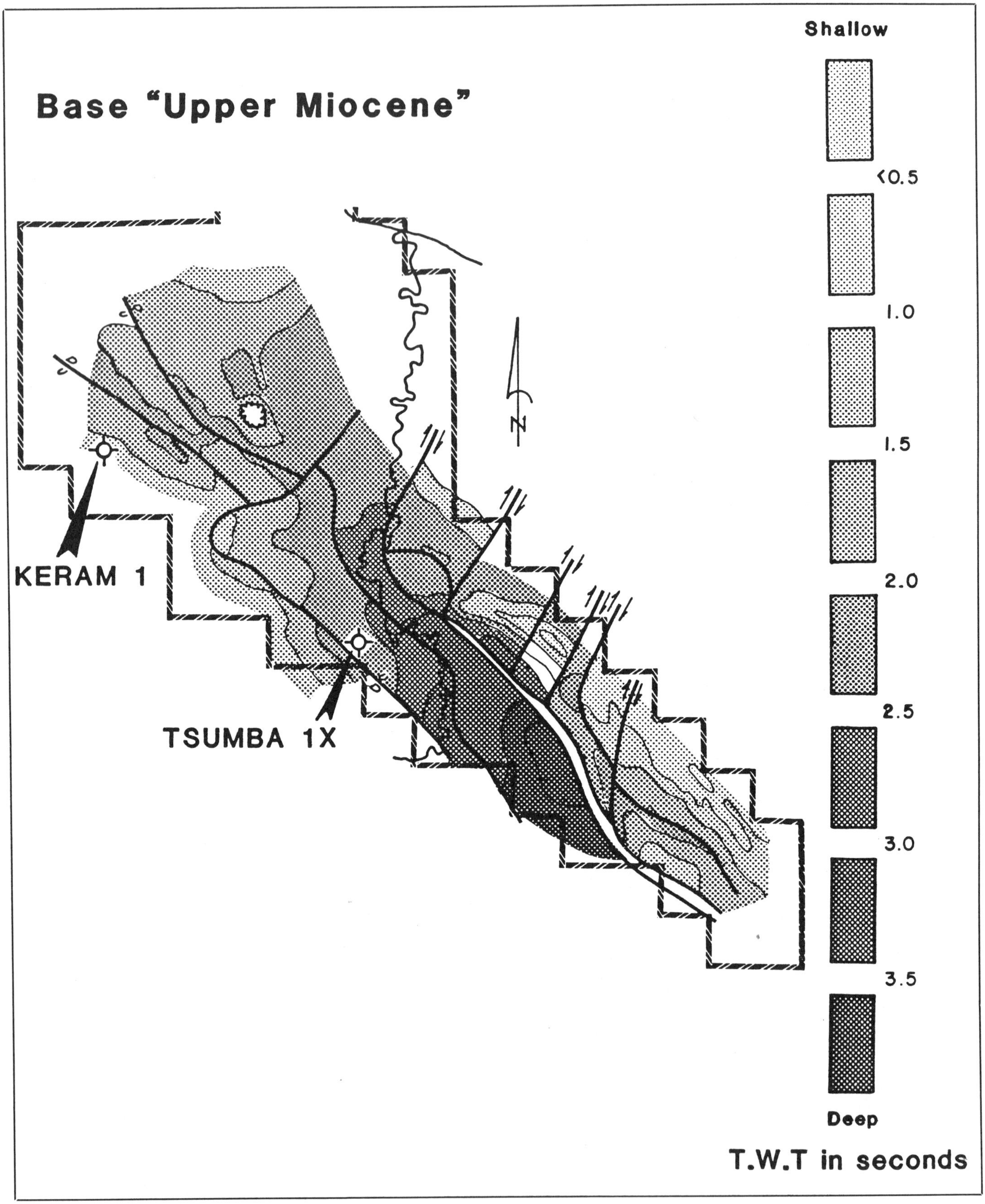


Figure 41. Simplified Coastal States seismic structural interpretation, Ramu Subbasin.

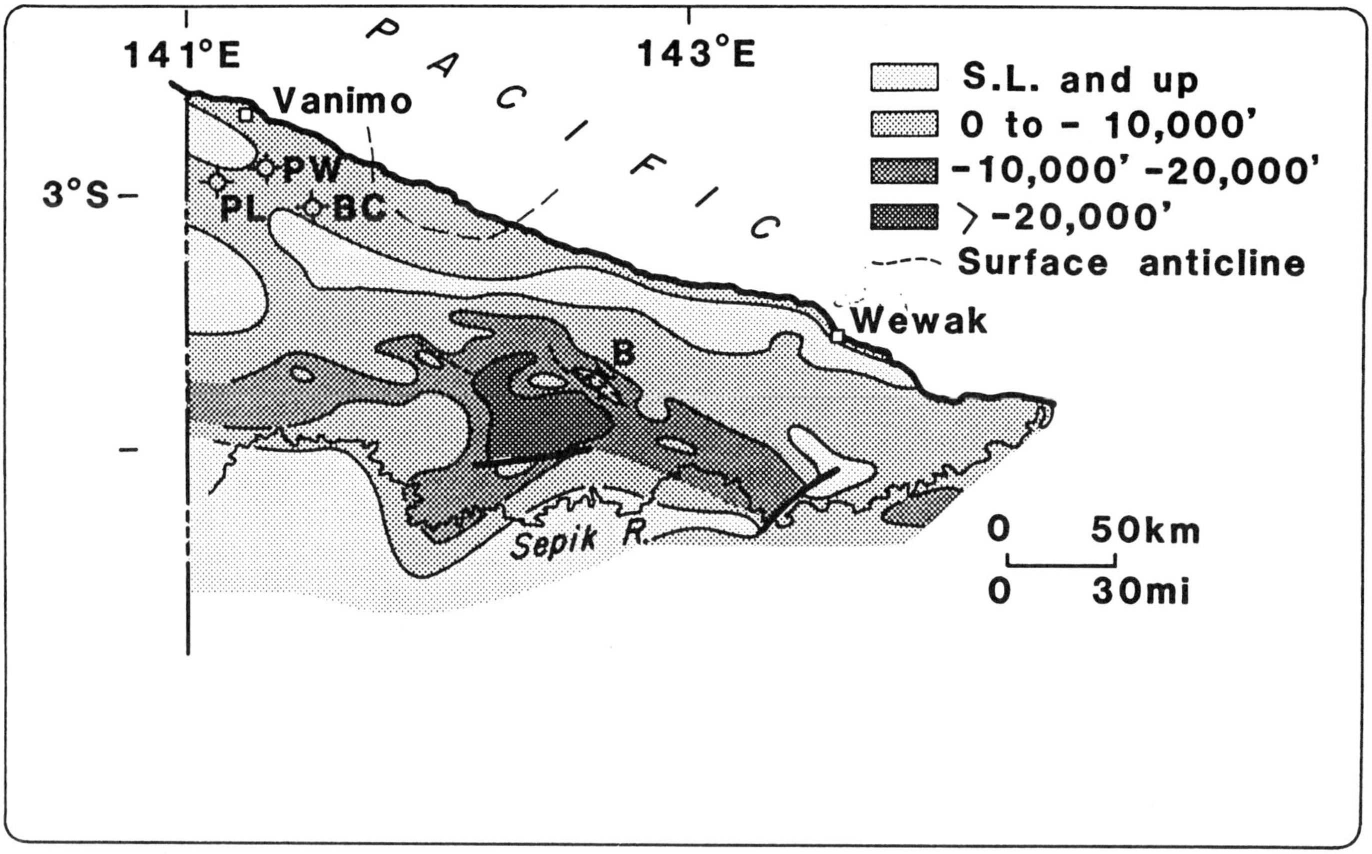


Figure 42. Aeromagnetic map of the Sepik Subbasin.

bonate reservoirs might be present, say north of lat. 4°30′S. Similarly, if clastic reservoirs of merit exist in the basin, they have not been tested on a significant structural closure. These might be found southeast of the Ramu River and east of the deeper part of the basin in which hydrocarbons might have been generated (the Josephstaal anticlinal trend). Earlier reconnaissance seismic work suggests untested structural closures occur 6 km southeast of Keram 1 and 18 km northeast of Keram 1. These closures are known as the Pankin and Kambot structures, respectively.

The Coastal States data and interpretations suggest that a number of other potential prospects appear to be available for further consideration and exploration.

*Sepik and Aitape subbasins*—Aeromagnetic data suggest appreciable thicknesses of Tertiary section in these subbasins. Although detailed seismic data are not available, one can probably assume that numerous structural traps could be found in such an active tectonic region (Figure 42).

In the eastern part of the Sepik Subbasin (PNG 1:250,000 sheets for Ambunti and Wewack) aeromagnetic interpretation and surface geology suggest that structures will probably have a northwesterly component. If surface geology is any criterion, (south flank of the Toricelli Range) then anticlines might occur in the central Sepik Subbasin with lengths of 10 to 30 km (6 to 18 mi) and widths of up to 5 km (3 mi). Vertical closures of a few tens of meters to hundreds of meters are possible.

*Aitape Subbasin*—Occupying the northwesternmost part of PNG, the Aitape Subbasin comprises an area of some 7850 km$^2$ (3030 mi$^2$). It appears to be asymmetric to the south basin. It has shallow basement along the coast to the north and passes southward some 45 km (27 mi) along a meridian through Vanimo (Figures 42, 43), into a deep structural and sedimentary trough near its southern margin.

The structurally higher coastal area contains two principal topographic and structural uplifts. These are the Oenake uplift to the west and the Serra Hills uplift to the east (Figure 42). The south margin of the subbasin is a pronounced east-west-trending fold belt and major fault zone. This is expressed by the Bewani Mountains to the west and the Torricelli Mountains to the east.

Across the Bewani-Torricelli fault zone, consisting of a series of east-trending anastomasing faults, largely vertical at the surface, there might be an overall vertical offset component of several thousand meters down to the north. Lower Cretaceous to Miocene basic intrusive and volcanic rocks form the core of the Bewani-Torricelli structural ridge. A few zones of serpentinite are reported (Norvick and Hutchinson, 1980).

## Oil/Gas Potential

Oil and gas have been generated in parts of the North New Guinea Basin. This is demonstrated by the presence of oil and gas seeps (Figure 38). Short-lived flows of oil are reported (Jacques and Robinson, 1977) from shallow bores drilled in the Matapau area, 50 km northwest of Wewak. These have been described as good-quality, mature, naptheno-aromatic oils. Gas seeps and minor gas shows (largely methane) have been identified throughout the area.

Limited geochemistry suggests the presence of oil/gas source rocks; however, maturity can be a problem. As such, areas with thicker section (deeper structures) are perhaps places which should be sought. Coincident with source poten-

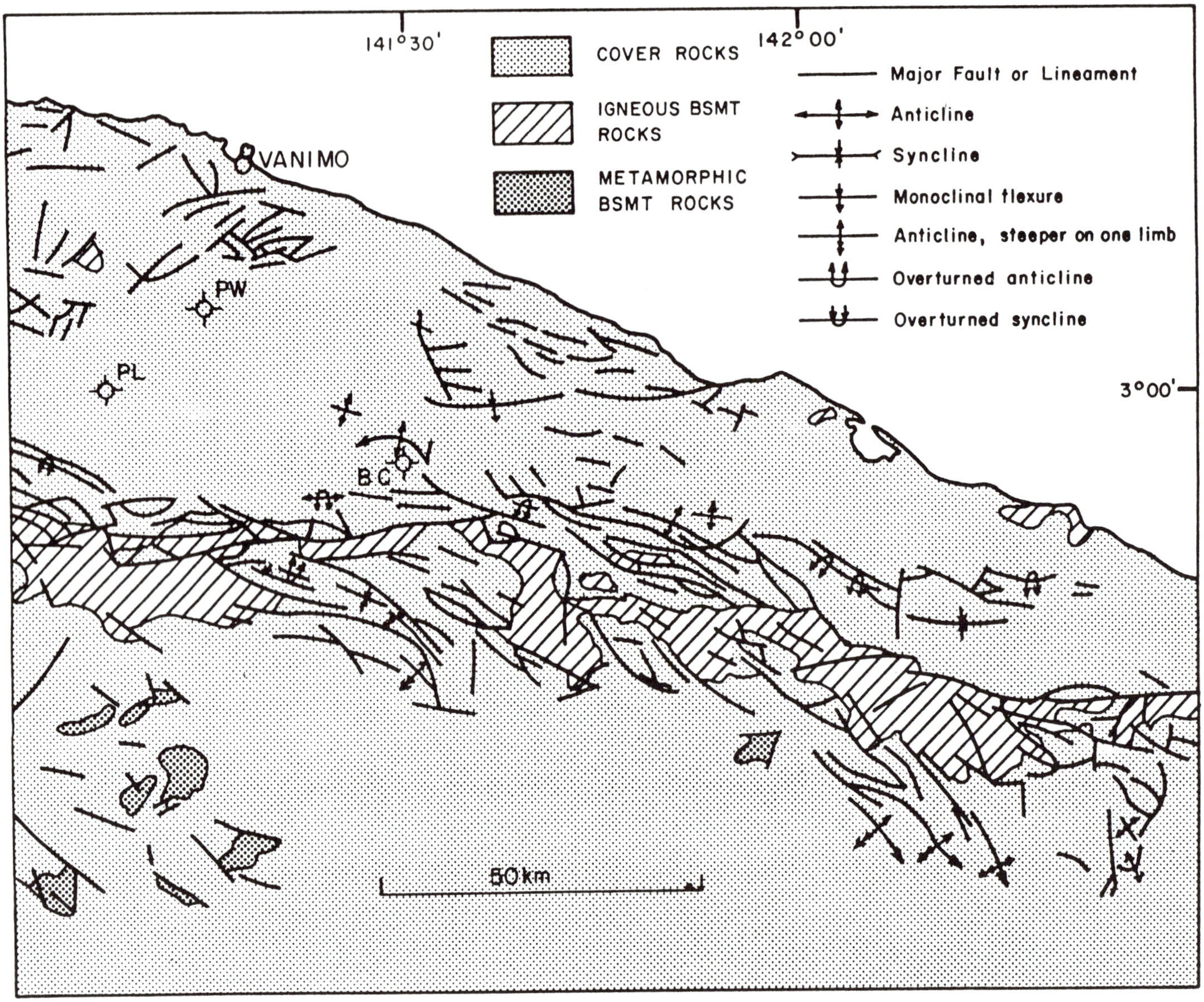


Figure 43. Aitape Basin, general geologic map.

tial, good reservoirs must also be found. For this reason, Miocene limestone reefs remain a significant objective wherever they face or are associated with a potential oil/gas-generating section.

## Play Concepts

*Aitape Subbasin*—As noted by Norvick and Hutchinson (1980), there are Miocene and Pliocene reef carbonates in the Oenaka and Serra Hills area along the north margin of the Aitage Subbasin. Where these pass southward into the deeper part of the subbasin, there might occur an oil and gas prospective east-west-trending reef front. This front facing a thicker, deeper clastic basin to the south might house potential oil accumulations generated in the deeper part of the basin.

Some oil and gas seeps have been identified in the south-central part of the Aitape Subbasin (Figure 44).

There are also numerous untested anticlines in the south part of the basin paralleling the Bewani-Torricelli Mountain front. Although many are structurally complex, more moderately folded structures might be found as one moves northward away from the mountains. Of the recent BHP tests (1984-1985), the Boap Creek 1 test was drilled several kilometers down plunge and 5 km south of the surface crest and axis of the Muru anticline (PNG Aitape-Vanimo 1:250,000 Geological Sheet, 1980). This is the only test drilled near the mapped surface expression of an anticline in the Aitape Subbasin.

*Ramu Subbasin*—In the Ramu Subbasin there are anticlinal structures that remain to be tested (Figure 45). The two wells, Keram 1 and Tsumba 1x, have not evaluated this potential. Keram 1 was not drilled on a structure and Tsumba 1x did not penetrate beds under closure until deep in the section, possibly the lower part of the basal 500 m drilled (Middle or Late Miocene).

Source rock analysis of Keram 1 samples suggest that the section below 920 m is favorable for oil/gas generation, whereas Tsumba 1x appears to be immature. In the southern part of the basin, structures within, or which are facing, more deeply buried potential oil source section should be evaluated. In the northern part of the basin (Keram region) maturity does not appear to be a major concern.

Perhaps the greatest potential to be evaluated in the Ramu

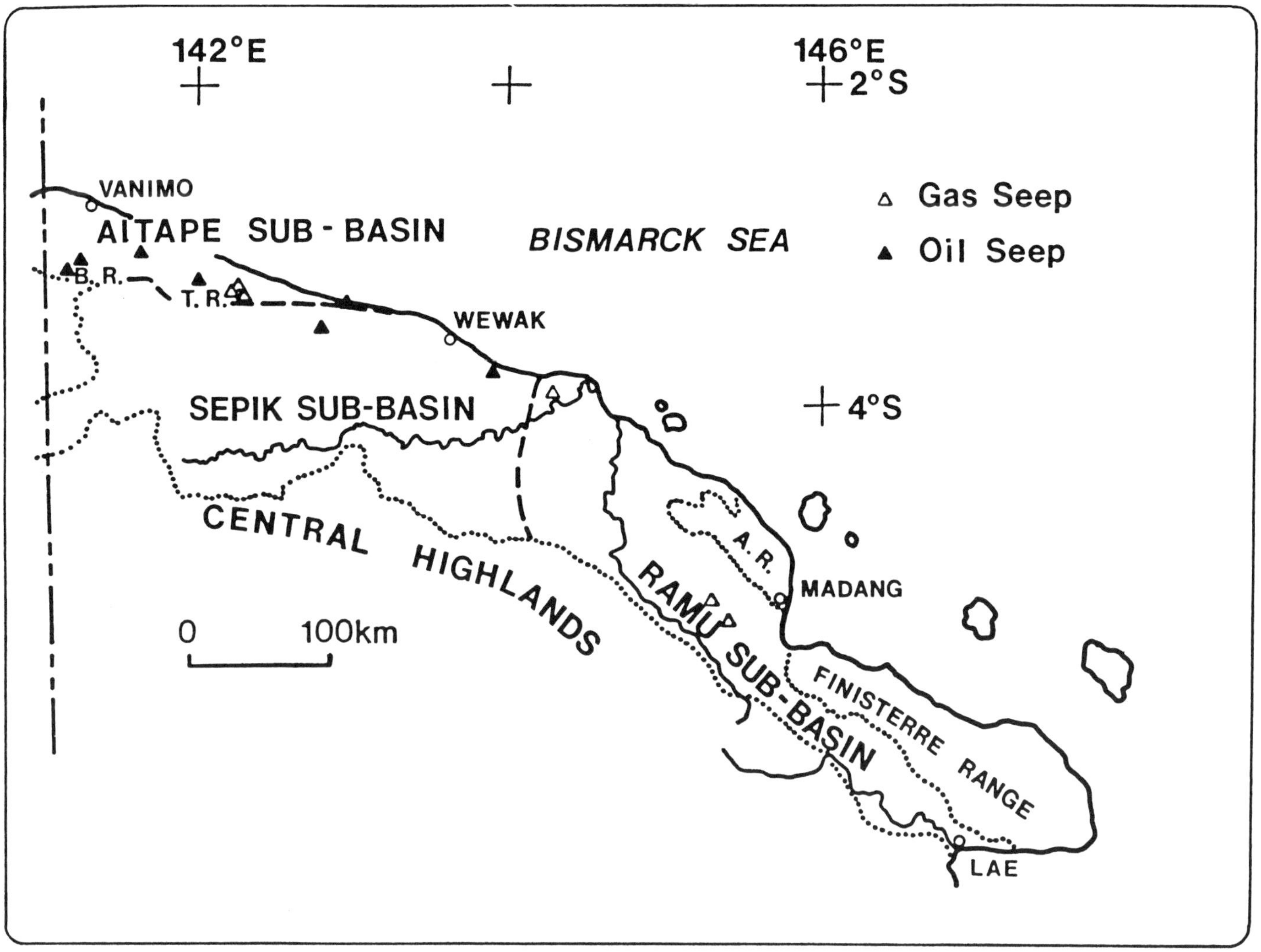


Figure 44. Oil and gas seeps in the North New Guinea Basin.

Subbasin relates to Miocene carbonate prospects. As shown on Figure 40 there is approximately 500 meters of limestone-dominated stratigraphic section in the Middle Miocene at Keram 1. This section was not present at Tsumba 1x. Figure 46 illustrates a potential area for oil and gas accumulation in Miocene carbonates and/or reefs in the northern Ramu Subbasin. This play concept needs further study and evaluation.

## CONCLUSIONS

The Papuan Basin offers a diverse array of exploration opportunities for oil and gas. The sedimentary history of the basin has been conducive for the accumulation of a wide variety of good-quality reservoirs in close association with organic-rich source rocks. A favorable temperature regime has been developed, allowing the attainment of suitable maturation levels for the generation of hydrocarbons. Finally, a varied structural history was led to the formation of a large number of trap types, many of which are capable of housing up to several hundred million barrels of liquid hydrocarbons.

The presence of large accumulations of both gas and condensate has already been proven by past exploration in the Papuan Fold Belt and on the Fly Platform. Excellent potential remains for extending successful play concepts into adjacent areas, such as along structural trend from the Juha and Iagifu discoveries and elsewhere in the central Fly Platform close to the Pasca discovery.

Significant accumulations of oil have, however, eluded explorationists in the past, despite strong evidence that oil has been generated in the Papuan Basin. Oil seeps in the 21° to 30° API range have long been known to occur in the Papuan Fold Belt and Aure Tectonic Belt. The large initial flow of 42° API oil from the Puri structure suggests that it and other structures along trend might contain significant quantities of oil.

The geochemical characteristics and maturation level of the Maril Shale and its lateral equivalents to the west constitute strong evidence for oil source potential on the central and western parts of the Fly Platform. Multiple Mesozoic and Tertiary reservoirs exist in these areas, offering attractive exploration targets.

Potential accumulations might exist in the virtually unexplored Morehead Subbasin, but lack of well control precludes firm conclusions about that area.

The Cape Vogel Basin offers varied opportunities for oil and gas exploration. Existing seismic-reflection coverage has revealed a thick sedimentary section, as well as diverse structural and stratigraphic trapping possibilities. The acquisition of modern, high-resolution seismic-reflection data in subsequent exploration programs will allow refinement of existing

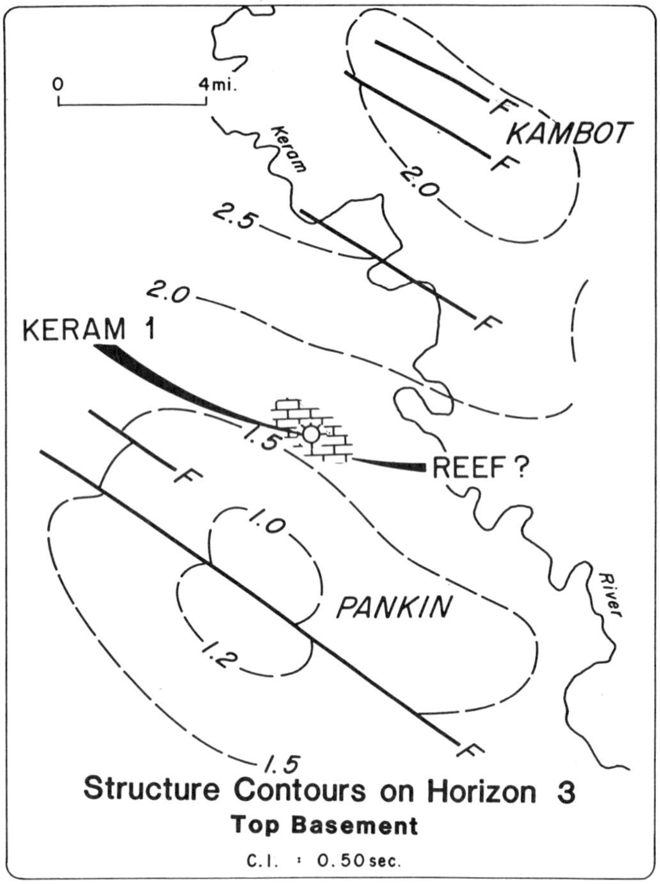


Figure 45. Untested structures in the Keram Area, Ramu Subbasin.

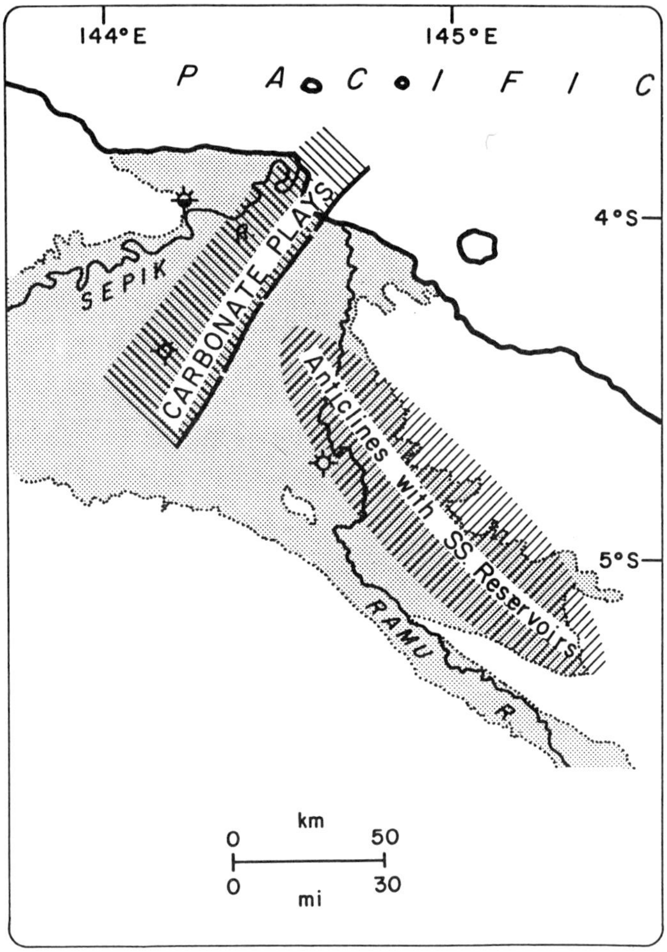


Figure 46. Northern Ramu Subbasin play concept areas.

play concepts, and will likely lead to the recognition of additional, more subtle trap types.

During the initial exploration phase offshore, the sedimentary sequence and structural configuration of the Cape Vogel Basin were virtually unknown. Integration and evaluation of existing data by RRA/FDB (1985) have contributed significantly to the development of viable play concepts, none of which have been tested. The availability of this new information provides opportunities for future exploration in the basin.

Available seismic-reflection coverage indicates that the Bougainville and New Ireland basins contain thick sedimentary sections with good reservoir potential. A wide variety of structural and stratigraphic trap types has been recognized, providing numerous leads for future exploration. Although little is known about the source potential of the two areas, present data suggest that a substantial volume of sediment probably lies within the oil generation zone in the major basin depocenters.

The RRA/FDB (1985) report contains valuable background geological and geophysical data for companies considering exploration in these frontier areas. The ultimate petroleum potential of the two basins cannot, however, be fully assessed until additional seismic data and well information are acquired.

The North New Guinea Basin appears to have most of the ingredients necessary for discovery of commercial oil/gas discoveries (oil and gas seeps, oil and gas shows in wells, reservoir and cap rocks, large structures). The area requires the drilling of many more wells to evaluate some of the possibilities alluded to previously in this paper.

Overall, there can be no better testimony to demonstrate the potential for oil and gas discoveries than the results of drill stem tests 5 and 6 as reported by Oil Search for the Iagifu 2 (Australian Stock Exchange Reports and Press Releases, April 1986).

## REFERENCES CITED

Exon, N.F., W.D. Stewart, M.J. Sandy, and D.L. Tiffin, in press, Geology of the eastern New Ireland Basin, northeastern Papua New Guinea: BMR Journal of Australian Geology and Geophysics, v. 10, no. 1.

Grund, R.B., 1976, North New Guinea Basin *in* R.B. Leslie et al., eds., Economic geology of Australia and Papua New Guinea: n 3 Petroleum, p. 499-506.

Hobson, D.M., 1986, A thin skinned model for the Papuan thrust belt and some implications for hydrocarbon exploration: Australian Petroleum Exploration Association Journal, v. 26, p. 214-224.

Jacques, A.L., and G.P. Robinson, 1977, The continent/island-arc collision in northern Papua New Guinea: Bureau of Mineral Resources, Journal of Australian Geology and Geophysics, p. 289-303.

Marlow, M.S., and N.F. Exon, 1985, Basin development and regional tectonics of the New Ireland and Manus regions, *in* Extended abstracts: Petroleum Geology of South Pacific Island Countries: 14th BMR Symposium, Canberra, Bureau of Mineral

Resources, Geology and Geophysics, Record 1985/32, p. 11-18.
Norvick, M., and D.S. Hutchinson, 1980, Explanatory notes on the Aitape-Vanimo geological sheet: Department of Minerals and Energy, Geological Survey of Papua New Guinea, 45 p.
Pigott, J.D., N. Trumbly, and M. O'Neal, 1983, Northern New Guinea wrench fault system; a manifestation of Late Cretaceous interactions between Australia and Pacific plates: Transactions of the Third Circum-Pacific Energy and Mineral Conference, Honolulu, p. 613-620.
Robertson Research Australia Ltd. and Flower Doery Buchan Ltd., 1984, Petroleum potential of the Papuan Basin, Papua New Guinea: Report prepared for the Geological Survey of Papua New Guinea, Department of Minerals and Energy. 4 volumes text and appendices, 2 volumes maps and seismic sections.
Robertson Research Australia Ltd. and Flower Doery Buchan Ltd., 1985, Petroleum potential of the Cape Vogel, Bougainville and New Ireland basins, Papua New Guinea: Report prepared for the Geological Survey of Papua New Guinea, Department of Minerals and Energy. 1 volume text and appendices, 4 volumes maps and seismic sections.

CHAPTER 9

# PHILIPPINE MICROPLATE TECTONICS AND HYDROCARBON EXPLORATION

**John J. Gallagher, Jr.**
*ARCO Oil & Gas Company*
*Plano, TX*

ABSTRACT

Hydrocarbon traps in the Philippine Islands developed during a long and complex history of microplate tectonics. Carbonate and clastic stratigraphic traps formed during Mesozoic and early Cenozoic rifting and drifting. Hydrocarbons, generated in deep rift basins, migrated to the traps during rifting and drifting. Later, Cenozoic compressional tectonic activity and concomitant faulting enhanced some traps and destroyed others. Seismic data offshore from Palawan Island reveal the early trap histories. Later trap histories can be interpreted from the seismic data combined with outcrop and remote sensing data. Understanding the microplate tectonic history of the Philippines is the key to interpreting trap histories.

## INTRODUCTION

The great complexity of Philippine geology and the consequent difficulty exploring for natural resources is demonstrated on a Landsat mosaic (Figure 1). The fragmented appearance of the islands suggests the fragmented nature for much of the geology. The key to successful future exploration for oil and gas in the Philippine Islands may depend on the explorer's ability to decipher what was where and when, and to figure out how this can help locate adequate reservoirs and traps. In other words, present geological knowledge may be sufficient to roughly specify the position of each tectonic element of the very complex Philippine Island tectonic collage during each step of its history. This knowledge may have significant favorable or unfavorable implications for exploration in the Philippine Islands as well as the Circum-South China Sea.

A main conclusion of this paper is that the tectonic source areas for each tectonic element can be identified with some degree of confidence (Figure 2). This paper suggests, in a general way, how the basinal areas in the Philippine Islands are associated with the tectonic elements. The paleopositions of both the tectonic elements and the basins are reconstructed using published data and concepts about movement of the tectonic elements (Figure 3).

The published literature disagrees about the paleoposition of certain specific components within tectonic elements. For example, the islands from Luzon southeastward to Mindanao may have moved northward (e.g., Karig, 1983), or they may have developed in-situ (McCabe, 1984). This paper combines the two views and suggests that the northward movement of the components of the tectonic elements along plate boundaries seems to best explain pre-mid-Miocene tectonics, and that the in-situ interpretation best explains later tectonics.

The components of the tectonic element in the Cuyo shelf area including Reed Bank, North Palawan, and parts of Mindoro—moved southeastward across the South China Sea and away from the South China margin during the Tertiary. There is general agreement in the published literature on this scenario, but reconstruction details differ (Ben-Avraham, 1978; Chen, Gallagher, and Kenyon, 1979; Taylor and Hayes, 1980; Holloway, 1981).

The components of the tectonic element which includes the Sulu Sea, southern Palawan Island, and the northeastern part of Borneo, moved northeastward and underwent some clockwise rotation during the Tertiary (Karig, 1983; Taylor and Hayes, 1983).

This paper roughly defines the limits of the three tectonic elements that appear to have distinctly separate geologic histories before mid-Miocene and briefly discusses the history of each tectonic element. The focal point of the paper is the discussion in some detail of sample play concepts for the basin offshore Palawan Island. Samples of play concepts for the other Philippine basins are shown by Saldivar-Sali et al. (1986).

The conclusion of this paper is that for the Palawan area exploration risk is high and risked returns on investment are low, because the size of the typical oil reservoir is probably small. There are potentially larger reservoirs farther offshore than has yet been explored.

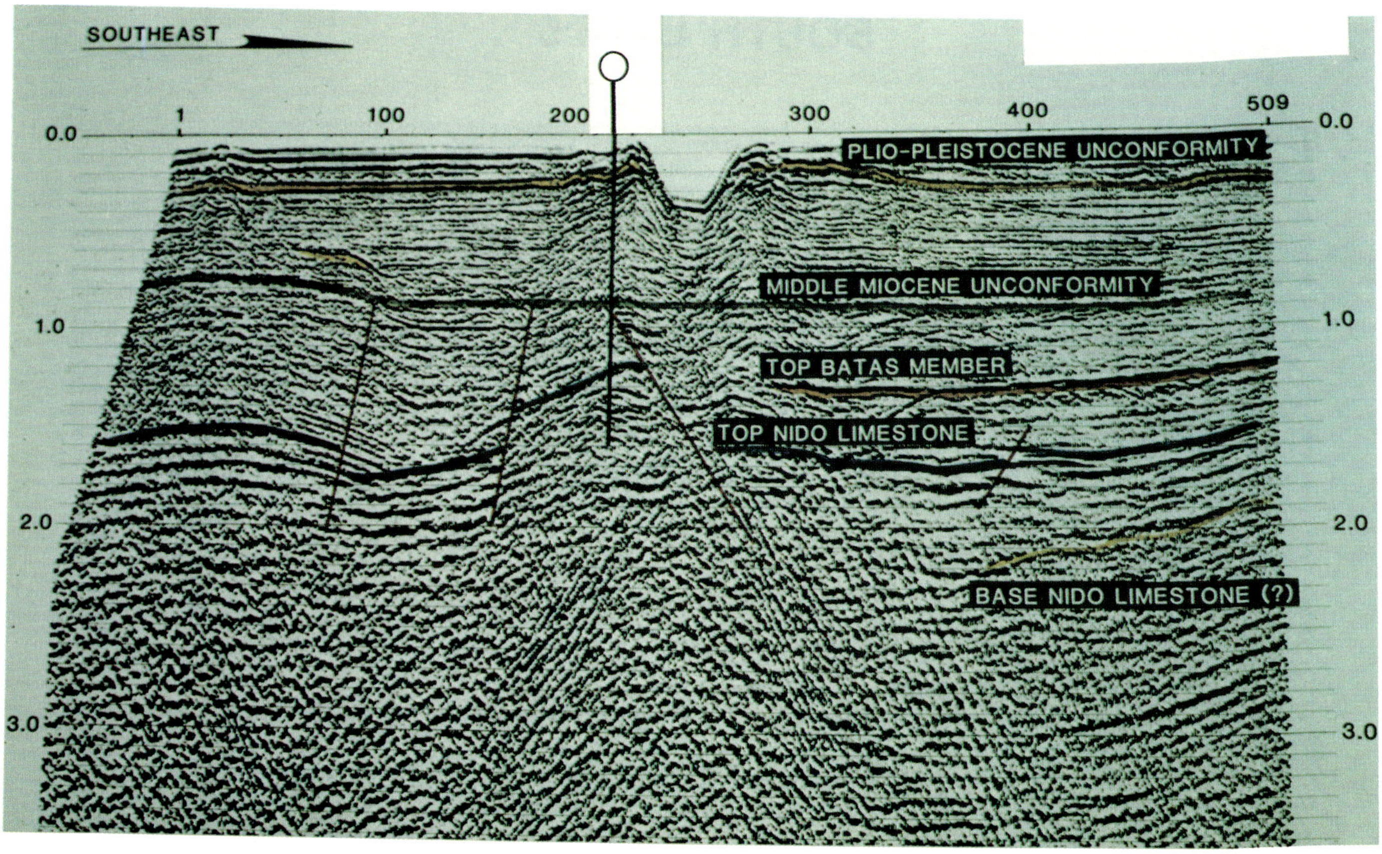


Figure 14. Migrated time section in the Bacuit area showing the offshore extension of the Ulugan Bay fault (UBF on Figure 3A).

above A on Figure 11 set up attractive-looking, potential trap structures.

In the Galoc area, faults below A and B on Figure 12 set up traps for clastic reservoir-prone units above them. These east-west-trending faults are related to the Paleocene-Eocene breakup of southeast China (Figure 3D). Younger clastic accumulations occur directly over the faults.

Above A on Figure 13 is the same strike-slip fault shown on Figure 11. This fault can be traced to the south and onshore Palawan Island. This fault is known to separate clastics to the east from carbonates to the west, both of which are proven reservoirs in this oil province.

The offshore extension of the Ulugan Bay fault (UBF Figure 3A) can be seen in the center of the seismic line on Figure 14. Pre-Mid-Miocene correlations across the fault are difficult, suggesting separation of the blocks on either side of the fault prior to that time. Block separations are shown in map view on Figures 3C to 3G. Younger correlations are possible, suggesting proximity from Mid-Miocene to present, as shown on Figure 3A and 3B. This may somehow explain why oil and gas are substantially more abundant northeast of the fault than southeast of the fault.

## CONCLUSIONS

The Philippine Islands have been looked at and rejected by many major oil exploration companies. Cities Service persisted and was rewarded with commercial oil discoveries. However, most major companies continue to resist exploring oil basins like those of the Philippine Islands, especially in times of depressed oil prices. The tectonic analysis in this paper shows that there are oil-trapping possibilities, especially in the basins of the Cuyo tectonic element. Some traps are located and others can be predicted using the rigid-indentor theory combined with the concept of continental-scale cataclasis. The challenge to explorers is to use these or other ideas to find oil traps that are large enough to justify the expense of modern exploration.

## ACKNOWLEDGMENTS

I am grateful to Occidental International Exploration and Production Company for releasing an earlier version of this report for publication and to Cities Service Company for permitting publication of the seismic data contained in this report. Thanks to Les Beddoes, Denis Brownlee, Allen Hatley, Hans Oesterle and Mike Roberts for informative discussions about Philippine geology.

## REFERENCES CITED

Beddoes, L.R., 1980, Hydrocarbon plays in Tertiary basins of Southeast Asia: Offshore Southeast Asia Conference, Singapore, 60 p.

Ben-Avraham, Z., 1978, The evolution of marginal basins and adjacent shelves in east and southeast Asia: Tectonophysics, v. 45, p. 269-288.

Ben-Avraham, Z., and S. Uyeda, 1973, The evolution of the China

basin and the Mesozoic paleogeography of Borneo: Earth and Planetary Science Letters, v. 18, p. 365-376.
Brownlee, D.N., and M.W. Longman, 1981, Depositional history of a lower Miocene pinnacle reef, Nido B. oil field: The Philippines, Proceedings 4th International Coral Reef Symposium, Manila, v. 1, p. 619-625.
Chen, P.H., J.J. Gallagher, Jr., and C.S. Kenyon, 1979, Paired rift-spread systems of China: American Geophysical Union Transactions, v. 60, p. 398.
Dewey, J.F., 1977, Suture zone complexities: A review: Tectonophysics, v. 40, p. 53-67.
Gallagher, J.J., Jr., 1981, Tectonics of China: Continental scale cataclastic flow, *in* Mechanical behavior of crustal rocks: American Geophysical Union Geophysical Monograph 24 (Handin Volume), p. 259-273.
Gallagher, J.J., Jr., 1984, Large-scale faults and the tectonics of China: Fourth International Conference on Basement Tectonics, Oslo, Norway, p. 173-185.
Gallagher, J.J., Jr., 1986, Philippine Islands: A tectonic railroad siding: AAPG Memoir 40 (Wallace Pratt Memorial Volume), p. 515-527.
Hamilton, W., 1979, Tectonics of the Indonesian region: U.S. Geological Survey Professional Paper 1078, 345 p.
Hatley, A.G., 1977, The Nido reef oil discovery in the Philippines, its significance: Proceedings, Association of Southeast Asian Nations Council on Petroleum, Jakarta, Indonesia, p. 263-277.
Hayes, D.E., and S.D. Lewis, 1984, A geophysical study of the Manila Trench,, Philippines, 1, Crustal structure, gravity, and regional tectonic evolution: Journal of Geophysical Research, v. 89, p. 9171-9195.
Holloway, N.H., 1981, The North Palawan block, Philippines: Its relation to the Asian mainland and its role in the evolution of the South China Sea: Geological Society of Malaysia Bulletin, v. 14, p. 19-58.
Holloway, N.H., 1982, North Palawan block, Philippines: Its relation to Asian mainland and role in evaluation South China Sea: AAPG Bulletin, v. 66, p. 1355-1382.
Hutchinson, C.S., 1975, Ophiolite in southeast Asia: GSA Bulletin, v. 86, p. 797-806.
Karig, D.E., 1983, Accreted terrains in the northern part of the Philippine archipelago: Tectonics, v. 2, p. 211-236.
Kenyon, C.S., P.H. Chen, and J.J. Gallagher, Jr., 1979, Tectonics of China's basement rocks: American Geophysical Union Transactions, v. 60, p. 398.
McCabe, R., 1984, Implications of paleomagnetic data on the collision related bending of island arcs: Tectonics, v. 3, p. 409-428.
McCabe, R., and J. Cole, 1987, Speculations on the late Mesozoic and Cenozoic evolution of the Southeast Asia margin, this volume.
Moore, G.F., and E.A. Silver, 1983, Collision processes in the northern Molucca Sea, *in* D.E. Hayes, ed., Tectonic and geologic evolution of southeast Asian seas and islands, Part 2: American Geophysical Union Geophysical Monograph 27, p. 360-372.
Murphy, R.W., 1975, Tertiary basins of southeast Asia: Proceedings Southeast Asia Petroleum Exploration Society, v. 2, p. 1-36.
Saldivar-Sali, A., H.G. Oesterle, and D.N. Brownlee, 1981, The geology of offshore northwest Palawan, Philippines: Proceedings, Second Association of Southeast Asian Council on Petroleum Conference and Exhibition, Manila, Philippines, p. 99-123.
Saldivar-Sali, A., L. Flower, A. Madrid, and R. Reyes, Jr., 1987, Tectonic setting and petroleum possibilities of Philippine sedimentary basins, this volume.
Tapponnier, P., and P. Molnar, 1976, Slip line field theory and large scale continental tectonics: Nature, v. 264, p. 319-324.
Taylor, B., and D.E. Hayes, 1980, The tectonic evolution of the South China basin, *in* D.E. Hayes, ed., Tectonic and geologic evolution of southeast Asian seas and islands: American Geophysical Union Geophysical Monograph Series, v. 2, p. 89-104.
Taylor, B., and D.E. Hayes, 1983, Origin and history of the South China Sea basin *in* D.E. Hayes, ed., Tectonic and geologic evolution of southeast Asian seas and islands, Part 2: American Geophysical Union Monograph 27, Washington, D.C., p. 23-56.
Uyeda, S., and Z. Ben-Avraham, 1972, Origin and development of the Philippine Sea: Nature, v. 240, p. 176-178.
Uyeda, S., and A. Miyashiro, 1975, Plate tectonics and the Japan Islands: A synthesis: GSA Bulletin, v. 85, p. 1159-1170.

CHAPTER 10

# OIL EXPLORATION IN AUSTRALIA

**Bryan R. Griffith**
*BHP Petroleum Pty. Ltd.*
*Melbourne, Australia*

## ABSTRACT

During recent years exploration activity in Australia has been at high levels. However, during a period of generally depressed economic conditions, this activity has been concentrated in the established areas where the risk-reward relationships have been judged the most favorable. Although oil discoveries have been at an encouraging level, Australia still finds itself with less oil reserves today than in 1979.

Without additional discoveries, indigenous oil production is projected to decline to a level 250,000 bbl/day below total demand by 1990 and 550,000 bbl/day below total demand by the year 2000.

Geological data would suggest that Australia's undiscovered oil resources have the potential to support a high level of self sufficiency. Given the proper business environment, the Australian exploration industry has the skills necessary to evaluate this potential effectively and efficiently. However, it is important, particularly in the light of industry's weakening expectations of profitability, that the Government pursue policies that encourage a high level of exploration activity in all areas, particularly in the high-risk, high-potential areas offshore. This will require recognition by Government of the critical balance which exists between risk and reward.

## RECENT EXPLORATION ACTIVITY IN AUSTRALIA

During the last few years, exploration activity in Australia, as indicated by the levels of exploration drilling, has been at record levels (Figure 1). The year 1984 was a record-breaking year for the industry. A total of 230 wildcat and appraisal wells were drilled, far exceeding the previous record of 156 wells drilled in 1965. Of the 230 wells drilled, 44 were offshore, surpassing the previous record of 30 set in 1972.

Despite the weakness of the world economies, and thus the general lack of investor confidence which was evident at the time, a high level of drilling activity has been maintained since that year. The year 1984 set new records when 247 wells were drilled, of which 204 were onshore and 43 offshore.

The latest calendar year survey of the Australian Petroleum Exploration Association indicates that, overall, the projected drilling activity for 1985 will continue the high trend of the previous 2 years. It is forecast that 1985 will see 251 wells drilled, 215 onshore and 36 offshore.

As shown by Figure 2, the same general trend of high activity can be seen in the number of wildcat wells drilled during the past three years and projected to be drilled during 1985. From the low level of activity seen during the mid 1970s wildcat drilling activity has increased significantly. Thus 146 wells were drilled in 1982, 164 wells in 1983, 183 wells in 1984, and a projected 186 wells are expected to be drilled in 1985. Perhaps as a move against this overall trend, wildcat activity offshore in 1984 was 35 wells, and the projection for 1985 is 21, somewhat below the levels experienced in 1982 and 1983.

In terms of the geographic distribution of exploration wells drilled (Figure 3), activity was concentrated in the established hydrocarbon provinces, in particular within the Cooper/Eromanga and Bowen/Surat basins. In 1982 and 1983 almost 50% of the total exploration wells drilled in Australia were located in these onshore areas. During 1984 about 55% of the total wells drilled were in these same areas.

The success ratio of new field wildcats is also being maintained at an encouraging level. As shown by Figure 4, the success ratios of new field wildcats during the period 1970 to 1977 were quite low and averaged about 11%. From 1978 onward, this ratio improved significantly, reaching levels in excess of 20% during the last 5 years.

As might be anticipated from such relatively high success ratios, oil discoveries made during recent years have been encouraging by Australian standards. In the period 1979 through 1983, at least 300 million barrels were added to the reserves inventory. As the drilling statistics would suggest, the bulk of these reserves has been discovered in the established hydrocarbon provinces both onshore and offshore.

The trend of encouraging oil discoveries continued through 1984 with Bodalla South in the Eromanga Basin and Tuna-4 in the Gippsland Basin. Together, these discoveries might add an additional 50 million barrels to the reserves inventory.

With the bulk of drilling activity concentrated in established areas, the oil discoveries made in 1983 by Turtle-1 in the Bonaparte Basin and Jabiru-1A in the northern Browse Basin are of particular interest and relevance. Although the commercial significance of these discoveries remains unknown at this time, they have established new areas of interest. With further exploration, these areas might develop into new, and possibly significant, hydrocarbon provinces.

The surge in exploration drilling activity during the past 3 years, and the relative success of that effort, has occurred during a period of economic recession in the free world. This per-

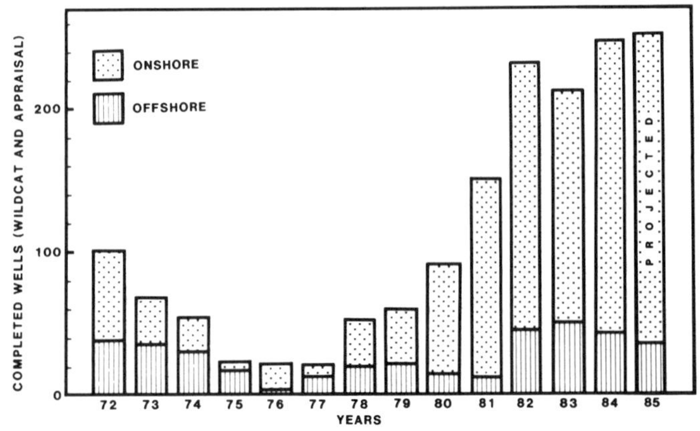


Figure 1. Exploration drilling activity.

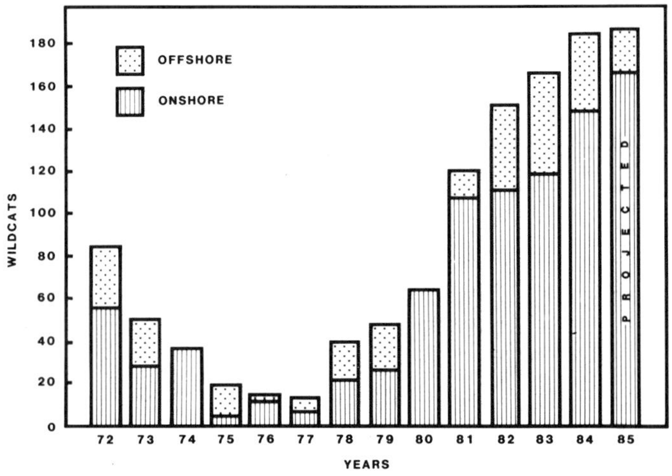


Figure 2. Wildcat drilling activity.

iod has been characterized by a decreased demand for oil products, weakening oil prices, and high interest rates—a combination of factors which elsewhere has resulted in a marked downturn in exploration activity.

Despite the high level of drilling activity, Australia has not been isolated from these worldwide economic events. In late 1982 and 1983 there was a shake out of the exploration industry, during which:

- exploration commitments were critically reviewed,
- many companies withdrew from the higher risk ventures, and
- companies highly dependent on the share market for exploration capital were either forced to reduce their activities or abandon participation in the industry.

Under these circumstances it is not surprising that most of the industry's exploration drilling effort was concentrated in the established areas, where the risk-reward relationships were judged to be the most favorable. In this regard, the emergence of the Cooper/Eromanga Basin as a significant oil province, following the Jackson discovery in December 1981, must be regarded as of considerable significance. The new oil play, as defined by the Jackson well, was pursued with considerable success. In more recent times, Tintaburra, drilled in 1983, and Bodalla South, drilled in 1984, have extended the play considerably and by so doing have engendered further interest and activity in the area.

Effectively, during the last few years, the Cooper/Eromanga Basin has provided the basis for a self-sustaining exploration effort. As an increasing number of small companies active in the area became producers, and generated their own cash flow, so an increasing number of companies became increasingly independent of the generally depressed equity markets to fund further exploration activities. Even for companies lacking production, their interest in trend acreage provided a means of gaining access to cash through the market, by farmout or partial sale of interest.

By comparison with the onshore, offshore drilling activity is projected to show a decline during 1985. The trend to lower levels of activity might well have been evident earlier if it were not for the fact that many wells were drilled either to satisfy permit commitments or long-term drilling contracts made during the more economically buoyant times prior to 1982.

The same apparent trend can be seen from the levels of seismic activity in the offshore, which fell from 52,516 km in 1982

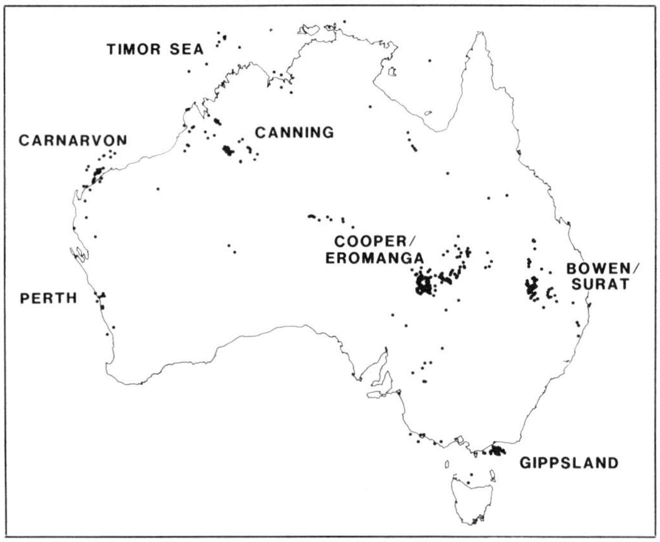


Figure 3. Distribution of exploration drilling activity, 1983 and 1984.

to 15,168 km in 1983. The year 1984 saw 19,224 km shot, and the projection for 1985 is 32,000 km. However, it should be noted that the 1985 projection includes 15,000 km of high-density 3-D seismic. The implications of these seismic statistics for future drilling activity need to be assessed cautiously. However, it is likely that without a new significant discovery or the release of additional acreage in the Timor Sea as a follow-up to the Jabiru and Turtle discoveries, lower levels of drilling activity in the offshore will continue for the foreseeable future.

## SUPPLY AND DEMAND FOR OIL AND CONDENSATE

The incentive and the need to maintain an intensive exploration effort is well illustrated by Figure 5, which shows a plot of forecast production of crude oil and condensate from known fields against projected total demand and local absorption capacity. The production curve shows a marked decline in the late 1980s, reflecting the decline of production from the major fields of the Gippsland Basin.

The plot illustrates two fundamental problems facing industry and Government alike:

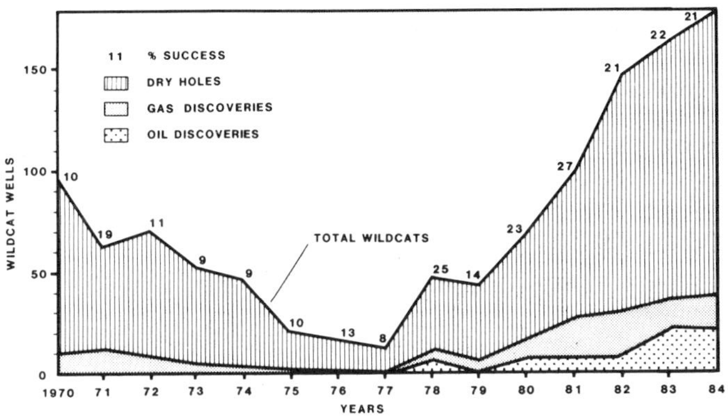


Figure 4. Wildcat discoveries.

(1) By the year 2000, without additional discoveries in the interim, the shortfall in production will amount to 550,000 bbl/day below total demand and 400,000 bbl/day below absorption level, and

(2) By the year 2000 "old" oil production will have declined to less than 20,000 bbl/day.

Thus, even to satisfy the projected levels of maximum local absorption in the year 2000, equivalent to about 70% self sufficiency, the exploration industry must bring on stream an additional 400,000 bbl/day from new discoveries. This production is equivalent to at least 1.5 billion barrels of additional developed reserves.

Such a shortfall is not a peculiar characteristic of the long term. By 1990 production from known fields will be almost 100,000 bbl/day short of local absorption and 250,000 bbl/day short of total demand.

The decline in old oil production is also significant, and something which must be viewed with concern for reasons other than loss of production. The Federal Government depends on the oil industry for well over 10% of budgeted revenue. Of this total contribution of almost $7 billion, by far the largest component of $3.4 billion is provided by the levy on old oil. With the projected decline in old oil production, of some 200,000 bbl/day by 1990, the loss to Government revenues is self evident.

Although recent exploration drilling activity has been at record levels, and the wildcat success ratios have been high, Australia still finds itself with less oil reserves today than in 1979 (Figure 6). Although the trend of progressive decline was halted in 1983, it is unlikely that this reversal can persist for any length of time if exploration activity is restricted to the established hydrocarbon provinces either offshore or onshore. These areas, although they will continue to yield significant volumes of additional oil reserves, are probably lacking in the large fields which are critical to oil self-sufficiency in future years.

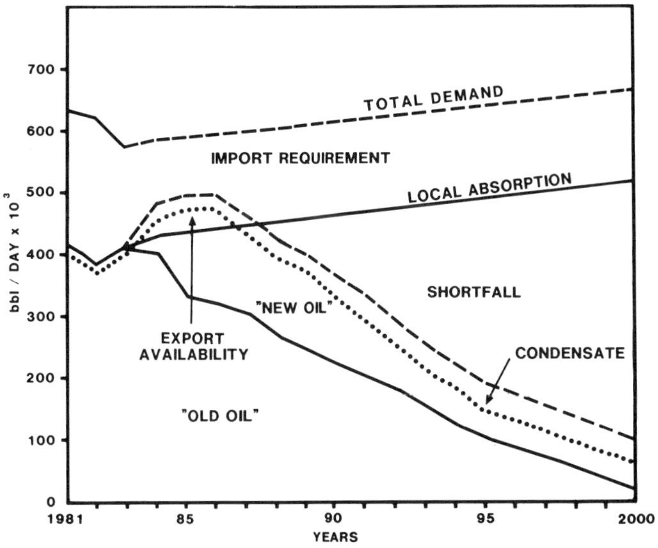


Figure 5. Crude oil and condensate—supply/demand.

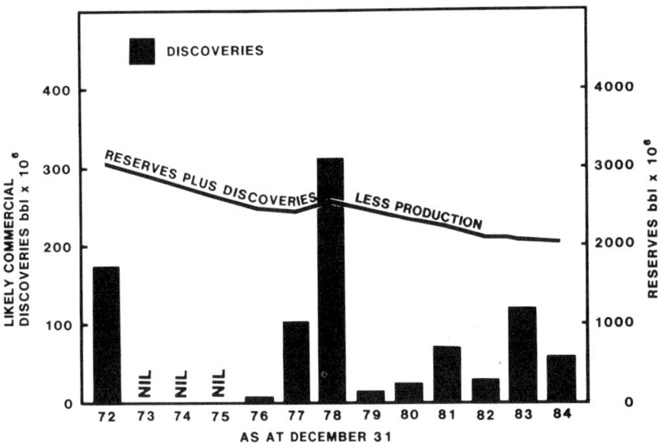


Figure 6. Australia's remaining oil reserves.

## AUSTRALIA'S UNDISCOVERED OIL POTENTIAL

Any discussion of prospects to the year 2000 would be lacking without at least some reference to the undiscovered oil potential of the continent. Because of the lack of adequate geological data, such assessments are fraught with uncertainty. Nevertheless, they are of considerable value in projecting, on the best data that are available, the likely range of oil production that can be anticipated from future discoveries.

In recent times both Esso and the Bureau of Mineral Resources have quoted their assessments of the continent's undiscovered oil resource base. For comparison these forecasts have been superimposed on Figure 7. Recognizing the uncertainties of the data base and the assessment techniques used, these two assessments should be considered as being in essential agreement.

On this basis, it can be concluded that there is a 50% probability that the undiscovered oil resources of the continent amount to approximately 3 billion barrels with an upside potential in excess of 5 billion barrels. Recognizing that about 4 billion barrels of oil have been discovered since 1950, most geologists would feel comfortable with this assessment.

The assessment indicates that, on the basis of current knowledge, there is potentially a sufficiently large undiscovered oil resource base to satisfy Australia's needs to the year 2000.

In providing further details of their analysis, the Bureau of Mineral Resources concluded that by far the greatest contribution to the undiscovered resource base would be provided by the offshore sedimentary basins and, of these, by far the greatest contribution would be provided by those basins off

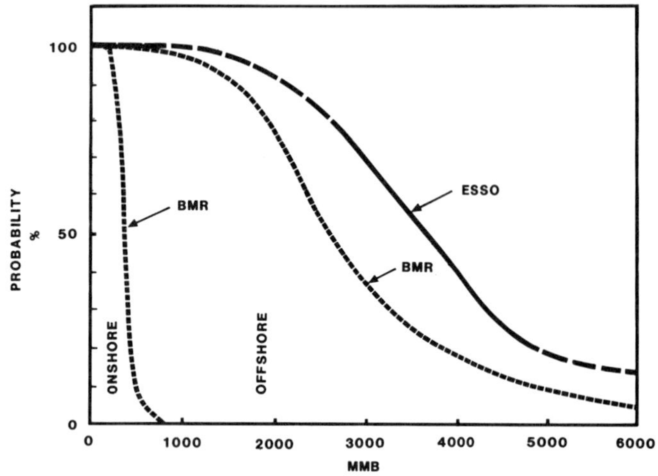


Figure 7. Australia's undiscovered oil resources.

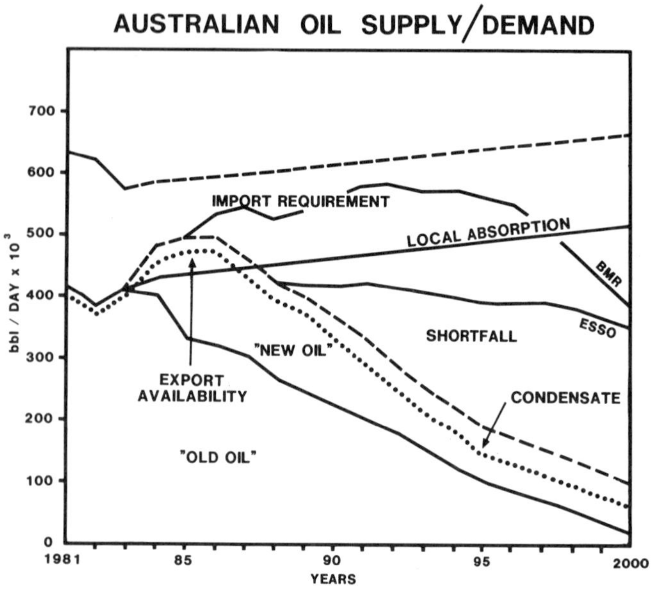


Figure 8. Possible future oil production profiles.

the northwest coast of Australia. Even before the significant discoveries at Jabiru and Turtle, most geologists, on the basis of past exploration results, would have supported this contention.

## POTENTIAL FUTURE OIL PRODUCTION LEVELS

The assessment of the resource base provides the first essential step in any determination of future oil production. However, conversion of that base to reserves, the levels of production and the timing of that production, will remain dependent on many factors, including:

(a) prevailing economic conditions,

(b) the nature and size of the individual fields making up the resource base—many small fields or possibly large fields of low productivity in the offshore could be noncommercial,

(c) the level of exploration activity,

(d) Government regulations and policy.

These variables defy prediction and, as a result, the range of possible production profiles to be derived from the assessed resource base becomes almost infinite.

The Bureau of Mineral Resources, assuming the discovery of 1 billion barrels of new reserves in the next 5 years, suggests that crude oil production might rise about 30% between now and 1989, and may plateau at this level before beginning to decline in 1996 (Figure 8). On this basis, Australia's production of crude oil and condensate could reach 90% of projected demand in the late 1980s and the early 1990s.

Esso, on the basis of probability analysis of all the variables involved, suggests that, in the most likely case, new discoveries will contribute only some 50,000 bbl/day by 1990, leaving Australia's total production at some 66% of self-sufficiency. The same projection concludes that by the year 2000 new discoveries will add 240,000 bbl/day to total production, leaving Australia 54% self sufficient in oil.

Such variation between the projections of two respected organizations only serves to highlight the uncertainties involved. The major variation between the two forecasts, and indeed the major source of uncertainty in any single production forecast, relates to the volume and timing of the contributions to be made from a small number of large pools.

In its projection, the Bureau of Mineral Resources assumes the rapid emergence of the Timor Sea area as a major hydrocarbon province. Although significant additions to any reserves inventory tend to come in cyclic "bursts," it can be misleading in any plan to project such an event as a single possible outcome.

In deciding its future policy options, it is important that the Government is not lulled into a sense of false security by the discoveries made during the last year in the Timor Sea. The Jabiru discovery is undoubtedly of considerable importance. However, the results of Jabiru-2 and Jabiru-3, and indeed the results of the total exploration effort that has been carried out in the general area since the early 1970s, would suggest the need for caution in any assessment of the hydrocarbon potential. Moreover, because of a complex of geological and geophysical factors, the oil resources of the Timor Sea will probably require a much greater effort and a considerably longer time to confirm than perhaps originally thought by many observers.

## REALIZATION OF AUSTRALIA'S OIL RESOURCE BASE

While stressing the need for caution in any assessment, the important fact remains that, on the basis of current knowledge, Australia's oil resource base has the potential to maintain Australia at a high level of self-sufficiency through the 1990s. On both economic and strategic grounds, the speed and effectiveness with which the resource base can be realized is of national concern.

To maximize the chances of maintaining a high level of self-sufficiency in oil, the industry must maintain a high exploration effort both onshore and offshore. Some 200-250 exploration wells a year will be required, involving the expenditure by industry of some $1000 million per year.

Such a massive commitment of both physical and financial resources, particularly in the high-cost and high-risk environment of the offshore, will be made only if the industry's perception of the potential reward in the event of success justifies the overall risks. Although the risk-reward relationship is affected by numerous factors, many outside the control of the Government, Government policies and regulations will continue to be of paramount importance.

Federal Government, generally, has recognized the importance of the oil exploration industry to the future well-being of the Australian economy. The introduction of the new oil pricing policy and the provision of a guaranteed market to local production have provided a significant boost to exploration activity. In more recent times such diverse Government initiatives as to allow the export of surplus crude production, and to introduce retention licenses to protect currently uneconomic resources, have been of direct benefit to the industry. Unfortunately, Governments of whatever persuasion, or whether State or Federal, are in constant need of additional revenue and apparently view the successful producer as a politically convenient means to satisfy this need. The need for stability of Government policy and the need to preserve a satisfactory relationship between risk and reward, so essential to a viable industry, can be too readily sacrificed for short-term financial gain.

At a time when lower projections of future oil prices are reducing the expected profitability of the exploration industry, the Federal Government has decided to introduce:

(a) a secondary tax on "new" oil, which, depending on area, will take the form of a rent or an excise levy, and

(b) a system of cash bonus bidding for "highly prospective" areas offshore.

Both forms of tax were opposed by the exploration industry as potentially damaging to the exploration effort. As ultimately defined, the additional taxation on new oil was acknowledged, at least by certain sections of the industry, as an acceptable compromise given the revenue objectives of the Government. However, almost by definition, the additional tax will have an adverse impact on exploration activity in the higher risk areas, both onshore and offshore. In the near term, at least, the impact on drilling statistics is likely to be small and difficult to measure. Unfortunately, the offshore probably carries the key to Australia's future self-sufficiency, and any reduction in rank wildcat activity in this area, however small, must be cause for some concern.

There remains the risk, well recognized by the industry, that the Government, having introduced a new tax vehicle and intent on maintaining its future revenues, will find it difficult to resist further taxation of the industry. With declining production of old oil, the Government will progressively lose some \$3. 4 billion of tax revenue per annum. If the Government looks to the oil industry to make up this deficit, by whatever means, it will face the prospect of setting in train a process that is likely to damage Australia's exploration effort, with all of its potential consequences for self-sufficiency in future years. Unfortunately, despite the realities of the situation, there is too often a tendency to place additional tax burdens on the existing producers, a facility now made easier in Australia by the existence of a number of different tax regimes in different areas. Such taxes on the producer not only reduce the cash flows necessary for further exploration but also reduce the confidence and expectations of the exploration industry as a whole.

Despite these real concerns, the vacant acreage in the Timor Sea area attracted considerable industry interest. However, there is no guarantee that this area will provide a solution, much less an easy solution, to the oil self-sufficiency problem. It is important that, for the future well-being of the industry, the Government recognize this fact in determining its future policy options.

## CONCLUSION

The maintenance of a high level of self-sufficiency through the 1990s is crucial to the Australian economy. The best geological data and expertise available would suggest that Australia's oil resource base has the potential to satisfy this objective. Given the appropriate business environment, the Australian oil exploration industry has the technical and managerial skills necessary to evaluate this potential effectively and efficiently.

In recent years, despite the weakness of the oil markets worldwide, exploration activity in Australia has been at record levels, and the discoveries made have been at an encouraging level. However, two comments are pertinent:

(1) Exploration activity has been concentrated in the established areas. These areas, although they will continue to yield significant discoveries, probably lack the large pools which are critical to self-sufficiency in future years.

(2) The Timor Sea area, although it has now emerged as an area of considerable interest, cannot be guaranteed to provide a solution, much less an easy solution, to the self sufficiency problem.

There remains the real prospect that, beginning in the late 1980s, Australia's oil production will drop progressively below the 70% self-sufficiency level. In deciding its future policy options, the Federal Government must allow for this fact. To maximize the chances of maintaining a high level of self-sufficiency it is critical that the Government pursues policies that encourage a high level of exploration activity in all areas, particularly in those areas which, although high risk, have the potential for large fields. This will require, among other things, stability of the fiscal regime and the recognition by Government of the critical balance which always exists between risk and reward.

CHAPTER 11

# DEEP SOURCE GAS POTENTIAL ALONG THE WEST COAST OF THE U.S. (An Overview of the U.S. Department of Energy's Deep Source Gas Project)

**W. J. Gwilliam**
**K. K. Cohen**
*U. S. Department of Energy*
*Morgantown Energy Technology Center*
*Morgantown, WV*

ABSTRACT

Deep Source Gas research is focused on the hypothesis that natural gas is generated in sediments carried to great depths at convergent plate boundaries in the earth's crust. These deeply emplaced sediments may source gas to shallower, drillable traps through deep fracture systems. Many areas of North America are believed to have experienced plate tectonic convergence. The western Cordilleran geologic province in particular appears to have thrust fault structures (associated with subduction and obduction) that enabled deep emplacement of hydrocarbon-generating sediments during more recent geologic ages (during the last 180 million years). The specific area of interest in this province encompasses approximately 1. 5 million $mi^2$ (3. 9 million $km^2$) of the western U. S. (including Alaska) and Canada; other portions of this same province extend southward into Mexico and Central and South America.

The ongoing research consists of basic studies of hydrocarbon generation, stability, and preservation at depths in excess of 30,000 ft (9150 m) in addition to a comprehensive evaluation of the geologic structures, stratigraphy, and geochemistry of the above region.

Results to date include geologic and geophysical evidence of deeply emplaced sedimentary rock units at depths exceeding 30,000 ft (9150 m) in western Washington and south-central Alaska, a new methodology for verifying deep methane stability via fluid inclusion studies, and a preliminary gas resource estimate of 3000 Tcf. If only a fraction of the gas thought to exist is found to be recoverable, it could have a profound influence upon the international energy industry.

## INTRODUCTION

There is general agreement that natural gas will continue to be an important alternative to other energy sources. Since 1976, the U. S. Department of Energy (DOE), Morgantown Energy Technology Center (METC) has been involved in unconventional gas recovery (UGR) research to characterize this resource and to enable the exploitation of natural gas from reservoirs that would otherwise be considered economically unrecoverable. The research has focused on low-porosity, low-permeability, or otherwise undeveloped resources whose sources and reservoirs are at conventional depths ($\leq$ 15,000 ft or 4575 m).

In 1982, a new effort emerged that was a natural outgrowth of DOE/METC's experience in UGR research. Natural gas resources are being investigated that required long-term, high-risk research and development, but offer a significant potential payoff. The level of risk is such that little or no private sector research is envisioned for the near future.

A workshop held in 1982 in Morgantown, West Virginia, addressed the subject of deep source gas from the perspective of three separate emplacement concepts: abiogenic gas, subducted organic-origin gas, and deep intracratonic sedimentary basin gas. Following the workshop, it was decided that METC research of deep source gas would focus on the subducted, organic-origin gas concept (Table 1). DOE/METC will continue to monitor the abiogenic gas research supported by the Gas Research Institute (GRI).

The existence of methane or methane-generating source rocks at great depth in the Earth's interior remains theoretical. If the existence of such gas was documented, it could radically influence the methods used to explore for new supplies. If

TABLE 1. DEEP SOURCE GAS RATIONALE

RESOURCE DEFINITION

Natural gas sourced at a depth of more than 30,000 feet

CONCEPTS OF ORIGIN

- Subducted organic matter
- Abiogenesis
- Deep sedimentary basin

RESOURCE EVALUATION

- Mode of source emplacement
- Methodology of migration
- Resource detection technology
- Mode of gas entrapment

POSSIBLE RESEARCH AREAS

- Plate tectonic theory
- Laboratory examination of rock and hydrocarbon samples taken from deep earth areas
- Carbon and helium isotope and rare earth gas studies of suspected deep source gas
- Field examination of suspected deep source gas seeps
- Concept development, laboratory experimentation, and field observation of hydrocarbon migration in the deep subsurface

even a fraction of the deep gas thought to exist proves to be recoverable, it could have a profound influence upon the international energy industry.

The nature of the deep source gas resource has yet to be identified. The primary emphasis of this research is to: establish the existence of natural gas formed at depths in excess of 30,000 ft (9150 m); relate such occurrences to conceptual models; define the limits of target areas; quantify the resource; and determine the significance of this gas to the nation's reserves.

Central to current deep gas research is the hypothesis that sediments carried to great depths source gas to shallower, drillable traps through deep fracture systems. Very early in the project, the Cordilleran geologic province of North America was a conspicuous candidate for study because of its modern (active) and suspect fossil (inactive) subduction zones. Although many areas of North America appear to have experienced convergent plate tectonic activity in past geologic times, the western Cordillera contains structures that enabled deep emplacement of hydrocarbon-generating sediments during more recent geologic ages (during the last 180 million years). For this reason, the Cordilleran geologic province has been designated as the prime study area for the project. The specific area of interest in this province encompasses approximately 1. 5 million mi$^2$ (3. 9 million km$^2$) of the western U. S. (including Alaska) and Canada; other portions of this same province extend southward into Mexico and Central and South America.

In order to focus on the subducted, organic-origin gas hypothesis, the initial emphasis has been on geological characterization. Geologic, geochemical, and geophysical studies will define target areas within regions of past and present subduction zones. Many disciplines in geology will play a role in unraveling what may become one of the most complex and fascinating geologic problems to emerge in natural gas exploration.

Although the emphasis of this research has been on the subducted, organic-origin gas hypothesis, DOE/METC continues to monitor worldwide deep drilling activities, mainly for geochemical (gas generation and preservation) data. Field verification of deep source gas occurrences will be difficult without the knowledge gained through this monitoring activity. Consequently, research in this field will not gain momentum until the target areas are better defined and a dependable geochemical technique can be developed for identifying and segregating deep-origin gas from near-surface sources. A subduction-zone deep-well test is planned for later-stage field verification if funds permit. The final planned phase of this project will consist of reserve studies of recoverable deep source gas, based on the newly developed exploration rationale.

Specific activities center on near-surface traps with deep gas sources in areas not previously considered prospective hydrocarbon resources. The success of this research plan will depend heavily on the demonstration of large-scale resource occurrences at economically drillable depths. Consequently, little emphasis will be placed on deep drilling technology.

## CIRCUM-PACIFIC GEOLOGIC CHARACTERIZATION OF CONVERGENT MARGINS

### Marine Studies

An attempt is being made to define the plate tectonics of convergence and sediment subduction at trench axes, to the first few tens of kilometers depths. Questions addressed include:

- How are sediments mechanically incorporated into a subduction zone?
- Is there a selective subduction process based on sediment composition or other factors?
- To what depth are sediments subducted?
- Under what conditions are sediments deposited on the overriding plate, injected into the mantle, or recycled to the surface from island-arc magmas?
- What are the effects of the angle of plate subduction, rate of subduction, volume of sediment influx, and its structure on the sediment subduction process?

Oceanographic studies have increased our understanding of what happens to sediments at deep sea trench locations.

Subduction sedimentary deposits beneath overriding crustal plates have been recognized at many different locations throughout the Circum-Pacific region. Seismic-reflection data, high-resolution sea-floor mapping methods, and other geophysical tools are used to visualize this phenomenon. Seismic-reflection profiles through the Japan Trench and Tonga Trench show the utility of these tools (Figures 1, 2). A structural interpretation of the Japan Trench and the Tonga Trench suggests that block faulting of the oceanic subduction plate has occurred.

Seismic data suggest that the geometric configuration of the descending oceanic plate consists of a series of smaller structures that serve to package the sediments. This allows their subduction to take place. Early models of subduction zones show the sediments are subducted through a rasping action of the block-faulted, underthrusting oceanic plate (Figure 3). The distribution of graben structures (downthrown, block-faulted valleys) and their orientations within the Circum-Pacific region is shown in Figure 4. When these investigations are completed, sediment subduction may be established as an important subduction process in the generation of large volumes of natural gas.

Trenches throughout the Circum-Pacific region are being actively investigated. It appears that the sediments arriving at a trench axis are being totally subducted in a least 70% of all trench regions (Hilde, 1983). These investigations are designed to model the geologic structure of subduction systems, and to show how sediment is actually carried deep into the earth's crust.

Seismic data from three grid lines located near Kodiak Island, Alaska, are an aid in estimating sediment subduction volumes. The data show conclusively that considerable sediment subduction is taking place along the southwestern coast of Alaska. An estimate will be made of the average amount of subducted sediment in each of the studied areas in North America. This will be the basis for estimating the total volume of sediment beneath the continental margin.

The Lesser Antilles island area (extending from the Virgin Islands through Grenada) may be the key to understanding hydrocarbon generation from both large-scale sediment subduction and deep, sedimentary basin formation. The sediments accumulating on the eastern side of the Lesser Antilles island area form a very wide and extremely thick sedimentary basin (as much as 64,000 ft or 19,520 m deep). This entire basin is apparently being slowly subducted beneath the Lesser Antilles island arc (Figure 5). The coincidence of a deep sedimentary basin of this magnitude adjacent to an active subduction system presents an unusual opportunity to study gas generation from both sources.

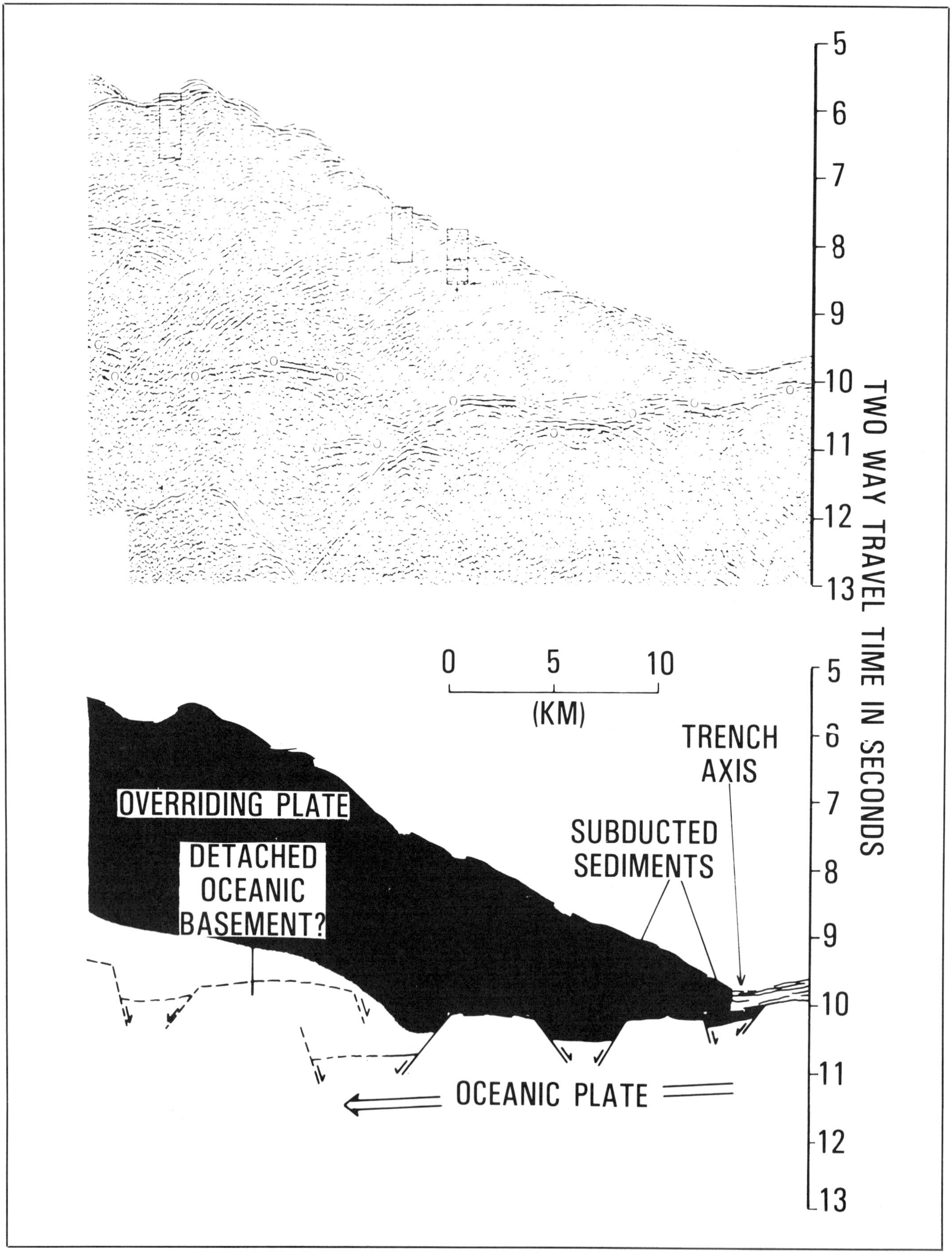


Figure 1. Seismic profile and interpretation of the Japan Trench.

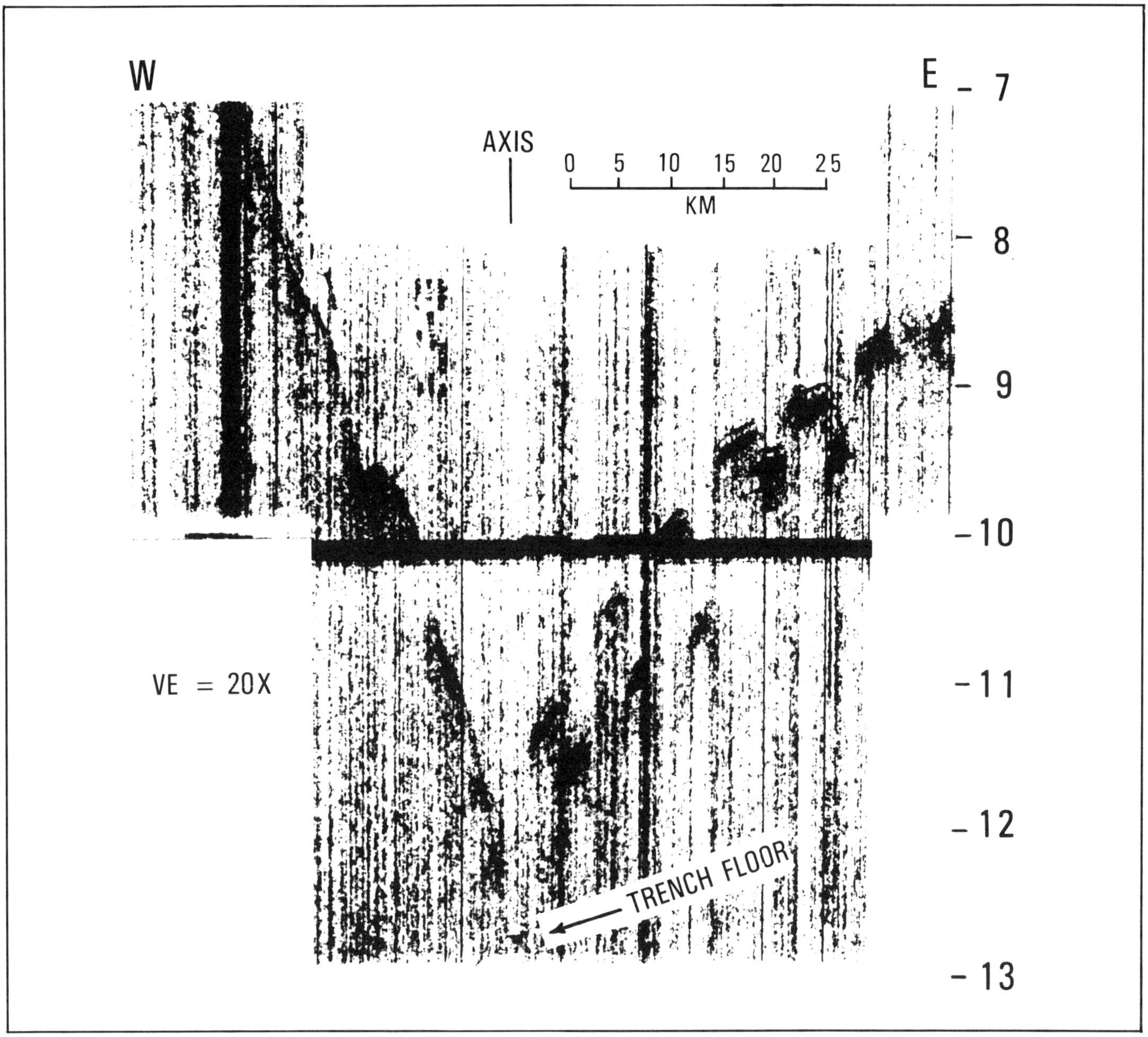


Figure 2. Seismic profile of the Tonga Trench showing horst-graben structures with accompanying sediment on the descending plate (right side of figure).

The objective of this effort is to investigate the plate tectonic structure of the Lesser Antilles (Barbados) accretionary prism and the occurrence of hydrocarbons. Specific project objectives include the following efforts:

- To detect the existence on Barbados of gas derived from underthrust or underplated sediments at the prism base, now at depths of 20 to 25 km.
- To understand the conditions and processes of organic maturation within and below the accretionary prism, and the evolution of migration paths and reservoirs for gas within this progressively deforming mass.
- To consider further strategies and methodologies by which gases generated at deep and shallow depths may be differentiated.
- To study accretion processes and structure in an active intraoceanic subduction zone (Lesser Antilles on Figure 5) with a large influx of terrigenous sediments.
- To provide a modern analog for the structural and hydrocarbon evolution of ancient Cordilleran accretionary complexes that contain an abundance of terrigenous sediment (such as Franciscan or Olympic peninsular terranes).

A major goal will be to develop methods by which gas that is generated from prism-base sources can be discriminated from gas that is generated from sediments at upper levels of the prism. Both theoretical and empirical methods will be developed, the latter method to be based on the characteristics and differences in the geochemistry and geology of sampled gas. There are three major elements in this study:

*Gas Characterization*—Gases are being collected from surface seeps (boiling springs) and wells on Barbados, and the thermal maturation and isotopic composition of these gases will be compared using proximal liquid and solid organics analyses. Gas samples will be collected in wells at and below gas and oil reservoirs, and in surface fault zones that probably extend deep into the subsurface and that are located as far as possible from reservoirs. The purpose of the comparison is to show whether the gases are derived from or cosourced with the oil and solid organics, or whether the gases are apparently derived from more mature sources. The latter case would suggest a deep gas source. It is possible that gases will be obtained that have different maturities at different lateral and vertical positions. This would indicate the existence of deep source gas pathways.

*Predictive Model*—Existing data (temperature, fluid pressure, rock composition, and dating) could be used to develop a model of growth for, and the pressure-temperature (PT) conditions at the base of, the Lesser Antilles accretionary prism. Deep Sea Drilling Project (DSDP) and other research data on the composition of organic precursors can be used to approximate changes in the composition of underthrusting sediments that occur with time. Organic and inorganic fluids and the solid components can be constructed into maturation models to indicate the most likely signatures of fluids from deep sources. Examples of such signatures could include (a) organic gas and oil composition and maturation indices; (b) chemical and isotopic composition of waters emerging from major fault zones, possibly indicating equilibrium with a higher PT regime; and (c) abundances and isotopic composition of gases such as argon (Ar) and helium (He), possibly indicating derivation from metamorphosed pelagic strata.

*Modern Accretionary Processes and Analogs for Ancient Complexes*—Detailed on-land work will be integrated with broader observations of marine geophysics and bottom sampling in the Barbados accretionary prism. This approach should advance the understanding of accretionary processes, and should provide modern analogs with which to interpret ancient accretionary complexes, such as those attached to the North American continent.

Modern accretionary-tectonics aspects of Barbados to be studied are:

- The mechanisms, structures, changes in physical properties, and defluidization of sediment transfer from ocean floor to toe of prism.
- Lateral changes in transfer characteristics along the prism toe that apparently depend on the thickness and rheology of the basin-plain sediment column.
- Partition of the basin-plain sediment column into an upper zone that is off-scraped at the toe and a lower zone that underthrusts the toe, and discovery of what controls the detachment level and how far below the prism the underthrust layer extends. (It has been shown to extend more than 80 km.)
- Progressive deformation within the prism between toe and structural high, changes in appression of folds and thrust wedges, and formation and deformation of upper slope basins. (The structural high at Barbados Island is near the culmination of such deformation.)
- Tectonic interaction between the structural high and oceanward side of the forearc basin.
- Development of a general theory on (a) major loci of organic-rich rocks in the Barbados prism, (b) thermal maturation as a function of position in the prism, (c) progressive development of gas migration paths, and (d) generation and destruction of gas reservoirs. (The deformation boundary of the structural high/forearc basin has a high reservoir potential and is easily studied on and near Barbados.)

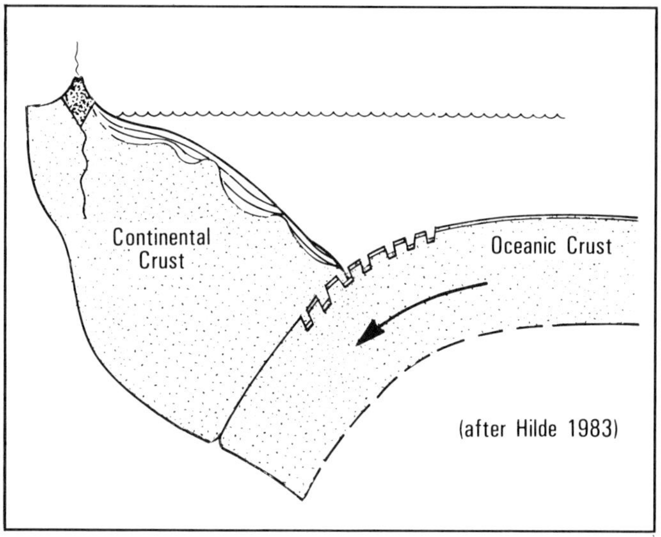


Figure 3. Continental-style sediment subduction.

## Continental Studies

In additional to marine investigations at active subduction zones, there has been a major effort to understand accretionary plate tectonics on the continents. Tectonostratigraphic terranes are defined as three-dimensional blocks of earth crustal material that have become accreted in a patchwork fashion to the margins of continents such as those of the Circum-Pacific region. The deep source gas hypothesis addresses how the terrane accretion concept provides a mechanism for deeply emplacing rocks that can generate hydrocarbons. Three principal activities that are being addressed in this research area:

*Gas Generation at Convergent Margins*—Concepts are being developed for gas generation, migration, and entrapment at convergent plate boundaries.

*Gas Generation at Fossil Subduction Zones*—Concepts are being developed that describe the generation, migration, and entrapment of deep gas resources from fossil subduction zones within the Circum-Pacific tectonostratigraphic terranes.

*Hydrocarbon Generation by Other Tectonic Processes*—Additional large- and small-scale plate tectonic processes deemed relevant to deep hydrocarbon generation, migration, and accumulation will be studied within the Cordilleran geologic province of western North America.

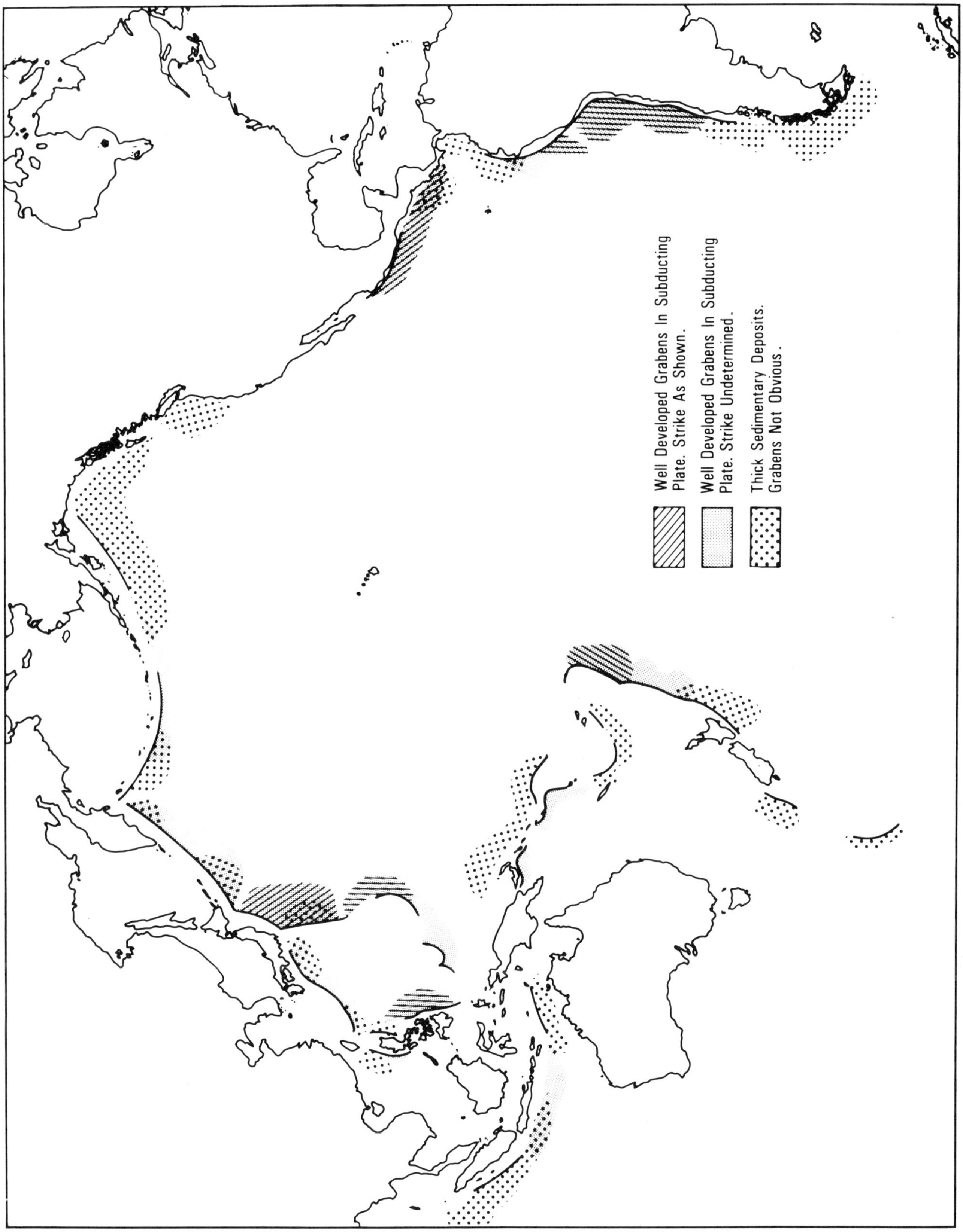


Figure 4. Circum-Pacific sediment subduction.

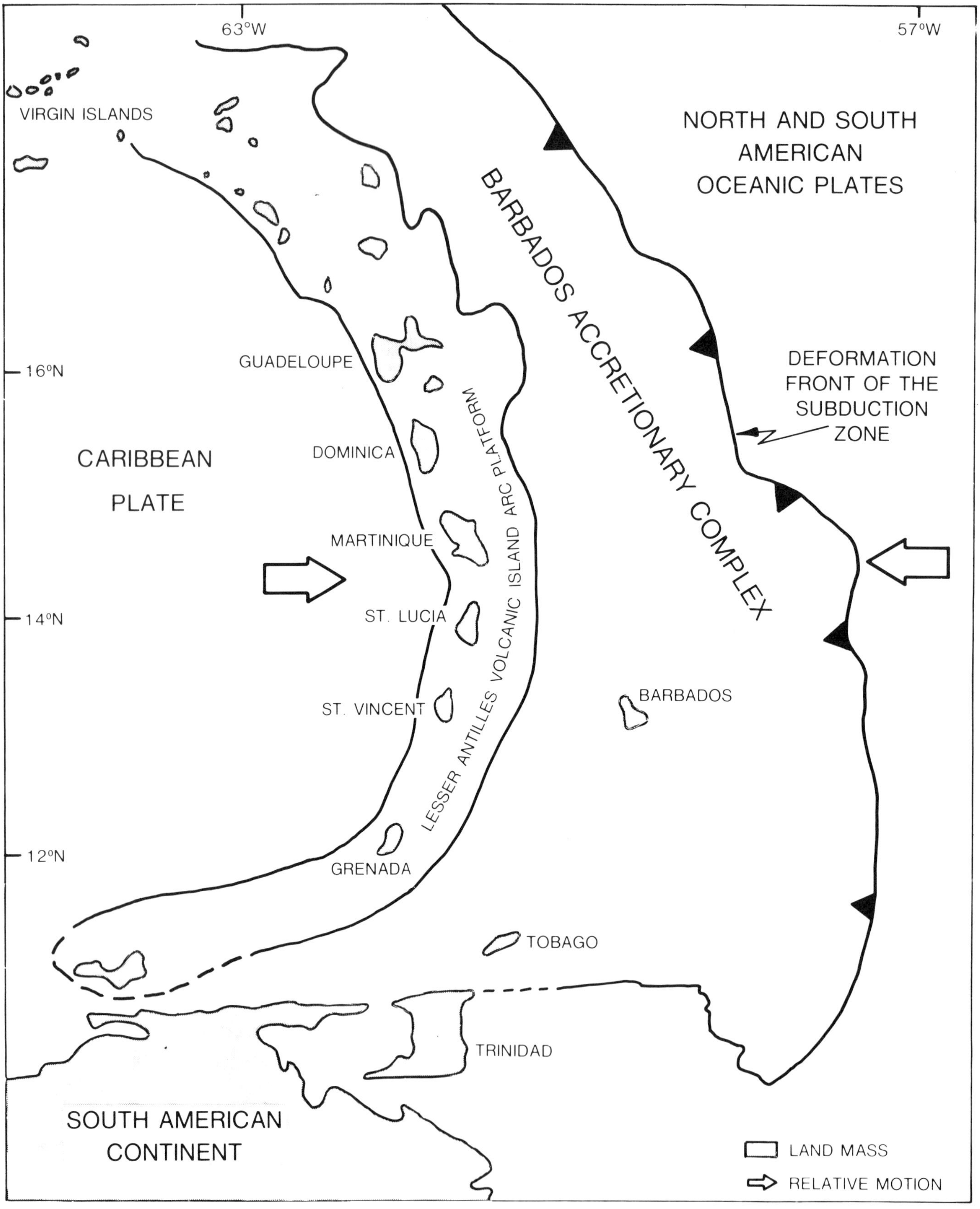


Figure 5. Plate and terrane distribution in the southeastern Caribbean.

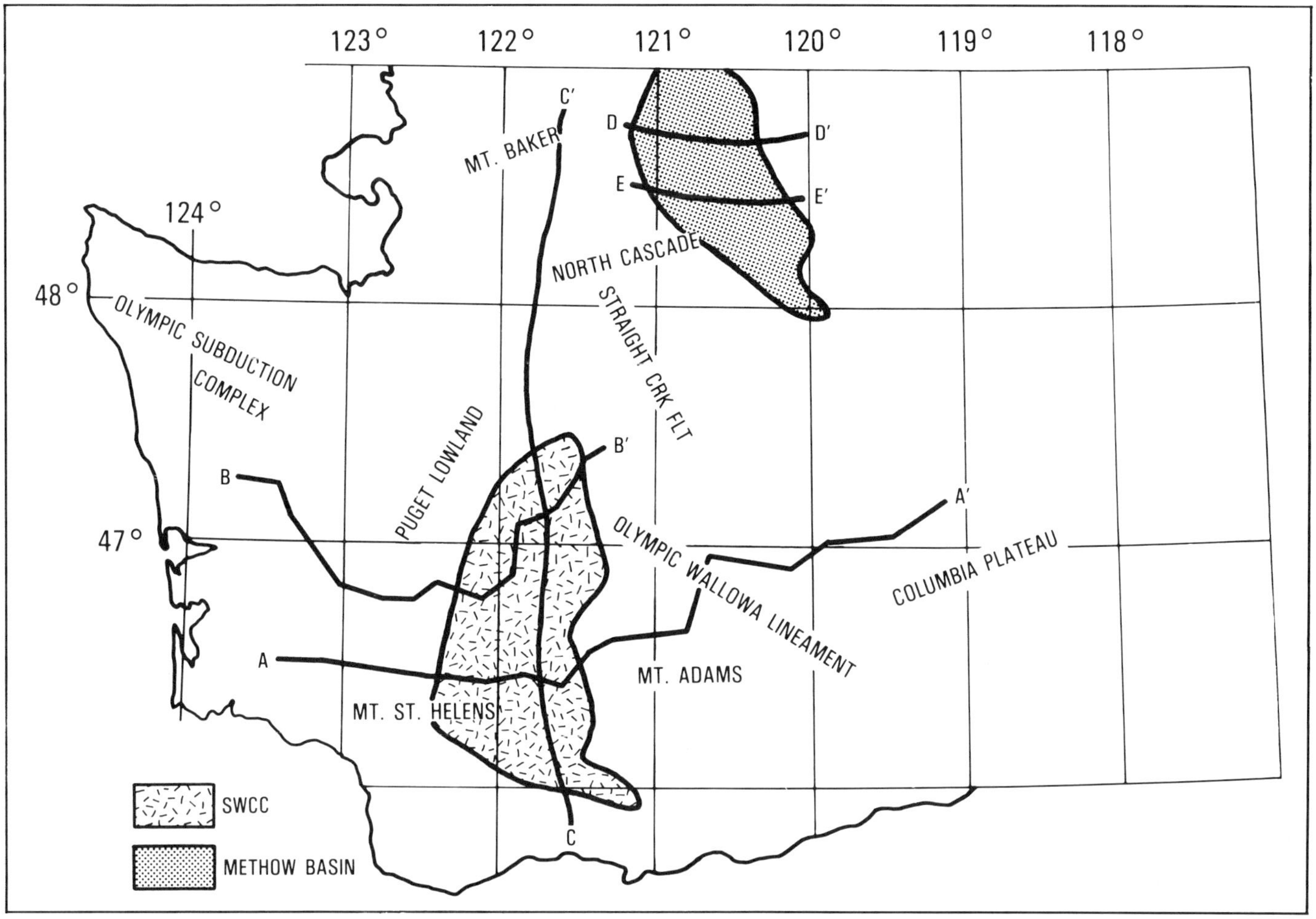


Figure 6. Locations of SWCC (Southern Washington Cascades Conductor), Methow Basin, and deep geophysical profiles in western Washington.

The object of this effort is to evaluate the potential for hydrocarbon generation as a result of sediment subduction, plate accretion, and other tectonic processes. These processes are believed to be, or to have been, active at the convergent plate margin boundary of western North America. The primary focus of these activities is to examine the rationale of the subducted, organic-origin gas hypothesis through field geology, geophysical data, and laboratory investigations.

Field geology studies in the Brooks Range of northern Alaska have also been completed. Results from this work are particularly encouraging in that many of the same hydrocarbon-rich producing formations of the Prudhoe Bay area of northern Alaska are now known to be deeply thrust faulted beneath the Brooks Range to the south. Several concepts have been developed for the generation of deep source gas. One concept involves the overthrusting (obduction) of oceanic rock units on top of known organic-bearing sedimentary rocks in the North Slope. Another concept involves the possible existence of a deep, compressed flysch basin in the central Alaska Range. This concept involves the subsidence of sedimentary rock units to great depth as a result of compressional forces from opposing thrust sheets. If these structural observations can be confirmed by future geophysical studies, DOE/METC's research will have significantly advanced current technology by demonstrating the existence of large-scale deep source target areas.

Finally, a detailed geologic field reconnaissance has been completed along a traverse extending from Cape Mendocino on the west, eastward across the northern Coast Ranges of the western margin of the Sacramento Valley. Information gathered from surface investigations such as this will aid in modeling the complex relationship of thrust-faulted rock units in the subsurface, including those believed to have been emplaced by subduction processes. A preliminary geologic map and a cross section of this area have been prepared. In addition, samples of the shallower rock units are being analyzed to determine their thermal maturity and the composition of the fluids they contain. The metamorphic and tectonic histories of this area have also been described. This study will aid in evaluating the gas-producing potential of rock units in this area that may have been emplaced at great depths by subduction.

## Deep Geophysical Studies of the Circum-Pacific

Sedimentary rock units of interest that occur in subduction systems can be identified on the basis of their physical properties. The most promising physical property for detection of subduction-related sediments is electrical resistivity. The usefulness of deep electrical sounding to the study of convergent margin sedimentary units has been demonstrated in studies in western Washington (Figure 6). Major sedimentary units have been found to be directly associated with convergent margin

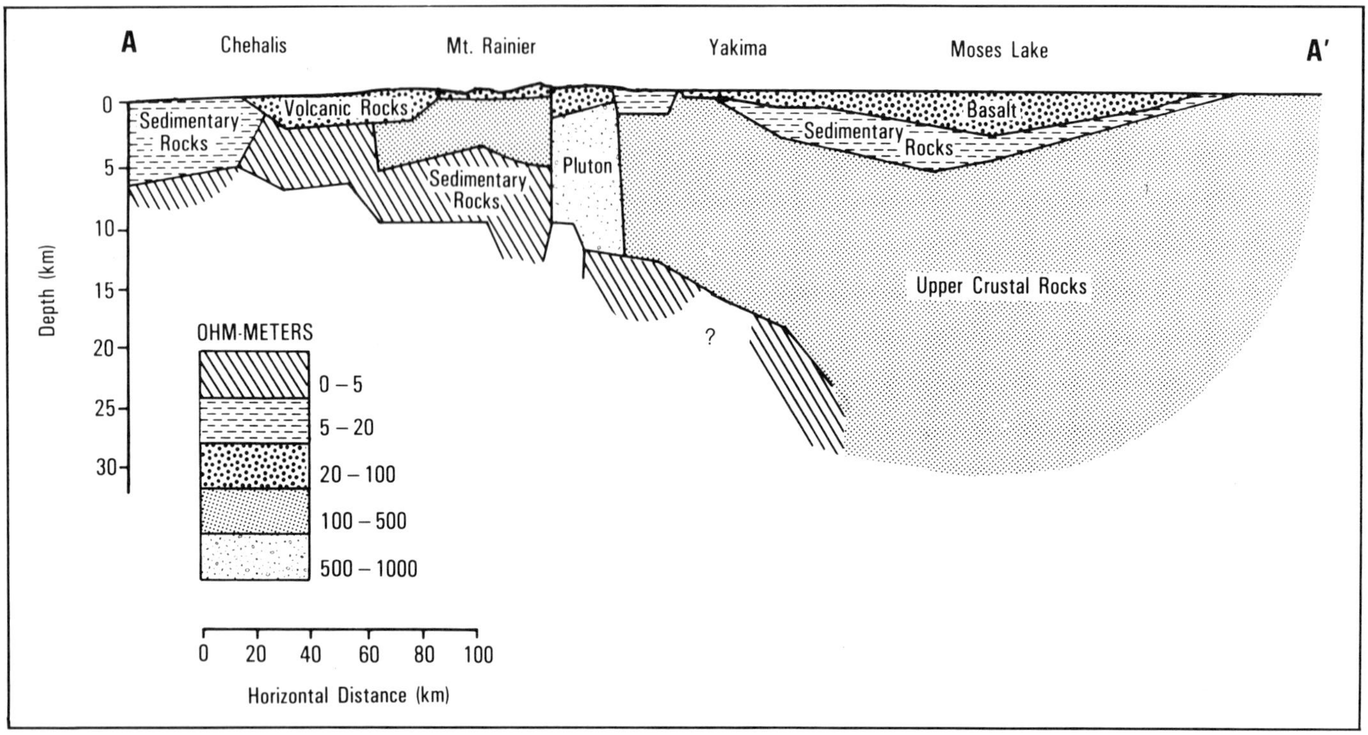


Figure 7. Deep geophysical MT profile A-A ′ through the SWCC (see Figure 6).

dynamics, and to extend to depths approaching those of the mantle. These features were found through the use of magnetotelluric (MT) studies, and it is possible that other such subduction-related sedimentary units can be located in the Cordillera. Although these deep electrical sounding methods were the main investigative tools, other geophysical and geological methods have been used.

The objective of this research effort has been to study deep, subduction-related sedimentary units in Washington, Oregon, and northern California using deep electrical soundings. MT soundings have been completed to depths of 20 to 40 km. Concurrent study of recent gravity and magnetic data from the Pacific Northwest has also been used as an aid in deep-sounding location and interpretation. The research on sediment subduction has consisted of a broad regional evaluation as follows:

- Analysis of surface geology for the most likely places in the accretionary terranes of the Cordillera to locate subduction-related sedimentary units using geophysical techniques.
- Analysis of available gravity, seismic, and aeromagnetic data for subduction characteristics; and investigation of geophysical anomalies that might have been caused by thick sedimentary units whose geometric configurations and dip angles are suggestive of subduction systems.
- Analysis of fault systems and seismicity for areas of strike-slip faulting unique to the regional setting (and possibly related to the accumulation of large volumes of sediment).

This research is being performed using the following methodology:

- Investigation of the area with deep electromagnetic soundings, using the MT method.
- Localized study of electrical properties to enable a detailed description of the lithology, using relative resistivity measurements from deep soundings.
- Interpretation of gravity and seismic data with constraints provided by deep electrical data, and vice versa.
- Incorporation of existing or planned major seismic refraction and reflection surveys across convergent margins using deep electromagnetic soundings. (An example would be the forthcoming Trans-Alaskan Crustal Transect [TACT] survey.)
- Incorporation of accretionary terrane and plate tectonic studies to refine geologic models suggested by geophysical data.

The MT soundings in western Washington suggest the presence of a fossil subduction zone, similar to that hypothesized by DOE/METC. The identified fossil subduction zone lies beneath surface volcanic rocks and a crystalline upper crust. Thick sequences of sedimentary rock are interpreted to extend within a descending plane to depths in excess of 25 km. These rocks were modeled as having anomalously low electrical resistivities (0 to 5 ohm-m). One of the surprising features of this occurrence is the apparent 3 to 15 km thickness of the sedimentary rocks.

The fossil subduction zone in Washington has been described as the Southern Washington Cascades Conductor (SWCC) (Figure 6). Three MT profiles completed through the SWCC provide a three-dimensional view of the SWCC (Figures 7, 8, 9). Abrupt or blocky contacts are shown between different lithologies as a consequence of MT modeling. Only major contrasts in conductivity can be modeled using the MT technique.

In an attempt to draw analogies between the SWCC and other areas, MT soundings have been completed in the

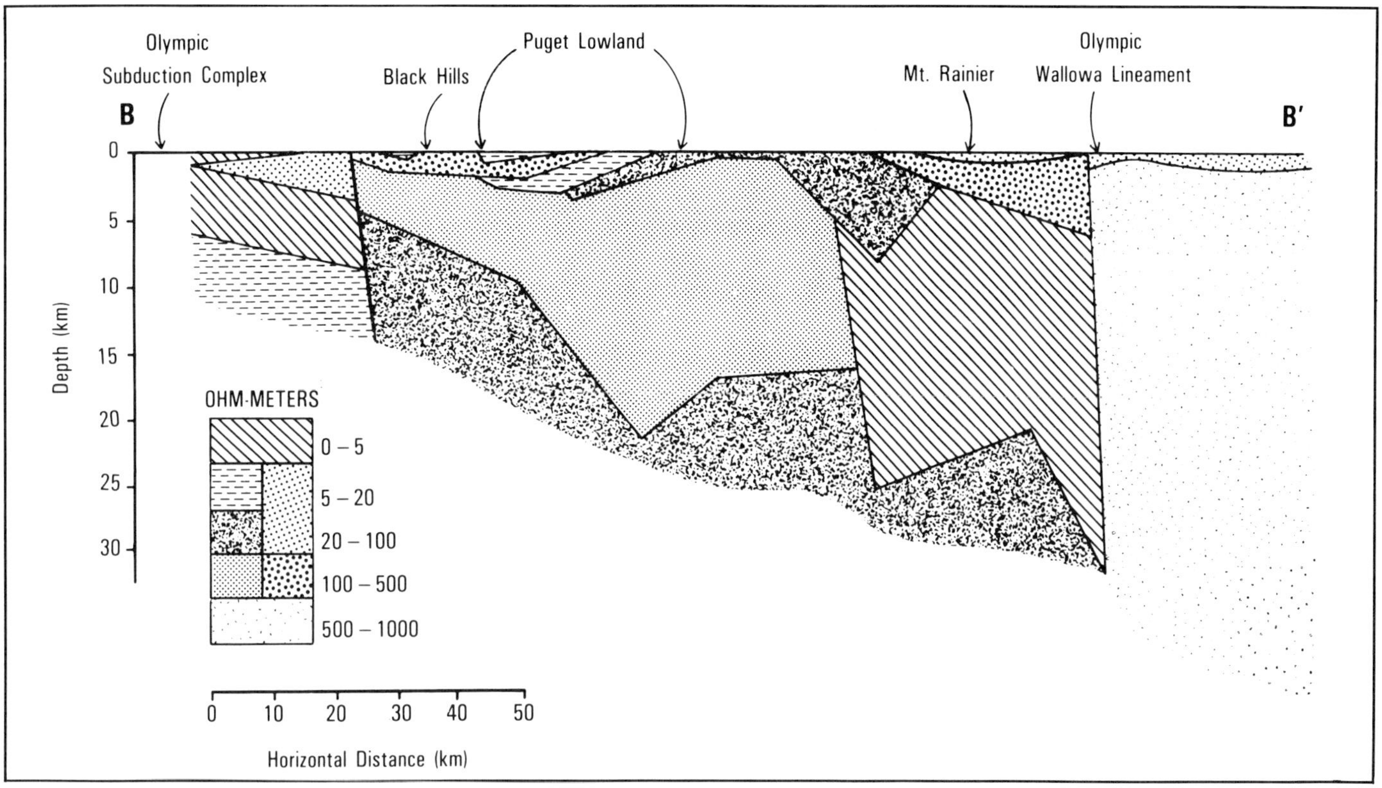


Figure 8. Deep geophysical MT profile B-B ′ through the SWCC (see Figure 6).

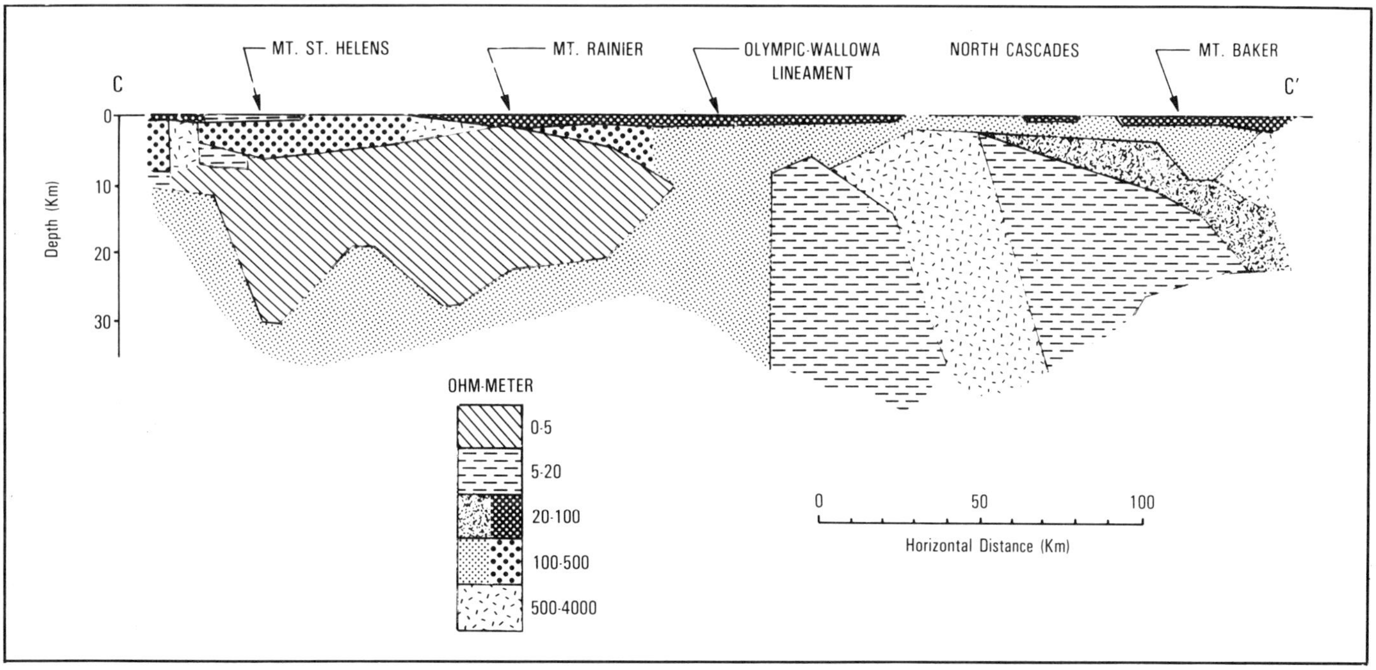


Figure 9. Deep geophysical MT profile C-C ′ through the SWCC (see Figure 6).

Methow Basin in north-central Washington (Figures 6, 10). The Methow Basin contains 15 to 20 km of marine and terrestrial sediments that were associated with a subduction zone. Magnetotelluric models suggest that a thick sequence (about 15 km) of low-resistivity sedimentary rocks has been emplaced to the depths of interest in deep source gas.

Soundings have also been conducted in Alaska, south of the Denali fault. Preliminary modeling of MT profiles shows that this area may also contain a thick sequence of deeply emplaced sedimentary rocks.

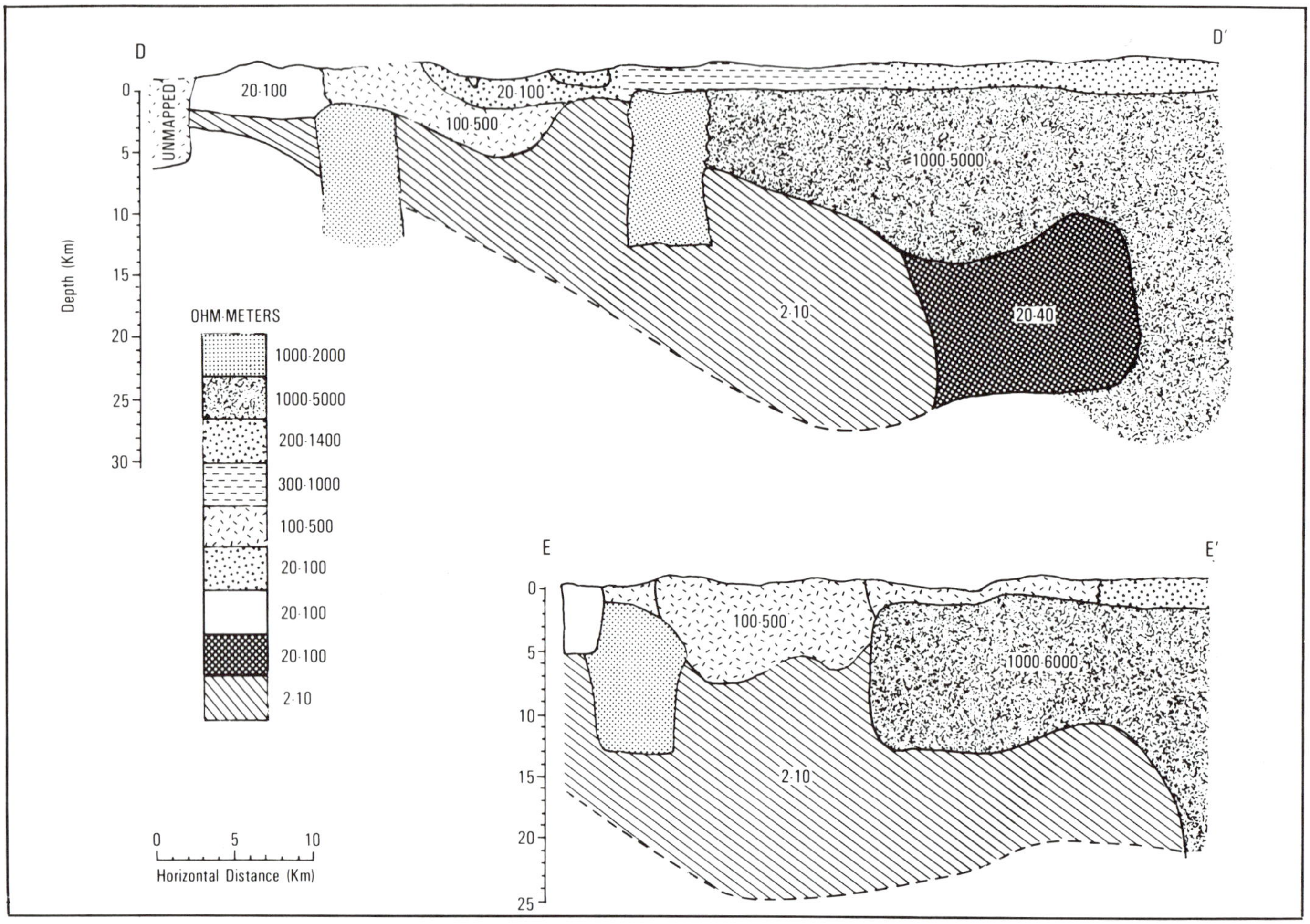


Figure 10. Deep geophysical MT profiles D-D′ and E-E′ through the Methow Basin (see Figure 6).

## THEORETICAL AND EXPERIMENTAL STUDIES OF DEEP SOURCE GAS COMPOSITION

In deep gas research there is a need to understand the depth to which hydrocarbons can be preserved, the depth at which free gas can be found, and the composition of this gas.

An analytical technique to detect the existence and stability of methane gas at extreme depths has been developed at the University of Tulsa. The existence of deep subducted gas as methane or of methane source rocks at great depths in the earth's interior has always been speculative. This new technique uses a combination of theoretical modeling and laboratory experimentation, and has helped determine that methane gas can exist at extreme depths. The reported research was concerned with only one aspect of the problem of commercial methane accumulation at great depth—the ability of methane to survive in contact with water and the rock matrix under projected conditions of temperature and pressure. Conditions of reservoir quality were not considered, nor was the problem of gas mobility.

A computer program was developed that can calculate the stabilities of natural gas components in reservoirs of various mineralogies. The computer program finds the minimum free energy in multicomponent (up to 70), multiphase (up to 20) systems for conditions corresponding to temperatures and pressures down to 40,000 ft (12,200 m), and does not assume ideal gas conditions. The minimum free energy is the equilibrium composition of the system.

Thermodynamic data were accumulated for relevant minerals, and suitable pressure and temperature input data were selected by using existing continental and oceanic gradient measurements. Normal or hydrostatic pressure was assumed to vary with depth at a rate of 0. 475 psi/ft (10. 74 kPa/m); abnormal or lithostatic pressure was assumed to vary with depth at a rate of 1. 0 psi/ft (22. 62 kPa/m). Subsurface temperatures are related to depth by the geothermal gradient, with values that range from 50 °C/km in tectonically active areas to 10 °C/km in ancient shields. Average surface temperature was assumed to be 20 °C.

The calculations illustrated that methane has considerable stability in sandstone reservoirs, even in the presence of feldspars and their clay breakdown products. However, calcite cement tends to thermally decompose at high temperatures, and the evolved carbon dioxide then dilutes the methane. The result would then be overall gas compositions that are unlikely to be economic. In carbonate systems, the main factor influencing methane stability is the presence of sulfur-bearing phases such as anhydrite. The sulfur compounds lead to hydrogen sulfide production, and methane is eliminated at

40,000 ft (12,200 m). The predicted gas composition then becomes $CO_2$-$H_2S$. However, iron-bearing phases can tie up some of the sulfur as sulfides, and methane then survives.

Predictions from the computer calculations were compared with actual analyses. Laboratory experiments were performed on test rock samples that contain trapped minute quantities of liquids and gas. These fluid inclusions provide samples of gas trapped within sealed mineral "bottles" that are not modified by gas loss during sample retrieval and are not contaminated by drilling mud. In addition, these gases formed at the same geologic time as the rocks and therefore serve as time capsules.

A computer-controlled, fast-scanning mass spectrometer system has been developed that can analyze individual fluid inclusions. Rock specimens containing fluid inclusions are heated in a vacuum to temperatures of more than 1800° F (982 °C) in special test devices. As the temperature increases, expansion of the fluids causes the rock to break. The gases are released and are analyzed for their chemical composition. The computer program recognizes the abrupt rise in partial pressures for any, or all, of six selected components.

Fluid inclusions have been analyzed successfully in quartz, calcite, feldspars, galena, and other minerals. Calcite cements from the deep Smackover Formation of the Gulf Coast (depth of approximately 20,000 ft or 6100 m) were analyzed in this manner and showed methane, carbon dioxide, and hydrogen sulfide compositions that were within the range calculated theoretically. Table 2 illustrates the expected gas composition in the monomineralic environment of a fluid inclusion in a silica or calcite host, as calculated by the equilibrium computer program.

## DETECTION AND DIFFERENTIATION SYSTEM FOR DEEP SOURCE GASES

The focus of this research on detection and differentiation is to develop a test methodology for (1) detecting the existence of deep subducted, organic-origin gas; and (2) segregating it from gases of suspected shallow-crustal organic origin and abiogenic origin.

Very little work has been published in the U. S. or European literature on the nature of hydrocarbons associated with subduction zones. In order to develop criteria for distinguishing gas origins, a reliable data base with a wide variety of potentially important measurements must be established. Helium isotope and carbon isotope analyses will probably help in distinguishing mantle, subducted organic, and conventional biogenic or thermogenic methane. However, these two measurements alone may not allow adequate source resolution.

Accordingly, other parameters must be examined, such as $CH_4/CO_2$ or nitrogen/helium ($N_2/He$) ratios, the concentration and isotopic composition of $C_2+$ hydrocarbon gases, and N and S isotope compositions. Boron, arsenic, selenium, and mercury may be additional source indicators. A general scheme for differentiating mantle, subducted organic, and conventional organic origins of natural gas has been developed (Table 3). This scheme will be refined as more data become available for the parameters listed, and as additional parameters are determined, such as $\delta^{15}N$, $\delta^{34}S$, and neon (Ne), argon (Ar), and xenon (Xe) stable isotopic compositions.

Substantial variability may exist in gases originating from the same source. For example, conventional thermogenic gases vary considerably in wetness, carbon and hydrogen isotope composition, and $CO_2$ content. Similar variations presumably exist in gases of mantle and subducted, organic origin. These variations may reflect mantle inhomogeneity, rate of subduction, heat flow, richness and type of organic matter in subducted sediments, and secondary processes such as migration, alteration, and mixing. Accordingly, the geological controls operative in each study area must be fully understood, and their possible influence on the measured gas and fluid parameters must be assessed.

## RESOURCE ESTIMATES

Geochemical and geophysical studies of the Aleutian Trench area of southern Alaska suggest that very large quantities of natural gas can be produced from subducted sediments. The research indicates that as much as 212 Tcf of natural gas could have been generated within the eastern Aleutian Trench area and held within the subduction system during the last 20 million years (Kvenvolden and von Huene, 1984). This figure represents approximately 9% of the currently known world resources of natural gas. The estimate is significant given the fact that the organic carbon content of this basin is relatively low (averaging about 0. 4%). However, the volume of sediment involved is very large.

Early modeling efforts indicate that the potential magnitude of deep gas resources in the western Cordillera may exceed 3000 Tcf of methane (Table 4). This resource estimate was derived from the rationale that a quantity of sediments was subducted along the western Cordilleran subduction zone in the last 180 million years. The volume was calculated by using a reasonable convergence rate, thickness of sediment column subducted, and length of subduction zone. A methane generation rate per unit volume of sediments and methane preservation factor were then multiplied by the volume of sediment in place to arrive at the gas reservoired.

## FUTURE WORK

Future activities will emphasize an increased range in earth science studies that are designed to further delineate the existence of deeply emplaced, hydrocarbon-generating source rocks. Thermal maturation history appears to be very important in modeling the generation of hydrocarbons in subduction systems. This technique will be a tremendous aid in estimating the deep source gas resource potential of areas such as western Washington.

Although much has been learned about the plate tectonic framework of the western Cordillera, the details of terrane accretion and how sediments are incorporated into the deeper portions of the earth's crust remain poorly understood. Geophysics will play an increasingly important role in locating areas like western Washington and south-central Alaska that appear promising. As with any effort in petroleum geology, the delineations of prospective regions is accomplished by combining surface geological data, well information, and remote sensing with geophysical leads.

It is hoped that the Deep Observation and Sampling of the Earth's Continental Crust (DOSECC) effort will provide an opportunity to field test the geophysical anomalies. Other

TABLE 2. GAS INCLUSION COMPOSITIONS IN CALCITE AND SILICA HOSTS

| DEPTH (feet) | PERCENT OF GASES PRESENT $CH_4$ | $H_2$ | $C_2H_6$ | $CO_2$ |
|---|---|---|---|---|
| | | Calcite Host | | |
| 10,000 | 99.84 | 0.16 | | |
| 15,000 | 99.75 | 0.25 | | |
| 20,000 | 99.40 | 0.60 | | |
| 25,000 | 99.60 | 0.40 | Trace | |
| 30,000 | 99.72 | 0.26 | 0.01 | 0.01 |
| 35,000 | 99.50 | 0.44 | 0.04 | 0.02 |
| 40,000 | 99.12 | 0.74 | 0.10 | 0.04 |
| | | Silica Host | | |
| 10,000 | 98.89 | 1.11 | | |
| 15,000 | 98.89 | 1.11 | | |
| 20,000 | 98.88 | 1.12 | | |
| 25,000 | 98.86 | 1.14 | | |
| 30,000 | 98.81 | 1.19 | Trace | |
| 35,000 | 99.07 | 0.92 | 0.01 | Trace |
| 40,000 | 98.91 | 1.03 | 0.03 | 0.03 |

TABLE 3. PRELIMINARY CLASSIFICATION SCHEME FOR CRUSTAL GAS ORIGINS

| Gas Components | Gas Source: Mantle, Abiogenetic | Subducted, Biogenic | Conventional, Biogenic |
|---|---|---|---|
| $R/R_a$* | $> 8$ | $\leqq 8$, $> 0.1$ | $< 0.1$ |
| $\delta^{13}C$ ($CH_4$) | > -20‰ | ≦ -20‰, ≧ -30‰ | < -30‰ |
| $CH_4/^3He$ | $< 10^7$ | $> 10^7$, $> 10^9$ | $> 10^9$ |
| % $C_2$+ | Very Low | Low to High | High |
| % $CO_2$ | High | Intermediate | Low |
| % B, As, Se, Hg | High | High | Low |

* $R = {}^3He/{}^4He$; $R_a$ = R (atmosphere)

**TABLE 4. ESTIMATES OF DEEP GAS RESOURCES IN THE WESTERN CORDILLERA**

ESTIMATE MODEL NO. 1. CONSTANT SUBDUCTING SLAB

| Subducted Sediment Rock Volume | Methane Generation |
|---|---|
| - Width (W) 14,869,492 ft | - Methane Generation per Unit Rock Volume (M) .550 $ft^3$ $CH_4/ft^3$ SED |
| - Thickness (T) 1312.4 ft | - Percent Reservoired (P) 1/200 |
| - Convergence rate (C) .3281 ft/yr (10 cm/yr) | = M x P |
| - Duration (D) 180 my | |
| = W × T × C × D | |

POTENTIAL GAS IN PLACE

= Subducted Rock Volume x Methane Generation

= 3,169 Tcf Methane

areas of research that will be investigated include: hydrocarbon migration studies and surface soil gas geochemical studies to grade the prospective regions.

## ACKNOWLEDGMENTS

The following are acknowledged for their substantial contributions of data and information for this project overview paper: Dr. Thomas Hilde of Texas A&M University; Dr. Keith Kvenvolden, Dr. Roland von Huene, Dr. David Howell, Dr. David Jones, Dr. Parke Snavely, and Dr. M. Clarke Blake of the USGS in Menlo Park, California; and Dr. W. D. Stanley of the USGS in Denver, Colorado; Dr. Peter Coney and Dr. Tekla Harms of the University of Arizona; Dr. Colin Barker of the University of Tulsa; Dr. Robert Speed of Northwestern University; Dr. Isaac Kaplan and Dr. Alan Jeffrey of Global Geochemistry Corporation; and Dr. Harmon Craig of the Scripps Institution of Oceanography.

## REFERENCES CITED

Barker, C. , 1985, A theoretical and experimental study of the composition of deep source gas: University of Tulsa, Contract No. DE-AC21-84MC21229, 91 p.

Bode, G. W. , 1973a, Carbon and carbonate analyses, Leg 18, *in* L. D. Kulm, R. von Huene, et al., eds. , Initial reports of the Deep Sea Drilling Project: Washington, D. C. , U. S. Government Printing Office, v. 18, p. 1069-1076.

Bode, G. W. , 1973b, Carbon-carbonate, *in* J. S. Creager, D. W. Scholl, et al., eds. , Initial reports of the Deep Sea Drilling Project: Washington, D. C. , U.S. Government Printing Office, v. 19, p. 663-665.

Claypool, G. E. , B. J. Presley, and I. R. Kaplan, 1973, Gas analyses in sediment samples from Legs 10, 11, 13, 14, 15, 18, and 19, *in* J. S. Creager, D. W. Scholl, et al., eds. , Initial reports of the Deep Sea Drilling Project: Washington, D. C. , U. S. Government Printing Office, v. 19, p. 879-884.

Kaplan, I. R. , and A. W. Jeffrey, 1985, Detection differentiation system development for deep source gases (proposed work): Under Contract No. DE-AC21-84MC21131.

Kvenvolden, K. A. , and R. von Huene, 1984, Potential for generation of natural gas in sediments of the convergent margin of the Aleutian Trench area: U. S. Geological Survey Open-File Report 84-404.

Speed, R. C. , 1985, Deep gas in the Barbardos accretionary prism: Sources, migration, detection; *in* Proceedings of the U. S. Department of Energy Deep Source Gas Program (a peer review), p. 110-143.

Stanley, W. D. , 1984, Tectonic study of Cascade Range and Columbia Plateau in Washington State based upon magnetotelluric soundings: Journal of Geophysical Research, 89(B6): p. 4447-4460.

Stanley, W. D. , 1985, Electromagnetic geophysics applied to sediment subduction and deep source gas: U. S. Geological Survey Progress Reports for FY 85, Under Interagency Agreement No. DE-AI21-83MC20422.

von Huene, R. , and L. D. Kulm, 1973, Tectonic summary of Leg 18, *in* L. D. Kulm, R. von Huene, et al., eds. , Initial reports of the Deep Sea Drilling Project: Washington, D. C. , U. S. Government Printing Office, v. 18, p. 1069-1076.

von Huene, R., et al., 1980, Summary, Japan Trench transect, *in* Initial Reports of the Deep Sea Drilling Project: Washington, D.C., U.S. Government Printing Office, v. 56, no. 57, Part I, p. 473-488.

CHAPTER 12

# DISCOVERIES, DEVELOPMENT AND FUTURE PROSPECTS—EROMANGA BASIN, AUSTRALIA

**R.J.S. Hollingsworth**
*CSR Oil and Gas Division*
*Adelaide, South Australia*

## ABSTRACT

The Eromanga Basin covers an area of about 1 million km$^2$ and is the largest subbasin of the Great Artesian Basin which occupies about one-fifth of the Australian continent. The basin contains an Early Jurassic to Late Cretaceous clastic sequence which attains a maximum thickness of about 3000 m in depocenters related to underlying Paleozoic basins.

Although the first commercial hydrocarbon discovery in the Eromanga Basin in 1976 was gas, subsequent discoveries have been mainly oil, thereby establishing the basin as a predominantly oil province. Currently some 60 oil fields, 4 fields containing oil and gas, and 5 gas fields have been discovered. Oil fields are characteristically small, containing recoverable reserves of less than 9 MMSTB, and most fields have less than 1 MMSTB. Present estimates of proved and probable oil and sales gas reserves are about 115 MMSTB and 60 BCF respectively, with oil production currently about 53,500 BPD and sales gas production at about 7.5 MMCFD.

Exploitation of oil fields has been marked by rapid commencement of production, with the mode dependent on proximity to the pipeline facilities at Moomba and Jackson. Fields of sufficient productivity close to these facilities are connected by oil trunklines, whereas oil from more distant fields is brought to the appropriate facility by truck.

The Eromanga Basin is widely regarded as the most prospective onshore basin in Australia, yet exploration coverage is still relatively sparse. Numerous challenges are still considered to exist in understanding the factors controlling accumulation, and further exploration is anticipated to result in additional discoveries.

## INTRODUCTION

The Eromanga Basin covers an area of about 1 million km$^2$ within four states in central eastern Australia—Queensland, South Australia, New South Wales, and the Northern Territory (Figure 1). The basin contains a sequence of Early Jurassic to Late Cretaceous sediments which have a maximum thickness of approximately 3 km in depocenters related to pre-existing underlying basins. Underlying the Eromanga Basin are several Late Carboniferous-Late Triassic basins, as well as undeformed to strongly folded and overthrust Paleozoic and Proterozoic sediments and Precambrian crystalline basement. The most important of the underlying Late Carboniferous-Late Triassic basins, with respect to petroleum exploration, is the Cooper Basin, which contains a maximum of about 2 km of nonmarine sediments.

Currently, the Cooper Basin supplies about 40% of Australia's domestic natural gas requirements. Most of the production is from Permian coal-rich fluvial sediments. The exploration history and development in the Eromanga Basin has been strongly influenced by the discovery pattern in the Cooper Basin.

Although the Eromanga Basin had been explored since 1924, the first commercial hydrocarbon discovery was made in 1976, when gas was discovered at Namur 1 in sandstones of the Late Jurassic-Early Cretaceous Namur Sandstone Member. This discovery was made some 13 years after the first commercial discovery in the Cooper Basin. In 1978 the major breakthrough in the Eromanga Basin exploration occurred at Strzelecki 3. This well was drilled primarily to assess the Permian but discovered oil in the Middle Jurassic Hutton Sandstone, producing a flow of 2400 BOPD. This discovery firmly established the petroleum potential of the basin and led to a reassessment of the widely held belief that the Eromanga Basin section was flushed because of widespread artesian water drive.

Further discoveries were made after the Strzelecki 3 success, and in December 1981 the Jackson field was discovered in southwest Queensland. Apart from establishing the petroleum prospectivity of this sector of the Eromanga Basin, the field was large enough to justify building a 790-km oil pipeline to Moonie. This pipeline has enhanced the commercial viability of exploitation of discoveries in this region.

To the end of June 1986 some 11 gas pools and 115 oil pools have been discovered in the Eromanga Basin. About 600 exploration wells have been drilled to date through the Eromanga Basin section. About 70% of this drilling has occurred since 1959. Production from the Eromanga Basin commenced in December 1982 from the Strzelecki field in South Australia. Jackson field in Queensland was brought on stream in February 1984 following the completion of the Jackson-Moonie pipeline.

This paper reviews discoveries and subsequent development during a phase of high activity in what is regarded as the

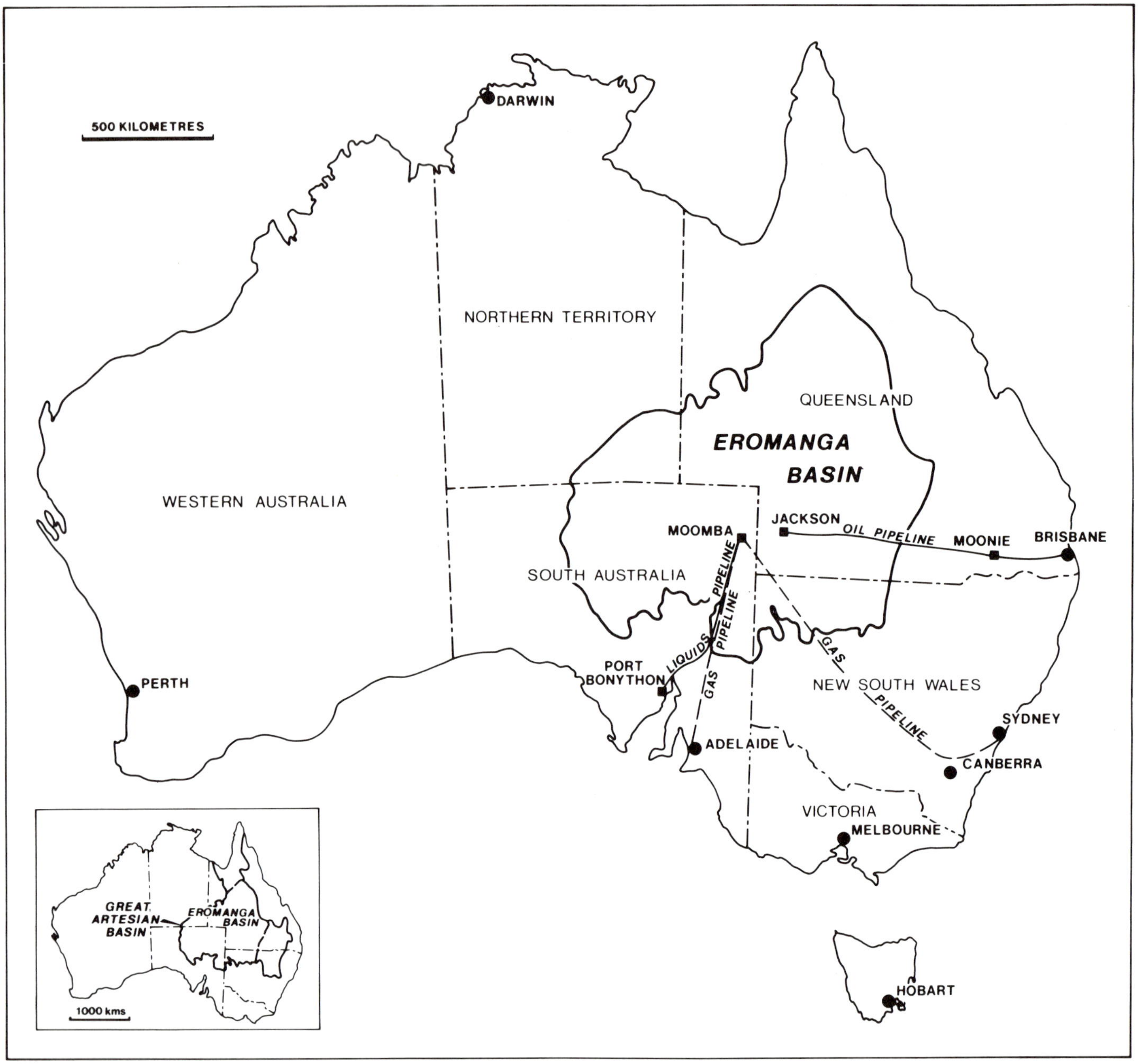


Figure 1. Location map.

most prospective onshore petroleum basin in Australia. An assessment of future exploration directions is also provided.

## GEOLOGY

The Eromanga Basin covers an area of about 1 million km$^2$. Together with the Carpentaria Basin and the Surat Basin, it forms part of the hydrogeological Great Artesian Basin, which occupies about 20% of the Australian continent and is one of the largest artesian basins in the world (Habermehl, 1980). Shallow basement ridges, the Euroka Arch and Nebine Ridge separate the Eromanga Basin from the Carpentaria Basin and Surat Basin to the north and east, respectively (Figure 2).

The basin is underlain by a series of Late Carboniferous-Late Triassic basins as well as older Paleozoic and Proterozoic sediments and Precambrian crystalline basement. From a petroleum perspective, the most important of the Late Carboniferous-Late Triassic basins is the centrally located Cooper Basin, which contains substantial Permian gas fields along with some oil fields. The older Paleozoic and Proterozoic sediments range from undeformed to strongly folded; although some discrete basins are recognized, other areas are poorly defined and understood.

The basin resulted from epeirogenic intracratonic downwarp and is consequently a very broad synclinal feature. It is characterized by a southwest tilt due to Cenozoic uplift of the eastern margins and subsidence in the central and southwestern parts (Figure 3). Major structures within the basin grade downward into faults and are generally the result of draping and differential compaction of sediments over basement faults and fault-bound basement blocks (Haber-

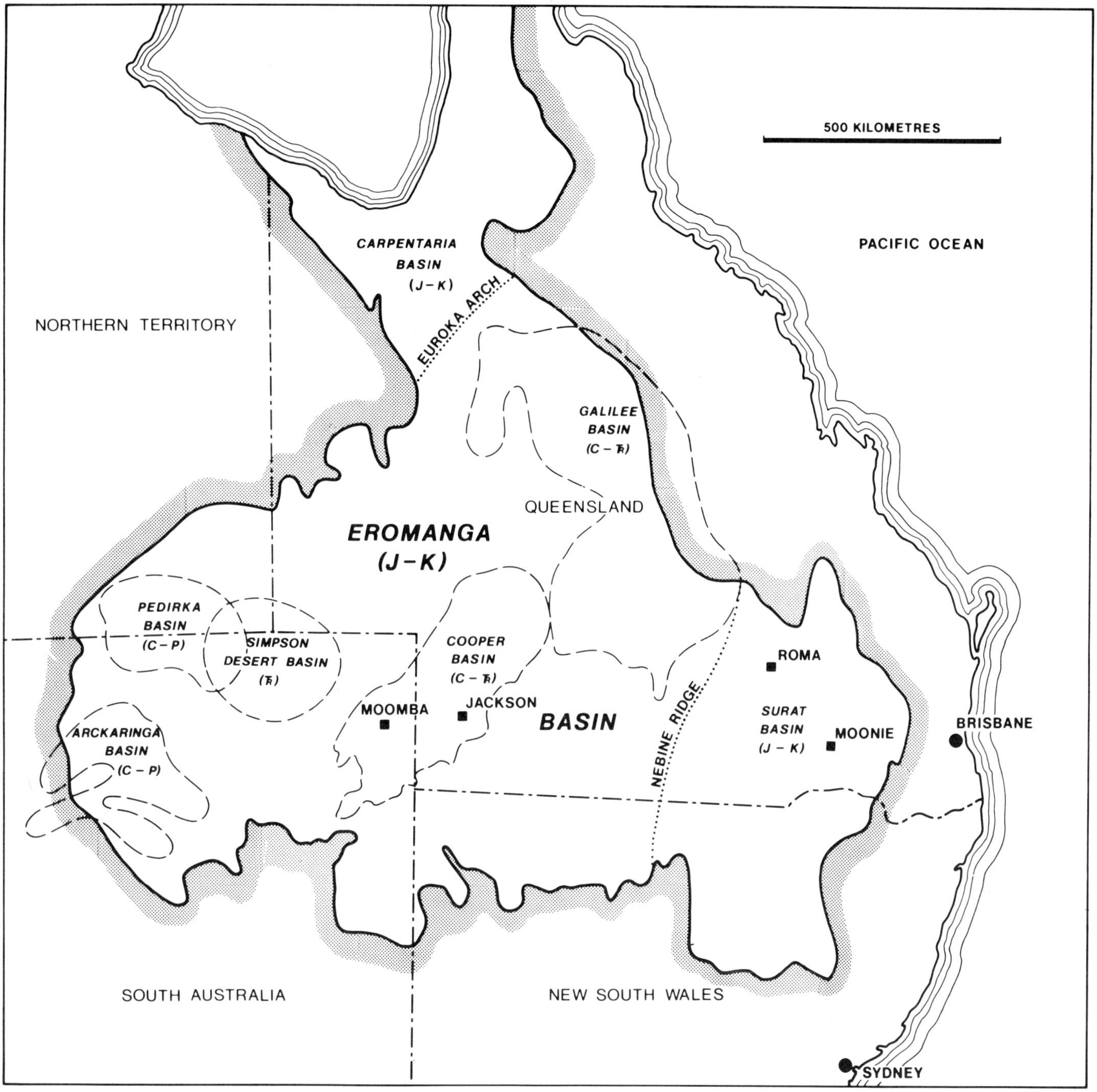


Figure 2. Adjacent basins and older basins relative to the Eromanga Basin.

mehl, 1986). In the Late Oligocene-Early Miocene a phase of left-lateral, wrench-induced folding appears to have created many of the large anticlines seen in the northern part of the basin and to have enhanced some structures in the southern part.

The sedimentary sequence in the Eromanga Basin consists of a conformable, almost horizontal sequence of sediments of Early Jurassic to Late Cretaceous age. The maximum thickness of about 3000 m is associated with older Paleozoic Basin depressions such as the Poolowanna Trough of the Pedirka Basin and the Windorah Trough of the Cooper Basin (Figure 3). The sequence thins toward the basement highs separating Eromanga Basin from Surat and Carpentaria, as well as toward the northern and southern margins. The stratigraphy of the Eromanga Basin is shown on Figure 4, along with that of the Surat Basin to the east. Reference has been made to the Surat Basin because many of the units outcrop in this area and have been studied in extensive detail, and as a result have provided understanding of the Eromanga Basin sequence.

The oldest Eromanga Basin sediments are Early Jurassic sandstones, siltstones, shales, and thin coals of the Poolowanna Formation deposited in the Poolowanna Trough (Moore, 1986). Sedimentation expanded beyond the Poolowanna Trough by late Early Jurassic, and during the rest of the Jurassic-earliest Cretaceous a sequence of fluvial-lacustrine-coal sediments was deposited. The Algebuckina

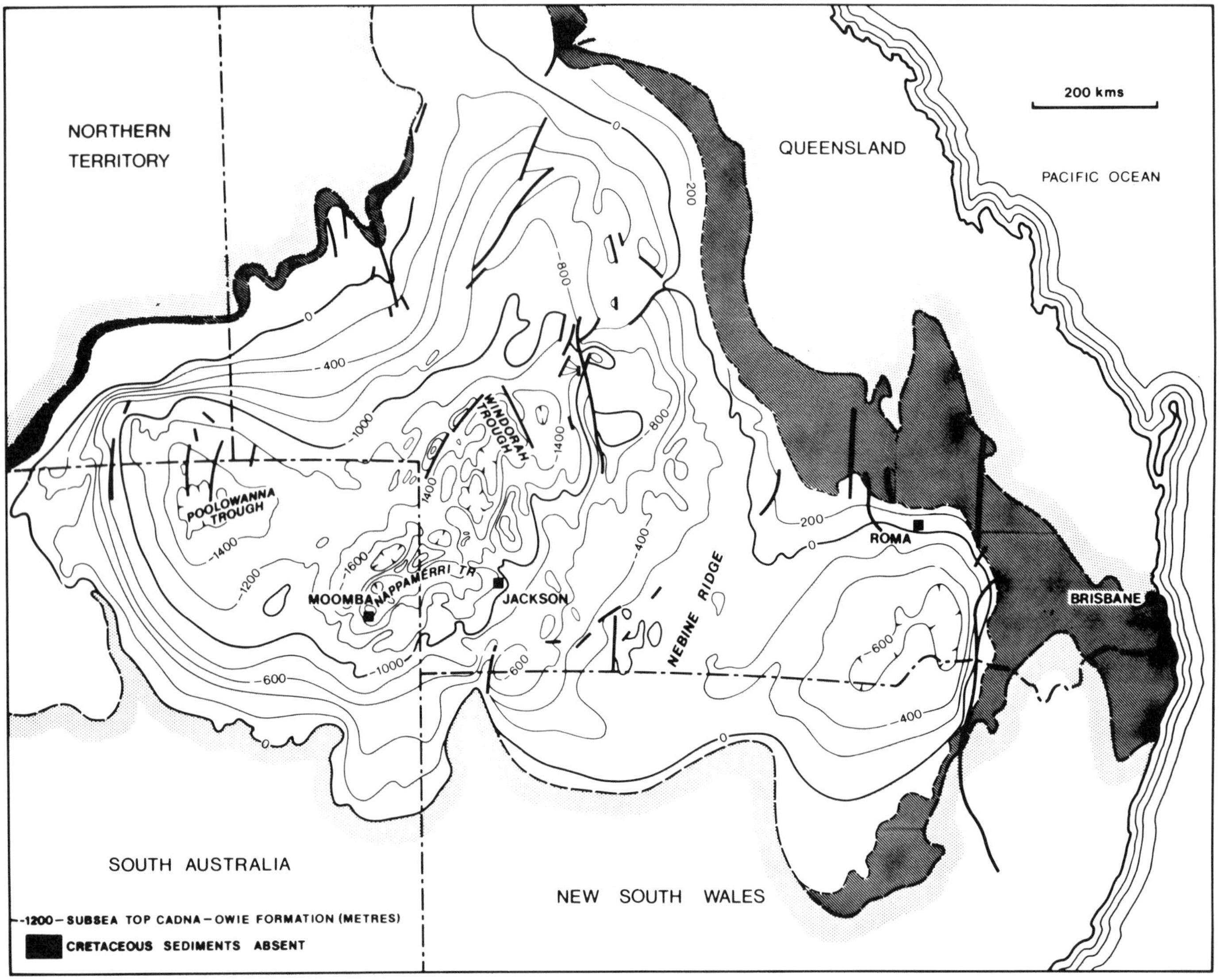


Figure 3. Structure contour map at C seismic horizon (top of Cadna-owie Formation).

Sandstone, Hutton Sandstone, Adori Sandstone, and Namur Sandstone Member represent extensive braided fluviatile sandstone units deposited during this phase. Flood-plain, swamp, and lacustrine environments are represented by the Birkhead Formation, Westbourne Formation, and Murta Member which consist of shales, siltstones, fine-grained sandstones, and minor coals. The great lateral extent of these environments can be observed by recognition of equivalent units in the Surat Basin and Carpentaria Basin.

Nonmarine sedimentation continued during the Neocomian, resulting in the deposition of mudstones, siltstones, and minor sandstones of the Cadna-owie Formation. A major marine transgression occurred during the Neocomian-Aptian, depositing mudstones of the Wallumbilla Formation and Bulldog Shale throughout the area. In southwest Queensland the Cadna-owie Formation and other Early Cretaceous units are locally dissected by prominent submarine channelling.

The Aptian to Albian sequence consists essentially of fine-grained marine sediments apart from two notable exceptions—locally developed oil shales of the Toolebuc Formation and thin shoreface sandstones of the Coorikiana Sandstone. Regression took place in the Albian, depositing the marginal marine sandy siltstones of the Mackunda Formation. Rapid subsidence during the Cenomanian resulted in deposition of a nonmarine section of fluvial/lacustrine sandstones, siltstones, shales, and minor coals of the Winton Formation.

In depocenters the Jurassic section reaches a maximum thickness of about 1000 m, whereas the Cretaceous section attains a maximum thickness of about 2000 m, of which about 1000 m consists of the Late Cretaceous (Cenomanian) Winton Formation.

## HYDROCARBON DISCOVERIES

Exploration for petroleum in the Eromanga Basin commenced in 1924 around the Longreach area in the northeastern part of the basin. However, despite often encouraging hydrocarbon shows, it was not until November 1976 that a commercial accumulation was discovered in the sequence when gas flowed at 13.9 MMCFD from the Namur Sandstone Member in the Namur 1 well. At the time of this discovery, it was thought that the gas was of Permian origin and had migrated upward into the Namur anticlinal trap via faults.

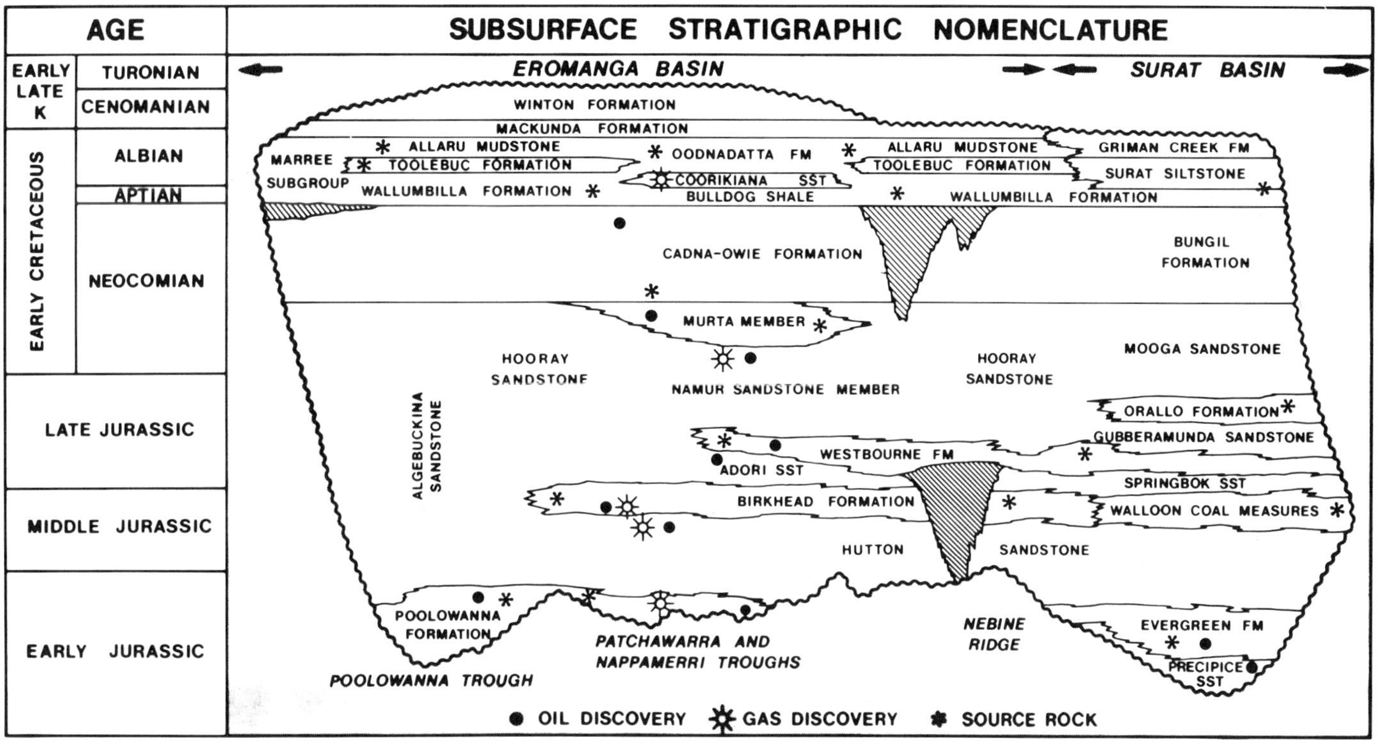


Figure 4. Major stratigraphic units of the Eromanga Basin.

An encouraging recovery of oil was made from the Poolowanna Formation at Poolowanna 1 in September 1977. This discovery located in the Poolowanna Trough is noncommercial and appears to be very small. Both Namur 1 and Poolowanna 1 now appear to be anomalous when compared with the later pattern of discoveries in the Eromanga Basin.

In October 1978, the Strzelecki 3 flowed at 2400 BPD from the Hutton Sandstone, thereby clearly establishing the petroleum prospectivity of the Eromanga Basin. From this discovery it became evident that the previously held belief that the Eromanga Basin section was nonprospective for hydrocarbon accumulations because of artesian flushing should be reconsidered. Following the success at Strzelecki, three key discoveries were made at Dullingari (Murta Member oil), Cuttapirrie (Early Jurassic oil), and Merrimelia (Namur Sandstone Member oil and Hutton Sandstone oil).

In December 1981, the Jackson 1 discovered oil in the Murta Member (338 BOPD), Westbourne Formation (2 tests at 188 BOPD and 1165 BOPD), and Hutton Sandstone (2616 BOPD). This well was the first oil discovery in the Queensland portion of the Eromanga Basin and by demonstrating the petroleum prospectivity of the area acted as an impetus for further exploration. In addition, its size allowed the building of a 790-km pipeline from Jackson to Moonie to transport oil to Brisbane for refining and/or shipping via an existing line from Moonie. This pipeline raised the economic viability of other discoveries in the immediate area of Jackson. Jackson is the largest discovery to date in the Eromanga Basin, containing some 43 MMSTB recoverable oil about 90% of which is within the Hutton Sandstone pool.

A high level of drilling activity followed the Jackson discovery and resulted in numerous discoveries in both South Australia and Queensland, thereby firmly establishing the Eromanga Basin as Australia's most prospective onshore area. At the end of June 1986, some 11 gas pools and 115 oil pools had been discovered in the Eromanga Basin sequence. Sixty fields containing oil, five containing gas, and four containing both oil and gas have been discovered (Table 1). Proved and probable initial recoverable reserves in the Eromanga Basin are currently estimated at about 115 MMSTB oil and about 60 BCF sales gas.

The majority of oil fields contain initial recoverable proved and probable reserves of less than 9 million barrels. The smallest fields comprise single pools containing significantly less than 1 million barrels of initial recoverable oil, and these reservoirs commonly occur in the Murta Member. Table 2 shows a classification of oil fields based on recoverable reserves. Eromanga Basin gas fields are typically small, with all fields containing less than 25 BCF initial recoverable proved and probable sales gas. The size distribution of fields is shown on Table 3.

Hydrocarbons occur in 10 stratigraphic units ranging in age from Early Jurassic to Early Cretaceous, as shown on Figure 4. Most of the traps are anticlinal structures (Figure 5), but some show a stratigraphic component particularly in interbedded sand/shale units such as the Murta Member, Westbourne Formation, and Birkhead Formation and recently discovered channel-fill seal plays in the Cadna-owie Formation. The average size of pools is generally small, the closure area is less than 1000 acres, and the average net pay is less than 20 ft or 6 m (Lavering, Passmore, and Paton, 1986). As a result, the definition of such pools needs a closely spaced seismic grid (1 km × 1 km) to clearly identify anticlinal crests.

Discoveries are characteristically single-pool or two-pool accumulations. However, some fields show marked contrast to this, such as Jackson, Merrimelia, and Talgaberry, where four commercial pools have been identified. The bulk of the oil reserves found to date (about 65%) are in the Hutton Sandstone. Although units such as the Murta Member have numerous discoveries, fields are small and do not contribute

TABLE 1

EROMANGA BASIN OIL AND GAS DISCOVERIES/FIELDS

| | Date | Discovery/Field | Reservoir(s) | Discovery/Field Type |
|---|---|---|---|---|
| OIL | Mar 1985 | Alwyn 1 | Murta | SW |
| | May 1982 | Big Lake | Namur, Hutton | MW |
| | Mar 1986 | Black Stump 1 | Basal Jurassic | SW |
| | Jun 1984 | Bodalla South | Hutton, Basal Jurassic | MW |
| | Jun 1984 | Bogala 1 | Murta, Birkhead | SW |
| | Nov 1984 | Bookabourdie | Birkhead, Hutton | MW |
| | Jun 1986 | Calamia West 1 | Hutton | SW |
| | Jun 1984 | Charo 1 | Hutton | SW |
| | Aug 1985 | Cook 1 | Hutton | SW |
| | Mar 1980 | Cuttapirrie 1 | Basal Jurassic | SW |
| | Feb 1986 | Dirkala 1 | Namur, Birkhead | SW |
| | Sep 1979 | Dullingari | Murta, Namur | MW |
| | Jan 1985 | Epsilon 3 | Birkhead | SW |
| | Sep 1985 | Glenvale 1 | Westbourne | SW |
| | Aug 1984 | Graham 1 | Birkhead | SW |
| | Jul 1983 | Gunna 1 | Murta | SW |
| | Jan 1986 | Ipundu | Cadna-owie, Murta | MW |
| | Dec 1981 | Jackson | Murta, Westbourne, Birkhead, Hutton | MW |
| | Apr 1982 | Jackson South | Westbourne, Birkhead | MW |
| | Jun 1985 | Jena 1 | Murta | SW |
| | Jun 1985 | Kenmore | Westbourne, Hutton, Basal Jurassic | MW |
| | Dec 1984 | Kercummurra 1 | Cadna-owie | SW |
| | Jul 1984 | Kerinna 1 | Hutton | SW |
| | May 1982 | Kihee | Murta | MW |
| | Aug 1984 | Kobari 1 | Murta | SW |
| | Sep 1985 | Koora | Adori | MW |
| | Jun 1985 | Kooroopa 1 | Adori | SW |
| | Jul 1984 | Limestone Creek/Biala | Murta, Namur | MW |
| | May 1981 | McKinlay | Namur | MW |
| | Aug 1981 | Merrimelia | Cadna-owie, Murta, Namur, Birkhead, Hutton | MW |
| | Apr 1985 | Mooliampah 1 | Murta, Namur, Adori | SW |
| | Feb 1981 | Moorari | Birkhead | MW |
| | Jun 1985 | Moothendella | Murta, Westbourne | MW |
| | Apr 1985 | Muteroo 1 | Hutton | SW |
| | Nov 1983 | Naccowlah South | Hutton | MW |
| | Dec 1983 | Naccowlah West | Murta, Hutton | MW |
| | Dec 1979 | Nappacoongee 2 | Birkhead | SW |
| | Sep 1983 | Narcoonowie | Murta, Namur, Birkhead, Hutton | MW |
| | Sep 1983 | Nockatunga | Murta, Birkhead | MW |
| | Apr 1986 | Nungeroo 1 | Namur | SW |

| | | | | |
|---|---|---|---|---|
| | Sep 1977 | Poolowanna 1 | Basal Jurassic | SW |
| | Dec 1983 | Sigma 1 | Murta, Namur | SW |
| | Jan 1986 | Spencer 3 | Namur, Birkhead | SW |
| | Mar 1986 | Spencer 4 | Namur, Birkhead | SW |
| | Oct 1978 | Strzelecki | Cadna-owie, Murta, Birkhead, Hutton | MW |
| | Nov 1985 | Takyah 1 | Murta, Birkhead | SW |
| | Jul 1985 | Talgaberry | Cadna-owie, Murta, Westbourne, Birkhead | MW |
| | Mar 1986 | Thungo 1 | Murta, Westbourne | SW |
| | Dec 1984 | Tickalara 2 | Namur | SW |
| | Dec 1983 | Tinpilla 1 | Murta | SW |
| | Dec 1983 | Tintaburra | Cadna-owie, Hutton | MW |
| | Nov 1985 | Toobunyah | Birkhead, Hutton | MW |
| | Mar 1985 | Ulandi 1 | Murta, Namur | SW |
| | Nov 1984 | Wancoocha | Birkhead, Hutton | MW |
| | Dec 1985 | Watson 1 | Hutton | SW |
| | Jun 1985 | Watson South | Hutton | MW |
| | Nov 1983 | Wilson | Murta, Namur, Westbourne, Hutton | MW |
| | Oct 1985 | Winna 1 | Murta | SW |
| | Mar 1984 | Yanda | Murta | MW |
| GAS | Nov 1976 | Namur | Namur | MW |
| | Aug 1978 | Wackett 1 | Birkhead | SW |
| | Jan 1981 | Marabooka | Coorikiana/Namur | MW |
| | Jun 1985 | Mina 1 | Coorikiana | SW |
| | Apr 1986 | Nanima 1 | Basal Jurassic | SW |
| OIL AND GAS | Dec 1983 | Challum | Murta (oil), Hutton (gas) | MW |
| | Aug 1983 | Chookoo | Hutton | MW |
| | Aug 1984 | Gidgealpa | Namur, Birkhead, Hutton (oil), Basal Jurassic (gas) | MW |
| | May 1984 | Mudera | Coorikiana (gas), Birkhead (oil) | MW |

SW - Single well discovery/field
MW - Multi well field

TABLE 2

EROMANGA BASIN OIL FIELDS

INITIAL RECOVERABLE PROVED AND PROBABLE RESERVES (MMSTB)

| < 1 | | 1-9 | 9-15 | 15-50 |
|---|---|---|---|---|
| Alwyn 1 | Kooroopa 1 | Bodalla South | Strzelecki | Jackson |
| Big Lake | McKinlay | Chookoo | | |
| Black Stump 1 | Mooliampah 1 | Dullingari | | |
| Bogala 1 | Moothendella | Gidgealpa | | |
| Bookabourdie | Mudera | Jackson South | | |
| Calamia West 1 | Nappacoongee 2 | Kenmore | | |
| Charo 1 | Narcoonowie | Limestone Creek/Biala | | |
| Challum | Nockatunga | Meranji | | |
| Cook 1 | Nungeroo 1 | Merrimelia | | |
| Cuttapirrie 1 | Poolowanna 1 | Moorari | | |
| Dirkala 1 | Sigma 1 | Muteroo | | |
| Epsilon 3 | Spencer 3 | Naccowlah South | | |
| Glenvale 1 | Takyah 1 | Naccowlah West | | |
| Graham 1 | Talgaberry | Spencer 4 | | |
| Gunna 1 | Thungo 1 | Tintaburra | | |
| Ipundu | Tickalara 2 | Toobunyah | | |
| Jena 1 | Tinpilla 1 | Wancoocha | | |
| Kercumurra 1 | Ulandi 1 | Wilson | | |
| Kerinna 1 | Watson 1 | | | |
| Kihee | Watson South | | | |
| Kobari 1 | Winna 1 | | | |
| Koora | Yanda | | | |

TABLE 3

EROMANGA BASIN GAS FIELDS

INITIAL RECOVERABLE PROVED AND PROBABLE SALES GAS (BCF)

| < 1 | 1-15 | 15-25 |
|---|---|---|
| Gidgealpa | Challum | Marabooka |
| Mina 1 | Chookoo | |
| Mudéra | Namur | |
| Nanima 1 | Wackett 1 | |

extensive reserves.

Most of the fields discovered to date occur in structures flanking the Nappamerri Trough (see Figure 6). The bulk of the reserves discovered to date occur in structures along the Naccowlah-Jackson trend along the northern rim of the Nappamerri Trough. It has, however, been encouraging that recent discoveries such as Tintaburra, Bodalla South, Kenmore, and Talgaberry occur away from the Nappamerri Trough, thereby extending the area of prospectivity quite significantly.

## PETROLEUM TYPE AND ORIGIN

The discovery pattern to date indicates the Eromanga Basin to be an oil rather than a gas province. Eromanga Basin oil is typically light (41° to 51° API), low-sulphur paraffinic, with two broad types being recognized (Figure 7). Oils from the Jurassic reservoirs—namely, the Basal Jurassic, Hutton Sandstone, Birkhead Formation, and Westbourne Formation—tend to be waxy, with pour points ranging from 10° to 23°C, and tend to be rich in $C_{21-27}$ alkanes. In contrast, oils from the Namur Sandstone Member, Murta Member, and Cadna-owie Formation tend to be condensate-like, with low pour points (-5° to 0°C). These oils have a low wax content and are dominated by $C_{6-12}$ alkanes. The oils have relatively low gas/oil ratios.

Gas from the Eromanga Basin reservoirs has a low $CO_2$ content and is dry, with typically less than 10% $C_{3+}$. The carbon isotopic compositions of the methane show a similar range to that seen in underlying Permian gases.

Extensive geochemical analyses have been undertaken on rock and hydrocarbon samples to assess petroleum sourcing in the Eromanga Basin. Analyses indicate that the Murta-type oils reservoired in the Namur Sandstone Member, Murta Member, and Cadna-owie Formation were generated from predominantly algal-bacterial and degraded terrestrial organic matter contained in the lacustrine shales of the Murta Member and Cadna-owie Formation. This organic matter has matured at relatively low vitrinite reflectances of about 0.4 to 0.6.

The Hutton-type oils reservoired in the Westbourne Formation, Adori Sandstone, Birkhead Formation, Hutton Sandstone, and Basal Jurassic units have a mixed source predominantly of terrestrial land plant origin with a minor contribution from algal-bacterial matter. Maturation of the organic matter generating these oils occurs generally in the reflectance range of 0.6-0.8.

Mixed source type occurs throughout the Jurassic section as well as the underlying Permian section of the Cooper Basin, thereby making it difficult to clearly identify the oil source. In a detailed study of the origin of oils along the Naccowlah-Jackson trend (Vincent, Mortimore, and McKirdy, 1985), it was concluded that the Jurassic oils were sourced from the mature Jurassic sequence found deeper in the Nappamerri Trough. Similarly, intra-Eromanga Basin sourcing is suggested for Hutton oils found at Tintaburra (Newton, 1986).

Gases found in the Jurassic reservoirs in the Namur Sandstone Member at Namur and the Birkhead Formation at Wackett show isotopic compositions similar to Permian gases of the Cooper Basin, and it has been argued that they are Permian sourced and have leaked into Jurassic reservoirs via faults (Kanstler et al., 1983). The gases do, however, exhibit very low $CO_2$ contents, which contrast with the Permian gases, and commonly have a high $i\text{-}C_4/n\text{-}C_4$ ratio, indicating that intra-Eromanga Basin sourcing cannot be entirely precluded.

## HYDROCARBON GENERATION, MIGRATION, AND ENTRAPMENT

Murta-type oils reservoired in the Namur Sandstone Member, Murta Member, and Cadna-owie Formation are assumed

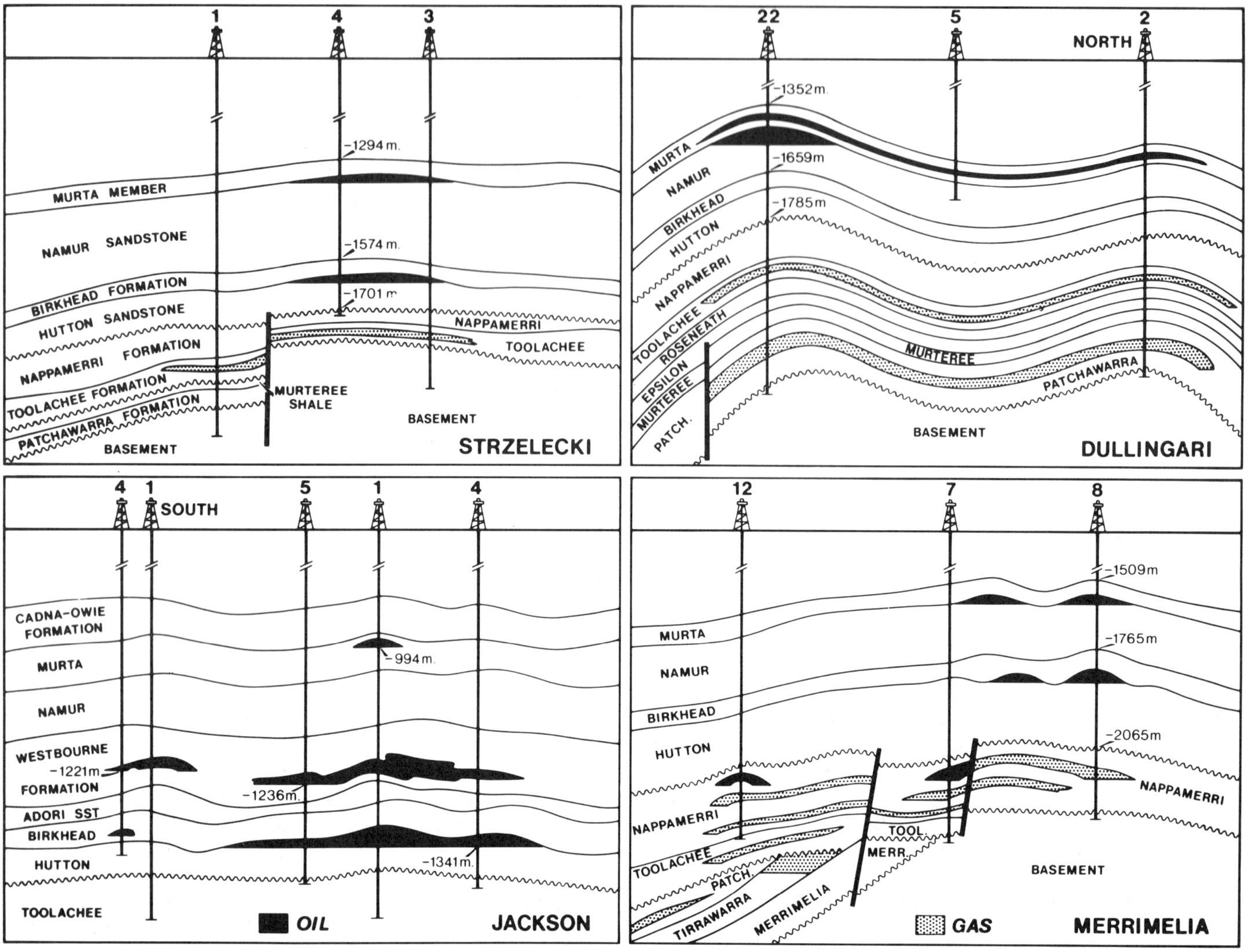


Figure 5. Geological cross sections of the Strzelecki, Dullingari, Jackson, and Merrimelia fields.

to have generated mainly from lacustrine shales of the Murta Member and Cadna-owie Formation. In most accumulations of this oil type it is geologically difficult to invoke sourcing from the Permian, although the possibility of such sourcing does exist for Namur Sandstone Member pools. The source of the Hutton-type oils reservoired in the Basal Jurassic, Hutton Sandstone, Birkhead, and Westbourne formations is more uncertain. Source rocks are contained in the Birkhead and Westbourne formations as well as the Basal Jurassic. The underlying Permian sequence can be readily invoked as providing at least some of the Hutton Sandstone oil where this unit directly overlies the Permian (as in the Jackson field).

The timing of generation is largely controlled by depth of burial variations and geothermal gradients. In areas such as the central part of the Nappamerri Trough, where optimum generation conditions are considered to exist, the lower part of the Jurassic sequence reached initial maturation in the early Late Cretaceous. Elsewhere, such levels were not attained until the Late Cretaceous or Tertiary.

The understanding of migration pathways in the Eromanga Basin is still poor. Originally it was thought that the extensive westward artesian water drive had flushed hydrocarbons from the basin. However, the absence of tilted oil-water contacts and the presence of several structures along the Naccowlah-Jackson trend approximately filled to spill point lend support to the view that the artesian system has had little flushing effect on trapped hydrocarbons.

Earliest discoveries in the Eromanga Basin were located around the flanks of the Nappamerri Trough, which led to the concept that the trough area was the source of the hydrocarbons, with short distance migration to flanking structures. However, recent discoveries such as those in the Tintaburra/Bodalla South area, Moothendella, and Charo confirm a much more extensive source area within the Eromanga Basin. The possibility of partial Permian sourcing cannot be entirely discounted in these discoveries also. In general, short-distance migration is thought to exist for most accumulations, and there is mounting evidence that many of the Hutton-type oils have been generated at depth and migrated into shallow reservoirs.

Most of the hydrocarbon discoveries found to date have been in structural traps. Consequently, the timing of structures and structural styles is vital. In the South Australian sector of the Eromanga Basin the bulk of structuring predates the Early Cretaceous, whereas many of the southwest Queensland features have had their main structural development in the

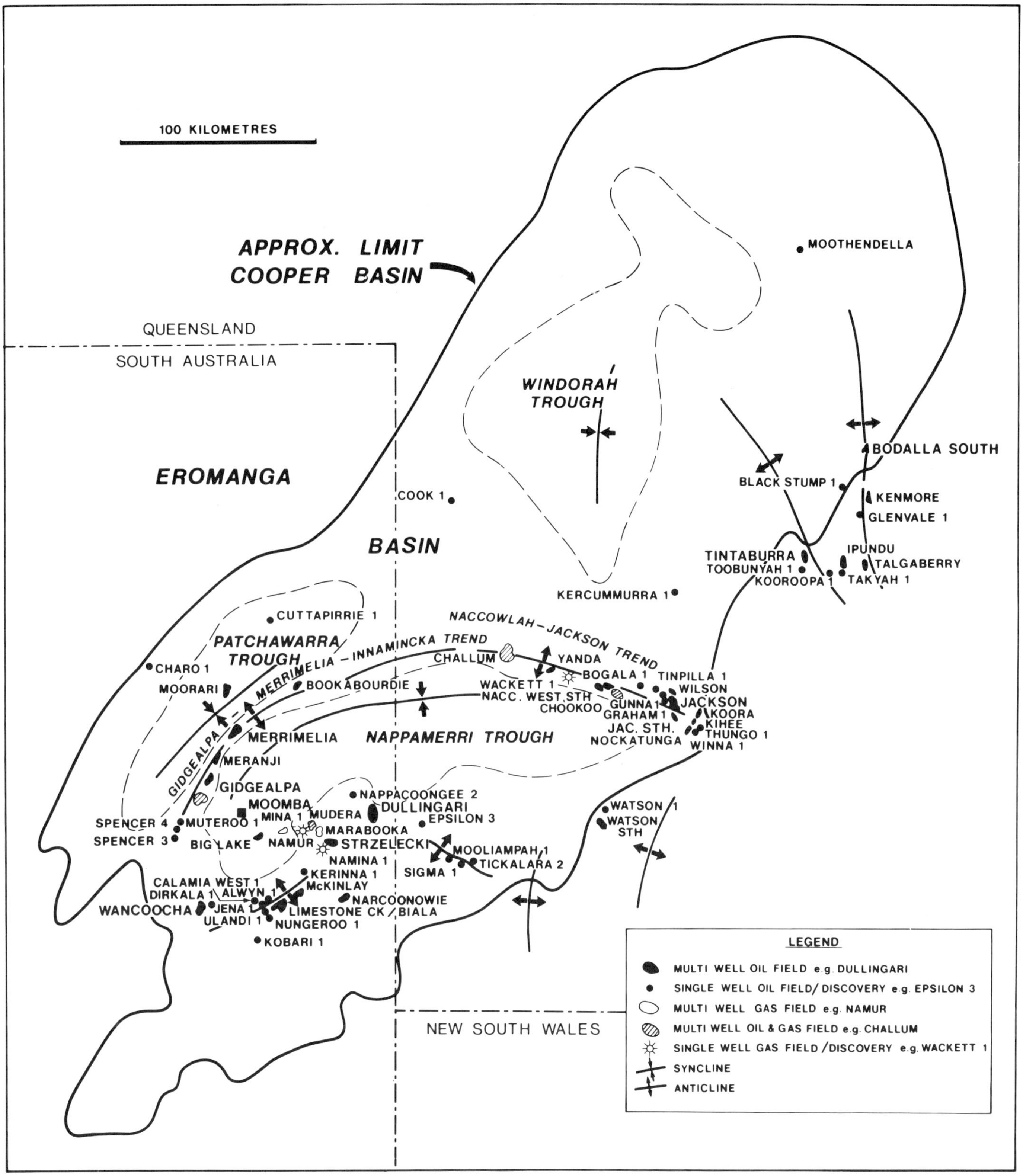


Figure 6. Location map of Eromanga Basin discoveries/fields.

Paleocene to Oligocene. Initially it was thought that "early" structuring (that is, pre-Aptian) was essential for hydrocarbon accumulations. Discoveries such as Tintaburra, where the main structuring phase occurred in the Tertiary, indicate that "late" structuring does not necessarily prelude accumulation.

It is concluded that a complex interplay exists between structuring, generation, and migration in the Eromanga Basin. Late structuring (Tertiary) does not prevent accumulation in that late generation and migration have definitely occurred in some areas.

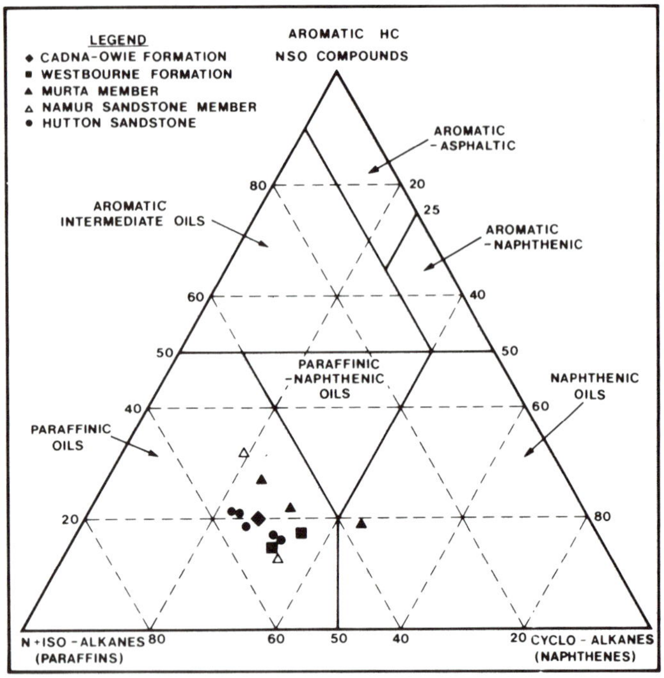


Figure 7. Gross composition of some Eromanga Basin oils.

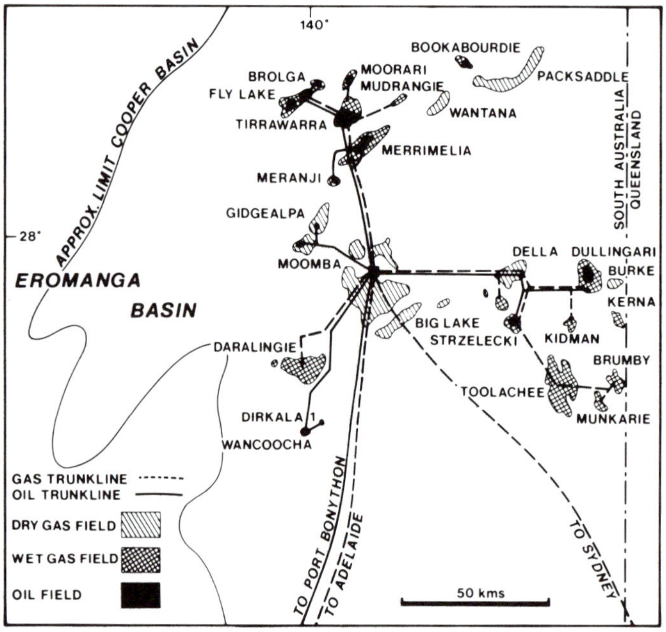


Figure 8. Location map of oil fields and wet gas fields associated with "Liquids Project" in South Australia.

## DEVELOPMENT—SOUTH AUSTRALIA

Oil produced from the Eromanga Basin within South Australia is transported via a liquids pipeline from Moomba to Port Bonython for marketing. This pipeline was built as part of the "Cooper Basin Liquids Project" and was completed in June 1984.

The need to sell crude oil and gas liquids from the Cooper Basin was recognized in the early 1970s with the discovery of the Permian Tirrawarra oil field, several smaller Permian oil fields, and numerous fields containing wet gas. However, the need to exploit wet gas fields became more urgent when gas contracts were signed in 1976 to supply the New South Wales market. Eromanga Basin oil discoveries at Strzelecki in 1978, Dullingari in 1979, and Merrimelia in 1981, further highlighted the necessity for a project to bring petroleum liquids from the Cooper/Eromanga Basins on stream and to market. The locations of Cooper Basin oil and wet gas fields and Eromanga Basin oil fields constituting the liquids project are shown on Figure 8. Eromanga Basin discoveries post-1981 which are connected by pipeline to Moomba are also shown.

Project design for the Cooper Basin Liquids Project commenced in late 1981 with construction beginning in 1982. The project was completed in June 1984 at a cost of about $1.4 billion. It involved the building of a 650-km, 355-mm-diameter liquids pipeline from Moomba to Port Bonython as well as substantial liquids processing plants at Moomba and storage and loading facilities at Port Bonython.

At the end of June 1986 in the South Australian sector of the Eromanga Basin oil production came from 18 fields. Seven of these fields are connected by pipeline to Moomba, and oil from the rest is trucked to Moomba. Total production rate is about 23,800 BOPD with the main fields being Strzelecki at 7200 BOPD and Dullingari at 4400 BOPD.

Well spacing is characteristically at 80 acres for the more productive reservoirs (Hutton Sandstone, Namur Sandstone Member) and 40-80 acres (Murta Member, Westbourne Formation) for less productive reservoirs. Lowest production rates are from fields like Kerinna, which was a single well on a Murta oil pool producing at about 30 BOPD.

The development facilities and field outline for the Dullingari Field, located about 65 km east of Moomba, are illustrated in Figure 9. The field produces gas (from 12 wells) from the Permian Cooper Basin section as well as oil from the Murta Member and Namur Sandstone Member in the Eromanga Basin sequence. Twelve wells, all of which are on artificial lift, produce from the Murta. Three potential Murta producers are currently shut in for pressure observations, and 4 Murta wells near the field limits are being used for water injection. One Murta well has been fraced to improve productivity. The Namur pool is produced from 3 wells, one of which is on artificial lift. Murta wells produce at rates ranging from 30-800 BOPD, and Namur wells produce at rates ranging from 150-800 BOPD.

The type of surface installations differ from field to field. Major fields such as Strzelecki, Dullingari, Merrimelia, and Wancoocha have crude-oil-gathering, treatment, and storage facilities including a pipeline to Moomba. Meranji, located about 9 km south-southwest of Merrimelia, and Dirkala 1, about 5 km northeast of Wancoocha, are connected by oil pipelines to the facilities at Merrimelia and Wancoocha, respectively. Fields such as Big Lake, Limestone Creek, and Narcoonowie have centralized gathering with trucking facilities, whereas fields which have low production or are remote from Moomba have individual wellsite trucking facilities. As can be seen, the type of development adopted is contingent on size of reserves, deliverability, and distance from existing facilities. As a generalization, small fields with production of less than 800 BOPD tend to favor the trucking option, either through individual wellsite facilities or a centralized gathering system; fields with a capacity greater than 2000 BOPD favor pipeline options of either being connected to an existing facil-

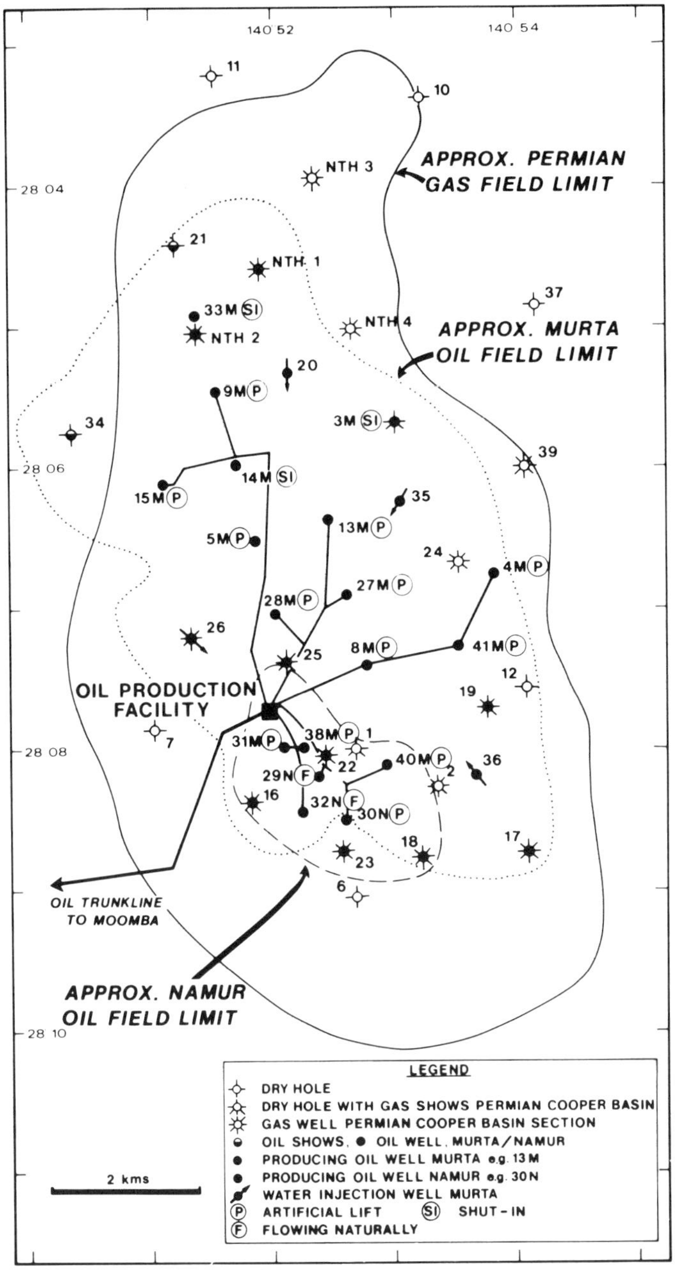


Figure 9. Dullingari field outline and development facilities.

ity or an individual pipeline system to Moomba (Scanlon and White, 1986).

Fluid gathered at the production facilities at Strzelecki, Dullingari, Merrimelia, Wancoocha, and Gidgealpa is allowed to cool and to separate into oil, water, and gas. The oil is then stored in tanks prior to pumping to Moomba through trunklines. For example, at Dullingari two 5000-barrel tanks are used, and the trunkline is 70 km long and 150 mm in diameter. At Gidgealpa, two 10,000-barrel tanks are installed, and the oil trunkline is 24 km long and 150 mm in diameter. Similarly, where oil is trucked to Moomba the fluids are stabilized at the wellsite or centralized gathering point and stored in tanks (usually only about 400-barrel capacity) before being loaded into trucks. A flow diagram outlining the Gidgealpa field production facilities is shown in Figure 10.

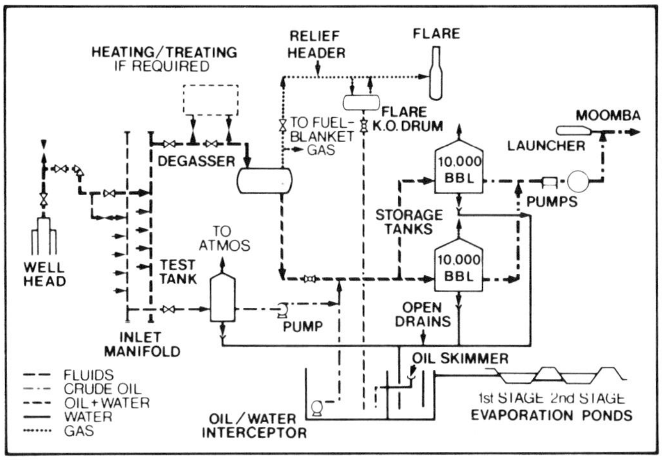


Figure 10. Flow diagram of Gidgealpa field production facilities.

The main treatment plant for Eromanga Basin oils produced in South Australia as well as Cooper Basin oils and gas liquids is at Moomba. Here, as part of the Cooper Basin Gas Liquids Project, a crude stabilization plant, a liquids recovery plant, and molecular sieve units were constructed. In addition, modifications were undertaken to the existing gas processing plant where natural gas is processed for the Adelaide and Sydney markets.

The Moomba complex has been designed to process 25.4 million $m^3$ of raw natural gas per day, 7250 kL of natural gas liquids per day, and 30,000 barrels of crude oil per day. Surface storage facilities consist of a 300,000-barrel crude oil storage tank and a 100,000-barrel condensate storage tank.

Ethane from the natural gas is being stored underground at Moomba in selected wells while its marketability and application to secondary recovery processes is investigated. The remaining natural gas liquids are comingled with the stabilized crude oil and piped to Port Bonython along a 355-mm-diameter, 650-km-long underground pipeline. This pipeline, which has a capacity of 80,000 barrels of liquids per day, has a pump station at Moomba and two additional along route.

The Port Bonython complex consists of a fractionation plant, storage tanks, and offshore loading facilities. The fractionation plant separates ethane (used as plant fuel), propane, and butane from condensate and oil. Five storage tanks, each of 250,000-barrel capacity, have been constructed. LPG (propane and butane), as well as condensate and crude oil, are pumped to tankers at the deepwater anchorage. Current annual through-put at Port Bonython is about 5.2 million barrels condensate, 8.2 million barrels crude oil, and about 50,000 tonnes LPG.

## DEVELOPMENT—QUEENSLAND

At the end of June 1986, there were 23 Eromanga Basin oil fields on production in southwest Queensland. The development methods are analogous to those in South Australia with Murta Member and Westbourne Formation oil pools being typically drained on 40-80 acre spacing. Pools in the more productive reservoirs, such as the Namur Sandstone Member, Hutton Sandstone, and Basal Jurassic, are drained on 80-acre spacing. Oil transportation from fields is dependent on field size and proximity to the production facilities at Jackson.

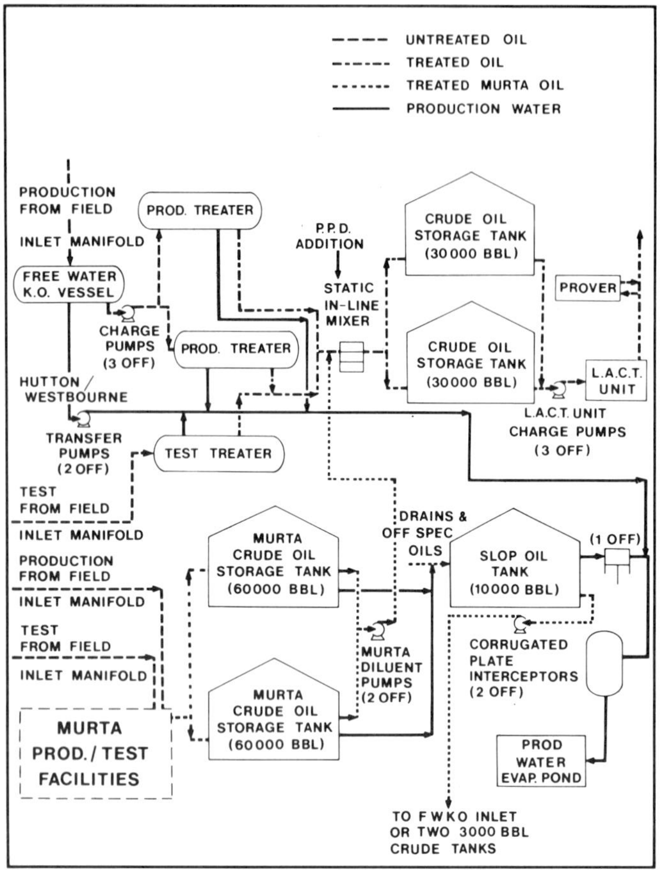


Figure 11. Flow diagram of Jackson production facilities.

Fields such as Jackson South, Wilson, Chookoo, Naccowlah South, and West proximal to Jackson have oil trunklines to Jackson; Naccowlah South and West have their own storage tanks. Smaller and more distant fields have either single-well or field-gathering and storage systems, with oil being trucked either to Jackson or other places in Queensland. At present about 90% of the trucked oil comes to Jackson.

The bulk of reserves found to date in Eromanga Basin in Queensland occur within a 50-km radius of Jackson. However, there have been some significant discoveries at and near the Tintaburra and Bodalla South fields about 100 km and 150 km, respectively, northeast of Jackson.

The Jackson field, the largest discovered to date in the Eromanga Basin, has played a crucial role in oil development in southwest Queensland. Following its discovery in December 1981, and discovery of the nearby Jackson South field in February 1982, a decision was made that reserves justified the building of field facilities at Jackson and construction of a 780-km, 324-mm-diameter underground oil pipeline from Jackson to Moonie to connect with an existing pipeline (307-km, 250-mm-diameter) from Moonie to Brisbane. The Jackson-Moonie pipeline was built during April 1983 to February 1984 at a cost of $125 million. The pipeline has an ultimate capacity of 55,000 BOPD via use of pumping facilities. The current pumping rate of about 27,000 BOPD is well short of capacity. The Jackson-Moonie oil pipeline, along with the allied infrastructure at Jackson, have acted as key factors in the commercial viability of Eromanga Basin oil fields in southwest Queensland.

At the Jackson field, oil is produced from the Murta Member, Westbourne Formation, and Hutton Sandstone. Twenty-nine wells have been drilled on the structure. At present, 4 crestally located wells produce from the areally small Murta pool at average rates of about 50 BOPD. Westbourne Formation production comes from 5 wells at rates ranging from 16-190 BOPD. An additional 10 wells have been completed in the Westbourne Formation. The main producing zone, the Hutton Sandstone, produces from 25 wells at rates generally in the order of 300-2000 BOPD but up to 3000 BOPD. Dual Westbourne/Hutton completions are common, and the bulk of the Westbourne production comes from dual-completed wells. Decrease in reservoir pressure and increase in water production have necessitated the use of artificial lift systems in about 95% of the wells in the Jackson field. Recently, electric submersible pumps have been installed in two Hutton wells to enhance production. These wells produce fluid at a rate of about 5000 barrels per day (3000 BOPD) and represent a significant advance in artificial lift techniques.

The Jackson Hutton and Westbourne crude oils contain negligible gas, but their high pour points (24 °C and 20 °C, respectively) and waxy nature cause significant problems during pipeline transportation. Pour point depressants are required during winter months, but during summer months no treatment is applied. At present oil from the Murta reservoir, with its low pour point of 0 °C, is added as a 5% diluent along with chemical treatment during colder months, and this increases production costs of Jackson crude. To date about 10 MMSTB oil have been produced from the Jackson field.

Surface installations at Jackson consist of treatment facilities to separate oil from water, and facilities to add chemical additives and diluent crudes for pour point management. There are two 30,000-barrel heated tanks to feed the pipeline and two 60,000-barrel tanks to stock Murta diluent crude (Figure 11).

How the proximity to Jackson influences the development mode is clearly demonstrated when comparing Naccowlah South/West fields with Bodalla South field.

Naccowlah South/West fields are located about 35 km west of Jackson. Naccowlah South contains a Hutton oil pool which is being produced at about 2500 BOPD from 6 wells (all wells are on artificial lift). Naccowlah West contains a small Murta oil pool and a significant Hutton oil pool. The Murta pool is exploited by 1 well (on artificial lift) which produces at about 45 BOPD. The Hutton pool is being produced at about 1600 BOPD from 6 wells, all of which are on artificial lift. Oil from these fields is gathered at crude storage facilities comprising two 30,000-barrel tanks for Murta crude with a single 32-km (250-mm-diameter) trunkline for transfer of both crudes to the Jackson plant. The Murta crude is a diluent source, as discussed above.

In contrast to these facilities, Bodalla South field located about 150 km northeast of Jackson only has trucking facilities. This field is currently producing at about 2600 BOPD from Hutton Sandstone and Basal Jurassic pools via 6 wells with artificial lift. The oil is gathered at a central processing area where water is removed and stored in tanks ahead of trucking to Moonie. Only about 10% of production is trucked to Jackson for transportation via pipeline.

## DEVELOPMENT SUMMARY

Total Eromanga Basin oil production is about 53,500 BOPD, of which about 45% is transported via the liquids line from

Moomba to Port Bonython in South Australia; 50% is transported via the Jackson to Moonie pipeline in Queensland; and 5% is trucked to market points in Queensland. As of June 30, 1986, production was coming from 41 fields. The Hutton Sandstone contains the majority of reserves (about 65%) and has the best production rates. To date about 28 MMSTB of oil has been produced from the Eromanga Basin. Current gas production is negligible at about 7.5 MMCFD sales gas from one field (Marabooka) in South Australia. A total of about 15 BCF sales gas has been produced to date from Eromanga Basin gas fields.

Fields of a suitable size and productivity near Moomba and Jackson are connected by oil trunklines to these respective centers. Other fields have gathering systems dependent on field size, and oil is trucked to either Moomba or Jackson or, in the case of about 5% of Queensland oil, to other market points. In South Australia the oil production from the Eromanga Basin is comingled with Permian oil and gas liquids from Cooper Basin fields as part of the Cooper Basin Liquids Project facilities.

Murta and Cadna-owie pools, which are generally the smallest in area, can often be exploited by 1-4 wells (Dullingari, however, is an exception to this with 13 Murta wells). On the average, oil pools in other formations are exploited by 4-6 wells. The largest pool in the basin, in the Hutton Sandstone at Jackson field, has some 25 producing wells.

The development history of Eromanga Basin oil fields has been characterized by rapid early production. This has been achieved by initially trucking oil and then, if field sizes warranted and infrastructure was proximal, the fields have been connected via oil trunklines to Moomba or Jackson. Fields close to infrastructure are often brought into production within a few months of discovery.

## FUTURE

The Eromanga Basin is currently widely rated as the most prospective onshore petroleum basin in Australia. The discovery pattern to date indicates the basin to be an oil rather than a gas province. The basin contains generally small fields, often multipooled, and most of the fields have less than 9 million barrels initial recoverable oil reserves. Discoveries to date have resulted in the construction of a substantial infrastructure of surface facilities and oil pipelines including two major pipelines to market points, namely, the Moomba to Port Bonython liquids line and the Jackson to Moonie oil line. Both of these lines have capacity for additional production from future discoveries. The Moomba to Port Bonython line currently carries about 80% of its 80,000 BOPD capacity, and the Jackson-Moonie line carries about 50% of its ultimate capacity of 55,000 BOPD. The commercial viability of future discoveries, particularly those within 50-70 km of Moomba or Jackson, is regarded as particularly attractive.

Exploration coverage in the Eromanga Basin is still relatively sparse. The most densely explored areas, proximal to Moomba and Jackson, have extensive 1 × 1 km seismic grids, whereas large areas have line density of 15 km. With the small size of potential traps in the Eromanga Basin sequence, it is obvious that a considerable amount of further seismic investigations will be required to adequately identify and define structures. Drilling density is similarly low basin-wide, with about 600 exploration wells in the whole basin (an area of about 1 million $km^2$). Highest drilling density is still only about 40 wells per 1000 $km^2$.

Current initial recoverable proved and probable oil reserves in the Eromanga Basin are estimated at about 110 million barrels. The inevitable question which arises is: What is the ultimate potential of the basin? This is naturally a difficult number to establish in view of the wide extent of the basin and attendant source rock and maturity variations. Calculations of the Jurassic potential based on sediment volume indicate at least 800 MMSTB original oil in place (OOIP) could be generated in the central part of the Eromanga Basin (the area roughly underlain by the Cooper Basin). This area is regarded as the most prospective to date and covers about 10% of the Eromanga Basin. Armstrong and Barr (1986) speculated about the potential of the Eromanga Basin by estimating the number of successful discoveries which could be made until the year 2000. They anticipated that reserves of 500 million barrels of oil could be discovered.

At this stage there is no shortage of undrilled prospects and leads in the Eromanga Basin sequence, and further exploration will undoubtedly discover many more. At least about 200-300 prospects and leads are currently defined.

Major challenges facing explorationists in the Eromanga Basin are seen as precise definition of small structures, finding discoveries away from currently known productive areas, and closer understanding of key factors controlling hydrocarbon accumulations. Another aspect which will require review is a better insight of the Cretaceous sequence, particularly its potential and influence on Jurassic accumulations.

Major challenges in the production area are artificial-lift applications, waterflooding and fracing applications, and the general capacity to respond rapidly to new discovery development.

Oil accumulations to date have generally been discovered at crests of anticlines, which need to be defined by 1 × 1 km grids, inasmuch as areal and vertical closures are small. As a result future exploration will require high-quality seismic to adequately identify features which commonly are very subtle. Although few structural/stratigraphic discoveries have been made to date, it is anticipated that more will become evident, further emphasizing the need for high-quality seismic.

Earliest Eromanga Basin discoveries were concentrated around the edges of the Nappamerri Trough. As exploration has progressed, discoveries have been made away from the trough rim. Discoveries in the Tintaburra and Bodalla South areas have been particularly valuable in extending the geographic spread of oil fields into a province some 100-150 km northeast of the Nappamerri Trough. Such breakthroughs are seen as vital thrusts in future exploration to expand the limits of the oil discovery in the basin.

As exploration has proceeded and discoveries have been made, more understanding has been gained about key parameters controlling Eromanga Basin accumulations. Some recent discoveries have invalidated previously held beliefs such as the need for early structuring and Permian sourcing for Jurassic/Cretaceous accumulations. Future exploration is anticipated to focus on the identification of key parameters.

Drilling to date has clearly established the petroleum potential of the Jurassic sequence in the Eromanga Basin. Similarly, the basal part of the Cretaceous sequence (pre-Aptian) with commercial oil pools has been identified as prospective. However, the rest of the Cretaceous sequence has so far only contained noncommercial hydrocarbon pools. Future drilling is expected to provide a better understanding of this section and will hopefully lead to commercial discoveries, as well as elucidating if the Cretaceous influences Jurassic accumulations.

In production activities, there is anticipated to be a strong emphasis on the need for application of artificial-lift techniques, which are already being applied extensively in many fields, notably Jackson and Dullingari. Such applications naturally increase production costs and impinge heavily on ultimate oil recovery. Similarly extensive studies are being undertaken on waterflooding and fracing possibilities in certain reservoirs. Successful development and application of such techniques on current and future discoveries will have a strong bearing on ultimate oil recoveries.

With the current infrastructure and level of experience it is anticipated that future discoveries will be brought on stream as rapidly as discoveries have been to date. This has been a hallmark of Eromanga Basin development and has acted as a positive impetus in the exploration and production in the basin.

The Eromanga Basin is seen as the most prospective onshore acreage in Australia for oil exploration and is anticipated to dominate the exploration activity in Australia for the next 5 years. The probability that most fields will be small will demand application of high-quality seismic along with a good understanding of parameters controlling oil accumulation. Attention will focus on recovery methods, and the current infrastructure is anticipated to permit the rapid exploitation of future discoveries.

## ACKNOWLEDGMENTS

The author wishes to thank colleagues in the CSR Oil and Gas Division, Adelaide, for their valued discussions and assistance in the preparation of this paper. This paper is published with the permission of the CSR Oil and Gas Division.

## REFERENCES CITED

Armstrong, J.L., and T.M. Barr, 1986, The Eromanga Basin, An overview of exploration and potential, *in* D.I. Gravestock, P.S. Moore, and G.M. Pitt, eds., Contributions to the geology and hydrocarbon potential of the Eromanga Basin: Geological Society of Australia Incorporated, Special Publication 12, p. 25-38.

Habermehl, M.A., 1980, The Great Artesian Basin, Australia: Bureau of Mineral Resources Journal of Australian Geology and Geophysics, v. 5, p. 9-38.

Habermehl, M.A., 1986, Regional groundwater movement, hydrochemistry and hydrocarbon migration in the Eromanga Basin, *in* D.I. Gravestock, P.S. Moore, and G.M. Pitt, eds., Contributions to the geology and hydrocarbon potential of the Eromanga Basin: Geological Society of Australia Incorporated, Special Publication 12, p. 353-376.

Kanstler, A.J., T.J.C. Prudence, A.C. Cook, and M. Zwigulis, 1983, Hydrocarbon habitat of the Cooper/Eromanga Basin, Australia: Australian Petroleum Exploration Association Journal, v. 23, n. 1, p. 75-92.

Lavering, I.H., V.L. Passmore, and I.M. Paton, 1986, Discovery and exploitation of new oil fields in the Cooper-Eromanga Basins: Australian Petroleum Exploration Association Journal, v. 26, n. 1, p. 250-259.

Moore, P.S., 1986, Jurassic and Triassic stratigraphy and hydrocarbon potential of the Poolowanna Trough (Simpson Desert region) northern South Australia, *in* D.I. Gravestock, P.S. Moore, and G.M. Pitt, eds., Contributions to the geology and hydrocarbon potential of the Eromanga Basin: Geological Society of Australia Incorporated, Special Publication 12, p. 39-51.

Newton, C.B., 1986, The Tintaburra oil field: Australian Petroleum Exploration Association Journal, v. 26, n. 1, p. 334-352.

Scanlon, R.J., and C.J. White, 1986, Surface facilities for oil fields in the South Australian sector of the Cooper Basin: Australian Petroleum Exploration Association Journal, v. 26, n. 1, p. 470-480.

Vincent, P.W., I.R. Mortimore, and D.M. McKirdy, 1985, Hydrocarbon generation, migration and entrapment in the Jackson-Naccowlah area ATP 259P, south-western Queensland: Australian Petroleum Exploration Association Journal, v. 25, n. 1, p. 62-84.

CHAPTER 13

# SOUTHEAST ASIA STRIVES FOR RAPID RESERVE ADDITIONS

**H.F. Keplinger**
*The Keplinger Companies*
*Houston, TX*

## ABSTRACT

Southeast Asia continues to be among the most active areas in oil and gas exploration worldwide. With economic growth continuing at rates which are higher than those in many industrial nations, there are pressures for continued energy development. Many countries still import substantial amounts of their energy. Realizing the difficulties that will be faced in the future, they have made heavy investments in energy development, particularly in natural gas and coal resources, over the last 5 years.

Although the results within individual countries have been mixed, substantial increases have been made in coal and natural gas production. This chapter will discuss energy and resource development in the framework of economic activity in the region; exploration, production, and reserves; energy self-sufficiency and diversification in the energy mix, and the implications for the future.

## INTRODUCTION

Southeast Asia continues be the most active oil and gas exploration region in the world outside of North America. It is the only area in the world which is currently increasing its well activity, which rose 8% to an estimated 8300 wells for 1986.

However, the 1986 increases are only one-half of the levels of increases during the last few years. The lower levels of activity are beginning to show up in the smaller reserve additions. During 1985, proven crude oil reserves increased by less than .5%, and proven natural gas reserves increased by 7.7%.

The outlook for future investment in exploration programs is not optimistic with the lowered oil prices. Even though production increased by more than 10% during 1985, the production increases did not offset the losses in oil revenues. As a result, such major exporting countries as Indonesia and Malaysia are experiencing declines in their exploration programs. The benefits of lower prices have gone to the oil-importing countries. If crude prices stay at the expected $15 per barrel during 1986, revenues for the region's oil exporters will drop to around $35 billion compared with $60 billion during 1985.

Many countries in Southeast Asia have had success in developing their natural gas and coal resources. Several oil and gas programs have been developed jointly between the national governments and private companies. With the impact of lower oil prices, countries such as China have offered incentive programs to attract capital for energy development. The need for capital will likely create many more incentive programs. Yet several trends are emerging in the development programs in the region:

(1) Coal resources are being developed more extensively in Australia and China, which therefore continue to receive the largest capital investments;
(2) Natural gas reserves, which are more widely distributed throughout the region than crude oil reserves, have been increased substantially, but future production scheduled for LNP exports may face competition with lower crude and coal prices;
(3) In efforts to diversify the energy mix, such oil-importing countries as Japan, South Korea, and Taiwan are moving rapidly toward nuclear power generation. The impacts of Chernobyl could slow the planned projects. But it appears that with the large shares already held by nuclear plants and prospective plants promising more competition, this will further suppress oil and gas prices, while putting the countries on a path of rising electricity rates in the future.

### Prospects for Continued Growth

The Southeast Asia countries are among the top performing economies in the world. While many industrial nations are struggling to achieve growth rates from 2 to 3% in the future, several Southeast Asia industrial economies are growing in the range from 5 to 10%. During 1986, South Korea's growth is expected to be about 9 to 10%, then slow to 8% in 1987.

However, some slowing is occurring because of problems associated with lower oil and commodity prices and the devaluation of the U.S. dollar against the yen. Japan's economy, now slowing from 4.4% in 1985, is expected to drop to a 2 to 3% growth rate during 1986 with the yen's appreciation. Taiwan, now growing in the range of 7 to 9%, is expected to slow to about 6.5% in the next 4 years. Hong Kong and Thailand are having difficulties, which will bring their growth rates down to the range of 4 to 5% by 1987. In most cases, the growth rates of 2 to 3% are expected with low unemployment, and low inflation.

### Impacts of Low Oil Prices

Lower oil prices have provided some impetus for the growth among oil importers. Japan, which imports about 85% of its energy requirements, may see its overall trade surplus grow to \$77 billion for this fiscal year compared with \$52.6 billion last year. However, Japan expects some slowdown in the level of its overall activity as Japanese goods become less competitive in worldwide markets with the appreciation of the yen. In addition, the decline in oil revenues among the producers means that they can order fewer Japanese goods.

South Korea is moving at higher-than-expected rates as a result of the lower oil prices, lower interest rates, and the lower U.S. dollar against the yen. Korea's industrial sector has been booming, with the sales of videotape recorders, autos, footwear, and textiles.

Lower oil prices have helped with 1986 trade deficits in several countries. The value of Thailand's oil imports has dropped from 24 to 16.5% of total imports during 1986, which lowered its trade deficit during the first quarter. New Zealand has had some improvement in its foreign trade deficits, which were \$1.9 billion by March 1986 compared with \$2.7 billion by March 1985. Many New Zealand commodities have been hurt as the U.S. dollar floated down.

The two countries hurt most with the oil price declines are Malaysia and Indonesia. Malaysia is expected to have a budget deficit for the first time in 10 years as the result of the lower oil prices coupled with a decline in prices of its other main commodity exports—palm oil, rubber, and tea. Indonesia has similar problems, because oil exports constitute 70% of its export income. It is likely to suffer a loss of about \$39 billion in foreign income, which will contribute to a negative growth rate this year.

While initially benefiting from lower oil prices, Singapore refiners hit a production high above 800,000 barrels per day of raw material intakes during March, but buyers began holding back in anticipation of further price declines. As a result of the buildup in inventories, refinery intakes fell rapidly to around 100,000 to 200,000 barrels per day during the summer.

### Currency Adjustments

The devaluation in the U.S. dollar and the appreciation of the yen continue to create difficulties for several countries. The yen appreciation has hurt the current account balances of Japan's major trading partners including Malaysia, Singapore, and Thailand. By May 1986, the Singapore dollar dropped by 30% against the yen from its position last September. As a result, first quarter trading results showed a 13% decline in Japanese imports.

### Adjustments With Trade Deficits

The trade deficits are affecting the rates of growth. For instance, China has clamped down on lending to its enterprises and has sharply reduced imports after recording a trade deficit of \$6.4 billion by mid-year. The largest portion of the deficit was \$3.48 billion with Japan.

The cost of servicing a \$29.4 billion public debt in Indonesia has risen to nearly 22% in 1985-86 from 16% in 1984-85. Indonesia also suffers from the depreciation of the yen, since 30% of its external debt is denominated in yen.

## ENERGY RESOURCES AND THEIR GROWTH

Southeast Asia is now reaping the benefits of exploration investments during the 1980s with additions to their coal and natural gas reserves. During 1985, Australia increased its crude reserves by about 10% and Malaysia gained about 3%. Smaller crude reserve additions were made in Japan, Malaysia, New Zealand, and Thailand, among others.

China, Indonesia, Malaysia, and Australia have made substantial natural gas reserve additions over the last 5 years. Australia has done well with its coal reserve additions. While China has been developing its vast coal reserves, its reserve position has not changed substantially since 1980 according to estimates in the International Energy Annual of the U.S. Department of Energy.

Several countries are showing declines in their reserves. The Philippines had a sharp 30% decline in reported crude and natural gas reserves from 1984 to 1985. Australia's crude reserves have dropped substantially over the last 5 years.

The region has an estimated 5 to 6% of the world's proven crude oil and natural gas reserves. The share in the world's proven coal reserves is about 20%, with most of the potential for further development in China.

### Energy Production

A comparison of production during the first four months in 1986 with the same period in 1985 indicates that crude energy production is about 5% higher than the previous year. Small gains were made in almost all of the countries except Australia, Taiwan, and the Philippines.

Natural gas production was about 50% higher in China and Indonesia for the first 4 months in 1986 than in the same period in the previous year.

Coal production increased in China by about 50 million tonnes and in Australia, by another 20 million tonnes.

### Exploration Activities

Though slowing, the pace of exploration activity exceeds that of any other region in the world. During 1986, countries in the Southeast Asia area are expected to drill about 8300 wells. Most of these wells (about 6900) are in China, and about one-half are dedicated to the search for crude oil. The footage drilled by China increased by more than 25% during 1985. There is a renewed interest in onshore exploration. In contrast, Australian well activity has declined by almost one-half with the lower oil prices and taxes on offshore production.

## ENERGY DEMAND AND FUTURE FORECASTS

In their energy resource development programs, countries in this region have tried to achieve a more diversified energy mix, to develop their own indigenous energy supplies, and to move away from heavy dependence on imported crude oil.

China is making great strides toward a full development of its own energy resources. Australia is in a favorable position with one of the more diversified energy mixes in the region.

The more industrialized nations—Hong Kong, Japan, Taiwan, and Singapore—have little or no natural resources. Their economies are highly vulnerable to shifts in the prices of imported energy, as well as supply disruptions. As a results, these countries are much more actively pursuing nuclear development.

### Forecasts on Energy Demand

In line with slowing rates for industrial development in the future, the forecasts for energy demand are around 2 to 3% for Australia and Japan. Higher rates are still forecasted for New Zealand, 4 to 5%; Philippines, 5 to 6%; Korea and Thailand, 6 to 7% over the next decade.

A United Nations survey of the energy development plans of some selected countries indicates a continued trend toward greater utilization of natural gas, coal, and nuclear energy. In about 5 to 6 years, Japan and Korea plan to have nuclear supplying more than 10% of their energy.

## CAPITAL INVESTMENTS IN ENERGY PROJECTS

China has made the most substantial investments in its energy development among the Southeast Asia countries. Many of its oil exploration programs continue with participation from the international companies. Foreign participation is seen as an essential element to lift China's oil production 60% by the year 2000 from its current 2.4 million b/d. Western groups have invested more than $1.7 billion in exploration since China began operations in 1979, according to the February 1986 Petroleum Economist. Many other countries are looking at incentive programs to attract foreign investment.

### Major Energy Projects

Heavy investments in export coal mines in 1980-81 increased Australia's coal reserves and put it in the position of being the world's largest coal exporter. Natural gas development has been increased from the northwest shelf over the last 5 years.

By mid-1985, China had 130 mines under construction with seven open cast mines in the northwest developed with foreign assistance, including a $250 million loan arrangement with the World Bank. More than 23 foreign operations have participated in exploration in the South China and Yellow seas and for natural gas deposits off of Hainan Island. Seven thermal power plants are under construction, with plans for an additional 5820 mW of capacity over the next 5 to 10 years. Six nuclear power plants will be built with a combined generating capacity of 10,000 mW over the next 15 years.

Many large redevelopment projects in Indonesia have been rescheduled because of declines in oil revenues. Additional financing of $300 million has been arranged for its tenth power plant.

South Korea's power plants are the largest budgeted projects, including 17 power plants to be built through 1996. Six of the plants will be nuclear, bringing Korea's total to 10 plants. A massive liquefied natural gas storage facility is also under construction.

New Zealand has the world's first commercial-size gas-to-gasoline plant. The plant will meet one-third of New Zealand's forecasted gasoline consumption in 1986. The investment is approximately $1.5 billion.

Oil and gas projects in Thailand have drawn the highest investment funding, with a 350 million $ft^3$/day gas separation plant. Thailand plans to double its present refining capacity in 3 years and to rehabilitate the Bangchak Oil Refinery.

## CONCLUSION

The future for oil and gas exploration should continue to be bright in the Southeast Asia area. The trends toward more widely diversified programs for energy utilization will engender heavier capital investments.

The only drawbacks are current economic conditions affecting many of the nations and the impact of lower oil prices on funds that have been traditionally available for oil and gas exploration. There are several factors which will continue to influence the environment:

(1) Many major oil companies have cut back on their exploration programs;
(2) The current growth in debt in these countries could impact the availability of capital in the future, particularly in those countries with high debt service ratios;
(3) Many major energy projects are now in electrification, nuclear projects and refineries, rather than energy exploration and development.

On the more favorable side, these countries need more energy exploration to meet the demands with their economic growth. As they experience the growth, the capital should be available for further expansion, since each must meet higher-than-average growth rates. As industrialization continues, the population in the area will have increasingly higher demands for consumer products, particularly automobiles, which will only add to their energy requirements in the future.

CHAPTER 14

# GEOTECTONIC REGIONALIZATION AND EVOLUTION OF THE EAST CHINA SEA AND ITS ENVIRONS

**Li Yinhuai**
*Institute of Geology*
*Academia Sinica, Beijing, People's Republic of China*

## INTRODUCTION

The East China Sea (ECS) and its adjacent sea areas is one of several major marginal basins developed in back of Pacific island arcs. Great attention has been paid by Chinese and foreign geoscientists to the tectonic characteristics and evolution of this region. It seems to be one of the key problems in our understanding the relationship between East Asia continent and West Pacific tectonics.

Since the 1970s, extensive geophysical and geological surveys have been made in the ECS and its adjacent areas, large amounts of valuable data have been obtained, and a division of tectonic units of this region has been made (Wageman et al., 1979; Jin and Yu, 1981; Feng and Zheng, 1982). These studies provided a useful scientific basis for the exploration and assessment of oil and gas resources. Nevertheless, there are many debates and poorly understood problems on the tectonic characteristics and dynamic mechanism of this region. The purpose of this paper is to provide a general review and preliminary discussion of these problems and to draw further attention to them.

## REGIONAL GEOLOGICAL AND GEOPHYSICAL SETTINGS

### Outline of Inland Geology

ECS and its adjacent sea areas are surrounded by the southeast China continent and Taiwan Island, the southern part of the Korea peninsula, southwest Japan, and the Ryukyu Islands. The formation and evolution of the geotectonic framework of the studied sea areas are closely related to that of the on-land tectonics.

In southeast China, the oldest rocks known are the Jian'ou Group. Its contact with Sinian metamorphic rocks is disconformable in the Sanming area of the Fujian province. The latter are conformably related to early Paleozoic low-grade metamorphic rocks called the Luofengxi Group. In the Ninghua and Qingliu areas, Devonian deposits of platform-cover type, the Nanjin Group, overlie the Luofengxi Group with angular unconformity. A similar situation is also seen on Hainan Island, where Sinian metamorphic rocks and the early Paleozoic Shilu Group of low-grade metamorphic phase are overlain unconformably by Devonian quartzite. Apparently, these unconformities are all the results of the Guangxi orogeny. In the Yunkai Mountain area, Eozoic-early Paleozoic geosynclinal flysch sediments are intercalated with basic, intermediate, and acidic volcanic rocks. These rocks have been strongly deformed and folded. This region was affected by the Yunan orogeny at the end of the Cambrian and by the Guangxi orogeny at the end of the Silurian.

In the southern part of the Korea peninsula (Sung, 1983), there are various schists and gneisses of the Ryong Nam Group and the Ryul Ri Group which have K-Ar isotopic ages of 1880 Ma. They are presumed to be of pre-Eozoic age.

The Ogcheon System in Korea is roughly equivalent to the Sinian System in China. The lower part of the Ogcheon System consists mainly of schists, and the upper part consists of glacioaqueous conglomerate-slate and flaglimestones. The Joseon System of early Paleozoic age is mainly limestones. The Pyeongan System of late Paleozoic age is composed of terrigenous clastics. The absence of Silurian and early Carboniferous sediments between the two above-mentioned systems indicates the effect of the Caledonian epeirogenetic event in that area.

In southwest Japan (Kizaki and Oshiro, 1977), Permian conglomerates contain gravels originally from the Hida Gneiss which have a K-Ar isotopic age of 1500-1700 Ma. The oldest fossiliferous strata are Silurian deposits of the platform type. Thick Devonian to Triassic deposits are of the geosyncline type. They are commonly associated with basic and ultrabasic rocks between the Median Tectonic Line and the Butsuzo Tectonic Line.

On both sides of the Median Tectonic Line, the post-Triassic sedimentary facies were greatly changed. On the northern side, the late Triassic to Jurassic sediments are the shallow, enclosed bay facies and nonmarine coal-bearing clastic deposits. The coal-bearing deposits lie unconformably on the older strata. This unconformity is a result of the Honshu orogeny. On the southern side, the late Triassic to early Miocene strata are of the geosyncline type, called the Shimando Group, which consist of turbidite sediments, cherts, limestone, basalt, and pillow lava. The Miyazaki Group of middle-late Miocene age unconformably overlies the Shimando Group. This unconformity is named the Takachiho orogeny (Ishiwada, 1981).

The basement rocks of the Ryukyu Islands are composed of green schists and low-grade metamorphic rocks such as clay slate, chert, slightly metamorphosed sandstone, limestone, and tuff. The low-grade metamorphic rocks contain fossils which indicate the existence of Permian, Triassic, and Cretaceous strata.

The Danan'ao Group in the central Taiwan Mountains yields Permian fossils (Chai, 1972). This group is composed mainly of schists believed to be of late Paleozoic to Triassic age. The Pihou Formation occurs on both sides of the Central Mountains. Lithologically, it is similar to the early Cretaceous to Paleocene fossiliferous strata found in boreholes in the Beigong area. On the eastern side of the Central Mountains, an extensively distributed Eocene conglomerate formation unconformably overlies the Pihou formation. This unconformity is called the Taiping orogeny.

The ridge and most of western flank of the Central Mountains are covered by an Eocene and early Miocene clay slate formation. Neogene strata are extensively distributed west of the Central Mountain ridge. This formation is composed of nonmetamorphic marine sandstones interlayered with shales. Its thickness reaches more than 8000 m.

The Taitung coast range is composed of Neogene sediments of the eugeosyncline type. According to Chai (1972), the lower Miocene Tulanshan Formation, which consists mainly of andesite lava and volcanic clastics, is the oldest deposit that outcrops. The Upper Miocene Takangkou Formation, which conformably overlies the Tulanshan, is composed of limestones and conglomerates. The Pliocene Chimei Formation is a series of typical turbidite deposits consisting of alternating sandstones and shales. In the southwest Taitung coast range, the Pliocene Lichi Formation is an ophiolite-mélange complex composed of unstratified mudstones and exotic blocks. This unconformity is termed the Suiren orogeny.

## GEOPHYSICAL CHARACTERISTICS OF THE SEA AREA

Wageman et al. (1979) and Bosum et al. (1979) have presented geomagnetic data indicating that the geomagnetic features west of the western boundary fault on the East China Sea shelf (ECSS) are consistent with those in the South China region and the southern part of the Korea peninsula. It is therefore suggested that this area is an offshore extension of the latter two areas. To the east of the western boundary fault on the ECSS, is a geomagnetic feature representing a gentle negative anomaly belt. The average depth of magnetic basement ranges from 4 to 6 km, and the deepest part may reach 10 km. On the outer margin of the ECSS, however, the positive and negative magnetic anomalies vary sharply, indicating that, in addition to folded sedimentary rocks, there were lava eruption and granite intrusions. The geomagnetic field in the Okinawa trough is characterized by more uniform positive and negative anomalies.

According to Chen et al. (1981), in the northern South China Sea Shelf (SCSS) north of the Beiwei-Shenhu fault zone, the geomagnetic field is characterized by gentle magnetic low anomalies that trend northeastward. This indicates that the orientation of major structures in this area is similar to that in southeast China.

On a seismic-refraction profile across the East China Sea Shelf basin (ECSSB) it can be seen that (Figure 1) the acoustic basement has a P wave velocity ranging from 5.8-6.8 km/sec. The sedimentary layers with P wave velocity of 5.2-5.4 km/sec and 3.6-4.4 km/sec have been compressed and folded on the buried uplifts. Sedimentary layers with a P wave velocity

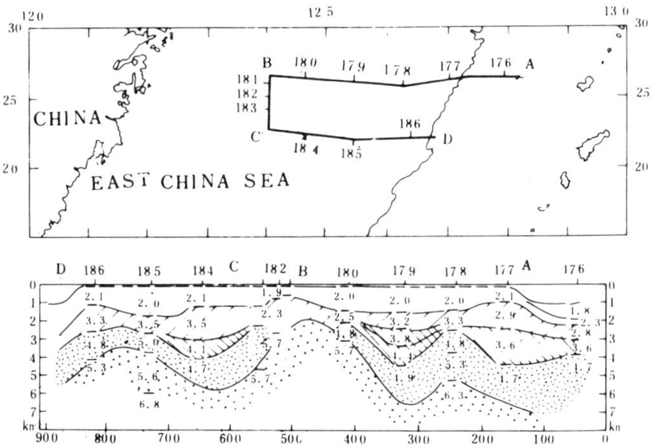


Figure 1. Seismic-refraction profile across East China Sea ECS (after Leyden and Murauchi, 1973).

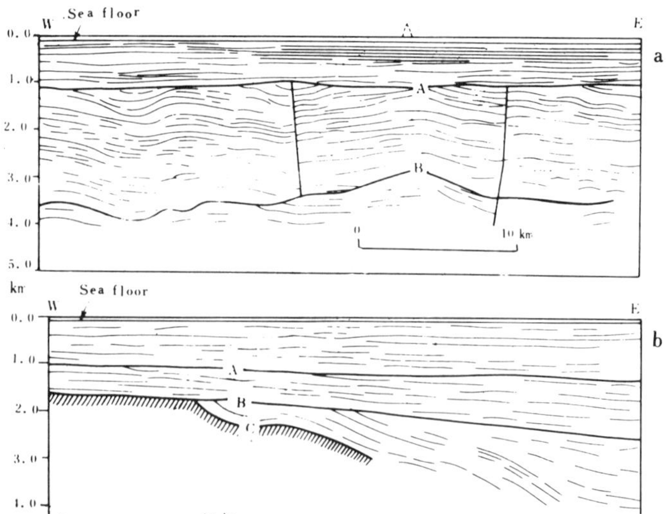


Figure 2. Seismic profile across an area (near lat. 30°33 ′N) south to Cheju Island (after Ishiwada, 1981).

of 3.6-4.4 km/sec are usually absent. All the above-mentioned layers are overlain unconformably by the layers with P wave velocity of 1.8-3.5 km/sec.

From seismic-reflection data the sedimentary layer with a P wave velocity of 1.8-3.5 km/sec can be subdivided into layers separated by an angular unconformity, as Leyden and Murauchi (1973) and Ludwig et al. (1973) suggested. The lower layer, with a P wave velocity of 2.2-3.6 km/sec, has been strongly compressed and folded. Its thickness may reach more than 2000 m at many locations. The upper layer shows a P wave velocity of 1.8-2.1 km/sec. Its thickness is about 1200 m. This layer is essentially undeformed and almost horizontally overlies the older sediments. Its thickness is gradually decreased landward, but the synchronous sediments in the Okinawa trough and on its slopes have been deformed and ruptured by faults.

By comparing the above-mentioned velocity layers with drill data from Taiwan Island, the age of these velocity layers may be determined roughly as follows: The formation with a P wave velocity of 5.8-6.8 km/sec is of pre-Eocene age. The layers with a P wave velocity of 5.2-5.4 km/sec and 3.6-4.4 km/sec are of Eocene-Oligocene and early Miocene age, respectively. The layers with a P wave velocity of 2.2-3.5 km/sec and 1.8-2.1 km/sec are of middle-late Miocene and Pliocene-Pleistocene age, respectively.

## NEOGENE OROGENY IN EAST CHINA SEA

Seismic-reflection data (Figure 2) show clearly that there are three unconformities (A, B, and C). The sediments (3rd layer) above unconformity A are of Pliocene-Pleistocene age. Sediments (2nd layer) between A and B, as well as sediments (1st layer) between B and C, are of middle-upper Miocene and Eocene-lower Miocene age, respectively.

The lower part of the 3rd layer has been subjected to weak deformation and tilting, but the upper part is an almost undisturbed horizontal layer.

The 2nd layer has been relatively strongly deformed and folded (Figure 2a). Figure 2b shows an area west of Figure 2a, where the basement outside the continent is rigid. The 2nd layer between boundaries A and B has an almost even-layered structure and does not differ structurally from the 3rd layer.

The 2nd layer in the central part of ECS has been strongly compressed and deformed (Figure 3) and is usually formed into a series of accompanying bar-type folds and reverse faults.

Unconformity A on the ECSS is a nearly horizontal and even surface, and unconformity B is also an eroded to different extent surface. It indicates that there was a peneplanation after these two orogenies. Geological data from the southwest Japan islands indicate that the tectogenesis reflected by unconformity B corresponds approximately to the Takachiho orogeny along the Shimando belt. The tectogenesis reflected by unconformity A is named the Longjing orogeny.

Results of the Takachiho orogeny are obvious along the northwestern coast of Kyushu Island. Here, the early Miocene Sasebo Formation consists of shallow-marine interlayers of sandstones and shales intercalated with many coal seams, and overlain unconformably by the late Miocene Noshima Formation in the western part of the Katamatsu coal field.

The Shimajiri Group distributed on the Kikai, Okinawa, and Miyako Islands corresponds to the Miyazaki Group. The

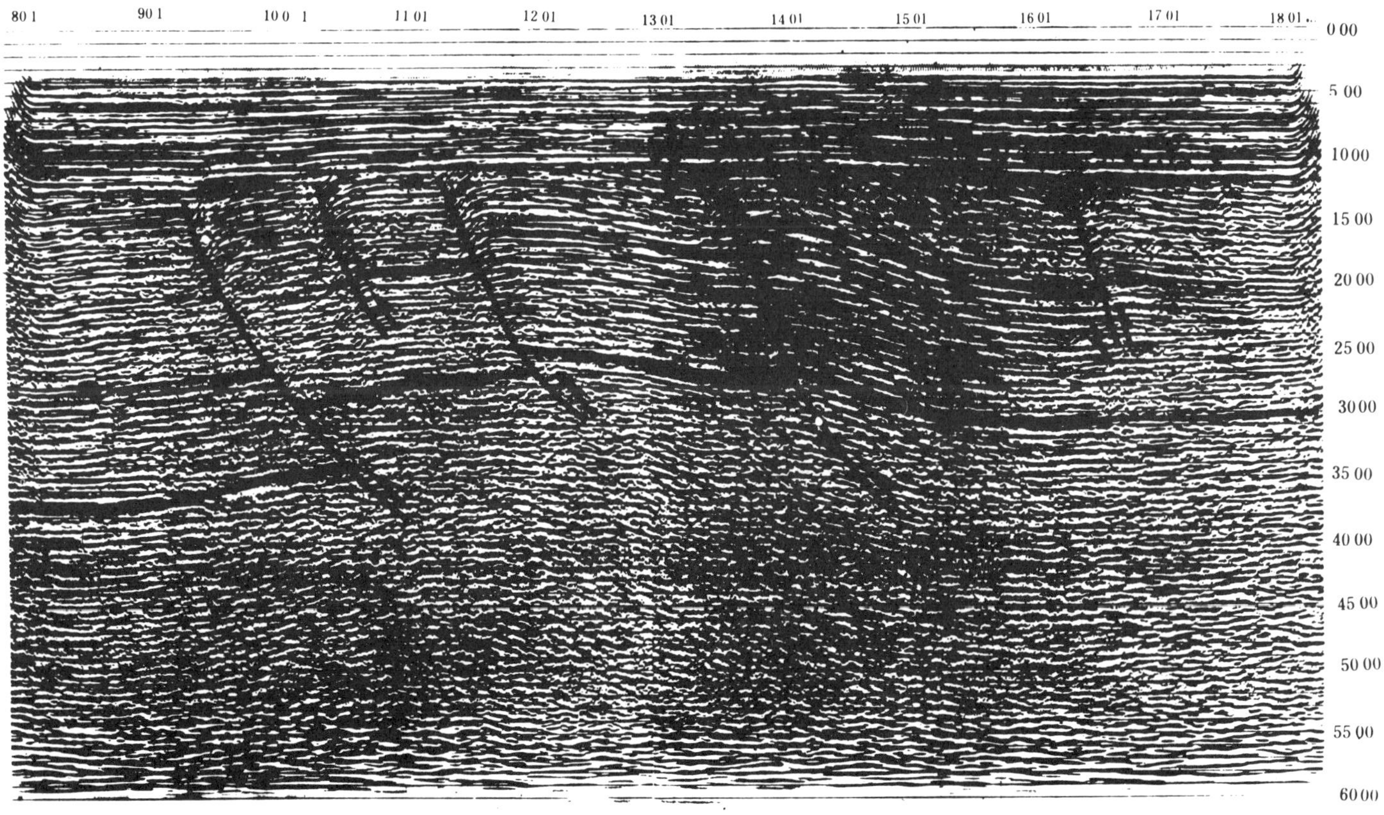


Figure 3. Seismic profile across bar D 692 in eastern Zhejian Province.

Okinawa hole No. 1 in the Shimajiri basin reveals an unconformable contact between the Shimajiri Group and the underlying Eocene-early Miocene sediments (Aiba and Sekiya, 1979). This unconformity must be a reflection of the Takachiho orogeny.

The Longjing orogeny was also intense on the Ryukyu Islands and southwest Japan. The absence of post-Miocene deposits on the Ryukyu Islands indicates that all of the Islands were land at that time. The formation of the old lands was accompanied by the eruption of a large amount of tuffs and by granite intrusions. The age of the Kume-Jima green tuffs determined by K-Ar dating is 17.7, 12.6 and 6.5 m.y., and the granites from Ishigaki-Shima and Osumi Islands have a K-Ar age of 12-22 m.y. They are of late Miocene age. The green tuffs and granites of this period are also widely distributed in the northwestern part of Kyushu, Goto Retto, and the Tsushima Islands.

The above-described data sufficiently indicate that the Takachiho and Longjing orogenies have occurred continuously in ECS, Ryukyu, and southwest Japan, and that their extent and intensity were quite large.

## GENERAL OUTLINE OF THE GEOTECTONIC REGIONALIZATION

ECS and its environs may be geotectonically divided into several tectonic belts based on the above-mentioned geological and geophysical data and in accordance with principles of geotectonic regionalization adopted by the authors for the period of intense folding by transformation of geosyncline into platform (Figure 4). A brief description of these belts follows.

### South China-Ogcheon Tectonic Belt

This belt extends northeasterly along the southeastern coast from Guangdong, Fujian, Zhejian, through Hupijiao and Cheju Island, to the southern part of the Korea peninsula. It includes part of ECSS west of the fault belt along the western border of ECSS and a wide shelf in the northern part of SCS north of the Beiwei-Shenhu fault belt.

This belt consists mainly of Caledonian folded basement and Devonian to Triassic sedimentary cover. The belt entered in a new geotectonic stage at the end of the Triassic. It is that the Indosinian and Yenshanian orogenies have succeeded Songrin and Daebo orogenies in southern part of Korea. These orogenies caused a series of NE-NNE-trending faults and downfaulted basins to be formed, and a large amount of intermediate-acidic volcanic eruptions and granitic intrusions have taken place.

Late Cretaceous-early Tertiary red mollasse deposits commonly have developed in the downfaulted basins. A set of Cretaceous terrigenous sediments and volcanic rocks 10 km thick was deposited in the Gyeongsang downfaulted basin (Inoue, 1981). The magmatism in the Pusan downfaulted basin took place from the Cretaceous to the late Tertiary-Quaternary.

### Southwest Japan Tectonic Belt

This belt has been bisected by the Median Tectonic Line into two zones. The inner zone (close to the Japan Sea) consists of metamorphic basement composed of relict Precambrian continental block and metamorphic basement of Indosinian folding. This basement is overlain unconformably by late Triassic and Jurassic coal-bearing deposits. The Honshu orogeny was accompanied by the formation of late Triassic and early Jurassic granite batholiths. Muscovite from the Hide Gneiss gives its K-Ar age of 185-214 m.y. This age may be the age of metamorphism caused by this orogeny.

The early Cretaceous volcanic eruptions and granite intrusions of 60-120 m.y. and the unconformity between late Cretaceous molasse and its underlying older deposits all indicate that the Sambakawa orogeny was considerably strong, and resulted in the formation of the Ryoke high-T and low-P metamorphic belt and the Sambakawa low-T and high-P metamorphic belt.

The outer zone (close to the Pacific Ocean) is composed of the Himalayan (Takachiho orogeny) fold basement of the Shimando Group. The basement has been intensely compressed, deformed, and faulted to form a series of reverse folds toward the Pacific Ocean and imbricate thrust faults.

### Ryukyu Arc Tectonic Belt

The Ryukyu arc tectonic belt extends for 1200 km from the southern end of Kyushu Island in the north, to the northern end of Taiwan Island in the south. It is a geomorphically sharp arcuate zone convex to the southeast and is located between the Okinawa Trough and the Ryukyu Trench. It consists of several tectonogeomorphic units, such as a parallel-oriented inner arc (Satsun-Shoto volcanic arc), an outer arc (Ryukyu front arc), and the Ryukyu Trench.

The inner arc is a chain of volcanic islands, such as Satsum-Shoto, Okinawa, and Iojima Islands, which are composed of early Quaternary pyroxene andesites.

The outer arc has been cut into two zones by the Tengan fault. The inner zone has been cut into three subzones—Ishigaki-Shima, Motobu, and Kunigami—by the Omoto tectonic line and the Hento fault. Late Paleozoic low-grade metamorphic rocks are exposed mainly in the Ishigaki-Shima and Motobu subzones, but the Mesozoic deposits are distributed in the Kunigami subzone. The outer zone, named the Shimaziri zone, has a folded basement composed of Eocene-early Miocene sandstones, shales, and basic lava. The basement is overlain unconformably by the upper Miocene-Pliocene Shimaziri Group. This unconformity must be a reflection of the Takachiho orogeny.

### Taiwan Tectonic Belt

This belt consists of the Taiwan Central Mountain west of the Taitung longitudinal valley and the western Taiwan plain. This is a Himalayan fold belt composed of tectonic layers of metamorphic basement of the Tanan'ao Schist and the Bihou Formation, and a Tertiary fold tectonic layer.

The Central Mountain consists of thick Eocene-early Miocene argillaceous slates. The Eocene sediments are absent in the western plain, but the thick Neogene sediments are overlain unconformably by Paleocene or Cretaceous low-grade metamorphic rocks which thin uniformly from east to west (Chai, 1972). This indicates that the metamorphic basement formed during Taiping orogeny is characterized by large relief, with uplift in the west and depression in the east.

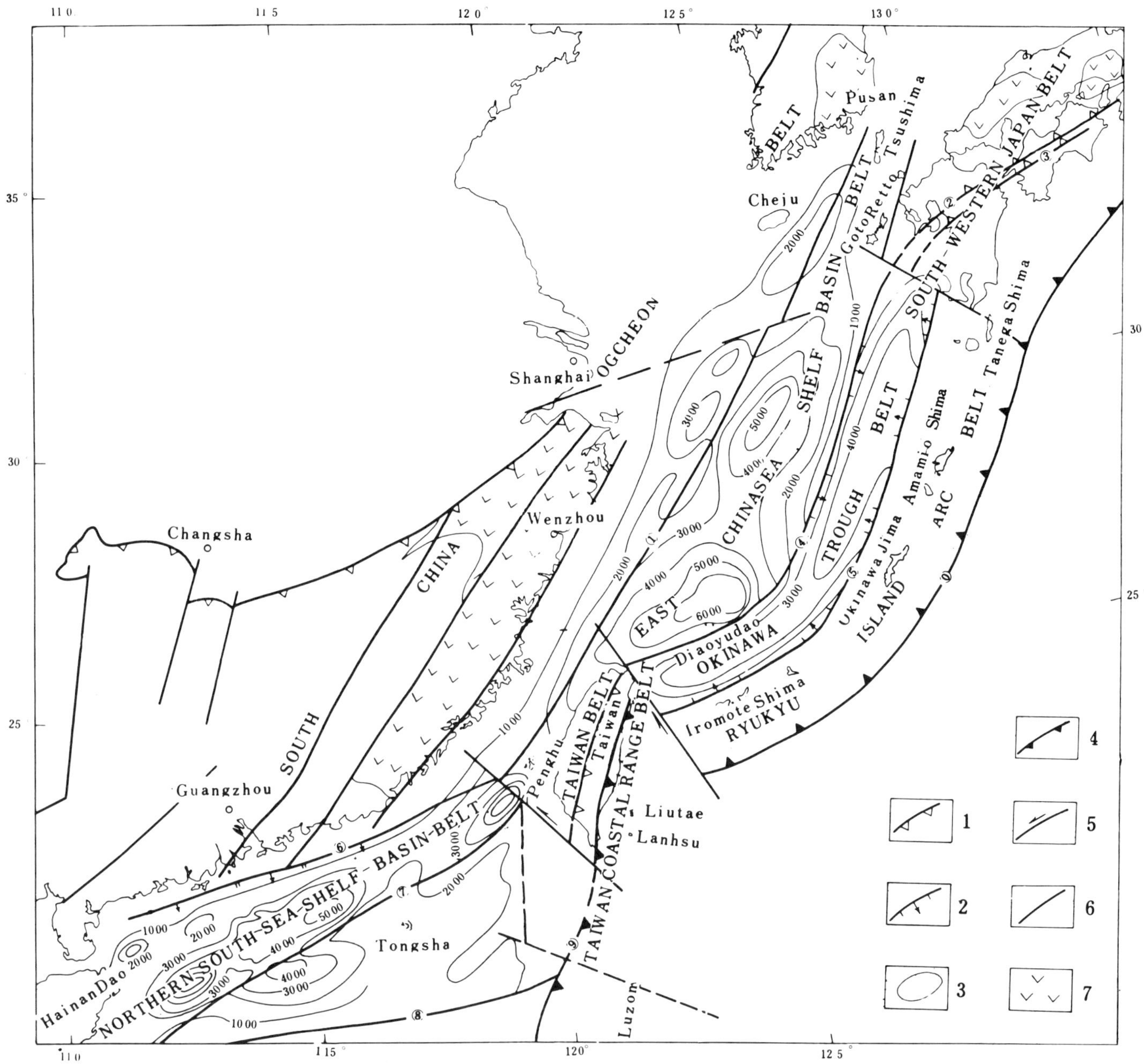


Figure 4. Geotectonic regionalization of the East China Sea and its environs (1) Fault belt along the western border of the East China Sea shelf; (2) Median tectonic line; (3) Butsuzo tectonic line; (4) Fault belt along the outer border of the East China Sea shelf; (5) Fault belt along the eastern border of the Okinawa Trough; (6) Great fault along the northern border of South China Sea Shelf Basin (SCSSB); (7) Beiwei-Shenhu fault belt; (8) Great fault along the southern border of SCSSB; (9) "A-type" subduction zone along the Taitung longitudinal valley-Manila Trench; (10) "B-type" subduction zone along the Ryukyu Trench.

The Toukoshan Formation on western Taiwan is a series of terrigenous piedmont molasse cumulus. It reflects the folding reversion of the Tertiary geosyncline on Taiwan and the start of regression from western Taiwan. The accumulation of the molasse deposits was followed by the onset of the Taiwan orogeny. This is confirmed by a general unconformity between late Pleistocene laterite and the underlying Toukoshan Formation.

## Taitung Coast Range Tectonic Belt

The Taitung belt is an independent tectonic unit adjacent to the western Pacific basin. A narrow Taitung longitudinal valley separates it from the Taiwan tectonic belt. Geological and geophysical data for Taiwan suggest that the Taitung coast range consists of oceanic crust (Chai, 1972). The Liji Formation is a relict fragment of ruptured oceanic crust formed by tectonism. Analogues to this relict fragment of oceanic crust are distributed along the Zambales Mountains on the western coast of Lusan Island.

Seismological data (Katsumata and Sykes, 1969) also show that the Taitung longitudinal valley separating the Taiwan tectonic belt with continental crust from the Taitung coast range with oceanic crust is a Neogene "A-type" subduction zone underthrust toward the Pacific Ocean.

The topographic map of the sea bottom from Taiwan to

Lusan shows that this "A-type" subduction zone extends southward and is approximately connected with recent eastward underthrusting of the Manila Trench. These areas must have been connected by a subduction zone in Neogene time.

## Tectonic Belt of East China Sea Shelf Basin (ECSSB)

This belt lies between the western border fault and the outer border fault of ECSS (Jin and Yu, 1981). It extends from Taiwan Beach in the southwest to Tsushima Island in the Korea strait in the northeast.

Drill data from the Penghu Islands indicate that a series of metamorphic clastics occurs under the Miocene unconformity. It is lithologically similar to the Cretaceous-Paleocene sediments in the Beigan area of Taiwan (Chou, 1969). On Diaoyudo Island, gravel sandstones with coal layers overlain unconformably by Ryukyu limestone are similar to the lower Miocene Yaeyama Group on the Yaeyama Islands of Japan and can be comparable with the latter. Seismic sounding suggests that the Diaoyudao Island and its neighboring areas are a gently tilted and extensive fold belt (Aiba and Sekiya, 1979).

On Tsushima Island in the Korea strait, the Eocene-early Miocene Tomishiro Group consists of thickly interlayed sandstone and shales, intercalated locally with carbonaceous shales and volcanic clastics, which have been intensely compressed and deformed to form a series of NE-trending folds (Inoue, 1981).

The above-described geological and seismological data show that this is an extensive Himalayan fold belt composed of tectonic layers of Cretaceous-Paleocene low-grade metamorphic basement, fold tectonic layers of Eocene-lower Miocene basement, and fold tectonic layers of middle-upper Miocene cover. The belt extend southwesterly, connects with the Taiwan tectonic belt, and then extends northerly through Gotc Reto into the area of Tsushima Island.

## Okinawa Trough Tectonic Belt

The Okinawa Trough is a geomorphologically valley-shaped trough which is deep in the south and shallow in the north. It is composed of complex metamorphic rocks which have a seismic velocity of 3.5-5.5 km/sec. Its age is presumably Eocene-early Miocene. Late Miocene and recent volcanic clastics with their seismic velocity of 1.8-3.1 km/sec unconformably overlie the basement (Leyden and Murauchi, 1973).

Geological data from hole To-KA1 (Aiba and Sekiya, 1979) on the northern segment of the Okinawa Trough suggest that the sediments and volcanic clastics and lava (K-Ar age of 6 m.y.) on the Eocene-early Miocene acoustic basement are overlain unconformably by a series of Pliocene-Pleistocene clastic sediments.

Several recent normal faults parallel to the trough have been found along both the eastern and western flanks of the entire Okinawa trough. The fault on the eastern flanks has caused striking offset and tilt of the strata. At some locations, the overlying sediments have been folded and warped by intrusion of magmatic bodies (Wageman et al., 1979). In the central part of the trough, a large ridge of biotite-quartz diorite and metamorphosed pillow lava basalts has been determined to be of late Miocene-Pleistocene age (Herman et al., 1979). It indicates that the Okinawa trough began to be pulled apart in late Miocene, and continues until now.

## Northern South China Sea Shelf Basin Belt

This belt is located in an area of northern SCSS north of the Beiwei-Shenhu fault belt, from Hainan Island in the southwest to Taiwan Beach in the northeast.

The previously described gravity and magnetic data indicate that the features of the magnetic field here are similar to those on the southeastern China coast. It can be seen on the isopach map of upper Cretaceous-lower Tertiary deposits in northern SCS compiled by Zhang (1981) that the northeast-trending Cathysian tectonism occurred not only in the south China Caledonian tectonic belt, but also in the northern SCS. These must be one geotectonic unit—a Caledonian fold tectonic belt.

A large amount of drilling and geophysical data obtained in recent years suggest that the northern SCSSB is a large Cenozoic sedimentary basin formed on basement of the Caledonian fold belt, which has been intensely reworked by the Yanshanian movement. The late Cretaceous-Tertiary deposits have completely developed in the basin. They are generally continuous sediments except for a sharp sedimentary hiatus or overlap unconformity between the upper and lower Tertiary sediments. This is quite different from the orogeny that occurred successively in Miocene time on the ECSS. Here, the late Cretaceous-early Tertiary fault activity was intensive, and commonly led to the formation of a series of northeast-trending downfaulted basins coupled with uplifts, which are characteristic of tectonic development of a continental rift pattern. The thick late Cretaceous-Paleocene sediments in the downfaulted basins are mainly terrigenous; these sediments are thin or absent on the uplifts. Since the Neogene, the basin sediments gradually changed from early local terrigenous to extensive marine sediments, forming three distinctive sedimentary stages of filling-overlapping-covering.

# MODEL OF GEOTECTONIC EVOLUTION

Deep crustal structure in the ECS and its environs is characterized by no large change in Moho relief. Under the Ryukyu arc, the Moho surface has been determined to be 30 km deep; it rises easterly to 19 km under the Ryukyu Trench. The Moho surface under the Okinawa Trough is 28 km deep; westward, under the ECSS, it again lowers to a depth of 30 km, and gradually deepens landward. The correlation of uneven structure of the Moho relief and superficial structure indicates a close genetic relationship between them.

A model of the formation and evolution of the geotectonic framework of the ECS, which has been constructed in accordance with the geotectonic dynamic model of pull-apart ocean and compression-formed continent (Li, 1984), is presented in Figure 5.

Recent drilling data suggest that plagioclase gneiss which unconformably occurs under the Paleocene marine sediments in the ECS east of Wenzhou city has a metamorphism age of 1680 ± 180 m.y. This age is close to the metamorphism age of Hida gneisses (1500-1700 m.y.) in Japan and to that of basement (1880 m.y.) under the Sobaeksan uplift in southern Korea, and their metamorphic facies are very similar. It is

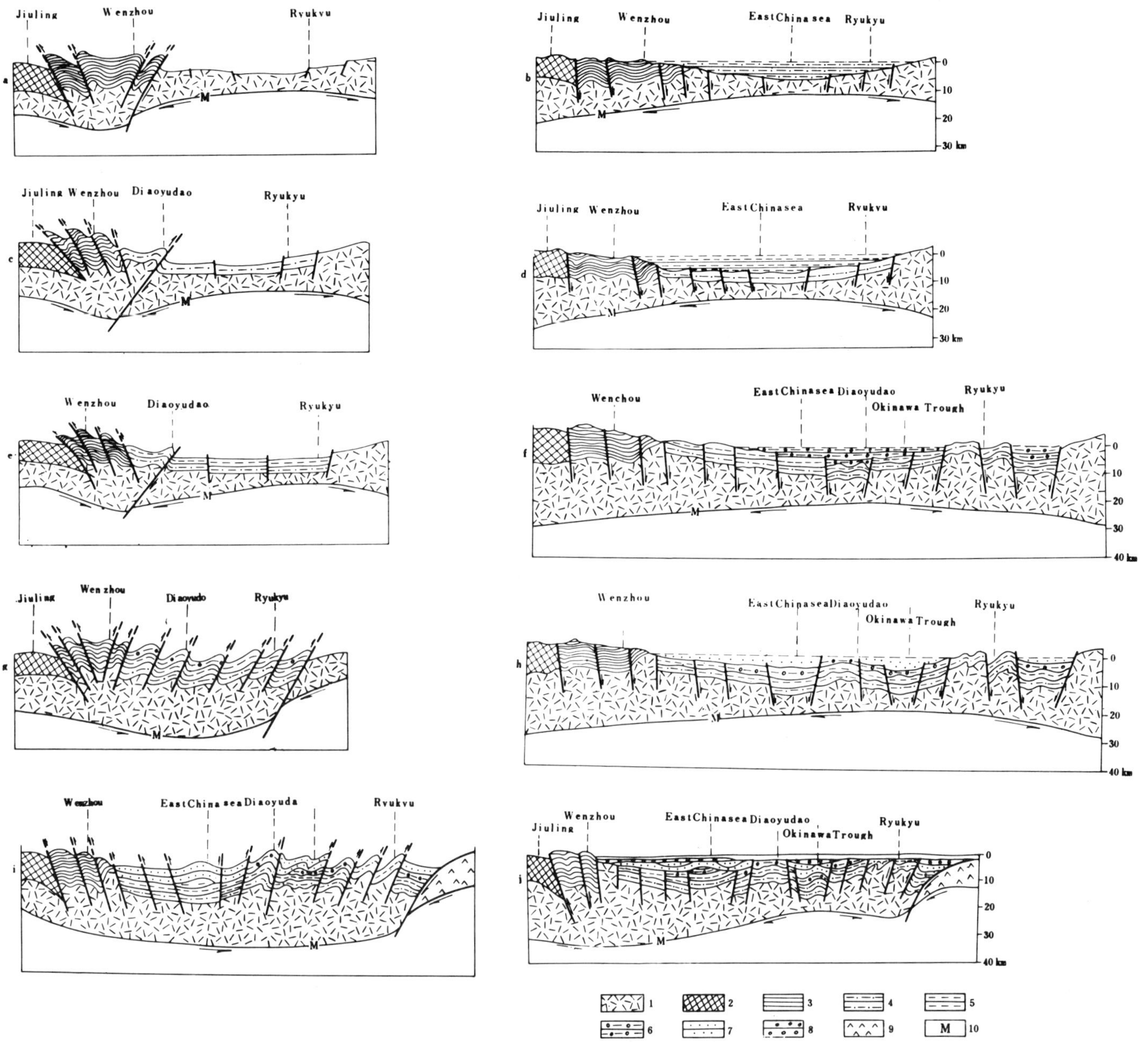


Figure 5. Model of formation and evolution of geotectonic framework of East China Sea and its environs; a) Caledonian orogenic period; b) Devonian-Triassic tectonic stage; c) Indosinian orogenic period; d) Late Triassic-early Cretaceous tectonic stage; e) Taiping orogenic period; f) Eocene-early Miocene tectonic stage; g) Takachiho orogenic period; h) Middle-late Miocene tectonic stage; i) Longjing orogenic period; j) Pliocene-Pleistocene tectonic stage. 1) Metamorphic basement of Zhongtiao period (1700-01800 m.y.); 2) Tectonic layer of Jiuling cycle; 3) Tectonic layer of Caledonian cycle; 4) Indosinian fold tectonic layer; 5) Yenshanian fold tectonic layer; 6) Early Himalayan fold tectonic layer; 7) Late Himalayan fold tectonic layer; 8) Pliocene-Pleistocene sedimentary cover; 9) Oceanic crust; 10) Moho surface.

inferred that the gneisses may constitute a unified metamorphic basement of the Zhongtiao period (Figure 5a). The basement has disintegrated since the Devonian; subsequently a wide area between the fault belt along the western border of the ECSS and the Ryukyu Trench-Taitung longitudinal valley was developed into a small ocean basin of the marginal rift type, with an extension pattern in the late Paleozoic, Mesozoic, and Cenozoic stages (Figure 5b, d, f, h,j). It has been developed by the multistage compressional movement in the Pacific orogenic cycle (Honshu, Sambagawa, Taiping, Takachiho, Longjing, and Taiwan orogenies) and gradually closed to form a recent geotectonic framework (Figure 5c, e, g, i).

Analysis of the model of formation and evolution of the geotectonic framework of the ECS and its environs made it possible to further realize that there is a landward underthrust Indosinian and Yenshanian paleosubduction belt approximately along the fault belt on the outer border of the ECSS and the Beiwei-Shenhu fault belt. A Mesozoic volcanic belt in the southwest Japan inner belt, in the southern part of the Korea peninsula, and along the southeastern China coast may have a certain genetic relationship to the underthrust and consumption of the paleosubduction belt.

Moreover, another Himalayan recent active subduction belt runs along Ryukyu Trench, Taidong longitudinal valley, and

Manila Trench. Seismological and geological data suggest that the subduction belt at Ryukyu Trench is a landward underthrust B-type subduction belt which caused the southwest Japan outer belt and Ryukyu Island arc to form a series of reverse folds and imbricate thrust faults coincident with the underthrust direction. Late Miocene green tuffs and the Quaternary volcanic belt developed widely on the Ryukyu Islands may have a close genetic relationship with the underthrust and consumption of the B-type subduction belt. But the A-type subduction belt underthrust toward the Pacific Ocean occurs along the Taitung longitudinal valley-Manila Trench. The formation of the Quaternary volcanic belt on Liutae and Lanyu Islands east of Taiwan may have resulted from the underthrust and consumption of the A-type subduction belt.

Two tectonic units with different types of the crust on Taiwan have fitted in Pliocene time, the buoyancy of Taiwan continental crust resisted the A-type subduction belt to continuously underthrust, resulting in fold reversion of Tertiary geosyncline on Taiwan, and in formation of a series of reverse folds and imbricate thrust faults coincident with the underthrust direction.

The crustal movement pattern with alternation of multistage extensional subsidence and compressional underthrust in the Pacific geotectonic cycle is a significant characteristic in geotectonic evolution of the East China Sea and its environs. The characteristic of geotectonic evolution has been controlled by deep structure; that is, a correlation of extensional subsidence and the mantle uplift and an association of compressional underthrust and mantle depression, which were the main factors for tectonic movement. The author considers that the geotectonic framework of the region is a result of the uplift and depression of the mantle.

## ACKNOWLEDGMENTS

I would like to express my cordial thanks to Prof. Zhong Dalai and Dr. Zhao Xitao for valuable discussion.

## REFERENCES

Aiba, J.I., and E. Sekiya, 1979, Journal of the Japanese Association of Petroleum Technologists, v. 44, n. 5, p. 329-340.

Bosum, W., et al., 1979, CCOP Technical Bulletin, ECAFE, 3: 1-14.

Chai, B.H.T., 1972, American Journal of Science, v. 272, p. 389-422.

Chen, S.Q., et al., 1981, Scientia Sinica, v. 24, n. 9, p. 1271-1284.

Chou, J.T., 1969, Technical Bulletin of CCOP, v. 2, p. 97-116.

Feng, Z.Q., and W.J. Zheng, 1982, Acta Geologica Sinica, n. 3, p. 213-222.

Hayes, D.E., 1983, The tectonic and geologic evolution of southeast Asian seas and islands, p. 124-138, American Geophysical Union, Washington, D.C.

Herman, B.M., et al., 1979, Geological and geophysical investigations of continental margins: AAPG Memoir 29, p. 199-208.

Inoue, E., 1981, Chishitsu New, v. 328, n. 12, p. 23-36.

Ishiwada, Y., 1981, Marine Sciences, v. 13, n. 3, p. 175-179.

Jin, X.L., and P.Z. Yu, 1981, Collection of Structural Geology, (no. 1): Beijing, China, Publishing House of Geology, p. 10-27.

Katsumata, M., and L.R. Sykes, 1969, Journal of Geophysical Research, v. 74, p. 5923-5948.

Kizaki, K., and I. Oshiro, 1977, Marine Sciences, v. 9, n. 8, p. 38-45.

Leyden, R.E.M., and S. Murauchi, 1973, AAPG Bulletin, v. 57, p. 2396-2403.

Li, Y.H., 1984, Scientia Geological Sinica, n. 2, p. 127-138.

Ludwig, W.J., et al., 1973, Journal of Geophysical Research, v. 78, p. 2508-2515.

Sung, K.C., 1983, Marine Geology of Korean Seas, D. Reidel Publishing Company, p. 7-13.

Wageman, J.M., et al., 1979, AAPG Bulletin, v. 54, n. 9, p. 1611-1643.

Zhang, Y.X., 1981, Collection of structural geology, (no. 1): Beijing, China, Publishing House of Geology, p. 28-36.

CHAPTER 15

# EXPLORATION AND DEVELOPMENT OF NATURAL GAS, PATTANI BASIN, GULF OF THAILAND

**Harold M. Lian**
**Ken Bradley**
*Unocal Thailand, Inc.*
*Bangkok, Thailand*

## ABSTRACT

The geology of the Gulf of Thailand features a series of north-south-trending ridges and linear, fault-bounded basins with a sedimentary section predominantly of upper Tertiary sediments. The Pattani Basin, located near the geographic center of the Gulf of Thailand, contains as much as 8 km of almost entirely nonmarine fluvio-deltaic sediments. The gas/condensate fields described in this paper are on the west flank of the Pattani Basin.

Gas reservoirs are thin, randomly distributed sandstone beds occurring from 1200 to 3000 m below sea level. At greater depths very high temperatures have caused a degradation of reservoir properties.

The gas fields are found on intensely faulted structures. The high fault density superimposed on the stratigraphic model limits the size of individual gas accumulations. Extensive 3-D seismic surveys have proven to be essential for delineating and developing these complex fields. An interactive computer system is used to interpret the 25,000 line-kilometers of 3-D data.

A new era in Thailand began when gas production commenced from the Erawan field in August 1981. The Baanpot, Satun, and Platong fields came on production between October 1983 and March 1985.

Two hundred forty-four development wells have been drilled from 23 platforms in the four fields. The productive limits of each field have yet to be established. Development work is continuing. These wells are capable of producing 13.45 $M^3$ (475 MMCF) of gas per day, which is considerably in excess of Thailand's current requirements. Condensate ratios average 40 barrels per million cubic feet of gas.

The advent of gas production has created a new industry in Thailand, with significant social and economic benefits to the Kingdom. Gas is used primarily for electrical generation. A lesser amount is used for industrial fuel. LPG is used extensively in the transportation sector.

## INTRODUCTION

In 1962 Union Oil Company of Thailand (now Unocal Thailand, Inc.) became the first oil company to be awarded exploration rights in the kingdom of Thailand. Unocal, operating for its co-venturers Mitsui Oil Exploration Company, Ltd., and Southeast Asia Petroleum Exploration Company, Ltd., is currently producing gas and condensate from four fields located in the Gulf of Thailand (Figure 1).

The fields are operated under concession agreements with the Royal Thai government. The concession area consists of four blocks totalling 8235 $km^2$, which is the area remaining after the final exploration period relinquishment in April 1982.

The discovery well was drilled by Unocal in 1972. The first sales contract to supply gas to the domestic market was signed in 1978. Development of the Erawan field began immediately thereafter, and production commenced in August 1981. A second gas sales contract covering the Platong, Satun, and Baanpot gas fields was signed in May 1982, and development of these fields proceeded concurrently with activities at Erawan. Production from these fields commenced between October 1983 and March 1985.

Gas production is currently limited by demand and is now averaging 8.5 million $M^3$ per day together with 2,385,000 L of condensate. The gas is delivered to the Petroleum Authority of Thailand's pipeline inlets at the offshore processing platforms. The condensate is pumped to a permanently moored storage tanker near Erawan field, from which liftings are made for both domestic and export markets.

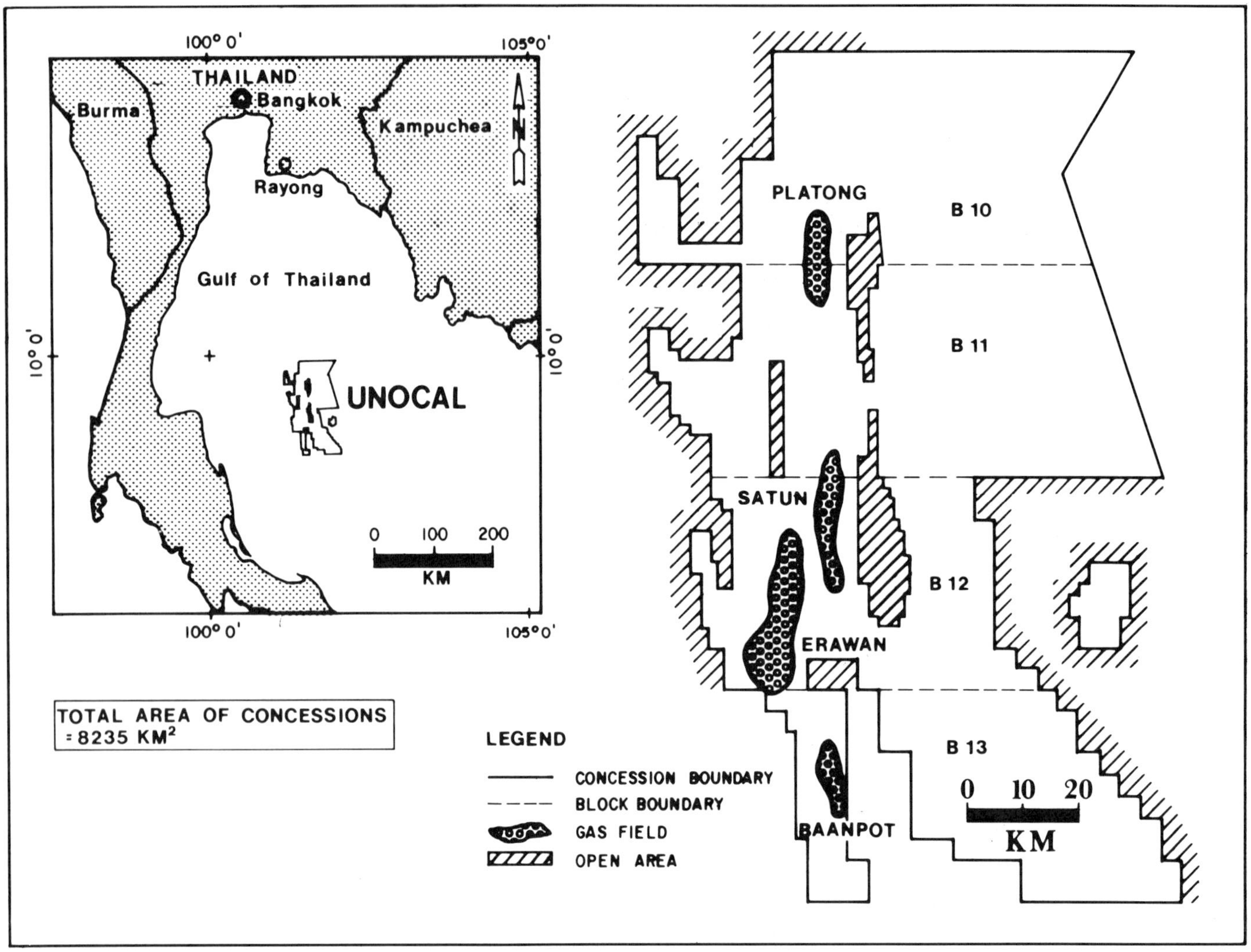


Figure 1. Location map of Unocal's gas fields.

The gas reservoirs are subquartzose sandstones of Miocene age occurring at depths between 1200 and 3000 m. Most gas production comes from reservoirs above 2650 m.

Two hundred forty-four development wells have been drilled from 23 platforms located in the four fields. Twelve of these platforms are sited in the Erawan field. The total daily gas production capacity at present is about 13.45 million $M^3$.

Economic exploitation of the gas resources of the Pattani basin has required state-of-the-art technology. The combination of high fault density, an abnormally high geothermal gradient, thin, narrow, and unpredictable sandstone reservoirs, and an overpressured section have been a continuing challenge to well planning, drilling, and completion methods.

## REGIONAL SETTING

The Gulf of Thailand is a northwest-trending reentrant into Sundaland, the southeasternmost extremity of the Eurasian plate. It has existed as a tectonic feature since the early Tertiary and at present is a shallow epicontinental sea with a maximum depth of about 86 m.

The region has a complex pre-Tertiary tectonic history involving a relatively stable Paleozoic period followed by complex and severe tectonics in the Mesozoic resulting from interaction of tectonic plates. Volcanism, granitic intrusion, local uplift, and subsidence produced the pre-Tertiary terrane on which the Tertiary sediments were deposited.

Prior to the middle Tertiary, rifting-initiated subsidence formed the proto-Gulf of Thailand. The precise timing of subsidence is unknown, as the oldest Tertiary sedimentary rocks penetrated by Unocal wells to date are late Oligocene. An older undrilled section is present in the basin center.

Early Tertiary sediments in the Gulf of Thailand are continental deposits, which are overlain by marginal marine sediments, the result of a regional transgression. In the Miocene, deltaic sedimentation and associated subaerial deposition became dominant in the Gulf during a basin-wide regression. A major realignment of plate movement in the mid-Miocene is recorded in the sedimentary sequence by a widespread unconformity.

When deposition resumed after the hiatus, it was marginal marine. This marine influence remained dominant to the present day apart from minor regressions. During much of the Tertiary, the surface was at or near sea level, and sedimentation appears to have kept pace with subsidence.

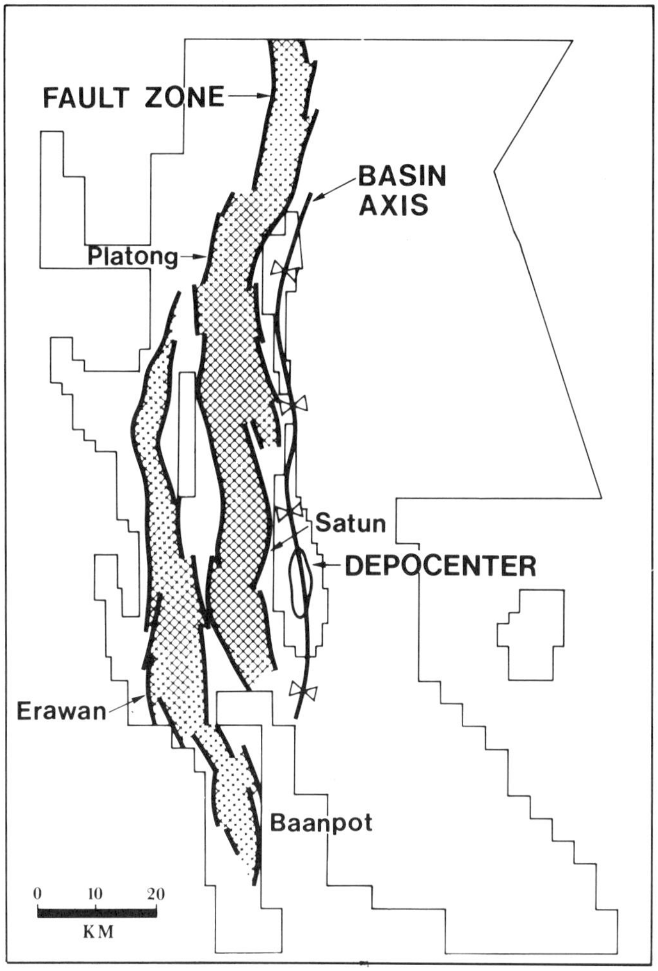


Figure 2. Regional tectonic setting of the Pattani Basin (after Burton, 1984).

## Regional Structure

The Gulf of Thailand appears to lie near the intersection of two major transcurrent fault systems (Figure 2). The southeast-trending Kanchanaburi fault system runs from the Burmese border through the Three Pagodas Pass and may extend along the northern and northeastern shore of the Gulf. The north-northeast-trending Phang Nga-Surat Thani and Ranong faults cut across the Malay Peninsula and into the western part of the Gulf. Direction of movement along both of these faults has been described as both sinistral and dextral by various authors. Structures in the Tertiary section indicate that the Gulf formed in an extensional tectonic environment; normal faulting dominates observable structural style. Whether the opening of the Gulf of Thailand results only from tensional movements or from combined tensional and strike-slip movements cannot be determined, as strike-slip faults have yet to be recognized in the Gulf.

The granitic belts of Peninsula Malaya extend offshore as elongate basement highs, such as the Ko Kra Ridge, which roughly divides the Gulf of Thailand into eastern and western sectors. West of the Ko Kra Ridge many narrow, elongate basins are separated by narrow north-south-trending horst blocks. East of the Ko Kra Ridge and west of the Khmer Shelf basement high are two basins with thick sedimentary accumulations—the Pattani Basin and the Malay Basin (Woollands and Haw, 1976) (Figure 3). Unocal's concessions are located within the Pattani Basin, which is approximately 400 km long in the north-south direction and 70 km wide, and contains as much as 8000 m of Tertiary sediments.

# STRUCTURES WITHIN THE PATTANI BASIN

Sediments in the Pattani Basin are cut by a multitude of closely spaced north-south-trending normal faults. The most complex structures have developed where basement-penetrating normal faults occur in an area of changing basement dip. This situation commonly occurs along depocenter margins, although it is not restricted to these areas. An idealized cross section of this structural style is illustrated in Figure 4.

West of the Pattani Basin's depocenter, where dense seismic coverage allows detailing of the presently producing fields, major en echelon, basement-penetrating normal faults trend north-northeast and dip toward the basin depocenter (down to the east). Ancillary faults are both subparallel and antithetic to the major normal faults and tend to trend due north in alignment with depocenter axes. These fault sets combine to form long, narrow zones as much as 20 km long which are in turn arranged in the en echelon fashion (Figure 5). It is possible that the major faults are controlled by zones of weakness in the basement, whereas ancillary faults are controlled by tensional stress and load-induced subsidence in the Tertiary section.

Field structure closely controls reservoir occurrence. Maximum gas accumulation occurs in the west flanks of the fields and is controlled by faults dipping toward the basin depocenter. These faults tend to penetrate deeper than the antithetic faults, and in addition to providing conduits for upward-migrating hydrocarbons, they appear to have provided trapping conditions at a critical stage. A characteristic of the west flank of the fields is that the beds in each fault block are rotated so that they are flat or dip away from the basin's depocenter with antiregional dip. Hydrocarbon accumulations on the east flank, which is dominated by the west-dipping, antithetic faults, are less significant.

These features are illustrated in Figure 4, a section across northern Erawan. The section crosses two structural trends; the western part of the line is in the Erawan field, and the eastern part of the section crosses the Satun field. At Erawan the basement-penetrating east-dipping normal faults are located on the west flank of the field. Closer to the graben are subsidiary down-to-the-basin faults which do not displace basement. Differing strata thickness across the major basement-penetrating faults clearly shows that fault movement was penecontemporaneous with sedimentation. Faulting antithetic to these faults is well developed and is restricted to the east flank of the structure. The throw across these faults is as much as 200 m at the base of Unit II, and this is significant enough to truncate some of the subsidiary east-dipping normal faults.

Small differences in basic structural style exist from field to field. Density of faulting is much greater in the Platong field than in the Erawan field. Throw on individual faults is sufficient to isolate the reservoirs so that they are pressure- and fluid-separated.

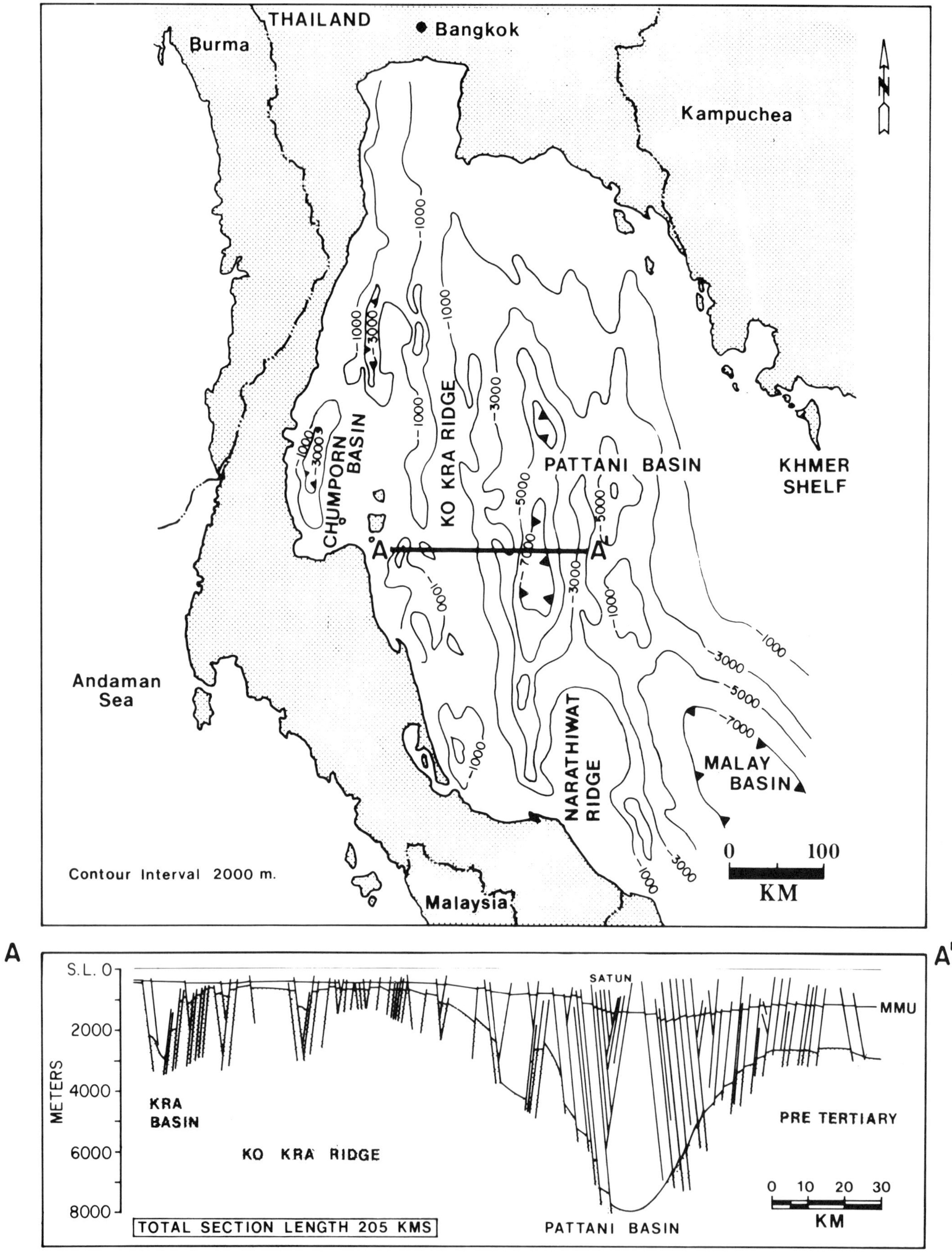


Figure 3. Gulf of Thailand base Tertiary structure and cross section.

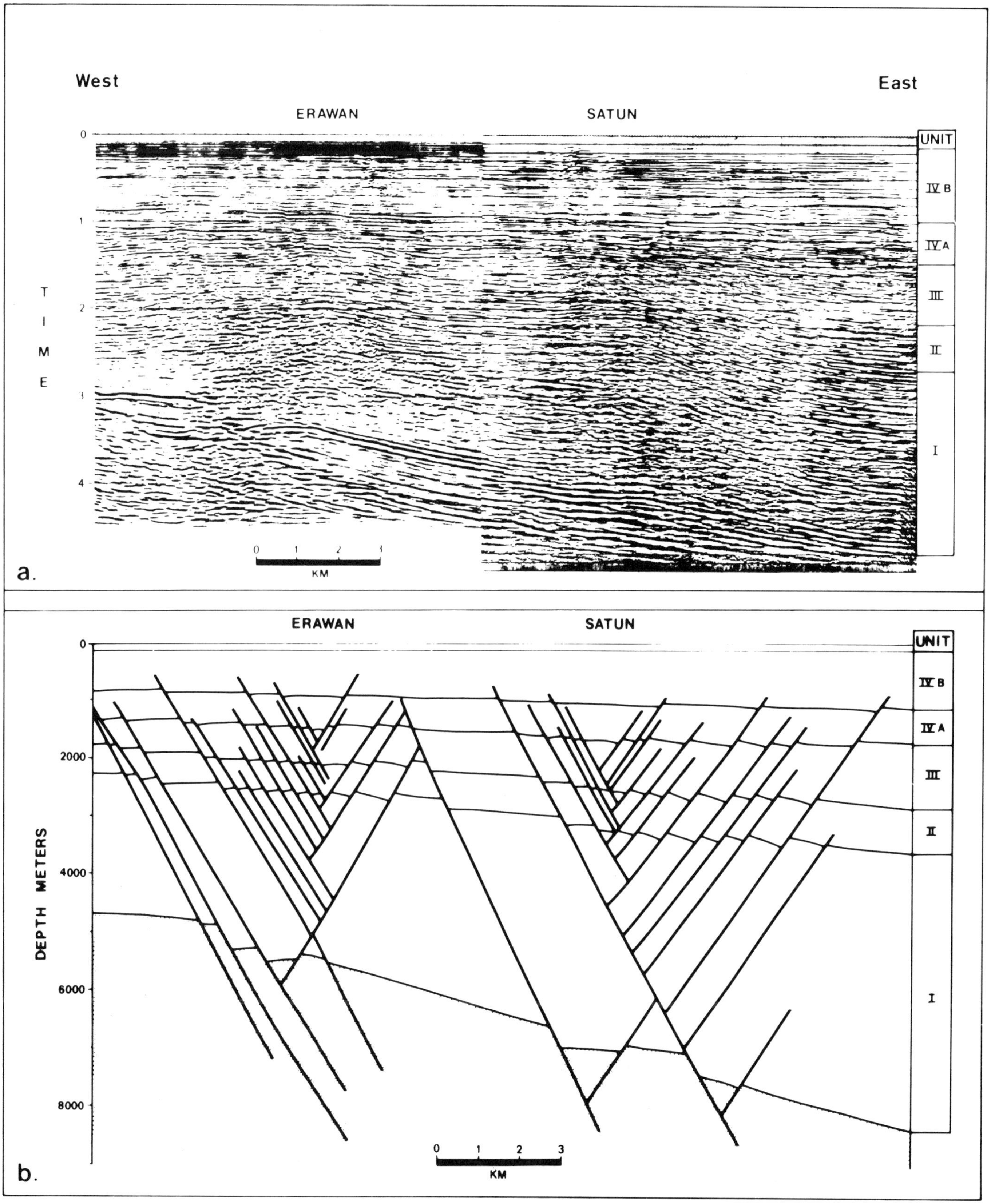


Figure 4. Seismic profile through Erawan and Satun fields and interpreted depth section.

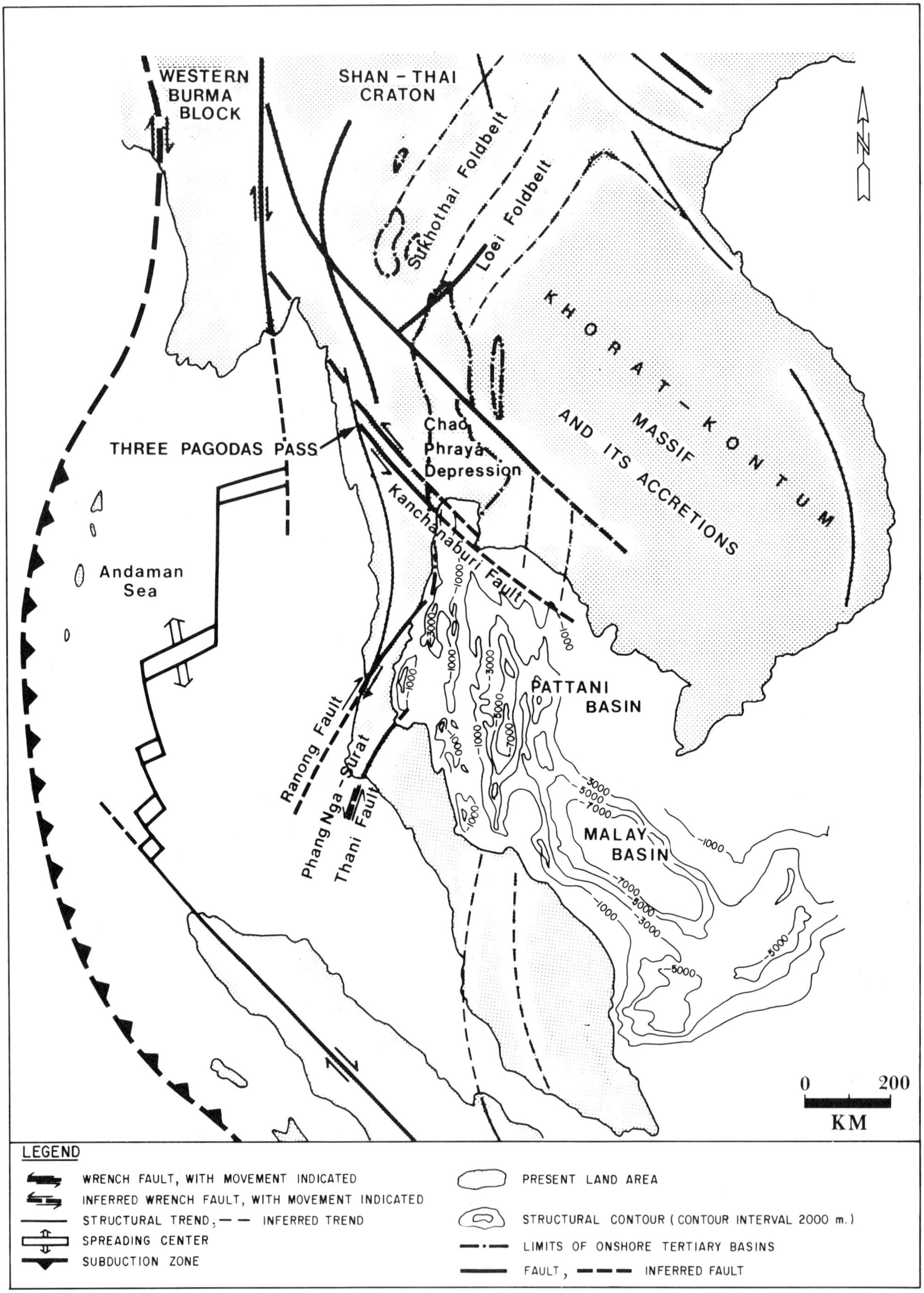


Figure 5. Trends of fault zones that contain gas fields.

## GENERAL STRATIGRAPHY—PATTANI BASIN

Production from Unocal's concession is entirely from the Tertiary section, which unconformably overlies a pre-Tertiary complex of plutonic, metamorphic, and sedimentary rocks. Pre-Tertiary rocks have been penetrated by only a few wells drilled on the ridges. They include Cretaceous granite and low-grade metasedimentary rocks of Paleozoic age, similar to those observed onshore.

In general it is practical to divide the Cenozoic section into four units, all of which appear to be time-transgressive to some extent. The four units represent periods of transgression and regression in the Gulf of Thailand. They are described in detail in Figure 6. Deposition in Unit I (the oldest) was of a fluvial-deltaic nature. Unit II sedimentary rocks represent a cycle of restricted marine transgression. Unit III is composed of nonmarine deltaic and fluvial sediments topped by the mid-Miocene unconformity. The Neogene sediments above the unconformity were deposited in predominantly coastal marine environments, ultimately leading to the modern day marine conditions.

Biostratigraphic controls throughout the Tertiary are very limited because of the generally nonmarine nature of deposition. Sandstone beds in all units are thin, generally averaging less than 5.5 m, narrow, and sinuous. The restricted and random position of sandstone beds is severe in Units I and III, but beds have somewhat greater lateral extent in Unit II. These beds frequently show upward-fining sequences, identified on wireline logs by steadily increasing gamma ray readings. Many are clearly identified seismically as meandering river channels. Sandstone grain size tends to be fine, but locally ranges from very-fine-grained to very-coarse-grained and granular.

## SANDSTONE COMPOSITION AND DIAGENESIS

Apart from depth-related diagenetic changes, sandstone composition is consistent throughout the concession area. A QFL plot (Figure 7) shows the range in composition of framework grains for these subquartzose sandstones. Lithic fragments commonly compose from 2 to 19% of the sandstones. Mudstone and siltstone rip-up clast are typically the most common fragments. Other lithic fragments include chert, carbonates, and quartz-muscovite schist. Detrital micas can locally compose up to 2% of the rock.

The amount and composition of cement changes with the progress of diagenesis, and is related to depth. Kaolinite cement is common in the depth zone from 1980 to 2285 m, together with well-developed quartz overgrowths. Below 2285 m quartz cement is abundant and the remaining pores are partly filled by kaolinite and illite. Carbonate cements are only locally significant.

Diagenesis, grain size, and sorting determine the reservoir properties. Porosity and permeability have been decreased by mechanical compaction, kaolinite cementation, and quartz overgrowths. Dissolution of detrital feldspar has locally enhanced porosity.

Porosity and permeability of sandstones in the Pattani Basin decrease rapidly with depth. Mean density log porosities decline from about 30% at 1220 m to 15% at 2440 m. At shallow depths porosity originates predominantly from relatively large spaces between sand grains where not inhibited by cement. At greater depth, pore spaces are ultimately the micropores created by the dissolution of feldspars. There are some exceptions to the depth-related decline in porosity, where some deeply buried medium- and coarse-grained sandstones have significant secondary porosity development. In general the coarse-grained, well-sorted sands have the highest porosities and permeabilities.

## PETROLEUM GEOLOGY

Economic gas production in the Pattani Basin is determined by the combination of pressure, temperature, and degradation of formation permeability. The economic basement occurs near a depth of 2740 m.

Geothermal gradients within the producing zones of the fields are exceptionally high and range from 2.0° to 5.8° C/100 m (2.2° to 3.2° F/100 ft).

### Correlation of Markers

Because of the laterally discontinuous nature of most of the sedimentation, correlation of markers among even closely spaced wells is difficult. In Erawan, a number of electric log markers are recognized in the section. These are either low-resistivity shales or coals. The majority of markers recognizable throughout the region occur in the marine-influenced Unit II. Correlation of the Platong wells is particularly difficult because of the greater fault density in that area. Reliable correlation is only possible when control is afforded by many wells, and even then correlation of sandstone beds can be uncertain.

### Source Rocks

Source rock studies are available for a limited number of Pattani Basin wells. Unit II is mature and appears to contain the richest oil and gas source rocks. The leanest source rocks are found in Unit III, although gas is produced from reservoirs within the unit, implying that migration has occurred. Samples of Satun and Baanpot condensates show good correlation to the mature section beneath the zone of reservoirs. In other wells analyzed, however, the correlation is doubtful. Migration may have occurred from depths below 2740 m or from laterally distant areas. Faulting may have provided conduits for hydrocarbon migration from depth.

High pristane/phytane ratios and characteristic isoprenoid distributions indicate that hydrocarbon were most likely generated from land-derived organic matter such as that found in the Tertiary section.

### Gas Composition

The gas varies slightly in composition from field to field. In Erawan, the gas contains 65 mole % methane, which is about the average composition for the Basin. Other components include 19% higher hydrocarbon group ($C_2$ to $C_7+$), 14% $CO_2$, and 2% nitrogen. Satun field gas consists of 72% methane, 17% higher hydrocarbon components, 11% $CO_2$, and a

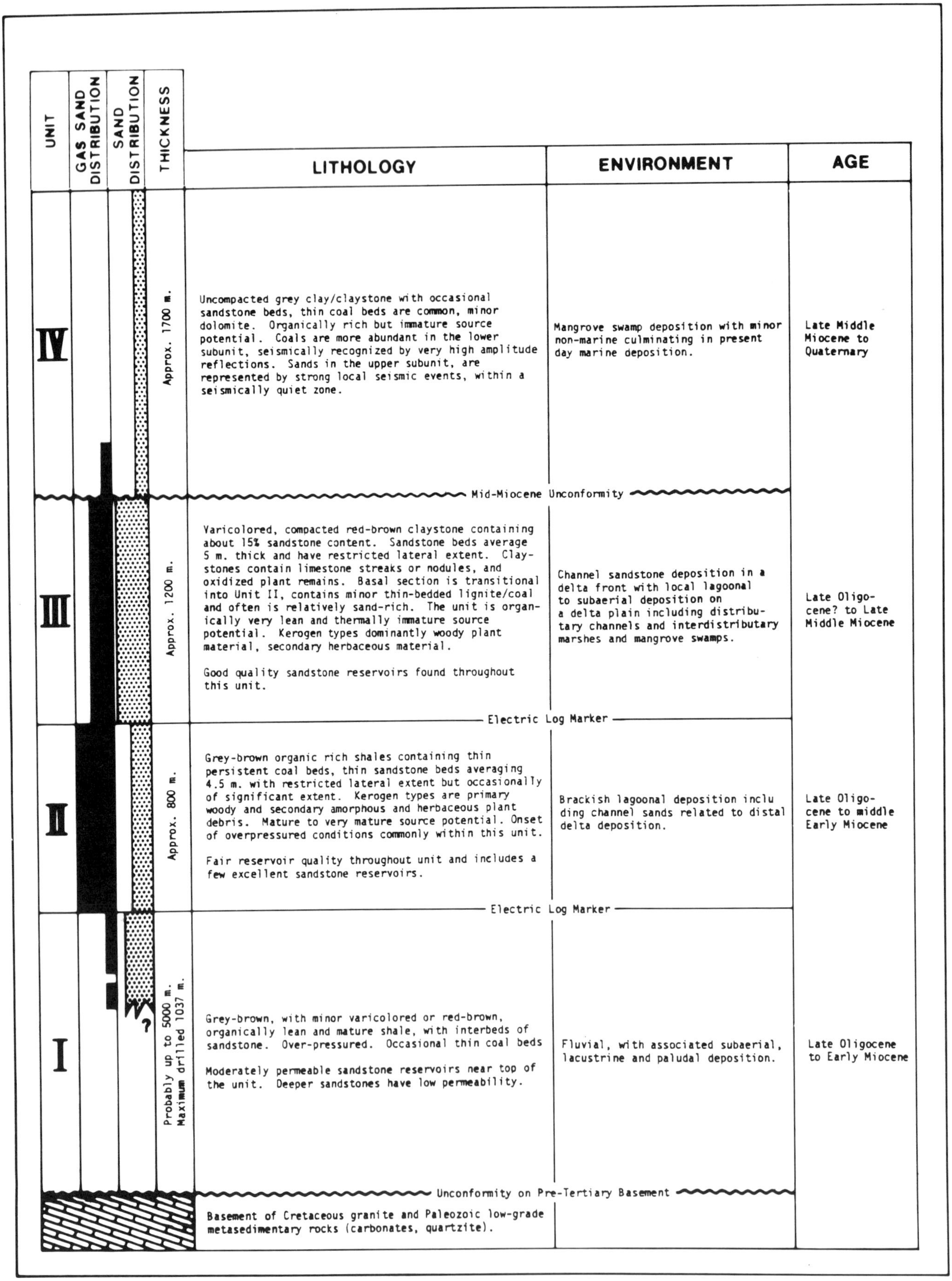


Figure 6. Pattani Basin stratigraphy.

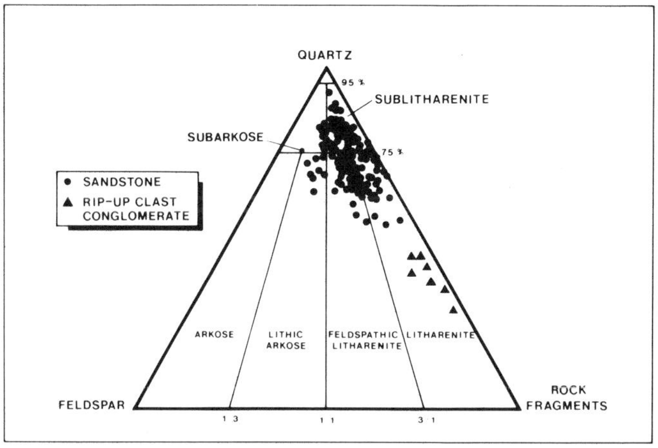


Figure 7. Composition of Miocene sandstones, Pattani Basin (Trevena and Clark, 1986).

trace of nitrogen. In the Platong field, higher hydrocarbon components are relatively richer, representing as much as 20% of the total. Carbon dioxide in significant amounts (65%) occurs locally in the north Erawan area, west of Erawan, and in the northwestern portion of Platong. The heat value of the gas and the gas-condensate ratio also differ from field to field. Heat value in BTU is 1080 in Erawan, and 1030 in Platong. The average gas-condensate ratios range from 38 barrels per million cubic feet of gas in Erawan to 70 in Platong. The condensate composition is 49% $C_7+$ and higher components, and 51% $C_2$ - $C_6$ with a trace of nitrogen. Basic properties of the condensate are 44.9° API gravity, 0.8014 gm/cc, and 130 molecular weight at 60°F. Minor amounts of high-gravity oil are produced from a few isolated reservoirs.

## OVERPRESSURE

A moderately overpressured section is commonly encountered in Pattani Basin wells, and it is most strongly developed in the organic-rich Unit II section. The most severe pressures, 7400 psig at 2740 m, have been encountered at Baanpot. Excess water has probably been generated by the alteration of clay minerals or by the alteration of organic matter in shales. These reactions take place at the elevated temperatures common in the Gulf and release water bound in the chemistry of minerals and organic matter. The top of the overpressured zone often corresponds to a zone of well-developed secondary porosity.

## SEISMIC TECHNIQUES

The full extent of the complex structural and stratigraphic problems became apparent as the Erawan field was developed. Advances in seismic technology in recent years have permitted these problems to be largely overcome.

Detailed 2-D seismic grids were initially used. It was soon realized that 3-D coverage was required for proper development. Unocal acquired its first 3-D survey in 1979, over the Erawan field. Subsequently, all potentially productive acreage within the contract areas has been covered by 3-D surveys totalling 25,000 line-km. 3-D seismic is essential for the delineation phase and ultimately for siting well platforms and individual development wells. Exhaustive processing tests have been examined during the past 6 years and, with appropriate processing, high-resolution data have been obtained with a good signal-to-noise ratio. This permits accurate structural and stratigraphic mapping. Unocal uses a dedicated interactive computer interpretation system, which enables the mass of data to be analyzed quickly.

### Seismic Stratigraphy

The seismic response follows the gross stratigraphic units (Figure 8).

Pre-Tertiary—the base Tertiary reflection is generally very strong and marks a major unconformity. Faulted and eroded pre-Tertiary section is evident in deeper parts of the basin.

Unit I—There are few control points for tying seismic data because of the small amount of deep well data. This unit has some extensive deep reflections among the common discontinuous events.

Unit II—A zone of widespread reflections associated with correlatable coals and low-velocity shales. Sands show little impedance contrast with the surrounding shales and rarely have a seismic response.

Unit III—A sequence of numerous, strong and discontinuous reflections. These events are correlatable with sands from log data which have low impedance relative to the shales. The unit is bounded by the continuous reflections of Unit II and the middle Miocene unconformity (MMU).

Unit IVa—Overlying the MMU is a package of very high amplitude reflections associated with widespread coal beds. This is the most easily correlatable unit from seismic, and gives excellent control of fault orientation.

Unit IVb—Generally a seismically quiet zone but with local very strong events that can cause severe degradation of the deeper data because of multiples. Recent channels at or near the surface are mappable and these give rise to noise trains that impair deep-data quality.

### Seismic Mapping

Good structural maps can be produced from the coal reflections of Unit IV and from the coals and low-velocity shales of Unit II. The mid-Miocene unconformity is poorly defined seismically because of the abundance of reflections from coals above and sands both above and below the unconformity. Faults can be defined accurately with 3-D migrated data. Stratigraphic mapping is essentially restricted to the redbeds of Unit III, where reflections are associated with low-impedance sands.

The sands all fall within the tuning thickness range; hence resolution of the top and base is not possible. It is relatively easy to map the areal extent of these discontinuous sands using an interactive system. Figure 8a shows the seismic amplitude of a narrow channel sand in the Erawan field. Faults are superimposed to show the structural relationship. A line generated along this sand body (Figure 8c) illustrates the excellent correlation compared with the discontinuous nature on a line perpendicular to the sand (Figure 8b). Synthetic seismic ties in the redbed sequence are excellent (Figure 8d), and sand bodies as thin as 15 to 20 ft (4 1/2 to 6 m) can be accurately mapped.

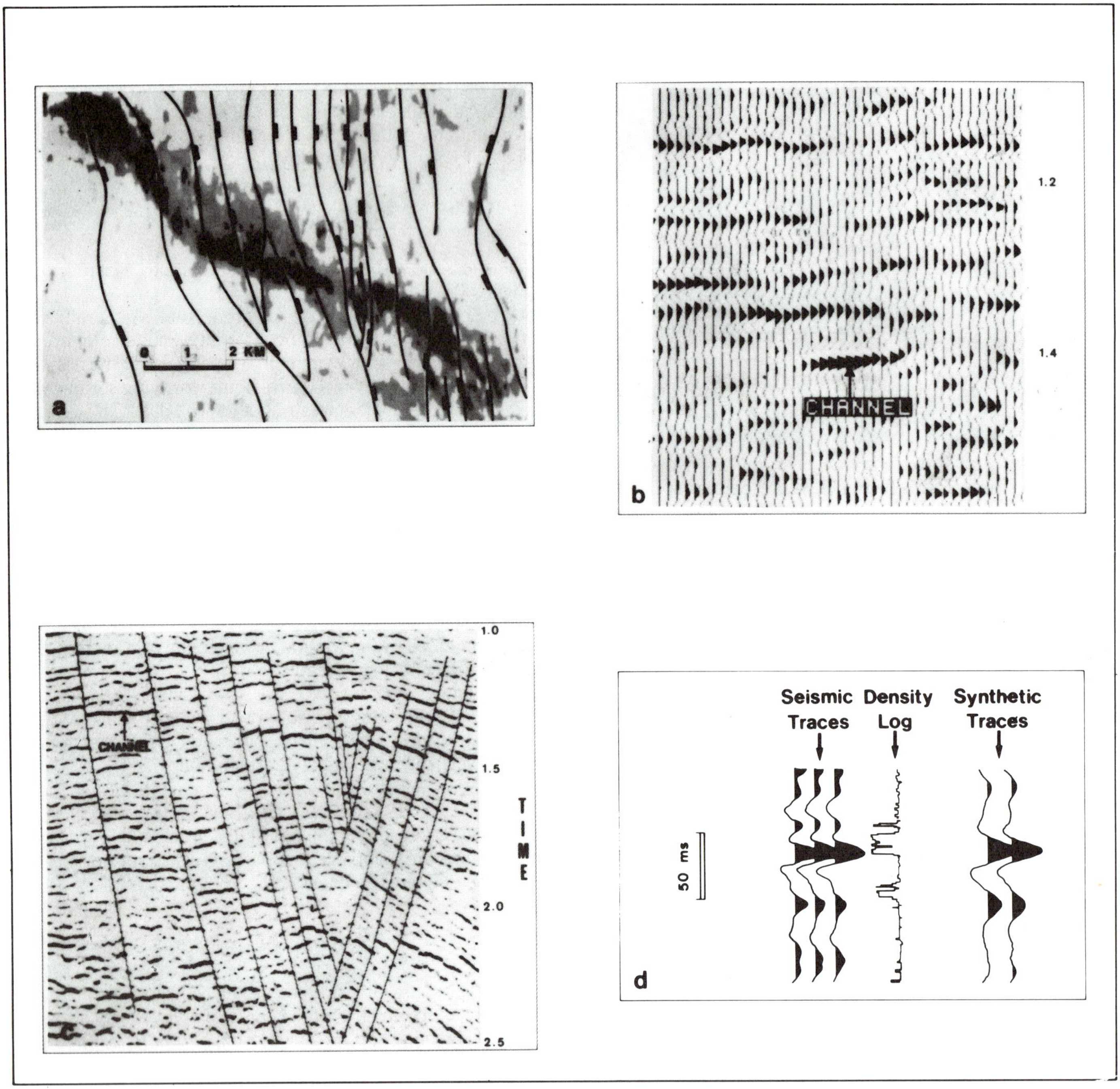


Figure 8. Seismic response in Unit III.

### Direct Hydrocarbon Detection

Numerous attempts have been made to relate seismic attributes, including offset dependent reflection, to gas accumulation. To date these have not been successful because of the complexities of tuning effects and particularly the presence of low gas saturations in wet sands.

## UTILIZATION OF NATURAL GAS

All of the gas production is sold to the Petroleum Authority of Thailand, a State enterprise, which operates the 425-km pipeline to shore. The gas is then processed at a gas separation plant, where carbon dioxide is removed and liquefied petroleum gas is produced for domestic demand.

The gas then flows through a 169-km onshore pipeline to Bangkok and its environs. The principal user is the Electricity Generating Authority of Thailand. At present natural gas generates 42% of Thailand's electricity. Gas is also used as fuel in a cement factory and in a ceramics factory.

## BENEFITS TO THAILAND

Even at this early stage in the development of its natural gas resources, the benefits to the Kingdom are becoming a significant factor in its economy. They will become increasingly important as gas production increases in future years.

Only a few years ago Thailand was almost completely dependent on imported energy. A concerted effort during the past 10 years, by both the private sector and the public sector, has reduced that dependence significantly by the large-scale development of Thailand's own energy resources. The public sector has exploited hydroelectric and lignite resources. The private sector has discovered and developed petroleum resources. Thailand's domestic energy production now accounts for nearly 50% of its energy demand.

At the start of this decade, Thailand relied on foreign sources for about 70% of its energy requirements. The picture started to improve in 1982, which was the first full year of natural gas production.

In the 10 years from 1980 to 1990, it is projected that Thailand will go from a nation importing 70% of its energy to one importing only 40% of its energy needs—and this in spite of a 75% increase in its total energy consumption during this period. That will be a remarkable achievement, and in large part it will be due to the increasing availability of natural gas from the fields in the Gulf of Thailand.

Natural gas has significantly reduced Thailand's dependence on imported oil, and in so doing has saved much foreign exchange and provided lower cost energy. For the first 4 1/2 years of gas production, it was priced at less than 50% of the cost of imported oil. With the drop in global oil prices, natural gas and oil are now at parity. Gas has created value-added downstream industries. It is the key to Thailand's planned Eastern Seaboard Development Project, which includes a projected fertilizer plant and a petrochemical plant.

Not the least of the contributions natural gas is making are the social benefits—the training of hundreds of Thais for responsible offshore production jobs, and the transfer of technology in the upstream operations of a new industry.

## ACKNOWLEDGMENTS

The writers wish to thank Unocal Corporation and its coventurers, Southeast Asia Petroleum Exploration Company, Ltd., and Mitsui Oil Exploration Company, Ltd., for permission to publish this paper. They acknowledge with gratitude the assistance of the Unocal Thailand, Inc., Exploration Department and particularly the contributions of L.J. Edwards, W.D.T. Gray, R. Risley, and Kosit Fuangswasdi.

## REFERENCES CITED

Burton, C.K., 1984, The tectonic framework of mainland southeast Asia: Proceedings of the Conference on Applications of Geology and the National Development, Bangkok, November 19-22, 1982, p. 255-268.

DuBois, E.P., 1980, Review of principal hydrocarbon-bearing basins of the South China Sea area: Proceedings of the EAPI/CCOP Workshop East-West Center, Honolulu, Hawaii, August 5-12, 1980, p. 1113-1140 (also in the CCOP Proceedings of the Seventeenth Session, Bangkok, 1980 p. 245-285).

DuBois, E.P., ed., 1981, Tertiary sedimentary basins of the Gulf of Thailand and South China Sea: Stratigraphy, structure and hydrocarbon occurrences: ASCOPE/CCOP Committees Paper, Jakarta, Indonesia, 72 p.

Knox, G.J., and L.L. Wakefield, 1983, An introduction to the geology of the Phitsanulok Basin: Proceedings of the Conference on Geology and Mineral Resources of Thailand, Department of Mineral Resources, Bangkok, November 19-28, 1983.

Parke, M.L., Jr., K.O. Emery, R. Szymankiewicz, and L.M. Reynolds, 1971, Structural framework of the continental margin in the South China Sea: United Nations ECAFE, CCOP Technical Bulletin, June 4, 1971, p. 103-142.

Trevena, A.S., and R.A. Clark, 1986, Diagenesis of sandstone reservoirs of Pattani Basin, Gulf of Thailand: AAPG Bulletin, v. 70, p. 299-308.

Woollands, M.A., and D. Haw, 1976, Tertiary stratigraphy and sedimentation in the Gulf of Thailand: Southeast Asia Petroleum Exploration Society, Offshore S.E. Asia Conference, Singapore, February 17-20, 1976, 22 p.

CHAPTER 16

# RECENT DEVELOPMENTS IN THE AUSTRALIAN PETROLEUM AND MINERALS INDUSTRIES

**T.J. McMahon**
*Petroleum Division*
*Department of Resources and Energy*
*Canberra, Australia*

## INTRODUCTION

I am pleased to have the opportunity to address this Conference on recent developments in the Australian petroleum and minerals industries, including recent Government actions taken in response to changes in market conditions.

In Australia, exploitation of mineral and energy resources is undertaken by the private sector, except for gas and electricity utilities, which are either semigovernment or closely regulated private sector bodies. Foreign companies have played an important part in the development of our mineral resources, and their participation is encouraged by the Australian Government.

The Government's influence on the resources sector is established mainly through its broad economic policy measures, which are supplemented by policies specific to particular industries. Onshore, the governments of the Australian States have jurisdiction over many important activities. Offshore, the Commonwealth (central) government has exclusive jurisdiction, but some legislation is administered jointly by the Commonwealth and the States. Wherever practicable, the Government is removing or reducing regulation of business operations.

## PETROLEUM

### Impact of Low Prices on Exploration, Development, Production, and Exports

Delegates will recall that early in December 1985, representative crude prices were around $US26 to 28 per barrel. By the end of July of this year, prices of internationally traded oils had fallen to around $US8 to 10 per barrel. There are no obvious signs that there will be a speedy return to 1985 price levels. Indeed, sustained prices of even $US15 to 20 per barrel might be some way off, although the short-term OPEC production agreement reached in the first week of August has brought some welcome relief from very low prices.

The sharp falls earlier in 1986 have had a significant impact on the exploration, development, and production sectors of the petroleum industry. Australian companies, like those elsewhere, have responded to the adverse impact of low prices by cutting down activities and expenditures very sharply.

The Australian Bureau of Mineral Resources has estimated that should current prices be maintained throughout 1986, petroleum exploration activity could decline by around 50% compared with the record level achieved in 1985. The Bureau estimates a decline from 226 wells in 1985 to 111 wells in 1986 for onshore exploration activity, and a reduction from 43 wells in 1985 to 32 wells in 1986 for offshore activity.

The figures for the development sector of the industry are even starker. While offshore development drilling may be sustained (17 wells), reflecting activity on projects already commenced in Bass Strait and the North West Shelf, onshore development drilling is predicted to fall from 75 wells in 1985 to 15 wells in 1986. Esso/BHP have announced deferral of $A1.0 billion in development work in Bass Strait (from a 4-year program of $A1.8 billion).

A big reduction in seismic activity is also expected, from about 91,000 km last year to 40,000 km in 1986.

Oil production onshore, which is all sold to Australian refiners under a Government-supervised allocation system, has not been greatly affected by low prices, although some small producers have shut-in. Offshore, a certain proportion of Bass Strait production was shut-in for some months, mainly as a result of difficulties in making profitable export or domestic free market sales. During August, however, export sales have resumed following policy changes made by the Government late in July.

The fall in prices has had a marked effect on profitability, the most severe impact being felt by companies with high ratios of debt to equity. Companies with foreign debt have also suffered from the sharp devaluation of the $A. Some selling off of interests to financially stronger companies has occurred.

Australia is currently experiencing a serious balance-of-trade deficit, caused in large part by declining prices for a number of our rural and mineral exports. Consequently, the loss of export income from crude oil sales is causing the Australian Government considerable concern. In the first 7 months of 1986, only 10 cargoes (6.55 million barrels) valued at $A203 million were exported. By contrast, in the whole of 1985, exports of crude oil and other refinery feedstocks earned $A1.6 billion in foreign exchange (64 shipments totalling 42.5 million barrels). The resumption of exports during August is therefore welcome, even though it has been on a smaller scale and at lower prices than before.

The adverse effects on profitability and cash flows have inevitably led to unemployment in the industry itself and in service industries. The Australian Petroleum Exploration Association has estimated that some 3000 jobs have been lost so far, and believes that further losses are likely.

## Government Responses: Profits-Related Taxes and Excise Concessions

Most Australian oil production is sold in the domestic market at officially calculated import parity prices, which track international market movements quite closely. The remainder is sold either domestically or on the international market at commercially negotiated prices. Over the years, these pricing arrangements have proved effective to encourage exploration and production of indigenous crude oil, more efficient end-use of petroleum products and, in appropriate circumstances, substitution of other energy forms for crude oil.

The year 1983 saw the emergence of an export surplus which grew rapidly until early in 1986, when the price collapse rendered the trade unprofitable for the Bass Strait producing companies.

The current situation has understandably led to calls by different sections of the industry for Government assistance. In responding to these calls, the Government has given particular attention to possible support for the industry through the taxation system.

The Government wanted to maintain a viable exploration and development industry, albeit somewhat smaller than before, able to take advantage of economic opportunities as they become available when oil prices recover. This requires Australia to maintain its relative attractiveness as a place to invest in petroleum exploration and development.

It was recognized that the combined effect of low prices and the level of taxation (crude oil excise) had discouraged production for the domestic free market and for export markets. The after-tax return to producers from free market sales, which are made from high-cost production from the marginal fields in Bass Strait, provided no incentive to pursue such sales.

In July, the Government reduced the effective rate of excise payable on oil which is sold on the free market—both domestic and export—to bring the net returns to producers closer to the levels that were being achieved before the sharp price falls earlier in 1986.

Following yet further falls in oil prices during July, the Government also reduced, with effect from 1 August 1986, excise rates on oil produced and sold in Australia under the allocation scheme. At the margin, some of these sales had also become unprofitable for the companies.

For onshore producers the excise liability has been completely removed until 30 June 1987, unless the import parity price (IPP) of Bass Strait oil reaches $A20 per barrel (on 1 August 1986 it was fixed at $A15.75). If the IPP recovers to $A20, there will be a review of these arrangements. For the offshore producers, the Government has reduced the top marginal rate of excise from 87% of import parity price to 80% until 30 June 1987, subject to the same review provision.

The Government is continuing to monitor developments in the world oil market, their impact on the oil industry in Australia, and the effects of the recent taxation reductions. At the present time, the companies are hopeful that those reductions will assist them to resume exporting in modest volumes, as well as to meet their obligations to supply the domestic market.

It is essential that Government policies remain relevant to the changing circumstances of the industry. There is close and continuous consultation with the companies to that end. The Government has already announced that a further review of the crude oil marketing arrangements within Australia will be carried out in 1987 with a view to the possible introduction of a free market from 1988.

In addition to these responses to immediate problems, the Government has reaffirmed an earlier policy stance in favor of replacing production-based excise and royalty imposts with profits-related taxes.

The Government believes that our traditional approach to revenue raising on the basis of production levels rather than profitability inevitably leads to problems as prices and profitability decline. Production-based revenue measures can make marginal developments unattractive and lead to the premature cessation of operations.

The Government had already introduced a profits-based Resource Rent Tax (RRT) for "greenfields" projects in the offshore area where it has exclusive jurisdiction. The first project to proceed under this arrangement was the Jabiru project in the Timor Sea, which has just commenced production for export—a project which almost certainly would not have proceeded without the stimulus provided by RRT.

In the onshore area, reform of the petroleum taxation arrangements can proceed only with the cooperation of the States and the Northern Territory. The Government recently renewed an earlier invitation to the States and the Northern Territory to introduce a Resource Rent Royalty (RRR), which is very similar to RRT, for all onshore production. This would require the States to remove royalty imposts as well. An important step had already been taken in cooperation with the State of Western Australia, with the introduction of RRR to oil production at Barrow Island. Discussions are also under way in relation to project in the Northern Territory. However, for most onshore projects, the present excise and royalty regime will continue to apply for some time at least.

## Self-Sufficiency Outlook

Having referred to Australia's oil exports between 1983 and early 1986, and to recent attempts to get exports moving again, I should say a little more about our prospects for oil self-sufficiency, if only to correct any misunderstanding that we could be significant exporters in the future.

Our level of crude oil self-sufficiency is forecast to decline from 95% in 1984-85 to about 52% in 1994-95. The Bass Strait producing fields are expected to go into decline very soon. The prospects of further modest discoveries of petroleum, however, are fairly good. Before the 1986 slump in exploration, the Bureau of Mineral Resources had estimated that there was a 50% chance that reserve additions during the next 10 years would be 950 million barrels. (At present, reserves stand at about 1.7 billion barrels.) Even if that rate of discovery could still be achieved, our self-sufficiency level would fall to about 60% in 1994-95 and 44% by the year 2000.

To the extent that future discoveries are made in areas remote from domestic markets, e.g., the Timor Sea, it may well be appropriate on economic grounds to export the production rather than transport it to Australian refineries. This is happening, for example, with production from the Jabiru discovery.

Nevertheless, Australia seems unlikely to be a substantial net exporter of crude. On the other hand, the recent discovery of a large additional deposit of condensate in the Goodwyn field, still being appraised, may lead to exports of up to 80 kbd of condensate during the 1990s.

I should add that even in periods when Australia is a net exporter, we will need to import some refinery feedstocks for production of lubricants, bitumen, and certain grades of fuel oil—Australian crudes are unsuitable for refining of these products.

## Recent Acreage Releases and Future Releases

Applications will shortly be invited for exploration permits in 10 areas in the Gippsland Basin in Bass Strait off the Victorian coast. The areas are in close proximity to producing fields and are likely to attract interest from local and overseas exploration companies. Information on the release will be disseminated widely among oil exploration companies. Applications will close early in 1987 and will be considered on the basis of the recently revised work program system.

Earlier this year, permits were issued in the Timor Sea area under a cash bid system. Twelve permits are current in the Timor Sea. Offers of permits will be made during the next few weeks for six areas offshore Western Australia.

The Government intends to maintain a program of regular releases of further exploration acreage to ensure that prospective explorers will not be hampered by lack of access to areas of interest. Bids will usually be invited on the basis of competitive work programs, but for areas of greater prospectivity, cash bids may be invited. The next release is tentatively planned for November or December 1986, and is likely to include areas off Western Australia, South Australia, Tasmania and Queensland.

## Australia-Indonesia Maritime Boundary

The maritime delimitation talks between Australia and Indonesia have continued constructively in recent months. So far as the area opposite East Timor is concerned, consideration is currently being given to the possibility of establishing a Joint Development Zone pending final resolution of the boundary. The JDZ would encompass the Kelp structure, which is considered to be prospective for petroleum.

## Natural Gas

The Bureau of Mineral Resources has estimated that, as of 30 June 1985, Australia's recoverable commercial reserves of natural gas totalled 531 billion cubic metres. Recoverable reserves which have not yet been declared commercially viable have been estimated at 991 billion cubic metres. These reserves would supply Australian requirements for gas for well over a century at present rates of usage.

The development of the natural gas industry has been one of the success stories on the Australian scene during the past 20 years. Demand has grown at an average rate of 11% per annum in the 10 years to 1983-84, and the share of gas in total energy consumption increased from 7% to 15%. Major pipelines have been constructed to bring gas from remote deposits to markets, e.g., from the Cooper Basin to Sydney and Adelaide, from Bass Strait to Melbourne, and from the North West Shelf to Perth. Another long-distance pipeline is now under construction between Palm Valley and Darwin. Gas has displaced oil on a major scale, and this is continuing.

## North West Shelf Project

The North West Shelf Project is not only the largest natural gas project but the largest resource development project ever undertaken in Australia. The fields contain 379 bcm of natural gas, 312 million barrels of gas condensate, 169 million barrels of LPG, and crude oil reserves of 29.7 million barrels. Current testing seems likely to prove up even larger reserves of condensate.

The first stage of the project, already completed, supplies natural gas to the domestic markets of Western Australia via a 1500-km pipeline. Supply of up to 385 mm cfd is contracted for a 20-year period. The second stage, now under construction, will supply as much as 6 million tonnes a year of liquified natural gas to a consortium of eight Japanese electricity and gas utilities for 19 years from October 1989. The capital cost of this stage in dollars of the day is estimated to be $A9.8 billion, excluding the cost of LNG tankers. Production of 1.4 million tonnes of condensate annually is also expected.

Foreign exchange earnings from the project, estimated at nearly $A50 billion over 19 years, will be doubly welcome in the 1990s and into the next century, as Australia's oil imports are expected to increase.

The Commonwealth and Western Australian Governments have sought to improve the opportunities for Australian industry to participate in the design and construction of the export phase of the North West Shelf Project. A National Liaison Group was established in May 1985 to promote Australian participation in the supply of equipment and services, provided price, quality, and performance criteria can be met by Australian firms. The fundamental principle is the Australian industry should have "full and fair opportunity" to compete in tenders for the project. Obviously there is still scope—indeed necessity—for very substantial foreign participation.

## Petroleum Exploration in the Asia-Pacific Region

Australian companies have had a limited involvement in the search for oil and gas in the Asia/Pacific region. For example, BHP has a 20% equity in five contracts in the Pearl River basin; CSR (through its subsidiary, AAR) and Ampol each have a 10% equity in a South China Sea contract area; and Ampol has a further 10% equity in two blocks in the Gulf of Beibu. In April 1986, the South China Sea consortium found oil, but further evaluation is necessary to assess the commercial potential of the discovery.

In May 1985, a consortium of Australian companies, led by CSR, was awarded a production-sharing contract for onshore petroleum exploration on Hainan Island.

Australian companies are also involved in the second round of bidding in China for offshore acreage, as well as in oil exploration in Indonesia and Papua New Guinea.

## COAL AND OTHER MINERALS

I turn now to a consideration of some of Australia's non-petroleum mineral resources, and their prospects and problems.

Australia is generously endowed with many high-grade and readily accessible mineral resources. Australian coal resources include all coal types except pure anthracite. Recoverable economic black coal reserves have been estimated at nearly 34 billion tonnes. Inferred resources are much larger. Demonstrated economic resources of iron ore are estimated at 16.2 billion tonnes, second only to Brazil in high-grade iron ore reserves.

Known bauxite reserves are about 5.265 billion tonnes, and demonstrated economic reserves are 2.9 billion tonnes. In 1985 Australia's demonstrated and inferred economic reserves of lead were 14.4 million tonnes, and those for zinc were 21.2 million tonnes.

Australia also has about 30% of the world's proven reserves of low-cost uranium. Production is confined to two modern open-cut mines operating in the Northern Territory (Ranger and Nabarlek) and a third underground mine in South Australia (Roxby Downs), which is scheduled to start production of copper, uranium, and gold during 1988. This is the world's largest uranium deposit, and among the largest copper and gold deposits.

Understandably, the minerals sector (including petroleum) has played an important role in the development of the Australian economy, and will continue to do so.

The sector has grown at almost double the rate of the national economy since the 1960s. In the 1980s its contribution to GDP, including minerals processing, had grown to around 8%. However, because of the large scale and capital-intensive nature of the industry, it accounts for only 2% of the workforce.

Investment in new mining and processing plants, although subject to cyclical fluctuation, is still providing considerable support to the Australian economy. At present, investment is being concentrated in a relatively narrow range of large projects—Roxby Downs, the Portland aluminum smelter, the Goonyella coal project, mineral sands processing plants, and a number of gold mining projects. When petroleum industry investment is included, new fixed capital expenditure totalled $A2.9 billion in the year ended June 1986 (41% more than the previous year). It is expected to rise again by 9% during the year ending June 1987.

Australia has traditionally relied on exports to underpin economic growth and provide a driving force for development. In 1985-86, mineral- and energy-related exports are estimated to have reached almost $A17 billion, more than half of total merchandise exports.

Australia is a major exporter of coal, iron ore, bauxite/alumina, lead, mineral sands, and more recently diamonds, and expects to become a leading exporter of gold within 2 to 3 years. But the big resource exports continue to be coal and iron ore.

Australia became the world's largest coal exporter in 1984 and retained that position in 1985. Japan was Australia's largest customer in 1985, taking 50% of the exports. Coal export volumes have continued to grow, on a year-on-year basis, reaching a new record of 89.9 million tonnes in 1985-86 (up 7.3% from the previous year).

Exports of iron ore in 1985-86 were valued at about $1948 million, making Australia the world's second largest exporter. Japan is our major export customer, taking 49.4 million tonnes of the total exports of 82.3 million tonnes.

Export competitiveness is essential to the viability of the Australian resources sector. And it is in this area that important challenges must be met. In recent years the sector has faced difficult international market conditions with weak demand and oversupply depressing prices for many commodities. The decline in commodity prices generally, including key items such as coal and iron ore, has accelerated during 1986, putting severe strains on Australia's balance of payments. The protracted recession in world mineral markets—now over 5 years long—has led to reduced exploration for almost all minerals except gold, mineral sands, diamonds, and platinum. It has also stimulated the industry to undertake a major effort to cut costs and increase productivity to remain internationally competitive.

It is worth noting that the present difficulties with mineral exports have emphasized the need for Australia to broaden the range of its major exports beyond the rural and minerals sectors. Those sectors have been magnificent export performers, but the time has come for manufactured goods and services to play a larger role. As market opportunities permit, it is also highly desirable that an increased proportion of our mineral resources themselves be processed in Australia, thereby adding value to our export earnings from the nation's resource endowments.

The competitiveness and profitability of Australia's mineral industries have been significantly helped by the substantial depreciation in the value of the $A—more than 30% in trade weighted terms since January 1985. Although the depreciation had some negative effects, including exchange losses, increased interest charges and higher priced imports, overall it has been highly beneficialy by raising the $A returns for most of our mineral exports. Without the depreciation there would have been widespread losses and mine closures; even with it, rates of return on capital employed are still low (less than 3% for about half of the respondents to a 1985 survey), and the overall profit position (5.7% for the industry as a whole) masked a considerable number of loss-making companies.

Some closures have occurred, especially in the nickel industry, and the sector as a whole has been exploring and implementing various ways of reducing costs and improving operating efficiency. Containment of labour costs is a vital and never-ending goal in this context. The protracted dispute at Broken Hill, which ended only last month, shows how painful this process of adjustment can be when long-established work practices are being overturned in the interest of increasing productive efficiency.

Industrial disputation in the mining industry has generally remained low over the past 3 years, although some problems this year have caused a temporary worsening. In an attempt to enhance the industry's competitiveness through both increased productivity and greater reliability as a supplier, the Government has placed great emphasis on developing a more harmonious industrial relations climate. Major initiatives have been the Prices and Incomes Accord (an economy-wide agreement between the Government and the union movement) and the establishment of mechanisms in the coal and iron ore industries for consultation between business, unions, and government on occupational, technological, and economic questions.

Many of the problems currently facing the resources sector are exemplified in the coal industry.

The international coal market is increasingly characterized by substantial oversupply, in conjunction with reduced demand for coking coal in some major export markets and lower growth in steaming coal demand. The dramatic fall in world oil prices in recent months has changed the relative competitiveness of other energy sources. However, the effect of low oil prices is still unclear, and will depend in part on perceptions about the outlook for security and price of oil supply in the longer term. The immediate concern is interfuel substitution by those consumers—in power generation, cement and steel making—who can readily change their energy mix without installing new equipment.

This has not yet seriously affected Australia's coal exports, but spot oil prices are certainly in the substitution zone, and there may be real downward pressures ahead. Against this, increased activity in the industrial nations flowing from lower oil prices should operate over time to increase the demand for both steaming and coking coal. It is too early to calculate how these various pressures will balance out.

Despite the current difficulties, it would be wrong to conclude that the prospect for the Australian resources sector is one of unmitigated gloom.

In the first place, the Asia-Pacific region, while not immune from adverse developments in the world economy, is still forecast to grow at a faster pace than the world generally. Some countries in the region may continue their strong growth patterns; for example, China, Thailand, and the Republic of Korea.

This suggests strengthening demand for raw and processed minerals, although this will be tempered by declining metals intensity, particularly in developed economies as their industrial sectors decline in importance and substitution of advanced materials proceeds.

A further factor which might increase mineral demand in the 1990s is the likelihood, raised recently by the "Group of Thirty," that a number of industrialized countries such as Japan will need to reinvest heavily in replacement of national infrastructure (e.g., bridges, urban railways, and residential buildings) during the 1990s.

The immediate outlook for some individual Australian resources is reasonably promising—the prices for gold, diamonds and mineral sands have kept up, against the general trend in minerals prices, and they continue to have sound prospects for the foreseeable future. Ilmenite, rutile, monazite, and zircon prices increased strongly in 1985 and seem likely to at least maintain these levels in the short term. Coal and iron ore production and export volumes are also holding up, though prices are not. The recent partial recovery in oil prices is obviously useful, although it would be unwise to put much reliance on an upward price path for the next few years.

Looking a little further ahead, Australia has some massive development projects under construction at the moment which should prove to be major new export earners. I have mentioned most of them already. Pre-eminent among them are the North West Shelf LNG project (whose profitability seems assured even on pessimistic oil and gas price assumptions) and the Roxby Downs/Olympic Dam copper, uranium, and gold project. Nuclear power will continue to be a competitive source of electricity in many countries. Its share could be boosted as economies expand to take advantage of the opportunities arising from lower oil prices. Thus the Australian uranium industry, as an efficient, unsubsidized producer looks forward with some confidence to the medium and long term. Current production capacity could rise to about 10,000 tonnes per annum during the next decade.

The Australian Government has a strict policy on uranium, limiting production to the Ranger, Nabarlek, and Roxby Downs mines and ensuring the exports of yellowcake are confined to countries which have agreed to meet stringent safeguard requirements.

Other current or prospective projects include:

- The Alcoa-led consortium's Portland aluminum smelter in Victoria ($1150 m), still on schedule for first production in November 1986.
- Several new satellite iron ore developments being considered in the Philbara region, all of which are planned to utilize existing infrastructure.
- Several new gold mine developments already entrain, and some being reopened. There is an emphasis on development of large, low-cost, open-pit operations.

## CONCLUSION

It is obvious that most of Australia's resource-based industries are going through a difficult time as they confront oversupply and declining prices in export markets. In this, they are not, of course, alone or even in a minority among world producers. Nor is this the first time—except for the petroleum producers—that they have endured hard times.

While there will be losses, and some mine closures and well shut-ins, the remedial measures already taken or in course should keep most Australian producers competitive. As I have noted, there are some bright prospects among the nonenergy minerals, and coal will continue to be sold in very large quantities even if prices are less than satisfactory.

We have:

- large and diverse resources,
- good technology,
- a skilled workforce,
- good infrastructure for exports, and
- a sophisticated deregulated financial system to marshal capital;

and we are therefore well placed to grow rapidly as the economies of our trading partners expand and markets grow again.

The resources sector, including minerals processing, will continue to make vital contributions to GDP and the balance of payments, despite current efforts to increase the range and value of exports of manufactures and services. The sector will continue to be largely the preserve of the private sector.

The Australian Government intends to maintain policies which are conducive to profitable investment in petroleum and minerals projects. Participation by foreign companies will continue to be welcomed, especially in association with Australian interests. The foreign investment rules will be administered in a way which ensures that no worthwhile enterprise fails to get off the ground for lack of Australian capital on reasonable terms and conditions. The progressive removal of unnecessary business regulation will be pursued. Winding back the rate of inflation, and improving of our international competitiveness, will remain major goals of policy.

CHAPTER 17

# CIRCUM-PACIFIC DIATOMITE DEPOSITS

F.K. North
*Carleton University*
*Ottawa, Ontario, Canada*

## ABSTRACT

Deformed marine diatomites of assured identification are all Oligocene or younger. They are not to be interpreted with oceanic diatom oozes as analogs, nor with California's Monterey Formation as prototype.

All examples apart from the unique Monterey are shallow-water deposits at convergent plate or microplate boundaries—in arc-trench gaps or immediate back-arc belts. Tethyan examples, along a collision boundary, are now slivers in late external flysch along Alpine thrust fronts. Circum-Pacific examples, at ocean-continent subduction boundaries, are exposed only on mountainous islands or peninsulas, the uplift (not folding) of which has protected the diatomites and their overlying evaporites from subduction. The control is tectonic and volcanic, not by water temperature or eustatism.

The unique Monterey Formation owes its spectacular development and preservation to the conversion of an arc-trench boundary to a transform boundary, at a triple junction, *before* the diatomite was deposited. The Monterey's capacity as an oil-source sediment does not stem from its true diatomite component. Other Circum-Pacific diatomites are of negligible significance to the petroleum geologist but are potentially minable for other uses in Japan, the Philippines, and Peru, and possibly in Chile.

## ENVIRONMENTS OF DIATOMITE DEPOSITION

Diatom-dominated sediments were originally recognized as one of the modern oceanic oozes (Figure 1). A few fossil examples were identified early, one of them becoming world-famous. More numerous fossil examples are now known to represent four distinct depositional environments:

a. Truly oceanic, but *not* on abyssal plains;
b. Marginal-oceanic, most of them now in strongly deformed belts and nearly all Neogene in age;
c. Epicontinental marine (especially polar), essentially undeformed; mostly late Mesozoic or early Tertiary;
d. Freshwater lacustrine, including glacial, interglacial, and postglacial lacustrine; mostly Pliocene or younger.

Lacustrine diatomites of group (d) are found in most continents and include some economically valuable deposits, especially in Europe; but they are of no concern to our present enquiry. Truly oceanic diatomites of group (a) are restricted to the flanks of oceanic ridges and fracture zones, or to seamounts, especially during the Miocene and Pleistocene. Even with this restriction, they are principally a Pacific and Circum-Antarctic phenomenon. They are essentially or totally unknown in such other ridge or fracture zone regions as the central Atlantic, the Gulf of Mexico, the Caribbean, the Arabian Sea, the Gulf of Aden, the Red Sea, and the Gulf of Suez. These oceanic deposits contain diatom species living in deep (but not necessarily cold) waters. They probably have no direct equivalents on the continents and we will not consider them further.

Diatomites of group (c) represent those of group (b) extended over continental platforms during high stands of boreal seas. They are coeval with the black, organic-rich shales attributed to anoxic events, and they are directly associated with phosphorites, though these two sediment types are much more common than diatomites. The ages of marine diatomites younger than the onset of Mesozoic drift, on continental platforms, are therefore forecastable by petroleum geologists, because they are of essentially the same ages as the principal young oil-source sediments—Kimmeridgian (in the Jura Mountains, coeval with the North Sea source sediment); Cenomanian-Turonian (especially in the Ukraine, coeval with Oceanic Horizon B); Santonian-Campanian (on the Russian platform and in the West Siberian Basin, coincident with the peak of phosphorite occurrence); Paleocene through Lower Eocene (in Denmark and the Caspian, Volga, and West Siberian regions of the USSR, essentially coincident with Oceanic Horizon A); and from the Middle-Upper Miocene boundary through the Lower Pliocene (in the U.S. Atlantic states, coincident with the principal Cenozoic source sediments).

But a conclusion that marine diatomite extensions over the continental platforms should be areally comparable with those of oil-source sediments would misinterpret the instigating control of diatomite deposition. Platformal diatomites are essentially *absent* from the onshore stratigraphic records of

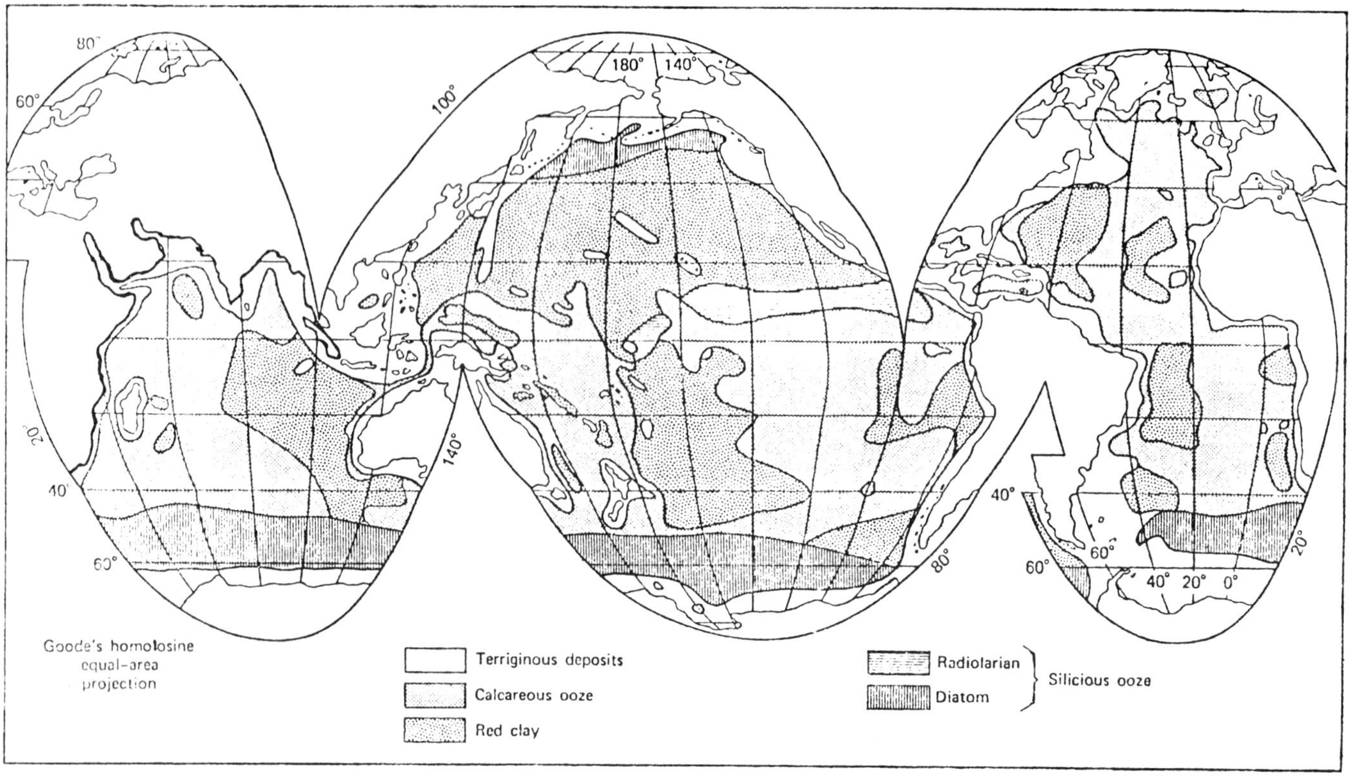


Figure 1. Distribution of sediments in modern oceans. Diatom oozes are essentially restricted to the northern Pacific rim and Circum-Antarctica. DSDP data have shown that radiolarian oozes are much more widespread than indicated. (From Menzies et al., 1973; after Kuenen. 1950).

Canada, Mexico, South America, Africa south of the equator, the Middle East (including Egypt), India, and Australia (and possibly from Antarctica also). China is unknown diatomite territory, to this author at least.

The Circum-Pacific diatomites, in fact, fall entirely within group (b), which therefore provides the chief concern of this paper. The diatomites are still associated with phosphorites, and their deposition was probably influenced by cold, upwelling waters, but they are not themselves cold-water deposits; they are by no means restricted to high latitudes; and they are not oceanic deposits. Their provenance may be elucidated if we consider first their better-known counterparts, of the same Neogene ages, in the European Tethyan belt.

## TETHYAN DIATOMITES

Oligocene or Miocene marine diatomites almost completely circumscribe the western Mediterranean, from northern Algeria through southern Spain, southernmost France, and around the northern and eastern rims of the Italian Apennines into southern Sicily (the type locality of the diatomite called tripolite). From the Adriatic Sea, mid-Tertiary diatomites continue through the northern Dinarides, around the front of the Carpathians and the Balkan Ranges to the shore of the Black Sea, across to the Kerch Peninsula and through the eastern Caucasus into the Caspian Sea (Figure 2).

The occurrences are clearly restricted to the Alpine fold belt, nearly all of them lying along the thrust-faulted frontal (external) zones. They reveal no association with either ocean floors (represented by ophiolites) or cold-water faunas. The European Neogene was split between two distinct marine provinces, separated by the belt of Hercynian massifs (Figure 3). The northwest province was boreal, and included the North Sea and the European Arctic. During the late Jurassic and Cretaceous, the predecessor of this basin received the sediments containing the familiar boreal ammonite and belemnite faunas; it also received the only Cretaceous and Paleogene diatomites known in Eurasia (Figure 4). During the Neogene, in contrast, all known diatomites are found in the southern basin (Figure 3). The deposits contain warm-water foraminiferal and molluscan faunas, and major evaporites (including the Messinian evaporite of the Mediterranean).

The diatomites of the Tethyan Tertiary are unquestionably the deposits of shallow lagoons, containing warm waters driven into embayments within rising mountain arcs during tectonically-instigated transgressions. They are the final deposits of external flysch successions, and are themselves commonly capped by evaporites.

The western Tethys was closed during the collision of two major continental plates, between which lay a mosaic of microplates (Figure 5). The preserved diatomites, originally the deposits of the arc-trench gaps and the immediate back-arc basins, now appear as thin, discontinuous slivers in folded and faulted flysch successions partially subducted between plates and accreting microplates. Their deposition was controlled by tectonism and volcanism, not by eustatism, water depth, or latitude.

## PACIFIC DIATOMITES

Pacific diatomites have been thought of in quite different terms. Because the present "home" of diatoms in the northern hemisphere is the Bering Sea, and the two best developed onshore Neogene diatomite successions are those of Japan

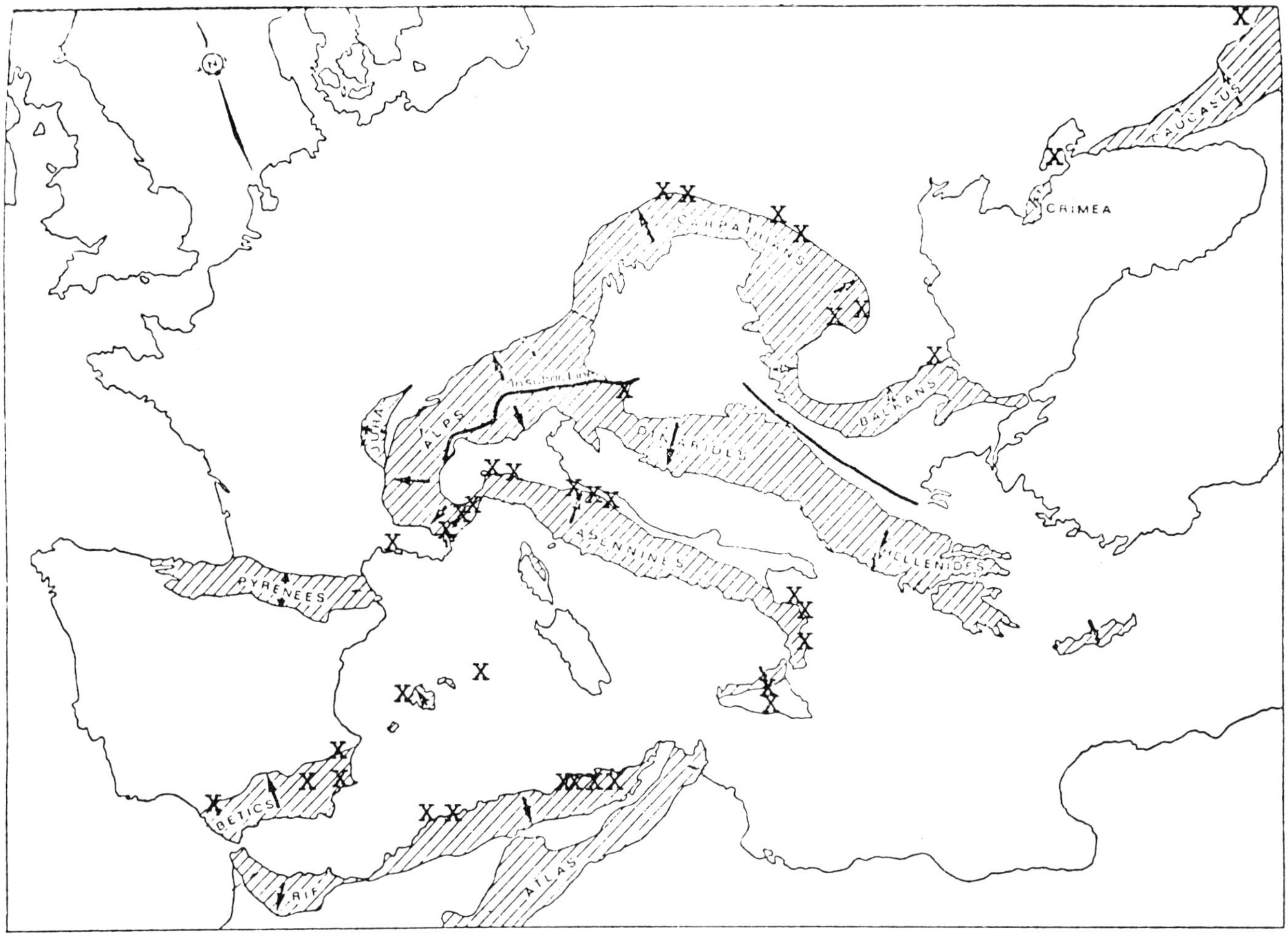


Figure 2. Alpine fold belts of Europe, showing directions of main deformational movements. Mid-Tertiary diatomite deposits marked schematically by x. (Base map after Ager, 1975).

and California, the impression has been created that the cold North Pacific, at least, is largely underlain by diatom oozes extending back in time at least to the Middle Miocene (Figure 6). Where these oozes are accessible to the drill (onshore, or in shallow coastal waters), they are thought to be the source rocks of prolific oil and may, in addition (where fractured), be the reservoir rocks also.

In fact, the diatomite occurrence principally responsible for this impression—the Monterey Formation of California—can easily be shown to be unique. All other Circum-Pacific diatomites of Miocene age are closely comparable with those of the Tethyan belt. The only substantial difference is in the framework of the tectonic control. Whereas the arcs and trenches of the Tethyan belt became involved in the Alpine collisional orogeny, those of the Circum-Pacific have been continuously involved in Andean or Aleutian subduction processes.

In the western and northern Pacific rim (Figure 7), the Aleutian-type margins are formed by mature island arcs and accompanying trenches, separated from the continents by wide back-arc basins. All exposed Neogene diatomites are on islands or rocky peninsulas within the back-arc basins or flanking them (in the seas of Japan and Okhotsk and the Bering Sea) or on isolated islands adjacent to the trenches in front of the arcs. They are not parts of the normal coastal plain successions, and they nowhere extend into the interior of any landmass even where marine Neogene strata are well developed. Along the eastern Pacific margin, the arcs have been added to the continents in the Andean pattern, and the back-arc basins have become petroliferous foredeeps between the cordillera and the shields. The only exposed diatomites are again on mountainous islands or peninsulas—except, of course, for our old friend the Monterey diatomite.

Coastal Central and South America provides an instructive example. From the Clarion fracture zone to the Gulf of Guayaquil, Central America and coastal Colombia and Ecuador border the Cocos Plate. The rapid spreading of this plate may reflect a constitution wholly oceanic. The rocks exposed along its eastern and northeastern borders are of oceanic crustal derivation, backed by a festoon of young volcanoes. There is no arc in the tectonic sense, and diatomaceous sediments lie offshore along the edges of the trench. No onshore occurrences of diatomite are known to me. From the Gulf of Guayaquil to beyond Isla Chiloe at 43 °S, the linear coastlines of Peru and Chile border the Nazca Plate. The linearity of the coasts is punctuated by a series of small peninsulas consisting of old, resistant rock (everywhere igneous or metamorphic, and possibly representing allochthonous fragments accreted to the continent during its westward drift). Every single penin-

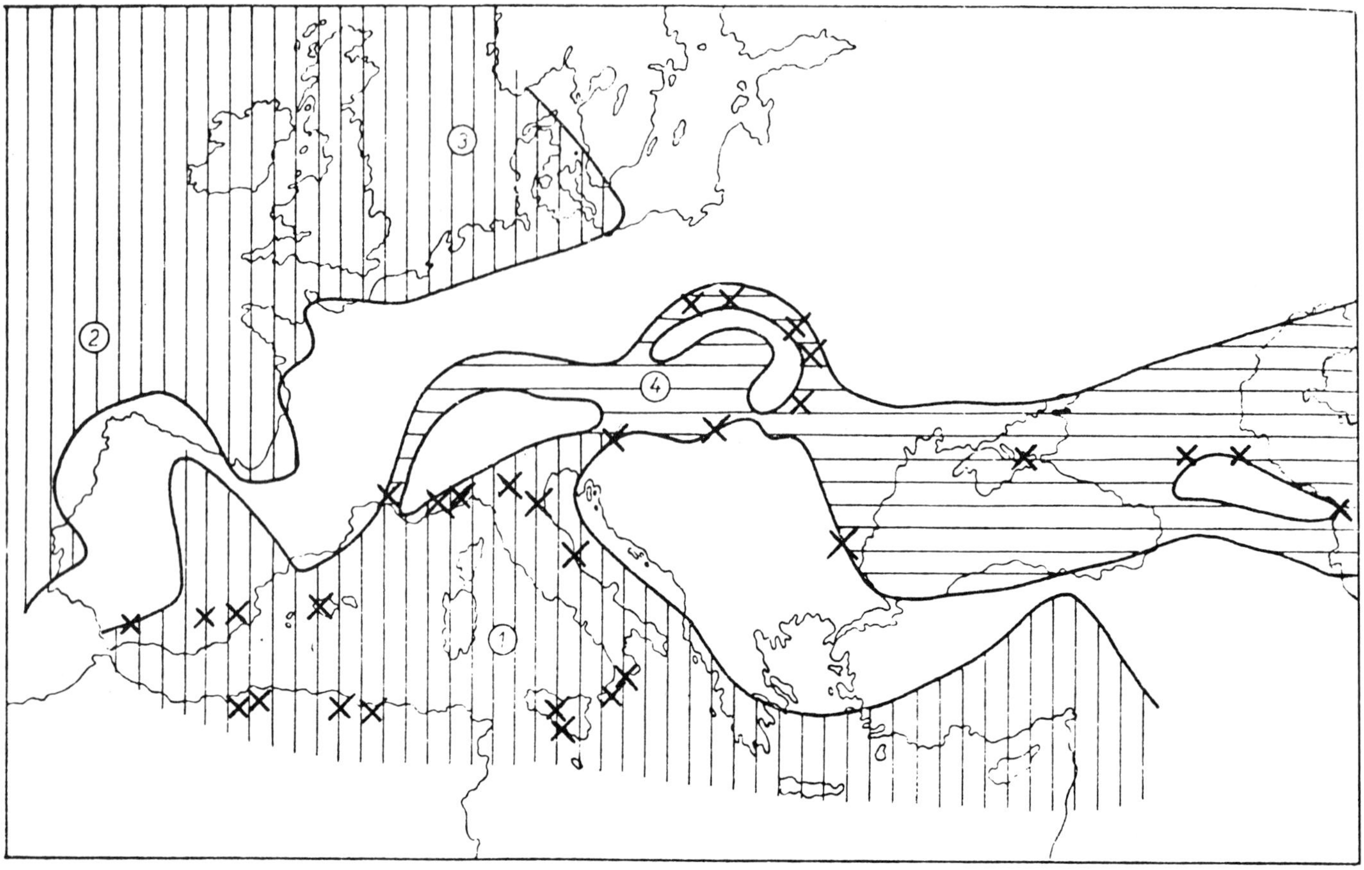


Figure 3. Neogene bioprovinces in Europe. Diatomite deposits marked schematically by x. (Base map from Cicha, 1970).

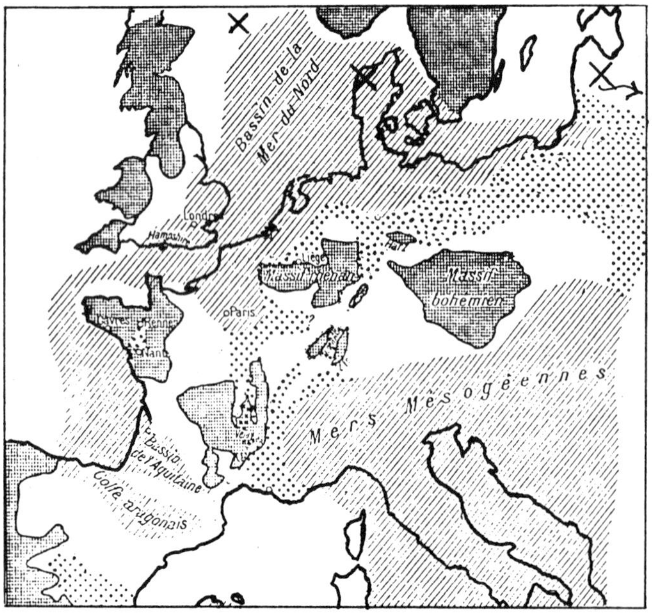


Figure 4. Paleogeography of Europe during the Paleogene, showing Eocene seaways (diagonal ruling) and Oligocene extensions of these (dots). Emergent Hercynian massifs (cross-hatched) separate a northern (boreal) basin, containing the only known diatomites of pre-Oligocene age (x), from a southern (Tethyan) basin with no known diatomites. (Base map after Gignoux, 1936).

sula is backed by a small area of marine Neogene strata, most if not all of them including diatomites—at Sechura, Peru (5 °S; Figure 8); along the Paracas Peninsula between Pisco and Yauca (15 °S; Figure 9); at Mejillones, Chile (23 °S; Figure 10); at Caldera (27 °S); at Tongoy, near Coquimbo (30 °S; Figure 11); at Tiltil, near Valparaiso (33 °S); possibly at Arauco and Lebu (37-38 °S); and at Taytao (46 °S). The diatomites found behind these peninsulas are otherwise unknown in the entire continent.

Like their Tethyan equivalents, the Circum-Pacific Neogene diatomites are characteristically underlain by phosphorites and overlain by evaporites. The phosphorite component is no doubt attributable to upwelling of Antarctic waters at the continental margin, but the sediments would have been subducted beneath that margin if oceanic crust were all that arrived at the trenches. Where fragments of older, resistant sialic crust arrived there, the narrow waters between these fragments and the arcs (the arc-trench gaps, in fact) shallowed through lagoonal phases until they were raised along steep faults during the final stages of Andean uplift. The Neogene strata are not moderately tilted or warped, but they have not been folded, overthrust, or dismembered as their Tethyan equivalents have been.

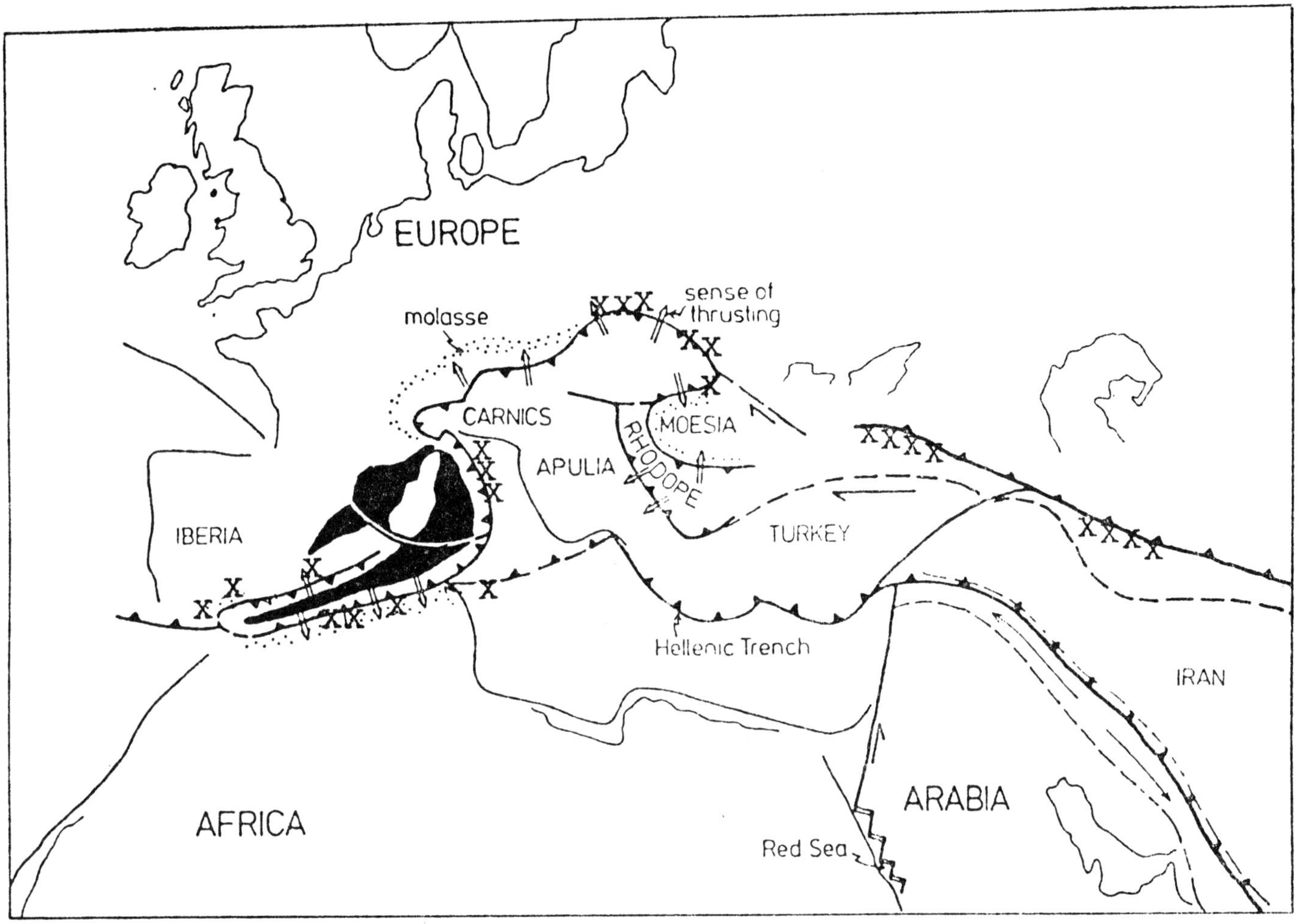


Figure 5. Proposed plate-boundary scheme for the Alpine-Mediterranean region of Europe during early Miocene time. Note relationship of Neogene diatomite deposits (x) to boundaries between plates and microplates. (Base map from Dewey et al., 1973).

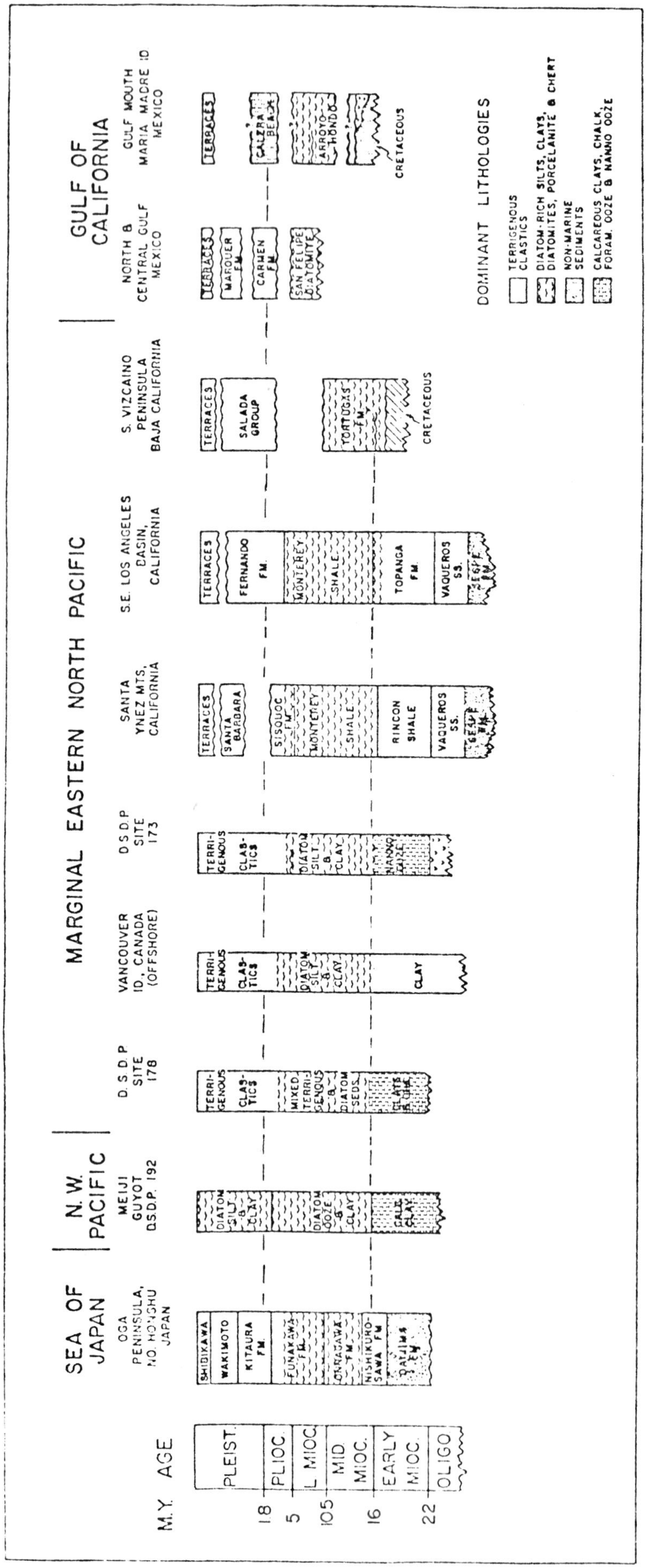


Figure 6. Correlation of diatom-bearing sequences in the Miocene of the North Pacific rim. Note that the vertical scale for the sections is in time (m.y.) and not in thickness. (From Ingle, 1980).

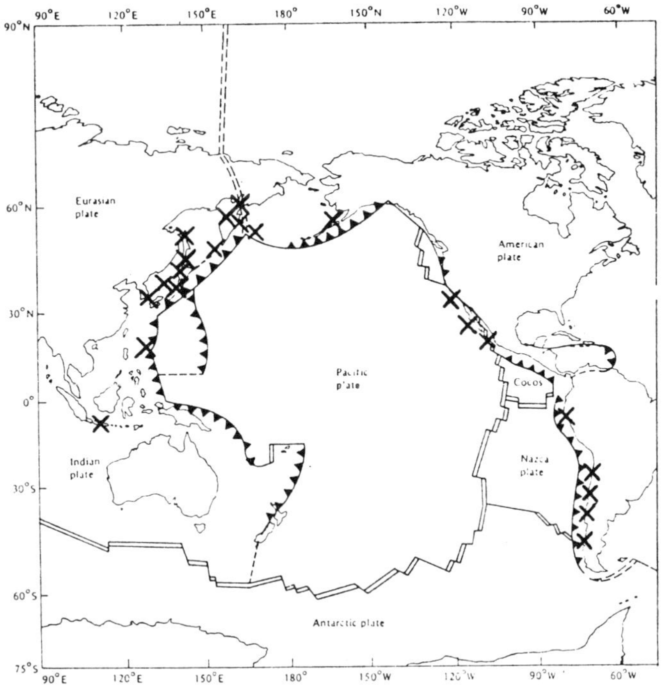


Figure 7. Boundaries of the Pacific plate and related deposits of Miocene diatomite (schematically indicated by x).

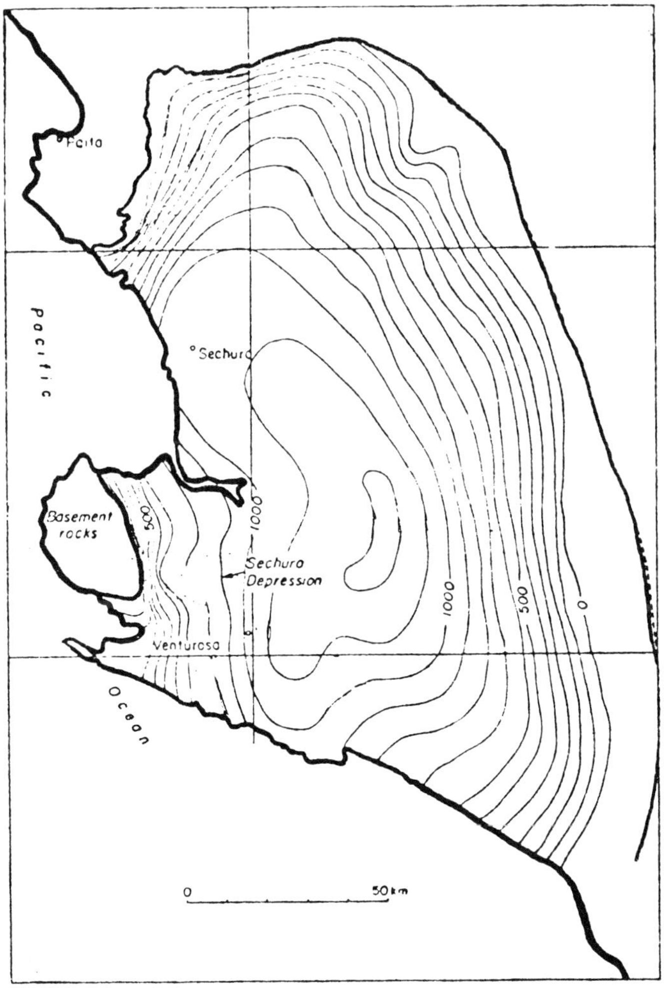


Figure 8. Isopachs of diatomite-bearing Miocene strata in the Sechura Desert, northwestern Peru. The strata occupy a small basin between uplifted basement outcrops (From Cheney, McClellan, and Montgomery, 1979).

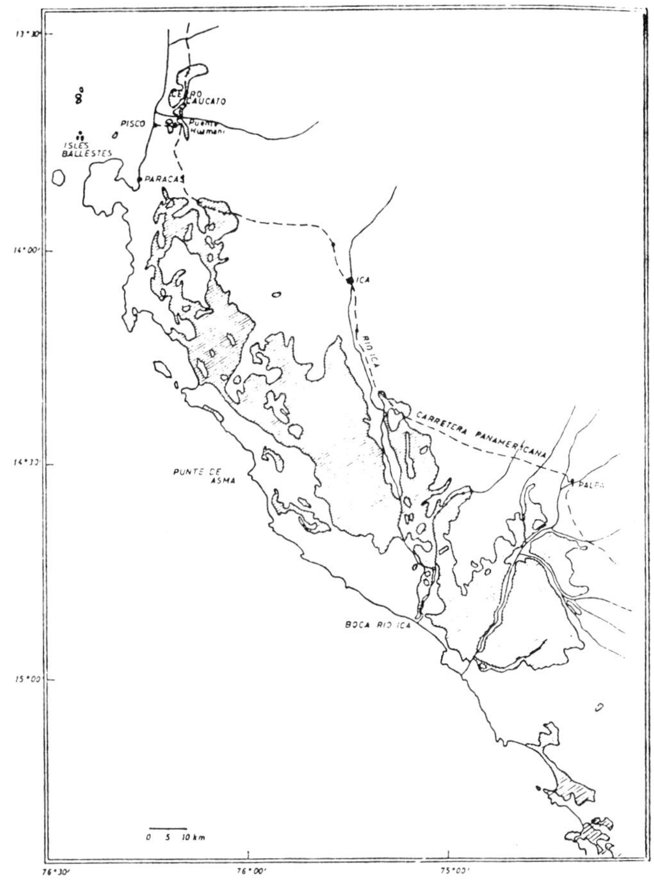


Figure 9. Distribution of Mio-Pliocene marine sediments (shaded) in the Paracas Peninsula area of Peru. The interval mapped includes the Miocene diatomite-bearing Pisco Formation, which is best exposed at Puente Huamani near the northern end of the outcrops (From Mertz, 1966).

## MONTEREY DIATOMITE

We may now consider the reason for the uniqueness of the Monterey Formation. From the Gulf of California to Cape Mendocino, there is no longer either a trench or a ridge. The arc-trench system of the early drift regime impinged upon a triple junction and was converted into a transform margin *before the onset* of diatomite deposition (Figure 12). The sediments were thereby preserved from subduction, dismemberment, or even uplift until they became folded during the Pleistocene deformation.

It is this fortuitous history that has led to the association of the Monterey Formation with prolific oil accumulations. The formation is world-famous as a source sediment, but only part of it is in any sense a diatomite, and it is unlikely that this part fulfills the source function. The more likely source sediment is the phosphatic member of the formation. In the Tethyan cases in which petroliferous successions include diatomite members (in the Carpathian and Caucasian basins most notably), those members are widely agreed not to be the source sediments.

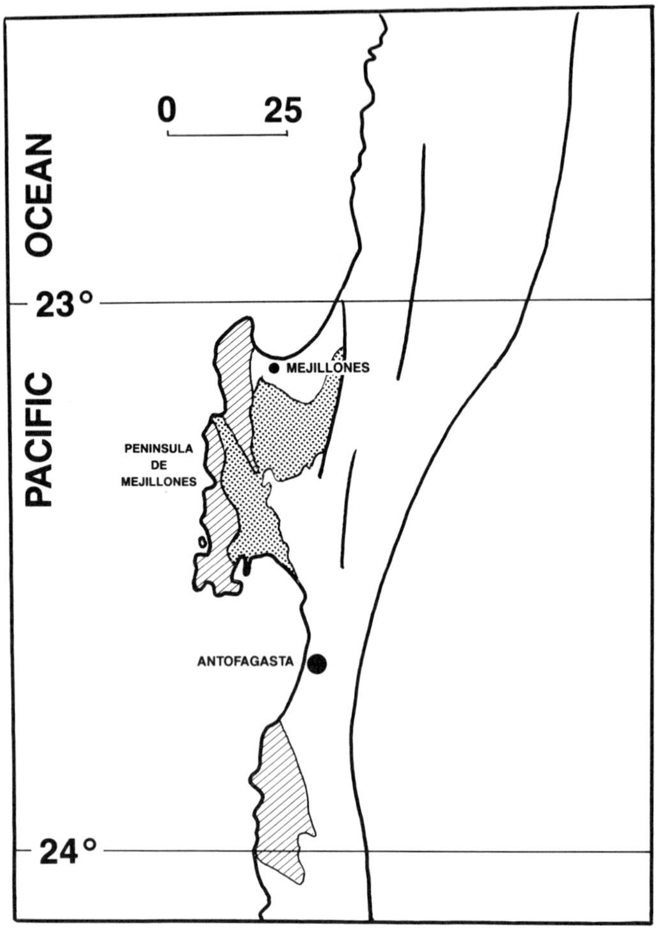


Figure 10. Miocene diatomite deposit (x) behind the basement uplift of the Mejillones Peninsula, Chile. (From the Geological Map of Chile).

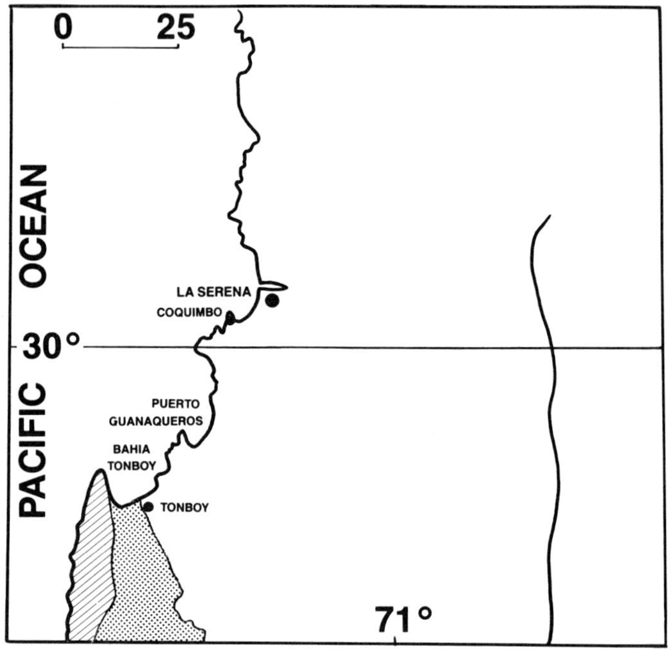


Figure 11. Miocene diatomite deposit (x) behind the basement uplift of the Tongoy Peninsula, Chile. (From the Geological Map of Chile).

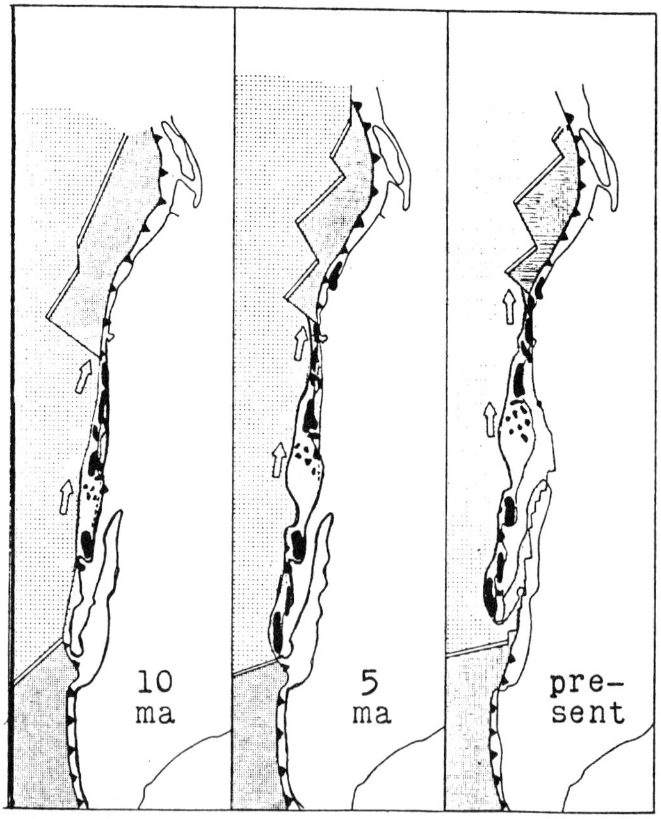


Figure 12. Spatial coincidence of Cenozoic marine basins of California and Baja California (containing diatomites) with transform margin at a converted triple junction. (From Blake et al., 1978).

## ECONOMIC EXPECTATIONS

We can have little expectation that any other Circum-Pacific diatomite occurrences will be associated with oil as rich as that associated with the Monterey. The great thickness of that formation and its preservation from dismemberment, plus the rapid accumulation of even thicker overburdens of turbidite sandstones in post-Monterey basins deeply depressed by oblique transpression, have provided a favorable geologic history not matched for diatomites along subduction boundaries.

Diatomite, however, is a versatile industrial material, and deposits in Japan, the Philippines, and Peru, at least, are sufficiently thick and pure to be minable if local demand were to justify this. If those countries have not maintained up-to-date inventories of their diatomite resources, I think they should do so.

## REFERENCES

Applegate, S.P., I. Ferrusquia-Villafranca, and L. Espinosa-Arrubarena, 1979, Preliminary observations on the geology and paleontology of the Arroyo Tiburon area, Bahia de Asuncion, Baja California Sur, Mexico, *in* Baja California Geology: GSA, p. 113-115.

Barron, J.A., 1975, Marine diatom biostratigraphy of Upper Miocene-Lower Pliocene strata of southern California: Journal of Paleontology, v. 49, p. 619-632.

Blake, M.C., Jr., R.H. Campbell, T.W. Dibblee, Jr., D.G. Howell, T.H. Nilsen, W.R. Normark, J.G. Vedder, and E.A. Silver, 1978, Neogene basin formation in relation to plate-tectonic evolution of San Andreas fault system, California: AAPG Bulletin, v. 62, p. 344-372.

Cheney, T.M., G.H. McClellan, and E.S. Montgomery, 1979, Sechura phosphate deposits, their stratigraphy, origin, and composition: Economic Geology, v. 74, p. 232-259.

Distanov, U.G., V.A. Kopeykin, T.A. Kuznetsova, and V.N. Silant'yev, 1971, Chert accumulation in Mesozoic and Cenozoic marine platform basins: Doklady Academy of Sciences U.S.S.R., v. 201, p. 83-86.

Fowler, G.A., W.N. Orr, and L.D. Kulm, 1971, An Upper Miocene diatomaceous rock unit on the Oregon continental shelf: Journal of Geology, v. 79, p. 603-608.

Frenguelli, J., 1949, Diatomeas fosiles de los yacimientos chilenos de Tiltil y Mejillones: Darwiniana, v. 9, p. 97-156.

Hanna, G.D., 1970, Fossil diatoms from the Pribilof Islands, Bering Sea, Alaska: Proceedings of the California Academy of Sciences, Ser. 4, v. 37, p. 167-233.

Ingle, J.C., 1980, Origin of Neogene diatomites around the North Pacific rim, *in* R.E. Garrison, ed., The Monterey Formation and related siliceous rocks of California: Pacific Section, SEPM, p. 159-179.

Isaacs, C.M., 1984, The Monterey—Key to offshore California boom: Oil and Gas Journal, p. 75-81.

Kanaya, T., and I. Koizumi, 1966, Interpretation of diatom thanatocoenoses from the North Pacific applied to a study of core V20-130: Tohoku University Science Reports, Ser. 2, v. 37, p. 80-130.

Koizumi, I., 1969, Tertiary diatom flora of Oga Peninsula, Akita Prefecture, northeast Japan: Tohoku University Science Reports, Ser. 2, v. 40, p. 171-225.

Loeblich, A.R., III, L.A. Loeblich, H. Tappan, and A.R. Loeblich, Jr., 1968, Annotated index of fossil and recent silicoflagellates and ebridians with descriptions and illustrations of validly proposed taxa: GSA Memoir 106, 319 p. (especially p. 15-23).

Mandra, Y.T., A.L. Brigger, and H. Mandra, 1973, Preliminary report on a study of fossil silicoflagellates from Oamaru, New Zealand: California Academy of Sciences Occasional Papers 107, p. 1-11.

Martinez-Pardo, R., and R. Caro Perez, 1980, Microfosiles siliceos de las diatomitas de Tongoy, Provincia de Coquimbo, Chile: Su significado biocronoestratigrafico, biocronogeologico, paleoecologico y paleogeografico: Revista Geologica de Chile, n. 10, p. 33-53.

Mertz, D., 1966, Mikropaläontologische und sedimentologische Untersuchungen der Pisco-Formation Südperus: Palaeontographica, Abt. B, v. 118, p. 1-48.

Rodnikova, R.D., K.M. Sevost'yanov, and A.Y. Taboyakov, 1968, Structural formation connection between southern Sakhalin Island and Hokkaido Island in relation to hydrocarbon potential: International Geology Review, v. 10, n. 7, p. 749-760.

Shmidt, O.A., 1974, Tectonic development of Komandorskie Islands: Geotectonics, n. 6, p. 377-383.

United States Government Printing Office, Washington, D.C., 1973-1980, Initial reports of the Deep Sea Drilling Project; especially volumes 22, 24, 28, 29, 32, 35, 38, 67, 75, 85, and 86.

Viteri, A.E., 1979, Geologia y recursos minerales de la Formacion Huaylas en el Altiplano de Arica: Actas 2nd Congreso Geologico Chileno, Arica, v. 2, p. C57-C68.

CHAPTER 18

# BIMA FIELD, INDONESIA, A SLEEPING GIANT

**Scott W. Prior**
*ARCO International Oil and Gas Company*
*Los Angeles, CA*

ABSTRACT

The Pertamina/ARCO Bima field, 50 miles (80 km) north of Jakarta, Indonesia, traps over 1 billion barrels of oil in Neogene carbonate and clastic reservoirs. The Bima structure was first drilled in 1974, but ZZZ-1 tested poorly, was abandoned, and the field was considered to be a noncommercial stratigraphic accumulation. Numerous reinterpretations finally provided reason to retest the structure, and ZU-1 became the discovery well 9 years after the first test.

About two-thirds of the field's 100 million barrels of primary recoverable reserves are within a northward-thickening wedge of Miocene Batu Raja Limestone. Additional pay is in the deeper, fractured portion of the Batu Raja and in thick, clean Talang Akar sands above basement gullies.

The northern third of the field is being developed from seven, nine-slot platforms. Development will include several horizontal wells to increase productivity and reduce the chance of gas coning. Production start-up is expected in late 1986 and phase one development is to be completed by July 1987. Peak production should be about 55,000 BOPD. The crude will move to the Pertamina/ARCO P flow station via a 45-mi (72-km) pipeline.

## INTRODUCTION

The Pertamina/ARCO Bima field is a shallow subgiant at the north end of the Seribu Platform, 50 to 60 mi (80 to 96 km) north of Jakarta, Indonesia (Figure 1). The Seribu Platform is a broad, shallow basement high that dips gently northward from west Java. It adjoins the Sunda Basin to the west, where a maximum of 14,000 ft (4270 m) of post-middle Oligocene sedimentary rocks were deposited. The contact between these two provinces is the Thousand Islands fault zone. Northeast of Bima the Seribu Platform rapidly dives into the North Seribu Trough, an Oligocene/Miocene half-graben containing 9000 ft (2745 m) to 11,000 ft (3355 m) of sedimentary rocks. This geological setting was conducive to the generation, migration, and entrapment of several billion barrels of oil and a significant amount of gas.

Bima field is one of the larger discoveries in the Pertamina/ARCO Offshore Northwest Java Production Sharing Contract. About 100 million barrels of oil and 70 billion cubic feet of gas are thought to be recoverable by primary production methods. The field is about 10 mi (16 km) long, has a productive area of 25 $mi^2$ (65 $km^2$), and is full to spill point.

Wildcat well ZZZ-1, drilled in 1974, was the first well to test the Bima high and might have been the discovery well had the structure, stratigraphy, and fluid characteristics of the reservoired hydrocarbons been understood at that time. ZZZ-1 just caught the downdip northern flank of the field, tested poorly from the Batu Raja Limestone, and was abandoned as a well with oil shows. The true significance of this well was unknown for another 9 years until the ZU-1 well "rediscovered" the field in November 1983.

This paper details the exploration history of the field and many factors that contributed to the delay between the drilling of ZZZ-1 and ZU-1. The geological history of the primary reservoirs is discussed, and development plans and drilling results through March 1986 are summarized.

## EXPLORATION HISTORY

The first Indonesian offshore production sharing contract was awarded in August 1966 to the Independent Indonesian American Petroleum Company (IIAPCO). The Offshore Northwest Java Contract area consisted of some 13.4 million acres in the Ardjuna, Jatibarang, and Sunda basins (Figure 2). IIAPCO assigned operatorship to Sinclair Oil in 1967 while retaining an interest in the acreage. Drilling began in 1968 and Atlantic Richfield Indonesia, Inc. became operator the following year upon merger with Sinclair Oil. The concession's first oil discovery, E-1, was drilled as the fifth test in the contract area in January 1969. The next 3 years saw an active exploration campaign, appraisal drilling, and development of the E and B discoveries.

Early ARCO successes in the Ardjuna Basin were primarily lower to middle Miocene Cibulakan Formation discoveries (Figure 3). IIAPCO, operator in the adjacent Southeast Sumatra Production Sharing Contract area in the Sunda Basin, was making significant discoveries at this time in the Oligocene to

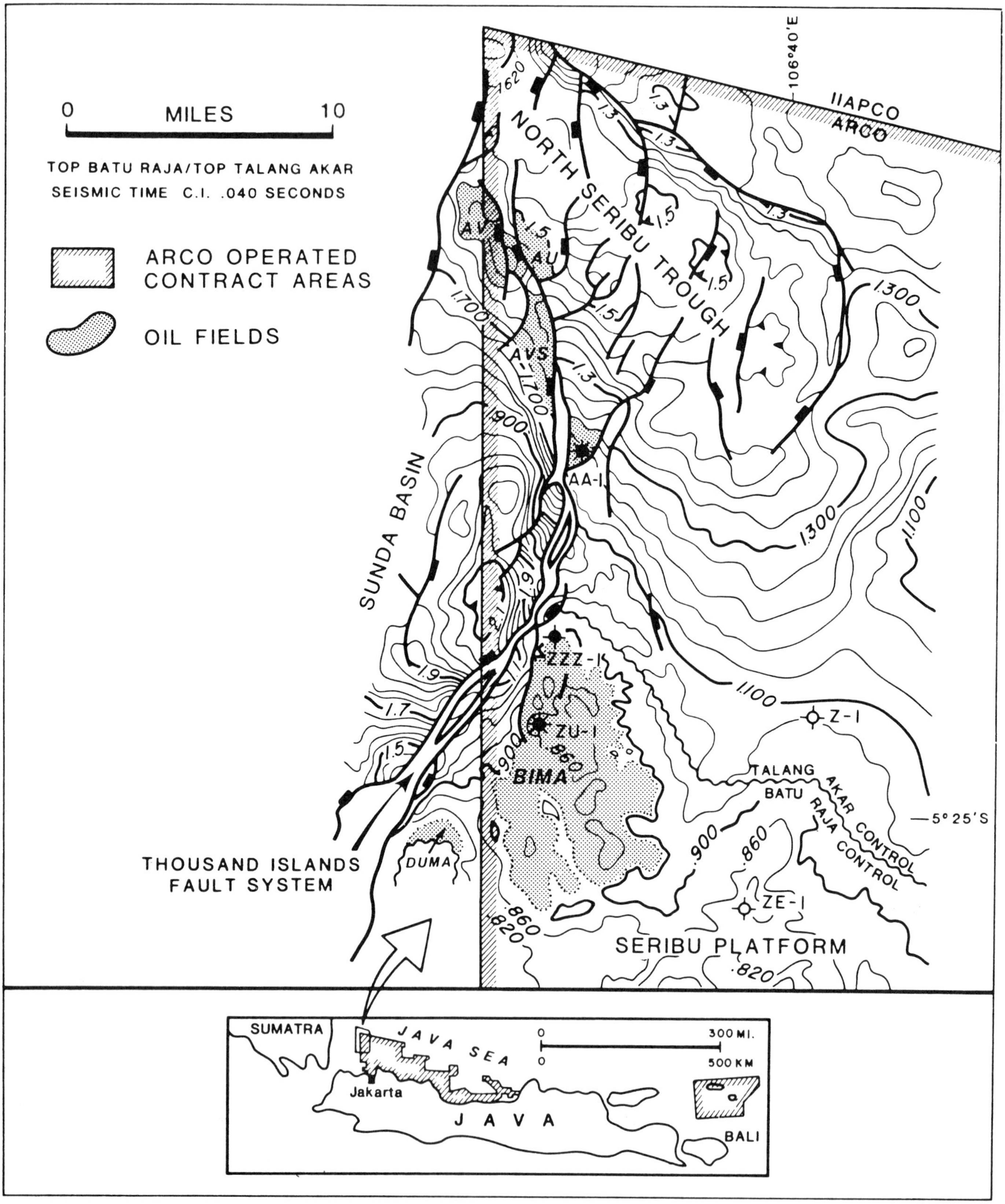


Figure 1. Index map locating the Bima field at the north end of the Seribu Platform in the Java Sea. Mapped seismic horizon is top Batu Raja Limestone or top Talang Akar Formation as noted.

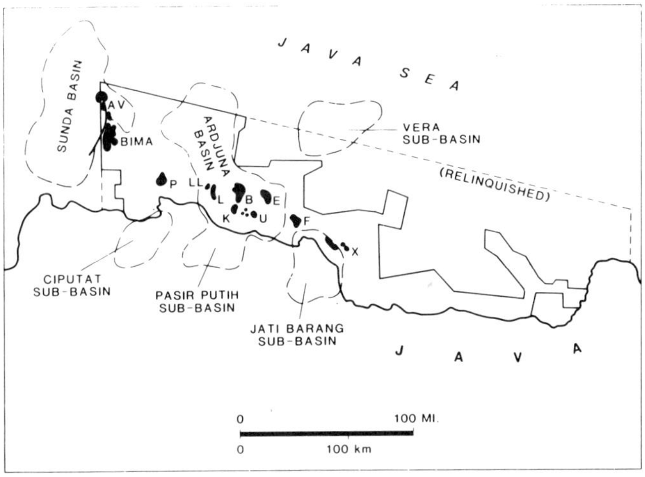


Figure 2. Location of major fields and sub-basins within the Offshore Northwest Java Contract area. Labeled fields, except for Bima, were found prior to drilling ZZZ-1 in 1974.

lower Miocene Talang Akar and Batu Raja formations (Figure 3). ARCO explorationists worked up several plays with Batu Raja and Talang Akar objectives along the east flank of the Sunda Basin in late 1973 and early 1974 prior to drilling ZZZ-1. One small Talang Akar discovery and one Batu Raja and Talang Akar discovery resulted from these efforts.

The ZZZ-1 test was proposed in June 1974 after a seismic interpretation (Figure 4) suggested the presence of a large, low-relief closure at basement and Batu Raja levels at the north end of the Seribu Platform. Primary targets were a possible reefal development of the Batu Raja Limestone and Talang Akar sandstone reservoirs expected to pinch out between basement and the Batu Raja Limestone. The well was positioned near the northern limit of the Batu Raja closure to test a maximum thickness of prospective Talang Akar section. The well came in 36 ft (11 m) and .020 seconds low to the expected top Batu Raja, but the bottom hole group of wireline logs showed a 49-ft (15-m) oil column on water in the Batu Raja and a possible thin oil pay zone in the Talang Akar (Figure 5). Although the Talang Akar Formation had very poor reservoir development, the exploration staff was encouraged by the Batu Raja possibilities and immediately requested permission to drill an updip stepout. Drill-stem testing of ZZZ-1 swabbed 2 barrels of oil from a thin Talang Akar sand, and the upper part of the Batu Raja flowed 87 BOPD (reported to be a 11° API oil-water emulsion). This certainly dampened enthusiasm for a second well, and ZZZ-1 was plugged and abandoned as a well with oil shows.

Water and oil analyses from the upper Batu Raja test were received after ZZZ-1 was abandoned. They showed the recovered fluid was indeed an oil-water emulsion and that the oil gravity was 24.4°. Analysis showed the recovered water to be mostly acid water and completion fluid; thus the reservoir had not cleaned up when the test was terminated. In retrospect, it is evident that the significance of the water analysis was not appreciated until after the ZU-1 (Bima) discovery well was drilled.

The ZZZ-1 test was immediately eclipsed by several new discoveries in the contract area. Throughout the remainder of 1974 and into 1975, six rigs were kept busy developing the X field and delineating the new AV and K discoveries (Figure 2). Substantial effort was channeled into the already identified productive trends in the Ardjuna Basin, since production facilities were already installed and additional discoveries could be tied in easily and economically.

By late 1975 almost all ARCO drilling ceased (Figure 6) as two important events took place. The first was a change to the production sharing contract, which made it difficult to determine future profitability based on new capital investments. The second was an unfavorable tax interpretation by the U.S. Internal Revenue Service, which did not allow the offset of Indonesian tax payments against US taxes (T.N. Machmud, personal communication, 1985).

Pertamina, in 1977, devised several incentives relating to accelerated cost recovery and rewrote the production sharing agreement to redress the IRS ruling. By 1978 ARCO had concluded that Indonesia again provided a favorable investment climate, and our backlog of prospects began to be drilled.

Seismic studies continued during the phase of slack drilling activity. A 1976 regional interpretation showed the Batu Raja with gentle north dip along the Seribu Platform and with only a single contour of closure updip from ZZZ-1. The study concluded that no step-out well to the ZZZ-1 test was warranted owing to the small size (300 acres) of the closure, reported 11° oil in ZZZ-1, and lack of Talang Akar sandstone development in the well.

In 1978, the ZZZ area was reinterpreted a third time using the same seismic data base as in the 1976 work. The top Batu Raja map was not significantly different, but a possible reef buildup was again interpreted on the north flank of the basement high (Figure 7). A stratigraphic model of oil entrapment within the reef was suggested to explain in the ZZZ-1 accumulation, since the lowest known oil fell outside the lowest interpreted closed contour. A step-out well was proposed one-half mile (805 m) south-southwest of ZZZ-1 on the top of the small closure. Pump testing the well was also recommended, since our engineering group had recently estimated ZZZ-1 could have been pumped at a rate of about 1000 BOPD with a drawdown of 500 psi.

Delineation well ZZZ-2 was proposed to management in July 1979. Approval was deferred since additional marine seismic data over the feature were scheduled to be acquired in a few months time. It was anticipated that the new data might better define the postulated carbonate buildup. When the 1979 seismic data were integrated, there was no significant change to the structural picture, but the Batu Raja was thought to become very thin as it approached the basement high to the south, and the "reefal" character of the Batu Raja was again questioned.

In July 1981 approval was again sought to drill ZZZ-2 at the previously proposed location to test the productivity of the Batu Raja reservoir on pump. The district office was asked for a well package with revised maps and economics. However, additional seismic data detailing the ZZZ feature were scheduled to be acquired in November 1981, and a seismic trade was being arranged with IIAPCO for data over the western flank of the Seribu Platform. It was deemed prudent to wait until these data were interpreted before resubmitting the well proposal and this resulted in another year delay.

The new seismic coverage was used in developing a Batu Raja interpretation in April 1982 showing two closures (Figure 8). The small northern closure was similar to that of previous

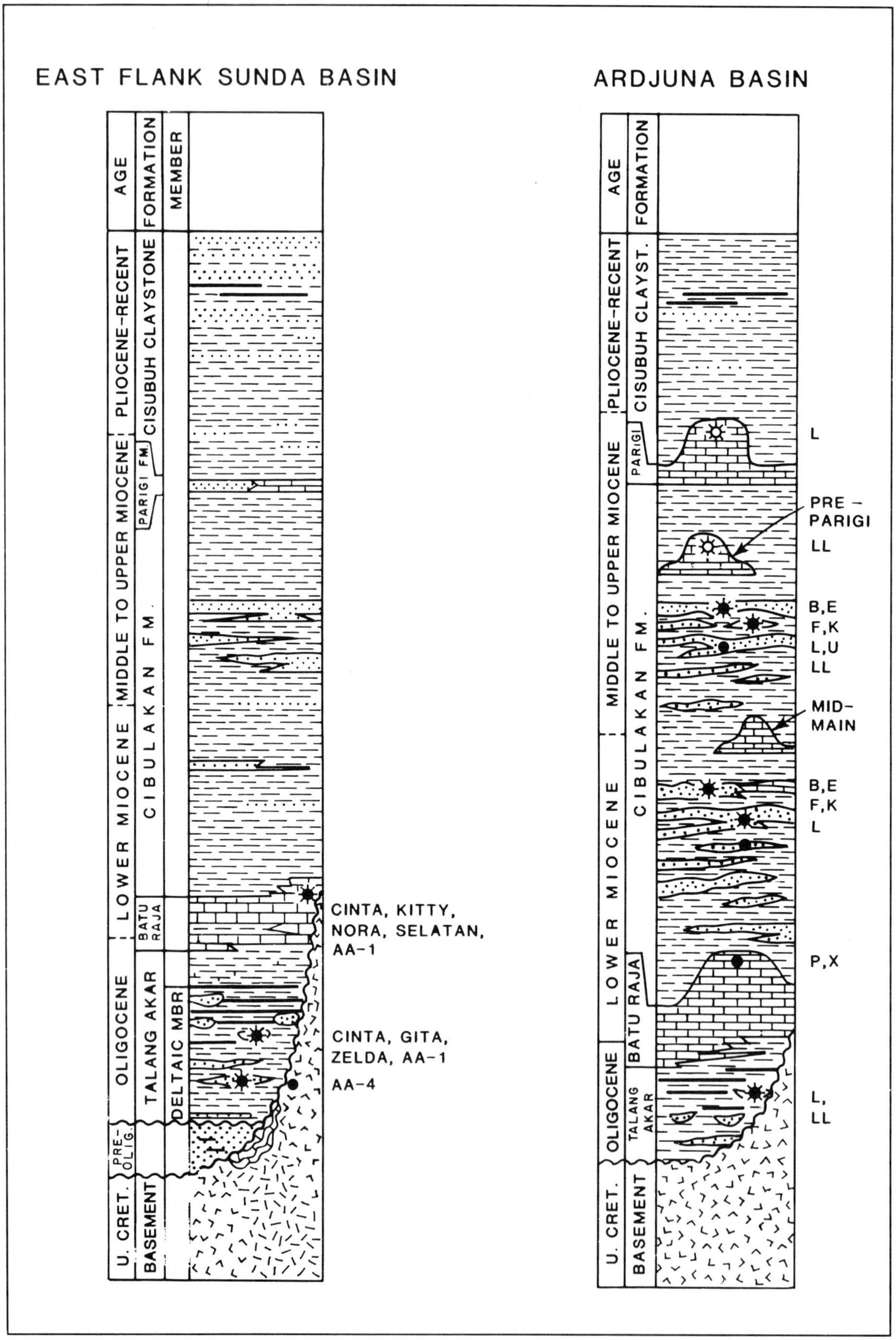


Figure 3. Generalized stratigraphy of the Sunda and Ardjuna basins. Listed fields were found prior to drilling ZZZ-1 in 1974.

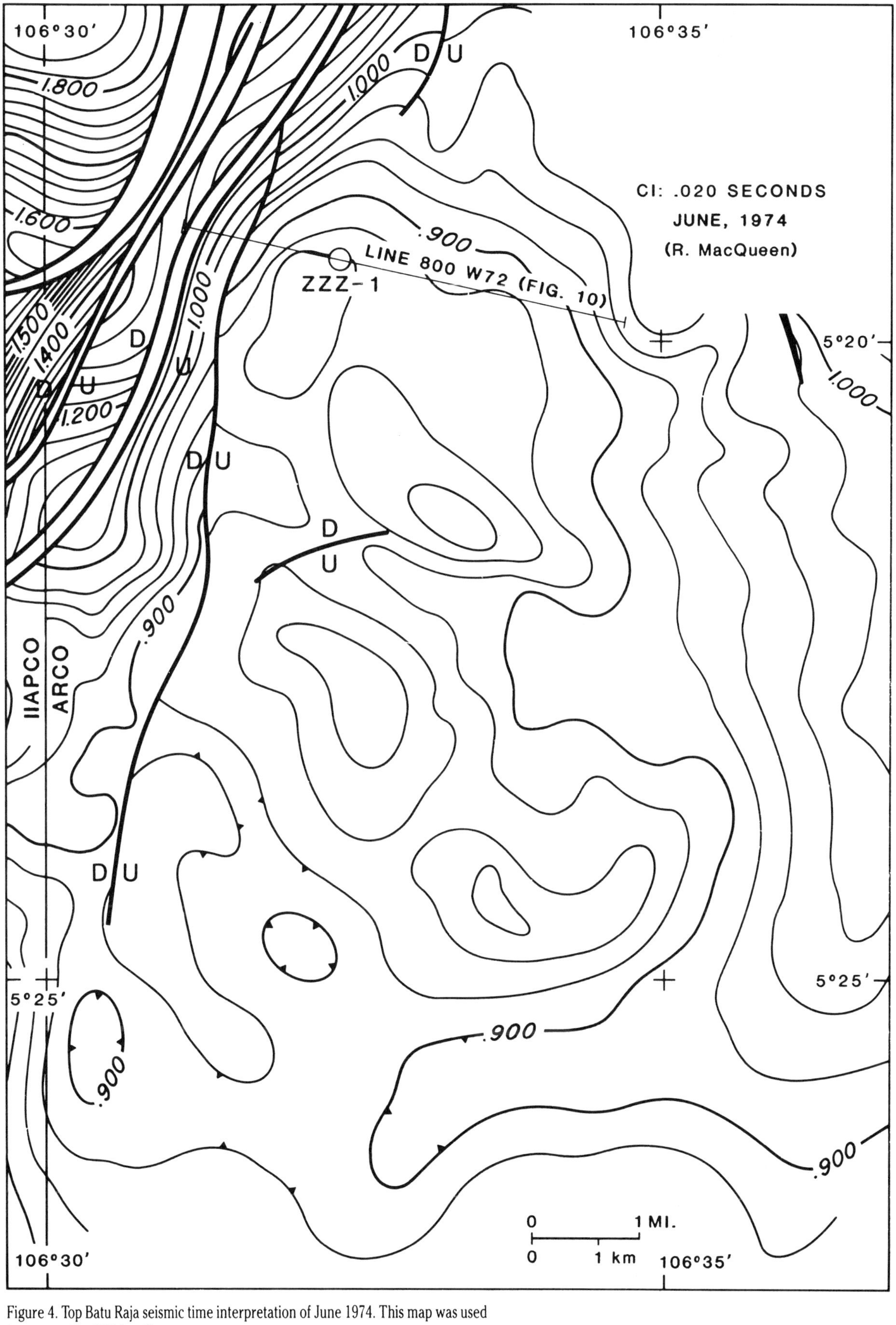


Figure 4. Top Batu Raja seismic time interpretation of June 1974. This map was used to propose the ZZZ-1 well.

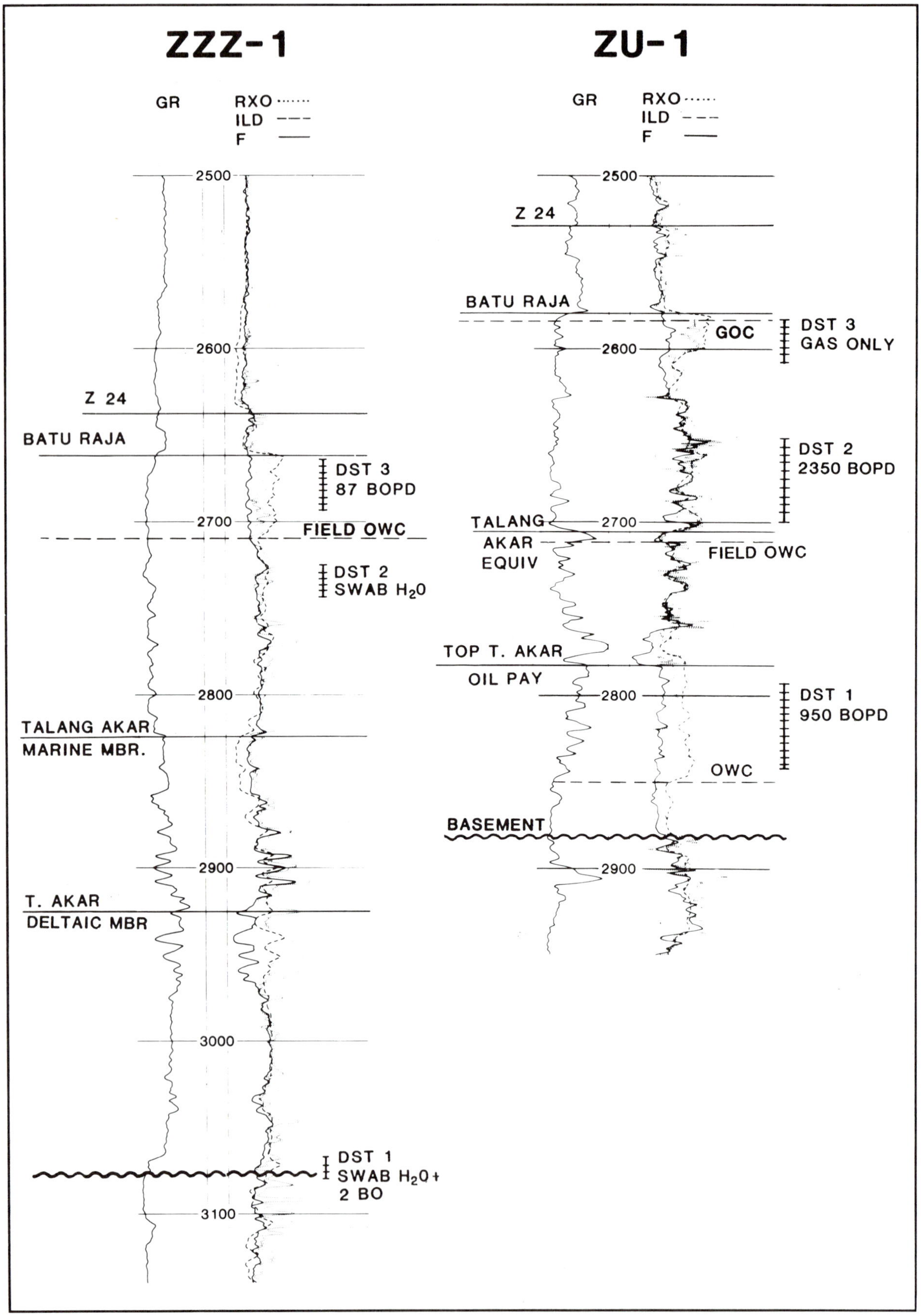


Figure 5. (A) ZZZ-1 merged F-overlay well log across the Batu Raja and Talang Akar intervals. There is a 40-ft (15-m) oil column on water in the upper Batu Raja and 4 ft (1.2 m) of possible oil indicated near the base of the Talang Akar.

(B) Merged F-overlay log of the ZU-1 Bima discovery well over the Batu Raja through basement interval. Note the spikes on the MSFL curve below 2550 ft (778 m). This interval is fractured.

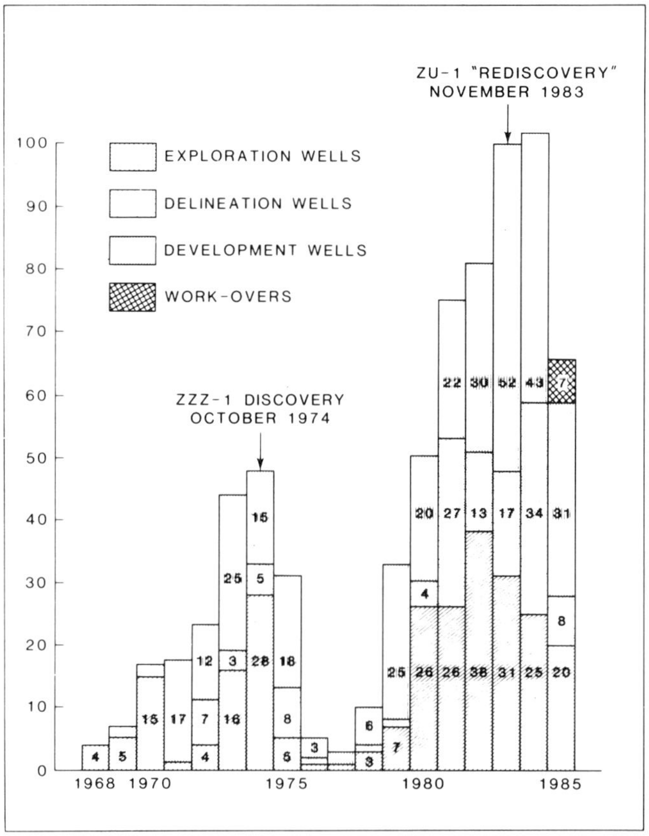


Figure 6. ARCO drilling activity in the Offshore Northwest Java Contract area from 1968 through 1985.

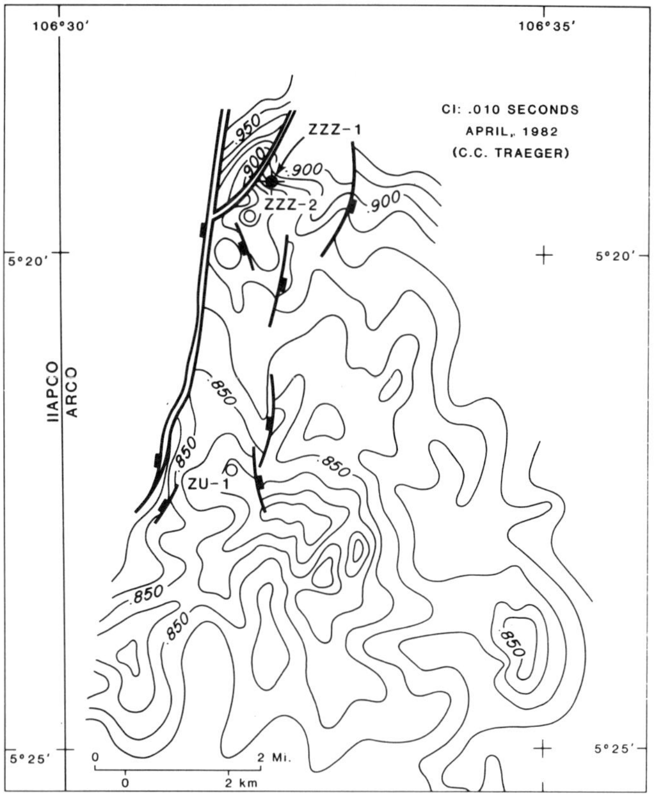


Figure 8. Top Batu Raja seismic time interpretation of April 1982. This interpreter did not recognize a reefal buildup, but he did define a significant closure 3 miles south of the ZZZ-1 well. This interpretation was used to propose ZU-1.

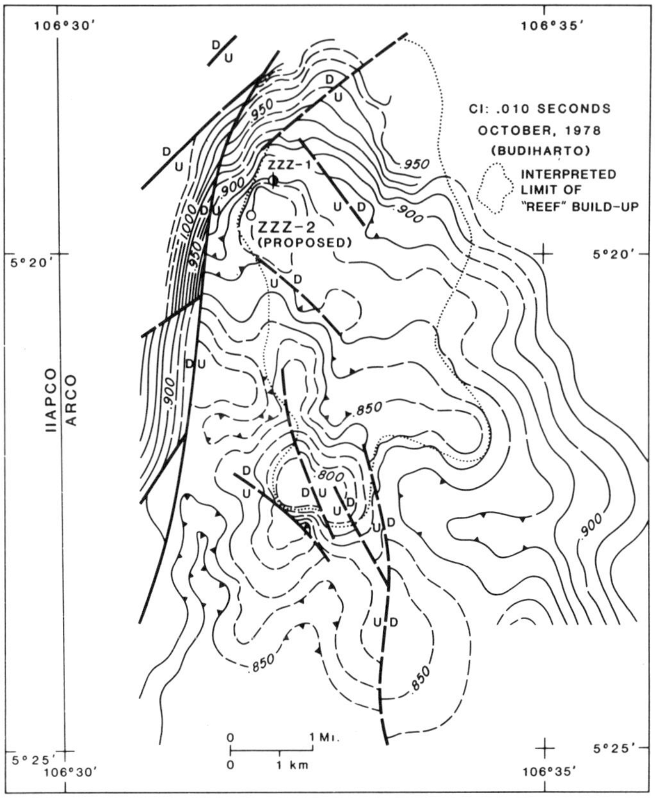


Figure 7. Top Batu Raja seismic time interpretation of October 1978. This interpretation stressed the "stratigraphic" nature of the trap owing to a poorly defined "reefal" buildup on the north flank of the basement arch.

interpretations. The southern closure, centered over the crest of the basement high, was larger in size and relief than shown on earlier interpretations. After integrating geological data, the district office agreed that the optimum test would be farther south than the original ZZZ-2 location and ZU-1 was recommended. The location was on the west flank of the southern Batu Raja closure and was above an east-facing basement reentrant (refer to Figure 13). This position was thought optimal to find stratigraphically trapped Talang Akar sands charged with oil from the Sunda Basin to the immediate west. Oil-bearing conglomeratic sands above basement gullies in the Playa del Rey field, California, were used as an analog to support this concept (Metzner, 1943).

The ZU-1 well proposal was submitted and approved in September 1983, and the well was drilled in November of that year. Wire-line logs showed at least 65 ft (20 m) net of Batu Raja oil over an interval 125 ft (38 m) thick (Figure 5). Drill-stem tests of the Batu Raja Limestone flowed at rates up to 2350 BOPD of 21° API oil. Tests of the uppermost Batu Raja were not successful owing to gas channeling behind pipe from the 4-ft-(1.2-m) thick gas cap at the top of the reservoir. Several cement squeezes were made to stop the gas channeling, but to no avail. A test of a lower Talang Akar sand flowed 950 BOPD of 20° API oil This sand body is 98 ft (30 m) thick and supported the proposed geological model.

A velocity survey was run with the bottom-hole group of logs. It suggested that the top Batu Raja reflection event was one cycle lower than previously interpreted. This was soon confirmed with a synthetic seismogram. Synthetic seismograms were not routinely made as part of the velocity survey on Java Sea wells drilled in the early 1970s. Thus, in ZZZ-1 a

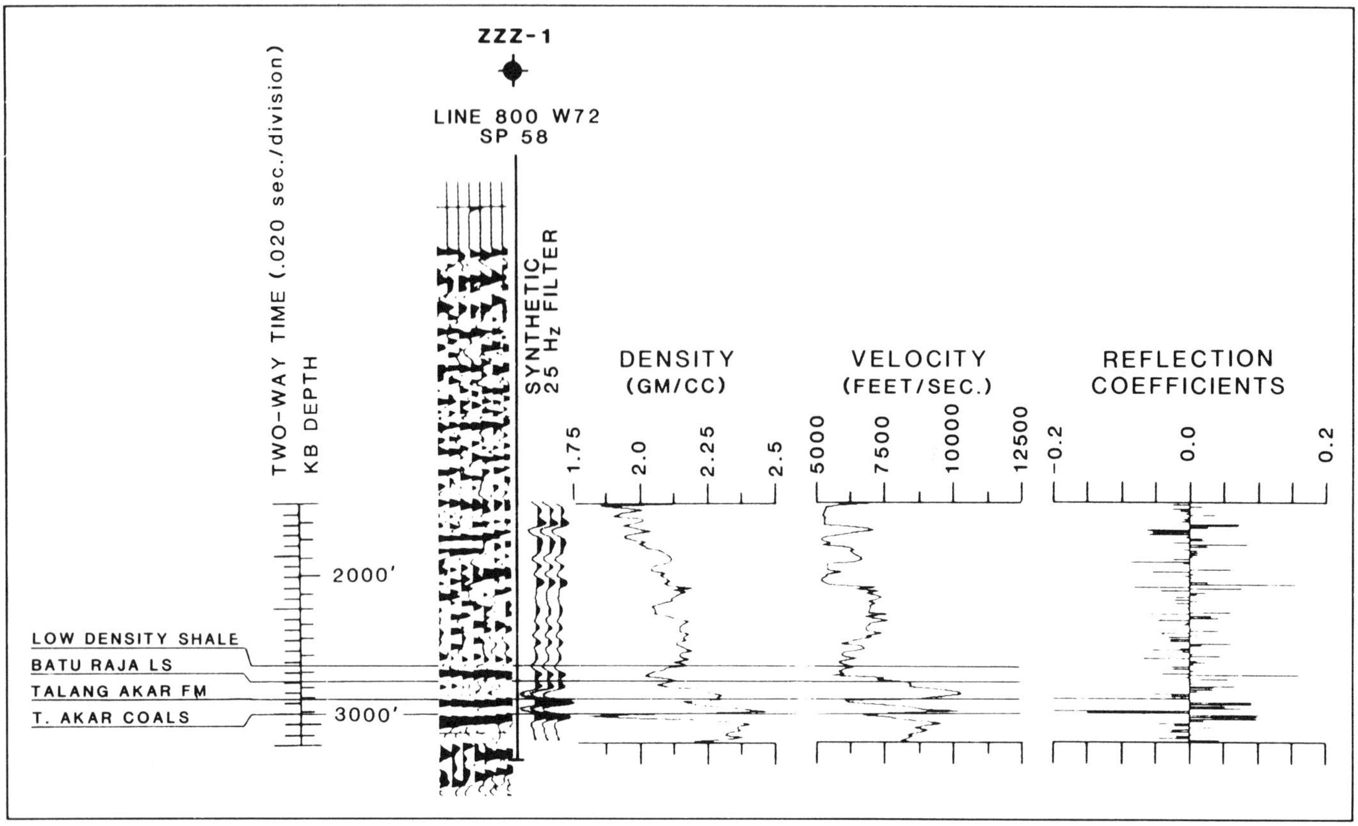


Figure 9. Synthetic seismogram constructed from ZZZ-1 check shot data and digitized well logs.

check shot survey was the only tie between well and seismic data. After the ZU-1 discovery, a synthetic seismogram was made from digitized ZZZ-1 density and sonic logs corrected with check shot data (Figure 9). It compares favorably with the trace on seismic line 800 W72, where ZZZ-1 was drilled. The upper event that was thought to be the Batu Raja was actually generated by a significant density contrast in the shale section about 80 ft (24 m) above the Batu Raja (Figure 5). Prior to ZU-1, seismic interpreters had recognized that the ZZZ-1 well-to-seismic tie was not ideal, but assumed that the event was correctly picked since it had the high-amplitude characteristics of a shale-carbonate interface. The real clincher was the apparent "reef edge" seen near shot point 113, line 800 W72 (Figure 10). Although there is a small break in slope here on the real top Batu Raja event, the upper event is actually faulted. This gives the appearance of a reef buildup. The fault was not recognized owing to the near-parallel orientation of the fault relative to line 800 W72. Northeast-trending lines acquired in 1983 clearly show this fault to be a compaction fault cutting the post-Batu Raja section and dying out downward or becoming a bedding plane fault on top of the limestone.

Seismic signature on the south flank of the field is also complicated by the Z24 limestone bed (Figure 5) which ranges in height above the Batu Raja from 25 ft (7.6 m) along the axis of the structure to about 110 ft (34 m) on the southwest flank. Before drilling ZU-1 this stringer was inadvertently picked as the top Batu Raja, since it generates a high-amplitude reflection event similar to that of the shale density contrast on the north flank of the structure.

Misidentifying the top Batu Raja and mapping on a shallower event did not permit the true extent of the closure to be discerned, since the lower event dips more steeply than the mapped horizon. This was especially critical on the south side of the high, where it had always been difficult to demonstrate any significant turnover.

As ZU-1 was being tested, a new Batu Raja structural interpretation was made using the correct seismic pick and included newly acquired seismic data over the southwest flank of the structure. These seismic lines were actually programmed to define a separate prospect, but their location and timeliness were well appreciated.

Correctly mapping the Batu Raja demonstrated that closure extended about 2 mi (3.2 km) farther south than the southern limit of most of the previous seismic maps (Figure 11). All earlier interpretations, except for the 1976 regional map, terminated at the southern edge of a seismic quadrangle base map. Most interpreters stopped mapping here because there were few seismic lines south of this area. The northern end of the Thousand Islands chain ends here, and the many offshore fish traps between the islands have made seismic acquisition difficult. The 1976 interpretation included the critical area, but not all 1968 seismic data were utilized, probably owing to the large fiducial shift and slight navigation problems with this vintage of data. The sharp, narrow syncline that closes off the south end of Bima is clearly defined on one of these lines. It is interesting to speculate that if all of the existing data had been integrated, then we might have discovered the field some 5 years earlier.

The ZU-2 and ZZZ-2 delineation wells were drilled immedi-

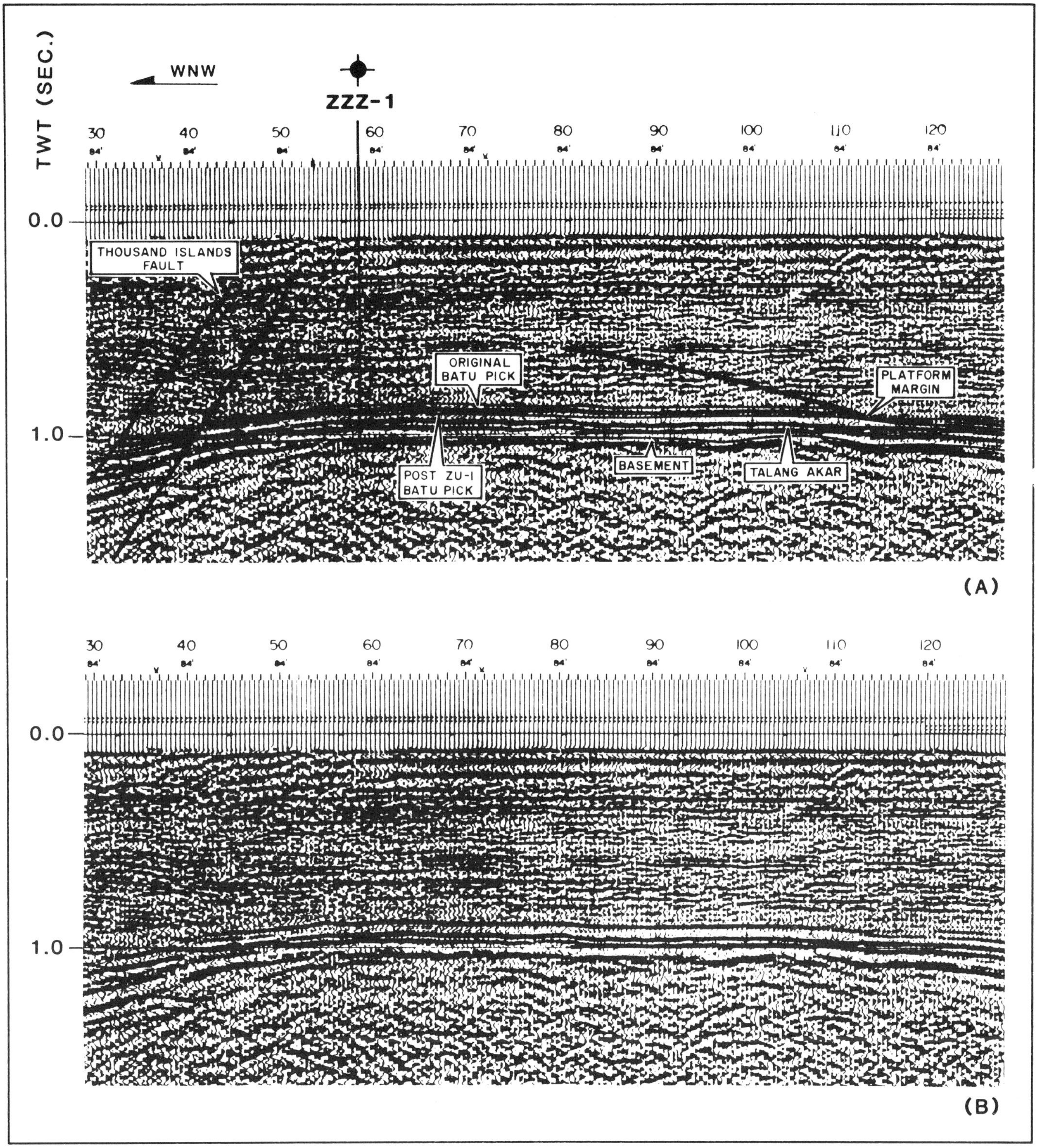


Figure 10. Seismic line 800 W72. (See Figure 4 for location). This line was used to locate ZZZ-1 and to support the reefal interpretation of the play. Note the apparent Batu Raja thinning to the right (east) of shot point 113. This is caused in part by a compaction fault above a low-relief platform margin or break in slope at the top of the Batu Raja. The low apparent dip angle results from the subparallel orientation of the fault and seismic section. (A) interpreted line, (B) uninterpreted line.

ately following ZU-1. ZU-2 was wholly within the gas cap, whereas ZZZ-2 was in the oil leg of the Batu Raja and flowed 948 BOPD. A basal Talang Akar sand with 41 ft (12.5 m) of oil on 3 ft (1 m) of water flowed 1290 BOPD of 26.7° API oil. This sand is stratigraphically separate from the ZU-1 sand as the oil-water contacts differ by about 150 ft (45 m).

The first three wells suggested the presence of oil and gas columns each about 130 ft (40 m) high. It was apparent that a significant discovery had been made.

An extensive delineation program was planned and by August 1984 fifteen additional delineation wells had been drilled (Figure 12). Twelve successful wells were suspended as

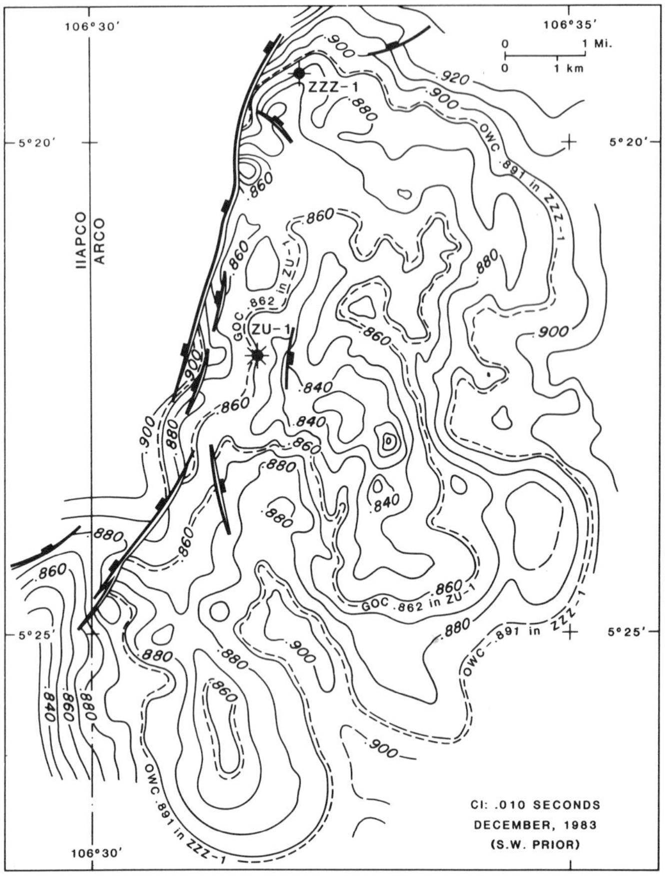


Figure 11. Top Batu Raja seismic time interpretation of December 1983. This was the first interpretation to correctly identify the top Batu Raja reflection event and to suggest that the ZZZ-1 and ZU-1 wells were within the same large structural closure.

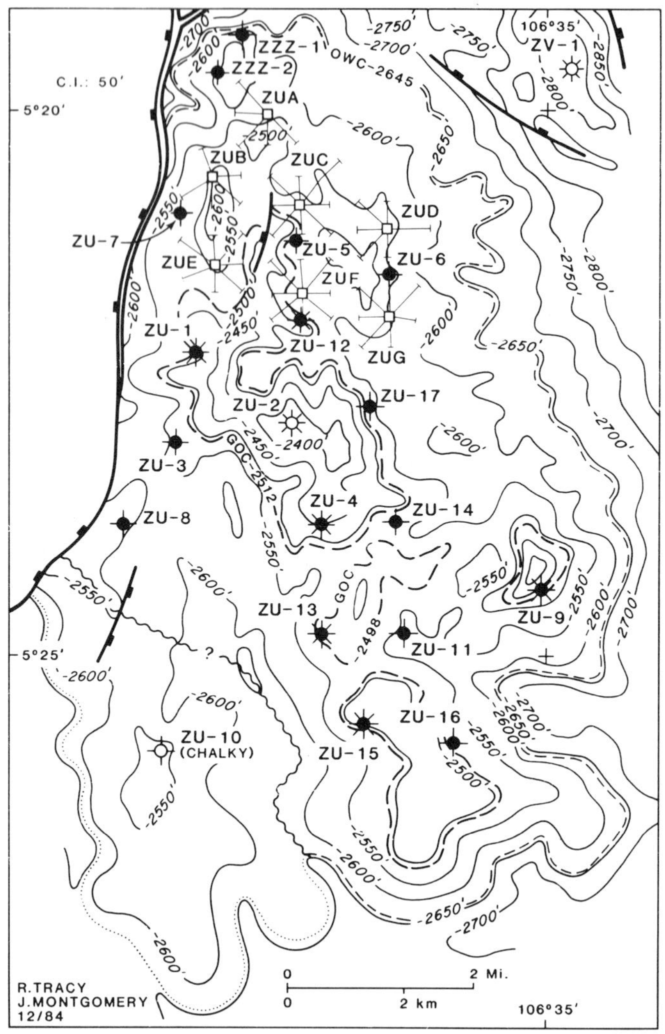


Figure 12. Top Batu Raja depth interpretation of December 1984 based on the delineation wells and the 1983 seismic data. Proposed phase one development well locations are shown.

possible future producers. ZU-3 was abandoned owing to thinness of the productive zone, ZU-7 was junked, and ZU-10 had no productive intervals owing to diagenetic chalkification of the Batu Raja reservoir. The extent of the Batu Raja chalky facies at the south end of the field has not been defined, and further drilling is warranted.

The delineation wells suggest the field oil-water contact has a 10 ft (3 m) south-to-north tilt. However, it is not possible to pick the contact in all wells, as several wells had no reservoir development at the depth of the oil-water contact. Drill-stem test results of the delineation wells show a progressive southward decrease in Batu Raja oil gravity from 24° API in the north to about 16° to 17° API in the south. It is concluded that an active Batu Raja Limestone aquifer, probably recharged on Java, tilts the oil-water contact to the north and is responsible for the increasing biodegration of the Batu Raja oil toward the south.

## GEOLOGICAL HISTORY

Basement beneath the Seribu Platform consists of diorite, quartzite, chlorite schist, and metamorphosed limestone. Radiometric age dates of basement cores samples from Seribu Platform wells range from about 66 m.y. to 84 m.y., or Late Cretaceous.

By the late Oligocene the north-dipping Seribu Platform basement surface had been eroded into a series of hills and ridges with about 400 ft (122 m) to 600 ft (183 m) of vertical relief (Figure 13). Topography was controlled by weathering characteristics of the basement lithotopes and by a pre-late Oligocene system of northwest-trending faults. These faults are reflected in the northwest-trending gullies incised into the basement surface.

The oldest unmetamorphosed sedimentary rocks overlying basement are alluvial, fluvial, and possibly shallow-marine rocks of the upper Oligocene Talang Akar Formation. Talang Akar rocks are erosional products of basement highs such as Bima, Duma, and the upthrown basement block north of the North Seribu Trough (refer to Figure 1).

Several discontinuous Talang Akar sandstone reservoirs have been drilled in other Bima field wells on the west, north, and northeast flanks of the field (Figure 14). These sands are relatively clean, fine-grained, moderately well sorted and tend to be thickest above basement gullies. The relative maturity of these sands and their stratigraphic position suggests they may be transgressive marine sands. Well control in the ZUD platform area suggests a reservoir geometry compatible with a strand-line depositional model.

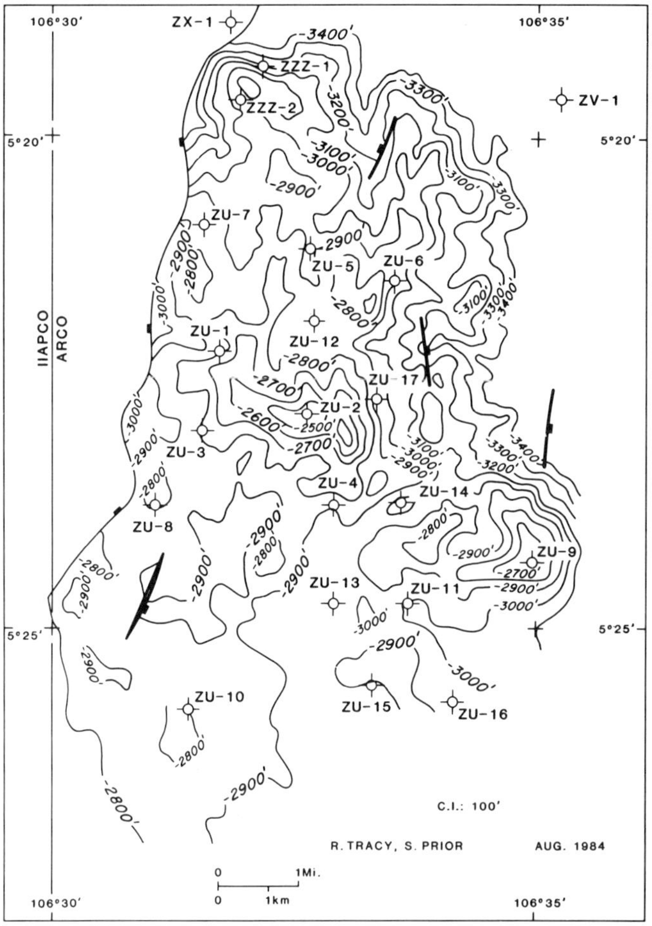

Figure 13. Base of Talang Akar/top of basement depth map.

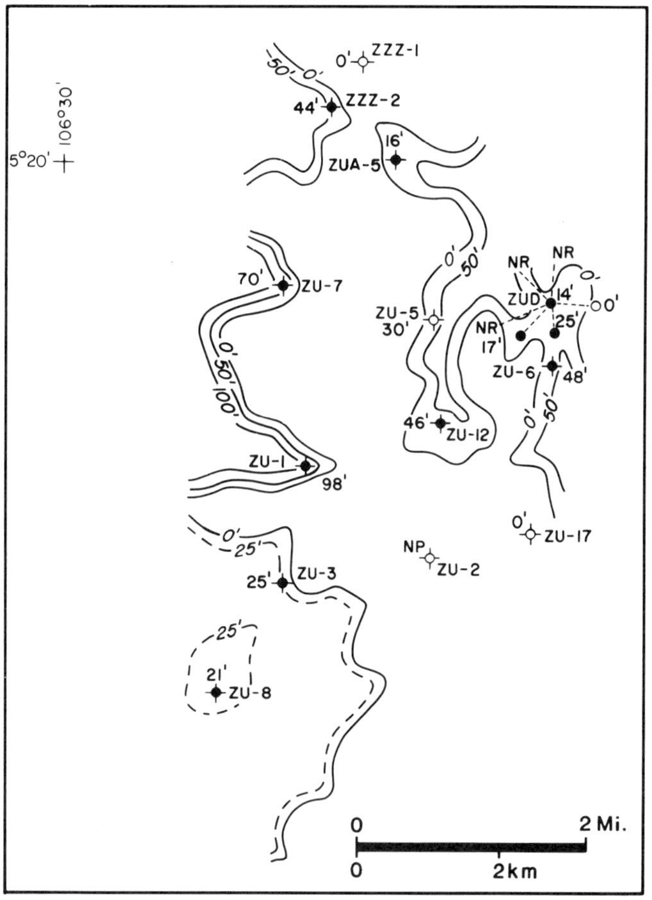

Figure 14. Lower Talang Akar net sand isopach map. Interpretation is based on well control and the basement configuration shown in Figure 13.

The upper 60 ft (18 m) to 100 ft (30 m) of the Talang Akar Formation belongs to the dominantly shaly marine member. Within this member, a variable development of thin limestone stringers interfingers with the stratigraphically oldest Batu Raja limestone in the ZU-4, 6, and 13 area. This facies change is observable on reflection seismic data and with wire-line logs (Figure 15).

Batu Raja Limestone deposition began in the latest Oligocene as a series of fringing reefs along the east and southeast flank of the Bima basement high. Similar east-facing basal Batu Raja fringing reefs are documented in the Sunda Basin west of the Bima area (Ardila, 1983). Several of the fringing reefs died out, were encased in Talang Akar shale, and are now gas-bearing (ZU-6, AA-1). However, the lower Batu Raja reefs became firmly established at ZU-4 and ZU-13. By the early Miocene the fringing reef complex had built upward and coalesced laterally with the slightly younger regional carbonate shelf.

Carbonate deposition continued with additional coral patch reefs becoming established on bathymetric highs. Intra-reef areas include typical lagoonal facies of foraminiferal wackestone, skeletal wackestone and packstone, and deeper water platy coral wackestone. Except for the coral boundstone and related proximal reef facies, the lower part of the Batu Raja formation contains a small, but significant percentage of clay fines (Figure 15). Presence of these clay fines has had a detrimental effect on the reservoir by greatly reducing permeability. Diagenetic processes acting on the argillaceous facies have also lowered the porosity. Embrittlement of the lower Batu Raja caused by secondary cementation has facilitated the propagation of compaction and tectonic fractures throughout this member, especially on the west side of the field.

The upper part of the Batu Raja is a northeast-thickening wedge of almost clay-free limestone that ranges in thickness from about 18 ft (5.5 m) in ZU-3 to 108 ft (33 m) in ZU-5. Virtual absence of clay fines differentiates this unit from the various underlying Batu Raja facies. The argillaceous-nonargillaceous limestone contact is well defined on natural gamma spectroscopy logs (Figure 15). The contact is time transgressive and becomes younger to the southwest. This relationship suggests that the clay provenance was to the southwest and retreated southward during transgression by the Batu Raja sea. The clay source could have been the Duma basement high that was exposed at this time.

Most of the upper Batu Raja interval consists of numerous small patch reefs within extensive lagoonal carbonates. Secondary dissolution of aragonite particles within the upper Batu Raja has locally increased porosity to 40%, although average porosity is about 28%. Dolomitization of some of the lime mud-rich intervals has taken place and drastically reduced the permeability. Dolomite occurs as a tight mosaic of anhedral crystals.

By the late early Miocene the rate of relative rise in sea level

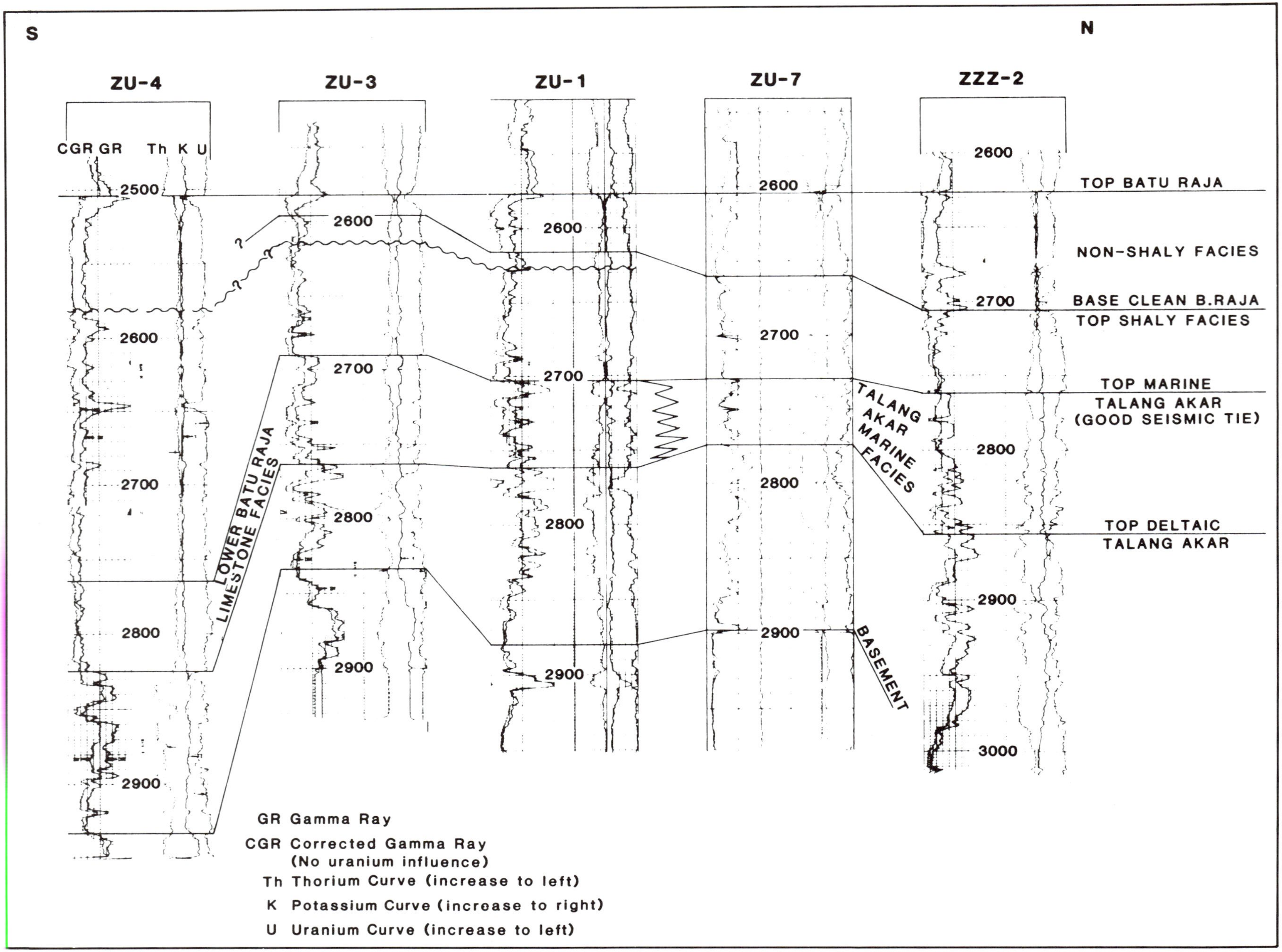


Figure 15. Natural gamma spectroscopy log cross section from ZU-4 to ZZZ-2. Note the northward thickening of the upper Batu Raja nonshaly facies. This is the primary reservoir interval in the field. Lack of clay in ZU-4 is attributed to almost continuous vertical reef growth on the high-energy, eastward-facing side of the Bima high.

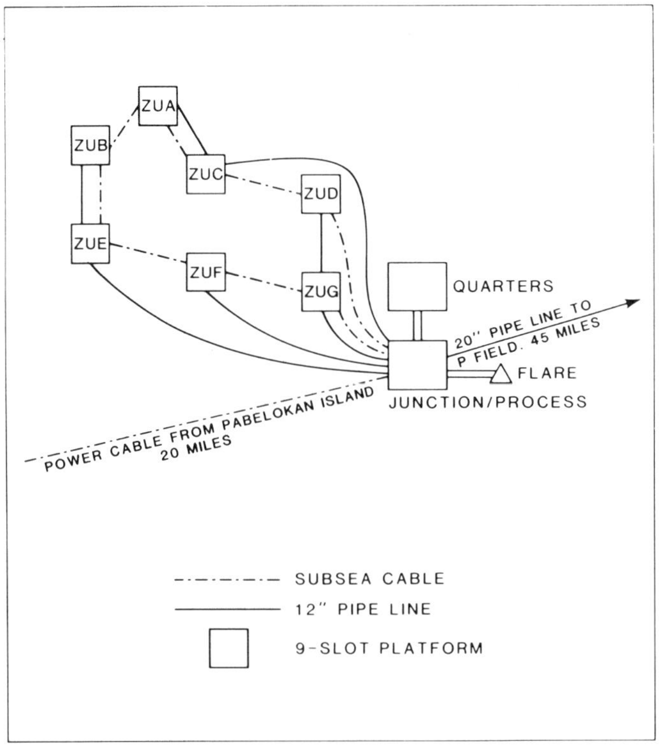


Figure 16. Bima phase one development plans.

was sufficient to drown the carbonate complex at Bima. Open marine conditions were established and water depths were almost certainly deeper than fair weather wave base. Cibulakan Formation clastic fines containing abundant planktonic forams and phosphatic pellets mantle the Batu Raja Limestone.

About 25 ft (7.6 m) to 125 ft (38 m) above the Batu Raja Limestone is a sandy limestone bed about 12 ft (3.6 m) to 25 ft (7.6 m) thick. This unit is named the Z24 zone and is within the lower part of the Cibulakan Formation (Figure 3). It is gas-bearing over most of the Bima field and is absent in wells drilled to the north and south of Bima. The bed contains very fine grained quartz sand, abundant planktonic forams, fine carbonate skeletal debris, and minor clay/marl matrix. The composition and stratigraphic position of this bed suggest the possibility of emplacement by some type of gravity-flow mechanism, perhaps storm induced. One possible source for this material would be the Duma area 7 mi (11 km) southwest of Bima, where a fringing reef complex still grew on the seaward flank of this partially emergent basement high.

The Z24 limestone is overlain by marine shale and shelf sandstones of the broadly regressive upper part of the Cibulakan Formation. Reservoirs within this unit are productive in the Ardjuna Basin to the east, but have not been significant from a petroleum standpoint in the Seribu Platform or Sunda Basin provinces.

## BIMA FIELD DEVELOPMENT

As the Bima delineation wells were assessed, it became apparent that as much as three-fourths of the primary recoverable Batu Raja reserves were in the upper porous wedge. This led to a two-phase development proposal. Phase one would develop most of the upper Batu Raja reserves plus the Talang Akar reserves in the northeastern third of the field. Phase two would develop the fractured part of the lower Batu Raja reservoir on the south side of the field, the remainder of the upper Batu Raja reserves, and the Talang Akar oil in the ZU-1/ZU-3 area.

Phase one development was originally planned with seven, nine-slot production platforms, 53 development wells, quarters and processing platforms, and a 45-mi (72-km), 20-in-(51-cm)-diameter pipeline to P field (Figure 16). Existing delineation wells in the phase one area were not suitably located for platform sites, owing to the gas channeling problem we experienced when drill-stem testing the upper Batu Raja reservoir in wells that pierced the gas cap. Phase one platform locations were sited downdip from the gas cap, so that the maximum number of wells could be drilled on approximately 100-acre spacing without encountering the gas cap.

Phase one development plans were submitted to Pertimina for review in November 1984. Approval was obtained in February 1985 with the proviso that one successful confirmation well be drilled and suspended at each of the proposed ZUA, ZUC, and ZUD platform locations (Figure 12). They were successfully drilled in April/May 1985.

At the time of initial phase one planning, a separate study group was investigating the feasibility of using drain-hole wells as part of the development. These wells would have deviation angles of 85° to 90° compared with conventional deviation angles of 45° to 50°. The purpose of a drain-hole well is to increase the length of the completion, hence productivity, and to lessen the chance of gas channeling and coning. The drain-hole feasibility study was not completed until after the development plans had been approved and several conventional development wells had been drilled. After review, Pertamina approved plans to drill an 85° well from the ZUD platform as a test case. ZUD-3 was drilled, logged, and 7-inch slotted liner set in 15 days without complications. The well was kicked off at 1260 ft (385 m) RKB and angle was built at a rate of 4.25° per hundred ft (30 m). An 85° angle was reached. The nearly horizontal portion of the well extends 400 ft (122 m) through the upper Batu Raja pay zone. Logging was accomplished by running a full suite of logging tools on the end of drill pipe. The well flowed 1400 BOPD after a small 3000-gal (11,356-L) HC1 acid job. This test was better than the delineation well average of about 1000 BOPD even though the well intersected below-average reservoir conditions. A larger acid job is envisioned when ZUD-3 is completed. Future drain-hole wells will have at least 800 ft (244 m) of completion behind slotted liner.

At the time of writing (March 1986) permission is being sought to modify our development plan to include a total of 16 drain-hole wells and 26 conventional wells. This will result in a net reduction of 11 conventional development wells in the field and will significantly reduce development costs. Field productivity and drainage efficiency are expected to be similar with either development plan.

Four rigs are scheduled to operate in the Bima field by April 1986. Field start-up at ZUC, ZUD, ZUE, ZUF, and ZUG is expected to commence in December 1986, and phase one development should be complete by July 1987. Phase one development is expected to exploit about 60 million barrels of reserves at a peak production rate of about 50,000 to 55,000

BOPD. Phase-two plans to develop up to 40 million barrels of reserves in the remainder of the field are being reviewed.

## CONCLUSIONS

A number of factors joined forces, as it were, to delay the discovery of the Bima field. These include the continued reporting of the erroneous 11° API oil value and failure to recognize the large size of the closure. It was a matter of not seeing the forest for the trees. This resulted from mispicking the top Batu Raja seismic event and thereby mapping a reflector with less critical dip than the Batu Raja reservoir. The piecemeal nature of in-fill seismic-data acquisition led to several approval delays. The repeated, relatively small acquisition programs were considered valid at the time because we were looking for a smaller field than that which was subsequently found. That better prospects were perceived to be present in the Ardjuna Basin caused us to devote most of our exploration efforts in this area. That this philosophy was correct can not really be questioned, since we have now produced more than 500 million barrels of oil from the Northwest Java Contract area and our discovery rate remains relatively high after 17 years of exploration. Had the Ardjuna Basin been less prolific, then the Bima field probably would have been found much earlier.

## ACKNOWLEDGMENTS

The author wishes to thank the management of ARCO and Pertamina for encouragement to write and permission to publish this paper. I also thank the many reviewers who made helpful suggestions.

## REFERENCES CITED

Ardila, L.E., 1983, The Krisna high, its geologic setting and related hydrocarbon accumulations: Southeast Asia Petroleum Exploration Society Proceedings, v. 6, p. 10-23.

Metzner, L.H., 1943, Playa del Ray oil field, *in* Geologic formations and economic development of the oil and gas fields of California: California State Division of Mines Bulletin 118, p. 292-294.

CHAPTER 19

# THE HOKKAIDO CORNER—HYDROCARBON EXPLORATION IN A SUBDUCTION SYSTEM

**Al Sabitay**
*Al Sabitay Consultants*
*Perth, Australia*

**John E. Shirley**
*Consultant*
*Melbourne, Australia*

## ABSTRACT

The area offshore from south-central Hokkaido in northern Japan is generally considered to be prospective for hydrocarbons. The prospective sequence is in a fore-arc basin of the Japan-Kurile subduction system and is here termed the "Hokkaido Corner."

At the junction of the Japan and Kurile trenches there is a direction change of the trench, and this is associated with a major transverse fault. Earthquake and focal mechanism studies show that the transverse fault disrupts the subducting slab below the prospective fore-arc basin. The history of deformation in the basin was controlled by the northwesterly movement of the Pacific Oceanic Plate and the subduction of this slab beneath the Japanese Island Arc. Seaward-dipping stacked thrust sheets recorded in wells are the result of gravity sliding from mountains at the eastern basin margin. These mountains were an uplifted belt at the "toe" of the continental mass overlying the subducting slab west of the Japan-Kurile trenches.

Further complexity results from a history of submarine canyon erosion, wrench faulting, volcanic activity, and rapid uplift and subsidence with the consequent complex sedimentation. The conventional integration of gravity, magnetics, seismic, well and petrophysics information together with substantial imagination are vital to the interpretation of this area.

The Hokkaido Corner is the term used here for the region at the junction of the Japan and Kurile trenches. This area, shown in Figure 1, is located where the Pacific Plate, moving in a general northwesterly direction, is deformed as it is subducted beneath the Japanese and Kurile Island arcs.

Petroleum exploration has been carried out in this area for some years. Prior to this exploration, earthquake activity had been monitored over a long period by Japanese observatories and, together with other regional geological and geophysical work, this work has contributed to a significant understanding of the geology. This full range of geological work has resulted in the compilation by Japanese organizations of a major data base. Exploration drilling that has been reported has included data from at least one well which had significant hydrocarbon shows in the volcanic "Green Tuffs," and all wells have confirmed the geologic complexity of the area.

The Hokkaido Corner includes a fore-arc basin within the Japan-Kurile Arc-Trench Systems. South of this area the Japan Arc is a typical detached arc with normal subduction (Dickinson and Seely, 1979). Here the Pacific Plate is moving generally north-west and is being subducted beneath the Japanese and Kurile Island Arcs. The basin straddles the major direction change between the Japan and Kurile Arc-Trench Systems. The southeast corner of the Island of Hokkaido lies about half way along a transverse fault zone joining the subsea expression of the convergence of the trenches and the associated and parallel Quaternary-Recent line of volcanoes. The Hokkaido Corner extends from the trench system west through the volcanic lineation located at the position where the descending plate has reached a depth of about 110 km and where significant crustal melting could have commenced.

The subduction system in the Hokkaido Corner is illustrated in Figure 1. Near the "toe" of the continental plate, landward-dipping thrusts consist of a combination of continental sediments and marine sediments scraped from the descending slab. An uplifted zone seen to the left of these thrust sheets consists of a melange of both volcanic and continental rocks. Gravity slides occurred on both sides of the uplifted zone as

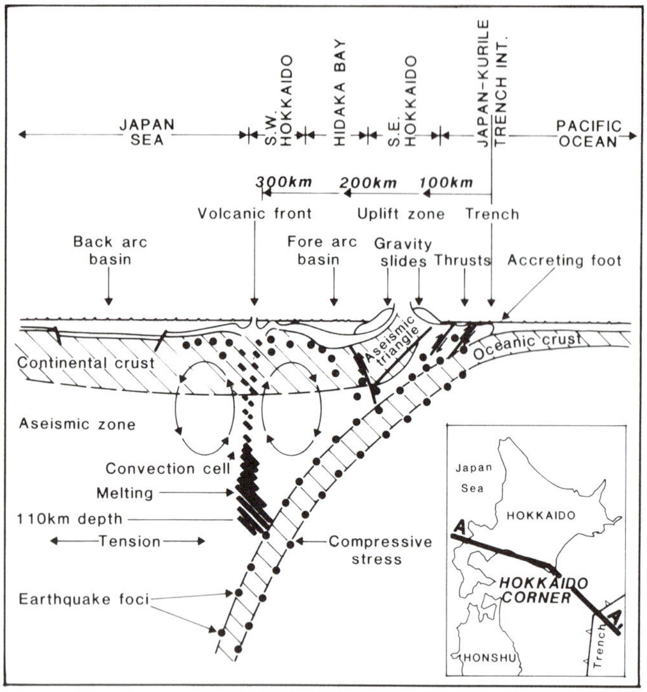


Figure 1. Generalized subduction system—Hokkaido Corner, Japan.

the highlands became unstable. Seaward-dipping thrusts have developed below the fore-arc basin. Landward, the volcanic front is located in present-day continental crust now overlying the subducting oceanic crust at a position where it is deeper than 100 km. Farther west, in a tensional stress regime, is a back-arc basin—the present site of the Sea of Japan.

Earthquake studies have provided much of the fundamental evidence for the basic delineation of the Hokkaido Corner. Murauchi and Ludwig (1981) have outlined the aseismic zones which underlie the uplifted area and the volcanic front considered to exist as a part of the magmatic volcanic cell. Shallow earthquakes are located above the convection cell in the area where thrusts, associated with the uplifted zone, have developed. The descending slab can be mapped using earthquake focal depths determined for both the top and the bottom of the slab. This is the Benioff zone, and Figure 2 shows the depth to the top of the subducting slab determined from the recent earthquake foci (Utsu, Ishikawa, and Katsui, 1973; Stauder and Mualchin, 1976). A typical cross section (A-B) of focal depth data is also shown in Figure 2.

The transcurrent fault zone at the Hokkaido Corner is inferred from the many shallow earthquakes along its length. Fault mechanisms, interpreted from these earthquake data and from those delineating the Benioff zone, show that the Japan subducting segment is descending at an angle of about 30°, whereas the Kurile segment dips more steeply (60°). At the Hokkaido Corner the Japan Arc segment is thought to override the Kurile segment with a large component of sinistral strike slip.

Low-angle thrusts penetrated in wells drilled in the vicinity of the Hokkaido Corner dip seaward instead of landward, as may be expected from the compressional stress directions inferred from an analysis of the subducting slab. There are alternate explanations for this. The thrusts may be the end result of landward tilting of the uplifted area with continued understuffing and seaward accretion of the overriding continental crust. Alternatively, the seaward-dipping thrusts may be the result of gravity slides descending from the unstable uplifted area as it continued to build up. Von Huene and Lee (1983) have suggested a number of reasons why gravity slides can be expected:

(1) Not all of the oceanic sediments on top of the subducting oceanic crust descend to mantle depths. Mass calculations indicate that about 70% of these sediments accrete on the leading edge of the overlying continental slab.

(2) Most of these sediments would be unconsolidated and thus unlikely to competently carry any stress.

(3) The sediments that do descend would dewater. The ascending fluids would help lubricate the basal part of the uplifted zone.

(4) Increases in pore pressure within the fore-arc sediments and the sediments attached to descending rocks would tend to decouple the overlying (fore-arc) sediments from the descending slab.

The zone of rapid transition from land and shelf depositional environments to that of deep-sea trenches is a common environment for submarine canyons and channels. Two major channels, the Kushiro Channel and the Hidaka Channel, as well as several minor channels, have been mapped near the Hokkaido Corner (von Huene and Arthur, 1982). Figure 3 shows the results of an extension of this study based on bathymetric contours and onshore drainage for the present channel system. Fossil channels are indicated in the area by turbidities in the well E-KA-1 (Figure 3) and pebble mudstones located onshore in Hokkaido (Takai, 1963). Other factors indicative of a turbidity regime—such as rapid uplift and subsidence, earthquakes, wrench faults and gravity slides—are also evident in the area. The fossil submarine channels could contribute to trap creation in the fore-arc basin; the Gippsland Basin in Australia is an excellent example of this.

The interpretation of magnetic data (Ogawa and Suyama, 1976) has delineated a series of northwest-southeast magnetic belts clearly related to changes in basement (Figure 3). The magnetics are complex and are considered to be only of minor interest in the study of hydrocarbon potential. Much of the magnetic response is from near-surface volcanics which have little or no bearing on the petroleum geology. The near-surface response generally masks the response from intermediate depths. The basement complexes are not discussed further except in the description of possible movements of the segments of the magmatic arcs.

The regional tectonic elements of the Hokkaido Corner are shown in Figure 4 and the regional Bouguer anomalies (Tomoda, n.d.) clearly confirm the principal features. A large positive gravity anomaly with an axis trending N20°W corresponds to the Kitakami Block of Northern Honshu. The Kitakami Block contains Permo-Triassic and older basement rocks extensively deformed and intruded by Cretaceous and Tertiary granites and ultrabasic rocks. The feature terminates to the north at the Tomakomai Ridge and is probably downfaulted on the eastern flank.

A major gravity low, which is not isostatically compensated, dominates the Hokkadio Corner. The principal feature is a fore-arc sedimentary basin composed of thick Tertiary sediments overlying Cretaceous metasediments and older basement rock. The basement has been downfaulted and highly

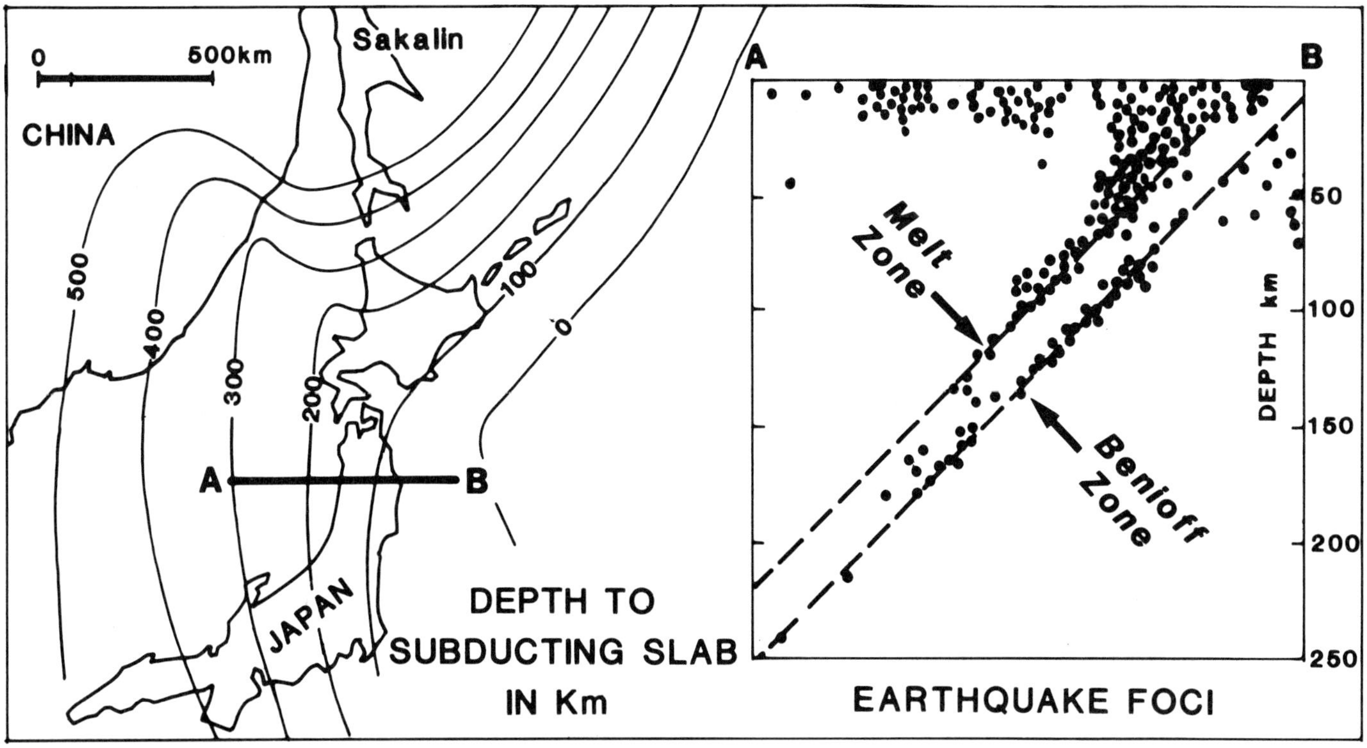


Figure 2. Delineation of the subducting slab at the Hokkaido Corner using earthquake foci and epicenters (after Utsu, Ishikawa, and Katsui, 1973; Stauder and Mualchin, 1976).

deformed as a result of the deep crustal and upper mantle tectonics of the Hokkaido Corner. The Bouguer anomaly trends N20 °W, paralleling the Kitakami High in the southwest. It terminates to the northwest in Central Hokkaido and has steep gradients on both flanks, which suggests major faulting. The Hidaka Mountains, a thrust-bounded belt of metamorphic rocks intruded by ultrabasics and granites, correspond to a positive elongate gravity anomaly which is again oriented N20 °W.

Figure 5 illustrates the gross geologic section at the Hokkaido Corner, and the location of the section line is shown on Figure 4 ($A_1$-A). Examination of this section reveals the structure of the fore-arc basin, the focus for hydrocarbon search. The sediments with hydrocarbon potential are predominantly in the Paleogene sequence. Sandstones of Eocene and Oligocene age are interbedded with argillaceous sediments, volcaniclastics, and coal measures. Turbidities are a common member of the suite.

The changing tectonic aspects of the subduction system result in many different environments of deposition. Basement consists of pre-Paleogene metasediments of mostly Cretaceous age. Trap types that can be expected are:

(1) Turnover at the leading edge of thrusts and gravity slides.
(2) Drape anticlines on pre-Paleogene horst blocks.
(3) Fault traps, including sandstones on footwalls juxtaposed to mudstones or with a fault seal at the hanging wall.
(4) Stratigraphic pinchouts.
(5) Keystone highs due to late, high-angle reverse faults.
(6) Channel sands with permeability barriers created by

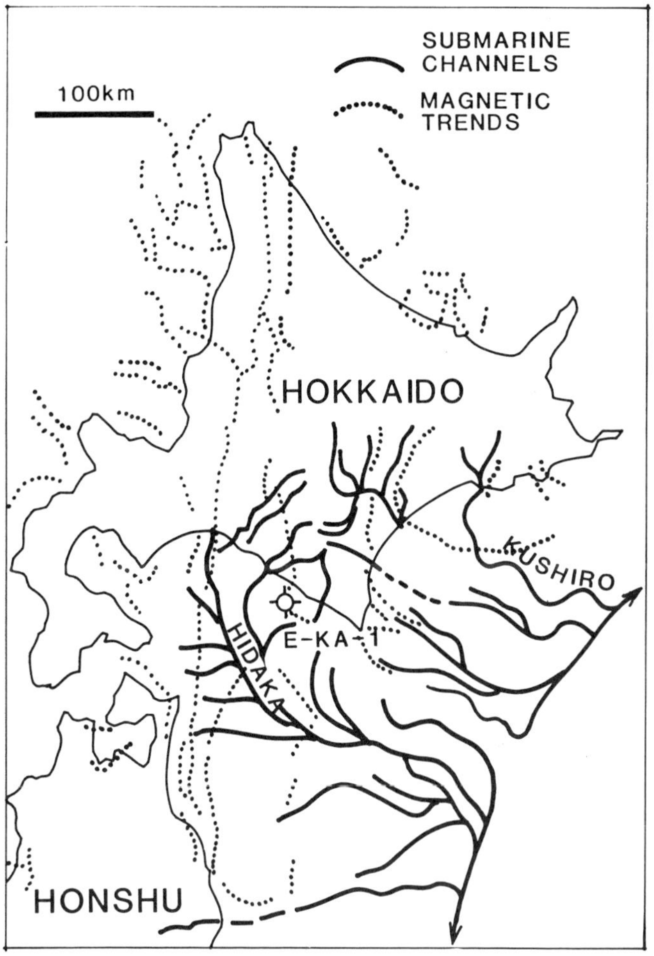


Figure 3. Submarine channels and magnetic trends in the vicinity of the Hokkaido Corner (after von Huene and Arthur, 1982; Ogawa and Suyama, 1976).

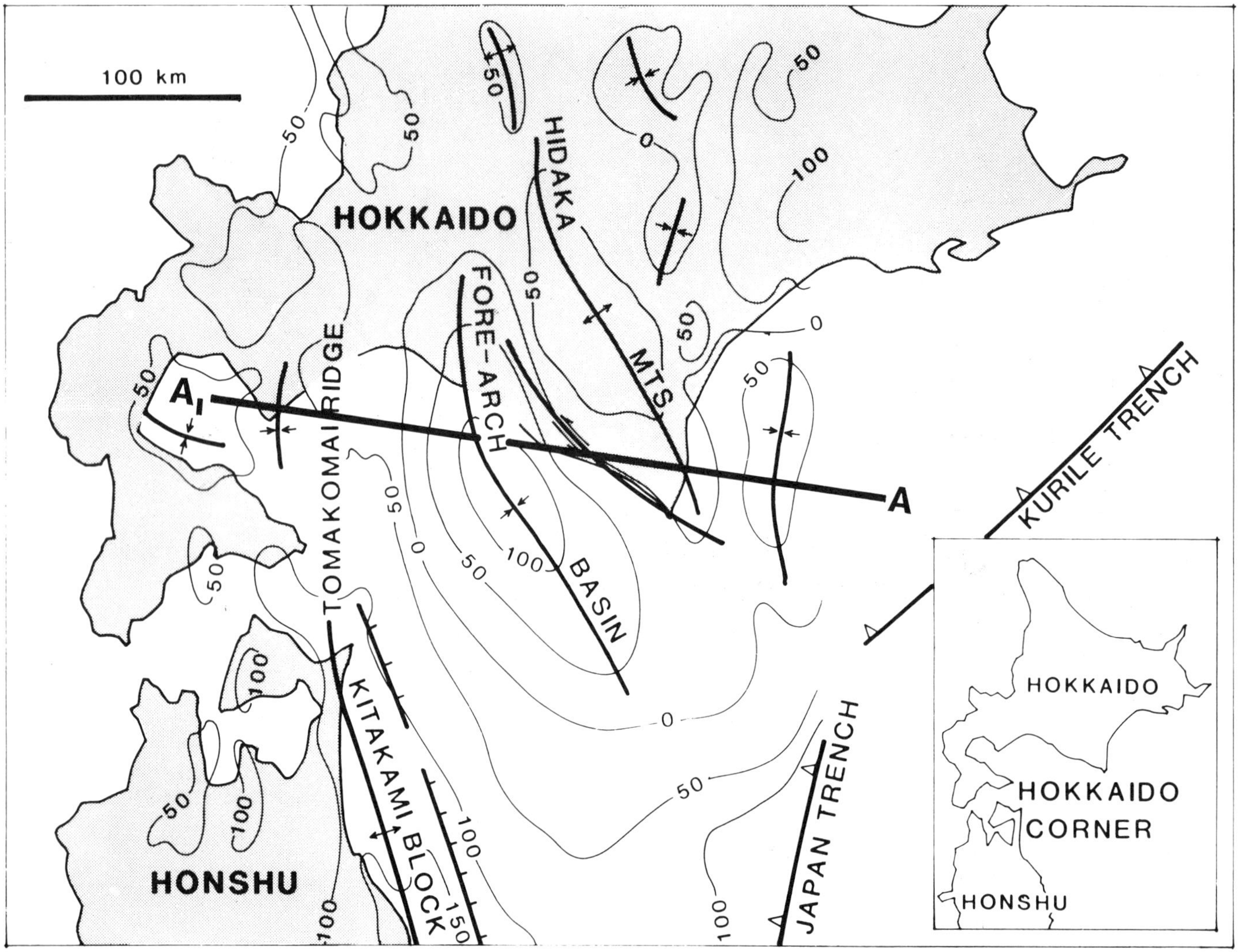


Figure 4. Bouguer anomalies and regional tectonic elements, Hokkaido Corner, Japan (after Tomoda).

argillaceous infill subsequent to submarine channel erosion.

(7) En echelon anticlines formed by wrenching. The southern coastline of Hokkaido probably marks the position of a primary wrench with left-lateral displacement.

The Hokkaido Corner has been in a nearly continuous compressive stress regime for at least the past 80 million years. During the first half of this time the stress, derived from the Pacific Plate movement, was directed approximately N10 °W. From 40 MYBP the stress direction was N60 °W, assuming the Eurasian Plate had been stationary. During some part of the more recent period, a transcurrent fracture zone (at the Hokkaido Corner) developed, inducing sinistral strike-slip movement, northeast side to the northwest. The evidence for this includes:

(1) Lateral offset of faults and fold axes onshore in Hokkaido.
(2) En echelon fault and fold patterns.
(3) Major faults parallel to the postulated transcurrent fault.

The geologic history in the Hokkaido Corner was developed using the strain ellipsoid concept and the wrench tectonics outlined in Wilcox, Harding, and Seely (1973). Some coupling between the subducting slab and the sediments in the fore-arc basin has been assumed. A left-lateral wrench model was used, with the compressional stress axis of the strain ellipsoid aligned with the plate tectonic stress field. By using this model at the Hokkaido Corner, the following comments can be made:

(1) The southwest-facing coastline of southeastern Hokkaido probably coincides with a primary wrench or synthetic left-lateral strike-slip fault (see Figure 4).
(2) Fold axes in the Hokkaido Corner should trend ENE-WSW to N-S depending on their age and proximity to sites of wrenching deformation. Normal faults should be oriented approximately E-W.
(3) Counterclockwise rotation of structural features can be expected from examination of the stress field history of the Hokkaido Corner.

The rotational aspect of major arc-trench system is illustrated by the volcanic arc segments. Figure 6 shows the results of movement of the volcanic arc segments in the vicinity of the Hokkaido Corner. Renneft (1979) has proposed that island arcs are rarely arcuate except in a gross sense but rather consist of a series of straight segments which, on a larger scale, fit a general arcuate form. Honza, Kagami, and

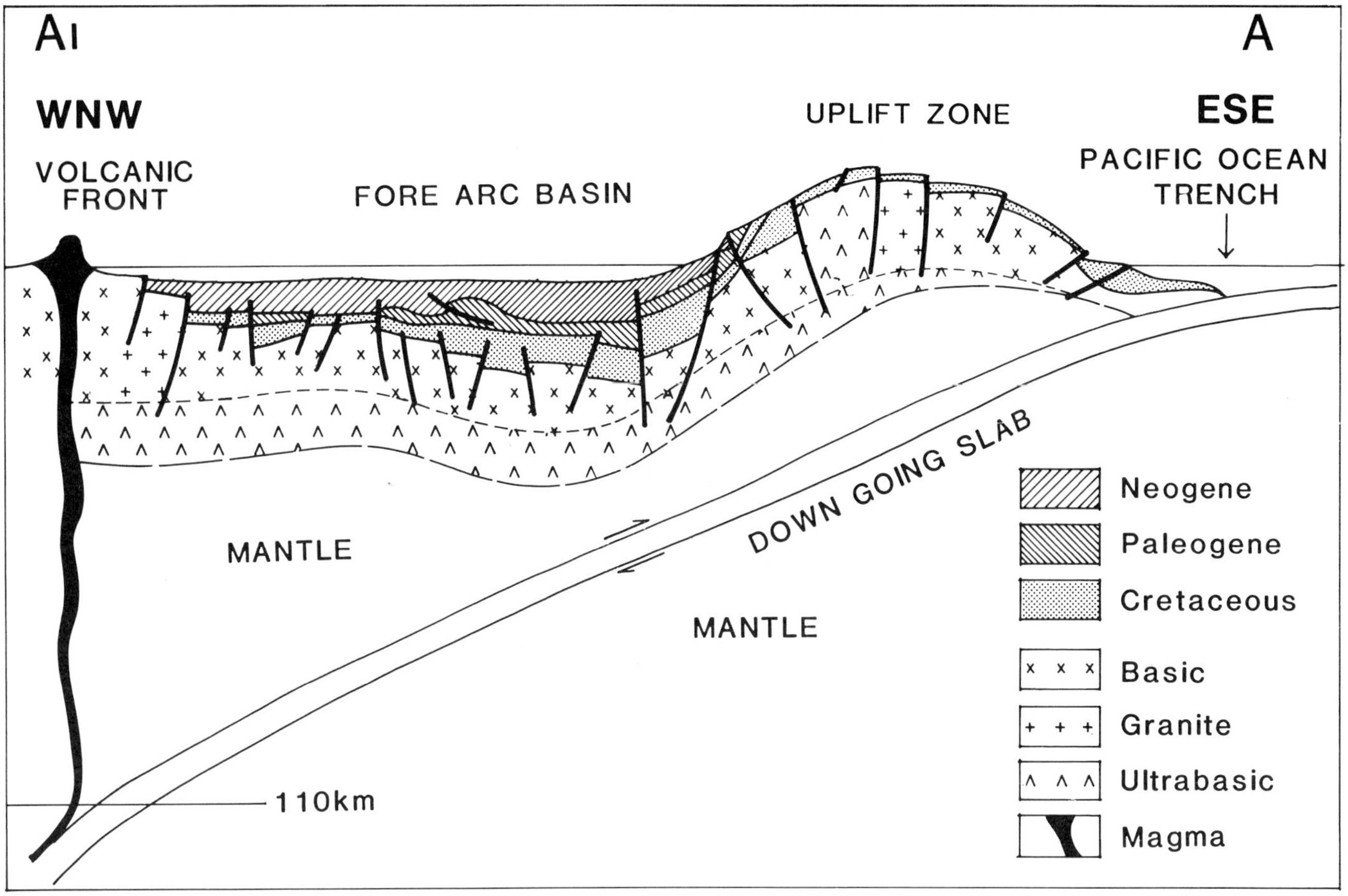


Figure 5. Geologic cross section typical of the Hokkaido Corner. Location of section $A_1$-A shown on Figure 4.

Nasu (1977) have delineated a series of volcanic arc segments in northeast Honshu, the positions of which are supported by earthquake data. Hokkaido and the southern Kurile Islands have been segmented using the location of volcanoes and the bathymetry of Hirokawa (1978). This segmentation can then be compared with the orientation of the the Japan and Kurile trenches; parallelism is expected considering the normal subduction system profile (see Figures 1, 5). The transcurrent faults interpreted at the extremities of each arc segment are normal to the trenches except in the vicinity of the Hokkaido Corner. The direction of strike-slip conforms to the apparent offset of each volcanic arc segment.

In the southern Kurile Arc the volcanic arc segments have apparently rotated counterclockwise with the amount of rotation increasing to the southwest. The segment of volcanic arc through Central Hokkaido appears to be part of the Kurile Arc, it having apparently rotated approximately 70° counterclockwise. Just south of Hokkaido the transcurrent fault indicates the hinge zone at the Hokkaido Corner. This zone has been the subject of detailed study of earthquake focal mechanism. Kanamori (1971), in a study of the 1968 Tokachi-Oki earthquake and its aftershocks, concluded that the movements were consistent with that of the Pacific Plate (N60 °W)

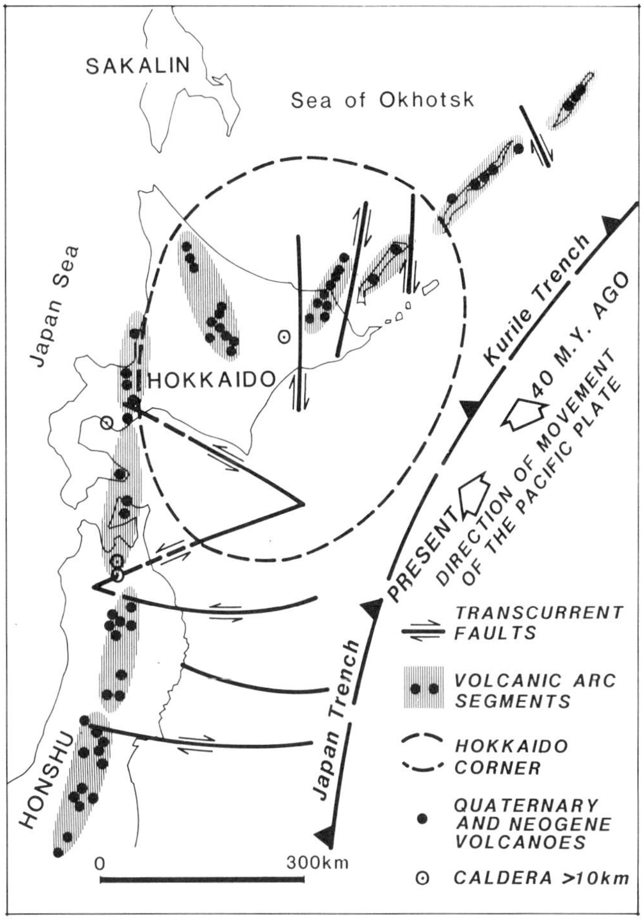


Figure 6. Volcanic arc segments and their suggested movement to give a general arcuate form (after Honza, Kagami, and Nasu, 1977).

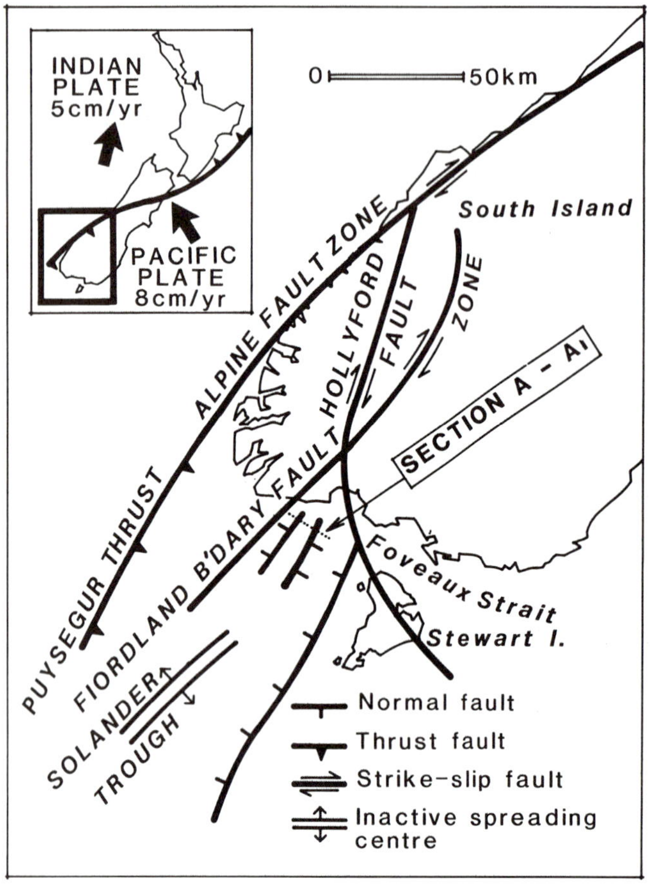


Figure 7. Tectonic elements of the South Island of New Zealand.

and had a significant strike-slip component. Stauder and Mualchin (1976), in a study of 65 earthquakes with magnitude greater than 5.7 in the Kurile Arc, show that the Pacific Plate moving N55 °W is consistent with the interpreted focal mechanisms. For eight earthquakes at the Hokkaido Corner, they interpret steeply dipping east-west fault planes with the north side downthrown. These faults reflect the rupture of the subducting oceanic slab in the contorted region of the Hokkaido Corner. The steeply dipping (60 °) Kurile segment is being overridden by the shallow-dipping (30 °) Japan segment (see Figure 2).

Inspection of some areas in the southwest Pacific illustrates where the understanding of these aspects of plate tectonics contributes to the interpretation of these areas of complex geology.

At the south end of the South Island of New Zealand (Figure 7) the Indian Plate is being subducted by the Pacific Plate except in the vicinity of the Puysegur Thrust, where the plate movements are mostly horizontal. The Pacific Plate is moving NNW at a rate of about 8 cm/year, whereas the Indian Plate is moving NNE at about 5 cm/year; hence the relative movement of the Pacific plate is about 3.5 cm/year NE. A consequence of these movements is the development of several

Figure 8. Seismic section across a thrust anticline in the offshore Solander Trough, New Zealand.

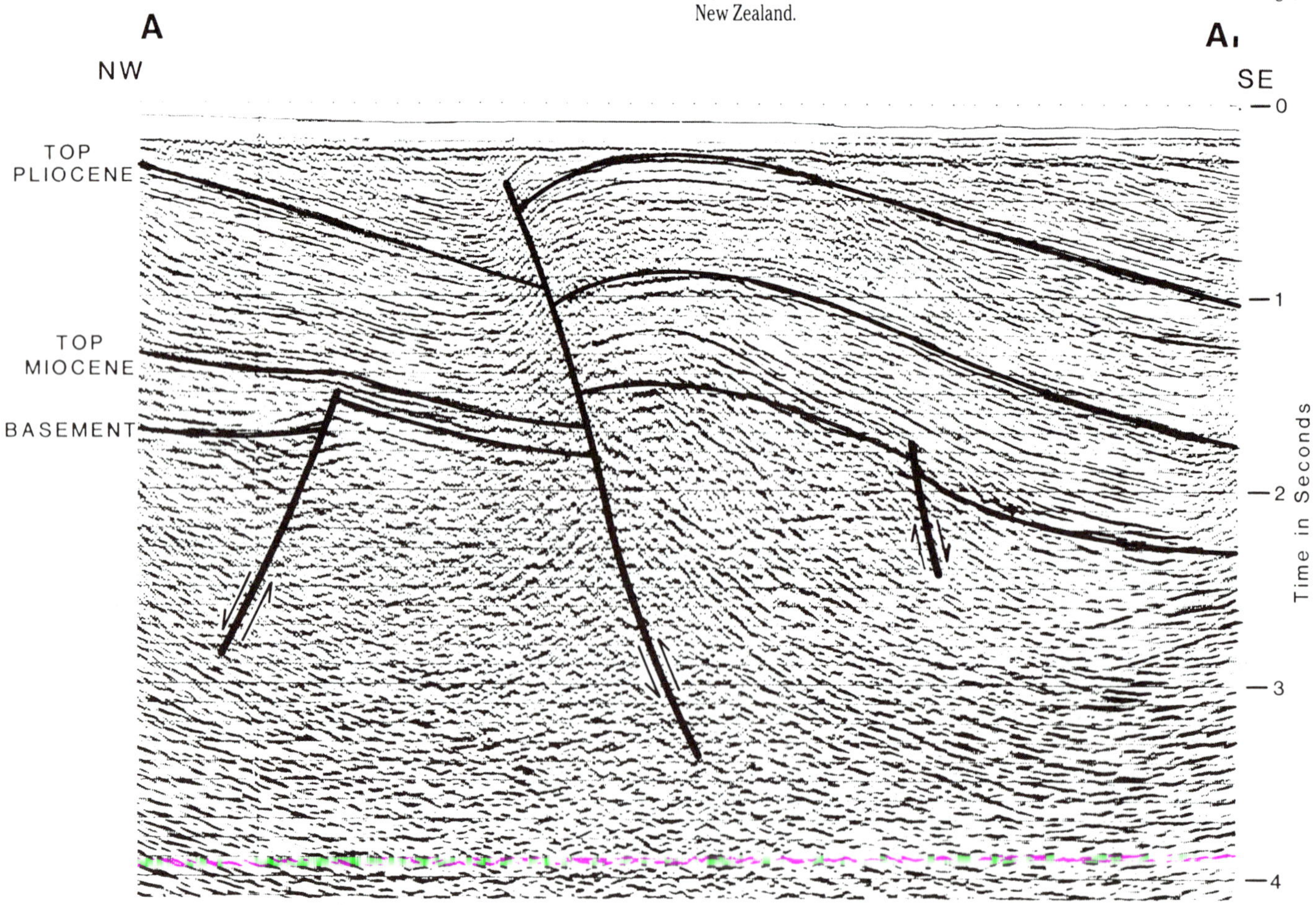

major strike-slip faults with associated earthquakes. The faults include the Alpine, Fiordland, and Hollyford dextral fault zones. The Solander Trough is the site of an inactive spreading center. One of the results of this complex of tectonic elements is the thrust anticline shown in Figure 8.

The northwest continental shelf of Australia (Figure 9) is the site of an aborted rift (Canning and Browse basins), continental drifting (India from the Browse Basin), transcurrent faults (Hall's Creek and King Leopold mobile zones), and present subduction (Timor). Wilcox, Harding, and Seely (1973) consider the Fitzroy Trough in the Canning Basin to be a "text book example" of a dextral divergent wrench graben. The consequences of wrench deformation are documented by Bentley (1984), Craig et al. (1984), and Smith (1984). Oil and gas fields near the Timor subduction zone, possibly related to plate tectonic processes, are located in Figure 9. The Australian-Indian Plate is currently moving NNE at about 8 cm/year, and Figure 10 shows the seismic expression of the probable resultant strike-slip movement on the eastern margin of the Browse Basin. The flower structure fault pattern, shown here at the pre-breakup unconformity, is commonly evident both in this area and in the Canning Basin (Purcell, 1984).

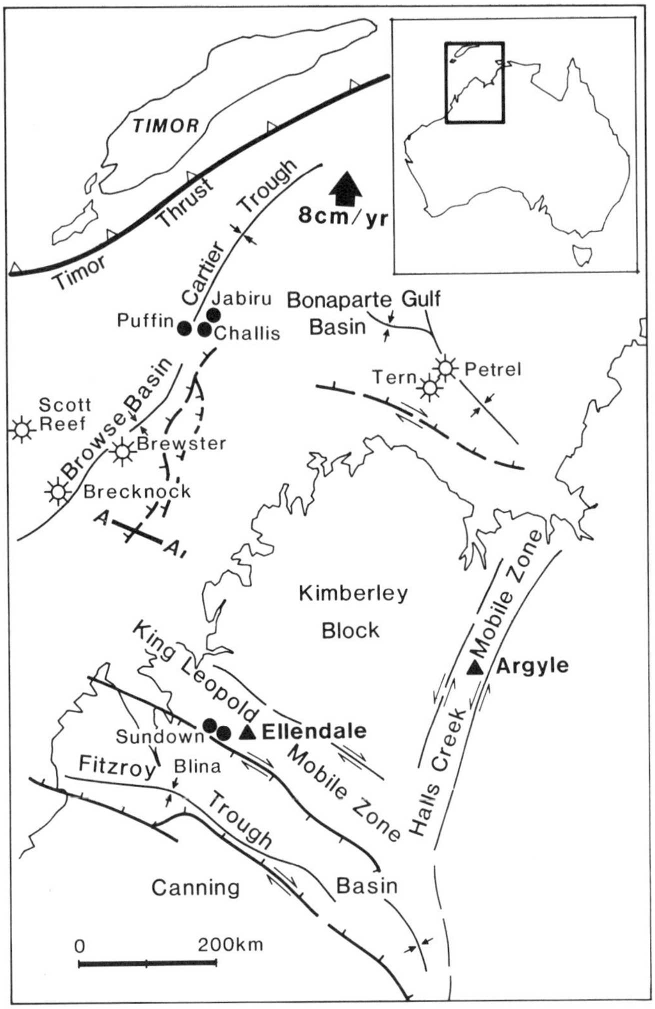


Figure 9. Tectonic elements of the northern North West Shelf, offshore Australia.

Bally (1983) has used the Andaman Sea area to illustrate negative flower structure in a zone of complex plate movements. Figure 11 shows an interpreted seismic section from Bally's Seismic Atlas, which is comparable with Figure 10 and the examples from the Canning Basin.

The recently increased activity in Papua, due to the discovery of probable commercial oil at Juha, directs attention to the position of this area in the plate tectonics history of the region. A cursory examination indicates that the thrusted anticlines (gravity slides) being drilled may lie in a fore-arc basin of the south-plunging New Guinea subduction system. Application of some of the concepts developed at the Hokkaido Corner may be of use in the interpretation of this area.

## ACKNOWLEDGMENT

The authors acknowledge the support of CRA Exploration Pty. Limited in the preparation of the figures for this paper.

## REFERENCES CITED

Bally, A.W., 1983 (ed.), Seismic expressions of structural styles: AAPG Studies in Geology 15, v. 3, p. 4.2-1–4.2.-12.

Bentley, J., 1984, Petroleum geology of Central Broome Platform, *in* P.G. Purcell, ed., The Canning Basin, W.A.: Perth, Proceedings of Geological Society of Australia Incorporated/Petroleum Exploration Society of Australia Symposium, p. 157-168.

Craig, J., J.W. Downey, A.D. Gibbs, and J.R. Russell, 1984, The application of LANDSAT imagery in structural interpretation of the Canning Basin, W.A., *in* P.G. Purcell, ed., The Canning Basin, W.A.: Perth, Proceedings of Geological Society of Australia Incorporated/Petroleum Exploration Society of Australia Symposium, p. 57-72.

Dickinson, W.R., and D.R. Seely, 1979, Structure and stratigraphy of forearc regions: AAPG Bulletin, v. 63, p. 2-41.

Hirokawa, O., (ed.), 1978, Geological map of Japan (2nd ed.): Geological Survey of Japan, scale 1:100,000.

Honza, E., H. Kagami, and N. Nasu, 1977, Neogene geological history of the Tohoku Island arc system: Journal of the Oceanographical Society of Japan, v. 33, n. 6, p. 297-310.

Kanamori, H., 1971, Focal mechanism of the Tokachi Oki earthquake of May 16, 1968: Tectonophysics, v. 12, p. 1-13.

Murauchi, S., and W.J. Ludwig, 1981, Crustal structure of the Japan trench, *in* Initial Reports of the Deep Sea Drilling Project: Washington, D.C., U.S. Government Printing Office, ,v. 57, p. 463-470.

Ogawa, K., and J. Suyama, 1976, Distribution of aeromagnetic anomalies, Hokkaido, Japan and its geologic implication in volcanoes and tectonosphere, Aoki, H. and Isuka, I.: Tokyo, Tokai University Press, p. 207-215.

Purcell, P.G., (ed.), 1984, The Canning Basin, Western Australia: Perth, Proceedings of Geological Society of Australia/Petroleum Exploration Society of Australia Symposium.

Renneft, T.S.M., 1979, Segmentation of island arcs: Journal of Petroleum Geology, v. 1, p. 35-53.

Smith, G., 1984, The tectonic development of the Gregory subbasin and adjacent areas, northeastern Canning Basin, *in* P.G. Purcell, ed., The Canning Basin, W.A.: Perth, Proceedings of Geological Society of Australia Incorporated/Petroleum Exploration Society of Australia Symposium, p. 109-120.

Stauder, W., and L. Mualchin, 1976, Fault motion in the larger earthquakes of the Kurile-Kamchatka arc and of the Hokkaido Corner area: Journal of Geophysical Research, v. 81, p. 297-308.

Takai, F., (ed.), 1963, Geology of Japan: Tokyo, University of Tokyo Press.

Tanaka, K., and T. Nozawa (eds.), 1977, Geology and mineral resources of Japan: Geological Survey of Japan, v. 1 (3rd ed.).

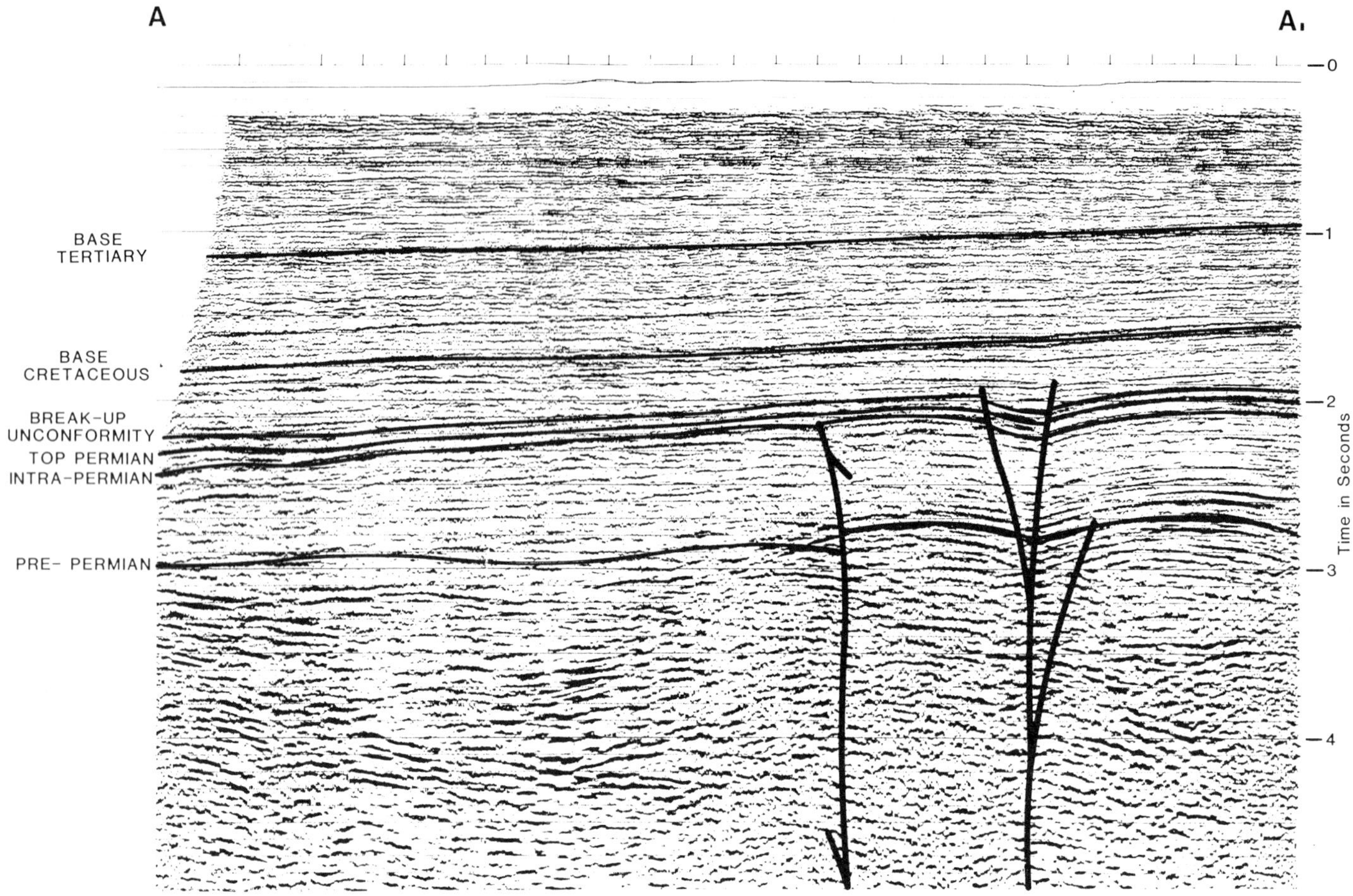


Figure 10. Seismic section illustrating flower structure in the Browse Basin, offshore Australia.

Tomoda, Y., n.d., Free air and Bouguer anomalies in and around Japan: Ocean Research Institute, University of Tokyo.

Utsu, T., T. Ishikawa, and Y. Katsui, 1973, In mobile belts in the world: Tokyo, Iwanami Publishing Co., p. 266.

von Huene, R., and M.A. Arthur, 1982, Sedimentation across the Japan trench: GSA Special Publication 10, p. 27-48.

von Huene, R., and H. Lee, 1983, The possible significance of pore fluid pressures in subduction zones: *in* Studies in continental margin geology: AAPG Memoir 34, p. 781-792.

Wilcox, R.E., T.P. Harding, and D.R. Seely, 1973, Basin wrench tectonics: AAPG Bulletin, v. 57, p. 74-86.

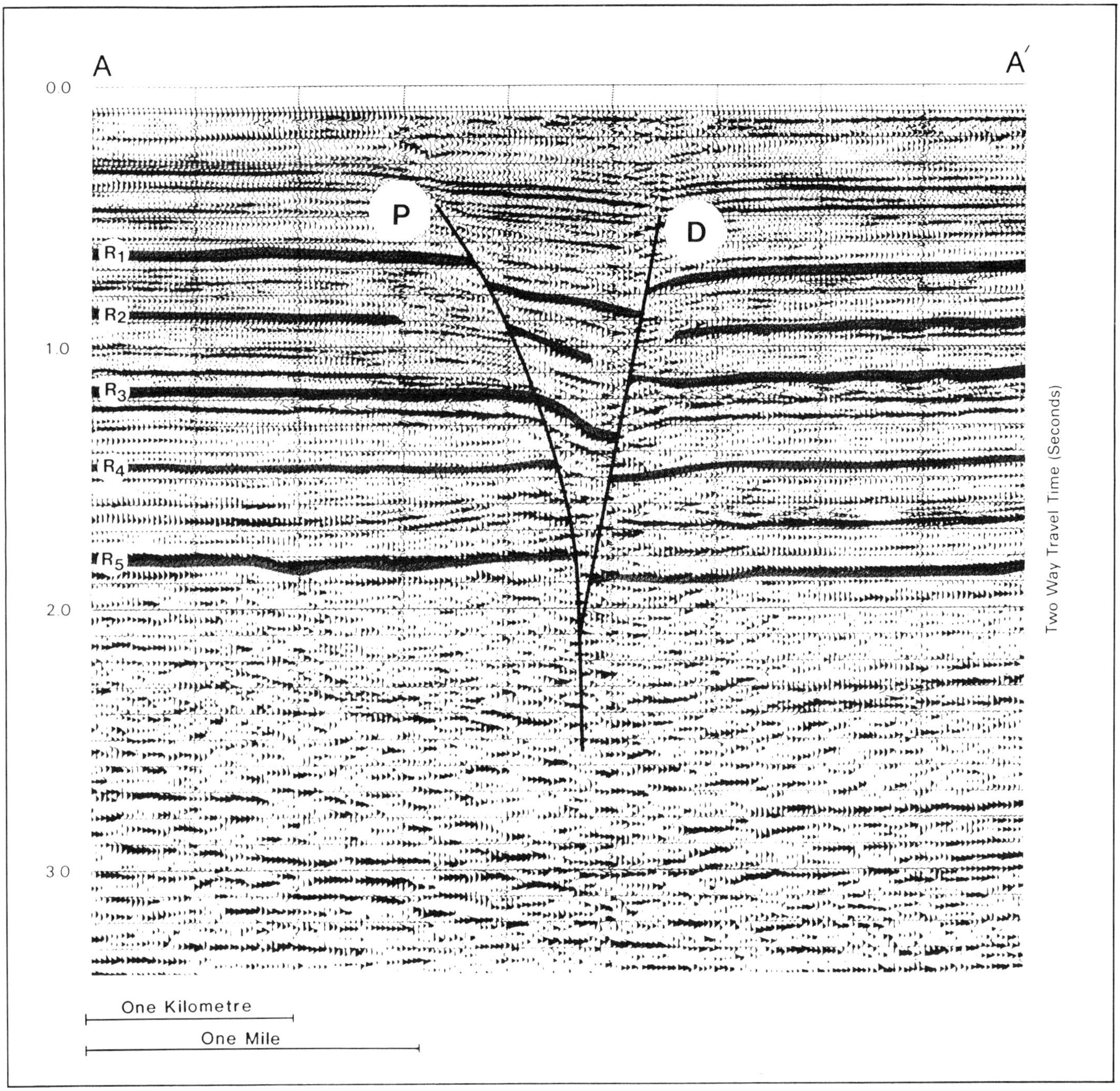


Figure 11. Seismic section illustrating flower structure in the Andaman Sea, Southwest Pacific (from Bally, 1983).

CHAPTER 20

# TECTONIC SETTING AND PETROLEUM POSSIBILITIES OF PHILIPPINE SEDIMENTARY BASINS

**Arthur Saldivar-Sali**
**Apollo P. Madrid**
**Raymundo A. Reyes, Jr.**
*Bureau of Energy Development, Philippine Ministry of Energy*
*Manila, Philippines*

**Louis G. Flower**
*Flower Doery Buchan Pty. Ltd.*
*Sydney, Australia*

ABSTRACT

The Philippines consists of two gross tectonic units: the island arc system and the microcontinental plate. The widely contrasting tectonic histories of these terranes strongly influenced the development and petroleum potential of 13 associated sedimentary basins. These basins can be broadly classified as forearc, backarc, or rift basins; the first two are identified with the island arc system and the last, with the microcontinental plate. Source rocks, reservoirs, traps, and seals evolved to different degrees in all basin types. Numerous hydrocarbon plates are conceived, of which only a few have been explored by drilling and/or validly tested to date. The basins, therefore, remain essentially frontier or marginally explored areas which merit further exploration efforts.

## INTRODUCTION

The Philippines comprises a region of complex tectonics that is progressively becoming better understood (Figure 1). The more recent regional studies include Hamilton (1979), Taylor and Hayes (1983), Holloway (1982), Karig (1983), and Mitchell et al. (in press).

The main north-trending sigmoidal-shaped Philippine archipelago is a 400-km-wide zone of active tectonism between the Eurasian and Philippine Sea plates, as evidenced by widespread volcanism, shallow crustal seismicity and generally rugged topography (Figure 2). The eastern limit of active deformation is marked by the Philippine and East Luzon trenches and an interconnecting transform zone east of central Luzon. No eastern trench exists between Luzon and Taiwan, but the East Luzon Trench is believed to be propagating northward (Lin and Tsai, 1981; Lewis and Hayes, 1983).

The western limit of active deformation is represented by the Manila, Negros, and Cotabato trenches, This otherwise continuous western trench system is disrupted where the main archipelago impinges on two northeast-trending ridges in the Eurasian plate—the Palawan-Mindoro Ridge and the Sulu-Zamboanga Ridge.

The active archipelagic zone is undergoing overall compressive deformation between the two major plates, but in the central Philippines there is a component of shear best known along the 1200-km-long Philippine Fault zone (Acharya, 1980). The overall present-day plate convergence that is being taken up along the main Philippine archipelago is in a northwesterly direction at about 7 to 10 cm/year (Seno, 1977; Minster and Jordan, 1979; Rankin et al., 1984).

The total convergence that has taken place between the Philippine Sea and Eurasian plates since the early Cenozoic is substantial. Available paleomagnetic data suggest a substantial northward component of motion of the Philippine Sea plate, apparently 15° to 25° since about 50 m.y. before present (Jarrard and Sasajima, 1980; Fuller et al., 1983). As a result of this cumulative convergence, a number of terranes—each with unique stratigraphy and geologic history—have been amalgamated into the main north-south archipelago. These include rifted continental margin, volcanic arcs, nonvolcanic arcs, and oceanic crusts. Only parts of the complex history of these terranes have been unraveled.

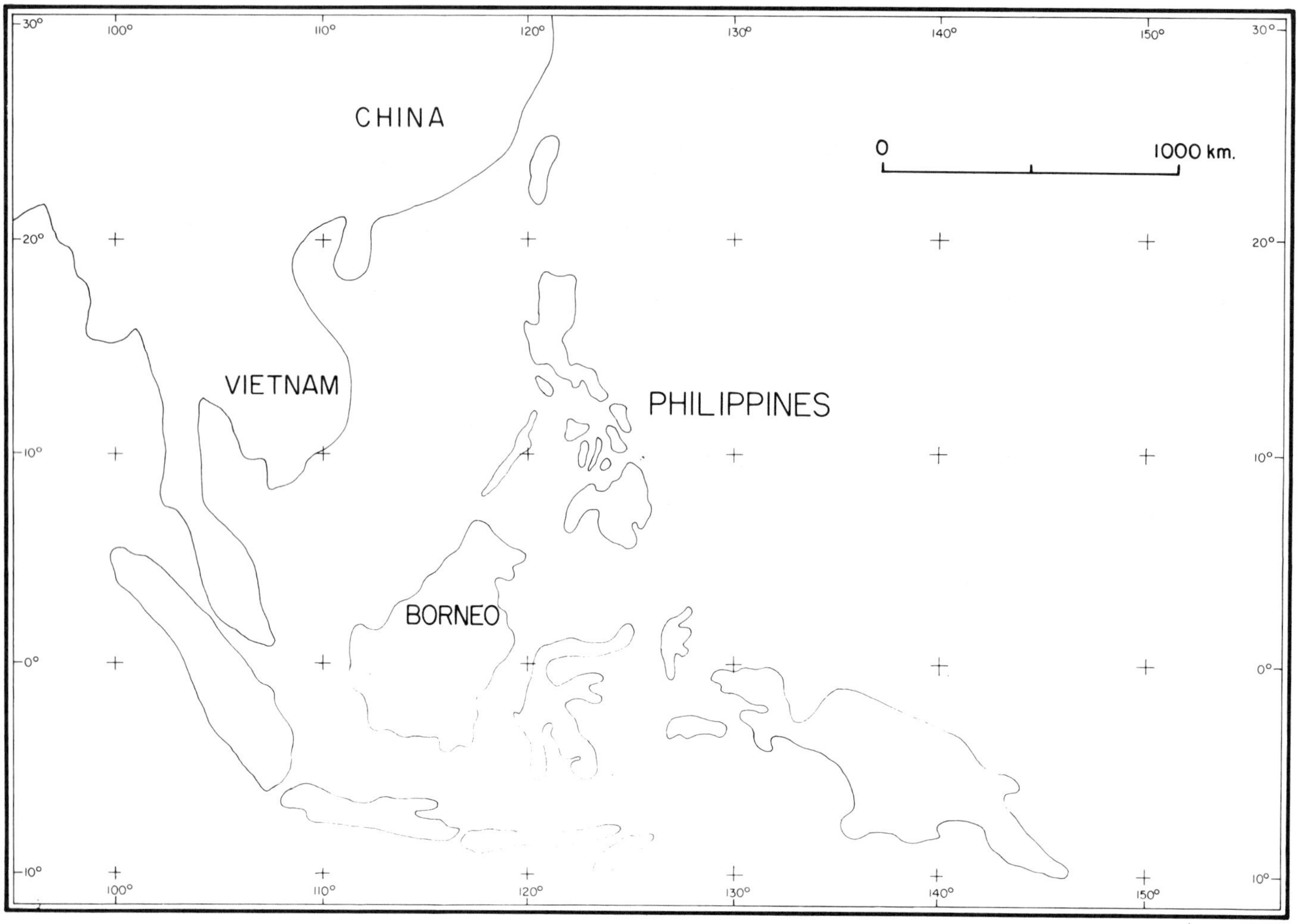


Figure 1. Index map of the Philippines.

## TECTONIC SETTING AND CLASSIFICATION OF SEDIMENTARY BASINS

The Philippines can be divided into two gross tectonic units (Figure 3). The first, which corresponds to the island arc system, consists of the main sigmoidal composite arc termed Philippine Archipelago and the northeast-trending series of minor arcs and basins in the Eurasian Plate. The second is equivalent to the rifted continental block or microplate found in the Eurasian Plate.

The island arc system developed in a convergent margin setting since its inception in the Pacific Plate (prior to formation of the Philippine Sea Plate) in the Late Mesozoic, whereas the microplate evolved in a convergent continental margin setting for much of the Mesozoic (in the southeast China margin), thence in a divergent margin setting. The two units were not contiguous until the late Tertiary.

Thirteen major polyhistory basins—11 in the island arc system and 2 in the microplate—occur within Philippine territorial limits (Figure 4). The island arc basins are generally deep and narrow linear troughs floored by intermediate crust. The microplate basins are broad, fault-dissected platforms or prisms associated with attenuated continental crust.

The precise classification of Philippine sedimentary basins according to tectonic setting is quite difficult in a number of cases because of tectonic overprinting and/or yet unclear tectonic evolution. Nevertheless, classification is attempted inasmuch as known regularities of petroleum prospects in analogous settings may be helpful in finding oil plays in both the frontier and marginally explored basins of the country.

The Ilocos Trough, Central Luzon Basin, West Luzon Platform, Iloilo Basin, Agusan-Davao Basin, Bicol Shelf-Lamon Bay, and the Balabac and Bancauan subbasins in the Sulu Sea fall under the category of forearc basins.

The Visayan Basin and Southeast Luzon Basin are considered as backarc basins modified by wrenching. The Cotabato Basin and Cagayan Basin, which are known inter-arc basins, may likewise be regarded as backarc basins during most of their sedimentation history. The Sandakan Subbasin in the Sulu Sea may be classified as a Tertiary delta superimposed on a backarc basin.

The North Palawan-Mindoro and South Palawan basins, including Reed Bank area, are rift basins modified by collision in their easternmost margins.

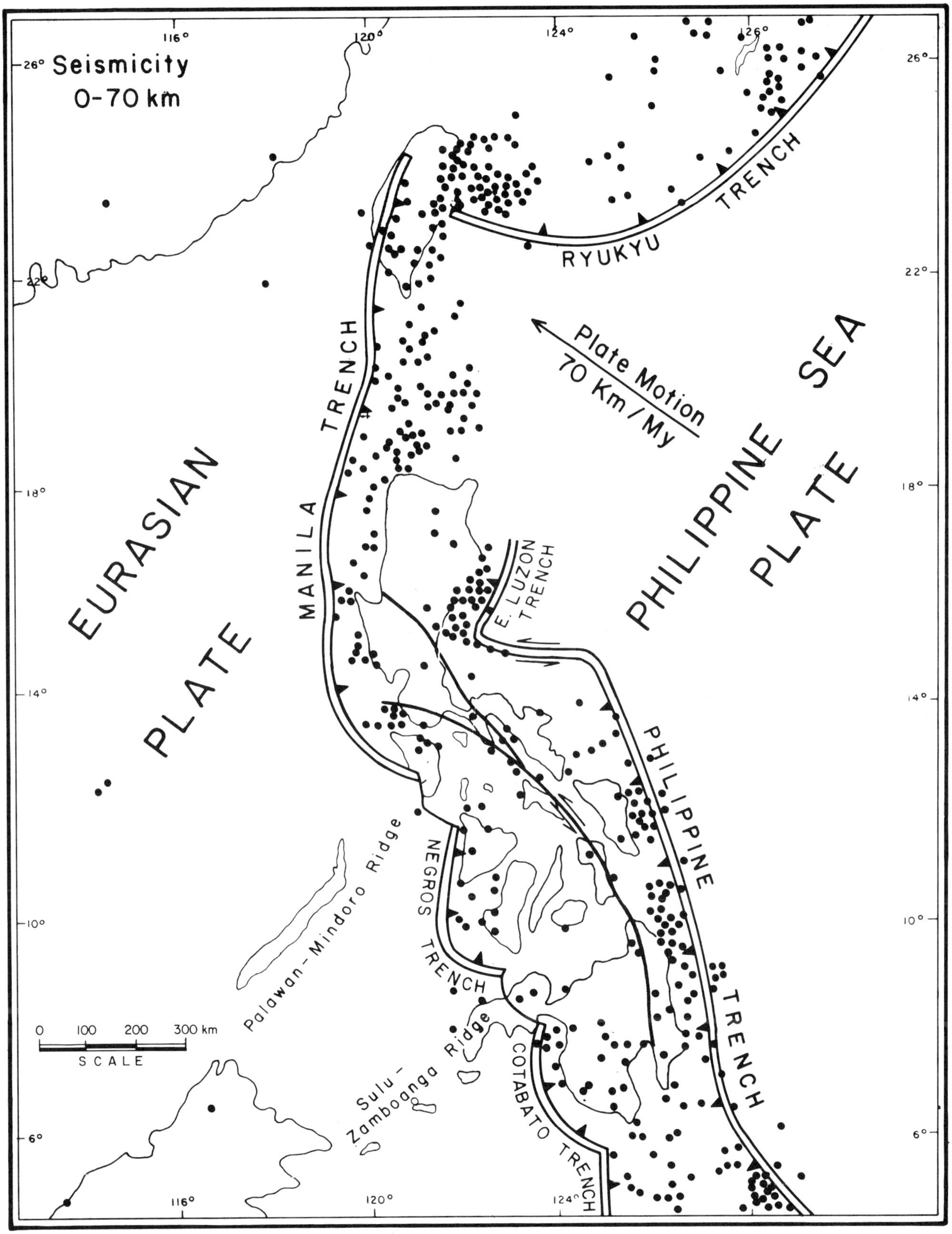


Figure 2. Present-day tectonics and shallow seismicity of the Philippines.

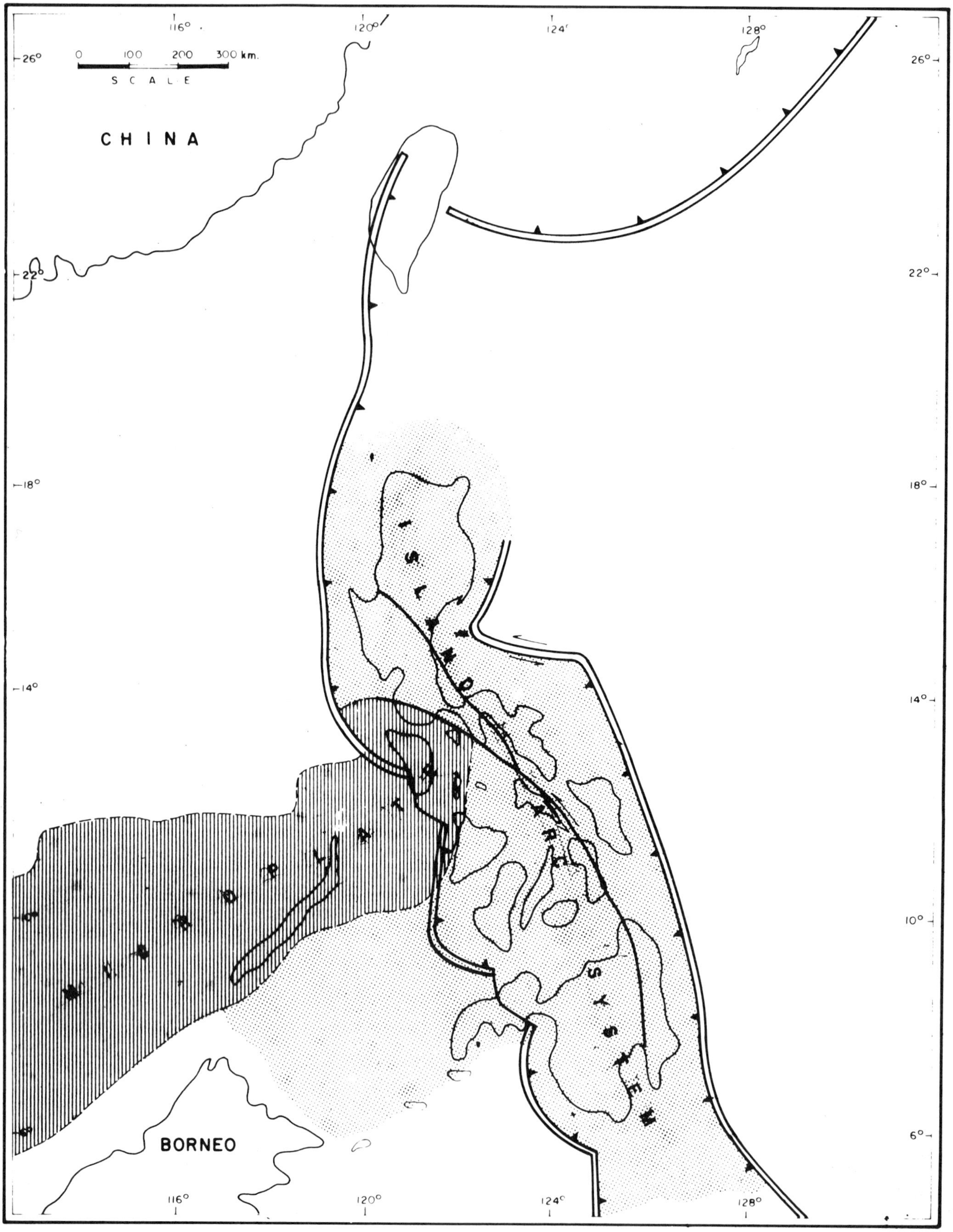


Figure 3. Megatectonic units of the Philippines.

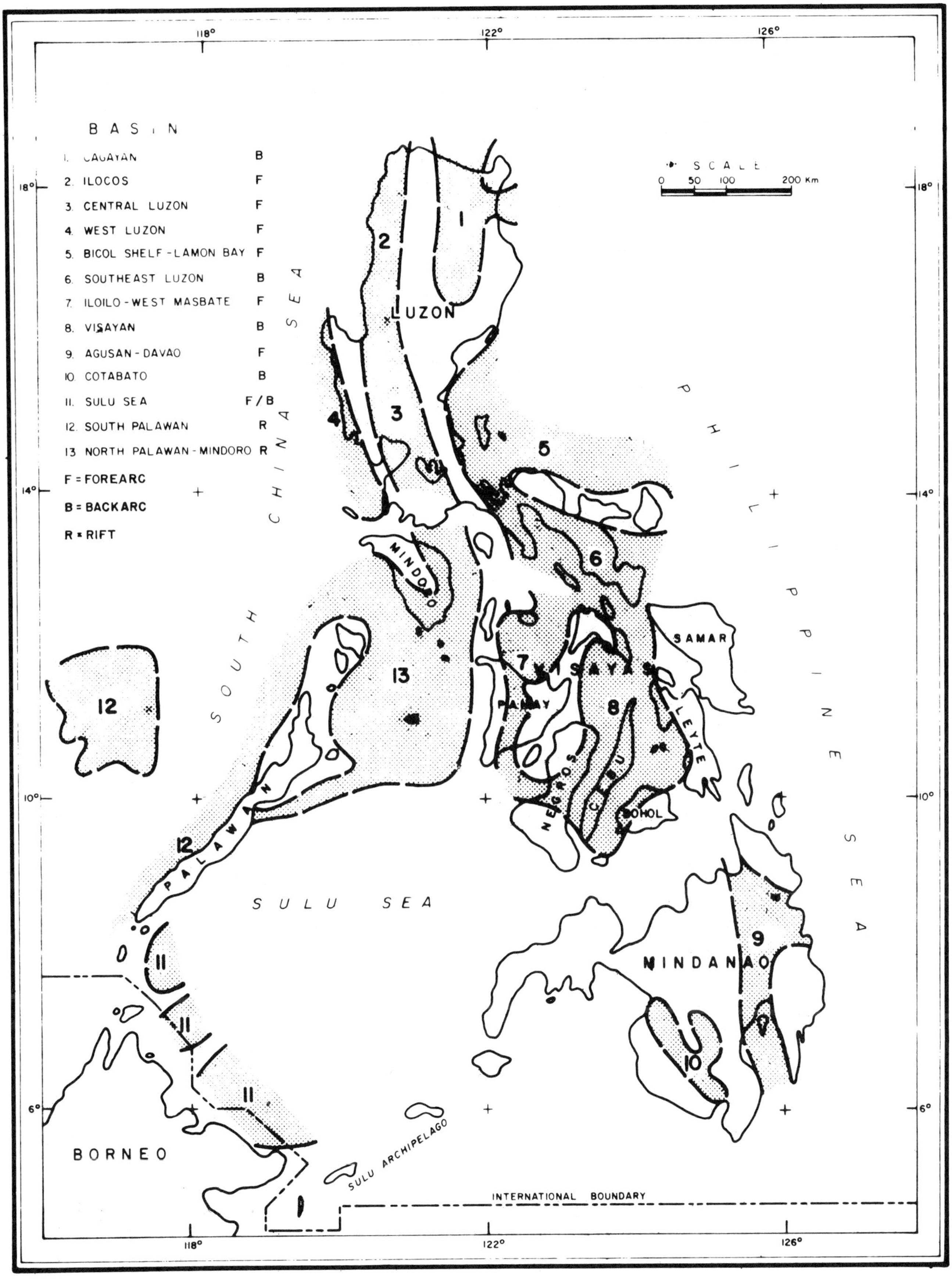


Figure 4. Sedimentary basins of the Philippines and their classification.

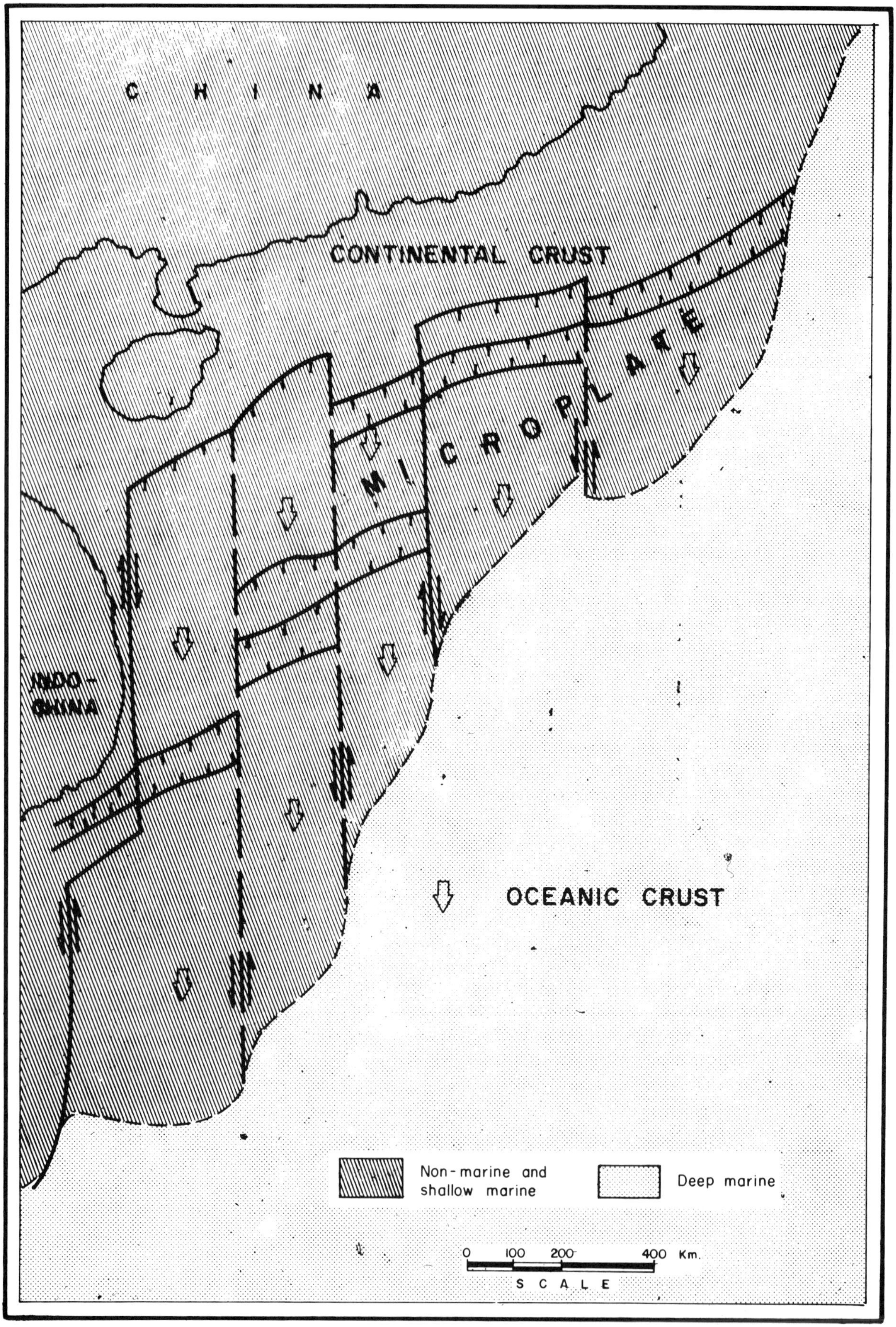


Figure 5. Rift phase (Paleocene to Early Oligocene) paleogeography of microplate basins.

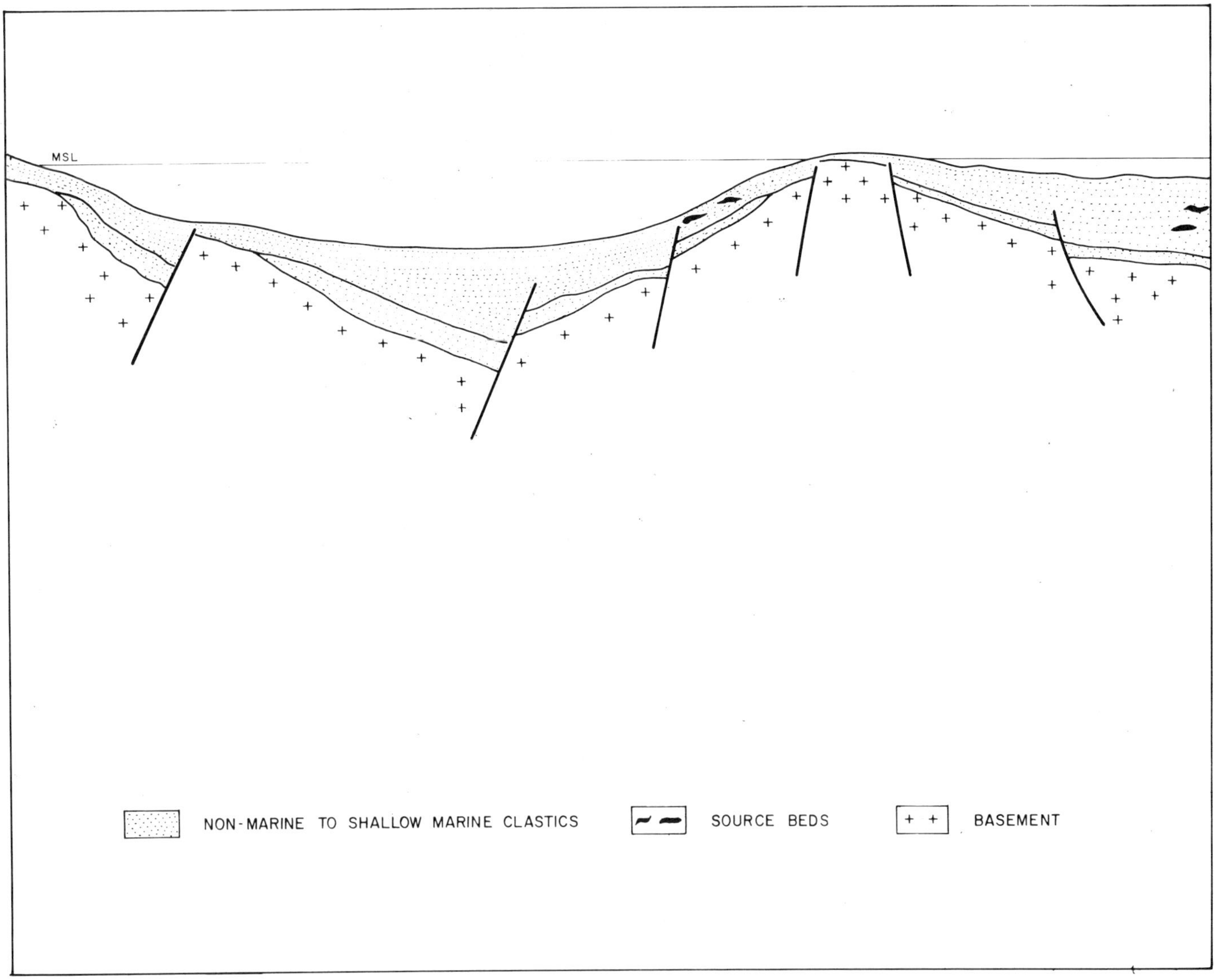


Figure 6. Rift phase (Paleocene to Early Oligocene) schematic profile of microplate basins.

## EVOLUTION OF THE BASINS AND DEVELOPMENT OF PETROLEUM PLAY ATTRIBUTES

Regional tectonics and the resultant structures controlled the evolution of Philippine sedimentary basins. Sedimentation in the island arc basins essentially began in the early Cenozoic and is characterized by the preponderance of volcaniclastics and reefal carbonates. In the microplate, on the other hand, sedimentation started in the Mesozoic and is dominated by nonvolcanic quartzose sediments.

### Microplate Basins

*Pre-Rift Phase (Late Paleozoic to Mesozoic)*—The microplate was originally an integral part of the southeast China continental massif, accumulating nonmarine and shallow-marine clastics and minor shallow-marine carbonates. Three major orogenic episodes related to plate convergence along the southeast craton margin—end of Paleozoic, Late Triassic, and Late Jurassic—caused regional uplift/erosion and magmatism in the area. Early Cretaceous sandstone reservoirs developed after the last uplift.

*Rift Phase (Paleocene to Early Oligocene)*—Continental crust attenuation occurred in the southeast China margin along a northeast-southwest trend at the end of the Cretaceous (Figures 5, 6). Tensional block faulting induced rift sedimentation, whereby nonmarine to shallow-marine clastics and shallow-marine carbonates filled half grabens.

Paleocene to Early Eocene sandstone reservoirs, as well as Eocene oil and gas source facies, developed in the basins of the microplate. Structural and combination traps in the form of truncated or nontruncated tilted fault blocks bearing transitional to shallow-marine sand bodies were common.

*Drift Phase (Late Oligocene to Middle Miocene)*—The mid-Oligocene crustal breakup and north-south opening of the South China Sea propelled the microplate southward (Figures 7, 8). During the Late Oligocene to Early Miocene, transgres-

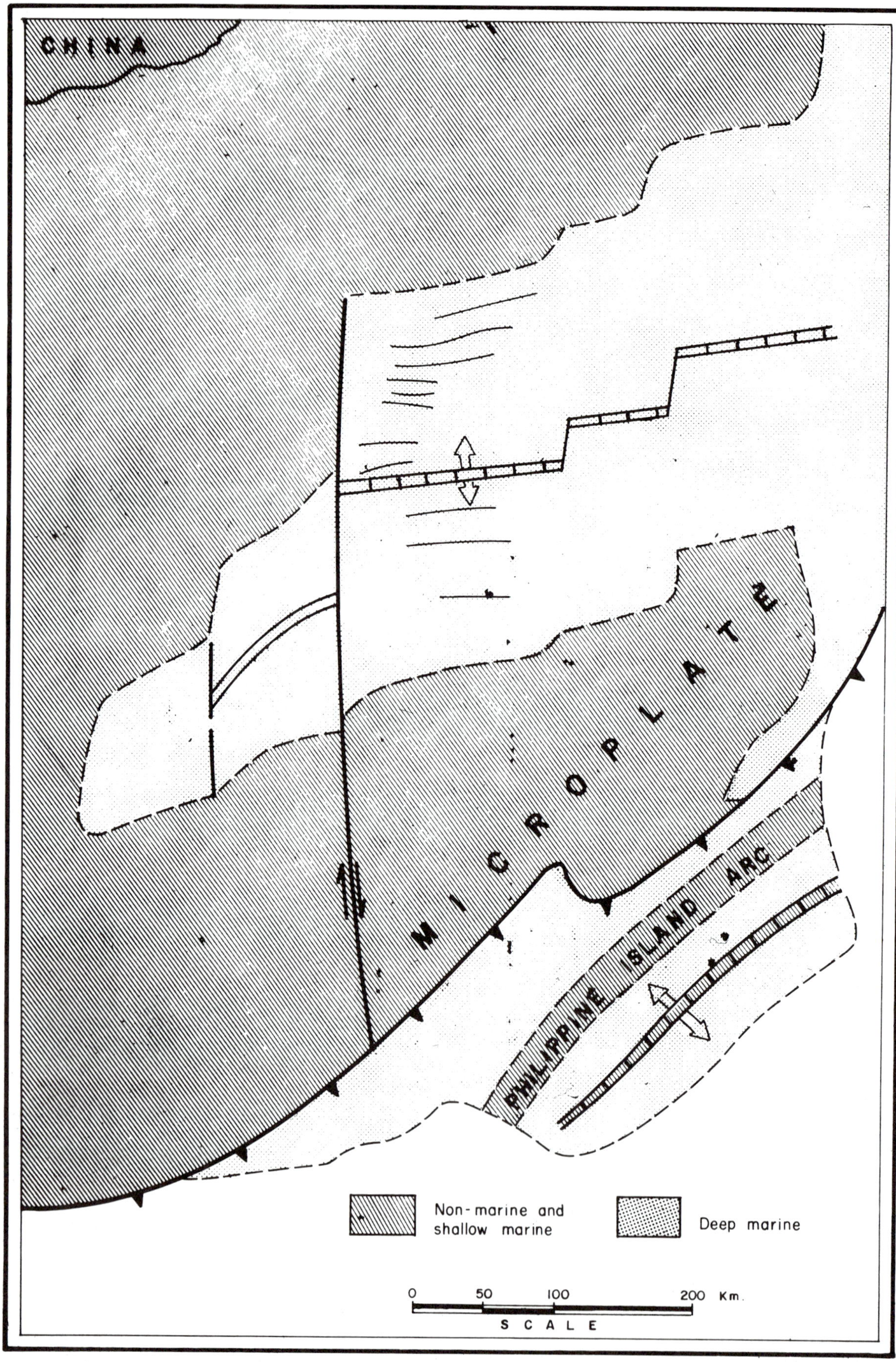


Figure 7. Drift phase (Late Oligocene to Middle Miocene) paleogeography of microplate basins.

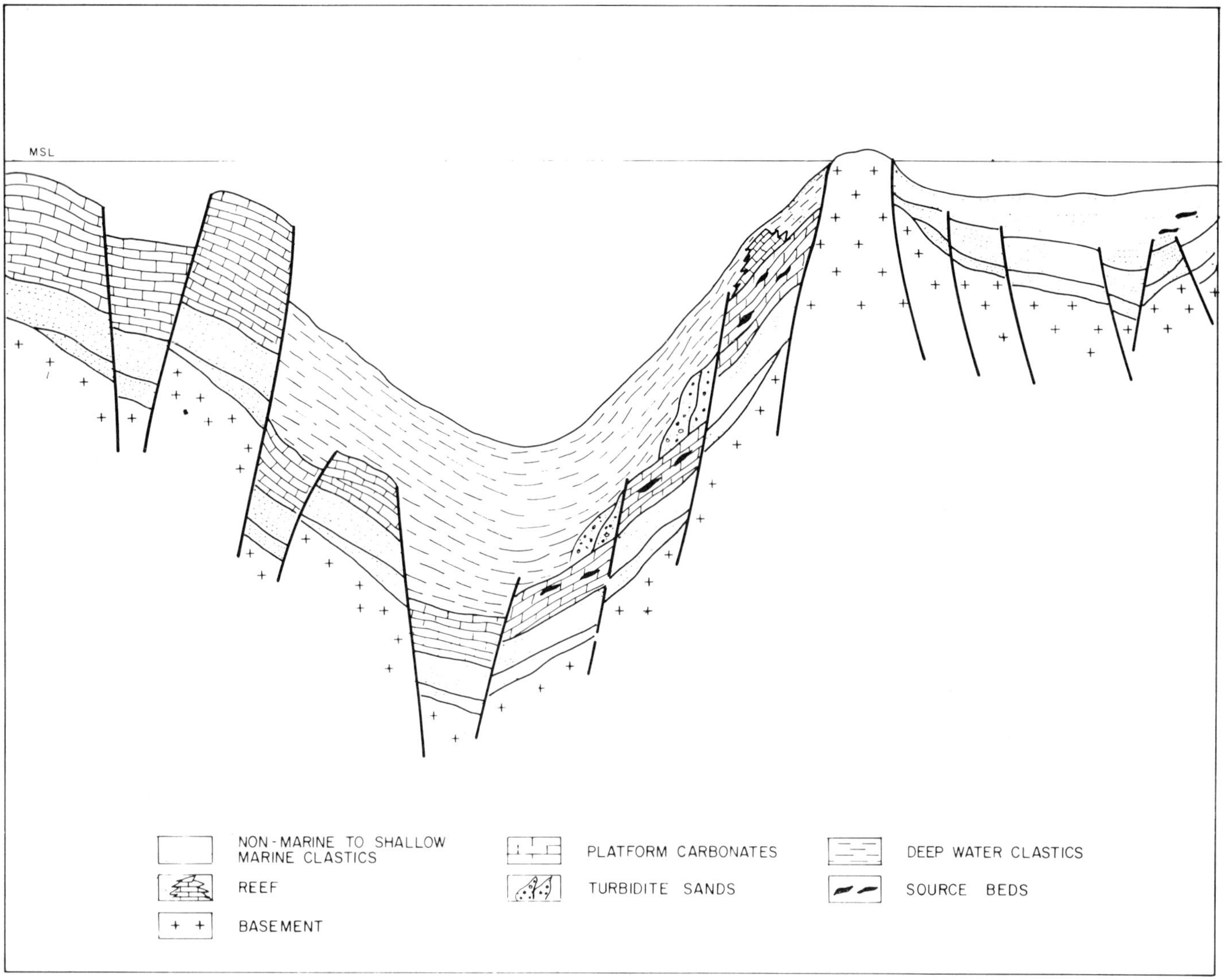


Figure 8. Drift phase (Late Oligocene to Middle Miocene) schematic profile of microplate basins.

sive carbonates blanketed the northwest-trailing part of the rifted block (Reed Bank-West Palawan region). Nonmarine and transitional marine clastics covered the southeast or leading half (Cuyo Platform-Mindoro region). In the late Early Miocene, the central part of the microplate (west of Palawan) subsided rapidly and accumulated turbiditic sandstones and carbonates. Middle Miocene saw continued deposition of shallow-marine carbonates in the Reed Bank area, deep-marine fine clastics west of Palawan, and transitional to shallow-marine clastics in the Cuyo Platform-Mindoro region.

During the period, the following reservoirs developed in various areas of the microplate: Late Oligocene platform carbonates, Early Miocene reefal carbonates and Early Miocene turbiditic sandstones and carbonates, and Early to Middle Miocene transitional to shallow-marine sandstones. Late Oligocene to Early Miocene carbonate oil source beds, Late Oligocene carbonate gas source facies, and Early/Middle Miocene oil source beds were likewise formed. The Middle Miocene regional mudstone seal was deposited west of Palawan.

Predominantly stratigraphic traps involving reefal buildups and turbiditic sandstone/limestone bodies were formed. Structural traps consisted mostly of drapes over Paleogene paleoridges or horsts.

*Collision Phase (End of Middle Miocene)*—The rifted microplate collided with the counterclockwise-rotating island arc system, elevating the basins on the former to erosional level (Figure 9). The attendant compressive stress was strongest in the eastern margin of the block, and structural traps such as trust and strike-slip fault structures and related folds were produced.

*Post-Collision Phase (Late Miocene)*—The entire Philippines was amalgamated into one tectonostratigraphic terrane and was situated more or less in its present-day geographic position (Figures 10, 11).

Shallow-marine carbonates were laid down in Reed Bank and South Palawan, shallow-marine carbonates and transitional to shallow-marine clastics in the Northwest Palawan Shelf and Cuyo Platform areas, and transitional to shallow-marine clastics east and south of Mindoro. The intervening areas were loci of deep-marine clastic sedimentation.

During this period, shallow-marine sandstone reservoirs were developed.

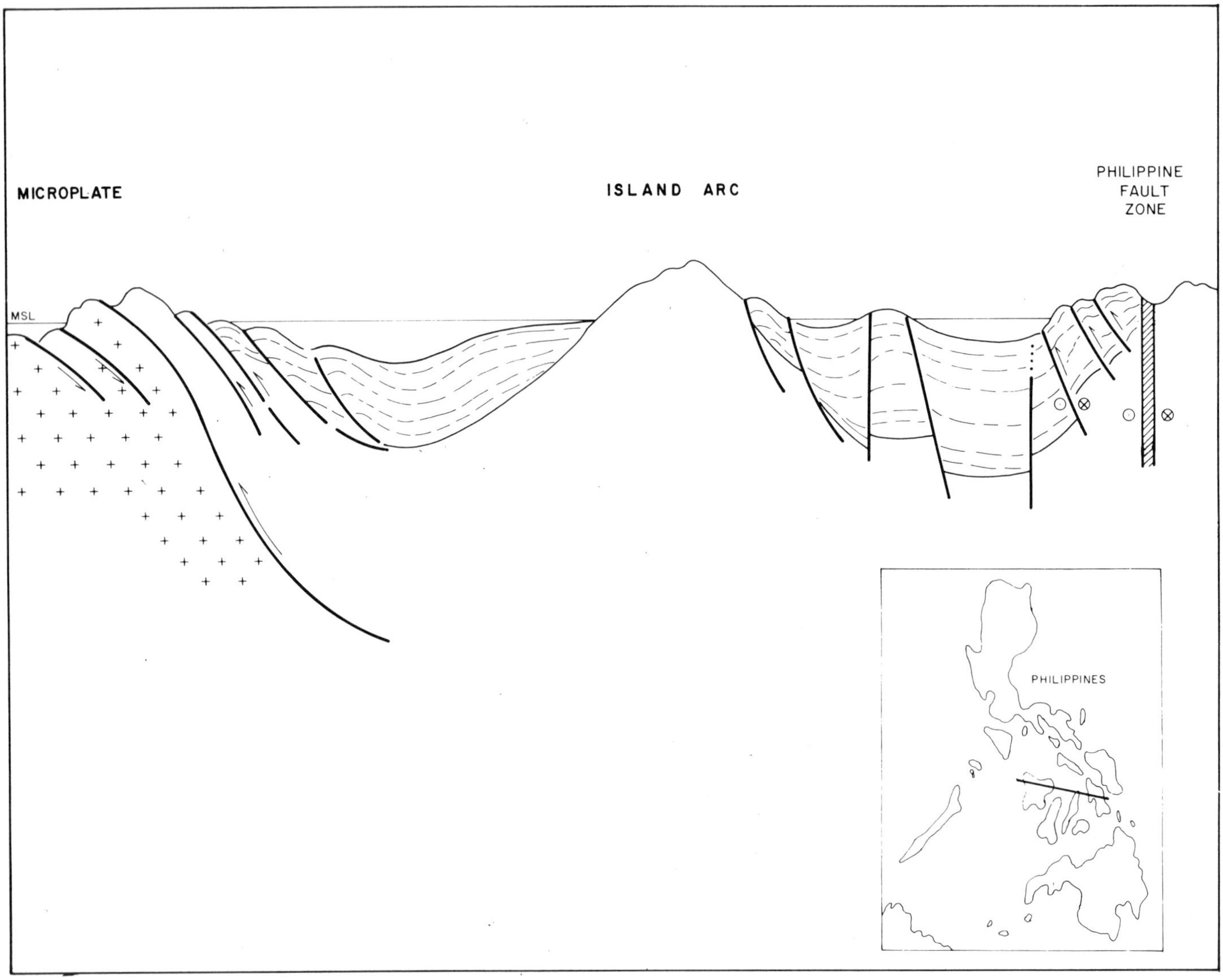


Figure 9. End of Middle Miocene collision schematic profile.

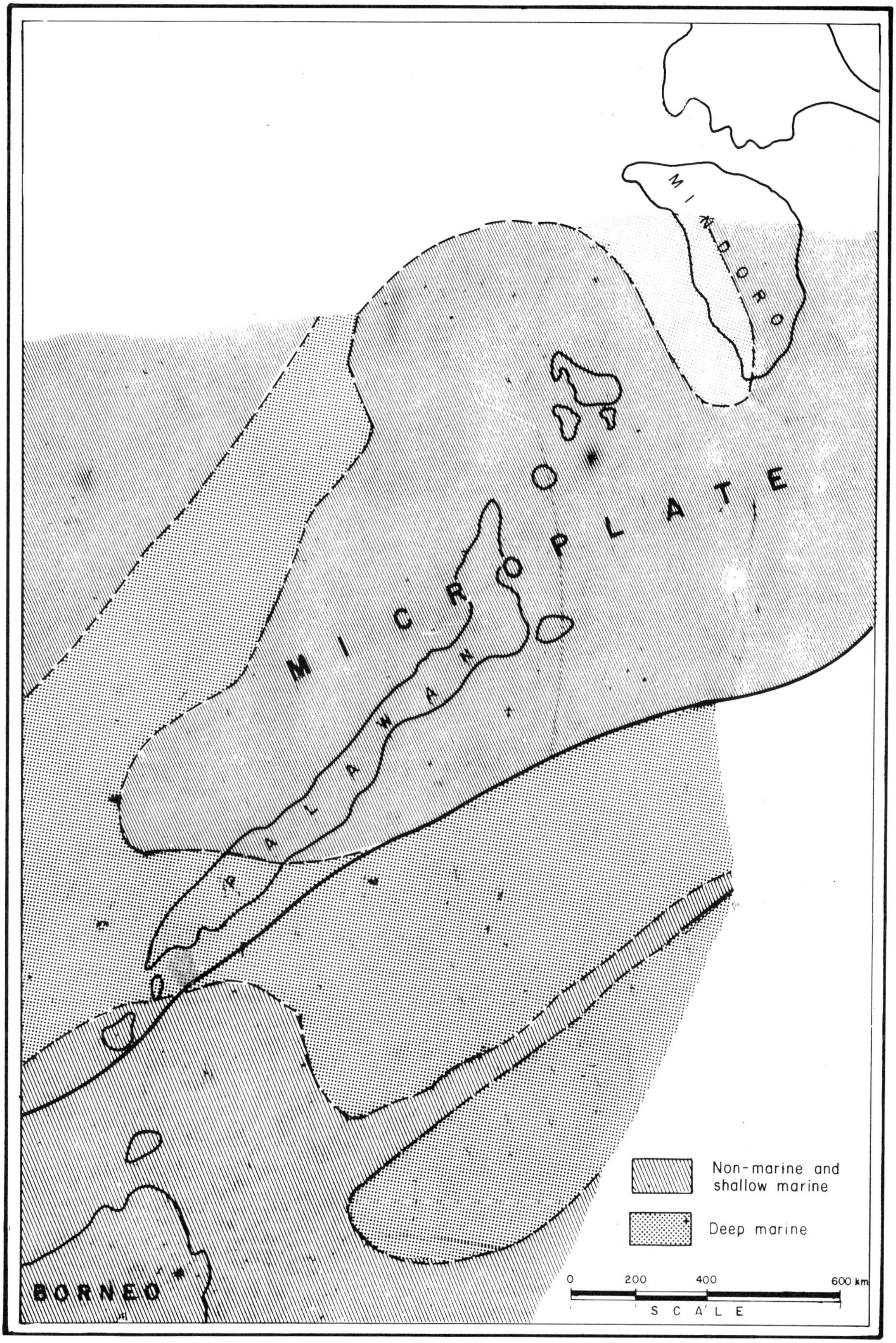


Figure 10. Late Miocene paleogeography of microplate basins.

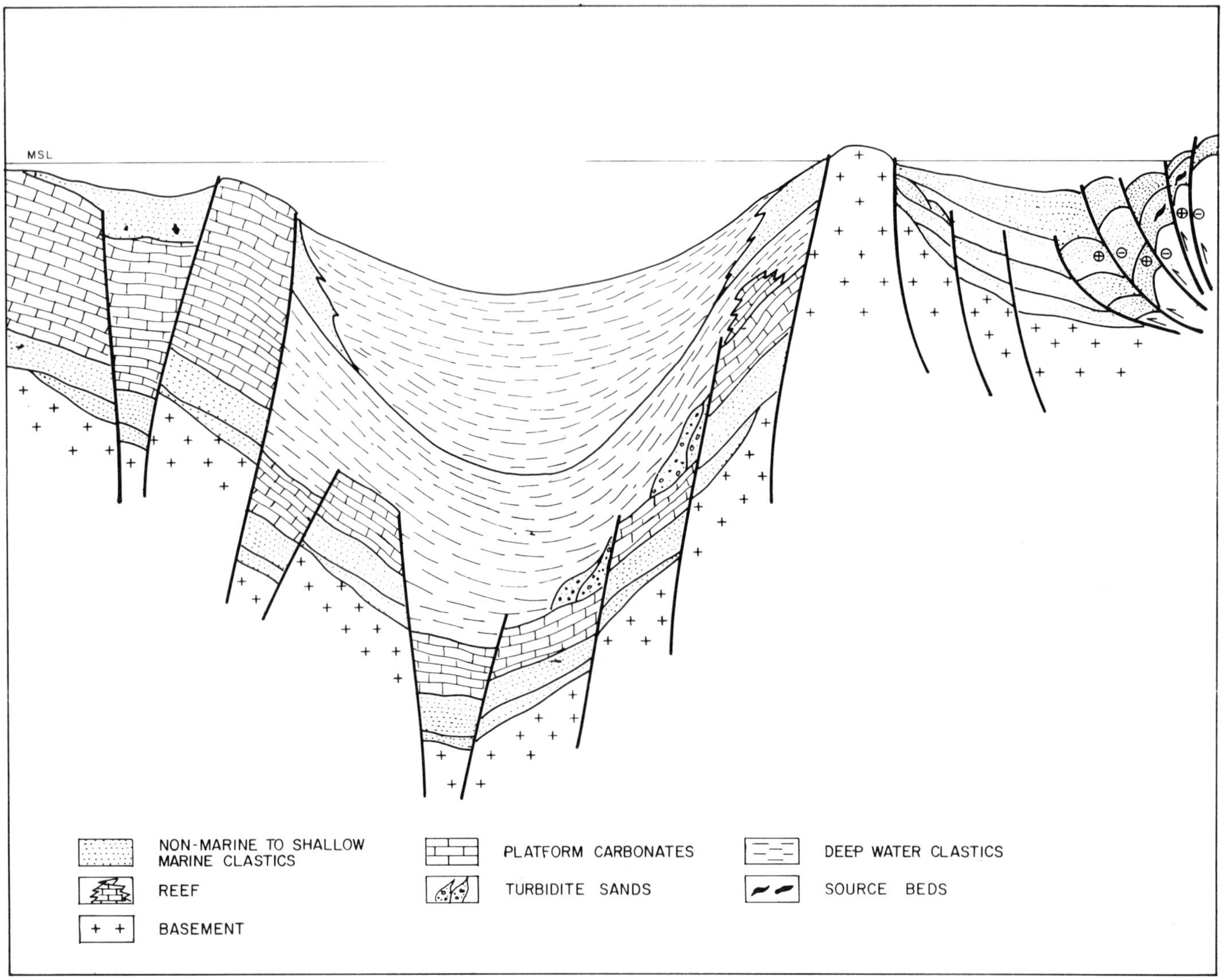


Figure 11. Late Miocene schematic profile of microplate basins.

## Island Arc Basins

*Late Cretaceous to Paleogene*—Much of the composite island arc system probably originated as part of an intraceanic Late Cretaceous protoarc/trench system. Consistent with the early stage of island arc development, sedimentation was subordinate to volcanism during the Paleogene. Volcaniclastics and minor carbonates interspersed with mafic/intermediate extrusives were deposited in a deep but progressively shallowing marine environment. The Paleogene sequence, with few exceptions, comprises the economic basement as it was weakly metamorphosed and/or hydrothermally altered by subsequent tectonism and magmatism.

Initiated as an apparent east-west trend, the protoisland arc system began its counterclockwise rotation in mid-Paleogene (Hsu and Sharon, 1970).

*Late Oligocene*—This period generally represents the beginning of continuous subsidence and infilling of the island arc basins. The basins started out as shallow marine depressions or sags which received clastic detrita from subdued land masses (Figures 12, 13). Neritic limestones with scattered reef growth developed in the periphery of positive areas and over shallow bathymetric highs. Coals and coaly sediments were deposited in restricted lagoons and deltaic embayments.

Potential reservoirs consist of transitional to shallow-marine sandstones. Potential oil source rocks, specifically coals and coaly shales, are present, particularly in the backarc basins. Stratigraphic traps such as onlaps and pinchouts were formed.

*Early Miocene*—Sedimentation proceeded as in the Late Oligocene as paleoenvironments essentially remained unchanged (Figure 12). Reefal carbonates grew over low-relief, block-faulted platform areas (horst features) of backarc basins and on narrow shelves of forearc regions (Figure 14). Transitional to shallow-marine clastics, commonly containing coals, were contemporaneously deposited.

Important reservoirs are represented by reefal carbonates and transitional to shallow-marine sandstones. Coals and coaly sediments developed as source facies in the backarc region. The principal stratigraphic traps consisted of reefal buildups and shelfal sand bodies. Fault blocks and drapes over paleohighs constituted the more important structural traps.

*Middle Miocene*—The beginning of this period marked the definition of shelf-slope hingelines and/or margin faults (Figures 15, 16). Fine clastics were deposited toward the basinal

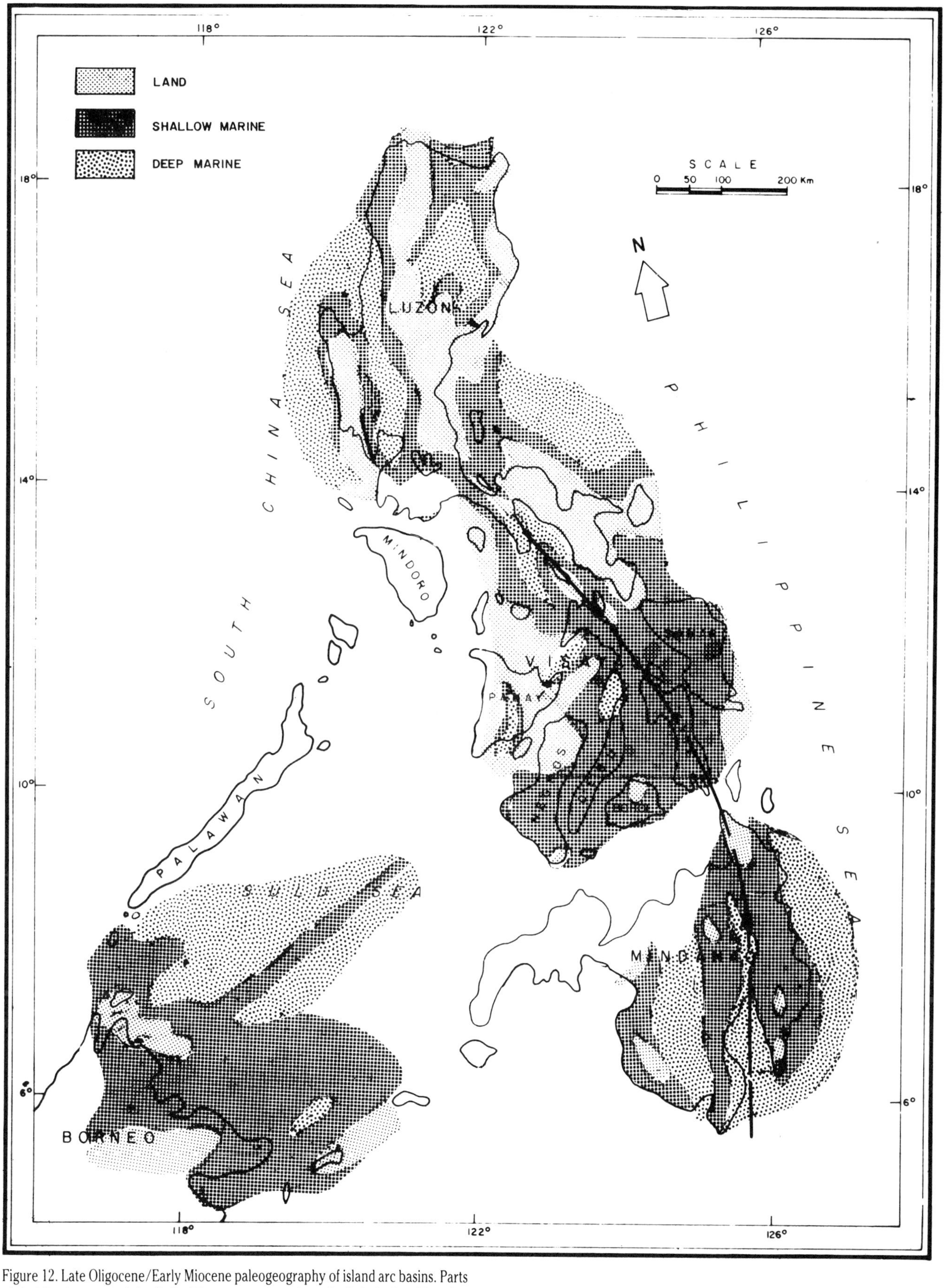


Figure 12. Late Oligocene/Early Miocene paleogeography of island arc basins. Parts of the island arc east of the dashed line may not have been in their depicted locations.

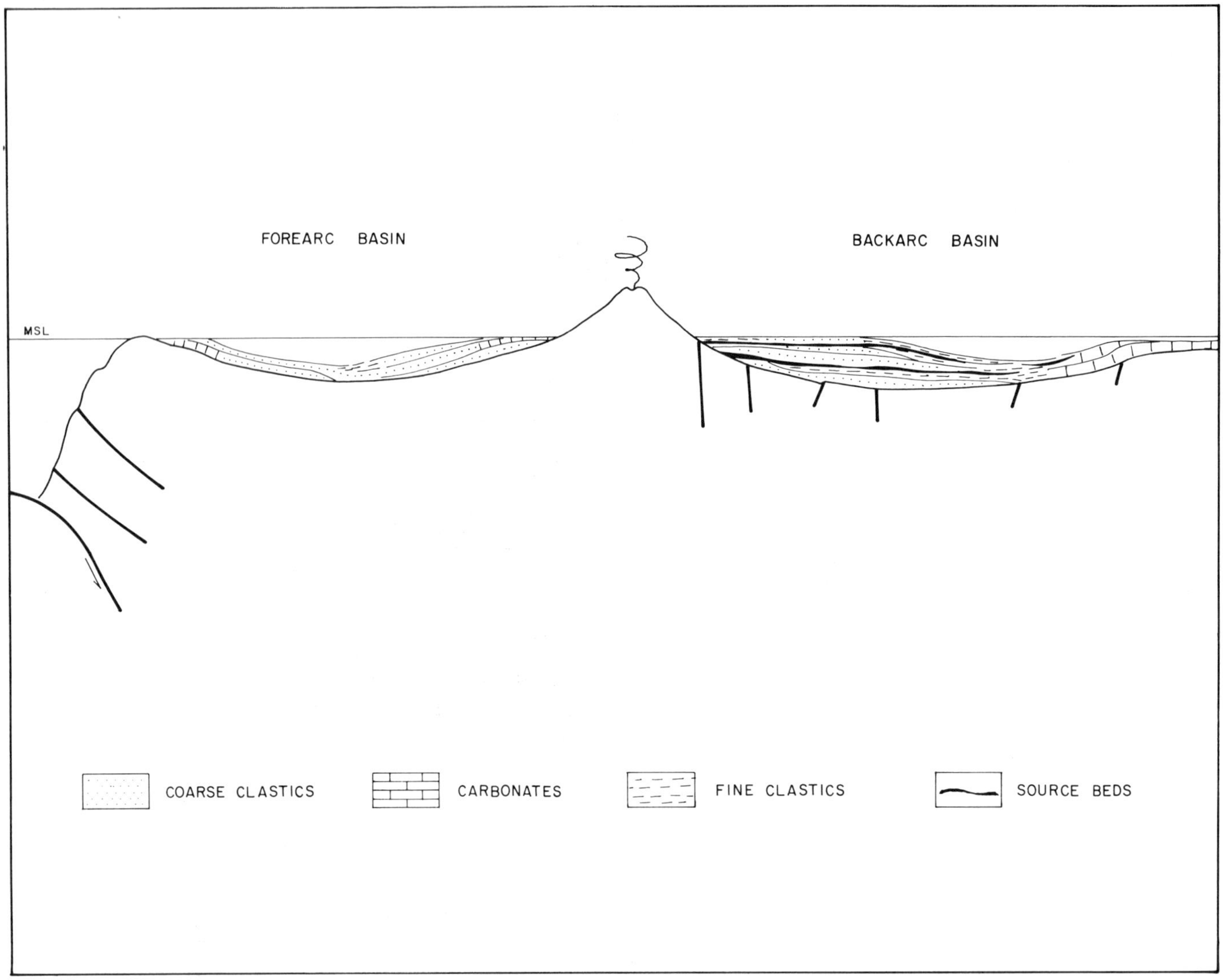


Figure 13. Late Oligocene schematic profile of island arc basins.

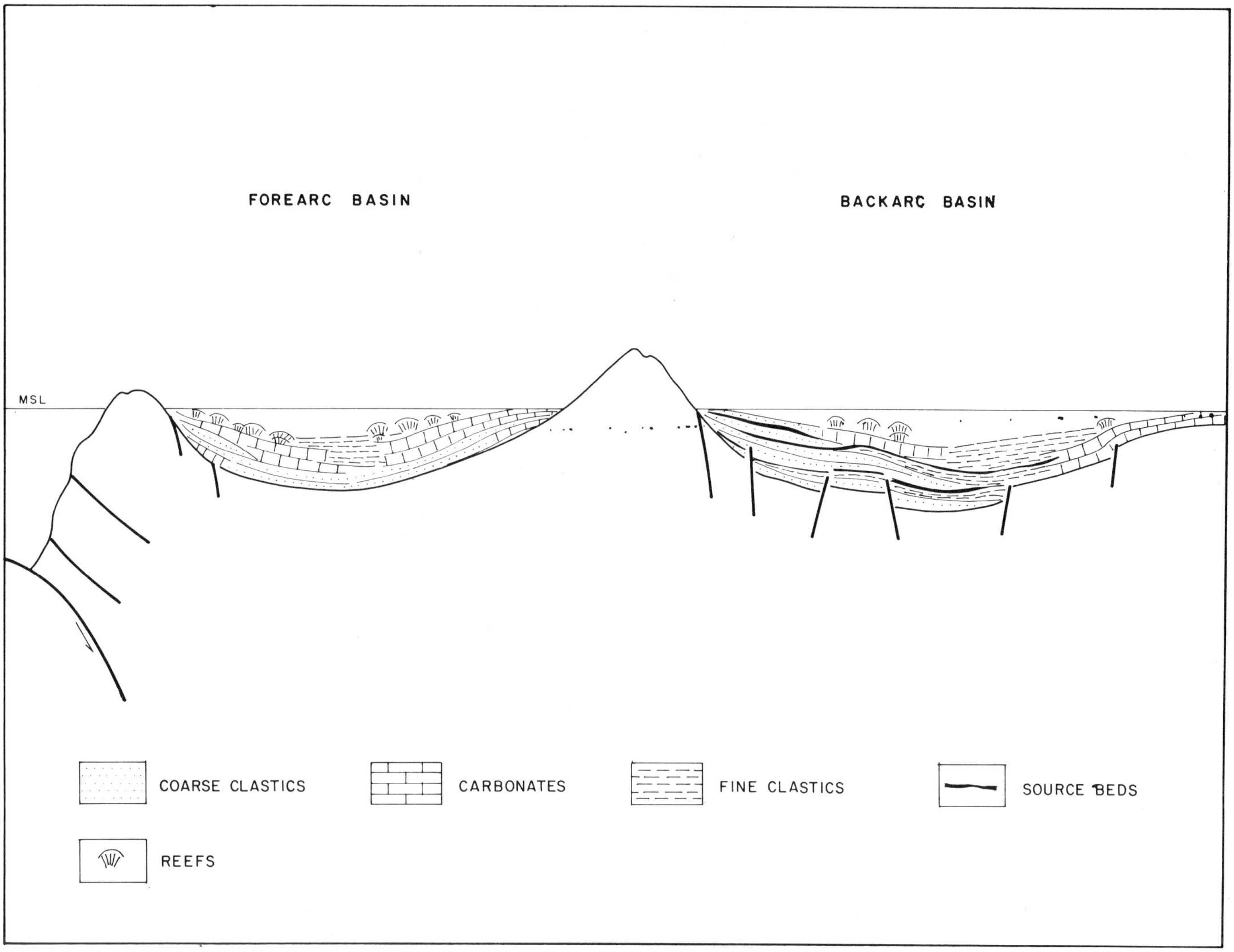


Figure 14. Early Miocene schematic profile of island arc basins.

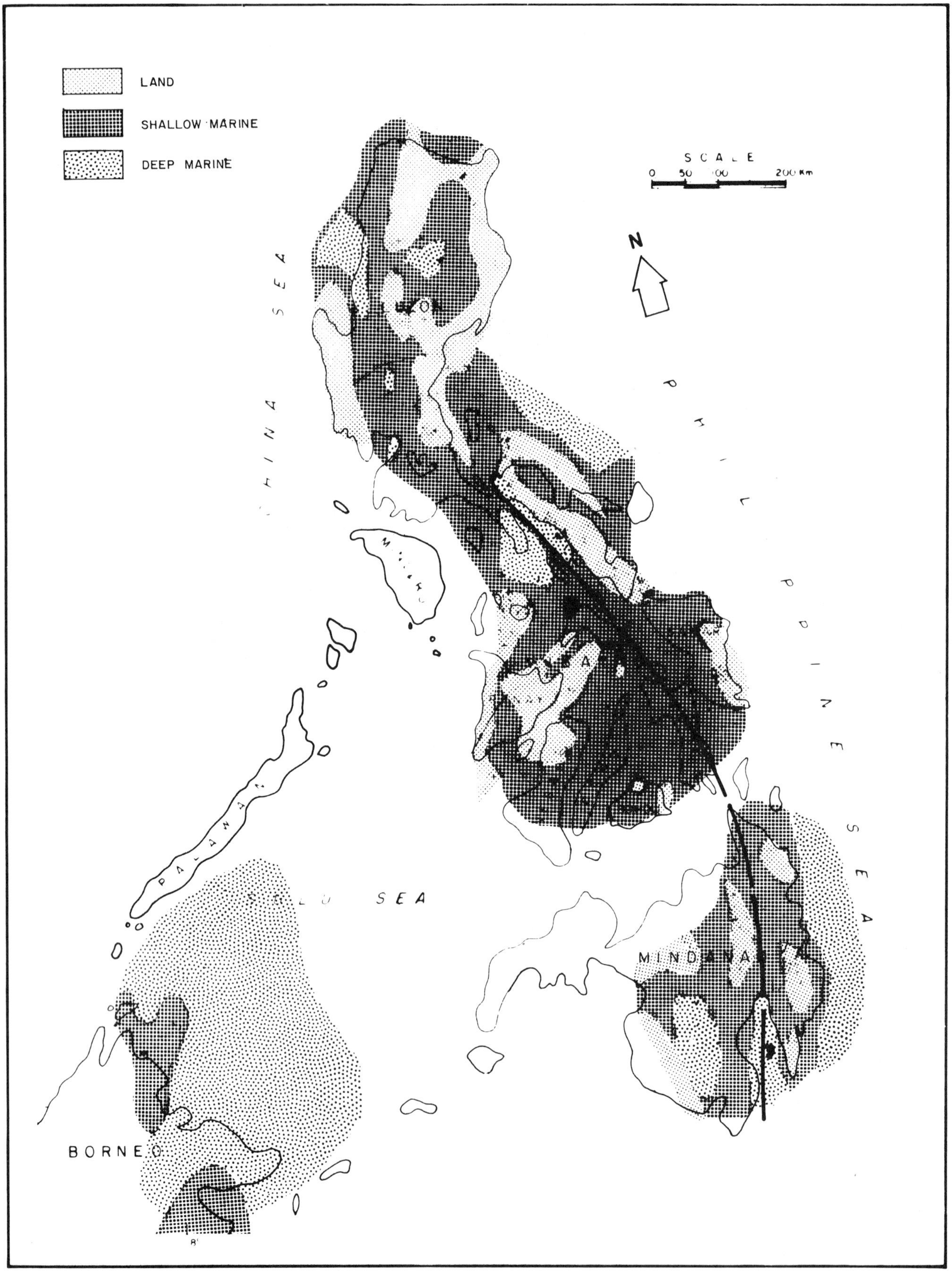


Figure 15. Middle Miocene paleogeography of island arc basins. Parts of the island arc east of the dashed line may not have been in their depicted locations.

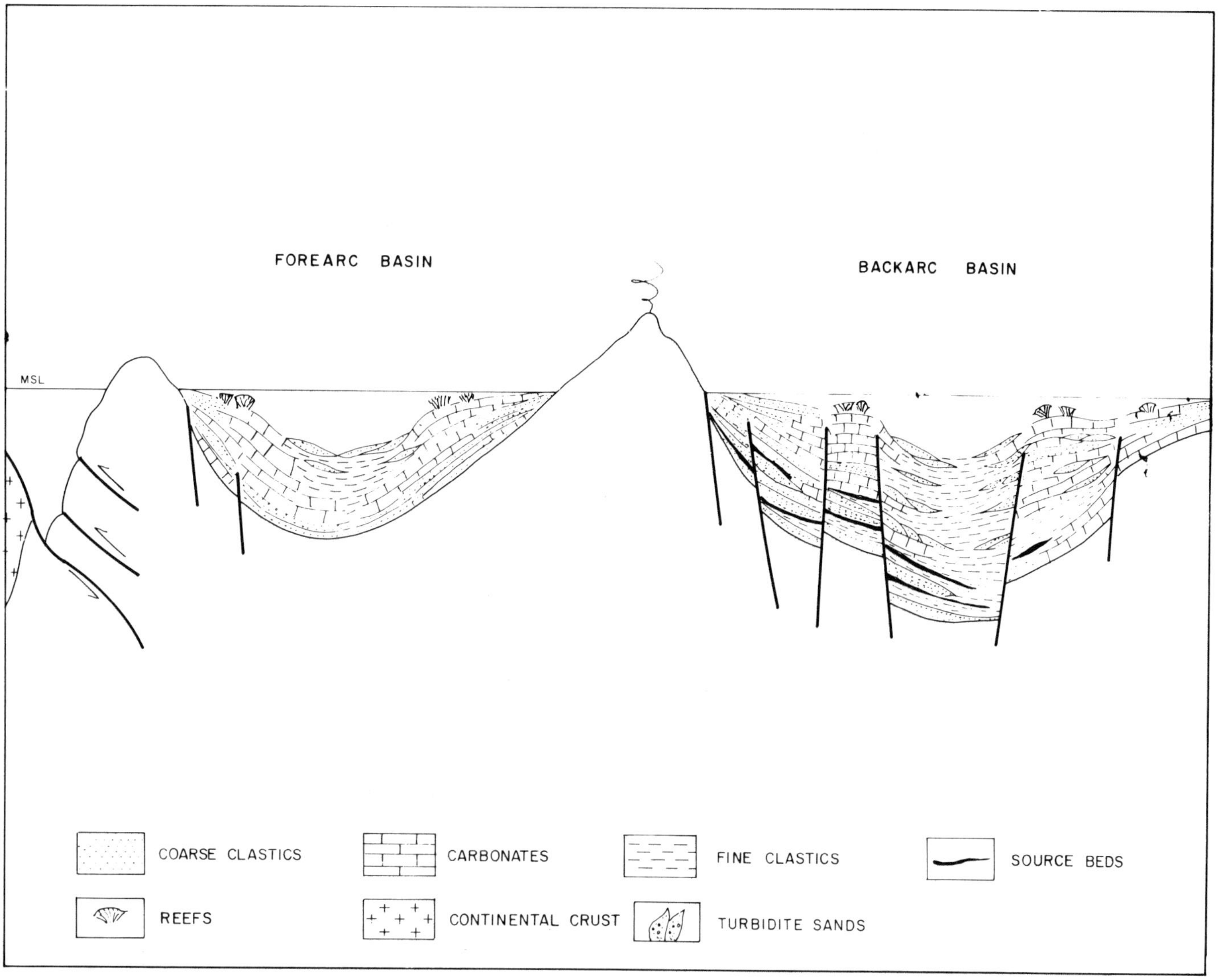


Figure 16. Middle Miocene schematic profile of island arc basins.

axes, whereas coarse clastics and carbonates, a few reefal, were accumulated on the shelves. Shallow-marine carbonates were also deposited on paleoridges established earlier. Mass flow deposits such as submarine fans and turbidite mounds formed on slopes and fault blocks, respectively. Deltaic distributary channels and submarine canyons were incised on the shelves.

Middle Miocene reservoirs include coarse shelfal clastics, turbidite sands, and reefal carbonates. Minor oil and gas source rocks occur in a few of the island arc basins. Traps formed were mostly stratigraphic and consisted of reefal buildups, turbidite mounds/lobes, and shelfal sand bodies (onlaps, pinchouts, channel deposits). Structural traps include drapes and simple anticlines.

*End of Middle Miocene (Collision)*—Convergent and divergent strike-slip motion occurred along the Philippine fault zone. This motion induced structural traps (or the structural component of combination traps). Thrust or wrench fault structures and related compressional folds, as well as normal fault block structures, were formed.

*Late Miocene (Post-Collision)*—Maximum subsidence of the basins was achieved through postcollision downfaulting and/or wrenching (Figures 17, 18). Accordingly, trough and graben geometries became pronounced in the depocenters. Mixed clastic-carbonate sedimentation occurred in well-defined shelf, slope, and basinal zones. Mostly fine clastics were deposited in the slope-basinal areas. Shallow-marine clastics and carbonates marked by prolific reef growth dominated sedimentation in the shelf/platform areas and paleohighs/horsts.

Proven reservoirs of this age include reefal carbonates and shallow-marine sandstones. Stratigraphic trap types, among them reefal buildups, fluvial channel sands, onlaps and pinchouts, are abundant. Folds, especially those associated with wrenching, comprise the structural traps. The Late Miocene deep-marine mudstone-dominated sequence acts as the regional seal for many of the forearc and backarc basins.

## PAST EXPLORATION PHILOSOPHIES

About 300 of the total of 450 exploratory wells in the Philippines were drilled between 1890 and 1972. The majority of these earlier wells were based on limited geoscientific data, and as in the early days of petroleum exploration, a substantial percentage chased surface oil and gas seepages. Many of

Figure 17. Late Miocene paleogeography of island arc basins.

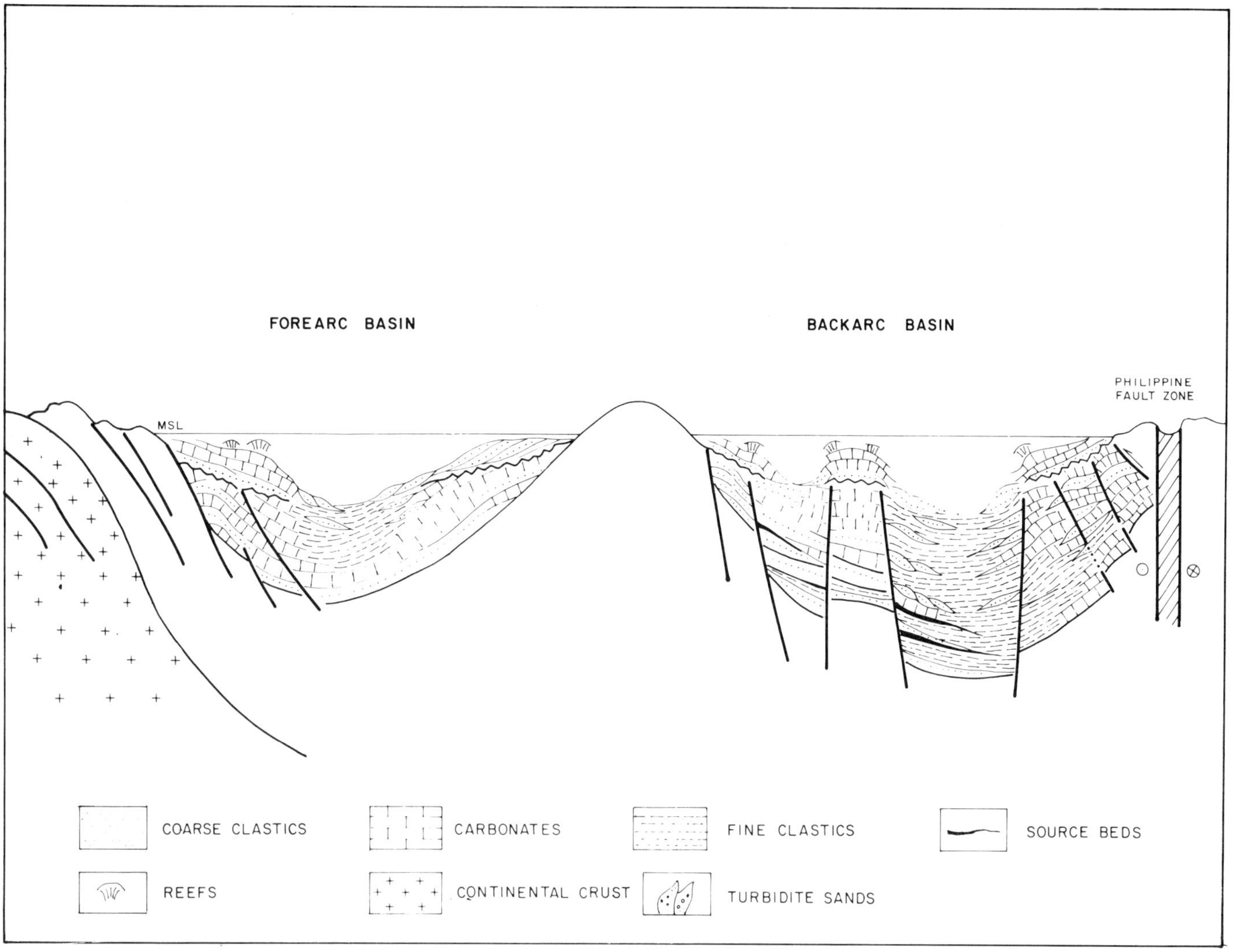


Figure 18. Late Miocene schematic profile of island arc basins.

the well locations utilized the anticlinal approach but were mainly based on surface geology occasionally supplemented with limited gravity data. In Southeast Luzon Basin, for example, 75% of the total of 70 wells were located solely on the basis of surface mapping.

The wells were also, in most cases, drilled to relatively shallow depths. In the Visayan Basin, 70% of the total of 216 wells were less than 700 m in total depth. A significant number of wells, mostly onshore, contributed little or no new information of regional or local geological value because of repeated drilling to shallow depths over the same structure. In the case of Bondoc Peninsula in Southeast Luzon Basin, 70% of the 47 onshore wells were drilled on only 2 anticlinal folds. In addition, in most areas, very little technical documentation on the older shallow wells is available today.

In summary, the strategy for this exploration phase up to 1972 basically involved: (1) definition of structural traps—usually Miocene anticlinal folds—mapped by surface geology and occasionally confirmed by geophysical surveys, and (2) exploration or step-out drilling with respect to oil and gas seeps.

The year 1973 saw the upsurge of offshore seismic and drilling activities and the widespread use of modern oil and gas exploration techniques. With the discovery of commercial oil offshore west of Palawan, the targets shifted from mainly structural to stratigraphic traps. This change is reflected in the statistics of exploratory wells for the period 1973 through 1984. Of 133 exploratory wells drilled during this exploration phase, 67% attempted to test reef stratigraphic traps, 3% turbidite stratigraphic traps, and the rest shelfal clastic reservoirs mostly in anticlinal traps.

It is clear that the traps which have been prognosticated and explored in previous wells represent only a fraction of the potential trapping mechanisms recognized or inferred from the evolution of the various basins. Furthermore, recent high-quality detailed data and analysis indicate that many of the wells were not valid tests of traps or prognosticated plays. Consequently, numerous play concepts remain to be tested by drilling.

Of the 13 basins defined in this study, 1 has no oil exploration well or even a shallow borehole, 1 has only one well, and 6 have not more than 12 wells. Numerous subbasins also have fewer than two wells. Thus a significant number of sedimentary depocenters must be considered as truly frontier areas.

## PLAY CONCEPTS

The main ingredients necessary for developing petroleum plays do occur in Philippine sedimentary basins. Play attributes such as oil and gas source rocks, clastic and carbonate reservoirs, regional and local seals, and a wide variety of types of traps can be demonstrated or reasonably expected in both the island arc-related and rift basins.

Numerous play types of different geologic risk and confidence levels are conceived for the basins. Play concepts which have been validly tested constitute a minor percentage of the total number of play types proposed.

### Forearc Basins

Gas and minor oil source rocks, carbonate reef reservoirs, volcaniclastic reservoirs which are quartz enriched in some places, regional and/or local seals, and stratigraphic traps (dominant) are expected in these basins.

The plays proposed include submarine fan complexes, carbonate reefs, delta fan complexes, shelfal sand pinchouts and onlaps, fluvial channel infills, growth structures, drapes over paleohighs, and wrench- or thrust-fault structures.

### Backarc Basins

Oil and minor gas source rocks, carbonate reef reservoirs, volcaniclastic reservoirs which are quartz enriched in some places, and regional and/or local seals occur. Structural traps resulting from wrench deformation are present in addition to stratigraphic traps, and these may form independent or combination traps.

The plays proposed are basically those enumerated for forearc basins plus compressional folds, either thrust- or strike-slip-related, reef talus in downthrown fault blocks, intrusion-generated folds, tilted fault blocks, carbonates on paleotopographic highs, and erosional truncation.

### Rift Basins

Oil and gas source rocks, quartzose sandstone and carbonate reservoirs, regional and/or local seals, and stratigraphic and structural traps are proven in these basins.

Among the plays proposed are reef and platform carbonates, turbidite pinchouts and drapes, compressional folds related to thrusting or wrenching, tilted fault blocks, simple anticlines/drapes, horst/ridge structures, onlap/pinchouts against paleohighs, and slope or fan wedges.

## CONCLUSIONS

1. The Philippines can be divided into two gross tectonic units—the island arc system and the microcontinental plate.
2. These two units have widely different tectonic histories, which are reflected in the stratigraphic and structural record of the associated sedimentary basins.
3. Philippine sedimentary basins can be broadly classified as forearc, backarc, or rift basins; the first two types are identified with the island arc system, and the last is identified with the microcontinental plate.
4. Source rocks, reservoirs, traps, and seals evolved to different degrees in all basin types.
5. Despite a century of exploration, it was only in the last 15 years that modern petroleum exploration commenced in these basins.
6. The basins, however, remain frontier or marginally explored areas, as only a few plays have been explored by drilling and/or validly tested to date.
7. Numerous untested play concepts in all three basin types enhance the possibilities for future hydrocarbon finds.

## ACKNOWLEDGMENTS

This paper was presented and published with the permission of the Ministry of Energy, Philippines.

The authors are indebted to Bureau of Energy Development geologists M.L. Alcala, J.J. Benavidez, and F.G. Rillera who sketched most of the geological illustrations. B.E. Guzman and G.E. Cabungcal supervised the preparation of the original slide artwork.

## REFERENCES CITED

Acharya, H.K., 1980, Seismic slip on the Philippine Fault and its tectonic implications: Geology, v. 8, p. 40-42.

Fuller, M., et al., 1983, Paleomagnetism of Luzon: American Geophysical Union Geophysical Monograph 29, p. 79-94.

Hamilton, W., 1979, Tectonics of the Indonesian region: U.S. Geological Survey Professional Paper 1078, 345 p.

Holloway, N.H., 1982, North Palawan block, Philippines—Its relation to Asian mainland and role in evolution of South China Sea: AAPG Bulletin, v. 66, p. 1355-1383.

Hsu, I., and L. Sharon, 1970, Paleomagnetism of Luzon igneous rocks: Paper presented at the U.S.-Japan Seminar on Paleomagnetism, Honolulu, Hawaii.

Jarrard, R.D., and S. Sasajima, 1980, Paleomagnetic synthesis for southeast Asia: Constraints on plate motion: American Geophysical Union Geophysical Monograph 23, p. 293-316.

Karig, D.E., 1983, Accreted terranes in the northern part of the Philippine Archipelago: Tectonics, v. 2, p. 211-236.

Lewis, S.D., and D.E. Hayes, 1983, The tectonics of northward propagating subduction along eastern Luzon, Philippine Islands: American Geophysical Union Geophysical Monograph 27, p. 57-78.

Lin, M.T., and Y.B. Tsai, 1981, Seismotectonics in Taiwan-Luzon area: Bulletin of the Institute of Earth Sciences, Academia Sinica, v. 1, p. 51-82.

Minster, J.B., and T.H. Jordan, 1979, Rotation vectors for the Philippine and Rivera plates: EOS American Geophysical Union Transactions, v. 60, p. 958.

Mitchell, A.H.G., et al., in press, Cenozoic evolution of the Philippine archipelago: Journal of Southeast Asian Earth Sciences, v. 1.

Rankin, B., et al., 1984, Kinematics of the Philippine Sea plate: Tectonics, v. 3.

Seno, T., 1977, The instantaneous rotation vector of the Philippine Sea plate relative to the Eurasian plate: Tectonophysics, v. 42, p. 209-226.

Taylor, B., and D.E. Hayes, 1983, Origin and history of the South China Sea basin: American Geophysical Union Geophysical Monograph 27, p. 23-56.

## NOT CITED

Philippine Bureau of Energy Development, 1986, Sedimentary basins of the Philippines—Their geology and hydrocarbon potential: Manila, 7 vols.

CHAPTER 21

# TWO DECADES OF HYDROCARBON EXPLORATION ACTIVITIES IN INDONESIA

**Atik Suardy**
**B. Simbolon**
**J. Taruno Ph.**
*PERTAMINA*
*Jakarta, Indonesia*

ABSTRACT

During the last two decades, the hydrocarbon exploration activities in Indonesia have been based on the Indonesian Energy Policy, which includes extensification and intensification of the hydrocarbon exploration programs. The extensification programs, which are mainly carried out in the offshore area, have led to the discovery of a new basin, prospects, and a new oil field; whereas the exploration intensification in the onshore area has also discovered new prospects, some of which were developed as oil fields. The first offshore hydrocarbon-producing fields in Indonesia came on stream in 1971, and since then the offshore area seems to be attractive.

New gas field discoveries such as Arun and Badak have been produced for LNG, and oil fields were found in the onshore area of Sumatra, Kalimantan, Java, Irian Jaya, and in the offshore of West Java, Madura, Natuna, and East Kalimantan. The exploration drilling success ratio within these two decades is approximately 32%, whereas the ratio of oil field development to discoveries is 54%. Because of technical and economic reasons, not all the discoveries can be developed into oil fields.

Recently, PERTAMINA Research and Development Division has organized a study of data that were contributed by PERTAMINA Exploration, Foreign Contractors, and the Science Institutes. Sixty basins have been discovered by this study, which covers the onshore and offshore areas including deep sea.

By using the PAUS ("Plan and Analysis of Uncertainty Situation") program of the Monte Carlo Simulation, the hydrocarbon resources within each basin have been estimated. Nevertheless, in the future, because of the development of the tectonic plate theory and computing techniques, and also because of the increase in exploration data, it is necessary to reassess Indonesian hydrocarbon reserves and resources.

## INTRODUCTION

Hydrocarbon exploration activities have been conducted in Indonesia since the 19th century. The Telaga Said oil field in North Sumatra was reported to be the first commercial oil field, which was brought into production in 1885; it was believed to be the beginning of the Indonesia oil industry.

During the period before 1966, exploration activities were not as bright as later because the government was still using inflexible old policies in hydrocarbon exploration. In that period there were three state oil companies—PERTAMIN, PERMIGAN, and PERMINA—which were later merged to form one large oil company, named PERTAMINA, the Indonesian State Oil Enterprise.

In the last two decades, the Indonesia government has instituted a new policy by establishing one state oil enterprise and inviting foreign investors to participate in the hydrocarbon exploration in this country. The terms of contracts and the attractive incentives have excited many investors during this period. Under the new policy, exploration became more active not only in the intensification programs onshore in areas of known fields but also offshore as extensification programs in virgin areas to define new prospects, sedimentary basins, and new oil/gas fields.

Offshore exploration activities were commenced in early 1966 and yielded the first production field of IIAPCO, named the "CINTA" field, which was discovered in 1971. Since then the offshore area has become attractive to many foreign oil companies which have offered to invest their capital for hydrocarbon in hunting in this country. These developments should mean a new bright era for the oil industry in Indonesia.

New oil and gas fields have been discovered in onshore Sumatra, Java, Kalimantan, and Irian Jaya; and in offshore West Java, Madura, Natuna, East Kalimantan, and Sulawesi.

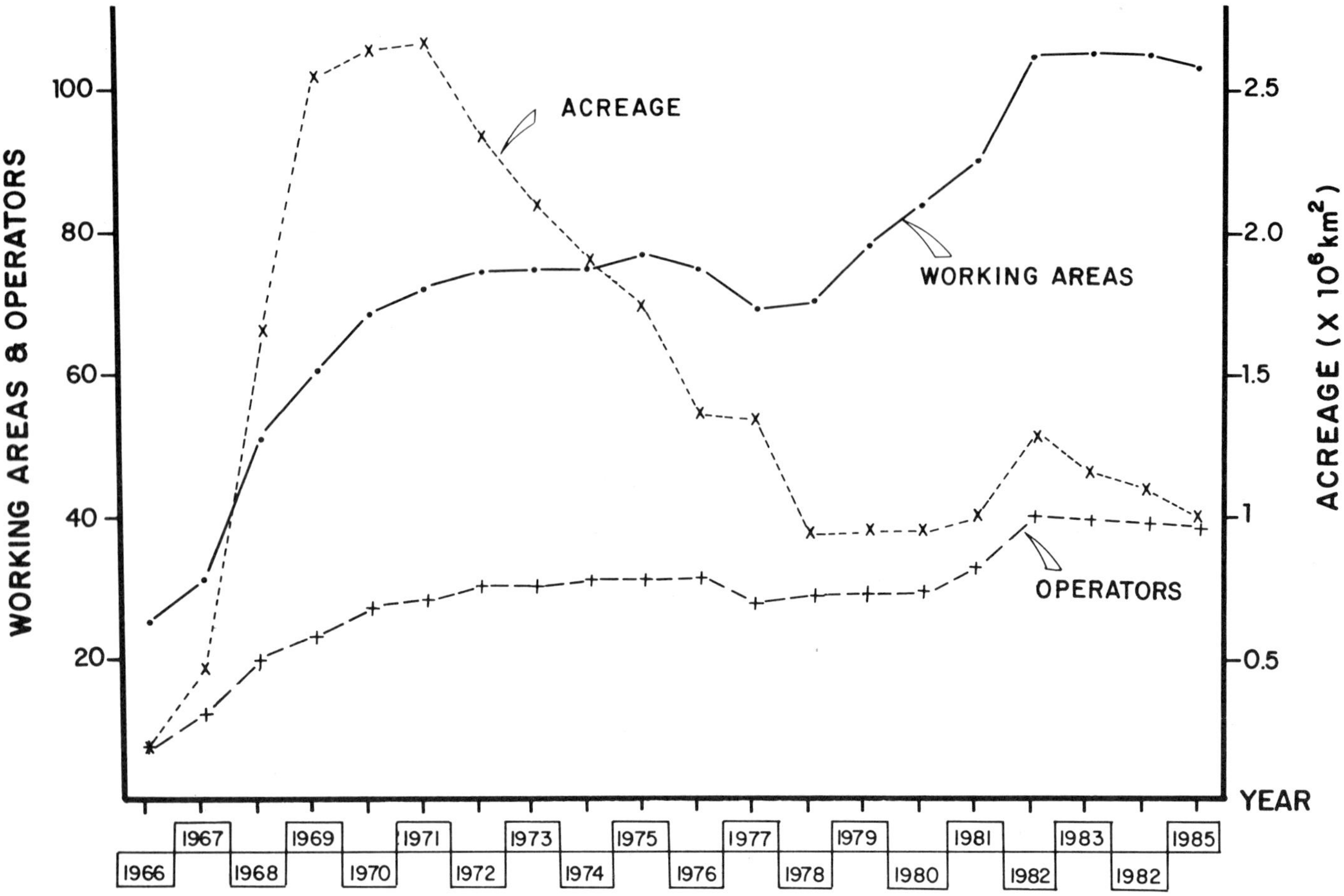


Figure 1. Statistics of acreage, working areas, and operators.

Oil production from the existing fields was sold as crude, or had been processed into other products as energy sources and industrial raw materials. Previously, gas production was used as gas lift or burned in the flare, but the discovery of giant gas fields such as Arun ('Mobil') and Badak ('Huffco') changed this situation. The gas has been processed into LPG and LNG for export since 1977.

## BACKGROUND

All the exploration stages that are usually used in oil industry, are also conducted in Indonesia. These exploration stages can be classified as follows:

(1) Data acquisition,
(2) Data processing,
(3) Data interpretation,
(4) Applications of the results of 1–3, and
(5) Data management.

Choosing of the exploration stage mainly depends on the exploration targets and the field conditions. These will determine the methods used in each stage.

Exploration methods that have been used during these last two decades in Indonesia are, among others:

(1) Remote sensing (SLAR, SAR, AERIAL PHOTO, LANDSAT, etc.)
(2) Geological methods: surface, mapping, geochemistry, paleontology, etc.
(3) Geophysical methods: gravity, magnetic, seismic, electric, etc.
(4) Drilling methods (shallow, deep, etc.)

The seismic reflection method is better known than the other methods because it is able to define the geometry of the subsurface continuously from bottom to top. Thus it can indicate the form, number of areal closures, and depth and location of structural or stratigraphic prospects.

Drilling is the only method that can be used directly to test and find oil/gas in the subsurface, and this will continue to be true in the future.

For these reasons, hydrocarbon activities during the last two decades were mostly oriented to seismic and drilling campaigns which generally yielding four aspects—sedimentary basins, prospects, resources, and reserves.

Lately, seismic and drilling methods have been the most important and have dominated the exploration activities because they define the structural prospects within the sedimentary basins used to estimate the hydrocarbon resources and reserves. It is certain that the assessment will change with evaluation of new information and data.

## WORKING AREAS, OPERATORS, AND TERMS OF CONTRACT

The working area and the exploration operators are two parameters that are closely related to each other in conducting hydrocarbon exploration efforts.

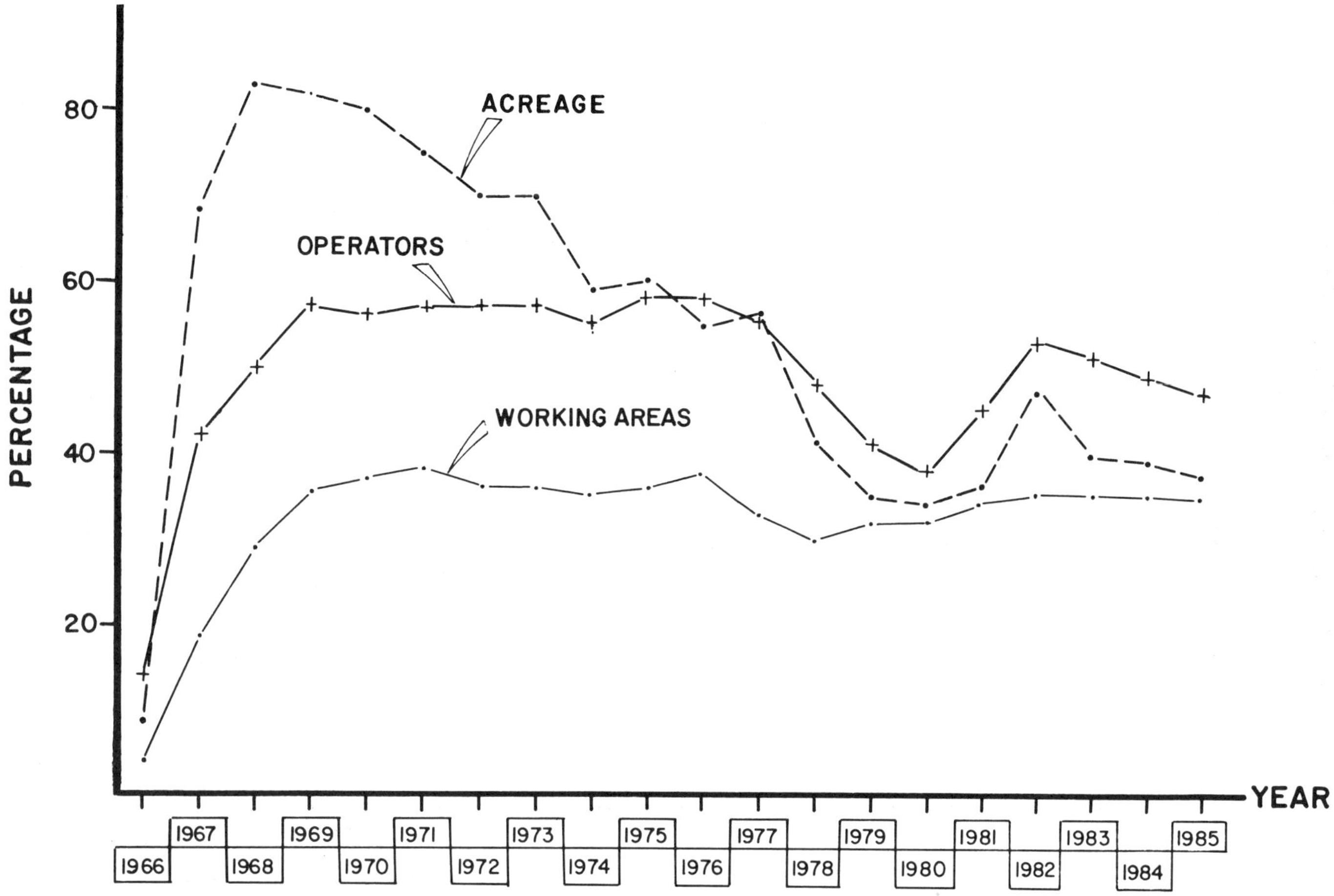


Figure 2. Percentage of offshore acreage, working areas, and operators.

The terms of contracts between the contractors and the Indonesian government cover the conditions of work and the shares of production each will receive if the working area is developed as a producing field.

According to the regulations, the whole area of Indonesia belongs to the Indonesian government, and contractors or investors can only get a permit for working in a certain area.

The Indonesia area is well known for its favorable geologic potential for oil/gas development because its strategic location can provide relatively easy access, and also because foreign oil companies had operated and produced oil/gas long before the period discussed in this chapter.

Long before 1966, there were two terms of contract—"contract of work" and "concession rights system"—which were applicable regardless of the role of oil/gas as a strategic commodity, either domestically or internationally. The "contract of work" came to an end in 1983.

In the post-1966 period, new government policies compelled the terms of contracts to be changed to what is known as a "Production Sharing Contract" (PSC), which is an agreement between the Indonesian government and foreign oil companies.

Before the Indonesian State Oil Enterprise PERTAMINA was born in 1968, there were two state oil companies in Indonesia named PERMINA and PERTAMIN. All the foreign oil companies had to make Production Sharing Contracts with these two state oil enterprises.

One of the conditions of the Production Sharing Contract requires the contractor to relinquish a part of the working area after 10 years, leaving the contractor with only a small part of the original contract area. Under this condition, large areas have been relinquished which were generally considered to be of low prospectivity.

On the other hand, because of the advancement of technology in exploration, especially seismic, the raising of oil prices, and the increasing demand for natural gas, these relinquished areas and new untested areas have excited the interest of many foreign oil companies.

Based on that condition, the Indonesian government in 1977 began to introduce the "Joint Operation Agreement System," giving a higher role to the Indonesian government as the owner of the area. The government invested 50% capital, whereas the contract conditions were the same as the standard term of the Production Sharing Contract.

A new contract form of the "Loan Agreement system" was introduced and developed in 1979 by PERTAMINA. By this system the investor does not have any certain specified working area, but all the exploration activities are conducted in the PERTAMINA working area. Warga Dalem, Pane, and Witoelar (1981) have discussed the terms of this contract in detail.

The relationship curves of acreage, working area and operators (Figures 1, 2, 3) show a "peak" from 1969-1973 and a "trough" from 1977-1980, though the number of working areas increased continuously, since the old contractors who had already much known about the hydrocarbon possibilities within the Indonesian basins were always interested in getting new working area.

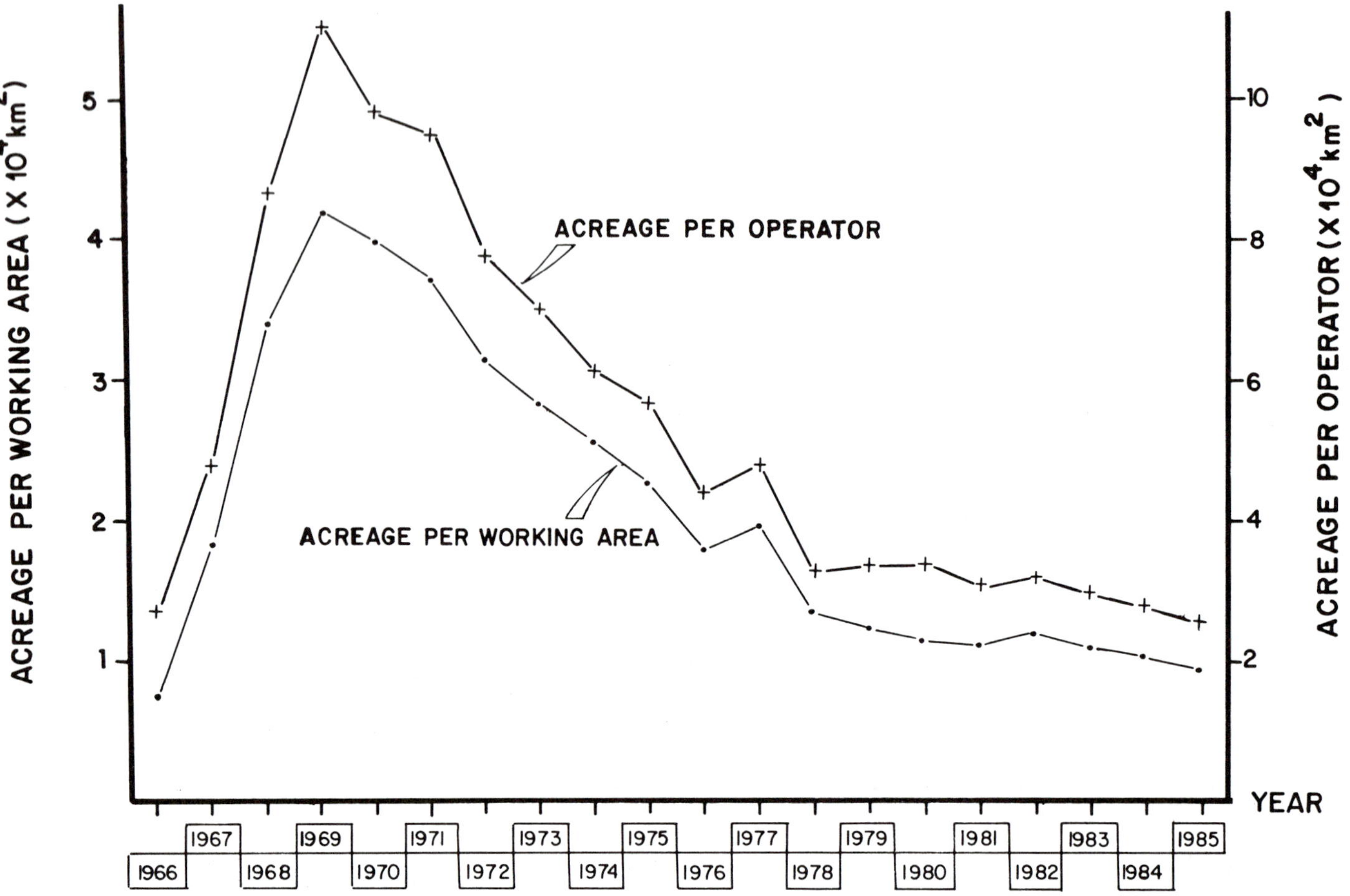


Figure 3. Acreage per operator versus acreage per working area.

## SEISMIC ACTIVITIES AND PROSPECTS

The seismic survey is still used in Indonesia as a well-known indirect method for hunting oil. The choice of the reflection seismic type rather than the refraction method is based on its ability for indicating subsurface geologic features.

Figure 4 shows the peak of seismic surveying in 1969, which decreased during the period 1977-1978. This was probably due to the impact of the 1974 world economic recession that was felt until 1978. During the period 1968-1974, the extensification program dominated the exploration activities. This was followed by a period of the intensification program of detailed and semidetailed surveys and studies which was conducted as a continuation of the previous program.

The second peak of seismic activities in 1980-1983 also related to the increasing of new working areas that were operated by both the new and old contractors (Figure 5). The offshore seismic surveys (Figure 6) have been more numerous than the onshore and also have yielded better quality data. This has been possible because of ease of data acquisition and data processing.

Offshore data acquisition is relatively faster and easier than onshore, because the offshore does not need an opening line or hole drilling, and also because of very easy transportation. Onshore data processing is hampered by "static correction" and also turnaround time on the processing sequence.

The number of refraction seismic surveys conducted during the last two decades is relatively very small (less than 1%) compared with the reflection method.

Many kinds of equipment and parameters are used in data acquisition, processing, and interpretation. The choosing of techniques and parameters depends on the exploration targets, the field conditions, and the year of execution. It seems that the exploration techniques from time to time always improve, becoming more accurate and more advanced.

The broad outlines of the exploration stages can be described as follows:

1. Data acquisition
   a. Recording equipment: analog to digitals of more than 100 channels
   b. Sampling rate: 8 ms to 1 ms and also 0.5 ms
   c. Profile space/distance: >20 km to < 1 km
   d. Trace space: > 100 m to < 10 m
   e. Recording system: manual to telemetry
   f. Coverage: singlefold to 3D
   g. Transportation: by man to helicopter
   h. Magnetic tapes: analog to digital 6250 BPI
   i. Energy sources:—onshore: explosives dynamite to minisosie—offshore/marine: explosives/dynamite to airgun, gas gun, flexcichoc, etc
   j. Drilling equipment, hand auger to rotary drill
   k. Other.
2. Data processing
   a. Computer: simple to sophisticated hard/softwares
   b. Processing: standard to acoustic impedance seismic stratigraphy modeling, slicing, etc

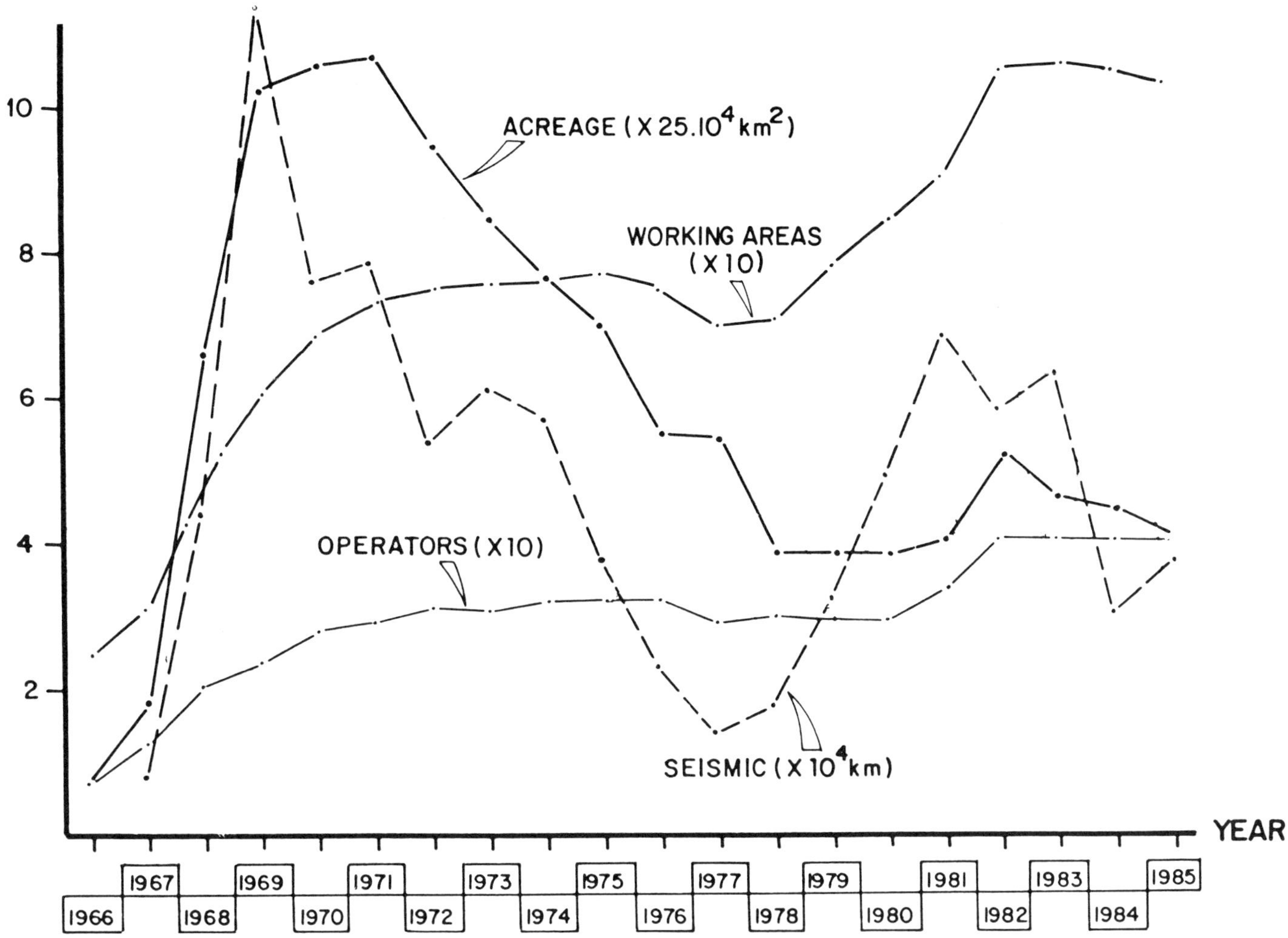


Figure 4. Statistics of acreage, operators, seismic coverage, and working areas.

3. Data interpretation
   Manual to automatic picking and computerized contouring
4. Data utilization
   Data can be used directly for locating the wildcat drilling (the exploration wellbore) and/or to anticipate hazard problems before and during the drilling
5. Data management
   This stage covers data storage either by manual or computerized systems. Data management is being improved by being able to handle a greater volume of data. The sophisticated retrieval system with terminal provides quick services.

The direct relationship between the progress of seismic surveys and the number of prospects discovered is very difficult to anticipate annually because of the data acquisition progress (Figure 7). This can be easily monitored each year but to define the number of prospects requires more time, since it is controlled by many kinds of aspects such as the data processing stage, interpretation, and detailed study of all geologic aspects.

For this reason, Figure 8 shows only the seismic survey progress, the annual number of prospects discovered, and the average kilometer profile for each prospect. The reconnaissance survey for every prospect requires about a 3000-km profile, whereas semidetailed and detailed surveys for prospects ready to drill require 300-500 km and 50-100 km, respectively, for each prospect.

Enhanced knowledge and experience is associated with the increase in ac tivity. This in turn is used to overcome problems and difficulties associated with exploration and drilling.

## EXPLORATION DRILLING AND DISCOVERIES

Drilling is the only direct method used to test the existence of hydrocarbon within any structural prospect.

The term exploration drilling is still debatable, for there are several interpretations, such as:

(1) Exploration drilling covers only "new field wildcat."
(2) Exploration drilling belongs to "new oil pool."
(3) Exploration drilling covers also the delineation or confirmation well.

In this chapter, the so-called exploration drilling is defined as "new field wildcat" or "prospect drilling"; that is, the first wildcat well drilled for testing a prospect. This definition is important for avoiding confusion in determining the success ratio.

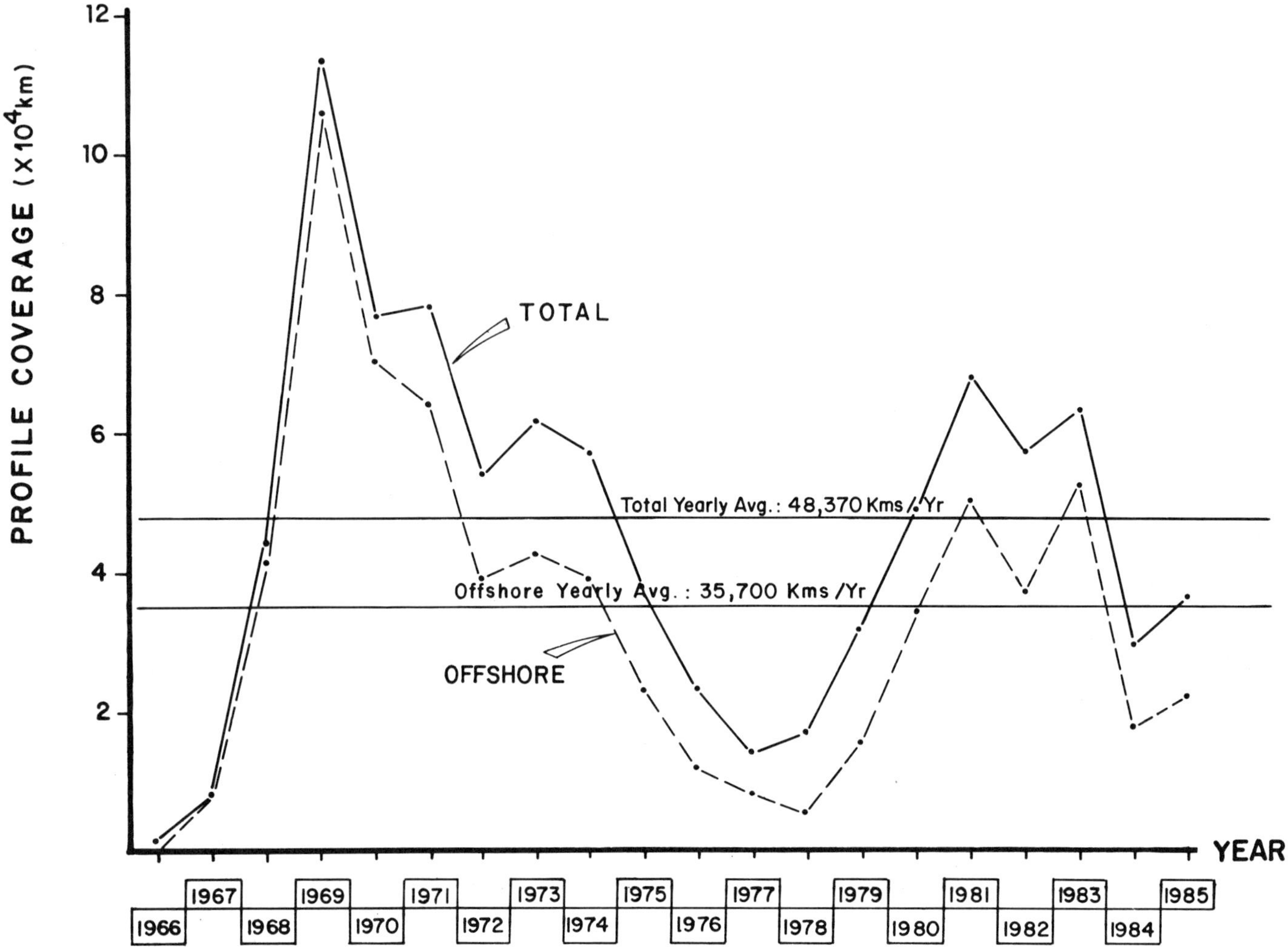


Figure 5. Statistics of seismic survey profile coverage.

The kind of equipment used in this method depends on the aim of the drilling, the depth target, and field conditions, and also on economic, efficiency, and effectivity considerations.

In general, it can be summarized as follows:

- on land (onshore): conventional drilling to helirig,
- marine/offshore: drillship, jack-up, barge,
- the system: straight and directional drilling.

Not all structural prospects known from seismic surveys can be drilled directly. Every structure should be studied carefully and in detail for finding the best choice.

Geologic surveys or basin studies must be conducted around a prospective area to determine reservoir characteristics, source rock maturity, age, environment of deposition, tectonic and sedimentation processes, and lateral-vertical migration possibilities, by using various kinds of methods such as stratigraphy, petrology/petrography, paleontology, petrophysics, geochemical analysis, and field mapping data.

Geochemistry is a relatively new method used in Indonesia that was known in early 1971 and is still a well-known method for source rock maturity detection and migration possibilities.

Priority for drilling a structural prospect must be based on a careful study of geologic factors. The anticipated resources—as defined by aspects of areal closure, prospective thickness, depth, and accessibility—must be great enough to justify drilling.

From the hydrocarbon potential and the economic risk point of view, the first priority is given to any structural prospect that has "high potential" with the lowest economic risk. This formula is also used for locating exploratory drilling in a prospective area. Drilling statistics (Figure 9) show a 33% discovery rate. These discovered prospects consist of 44% in offshore areas (Figure 10) and 56% in onshore areas.

The exploration peak reached during 1971-1975 was followed by a decrease during the period 1976-1977. This was possibly due to the world economic recession in 1974, the impact of which extended until 1976. The rise in world oil prices in 1980 resulted in an increase in drilling, which reached its second peak during the period 1982-1983.

Wildcat drilling carried out on a selected prospect is not always successful, even though the best factors have been considered. Quite often a prospect is drilled by more than 3 wells for testing its hydrocarbon content.

Sometimes unpredicted "technical trouble" (such as a blow out, etc.) increases the difficulty in drilling.

The statistic shows that the drilling compared to the discoveries is in proportion.

As has been mentioned, during the last two decades, of the average of 89.2 prospects, only 28.4 became oil/gas discov-

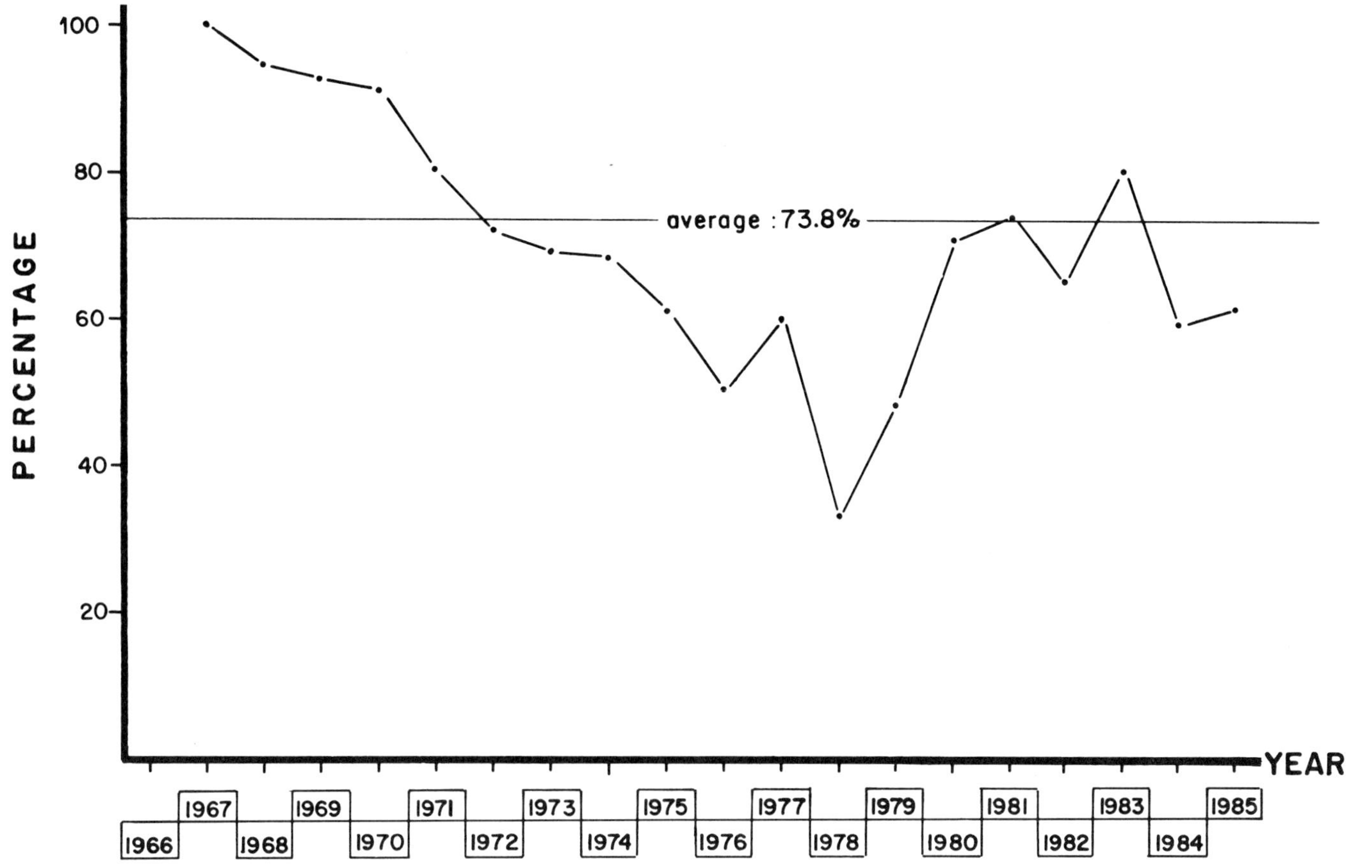


Figure 6. Percentage of offshore seismic surveys.

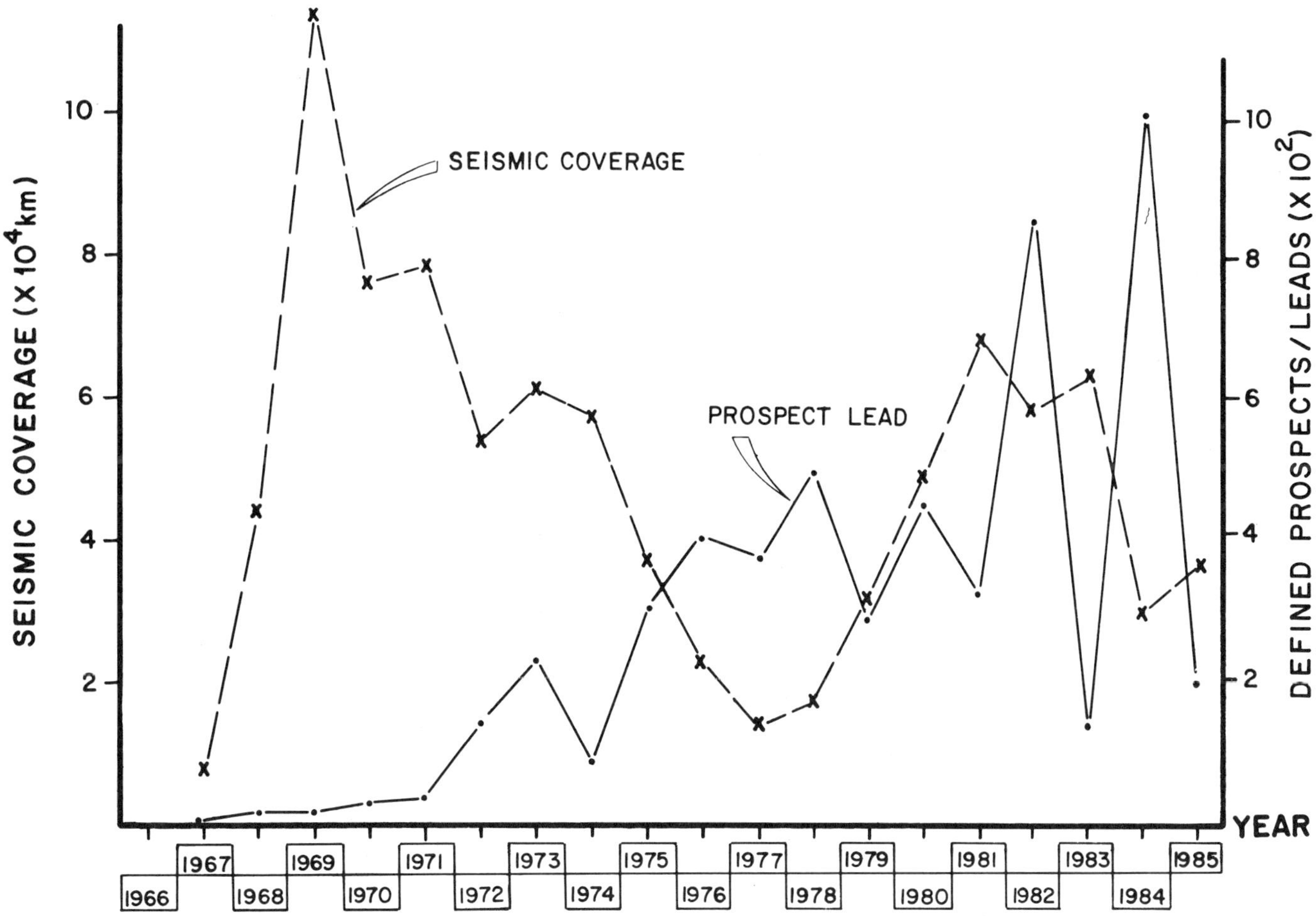


Figure 7. Statistics of seismic coverage and discovered prospects.

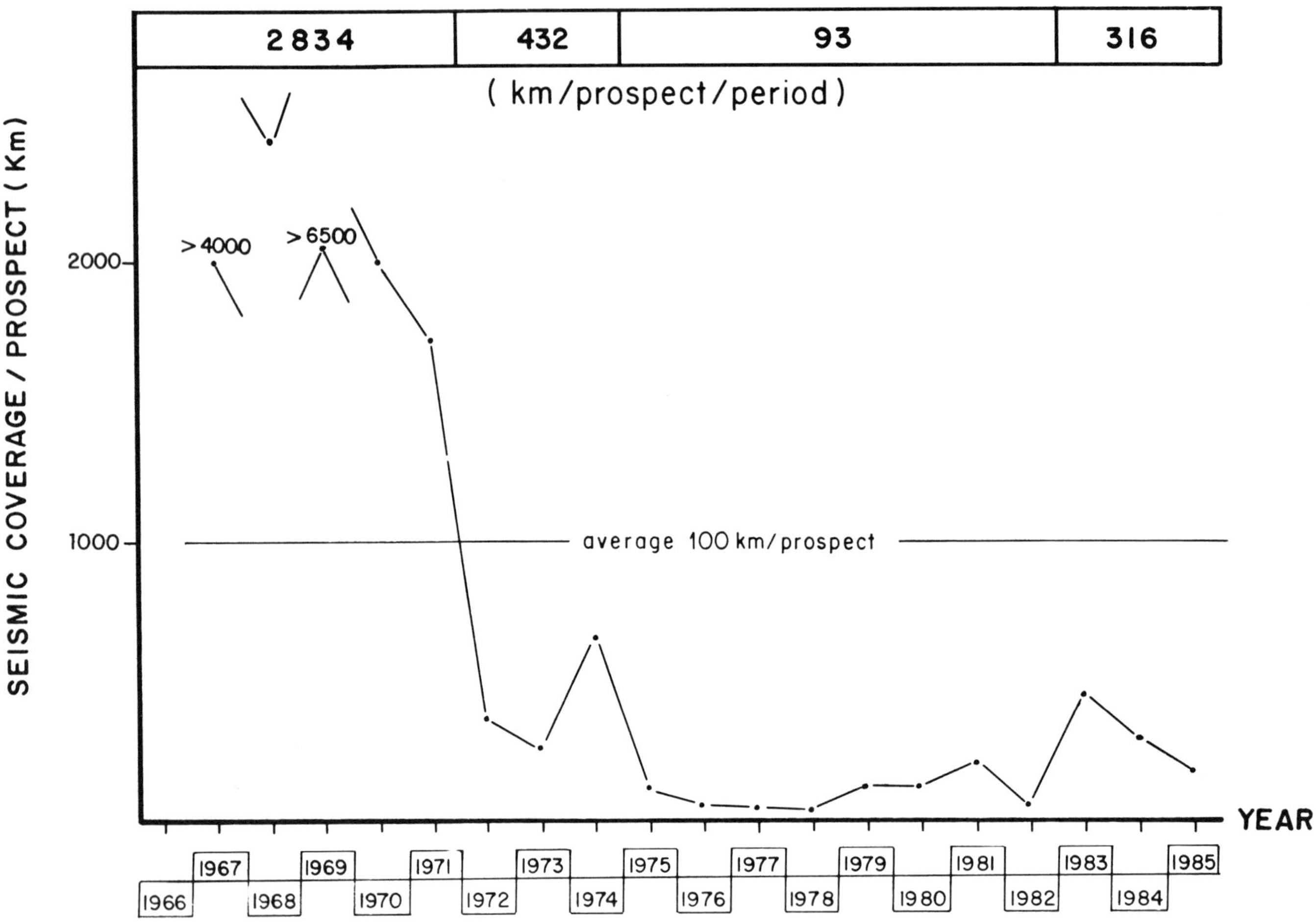


Figure 8. Seismic coverage by prospect/lead.

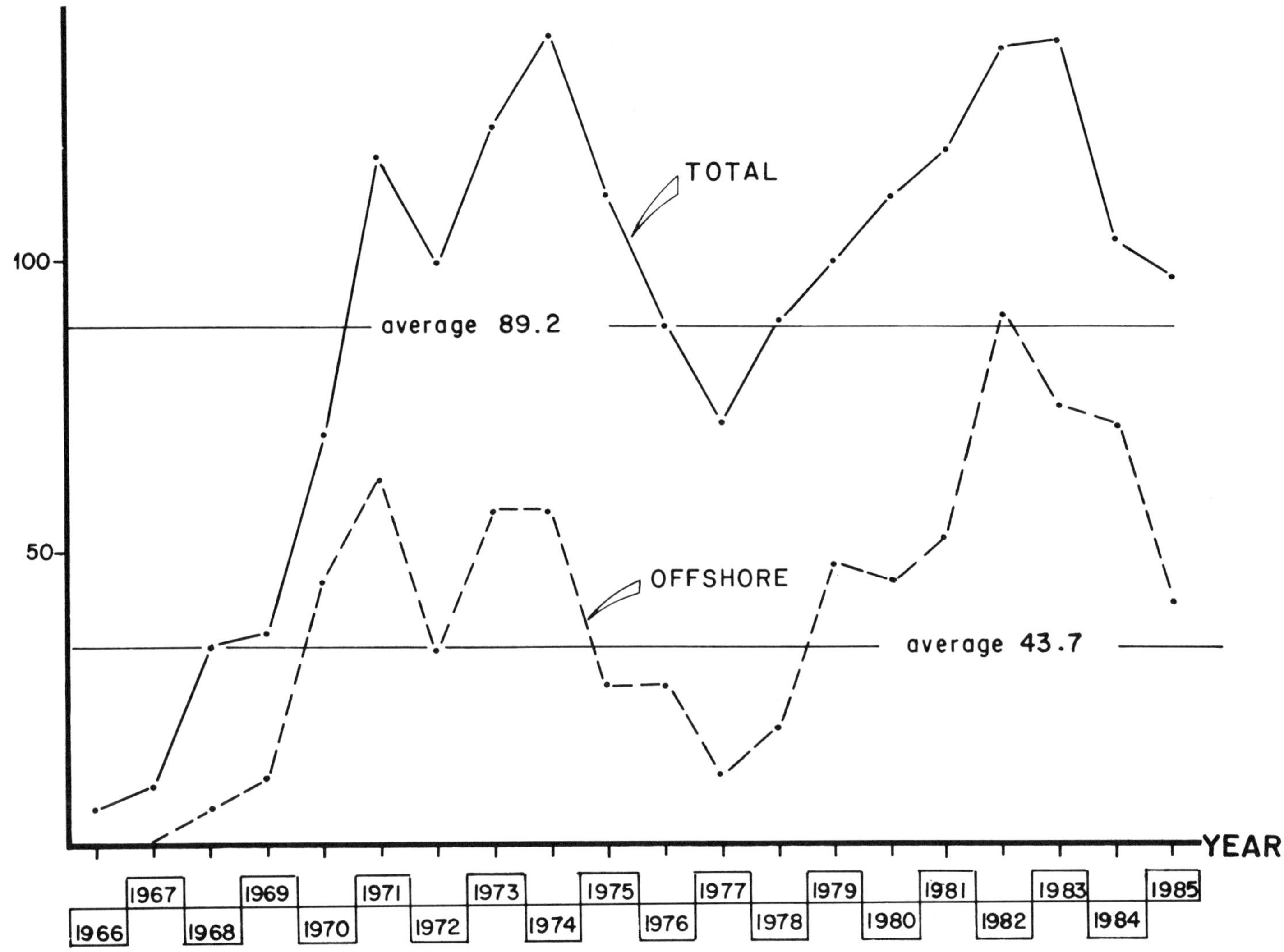


Figure 9. Statistics of total drilled prospects versus offshore drilled prospects.

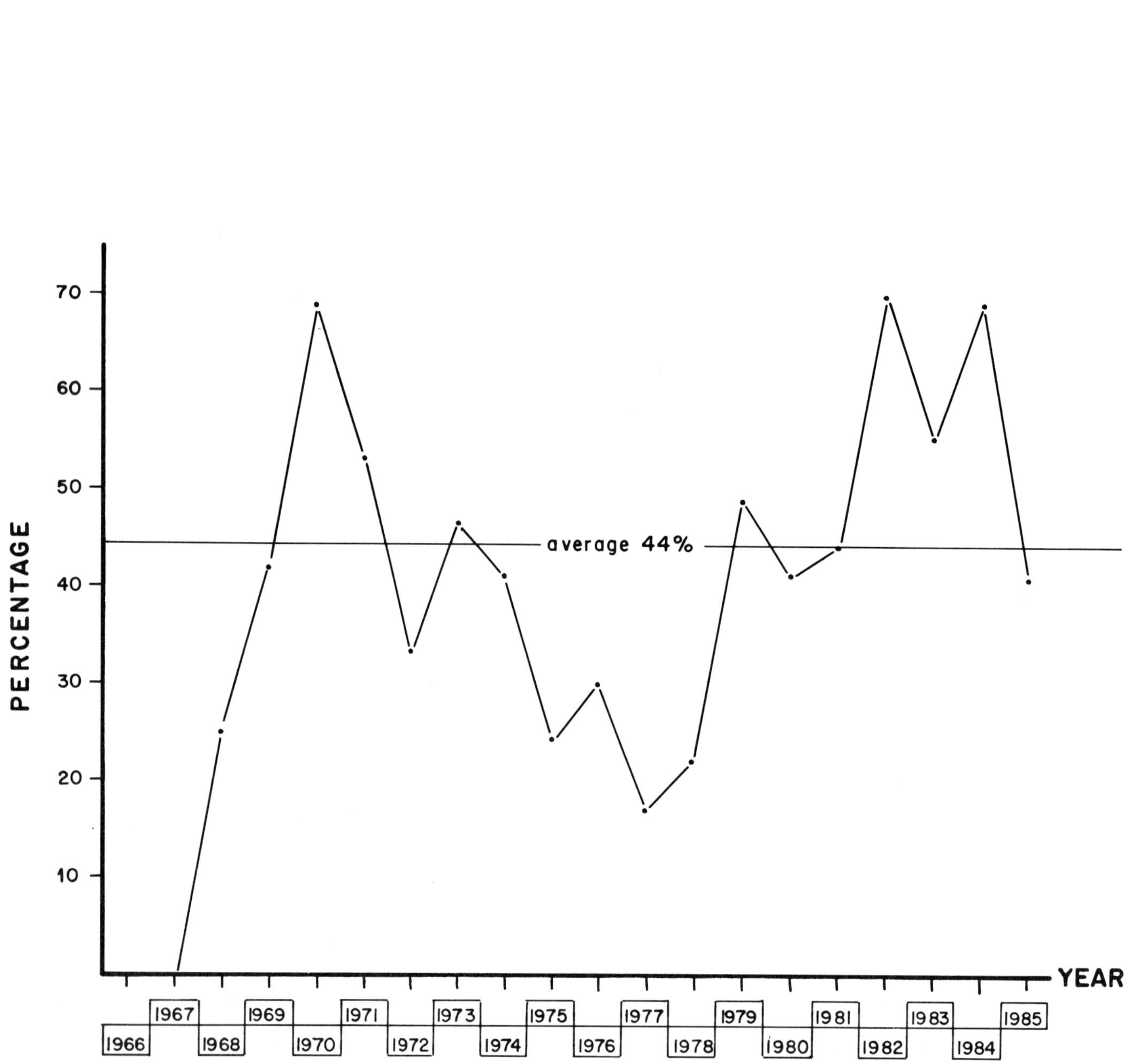


Figure 10. Percentage of drilled prospects offshore.

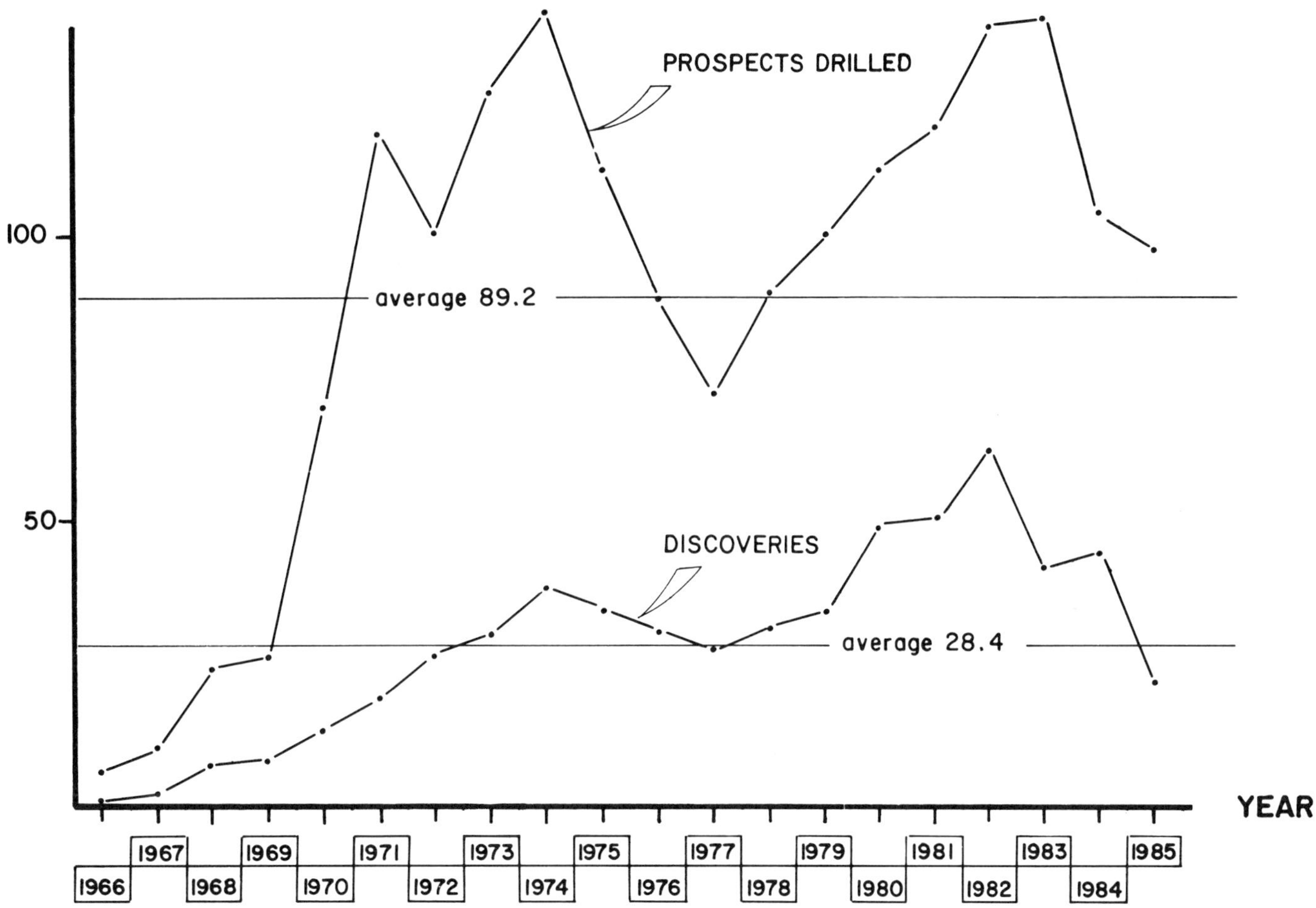


Figure 11. Statistics of drilled prospects and discoveries.

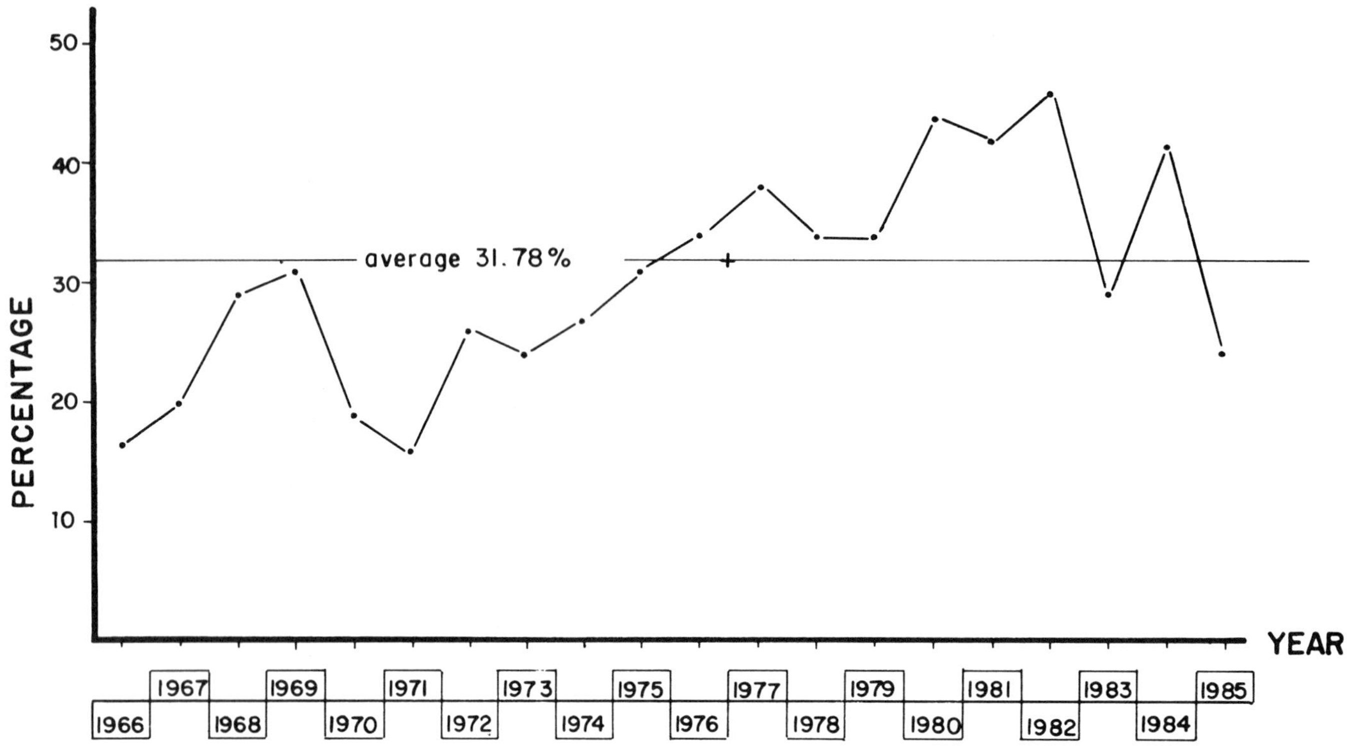


Figure 12. Success ratio by prospects (percentage).

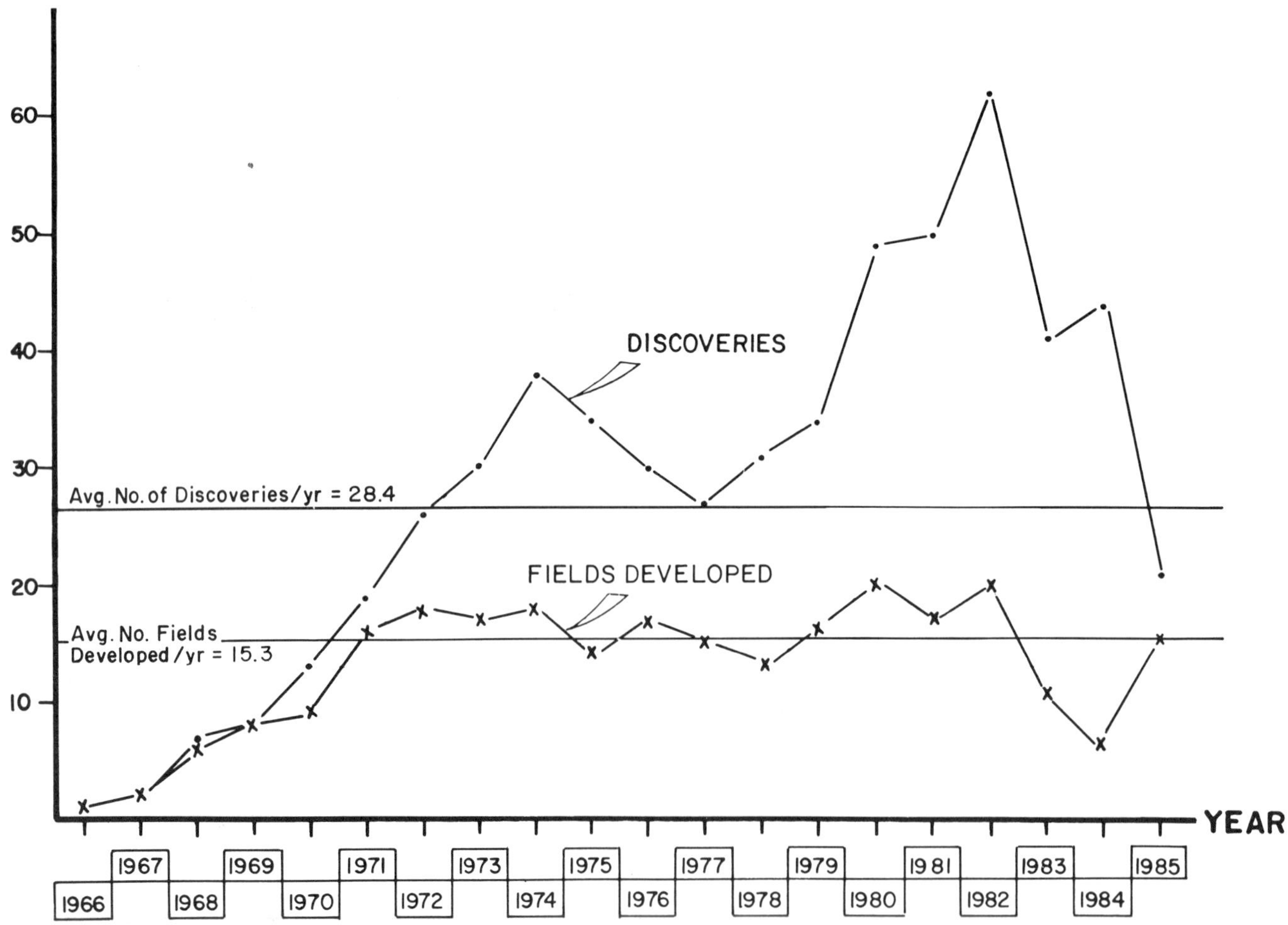


Figure 13. Statistics of discoveries and development fields.

Figure 14. Success ratio by field (percentage).

eries (or 31.78%) (Figures 11 and 12). Figure 12 shows that since 1975, the discoveries percentage tended to increase and reach the peak in 1980-1982.

The first onshore discovery during the last two decades was the Warukin Selatan prospect in Kalimantan Island, and the first offshore discovery was the Arjuna field in West Java. These discoveries were developed as oil/gas fields by drilling of delineation and development wells.

The discoveries, both in onshore and offshore areas, have proven that Indonesia has a great hydrocarbon potential. The success of these discoveries has stimulated investors to contract for new working areas, especially in offshore areas.

## DEVELOPMENT AND ADDITIONAL RESERVES

The development stage in the oil business is a continuation of the exploration stage. As has been stated, not all of the drilled prospects result in discovery of hydrocarbons. Likewise not all of the discoveries can be developed into oil/gas producing fields. The success ratio is influenced by economic factors, technological improvements, and government policy in regard to contract terms.

The time factor plays a very important role for foreign companies, because developing any discovered prospect into a producing field requires detailed feasibility studies of all aspects including field conditions and accessibility, world oil prices, suitable technology, etc. These considerations should provid maximum benefits to both the Indonesian government and the investors.

The relatively short time to be considered should be no longer than the time stated in the contracts. An ideal example is the case of the discoveries in the AL-1X structure in the Natuna Sea. The discoveries have been known for a long time and have been under lease by several companies, of which the first holder was AGIP and the last one is ESSO. This field is not yet developed, although there are many facilities with adequate technology, and the reserves are sufficient for profitable exploitation. However, the high $CO_2$ content of 72% tends to discourage efforts to develop this field.

The same case of a high $CO_2$ content was found in the Cilamaya Selatan prospect, onshore N.W. Java, in the PERTAMINA working area. Unlike the AL-1X, the Cilamaya can be developed because the $CO_2$ content can be directly sold to the Coca-Cola industry.

Statistically speaking, the discoveries, the development, and the reserves can be elucidated as follows. The curves plotted for discoveries and developments are nearly parallel.

The increase in field development during 1966-1970 was due to an increase in offshore discoveries with proven oil content.

During 1971-1982, field development was rather flat, and it decreased in 1983. Only 12.9, or 44%, of the 28.4 structural prospects were developed as oil/gas fields (Figure 13).

The curve of the success-ratio trend of developed fields versus discoveries tends to decrease with time (Figure 14), because of increasing difficulty in developing the discoveries into oil/gas fields due to critical factors—economic, technical, accessibility, etc. Discoveries are classified as uneconomic grade if the amount of resources or OOIP is relatively small, if

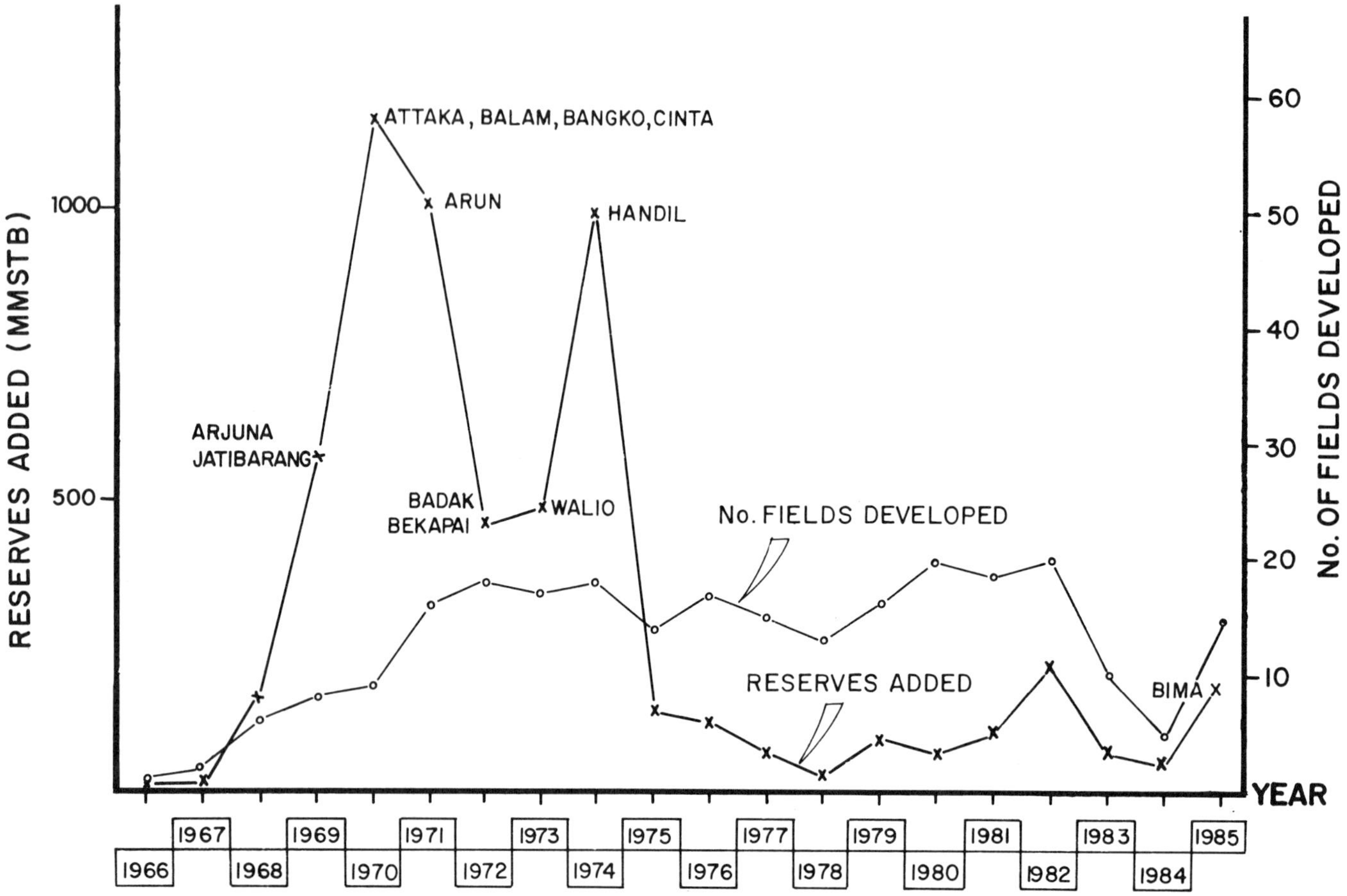


Figure 15. Statistics of development fields and additional reserves.

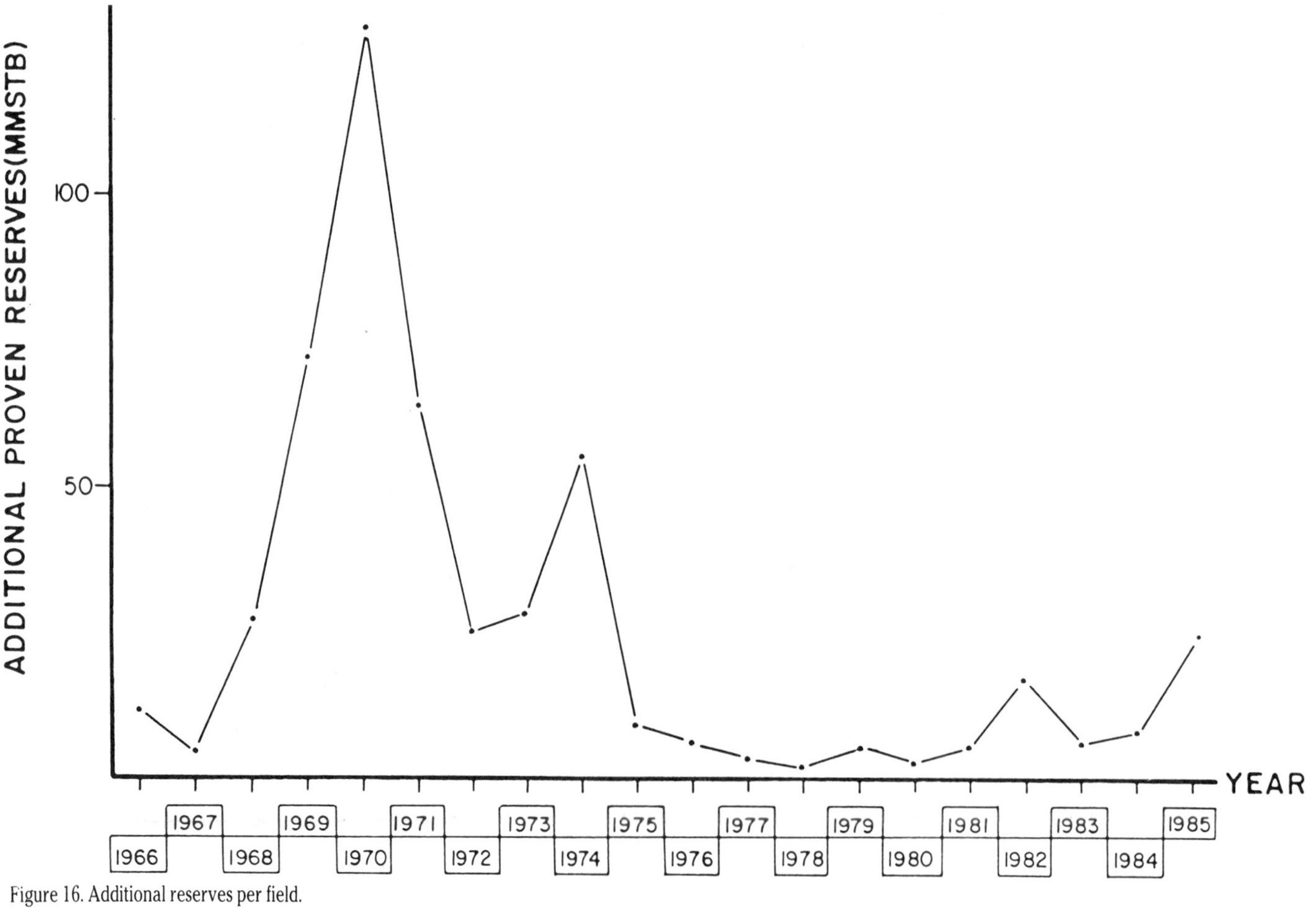


Figure 16. Additional reserves per field.

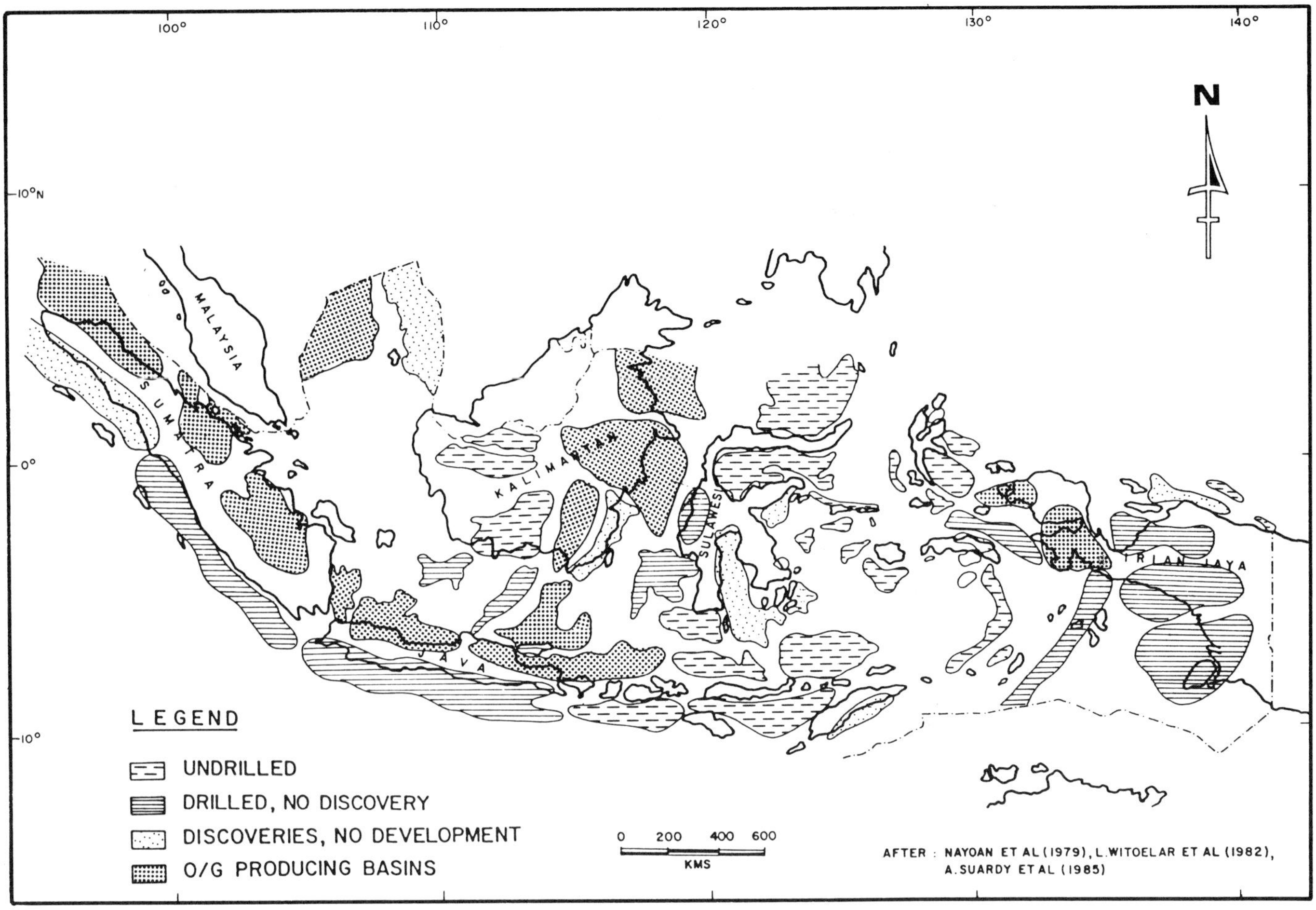


Figure 17. Sedimentary basin location map of Indonesia.

the location is very remote, and if there is a decline in world oil prices.

These conclusions are illustrated in Figure 15, which shows statistical curves of the number of development fields and the additional proven reserves as of January 1, 1986.

Oil/gas fields with more than 100 MMBO proven reserves were only discovered during 1969-1974, whereas before and after this period, oil/gas fields having more than 50 MMBO proven reserves were not discovered. The statistical curves of additional reserves per field (Figure 16) shows that oil fields with additional reserves of more than 25 MMBO were only found during the period 1968-1974, although the number of discovered prospects developed is nearly the same as in the following period of 1974-1985.

## SEDIMENTARY BASINS AND RESOURCES

As has been mentioned, sedimentary basins and resources are the result of exploration activities. The sedimentary basin is the main target of the continuation of exploration activities.

The end goal of finding commercial hydrocarbon deposits will not be achieved if there are no sedimentary basins that contain hydrocarbon resources. An increase in the amount of data can help define the sedimentary basin more accurately, and therefore the amount of hydrocarbon resources can be estimated more accurately. The assessment of resources becomes more accurate with the development and improvement of the computer technology or software, sophisticated hardware, and geologic data obtained from exploration activities, although the basic parameters have not changed much.

The sedimentary basins is usually defined by using a thickness of 1000 m of sedimentary rock or 2000 milliseconds two-way time on a seismic map as a boundary, combined with similarity of geologic characteristics.

Indonesian sedimentary basins and their resources have been discussed by several authors, and some changes have been noted. Prior to 1979 it was reported that there were 27 sedimentary basins in Indonesia, whereas Nayoan et al. (1980) noted 40 basins with total hydrocarbon resources of 89 BBOE. Witoelar, Achmad, and Reymondd (1982) discussed Indonesian deep-sea basins without noting their resources.

In the centennial commemoration of the Indonesian oil industry (1985), Atik Suardy et al., using the newest data from the latest exploration activities provided more complete information on Indonesian sedimentary basins.

They classified the Tertiary sedimentary basins of Indonesia into four groups, based on stages of exploration activities, as follows:

(1) Undrilled Tertiary sedimentary basins,

(2) Drilled Tertiary sedimentary basins without any discoveries,

(3) Tertiary sedimentary basins with undeveloped oil/gas discoveries, and

(4) Oil/gas-producing Tertiary sedimentary basins.

Atik Suardy et al. (1985) reported 60 Tertiary sedimentary basins in Indonesia offshore and onshore (Figure 17). Several

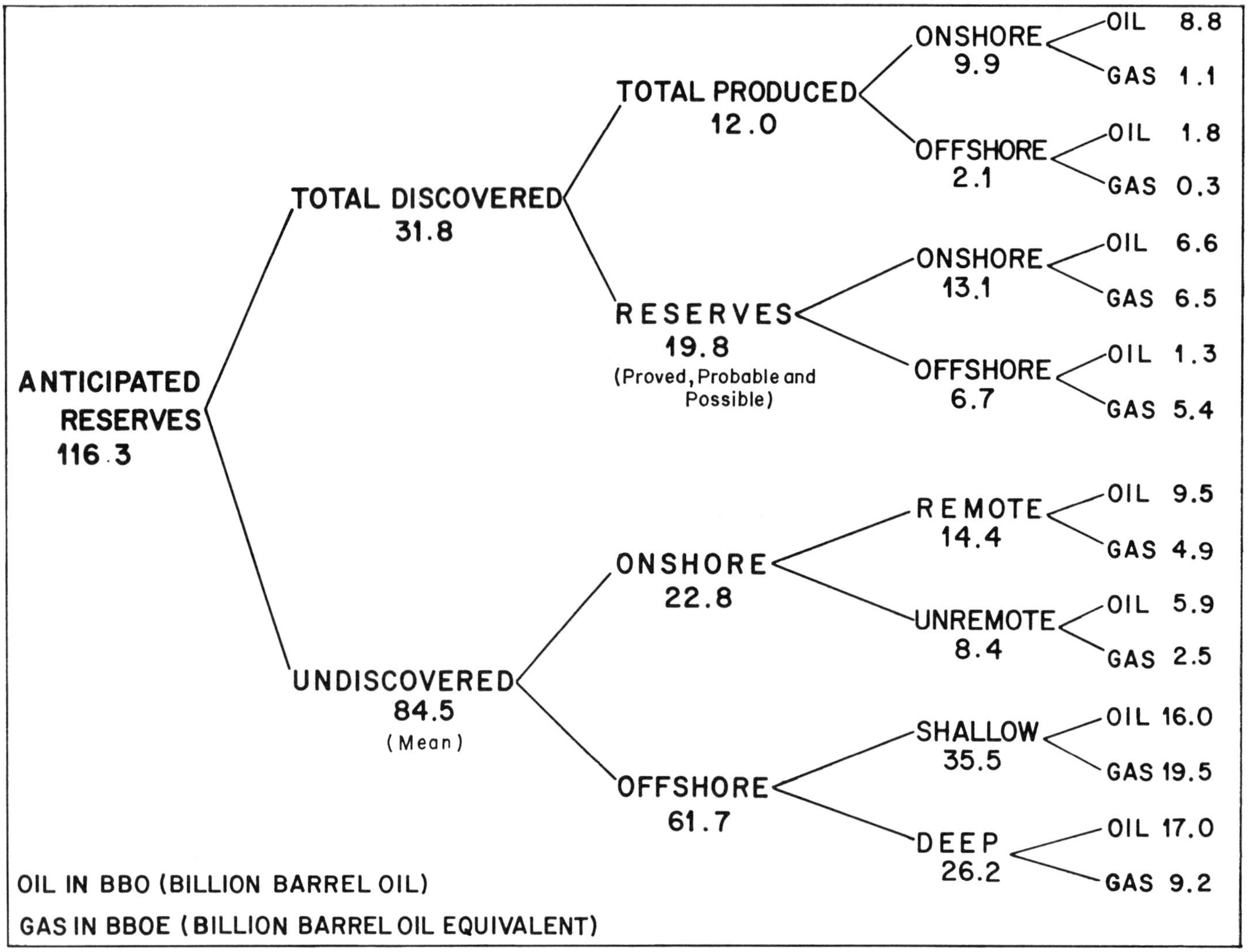


Figure 18. Anticipated hydrocarbon reserves of Indonesia (status 1985).

of the basins include both offshore and onshore areas in the same basin.

Data for offshore sedimentary basins were collected during 1966-1985, because during the previous period exploration was only carried out onshore.

Five of the potentially producing basins—Sunda, E. Natuna, N.E. Java Sea, W. Natuna, Bone basins—were discovered during 1966-1985, and the rest had already been defined before this period.

Besides finding a different number of sedimentary basins compared with previous studies, Atik Suardy et al. (1985) used the sophisticated computation techniques (hardware + software) in assessing the total hydrocarbon resources from each basin. The resources estimates were made by using IBM-3420 hardware and the Monte Carlo simulation technique with the PAUS (Plan and Analysis of Uncertainty Situation) program of PERTAMINA.

Though the tectonic plate theory has been introduced, it is not used yet in determining the estimated parameters of producing and potential basins. Conventional parameters are still used, because the available data are relatively small.

Estimates of the total hydrocarbon resources of every basin in general can be classified into three categories—minimum, maximum, and mean—because the basic parameters put into "triangle form" are "minimum, maximum, and most likely."

The result of this assessment and detailed descriptions of each area are shown in the Figure 18. These results conclude that Indonesian hydrocarbon resources are still enormous and hence still exciting for exploration, mainly in eastern Indonesia, an area that could be a challenge to investors and to oil industry experts, especially explorationists.

Assessments of hydrocarbon resources could change in the near future with further development of the application of the plate tectonic theory, the improvement of calculation techniques, and the increase in available data resulting from advanced exploration activities with sophisticated technology.

## SUMMARY AND CONCLUSION

Hydrocarbon exploration activities within the last two decades have been conducted in accordance with Indonesian government investment policies.

Exploration progress is closely related to and depends on the world situation, which is reflected in the increasing role of

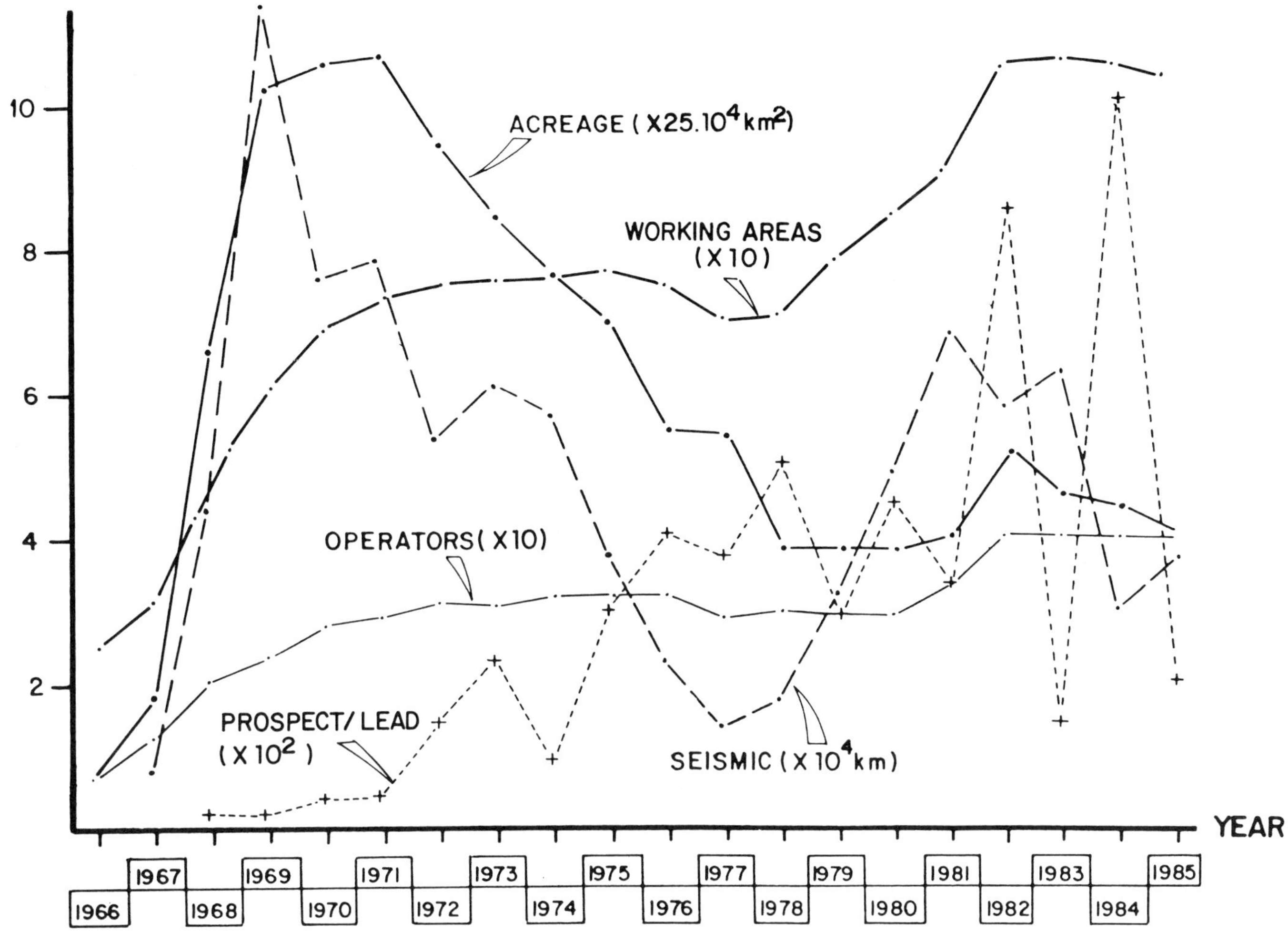


Figure 19. Statistics of acreage, working areas, operators, seismic coverage, and prospects.

oil/gas as a strategic, as well as a commercial, commodity. Exploration activity achieved its peak in the pre-1973 period. The investment policy of the Indonesian government was influenced by the Middle East crisis, which caused foreign investors to move their interests to Indonesia, where they could find a stable and hospitable environment. The trough of the activities curve was in 1974-1980, possibly caused mainly by the world energy crisis in 1973 (Figure 19).

Exploration activities tended to rise after 1980, in accordance with increasing oil prices, which encouraged foreign investors to conduct exploration activities as they had done prior to 1973. Though the observations and the statistics show that it was more and more difficult to find "giant fields" because of the remoteness of the area and the more complicated geologic conditions, government policies were more attractive to investors and technologic improvements caused an increase in exploration activities (Figure 20).

Exploration activities and the advancement of technology have had a direct impact on the results obtained and also have provided new information on the number of Indonesian sedimentary basins and their resources. To date, 60 Indonesian sedimentary basins are known, and with discovered resources have about 84 BBOE. This number, of course, would change in accordance with the availability of data and the improvement of computation techniques.

Estimated resources can only be used if exploration activities are conducted continuously by using advanced technology, which requires more time, manpower, and high costs to make the resources producible and reserves usable.

## ACKNOWLEDGMENT

The writers wish to thank the PERTAMINA management for allowing them to use their exploration data in writing this paper and for the opportunity to present it at The Fourth Circum-Pacific Energy and Mineral Resources Conference.

## SELECTED REFERENCES

Nayoan, G.A.S., A. Suardy, and (?) Suyanto, 1979, Nota tentage cadangan minyak dan gas bumi Indonesia: Suatu pemikiran ahli geologi: Majalah LAGI, jilid 7 No. 1, 1980.

Suardy, Atik, (?) Suyanto, and F.X. Hariadi, 1980, Note on assessment of undiscovered recoverable hydrocarbon resources in Indonesia, CCOP seminar, Kuala Lumpur, Malaysia.

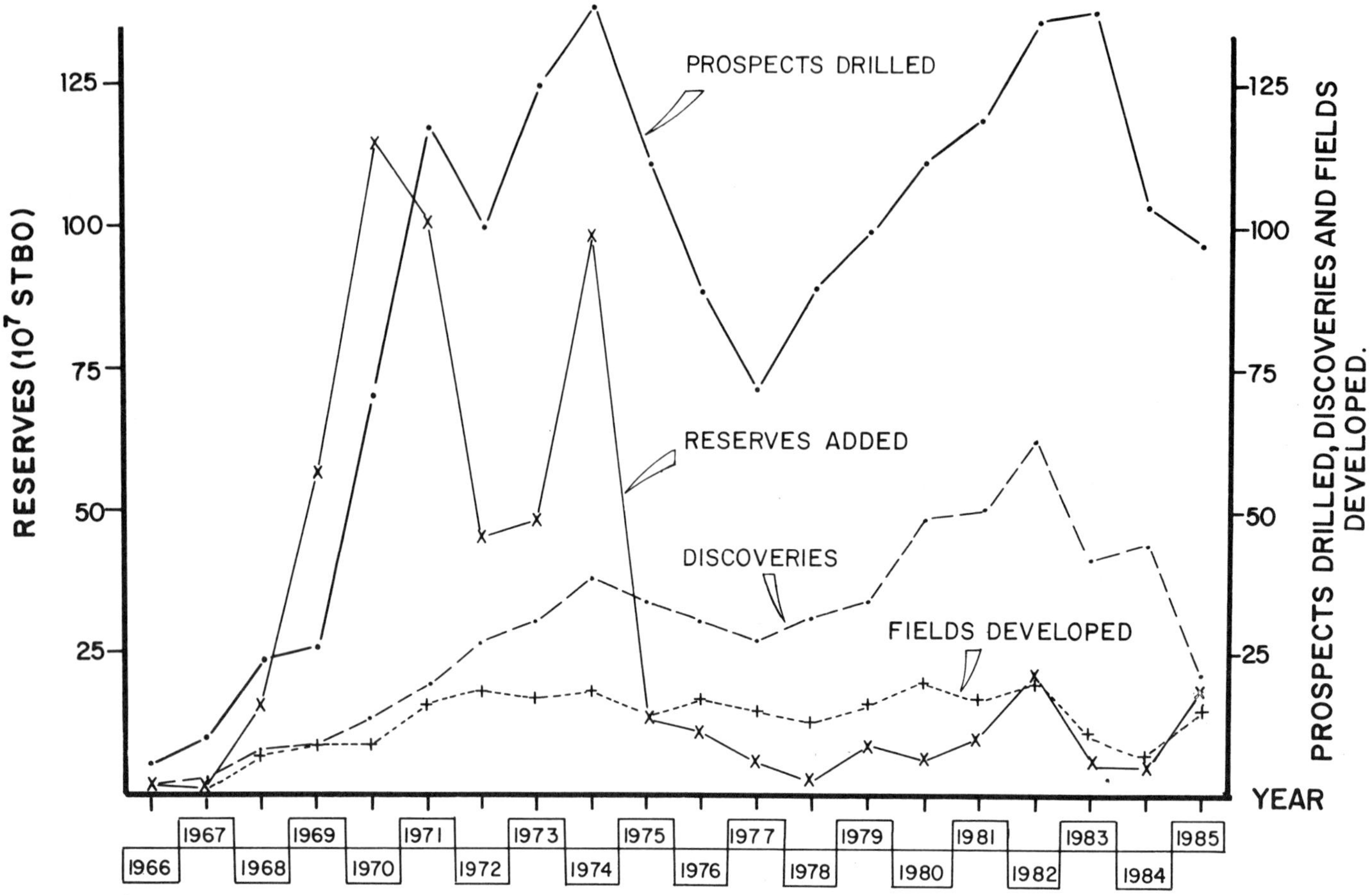


Figure 20. Statistics of prospects drilled, discoveries, development fields, and reserves.


Suardy, Atik, and J. Taruno P.H.,1985: Perkiraan Sumber Daya Hidrokarbon Indonesia (Revisi Peta Sumberdaya Energi Indonesia, 1980). Seminar Peringatan Seratus Tahun Usaha Perminyakan di Indonesia, Jakarta.

Warga Dalem, M.A., S.Z. Pane, and L. Witoelar, 1981, The Indonesian "formula" for maximizing petroleum exploration activities: Petroleum Exploration Strategies in Developing: Countries, The Hague, Netherlands.

Witoelar, L.K., Z. Achmad, and A. Reymondd, 1982, Deep sea basin in Indonesia: 11th Indon. Petr. Assoc. (IPA) Convention, Jakarta.

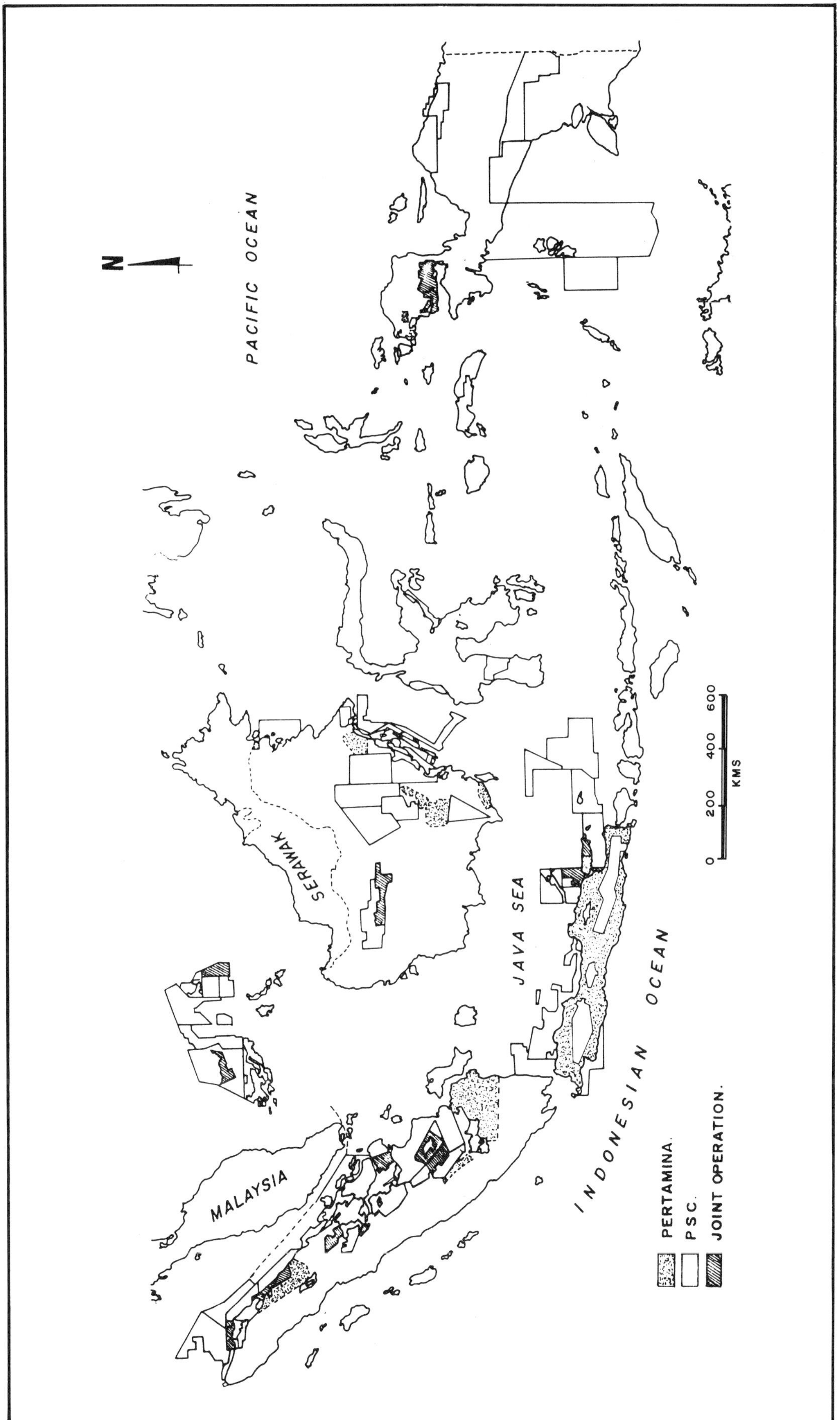


Plate 1. Map of Indonesia.

CHAPTER 22

# TERTIARY DEPOSITIONAL FEATURES, SOURCE BEDS, AND RESERVOIR BEDS IN THE NORTHERN PART OF THE SOUTH CHINA SEA

**Wang Shanshu**
*China National Offshore Oil Corporation*
*Zhuoxian, People's Republic of China*

## INTRODUCTION

The northern part of the South China Sea (SCS) lies between 16°40′–23° 00′N and 108°00′–120°00′E, including Beibu Gulf, a broad continental shelf and slope of the Guangdong province and Hainan, Taiwan islands. It is 1300 km long (east–west) and 200 to 400 km wide (north–south), covering an area of about 400,000 $km^2$. Geophysical reconnaissance and records for tens of wells indicate that this region contains widely distributed Tertiary deposits exceeding 10,000 m in thickness. The total area containing sedimentary rocks more than 1000 m thick is about 300,000 $km^2$. It consists of five sedimentary basins, such as Beibu Gulf basin, Ying Ge Hai basin, southwestern Taiwan basin (Figures 1, 2). Within the Beibu Gulf basin, the southeastern Hainan basin and the Pearl River Mouth basin, oil and gas accumulations have been discovered in Lower Tertiary Liushagang, Lingshui, Zhuhai formations and upper Tertiary Zhujiang, Jiaowei, Hanjiang formations, respectively; oil and gas production has also come from the Oligocene of the southwest Taiwan basin. From all these indications, it seems that this region has a very good oil and gas potential. On the basis of drilling, logging, and regional seismic stratigraphical data, the writer describes the depositional features of the Tertiary in the northern part of the SCS and the source-reservoir-cap assemblage controlled by these features.

## REGIONAL GEOLOGY

During the Cenozoic, the northern part of SCS was a passive continental margin in the northern part of the South China deep sea basin (Figure 3), which was deeply affected by the deep sea. It experienced three evolutionary phases—a rift phase in Paleocene to Early Oligocene, a pulling-apart phase in the Middle Oligocene to Early Miocene, and a delta to open-sea phase in the Middle Miocene to Quaternary. Three depositional structural sequences with their own characteristics and different depositional patterns were established.

### Rift Phase

In this phase, the development of a large number of tensile faults resulted in the formation of grabens and half-grabens in which 1000 to 5000 m of Paleocene, Eocene, and Early Oligocene sediments were deposited. In the early stages, the grabens were filled by slope wash and by fluvial and alluvial sediments, with smaller areas of lacustrine deposits. During the late stages, the rifted or faulted depressions became larger and deeper, forming a larger fluvio-lacustrine depositional system with fast sedimentation rate, multiple sources, and strong separation. Along the depressions a rift was separated from the marine basin by the southern margin.

### Pulling-Apart Phase

When the South China Sea expanded, 32–17 Ma, the changes in the ocean crust occurred. As the result of sea-floor spreading, the crust in the northern part of the SCS was thinned and became a transitional crust, which caused a gradual extension of the rifted depressions. The materials were mainly sourced from the north. Owing to the rise in sea level and the progressive transgression northward, the south marginal uplift was partly submerged as a barrier island. So the depressions were transformed to a semirestricted sea depositional system in which sediments accumulated to a thickness of 500 to 3000 m.

### Delta to Open-Sea Phase

From about the early Miocene, while the sea-floor spreading center of the SCS was much farther from the continental

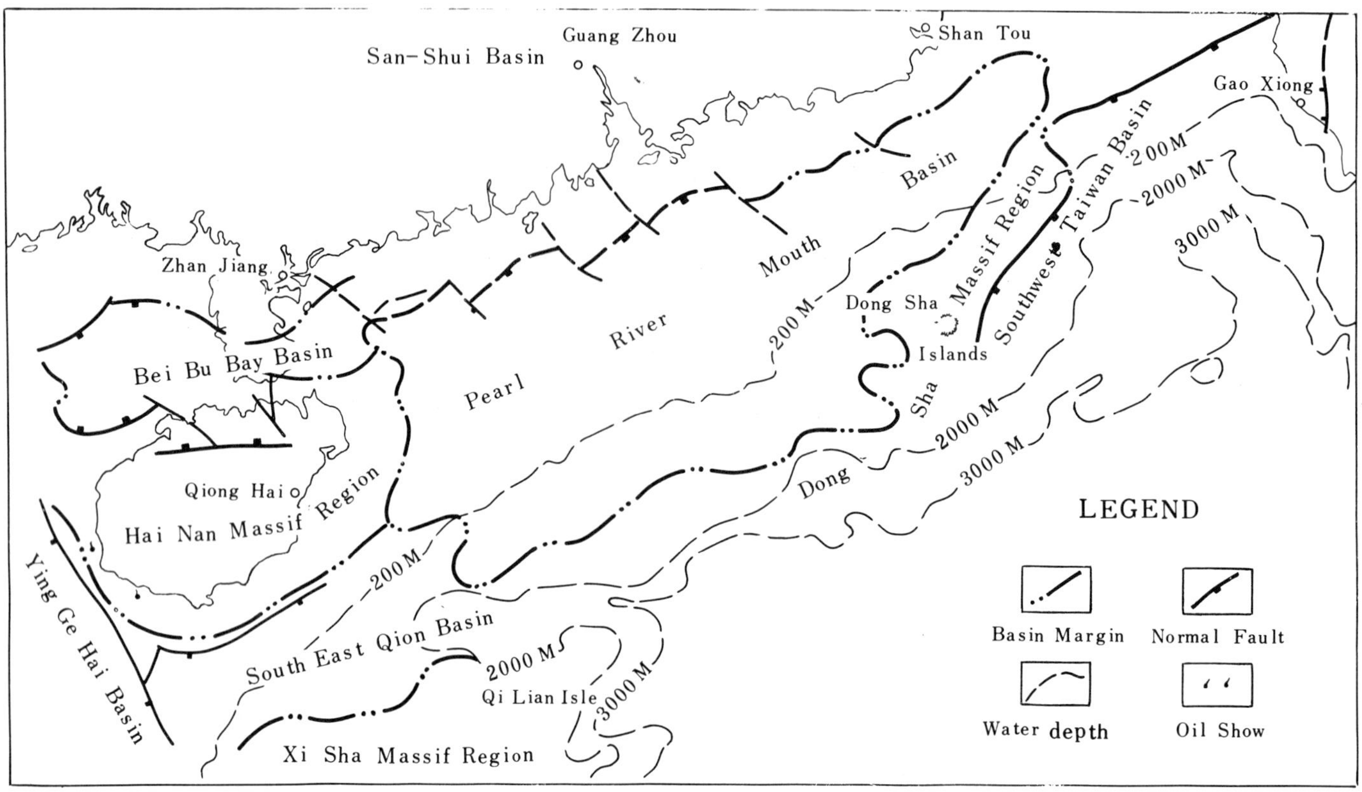


Figure 1. Sketch map of sedimentary basins in the northern part of South China Sea.

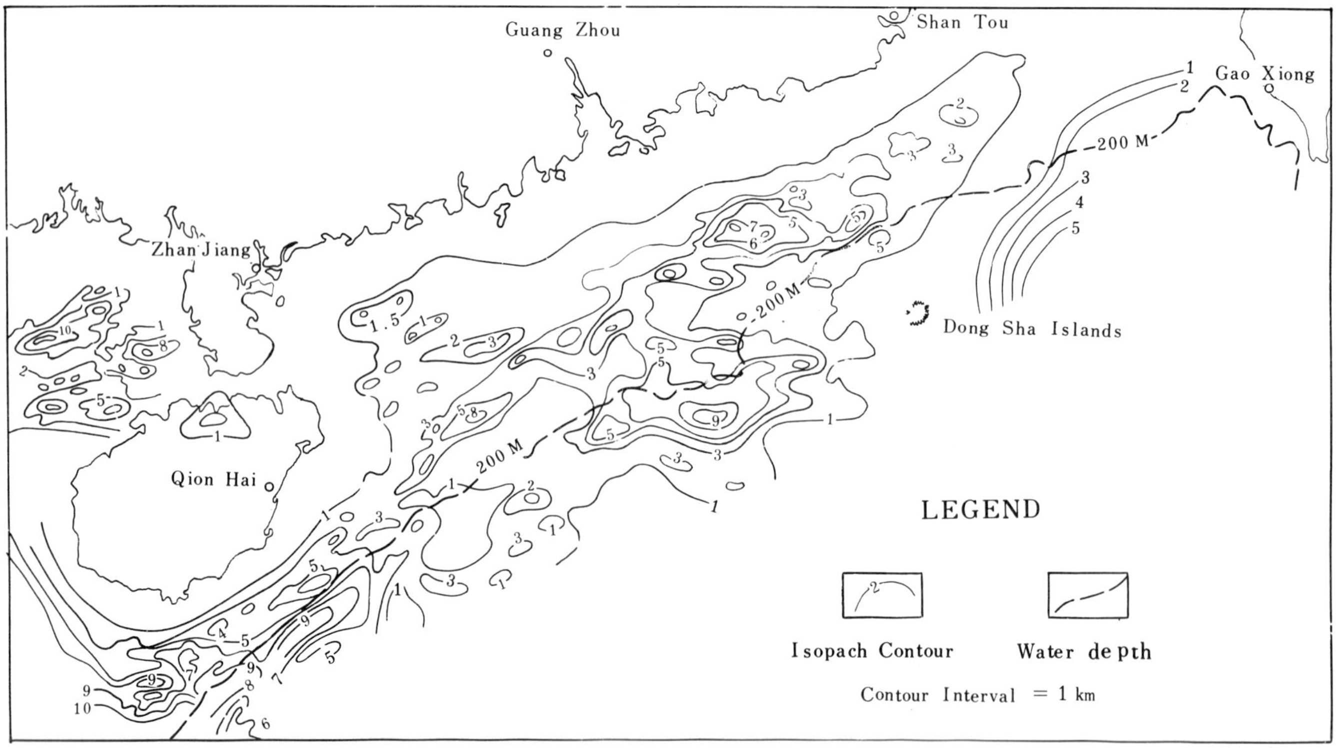


Figure 2. Isopach map of Cenozoic in the northern part of South China Sea.

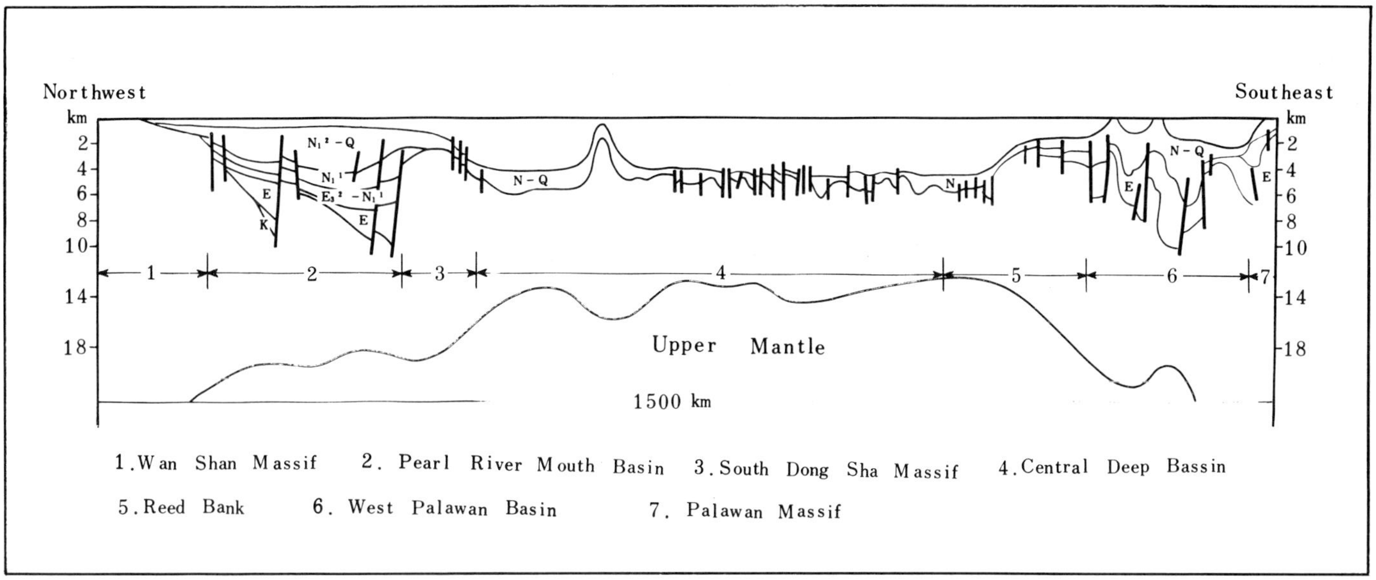


Figure 3. Section of geological structure of South China Sea.

margin, the transitional crust cooled and contracted gradually, the northward transgression intensified, and the marginal uplift was submerged. Thus, the depressions in the northern part of SCS were joined together with the SCS marine basin and became an open sea. During this period, the large rivers in the backland of Guangxi and Guangdong, with their strong transportation action, supplied sediments to the sea and formed a deltaic–open-sea depositional system prograding southward from 1000 to 5000 m in thickness.

## DEPOSITIONAL FEATURES

As a result of exploration, a Tertiary stratigraphical sequence on the northern part of the SCS has been established (Table 1). However, the exact age is still in question and needs to be determined by further work.

### Paleogene

The Paleogene is widely distributed and reaches a maximum thickness of more than 8000 m. The characteristics of each Paleogene formation are as follows:

The Changliu Formation of Paleocene age is developed at the base of rifted depressions in the Beibu Gulf basin, where its distribution is limited. A maximum of 800 m has been penetrated by drilling. The Changliu consists of nonmarine freshwater deposits, predominantly fluvial and alluvial facies. It is mainly brown and purple shale interbedded with fine conglomerate; the lower part is variegated shale interbedded with light-gray fine conglomerate, locally intercalated with dark-gray shale and sandstone.

The Shenhu Formation, which is developed in the Pearl River Mouth basin, consists of gray sandstone intercalated with multicolored mudstone in the lower part and gray sandstone intercalated with gray shale in the upper part. It is principally Paleocene but probably also contains some beds of late Cretaceous age.

The Liushagang Formation in the Beibu Gulf basin is Eocene to early Oligocene in age. It can be divided into three members from the top to bottom. The first and third members are dark-gray shale interbedded with grayish-white sandstone and conglomeratic sandstone; the second member is predominantly massive dark-gray shale and mudstone, with a few thin intercalated beds of sandstone. Oil shale is present at the top, and rich siderite in the middle. Large spinal fossils are also noted. The maximum thickness is more than 1500 m. The first member was deposited in a shallow lacustrine environment; the second member represents a medium to deep lacustrine environment with some turbidites; and the third member is a coastal lacustrine-marsh facies, probable with some sea water transgression.

The Wenchang Formation consists of Eocene lacustrine sediments in the Pearl River Mouth basin. The maximum penetrated interval is 912 m. It consists of dark-gray, gray and brown mudstone and shale, intercalated with grayish-white sandstone; more mudstone and shale are present in the middle of the formation.

The Enping Formation conformable overlies the Wenchang member in the Pearl River Mouth basin. The maximum thickness of 1140 m can be divided into a lower part consisting of grayish-white massive sandstone interbedded with coal layers, and an upper part consisting of gray and grayish-brown mudstone interbedded with grayish-white sandstone. The Enping is represented by shallow lacustrine and marsh facies of Early Oligocene age, probably including some beds of Late Eocene age.

A maximum thickness of 910 m of the Yacheng Formation was penetrated in the southeastern Hainan basin. It can be divided into three members from top to bottom. The first member is dark-gray mudstone interbedded with grayish-white and light-gray sandy conglomerate and intercalated with coal seams. The second member consists of dark-gray mudstone and silty mudstone. The third member is similar to the first member, but contains more coal intercalations, and the fine conglomerate and mudstone in the lower part are brownish red. The Yacheng represents a fluvial-marsh and lacustrine facies of Oligocene age.

The Weizhou, Lingshui, and Zhuhai formations are present in the Beibu Gulf, southeastern Qion, and Pearl River Mouth

Table 1A. Cenozoic stratigraphy.

| Geological Age | | Beibu Gulf Basin | Ying Ge-Hai Basin | Southeastern Qion Basin | Pearl River Mouth Basin | Southwestern Taiwan |
|---|---|---|---|---|---|---|
| Cenozoic: Quaternary | Q | (100 m) marine and continental | (130-200 m) neritic to deep marine | | (160-300 m) neritic and bathyal | Quaternary to upper Miocene |
| Cenozoic, Tertiary, Neogene: Pliocene | N2 | Wanglougan (300-400 m) littoral | Ying Ge-Hai (200-1800 m) neritic and bathyal | | Wanshan (400-600 m) neritic | (1500-2500 m) |
| Cenozoic, Tertiary, Neogene: Miocene | $N_1^3$ | Frow (150-600 m) littoral | Huang Liu (650 m) littoral and neritic | | Yeuhai (100-400 m) neritic | neritic to deep marine |
| Cenozoic, Tertiary, Neogene: Miocene | $N_1^2$ N | Jiaowei (400-600 m) neritic | Meishan (100-600 m) neritic | | Hanjiang (300-650 m) fluvial delta | Middle Miocene (200-500 m) neritic |
| Cenozoic, Tertiary, Neogene: Miocene | $N_1^1$ | Xia Yang (62-440 m) littoral | Sanya (430 m) neritic | Sanya (429 m) neritic | Zhujiang (377-1345 m) semi-closed marine | Lower Miocene (1275 m) neritic |

Table 1B. Cenozoic stratigraphy (continued).

| Geological Age | | Beibu Gulf Basin | Ying Ge-Hai Basin | Southeastern Qion Basin | Pearl River Mouth Basin | Southwestern Taiwan |
|---|---|---|---|---|---|---|
| Cenozoic, Tertiary, Paleogene: Upper Oligocene | $E_3^2$ | Weizhou (1000 m) fluvial | | Lingshui (582 m) littoral | Zhuhai (800 m) marine and continental | Oligocene (100-1000 m) neritic |
| Cenozoic, Tertiary, Paleogene: Lower Oligocene | $E_3^1$ | Liushagang ($E2 + E_3^1$ = 1500 m) lacustrine | | Yacheng (910 m) fluvio-lacustrine | Enping (1140 m) shallow-continental | |
| Cenozoic, Tertiary, Paleogene: Eocene | E2 | Liushang ($E2 + E_3^1$ = 1500 m) lacustrine | | | Wenchang (1140 m) shallow-continental | |
| Cenozoic, Tertiary, Paleogene: Paleocene | E1 | Changliu (800 m) continental | | | Shenhu (500 m?) continental | |

basins, respectively. They are late Oligocene in age and probably also contain Middle to Early Miocene sediments.

The Weizhou Formation of the Beibu Gulf basin is a predominantly fluvial facies which consists of purple, grayish-yellow, light-green, and variegated mudstone with grayish-white sandstone, gritstone, and conglomeratic gritstone. Glauconite and rhizophora pollen in the middle part and planktonic foraminifera in the upper part indicate marine intercalations, which are regarded as Late Oligocene to Early Miocene. Some geologists consider the Weizhou to be the same age as the Yacheng and Enping formations.

The Lingshui Formation was penetrated in the Ying No. 9 well in the southeastern Qion basin. It consists of grayish-white, gray, and light-gray sandstones, sandy gritstone and dark-gray mudstones; the basal part is shale interbedded with grayish-white, light-purplish-brown, and red bioclastic limestones which are rich in algae, sporopollen, foraminifera, and pelecypods. The formation is 582 m in thickness and represents a littoral to shallow-marine facies of Late Oligocene to Early Miocene age.

A maximum thickness of 800 m of the Zhuai Formation was penetrated in the Pearl River Mouth basin. The formation is a

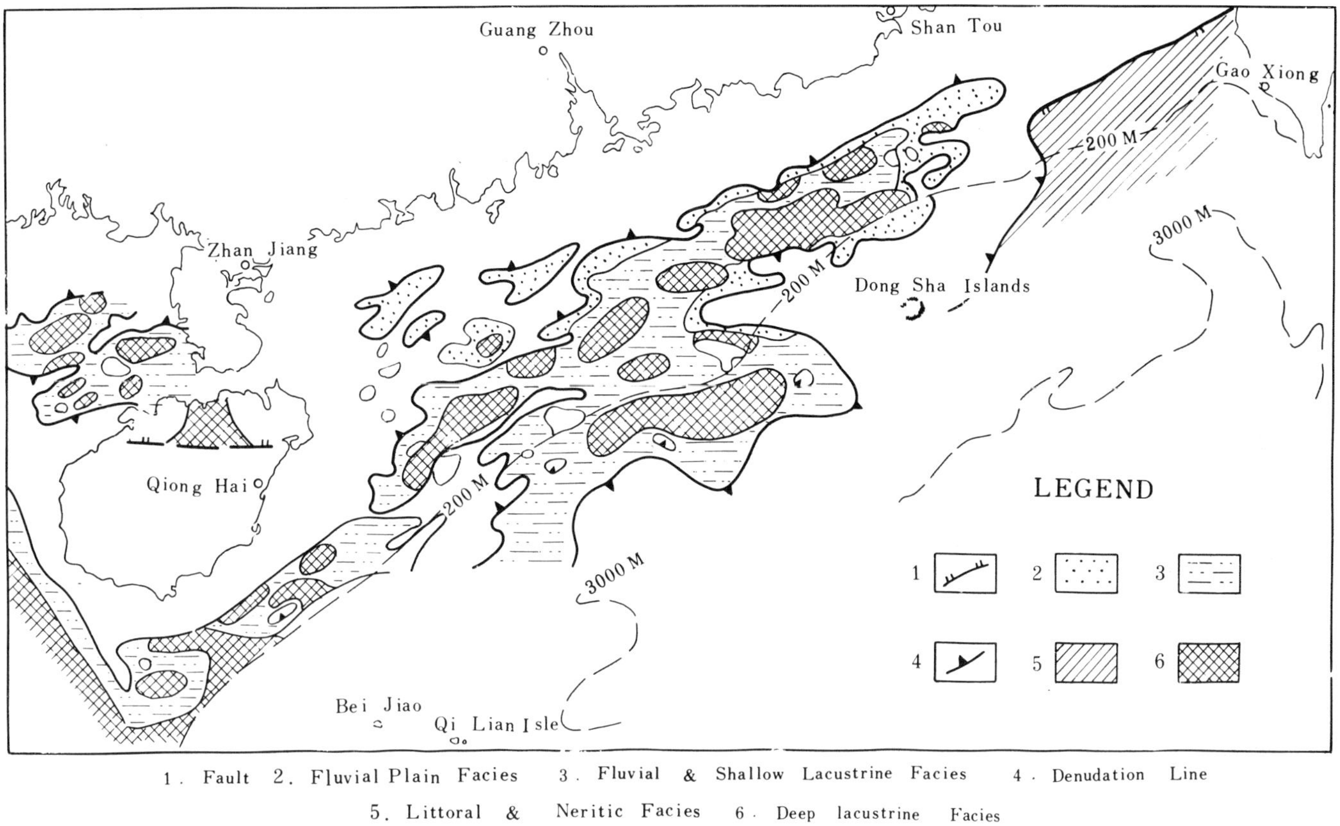


Figure 4. Lithofacies in the northern part of South China Sea from Eocene to Lower Oligocene.

normal sedimentary cycle. It consists of grayish-white massive sandstone intercalated with shale in the lower part and brownish-gray mudstone intercalated with brownish-gray sandstone in the upper part. The formation represents marine and nonmarine alternations of Late Oligocene age.

In the southwestern Taiwan basin, Oligocene shallow-marine sediments of 183 m were tested in the CFC-1 well, which consists of interbedded sandstone and mudstone and contacts unconformably with overlain and underlain sequences.

In general, the Early Tertiary in the northern part of the SCS was controlled by structural and depositional environments. The sediment characteristics are variable, such as the Paleocene comprising infilling deposits of rifted depressions, while the Eocene is fluvio-lacustrine deposits left when the rifted depressions were continuously spreading and water was deepening. The middle Eocene was the peak stage of the basin development and sea water transgression. During Early Oligocene, the faulted depressions spread further and the lacustrine deposits became more extensive (Figure 4). A regional transgression in the Middle Oligocene resulted in a littoral and shallow-marine environment in the Yinggehai, southeast Hainan, and southwest Taiwan basins.

## Neogene

The Neogene distribution was more continuous and wider in the northern part of the SCS than in the Paleogene, with a maximum thickness of 9000 m. From bottom to top, a semirestricted sea environment changed to a delta coastal and open marine environment.

The Xiayang, Sanya and Zhujiang formations of Early Miocene age are distributed in the Beibu Gulf, Yinggehai-southeast Hainan, and Pearl River Mouth basins, respectively.

The Xiayang Formation consists of greenish-gray and grayish-white fine conglomerate and conglomeratic sandstone intercalated with some gray mudstone and sandy mudstone. Littoral deposits in the basal part consist of thin layers of limestone or dolomite containing foraminifera and ostracods. The total thickness is 62 to 440 m.

The Sanya Formation in the Ying No. 9 well of the southeast Hainan basin consists of gray mudstone intercalated with gray and greenish-gray argillaceous siltstone. There are four calcareous quartzose sandstone layers in the upper part and a large quantity of foraminifera in the basal part. It is of 429 m in thickness and represents a shallow-marine facies.

The Zhujiang Formation is found in the Pearl River Mouth basin, where it ranges in thickness from 377 to 1345 m. The lower part consists of gray sandstone intercalated with mudstone; the upper part is a brownish-gray mudstone member intercalated with a few beds of light-gray silty and fine-grained sandstone. Bioreefs, banks, and platform carbonates are developed at the basal part of the mudstone member.

A Miocene sequence 1275 m thick was penetrated by drilling in the southwest Taiwan basin. It consists of intercalated light-gray sandstone and mudstone representing a shallow-marine facies.

On the basis of regional seismic stratigraphical analyses of sediments deposited since the Early Miocene, the source of material in the northern part of the SCS was mainly from the north, because of extensive transgressions of the sea in that

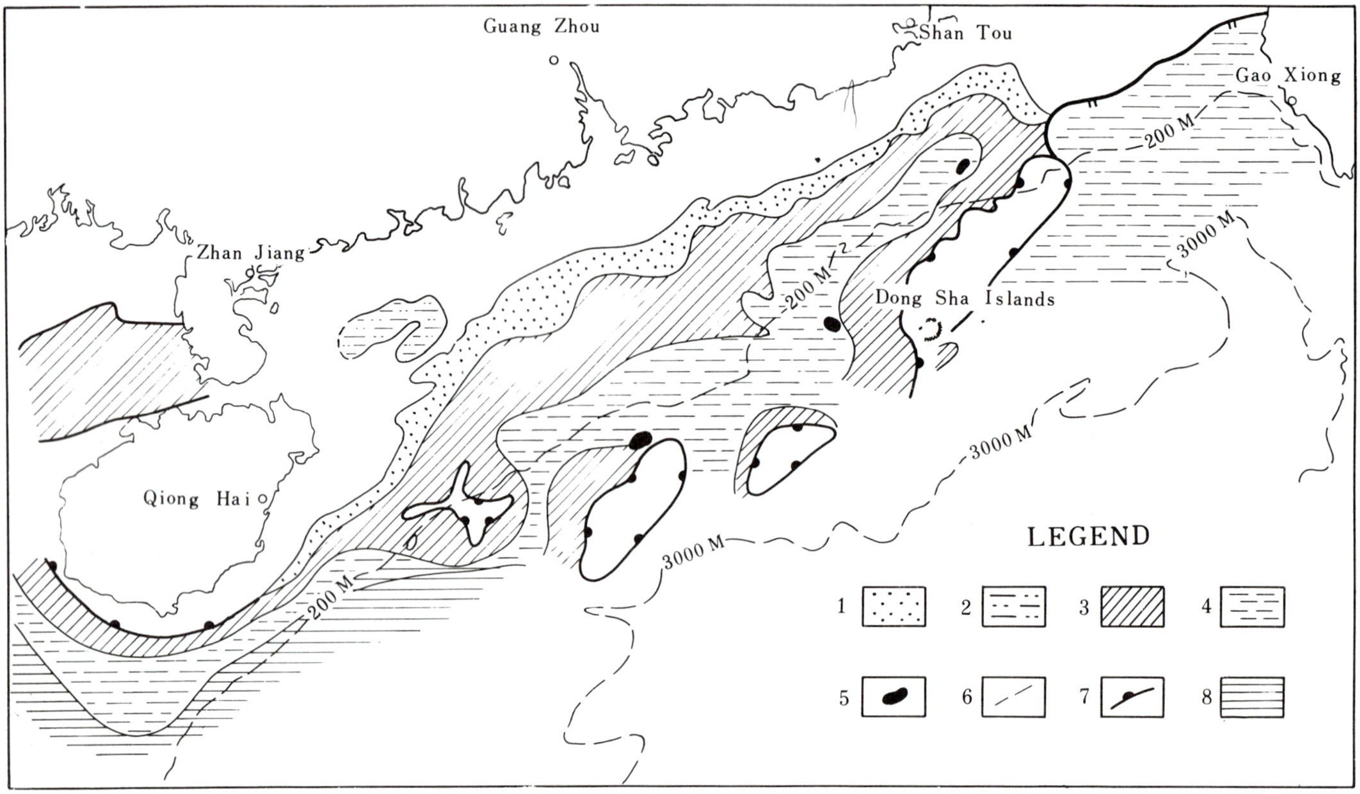


Figure 5. Lithofacies in the northern part of South China Sea from Upper Oligocene to Lower Miocene.

area. The sea water gradually flooded over the Pearl River Mouth basin, but an uplift that remained on the southern margin separated the basin from the open sea and provided a source of clastic material from the south. Deposition of these clastics formed a system of semiclosed bioreefs along the southern marginal uplift, beyond which to the south was an open sea (Figure 5).

The distribution of Middle Miocene is more extensive, and its maximum thickness is more than 5000 m. In the Beibu Gulf basin, it consists of interbedded sandstone, fine conglomerate, and mudstone with a thickness of 1000 to 2000 m. The Middle Miocene is a littoral facies which is characterized by its monolithology.

The Meishan, Huangliu, and Yinggehai formations in the Yinggehai and southeastern Hainan basins consist mainly of grayish-green and gray mudstone intercalated with sandstone and fine conglomerate. These formations are shallow-marine deposits, 1000 to 5000 m in thickness, which thicken and become finer grained southward. Bioreefs were found at the basal part of the Middle Miocene on the southwestern margin of the Hainan massif. During the Middle Miocene, the sedimentation gradually extended northward, and the delta developed progressively southward. Turbidites occurred in the prodelta areas.

The Middle Miocene and the overlying sequence penetrated by drilling in the Pearl River Mouth basin is 1700 m thick and consists of interbedded light-gray and grayish-green mudstone and sandstone and fine conglomerate. Seismic data indicate that these deposits thicken southward a maximum of 4000 m. The Hanjiang formation of Middle Miocene age outside the present Pearl River Mouth and Hanjiang Mouth formed a very large deltaic-open sea depositional system. The delta has prograded to the edge of the present continental slope (Figure 6).

## SOURCE BEDS AND RESERVOIR BEDS

Analysis of depositional and environmental features indicates that the Tertiary is the principal oil- and gas-bearing sequence in the northern part of the SCS.

### Source Beds

The sedimentation rate for the Tertiary in the northern part of the SCS is as fast as more than 1000 m/Ma, and the geothermal gradient is more than 3.6 °C/100 m. These factors are favorable for the accumulation, preservation and conversion of organic matter to hydrocarbon. There are five principal series of source beds:

A. The Liushagang Formation in the Beibu Gulf, especially the Liushagang II member, was deposited during the peak development period of the basin. It ranges in thickness from 400 to 800 m and consists predominantly of dark mudstone intercalated with a few beds of sand-

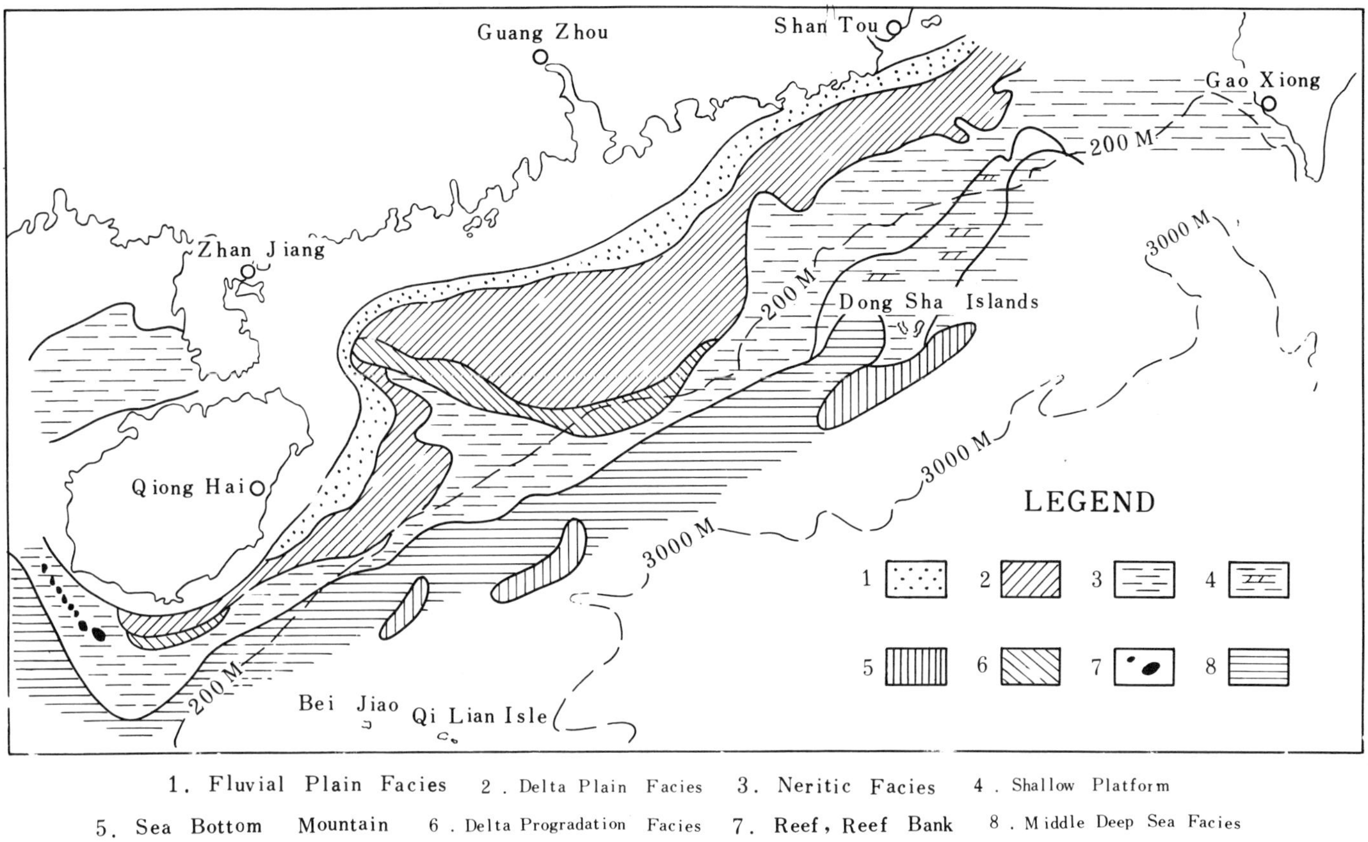


Figure 6. Lithofacies of Middle Miocene in the northern part of South China Sea.

stone. The average organic carbon content is 1.65%, the chloroform asphalt "A" content is 0.212% and the total hydrocarbons are 1373 ppm. Kerogens are mainly the humic type. This is one of the best source rocks ever found in the northern part of SCS.

B. The Wenchang Formation is a very good source rock, with an organic carbon content of 2.4%, a chloroform asphalt "A" content of 0.30%, and total hydrocarbons 1500 ppm. Kerogens are also mainly the humic type. The Wenchang can be correlated with the source beds of the Reed Bank basin.
C. The Enping Formation is a good source bed which is widely distributed. It has an organic content of 1.43%, a chloroform asphalt "A" content of 0.23%, and total hydrocarbons 1000 ppm.
D. The Yacheng Formation is an important source bed in the southern Hainan basin, where it consists of dark-gray mudstone intercalated with coal layers and carboniferous shales. The organic carbon content is more than 2% and the chloroform asphalt "A" content is from 0.326 to 0.19%. The kerogen is predominantly the humic type.
E. The Zhujiang Formation in the Pearl River Mouth basin is also a petroleum-generating sequence which was deposited in a semirestricted marine environment. In the center of the basin, it contains 0.90% organic carbon, 0.08% chloroform "A," and 300 ppm total hydrocarbons. The kerogens are mainly humic.

## Reservoir Beds

Studies on basin evolution, sedimentary characteristics in the northern part of the SCS, and data from well tests, indicate three Tertiary reservoir sequences.

A. The Liushagang Formation in the Beibu Gulf basin is the reservoir bed. Here many varieties of fluvio-lacustrine sand bodies are developed. In some wells, tests show productivities of individual sand bodies ranges from tens to hundreds of tons per day.
B. The Zhuhai and Lingshui formations are regional reservoir beds. Fluvio-lacustrine sand bodies and shallow-marine sandstone sheets are the principal reservoirs. The oil productivity per well is ranging from tens to hundreds of tons per day, and the gas productivity is high.
C. Many Neogene sand layers are good reservoirs, such as those found in the Sanya and Meishan formations. In these formations the porosities are 26 to 31%, the permeability is 2.6 to 4.8 md, and the thickness of a sand zone is from tens of meters to more than a hundred meters. Bioreefs are also found in the Meishan Formation.

The deltaic and littoral-shallow marine sandstones and bioreefs in the Zhujiang Formation of the Pearl River Mouth basin are very important reservoir beds. Their productivities range from a hundred to many hundred tons per day.

The delta formed from sandstones in the Hanjiang formation is widely distributed in the region. Porosity is 30 to 32% and permeability is 2 to 5 md. Heavy crude oil has been discovered in the Hanjiang Formation.

Further exploration is certain to reveal more oil and gas reservoir beds in the Tertiary.

## Source-Reservoir-Cap Association

Three Tertiary source-reservoir-cap associations can be identified in the northern part of the the SCS, in vertical order as indicated below.

A. Lower association—The Eocene in the Beibu Gulf basin is an oil-prone association that contains source rocks, reservoir rocks, and mudstones as cap rocks.

B. Middle association—An association rich in oil and gas is recognized in the Pearl River Mouth basin, southeast Qion basin, and southwest Taiwan basin. The source beds are believed to be Eocene and Oligocene; and the mudstones within the Oligocene or Lower Miocene act as cap rocks.

C. Upper association—High oil flows and gas zones have been discovered in the Pearl River Mouth basin. The Oligocene and Lower Miocene are believed to be the source beds; the lower part of the lower Miocene is considered to represent the reservoirs; and the upper part of the Lower Miocene acts as cap rocks.

The lower association predominates in the Beibu Gulf basin, the upper and middle association predominates in the southeast Qion basin, and possibly the upper association predominates in the Ying Ge Hai basin. However, all three associations are possibly developed in the Pearl River Mouth basin. In short, the northern part of the SCS in China is an area with multiple source beds and multiple reservoir beds, and great potential for oil prospecting.

# HYDROCARBONS ABSTRACTS

BUENO SALAZAR, RAFAEL, Empresa Colombiana de Petróleos, Bogotá, Colombia

## Energy Resources of Pacific Coast of Colombia

Despite failure of modest exploration efforts to yield commercial hydrocarbon production in the Choco-Pacific coastal basin of Colombia, recent geophysical, geochemical, and surface geologic investigations indicate a potential for petroleum accumulations, which could be related to fields located on the western basins of Ecuador that in fact constitute an extension of the Colombian Pacific geologic scheme.

The Choco-Pacific coastal basin of Colombia covers an area of approximately 70,000 $km^2$, of which 14,000 $km^2$ lies offshore. The structural style of this area corresponds to a convergent plate basin created over folded oceanic sediments and adjacent to the subduction zone. Such a framework could be conducive to an attractive array of potential hydrocarbon-bearing traps.

Geochemical knowledge of potential source rocks of Cretaceous and early Tertiary age confers an added attraction to the area. Most evaluations reveal kerogen-rich, gas-prone organic matter. Nevertheless, the existence of oil seeps from Cretaceous outcrops could indicate sufficient thermal maturity for oil generation. Adequate reservoirs could be found in sandy or calcareous rocks of late Eocene to Oligocene age, predominantly of marine origin with an estimated thickness exceeding 20,000 ft. Colombia has been one of the leading world producers of gold and platinum, mostly derived from the vast alluvial cover of the onshore area of the basin. In rocks cropping out in the Western Cordillera (eastern margin of the basin), deposits of potentially commercial value of porphyry copper and molybdenum, as well as massive sulfur, manganese, and bauxite, have been found.

LIU GUANGDING, Bureau of Petroleum Geology and Marine Geology, Ministry of Geology and Mineral Resources, Beijing, People's Republic of China

## Geology and Petroleum Prospects in East China Sea

Geologic studies of the petroleum potential of the East China Sea have progressed through three stages. (1) The tectonic divisions and evolution of the East China Sea were investigated using geophysical data, including gravimetric, magnetic, and seismic information. (2) Seismic data over the Shelf basin were interpreted and mapped, to understand the Cenozoic structural history of the East China Sea. (3) More detailed seismic shooting has allowed for detailed prospect mapping and seismo-stratigraphic studies. This stage has resulted in the mapping of numerous drillable prospects, several among them having been proved to contain commercial quantities of oil and gas.

Presently, results from the Lingfeng 1 well point to new and different directions for exploration in the basin.

SCRUTTON, MICHAEL E., Robertson Research, Inc., Houston, TX, and GREGORIO F. ESCALANTE, Consulting Geologist, San Jose, Costa Rica

## Petroleum Geology of Pacific Margin of Central America and Northern South America, from Guatemala to Ecuador

Exploration for hydrocarbons along the Pacific margin of Central America and northern South America has been limited and spasmodic. Less than 100 exploration wells have been drilled, with nearly 50 of these being in the Santa Elena, Progreso, and Guayas basins in Ecuador. Shows have been reported in some wells, and a few oil seeps are known. The only commercial production established to date has been from the Santa Elena Peninsula in Ecuador in the extreme south of the study area.

Understanding of the geology in this part of the continental margin is incomplete at best. This paper reviews present-day knowledge in an attempt to define the sedimentary basins better, to characterize their structure and stratigraphy, and to assess their petroleum prospects.

The area of continental margin reviewed is to the north, located northwest of the trench system where oceanic crust of the Cocos plate subducts under the Caribbean plate, and to the south, where the northern part of the Nazca plate collides with the South American plate. This plate tectonic setting forms the framework on which local structural and sedimentary events have created a series of relatively small trench-slope and forearc basins in what is now the coastal plain and adjacent offshore area of Central and South America, south or west of a line of mountain ranges with active volcanism. Sedimentary fill is generally of Tertiary age.

The basins and subbasins recognized and described include: in Ecuador—Guayas, Santa Elena, Progreso, Valdivia, Bajo Grande, Manta, Muisne-Esmeraldas, and Borbon; in Colombia—Choco-Pacific; in Panama—Gulf of Panama basin complex (Santiago, Tonosi, Sambu), and Burica-Chiriqui; in Costa Rica—Terraba and Coronado/Tempisque; in Nicaragua—San Juan del Sur; and in the Honduras, El Salvador, and Guatemala—the Pacific coastal basin.

ZHAI GUANGMING, Scientific Research Institute of the Ministry of the Petroleum Industry, Beijing, People's Republic of China

## Geologic Characteristics of Sedimentary Basins in China

The oil- and gas-bearing basins of China can be divided into four geographic areas, based on structural style. Eastern China contains the major producing fields, where oil and gas accumulations are found in combination structural-stratigraphic traps. In western China, oil and gas fields are distributed widely. Most traps are structural, with some stratigraphic controls on accumulations. The central part of China contains marine carbonate and nonmarine clastic sediments, which are prolific reservoirs. These traps show transitional structural features. The continental shelf offers excellent geologic conditions and a good potential for finding oil.

China is unique in the characteristics of its oil and gas basins. These include a multiple supply of source material, multicyclic deposition, multiple source-reservoir–cap-rock assemblages, and a variety of lithology and lithofacies.

# HYDROCARBONS (NEW FRONTIERS)

CHAPTER 23

# DEEP CONVERGENT MARGIN BASINS OF WASHINGTON STATE—PAST HYDROCARBON SOURCE?

**K. Kluger Cohen**
**W.J. Gwilliam**
*U.S. Department of Energy*
*Morgantown Energy Technology Center*
*Morgantown, WV*

**T.H. Anderson**
**E. Lidiak**
*Department of Geology and Planetary Science*
*University of Pittsburgh*
*Pittsburgh, PA*

## ABSTRACT

Magnetotelluric soundings conducted in western and north-central Washington have located large volumes of low-resistivity rocks (1 to 20 ohm-meters) extending to depths as great as 25 km (Stanley, 1984; Stanley and Plesha, 1985; and Stanley, Finn, and Plesha, 1986). These deep-seated packages have been interpreted as most likely representing marine sedimentary rocks. As part of the Unconventional Gas Recovery Program in the U.S. Department of Energy, we explored the hydrocarbon potential and tectonic histories of these deep-seated features.

Deep units in western Washington comprise a large volume of rock described previously as the "Southern Washington Cascades Conductor." Reconstruction models suggest they may record forearc or marginal basin sedimentation associated with Late Cretaceous and/or early Tertiary subduction. If analogies drawn between these deep rocks and the Late Cretaceous Hornbrook-Ochoco basin of southern Oregon and northern California are correct, preliminary source rock studies suggest that the deep rocks may have served as good source rocks. Analogies which compare the deep rocks with Early Tertiary sediments that crop out in the Pacific Northwest need to be investigated further.

Magnetotelluric data from north-central Washington also indicate sedimentary rocks lie deep under the Methow basin. If extrapolations of data obtained from exposed Methow sections in this basin are accurate, the deep rocks probably formed in forearc basins and are, at a minimum, thermally mature. Recent COCORP deep seismic surveys in the Pacific Northwest may add insight into these interpretations; however, our conclusions will remain speculative until deep drilling is undertaken.

## INTRODUCTION

Natural gas resources with unique geology and production mechanisms have been classified as unconventional by the U.S. Department of Energy. As part of the Unconventional Gas Recovery Program in the U.S. Department of Energy, we are investigating whether source rocks deposited along convergent margins emplaced to depths of greater than 30,000 ft (approximately 9 km) in the U.S. Pacific Northwest generated natural gas in their history. Currently, research is focused upon hydrocarbon occurrences and potential in active convergent margins of the Circum-Pacific, and sites of paleoconvergence in the North American Western Cordillera (Gwilliam and Cohen, this volume).

Petroleum occurrences in the Pacific Northwest are not uncommon; thus, hydrocarbon source rocks are believed to exist in this region. However, the Mist gas field is the only commercially producing field in the Pacific Northwest. Magnetotelluric soundings completed in Washington State (Stanley, 1984; Stanley and Plesha, 1985; Stanley, Finn, and Plesha,

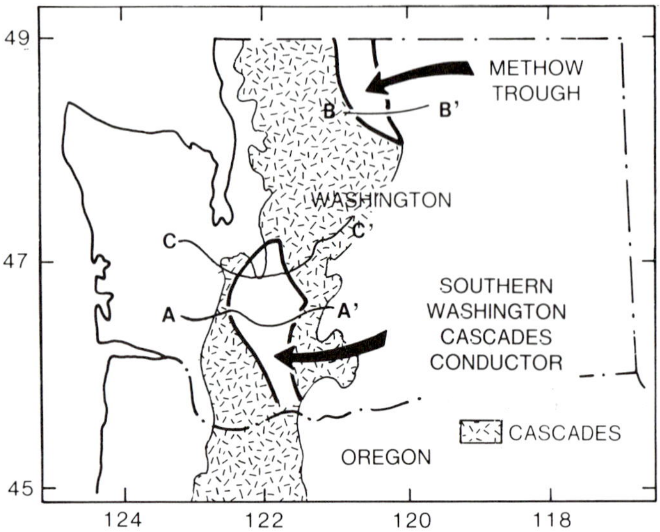


Figure 1. Map of Washington and Oregon with study areas. Areas include the Southern Washington Cascades Conductor area (SWCC) and the Methow basin. Cross sections (see Figure 2) are magnetotelluric sections from Stanley (1984), Stanley and Plesha (1985), and Stanley, Finn, and Plesha (1986).

1986) are interpreted to indicate that two areas contain deep marine sedimentary rocks: (1) a northwest-trending region beneath the southern Cascades of western Washington and; (2) the Methow basin (Figure 1). Modeled geoelectric cross sections suggest low-resistivity rocks (1 to 20 ohm-meters) range from depths of 5 to 25 km (Figure 2; Stanley and Plesha, 1985; Stanley, Finn, and Plesha, 1986). This area has long been a site of convergence, and subduction continues offshore Washington and Oregon. We consider it likely that these deep sedimentary packages were associated with paleoconvergence and may record hydrocarbon source rock deposition.

The goals of our investigations are to evaluate the hydrocarbon potential and tectonic history of these deep packages. Because data pertinent to these problems are sparse, indirect methods must be used to investigate the deep rocks. Magnetotelluric, COCORP (Consortium for Continental Reflection Profiling) deep seismic, aeromagnetic, gravity, earthquake seismicity, and deep drill hole data must be integrated to provide the most comprehensive interpretation of this mid-crustal region of interest (depths greater than 9 km). Source rock and thermal maturation studies of exposed rocks postulated to be equivalents of the deep rocks may provide insight into the hydrocarbon potential of the deep packages. Our plate tectonic reconstructions reveal constraints that should be considered during assessment of the geologic evolution of these rocks.

## WESTERN WASHINGTON

### Previous Geophysical Investigations

Magnetotelluric soundings conducted in western Washington (Figure 1) suggest that a large volume of low-resistivity rocks (1 to 5 ohm-meters) exists in the subsurface (Stanley, 1984; Stanley, Finn, and Plesha, 1986). This phenomenon has been described as the "Southern Washington Cascades Conductor" (SWCC; Stanley, Finn, and Plesha, 1986). Three east-west cross sections (Figure 2) have been modeled to date. The southernmost east-west cross section shows an east-dipping low-resistivity package ranging in depth from 2 to 15 km. The northerly east-west cross section shows a package of low-resistivity rocks extending from depths of 5 km, apparently compressed against more electrically resistant, crystalline rocks of the North Cascades. The low-resistivity rocks do not crop out, but are overlain by volcanic, plutonic, and other upper crustal rocks. The package has been interpreted to most likely represent marine sedimentary rocks of Cretaceous-Early Tertiary age based upon geologic and geophysical constraints.

Regional Bouguer gravity for the area, determined by fitting a second-order polynomial trend surface to the complete Bouguer gravity, is dominated by a large east-west gradient and the Cascade low (Phillips, 1983). These regional features of the Bouguer gravity can be correlated directly with the depth to the downgoing slab in this area, and the roots of the Cascade Mountains, respectively (Phillips, 1983). After removal of the regional field, no correlation is apparent between the residual gravity and the conductivity anomaly margins. Seismicity, aeromagnetic, and gravity data for the area have been summarized by Stanley, Finn, and Plesha (1986). Significant features interpreted from these data include a northwesterly trending seismic lineament which has seismic events that show right-lateral strike-slip focal mechanisms (Weaver and Smith, 1983). Linear or curvilinear magnetic lows dominate the area. The seismic lineament and the aeromagnetic low patterns are approximately coincident with the margins of the conductivity anomaly (Stanley, Finn, and Plesha, 1986).

### Tectonic Reconstructions

Four models are derived to evaluate how such a large volume of marine sedimentary rocks was deposited and emplaced to great depths in the Pacific Northwest. The primary assumptions are that the SWCC rocks are sediments that accumulated near the margin of North America during the Cretaceous-Early Tertiary. During this interval, the margin was dominated by convergence and transform faulting. The SWCC rocks may be autochthonous or allochthonous. In order to generate these reconstructions, some key tectonic elements in the Pacific Northwest were digitized so that accurate movements of these elements would be possible through computer modeling. The key elements are: SWCC, Methow basin, Oregon Coast Range, Klamath Mountains, Blue Mountains, and the Straight Creek fault (Figures 3, 4).

In the first reconstruction, the model of Magill, Cox, and Duncan (1981) for the Oregon Coast Range is considered (Figure 3). This model is based upon Tertiary paleomagnetic and geologic data. Eocene volcanics (Crescent, Siletz River, and Roseburg Formations) which form the basement rocks of the Oregon Coast Range have been shown to be oceanic in character (Duncan, 1982; Wells et al., 1984). In this model, these rocks formed as a seamount complex on the Farallon or Kula plate. Subsequently, this complex was then trapped between two active subduction zones, one south and one east of the complex. A wedge-shaped piece of oceanic crust carrying this seamount province was gradually subducted to the east, concurrently with clockwise rotation of the wedge between 50 to 42 Ma until it accreted to North America. The pole of rotation was near the Klamath Mountains. Between 20 Ma and present, and in response to the Basin and Range extension, the Oregon Coast Range was again rotated clockwise to its

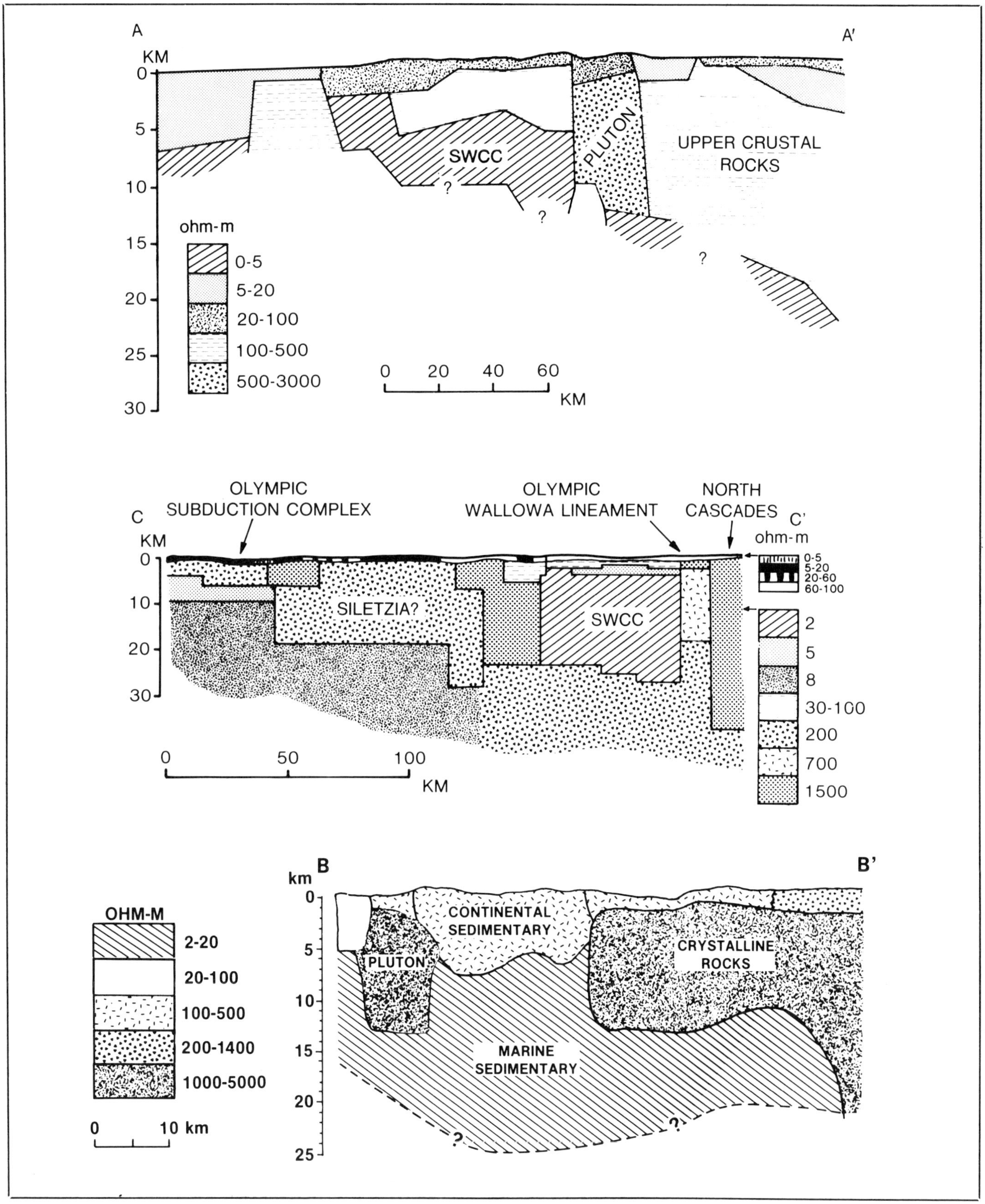


Figure 2. Magnetotelluric cross sections through study areas (see Figure 1). Data from Stanley (1984), Stanley and Plesha (1985), and Stanley, Finn, and Plesha (1986). Cross sections A–A′ and C–C′ are through the SWCC. Rocks of interest have been modeled with resistivities of 0 to 5 ohm-meters and extend from depths of 5 to 25 km. Cross section B–B′ is through the Methow basin. Rocks of interest have been modeled with resistivities of 2 to 20 ohm-meters and extend from depths of 5 to 25 km.

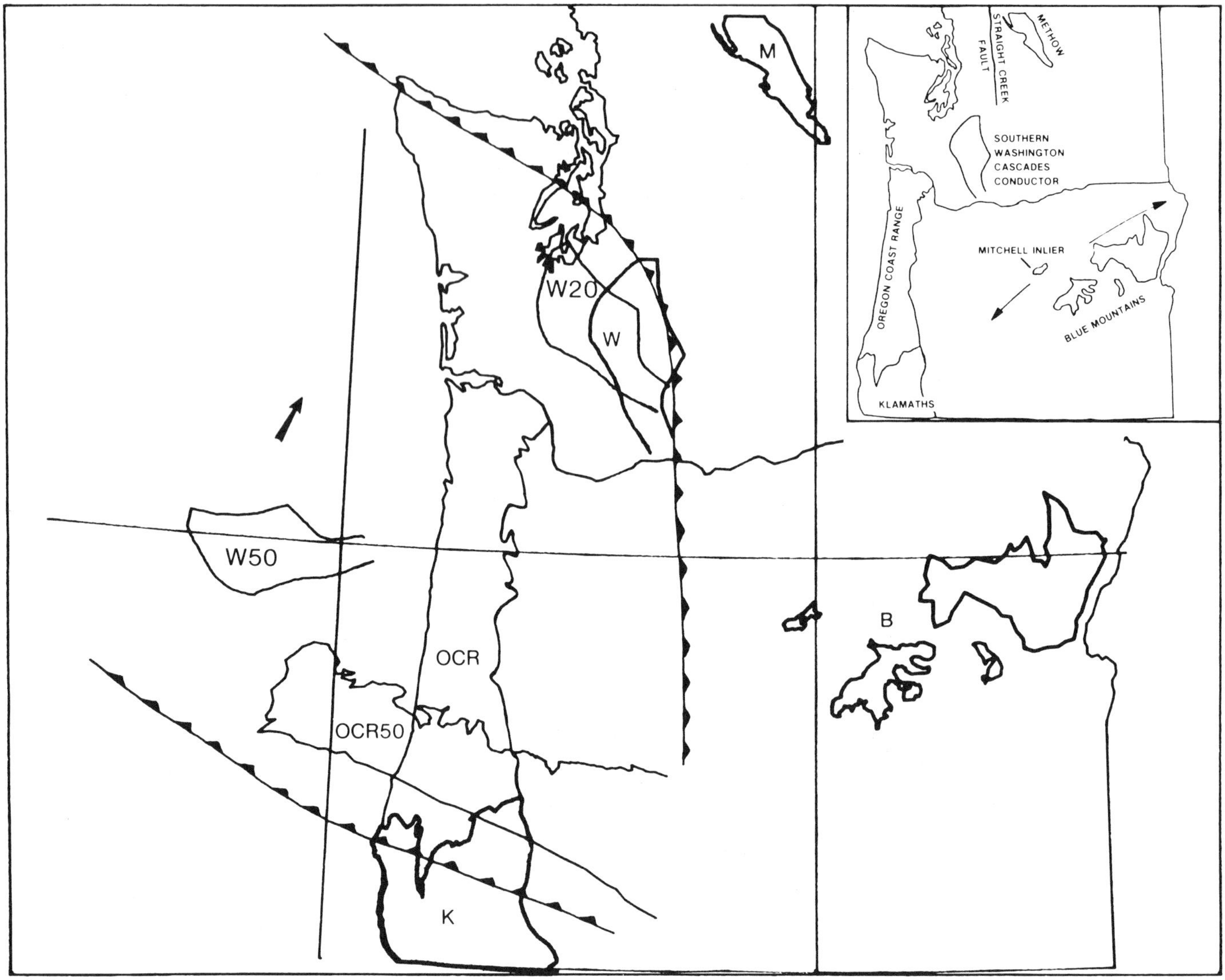


Figure 3. Early Tertiary reconstruction for western Washington study area (models 1 and 2). Lambert projection with key tectonic elements: M = Methow, B = Blue Mountains, K = Klamath Mountains, W = SWCC, W20 = SWCC 20 Ma, W50 = SWCC 50 Ma, OCR = Oregon Coast Range, and OCR50 = Oregon Coast Range 50 Ma. Inset shows present-day position of these elements. Microplate position carrying Oregon Coast Range and bound by subduction zones is from model of Magill, Cox, and Duncan (1981). Poles and angles of rotations used are also from Magill, Cox, and Duncan (1981). To restore SWCC to 20 Ma position, pole and angle of rotation used was 46.0° N, 122.5° W and 30.4°, respectively. Position of SWCC and Oregon Coast Range then rotated to 50 Ma position by concatenating previous move with an angle of 46.0° about a pole of rotation 43.0° N, 121.3° W.

present position, about a pole of rotation near Mount St. Helens.

The autochthonous position of the SWCC (W; Figure 3) lies along the postulated easternmost subduction zone associated with the Oregon Coast Range microplate at 50 Ma in this model. The thick volume of low-resistivity rocks would, therefore, record Early Tertiary accretionary wedge and/or fore-arc basin deposition of a short-lived subduction zone.

On the other hand, the SWCC may record marginal basin sedimentation dispersed on the wedge-shaped microplate carrying the Oregon Coast Range during the Early Tertiary. Thus, it can be considered allochthonous and experienced similar motion as the Oregon Coast Range. The two phases of rotations suggested by Magill, Cox, and Duncan (1981) for the Oregon Coast Range were used to restore the SWCC position at 50 and 20 Ma (Figure 3). If the SWCC experienced the same rotations as the Oregon Coast Range, then the positions of the SWCC package shown in Figure 3 (50 and 20 Ma) represent the maximum distances the SWCC may have been translated. The sediments would have been emplaced to their present depths during convergence.

Recently, Wells et al. (1984) suggested that the Oregon Coast Range basement volcanic rocks formed in an extensional basin. This leads to a third model for the origin of the SWCC as Early Tertiary rift basin deposition. A terrane once juxtaposed against western Oregon during the Early Tertiary was later displaced northward via rifting and dextral slip faults. As rifting became active, the Oregon Coast Range basement rocks formed as volcanics in this rift basin. Sedimentation in this basin may be recorded by the SWCC. Compression from the Farallon plate to the west may have closed this basin and subducted its eastern margin beneath North America. The terrane once west of Oregon was transported to Alaska by these events.

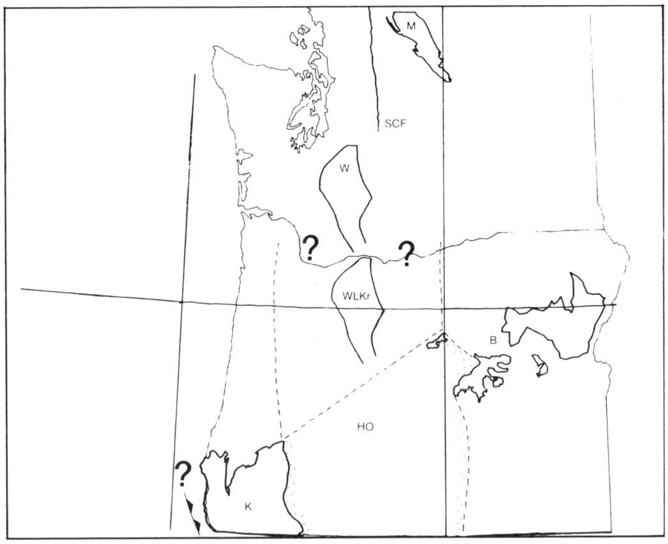


Figure 4. Late Cretaceous reconstruction for western Washington study area (model 4). Lambert projection with same notations used in Figure 3. New notations include: SCF = Straight Creek fault, WLKr = SWCC in Late Cretaceous, and HO = Hornbrook-Ochoco basin (east and west margins marked by stippled pattern). Position of Hornbrook-Ochoco basin from Nilsen (1984). Using a pole of rotation (35.4° N, 158.7° W) determined by Price and Carmichael (1986) for the Tintina Trench-Northern Rocky Mountain Trench and Fraser River-Straight Creek fault zones, the SWCC package is rotated 3° clockwise to restore displacement along the Straight Creek fault and move the SWCC package into its Late Cretaceous position.

In a fourth reconstruction, we investigate the possible roles of strike-slip and/or transform fault displacements in the Pacific Northwest (Davis, Monger, and Burchfiel, 1978; Johnson, 1984; Price and Carmichael, 1986; Stanley, Finn, and Plesha, 1986), and their displacement histories relative to the SWCC (Figure 4).

Reported right-lateral displacement along these faults varies. Values range from 450 km since the mid-Cretaceous (Price and Carmichael, 1986), to 700 km or >900 km since pre-mid Cretaceous (Gabrielse, 1985) for the Tintina trench-Northern Rocky Mountain trench. For the Straight Creek-Fraser River fault, postulated displacements range from 80 to 110 km since early Late Cretaceous (Price and Carmichael, 1986) to almost 200 km (Davis, Monger, and Burchfiel, 1978).

Price and Carmichael (1986) speculate that the Straight Creek-Fraser River fault in northern Washington-British Columbia was part of a large transform fault system in the Late Cretaceous-Paleogene. This system lies immediately northeast of the SWCC (Figure 4). These workers suggest that the Tintina trench-Northern Rocky Mountain trench fault zone and the Straight Creek-Fraser River fault zone are concentric, small circle fault segments that comprised an intercontinental transform fault. They calculate a pole of rotation for the faults from a loci of points located on them. Other workers have suggested that these faults represent transform faults which record motion between North America and the Kula or Farallon plate (Davis, Monger, and Burchfiel, 1978; Monger and Price, 1979).

In the fourth model, it is assumed that deposition in the SWCC package is, in part, pre-Straight Creek fault displacement. Using the pole of rotation calculated by Price and Carmichael (35.4° N, 158.7° W; 1986) for the Tintina trench-Northern Rocky Mountain trench and Straight Creek-Fraser River fault systems, combined with the speculative right-lateral displacement of 190 km for the Straight Creek fault (Davis, Monger, and Burchfiel, 1978), we compute an angle of rotation of 3° for the Straight Creek fault displacement and restored the SWCC to its pre-Late Cretaceous position, by rotating it clockwise about the specified pole of rotation (Figure 4).

Currently, we do not favor one of these models over the others; however, the fourth model leads to additional avenues of investigation. Are Late Mesozoic rocks in eastern and central Oregon related to the SWCC's depositional history? In its Late Cretaceous position, the SWCC is juxtaposed against Late Cretaceous sediments which crop out in the Mitchell inlier in the Blue Mountains. These sediments have also been sampled by drilling to the south (Nilsen, 1984).

Recent workers have speculated on the existence of the Hornbrook-Ochoco Late Cretaceous forearc basin (Nilsen, 1984; Alldredge and Meigs, 1984), which is postulated to extend northward from the Great Valley into central Oregon. The existence of this basin is based upon the Late Cretaceous Hornbrook Formation (1.3 km thick), which crops out east of the Klamath Mountains, and Late Cretaceous strata in the subsurface south of and cropping out in the Blue Mountains (4.3 and 2.7 km thick, respectively) (Nilsen, 1984; Dickinson and Thayer, 1978). The SWCC may be a northward extension of this Late Cretaceous basin which was later disrupted by the Straight Creek-Fraser River fault.

An alternative to this model is that the SWCC represents a northward extension of the Hornbrook-Ochoco basin, but it was not displaced by the Straight Creek fault. In this case, its present position represents the original site of deposition.

## Hydrocarbon Potential

The possibility of obtaining a physical sample of the low-resistivity rocks is remote at this time. However, hypothesized equivalent rocks can be sampled, and the data may be extrapolated to the deep rocks. In the fourth reconstruction model discussed, we speculated that the Hornbrook-Ochoco basin sediments may be equivalent to the SWCC sediments. We sampled the Late Cretaceous rocks in the Mitchell inlier (Hudspeth and Gable Creek Formations) of the Blue Mountains to complete source rock and thermal maturation studies. Jurassic units (Trowbridge and Lonesome Formations) were also sampled in the John Day inlier of the Blue Mountains.

Source rock studies using thermal hydrocarbon analyses of the samples are being completed with a ROCK-EVAL analyzer. Preliminary data are available only for the Jurassic samples, but data for the Cretaceous units have recently become available from previous workers (Sidle and Richers, 1985). Mean Cretaceous pyrolysis yields $S_1$, $S_2$, and TOC (total organic carbon) values (Table 1) suggest that these rocks would serve as fair to good source rocks. Furthermore, kerogen typing techniques indicate that Type II (oil/gas prone) and III (gas prone) kerogen is prevalent. The Tmax mean value indicates that these rocks are mature with respect to the oil window, and entering the gas window. Data from the Jurassic samples show lower mean $S_1$, $S_2$, and TOC values falling within the poor to good source rock range. Tmax value suggest the rocks are mature with respect to the oil window. If the extrapolation of these data to the SWCC rocks is valid, it suggests these rocks may have had hydrocarbon generative potential in their past.

Table 1. Pyrolysis source rock data from Jurassic and Cretaceous units in the Blue Mountains.[1]

Jurassic Sample Pyrolysis

| Number of Samples | $T_{max}$ (°C) | Mean: mgram/gram $S_1$ | $S_2$ | $S_3$ | TOC wt% | HI | OI |
|---|---|---|---|---|---|---|---|
| 18 | 450 | .07 | .48 | .57 | .88 | 54 | 70 |

Cretaceous Sample Pyrolysis (Sidle and Richers, 1985)

| Number of Samples | $T_{max}$ (°C) | Mean: mgram/gram $S_1$ | $S_2$ | $S_3$ | TOC wt% | HI | OI |
|---|---|---|---|---|---|---|---|
| 16 | 462 | 1.97 | 4.05 | 1.01 | 1.85 | Type II and III Kerogen | |

[1] Nomenclature: $T_{max}$ = Maximum temperature at which maximum generation of hydrocarbons from kerogen occurs; $S_1$ = mgram hydrocarbon/gram of rock; $S_2$ = mgram hydrocarbon/gram of rock; $S_3$ = mgram $CO_2$/gram of rock; TOC= total organic carbon; HO = hydrogen index; and OI = oxygen index.

Recent industry activity has occurred in northeast California in the Hornbrook basin. As of 1984, more than 40 oil and gas companies have acquired over 1.6 million acres of land based upon the speculation that the Late Cretaceous-Tertiary sediments of the Hornbrook basin have sourced and reservoired hydrocarbons (Alldredge and Meigs, 1984). This provides an encouraging note if our speculations that these sediments are equivalent to the SWCC basin rocks are valid.

## METHOW BASIN

Deep, low-resistivity rocks have also been postulated to underlie the Methow basin to depths of 25 km (Figure 2; Stanley and Plesha, 1985). The Methow basin is a fault-bounded basin complex composed of 15 to 20 km of marine and nonmarine clastics and volcanics (Barksdale, 1975; Tennyson and Cole, 1978; Trexler, 1984; Trexler and Bourgeois, 1985). The predominantly Jurassic and Cretaceous exposed section (Figure 5) is highly folded and faulted. The Methow basin has been interpreted as a forearc basin (Davis, Monger, and Burchfiel, 1978) or successor basin (Trexler, 1984), which may be either autochthonous (Davis, Monger, and Burchfiel, 1978) or allochthonous (Granirer, Burmester, and Beck, 1986).

The deep low-resistivity rocks have been interpreted as most likely representing sedimentary rocks, and to be most similar to the lowest black shale and volcaniclastic Jurassic units which crop out in the basin (Stanley and Plesha, 1985). If the exposed section is representative of the deep rocks, then analogies may be drawn between the surface and deep rocks. Black shales from the exposed Jurassic and Cretaceous sections were sampled for source rock and thermal maturation studies (Figure 5). Reconnaissance paleomagnetic studies were conducted with the red beds of the Midnight Peak Formation to provide constraints on the tectonic history and emplacement mechanism of the deep rocks (Figure 5).

### Paleomagnetism and Tectonic Reconstruction Models

Red beds from three sites in the Midnight Peak Formation in the Methow basin were subjected to thermal and alternating

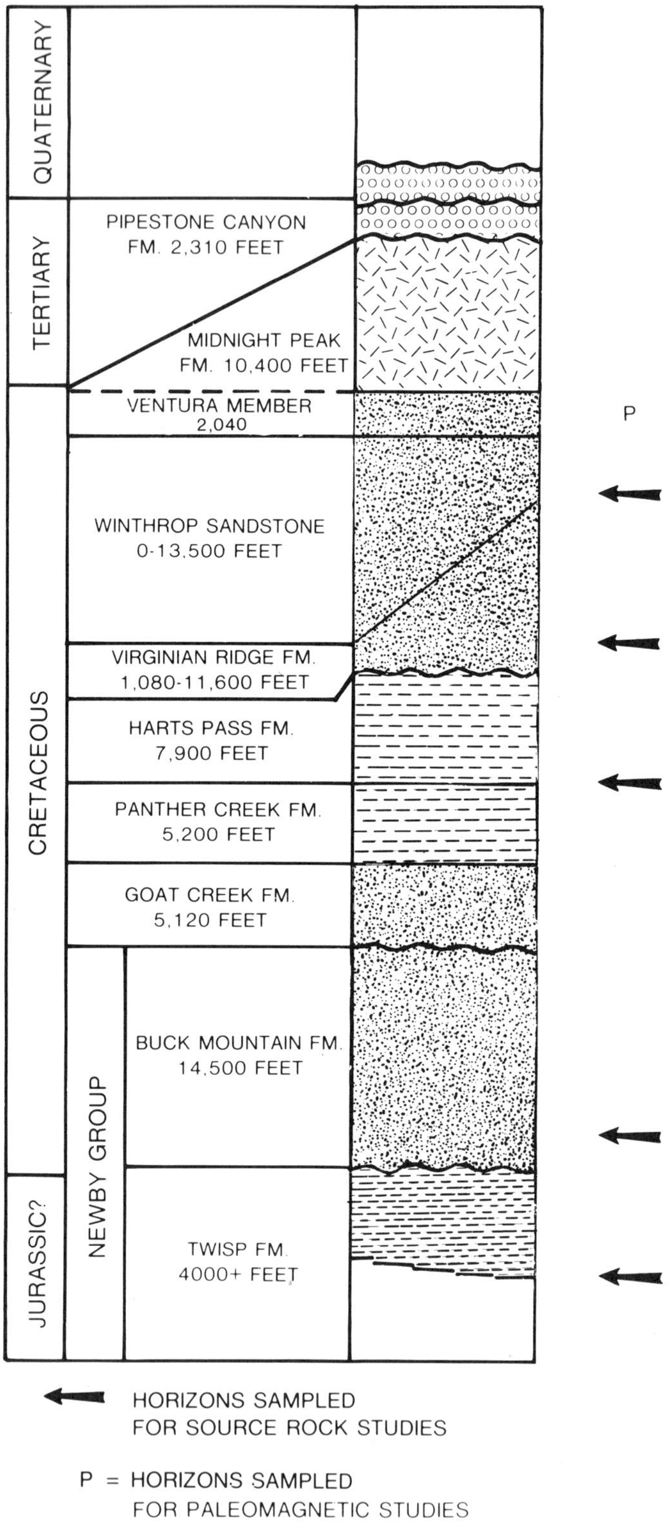


Figure 5. Stratigraphy and horizons sampled in the Methow basin. Stratigraphy from Barksdale (1975).

field demagnetization. Field directions were inspected both uncorrected and corrected for bedding attitudes. Although opposite limbs of a fold were not sampled for a fold test, bedding attitudes differed significantly at the sites; these data comprise a modified fold test. Field directions were in closest agreement after correction for folding. Field directions were determined by fitting a line (not constrained to pass through the origin) through points which formed a line segment on the Zijderveld diagram (Zijderveld, 1967), and by the minimum dispersion criteria. A mean field direction determined from the three sites (I = 72.8°, D = 24.5°; alpha-95 = 7.6°, k = 28.1; number of samples = 8; number of specimens = 14) has an inclination which is in good agreement with expected field directions as determined from Irving's (1979) Late Cretaceous North American pole. This mean field direction determined from reconnaissance data suggests that the basin has not been translated significantly northward since Late Cretaceous with respect to North America, but has been rotated clockwise 49° about a vertical axis within or near the basin. However, error associated with the mean field direction must be considered.

Paleomagnetic data which has recently become available from both the Midnight Peak and Winthrop Formations in the Methow basin have been interpreted differently (Granirer, Burmester, and Beck, 1986). These workers sampled opposite limbs of a plunging fold and obtained a mean direction (I = 61.1°, D = 12.0°, alpha-95 = 4.8°, k = 105, number of sites = 10) correcting for only 46% of bedding tilt. They suggest that (1) these rocks were remagnetized during folding, (2) the basin is allochthonous, and (3) the basin may have been transported as much as 1400 km poleward, in addition to being rotated 48° clockwise. Correcting for 100% of bedding gave essentially the same mean direction with more dispersion of field directions.

More work needs to be completed before the basin can be unequivocally shown to be autochthonous or allochthonous. In any event, both data sets suggest that the basin has rotated clockwise significantly. What role has this rotation played in the history of the basin, and is it related to the emplacement of the deep rocks?

If the basin is autochthonous, then the Methow sediments accumulated west of the Okanogan complex. An allochthonous interpretation is that the Methow sediments accumulated east of the North Cascade subcontinent (Cascadia), which was somewhere in the Pacific; these traveled together as a package until they accreted to North America. Stanley (personal communication, 1985) has suggested that emplacement of the deep Methow rocks may have been a direct result of the collisional event provided by the arrival of Cascadia.

The arrival of Cascadia during the Mesozoic has been postulated to be either a result of transform faulting (Davis, Monger, and Burchfiel, 1978) or convergence (Hamilton, 1978). In either scenario, the Methow sediments may be either autochthonous and placed west of the Okanogan complex, or allochthonous and attached to the east margin of Cascadia. As Cascadia converged against North America, sedimentation may have been compressed and emplaced to great depths. If Cascadia arrived by dextral transform faulting, then the clockwise rotation or ball-bearing tectonics observed in the Methow may be a result of this faulting. The nature of the Methow boundary faults (Chewack-Pasayten and Ross Lake fault zone), additional paleomagnetic data, and fossil assemblages in the basin may provide constraints upon the tectonic evolution of this region.

## Hydrocarbon Potential

Pyrolysis results for both the Cretaceous and Jurassic rocks show low $S_1$, $S_2$, $S_3$, and TOC (total organic carbon) values (Table 2). Results from thermal maturation studies indicate a possible explanation for these low values (Figure 6). Mean

Table 2. Pyrolysis source rock data from Jurassic and Cretaceous units in the Methow Basin.[1]

Cretaceous Sample Pyrolysis

| Number of Samples | Mean mgram/gram $S_1$ | Mean mgram/gram $S_2$ | Mean mgram/gram $S_3$ | TOC wt% |
|---|---|---|---|---|
| 43 | .03 | .03 | .05 | .21 |

Jurassic Sample Pyrolysis

| Number of Samples | Mean mgram/gram $S_1$ | Mean mgram/gram $S_2$ | Mean mgram/gram $S_3$ | TOC wt% |
|---|---|---|---|---|
| 14 | 0 | 0 | .02 | .43 |

[1] Nomenclature: $T_{max}$ = Maximum temperature at which maximum generation of hydrocarbons from kerogen occurs; $S_1$ = mgram hydrocarbon/gram of rock; $S_2$ = mgram hydrocarbon/gram of rock; $S_3$ = mgram $CO_2$/gram of rock; and TOC = total organic carbon.

vitrinite reflection values for these formations fall within the gas window. Thermal alteration index values had a wider range, but studies show these lower values may be a result of kerogen bleaching due to surface weathering. Vitrinite reflectance values are probably more representative of thermal maturation of these rocks.

These preliminary thermal maturation data indicate that the rocks were subjected to extensive temperatures (and depths) for a substantial period of time to reach the gas window. Hydrocarbons may have been generated earlier in the basin's history; however, there are no shows in the area, suggesting this is doubtful. If these data from surface rocks can be extrapolated to the rocks at great depth, we can conclude that they also have questionable hydrocarbon potential and are thermally mature or overmature.

# DISCUSSION AND TECTONIC SYNTHESIS

Several models have been explored to explain the origin of the SWCC deep rocks (Table 3), and models for the exposed Methow section suggest that the underlying deep rocks had a forearc basin history. A viable, integrated reconstruction for the evolution of both deep basins mapped in the Pacific Northwest is derived by combining models explored in this paper. In a Late Cretaceous paleogeography, North America is positioned at its correct paleolatitude using an oblique Mercator projection and a Late Cretaceous paleomagnetic pole (Irving, 1979) as the pole of projection (Figure 7A).

Subduction during the Late Cretaceous occurred west of the Hornbrook-Ochoco basin, and extended northward through the Straight Creek fault area (Figure 7B). Forearc deposition represented by the Great Valley sequence (south of map area)

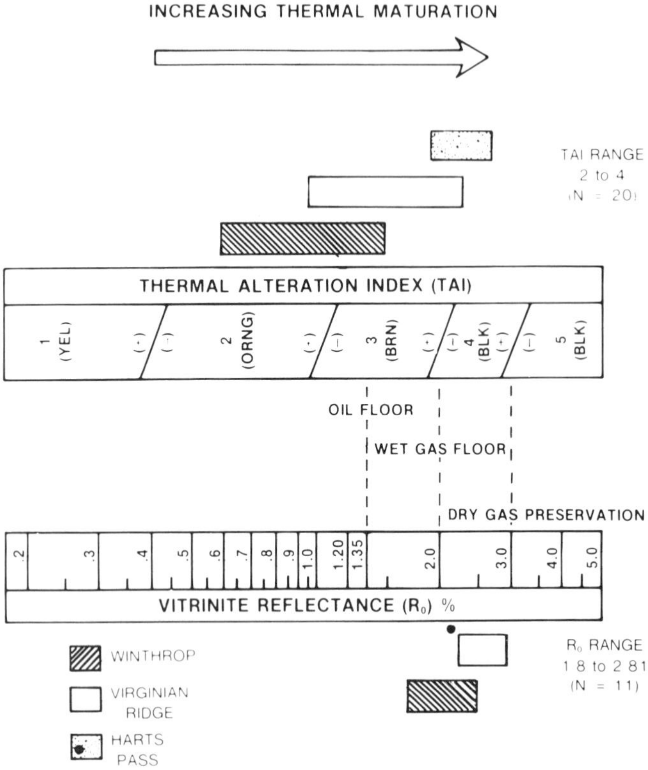


Figure 6. Thermal maturation data for the Methow basin. Zones of petroleum occurrence from Dow (1977).

Table 3. Summary of reconstruction models proposed for the SWCC.

| Model | Tectonic Setting | Tectonic Nature | Sources of Data |
|---|---|---|---|
| 1 | Tertiary forearc basin | Autochthonous | Magill and others, 1981 |
| 2 | Tertiary marginal basin | Allochthonous | Magill and others, 1981 |
| 3 | Tertiary rift basin | Autochthonous | Wells and others, 1984 |
| 4 | Mesozoic forearc basin | Allochthonous or Autochthonous | Price and Carmichael, 1986; Nilsen, 1984 |

and the proposed Hornbrook-Ochoco basin may document the position and timing of the subduction zone. Deposition in the Methow may also have been associated with this subduction. A comparison of ages of marine Cretaceous strata in the Methow, Ochoco, and Hornbrook basins (Figure 8) suggests that subduction may have been occurring as early as Aptian-Albian. We have restored the Straight Creek fault displacement (190 km), and position the SWCC package as the northern extension of the Hornbrook-Ochoco basin (Figure 7B). Then, perhaps because of oblique subduction, the slip component becomes dominant over convergence, and dextral displacement along the Straight Creek fault occurs. The SWCC package is transported northward during Late Cretaceous-Early Tertiary (Figure 7C). Alternatively, the SWCC was not displaced by the Straight Creek fault, but was autochthonous, Late Mesozoic deposition equivalent to the Hornbrook-Ochoco basin which extended further to the north.

The evolution of these two deep basins also involves the origin of Cascadia. The question of whether Cascadia arrived by convergence or transform faulting must not be neglected in models accepted for this area. Earlier in the Jurassic and Cretaceous, the trench may have been east of the Straight Creek fault, between the Methow and the North Cascades. Then, as Cascadia accreted by mid-Cretaceous, subduction jumped to the west to become roughly coincident with the present Straight Creek fault. The clockwise rotation apparently observed in the Methow basin (Figure 7C) may have been due to dextral movement along the Straight Creek fault or an unidentified fault associated with the arrival of Cascadia.

During the Early Tertiary, the basement rocks of the Oregon Coast Range may have recorded volcanism in a seamount complex (Duncan, 1982). As suggested by the model of Magill, Cox, and Duncan (1981), due to reorganization of plates, fragmentation of the Farallon plate may have occurred with subduction commencing to the south and east of the Oregon Coast Range basement rocks (Figure 7D). Rotation of the trapped oceanic plate followed (Figure 7D). A northern extension of the Oregon Coast Range seamount complex may have compressed the accumulation of sediments in the SWCC basin. Stanley, Finn, and Plesha (1986) speculate that a relatively electrically resistant package west of the SWCC area, which also produces gravity and magnetic highs, may represent Siletzia (Figure 2). The Oregon Coast Range jockeyed into its final position by Basin and Range extension (Magill, Cox, and Duncan, 1981). An alternative to this Tertiary history may have been the development of an extensional basin which is recorded by the Oregon Coast Range basement rocks (Wells et al., 1984). Subsequently, compression and subduction of the SWCC rocks occurred because of convergence.

This synthesis was an exercise in exploring a viable scenario for the tectonic history of the deep rocks of interest. We currently have no preference for any of these models discussed, and other models should be investigated in the future.

## CONCLUSIONS

Petroleum occurrences in the Pacific Northwest are not uncommon; however, the only commercially producing field is the Mist gas field in northwest Oregon. Ultradeep basins, such as those we are investigating in the Western Cordillera, have not previously been identified nor considered in hydrocarbon exploration. The sediments in these basins may have generated methane in the past. We have speculated about the hydrocarbon potential and tectonic history of deep, low-resistivity packages mapped by magnetotellurics in the Pacific Northwest. A major assumption has been that the deep, low-resistivity packages most likely record marine sedimentary deposition. Other possibilities for the conductive rocks are outlined in Stanley, Finn, and Plesha (1986). Deep drilling, such as that proposed by DOSECC (Deep Observation and Sampling of the Earth's Continental Crust), is necessary in order to confirm existing models.

The interpretation of these magnetotelluric anomalies as deep sedimentary basins leads to a new perspective on the tectonic history of the Pacific Northwest. Tectonic reconstructions suggest that the SWCC and Methow deep packages may be either allochthonous or autochthonous. Analogies drawn between the exposed rocks in the Methow basin and the deep, inaccessible rocks suggest that the deep rocks currently have marginal source rock potential and are, at minimum, thermally mature.

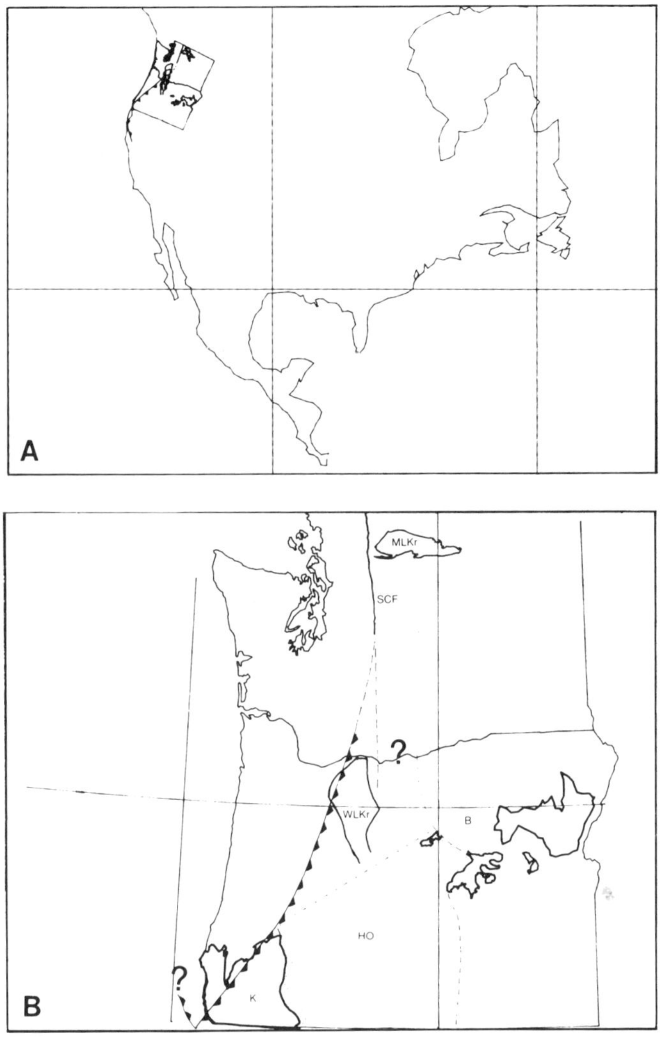


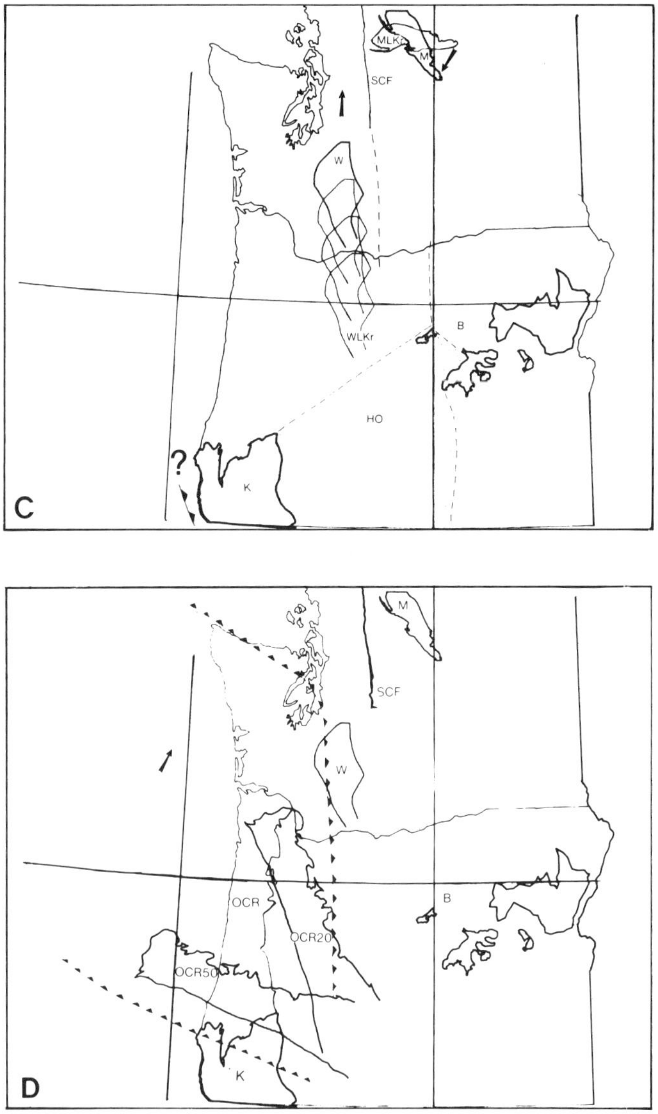


Figure 7. Late Cretaceous through present-day reconstructions for the areas of interest in the Pacific Northwest. Refer to Figures 3 and 4 for key elements digitized. Figures 7B through 7D are Lambert projections.
A. Oblique Mercator projection map for Late Cretaceous using Irving's (1979) Late Cretaceous paleomagnetic pole as map projection pole.
B. Late Cretaceous reconstruction in which displacement on Straight Creek fault has been restored. SWCC is in its Late Cretaceous position as the northern extension of the Hornbrook-Ochoco forearc basin. The Methow basin has been rotated counterclockwise, 49° about a pole of rotation in the basin to restore it to its Late Cretaceous position. Notation MLKr = Methow in Late Cretaceous.
C. Late Cretaceous-Early Tertiary reconstruction during which displacement along the Straight Creek fault occurs, moving SWCC northward to its present position. Methow basin is rotated clockwise in response to dextral, ball-bearing tectonics.
D. Early Tertiary to present-day reconstruction. The Oregon Coast Range, which has formed out in the Pacific, experienced two phases of rotations at 50 and 20 Ma as suggested by Magill, Cox, and Duncan (1981). The accretion of a northern segment to North American may have compressed the SWCC sediments.

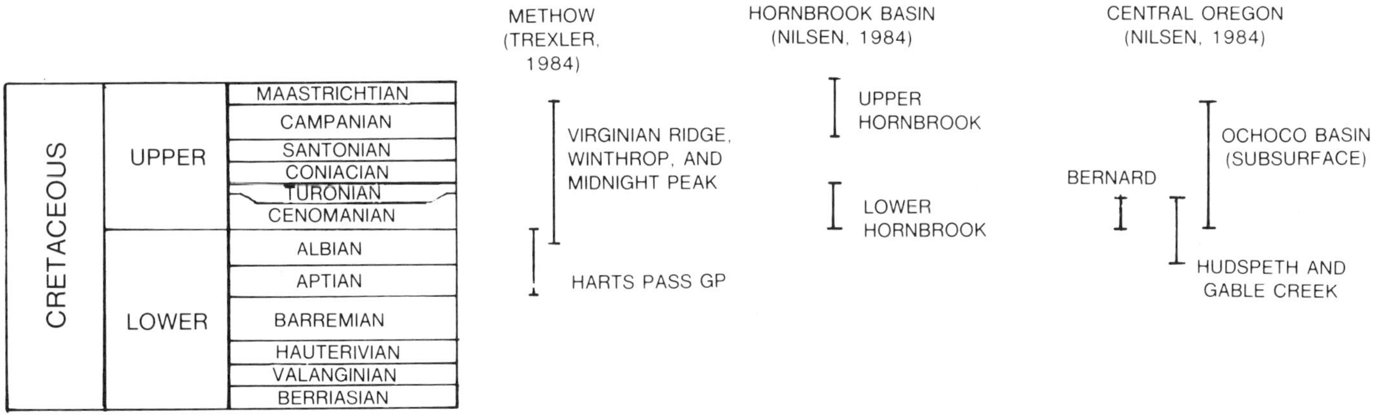


Figure 8. Chronostratigraphy of key Cretaceous formations of interest to this study.

The SWCC basin may be depositionally related to Late Cretaceous sediments which comprise the Hornbrook-Ochoco basin farther to the south in Oregon and California. This analogy suggests that the SWCC sediments once had good to fair hydrocarbon potential, as determined by source rock studies in the Blue Mountains. Additionally, the rocks are, at a minimum, thermally mature with respect to the oil window. Recent industry activity in the Hornbrook basin in northeastern California (Alldredge and Meigs, 1984) makes this an especially provocative scenario.

## ACKNOWLEDGMENTS

W.D. Stanley's magnetotelluric data in the Pacific Northwest provided the framework for our investigations in this area. Models explored in this paper grew out of discussions with him. We thank J. Reichelderfer for providing some of the geochemical analyses for this study. V. Schmidt and the University of Pittsburgh's Paleomagnetism Laboratory aided in the paleomagnetic analyses. We also are grateful to P.D. Snavely for discussions of the geology of the Pacific Northwest.

## REFERENCES CITED

Alldredge, M.H., and J.V. Meigs, 1984, N.E. California area drawing interest: Oil and Gas Journal, v. 82, p. 83-87.

Barksdale, J.D., 1975, Geology of the Methow Valley, Okanogan County, Washington: State of Washington, Division of Geology and Earth Resources Bulletin 68, 72 p.

Davis, G.A., J.W.H. Monger, and B. C. Burchfiel, 1978, Mesozoic construction of the Cordilleran "collage," central British Columbia to central California, *in* D.G. Howell and K.A. McDougall, eds., Mesozoic paleogeography of the western United States: SEPM, p. 1-32.

Dickinson, W.R., and T.P. Thayer, 1978, Paleogeographic and paleotectonic implications of Mesozoic stratigraphy and structure in the John Day inlier of central Oregon, *in* D.G. Howell and K.A. McDougall, eds., Mesozoic paleogeography of the western United States: SEPM, p. 147-162.

Dow, W. G., 1977, Kerogen studies and geological interpretations: Journal of Geochemical Exploration, v. 7, p. 79-99.

Duncan, R.A., 1982, A captured island chain in the Coast Range of Oregon and Washington: Journal of Geophysical Research, p. 10827-10837.

Gabrielse, H., 1985, Major dextral displacements along the northern Rocky Mountain Trench and related lineaments in north-central British Columbia: GSA Bulletin, v. 96, p. 1-14.

Granirer, J.L., R.F. Burmester, and M.E. Beck, Jr., 1986, Cretaceous paleomagnetism of the Methow-Pasayten belt, Washington: Geophysical Research Letters, v. 13, p. 733-736.

Gwilliam, W.J., and K.K. Cohen, 1987, Deep source gas potential along the west coast of the U.S.: this volume.

Hamilton, W., 1978, Mesozoic tectonics of the western United States, *in* D.G. Howell and K.A. McDougall, eds., Mesozoic Paleogeography of the western United States: SEPM, p. 33-70.

Irving, E., 1979, Paleopoles and paleolatitudes of North America and speculations about displaced terrains: Canadian Journal of Earth Science, v. 16, p. 669-694.

Johnson, S.Y., 1984, Evidence for a margin-truncating transcurrent fault (pre-late Eocene) in western Washington: Geology, v. 12, p. 538-541.

Magill, J., A. Cox, and R. Duncan, 1981, Tillamook Volcanic Series: Further evidence for tectonic rotation of the Oregon Coast Range: Journal of Geophysical Research, v. 86, p. 2953-2970.

Monger, J.W.H., and R.A. Price, 1979, Geodynamic evolution of the Canadian Cordillera—Progress and problems: Canadian Journal of Earth Science, v. 16, p. 770-791.

Nilsen, T.H., 1984, Tectonics and sedimentation of the upper Cretaceous Hornbrook Formation, Oregon and California, *in* J.K. Crouch and S.B. Bachman, eds., Tectonics and sedimentation along the California margin: SEPM, Pacific Section, v. 38, p. 101-118.

Phillips, W.M., 1983, Preliminary interpretation of regional gravity information from the southern Cascade Mountains of Washington, *in* M.A. Korosec, W.M. Phillips, and J.E. Schuster, eds., The 1980-1982 Geothermal Resource Assessment Program in Washington: State of Washington, Division of Geology and Earth Resources Open-File Report 83-7, 297 p.

Price, R.A., and D.M. Carmichael, 1986, Geometric test for late Cretaceous-Paleogene intracontinental transform faulting in the Canadian Cordillera: Geology, v. 14, p. 468-471.

Sidle, W.C., and D.M. Richers, 1985, Geochemical reconnaissance of Cretaceous inliers in north-central Oregon: AAPG Bulletin, v. 69, p. 412-421.

Snavely, P.D., Jr., N.S. MacLeod, and W.W. Rau, 1970, Summary of the Tillamook area, northern Oregon Coast Range: U.S. Geological Survey Professional Paper, 650-A, 47 p.

Stanley, W.D., 1984, Tectonic study of Cascade Range and Columbia Plateau in Washington State based upon magnetotelluric soundings: Journal of Geophysical Research, v. 89, p. 4447-4460.

Stanley, W.D., and J.L. Plesha, 1985, Progress Report on U.S. Geological Survey — Department of Energy Interagency Agreement DE-A121-83MC204, Task No. 4, Electromagnetic geophysics applied to sediment subduction and deep source gas: U.S. Geological Survey Open-File Report 85-252, 20 p.

Stanley, W.D., C. Finn, and J.L. Plesha, 1986, Tectonics and conductivity structures in the southern Washington Cascades: Journal of Geophysical Research, in review.

Tennyson, M.E., and M.R. Cole, 1978, Tectonic significance of upper Mesozoic Methow-Pasayten sequence, northeastern Cascade Range, Washington and British Columbia, *in* D.G. Howell and K.A. McDougall, eds., Mesozoic paleogeography of the western United States: SEPM, p. 499-508.

Trexler, J.H., Jr., 1984, Stratigraphy, sedimentology, and tectonic significance of the Upper Cretaceous Virginian Ridge Formation, Methow basin, Washington: Implications for tectonic history of the North Cascades: Seattle, Washington, University of Washington, unpublished Ph.D. dissertation, 173 p.

Trexler, J.H., Jr., and ? Bourgeois, 1985, Evidence for mid-Cretaceous wrench faulting in the Methow basin, Washington: Tectonostratigraphic setting of the Virginian Ridge Formation: Tectonics, v. 4, p. 379-394.

Weaver, C.S., and S.W. Smith, 1983, Regional tectonic and earthquake hazard implications of a crustal fault zone in southwestern Washington: Journal of Geophysical Research, v. 88, p. 10371-10383.

Wells, R.E., D.C. Engebretson, P.D. Snavely, and R.S. Coe, 1984, Cenozoic plate motions and the volcano-tectonic evolution of western Oregon and Washington: Tectonics, v. 3, p. 275-294.

Zijderveld, J.D.A., 1967, A.C. Demagnetization of rocks: Analysis of results, *in* D.W. Collinson, K.M. Creer, and S.K. Runcorn, eds., Methods in paleomagnetism, developments in solid earth geophysics, v. 3: Amsterdam, Elsevier, p. 254-286.

# CHAPTER 24

# NATURAL GAS HYDRATES OF THE OFFSHORE CIRCUM-PACIFIC MARGIN—A FUTURE ENERGY RESOURCE?

**Keith A. Kvenvolden and Alan K. Cooper**
*U.S. Geological Survey*
*Menlo Park, CA*

ABSTRACT

Natural gas hydrates are probably present within the uppermost 1100 m (3600 ft) of oceanic sediment in the following regions of outer continental margins rimming the Pacific Basin: (1) the landward slope of the Middle America Trench off Central America, (2) the continental slope off the western conterminous Untied States, (3) the landward slope of the western and eastern Aleutian Trench, (4) the continental slope of the Beringian and Koryak margins, (5) the landward slope of the Nankai Trough off Japan, (6) the Timor Trough northwest of Australia, (7) the continental slope east of the North Island of New Zealand, (8) the basinal sediment of the Ross Sea and the continental margin off Wilkes Land, Antarctica, and (9) the landward slope of the Peru-Chile Trench. In some places these gas hydrates may contain and cap significant quantities of methane. Geophysical evidence for gas hydrates rests mainly on the widespread occurrence on marine seismic records of an anomalous reflection event that apparently marks the base of the zone of gas hydrate. Geochemical evidence consists of analyses of gases and interstitial fluids obtained from drilling in offshore sedimentary deposits, particularly at eight Deep Sea Drilling Project (DSDP) sites cored adjacent to the Middle America Trench where gas hydrates were recovered.

Natural gas hydrates will likely be identified in many other Circum-Pacific regions as exploration for offshore petroleum moves into deeper waters over continental and island-arc slopes. Certainly future drilling in these regions will need to consider the geological and engineering consequences of gas hydrates. Gas hydrates are currently not considered economic hydrocarbon resources. Further testing, however, of gas hydrates and exploration for associated reservoirs may demonstrate that gas hydrates can be an important energy resource for the future.

## INTRODUCTION

Outer continental margins that rim the Pacific Basin may contain vast quantities of natural gas trapped in or beneath crystalline substances known as gas hydrates. These substances, which occupy the open spaces of sediment, are made up of rigid frameworks of water molecules in which natural gas is trapped and effectively condensed. Gas hydrates form under specific pressure-temperature conditions that are met in oceanic sediment at depths up to about 1100 m below the sea floor and in water depths greater than 300 to 500 m (reviewed by Kvenvolden and McMenamin, 1980). The exact location of gas hydrates in oceanic sediment is further influenced by gas and water chemistries, temperature and pressure gradients, and lithology.

The appropriate pressure-temperature conditions under which gas hydrates can theoretically form in a pure-methane and pure-water system is shown in Figure 1; however, gas in continental-margin sediments is not pure methane, and pore waters commonly contain some salt. Impurities in the methane include small concentrations of ethane and hydrocarbon gases of higher molecular weight, as well as $CO_2$ (McIver, 1974; Claypool, Presley, and Kaplan, 1973). The additional components in the gas mixture shift the phase boundary to the right in Figure 1, whereas impurities in the water such as salts shift the boundary to the left. Thus, the resulting shifts in the equilibrium phase boundary tend to cancel each other, and unless the gas composition and salinity are known, a pure-methane and pure-water system provides the best first approximation for establishing the pressure-temperature relationship of naturally occurring gas hydrates (Claypool and Kaplan, 1974). According to the phase diagram (Figure 1), gas hydrates are possible where pressures are greater than about 30 atmospheres corresponding to temperatures greater than 0 °C. In sediments, subsurface temperatures, which are dependent on geothermal gradients, and lithostatic and hydrostatic pressures control the depth range in which the gas hydrates can exist. The base of the gas hydrate zone coincides with a

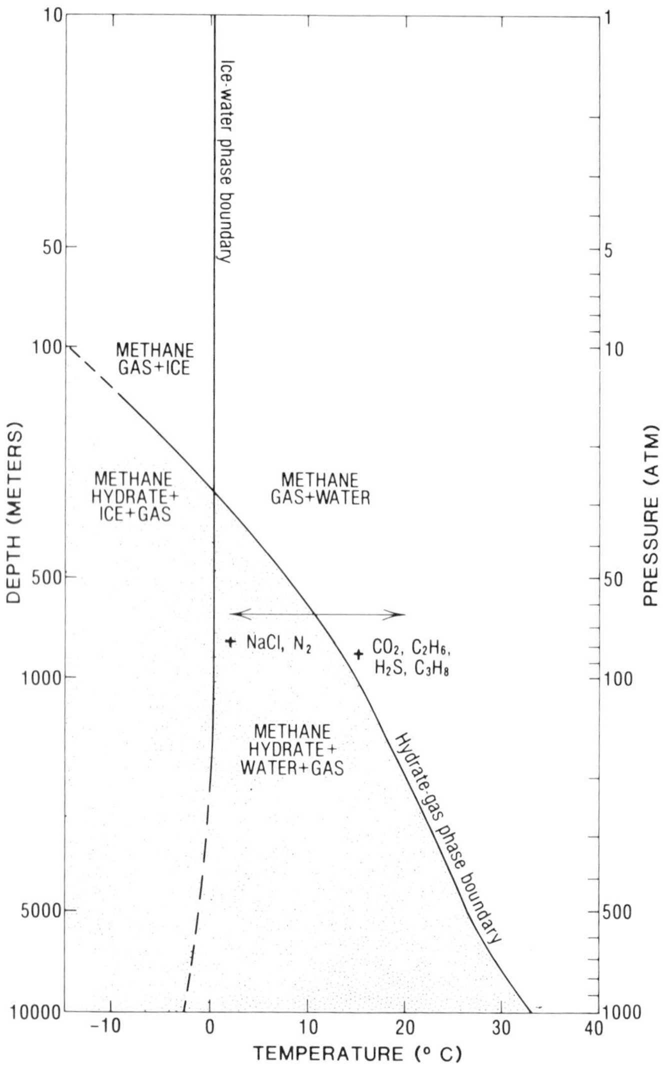


Figure 1. Phase diagram showing boundary between free methane gas (no pattern) and methane hydrate (pattern) for a pure-water and pure-methane system. Pressure is converted to depth assuming a lithostatic-hydrostatic pressure gradient of 0.1 atm/m. The diagram is useful for predicting depths and temperatures at which gas hydrates should occur (modified from Kvenvolden and McMenamin, 1980).

pressure-temperature boundary that represents the maximum depth at which the gas hydrate is stable.

An additional requirement for natural gas hydrate formation is that large volumes of methane be present to initiate and stabilize the gas-hydrate structure. Up to about 164 volumes of methane can be present for each equivalent volume of water (Davidson et al., 1978). Only under special geological circumstances involving high rates of formation and entrapment can methane approaching this large amount be found. The source of methane for gas-hydrate formation may be either biogenic or thermogenic. Biogenic methane results from the microbial breakdown of organic matter common in shallow sediment; thermogenic methane comes from the thermal alteration of organic matter in sediment at depth. In either case organic matter must be present. A large amount of organic matter can be expected to accumulate in continental-margin sediment, but it is not common in distal sediment of the ocean basins; therefore, gas hydrates are likely to be confined to continental slope and rise deposits underlying the margins of ocean basin.

The identification of gas hydrates in oceanic sediment has been based mainly on the appearance of an anomalous acoustic reflector that approximately parallels the sea floor on marine seismic records (Shipley et al., 1979). This bottom-simulating reflector, or BSR, is thought to result from a velocity contrast between the overlying, higher velocity sediment containing gas hydrates and the underlying, lower velocity sediment possibly containing free gas. This velocity contrast commonly produces large reflection coefficients and reversed polarities relative to the sea-floor reflector. The depth of this reflector generally correlates with the theoretical pressure-temperature-dependent depth for the base of the gas-hydrate zone. Thus, besides approximately paralleling the sea floor, the gas-hydrate-caused BSR should also increase in sub-bottom depth with increasing water depth (i.e., with increasing pressure). In addition, the depth to the BSR can, in theory, be used to estimate geothermal gradients within gas hydrate zones because of the pressure-temperature dependency of gas-hydrate occurrence. BSRs attributed to gas hydrates have been observed on marine seismic-reflection records from at least 19 locations along continental margins of the world (Kvenvolden and Barnard, 1983).

In the Circum-Pacific region, BSRs, inferred to be related to gas hydrates, were first noted on marine seismic-reflection profiles recorded along the Pacific coast of Central America from Panama to Acapulco, Mexico (Shipley et al., 1979). The existence of gas hydrates in these areas was later confirmed at eight sites by Deep Sea Drilling Project (DSDP) coring off Mexico (Shipley and Didyk, 1982), Guatemala (Harrison and Curiale, 1982; Kvenvolden and McDonald, 1985), and Costa Rica (Kvenvolden and McDonald, 1985). The Central America margin is the only continental margin of the Pacific Basin where gas hydrates have actually been sampled. Elsewhere in the Circum-Pacific region, the presence of gas hydrates has been inferred mainly on the basis of BSRs. This paper reviews the reported occurrences of gas hydrates in continental-margin sediments bordering the entire Pacific Basin (Figure 2) and attempts to assess their importance as a potential natural gas resource.

## OFFSHORE GAS HYDRATE OCCURRENCES

### Central America (Middle America Trench)

Shipley et al. (1979) noted that marine seismic-reflection records across the continental slopes of Panama, Costa Rica, Nicaragua, Guatemala, and Mexico commonly,, but not always, have BSRs, which they attributed to gas hydrates. Figure 3 illustrates one well-developed BSR on a marine seismic-reflection profile recorded over the landward slope of the Middle America Trench offshore from Mexico. This BSR behaves as predicted for a reflector from the base of gas hydrates in that it approximately parallels the sea floor, and the subbottom depth of the reflector increases with increasing water depth. Table 1 and Figure 4 summarize information on the seismically inferred gas hydrates of the Circum-Pacific margin including occurrences offshore from Central America. Where gas-hydrate-related BSRs have been observed, estimates of geothermal gradients have been made (Shipley et al., 1979; Yamano et al., 1982). Along the Middle America Trench,

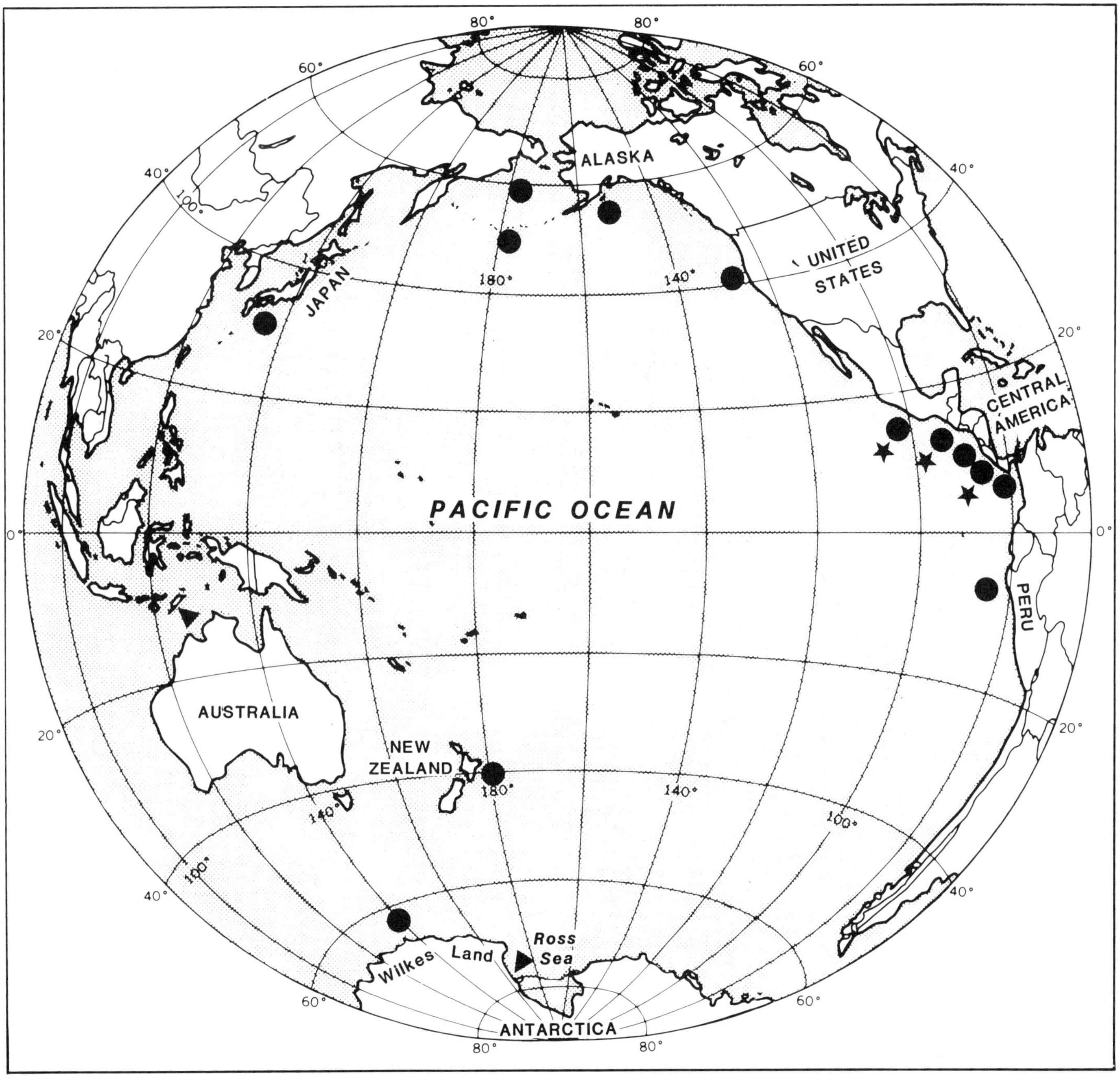


Figure 2. Location map of inferred and known gas hydrate occurrences in the Circum-Pacific region. Areas with gas hydrate inferred from BSRs are noted by dots; sites where gas hydrates have been recovered are shown by stars; and locations having only geochemical evidence for gas hydrates, such as unusually high gas concentrations, are marked by triangles.

the average geothermal gradients range from 22 °C/km off from Mexico to 66 °C/km off Panama (Table 1).

In 1979, coring on three DSDP legs recovered gas hydrates off Central America. Leg 66 drilled the landward slope of the Middle America Trench in an area near Mexico (Figure 5) where BSRs are common (Watkins et al., 1982). Solid gas hydrates were recovered from three holes (Shipley and Didyk, 1982); these occurred as icelike inclusions or, more commonly, as icelike substances occupying pore space and cementing volcanic ash and fine sand (Table 2). A large amount of methane was released during decomposition of the icelike material, demonstrating that these substances were gas hydrates.

On Leg 67 further drilling was done into the landward slope of the Middle America Trench near Guatemala (Figure 5) along a transect where no BSRs were originally observed (Aubouin et al., 1982); later reprocessing of the seismic records did, however, reveal a weak BSR there (von Huene, Ladd, and Norton, 1982). On the other hand, seismic line about 10 nautical miles (18 km) west of the drill sites showed a well-defined BSR (Ladd et al., 1982). At Site 497 a piece of solid gas hydrate was recovered from a core catcher near the bottom of the hole. Gas hydrates cementing vitric sands were recovered near the bottom of a hole at Site 498 (Table 2).

In 1982 DSDP Leg 84 again drilled into the landward slope of the Middle America Trench (Figure 5) at five sites (566

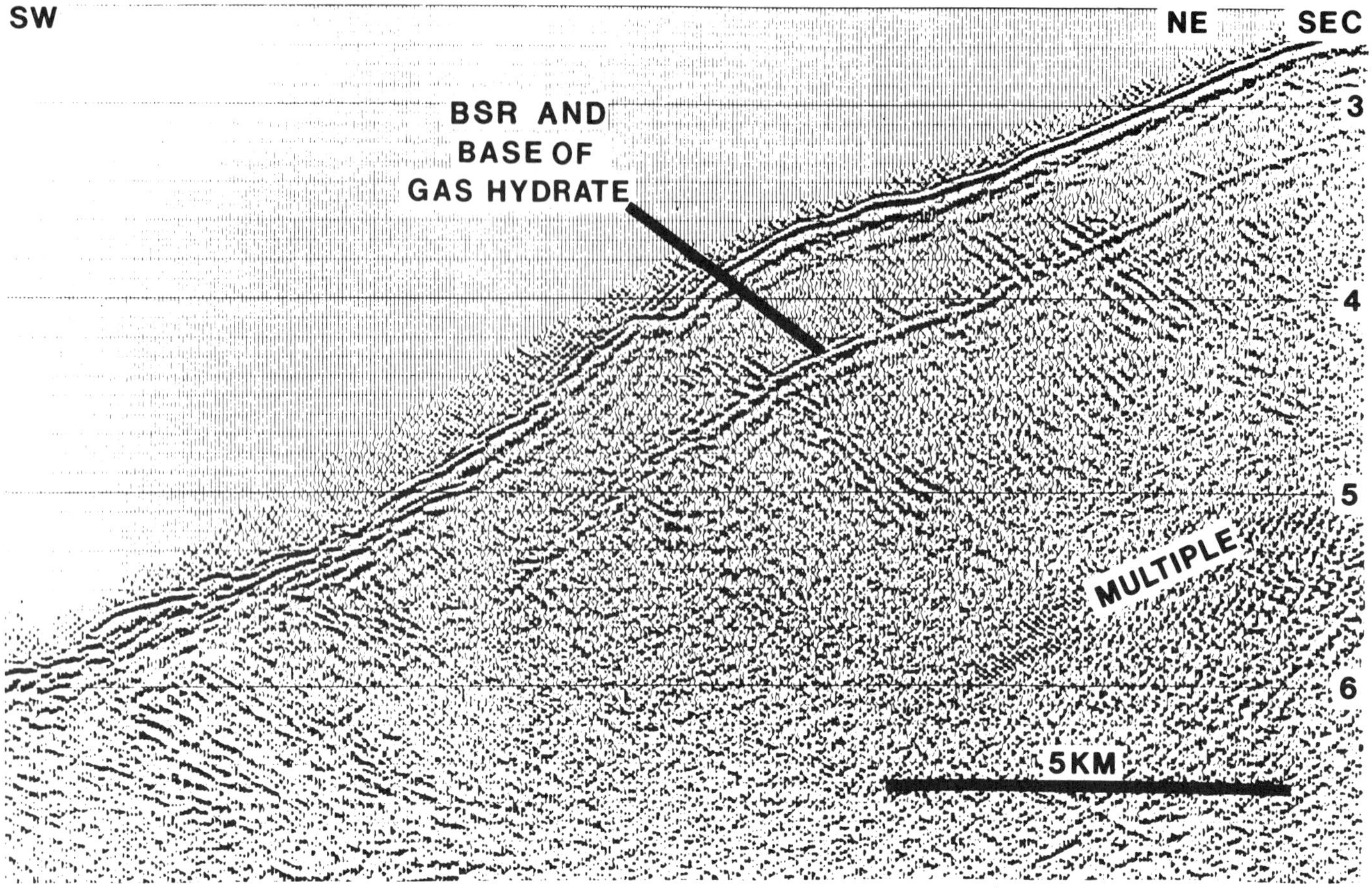


Figure 3. A 24-fold multichannel seismic-reflection profile of the landward slope of the Middle America Trench offshore from Mexico. The reflector at the base of the inferred gas hydrate (BSR) approximately follows the topography of the sea floor and deepens with increasing water depth (from Shipley et al., 1979).

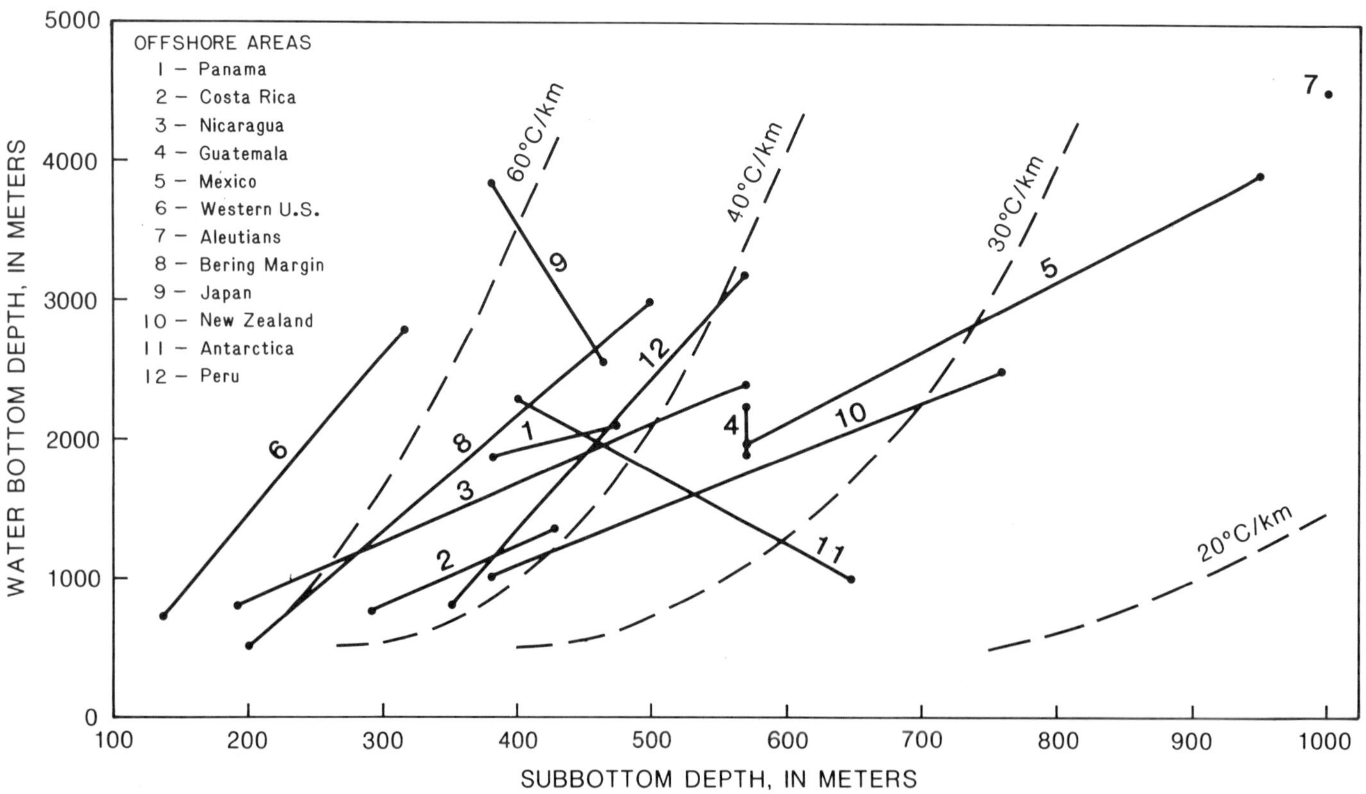

**Table 1. Seismically Inferred Gas Hydrates of the Circum-Pacific Margin**

| Location[1] | Water Depth(m) | Subbottom Depth of BSR(m) | Geothermal Gradient(°C/km)[2] min | max | ave | References |
|---|---|---|---|---|---|---|
| Central America | | | | | | |
| Panama | 1,875-2,100[3] | 380-475[4] | 66 | 82 | 77 | Shipley et al. (1979) Yamano et al. (1982) |
| Costa Rica | 750-1,350[3] | 290-430[4] | 30 | 44 | 35 | Shipley et al. (1979) Yamano et al. (1982) |
| Nicaragua | 800-2,400 | 190-570[4] | 30 | 35 | 33 | Shipley et al. (1979) Yamano et al. (1982) |
| Guatemala | 1,875-2,250[3] | 570[4] | 27 | 32 | 30 | von Huene, Ladd, and Norton (1982) |
| Mexico | 1,950-3,900[3] | 570-950[4] | 22 | 31 | 28 | Shipley et al. (1979) Yamano et al. (1982) |
| Western U.S. | 725-2,800 | 135-315 | 55 | 66 | 55 | Field and Kvenvolden (1985) |
| South Alaska | 500-2,300<br>1,600-2,300 | 300-800<br>570-670 | 31 | 24 | 30 | MacLeod (1982)<br>Kvenvolden and von Huene (1985) |
| Aleutian Islands | 4,500 | 1,000 | 23 | 21 | 22 | McCarthy et al. (1984) |
| Bering Sea Margin | 500-3,000 | 200-500 | Not determined | | | Cooper, Scholl, and Marlow (1986) |
| Japan | 2,535-3,850 | 465-380 | 42 | 66 | 54 | Yamano et al. (1982) |
| New Zealand | 1,000-2,500 | 380-760[4] | 20 (by inference) | | | Katz (1982) |
| Antarctica | 1,000-2,300 | 650-400[5] | 32 | 47 | 39 | Kvenvolden, Golan-Bac and Rapp (in press) |
| Peru | 800-3,200 | 350-570[4] | Not determined | | | Shepard (1979) |

1 Locations refer to offshore areas adjacent to the named places.

2 Geothermal gradients were calculated by the referenced authors using water depths and subbottom depths of BSRs.

3 Estimated using a water veolocity of 1,500 m/sec.

4 Estimated using a sediment velocity of 1,900 m/sec.

5 Estimated using a sediment velocity of 1,730 m/sec.

through 570) near Guatemala and at one site (565) off Costa Rica (von Huene et al., 1985). A weak BSR occurs on seismic records at Site 568 (von Huene, Ladd, and Norton, 1982). Gas hydrates recovered at three sites (565, 568, and 570) appeared as solid pieces of white, icelike material occupying fractures in mudstone; as rapidly outgassing, coarse-grained sediment, and as a core of massive gas hydrate (Figure 6 and Table 2).

Organic geochemical studies, particularly on samples recovered on Leg 84, have concluded that the methane in the gas hydrates is biogenic based on the carbon isotopic relationships and molecular compositions of the natural gas mixtures (Kvenvolden et al., 1984; Kvenvolden and McDonald, 1985; Claypool et al., 1985). Inorganic geochemical studies of pore waters in sediment associated with gas hydrates on Leg 67 showed that the salinity and chloride contents of these pore waters decrease with increasing depths (Hesse and Harrison, 1981; Harrison, Hesse, and Gieskes, 1982); these observations were further documented on Leg 84 (Kvenvolden and McDonald, 1985; Hesse, Level, and Gieskes, 1985). Gas hydrates decompose during drilling, thereby releasing fresh water (and methane) into the surrounding sediment and thus decreasing the salinity of the sediment pore water.

◄ Figure 4. Graph showing water-bottom depths and subbottom depths to BSRs for regions of the Circum-Pacific margin. Data are from Table 1. Lines connect maximum and minimum values from each area. Note that subbottom depths of BSRs normally increase with increasing water depths, except for areas off Japan and Antarctica, where geothermal gradients are believed to change laterally and increase with increasing water depths. Dashed curves represent geothermal gradients calculated from Figure 1 and based on assumed bottom-water temperature of 2° C.

## Western Conterminous United States

A well-defined BSR, inferred on the basis of its depth and pervasiveness to be associated with gas hydrates, has been

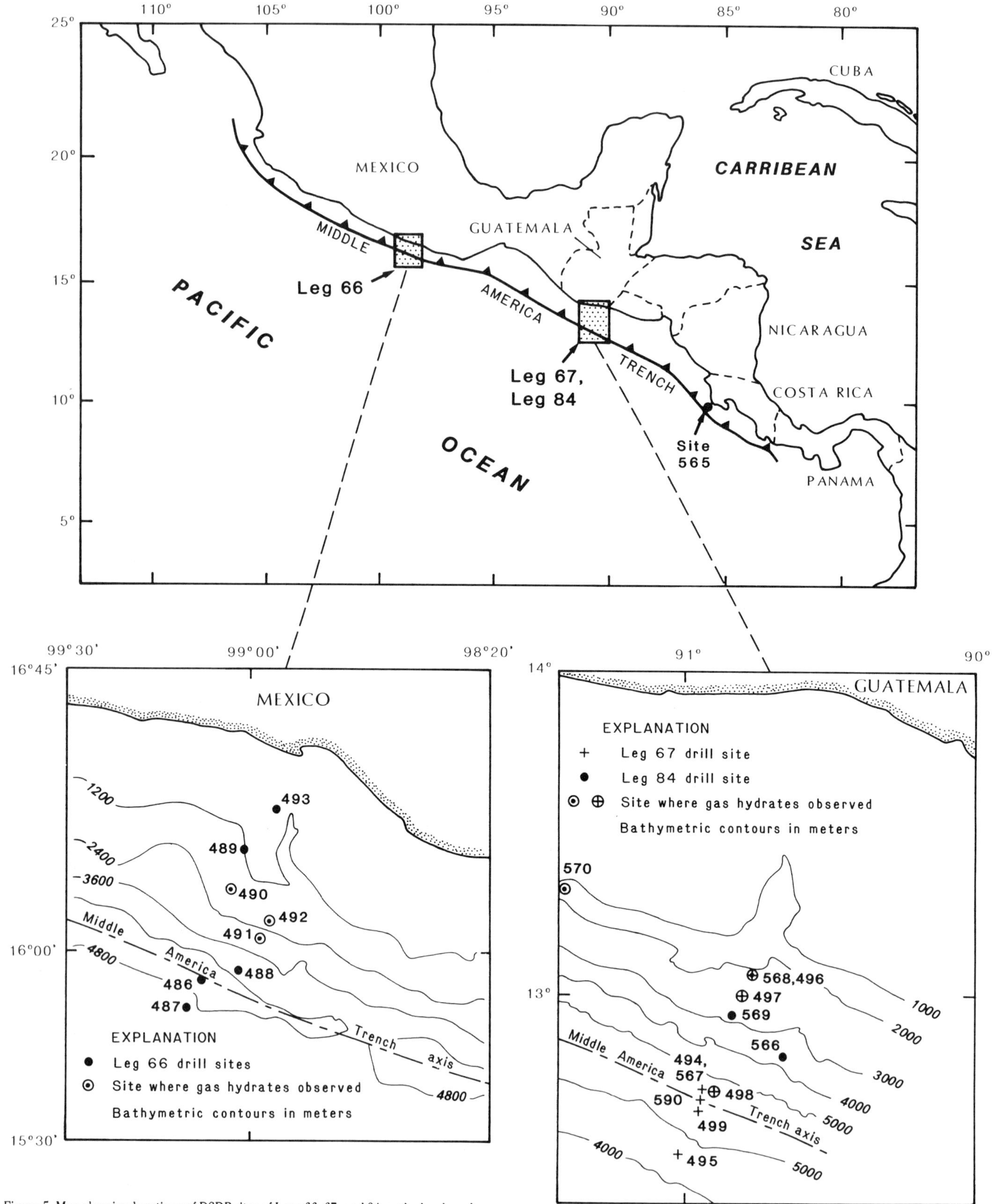


Figure 5. Map showing locations of DSDP sites of Legs 66, 67, and 84 on the landward slope of the Middle America Trench where gas hydrates were recovered during drilling. Legs 67 and 84 were drilled in the same region except for Site 565, which was drilled on Leg 84 offshore from Costa Rica.

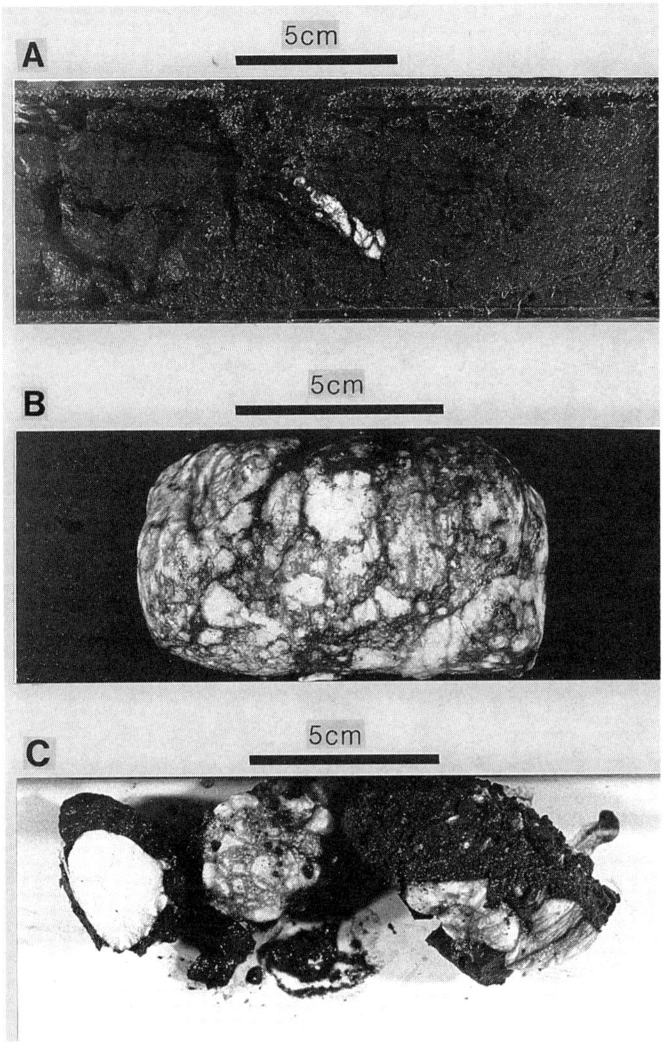


Figure 6. Photographs of examples of gas hydrates recovered on DSDP Leg 84 (scale in centimeters): (A) Site 570, 246 m subbottom; (B) Site 570, 249 m subbottom; (C) Site 568, 404 m subbottom (Table 2).

Table 2. Gas Hydrates recovered by DSDP off Central America

| Leg | Site | Water Depth(m) | Subbottom Depth(m) | Description |
|---|---|---|---|---|
| 66 | 490 | 1761 | 140 | Inclusion in mud |
| 66 | 490 | 1761 | 146 | Cementing ash |
| 66 | 490 | 1761 | 167 | Cementing ash |
| 66 | 490 | 1761 | 365 | Cementing ash |
| 66 | 491 | 2883 | 89 | Inclusion in mud |
| 66 | 491 | 2883 | 163 | Cementing sand |
| 66 | 491 | 2883 | 168 | Inclusion in mud |
| 66 | 492 | 1935 | 141 | Cementing ash |
| 66 | 492 | 1935 | 170 | Cementing ash |
| 67 | 497 | 2350 | 368 | Inclusion found in core catcher |
| 67 | 498 | 5480 | 310 | Cementing vitric sand |
| 84 | 565 | 3396 | 285 | Inclusions in mud |
| 84 | 565 | 3430 | 319 | Inclusion and associated with sand |
| 84 | 568 | 2434 | 404 | Inclusion in fractured mudstone |
| 84 | 570 | 1910 | 192 | Cementing laminated ash |
| 84 | 570 | 1965 | 246 | Inclusion in fractured mudstone |
| 84 | 570 | 1968 | 249 | Massive gas hydrate 1.05 cm long |
| 84 | 570 | 1982 | 264 | Inclusion in fractured mudstone |
| 84 | 570 | 1991 | 273 | Inclusion in fractured mudstone |
| 84 | 570 | 2021 | 303 | Cementing ash |
| 84 | 570 | 2055 | 338 | Cementing sand |

identified on marine seismic-reflection profiles recorded from the continental margin of northern California (Figure 2) by Field, Clarke, and White (1980), Biddle and Seely, (1983), and Field and Kvenvolden (1985). The BSR lies at subsurface depths of 135 to 315 m beneath the Klamath Plateau off northernmost California and probably southernmost Oregon in water depths of about 725 to at least 2800 (Table 1). The BSR deepens with increasing water depth (Figure 4), and the gas hydrate apparently covers an area of at least $3000^2$. The geothermal gradient in the area, as calculated from BSRs, averages about 55 °C/km (Table 1). This acoustically inferred gas hydrate is the only reported occurrence in the continental margin of the western conterminous United States.

## South Alaska and the Aleutian Islands (Aleutian Trench)

In the Gulf of Alaska (Figure 7), BSRs have been observed on seismic-reflection profiles across the upper continental slope bordering the Aleutian Trench southeast of Kodiak Island (Kvenvolden and von Huene, 1985). In this area, BSRs were observed at subbottom depths of 570 to 670 m in water depths of 1600 to 2300 m (Table 1). MacLeod (1982) also identified BSRs in sediments of the Gulf of Alaska, but he did not give specific locations; he showed that subbottom depths of the BSRs ranged from about 300 to 800 m for water depths of about 500 to 2300 m. In this area the average geothermal gradient is calculated to be about 30 °C/km (Table 1).

Evidence for gas hydrates has been noted in sediments of the Aleutian Terrace, which borders the Pacific side of the Aleutian Ridge (Scholl, Vallier, and Stevenson, 1983; McCarthy et al., 1984). Seismic-reflection profiles crossing the southern flank of the Atka Basin, south of Amlia Island, commonly show BSRs at subbottom depths of about 1000 m in water about 4500 m deep; the geothermal gradient here is estimated to be about 22 °C/km (Table 1). DSDP drilling at Site 186 near the Atka Basin encountered very high concentrations of methane, but not gas hydrates (Creager, Scholl, et al., 1973). The high methane concentrations may result from the decomposition of gas hydrates during drilling, and the BSRs may mark the base of the zone of gas hydrate occurrence. On Figure 7, we have extrapolated the area of possible occurrence of gas hydrates along the south flank of the Aleutian Ridge based on the similarity of geologic and tectonic framework between the areas where the BSRs have been identified.

## Beringian and Koryak Margins

BSRs have been observed on marine seismic-reflection profiles recorded in the Bering Sea on the continental slope and

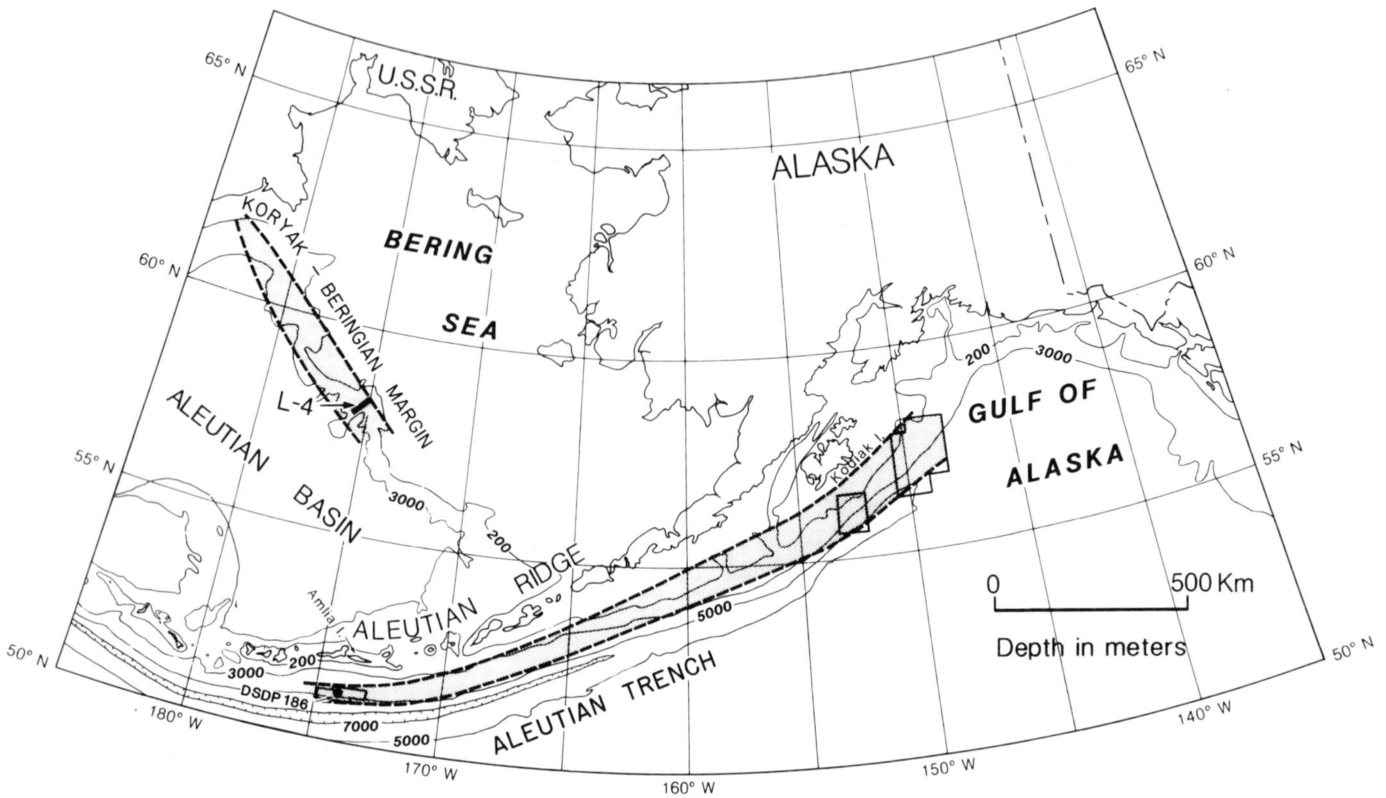


Figure 7. Map showing the locations of gas hydrates off Alaska, as inferred from the occurrence of BSRs on marine seismic-reflection profiles. The enclosed areas south of Kodiak Island and Amlia Island represent grids of seismic-reflection lines, some of which show BSRs.

rise of the Beringian and Koryak Margins (Marlow et al., 1981). In this area, BSRs have ben related to two sources—gas hydrates and diagenetic alteration of siliceous sediments. The acronym BSR was coined by Scholl and Creager (1973) for anomalous acoustic reflectors caused by the diagenesis of siliceous sediment, not by gas hydrates. More recent work, however, has identified two BSRs which often occur simultaneously (summarized in Cooper, Scholl, and Marlow, 1986). For example, in Figure 8 the more shallow of the two BSRs (BSR-GH) generally lies at 0.3 to 0.6 s subsurface and marks the base of the zone of gas hydrates. The deeper BSR (BSR-DB) is typically found at 0.9 to 1.2 s subbottom and is thought to follow the silica-diagenetic boundary between overlying and partially unconsolidated opal-A-rich sediment and underlying more indurated mudstone rich in opal-CT. The characteristics of these BSRs as documented by Cooper, Scholl, and Marlow (1986) differ significantly (Table 3).

The shallow BSR (BSR-GH) is of most importance to our discussion of gas hydrates, but knowledge of both types of BSRs is necessary to prevent misinterpretation of seismic records. The BSRs associated with gas hydrates of the Beringian and Koryak Margins are typically weak reflectors that are most easily observed beneath eroded knolls and areas of deformed strata underlying the continental slope and rise at water depths of 500 to 3000 m.

Figure 7 we have extrapolated the area of possible occurrence of seismically inferred gas hydrates along part of the Beringian and Koryak Margins based on these marine seismic data.

## Japan (Nankai Trough)

On the western side of the Pacific Basin (Figure 2), gas hydrates are believed to occur in sediments of the landward slope of the Nankai Trough near the Japanese islands of Kyushu and Shikoku (Aoki, Tamano, and Kato, 1983). Marine seismic-reflection records show a strong BSR, averaging about 0.5 s below the sea bottom, which is thought to correspond with the base of the zone of gas hydrates (Table 1). The BSRs show negative reflection polarities, as would be expected for gas hydrates; however, the BSRs do not increase in subbottom depth with increasing water depth, as theoretically required by pressure-temperature constraints on gas hydrate occurrence (Figure 4). Instead, the reflectors become more shallow in the deeper water possibly as a result of increasing heat flow and, consequently, increasing geothermal gradients down slope toward the Nankai Trough (Yamano et al., 1982).

## Australia (Timor Trough)

McKirdy and Cook (1980) suggested that gas hydrates may account for the higher than average concentrations of methane in sediment at DSDP Site 262, located on the axis of the Timor Trough northeast of Australia in water 2315 m deep (Figure 2). They based their conclusions on consideration of unusual relationships between alkalinity and salinity profiles and the high amounts of methane they observed. Pore-water

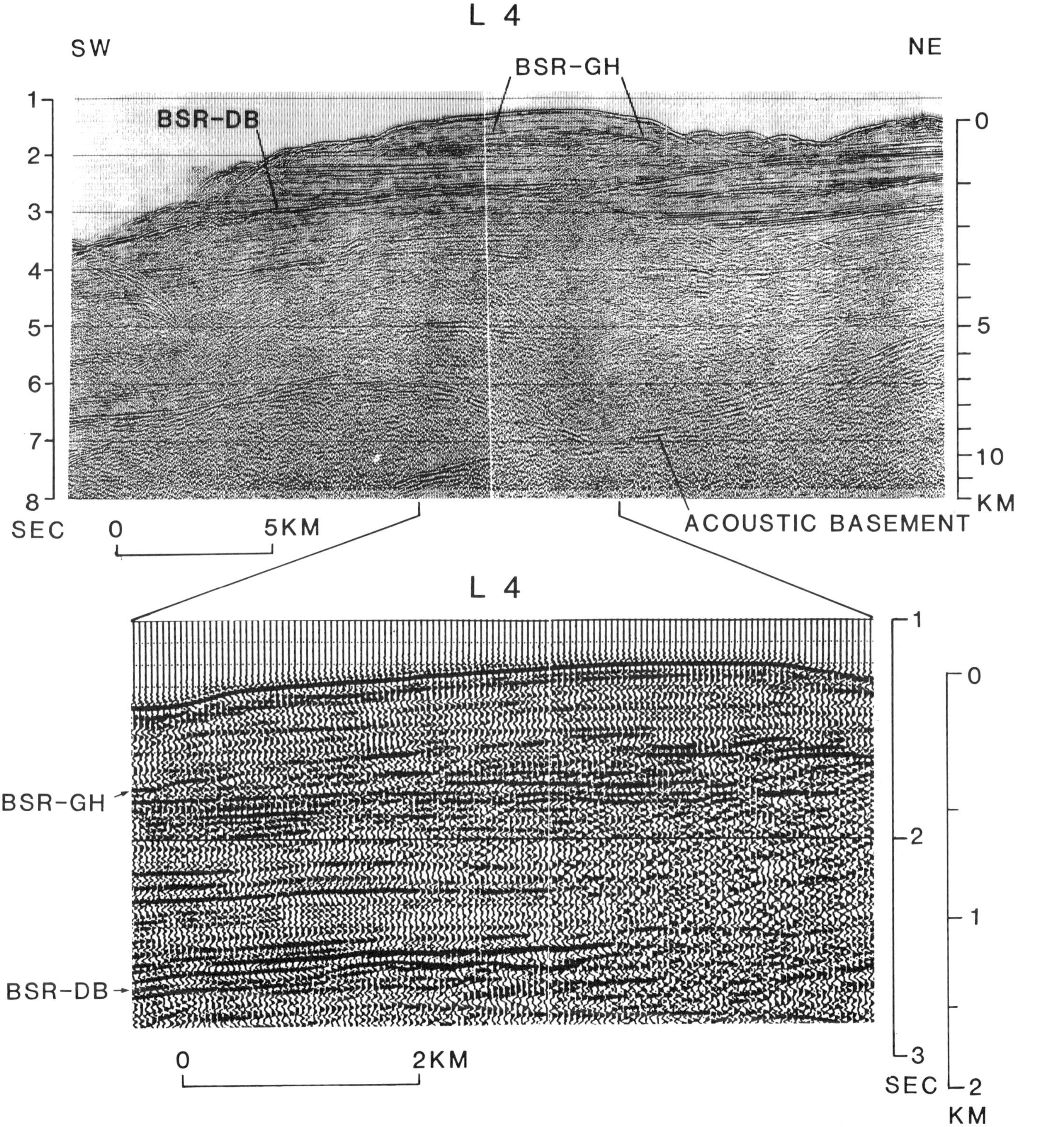


Figure 8. Seismic-reflection profile L4 across the Beringian Margin continental slope (Figure 7). The profile shows two BSRs, one at the base of an inferred gas hydrate (BSR-GH) and one at a diagenetic boundary (BSR-DB) (Cooper, Scholl, and Marlow, 1986).

salinity at Site 262 increases significantly with increasing depth, an effect attributed to an influx of highly saline water, from underlying or nearby evaporite deposits. This influx of saline water would mask any pore water freshening due to the presence of gas hydrates. Thus, the evidence for gas hydrate at this site is equivocal.

## New Zealand (East Coast Fold Belt)

Southeast of North Island, New Zealand (Figure 2), BSRs occur on a number of marine seismic-reflection profiles across the lower continental slope where water depths range from 1000 to 2500 m (Katz, 1981, 1982). The BSRs at subbottom depths between about 380 and 760 m cut across the bedding planes of the offshore portion of the East Coast Fold Belt (Table 1); seismic reflections are stronger in the core of anticlines and weaker or nonexistent in synclines. The BSRs suggest that gas hydrates occur within a belt 500 km long and several tens of kilometers wide, covering an area of 10,000 to 20,000 km$^2$. In contrast with the BSRs off Japan, the BSRs off New Zealand tend to increase in subbottom depth with increasing water depth (Figure 4). This is the expected behavior for a gas-hydrate-related BSR in an area with a uniform geothermal gradient, which is inferred by Katz (1982) to be about 20 °C/km (Table 1).

## Antarctica (Wilkes Land Margin and Ross Sea)

Gas hydrates have been inferred by Kvenvolden, Golan-Bac, and Rapp (in press) to be present in sediments offshore from Wilkes Land (Figure 2). A multichannel seismic-reflection survey (Eittreim et al., 1984) revealed a BSR (Figure 9) that had characteristics suggesting the base of a zone of gas hydrate: (1) the reflector is at a depth consistent with the pressure-temperature stability field of gas hydrate, and (2) the reflector shows a reversal of polarity. A third criterion that the subbottom depth of the reflector increases with increasing depth of water is not met (Figure 4), possibly because of ocean-ward increasing geothermal gradients as in the case of the inferred gas hydrate off Japan (Yamano et al., 1982).

Gas hydrates may exist beneath the Ross Sea continental shelf (Figure 2). In 1973 DSDP drilled four holes there, and hydrocarbon gases were detected at Sites 271, 272, and 273 (Hayes and Frakes, 1975). Shipboard observations of hydrocarbon gases were later augmented by the shore-based studies of McIver (1975), who suggested that the high concentrations of methane recovered at these sites may have been mobilized from gas hydrates. Studies of interstitial water for the DSDP Ross Sea sites showed that salinity values decrease with increasing depth (Mann and Gieskes, 1975). Initially this decrease was thought to indicate freshwater intrusion, but we now know that decreasing salinity of the sediment pore water can also be indicative of gas hydrates (Hesse and Harrison, 1981). We believe that the combination of decreased salinity and high concentrations of methane support McIver's original suggestion that gas hydrates exist in Ross Sea sediments. However, BSRs have not been reported in the extensive seismic-reflection data coverage of the Ross Sea continental shelf and margin (Hinz and Block, 1983; Cooper, Davey, and Behrendt, in press).

## Peru (Peru-Chile Trench)

Near Peru (Figure 2), BSRs inferred to indicate gas hydrates have been observed on most seismic-reflection profiles obtained over the landward slope of the Peru-Chile Trench (Shepard, 1979). Between about lat. 5° and 7 °S, these reflectors occur at subbottom depths between about 350 and 570 m, where water depths range from 800 to 3200 m (Table 1). The BSRs here become deeper with increasing water depth (Figure 4). Also at lat. 9 °S, von Huene, Kulm, and Miller (1985) pointed out a discontinuous BSR on another multichannel seismic-reflection record across the outer continental margin of central Peru.

Table 3. Characteristics of bottom simulating reflectors (BSRs) found for sediments of the Bering Sea

| | BSR-GH (Gas Hydrate) | BSR-DB (Diagenetic Boundary) |
|---|---|---|
| **I. General** | | |
| A. Description: | Reflection that marks the base of a gas hydrate zone - the transition from gas hydrate to free gas. | Reflector that marks a silica diagenetic boundary between overlying opal-rich (opal-A) sediment to underlying, indurated, cristobalite-bearing (opal-CT) mudstone. |
| B. Areal Distribution: | Presumed regional in extent, in water depths greater than 500 m. Identifiable most easily in areas with bathymetric relief and undeformed sediment. | Regional extent in deep-water (>200 m) areas where siliceous diatomaceous sediments are found or are presumed to be present. |
| C. Controlling factors: | Sufficient gas concentrations and suitable pressure-temperature conditions. | Appropriate conditions (pressure, temperature, geochemistry) for diagenesis of biogenic to crystalline silica. |
| **II. Geophysical Signature** | | |
| A. Geometry | Reflector that mimics seafloor bathymetry and commonly cuts across other layered sub-seafloor reflectors. | Same description as BSR-GH but found at greater sub-seafloor depths. |
| B. Sub-seafloor depth: | Depth to BSR-GH INCREASES with increasing water depth - usually found at 200-500 m (0.3-0.6 s) sub-seafloor. | Depth to BSR-DB DECREASES or is the same with increasing water depth - usually found 700-1000 m (0.9-1.2 s) sub-seafloor. |
| C. Sub-seafloor velocity: | Higher than typical sediment velocity ABOVE the BSR-GH. | Higher than typical sediment velocity BELOW the BSR-DB. |
| D. Phase of BSR reflector: | INVERTED (negative) phase. | NORMAL (positive) phase. |
| E. Amplitude of BSR reflector: | Highly variable, commonly faint in multi-channel seismic data. Hydrate zone above BSR-GH often associated with diffuse and faint reflectors. | Usually large amplitude and easily identifiable except where BSR-DB is conformable with other reflectors. |
| **III. Significance for Hydrocarbon Distribution** | | |
| A. General: | Gas hydrate zone may (1) contain significant included gas and (2) act as a seal that prevents upward migration of hydrocarbons. | Diagenetic boundary (1) has a major decrease in porosity and permeability beneath the boundary and (2) may influence migration path and hydrocarbon trapping. |
| B. Traps: | Hydrocarbons (liquid? and gas) may be trapped and accumulate beneath impermeable gas hydrate zone. | In areas of bathymetric relief, laterally migrating hydrocarbons may be trapped against diagenetic boundary in porous interbedded strata. |
| C. Migration of hydrocarbons: | Possible lateral migration of hydrocarbons along base of gas hydrate zone. | Possible migration beneath BSR-DB via fractures in highly indurated rocks or in unaltered strata originally poor in biogenic silica. |
| D. Operational hazards: | Pockets of overpressured gas trapped beneath gas hydrate zone. | BSR-DB may form a slip surface above which sediment bodies may slide. |

# RESOURCE IMPLICATIONS

This report shows the evidence for gas hydrates that probably exist beneath continental and island-arc margins around the Pacific Basin. As exploration for petroleum moves into deeper waters, and slope and rise sediments become targets for investigation, more gas hydrate occurrences are likely to be identified.

Although large volumes of gas can be included within gas

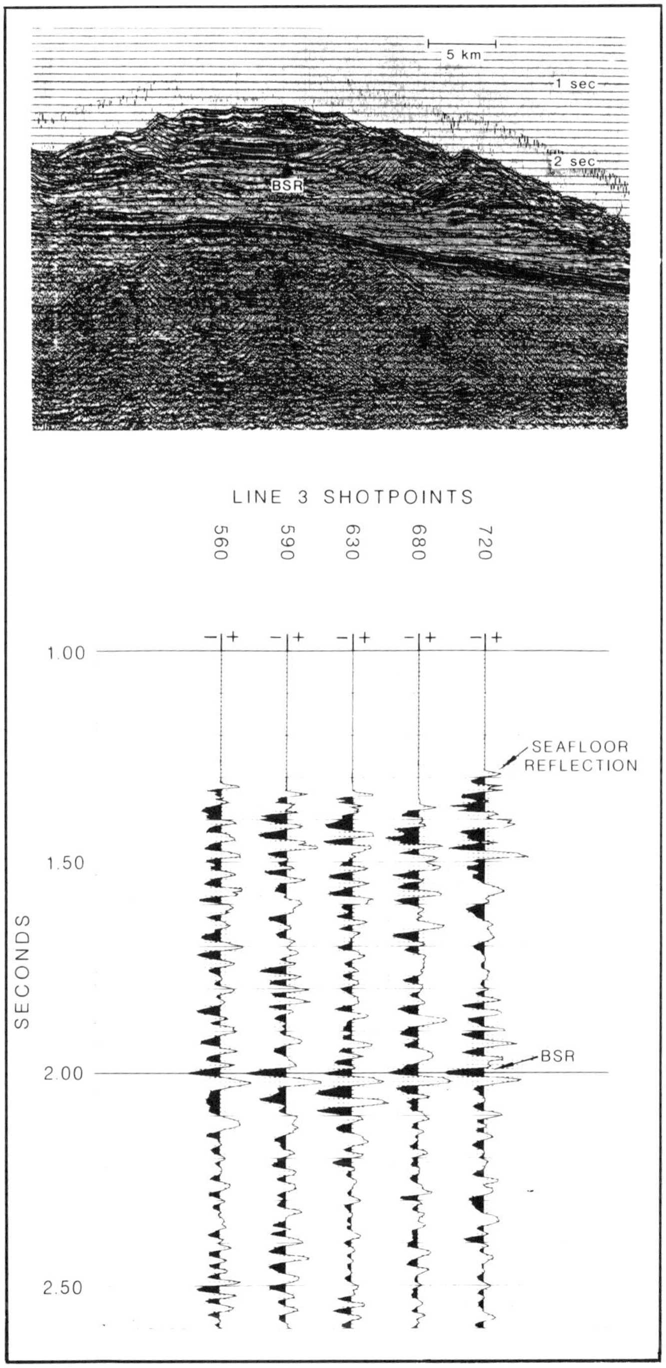


Figure 9. (A) Segment of multichannel seismic Line 3 from the Wilkes Land margin of Antarctica showing an anomalous acoustic reflector (BSR) which may mark the base of the zone of gas hydrates and (B) five single-trace seismic-reflection records from Line 3 showing reversal of reflection polarity at the BSR. Dark peaks are negative polarity. The sea-floor reflection and the BSR are of opposite polarity. (From Kvenvolden, Golan-Bac, and Rapp, in press).

hydrates, current technology cannot extract hydrated gas economically even if production wells are able to drill into the gas hydrates (Franklin, 1980). A main asset of marine gas hydrates is that they may form impermeable barriers or seals under which economic deposits of free natural gas and possibly oil can collect (Hedberg, 1980). At least three kinds of structural or stratigraphic traps are possible (Dillon, Grow, and Paull, 1980): (1) dome-shaped structures under closed topographic highs, where the base of the gas hydrate zone mimics the topography of the sea floor; (2) dome-shaped structures at the base of the gas hydrate, created by local increases in heat flow that are caused by local changes in thermal conductivity such as the occurrence of a salt dome; and (3) stratigraphic traps, where gas-containing beds dip toward the sea floor and are sealed at their updip ends by the gas-hydrate zone. Of course, porous and permeable reservoirs must be present in order for exploitable traps of natural gas to be formed in association with gas hydrates.

Any volumetric estimates for gas hydrates are highly speculative because of the incomplete knowledge of gas hydrate occurrences. Worldwide estimates for gas hydrates range from 110,000 to 270,000,000 trillion cubic feet (Tcf) or $3.1 \times 10^{15}$ to $7.6 \times 10^{18}$ $m^3$ in oceanic sediment (Potential Gas Committee, 1981). The upper limit of this estimate was made without regard for the distribution of sedimentary basins, the availability of natural gas, or the reservoir quality; the lower limit attempts to take these factors into account. If only 1% of gas-hydrate resources are within porous and permeable strata, then the most conservative estimate for recoverable gas from marine gas hydrates is still more than 1100 Tcf, which is half of the proved and currently recoverable conventional gas resources (Potential Gas Committee, 1981). These estimates do not consider possible accumulations of free gas beneath gas hydrate layers. Because the Circum-Pacific margin includes a significant portion of the continental margins of the world, about half of the gas hydrate resources, or 550 Tcf of natural gas, may be present there.

Any future drilling on continental slopes and rises will have to pass through the zone of gas hydrate stability. If gas hydrates are present, drilling and casing procedures will have to be designed to cope with their presence. If gas is trapped beneath the gas hydrates, then appropriate precautions will have to be taken to prevent its uncontrolled release.

## SUMMARY

The evidence for gas hydrates in sediments of the Circum-Pacific margin is based mainly on the occurrence of anomalous acoustic reflectors (BSRs) on marine seismic-reflection records. Drilling in some areas off Central America where BSRs have been mapped has proved that gas hydrates are actually present; finding gas hydrates there strengthens the case for the occurrence of gas hydrates in other regions, especially where BSRs have been observed. Gas hydrates will likely be found in many regions surrounding the Pacific Basin as exploration for offshore petroleum moves to the deeper water over slopes and rises of margins of continents and island arcs. The amount of potentially producible gas from gas hydrates of the Circum-Pacific margin is estimated to be 550 Tcf; this number is highly speculative and is calculated without information regarding reservoir quality. Future drilling on continental slopes and rises will have to penetrate the zone where gas hydrates are stable; free gas, trapped by gas hydrates, could constitute a drilling hazard. Therefore, gas hydrates will have to be considered in future exploration and production strategies.

## ACKNOWLEDGMENTS

This work has been partially supported under the U.S. Geological Survey/Department of Energy Interagency Agreement No. DE-AI21-83MC20422.

## REFERENCES CITED


Aoki, Y., T. Tamano, and S. Kato, 1983, Detailed structure of the Nankai Trough from migrated seismic sections, *in* J.S. Watkins and C.L. Drake, eds., Studies in continental margin geology: AAPG Memoir 34, p. 309-322.

Aubouin, J., R. von Huene, et al., 1982, Initial reports of the Deep Sea Drilling Project, Volume 67: Washington, D.C., U.S. Government Printing Office, 799 p.

Biddle, K.T., and D.R. Seely, 1983, Structure of subduction complex, *in* A.W. Bally, ed., Seismic expression of structural styles: AAPG Studies in Geology 15, v. 3, p. 3.4.2., p. 129-134.

Claypool, G.E., and I.R. Kaplan, 1974, The origin and distribution of methane in marine sediments, *in* I.R. Kaplan, ed., Natural gases in marine sediments: New York, Plenum Press, p. 94-129.

Claypool, G.E., B.J. Presley, and I.R. Kaplan, 1973, Gas analysis of sediment samples from legs 10, 11, 13, 14, 15, 18, and 19, *in* J.S. Creager, D.W. Scholl, et al., Initial reports of the Deep Sea Drilling Project, Volume 19: Washington, D.C., U.S. Government Printing Office, p. 879-884.

Claypool, G.E., C.N. Threlkeld, P.N. Mankiewicz, M.A. Arthur, and T.F. Anderson, 1985, Isotopic compositions of interstitial fluids and origin of methane in slope sediment of the Middle America Trench, Deep Sea Drilling Project Leg 84, *in* R. von Huene, J. Aubouin, et al., Initial reports of the Deep Sea Drilling Project, Volume 84: Washington, D.C., U.S. Government Printing Office, p. 683-691.

Cooper, A.K., D.W. Scholl, and M.S. Marlow, Structural framework, sedimentary sequences, and hydrocarbon potential of the Aleutian and Bowers Basins, Bering Sea, *in* D.W. Scholl, A. Grantz, and J.G. Vedder, eds., Geology and resource potential of the continental margin of western North American and adjacent ocean basins—Beaufort Sea to Baja California: AAPG, in press.

Cooper, A.K., F.J. Daveu, and J.C. Behrendt, 1987, Seismic stratigraphy and structure of the Victoria Land basin, western Ross Sea, Antarctica: *in* A. Cooper, S. Eittreim, M. Hampton, and F. Davey, eds., The Antarctic continental margin—Geology and geophysics of the Wilkes Land and the western Ross Sea: AAPG Earth Science Series volumes 5A and 5B.

Creager, J.S., D.W. Scholl, et al., 1973, Initial reports of the Deep Sea Drilling Project, Volume 19: Washington, D.C., U.S. Government Printing Office, 913 p.

Davidson, D.W., M.K. El-Defrawy, M.D. Fuglem, and A.S. Judge, 1978, Natural gas hydrates in northern Canada: Proceedings, 3rd International Conference on Permafrost, v. 1, p. 937-943.

Dillon, W.P., J.A. Grow, and C.K. Paull, 1980, Unconventional gas hydrate seals may trap gas off southeast U.S.: Oil and Gas Journal, v. 78, n. 1, p. 124-130.

Eittreim, S.L., A.K. Cooper, and Scientific Staff, 1984, Marine geological and geophysical investigations of the Antarctic continental margin, 19984: U.S. Geological Survey Circular 935, 12 p.

Field, M.E., and K.A. Kvenvolden, 1985, Gas hydrates on the northern California continental margin: Geology v. 13, p. 517-520.

Field, M.E., S.H. Clarke, and M.E. White, 1980, Geology and geologic hazards of offshore Eel River Basin, northern California continental margin: U.S. Geological Survey Open-File Report 80-1080, 80 p.

Franklin, L.J., 1980, In-situ hydrates—A potential gas source: Petroleum Engineer International, v. 52, n. 13, p. 112-122.

Harrison, W.E., and J.A. Curiale, 1982, Gas hydrates in sediments of holes 497 and 498A, Deep Sea Drilling Project Leg 67, *in* J. Aubouin, R. von Huene, et al., Initial reports of the Deep Sea Drilling Project, volume 67: Washington, D.C., U.S. Government Printing Office, p. 591-594.

Harrison, W.E., R. Hesse, and J.M. Gieskes, 1982, Relationship between sedimentary facies and interstitial water chemistry of slope, trench and Cocos Plate sites from the Middle America Trench transect, active margin off Guatemala, Deep Sea Drilling Project Leg 67, *in* J. Aubouin, R. von Huene, et al., Initial reports of the Deep Sea Drilling Project, Volume 67: Washington, D.C., U.S. Government Printing Office, p. 603-613.

Hayes, D.E., and L.A. Frakes, 1975, General synthesis, Deep Sea Drilling Project Leg 28, *in* D.E. Hayes, L.A. Frakes, et al., Initial reports of the Deep Sea Drilling Project, Volume 28: Washington, D.C., U.S. Government Printing Office, p. 919-942.

Hedberg, H.D., 1980, Methane generation and petroleum migration, *in* W.H. Roberts III and R.J. Cordell, eds., Problems of petroleum migration: AAPG Studies in Geology 10, p. 179-206.

Hesse, R., and W.E. Harrison, 1981, Gas hydrates (clathrates) causing pore-water freshening and oxygen isotope fractionation in deep-water sedimentary sections of terrigenous continental margins: Earth and Planetary Science Letters, v. 55, p. 453-462.

Hesse, R., J. Level, and J.M. Gieskes, 1985, Interstitial water chemistry of gas-hydrate-bearing sections on the Middle America Trench slope, Deep Sea Drilling Project leg 84, *in* R. von Huene, J. Aubouin, et al., Initial reports of the Deep Sea Drilling Project, Volume 84: Washington, D.C., U.S. Government Printing Office, p. 727-736.

Hinz, K., and M. Block, 1983, Results of geophysical investigations in the Weddell Sea and in the Ross Sea, Antarctica: Proceedings of 11th World Petroleum Congress (London), v. PD 2, n. 1, p. 1-13.

Katz, H.R., 1981, Probable gas hydrate in continental slope east of the North Island, New Zealand: Journal of Petroleum Geology, v. 3, p. 315-324.

Katz, H.R., 1982, Evidence for gas hydrates beneath the continental slope, East Coast, North Island, New Zealand: New Zealand Journal of Geology and Geophysics, v. 25, p. 193-199.

Kvenvolden, K.A., and M.A. McMenamin, 1980, Hydrates of natural gas: A review of their geologic occurrence: U.S. Geological Survey Circular 825, 11 p.

Kvenvolden, K.A., and L.A. Barnard, 1983, Hydrates of natural gas in continental margins, *in* J.S. Watkins and C.L. Drake, eds., Studies in continental margin geology: AAPG Memoir 34, p. 631-640.

Kvenvolden, K.A., and T.J. McDonald, 1985, Gas hydrates of the Middle America Trench—Deep Sea Drilling Project Leg 84, *in* R. von Huene, J. Aubouin, et al., Initial reports of the Deep Sea Drilling Project, Volume 84: Washington, D.C., U.S. Government Printing Office, p. 667-682.

Kvenvolden, K.A., and R. von Huene, 1985, Natural gas generation in sediments of the convergent margin of the eastern Aleutian Trench area, *in* D.G. Howell, ed., Tectonostratigraphic terranes of the Circum-Pacific region: Circum-Pacific Council for Energy and Minerals, Earth Sciences Series, n. 1, p. 31-49.

Kvenvolden, K.A., G.E. Claypool, C.N. Threlkeld, and E.D. Sloan, 1984, Geochemistry of a naturally occurring massive marine gas hydrate: Organic Geochemistry, v. 6, p. 703-713.

Kvenvolden, K.A., M. Golan-Bac, and J.B. Rapp, in press, Hydrocarbon geochemistry of sediments offshore from Antarctica: I. Wilkes Land Continental Margin, *in* S.L. Eittreim, M.A. Hampton, A.K. Cooper, and F.J. Davey, eds., The Antarctic continental margins: Geology and geophysics of offshore Wilkes Land and the western Ross Sea: Circum-Pacific Council for Energy and Mineral Resources, Earth Science Series, volumes 5A and 5B.

Ladd, J., A.K. Ibrahim, K.J. McMillen, G.V. Latham, and R.E. von Huene, 1982, Interpretation of seismic reflection data of the Middle America Trench offshore Guatemala, *in* J. Aubouin, R. von Huene, et al., Initial reports of the Deep Sea Drilling Project, Volume 67: Washington, D.C., U.S. Government Printing Office, p. 675-689.

MacLeod, M.K., 1982, Gas hydrates in ocean bottom sediments: AAPG Bulletin, v. 66, p. 2649-2662.

Mann, R., and J.M. Gieskes, 1975, Interstitial water studies, Leg 28, *in* D.E. Hayes, L.A. Frakes, et al., Initial reports of the Deep Sea Drilling Project, Volume 28, Washington, D.C., U.S. Government Printing Office, p. 805-814.

Marlow, M.S., P. Carlson, A.K. Cooper, H. Karl, H. McLean, R. McMullin, and M.B. Lynch, 1981, Hydrocarbon resource report for proposed OCS Sale No. (?) Navarin Basin, Alaska: U.S. Geological Survey Open-File Report 81-252, 83 p.

McCarthy, J., A.J. Stevenson, D.W. Scholl, and T.L. Vallier, 1984, Speculations on the petroleum geology of the accretionary body: An example from the central Aleutians: Marine and Petroleum Geology, v. 1, p. 151-167.

McIver, R.D., 1974, Hydrocarbon gas (methane) in canned deep sea drilling project core samples, *in* I.R. Kaplan, ed., Natural gases in marine sediment: New York, Plenum Press, p. 65-69.

McIver, R.D., 1975, Hydrocarbon gases in canned core samples from Leg 28 Sites 271, 272, and 273, Ross Sea, *in* D.E. Hayes, L.A. Frakes, et al., Initial reports of the Deep Sea Drilling Project, Volume 28: Washington, D.C., U.S. Government Printing Office, p. 815-817.

McKirdy, D.M., and P.J. Cook, 1980, Organic geochemistry of Pliocene-Pleistocene calcareous sediments, DSDP Site 262, Timor Trough: AAPG Bulletin, v. 64, p. 2118-2138.

Potential Gas Committee, 1981, Potential supply of natural gas in the United States (as of December 31, 1980): Potential Gas Agency, Colorado School of Mines, Golden, Colorado, 119 p.

Scholl, D.W., and J.S. Creager, 1973, Geologic synthesis of Leg 19 (DSDP) results: Far north Pacific, Aleutian Ridge, and Bering Sea, *in* J.S. Creager and D.W. Scholl, Initial reports of the Deep Sea Drilling Project, Volume 19: Washington, D.C., U.S. Government Printing Office, p. 897-913.

Scholl, D.W., T.L. Vallier, and A.J. Stevenson, 1983, Geologic evolution of the Aleutian Ridge—Implications for petroleum resources: Journal of the Alaska Geological Society, v. 3, p. 33-46.

Shepard, G.L., 1979, Shallow crustal structure and marine geology of convergence zone, northwest Peru and southwest Ecuador: Honolulu, University of Hawaii, unpublished, Ph.D. dissertation, p. 157-165.

Shipley, T.H., and B.M. Didyk, 1982, Occurrence of methane hydrates offshore southern Mexico, *in* J.S. Watkins, J.C. Moore, et al., Initial reports of the Deep Sea Drilling Project, Volume 66: Washington, D.C., U.S. Government Printing Office, p. 547-555.

Shipley, T.H., M.H. Houston, R.T. Buffler, F.J. Shaub, K.J. McMillen, J. Ladd, and J.L. Worzel, 1979, Seismic reflection evidence for widespread possible gas hydrate horizons on continental slopes and rises: AAPG Bulletin, v. 63, p. 2204-2213.

von Huene, R., J. Ladd, and I. Norton, 1982, Geophysical observations of slope deposits, Middle America Trench off Guatemala, Deep Sea Drilling Project Leg 67, *in* J. Aubouin, R. von Huene, et al., Initial reports of the Deep Sea Drilling Project, Volume 67: Washington, D.C., U.S. Government Printing Office, p. 719-732.

von Huene, R., J. Aubouin, et al., 1985, Initial reports of the Deep Sea Drilling Project, Volume 84: Washington, D.C., U.S. Government Printing Office, 967 p.

von Huene, R., L.D. Kulm, and J. Miller, 1985, Structure of the frontal part of the Andean convergent margin: Journal of Geophysical Research, v. 90, p. 5429-5442.

Watkins, J.S., J.C. Moore, et al., 1982,, Initial reports of the Deep Sea Drilling Project, Volume 66: Washington, D.C., U.S. Government Printing Office, 864 p.

Yamano, M., S. Uyeda, Y. Aoki, and T.H. Shipley, 1982, Estimates of heat flow derived from gas hydrates: Geology, v. 10, p. 339-343.

CHAPTER 25

# GEOLOGY OF THE SUPER GIANT CAÑO LIMON FIELD AND THE LLANOS BASIN, COLOMBIA

**C. N. McCollough**
*Occidental International Exploration and Production Co.*
*Bakersfield, CA*

## ABSTRACT

After 40 years of sporadic exploration, which yielded negative or marginal results, the Llanos Basin of eastern Colombia was thrust to the forefront of world attention by the discovery of the super-giant Caño Limon field in July 1983.

This discovery was the culmination of an intensive 3-year exploration effort by Occidental involving 4000 km of dynamite seismic, 20 stratigraphic tests from 1300 to 3500 ft deep, (397 to 1066 m) and 13 exploratory wells. Cost prior to discovery was $45 million.

The Llanos is an oily basin with abundant excellent reservoir sands. The problem has been to define traps. The basin deep lies along the foot of the Eastern Cordillera from which a sediment thickness of 30,000 ft (9150 m) thins gradually to the east on a homoclinal flank terminating at the Guyana shield. Except for the very young folding along the Andean front, structural traps are generally sparse and subtle.

An exception is the Caño Limon area, which is dominated by major early Tertiary NE-SW strike-slip faulting. Concurrent folding, in combination with fault sealing, formed the Caño Limon field and other smaller fields in the area.

The Caño Limon field along with its other half, the Guafita field in Venezuela, contains an estimated 3160 MM BOIP of which 1460 MM barrels is expected to be recovered.

The bulk of the oil is in deltaic sands of the Eocene Mirador, with additional reservoirs in the Upper Cretaceous. Average porosity of the Mirador is 0. 24, permeability 4 darcys, and SW 0. 24. The average initial flow rate of the first 15 wells is 10,000 BOPD.

Exploration opportunities exist in a deep unexplored Paleozoic rift, in stratigraphic traps of Cretaceous and Tertiary age, and in fault-related traps along the basin flanks.

## INTRODUCTION

The month of June 1983 was eventful for oil exploration in Colombia. An elite panel of eight Colombian geologists was called back from jobs overseas to assess the oil situation in the country. The focus was on the Llanos Basin where the reserves of Intercol's (Exxon's) deep Arauca field discovery were suspected to be closer to 3 million barrels than the 100 million barrels estimated initially, and Ecopetrol's Apiay discovery was being recognized as questionably commercial.

On June 21, the panel presented its findings to the President of Colombia. The next day El Tiempo, the leading newspaper, carried the story under the headline, "Petroleum Expectations in the Llanos Collapse."

Ironically, unknown to the panel, Occidental's Caño Limon-1 was at the time drilling ahead in the Cretaceous, after having penetrated 174 ft (53 m) of net oil pay in the Eocene Mirado sands in what was to become a super giant field (relative to Latin America), or at least a giant field on world-class scale. Oxy's current reserve estimates are 1. 4 billion barrels of 30° API oil—1. 0 billion in Colombia and 0. 4 billion in the Guafita portion on the Venezuelan side. Suddenly, the unpopular Llanos Basin became the hottest exploration play in the world.

Although the geological panel reached the logical conclusion based on the information available to it, this incident illustrates that new frontiers are new frontiers precisely because their potential has not been generally recognized, and that for those with imagination and a willingness to take risks, new frontiers and opportunities still exist.

## GEOGRAPHIC SETTING

The Llanos Basin covers an area of some 196,000 km$^2$ in the low-lying grassy plains east of the Eastern Cordillera (Andes) in Colombia (Figures 1, 2). The basin is cut by numerous easterly flowing jungle-lined rivers feeding the major Meta River, which flows northeasterly to join the Orinoco River. To the east the grasslands give way to Amazonian jungle, and in the north near Caño Limon large lily pad swamps remain wet even through the dry season.

Heavy rainfall virtually prohibits land activity for 8 months of the year. In the 4-month dry season roads become deep in

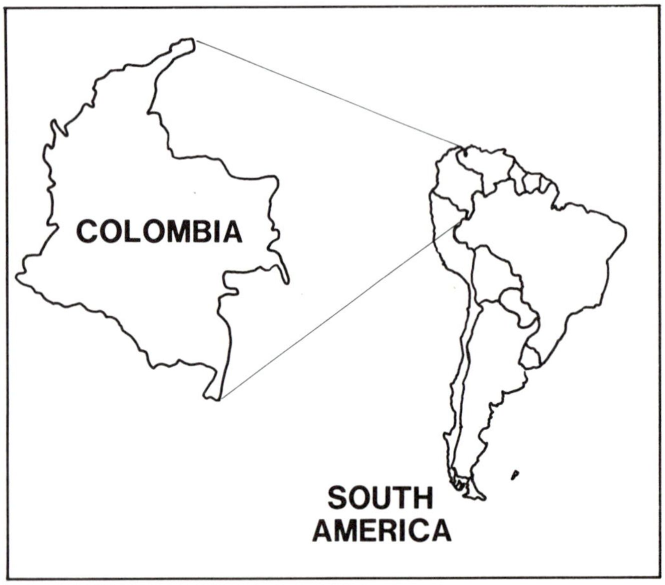


Figure 1.—Colombia's location in South America.

dust. The region is sparsely settled, with the two main towns being Villavicencio in the southwest and Arauca in the north. Cattle ranching is the only significant legal agricultural activity.

## EXPLORATION HISTORY

Llanos exploration has been very cyclical (Figure 3). The first well in the basin was the San Martin-1, drilled by Shell Oil along the southern basin edge in 1944.

The well tested heavy oil and was followed by the San Martin-2 through -6 wells, but the area proved subcommercial and was abandoned.

In the south-central part of the basin, Shell also drilled the wildcats Chaviva-1 in 1946 and Voragine-1 in 1947. Neither was successful, and exploratory drilling in the basin ceased for more than a decade.

In 1958, Intercol drilled five stratigraphic tests in the east-central part of the basin. During 1959-62, six wildcats were drilled in the Llanos, three of them by Texaco on the Guavio anticline in the foothills belt. The Guavio-1, drilled in 1960, reportedly flowed a total of 554 bbl of 39° gravity oil from the Tertiary on two drillstem tests. But two delineation wells,

Figure 2.—Index map showing the Llanos Basin in Colombia.

Guavio-2 and Guavio-3, failed to establish commerciality. The other tests during the period were dry holes, ending Llanos exploration's second phase in 1962.

Chevron launched the third phase of Llanos exploration in 1969 with its Castilla-1 wildcat, which tested heavy oil—about 14°—from the Cretaceous. Chevron subsequently has developed the field and is trucking production to Bogota. Ironically, the Castilla-1 is only 1.5 mi (2.4 km) from San Martin-1.

The Castilla strike sparked a surge of exploration lasting until 1978. During this period operators drilled 25 exploratory wells, mostly in the foothills fold-thrust province. Exploration in this area was not commercially successful, although some

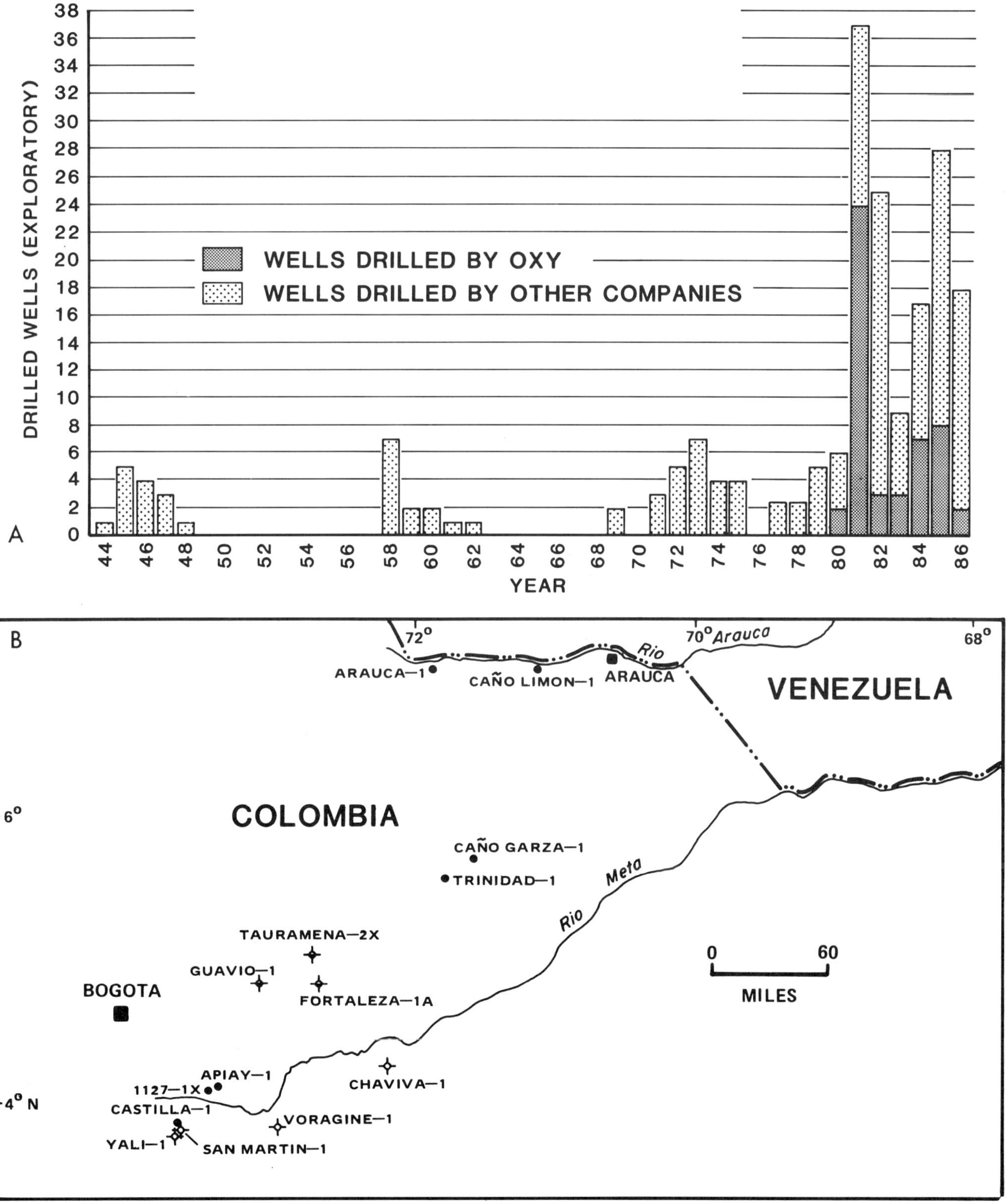


Figure 3.—A. Drilling history of the Llanos Basin. B. Location map of llanos exploration.

Figure 4.—South American sub-Andean basins.

wells—most notably Phillips' 1127-1X, Chevron's Yali-1, and Ecopetrol's Tauramena-2X—encountered significant shows. Exploratory drilling in this area was hindered by the depths of the objectives and problems created by swelling of the Leon shale.

During this period Elf Aquitaine began its exploration program in the central Llanos, drilling the Trinidad-1 discovery well in 1974 that tested about 1600 b/d of 32° gravity oil from the Mirador.

Exploration declined sharply during 1976-79, totaling only two wells in 1977 and one in 1978. None of the wells was successful, although Elf Aquitaine's Fortalexa-1A encountered subcommercial shows.

Exploration activity increased somewhat in 1979, with Elf accounting for three of the four wells spudded that year. Elf drilled and completed the Trinidad-2 confirmation well and discovered the Caño Garza field.

Intercol spudded the Arauca-1 well in 1979, completing the well as a discovery at depths of 18,090 to 18,140 ft (5517 to 5533 m) in 1980. The well tested 1941 b/d (41° gravity), and the Arauca-2 well tested a total of 4970 b/d (34° and 41° gravity) from Mirador and Cretaceous intervals in 1981. At the time of discovery and confirmation, the Arauca discovery was hailed as a major find, with recoverable reserves of more than 100 million barrels. Subsequent delineation drilling was disappointing, and reserve estimates have been downgraded to about 3. 5 million barrels.

Llanos exploration accelerated in 1980, primarily because of increased activity by Ecopetrol, Elf et al. , and Intercol, along with the start of Occidental's exploration program. A

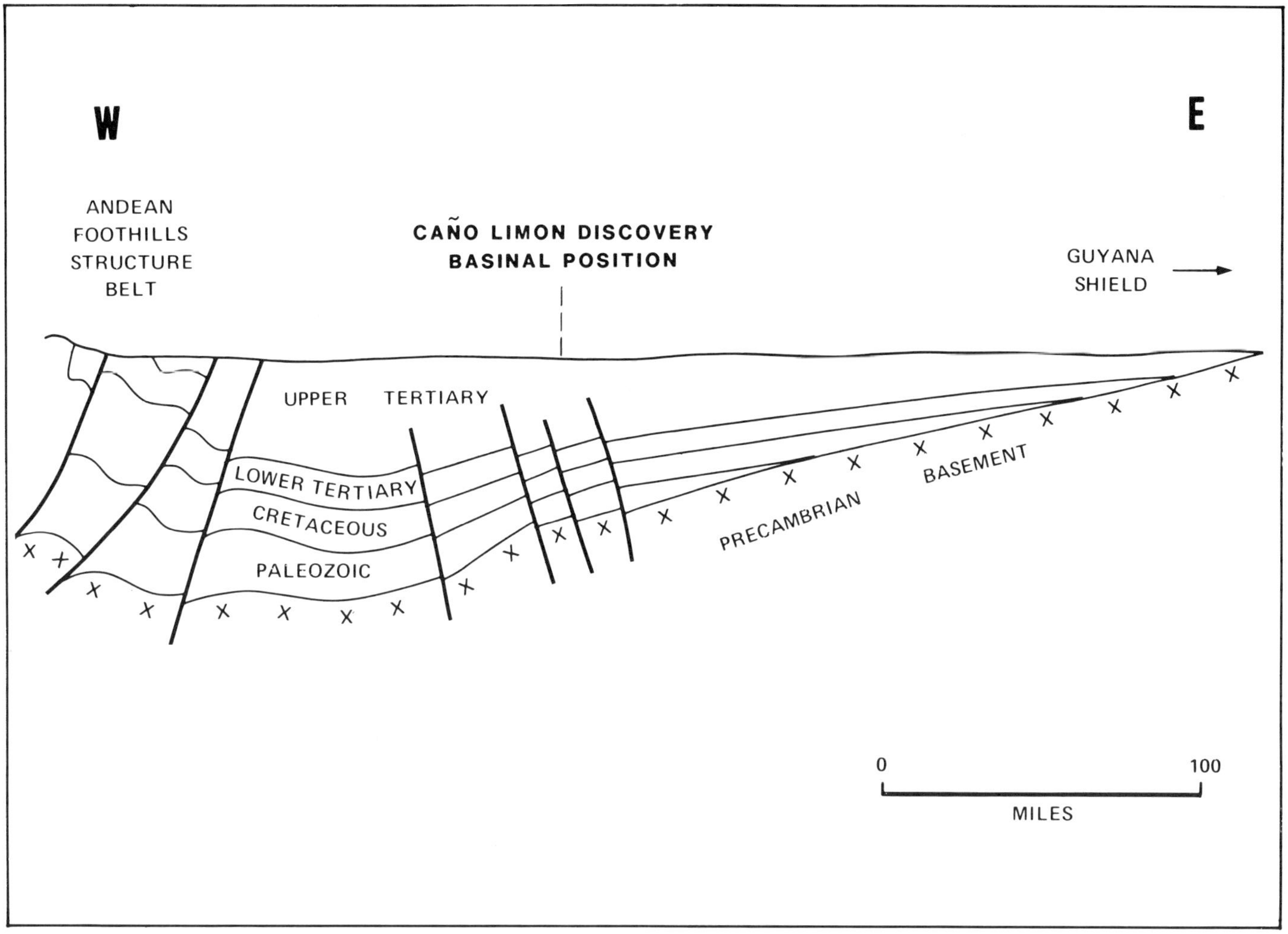


Figure 5.—Schematic cross section of the Llanos Basin

total of 35 exploratory and confirmation wells and 37 stratigraphic tests (20 by Oxy, 17 by Intercol) were drilled during 1980-82.

Much of Intercol's activity focused on a heavy-oil play in the southern part of the basin, which yielded four noncommercial discoveries during 1981-82. Ecopetrol discovered the Apiay field on the margins of the foothills belt in 1981, testing 4509 b/d of 24.6° gravity oil from the Cretaceous.

Oxy's Caño Limon-1 discovery, which tested oil from two Mirador zones at a rate of 19,690 b/d (about 31° average gravity), was announced in July of 1983. Subsequent drilling has established the discovery as a giant oil field, which Oxy estimates could ultimately produce more than 1 billion barrels of oil from the Colombian portion, and 1.4 billion barrels including the Venezuelan portion. The latter portion is called the Guafita field.

Elf Aquitaine has continued its success in the central basin, as has Ecopetrol. In addition, Ecopetrol has added two new discoveries near the Apiay field.

The accelerated exploration success focused greater industry attention on the basin, sparking an intense flurry of acreage acquisition during 1984 and 1985.

The commercial successes of the last few years clearly have brought the Llanos into a new era. After 40 years of nearly fruitless exploration, what once had been dismissed as a marginal petroleum province became one of the hottest exploration plays in the world.

## BASIN SETTING

The Llanos Basin is one of a long chain of sub-Andean basins extending from Venezuela to Argentina (Figure 4). These basins, which lie east of the Cordillera (Andes), are characterized by a fold-thrust belt at the mountain front, a basin deep immediately to the east, and a shallowing onto granitic shield at the eastern margin (Figures 5, 6).

The maximum depth of the Llanos Basin is estimated to be at least 30,000 ft (9150 m). The sediment package thins to zero 500 km to the east, where the granitic Guyana shield is exposed.

## BASIN EVOLUTION

Until the Andean Orogeny in late Miocene, the Llanos Basin and the Magdalena Basin to the west were one. The basin deep was in the area now occupied by the Eastern Cordillera. With the uplift of the Eastern Cordillera during the Andean orogeny of late Miocene to Recent age, the basin was bisected, and the larger eastern part became the Llanos Basin.

An earlier component of the basin evolution occurred during the Paleozoic and early Mesozoic. A large failed rift is now recognized trending northeast-southwest through the northern part of the basin. The depth of this graben is not known, but seismic reflectors are seen to at least five seconds, representing an estimated 28,000 ft (8540 m). The sediments in the

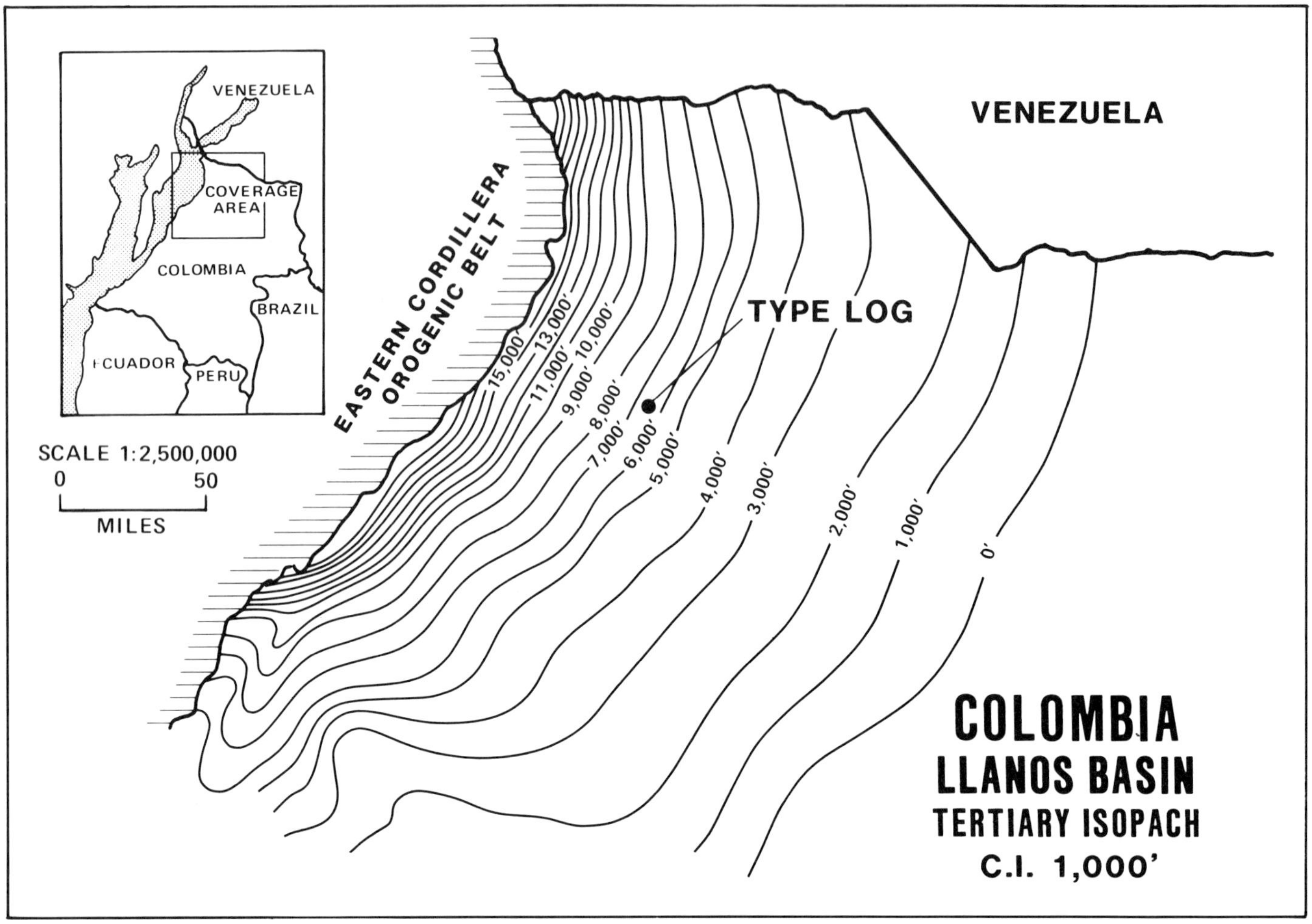


Figure 6.—Tertiary isopach of the Llanos Basin.

graben are mainly Cambro-Ordovician with possible Devonian in some places. In limited areas the Paleozoic is overlain by Jura-Triassic red beds. As yet no information to explain the Permo-Carboniferous hiatus is available.

A major unconformity at the base of the Cretaceous puts early Cretaceous on Paleozoic in the deeper western part of the basin, and progressively younger Cretaceous on older Paleozoic and ultimately on granite basement to the east. In the southeastern Llanos a Paleozoic subbasin preserved a considerable thickness of Cambro-Ordovician sandstones and shales overlain by Tertiary sediments rather than Cretaceous.

## STRATIGRAPHY

The stratigraphy of the Paleozoic is not well known (Figure 7). In the rift or graben of the northern area there are indications of carbonate buildups on seismic, but the age and nature are uncertain. The few wells drilled into Paleozoic have found mainly siltstones and shales with a few sandstones. Paleozoic sediments over the major part of the basin are well indurated but cannot be called metamorphic. Sandstones with porosities up to 22% have been seen, but the usual values are in the 10% to 12% range. Whether adequate oil source rocks exist is a question.

With the exception of minor carbonates in the Paleozoic and Cretaceous, the Llanos Basin is filled with sands, shales, and siltstones. Prior to the Andean orogeny, probably all sediments entering the basin were derived from the granitic Guyana shield to the east and southeast. A western source, the Central Cordillera, contributed sediments to the Magdalena part of the basin. After the onset of the Andean orogeny, a probably dual eastern and western source was present which later became a totally western source as basinal sediments, now exposed in the Eastern Cordillera, were eroded and carried eastward.

Cretaceous units ranging in age from Albian or possibly older to Campanian are predominantly alluvial, deltaic, turbiditic, and basal transgressive sands. Shales are a minor constituent to the east, but the basinal shaly equivalents to the continental/deltaic sands thicken into the basin and contain turbidites, as might be expected. Cretaceous sands are generally of very good reservoir quality.

The Paleocene Cuervos-Barco shale-sand units are present in the western part of the basin but terminate along an eastern pinchout, the position of which is controversial. The Barco sands are productive in Intercol's deep Arauca field.

Another controversial unit is the Eocene Mirador sand. It is widely believed that the Mirador sand in outcrop in the Eastern Cordillera and Merida Andes of Venezuela is not the same unit that is the main producing sand in Caño Limon and other Llanos fields. There is growing support for the idea that the "true Mirador" of the outcrops is stratigraphically deeper and pinches out to the east somewhere in the western part of the basin. The other Mirador, which is present over most of the Llanos, may be stratigraphically shallower and in some areas

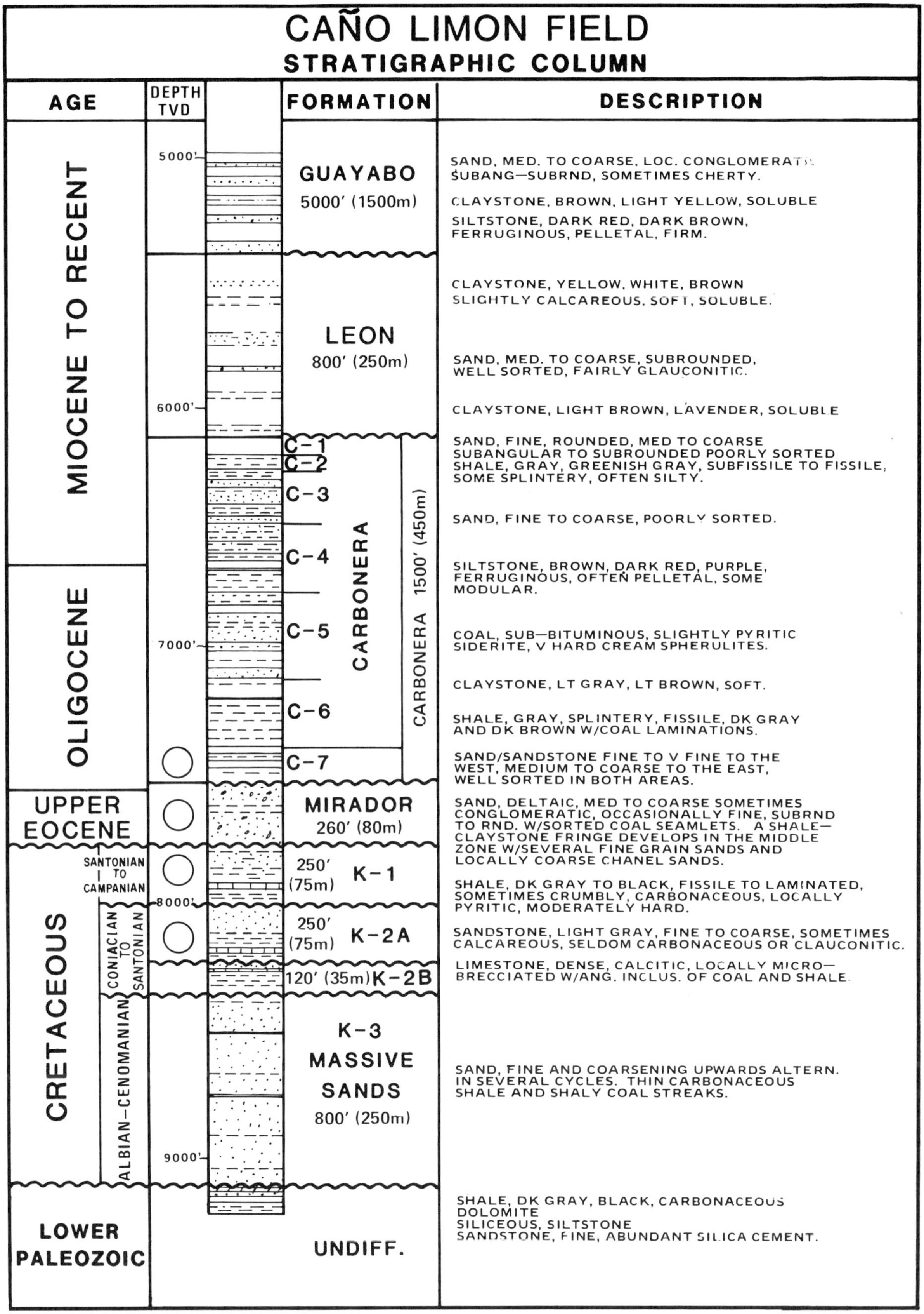


Figure 7.—Stratigraphic column of the Caño Limon field.

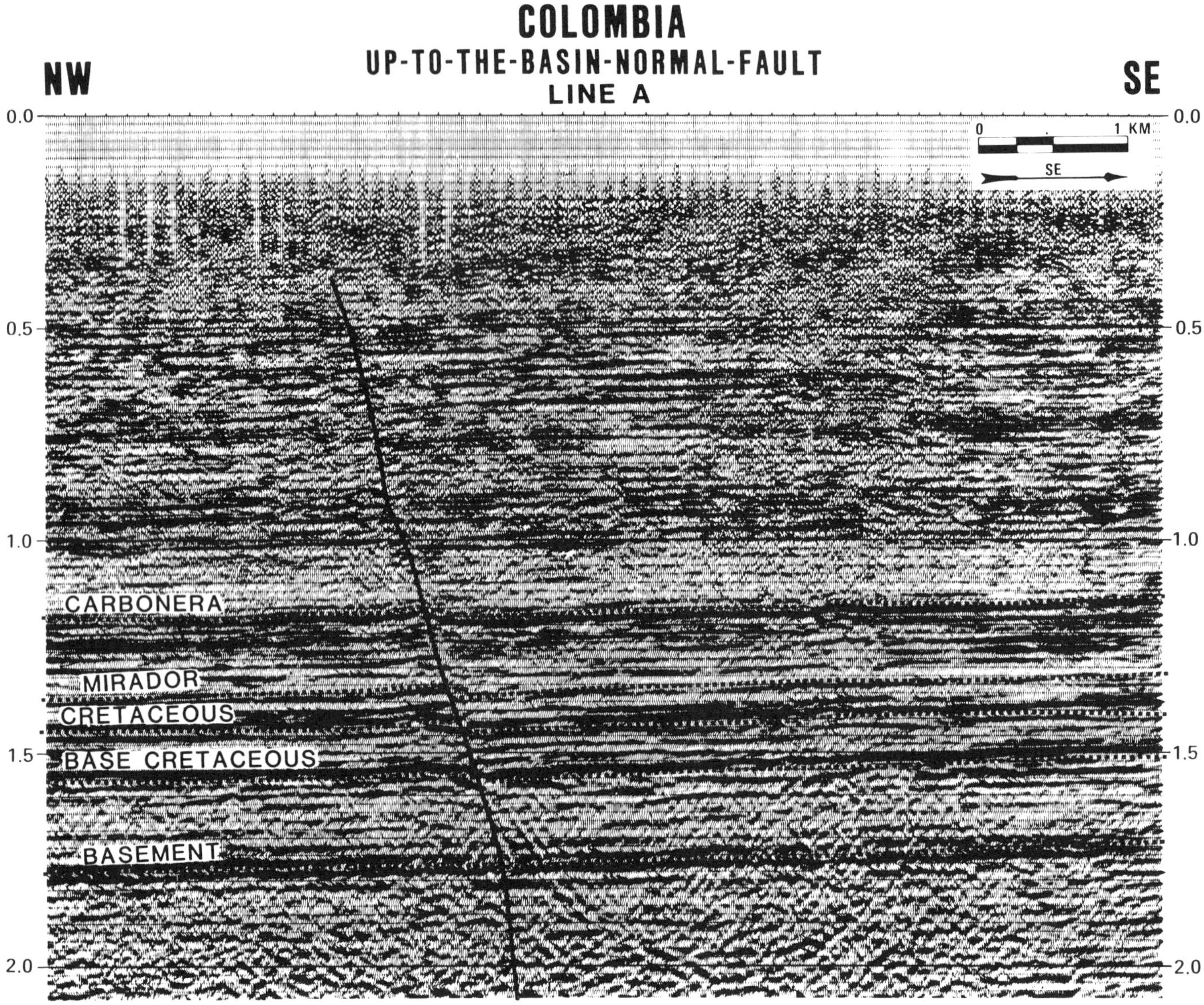


Figure 8.—Line A, showing an up-to-the basin (antithetic) fault. For location, see Figure 9.

may overlap the "true Mirador." At least a portion of the higher "Mirador" may more correctly be a basal sand member of the Carbonera Formation. The Oligocene-Eocene boundary occurs within the upper part of the "Mirador" Formation at Caño Limon on the basis of palynology. The distinction between "Mirador" and Carbonera at Caño Limon is more tied to the position of the unconformity/disconformity and structural growth related to Caño Limon fault displacement than to age.

In any event the "Mirador' is present over essentially all the Llanos Basin as a major regressive unit. It appears to be a series of stacked deltaic and alluvial sands with few shales. It is an excellent reservoir, which is about 200 to more than 500 ft (61 to 152 m) thick. In the remainder of this paper, "Mirador" refers to this unit.

The Carbonera Formation is Oligocene in the lower part and appears to be Miocene in the upper part. It incorporates three to four shaly transgressive units and the same number of regressive sandy units. The sandy regressive units are predominantly deltaic and contain coal swamp deposits and thin coal beds, which give the formation its name. Within the pro delta and basinal shales are some thin sands that are probably turbidites.

Oil has been found at several levels of the Carbonera, but mainly in the lower sands just above the Mirador.

Overlying the Carbonera is the Miocene Leon Shale, a greenish-gray claystone 560 to 785 ft (170 to 239 m) thick. In the northern Llanos near Caño Limon, the Leon contains numerous sand interbeds which make its upper boundary difficult to pick. This is the most ubiquitous shale in the basin, but to date it seems to have no economic significance as no oil is known to be trapped in the sands immediately beneath it.

Above the Leon shale is a thick, easterly thinning wedge of sands and shales of the Miocene to Recent Guayabo Formation. This formation has no known hydrocarbon potential.

## STRUCTURE

The primary deterrent to exploration in the Llanos Basin has been that, except for the mountain front, structural activity seemed to be virtually nonexistent. Early reconnaissance seismic and in some areas reasonably closely spaced seismic grids failed to define interesting structure. A number of wells were drilled along the fold-thrust mountain front belt, but without commercial success. Wells were generally deep and drilling was difficult and expensive. Several never reached their objective. Small amounts of oil were found, but often it was degraded. Fresh waters and porosity destruction by diagenesis were also common.

Subtle up-to-the-basin (antithetic) faults were recognized on seismic in the west-central part of the basin. The first potentially commercial light oil discovery, Elf Aquitaine et al. Trinidad-1, was made in 1974 on a structural closure against one of these faults. Since then more than 20 small oil fields with reserves of 2 to 15 million barrels each have been discovered on similar structures.

These antithetic faults are very young and on many seismic lines appear to come essentially to the surface (Figure 8). The writer believes they were formed too late to trap oil from the main migration from Cretaceous sources and are being sourced by younger shales, possibly still generating. Most of these traps do not appear to be filled to capacity, although it is difficult to determine what the capacity really is.

The major strike-slip fault system responsible for the trapping structures in the Caño Limon area is still being studied with regard to the nature and timing of the tectonic forces involved.

## OIL SOURCE

Until the discovery of Caño Limon, the Llanos Basin was believed to have a limited oil source. As suggested above, the problem may be timing rather than quantity of source. The main source of oil likely was marine shales of the Cretaceous, most of which are now uplifted in the Eastern Cordillera. Estimated sediment thickness and geothermal gradients indicate that the oil probably was generated and expelled from the Cretaceous source during the Oligocene. Any traps in the basin that were present at that time should be filled. Since there were apparently insufficient traps, much of this oil migrated to the margins of the basin and was degraded or lost. Heavy oil and oxidized oil residue in a large number of wells along the southern and southeastern margins of the basin testify to this thesis.

The antithetic fault traps were not present at this time. The oil in these accumulations is likely sourced from younger shales, possibly the basinal equivalents of the Mirador, the Cuervos, or even the Carbonera.

This difference in sources is supported by geochemical data showing that oils from early traps such as Caño Limon and Ecopetrol's Apiay field in the southern Llanos are from a strongly marine source, whereas oils from the antithetic fault traps have a much more terrestrial origin. This younger source may indeed be limited, or perhaps it is only that the young structures found so far have limited capacity.

## OCCIDENTAL'S EXPLORATION PROGRAM

When Oxy entered the Llanos Basin in 1980, much of what we know today was unknown. The factors in the decision to explore an unpopular basin were the following:

1. It was apparent that excellent reservoirs were everywhere.
2. Sealing shales appeared adequate in number and competency, although in the eastern parts of the basin this was a concern.
3. Contrary to some published reports, we deduced that there probably had been more than adequate mature source rocks in the pre-Andean basin deep.
4. Despite the indications at that time, it was reasoned that no basin of this size could completely escape structural deformation throughout its history.
5. Although not proven, the Arauca Arch was believed to exist at the Colombia-Venezuela border. It was postulated that this could provide the necessary structural and stratigraphic complications and migration focus for oil accumulations.
6. Some 90,000 $km^2$ were totally unexplored.
7. Ecopetrol, the national oil company, had an enlightened approach to attracting foreign exploration while protecting the national interests.

On these bases, Oxy signed association contracts on five blocks of 1 million hectares each—a total of 12. 4 million acres (Figure 9). Within the selected areas there existed a small amount of seismic, three stratigraphic tests, and one exploratory well (La Heliera-1) whose location was determined at least partly by how far up the Rio Casanare the barges transporting the rig were able to navigate.

The first contract year commitment was to drill four stratigraphic tests on each block. At the end of the first year, half of the acreage had to be relinquished.

In order to locate the stratigraphic tests most effectively and to prepare for the 50% relinquishment, we rushed to start recording a large seismic program. The effective date of the contracts was July 1, 1980, and by August a program of 1100 km of seismic on the major rivers was begun. Portable dynamite crews started the land program shortly after.

As one of the potential accumulation types was a heavy-oil belt such as the Orinoco belt in Venezuela, many of the stratigraphic tests were drilled along the shallow eastern parts of the blocks.

Information gained from the first year seismic and stratigraphic program was as follows:

1. Seven of the stratigraphic tests found shows of oil, including dead residuals along the shallow eastern margin of the basin.
2. Several antithetic faults for potential traps were found in the east-central part of the basin.
3. A new type of structure—drape folds over granite knobs—was found in the northeastern areas (Figure 10).
4. A new type of large early faulting was seen on seismic lines in the area southwest of Caño Limon (Figure 11).

In the second year additional seismic was recorded and a commitment of 10 exploratory wells was completed. Additional encouragement resulted; in particular, recognition that the faults in the Caño Limon area formed a major system and were probably strike-slip movements.

The possibility of folding associated with the major fault system began to spark enthusiasm, and a third small seismic program was recorded in the Caño Limon area to check several indications of folding. In particular, line 61. 7 was shot along the south (downthrown) side of the main NE-SW fault now called the Caño Limon fault (Figure 12). This was to check a very small rollover which later was drilled as the Caño Verde

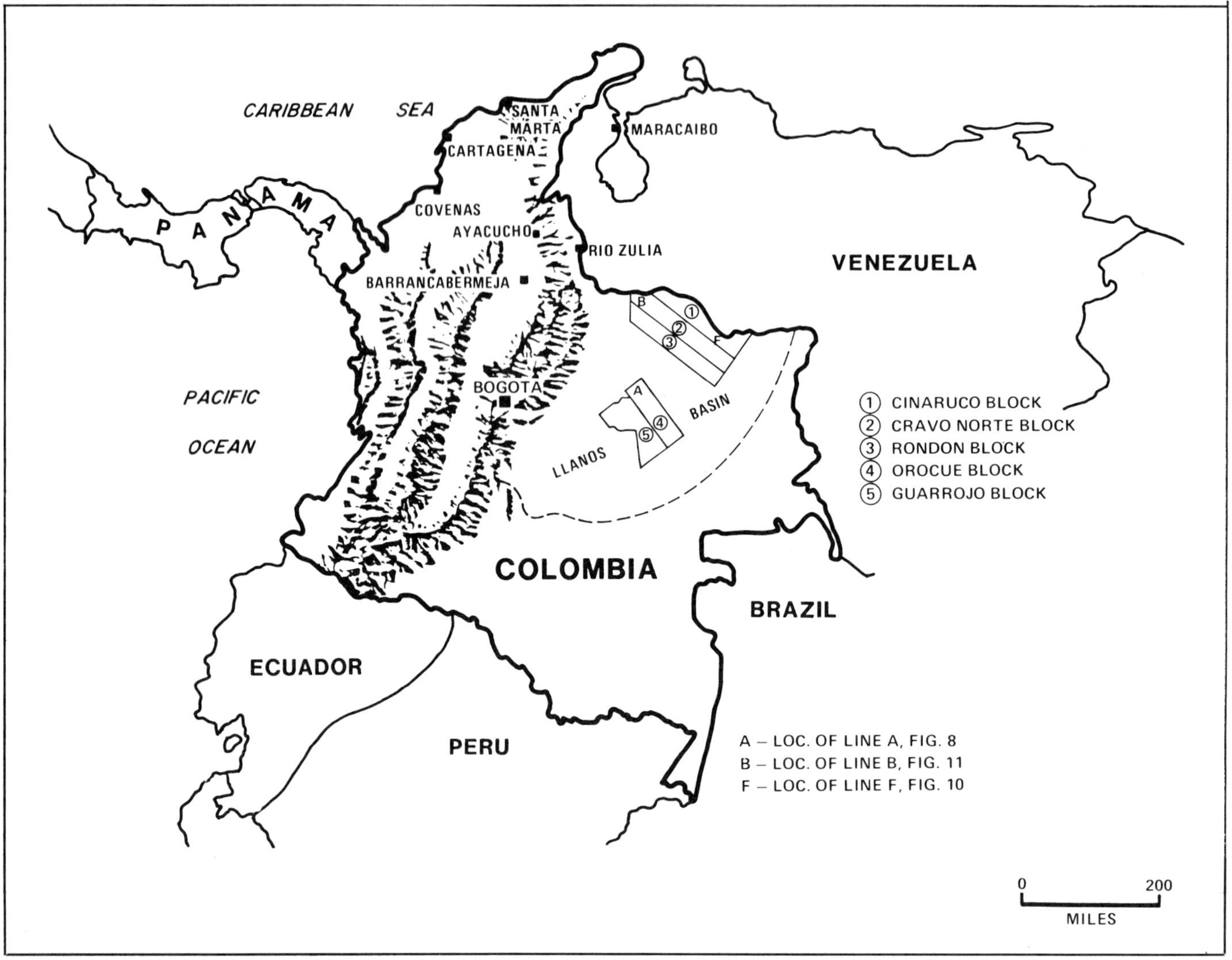


Figure 9.—Locations of original Oxy blocks, and of lines A, B, and F in Figures 8, 11, and 10, respectively.

discovery, but farther to the northeast a larger rollover was found which was to be the Caño Limon discovery.

This was late 1982 and the oil industry had begun to slump as a result of falling oil prices.

Oxy was in the process of buying Cities Service, and there was little top management enthusiasm for drilling another well in the Llanos. The decision was made to farm out 50% of the whole area for the drilling of the Caño Limon well. Exploration enthusiasm for the prospect was high, but from a business standpoint it appeared the options might be to farmout or drop.

Thus began Oxy's largest farmout effort. A total of 73 contacts were made with 45 companies with presentations to a majority of these.

In the meantime it was found that a well location could not be constructed in the swampy area overlying the prospect. A site was found on the natural levee of the Caño Agua de Limon, and a 43-km road from Troncal was started while equipment made its way to the drill site to raise the site elevation 2 m.

Inasmuch as a deviated well would have to be drilled, we realized that a seismic line from the drillsite to the apparent crest of the structure was necessary for proper well planning. Our interpretation at that point—based on a one-line rollover—was a small four-way dip closure; but it was apparent that we did not know what the structural configuration really was (Figure 13). So, in addition to a line NW-SE from the drillsite to the target, a second NNE-SSW line was shot across the apparent crest of the structure. These two lines showed that it was not a four-way closure, but was closed against the major Caño Limon fault (Figure 14). Thus it became necessary to believe that the fault would form a seal.

As the end of the dry season approached, it was clear that we needed to get a rig in before heavy rains closed the road. Management agreed to go ahead if farmout attempts were vigorously continued. The rig was trucked in as the rains began. The trucks carrying the last load, the main mud pit, became stuck short of the location; and the mud pit was dragged by bulldozers to the Caño Limon and onto a barge to be transported to the drill site. The well was spudded on April

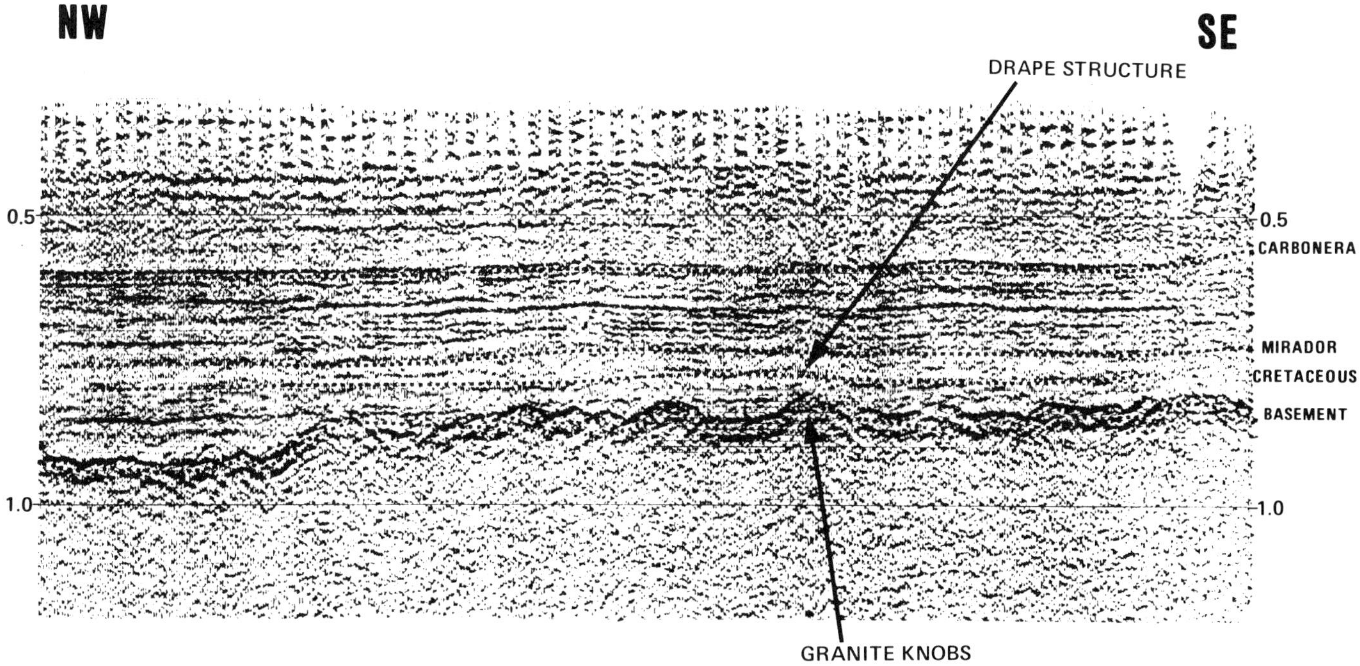


Figure 10.—Line F, showing drape folds over granite knobs. For location, see Figure 9.

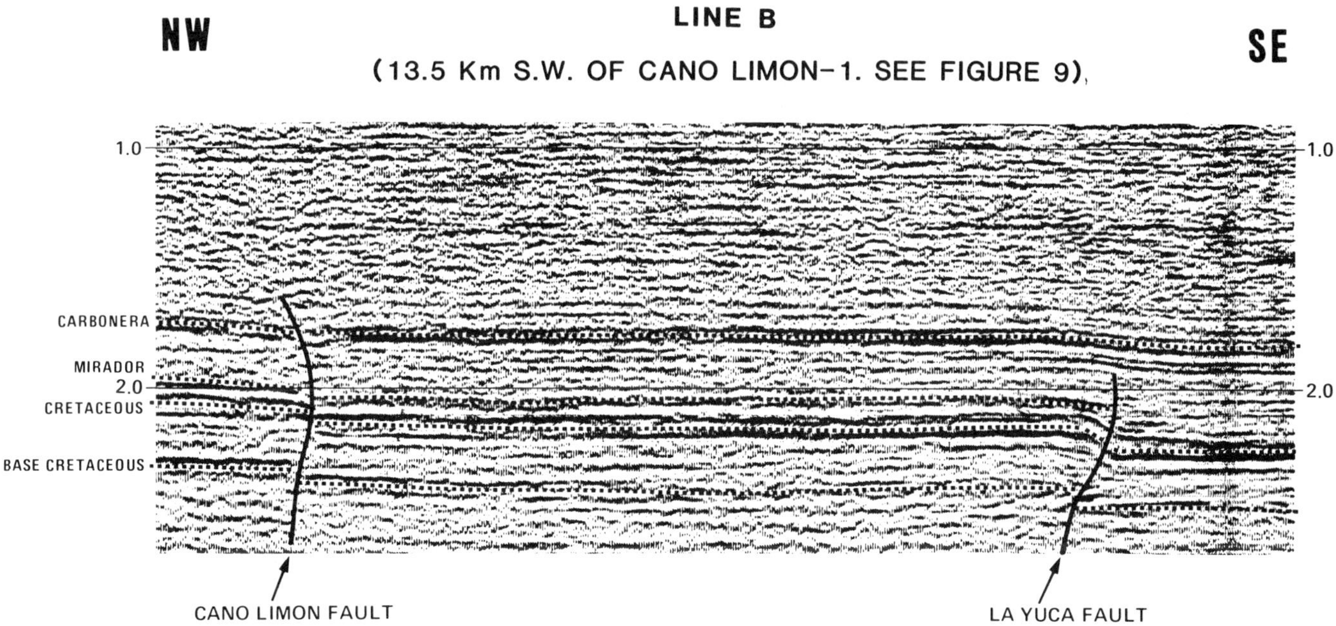


Figure 11.—Line B, showing strike slip faults in the Cravo Norte block. For location, see Figure 9.

20, 1983. Farmout attempts continued, and at least one major company was still considering the farmout when the Leon Shale was reached. At that point special conditions were telexed so this company could continue to try to convince their management as we approached the objective Mirador zone. Fortunately, they never did.

The Mirador sands were penetrated with no gas shows and only traces of fluorescence on a few sand grains. It did not look good.

We were stunned by the first log which showed 160 ft (49 m) in two massive sands with resistivities up to a mind-boggling 2500 ohm-meters. There was no doubt we had made a significant discovery. The well tested at a combined rate of 10,690 BOPD from the two zones; one zone on a 1-7/8 inch choke

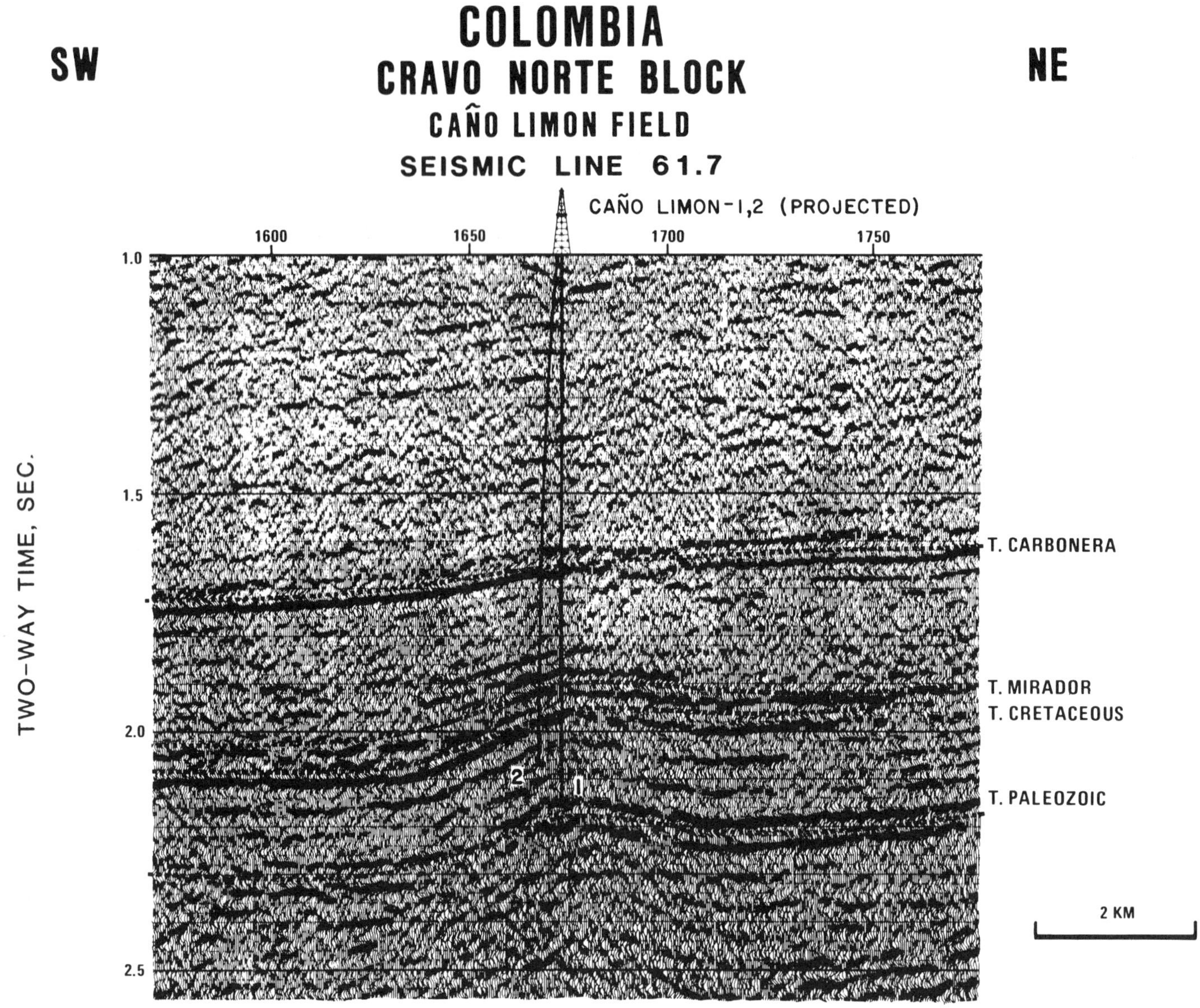


Figure 12.—Line 61.7, shot along the downthrown side of the Caño Limon fault. For location, see Figure 13.

and the other on a 1/2-inch choke. The API gravity varied from 33.4° at the top of the Mirador to 29° at the base. No oil-water contact was present in the Mirador in this well.

The rig was skidded on the same location and Caño Limon-2 was directionally drilled to a target location on the west flank of the structure. The oil-water contact was penetrated and commerciality was granted by Ecopetrol on a reserve estimate of 64 MMBO.

Drilling was suspended until the following dry season when equipment could again enter the area to build locations. Three locations were built—one for Caño Limon-3 on the southeast flank of the structure and two for wildcats on nearby structures updip to the northeast, La Yuca and Matanegra.

The first indication that a major oil field could be present came in Caño Limon-3, where the oil-water contact was found to be lower than in Caño Limon-2 and also lower than the syncline between the Caño Limon and La Yuca structures. The La Yuca-1 well soon added support to the idea that the two structures were in a common oil accumulation. Oil was found down to a major sealing shale in the Cretaceous K-1 zone, and pressure measurements indicated an oil-water contact at the same depth as in Caño Limon-3. The interconnection was subsequently proved by La Yuca-3, which found oil in the syncline.

Matanegra-1 found oil in the Mirador, K-1 and K-2 sands, adding a major fault-bounded segment to the Caño Limon field. This fault wedge is not in communication with the Caño Limon-La Yuca part of the field, as shown by lack of pressure response and different oil-water contacts across the Caño Limon fault.

An interesting aside is that the Caño Limon fault separating Caño Limon and Matanegra has now been accurately positioned with 3-D seismic. Had the deviation of the Caño Limon-1 been 300 m less, we would have been on the wrong side of the fault and would have missed the field. After the Caño Limon discovery, Corpoven drilled its Guafita-1X discovery across the border in Venezuela on a structure to the northeast

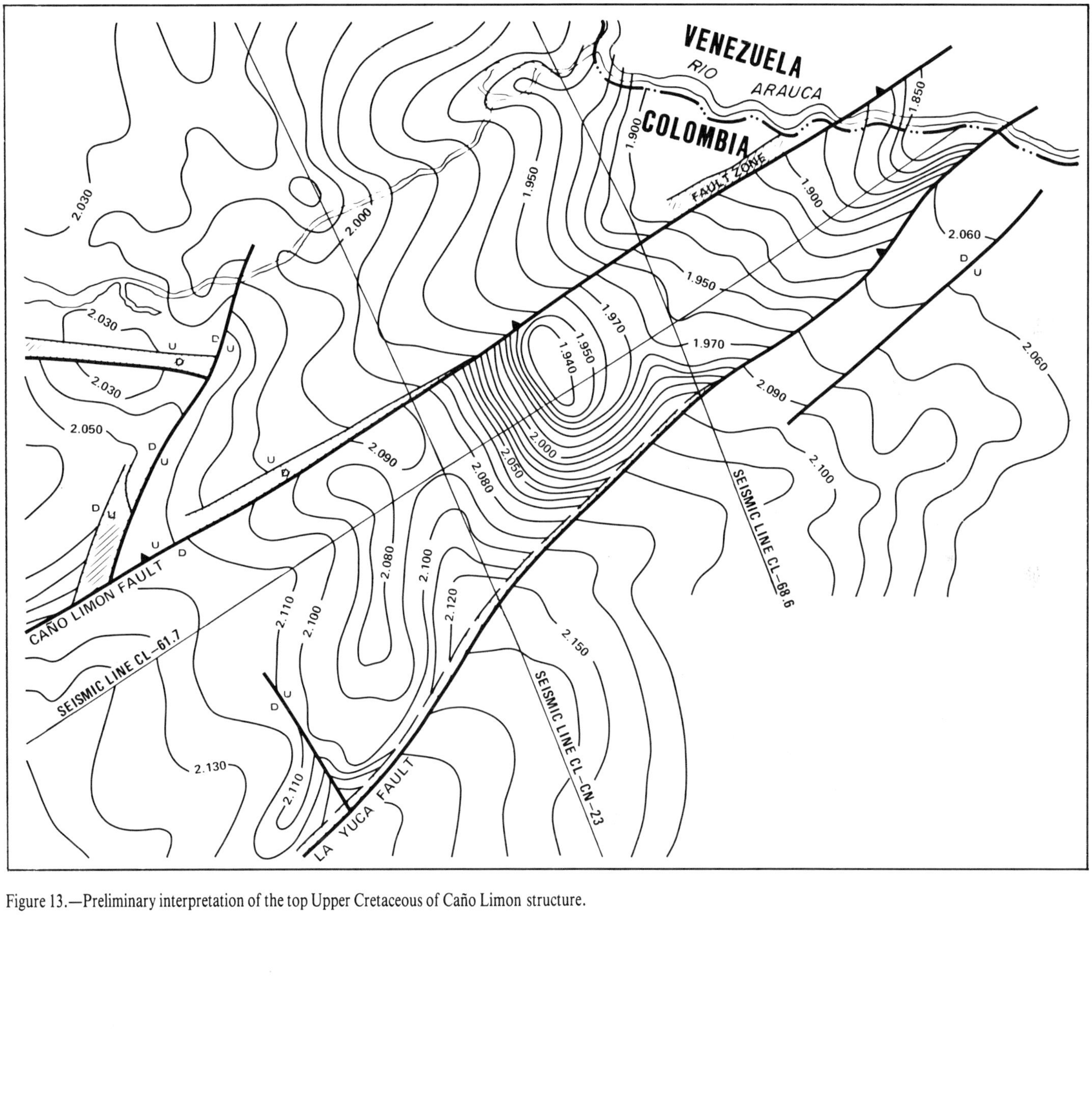


Figure 13.—Preliminary interpretation of the top Upper Cretaceous of Caño Limon structure.

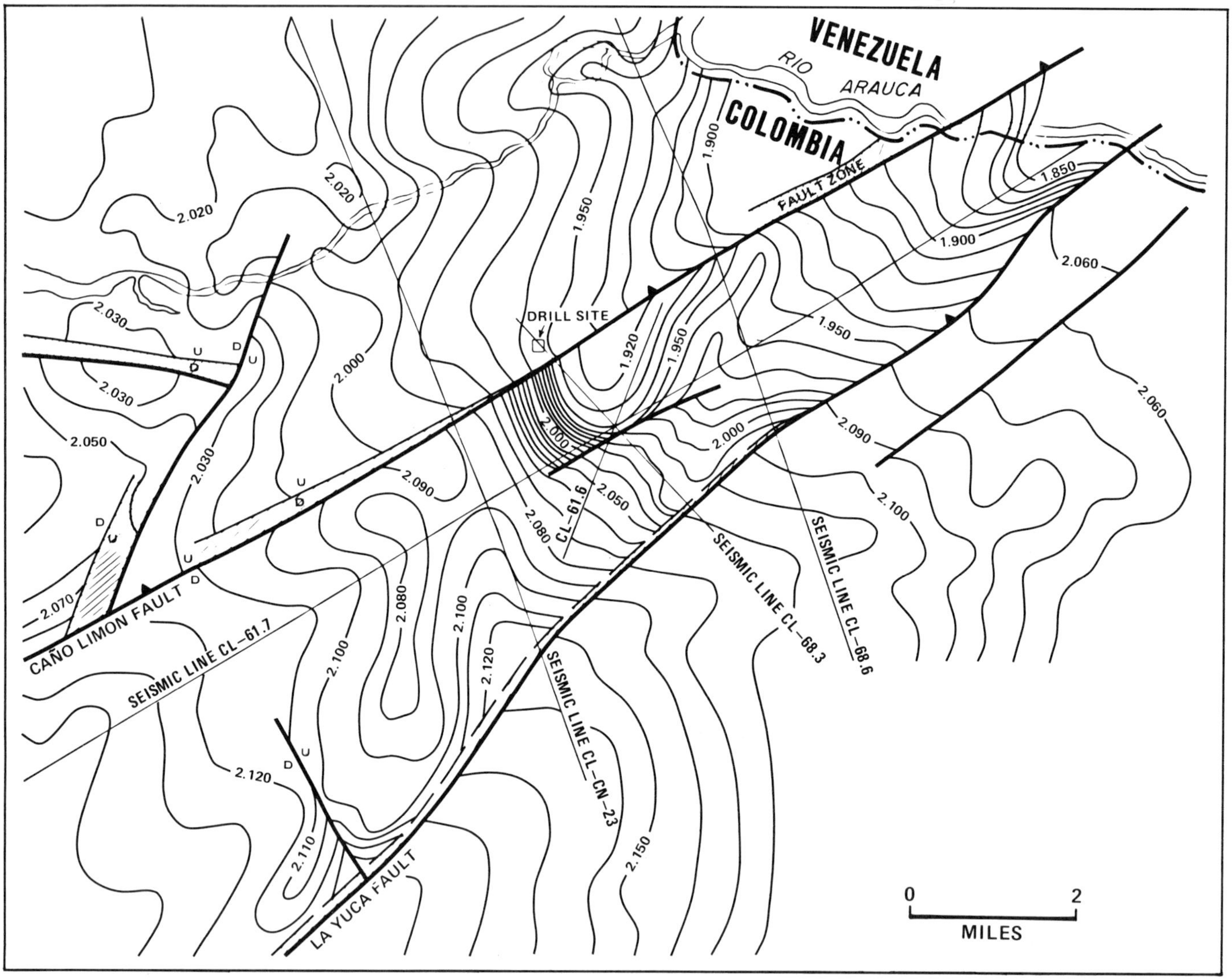


Figure 14.—Pre-drilling interpretation of the top Upper Cretaceous of Caño Limon structure.

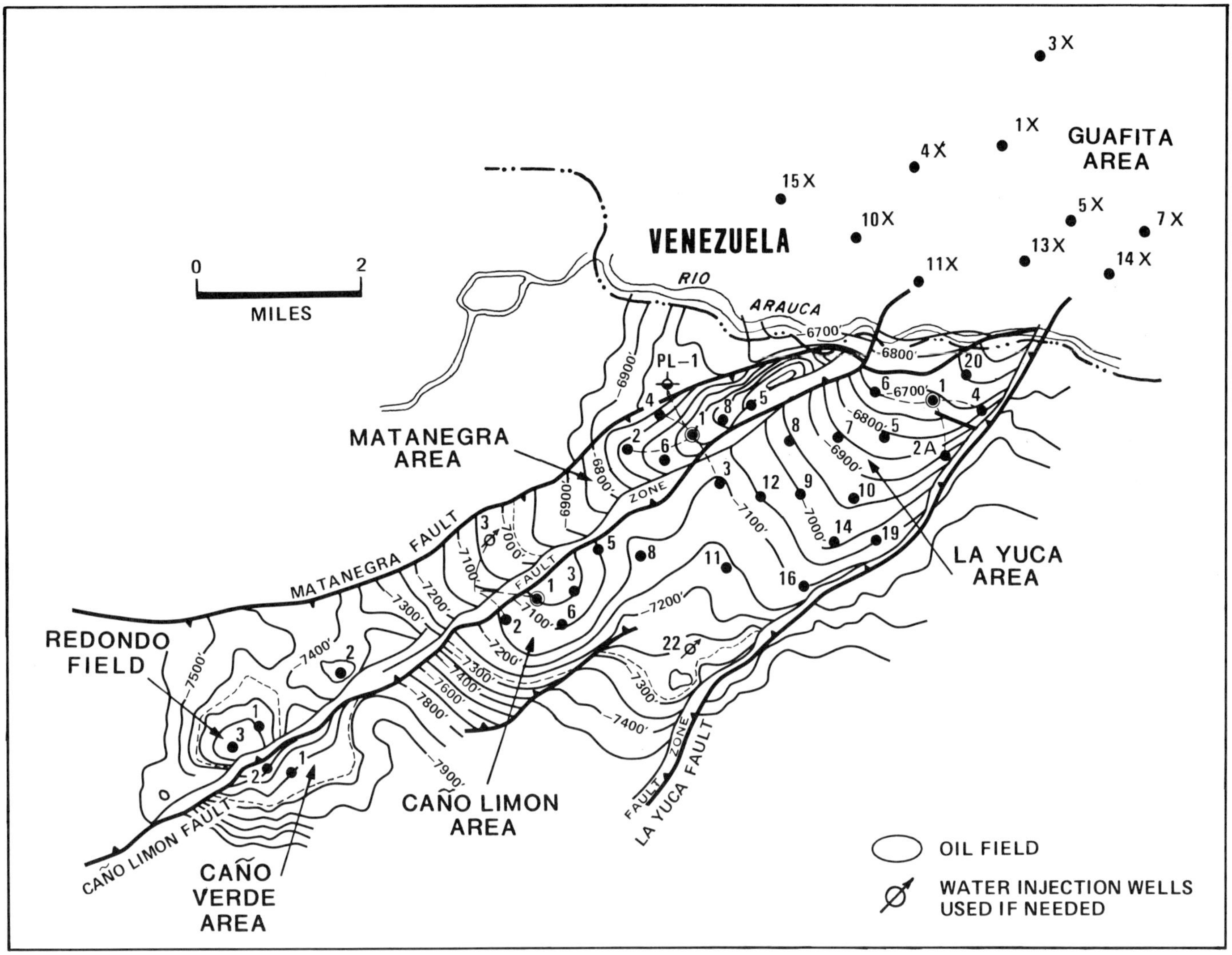


Figure 15.—Contours on top Mirador, Caño Limon field.

and on trend with the Caño Limon field. It is now clear that the Caño Limon and Guafita fields are in the same faulted structure, but it is not known if they are in fluid communication because of several large normal faults along the Colombia-Venezuela border (Figures 15, 16, 17).

## CAÑO LIMON FIELD

Caño Limon has been described as "the kind of field geologists and engineers dream of."

Some of the attributes that give it this reputation are:

1. In the Mirador, which holds 80% of the reserves, the sequence of deltaic channel and fringe sands results in excellent to good reservoirs, with several continuous shales to reduce major fluid crossflow and assist in reservoir management capability (Figure 18). Cretaceous reservoirs are also good to excellent.
2. Permeabilities differ widely depending on facies but average 1145 millidarcys from core analysis (under overburden pressure), and measure in the 2 to 15 darcy range on pressure buildup analyses.
3. Flowing well rates can exceed 20,000 BOPD.
4. The oil gravity averages 29. 5° API and is low sulfur (0. 4%).
5. The gas-oil ratio is about 7 cu ft per barrel, resulting in a formation volume factor of 1. 05, or virtually no shrinkage.
6. Formation waters are fresh (100-200 ppm NaCl) and noncorrosive in the water leg, and are only slightly more saline within the oil accumulation.
7. A strong natural water drive may allow minimal pressure support by water injection. Oxy reservoir engineers believe overall recovery should reach or perhaps exceed 50%.
8. An abundant supply of good-quality fresh water for pressure maintenance is present in high-permeability shallow sands.
9. Wells are very easy to drill. Depths are moderate, with the top of the reservoir averaging 7500 ft (2288 m) and total depths around 8000 ft (2440 m).
10. Seismic resolution is good. A 3-D survey has been recorded to detail structural and stratigraphic relationships.

A negative factor is that the lack of gas will require burning crude oil for all major fuel needs, and will require early artifi-

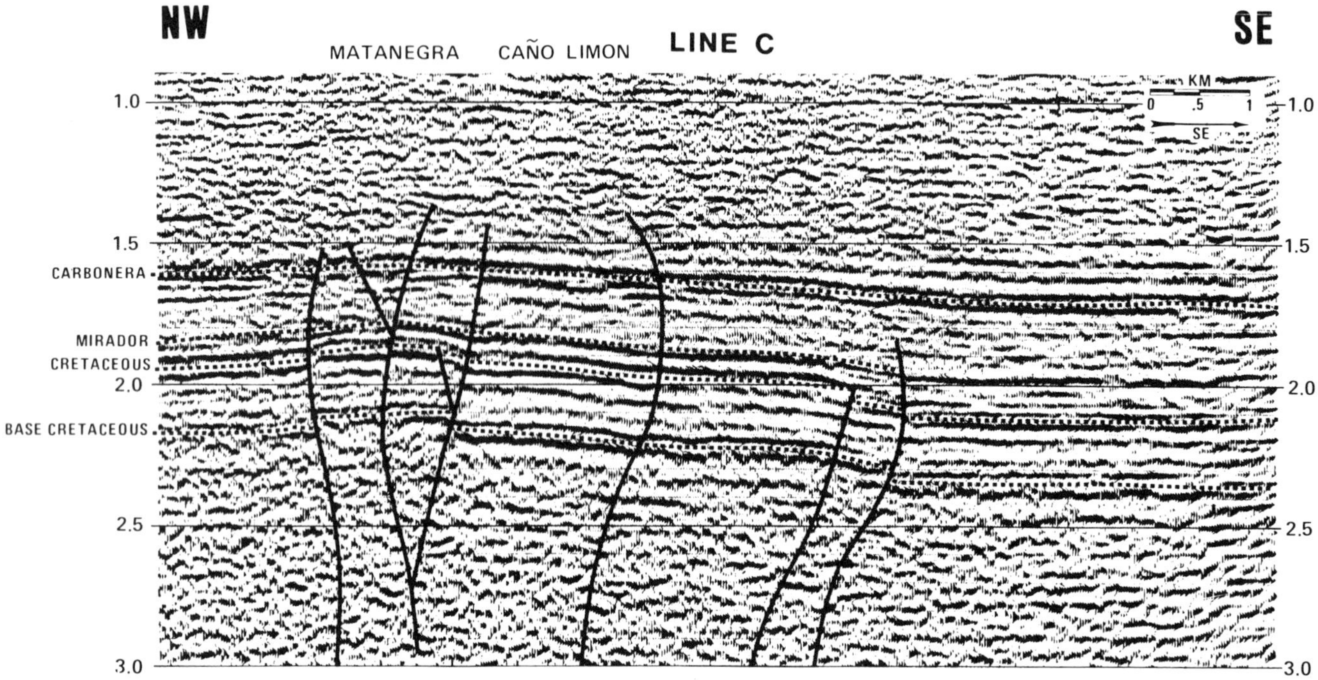


Figure 16.—Line C, northwest to southeast through Caño Limon field (Cravo Norte block).

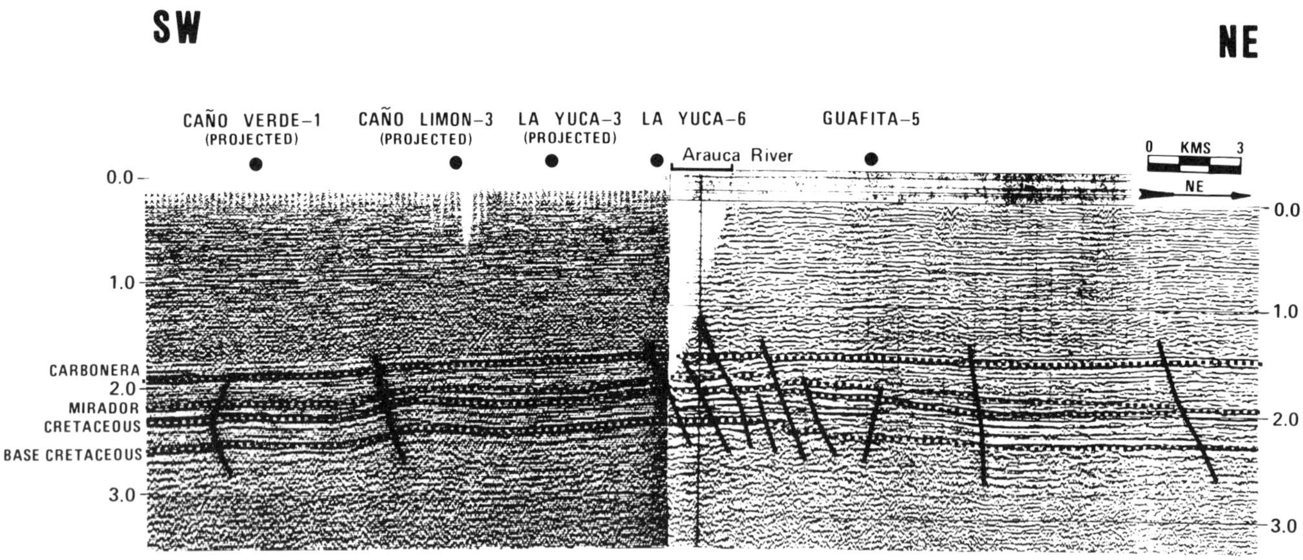


Figure 17.—Line D, southwest to northeast through Caño Limon and Guafita areas.

cial lift by electrical submersible pumps when water cuts start to increase.

Field characteristics are shown in tabular form on Figure 19.

## Structure

The Caño Limon-Guafita structure resulted from strike-slip movements of the Caño Limon and La Yuca faults during the Eocene and early Oligocene. Minor folding may have occurred just prior to Mirador deposition or during early Mirador time, as there is some thinning of the Mirador over the crestal areas. The main folding clearly took place after Mirador and before Carbonera, as indicated by marked thinning of basal Carbonera units over structural highs. Although folding of the Caño Limon structural complex terminated early in Carbonera time, movement on the major faults continued longer, as shown by thickening of higher Carbonera zones in downthrown blocks and fault cuts in some of these higher zones.

Lateral offset on the order of 2 to 4 km on the major faults is estimated from offsets in the thickness trends within specific

## CAÑO LIMON - CHARACTERISTICS

AVERAGE VALUES

| | |
|---|---|
| NET PAY | 192' |
| POROSITY | 24% |
| PERMEABILITY (Stressed Core) | 1145md |
| PERMEABILITY (DST) | 2 - 15 Darcys |
| WATER SATURATION | 23.6% |
| OIL GRAVITY | 29.5° API |
| GAS OIL RATIO | 7 CU. FT./bbl |
| SULFUR | 0.4% |
| EST. REC. EFF. | > 50% |

Figure 18.—Characteristics of the Caño Limon field.

Mirador and Carbonera units and offsets of certain areally limited Cretaceous marker beds.

## Stratigraphy

Cretaceous sands of Albian to Campanian age rest on Cambro-Ordovician sandstones, siltstones, and shales over most of the Caño Limon area (Figure 7). In a few places (Arauquita-1 well and Guafita area) Triassic sand-shale red beds are preserved beneath the base Cretaceous unconformity.

The Cretaceous has been divided into four members—K-1, K-2A, K-2B, and K-3—in the Caño Limon area (Figure 19), but it now appears that this K-1 is not the K-1 as defined by Ecopetrol on a regional basis but is the regional K-2. The regional K-1 is apparently not present. Whether it is missing by nondeposition or truncation is not yet known, but its absence confirms that the Arauca Arch or high was present in the late Cretaceous or early Tertiary.

The K-3 basal member is a massive sand unit of stacked continental braided stream deposits with an average thickness of 800 ft (355 m). Oil has been found in the K-3 only in the Caño Rondon-1 well, a small discovery 33 km southwest of Caño Limon.

The K-2 members of Santonian to Coniacian age are marine-dominated regressive sequences (deltaic) containing several thin calcareous sandstones and shelly limestones. The K-2A averages 250 ft (76 m) in thickness. It is a major reservoir sand in the Matanegra fault block, where it has produced at a rate of 10,784 BOPD. It has also produced oil in the Caño Rondon-1 well. The K-2B averages 120 ft (37 m) in thickness but has not yet been found to be productive.

The K-1 falls within the Santonian to Campanian age interval. It is a more marine sequence of pro-delta shales, thin limestones, and shelf sands with deltaic sands near the top. This interval is productive in the La Yuca and Matanegra areas and produced at rates of up to 8000 BOPD in La Yuca-4. The average thickness of the K-1 is 250 ft (76 m).

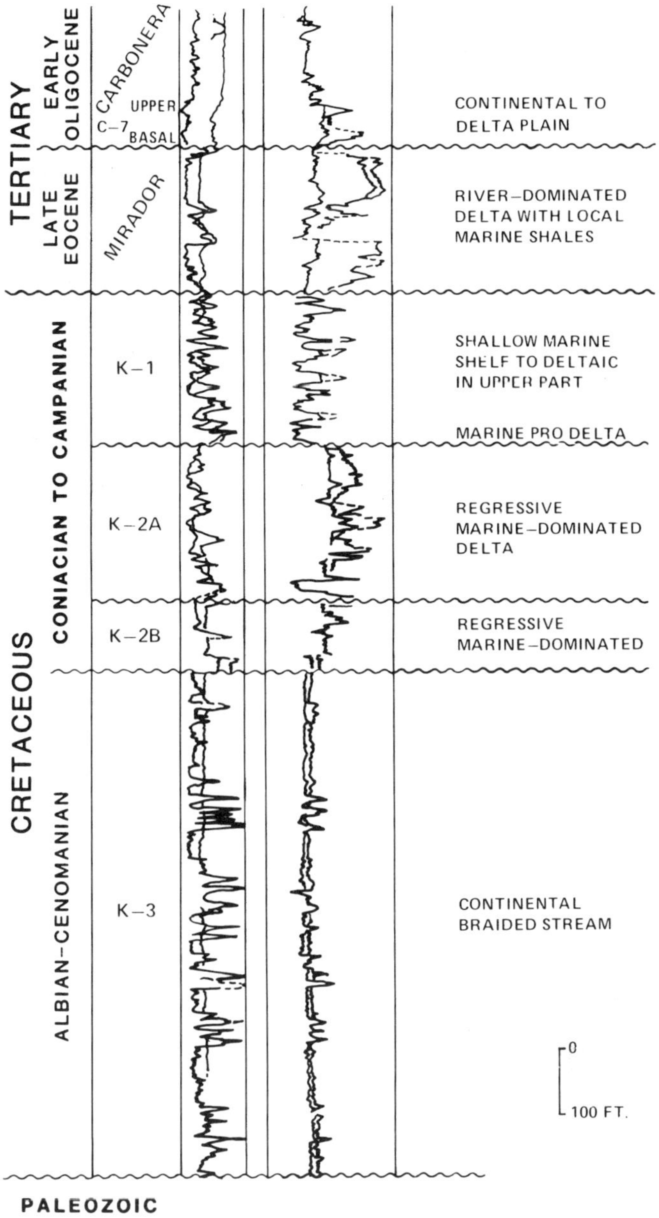


Figure 19.—Sedimentary sequence of the Cretaceous and Tertiary, Caño Limon field.

The Mirador zone of Eocene age is mainly a river-dominated delta system but has several marine shales, delta fringe sands, and probably wave-dominated delta sands. The river-dominated delta channel sequences are commonly massive sand bodies with excellent permeability and good continuity in the channel direction. Continuity normal to the channel direction is expected to vary. Pressure interference tests between wells have all shown rapid response when equivalent Mirador zones were open, and no response when the zones were not the same. An injectivity test into the La Yuca-22 water-injection well produced pressure responses in Caño Limon-6 after 7 hours of injection (a distance of 2. 4 km) and in La Yucua-11 after 11 hours of injection (a distance of 1. 6 km).

The average thickness of the Mirador zone is 260 ft (79 m).

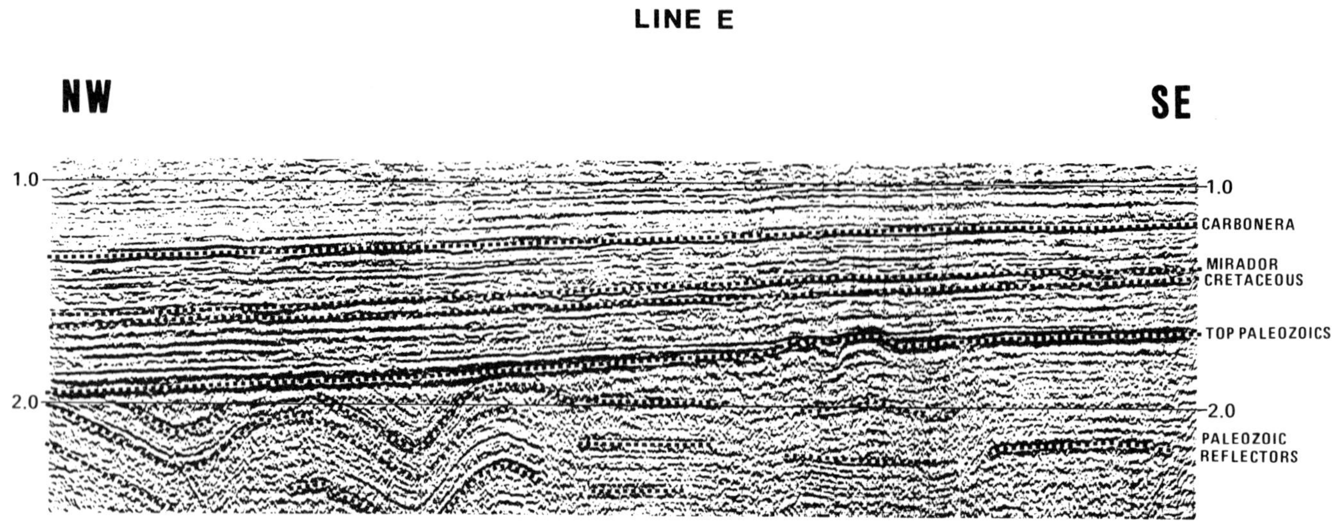


Figure 20.—Line E, showing a Paleozoic graben in the Llanos Basin.

Approximately 80% of the Caño Limon reserves are in this reservoir.

The basal sands of the Carbonera C-7 zone are productive on the flanks of the Caño Limon and La Yuca structures. The distribution and nature of these sands are not yet well understood.

## Future Potential

Although there may not be other fields as large as Caño Limon in the Llanos Basin, the normal distribution of field sizes typically found in producing basins would indicate that a number of fields 100 MMBO or larger should yet be found.

*Paleozoic.* This mainly Cambro-Ordovician zone is almost unexplored. Seismic shows extensive structural deformation of the Paleozoic in various parts of the basin (Figure 20), but available stratigraphic information is only modestly encouraging. Most of the sandstones are very well indurated, with porosities of reservoir quality in only a few wells. Potential source rock has been seen in a few wells, but little is known about quantity, quality, or distribution. Areas that need further testing are the apparent carbonate buildups in the northern graben and any area where seismic shows structures with zones of high seismic reflectivity.

*Cretaceous.* In addition to any structural traps that are found, there should be stratigraphic trapping potential in sands such as turbidites that are encased in basinal shales.

*Mirador and Carbonera.* At the present time there is considerable remaining potential for small (5 to 15 MMBO) discoveries in the antithetic fault traps. Additionally, stratigraphic traps at several levels in the Carbonera offer potential.

## ACKNOWLEDGMENTS

The writer thanks the management of Occidental International Exploration and Production Company, along with the managements of partner companies in Colombia, Ecopetrol and Shell, for permission to publish this paper.

I also wish to recognize the efforts of all those who participated in and supported the Llanos exploration program. This includes intended co-author, Earl Padfield, who was unable to participate in the writing of this paper, but whose earlier work and ideas have influenced our understanding of the Llanos and of the Caño Limon field.

## REFERENCES CITED

McCollough, C. N. , and E. P. Padfield, 1985, Petroleum geology of Colombia's Llanos Basin: A summary: Oil and Gas Journal, Reprint from April 15 and 22 issues, p. 9-13.

Robertson Research, 1983, The Northern Llanos of Colombia—Hydrocarbon potential and stratigraphic control, 3 volumes.

Valderrama, R. , 1982, Desarrollo de Facies de la Cuenca de los Llanos orientales de Colombia: Simposio "Exploracion petrolera en las cuencas subandinas de Venezuela, Colombia, Ecuador y Peru": Bogota, Asociacion Colombiana de Geologos y Geofisicos de Petroleos.

CHAPTER 26

# IDENTIFICATION AND EVOLUTION OF ANTARCTIC SEDIMENTARY BASINS

**Patrick G. Quilty**
*Antarctic Division, Department of Science*
*Kingston, Australia*

ABSTRACT

Antarctic sedimentary basin identification and evolution are researched by assembling data that are more fragmentary than data available for equivalent studies on other continents.

Traditional methods—geological and geophysical field work—have proved deficient in identifying onshore Antarctic sedimentary basins, because 98% of the continent is covered with an average thickness of 2.5 km of ice. Radio echo sounding combined with aeromagnetic surveys can identify basin location and depth to basement. Studies of coastal outcrop and the fossil record contained in offshore glacial debris give clues to the age and environment of deposition of sediments in onshore basins.

Offshore basins can be identified by standard marine geophysical and geological techniques, but specialized shipping and supply lines are needed. The evolution of the basins is assembled by using evidence from tectonics, shallow research drilling, offshore geophysics, analysis of microfossils in glacial debris, and whatever other piecemeal data become available.

## INTRODUCTION

Until recently, Antarctica has been seen as a remote, beautiful, forbidding, romantic place for the scientifically inclined and adventurous. It was perceived as extraordinary and unique.

Now, with the growth of the concept of "spaceship earth," the Antarctic is being recognized as just another region of a total global ecosystem. Its atmosphere and ocean water regularly interchange globally, and anthropogenic effects from elsewhere do influence it and are leading to some concern.

Increased knowledge of the continent has contributed to its demystification and has given rise to the possibility that it is a potential source of resources such as water (from ice), marine protein (from krill and fish), and minerals (particularly offshore hydrocarbons).

As in so many aspects of things Antarctic, the search for sedimentary basins has called for uniquely Antarctic techniques to solve uniquely Antarctic problems.

Antarctica is the coldest, windiest, and most isolated continent. It is characterized by long periods of winter darkness and summer light. It is inaccessible during winter, and at best, supply lines are long, fragile, and expensive.

Specialized shipping and aircraft are needed, but even these can be frozen in or can break down, as the early 1985 besetting of the icebreaker M/V *Mikhail Somov* and the late 1985 besetting of M/V *Nella Dan* amply demonstrate. The sinking in recent years of M/V *Gottland II* and *Southern Quest* shows the dangers of using inadequate shipping.

Because the area claimed by Australia is where its program is concentrated, I have focused my comments on it (Figure 1).

## DATA BASE

The search for Antarctic sedimentary basins depends on the application of many techniques and the integration of a variety of very inadequate data sets.

### Field Observations

For basin identification on other continents, it has been traditional to obtain the first indications from the results of geological and geophysical fieldwork.

The results of this approach in Antarctica are not very helpful. Approximately 98% of Antarctica is covered to an average thickness of some 2500 m by ice. Outcrop is very limited except in such places as the Transantarctic Mountains. Exposures are generally on the edge of a roughly saucer-shaped continental surface.

The Transantarctic Mountains constitute the best exposed area and contain magnificent exposures of essentially flat-lying nonmarine Triassic overlying nonmarine Permian sedimentary rocks, little disturbed other than by faulting and

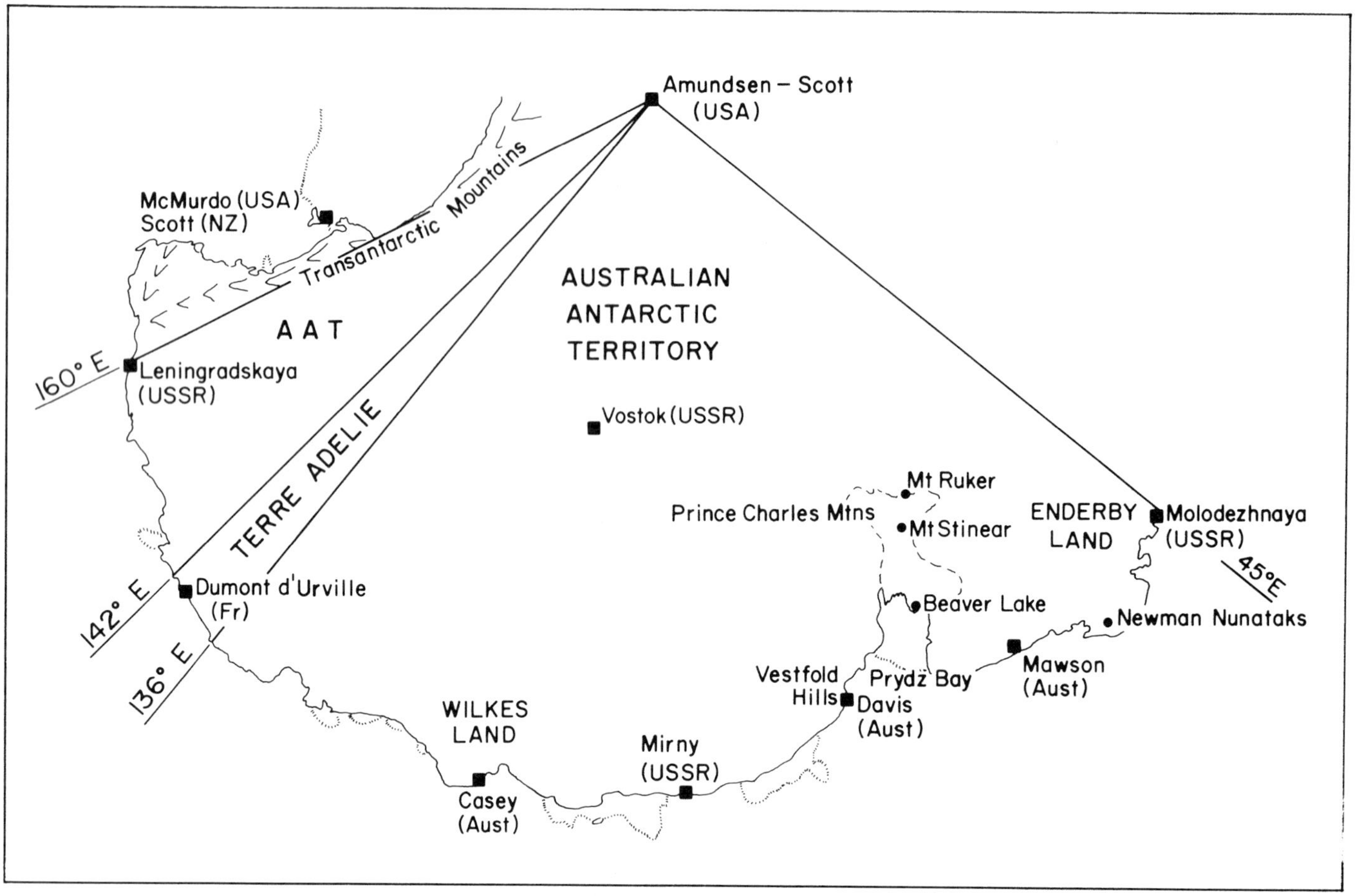


Figure 1. The area under consideration including key localities.

tilting, associated with intrusion of the mid-Jurassic Ferrar Dolerites. Sedimentation seems to have been essentially continuous, consisting dominantly of sandstone, with coal and glacial sediments in the Permian.

These sediments disconformably overlie little-disturbed Devonian and Cambrian marine sediments. The stratigraphic column and tectonic activity are shown diagrammatically on Figure 2.

Webb and collaborators (Webb et al., 1984) have shown the existence of faunas and floras of various Late Cretaceous and Tertiary ages in young glacial sediments at high altitudes. The provenance of this fossil material is the subject of considerable debate. The sediment ages seem to equate reasonably with high sea level peaks and possibly "warm" seas in the Tertiary, as suggested by Vail, Mitchum, and Thompson (1977), Quilty (1977, 1980), and McGowran (1979).

Elsewhere in the Australian Antarctic Territory (AAT), exposed rocks are very dominantly old Precambrian high-grade metamorphic assemblages affected by a continent-wide metamorphic/tectonic event about 500 million years ago. They are intruded by Phanerozoic basalts of various ages (James and Tingey, 1983).

In the Beaver Lake area of the Lambert Graben (Crohn, 1959; Mond, 1972), there are Permian nonmarine sediments with coal (Figure 3) in a large-scale Antarctic aulacogen.

Pliocene marine sediments a few meters thick have recently been identified in the Vestfold Hills (Pickard et al., in press), and this find suggests rocks of this age may be more widespread.

Many areas of Antarctica contain glacial sediments, both dated (for example, 5000 to 8000 years and younger in the Vestfold Hills; Adamson and Pickard, 1983) and undated (see Tingey, 1982).

Other than in the Transantarctic Mountains, sediment thickness is not known, from fieldwork studies, to be very great.

## Radio Echo Sounding

Perhaps the most spectacular technological development to provide answers to peculiarly Antarctic questions was the concept of radio echo sounding, which can be used to identify:

- the ice-atmosphere interface
- the ice-bedrock interface
- any major inhomogeneities in the ice

A flat reflection at the base of the ice has led to the mapping of large subice lakes, but whether fresh or saline water is not clear. Figure 4 is an illustration of a typical ice radar record.

Roughness of subice topography has been taken as a guide to bedrock type, whether hard basement or softer, smoother material such as sediment cover. Boundaries between various units can also be interpreted (for example, as faults).

Research based on a very extensive 30- to 50-km grid gives a rough approximation of subice structure and rock type. The

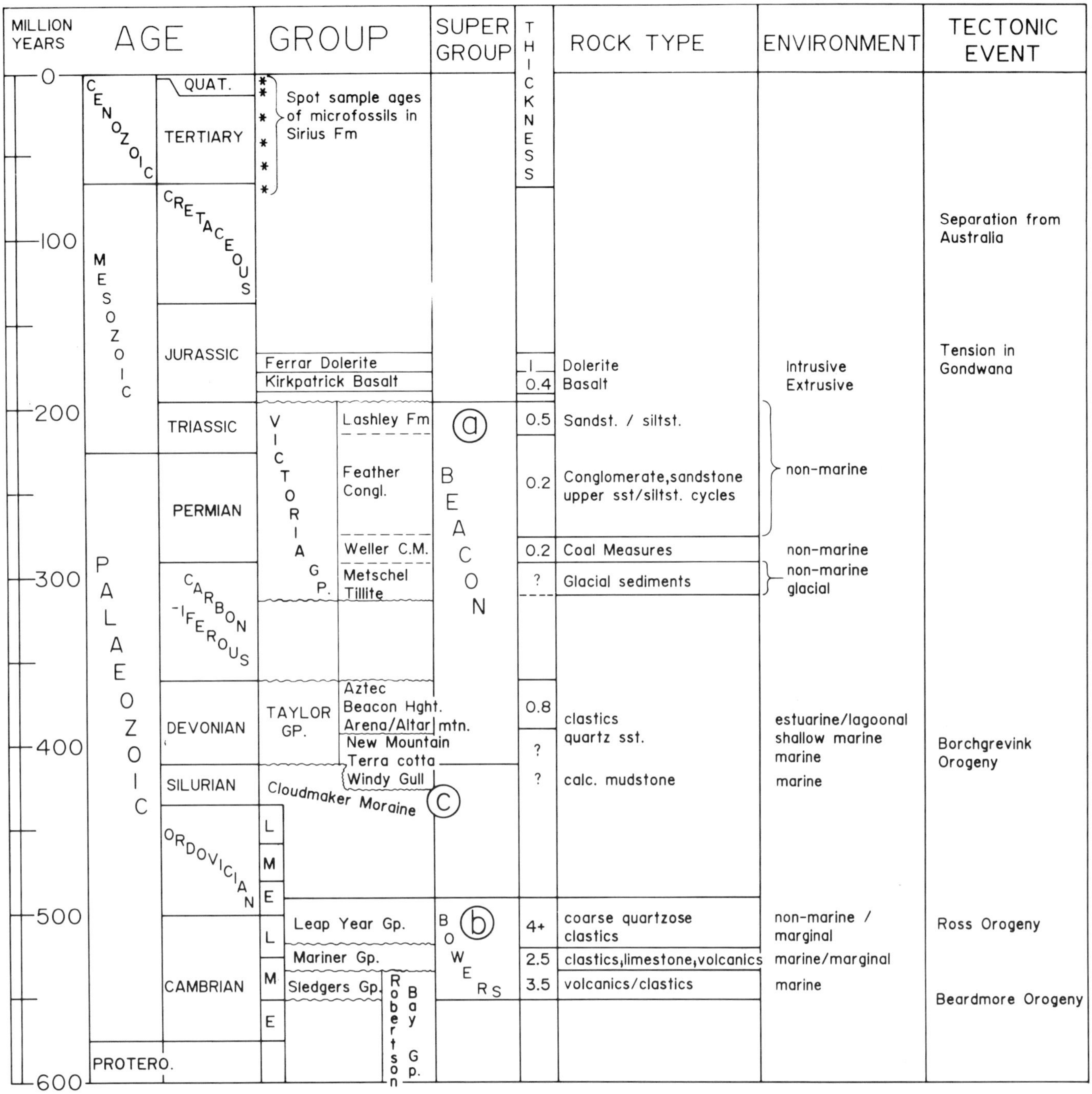


(a) Stratigraphy from Whitby, Rose & McElroy, 1983, S.Victoria Land & Kyle & Schopf, 1982.

(b) After Laird & Bradshaw, 1983, N.Victoria Land

(c) After Jell 1981

Figure 2. Stratigraphic/tectonic summary, Transantarctic Mountains.

grid covers more than half the Australian Antarctic Territory.

Rare airborne magnetometer information constrains hypotheses on depth to magnetic basement. Combined with echo sounding data, the program provides data on:

- ice surface elevation (Figure 5)
- ice thickness (Figure 6)
- internal ice stratigraphy
- areas of subice lakes
- subice topography (Figure 7)
- subice general geology
- thickness of subice sediment cover

A few gravity surveys, satellite magnetometer measurements, and seismic-refraction and seismic-reflection surveys supplement radio echo sounding to yield data on onshore Antarctic basins (Drewry, 1972, 1975, 1976; Steed, 1983; Steed and Drewry, 1982; Ritzwoller and Bentley, 1983).

This program has been used to identify the Wilkes and

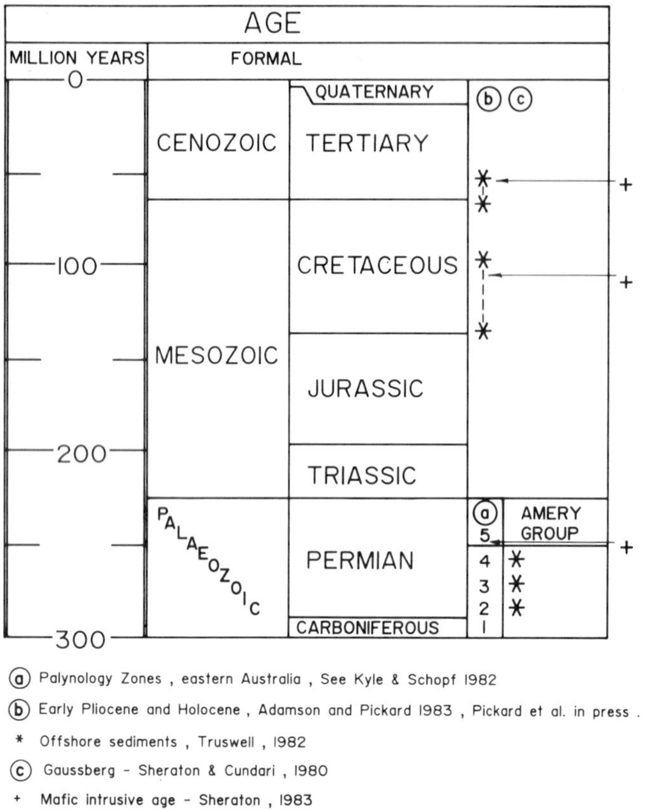


Figure 3. Stratigraphy of the Lambert Graben, east Antarctica.

Aurora Basins (Figure 8), the Gamburtsev Mountains, and Highlands A, B, C, and D. The Wilkes Basin is some 1400 km long with a north-south orientation and 200 to 600 km wide, widening to the north. The highlands separate it from the less well defined Aurora Basin to the west.

Sediment thicknesses in the Wilkes Basin, although normally about 3 km, may locally be as much as 6.5 km. In the Aurora Basin, 2 km seems a reasonable approximation. These are essentially spot thicknesses. The Wilkes Basin may extend to the coastline and onto the continental shelf in the vicinity of long. 147°-155°E.

## Continental Reconstruction

Antarctica has often been referred to as the "Keystone" of Gondwana (Figure 9), which broke up by a series of well-documented steps when various land masses moved away and left Antarctica close to the south geographic pole (Veevers, 1984). This evolution and its economic consequences were reviewed by Quilty (1984) and Tingey (1984).

India wrapped around Enderby Land (Figures 10, 11) with southern India and Sri Lanka resting to the east of Gunnerus Bank (Fedorov et al., 1982; Johnson and Veevers, 1984). India

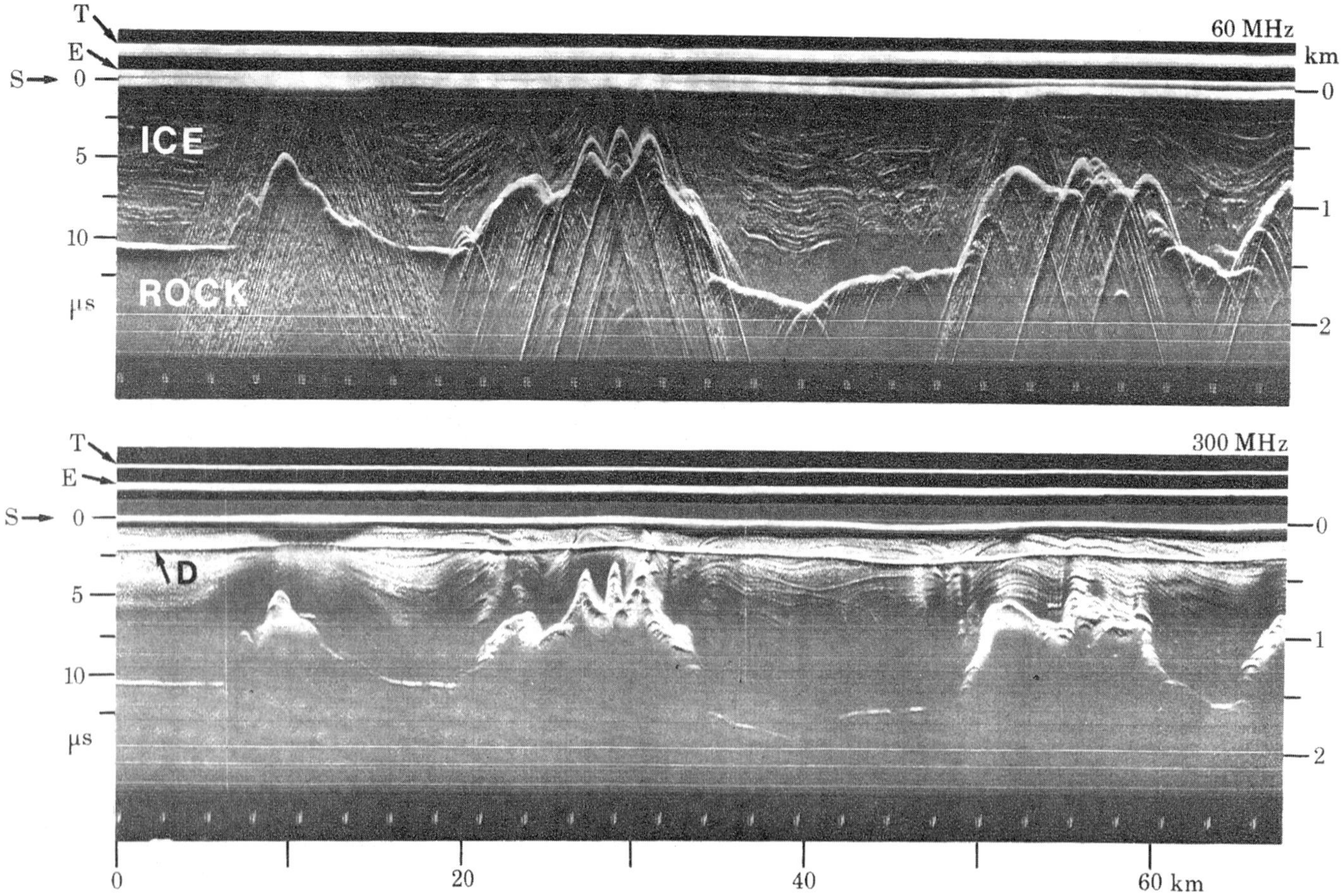


Figure 4. Typical ice radar records at 100 and 300 MHz (from Robin, Drewry, and Meldrum, 1977). Reproduced with approval of the Royal Society of London.

departed Gondwana in the Early Cretaceous (128 million years ago; Johnson and Veevers, 1984) allowing the development of marine conditions along this part of the AAT coast (Figure 12). The Lambert Graben may contain a significant nonmarine sequence. The stratigraphy probably is similar to that of the east coast basins of India (Sastri et al., 1974).

Australia lay against Antarctica well east of the Indian position. The southwestern tip of Australia may once have lain opposite the position of Casey station (Figure 1) and Tasmania off King George V Land.

Johnson, Powell, and Veevers (1976) envisaged Greater India filling the gap between the reconstructed positions of present-day India and Australia. Other reconstructions leave a gap as oceanic *sinus australis* (Dietz and Holden, 1971).

Australia departed Antarctica at or before 86 million years ago (Late Cretaceous) (Cande and Mutter, 1982) or 95 million years ago (Veevers, 1984), but movement was very slow until 55 million years ago (Late Paleocene) when the process accelerated (Weissel and Hayes, 1972, 1974; Cande and Mutter, 1982). Thus restricted marine conditions became established along this part of the AAT coast (Figure 13).

Tensional graben probably marked the continental rifts, and nonmarine infill, in structures parallel to the coast, can be expected. The mode is probably similar to that of the Ceduna Terrace (Bouef and Doust, 1975; Fraser and Tilbury, 1979; Bein and Taylor, 1981) or Otway Basin (Douglas and Ferguson, 1976).

There is a series of subice highs at approximately long. 135 °E. This corresponds to the position of the Flinders Ranges (which were not prominent when the continents were joined). Veevers (1982, 1984) examined this in detail and used it to predict that both Australia and Antarctica had complementary Mesozoic drainage systems (Figure 14) emptying into the Ceduna Terrace, Otway Basin, and Antarctic equivalent positions. Such drainage has implications for the volume of sediment to be expected off the Antarctic coast. Veevers and coauthors also took the Eastern Australian Highlands and the Transantarctic Mountains to be extensions of each other, to have a similar history, and to provide an eastern limit to the proposed drainage.

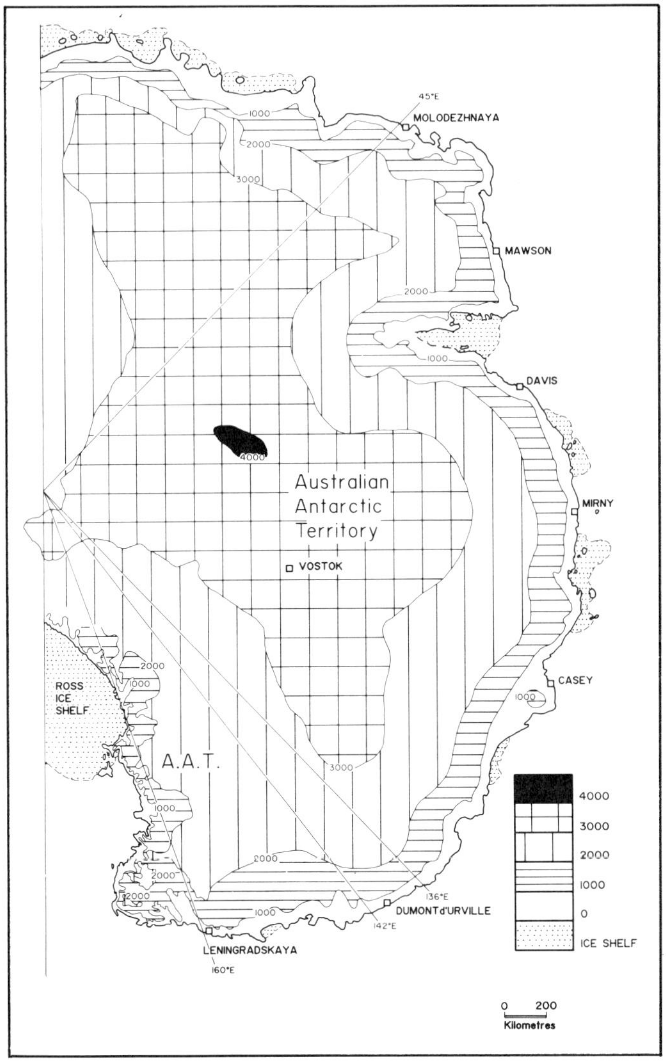


Figure 5. Ice surface elevation (in meters).

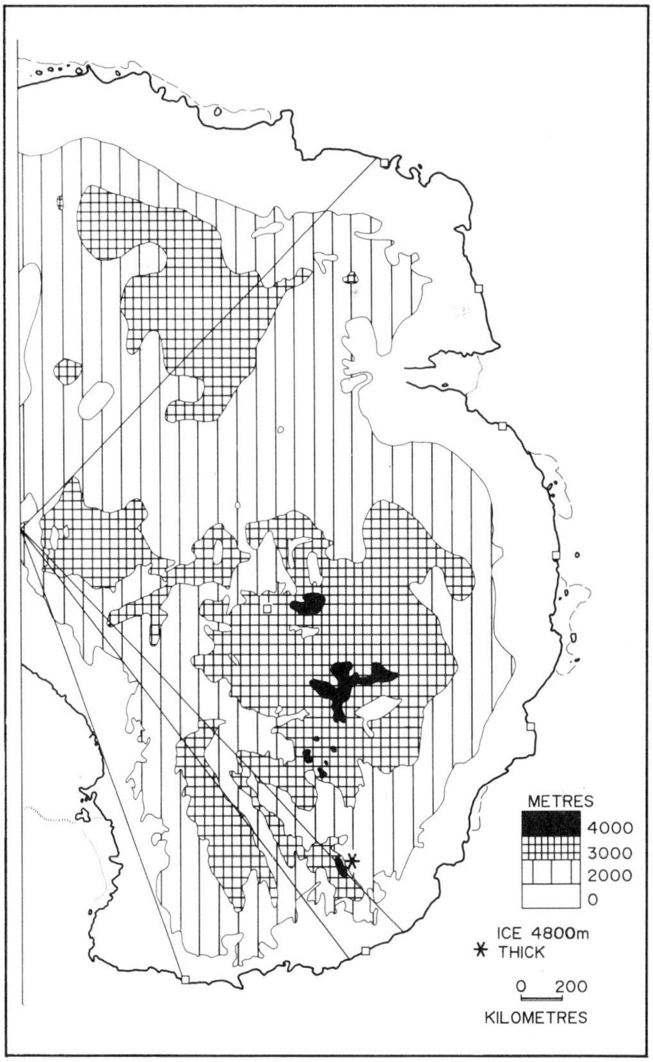


Figure 6. Ice thickness from ice radar studies (after Drewry, 1983).

Figure 7. Subice topography from ice radar studies (after Drewry, 1983).

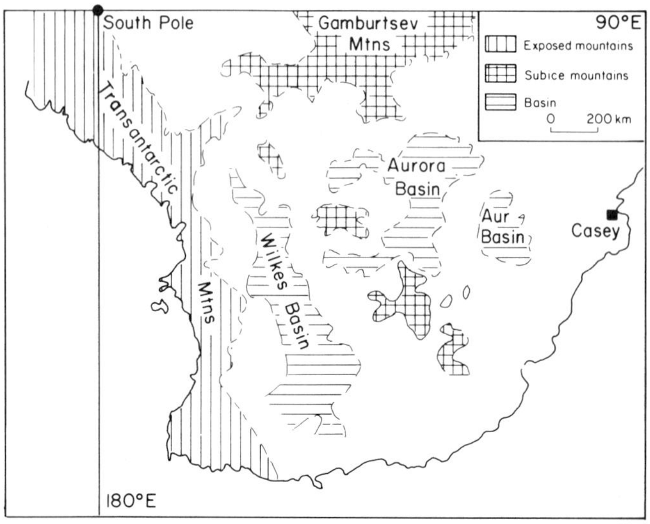


Figure 8. Wilkes and Aurora basins and Gamburtsev Mountains, east Antarctica (after Drewry, 1976).

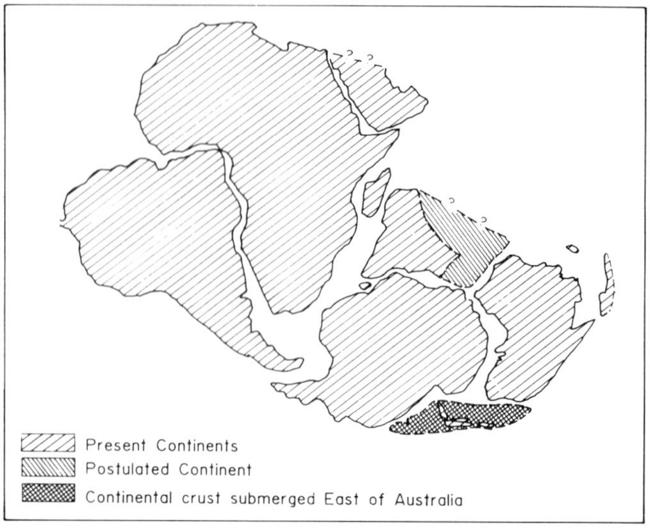


Figure 9. Gondwana until 128 million years ago.

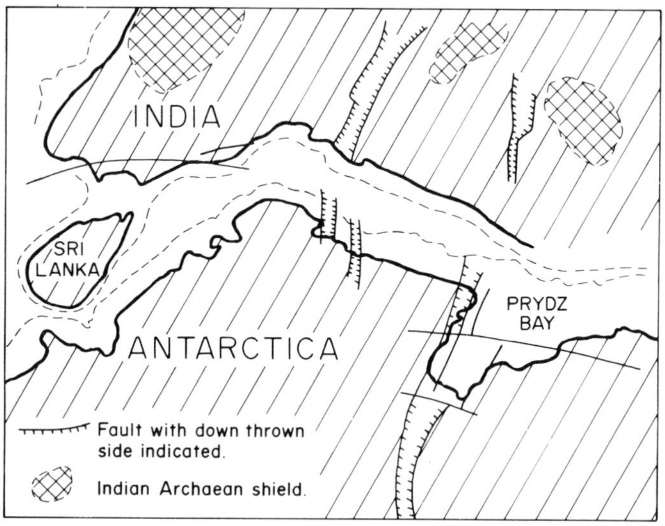


Figure 10. Relationship between India, Sri Lanka, and Antarctica Gondwana (after Fedorov, Ravich, and Hoffman, 1982).

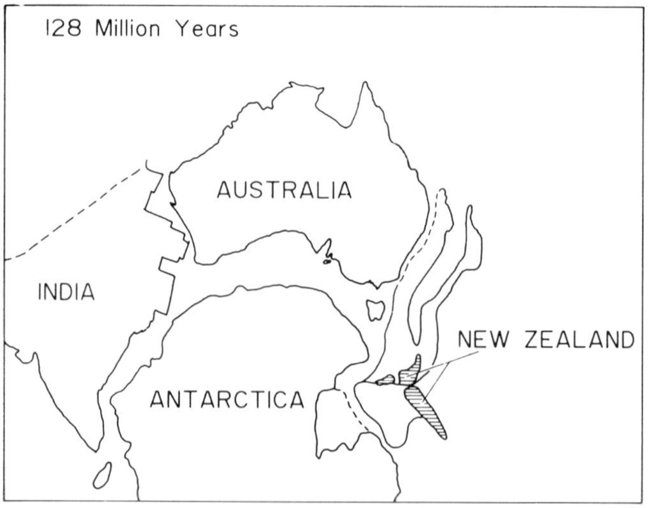


Figure 11. Gondwana (part only) at 128 million years (after Veevers, 1984).

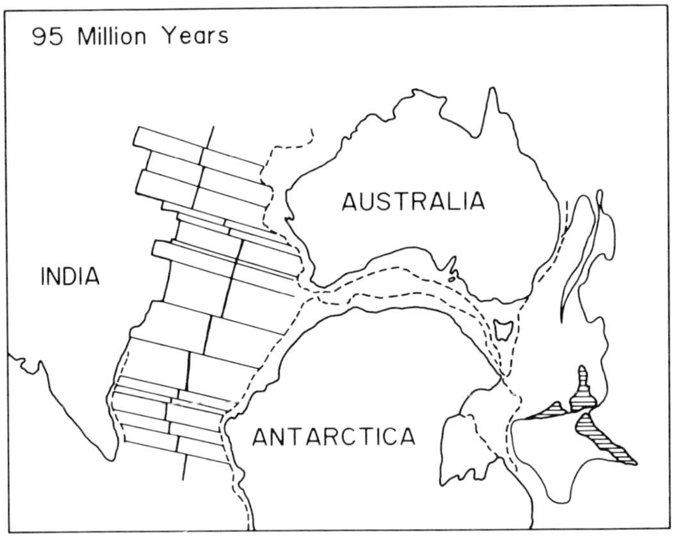


Figure 12. Gondwana (part only) at 95 million years (after Veevers, 1984).

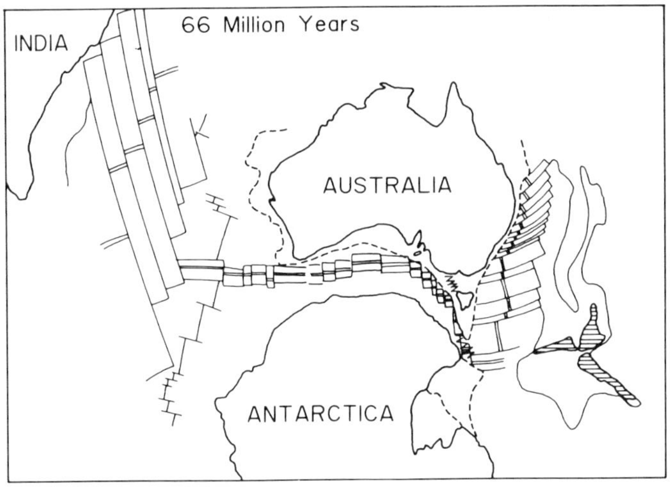


Figure 13. Gondwana (part only) at 66 million years (after Veevers, 1984).

## Marine Seismic Surveys

Marine seismic surveys in Antarctic waters are in their infancy, and vast areas have had no surveys at all as yet.

In the deep ocean basins, there have been numerous single-channel surveys but the sediment is generally thin and in a nonbasin mode. The results until 1971 were summarized by Houtz and Markl (1972) and are reproduced as Figure 15. Later surveys of this type have not changed the basic pattern outlined in that paper.

Inadequate seismic cover near the continent is a result of:

- inability to penetrate ice with traditional seismic survey vessels (minimal ice strengthening is not enough),
- the period available for surveying is short and not accurately predictable,
- the danger to expensive survey equipment,
- sea conditions which are harsher than experienced elsewhere,
- long and expensive supply lines.

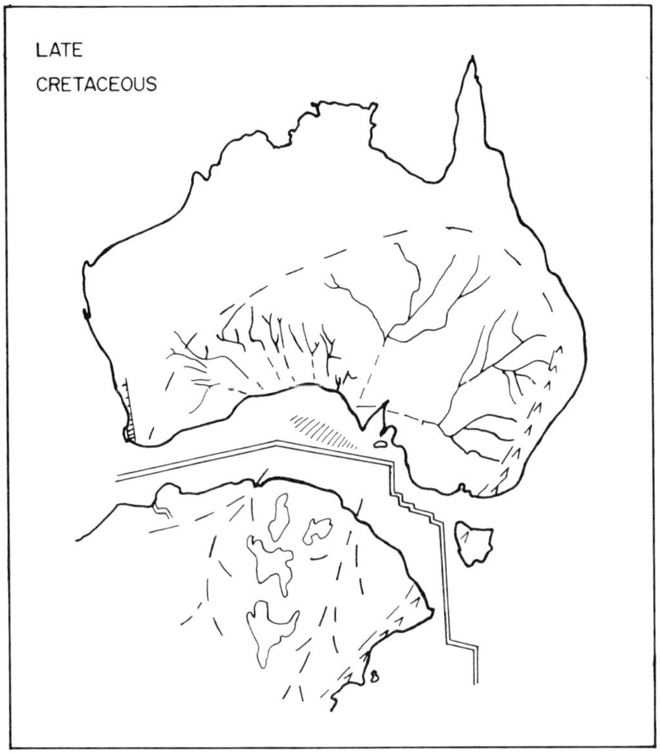


Figure 14. Complementary Late Cretaceous drainage systems in Australia and Antarctica (after Veevers, 1982).

Seismic programs off the subject area (Figure 16) are:

- 1981/82 Australian program in Prydz Bay (Stagg, Ramsay, and Whitworth, 1983; Stagg, 1985)—5000 km of 6-fold reflection data.
- 1981/82 Prakla-Seismos survey—3000 km of 48-fold reflection data off Terre Adelie and the eastern sector of the AAT (Behrendt 1983; Wannesson et al., 1985).
- 1982/83 Japanese program in the Ross Sea and off Terre Adelie (Sato et al., 1984; Tsurumaya et al., 1985)—820 km of seismic-reflection data, 680 km of multichannel data.
- 1983/84 United States program (Eittreim, Cooper, et al., 1984; Eittreim, Hampton, and Childs, 1985; Eittreim and Smith, 1986)—1800 km of 24-fold reflection data in approximately the same position as the French program but extending 5° west.
- 1984/85 Japanese program in the Prydz Bay region (Mizukoshi et al., in press)—2350 km of seismic-reflection data.
- 1985/86 program of USSR—1000 km of multichannel seismic, Prydz Bay region; details not yet available.

There were plans for a 1985/86 Japanese cruise, again using RV *Hakurei Maru* off Queen Maud and Enderby Lands, but details are not yet available.

The Australian program (Stagg, 1985) (Figure 17) identified a significant sedimentary basin in the Prydz Bay region and showed that the structural grain, previously identified onshore, continued offshore and controls deposition.

All is consistent with the failed-arm concept, perhaps a four-armed rift-rift-rift-rift, including the Mahanadi Graben of northeast India as an opposing failed arm, and the other two

Figure 15. Distribution of sediment thickness between Australia and Antarctica (from Houtz and Markl, 1972). Reproduced with the approval of the American Geophysical Union.

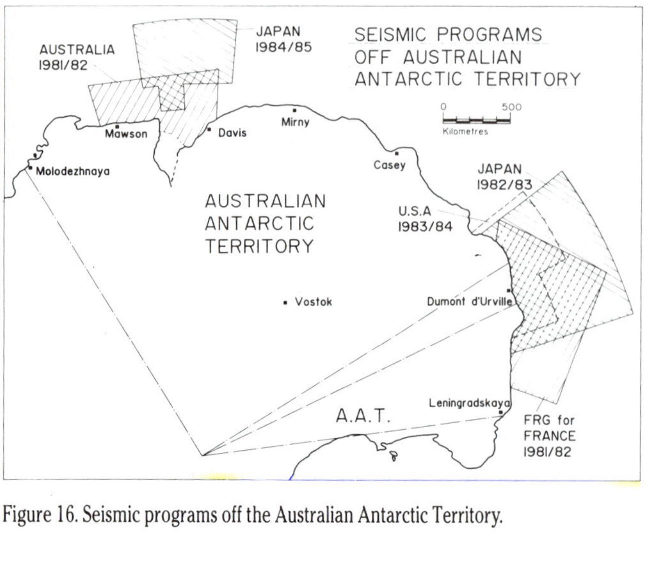


Figure 16. Seismic programs off the Australian Antarctic Territory.

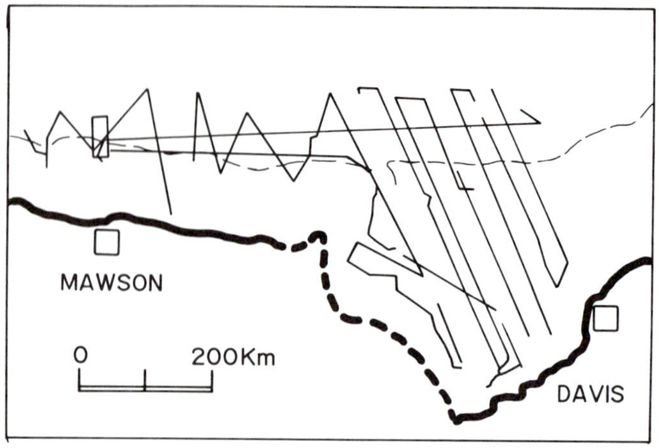


Figure 17. Track chart M/V *Nella Dan* seismic program, early 1982.

forming the rift along which the proto-Indian Ocean developed.

Stagg differentiated the thick sequence in Prydz Bay from that on MacRobertson Shelf which consists of thin carbonate-bearing sediments, probably everywhere overlying acoustic basement. He identified six shallow (PS1-6) and five deep (PD1-5) water sequences (Figure 18), the former on the continental shelf and the latter on the slope and rise.

PS1 and 2 are tentatively identified as postbreakup sediments and thus marine and younger than 125 to 130 million years. PS3-5 are taken to be Permian to Early Cretaceous and thus prebreakup and dominantly terrestrial. PS6 is thought to be Cambrian and older. PS1-5 are believed to be dominantly terrigenous. The results described by Mizukoshi et al. (in press) are generally in agreement with those of Stagg (1985).

Sediment thickness may be as great at 10 km, based on the informally reported USSR program.

The main depocentre in Prydz Bay is in the eastern half of the bay.

In the deep-water sequences, PD1 and 2 also seem to be postbreakup sequences, and PD3 seems to be a volcanic unit of limited distribution. PD4 and 5 are questionably identified as belonging to other Mesozoic units. At least 3 km can be identified.

Three major seismic surveys have been conducted off the coast of Terre Adelie and George V Land by France, Japan, and the U.S.A. The timing of the programs and publication coincided with an interval during which ideas of continental-margin evolution and Australia-Antarctica breakup were evolving rapidly.

Wannesson et al. (1985) summarized the Prakla-Seismos program, which included a pass over DSDP Site 269 and a profile from the Wilkes Abyssal Plain across the continental shelf.

The cover on the abyssal plain (Figure 19) consists of 1 to 1.5 km of sediments lying on oceanic basement. The upper sedimentary unit, which represents two-thirds of the total column, was taken to be composed of thin-bedded Neogene turbidites; this is consistent with the interpretations of Hayes and Connolly (1972) and Connolly and Payne (1972) for the abyssal plain morphology in this area.

Deeper units were interpreted as a mixed coarse sediment/volcanic assemblage overlain by a transparent, dominantly pelagic unit, which together represent Middle Eocene and Oligocene sedimentation. These correspond to units not reached

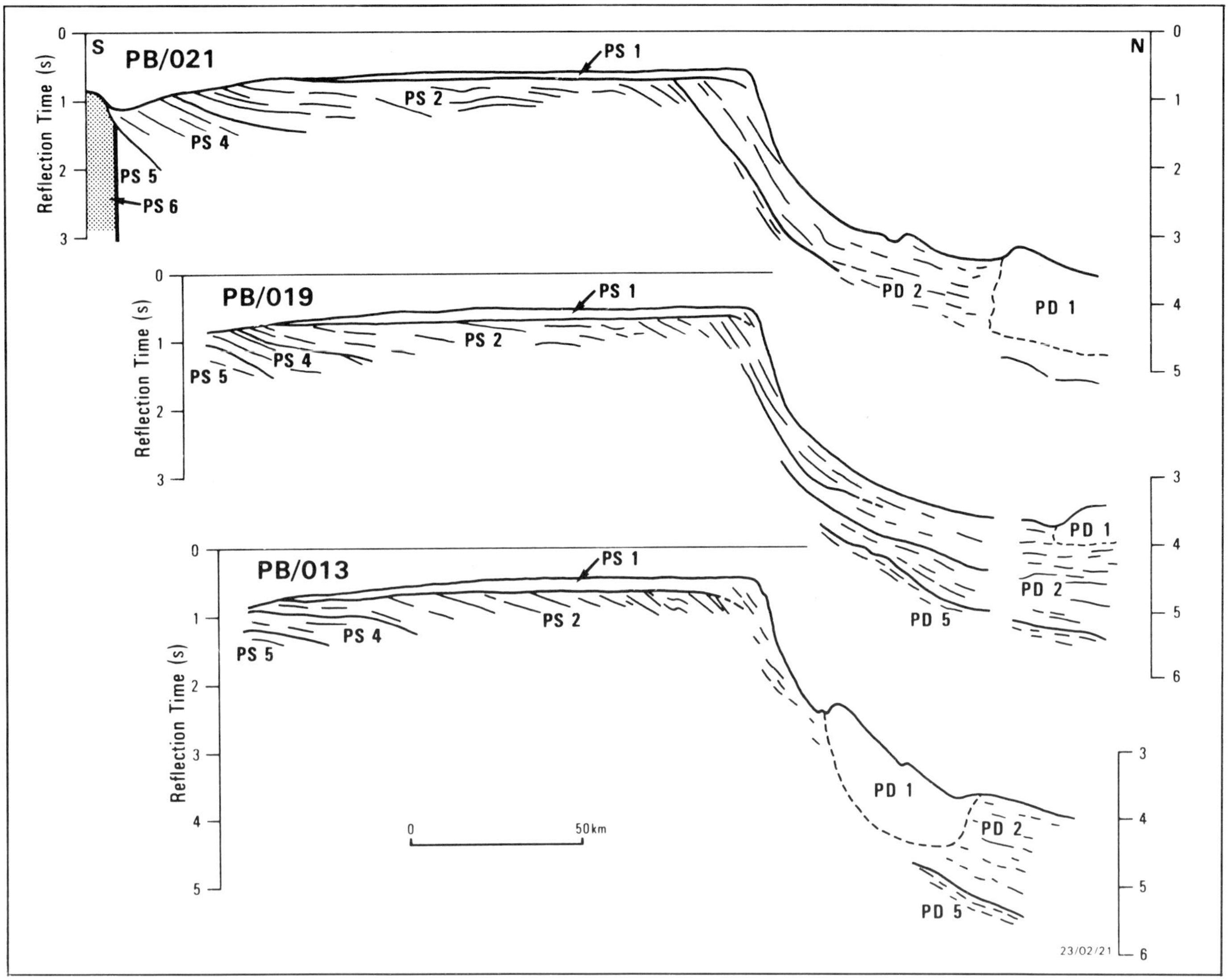


Figure 18. PS and PD sequences identified in Prydz Bay (from Stagg, 1985). Reproduced with the approval of *Tectonophysics*.

during the drilling of DSDP Site 269, which penetrated only 958 m of an estimated 1500-m section.

Interrupting the abyssal sequence is a 150-km-long "high" against which all three units discussed earlier thin markedly. It was taken as a volcanic unit, probably formed at the time of sea-floor generation, and may equate with the volcanic unit identified as PD3 by Stagg (1985).

The lower units (1 and 2) thicken onto the continental slope and rise. The total sequence thickens again onto the outer continental shelf, but the shelf thickening is caused by an expansion of the younger unit 2 at the expense of unit 1 which thins southwards by onlap onto basement (Figure 20). Maximum thickness on the continental shelf (here some 120-km wide) is 6 to 7 km.

On the slope, two unconformities are identified bounding Unit 2, which must then be dominantly Oligocene as the unconformities were interpreted as being Late Eocene and Late Oligocene by inference from oceanographic development in the region at the time. The Late Eocene unconformity is taken to correspond with the opening of the gap between Antarctica and the South Tasman Rise. The Late Oligocene unconformity is then due to the effects of development of major glaciation on Antarctica, so well documented in the Ross Sea (Hayes and Frakes, 1975).

The Japanese 1982/83 Terre Adelie cruise (Sato et al., 1984), like the French (Wannesson et al., 1985), also passed over DSDP site 269 (Figure 21).

Sato et al. (1984) recognized four stratigraphic units taken to represent an entire Tertiary section. Unit D3 of Sato et al. equates exactly with Unit 1 of Wannesson et al. (1985). The geological interpretation is different. Sato et al. take D3 to be Early Paleogene, in contrast with the Middle Eocene (at the base) of Wannesson et al. Wannesson et al. identified Unit 2 as extending from the top of 1-D3 to approximately the base of the DSDP Site 269 drilled section, and took it to be mainly Oligocene. Sato et al. took this interval to be Middle Paleogene, probably in conflict with Wannesson et al. Elsewhere, Sato et al. identified more than 5 km of Tertiary, not much in conflict with Wannesson et al.

The cruise also collected piston cores and bottom-sediment

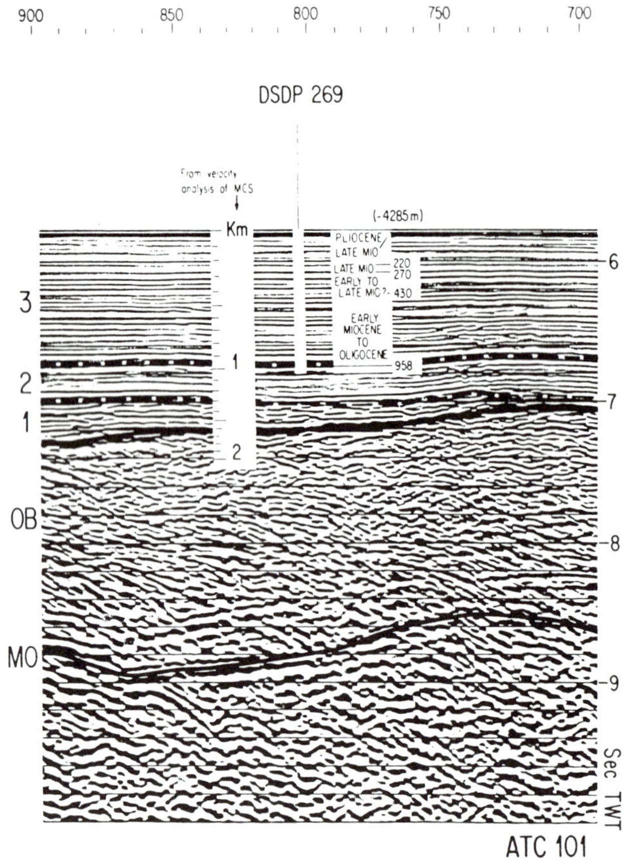


Figure 19. *M.V. Explora* record over DSDP Site 269, and its interpretation (from Wannesson et al., 1985). Reproduced from *Marine and Petroleum Geology* by permission of the publisher, Butterworth and Co. (Publishers), Ltd.

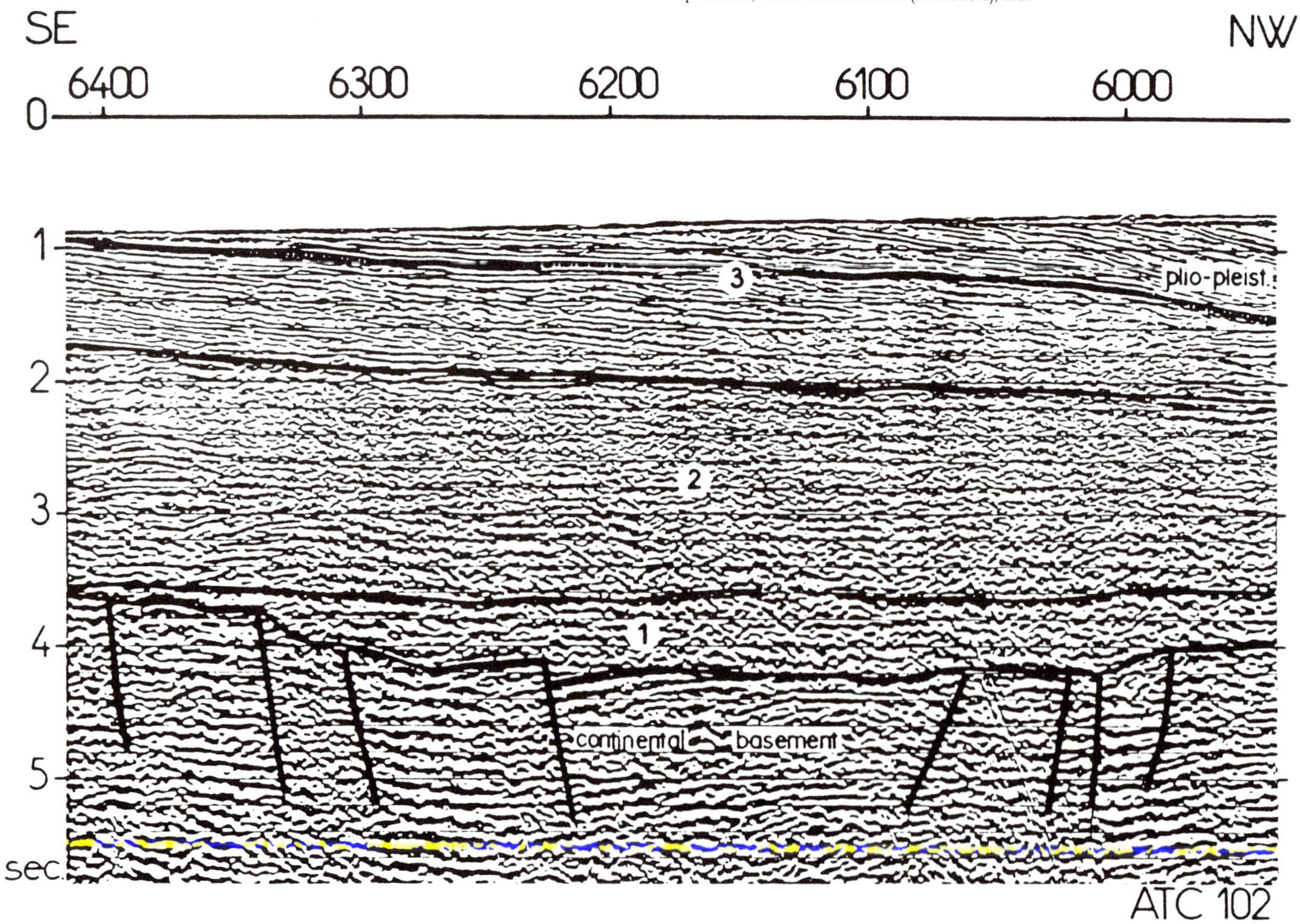


Figure 20. Variation in unit thickness on the continental shelf (from Wannesson et al., 1985). Reproduced from *Marine and Petroleum Geology* by permission of the publisher, Butterworth and Co. (Publishers), Ltd.

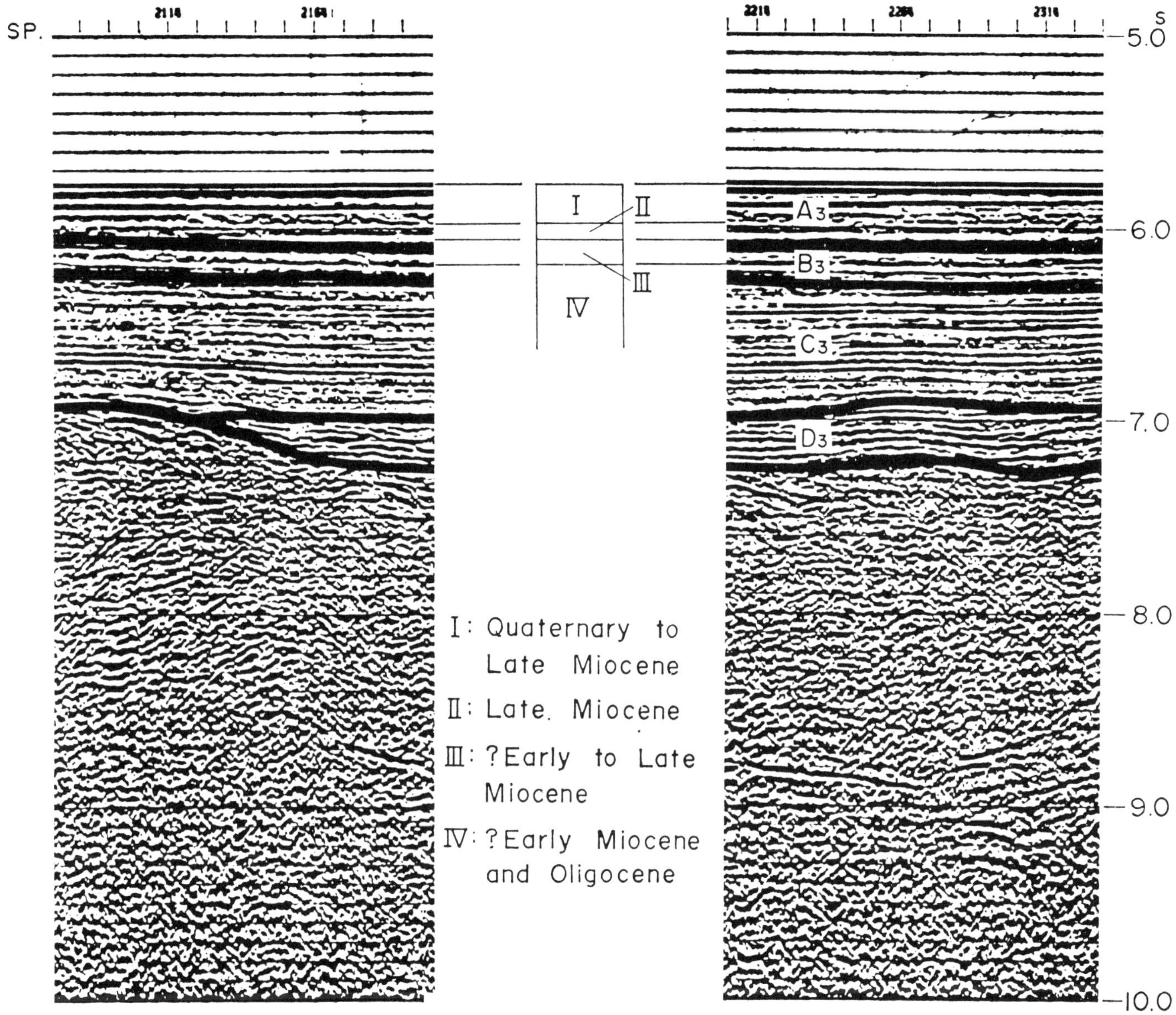


Figure 21. *R.V. Hakurei Maru* record over DSDP Site 269, and its interpretation (from Sato et al., 1984). Reproduced with the approval of the Japanese National Institute for Polar Research.

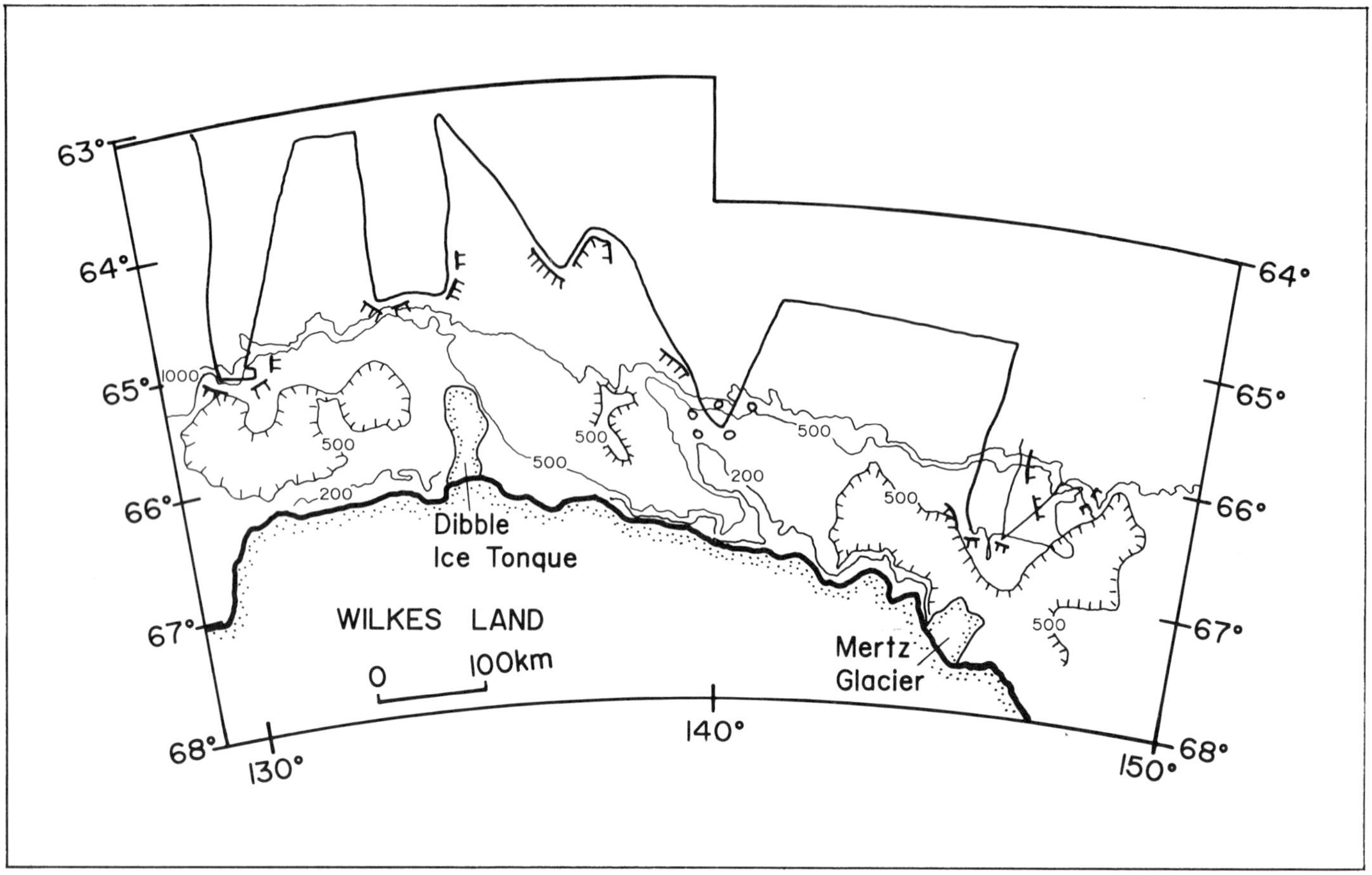


Figure 22. *R. V. S.P. Lee* tracks over east Antarctic continental margin, 1984 cruise (after Eittreim, Cooper, et al., 1984).

samples, and took heat-flow measurements at three stations. Quaternary, Miocene, and Late Oligocene sediments were recovered.

Eittreim and Smith (1987) present the results of the 1983/84 *S.P. Lee* cruise in the same region (Figure 22). Their interpretation proposes an earlier age for the various units identified in all seismic programs. Whereas Unit 1 of Wannesson et al. (1985) is taken as commencing approximately in the Middle Eocene and by Sato et al. (1984) as Early Paleogene, Eittreim and Smith would have it sitting on an unconformity surface (K2) probably Early Cretaceous in age and representing a rift-onset unconformity. The latter interpretation is more in keeping with more recent ideas (Cande and Mutter, 1982; Veevers, 1984) on the timing of and events in the separation of Australia from Antarctica.

## Direct Sampling of Marine Sediment

Echols and Kennett (1973) showed the distribution of seafloor sediment samples taken until that time. Figure 23 is a summary of the locations appropriate to this paper. They are dominantly from the British-Australian-New Zealand Research Expedition (BANZARE: 1929-31) and the Australasian Antarctic Expedition (AAE: 1911-14). Figures 24 and 25 show the locations of later samples by the 1982 Australian seismic program (Stagg, 1985; Quilty, 1985), the United States 1979 Operation Deep Freeze (Milam and Anderson, 1981), and the United States Geological Survey's 1983/84 seismic program (Eittreim, Cooper, et al., 1984).

The sediment samples tell, through their organic content, of the past and what may be under the ice.

Truswell (1982) studied the sediments collected by Mawson's 1911-1914 AAE, from various localities off the AAT, which yielded a variety of palynomorphs of several ages well known from Australian occurrences.

Early Cretaceous and Late Cretaceous-Eocene floras were identified, and the Eocene had a marine component. Similarities with Otway Basin equivalents are strong.

Domack, Fairchild, and Anderson (1980) recorded an in situ Early Cretaceous nonmarine flora from a single 40-cm-long core (Figure 25).

These studies suggest that the Wilkes Basin contains a significant Cretaceous and perhaps Tertiary sediment column and that the specimens recovered (other than those recorded by Domack, Fairchild, and Anderson) made their way to the present continental shelf by glacial transport.

Farther west, Truswell recognized abundant Early to mid-Cretaceous and Late Cretaceous to Eocene assemblages, the latter again with a marine component.

The major area sampled by the AAE was immediately west of the Shackleton Ice Shelf. The assemblages suggest a nearby almost complete nonmarine Permian sequence, different from and more complete than that known in the Beaver Lake area of the Lambert Graben. Also present are Late Jurassic and Albian floras.

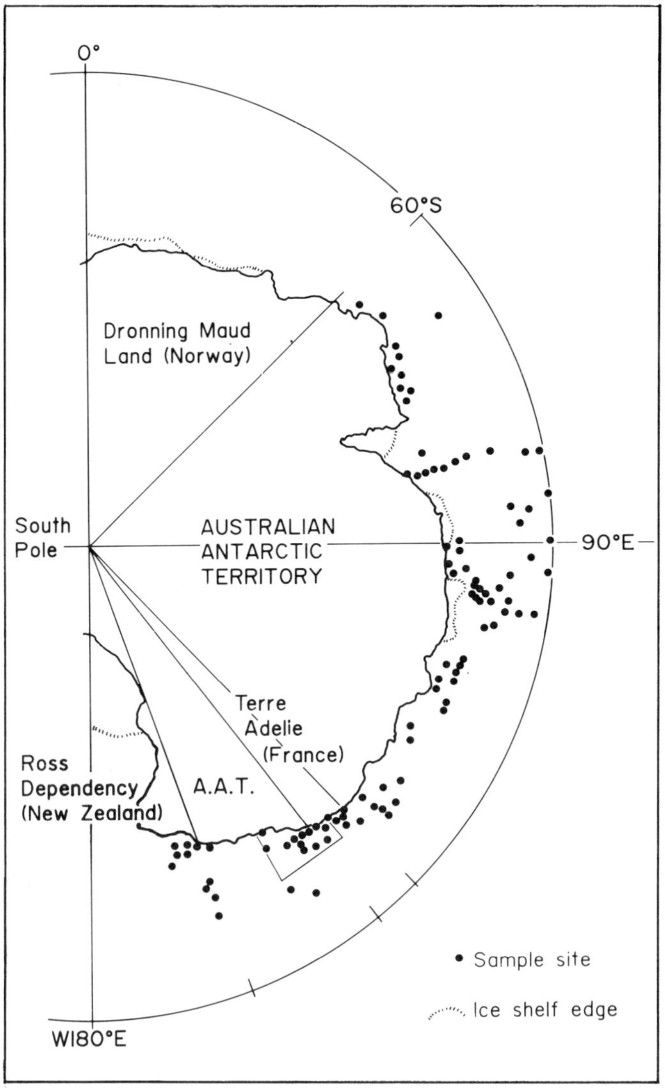


Figure 23. Sediment sample locations prior to 1972 (after Echols and Kennett, 1973).

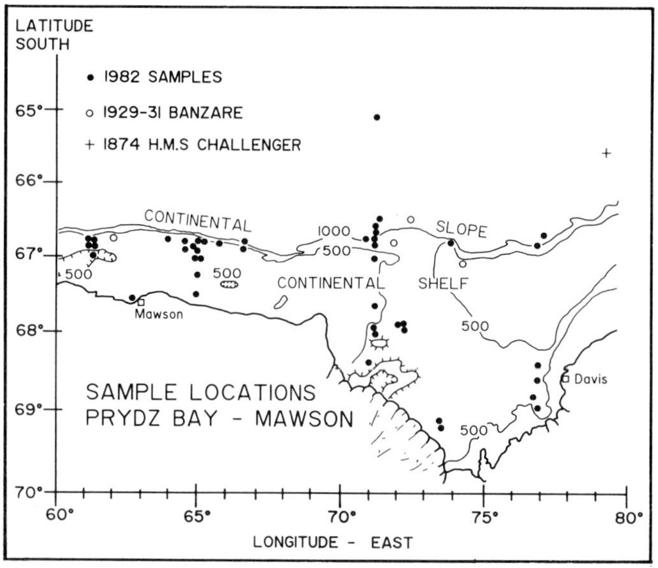


Figure 24. Sediment sample locations Prydz Bay, early 1982.

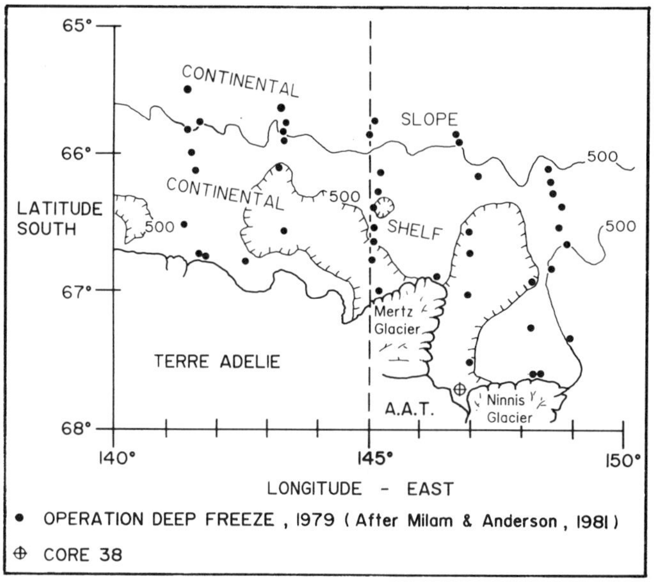


Figure 25. Sediment sample locations of Terre Adelie and eastern Australian Antarctic Territory.

In addition, both Late Cretaceous and Paleocene nonmarine elements are identifiable. A marine dinoflagellate component is closely allied with well-known Late Paleocene-Early Eocene marine floras documented elsewhere in polar regions.

The above studies have shed valuable information on what may lie beneath the ice cover, and the results are generally in agreement with predictions based on tectonic history.

## Drilling

There has been no prospecting or exploration drilling in the Antarctic.

Scientific drilling under the auspices of the Deep Sea Drilling Project (DSDP) on Legs 28 and 35 of that program (Hayes and Frakes, 1975; Hollister, Craddock, et al., 1976) has already been mentioned. Figures 26 and 27 show the locations of drill sites of the Deep Sea Drilling Project and of various project localities.

There has been a series of shallow drilling programs in the McMurdo Sound-Dry Valleys area in an attempt to answer glacial-geological stratigraphic, volcanic, and geochemical questions posed by field work in the region. The results help fill gaps in the Cenozoic geological history, which cannot be filled from studies in the area to the west.

The Dry Valleys Drilling Project (DVDP) drilled 15 holes (see Figure 27 for localities), 3 of which were more than 300 m deep (Torii, 1981; McGinnis, 1981). The results of this drilling provided data on the Miocene to recent history of the region.

Since the DVDP was completed, additional programs with similar aims have been developed for different localities. These programs are the Ross Ice Shelf Project (RISP), the collaborative United States-New Zealand McMurdo Sound Sedimentary and Tectonic Study, or MSSTS (1979), and the U.S.-New Zealand Cenozoic Investigations of the western Ross Sea CIROS (1984-1986).

The MSSTS program drilled only a single hole, which did not reach basement but penetrated 226 m of Middle Miocene and younger glacial sediment.

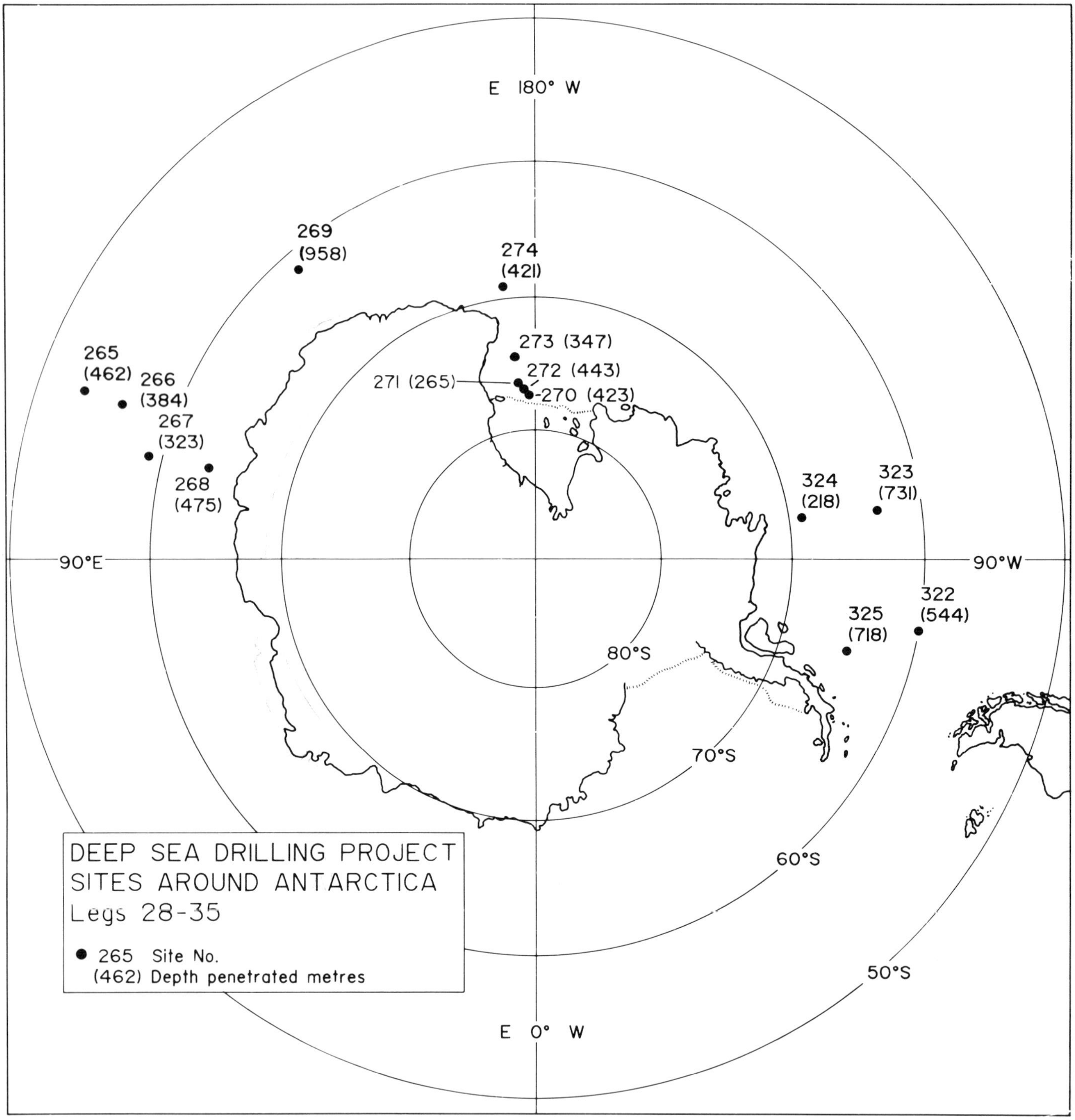


Figure 26. DSDP sites around Antarctica, legs 28–35.

## HYDROCARBON INDICATIONS

Deep Sea Drilling Project Leg 28 (Hayes and Frakes, 1975) drilled several holes in the Ross Sea area and off the AAT. Sites 271, 272, and 273 (Figures 26, 28) yielded hydrocarbon gases probably generated by thermogenic means (McIver, 1975), although the amount is very minor and its significance is unclear.

More recently, Whiticar, Suess, and Wehner (1985) have documented the occurrence of thermogenic hydrocarbons in a gravity core recovered from Bransfield Strait (Figure 28). Whiticar, Suess, and Wehner stated, "the data provide unambiguous geochemical evidence of active petroleum source rocks along the Antarctic continent." This discovery occurs in an area of high heat flow, probably from a nearby volcanic source, so generation takes place at a shallower depth than normally expected. However, this means very little, as the basin is only some 2 million years old.

## ECONOMIC POTENTIAL

Over recent years, there has been a series of reviews such as those by Behrendt (1983; in press), Bergsager (1983),

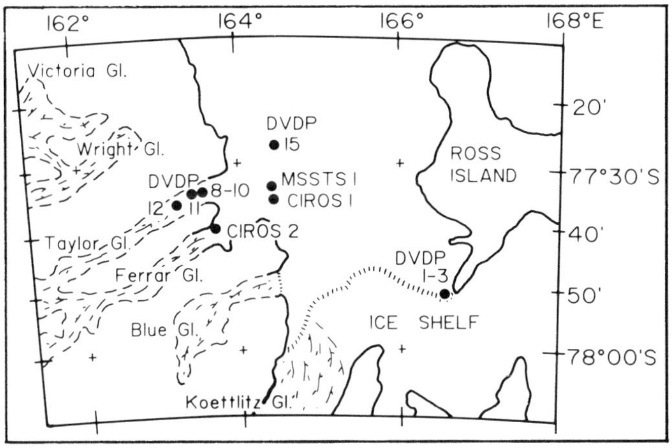


Figure 27. Drilling locations, Ross Sea—DVDP, CIROS, MSSTS (after Barrett, 1985).

Cameron (1981), Gjelsvik (1982), Roland (1983), Tessensohn (1984), and Quilty (1984), each adding a little to those written before and each incorporating new data as scientific research results progressively become available.

Almost all seismic surveys have indicated that sedimentary sequences are thick enough, and probably of an appropriate age to yield hydrocarbons. There is nothing to suggest that Antarctica is any less hydrocarbon prone than any other continent.

Several factors, discussed in the references cited above, militate against any attempts in the near future to search actively for or attempt to develop any Antarctic hydrocarbons. In the case of the AAT region, the major problems can be summarized as follows:

(1) Legal—There is no legal system that is unanimously accepted internationally under which an explorer can gain an assured title to an exploration permit in Antarctica.

At present, Antarctic Treaty nations are working toward developing a regime to control any mineral resource (which includes petroleum) activity in Antarctica. Until a regime is developed and formalized, Antarctic Treaty nations are not free to explore, or even to undertake prospecting. Only scientific study with the obligation of free exchange of information is permitted.

(2) Environmental—Antarctica is an undeveloped continent, surrounded by an undeveloped ocean which has suffered virtually no environmental impact by virtue of man's activities, other than the depradations of whaling. This couplet of undeveloped continent and surrounding ocean is unique.

In addition to the emotional/philosophical call for conservation, the fact that Antarctica controls much of the world's weather and is the key to sea level are practical reasons why it must be preserved from locally caused changes. The potential "greenhouse" effect caused by humanity's actions elsewhere is still probably the major threat.

(3) Technological—As discussed under marine seismic activity, the Antarctic environment is the most hostile on earth, and entirely new technological developments will be needed before any economic extraction can be contemplated. Arctic technology is generally not appropriate.

(4) Economic—This is, as I see it, the most effective barrier to exploration in the Antarctic context. The cost of new technology, the length and complexity of supply lines, the absence of a local workforce, the enormous cost of any infrastructure, the cost compared with alternative energy sources, and the cost of environmental safeguards will, in my opinion, prevent Antarctic development, certainly for the foreseeable future. The economic prospects for Antarctic oil are forcefully discussed by Garrett (1985), who compares them with the exploitation of lunar gold.

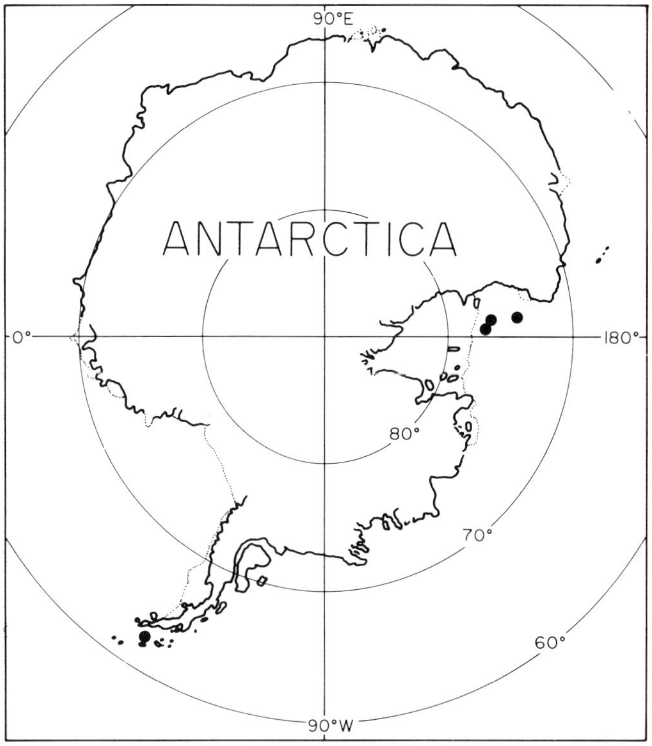


Figure 28. Locations known to have yielded thermogenic hydrocarbon traces.

How much oil is in Antarctica? No one knows and any estimates are barely short of pure speculation. St. John (1984) estimated the sediment volume in most of the Antarctic basins and suggested that the Enderby, Queen Maud, and Scott basins could contain speculated 74 BOE, some 30% of the total estimated for Antarctica.

## EPILOGUE

Antarctica is a place of incredible beauty. It is pristine. It is unknown.

I would be very disappointed if, as a result of man's unnecessary activities or as a result of his carelessness or greed, its delicate balance was disturbed. I have no personal wish or aim to prevent legitimate aspirations for Antarctica, but please let's look after it; it's the only continent we have not changed irreparably.

## ACKNOWLEDGMENTS

I thank the Petroleum Exploration Society of Australia for choosing me as its 1985 Distinguished Lecturer, the focus for preparing an earlier, larger version of this paper, published in its journal.

Dr. John Behrendt helped me identify some sources of data and Drs. S. Asakura and J. Wannesson provided useful data on request. Mr. R. Tingey reviewed an earlier version of the paper. To all I owe a great debt.

The paper was typed by Lesley Manson and Betty Willemer and the drafting was performed by John Cox.

## REFERENCES CITED

Adamson, D., and J. Pickard, 1983, Late Quaternary ice movement across the Vestfold Hills, East Antarctica, *in* R.L. Oliver et al., eds., Antarctic earth science: Australian Academy of Science, p. 465-469.

Barrett, P.J. 1985, Plio-Pleistocene glacial sequence as cored at CIROS 2, Ferrar Fjord, Western McMurdo Sound: NZ Antarctic Record, v. 6, n. 2, p. 8-19.

Behrendt, J.C., (ed.), 1983, Petroleum and mineral resources of Antarctica: U.S. Geological Survey Circular 909, p. 1-75.

Behrendt, J.C., in press, Scientific studies on the question of petroleum resources of Antarctica, *in* J. Splettstoesser, ed., Mineral resource potential of Antarctica: University of Texas Press.

Bein, J., and M.L. Taylor, 1981, The Eyre sub-basin: Recent exploration results: Journal of the Australian Petroleum Exploration Association, v. 21, p. 91-98.

Bergsager, A., 1983, Basic conditions for the exploration and exploitation of mineral resources in the Antarctic: Options and precedents, *in* F. Orrego Vicuna, ed., Antarctic resources policy: Scientific, legal and political issues: London, Cambridge University Press, p. 167-183.

Boeuf, M.G., and H. Doust, 1975, Structure and development of the southern margin of Australia: Journal of the Australian Petroleum Exploration Association, v. 15, p. 33-43.

Cameron, P.J., 1981, The petroleum potential of Antarctica and its continental margin: Journal of the Australian Petroleum Exploration Association, v. 21, p. 99-111.

Cande, S.C., and J.C. Mutter, 1982, A revised identification of the oldest seafloor spreading anomalies between Australia and Antarctica: Earth and Planet Science Letters, v. 58, p. 151-160.

Connolly, J.R., and R.R. Payne, 1972, Sedimentary patterns within a continent - Mid-oceanic ridge-continent profile: Indian Ocean south of Australia, *in* D.E. Hayes, ed., Antarctic oceanology II: The Australian-New Zealand Sector: American Geophysical Union, Antarctic Research Series, v. 19, p. 295-315.

Crohn, P.W., 1959, A contribution to the geology and glaciology of the western part of the Australian Antarctic Territory: Bulletin of the Bureau of Mineral Resources, Geology, and Geophysics of Australia, v. 52, p. 43, 44.

Dietz, R.S., and J.C. Holden, 1971, Pre-Mesozoic oceanic crust in the eastern Indian Ocean (Wharton Basin?): Nature, v. 229, p. 309-312.

Domack, E.W., W.W. Fairchild, and J.B. Anderson, 1980, Lower Cretaceous sediment from the East Antarctic continental shelf: Nature, v. 287, p. 625, 626.

Douglas, J.G., and J.A. Ferguson, (eds.), 1976, Geology of Victoria: Geological Society of Australia Special Publication 5, p. 1.

Drewry, D.J., 1972, Subglacial morphology between the Transantarctic Mountains and the South Pole, *in* R.J. Adie, ed., Antarctic geology and geophysics: Oslo, Universitetsforlaget, p. 693-703.

Drewry, D.J., 1975, Radio echo sounding map of Antarctica: Polar Record, v. 17, p. 359-374.

Drewry, D.J., 1976, Sedimentary basins of the East Antarctic craton from geophysical evidence: Tectonophysics, v. 36, p. 301-314.

Drewry, D.J., (ed.), 1983, Antarctica: Glaciological and geophysical folio: Cambridge, Scott Polar Research Institute, Sheets 1-9.

Echols, R.J., and J.P. Kennett, 1973, Distribution of foraminifera in the surface sediments, *in* V. Bushnell, ed., American Geographical Society Antarctic Map Folio Series, v. 17, p. 13-17.

Eittreim, S.L., and G.L. Smith, 1987, Seismic sequences and their distribution on the Wilkes Land margin, *in* S.L. Eittreim, and M.A. Hampton, eds., The Antarctic continental margin: Geology and geophysics of offshore Wilkes Land: Houston, Texas, Circum-Pacific Council for Energy and Mineral Resources.

Eittreim, S.L., A.K. Cooper, et al., 1984, Marine geological and geophysical investigations of the Antarctic continental margin, 1984: U.S. Geological Survey Circular 935, p. 172.

Eittreim, S.L., M.A. Hampton, and J.R. Childs, 1985, Seismic-reflection signature of Cretaceous continental breakup on the Wilkes Land margin, Antarctica: Science, v. 229, p. 1082-1084.

Fedorov, L.V., M.G. Ravich, and J. Hoffman, 1982, Geologic comparison of southeastern peninsular India and Sri Lanka with a part of East Antarctica (Enderby Land, MacRobertson Land and Princess Elizabeth Land), *in* C. Craddock, ed., Antarctic Geoscience: Madison, University of Wisconsin Press, p. 65-71.

Fedorov, L.V., G.E. Grikurov, R.G. Kurinin, and V.N. Masolov, 1982, Crustal structure of the Lambert Glacier area from geophysical data, *in* C. Craddock, ed., Antarctic Geoscience: Madison, University of Wisconsin Press, p. 931-936.

Fraser, A.R., and L.A. Tilbury, 1979, Structure and stratigraphy of the Ceduna Terrace region: Journal of the Australian Petroleum Exploration Association, v. 19, p. 53-65.

Garrett, J.N., 1985, The economics of Antarctic oil, *in* L.M. Alexander, and L.C. Hanson, eds., Antarctic policies and marine resources: Critical choices for the 1980s: Center for Ocean Management Studies, University of Rhode Island, p. 185-190.

Gjelsvik, T., 1982, The mineral resources of Antarctica: Progress in their identification, *in* F. Orrego Vicuna, ed., Antarctic resources policy: Scientific, legal and political issues: London, Cambridge University Press, p. 61-76.

Hayes, D.E., and J.R. Connolly, 1972, Morphology of the southeast Indian Ocean, *in* D.E. Hayes, ed., Antarctic oceanology II: The Australian-New Zealand sector: American Geophysical Union, Antarctic Research Series, v. 19, p. 125-145.

Hayes, D.E., and L.A. Frakes, 1975, General synthesis Deep Sea Drilling Project Leg 28, *in* D.E. Hayes, L.A. Frakes, et al., Initial Reports Deep Sea Drilling Project: Washington, D.C., U.S. Government Printing Office, 28, p. 919-942.

Hollister, C.D., C. Craddock, et al., 1976, Initial Reports Deep Sea Drilling Project, Leg. 35: Washington, D.C., U.S. Government Printing Office, p. 1-929.

Houtz, R.E., and R.G. Markl, 1972, Seismic profiler data between Antarctica and Australia, *in* D.E. Hayes, ed., Antarctic oceanology II: The Australian-New Zealand Sector: American Geophysical Union, Antarctic Research Series, v. 19, p. 147-164.

James, P.R., and R.J. Tingey, 1983, The Precambrian geological evolution of the East Antarctic metamorphic shield—A review, *in* R.L. Oliver, et al., eds., Antarctic earth science: Australian Academy of Science, p. 5-10.

Jell, J.S., 1981, Silurian tabulate corals from the Cloudmaker Moraine, Beardmore Glacier, Antarctica: Alcheringa, v. 5, p. 311-316.

Johnson, B.D., and J.J. Veevers, 1984, Oceanic palaeomagnetism, *in* J.J. Veevers, ed., Phanerozoic earth history of Australia: Oxford, Clarendon Press, p. 17-38.

Johnson, B.D., C.McA. Powell, and J.J. Veevers, 1976, Spreading history of the Eastern Indian Ocean and Greater India's northward flight from Antarctica and Australia: GSA Bulletin, v. 87, p. 1560-1566.

Kyle, R.A., and J.M. Schopf, 1982, Permian and Triassic palynostratigraphy of the Victoria Group, Transantarctic Mountains, *in* C. Craddock, ed., Antarctic Geoscience: Madison, University of Wisconsin Press, p. 649-659.

Laird, M.G., and J.D. Bradshaw, 1983, New data on the Lower Palaeozoic Bowers Supergroup, Northern Victoria Land, *in* Oliver et al., eds., Antarctic earth science: Canberra, Australian Academy of Science, p. 123-126.

McGinnis, L.D., (ed.), 1981, Dry Valley Drilling Project: American Geophysical Union Antarctic Research Series 33.

McGowran, B., 1979, The Tertiary of Australia: Foraminiferal overview, Mar. Micropaleont. 4: p. 235-264.

McIver, R.D., 1975, Hydrocarbon gases in canned core samples from Leg 28 Sites 271, 272 and 273, Ross Sea, *in* D.E. Hayes, L.A. Frakes, et al., Initial Reports of the Deep Sea Drilling Project: Washington, D.C., U.S. Government Printing Office, v. 28, p. 815-817.

Milam, R.W., and J.B. Anderson, 1981, Distribution and ecology of recent bethonic foraminifera off the Adelie-George V continental shelf and slope Antarctica. Mar. Micropaleont., v. 6, p. 287-325.

Mizukoshi, I., H. Sunouchi, T. Saki, and S. Sato, in press, Preliminary report of geological and geophysical surveys off Amery Ice Shelf, East Antarctica.

Mond, A., 1972, Permian sediments of the Beaver Lake area, Prince Charles Mountains, *in* R.J. Aide, ed., Antarctic geology and geophysics: Oslo, Universitetsforlaget, p. 585-589.

Norton, I.O., and P. Molnar, 1977, Implications of a revised fit between Australia and Antarctica for the evolution of the eastern Indian Ocean: Nature, v. 267, p. 338-339.

Pickard, J., D.A. Adamson, D.M. Harwood, G.H. Miller, P.G. Quilty, and R.K. Dell, in press, Late Tertiary marine sediments and coastline of East Antarctica: South African Journal of Science.

Quilty, P.G., 1977, Cenozoic sedimentation cycles in Western Australia: Geology, v. 5, p. 336-340.

Quilty, P.G., 1980, Sedimentation cycles in the Cretaceous and Cenozoic of Western Australia: Tectonophysics, v. 63, p. 349-366.

Quilty, P.G., 1984, Mineral resources of the Australian Antarctic Territory, *in* S. Harris, ed., Australia's Antarctic Policy Options: CRES Monograph 11, p. 165-203.

Quilty, P.G., 1985, Distribution of foraminiferids in sediments of Prydz Bay, Antarctica: South Australia Department of Mines and Energy Special Publication 5 (Ludbrook Honour Volume), p. 329-340.

Ritzwoller, M.H., and C.R. Bentley, 1983, Magnetic anomalies over Antarctica measured from MAGSAT, *in* R.L. Oliver, et al., eds., Antarctic earth science: Australian Academy of Science, p. 504-507.

Robin, G. de Q., D.J. Drewry, and D.T. Meldrum, 1977, International studies of ice sheet and bedrock: Phil. Trans. Roy. Soc. London B., v. 279, p. 185-196.

Roland, N.W., 1983, Mineralische resourcen der Antarktis: Geogologische Rundschau, v. 35, p. 121-122.

Sastri, V.V., A.T.R. Raju, R.N. Sinha, and B.S. Venkatachala, 1974, Evolution of the Mesozoic sedimentary basins on the east coast of India: Journal of the Australian Petroleum Exploration Association, v. 14, p. 29-41.

Sato, S., N. Asakura, N. Saki, N. Oikawa, and Y. Kaneda, 1984, Preliminary results of geological and geophysical surveys in the Ross Sea and in the Dumont d'Urville Sea, off Antarctica: National Institute for Polar Research Memoir 33: Proceedings of the 4th Symp. Antarctic Geoscience, p. 66-92.

Sheraton, J.W., 1983, Geochemistry of mafic igneous rocks of the northern Prince Charles Mountains, Antarctica: Journal of the Geological Society of Australia, v. 30, p. 295-304.

Sheraton, J.W., and A. Cundari, 1980, Leucities from Gaussberg, Antarctica: Contributions to Mineralogy and Petrology, v. 71, p. 417-427.

Stagg, H.M.J., 1985, The structure and origin of Prydz Bay and Mac.Robertson Shelf, East Antarctica: Tectonophysics, v. 114, p. 315-340.

Stagg, H.M.J., D.C. Ramsay, and R. Whitworth, 1983, Preliminary report of a marine geophysical survey between Davis and Mawson stations 1982, *in* R.L. Oliver et al., eds., Antarctic earth science: Australian Academy of Science, p. 527-532.

Steed, R.H.N., 1983, Structural interpretations of Wilkes Land, Antarctica, *in* R.L. Oliver, et al., eds., Antarctic earth science: Australian Academy of Science, p. 567-572.

Steed, R.H.N., and D.J. Drewry, 1982, Radio echo sounding investigations of Wilkes Land, Antarctica, *in* C. Craddock, ed., Antarctic geoscience: Madison, University of Wisconsin Press, p. 969-976.

Tessensohn, F., 1984, Present knowledge of non-living resource in the Antarctic, possibilities for their exploitation and scientific perspectives, *in* R. Wolfrum, ed., Antarctic challenge: Berlin, Duncker & Humblot, p. 189-210.

Tingey, R.J., 1982, The geologic evolution of the Prince Charles Mountains—An Antarctic Archaean cratonic block, *in* C. Craddock, ed., Antarctic geoscience: Madison, University of Wisconsin Press, p. 455-464.

Tingey, R.J., 1984, Comments (on preceding paper by Quilty), *in* S. Harris, ed., Australia's Antarctic policy options: CRES Monograph 11, p. 204-215.

Torii, T., 1981, A review of the Dry Valleys Drilling Project, 1971-1976: Polar Record, v. 20, p. 533-541.

Truswell,, E.M., 1982, Palynology of seafloor samples collected by the 1911-14 Australasian Antarctic Expedition: Implications for the geology of East Antarctica: Journal of the Geological Society of Australia, v. 29, p. 343-356.

Tsurumaya, Y., M. Tanahashi, T. Saki, T. Machihara, and N. Asakura, 1985, Preliminary report of the marine geophysical and geological surveys off Wilkes Land, Antarctica in 1983-1984: National Institute for Polar Research Memoir 37: Proceedings of the 5th Symposium on Antarctic Geoscience, p. 48-62.

Vail, P.R., R.M. Mitchum, and S. Thompson, 1977, Seismic stratigraphy and global changes of sea level: AAPG Memoir 26, p. 83-97.

Veevers, J.J., 1982, Australian-Antarctic depression from the mid-ocean ridge to adjacent continents: Nature, v. 295, p. 315-317.

Veevers, J.J., (ed.), 1984, Phanerozoic earth history of Australia: Oxford, Clarendon Press, p. 1-418.

Wannesson, J., M. Pelras, B. Petitperrin, M. Perret, and J. Segoufin, 1985, A geophysical transect of the Adelie margin, East Antarctica: Marine and Petroleum Geology.

Webb, P.N., D.M. Harwood, B.C. McKelvey, J.H. Mercer, and L.D. Stott, 1984, Cenozoic marine sedimentation and ice-volume variation on the East Antarctic craton: Geology, v. 12, p. 287-291.

Weissel, J.K., and D.E. Hayes, 1972, Magnetic anomalies in the southeast Indian Ocean: American Geophysical Union, Antarctic Research Series, v. 19, p. 165-196.

Weissel, J.K., and D.E. Hayes, 1974, The Australian-Antarctic discordance: New routes and implications: Journal of Geophysical Research, v. 79, p. 2579-2587.

Wellman, P., and R.J. Tingey, 1976, Gravity evidence for a major crustal fracture in eastern Antarctica: BMR Journal 1, p. 105-108.

Whitby, K.J., G. Rose, and C.T. McElroy, 1983, Formational mapping of the Beacon Supergroup type area with special reference to the Weller Coal Measures, South Victoria Land, Antarctica, *in* R.L. Oliver, et al., eds., Antarctic earth science: Canberra, Australian Academy of Science, p. 228-232.

Whiticar, M.J., E. Suess, and H. Wehner, 1985, Thermogenic hydrocarbons in surface sediments of the Bransfield Strait, Antarctic Peninsula: Nature, v. 314, p. 87-90.

# HYDROCARBONS (NEW FRONTIERS) ABSTRACTS

MAYCOCK, IAN D., Hunt Oil Company, Dallas, TX

## Oil Exploration and Development in Marib/Al Jawf Basin, Yemen Arab Republic

In 1981, Yemen Hunt Oil Company (YHOC) negotiated a production-sharing agreement covering 12,600 $km^2$ in the northeast part of the Yemen Arab Republic. A reconnaissance seismic program of 1,864 km acquired in 1982 revealed the presence of a major half graben, designated the Marib/Al Jawf basin by YHOC. A sedimentary section up to 18,000 ft thick has been recognized. Geologic field mapping identified Jurassic carbonates covered by Cretaceous sands overlying Permian glaciolacustrine sediments, Paleozoic sandstones, or Precambrian basement.

The first well drilled in 1984, aimed at a possible Jurassic carbonate objective, encountered hydrocarbon-bearing sands in the Jurassic-Cretaceous transition between 5,000 and 6,000 ft. A successful appraisal drilling program has demonstrated satisfactory lateral reservoir continuity. Further wildcat drilling demonstrates macro-unit correlation within the eastern part of the basin. Rapid basin development apparently commenced in the late Kimmeridgian, culminating with the deposition of Tithonian evaporites.

Available geochemical analysis indicates sourcing from restricted-basin sediments. Excellent traps, reservoirs, and source beds underlying the Tithonian evaporites indicate that a significant new petroliferous province is present.

# SEDIMENTARY BASIN DEVELOPMENT

CHAPTER 27

# A CRUSTAL CLASSIFICATION FOR HYDROCARBON EXPLORATION

**Dave R. Kingston**
**Carr P. Dishroon**
*Esso Exploration, Inc.*
*Houston, TX*

**Phillip A. Williams**
*Esso Exploration, Inc.*
*Surfside, TX*

ABSTRACT

A system is proposed which classifies the earth's crust, worldwide, into three general categories—continental, transitional, and oceanic. These are in turn subdivided into eight crustal types using the parameters of genetic origin, crustal composition, and thickness. Oil and gas accumulations worldwide may be grouped according to the type of crust underlying the basin. Certain crustal types are commonly overlain by petroliferous basins, others rarely, and still others have no known hydrocarbon association. This chapter gives a description of the various types of crust and how each influences hydrocarbon basin development.

## INTRODUCTION

Three general categories of crust—continental, transitional and oceanic—have been used by various authors for many years. Fyfe (1978), Green (1977), Dewey and Windley (1981), and others have demonstrated subdivisions through various types of crustal evolution. The authors propose that these three major categories be divided into eight subtypes that seem to occupy fairly prominent positions in the wide range of crustal compositions and thicknesses. Further splitting of these crustal types may be done, but this appears to needlessly complicate the system. This crustal division scheme is best utilized in hydrocarbon exploration when used in conjunction with the Global Basin Classification System described by Kingston, Dishroon, and Williams (1983). In this chapter we will attempt to show how the type of underlying crust influences the type of basin formation, sedimentary fill, type of structuring, and ultimately hydrocarbon plays.

Figure 1 shows a simplified diagram of near-surface earth layers. A plate of lithosphere includes the crust and the upper part of the mantle down into the low-velocity zone (LVZ). This classification concerns the crustal portion of the lithosphere.

## CRUSTAL CLASSIFICATION

Figure 2 shows the three main crustal divisions—continental, oceanic, and transitional—and the eight subtypes of each. From this diagram, it may be seen that the crustal classification system is based upon three main parameters—genetic origin, crustal composition, and thickness. Age is also a minor factor, older crust being more brittle than younger crust. In simplified terms, the earth's crust has two end members—the thick, lighter continental crust and the thin, denser oceanic crust. Between these two are various types of intermediate or transitional crust. Continental crust consists of an upper, lighter acid igneous (granitic) layer, and a lower, heavier basic igneous layer. Oceanic crust consists of basic igneous rocks and overlying oceanic basalts. The granitic layer shows a range in thickness under continents depending upon its origin (Soller, Ray, and Brown, 1982). In recent years the contributions of COCORP (Consortium of Continental Reflection Profiles) have revolutionized our ideas about the origin and tectonic history of the crust, primarily in orogenic areas (Oliver, 1985).

The crustal types found within a given area are not static but can thicken (become more continental) or thin (become more oceanic), depending upon the local type of crustal evolution. Oceanic crust may start out normally as a very thin basaltic layer and, by being cycled through various convergent processes, thicken and receive lighter components until it becomes continental in thickness and composition. Alternately, continental crust, possibly starting out as a very thick active orogen, may, by passing through various cycles of divergence, thin down and become almost oceanic in nature. There can be many intermediate positions between the two extremes of thick continental and thin oceanic crust.

The genetic origin and evolution of crust is controlled by various thickening or thinning processes, as shown in Figure 3. Crustal thickening may be caused by any one of several processes: (1) converging plate interaction, such as subduction, and plate collision, or by (2) load tectonism, where great

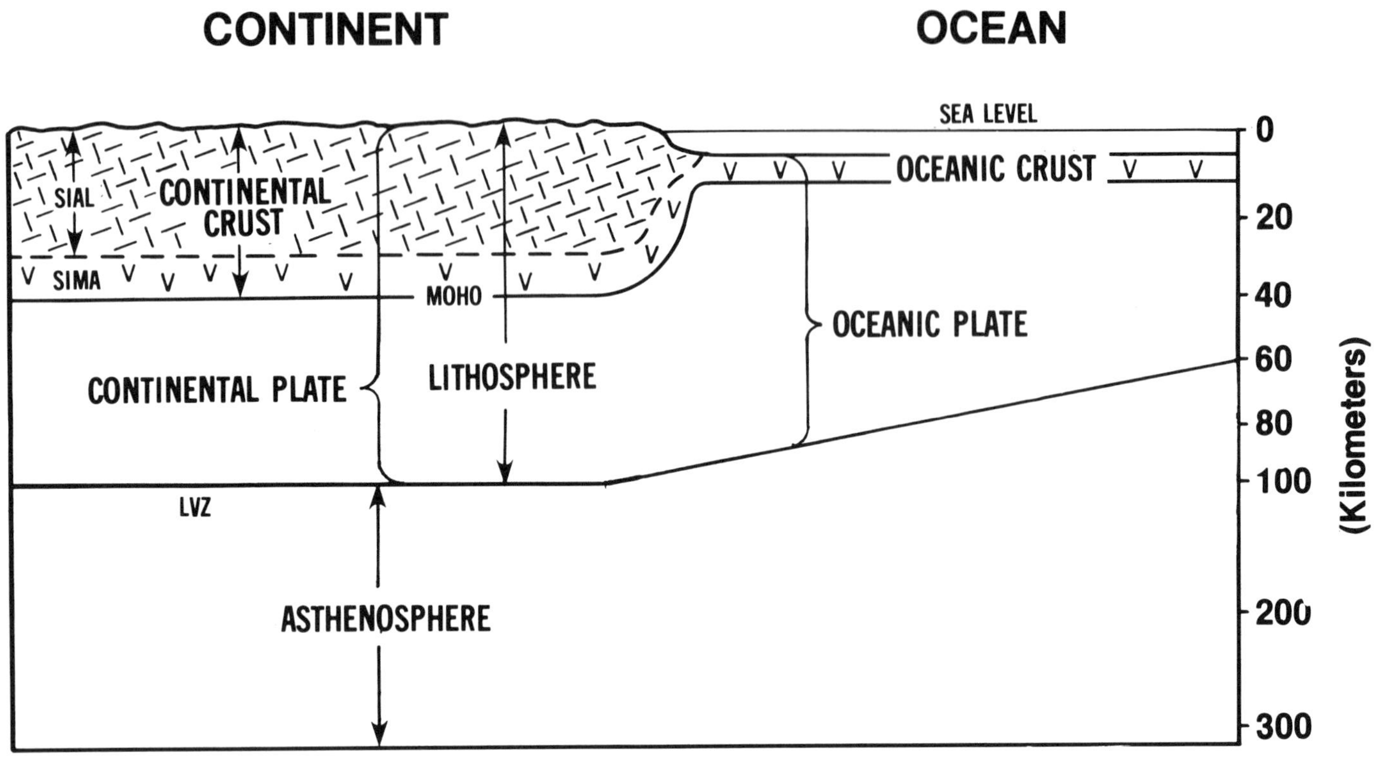


Figure 1.

thicknesses of sediments can create new crust at depth. Subduction is the convergence of two plates, whereby one plate overrides the other (Roeder, 1973). This can be either two oceanic plates, one overriding the other, or continental plate overriding a plate with oceanic or mixed oceanic/continental crust. New continental crust is formed and/or old continental crust is thickened by the incorporation of subduction-involved trench-slope sediments and associated igneous intrusions (Karig and Kay, 1981; Thorpe, Francis, and Harmon, 1981).

If both plates involved in the subduction have oceanic crust, a thicker or supraoceanic crust is formed (Figure 2). Supraoceanic crust may be considered as a first step in oceanic crustal thickening. If subduction carries a plate of supraoceanic crust into another, the resulting collision could form an even thicker supraoceanic plate, or one consisting of protocontinental crust. The subduction process can create three different types of new crust: (1) supraoceanic, (2) protocontinental, and (3) thickened continental crust in an active orogen, as seen in Figure 2.

Plate collision between converging continental plates forms major orogenies and thick continental crust. The direction of convergence can range from almost parallel to head-on. The plates themselves may buckle and thicken up with an accordianlike action or by stacking of thrust sheets. Convergence of two oceanic crustal plates with differential rates, or direction of movement, will form supraoceanic crust. Some oceanic ridges (such as the 90 East Ridge in the Indian Ocean) may be formed in this way.

Load tectonism may be another method of nonorogenic crustal thickening. When oceanic crust is loaded with thick sediments, it is depressed. If the sediments are thick enough (10 to 12 km), the increasing load alters the basal sediment to form protocontinental crust. A high heat flow may speed up the process. Sediments involved can be volcanic basalts, tuff, volcanogenics, graywackes, turbidites, carbonates, and evaporties (Dickenson, 1974). One example of this is the north Caspian basin in Russia, where sediments more than 20 km thick have been recorded. The weight of this thick sediment accumulation would cause low- to medium-grade metamorphism to occur at the base. The new continental crust formed probably consists of 6 to 10 km of basic igneous rock, 6 to 10 km of metamorphics at the base of the sediments, and the rest the unmetamorphosed sediments themselves. Some of these areas may start out as oceanic and end up as a transitional crust. Crustal thickening may also occur through magmatic processes such as extrusion of flood basalts or basaltic underplating.

Crustal thinning, as shown in Figure 3, can be caused by four different processes: (1) surface erosion, (2) subcrustal melting, (3) crustal fracture and divergence, and (4) divergent wrenching. Surface erosion is a long-term crustal thinning and redistribution process. After continental crust has been formed and convergence and thickening have ceased, erosion begins to cut the top off the active orogen. Gradually, the orogen core becomes cooler and more brittle and rises as the upper mantle discontinuity, or Moho, seeks a more stable position. Eventually, the mountains are removed and the sediments redistributed to other areas. This isostatic uplift and removal by erosion continues more slowly throughout the old orogen and low-lying shield stages. As the metamorphosed roots of the old orogen are exposed, the isostatic uplift almost ceases and, in turn, so do the erosional processes.

Subcrustal melting, or anatexis, is a process which thins the crust from the bottom. Mantle plumes or other deep vertical phenomena bring the mantle upward, thinning the crust (Rezanov and Faytl'son, 1974). Fracture or divergent rifting

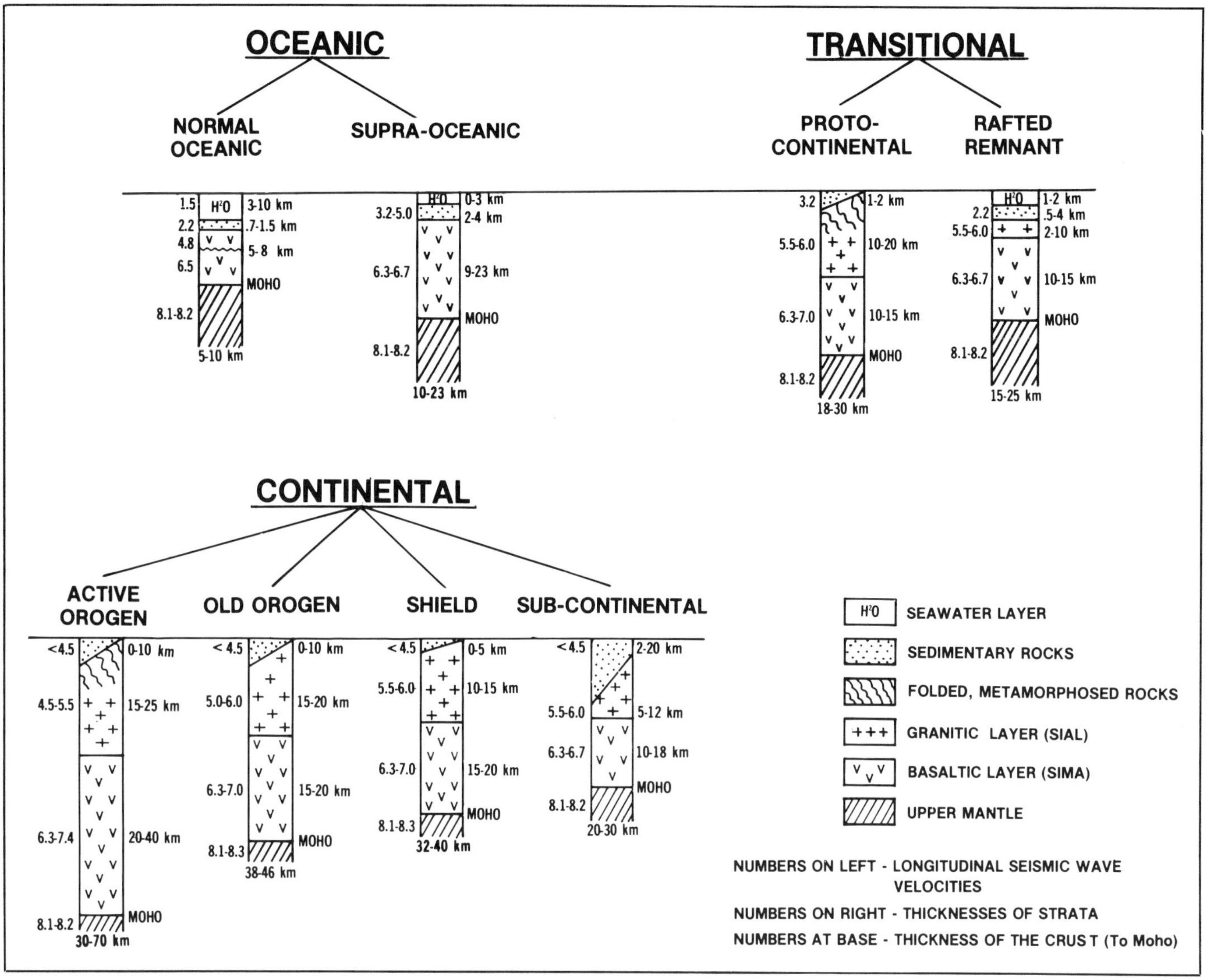


Figure 2.

causes continental crust to thin by breaking up. Divergent forces may cause fracture basins to form with attendant crustal thinning. The process may continue until continental separation takes place.

Divergent wrenching causes thinning or splitting of the crust by divergent lateral plate movements. When this happens in the continental plates, crustal thinning occurs. The shearing may break completely through the crust, creating a gap which fills with basaltic material from below. This shearing action may result from major wrenching with some divergence, wrench coupling, or differential plate rotation, with one side outrunning the other. It is believed that this shearing movement has formed, at least initially, some of the recognized "small ocean basins" (Gulf of California, South China Sea) which are floored with basaltic crust of oceanic composition. Eventually, after considerable divergent shearing separation in a young ocean, a spread center may develop, linear magnetic anomalies are formed, and normal oceanic crust is laid down.

The crustal classification resulting from these thickening, thinning, and compositional changes is summarized in Figures 2, 3, and 4. Note that there can be many intermediate positions between the individually named crustal types in Figure 2. To simplify the numerous possible crustal variations, we have selected the eight shown as representing the diagnostic types that are most useful in hydrocarbon exploration. These have been consolidated into the three main groups—continental, transitional, and oceanic.

In addition to crustal differences between the eight selected types, there are also alterations and recombinations. For example, broken-up fragments of continental crust (rafted remnants) may be trapped in a zone of convergence and emerge as foldbelt microplates with entirely different crustal characteristics and exploration potentials than they previously possessed. Crust underlying many hydrocarbon-producing basins show various modifications or combinations of the eight classified types. These may be referred to as mixed crustal types (not classified) such as protocontinental crust containing rafted remnants, or rafted remnants injected by oceanic basalts.

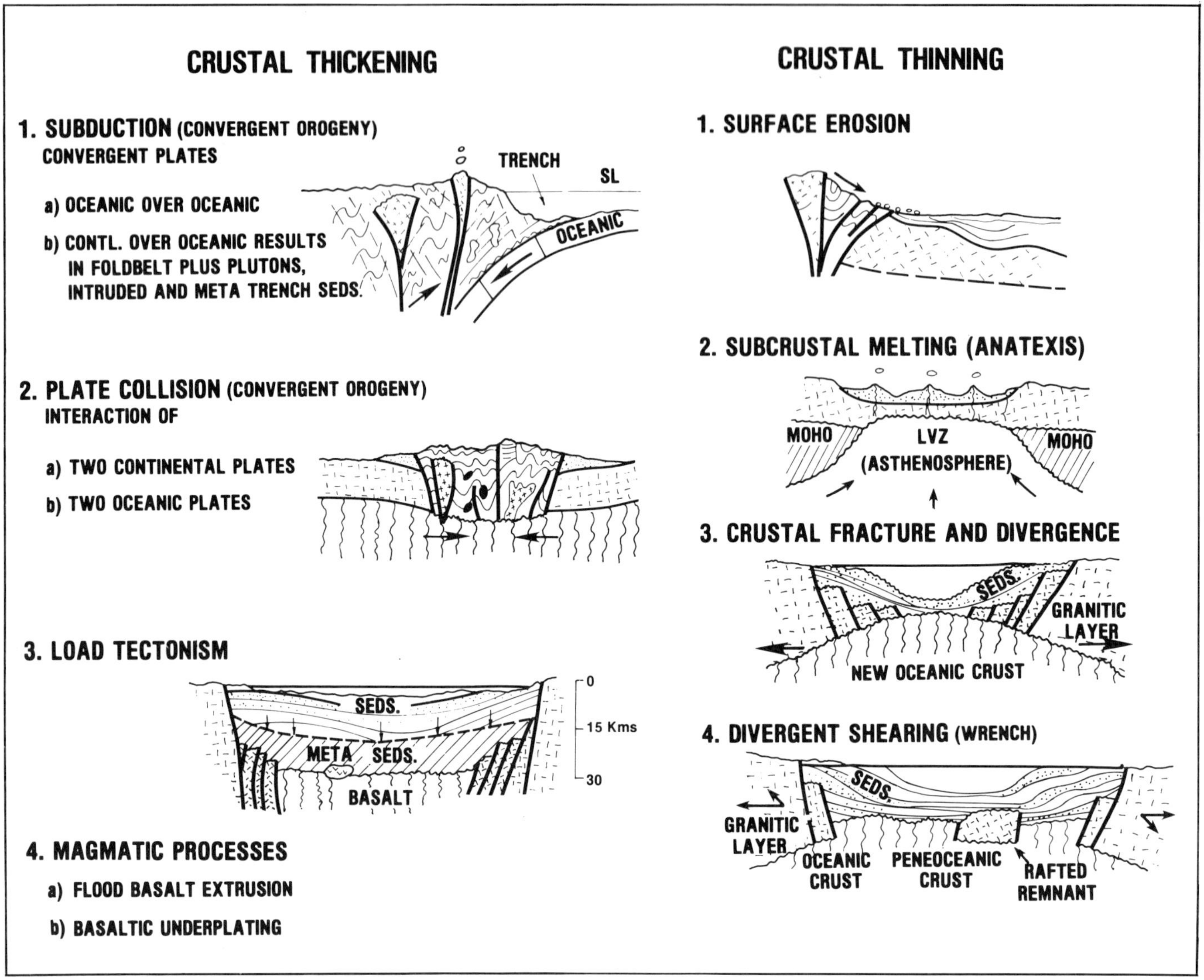


Figure 3.

## INFLUENCE OF CRUSTAL TYPES ON HYDROCARBON POTENTIAL

The main purpose of this crustal classification is to help evaluate the hydrocarbon potential of sedimentary basins. The system may be used by the explorationist in two ways. First, it can be documented that certain crustal types are commonly overlain by hydrocarbon-bearing basins. Other types of crust rarely are associated with hydrocarbon-producing basins. Therefore, it is important to know the type of crust in a given area before initiating an exploration venture. Second, once the crustal type is known, some of the physical characteristics of overlying basins may be predicted. For example, it is possible to predict the probable basin-forming tectonic origin, type of sedimentary fill, structure size and reservoir quality, thermal history for hydrocarbon maturation and diagenesis, plus other key factors (Kingston, Dishroon, and Williams, 1983). Crustal classification is a necessary first step in basin classification and play definition, and is primarily used to evaluate or downgrade unprospective areas. Figure 5 is a chart summarizing the various physical properties of each of the crustal types, including crustal examples and associated basin types. Figure 6 illustrates the association of basin and crustal type. Figure 7 shows the crustal classification utilized in Southeast Asia cross sections.

### Oceanic Crustal Types

Normal oceanic crust consists of oceanic basalt plus sima. It is extruded along divergent spread centers and therefore has magnetic stripes. It is normally 5 to 8 km thick. Older normal oceanic crust which has cooled and subsided (Sleep, 1971), may be located in water 4 to 5 km deep and is commonly very stable. There are several unclassified variations of normal oceanic crust which may be mentioned. A thinner type of oceanic crust is located in spreading centers. This type of crust is not preserved, and as soon as it cools and moves away from the spread center ridge, it becomes normal oceanic basalt.

"Peneoceanic crust" is a term that may be used for slightly thicker variations of normal oceanic crust. It is formed as a result of plate separation by oblique divergent wrenching. This may be emplaced either below sea level or above, as in

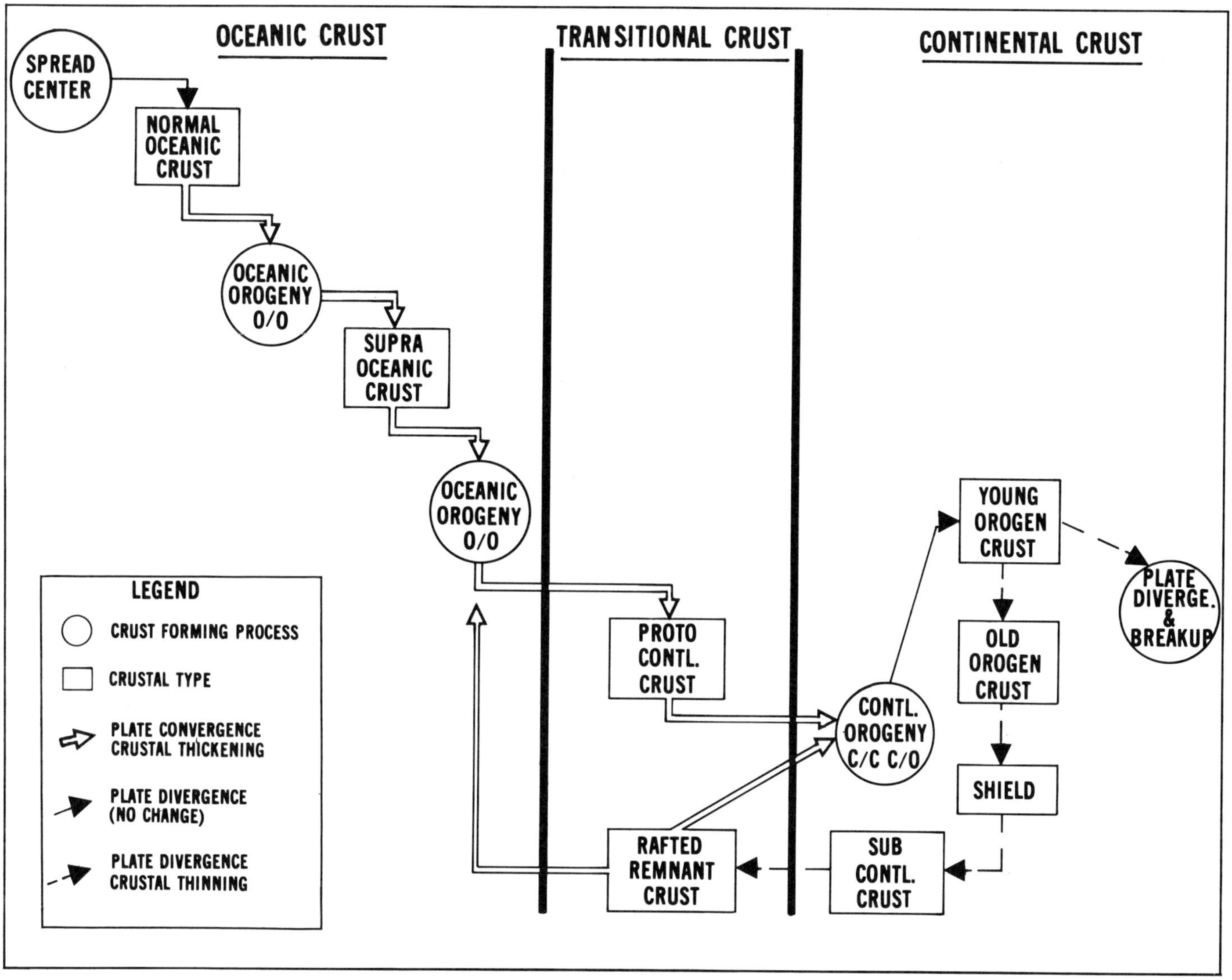


Figure 4.

the Afar Triangle of east Africa (Morton and Black, 1975). Peneoceanic crust consists of oceanic-type basic igneous rocks plus basalt, and appears to be somewhat thicker than normal oceanic crust. Along the edges of continental crustal blocks it can be injected into diverging faults or fractures, causing the formation of complex mixed crustal types, such as partly melted or oceanized continental crust. Peneoceanic crust may have no magnetic stripe pattern, or may have one that is poorly defined, and may be the cause of the magnetic quiet zone found adjacent to many continents or in small ocean basins. Peneoceanic crust may subside very rapidly after formation, accumulating a stack of volcanoclastics, tuffs, oceanic basalts, and graywackes, which make the crust appear to be very thick though still oceanic in composition. Examples are parts of the Caribbean Sea, or some of the small ocean basins in Southeast Asia (Celebes Sea, South China Sea). Both the spread center and peneoceanic crustal types are interesting variations of the normal oceanic crust that is found worldwide; however, they have no particular exploration significance and, as a result, only the term oceanic crust is used in this classification.

Supraoceanic crust (Figure 2) is oceanic crust that has become thicker than normal by passing through some crustal thickening process, such as subduction, plate collision, or basalt underplating (mantle plumes). Supraoceanic crust may have a thickness of 9 to 32 km (commonly 16 to 20 km) and is composed of oceanic-type basic rocks (gabbro/basalt and ultramafics). It should contain no acidic or continental-type rocks such as diorite/andesite. Examples of supraoceanic crust are found in seamount chains and oceanic ridges such as 90 East, Walvis, and parts of the Isthmus of Panama. Supraoceanic crust is most commonly formed in major oceanic/oceanic-subduction zones, where the crust is thickening. Supraoceanic crust may be the result of a single subduction event or repeated subduction phases, so long as the composition remains basic (oceanic).

Three types of sedimentary basins are found overlying oceanic crust—Trenches (T), Oceanic Sags (OS), and the outer edges of Margin Sag (MS) basins as shown in Figure 6. Some examples of these basins are the Java, Japan, and Cayman Trenches (T), the oceanic sags (OS) of the South China Sea, Celebes Sea, and Sea of Japan, and the outer edges of Margin

| CRUSTAL TYPES | NORMAL OCEANIC | SUPRA OCEANIC | PROTO CONTINENT | ACTIVE OROGEN | OLD OROGEN | SHIELD | SUB CONTINENTAL | RAFTED REMNANT |
|---|---|---|---|---|---|---|---|---|
| MODE OF FORMATION | Cooled Crust From Spread Center | First Order Oceanic Over Oceanic<br>Subduction Or Convergent Plate Collision | Collision Of Supra-Oceanic Arcs<br>Second Order Subduction<br>Intersect Of Spread Center And (Wrench) Plate Margin | Orogeny Involving Two Plates (One Continental) Converging<br>–Plate Collision<br>–Subduction | Erosion Of Active Orogens | Erosion Of Old Orogens | Erosion Of Shields Plus Divergence Wrench Or Interior Fracture | Continental Fragment Left Behind By Divergent Plate Separation |
| CRUSTAL THICKNESS DOWN TO MOHO | 5 10 Kms | 10 23 Kms (Usually 20 Kms) | 18 30 Kms | 30 70 Kms (Sometimes Thinner) | 38 - 46 Kms (Usually 38 42 Kms) | 32 - 40 Kms (Usually 34 36 Kms) | 20 - 32 Kms (Average 25 Kms) | 15 - 25 Kms |
| COMPOSITION | Oceanic Basalt Sima | Oceanic Basalt Volcanics Sima | Acidic and Basic Volcanics, Intrusions Metamorphics | Volcanics Igneous Plutons Metamorphic Sediments | (Igneous Plutons) Metamorphic Sediments Volcanics | Igneous Metamorphics (Roots of Mountains) | Same as Shield | Shield Composition Sometimes with Basalt Dikes |
| STABILITY | Stable | Unstable | Strongly Unstable | Strongly Ustable | Stable | Very Stable | Stable | Stable to Unstable |
| GEOLOGIC AGE OF PRESENT CRUSTAL TYPES | Upper Ju ssic to Recent | Upper Cretaceous to Recent | Jurassic to Recent | Upper Jurassic to Tertiary | Upper Jurassic And Older | Upper Jurassic And Older | Upper Jurassic And Older | Sub Continental Crust Modified By Tertiary Volcanics |
| ASSOCIATED BASIN TYPE | Oceanic Sags<br>Accretionary Prisms | Trenches<br>Oceanic<br>Wrench Basins | Trenches<br>Wrench Basins | Foldbelts<br>Foreland Basin<br>Wrench Basins<br>Foldbelt Microplates | Interior Fracture<br>Interior Sag<br>Foreland Basin | Interior Sag<br>Interior Fracture | Interior Fracture<br>Interior Sag,<br>Margin Sag | Interior Fracture<br>Margin Sags |
| SPECIAL CHARACTERISTICS | Magnetically Reversed Stripes | Thickened Basalic Layer | Mainly Oceanic Composition But Containing Some Acid Igneous Rocks | Thick Crust or "Root" Under Mountain Range<br>Young, Hot, Ductile | Areas Of Thickened Crust Origin Unknown | Cold And Old –The Roots Of Old Mountains | Thinned Continental Crust | Sometimes Injected By Basalt Dikes |
| EXAMPLES | North Atlantic | Marshalls And Marianas Island Chains<br>90 E Ridge | Philippines | Andes Mountains, South America<br>Sierra Nevada, California | Appalachians<br>Urals | Canadian Shield<br>Russian Platform | Gulf Coast, U.S.A.<br>Persian Gulf<br>North Sea<br>Gabon, Sirte | Rockall Banks<br>Campbell and Exmouth Plateaus |

Figure 5.

Sag (MS) basins such as the Ganges and Indus river cones. Future oil prospects in these oceanic basin types (Kingston, Dishroon, and Williams, 1983) do not look very encouraging for the following reasons. First, these basins are covered by deep water, which would make exploration difficult and expensive. Second, the sediment cover in most oceanic basins is thin, consisting of fine-grained pelagic and volcanic-derived siltstones and muds. Where sediments are thicker (trenches, accretionary prisms) they are still fine-grained, and commonly are slump-deposits. As a result, potential reservoir rocks typically are missing. Third, source rocks in the open ocean commonly are quite lean, and have TOCs too low to generate hydrocarbons. Even worse, oceanic basins are cold, with very low thermal histories (after crustal emplacement). These temperatures would require an exceptionally great thickness of sediments for hydrocarbon maturation. This thick sediment column would probably ensure that any porosity would be destroyed by diagenesis before peak maturation temperatures were achieved. To our knowledge to date, no commercial hydrocarbons have been found in sedimentary basins overlying true oceanic crust.

## Transitional Crustal Types

Two types of crust are included in this category. One is called protocontinental crust, and is intermediate in composition and thickness between oceanic and continental. It is newly formed crust that once was oceanic but has thickened up and contains some acid igneous intrusions (becoming more continental). The second type of transitional crust, called a rafted remnant, is just the reverse. It is a fragment of old continental crust that is being thinned down by stretching or faulting; it may have basaltic intrusions (becoming more oceanic), and has been "left behind" as the crustal plates diverge.

Rafted remnants are pieces of old continental crust rafted out into the ocean or left behind after divergent continental breakup and separation. These continental blocks, in order to be labeled as rafted remnants, should be surrounded on at least three sides by younger oceanic crust. They may be large, such as the Demerara Rise off South America and the Exmouth plateau northwest of Australia, or they may be small, like the Seychelles in the Indian Ocean. The average crustal thickness of rafted remnants is 16 to 25 km. During the

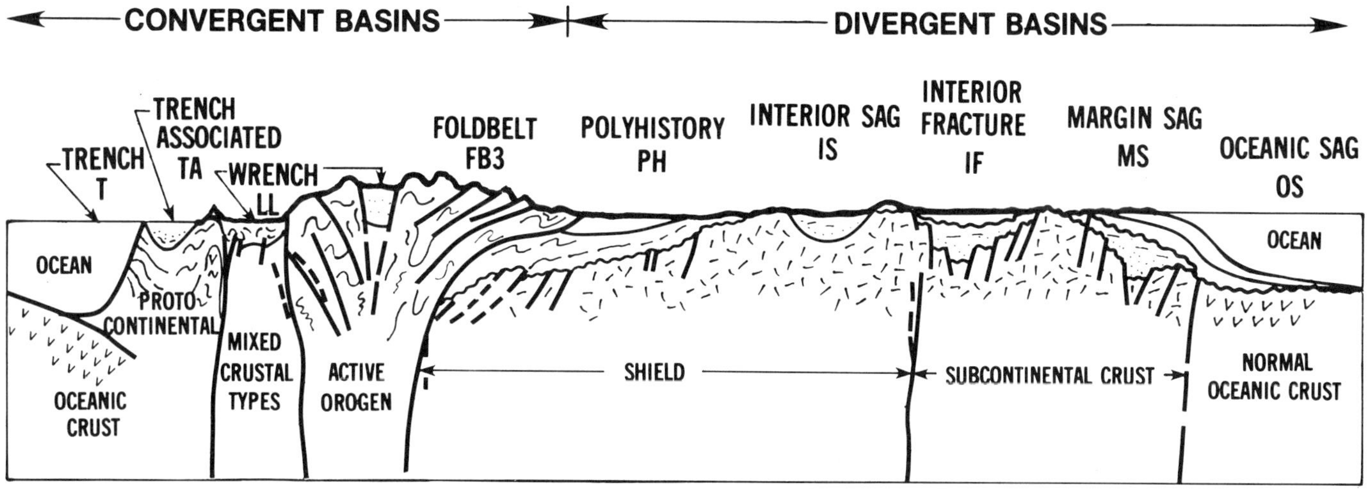


Figure 6.

period of breakup and separation, some rafted remnants are intensively intruded by basalt dikes, as may be seen on the island of Lanzarote in the Canary Islands. As a result, they may be more deeply subsided than most thinned subcontinental crust. The process of fracturing and basalt injection may destroy the older sedimentary section, including the hydrocarbon prospects therein. If the continental microplate is big enough, it may act as a microcontinent, and be called subcontinental crust, rather than a rafted remnant. Examples of minicontinents are Madagascar, Sicily, Sardinia, and the Campbell Plateau in the South Pacific.

Rafted remnants are no longer true continental blocks, but not really oceanic either, although they may subside to moderate oceanic depths (1000 to 2000 m). Other examples of rafted remnants of various sizes are the Rockall Bank in the North Atlantic, Chatham Rise in the South Pacific (Figure 7a) and Reed Bank in the South China Sea (Figure 7b). Rafted remnants may remain far from land, collecting either thin-bedded pelagic sediments or accumulating carbonate banks, as have the Bahamas. In some places these features are basified by some oceanizing process, injected by basic dike swarms, or melted by a rising plume from the asthenosphere, possibly forming deep oceanic rises or ridges. Some rafted remnant blocks are moved to convergent plate margins and destroyed or incorporated into fold-belts (microplate accretion) by subduction or plate collision (Churkin, Carter, and Trexler, 1980). Some of these mixed crustal areas may be seen in the Sulawesi-Halmahera region in southeast Asia (Figure 8).

Interior fracture (IF) and margin sag (MS) basin types (Kingston, Dishroon, and Williams, 1983) are associated with rafted remnant crustal features. Many exploration wells have been drilled on some substantial rafted remnant features, with poor success to date. Some of these are: the Seychelles, Exmouth Plateau, Demerara Rise, and Campbell Plateau. These ventures have all encountered similar problems such as low thermal gradients, lean source rocks, poor reservoirs, and difficult drilling (deep water). Exploration ventures conducted on present-day rafted remnant crustal features appear to be very high risk.

Protocontinental crust is commonly the result of crustal thickening due to repeated subduction, or the collision of two supraoceanic arcs. It is mainly formed along convergent oceanic plate margins. Protocontinents themselves may be located in the oceanic realm some distance away from established continents, as in the case of Mauritius, the Philippines, Iceland, and Fiji. Protocontinental bodies contain both basic and acidic igneous rocks, such as basalts, diorite (andesite), granodiorite or granite (Gastil, 1979; Pitcher, 1982). Metamorphic rocks derived from load tectonism or previous orogenies may also be found. Static load tectonism (Figure 3), where oceanic crust is covered by thick sediments and depressed, may also produce a form of protocontinental crust. In this case, sediments thicken to ± 15 km, and the increased load may alter the basal sediments to low-grade metamorphics without the involvement of convergent plate movements or igneous activity. Basement under the Niger River delta or the North Caspian Basin in the USSR may have formed in such a manner. Tectonized accretionary prisms or wedges (Karig and Sharman, 1975) are common ingredients in protocontinental crust. These forms are typically associated with continental/oceanic or oceanic/oceanic subduction zones, which include shearing or thrusting structural movements stacking and deforming the sediment pile, as found in the Mentawi Islands of Indonesia.

Protocontinental crust is the first stage of evolution that we could call transitional rather than thickened oceanic. The thickness of protocontinental crust may have a wide range but is typically about 18 to 26 km. Basin types associated with protocontinental crust are those associated with convergent plate margins; wrench (LL), such as the Cagayan and Manila basins (Philippines), the Taranaki (NZ), and the Pacifico and Jipijapa basins of Colombia and Ecuador; and trench-associated basins (TA), such as the Gulf of Alaska and the Abukuma-Kitikama basins of Japan. Wrenched foldbelts (L3FB) such as East Coast (Hawkes Bay) in New Zealand are also found on convergent plate margins, and it is often difficult to ascertain whether the crust should be called "protocontinental" or "active orogen."

Continental growth around much of the Pacific Ocean margin has been referred to as "allochthonous or accreted terraines," which are really protocontinental crustal types or young active orogens (Coney, Jones, and Monger, 1980; Ben-Avraham et al., 1981). Basins overlying accretionary prism or protocontinental crust have generally had very poor exploration success for either oil or gas. This has been due mainly to high lithics-low quartz sandstone reservoirs derived from

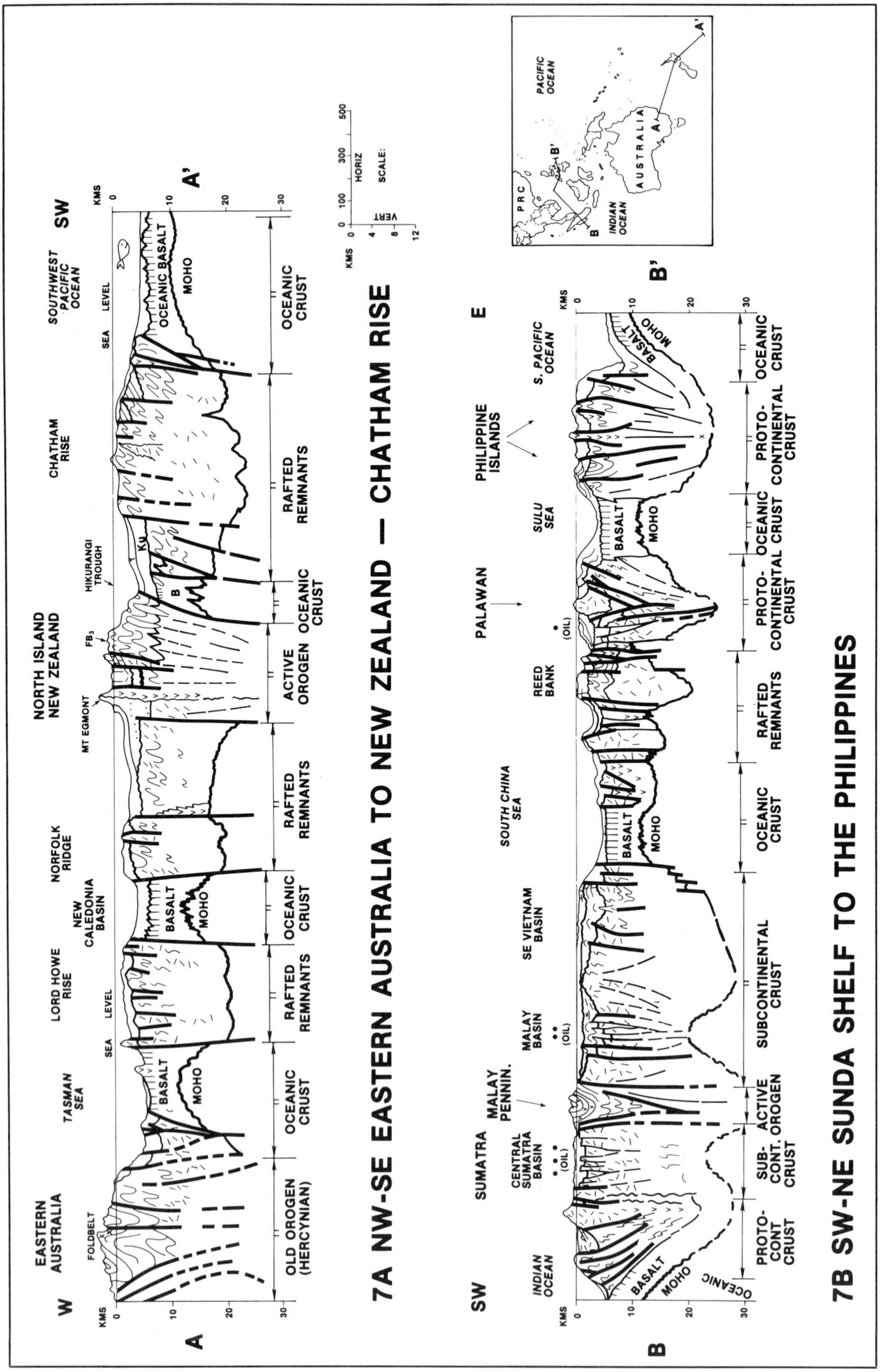
7A NW-SE EASTERN AUSTRALIA TO NEW ZEALAND — CHATHAM RISE
A
A'
W
SW
KMS
EASTERN AUSTRALIA
FOLDBELT
TASMAN SEA
SEA LEVEL
LORD HOWE RISE
NEW CALEDONIA BASIN
NORFOLK RIDGE
MT EGMONT
NORTH ISLAND NEW ZEALAND
HIKURANGI TROUGH
CHATHAM RISE
SOUTHWEST PACIFIC OCEAN
BASALT
MOHO
OCEANIC BASALT
OLD OROGEN (HERCYNIAN)
OCEANIC CRUST
RAFTED REMNANTS
ACTIVE OROGEN
7B SW-NE SUNDA SHELF TO THE PHILIPPINES
B
B'
SW
E
INDIAN OCEAN
SUMATRA
CENTRAL SUMATRA BASIN
(OIL)
MALAY PENNIN.
MALAY BASIN
SE VIETNAM BASIN
SOUTH CHINA SEA
REED BANK
PALAWAN
SULU SEA
PHILIPPINE ISLANDS
S. PACIFIC OCEAN
OCEANIC
PROTO-CONT CRUST
SUB-CONT. CRUST
ACTIVE OROGEN
SUBCONTINENTAL CRUST
OCEANIC CRUST
RAFTED REMNANTS
PROTO-CONTINENTAL CRUST
SCALE:
HORIZ
VERT
PRC
PACIFIC OCEAN
INDIAN OCEAN
AUSTRALIA

Figure 7.

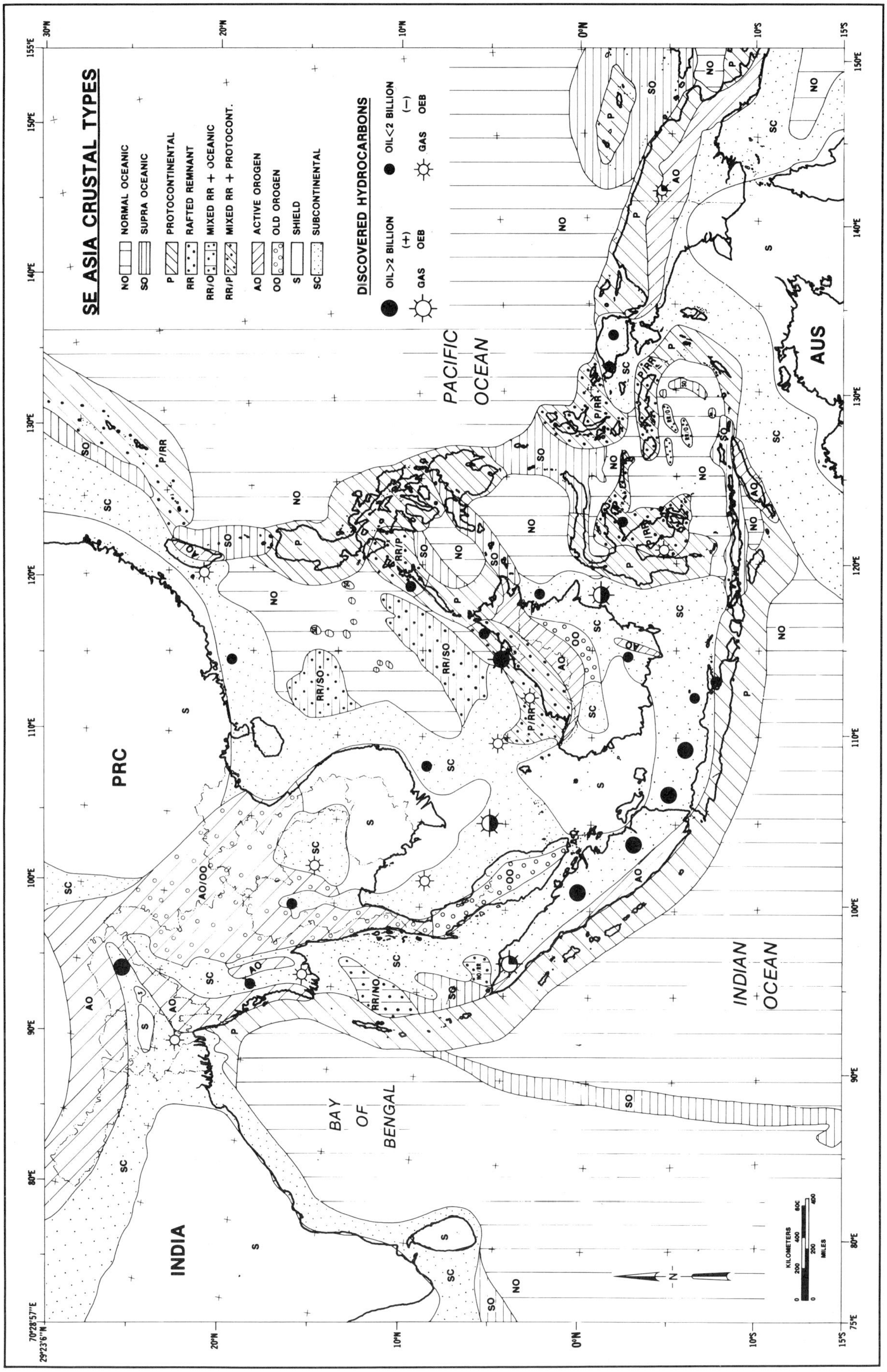
SE ASIA CRUSTAL TYPES
NO NORMAL OCEANIC
SO SUPRA OCEANIC
P PROTOCONTINENTAL
RR RAFTED REMNANT
RR/O MIXED RR + OCEANIC
RR/P MIXED RR + PROTOCONT.
AO ACTIVE OROGEN
OO OLD OROGEN
S SHIELD
SC SUBCONTINENTAL
DISCOVERED HYDROCARBONS
OIL>2 BILLION (+)
OIL<2 BILLION (−)
GAS
OEB
PACIFIC OCEAN
INDIAN OCEAN
BAY OF BENGAL
PRC
INDIA
AUS
KILOMETERS
MILES

Figure 8.

these crustal types, low geothermal temperatures affecting hydrocarbon maturation, and a host of other problems that are associated with oceanic/volcanic-derived crustal materials. Most coastal Circum-Pacific basins are underlain by protocontinental crust, and most of them are poorly prospective as a result. A few exceptions exist where accreted terraines of protocontinental crust contain rafted remnants of older acidic crust, or have numerous young granitic plutons. These mixed crustal elements appear to provide quartz sand for improved reservoir rocks. They also have a higher heat flow for hydrocarbon maturation. Many of the hydrocarbon-producing Circum-Pacific basins such as Talara (Peru), Cook Inlet (Alaska), Niigata (Japan), and the California coastal basins are resting directly on, or adjacent to, older continental crustal elements containing acid igneous or metamorphic rocks. Another example is seen in North Palawan (Figures 7b, 8) which is a local hydrocarbon-producing area resting on mixed crust—protocontinental with rafted remnants. Other Circum-Pacific margin basins, formed over protocontinental crust without these acidic-rock elements, have had almost no hydrocarbon discoveries, despite extensive drilling.

## Continental Crustal Types

Active orogens are young mountain systems formed by the convergent interaction of two plates in subduction or plate collision (Molnar and Chen, 1982). One or both plates must be continental. The crust thickens and the orogen develops a root of thickened crust which can extend 40 to 70 km down to the Moho, as shown by deep seismic-reflection profiling (Ando et al., 1984). Active orogens are very unstable, have high temperatures, and commonly contain soft schist with folded sediments on top (Hsu, 1982). They deform easily by shearing and compressional movements. Present-day, active orogens are typically quite young (Tertiary-Cretaceous) and have a wide diversity ranging from high mountains with a deep root (Himalayas, Andes) to lower, shallower features as shown in Figure 7 (New Zealand). Smaller active orogens can merge directly into protocontinental crust. Parts of southern Alaska and eastern Siberia consist of such active orogens and protocontinental crust containing large and small (rafted remnant) microplates of Mesozoic, Paleozoic, and Precambrian continental crust. Associated with active orogens are wrench (LL) basins such as the upper and middle Magdalena in Colombia, the Cankiri and Sivas in Turkey, and the Hodna in Algeria; trench-associated (TA) basins such as Mamberamo in Irian Jaya, and some complex polyhistory intermontaine basins. Of the six types of active orogen foldbelts listed in the Global Basin Classification system (Kingston, Dishroon, and Williams, 1983, Figure 10) only three produce commercial hydrocarbons. Worldwide, with the exception of the Zagros foldbelt in Iran, past oil and gas discoveries in basins associated with active orogens can be rated as only fair to good.

Old orogens are worn-down mountain ranges, smoothed remnants of erstwhile active orogenic belts. Old orogens (Appalachians and Urals) generally have crustal thicknesses from 38 to 46 km, some of it tectonically stacked as described in various COCORP papers (Lillie, 1984; Oliver, 1982). Erosion exposes the cores of the orogens that were formed at depth. Old orogens, not being active, have cooled somewhat and the crust is more brittle, or less mobile than active orogens, with more of a tendency to fracture rather than fold. Interior fracture, or interior sag basins, rather than wrench basins are normally associated with this type of crust. Worldwide, we have found that most documented old orogens are pre-Jurassic. Some are as old as Precambrian. Hydrocarbon-bearing basins associated with the flanks of old orogens include interior fractures (IF), interior sags (IS), margin sag (MS), and complex "foreland' polyhistory basins such as the Appalachian (USA), and the Volga-Urals and Pechora (USSR). Old orogens have the same associated plays as active (young) orogens and produce hydrocarbons from buried thrustbelts as found along the western flanks of the Appalachians and Urals, as well as wrenched foldbelts (Ardmore). One of the main problems in exploring hydrocarbon traps in either young or old orogens is to find those with seals unbroken over the long period of time since folding-formation. Even with these constraints worldwide, the record of past oil and gas discoveries in basins associated with old orogen crust can be rated as fair to good.

Shields are formed when orogens are completely worn down to their exposed roots. A shield area may also contain very old Precambrian crust of undetermined orogen. They are tectonically stable, with no active crustal generation, form the cores of the continental massifs, and are at or near isostatic equilibrium. The crust is more stable and brittle than old orogens and may respond to plate rotational stresses with long episodic wrench faults or fractures. When these features pass through sedimentary basins they may produce large, potentially prospective anticlines, such as the Nesson (Williston Basin, USA) or the Bray (Paris Basin, France). Shield areas characteristically have a crust ranging from 32 to 40 km in thickness. Present-day shield areas are commonly made up of old Precambrian continental crust, such as the Canadian, African, or Siberian massifs. Only a few younger elements could be classed as shields, like the Piedmont of North America, or parts of the Hercynian terrane (Caledonian-Hercynian orogenies) of Europe. Classic shields have the old basement rocks outcropping at the surface, whereas "platforms" are shields covered by flat-lying platform-type sediments.

Basin types formed over shield areas are commonly interior sags or interior fracture grabens. These basins can either be simple or complex poly-history types, as seen in Figure 6. Many hydrocarbon-bearing basins overlie shield areas. They normally contain many small fields with only a few giants. The basins of the north-central United States such as Salina-Forest City, Illinois, and the Cincinnati Arch, fall into this category. Prospects over this kind of crust must be rated as good, but not too prolific.

Subcontinental crust is formed in areas of plate divergence, commonly Atlantic-type divergent continental margins. It can, however, be found in the interior of continents. Subcontinental crust is pre-existing crust (typically old orogen or shield) that has been thinned down by stretching, normal faulting, divergent wrenching, or subcrustal melting. Thicknesses of subcontinental crust range from about 20 to 32 km, with an average of about 25 km. Most major sedimentary basins are underlain by this type of crust.

The basin types formed over subcontinental crust are the most varied and important from an oil-generating viewpoint. Hydrocarbon formation and accumulation conditions are enhanced because subcontinental crust supplies quartz sand for reservoirs, restricted depositional conditions for rich source rocks, and good temperature gradients for maturation. Because the crust is thinner than shields/platforms, the sedimentary basins can be deeper and have larger prospects. Interior fracture basins (Sirte-Libya, Gulf of Suez-Egypt, Pripyat, Dnepr-Donetz-USSR), interior sags (Ghadames-Libya, Michigan), and assymetric (foreland) polyhistory basins (Anadarko, Persian Gulf) commonly overlie subcontinental crust. Thinned subcontinental crust on exterior margins can be overlain by fairly simple margin sags such as the Gabon, Congo, or Niger Delta basins. Past oil and gas discoveries associated with basins overlying subcontinental crust include most of the major hydrocarbon basins of the world and a vast majority of the giant fields. Figure 8 shows a crustal map of southeast Asia. Most of the major hydrocarbon accumulations are located primarily on subcontinental crust around the sundaland shield. Smaller accumulations are found on other types of continental, or mixed continental/transitional crust. It may be noted that in new crustal assembly areas (as in southeast Asia) that the key regional map for locating hydrocarbon-bearing basins is the map of crustal types. This is also true for the whole Circum-Pacific region.

## MINERALS AND CRUSTAL TYPE

In addition to hydrocarbons, the crustal classification may be utilized in exploration for mineral deposits. There appears to be a correlation between certain groups of ore deposits and crustal type. For example, heavy mineral deposits such as chromium and nickel are associated with basic and ultrabasic crustal types. Porphyry copper and vein mineral deposits are commonly found in protocontinental crust. The division of continental crust into "new" and "old" categories is useful in minerals exploration. Young, active orogens may contain a wide variety of low-temperature vein mineral deposits. The roots of old orogens and shields, on the other hand, may contain more intermediate and higher temperature mineral deposits. Useful predictions as to the location of sedimentary ore deposits may be made by studying the types of crust surrounding basin areas. Geothermal studies may also find this crustal classification of use.

## SUMMARY

In the past, worldwide, most of the hydrocarbons found have been located in basins overlying or adjacent to some type of continental crust. Basins located over subcontinental crust contain most of the oil and gas and are rated best, shields and old orogens are rated second, and active orogens third. The transitional crustal types, located in the open ocean or on continental margins, have in the past had poor exploration success, despite determined drilling efforts by industry. Basins overlying mixed crustal types along convergent margins or within foldbelts have a spotty record; most are poor but some are good. Basins overlying oceanic crustal types have had virtually no exploration success as of this date.

## ACKNOWLEDGMENTS

The authors would like to thank the management of Esso Exploration, Inc., for permission to publish this paper; T.H. Nelson, J.B. Sangree, R.C. Vierbuchen, Jr., and L.E. Brow for comments and assistance in revisions; and R.T. Schweisberger for translations.

## REFERENCES CITED

Ando, C.J., B.L. Czuchra, S.L. Klemperer, L.D. Brown, M.J. Cheadle, F.A. Cook, J.E. Oliver, S. Kaufman, T. Walsh, J.B. Thompson, Jr., J.B. Lyons, and J.L. Rosenfeld, 1984, Crustal profile of mountain belt, COCORP deep seismic reflection profiling in New England Appalachians and implications for architecture of convergent mountain chains: AAPG Bulletin, v. 68, p. 819-837.

Ben-Avraham, Z., A. Nur, D. Jones, and A. Cox, 1981, Continental accretion: From oceanic plateaus to allochthonous terranes: Science, v. 213, p. 47-54.

Churkin, M., Jr., C. Carter, J.H. Trexler, Jr., 1980, Collision-deformed Paleozoic continental margin of Alaska—Foundation for microplate accretion: GSA Bulletin, v. 91, pt. 1, p. 648-654.

Coney, P.J., D.L. Jones, and J.W.H. Monger, 1980, Cordilleran suspect terranes: Nature, v. 228, p. 329-333.

Cummings, D., and G.I. Shiller, 1971, Isopach map of the earth's crust: Earth Science Review, v. 7, p. 97-125.

Dewey, J.F., and J.M. Bird, 1970, Mountains belts and the new global tectonics: Journal of Geophysical Research, v. 75, p. 2625-2647.

Dewey, J.F., and B.F. Windley, 1981, Growth and differentiation of the continental crust: Philosophical Transactions of the Royal Society of London, Ser. A, A301, p. 189-206.

Dickenson, W.R., 1974, Plate tectonics and sedimentation, *in* Tectonics and sedimentation: SEPM Special Publication 22, p. 1-27.

Fyfe, W.S., 1978, The evolution of the earth's crust: Modern plate tectonics to ancient hot spot tectonics?: Chemical Geology, v. 23, p. 89-114.

Gastil, R.G., 1979, A conceptual hypothesis for the relation of differing tectonic terrane to plutonic emplacement: Geology, v. 7, p. 542-544.

Green, A.R., 1977, Evolution of earth's crust and sedimentary basin development: AGU Monograph 20, p. 1-17.

Hsu, K.J., 1982, Geosynclines in plate-tectonic settings: Sediments in mountains, *in* K.J. Hsu, ed., Mountain building processes: New York, Academic Press, ch. 1-1, p. 3-12.

Jarvis, G.T., and D.P. McKenzie, 1980, Sedimentary basin formation with finite extension rates: Earth and Planetary Science Letters, v. 48, n. 1, p. 42-52.

Karig, D.E., and G.F. Sharman, III, 1975, Subduction and accretion in trenches: GSA Bulletin, v. 86, p. 377-389.

Karig, D.E., and R.W. Kay, 1981, Fate of sediments on the descending plate at convergent margins, Philosophical Transactions of the Royal Society of London, Ser. A., A301, p. 233-251.

Kingston, D.R., C.P. Dishroon, and P.A. Williams, 1983, A global basin classification system: AAPG Bulletin, v. 67, p. 2175-2193.

Lillie, R.J., 1984, Tectonic implications of subthrust structures revealed by seismic profiling of Appalachian-Ouachita belt: Tectonics, v. 3, n. 6, p. 619-646.

Molnar, P., and Wang-Ping Chen, 1982, Seismicity and mountain building, *in* K.J. Hsu, ed., Mountain building processes: New York, Academic Press, ch. 1-4, p. 41-57.

Morton, W.H., and R. Black, 1975, Crustal attenuation in Afar, Afar depression of Ethiopia, v. 1, Inter-union Commission on Geodynamics Sci. Re. 14: Stuttgart, E. Scheizerbart'sche Verlagsbuchhandlung, p. 55-65.

Oliver, J., 1982, Probing the structure of the deep continental crust: Science, v. 216, n. 4547, p. 689-695.

Oliver, J., 1985, Continental tectonics and the continental lithosphere: Testimony before Subcommittee on Science, Research, and Technology, Committee on Science and Technology, U.S. House of Representatives, 19 March 1985.

Pitcher, W.S., 1982, Granite type and tectonic environment, *in* K.J. Hsu, ed., Mountain building processes: New York, Academic Press, ch. 1-3, p. 19-40.

Rezanov, I.A., and A. Sh. Faytl'son, 1974, Mekhanizm Obrazovaniya Okeanicheskoy Kory, Vyssh Ucheb Zavedeniy Izvestiya: Geologiya i Razvedka, n. 2, p. 13-15.

Roeder, D.H., 1973, Subduction and orogeny: Journal of Geophysical Research, v. 78, p. 5005-5024.

Sleep, N.H., 1971, Thermal effects of the formation of Atlantic continental margins by continental breakup: Geophysical Journal of the Royal Astronomical Society, v. 24, p. 325-350.

Soller, D.R., R.D. Ray, and R.D. Brown, 1982, A new global crustal thickness map: Tectonics, v. 1, p. 125-149.

Thorpe, R.S., P.W. Francis, and R.S. Harmon, 1981, Andean andesites and crustal growth: Philosophical Transactions of the Royal Society of London, Ser. A., A301, p. 305-320.

CHAPTER 28

# GLOBAL BASIN CLASSIFICATION AND HYDROCARBON PLAYS

**D.R. Kingston**
*Esso Exploration Inc.*
*Houston, TX*

**C.P. Dishroon**
*Esso Exploration Inc.*
*Surrey, U.K.*

**P.A. Williams**
*Esso Exploration Inc.*
*Surfside, TX*

ABSTRACT

This proposed system classifies sedimentary basins, worldwide, into specific as well as general categories. The geologic history of each basin may be subdivided into cycles using three parameters—basin-forming tectonics, depositional sequences, and basin-modifying tectonics. Sedimentary basins may be simple, with one or two tectonic/sedimentary cycles, or they may be complex polyhistory basins with many different cycles and events. There are eight basic cycle types in this classification, which cover continental, continental-margin, and oceanic areas. The eight basic cycle types, their depositional fills, and tectonic modifiers have been given letter and number symbols so that the specific geologic history of each basin may be written as a formula. These formulas may then be compared and similarities or differences between basins noted. Prolific hydrocarbon basins are categorized, and the plays are compared to specific tectonic and depositional events. Some plays are controlled by basin-initiating tectonic events, such as interior fracture or interior sag, and the type of sedimentation inherent in each basin type. In more complex polyhistory basins, oil and gas plays are commonly associated with combinations of various cycles, or by basin-modifying tectonics such as episodic wrenching, subsidence, foldbelts, and basin tilt.

## INTRODUCTION

Various basin classification systems have been proposed in recent years, e.g., Weeks (1952), Knebel and Rodriguez-Eraso (1956), Uspenskaya (1967), Halbouty et al. (1970a, b), Perrodon (1971), Klemme (1971a, b, 1975), Bally (1975), Huff (1978), Bally and Snelson (1980), and Bois, Bouche, and Pelet (1982). The geologic history of two continental-margin basins or two cratonic basins may be similar in general aspects, but will show important differences in detail; consequently, in the past, two basins could be compared only in very general terms. An alternative system proposed herein compares basins in both general and specific terms.

Esso Exploration began plate tectonic restorations on a global scale in the late 1960s. The result was the development of a basin classification system, whereby all sedimentary basins, worldwide, could be classified according to their structural genesis and evolutionary history. Contributions from Exxon domestic and overseas affiliates provided data on all 650 identifiable sedimentary basins worldwide. We were able to classify all of these basins within the systems, the accuracy depending on the quality of the data and our collective knowledge of the regional geology.

The basic unit in this classification is the cycle, which consists of the sediments deposited during one tectonic episode. Some basins have only one sedimentary or tectonic cycle. These are called simple basins. Most basins, however, contain more than one tectonic/sedimentary cycle, and are called polyhistory basins. Figure 1 is a chart for the classification of simple basins and the identification of cycles for polyhistory basins. It should be noted that the terms "basin" and "cycle" may be used interchangeably in this system as a unit formed by one structural mode of basin formation. Basins, both simple and complex, may be classified by analyzing their geologic history in the context of plate tectonics. The major elements of this history are (1) depositional cycle or sequences, (2) basin-forming tectonics, and (3) basin-modifying tectonics.

## DEPOSITIONAL SEQUENCES

The first major elements used in the basin classification system are depositional cycles and stages. A cycle is defined as the sediments deposited during one tectonic period. The minimum stratigraphic unit that can be called a cycle must have

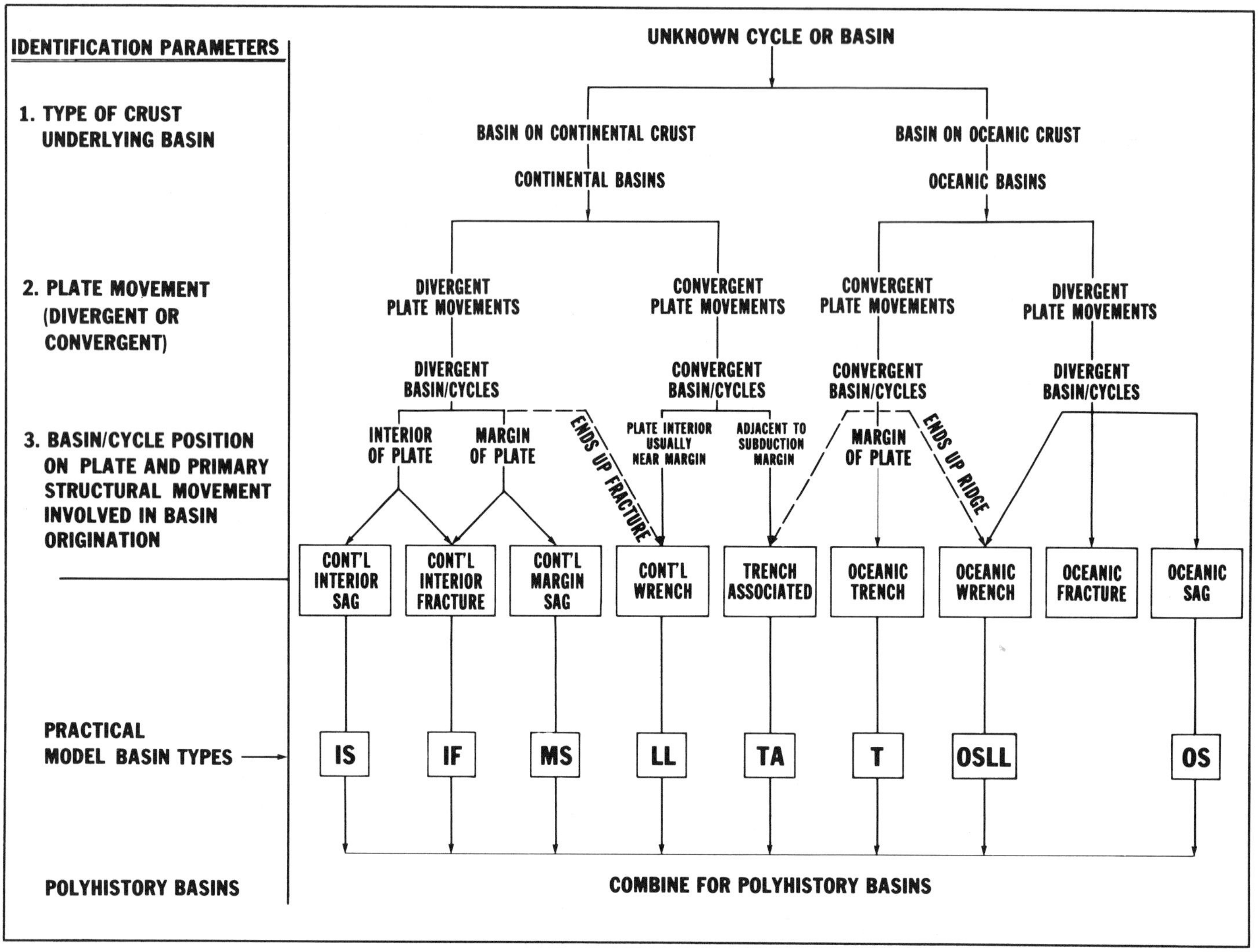


Figure 1. Basin/cycle identification key for basin-forming tectonics.

significance in the development of a basin, either in thickness or span of geologic time. This allows us to lump thin units of high-shelf or wedge-edge deposits, which may form over long periods of time, into just a few cycles and to split thick prograding deposits into identifiable units.

Figure 2 shows the relationship of the depositional stages to the tectonic cycle. One can think on the concept of wedge base, wedge middle, and wedge top as one sedimentary cycle with the three stages representing the three elements of one major transgressive-regressive wedge (White, 1980). Stage 1 of the cycle corresponds to a nonmarine wedge base. This includes primarily nonmarine flood-plain, lagoonal, and beach deposits, if they can be distinguished. Sedimentary types normally present are nonmarine conglomerates, sandstones, and shales. Other lithologic characteristics less commonly found are red beds, coals, volcanics and fresh-water limestones.

Stage 2 is the marine wedge middle. Lithologic types most commonly found here are marine shales, limestones, and sandstones. All massive salt is included here on the theory that thick evaporites generally indicate a marine connection or at least the drying of a marine-connected tongue. Other less common lithologic characteristics found in stage 2 are volcanics, marine coals, flysch and other turbidites, and deep-water marls and pelagic deposits. Stage 2 may also contain the distal ends of nonmarine tongues provided these do not exceed 50% of the total.

Stage 3 is the nonmarine wedge top and the associated regional unconformity. Lithologically, it typically resembles the wedge base with more than 50% nonmarine conglomerates, sandstones, shales, red beds, coals, freshwater limestones, and minor evaporites. Postwedge-top unconformities are included in stage 3.

The sedimentary stages should be described from the center of the depositional cycle in enclosed basins or from the thickest part of the wedge in a margin basin open to the sea on one side. Referring to a cross section of the wedge concept shown in Figure 2A, it is evident that if the portion of the basin studied is too far updip, past the pinch-out of the marine wedge middle at c, only the nonmarine rocks of stages 3 and 1 will be described. Conversely, section a of Figure 2A shows that by going too far downdip, only the stage 2 marine wedge may be described.

## BASIN-FORMING TECTONICS

Basin-forming tectonic style is the second major element used in classification. Figure 1 shows the identification key

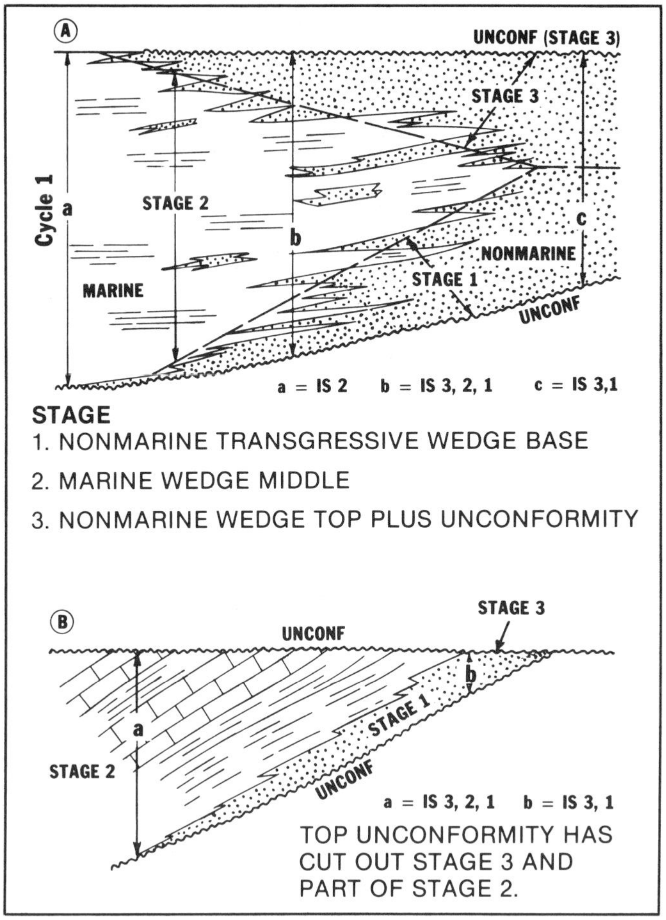


Figure 2. Relationship of depositional stages to sedimentary wedges.

used in distinguishing the various basin- or cycle-forming tectonic styles. The three parameters used in the general identification of these cycle types are shown on the left in Figure 1. The composition of the crust underlying the basin is the first parameter necessary for identification (see Figure 3). There should be no difficulty in determining crustal composition if dealing with true continental craton or true oceanic crust. Intermediate crustal composition may present a problem that can generally be resolved (Kingston, Dishroon, and Williams, 1985). For the purpose of cycle identification, it need only be known whether the basin lies on continental or oceanic crust.

The type of past plate movement involved during the cycle or basin formation is the second parameter used in Figure 1. Fundamentally, there are two types of plate movements affecting cycle or basin formation: (1) divergence and (2) convergence. It has been argued that transform movements are a third type, but these rarely show perfect side-by-side motion and generally exhibit some divergence or convergence. Small angles of convergence show up as wrenching or foldbelts, and small angles of divergence appear as normal faulting or sagging. Consequently, transform movements are not specified in the basin classification.

Convergence normally affects the margins of actively colliding plates, particularly of overriding plates. Strong convergence may be transmitted into the interior of cratonic plates along major shear zones (episodic wrenches), deforming basins well into the interior and away from the convergent margins. Convergent and divergent margins are found on both continental and oceanic crustal plates.

The third parameter used in Figure 1 is basin position on the plate (continental interior or margin) and primary structural movement (sagging, normal faulting, or wrench faulting). These combinations give rise to the theoretical model or 10 simple basin or cycle types (shown in parentheses on Figure 1). Two of these (OTA and OF) are not considered at this time to be prospective for hydrocarbons and are omitted from the practical model, which has eight simple basin types. These eight cycle or basin types are the ones used in this basin classification. A cycle or simple basin is normally referred to by its abbreviated letter; thus, margin sag cycles are MS, wrench or shear are LL (lateral), oceanic sags are OS, and so forth.

From the standpoint of petroleum exploration, the greatest number of cycles fall into four major categories and the rest into four minor categories. The major categories are interior sag (IS), interior fracture (IF), margin sag (MS), and wrench (LL). The minor categories are trench (T), trench associated (TA), oceanic sag (OS), and oceanic wrench (OSLL).

Globally, more than 98% of the hydrocarbons discovered to date have been found in the four major cycle categories which are associated with continental crust. All the basins within this classification consist of one or more of these cycles, plus modifying structural events. For example, a foldbelt, which is complete folding of a basin or part of one, is not considered to be a basin or cycle type, but is treated as a tectonic event that affects pre-existing basins. Some of these cycle types may be split into variations for play analysis and evaluation.

## Divergent Cycle Types

Divergent plate movements are the underlying cause of several important basin types. These may be single-cycle (simple) basins or multicycle polyhistory basins.

*Interior Sag Cycles/Basins (IS)*—These cycles or basins are defined as being located entirely on continental crust in areas of divergence. They are found in the interior of continental masses, not at the plate margin, and if near the edge their axes are generally at a significant angle to the margin. Interior sag basins (IS) are normally more or less circular or oval in shape and generally do not accumulate as great a thickness of sediments as continental margin basins. They are formed by simple sagging of the continental crust, as shown in Figure 4. Many interior sag basins originated in the Paleozoic. Some are simple or single-cycle basins, whereas others contain several repeated sag cycles. Figure 4 shows the evolution of a typical interior sag (IS) basin. The basin simply sags with minor or no faulting. Interior sag cycles are commonly found in polyhistory or multiple-cycle basins.

*Interior Fracture Cycles/Basins (IF)*—This basin/cycle type is defined as being found on continental crust, either in the interior of present plates or at the crustal margins of old continental plates. Interior fracture basins are caused by divergence and tension within the continental block. Vertical horst-and-graben faulting and subsidence are the dominant features.

Figure 5 is a series of restored cross sections showing the Gulf of Suez, a typical interior fracture basin. Cycle 1 shows basin formation in the Early Cretaceous, with tension block faulting and subsidence. Stage 1 nonmarine sandstone and shales are deposited. During stage 2, marine rocks fill the Suez graben as block faulting and subsidence continues. Reservoir sands and carbonates are deposited over the highs while shales fill the lows. Continued subsidence and compaction

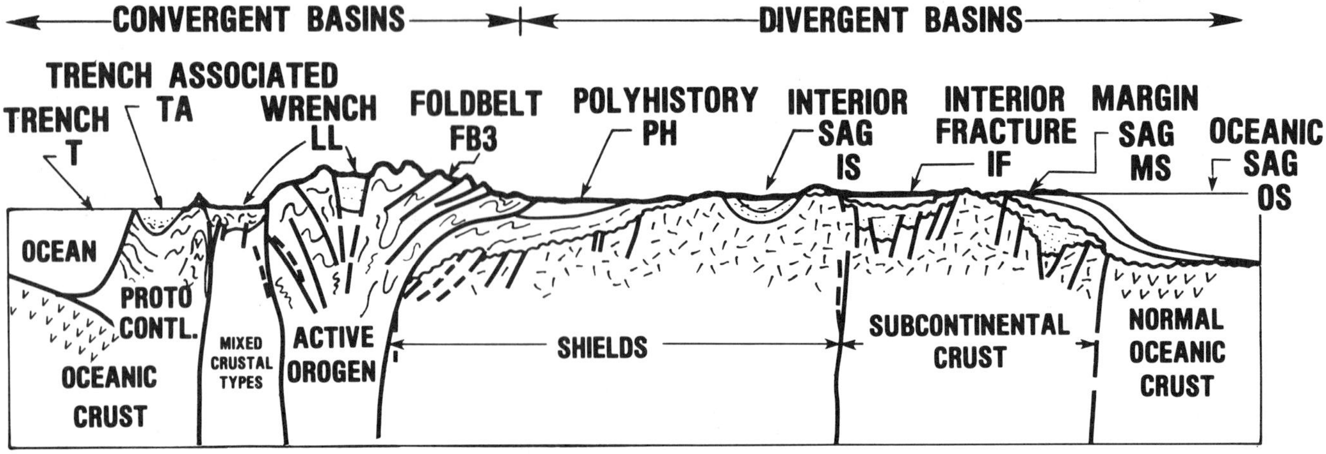


Figure 3. Basin types and crustal association.

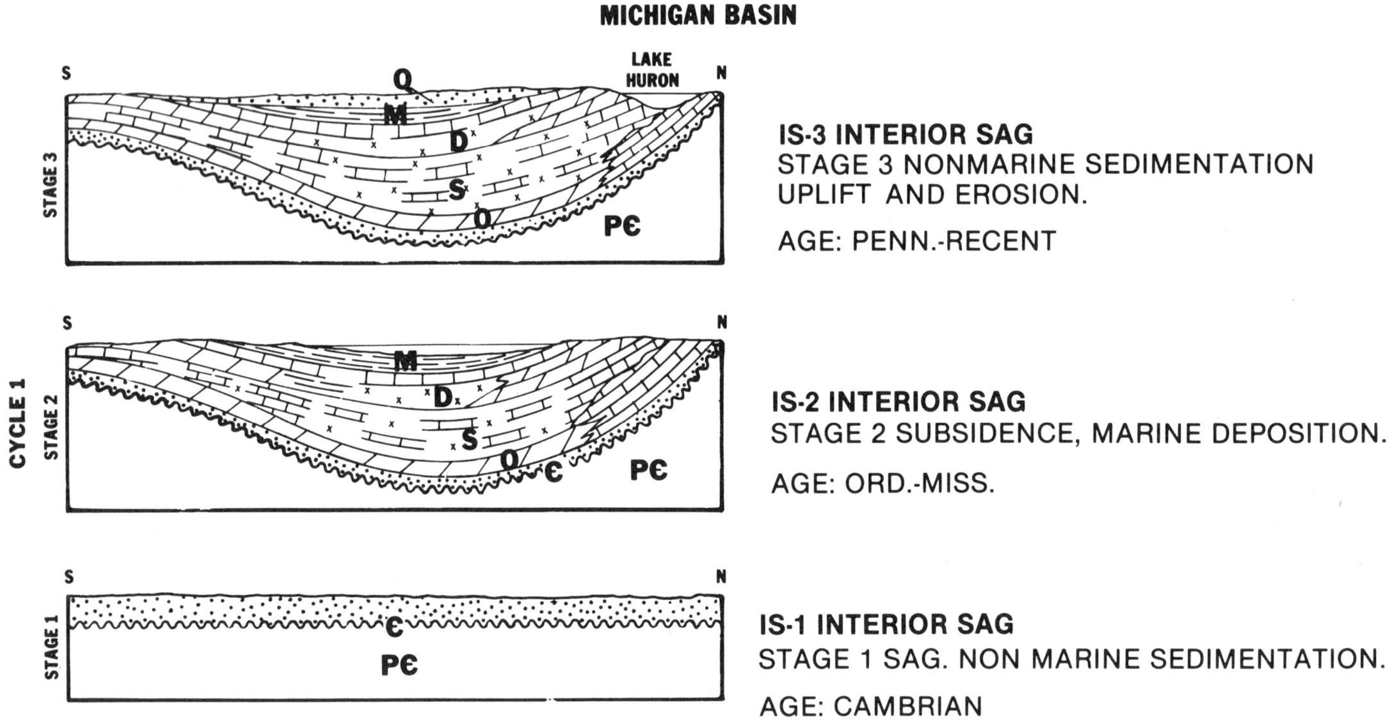


Figure 4. Evolution of Interior Sag basin.

cause drape of beds over high-standing fault blocks. During the third or final stage of interior fracture the basin fills or is uplifted, allowing nonmarine wedge-top sediments to be deposited; unconformities may truncate the top of the section.

*Margin Sag Cycles/Basins (MS)*—Margin sag cycles/basins are located on the outer edges of continental crust blocks in areas of divergence. The basin axes lie parallel to the continental/oceanic crust boundary, and the sediments may overlap onto oceanic crust. Such basins are referred to as being located on "Atlantic-type" margins. All margin sag basins have at least two basin-forming tectonic origins and are polyhistory basins.

The evolution of a typical margin sag basin is shown in Figure 6. The basin begins with the cracking of a cratonic mass by divergence. This first phase (stage 1 of cycle 1), as previously outlined, is called interior fracture and may resemble the present-day rift valleys of Africa. These grabens generally are filled with nonmarine sediments. During graben formation, basal block fault structures are formed and buried by sediments. Probably no continental separation occurs at this time, though deeper fracture zones may have extended through the crust. There is generally no stage 2 in this cycle, inasmuch as the continents have not separated enough to allow a marine incursion.

The next phase of basin formation, stage 3 of cycle 1, is the end of interior fracture basin development. Continued continental divergence and graben subsidence by block faulting are typically accompanied by nonmarine deposition. One or more deep fracture zones may form as potential separation centers. These fracture zones may fill with basalt intrusions and dike swarms. Stage 3 of cycle 1 normally ends with a major unconformity. Basement block faults cease differential

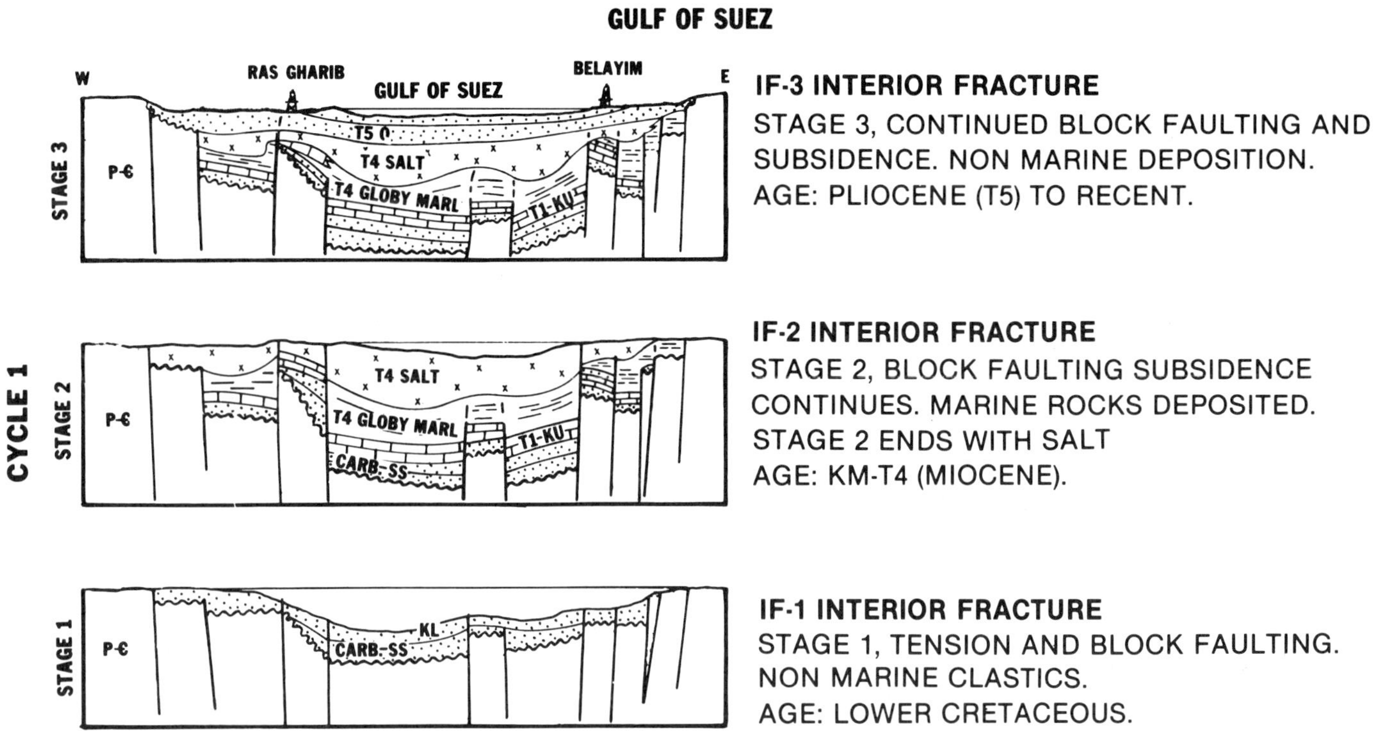


Figure 5. Evolution of Interior Fracture basin.

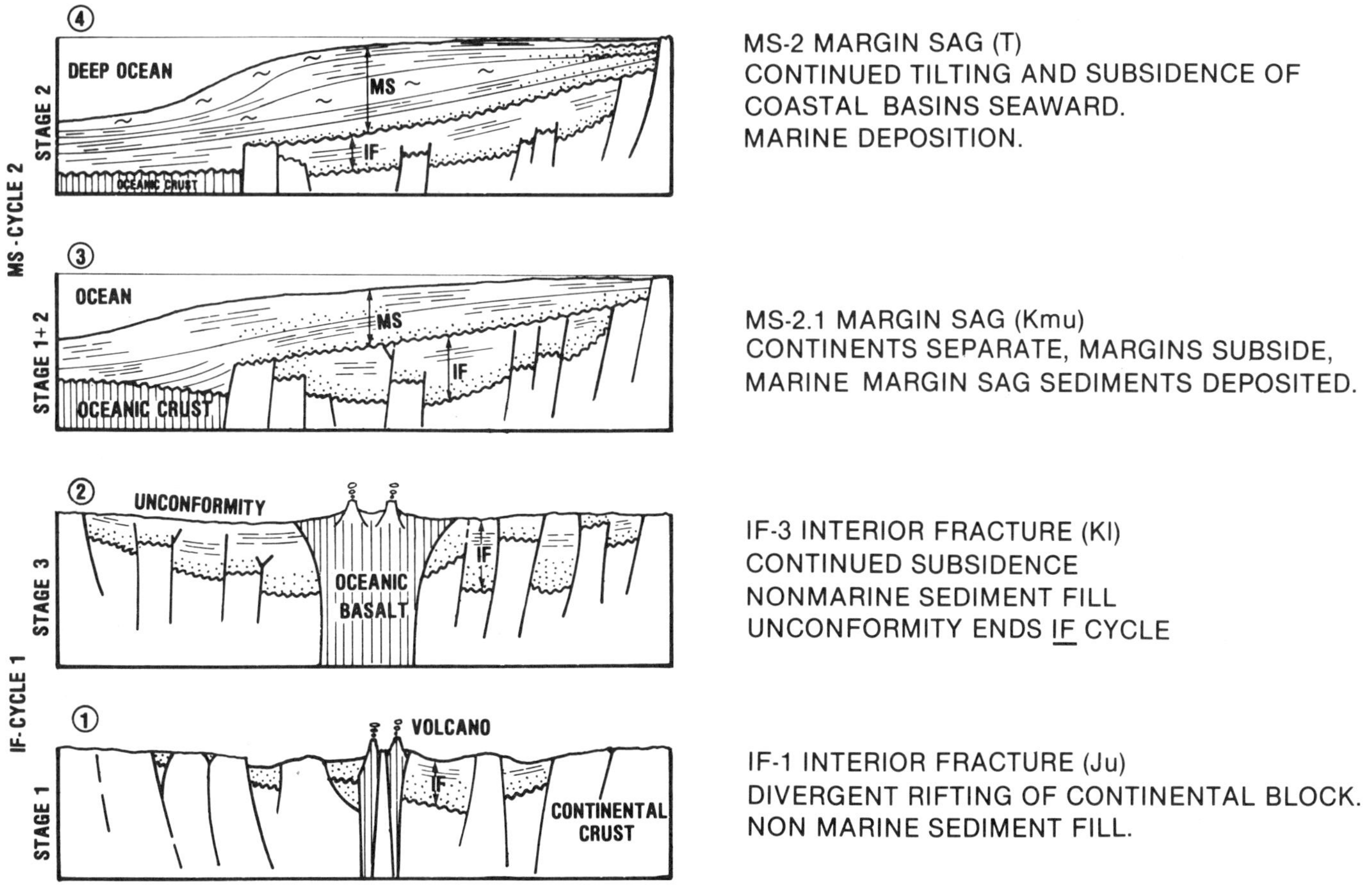


Figure 6. Evolution of Continental Margin Sag basin.

movement near the end of the interior fracture cycle, and overlying beds generally do not show any rejuvenated movement or structure. It may be that continental rifting freezes the basement block faults on either side but whatever the cause, the type of structural origin changes from interior fracture to margin sag at this time.

Margin sag is initiated as the spreading center in the interior fracture grabens is activated and begins to grow. The continents separate and begin to move apart. Basement faults are no longer independently active, and basement begins to subside as one block. The entire edge of the continent sinks. Simultaneously, stage 1 of margin sag cycle 2 commonly begins with deposition of nonmarine beds and a few minor evaporites.

Stage 2 of margin sag includes continued subsidence and continental separation, and is identified by marine deposition. Marine waters invade the infilled graben system for the first time, depositing clastics, carbonates, or massive salt at the base of the series. As the oceanic spreading center expands, the continents separate. New layers of oceanic crust form and the older ones cool.

On many present-day divergent continental margins, there is evidence of very rapid subsidence and oceanward tilting of the basins, beginning in the Early Cretaceous and accelerating in the Tertiary. Marine sediments on the open shelf mark periods of highstands of sea, and nonmarine sediments and paleoslope unconformities mark the lowstands. As the ocean basins become deeper, clinoform deposition into the deep water dominates all previous patterns.

Margin sag basins have been divided into four general types by Beck and Lehner (1974). These are normal clastic, carbonate bank, major delta, and salt tectonics (Figure 7). Most of the more than 100 margin sag basins we have identified can be placed in one of these four groups. These four margin sag basin types, however, are not subdivided from the margin sag category; all are referred to as margin sag (MS) basins or cycles. The advantage of these groupings is in their oil play association, as will be discussed later in this paper.

One variation of the margin sag cycle should be noted separately—it consists of "old" continental margin basins that, as suggested by depositional evidence, have been converted into interior basins by plate collision and the subsequent formation of a foldbelt on the seaward side. They were margin sag cycles transformed into interior sag cycles by orogeny, and are given the designation MSIS. The main reason for the separation and identification of margin sag-interior sag (MSIS) cycles is that they are the classic asymmetrical or foreland basins and, world-wide, many are prolific producers of hydrocarbons.

## Convergent Cycle Types

Basins classified in this category are those formed on margins or nearby interiors of two or more plates converging toward one another. Most basins on converging plates mainly exhibit tensional features. Figure 8 shows a simplified version of the relative positions and development of basins on convergent plate margins.

*Wrench or Shear Cycles/Basins (LL)*—These strike-slip, shear, or wrench basins are referred to here as double "L" cycles (LL), for lateral movement. These cycles are found on continental or intermediate crust. For this classification, hydrocarbon-prospective wrench basins are restricted to those found on or directly adjacent to continental crust.

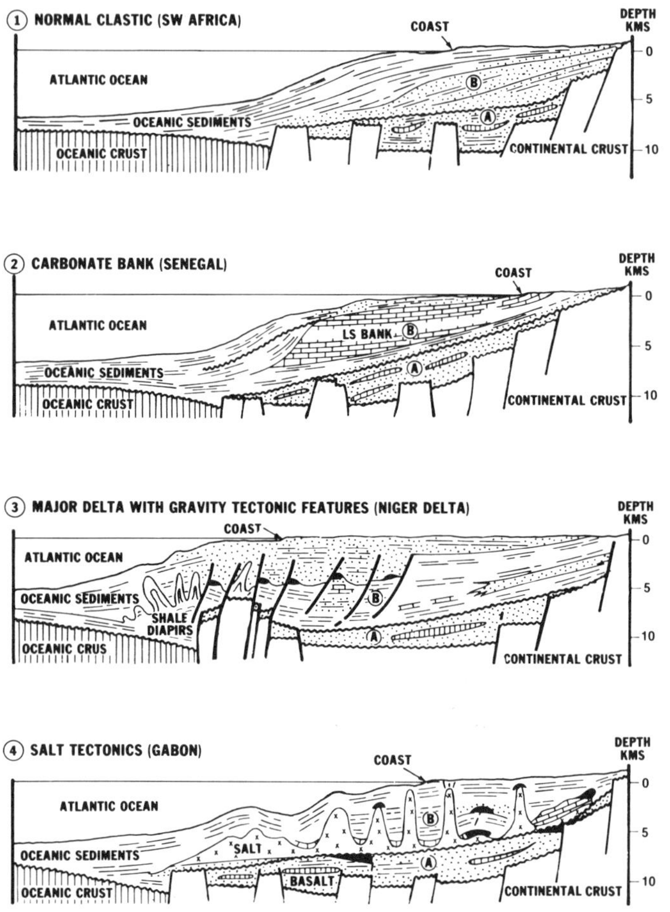


Modified after Beck (AAPG 1974)

Figure 7. Types of Continental Margin Sag basins, showing oil plays.

Wrench basins/cycles are found in areas of two or more converging plates. They are formed by a divergent wrench couplet with strike-slip faults along two or more sides, as shown in Figure 9.

Most wrench couplet or shear basins are found in the areas of present-day, or Tertiary, plate convergence. Typical areas where these basins occur are: (1) the periphery of the Pacific Ocean, including Antarctica; (2) Southeast Asia; (3) the Himalayan-Alpine chain from the Solomon Islands to Spain; and (4) around the Caribbean (Antilles arc). For most of these shear basins, stage 1 (basin initiation by divergent wrenching) appears similar to an interior fracture (IF) basin, with tensional block faulting and little or no evidence of wrenching. Nonmarine wedge-base sediments are deposited, unless the basin is initiated underwater.

Sometime after the basin is initiated in stage 1 or 2, wrench deformation of the basin begins. Wrench structures may form along the flanks or within the basin. Major wrenches outline the basin, and minor wrenches form structures within the basin. If the basin is near enough to the ocean, marine sediments may be deposited; many wrench basins have no marine fill, because they are too far from the sea to have had invasions of marine waters.

In stage 3 of the wrench (LL) cycle, the basin is uplifted and eroded, and the continued shearing may begin to destroy the structures and parts of the basin. The term "L3FB" is used for the final stage of a basin completely wrenched to a foldbelt. Continued plate convergence may result in orogeny.

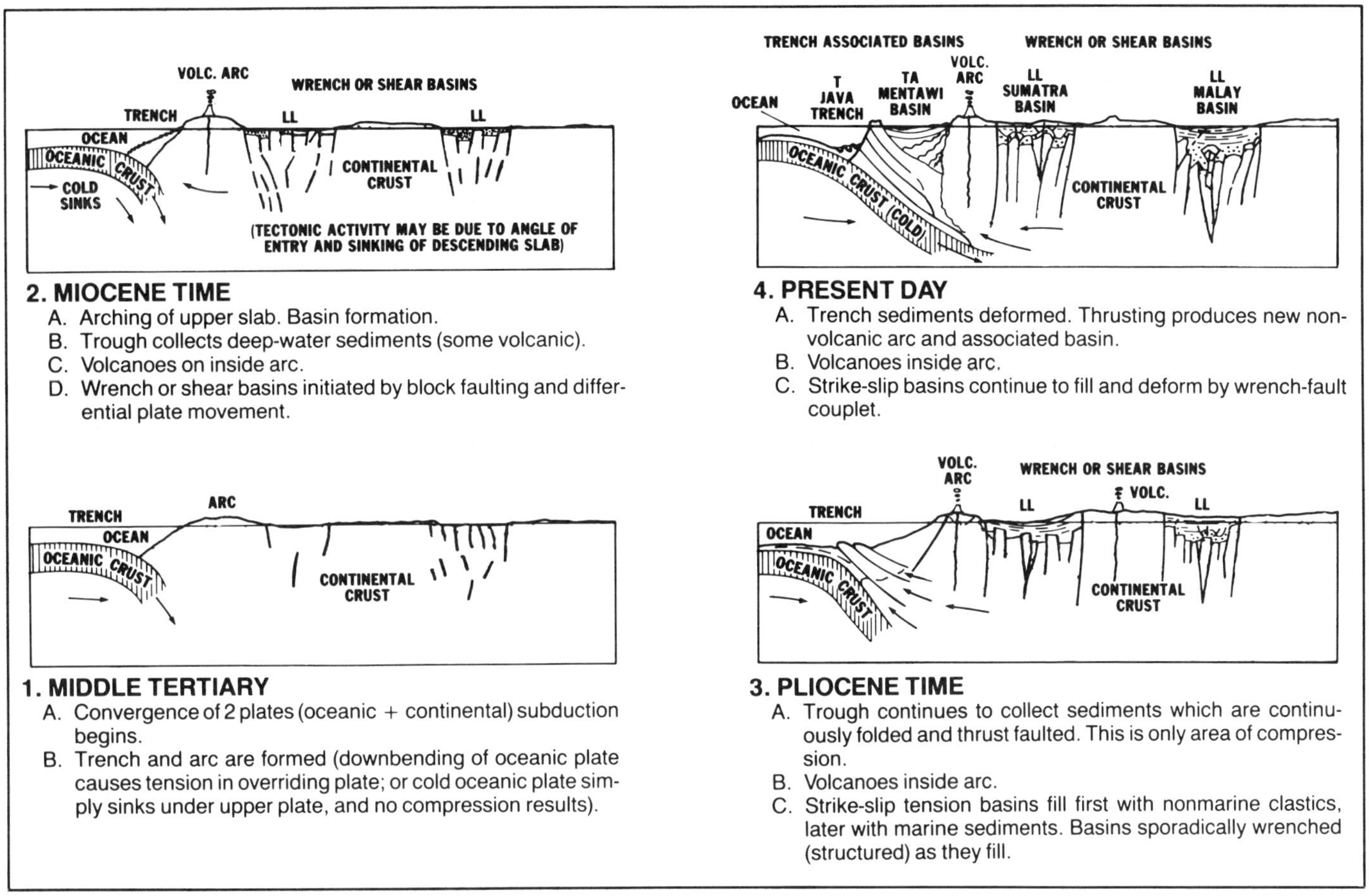


Figure 8. Evolution of Convergent basins.

If the wrench-faulting process continues over a long enough period, it will eventually destroy the LL basin. As a result, few examples are found of preserved wrench basins older than Tertiary. Typically, pre-Tertiary wrench (LL) basins occur as wrenched foldbelts, with only fragments of the destroyed basins being recognized. However, if convergence between the two plates and, therefore, the engine driving the mechanism has stopped at some intermediate interval, the basins existing at that time may be preserved. Some portions of the wrench basins of Oklahoma (Ardmore) are believed to represent different stages of arrested wrenching. Sedimentary fill in wrench (LL) basins is extremely varied. Marine clastics, carbonates and evaporites, nonmarine clastics, volcaniclastics, and flysch-chert-ophiolites are all found, depending on depositional conditions.

*Trenches (T) and Trench-Associated (TA) Basins*—To date, very few commercial oil or gas fields have been found in these basin types. Trench-associated cycles/basins (TA), as shown in Figure 8, are located on convergent continental plate margins, landward of the trench or nonvolcanic arc, if one is present. The basins generally are built on folded trench sediments, not continental crust, and are formed by a simple sag, typically deformed by contemporaneous wrenching. Trench-associated (TA) basins are likely to be filled with a high percentage of volcaniclastic sediments, though quartz or arkosic sands may be found, given the proper nearby sediment source area. Sediments in the these basins generally are marine to deep marine, though some have been found to contain nonmarine materials in the lower stage (Abukuma basin of Japan). Trench-associated (TA) basins easily may be confused with wrench (LL) basins. Forearc basins are included in this trench-associated category. In our global basin classification system, we find many "forearc basins" with no arc association (Sabah, Palawan, Philippines, etc.). As a consequence, we prefer to use the term "trench-associated basins."

Trenches (T) are located (1) on oceanic crust, and (2) at the margins of two or more converging plates. Present-day trenches are relatively narrow downwarps located over subduction zones, as shown in Figure 8. Active trenches are areas of low heat flow and are Tertiary in age; older trenches have been converted into foldbelts.

We recognize two types of trenches. The first involves one oceanic plate overriding another, forming a midocean trench such as the Mariana, Aleutian, and Philippine trenches. These features normally have little sediment fill, and the amount they do have primarily is volcanogenic and deep-water pelagic. The second type involves an oceanic plate overridden by a continental plate. This trench can receive oceanic pelagic sediments and volcanics, as well as land-derived fine clastics. These ocean-margin trenches accumulate very thick deep-water deposits. Continued convergence, with its attendant compression and shearing, eventually results in a foldbelt. Most trenches and trench-associated basins end up as foldbelts, and such folded trench sediments are given the designation "FB3T."

*Oceanic Sags (OS)*—Oceanic sags are areas where oceanic

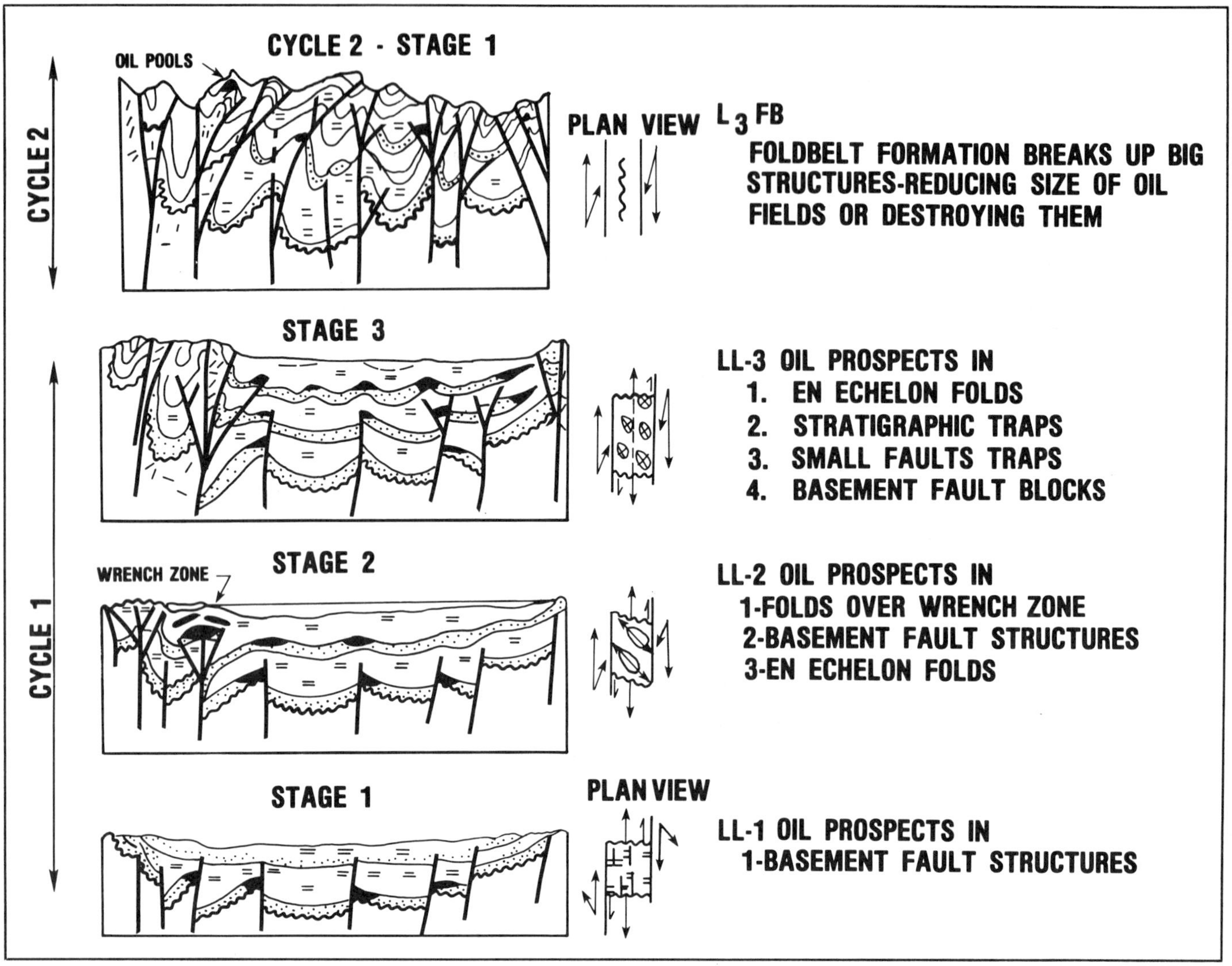


Figure 9. Evolution of Wrench or Shear basins, showing oil plays.

crust has been formed by continental separation and the formation of a spreading center. Subsequent cooling and subsidence of the ocean crust have caused it to sink. The sag areas that have accumulated a significant thickness of sediments are called oceanic sag (OS) basins. Most of these areas of thick sediment are near continental blocks or island arcs and may be associated with either convergence or divergence. The sediments consist of pelagic material, volcaniclastics, and distal turbidities.

*Oceanic Sags–Wrench Couplet Cycles/Basins (OSLL)*— These are the minioceans or small ocean basins floored by oceanic or transitional crust. Such basins are believed to have been formed by a large divergent wrench couplet (rather than mantle upwelling and sea-floor spreading) that opened a miniocean through separation of the sialic layer. These basins may be filled with pelagic materials, volcaniclastics, or distal turbidites, as in oceanic sag (OS) basins.

## BASIN-MODIFYING TECTONICS

The third major element used to classify basins is basin-modifying tectonics. Basins or cycles formed by one type of tectonic movement may be changed during their history by other structural events. There are three types of basin-modifying tectonics—episodic wrenches (L), adjacent foldbelts (FB), and complete folding of a basin area (FB3) which is foldbelt formation.

### Definition of Episodic Wrenches and Foldbelts

Episodic wrenches are designated by the single letter "L," and represent a wide variety of lateral movements not connected with basin/cycle origin. Episodic wrenches modify basins formed by other means and are found in basins with all ages of basement rocks. It is believed that old zones of weakness in the basement, such as old sutures, interior fracture zones, plate boundaries, etc., move periodically or episodically in response to plate movements. These plate movements are manifested at the surface by plate collisions, rotations, fragmentation, and by subduction zones. Wrench faulting or rotational movements are prevalent in basin histories on all continents and react in different ways at different times. Crustal blocks floating on the asthenosphere may be similar to the jostling of ice flows on the polar seas.

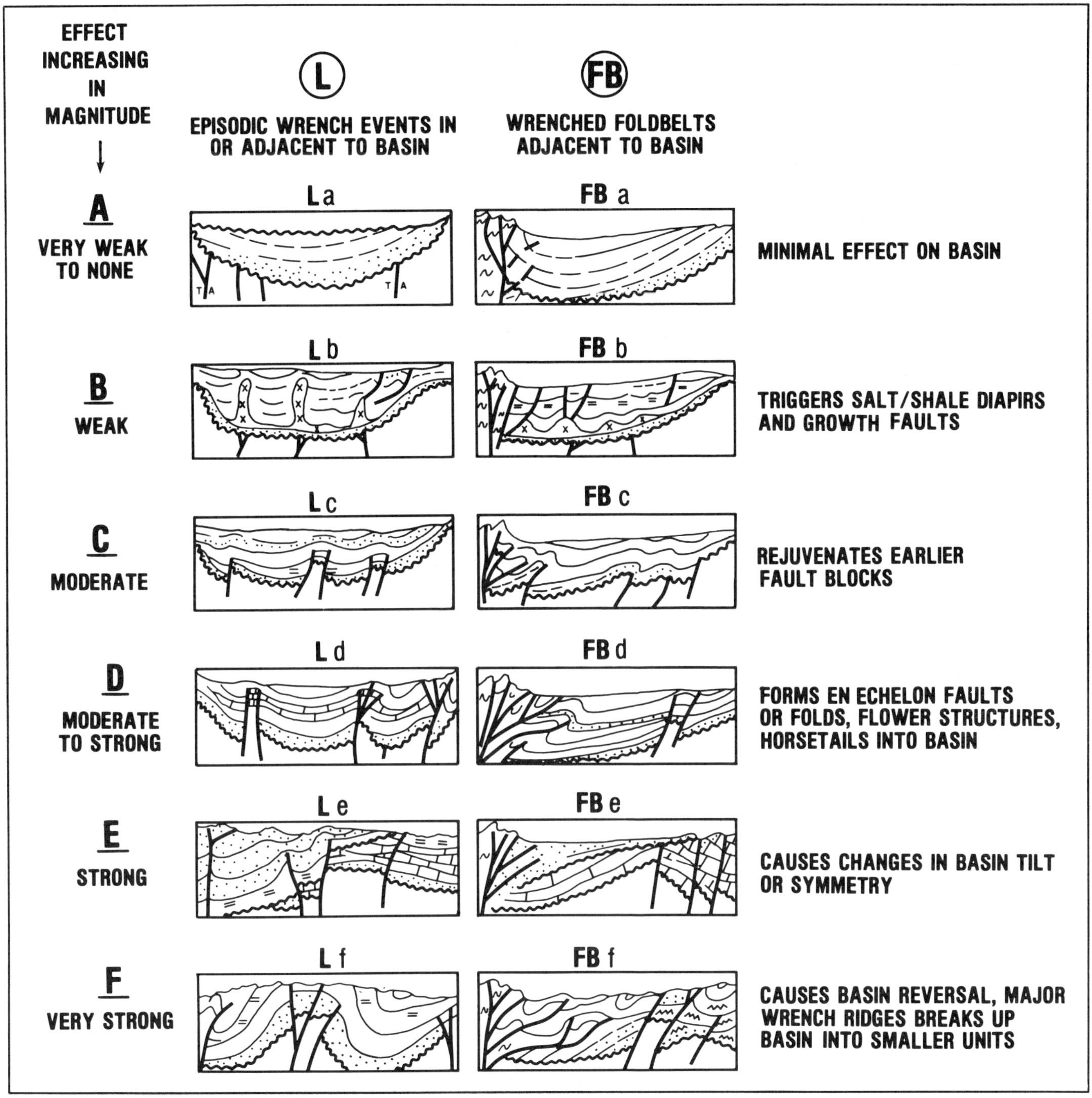


Figure 10. Tectonic modifiers of primary basin types (Wrench and adjacent foldbelts).

Generally, the origin of an episodic wrenching or lateral L movement is fairly easy to ascertain, given good plate-tectonic reconstructions. The event that triggers isolated wrench movements may be on some other side of the plate in question, or even on some other part of the earth. The effect of different intensities of L movements is shown in Figure 10.

Foldbelts are caused by convergence of two or more plates. Basin areas caught in this convergence may be completely or only partly folded. Basins not completely folded are not considered to be foldbelts and are said to have been episodically wrenched. Basins completely folded are called foldbelts (FB3). Foldbelt formulas are derived from the basin that was folded. If the edge of a large basin is folded, the formula for the foldbelt is the same as that of the basin except that the youngest event will be some type of folding. Completely folded rocks, or foldbelts (FB3), are commonly found adjacent to relatively unfolded basins and, in fact, may grade into basins. The expression used in the basin formula to denote an adjacent foldbelt is "FB." Foldbelts have various effects on adjacent basins as shown in Figure 10.

## Tectonic Modifiers

The tectonic modifiers of primary basin types are listed in Figure 10 in order of increasing magnitude, downward from "a" to "f." Each of these effects is found associated with both episodic wrenches (L) and adjacent foldbelts (FB). The very weak "a" effect is known to occur within or adjacent to a basin with minimal structural effect. La would mean that a wrench passed through or adjacent to a basin but caused no faulting or folding visible at the surface or on seismic data. Porosity and permeability on structure, however, may be affected.

Lb and FBb illustrate the "b" effect—still very weak on the scale. The "b" effect triggers salt or shale diapirs and growth faults within the basin and can cause open folds in basins adjacent to FBb. We believe that without a tectonic event of "b" intensity or stronger, salt and shale may not be triggered to flow. The movement could be described best as "jiggling." There are numerous examples, worldwide, of basins with thick salt layers and plenty of load that have never flowed to produce domes (e.g., Yakutsk, Touggourt Permian basins), nor do they exhibit any other evidence of postsalt deposition structuring. It can be concluded that slight plate motions or jiggling are required to initiate or trigger salt and shale diapirs and growth-fault movements. Salt and shale diapirs, once triggered by an Lb event, could continue to grow by static load without further jiggling to keep them moving.

Lc and FBc illustrate the "c" effect which is rejuvenation of pre-existing blocks, either interior fractures or basement. This jostling of older blocks can cause structural growth which may or may not reach the surface. Many of the world's giant fields owe their structural growth to "c" effects. It is important to note that the "a," "b," and "c" effects of domes or arches may not reach the top of the structured cycle, which is also the old ground surface. The first modifier to reach the old ground surface as wrench-generated faults and folds is the "d" effect, which is rated as a moderate to strong event. It is convenient to divide the list of modifiers into the ones causing "weak" effects (a, b, c) and the ones causing "strong" effects (d, e, f).

Ld and FBd are examples of classic wrenching. Relative plate movement is enough to cause en echelon faults or folds to be well developed. Here are found the first flower structures recognizable on seismic records or visible on the surface. Horsetails (a series of en echelon faults or folds) may be seen as fanning out of foldbelts or wrench zones into basins. The Ld flower structures are fairly modest ones and do not bring basement to the surface, as they do in stronger wrench events. Le and FBe are strong plate tectonic effects that significantly alter basin tilt, causing marked basin asymmetry. Tilt in one direction, or change of tilt direction, is an "e" effect.

Lf and FBf effects are the strongest episodic wrench events we record in a basin. Basins are turned inside out or "reversed," with synclines becoming anticlines. Flower structures bring basement or very old rocks to the surface. The "f" effects also see the formation of major ridges or arches in a basin, the breakup of plates under the basins, and consequent formation of new smaller basins out of the old megabasin. The basin can be ripped up extensively and still be called an "f" effect. However, if the basin is completely folded, destroyed, or altered by faulting we would go one step further than "f" and call it a foldbelt (FB3).

It should be noted that any of the characteristic tectonics of the weaker modifiers may be found in any stronger one. For example, salt and shale diapirs and growth faults (Lb) may be found along with rejuvenated block movements (Lc). All of the tectonic parameters, salt domes, jostled blocks, flower structures, basin tilts, etc., may be found in Lf.

Several important points need to be emphasized concerning L and FB events. First, the tectonic modifiers affect, in varying degrees, basins already formed by other processes. Second, an episodic wrenching zone (L) can turn into a wrenched foldbelt (FB) along its length, as a matter of degree of wrenching. We believe that most foldbelts are caused by wrenching movements or convergence at some oblique angle (other than 180°). Third, the modifiers are described as to what effect they have on the basin, not on the wrench or foldbelt. For example, a wrench alongside a basin may be very disruptive locally but have little or no effect on the basin itself. Similarly, a foldbelt alongside a basin may be vaulted or highly deformed mountains but have little structural effect on the adjacent basin. It is the structural effect on the basin that is described in Figure 10. If the foldbelt is hydrocarbon-prospective, it will have its own description.

### Tectonic Modifiers–Foldbelts

Foldbelts represent sutures where past plates have converged or are still converging. This convergence results in compression and shearing motions that cause the rocks to be wrenched and folded. If ultrabasic rocks, serpentine, chert, volcanic flysch, and other oceanic sediments are found in foldbelts, it is assumed that an old area of oceanic crust was destroyed by subduction or plate collision, and the foldbelt suture is all that remains of the vanished oceanic plate.

The five different types of foldbelts (FB3) are shown in Figure 11. They are (1) FB3B, the topographically low, open folds, which may be partly or almost completely buried by younger sediments, as in part of the Iranian foldbelt; (2) FB3U, the uplifted foldbelts or high-mountain ranges, generally having imbricate structure, such as the Canadian Rockies; (3) FB3F, the topographically low but complex folds with imbricate structure (thrustbelts), which may be partly or almost completely buried by younger sediments, as in the Vienna basin; (4) FB3T, folded trench sediments, found on converging plate margins; and (5) L3FB, wrenched to a foldbelt (can have horsetails), invariably a former LL basin. Of the five foldbelt types listed, only three—FB3B, FB3F, and L3FB—presently produce commercial hydrocarbons.

## BASIN CLASSIFICATION

All basins, worldwide, may be classified by using the structural and stratigraphic elements previously discussed. It is possible to combine these elements to make a formula for each of the basins within the system. The formula is a general expression of the basin's structural and stratigraphic history. It does not describe unit thickness, rock color, source, reservoir, grain sizes, and many other factors necessary for basin analysis. Therefore, there is no magic formula which can separate sedimentary basins into oil-and-gas-prone versus barren. Formulas are simply useful means for summarizing the important points in a basin's structural and depositional history. In classifying a basin and writing its formula, it is important to outline only the main events in the basin history, and not to attempt to describe all detail. Too much detail results in long and needlessly complicated formulas, which are difficult to use.

Basins may be classified by comparing basin parameters, as shown in Figure 12. The data needed are regional maps and cross sections, well logs or surface sections, and regional seismic lines, if available. From the maps and cross sections, the geologic and plate tectonic history of the basin may be

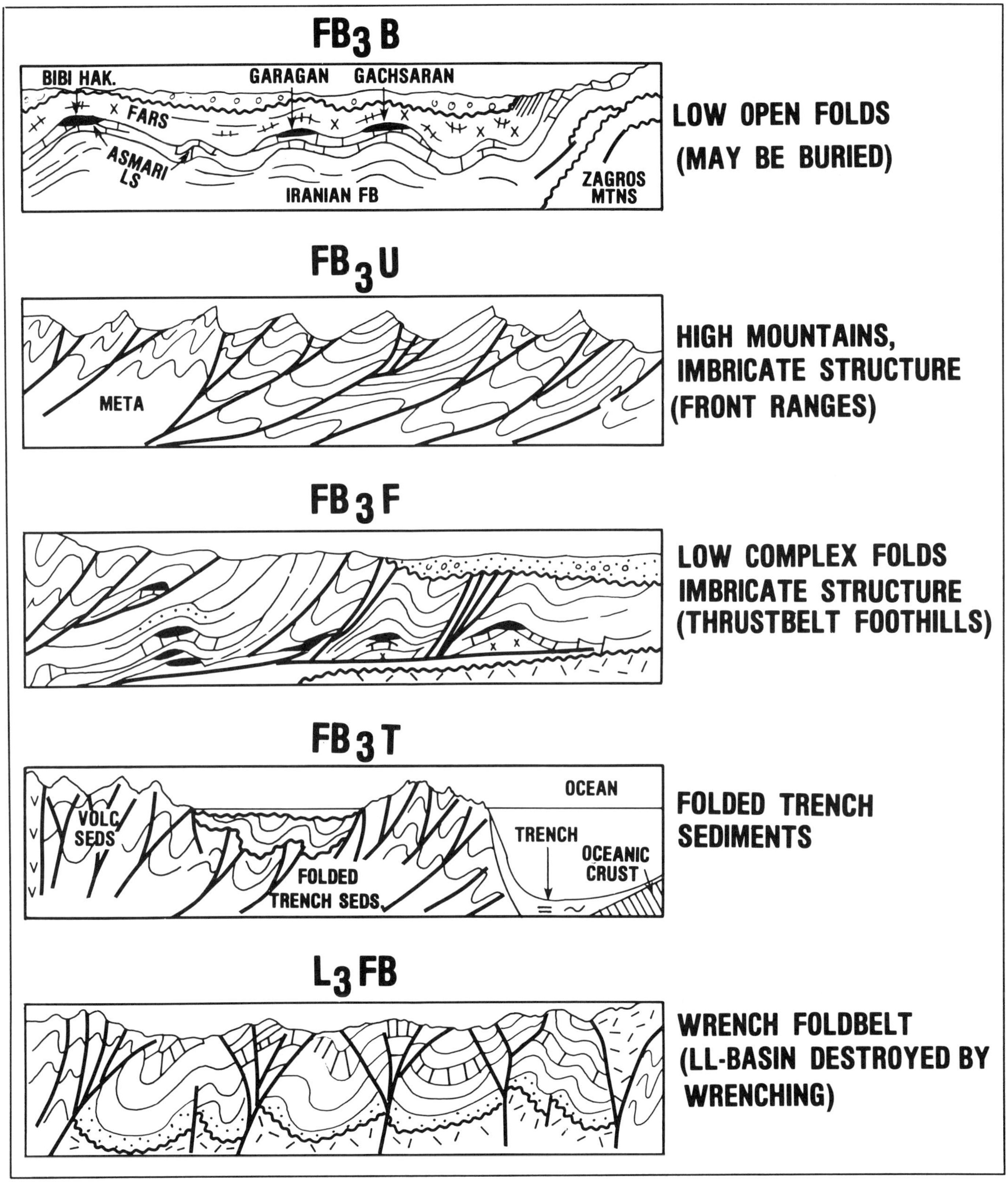


Figure 11. Foldbelt types ($FB_3$).

deduced. Major unconformity breaks within the cross sections should be restored to the old paleosurface, and a series of historically restored cross sections should be made. From these, the basin history can be broken down into cycles, stages and tectonic events. These may be compiled as shown in Figure 12 to derive the proper basin classification and formula.

Figure 12 shows how the (1) cycles, (2) stages, and (3) tectonic events of a polyhistory basin combine to create the basin formula. Figure 12A is a cross section of the basin to be formulated. Figure 12B translates the cross section events into the

## A. CROSS SECTION OF AN ATLANTIC MARGIN SAG BASIN SHOWING CYCLE POSITIONS, BURIAL HISTORY AND OIL MATURATION

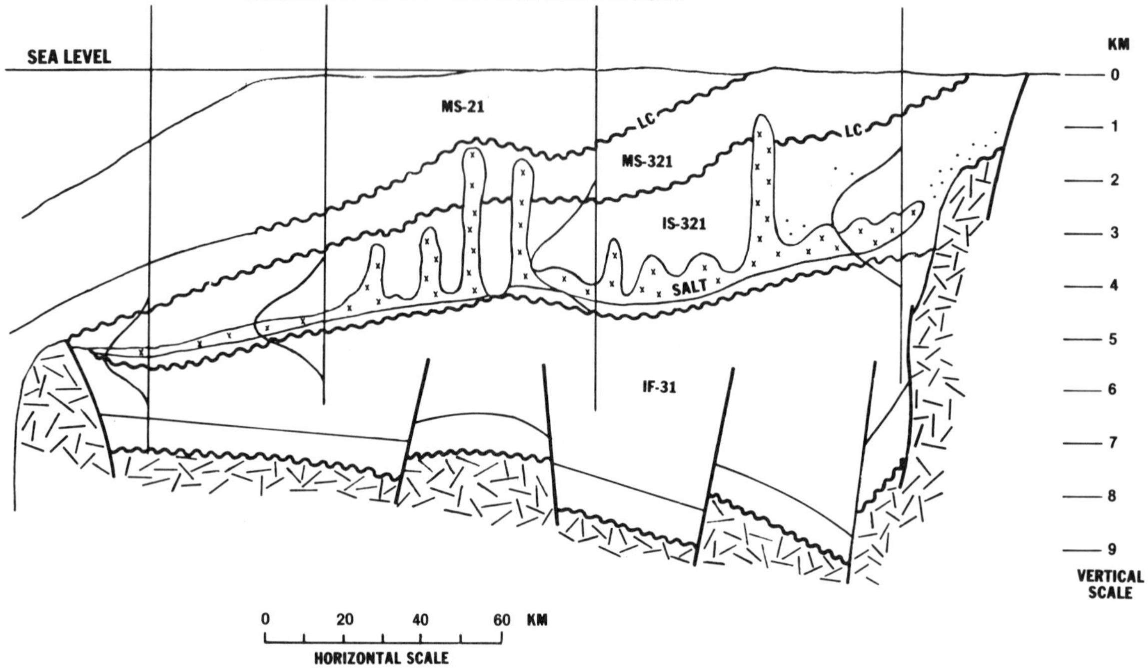


## B. COMPARISON OF BASIN PARAMETERS

BASIN: "ATLANTIC MARGIN"

Q-Tm Tm Tm-Km Km Km-Kl Kl-Ju

FORMULA: MS - 21/LC/MS - 321/LC/IS - 321/IF - 31

| CYCLE NO. | SEQ. | AGE | BASIN/CYCLE TYPE | TECTONIC EVENTS | DEPOSITIONAL STAGE | PROD. PLAYS |
|---|---|---|---|---|---|---|
| 4 | A | QUAT. TO L. MIO. | MARGIN SAG MS | | 2 (MARINE, WEDGE MIDDLE)<br>1 (NON MARINE, TRANSGRESSIVE) | |
| | | | | LC | | |
| 3 | B | L. MIO. TO M. CRET. | MARGIN SAG MS | | 3 (NON MARINE, REGRESSIVE)<br>2<br>1 | SALT STRUCTURES CLASTIC RESERVOIRS |
| | | | | LC | | |
| 2 | C | M. CRET. TO L. CRET. | INTERIOR SAG IS | | 3<br>2 (SALT)<br>1 | DRAPE OVER BASEMENT FAULT BLOCK, CLASTIC RESERVOIR |
| 1 | D | L. CRET. TO U. JUR. | INTERIOR FRACTURE IF | | 3–1 ALL NON MARINE | SUB-UNCONFORMITY ON BASEMENT FAULT BLOCK STRUCTURE, CLASTIC RESERVOIR |

Figure 12. Comparison of basin parameters.

formula. It should be noted that cycles are ended by any of four events: (1) a change in cycle type (basin-forming tectonics); (2) the occurrence of a significant wrench or folding event, such as L, FB, or FB3 (tectonics affecting basins); (3) a major sedimentary transgression and regression; or (4) a regional unconformity which may be caused by L, FB, FB3 events but commonly occurs without them.

Viewing the events in the example shown on Figure 12, as they would become apparent in drilling a well, the youngest cycle is a margin sag (MS) which is still in stage 2 because the continental margin is under water. The marine wedge overlies a wedge base of Early Miocene age and an unconformity; this unconformity corresponds to an episodic wrench which may have, in part, caused the unconformity. This Lc event reactivated the salt domes. The next older event is another margin sag (MS) cycle with wedge top, middle, and base (3, 2, 1) followed by another Lc event in the middle Cretaceous. This event initiated the salt dome movement in the basin and probably affected the unconformity. The next cycle down is an interior sag, IS-321, with thick salt in the stage 2 portion. The bottom cycle, an interior fracture, IF-31, has an unconformity at the top which is not related to L or FB events. The oldest cycle is entirely nonmarine. Combination of these cycles, stages, and events results in the formula being written youngest to oldest as follows: MS-21/Lc/MS-321/Lc/IS-321/IF-31. The abbreviated ages of each cycle may be written above the formula, as shown in Figure 12B.

St. John (1980) has published a map showing the locations of world sedimentary basins. We have included approximately 650 of these basins in the global classification system. It is obviously too complicated to refer to each one by its specific formula, so another description must be available for more general use. The writers have found it convenient to use the following categories. On a global basis, about two-thirds of the basins in the system may be called simple or single-cycle basins. These have only one basin-forming tectonic cycle or, if they have other cycles and tectonic events, are dominated by one type. These basins are grouped under the name of this dominant cycle such as interior fracture basins, interior sags, margin sags, wrench or shear basins, oceanic sags, oceanic wrenches, trenches, and trench-associated basins. The more complex polyhistory basins that cannot be categorized with the eight basic cycle types make up the remaining one-third of the basins classified globally. These are referred to simply as complex polyhistory basins, and further subdivision is not proposed at this time. Three regional sections, drawn across a variety of polyhistory basins, are presented in Figure 13 to illustrate examples of how the more complex basins are classified. The formula for each basin is constructed by using the information available on the section (Kingston et al., 1983a and 1983b).

## OIL-PLAY PREDICTION

After a basin is classified, it is possible to identify the productive cycle or cycles within it. Discovered hydrocarbon reserves for each of these productive cycles, worldwide, may then be tabulated, and evaluations of the various cycle types can be made. Productive cycles in explored basins may be compared with similar cycles and tectonic events in unexplored areas. Assessment procedures for new basin areas are not discussed here. However, we point out general oil-play controls for each of the cycle types that we have found to be of value in basin play predictions.

### Interior Sag (IS) Cycles/Basins

Interior sags are stable cycles, commonly remnants of larger Paleozoic basins. These cycles have an interior, restricted depositional environment, normally good for source rock development. Shallow-marine and nonmarine conditions also promote the widespread extent of blanket-type reservoirs and evaporite seals. Poor structural focus is a common problem in interior sag cycles. Stratigraphic traps such as carbonate reefs (Michigan) and pinchout sands are commonly found in interior sag (IS) cycles. The oil plays in these cycles generally are associated with basement arches, noses, and topographic features within the basin (Figure 14).

Larger interior sag basins generally are more prospective, whereas small ones commonly are too shallow to generate hydrocarbons, with too little burial for hydrocarbon maturation, especially for Mesozoic and Tertiary rocks. Interior sag cycles/basins, though commonly large, may be very shallow. The larger and deeper Paleozoic basins generally are the most prospective for hydrocarbons. Most simple interior sag (IS) basins contain poor structural prospects or few prospects in small fields. Polyhistory basins that contain interior sag (IS) cycles affected by later structuring may become very prospective. Of all interior sag (IS) cycles studied worldwide, 49% were found to produce commercial quantities of hydrocarbons.

### Interior Fracture (IF) Cycles/Basins

Interior fracture basins include some of the most prolific oil basins in the world, such as Sirte in Libya, west Siberia in the USSR, and the Gulf of Suez in Egypt. Normally there are four main plays found in interior fracture basins (Figure 15). These are all reservoirs associated in some way with major block-fault structures. There is a pre-IF reservoir such as the Gargaf sand in Libya, a basal IF reservoir like the Nubian sand in Egypt and Libya, younger marine reservoirs draped over fault blocks (Nasser field, Libya), and post-IF-cycle reservoirs where hydrocarbons have migrated vertically into younger beds.

Active block faulting normally continues throughout the history of the basin, thereby constantly rejuvenating structural growth. Episodic wrench movements may also occur. Better reservoirs are associated with structural highs, and source rocks with the lows throughout the development of the cycle, ideal conditions for local migration and entrapment. The cycle 1 nonmarine play, with its built-in source, reservoir and structure, attains its best development in interior fracture (IF) cycles.

A semirestricted depositional environment, commonly found in basal interior fracture (IF) cycles, may produce widespread evaporites, or blanket sand, and rich oil source rocks such as the Socna Shale in Libya. Size is an important factor in interior fracture basins. Generally, the larger basins have a better chance for exploration success, although a few smaller basins, like Vienna in Austria, and the Gulf of Suez, have good production.

Interior fracture cycles are found commonly as simple basins. They may also be found as cycles within polyhistory basins, particularly the initiating cycle. Interior fracture

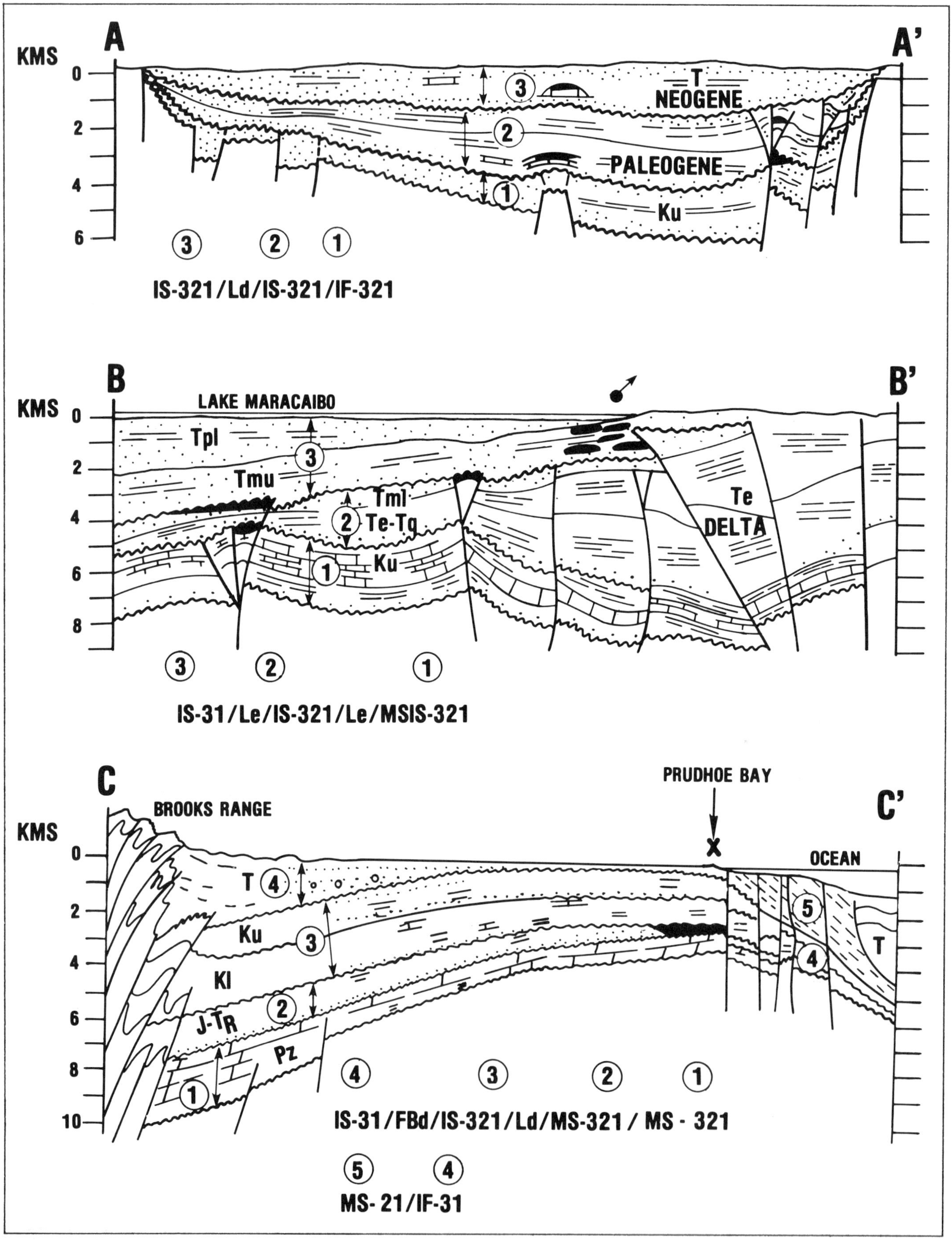


Figure 13. Regional cross sections, A–A′ to C–C′.

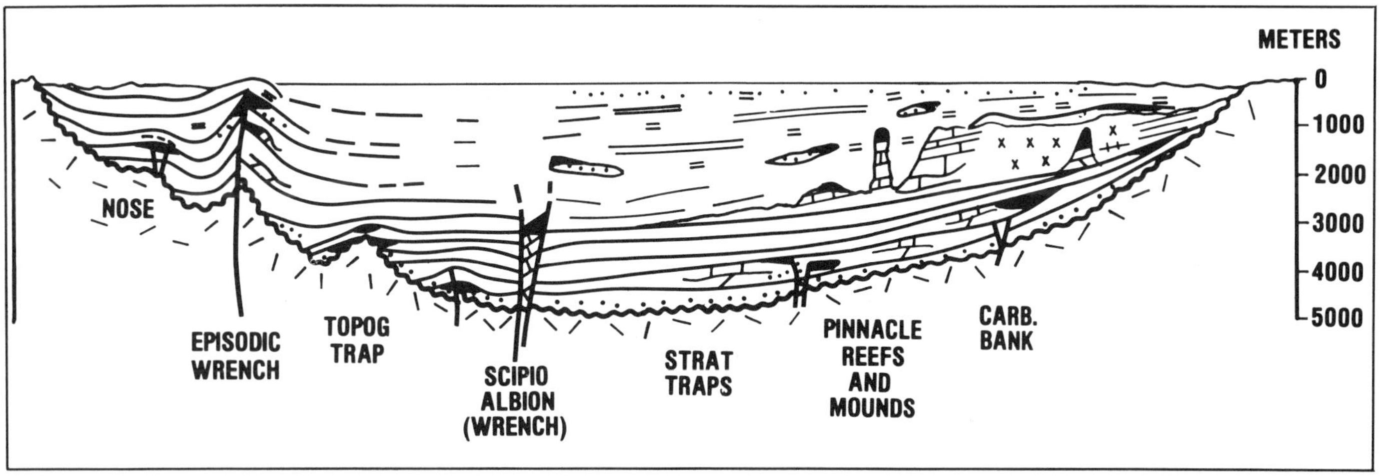


Figure 14. Interior Sag (IS) basin, oil play types.

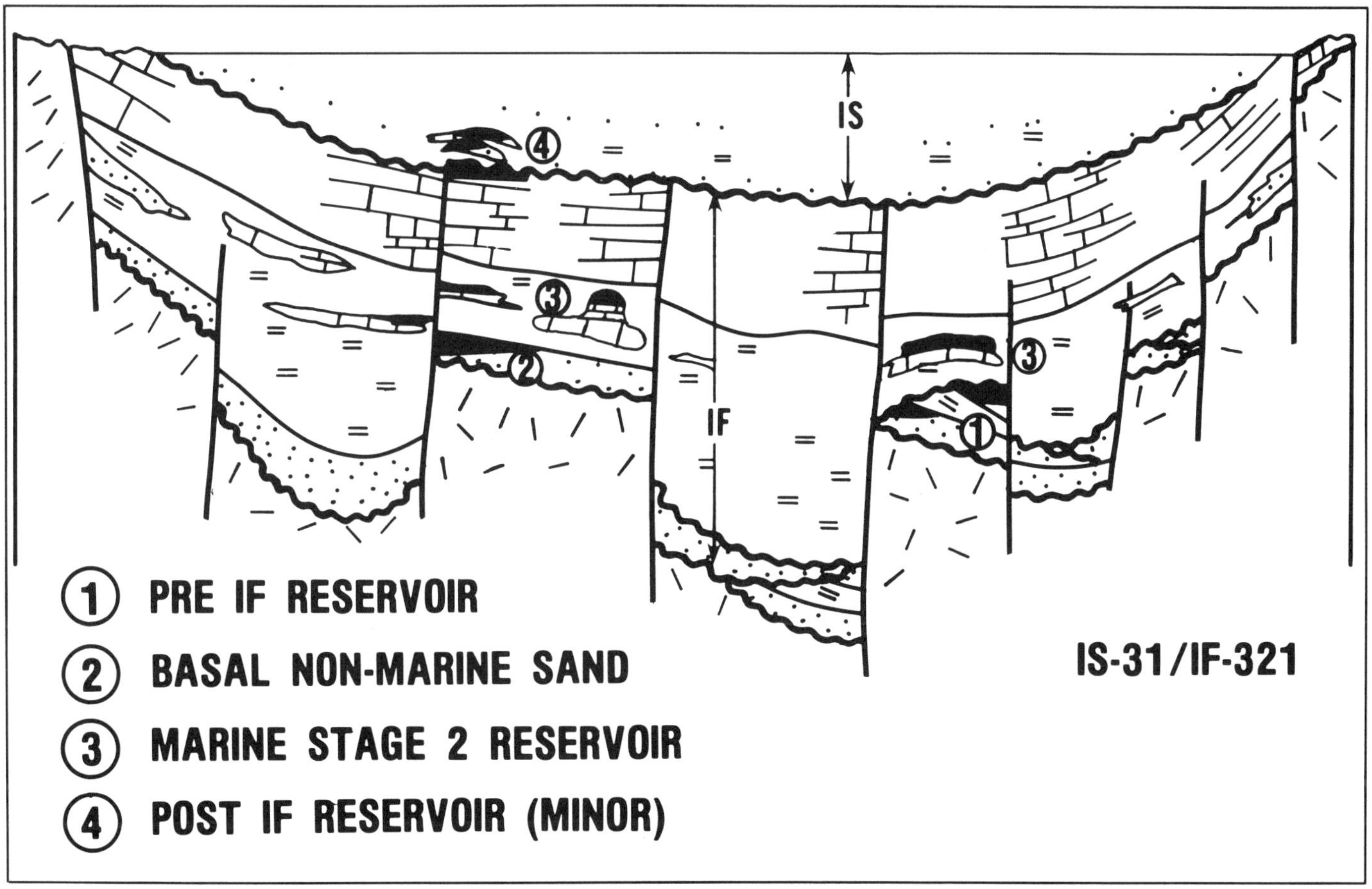


Figure 15. Interior Fracture (IF) basin, major oil play types.

basins, which have been separated by continental "pull-apart" and now comprise the bottom cycle in margin sag basins, are largely unexplored at present. Some production has been established in these continental margin interior fracture cycles in Gabon, Angola, and Brazil.

The major plays in interior fracture basins are all related to the horst and graben features that dominate the structure and stratigraphy. The largest plays are found in reservoirs directly overlying high basement blocks; examples are sandstones (Samotlor in west Siberia) and carbonates (Nasser in Libya). Of all interior fracture basins studied worldwide, 35% of them produce commercial quantities of hydrocarbons.

## Margin Sag (MS) Cycles/Basins

The divergent continental margin basins are a complex group (Figure 7). They have a variety of potential plays, including basal fault blocks (Brazil, Angola), salt domes (Gabon), and a few major deltas with gravity tectonic features

(Niger delta). The combination of basement block-fault structure, good reservoir sand, and source shale may be the best oil play in margin basins. Unfortunately, it is generally not at optimum drilling depth and is commonly too deep (overmature) or too shallow (immature). To date, not much drilling has been done for basal cycle interior fracture (IF) plays in margin sag (MS) basins.

Most cycle 1 fault blocks, the best potential structures, freeze or cease differential movements after continental separation. Because there is no structural growth, there is normally no drape or closure of cycle 2 younger beds over the old blocks, and the play must be downgraded as a result.

Above the basal interior fracture (IF) cycle, younger units may consist of interior sag or margin sag cycles. Simple margin deposition and sag with no other tectonic movements appear to be detrimental to preservation of hydrocarbons in these basins. Contemporaneous or younger salt dome uplift, growth faults, and wrench faulting appear to increase the incidence of oil discovery in these basins.

The size of margin sag basins/cycles is an important factor. If the continental shelf is narrow and small, the cycle will be highly tilted. Most young Tertiary margin sag cycles are deposited as thick sediment pods or clinoform wedges just off the outer shelf. The subsidence and seaward tilt are perhaps the most important negative factors. In margin sag cycles, marine shales are located seaward, downdip, and with porosity (sand) increasing updip to the outcrop. In the Late Cretaceous and Tertiary, rapid subsidence of the outer margins of the basins tilted many gentle to moderate structural closures, and the oil has been lost updip. The updip flanks of margin sag basins in Nigeria, Gabon, and Angola contain extensive tar deposits at the outcrop, evidence of oil lost by basin tilting.

Some margin sag cycles contain major river deltas which most explorationists believe to be good places to look for hydrocarbons. The writers have analyzed the 60 present-day major river delta systems around the world, and only five are able to produce more than marginally commercial quantities of hydrocarbons. The five major deltas are the Mississippi, Niger, Mahakam, Baram, and Mackenzie. The other 55 deltas (most have been drilled at least once) have produced only very small amounts. Older margin sag delta cycles, perhaps buried or structured by younger cycles or events, generally are prolific producers. Overpressured shale sections are common features of margin sag river delta cycles.

Extensive carbonate banks with reefs are found in many margin sag cycles; however, these carbonate bank edges have had minimal success when drilled. Several broad carbonate platforms that have been structured by younger tectonic events occasionally may provide reservoirs for substantial oil plays, as in the Tabasco-Salinas basin of Mexico. The percent of productive divergent margin sag cycles worldwide, compared with those nonproductive, is approximately 20%. Most margin sag basins should be considered high-risk exploration ventures.

Margin sag-interior sag (MSIS) cycles/basins are variations of margin sags that have many attributes of interior sag cycles. Most of the MSIS cycles were deposited on older continental plate margins which appear to have been broad, flat, and only slightly tilted. These stable shelves accumulated extensive reservoirs, such as blanket sands and carbonates. Thick source rocks and seals commonly were deposited as well. Subsequent structuring produced the world's most prolific basin/cycle type in the Persian Gulf, the MSIS cycle with wrench anticlines, blanket carbonate reservoirs, and evaporite seals. Approximately 50% of the MSIS cycles worldwide produce commercial hydrocarbons, one of the highest ratios of cycles we have studied. The key parameters for giant field potential in these MSIS cycles, sometimes called foreland basins, are large structures and extensive reservoirs with effective seals. In the MSIS cycles of the Permian basin of Texas, the blanket reservoirs are missing, but the source rocks are prolific. Good evaporite seals are present, and although the MSIS cycle has almost 6 billion bbl of oil equivalent, most of the fields are small. In western Canada, the Alberta basin has extensive reservoirs and source rocks, but the basin is primarily unstructured. Carbonate reefs, which are really stratigraphic traps, are the main traps available in the cycle and are small. However, MSIS cycles should be examined carefully, for they have an enviable record of productivity.

## Wrench or Shear (LL) Cycles/Basins

Plays in wrench (LL) basins are summarized in Figure 9. Probably the most important factor affecting wrench basins is timing. Many LL basins are formed, filled, structured, and destroyed in a relatively short geologic time. The optimum combination of these short-term factors can result in highly productive basins, such as the Los Angeles basin in California. However, optimum factors are not always present, and basins may be too young and immature or too old and overstructured. In these situations, the basins must be considered to be very high risk.

A critical factor in shear basins is the presence or absence of quartz sand for reservoir rocks. Mesozoic and older basement usually contains quartz sand sources. Tertiary crust, however, is normally composed of oceanic basalt and volcanogenic sediments, which substantially degrade reservoir quality. Wrench basins must be situated on or adjacent to good quartz-sand-yielding areas to insure the existence of predictable quartz sand reservoirs; this generally means Mesozoic or older crust. Carbonate rocks and chert, as in California, are also found occasionally as reservoirs in this type of basin.

Normally a high heat flow exists in active wrench (LL) basins. Some cycles may be too hot, and therefore gas-prone or overmature in the deeper parts. Because most wrench basins are very young, the source shales generally are immature, even at intermediate depths (8000 to 10,000 ft or 2500 to 3000 m). One of the exploration pitfalls of this type of basin is the potential immaturity of the upper source shales, whereas the deep portion of the basin may be overmature, with shales cooked and sandstone porosity lowered by diagenesis. This results in a very thin optimum zone between rocks that are too shallow or too deep. Unless reservoir sands and structural closures are found in this optimum zone, dry holes or gas may be the result.

Geologically speaking, wrench basins rapidly evolve from youth to maturity to old age. The type of oil trap found in each of these stages can change, also. Referring to Figure 9, we see that in stage 1 the main oil prospects are in basement fault structures. In stage 2 of cycle 1, an early mature phase, we find hydrocarbon prospects in folds over wrench zones (flower structures), en echelon folds, and basement fault blocks. In stage 3, a late mature phase, oil prospects are in en echelon folds, stratigraphic traps, small fault traps, and some basement fault blocks. Finally, in cycle 2, after the basin has been completely folded, most of the fields have been progressively destroyed by tectonic activity.

In prolific shear basins, oil migrates from older to younger structures as the wrenching progresses. Older structures are deformed and destroyed, and the oil migrates or is lost in the process. If convergence between the two plates stops, the wrench faulting will cease, and the basins and potential plays could remain frozen in their last evolutionary stage. Size is not a critical factor in LL shear basins, as most of them are relatively small.

LL shear basins are moderate-risk prospects. They have many constructive and destructive features to consider, and they are short-lived. The main plays in LL shear basins are sandstone reservoirs and wrench-generated block faults or anticlines. Most fields are moderate in size; the largest we have listed is Minas in central Sumatra, with almost 5 billion bbl of oil equivalent. Some LL shear basins have prolific source rocks, such as those in California and Sumatra; producible stratigraphic traps are found, as well as structures, as the oil seeks to fill all available porosity. Approximately 47% of all wrench (LL) cycles studied worldwide were found to produce commercial hydrocarbons.

## Trench-Associated (TA) Cycles/Basins

Trench-associated (TA) cycles/basins are found along ancient or modern subduction zones at convergent plate boundaries. They are commonly developed on the continental or high side of the oceanic trench, and oceanward of the first volcanic or magmatic arc. The source of the sediment is mainly the landward side with shallow-water clastics or carbonates, and some deep-marine deposition offshore. Volcanogenic sediments comprise the bulk of the clastic rock, and porosity/permeability is likely to be destroyed by diagenesis at depths above the hydrocarbon maturation zone. Only a few of the trench-associated cycles, worldwide, contained any quartz-prone reservoirs (e.g., Sabah, Malaysia). Carbonates and chert are also found in places. Structuring, if present, is related to subduction, and wrench fault features are commonly present. The crust under these basins is composed of tectonized or folded trench sediments ($FB_3T$). Trench-associated cycles will eventually self-destruct and become $FB_3T$ folded trench foldbelts. Exploration drilling in basins such as Mentawi and south Java in Indonesia, Abukuma in Japan, the Gulf of Alaska, and others have produced a poor record of discoveries to date. As potential hydrocarbon prospects, the basins must be considered minor and very high-risk areas.

## Oceanic Sags (OS) and Oceanic Wrench Basins (OSLL)

The chances are probably small that oil plays will be found in oceanic basins filled with volcanogenic/deep-water pelagic sediments. Questionable source rocks of deep-marine pelagic sediments, thermal histories of low temperatures, poor volcanogenic reservoir rocks, and probable lack of structure combine with the presence of deep water to make these basins extremely high risk. Some of the oceanic sags contain deep-water fans from river deltas that have built out over oceanic crust, as are found in the south Caspian and Black seas in the USSR. One oil-prospective area, possibly underlain by oceanic crust and which has these features, is offshore from the Niger delta. Its seismic signature, indicating oceanic crust, is one of flat, featureless overlying sediments in deep water. Because of this lack of structure, and thermal histories of low temperatures, most oceanic basin areas associated with continental blocks must be considered to be very high risk. Most sediment wedges filling oceanic sags are classified with margins sags (basin flank) and not with the basin center.

## Polyhistory Basins

Most large hydrocarbon-producing basins are the polyhistory type, containing numerous and perhaps varied cycles and tectonic events. The oil-play controls of single cycles apply in a general way to polyhistory basins, but they are commonly modified later by tectonics or subsidence. Classification and the writing of formulas are very useful in the evaluating of complex polyhistory basins. The formulas provide a shorthand history of each basin, permitting identification of the geologic events that form the basis for the oil plays. Once the plays are identified, they may be compared with those in other similar polyhistory basins. Change in structural genesis appears to enhance oil prospects. In Kingston et al. (1983a and 1983b, Figure 13), a series of restored cross sections shows the evolution of the Persian Gulf. In this diagram the main events are: (1) basin formation, (2) subsidence and fill, (3) structuring, and (4) renewed subsidence and "loading" of the structures. This sequence is common to many productive polyhistory basins. The loading of the structures by more subsidence and fill often appears to be critical. Polyhistory basins ending with interior sag (IS) appear to be the best situated for the generation and preservation of oil and gas. The final interior sag generally does not cause tilting of the structures, but commonly provides the sediment loading which triggers maturation and migration of hydrocarbons. Too much structuring at the end of the polyhistory basin formation is generally more destructive than constructive to the pooling and preserving of hydrocarbons.

## CONCLUSIONS

This global basin classification system identifies and compares basins in specific as well as general terms. The system is based on the genesis and evolution of basins in the context of their geologic history. The main elements used to classify basins are basin-forming tectonics, depositional cycles, and basin-modifying tectonics. Basin-forming tectonics are deduced by knowledge of the type of underlying crust, past plate tectonic history, basin location on the plate, and type of primary structural movement involved in the basin formation, such as sagging or faulting. The result is eight single tectonic-cycle or simple basin types that are termed interior sag, margin sag, interior fracture, wrench, trench, trench-associated, oceanic sag, and oceanic wrench.

Basin-modifying tectonics include episodic wrenches, basin-adjacent foldbelts, and completely folded basins. These have been identified and placed on a scale of increasing magnitude, from movements of slight to major structural effects. More complex basins may contain several different tectonic cycles, plus basin-modifying tectonic events. These are called polyhistory basins. The eight simple basin types, their depositional fills, and tectonic modifiers have been given letter and number symbols so that the specific geologic history of each basin may be written as a formula. These formulas can be compared to show similarities or differences between basins.

The primary uses of the global basin classification are summarized as follows: (1) to locate and identify all basins of the world in one framework (the system can expand the explorationist's viewpoint to include all possible basin types, and not just those with which he or she has had personal experience—it is an aid to exploration thinking); (2) to permit the separation of complex basins into their simple component parts, for analysis as simple units; (3) to compare plays within one or two basins of the same type, or two or more basins classified as different types; (4) to provide a system for evaluating favorable plays and risks for each basin type, because risks should be understood before venture decision; (5) to predict what geologic events must be found in a basin to improve oil prospectiveness; (6) to enhance the prediction of oil potential in unknown or poorly known basins by referring to known basins of the same classified type; (7) to provide a system where the paleontology, seismic stratigraphy, geochemistry, and sedimentary history of like (and different) basins can be compared and evaluated; (8) to permit location and assessment to the best specific-play areas in a basin, not just total sediment volume; and (9) to act as a vehicle for comparative assessment of hydrocarbon basins worldwide.

## ACKNOWLEDGMENTS

The authors would like to thank the management of Esso Exploration, Inc., for permission to publish this paper; T.A. Fitzgerald, H.M. Gehman, Jr., D.S. McPherron, T.H. Nelson, and J.B. Sangree for comments and assistance in revisions. This paper is a revised version of two papers first published in the AAPG Bulletin, v. 67/12, December 1983.

## REFERENCES CITED

Bally, A.W., 1975, A geodynamic scenario for hydrocarbon occurrences: 9th World Petroleum Congress Proceedings, v. 2, p. 33-44.

Bally, A.W., and S. Snelson, 1980, Realms of subsidence, *in* A.D. Miall, ed., Facts and principles of world petroleum occurrence: Canadian Society of Petroleum Geologists Memoir 6, p. 9-75.

Bois, C., P. Bouche, and R. Pelet, 1982, Global geologic history and distribution of hydrocarbon reserves: AAPG Bulletin, v. 66, p. 1248-1270.

Halbouty, M.T., A.A. Meyerhoff, R.E. King, R.H. Dott, Sr., H.D. Klemme, and T. Shahad, 1970a, World's giant oil and gas fields, geologic factors affecting their formation, and basin classification; Part I, Giant oil and gas fields, *in* Geology of giant petroleum fields: AAPG Memoir 14, p. 502-528.

Halbouty, M.T., R.E. King, H.D. Klemme, R.H. Dott, Sr., and A.A. Meyerhoff, 1970b, World's giant oil and gas fields, geologic factors affecting their formation, and basin classification; Part II, Factors affecting formation of giant oil and gas fields, and basin classification, *in* Geology of giant petroleum fields: AAPG Memoir 14, p. 528-555.

Huff, K.F., 1978, Frontiers of world oil exploration: Oil and Gas Journal, v. 76, n. 40, p. 214-220.

Kingston, D.R., C.P. Dishroon, and P.A. Williams, 1983a, Global basin classification system: AAPG Bulletin, v. 67, p. 2175-2193.

Kingston, D.R., C.P. Dishroon, and P.A. Williams, 1983b, Hydrocarbon plays and global basin classification: AAPG Bulletin, v. 67, p. 2194-2198.

Kingston, D.R., C.P. Dishroon, and P.A. Williams, 1985, A crustal classification for hydrocarbon exploration: Oil and Gas Journal, v. 83, n. 10, p. 146-156.

Klemme, H.D., 1971a, What giants and their basins have in common: Oil and Gas Journal, v. 69, n. 9, p. 85-90.

Klemme, H.D., 1971b, To find a giant, find the right basin: Oil and Gas Journal, v. 69, n. 10, p. 103-110.

Klemme, H.D., 1975, Giant oil fields related to their geologic setting—A possible guide to exploration: Bulletin of Canadian Petroleum Geology, v. 23, p. 30-66.

Knebel, G.M., and G. Rodriguez-Eraso, 1956, Habitat of some oil: AAPG Bulletin, v. 40, p. 547-561.

Nehring, R., 1978, Giant oil fields and world resources (prepared for the Central Intelligence Agency): Santa Monica, CA, Rand Corporation, R 2284-CIA, 162 p.

Perrodon, A., 1971, Essai de classification des bassins sedementaires: Nancy, France, Science de la Terre, v. 16, p. 195-227.

St. John, B., 1980, Sedimentary basins of the world and giant hydrocarbon accumulations: AAPG, 23 p., 1 map.

Uspenskaya, N. Yu., 1967, Principles of oil and gas territories, subdivisions, and the classification of oil and gas accumulations: Amsterdam, Elsevier Publishing Co., 7th World Petroleum Congress Proceedings, v. 2, p. 961-969.

Weeks, L.G., 1952, Factors of sedimentary basin development that control oil occurrence: AAPG Bulletin, v. 36, p. 2071-2124.

White, D.A., 1980, Assessing oil and gas plays in facies-cycle wedges: AAPG Bulletin, v. 64, p. 1158-1178.

Zeigler, P.A., 1975, North Sea basin history in the tectonic framework of north-western Europe, *in* A.W. Woodland, ed., Petroleum and the continental shelf of north-west Europe: New York, John Wiley and Sons, v. 1, p. 131-150.

CHAPTER 29

# COMPLEX PROGRAM OF DEEP STRUCTURE INVESTIGATIONS IN THE USSR TERRITORY

**E.A. Kozlovsky**
**N.A. Bogdanov**
*Institute of the Lithosphere*
*U.S.S.R. Academy of Sciences*
*Moscow, U.S.S.R.*

## INTRODUCTION

The investigation of the deep-seated structure of the earth's interior and the geodynamic regimes of the lithosphere promotes the development of fundamental geological sciences of the earth, establishing theoretical bases for the solution of the applied targets of geology on the expansion of the mineral raw material base of the USSR and national-economic development of the geological environment. These investigations are carried out in the USSR under the supervision of the Interdepartmental Scientific Council of the State Committee on Science and Technology on the problem of "The investigation of the earth's interior and super-deep drilling," which includes about 200 prominent scientists and specialists.

## TARGETS OF COMPLEX INVESTIGATIONS OF THE DEEP INTERIOR OF THE EARTH

The most important targets specified by the Scientific Council included the solution of fundamental and applied problems of geology to reveal new regularities of mineral distribution, in order to solve problems of uncertainty, discovery, and mineral resource estimation (Kozlovsky, 1982a, 1984a).

The above targets are solved within the bounds of a scientific-engineering programme approved by the Scientific Council. These consist of studies of deep structures in the earth's crust based on regional and deep geologic-geophysical investigations together with drilling of deep and super-deep boreholes. As a result, a number of type models of tectonospheres for the regions having different geodynamic environments for the main oil- and gas-bearing areas of the country has been established (Figure 1).

The network of interrelated geophysical profiles of deep seismic sounding, distributed on the USSR territory and based on deep and superdeep boreholes, drilled at the points of profile intersections appears to be the basis for deep geologic-geophysical investigations. The system will also include prognosticated geophysical proving grounds of the USSR Academy of Sciences where variations of geophysical fields are studied (Figure 2).

The main target of investigations along the extensive framework profiles of the first class, the so-called geotraverses, is to study principal differences in the internal structure and in the state of lithospheric matter in the regions with different geodynamic regime of development. Profiles of the second class are made within the uniform tectonic blocks or in a certain system of the latter, as well as their boundaries (conjugation lines, faults) and should assist in solving structure-tectonic, lithological, and other regional tasks. The main purpose of work inside the framework proving grounds of the first and second classes is to solve targets related to medium- and large-scale prognostication and prospecting for mineral deposits.

The investigation of the upper mantle structure in different geotectonic zones has been envisaged by the program. The programme includes studying the peculiarities of modern geodynamic processes occurring in the Urals and making a geomagnetic model of this folded structure. Complex geologic-geophysical investigations of the crust and upper mantle and heat flows will be carried out in the Baikal rift zone and adjacent areas of Cenozoic activity of Inner Asia and also between the Siberian Platform and Verkhoyan-Kolima folded zone. Investigations will also be done to study factors that may control mineralization in the Norilsk region, including research on deep-seated structures in the earth's crust and upper mantle and associated magmatic formations. The investigations of deep sedimentary complexes and underlying basement formations and the crust on the whole within the shelf areas of the Barents and the Kara Seas, the transitional zones from the Asian continent to the Pacific Ocean, prove to be of great practical importance. The reference section of the earth's crust on the international geotraverse within the Soviet part of the Baltic Shield to reach the Kola superdeep borehole is also to be compiled. According to the

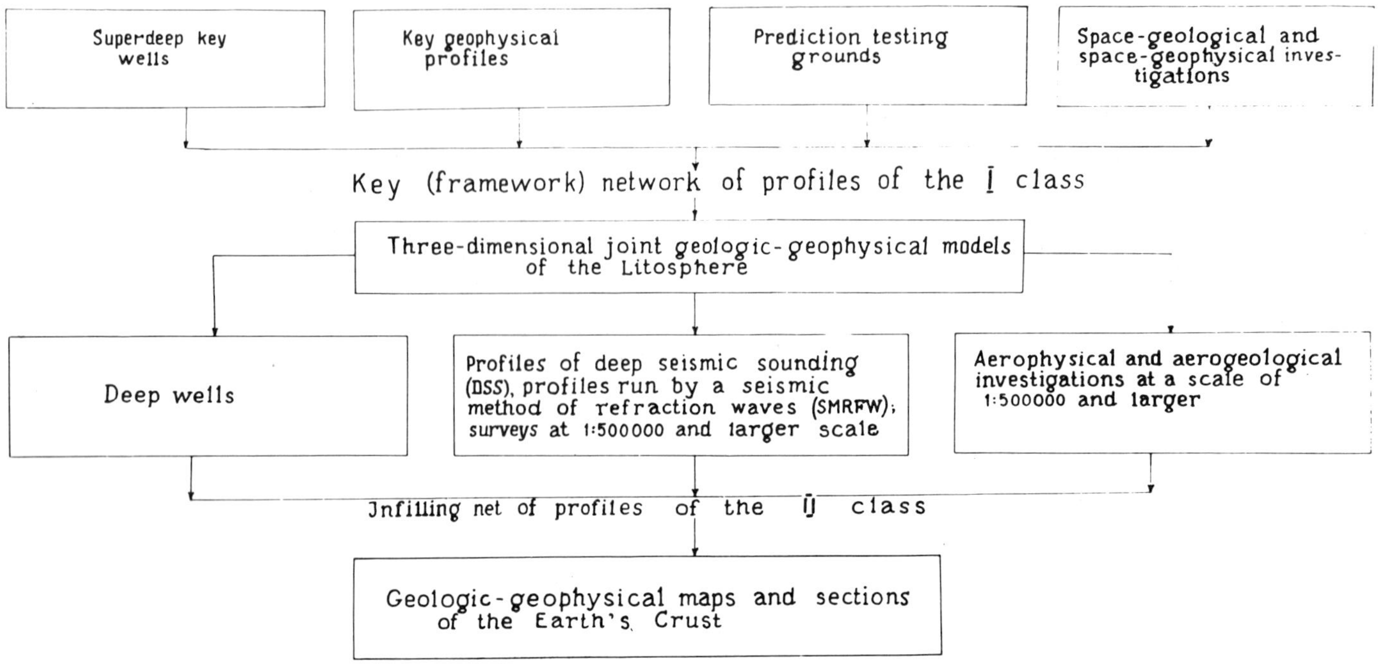


Figure 1. Scheme of the regional study of the earth's crust and mantle within the limits of the USSR territory.

programme it is envisaged to make models of the earth's crust and upper mantle structure, as well as new exploration methods for mineral deposits, to compile exploration maps with quantitative estimation of resources, and to determine trends for prospecting and exploration for the main types of minerals in the promising areas of the country. This programme includes the assignments aimed at the solution of a number of applied targets, closely related to the fundamental theoretical elaborations in the field of deep structure and development of the earth's crust, including further study of processes of formation and regularities of mineral distribution.

Three stages can be distinguished in the investigation of deep structure of the USSR territory. The first stage includes the 1960s, when the tasks of these investigations were formulated, scientific preparation was made, and national technical facilities for superdeep drilling and geologic-geophysical investigations in boreholes as much as 10 to 15 km deep were provided. The second stage was in the 1970s, during which the experimental drilling of the Kola and Saatlin super-deep boreholes, as well as investigations of individual regions using depth geophysical methods were made. The third stage, started in 1981, is characterized by the transition to the systematic study of the earth's crust and upper mantle.

## Main Results of Work From 1970-1985

In the 1970s and 1980s, complex geologic-geophysical investigations of the most important regions were performed. These included a total of 18 km of profiles of deep seismic sounding, which aided in interpreting the deep structure of the earth crust—the gypsometrical position and the relief of Mohorovičić discontinuity, Pre-Riphean basement in the Eastern Siberia, and a number of intermediate boundaries in the consolidated crust and sedimentary cover were traced; fault zones and their vertical extension were established; regional boundaries and deep-seated structures of positive and negative tectonic elements, which can be related to endogenic mineralization and to areas of oil and gas accumulation, were specified. New data on the structure and physical parameters of the upper mantle of the earth were obtained. Velocity sections to a depth of about 400 km were compiled, absorbing rock properties were studied, and generalized geological interpretation of the observed areal velocity anomalies of longitudinal seismic waves was given. For the first time volumetrical tectonic and geodynamic models of the structure of the upper layer of the earth's crust and upper mantle were compiled, and depth criteria for prognostication and mineral prospecting were specified.

Complex studies of geophysical data made it possible to establish considerable vertical and horizontal heterogeneity in the structure of the earth's crust and lithosphere; to reveal complicated relationships of deep structure with near-surface structures; to trace the boundaries of different geodynamic situations of the environment to reveal heterogeneity of the velocity properties of the earth's crust and the mantle, and their variations in space and in time.

The above studies permitted us to continue establishing the framework grid of the first class regional profiles within the limits of the Siberian platform and in the rest of the USSR in order to correlate the deep and superdeep boreholes that had been drilled and those that were being drilled.

On the basis of the new data obtained a number of 1:10,000,000 scale geologic-geophysical maps of the USSR territory were compiled, including the map of metallogenic zoning of the USSR territory; international map of heat flow of Europe (the Soviet part); geological map of the USSR, scale 1:2,500,000; structural map on the reference horizons of the sedimentary cover of the USSR platform territories; maps of crystalline basement surface of the platforms; relief of the Mohorovičić discontinuity; structure and types of the consolidated crust; geodynamics of the earth's crust and upper man-

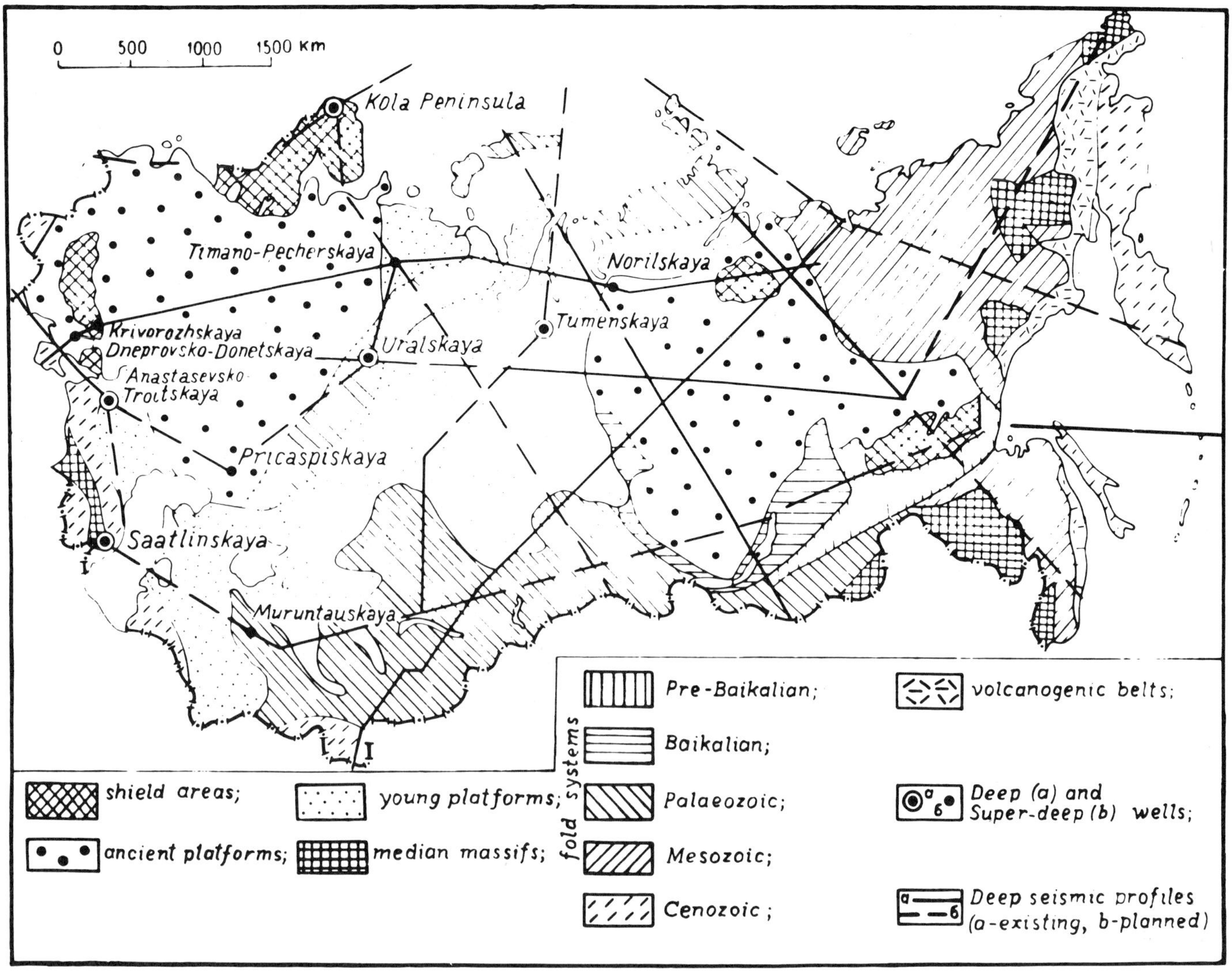


Figure 2. Major structural divisions of the USSR, showing the location of seismic profiles and deep drillholes.

tle; physical parameters of the upper mantle; oil and gas potential by depth criteria; and many others.

The above maps prove to be of the utmost importance for study of geological peculiarities of the most important oil- and gas-bearing and metalliferous provinces, and for appraisal of deposit discovery, as well as for revealing new criteria which can be used in mineral prospecting in poorly investigated areas. A system of complex geologic-geophysical sections of the earth's crust and upper mantle was compiled for the main tectonic elements, oil- and gas-bearing and metalliferous provinces.

The system of regional depth study of the earth's crust and upper mantle of the USSR territory and main instructions for carrying out geologic-geophysical investigations on the reference geophysical profiles was elaborated and put into practice.

The most important element of the program is the complex investigation of the deep continental structure of the earth's crust by drilling superdeep boreholes. The first of these boreholes was drilled at the Kola peninsula, in the Pechenga copper-nickel mineralized area of the Baltic shield. This area is composed of old crystalline rocks, which can serve as a reference shield for obtaining knowledge of geology and ore formation not only of the Baltic Sea, but to a certain extent, of the ancient continental crust and the Precambrian of the world.

Drilling of the Kola boreholes has provided important information on the geological structure, and on physical–chemical and mechanical properties in the deep zone of the earth's crust, essentially changing the theoretical notions that prevailed before the drilling (Kozlovsky, 1982b; Kozlovsky and Yanshin, 1984). The most important results of the Kola borehole drilling, which reached a depth of 12,066 m, are summarized in the following discussion.

The borehole intersected deposits characterizing the interval of geological history from 3 to 1.6 billion years, represented by Proterozoic and Archean rocks. For the first time in one continuous section, the metamorphic zonation was studied in detail from prinite-pumpellyite to amphibolite facies. The first reliable vertical geochemical section of the Precambrian earth crust was compiled on the basis of voluminous analytical data. For the first time mineralized underground fracture waters were found in deep zones of continental earth

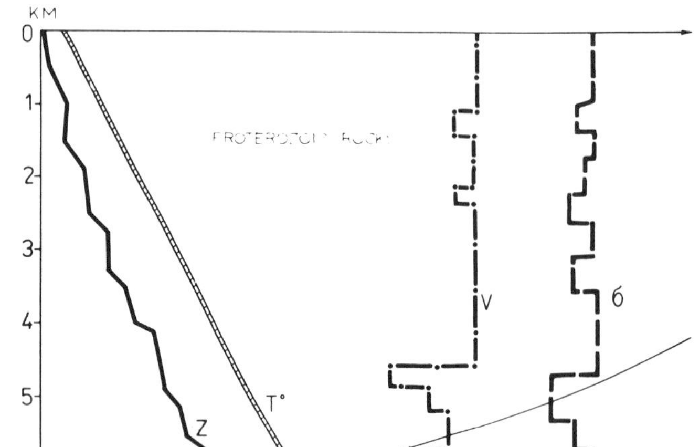


Figure 3. Kola superdeep well. Thermodynamic state of rocks at great depth.

crust and their hydrogeological zonation was established. Gases were established to play an important part in the development of structure-formational zones of the earth crust; the content of hydrogen and helium increases with depth, whereas that of hydrocarbon gas, confined mainly to the sedimentary sequences of the Proterozoic complex, decrease.

The possibility that ore minerals formed in economic accumulations at great depths has been proved in principle. This conclusion is of the utmost importance for further development of the science of mineralogy, and for prospecting for new mineral deposits at depth. The conclusion of the initial horizontal occurrence of nickel-bearing intrusions and their fragmenting due to flaky displacements of tectonic blocks appears to be especially important; it enhances the possibilities for discovery of copper-nickel ores in the Pechenga structure. Additional data obtained on the composition and physical properties of certain rocks may be very important in evaluating the potential for undiscovered deposits of phosphate raw material, ferruginous ores, copper, mica, and rare metals.

Our knowledge of the character and the nature of physical fields, i.e., electromagnetic, acoustic, and radiation—as well as their relationship to material composition, structure-textural peculiarities, and thermodynamic rock properties—has also been expanded.

For the first time, theoretical notions on the change of temperature regime at greater depths have been experimentally tested, and modern and ancient geothermal gradients in the crystalline shield have been established.

These studies have established that the character of physical rock property alternations and the development of geophysical boundaries in the earth's crust depends, to a greater extent, on a strain state of rocks (its distribution in depth). The boundaries of sharp change of rock strain state correspond to steplike changes of temperature gradient and heat flow (Figure 3). Strain redistribution is most likely caused by the internal processes in the earth crust. The analyses of different rank strain fields permits us to make a conclusion on the distinctly pronounced horizontal stratification of the earth's crust, related to the alternation of local conditions and deformation mechanisms of different deep horizons.

The generalization of complex geological and geophysical investigation results permitted us to make a conclusion that most of the inclined seismic boundaries known for the upper part of the section coincide with the boundaries of geological bodies, i.e. it has a material nature. Gentle seismic boundaries in the lower part of the section are related to cataclasis zones, having probably physical nature.

When drilling the Kola borehole, the basaltic layer expected from geophysical data was not intersected. Thus, the boundary of sharp change of seismic wave velocities intersected by the borehole proved to be related not to the surface of granitic and basaltic layer separation, but to the zone or rock discompaction in the process of progressive metamorphism; this zone represents a "closed" system, in which liquid and volatile components cannot be removed, and which results in the formation of fissures and discompaction of rocks. The established phenomenon of "hydrogenic" discompaction of metamorphic rocks helps us to understand the geological nature of the reflecting boundaries and waveguides in the earth's crust.

The data obtained make it possible to interpret the information on regional superficial and borehole geophysical investigations from a new viewpoint, and to provide increased accuracy in geophysical investigations, particularly deep seismic sounding.

The formulated conclusions on the evolution of the Precambrian continental earth crust in the interval from 3 to 1.6 billion years and compilation of its geologic content can be considered to be grand total of the data studies obtained from the Kola borehole drilling. According to the new model, the Archean continental crust consists of three layers: granitogneissic (0-15 km), granulitogneissic (15-30 km), and a lower layer (30-40 km) represented by protocrust.

The main drilling results of the Kola borehole were widely discussed at the XXVII Session of the International Geological Congress in Moscow in 1984 and were published in the monograph *Superdeep well of the Kola Peninsula* (Kozlovsky, 1984b).

The high yields of results from these scientific investigations made on the basis of the borehole data is attributed to the creation and successful practical application of a number of new technical facilities and progressive technological drilling processes, as well as to the development and utilization of highly efficient geophysical apparatus.

A unique drilling unit BU-15000 of 400-t load-carrying

capacity at 400 kg/cm$^2$ discharge pressure with maximum automatization of drilling processes (descend and hoisting of drilling instrument, drill-bit feed) and variable-speed d.c. drive-control of the main technical processes was created. The unit is designed for borehole drilling to a depth of 15 km. The method of advance open shaft drilling of optimum diameter was developed and put into practice (Figure 4). This eliminated wear of cemented casings and prevented complications, thus providing for drilling to a greater depth; increased the yield of information from intraborehole investigations; corrected all drilling processes; and made it possible to make decisions on further deepening of the borehole on the basis of actual geologic–engineering data.

Rock destructive tools and bottom engines with the appropriate specifications for deep conditions, as well as efficient means of turbodrill control at the bottom, were designed and put into production. Without the above facilities there is no possibility of drilling with the bottom engine to the depth of 8-9 km with the process controlled by sensing devices.

New types of core barrels have been designed and introduced into practice, thus enabling an increase in core recovery.

The Saatly borehole is the second superdeep hole aiming at the study of petroleum potential of Mesozoic and Paleozoic sediments within the Kura depression in the Transcaucasus; now it is being drilled below 8 km. The borehole has intersected Mesozoic volcanic beds and, when drilled deeper, should penetrate the entire sequence and reach Paleozoic formations. The drilling of six boreholes is in progress; at the end of July 1986 the depth of the boreholes was as follows: the Dnieper-Donats borehole—4655 m, Precaspian—3585 m, Krivoi Rog—1970 m, Muruntan—1020 m, Timan-Pechora—1616 m, Ural—2553 m.

So, during 1981-1985 in the USSR, the first stage of qualitatively new studies of the interior of the earth was successfully completed. It consisted of complex studies of all the structural levels of the earth's crust and mantle for predictive purposes as well as to search for concealed mineralization and deep zones of oil and gas accumulation.

A principally new scientific solution to global problems consists of a systematic approach to the collection, processing, and integrated analysis of geological and geophysical data at all the levels attained by up-to-date technical means (space-air-earth-superdeep borehole).

## Trends and Tasks of Complex Studies of a Deep Part of the Earth's Interior in the USSR During 1986-1990 and Up To 2000

The State Committee on Science and Technology and the State Planning Committee (Gosplan) of the USSR have approved a work scheme according to the new All-Union Programme of complex studies of the earth's interior which contemplates, along with the work now in progress, the solution of the following problems:

ascertaining the regularities of structure and evolution of the earth's crust within continents and on shelf zones, as well as detecting the sources of geodynamic processes;

studying the processes of sedimentation, differentiation of magmatic chambers, genesis and migration of fluids (solutions and gases) in the earth's crust, and present-day geothermal and geodynamic conditions.

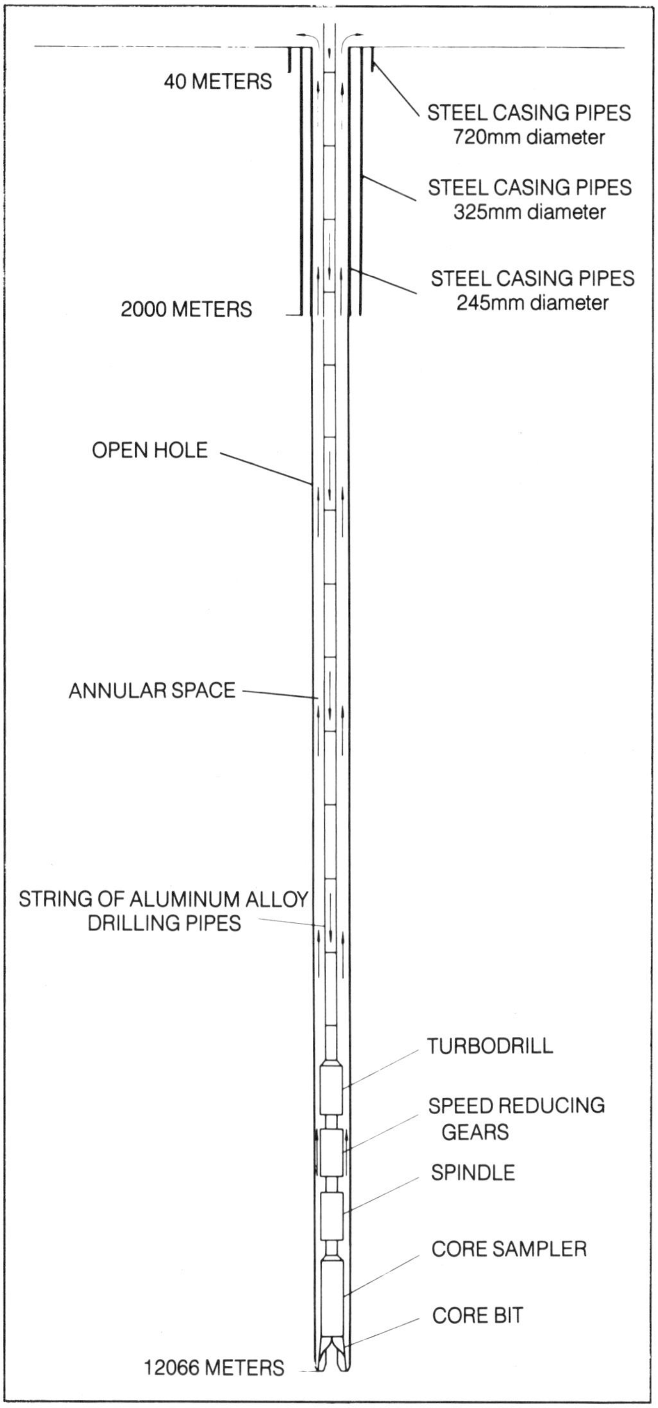

Figure 4. Schematic of the Kola well.

Of principal importance is setting up the peculiar depth laboratories on the basis of boreholes having been drilled to feel the pulse of the planet at various depth levels. Recording a number of physical, thermal, seismic, acoustic, hydrogeological, and some other data directly from well holes provides a unique means for observing the present-day geodynamic processes and representing a true life of the deep portion of the earth's crust. Such information will serve as an objective basis for reconstruction of the geological past and will provide a number of quantitative characteristics of geological processes.

The programme calls for the further development of a network of regional geotraverses and the drilling of deep and superdeep boreholes. In contrast to the work done between 1981 and 1985, complex investigations within large geodynamic test areas should be carried out during 1986-1990 in close connection with drilling deep holes in the most important mineral areas. A network of interrelated regional geotraverses, deep and super-deep boreholes to be drilled in different geological conditions, geodynamic test areas, and deep boreholes within mineral areas will provide volumetric three-dimensional data on several depth levels of the earth's crust.

It is also planned to intensify efforts on improvement of the computer methods of interpreting geophysical data as related to the new concepts of seismic wave fields and to study the nature of physical boundaries in the earth's crust and upper mantle.

During 1986-2000, regional geological and geophysical studies will be promoted to carry out investigations along profiles mainly in areas of the Far East, Western and Eastern Siberia, Russian platform, Central Asia, and shelf areas of the Arctic and Pacific Oceans.

Complex studies along geotraverses are carried out ahead of drilling deep and superdeep boreholes. Drilling data, in turn, will be used to corroborate the results of geophysical interpretation.

Direct geological information will be obtained on the structure of the lithosphere at depth, with evaluation of petroleum and mineral potential from test grounds, deep and superdeep boreholes, which are intended for solving two principal groups of tasks:

study of deep levels of the main geostructures corresponding to various geodynamic conditions and evaluation of their petroleum and mineral potential;

study of the subsurface, evaluation of ore-bearing capacity along the vertical, and ascertaining the factors for prediction of concealed mineralization in the typical mineral areas.

The boreholes intended for solving the task of the first group are located in such a way that they penetrate into deep layers of geostructures of different types, thus permitting the studied intervals as if to build up a sequence according to the age of deposits. This will make it possible to construct a true geological section over a wide depth range.

The Kola and Krivoi Rog superdeep boreholes make it possible to study old formations of shields and crystalline massifs and to gain comprehensive information on their structure as well as on composition and mineral potential of granulo-basaltic and gneissic granite protogeosynclinal units. The Kola superdeep borehole should reach 13 km by 1990, the designed depth being 15 km. The Krivoi Rog superdeep hole will be drilled to 8 km, with designed depth of 12 km. This borehole is aimed at ascertaining the nature of distribution of rich iron ore downward. The penetration through the whole Krivoi Rog sequence of Lower Proterozoic age will make it possible to clarify the existing idea of its ore-bearing capacity, the latter being of importance for general evaluation of mineral potential of the Krivoi Rog basin.

The tasks of studying deep levels of eugeosynclines, ascertaining the nature of their basements, and estimating mineral potential of the eugeosynclinal units at depth are being solved with the support of drilling data from the Ural borehole. By 1990 the borehole should reach 7 km, its designed depth being 15 km. The parametric data on the physical conditions of the interior of the Urals should be obtained, these being required for the reinterpretation and specifying of the previously compiled seismic sections. It is anticipated that drilling data from the Ural borehole would make it possible to solve a number of principal theoretical problems concerning tectonic evaluation of the earth, primarily these unified by the concept of "plate tectonics."

Evaluation of gold-bearing capacity of deep layers of the miogeodynclinal complexes in Uzbekistan and interpretation of the nature of their basements should be performed when drilling the Muruntau borehole, which should reach the designed depth of 7 km by 1990. The drilling is aimed at ascertaining the position of ore-bearing layers in the stratigraphic section, as well as the regularities and mechanism of mineralization concentration in ore-bearing horizons and the nature of distribution of ore mineralization downward to the designed depth.

A group of wells make it possible to reveal and estimate prospects of oil and gas potential in deep horizons of platform sedimentary strata, among them the Caspian well drilling in subsalt deposits. This well will penetrate to the projected depth of 7 km.

The Dnieper-Donets well, predetermined for drilling in Carboniferous oil and gas deposits with a possible penetration into the base rocks, should be completed in 1990 (7 km). In 1990 the Timano-Pechora well, which will estimate oil- and gas-bearing capacity in Devonian and pre-Devonian deposits, will reach the planned depth of 7 km. The projected depth of the Tyumen well is 8 km; it aims at drilling in potentially petroleum-bearing pre-Jurassic deposits and in the underlying basement. By 1991 it is scheduled to drill 6 km. The Novo-Yelkhovskaya well, projected to be 7 km deep, provides an estimate of oil and gas presence within deep strata in Tatar Autonomous Soviet Socialist Republic. The Saatlinskaya well will penetrate to a depth of 11.5 km. The Kuban well, with a planned depth of 12 km, is to drill in the Mesozoic-Cenozoic section of the West Kuban Trough to estimate its oil and gas potential. In 1986-1990 the wells will penetrate to depths of 3200 and 4500 m. Wells are contemplated to penetrate the crustal sequence to an ultimate thickness of about 60 km; they will give information about the rock composition of a wide age range and different regimes of origin.

The second group of problems (subsurface study of ore areas) is solved by drilling wells to projected depth of 3-5 km and by integrated geological and geophysical studies. For 1986-1990, holes are scheduled to be drilled in many ore regions including Norilsk (3.5 km), Pechenga (4 km), Khibiny (2 km), Lesser Caucasus (2 km), etc. All the wells are allocated to regional geotraverses. Operations are planned to determine whether additional holes should be drilled in other ore areas, including diamond- and gold-containing regions, and the possible sites for such drilling.

Besides applied problems solved by efforts in the Programme, its realization will make it possible to obtain new data for constructing global models of geotectonics, oil and gas potential, and metallogeny supported by actual data. In this respect the following trends are the most interesting:

- horizontal stratification of lithosphere and its nature
- specific features in the Precambrian history of the earth
- heat and radiation flow
- present geodynamic activity

- differences in rock properties at various depth
- distribution and rearrangement of hydrocarbons and ore-forming elements along different deep horizons
- role of metamorphism in the formation of oil, and gas, and ores
- water migration at great depths

All the above suggest that scientific efforts for 1986-1990 possess not only a qualitative but also a quantitative novelty. It will be necessary to drill deep and ultradeep wells to more than 30 km. A considerable part of these wells will penetrate in depth ranges for which reliable engineering facilities and technological processes are still lacking. It is contemplated to drill holes to a depth of more than 22 km in ore regions.

In conclusion we should like to emphasize that the realization of planned studies will make available unique integrated data about deep structure of the lithosphere. In addition, the presence of ores and hydrocarbons within the main types of the world's geostructures will be linked to different geodynamic situations. In the process of implementing the programme for ore regions, complex models will be constructed reflecting regularities in the stage localization of the mineralization and determining methods of prediction and search for buried mineralization at depths accessible for future development. The fulfillment of the programme for the study of subsurface horizons on the USSR territory, in our opinion, will provide the opportunity not only to understand better the fundamental theoretical problems but also to solve a wide range of applied problems in geology. Eventually, the progress of the earth sciences, as a whole, will benefit from its realization.

## REFERENCES CITED

Kozlovsky, Y.A., 1982a, Complex programme of the investigation of deep structure of the Earth's interior: Sovetskaya Geologiya, n. 9, p. 3-13.

Kozlovsky, Y.A., 1982b, Kola super-deep interim results and prospects: Episodes, n. 4, p. 9-11.

Kozlovsky, Y.A., 1984a, Study of the deep Earth's structure: Achievements and problems: Razvedka i okhrana nedr, n. 7, p. 12-18 and n. 8, p. 17-21.

Kozlovsky, Y.A., (ed.), 1984b, Superdeep well of the Kola Peninsula: Moscow, "Nedra," 490 p.

Kozlovsky, Y.A., and A.L. Yanshin, 1984, Soviet geology: Achievements and prospects: Episodes, v. 7, n. 1, p. 3-8.

CHAPTER 30

# SPECULATIONS ON THE LATE MESOZOIC AND CENOZOIC EVOLUTION OF THE SOUTHEAST ASIAN MARGIN

**Robert McCabe**
**Jay Cole**
*Texas A&M University*
*College Station, TX*

## ABSTRACT

The Sulu-Celebes-Banda Basins are three poorly understood marginal basins located between northwest Australia and southeast Asia. Analysis of magnetic anomalies, heat-flow data, bathymetry, and on-land geology has led to the conclusion that these three basins are remnants of a once-continuous ocean basin (Lee and McCabe, in press). Identified east-west-trending anomalies in all three of these basins suggest the following ages: Banda (119-136 Ma; Lapouille et al., recently submitted manuscript); and Sulu (41-47 Ma; Lee and McCabe, in press). The anomalies identified in each of these basins become progressively younger toward the north.

The on-land geology of this region is complicated but stratigraphic and paleomagnetic studies on pre-Oligocene rocks by numerous workers are consistent with the interpretation that the older landmasses that presently dissect the basin were translated into their present position before late Miocene. Stratigraphic studies from the Sula Microcontinent, Buru, Ceram, and Timor show close correlation with the stratigraphy of northwest Australia or New Guinea. Paleomagnetic studies from Timor suggest that a portion of the island was part of Australia since early Mesozoic. Paleomagnetic and stratigraphic data from Borneo and Sulawesi suggest that these landmasses share a common origin and that Sulawesi rifted eastward off Borneo during the Tertiary. Paleomagnetic and stratigraphic data from the Philippines suggest that the Philippine Arc is a composite of early Cretaceous to Recent arcs and oceanic crustal fragments that were translated clockwise from the southeast. The clockwise rotation of the Philippines, coupled with paleoinclination studies, suggests that this arc shares a common middle Tertiary history with the West Philippine Sea. Stratigraphic and paleomagnetic data from the North Palawan Continental Terrane are consistent with the interpretation that this terrane rifted away from China during the middle Tertiary.

## INTRODUCTION

The Banda, Celebes, and Sulu Seas (Figure 1) are three poorly understood marginal seas located in the complex region between the Australian and Eurasian plates. Because of the limited data that exist for these basins and the complex middle to late Tertiary tectonics, no overall interpretation of the region has been proposed. Lee and McCabe (1986) are believed to have identified east- to northeast-trending anomalies in each of these basins, which are consistently younger northward. The age Lee and McCabe assign to these anomalies are: (1) early Cretaceous, anomalies M1-M10 in the Banda Basin; (2) latest Cretaceous, anomalies 30-33 in the Celebes Sea; and (3) Eocene, anomalies 17-20 in the Sulu Sea. These ages suggest that the three basins are remnants of a once-continuous early Cretaceous to Eocene ocean basin which is now dissected and trapped. Their conclusion is also supported by heat-flow data and the overall bathymetry of each of these basins.

An examination of Figure 1 shows that this region is presently composed of a hodgepodge collection of landmasses. Many of these landmasses contain known sequences of pre-Tertiary rocks (van Bemmelen, 1949; Hamilton, 1978, 1979; Silver and Smith, 1983; Pigram et al., 1985). If the above model—which suggests that the present configuration of this region is the result of middle to late Tertiary tectonic movements which dissected this older oceanic basin—is correct, then each of the landmasses that currently occupy the region must have originated elsewhere.

Here, we will examine existing paleomagnetic data and stratigraphic data from the various landmasses between northwest Australia and southern China. The data will be used to show that the landmasses presently located in this region were either built in place by late Tertiary to Recent volcanic

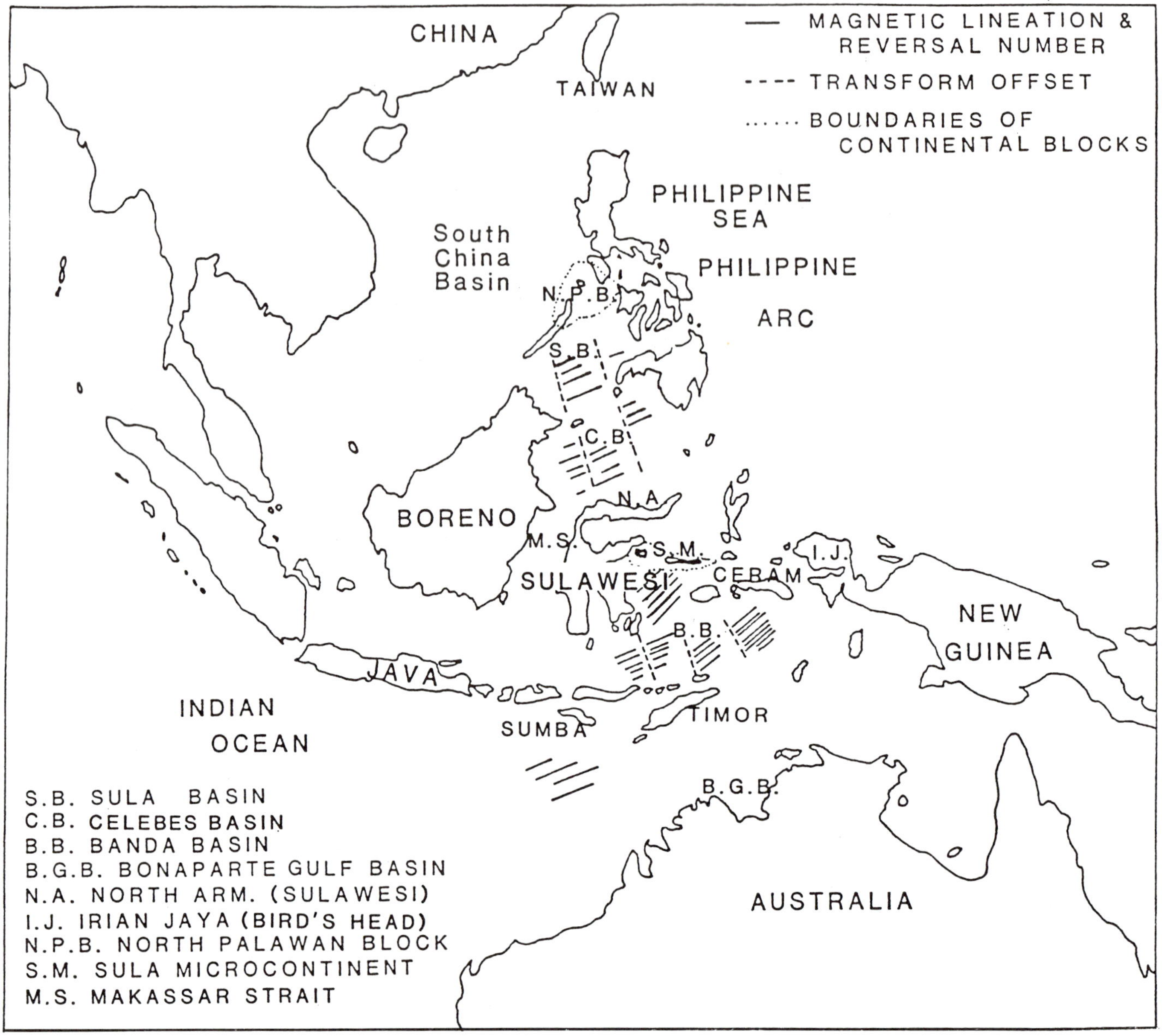


Figure 1—Generalized tectonic map of the southeast Asian region. The magnetic anomalies shown in the figure are from Lapouille et al. (recently submitted manuscript) for the Banda basin and from Lee and McCabe (in press) for the Celebes and Sulu basins.

activity or were translated into their present position by Tertiary tectonic motions.

## PALEOMAGNETIC DATA

Below, we will examine paleomagnetic data from the Timor, Borneo, Sulawesi, and the Philippines (Table 1). The data will be used to try to constrain the pre-middle Tertiary affinity of each of these regions. It should be pointed out that, despite the success of paleomagnetic studies in other parts of the world, paleomagnetic data from southeast Asia are somewhat limited and the reported data are hampered by a number of factors. These factors include limited access to exposures, poor geological mapping, limited age constraints, and severe tropical weathering. The combined effect of these factors has limited the overall density of sites in this region. Although it will take years before we have a reliable paleomagnetic data base for the region, enough results now exist to try to constrain the possible affinity of some of the landmasses between Australia and South China. Below, we will discuss the paleomagnetic results from this region in terms of three independent groups of islands: (1) Timor-Australia; (2) Borneo-Sulawesi; and (3) the Philippines. This discussion will attempt to show that, although the relative movements between these three island groups are poorly known, the paleomagnetic data from the region show that all three of these groups underwent independent Tertiary movements. The paleomagnetic data discussed below are consistent with the interpretation that these island groups arrived in their present position during the late Tertiary.

## Timor-Australia

The early to middle Mesozoic rocks of the northwest Australia shelf (Figure 2) are characterized by fluviodeltaic and shallow-marine sequences (Powell, 1976). These are in turn overlain by an upper Jurassic to lower Cretaceous sequence of clastic sediments deposited within marginal to restricted marine environments (Powell, 1976; Crostella, 1977). Open marine marlstones and claystones comprise the upper Cretaceous to lower Eocene sequence, which is overlain by a middle Eocene to Recent carbonate shelf that prograded northwestward throughout this time (Kaye, 1972; Crostella, 1977; Allen, Pearce, and Gardner, 1978). The geology of this shelf and the recognition of late Jurassic ENE-trending magnetic anomalies in the Argo Basin suggest that the northwestern shelf of Australia exemplifies an Atlantic-type rifted margin.

Timor occurs along the northern edge of the Australian margin. Many of the Permian, Mesozoic, and lower Tertiary rocks of Timor (Figure 2) reflect shallow-water facies and faunas similar to those of the northwest Australian shelf (Audley-Charles, 1965, 1968; Veevers, 1969; Teichert, 1970; Waterhouse, 1972; Hamilton, 1979). Although strong differences of opinion exist as to whether Timor represents a "rootless" accretionary prism (Hamilton, 1978, 1979; Von der Borch, 1979) or a series of exotic nappes and a "para-autochthon/autochthon/allochthon" association (Audley-Charles, 1983, 1986), most workers in the region generally agree that the island of Timor is merely the chaotic product of the collision of the Australian shelf with the Banda Arc.

Paleomagnetic studies on Timor are reported for rocks that range in age from Permo-Triassic to Tertiary (Chamalaun, 1977a,b; Chamalaun and Sunata, 1982). The paleomagnetic poles for the rocks of various ages are shown in Figure 3a. Although Chamalaun and Sunata point out that there is quite a bit of uncertainty as to the ages of some of the formations, when these Timor poles are plotted against the Mesozoic through Cenozoic polar wander curve of Australia (Embleton, 1981), the agreement is good. This good correlation between Timor and Australia polar wander paths suggest that portions of Timor are part of the Australian margin and that there has been no significant drift between Timor and Australia since Permo-Triassic. The affinity of Timor to the northwest Australian shelf is also supported by Chamalaun and Sunata's paleoinclination data from Timor. These data show that Timor underwent northward drift during the Cenozoic. This motion is consistent with the Tertiary northward translation of Australia resulting from the opening of the southeast Indian Ocean.

An examination of Figure 3a shows two poles from Timor (the Cretaceous Wai Formation and the Pliocene(?) Oecussi Volcanics) that do not conform to the Australian margin origin. Chamalaun and Sunata (1982) suggest that these two anomalous results are from formations located on thrust sheets which were emplaced during the Neogene collision between the eastern Banda Arc and Australia.

On the northern edge of the Banda Sea are a series of islands which are also stratigraphically related to the northern Australian-New Guinea shelf region. At present, no pre-Neogene paleomagnetic data are reported from this region. One of the largest landmasses in this region is the Sula Platform. This platform is regarded as being a microcontinental fragment (Hamilton, 1979; McCaffrey, Silver, and Raitt, 1981) that has moved westward from New Guinea via the Sorong fault zone (Visser and Hermes, 1962; Krause, 1965; Hermes, 1968, 1974; Gribi, 1973; Hamilton, 1978, 1979; Norvick, 1979; Silver and Smith, 1983; Silver, McCaffrey, and Smith, 1983; Pigram et al., 1985; Silver et al., 1985). This westward translation is believed to have occurred during the Neogene and was eventually terminated by the late Miocene collision of this fragment with eastern Sulawesi (Hamilton, 1979). Through stratigraphic correlation it has been suggested that the Sula Platform originated from either the area north of Misool (Audley-Charles, Carter, and Milson, 1972), the Bird's Head of Irian Jaya (Visser and Hermes, 1962; Hermes, 1968; Norvick, 1979), or the Bird's Neck of Irian Jaya (Hamilton, 1979). Pigram et al. (1985), on the other hand, believe the stratigraphy of the Sula Platform correlates better with that of central Papua New Guinea (Figure 4), and thus suggest a more easterly origin for this microcontinental fragment. Whatever the case, it is generally agreed that the Sula microcontinent originated from somewhere in the western two-thirds of New Guinea before being displaced westward (via strike-slip movements) and subsequently colliding with eastern Sulawesi. Silver et al. (1985) further suggest the Sula Microcontinent and other submerged ridges of the Banda Sea represent slivers of Irian Jaya that have been sinistrally strike-slip faulted during the Neogene into their current location separating the north and south Banda basins.

## Borneo and Sulawesi

The island of Borneo, located in the central portion of the Sunda region, is a key to any regional tectonic understanding of southeast Asia. The eastern half of Borneo (Figure 5) consists of Cretaceous clastic sediments and volcanic rocks and possibly older metasedimentary rocks (van Bemmelen, 1949; Hamilton, 1979). These rocks are intruded by granitoid plutonic rocks which have yielded early to late Cretaceous ages on individual mineral separates (Haile, McElhinny, and McDougall, 1977). Oil exploration wells in the Java Sea (south of Borneo) have encountered similar metasedimentary rocks, lower Cretaceous limestones and clastic sediments, and granitiod intrusions (Hamilton, 1979), suggesting that the Java Sea is underlain by the same pre-Cenozoic basement. These Mesozoic sequences are overlain in eastern Kalimantan by Eocene coarse clastic sedimentary rocks, Tertiary siliceous volcanics, and transgressive marine clastic sedimentary rocks (van Bemmelen, 1949). These shallow-marine sequences are overlain by Oligocene and early Miocene shelf limestones and marine clastics which prograde eastward. The Cenozoic sedimentary sequences reported in eastern Kalimantan suggested to Hamilton (1979) that these rocks were deposited in response to the middle Tertiary opening of the Makassar Strait. The occurrence of early to middle Tertiary volcanic rocks on the South Arm of Sulawesi (Figure 5) (van Bemmelen, 1949; van Leeuwen, 1975; Katili, 1978; Sukamto, 1975a,b; Hamilton, 1979), subduction melange in eastern Sulawesi (Hamilton, 1979; Katili, 1978), and middle Tertiary north-trending extensional faulting in the eastern margin of the Java Sea, all support the suggestion of Hamilton for the back-arc-related eastward migration of Sulawesi with respect to Borneo in the middle Tertiary.

The results from paleomagnetic studies on Borneo rocks have been reported from Kalimantan and Sarawak (Haile,

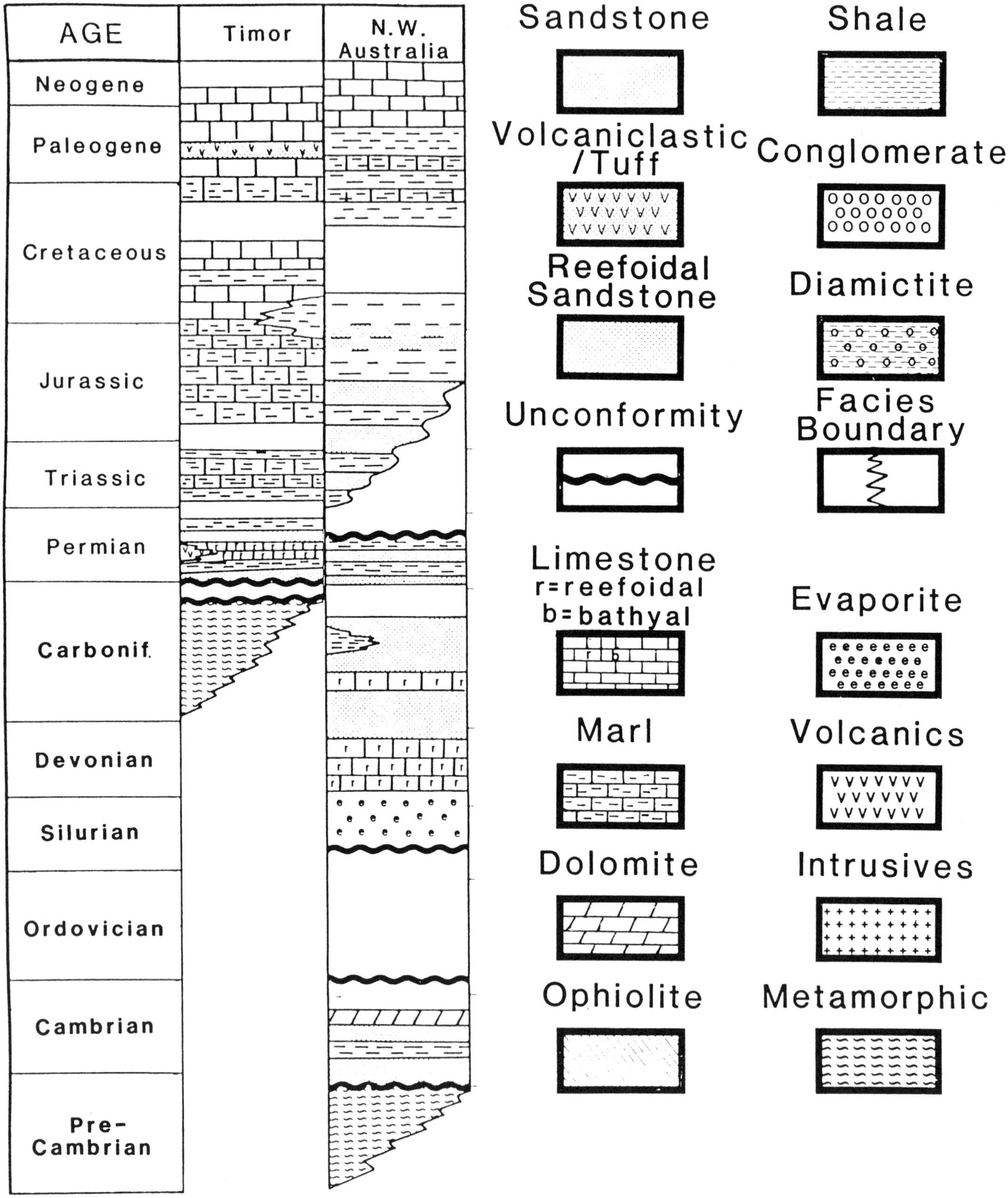


Figure 2—Generalized stratigraphic columns for the northwest Australian shelf and the island of Timor. The Mesozoic and Cenozoic stratigraphies have suggested to past workers in the region that portions of Timor exhibit depositional affinities to the northwest Australian shelf. The stratigraphic legend on the edge of the figure is the same for Figures 4–7.

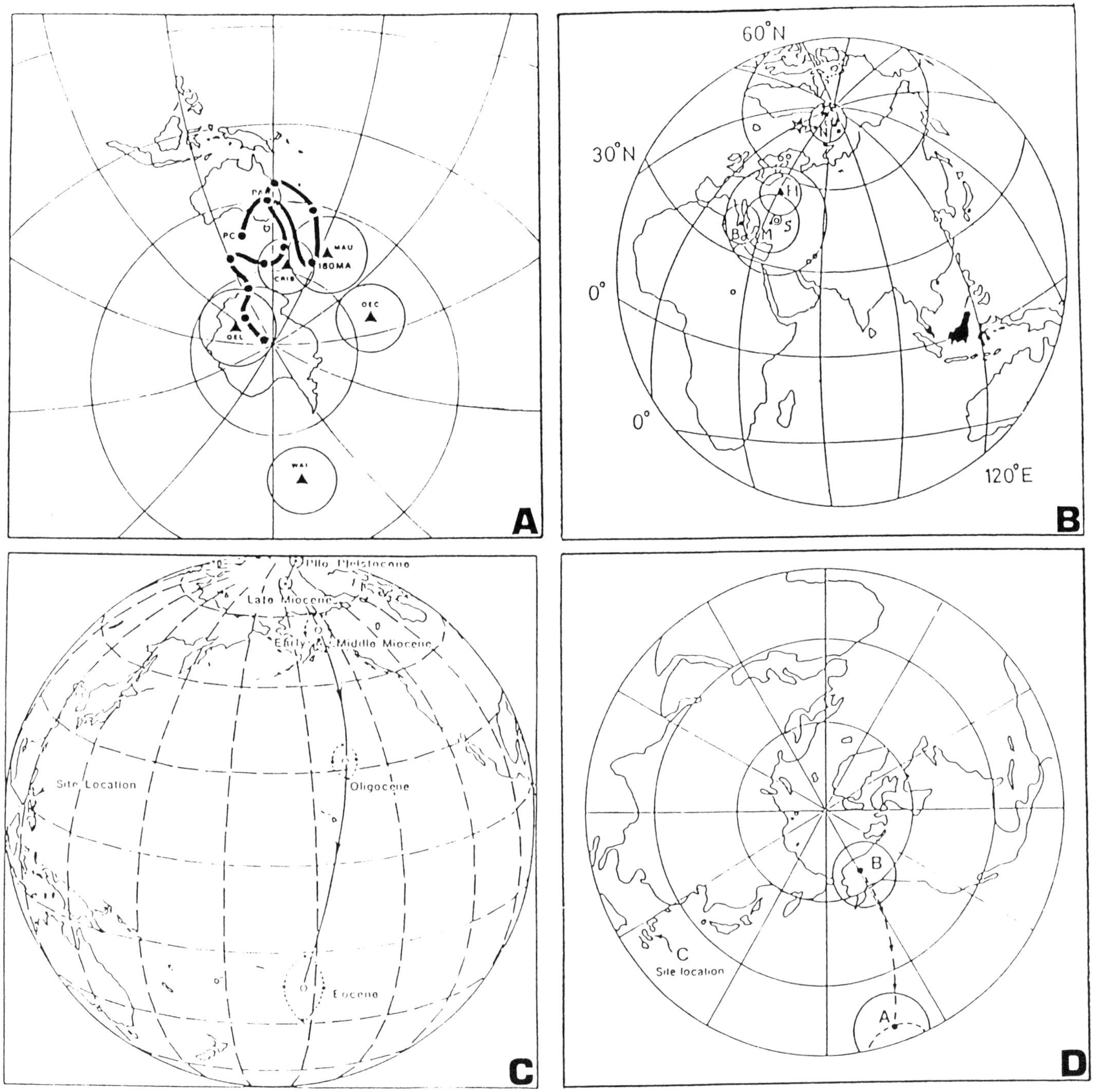


Figure 3—(a) Comparison of Permian to Tertiary polar wander curves of Australia (Embleton, 1981) and Timor (Chamalaun and Sunata, 1982). PC and PA correspond to the late Permian and early Triassic Australian poles, respectively. Oec and Qel correspond to Tertiary poles from Timor, WAI, corresponds to a Cretaceous pole. MAU and CRIB correspond to Permian to early Mesozoic poles. Chamaluan and Sunata argue that these pre-Cretaceous poles were remagnetized around 180 MA. The coincidence between the timing of this possible remagnetization and initiation of rifting in NW Australia suggest that this remagnetization could correspond with the break-up of the Australian margin. Two poles (WAI and Oec) show marked divergence from the Australian polar wander curve. Chamaluan and Sunata argue that these two poles are from formations that are on thrust sheets, and hence the divergence is related to structural complications. Note that the close proximity of the Timor poles (excluding WAI and Oec) to the Australian polar wander curve suggests a common Phanerozoic translational history for Timor and Australia. This suggests that a portion of Timor was part of the Australian margin since the Paleozoic. (b) Comparison of paleomagnetic pole positions for the Sunda region. M: late Cretaceous pole from the Malay Peninsula (McElhinny, Haile, and Crawford, 1974; Haile, McElhinny, and McDougall, 1977). B: Late Cretaceous pole from west Kalimantan, Borneo (Haile, McElhinny, and McDougall, 1977). H: Late Jurassic to Early Cretaceous pole from the South Arm of Sulawesi (Haile, 1978). S: Paleogene to early Miocene pole from the South Arm (Sasajima et al., 1980); pole positions for individual sites are shown as small open circles. N: post-middle Miocene pole from the South Arm (Sasajima et al., 1980); pole positions for individual sites are shown as small solid circles. Star: late Neogene pole of Sabah, Borneo (Schmidtke et al., 1985). The present position of Borneo, Malay, and Sulawesi is indicated by dark shading. Note that the pre-middle Miocene poles and the post-middle Miocene poles form two distinct populations and suggest that, within the limits of paleomagnetic data, the entire region shares a common Tertiary translation history. Figure is modified after Figure 4 of Jarrard and Sasajima, 1980. (c) Cenozoic apparent polar wander path of the western Philippines, excluding the North Palawan Continental Terrane. The data used for this path (McCabe et al., in press) excludes the data from the portions of the Philippines where the observed declinations are believed to be collision related (e.g., early-middle Miocene data from Marinduque, and late Miocene data from Central Luzon); (d) Location of late Jurassic paleopole positions of Mindoro (a) (McCabe et al., in press) and South China (b). The position of the Miondoro pole is consistent with the southward migration of the North Palawan Block as a result of the Oligocene to Miocene opening of the South China Sea (Taylor and Hayes, 1980).

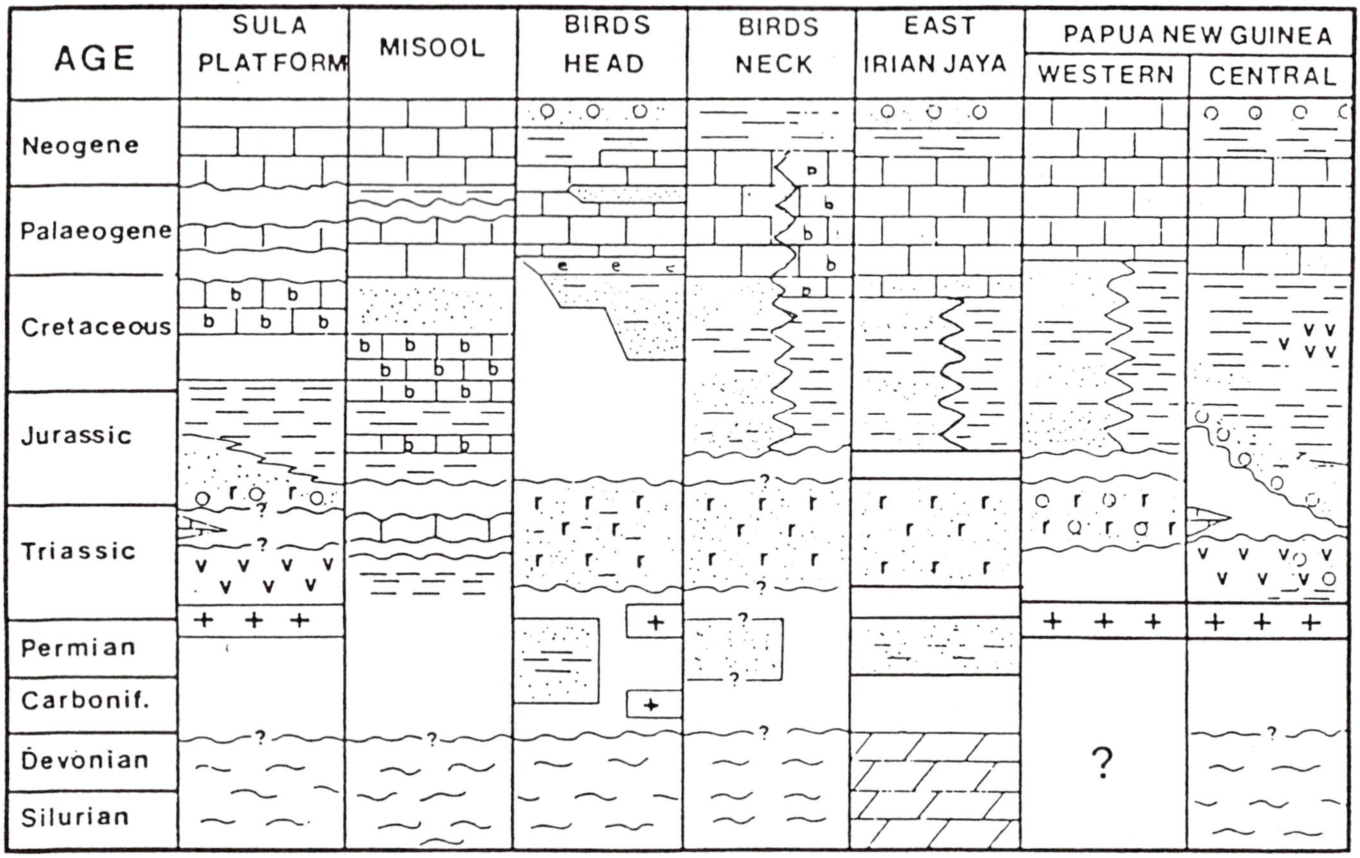


Figure 4—Generalized stratigraphic columns for the Sula Platform, Misool Archipelago, Irian Jaya, and Papua New Guinea (after Pigram et al., 1985). This figure was used by Pigram et al. to suggest that the stratigraphy of the Sula Platform best correlates with that of central Papua New Guinea. The stratigraphic legend for the figure is on Figure 2.

McElhinny, and McDougall, 1977; Haile, 1979) and Sabah. These results are summarized in Table 1 and Figure 3B. Haile, McElhinny, and McDougall (1977) reported the results of a paleomagnetic study of late Cretaceous rocks from the basement of western Borneo. Their study suggested 50° of counterclockwise rotation but no latitudinal change of the western portion of Borneo since the Cretaceous. The late Cretaceous pole for Borneo is in good agreement with the Cretaceous pole for the Malay Peninsula (McElhinny, Haile, and Crawford, 1974) (Figure 3b), which suggests that there has been little to no differential movement between Borneo and Malaysia since the Cretaceous.

Tertiary data from Borneo are limited to a small number of sites that are separated by rather large distances. Preliminary results on Oligocene/Miocene red beds from Sarawak were reported by Haile (1979). The results indicate that these red beds are not distinguishable from the present field, suggesting that the counterclockwise rotation of Borneo was complete by the late Tertiary. Paleomagnetic results from the middle Tertiary Crocker Formation in Sabah, northeast of Sarawak, show evidence for greater than 60° of counterclockwise rotation (Table 1), whereas results from one site from a late Miocene intrusion are indistinguishable from the present field (Schmidtke et al., 1985). If we assume that these two sets of limited data (Haile, 1979; Schmidtke et al., 1985) are reliable and that the island of Boreno has behaved as a single tectonic unit during the Cenozoic (at least at the resolution of paleomagnetic data), the results suggest that Borneo underwent a large counterclockwise rotation that was completed by late Tertiary. It must be pointed out that, although the late Cretaceous data and limited Tertiary data from Borneo suggest that the island has behaved as a single tectonic unit, the northern part of the island is composed of numerous ophiolitic fragments, which Hamilton (1979) has suggested represent Cretaceous and Tertiary melange belts. This structural complexity, combined with conflicting paleomagnetic results from Sabah (Schmidtke et al., 1985), argues that more work is needed on the structure of the region. Until better regional geologic mapping is carried out, it is impossible to ascertain if Borneo is behaving as a single tectonic unit or is composed of numerous unrelated tectonostratigraphic terranes.

Sulawesi is a peculiar K-shaped island just east of Borneo. The distinctive shape of the island, the position of the Sula microcontinent, and eastward-thrusted ophiolites on the eastern arms of Sulawesi led workers (Hamilton, 1973, 1978, 1979; Katili, 1978; Silver, McCaffrey, and Smith, 1983) to interpret Sulawesi as the site of a Miocene collision between an east-facing island arc and a passive Paleozoic microcontinent (Sula microcontinent) situated on the downgoing plate.

Three separate groups have reported paleomagnetic results from Sulawesi. Haile (1978) was the first to do a paleomagnetic study of the island. His study of Jurassic to early Cretaceous cherts from the South Arm showed that the paleomagnetic pole position is close to the Cretaceous pole positions obtained from Borneo (Haile, McElhinny, and McDougall, 1977) and the Malay peninsula (McElhinny, Haile,

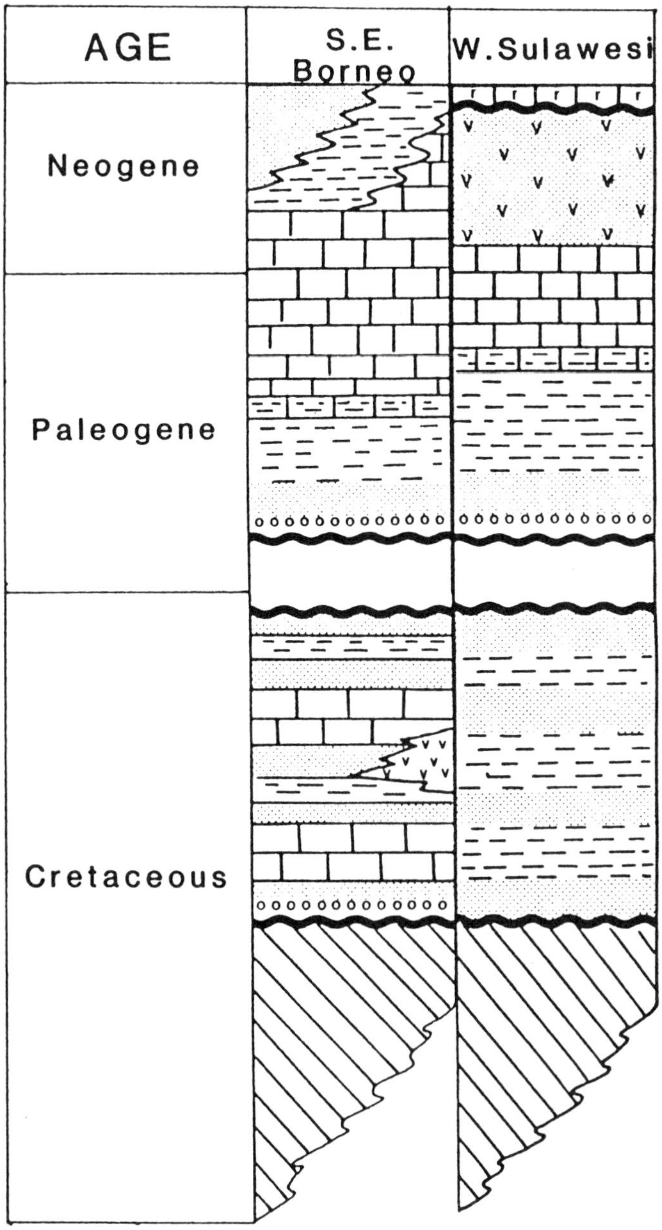


Figure 5—Generalized stratigraphic columns for southeastern Borneo and western Sulawesi. The divergence in their stratigraphical similarities during the early to middle Neogene is consistent with Hamilton's (1979) eastward migration of western Sulawesi from eastern Borneo during the middle Tertiary. The stratigraphic legend is on Figure 2.

and Crawford, 1974). These results (Figure 3b) suggest that the South Arm of Sulawesi probably occupied a position near Borneo and the Malay Peninsula. Besides the Mesozoic pole positions, Haile also found that the paleoinclinations from the Mesozoic sites on Sulawesi were shallow and consistent with the inclination data from Borneo. This similarity of paleopole positions and shallow inclinations for Sulawesi and Borneo argues that the South Arm of Sulawesi probably lay close to Borneo since the Cretaceous.

In contrast with the results from the South Arm, preliminary paleoinclination data from Mesozoic cherts from the Southeast Arm (Haile, 1978) argue that these rocks were formed at much higher latitudes. Such high latitudes are consistent with the theory that part of the rocks from the Southeast Arm were formed in a southern Mesozoic ocean basin and traveled northward by plate motions until they were eventually accreted onto the equatorial Sulawesi island arc.

During his investigation of Mesozoic sites, Haile also collected four sites from late Cenozoic volcanic rocks from the South Arm. The results of this study (Table 1, Figure 3b) suggested that the South Arm has not rotated during the Tertiary. In addition to Haile's Cenozoic results from the South Arm, Sasajima et al. (1980) reported results from an additional 15 sites from the South Arm. Although the late Miocene to Recent results of Sasajima et al. are consistent with Haile's conclusions, all of their pre-late Miocene samples are rotated nearly 40° counterclockwise from the present field direction. Although there are differences between the pre-Neogene data sets of Sasajima et al. and those of Haile for this region, both data sets are statistically indistinguishable from the pre-Neogene pole position of Borneo and support the conclusion that both Borneo and Sulawesi have been part of the same tectonic unit since the Cretaceous.

Paleomagnetic work from the North Arm is reported by Otofuji et al. (1981). In contrast with the South Arm, they suggested that the North Arm rotated more than 90° between Eocene and early Miocene, whereas Plio-Pleistocene directions from the North Arm are indistinguishable from the present dipole direction.

On the basis of paleomagnetic data, it appears that both the South Arm of Sulawesi and Borneo have been part of the same block since the Cretaceous, and both of these areas underwent an early to middle Tertiary counterclockwise rotation which was completed by the Miocene. Paleomagnetic data from the North Arm of Sulawesi suggest that this area has rotated with respect to the South Arm during latest Tertiary. It has been suggested that this differential rotation resulted from the collision between the Sula Microcontinent and Sulawesi during the Miocene (Silver, McCaffrey, and Smith, 1983; McCabe, 1984a).

## Philippines

With the exception of the North Palawan Continental Terrane, the remainder of the Philippine Archipelago (Figure 6) consists of a series of Cretaceous to Recent ophiolites and associated metasediments, basaltic and andesitic lavas, grainitiod intrusions, volcanogenic sandstones and shales, as well as intercalated bedded and reefal limestones (Hamilton, 1979; Hashimoto, 1981; Balce et al., 1981; Karig, 1983; McCabe, Almasco, and Yumul, 1985; Hawkins et al., 1985). These sequences imply an island-arc-related history for the Philippines since Cretaceous without continentally influenced sedimentation.

Stratigraphic data from the North Palawan Continental Terrane (Figure 7) suggest that this terrane rifted off South China beginning in the mid-Oligocene (Holloway, 1982). This rifting scenario is also supported by marine magnetic anomaly studies in the South China Sea (Taylor and Hayes, 1980, 1983). The middle Miocene and older (through upper Paleozoic) rocks of the South China shelf and North Palawan include nonmarine to paralic feldspathic sandstones and conglomerates with intercalated siltstones, shales, coals, and a few granitiod intrusions (Hashimoto and Sato, 1973), as well as four correlatable unconformities (Holloway, 1982; Taylor and Hayes, 1980).

TABLE 1 PALEOMAGNETIC RESULTS FROM TIMOR, SULAWESI, BORNEO, MALAYSIA, AND PHILIPPINES

| Age | Number of Sites | Number of Samples | Stability | Declination | Inclination | α95 | Pole latitude | Pole longitude | dp | dm | reference number |
|---|---|---|---|---|---|---|---|---|---|---|---|
| **TIMOR** | | | | | | | | | | | |
| Tertiary | -- | 9 | TA | 190.6 | 35.2 | 15.6 | 75.0 | 264.4E | -- | -- | 1 |
| Tertiary | -- | 20 | TAP | 320.4 | 7.1 | 12.0 | 54.0E | 51.0E | -- | -- | 1 |
| Cretaceous | -- | 12 | T | 12.1 | 51.1 | 11.1 | 47.6N | 140.8E | -- | -- | 1 |
| Upper Permian | -- | 32 | TRF | 158.6 | 53.3 | 7.1 | 58.3N | 341.3E | -- | -- | 2 |
| Permian/Jurassic (?) | -- | 10 | TF | 140.0 | 45.0 | 13.0 | 48.0N | 5.0E | -- | -- | 3 |
| **BORNEO** | | | | | | | | | | | |
| late Miocene | 1 | -- | ? | 349.8 | 0.1 | 2.4 | -- | -- | -- | -- | 4 |
| Oligo-Miocene | 2 | 10 | T | 178.0 | -3.0 | 4.0 | 88.0N | 10.0E | 2.0 | 4.0 | 5 |
| middle Tertiary | 2 | -- | ? | 286.0 | -1.3 | -- | 15.7N | 24.0E | -- | -- | 4 |
| Upper Cretaceous | -- | 39 | TR | 311.0 | 0.0 | 8.3 | 41.0N | 21.0E | -- | -- | 6 |
| **SULAWESI** | | | | | | | | | | | |
| South Arm | | | | | | | | | | | |
| Quaternary | 4 | 13 | A | 358.0 | -18.0 | 12.0 | 81.0N | 359.0E | 6.0 | 12.0 | 7 |
| Plio-mid Miocene | 5 | 47 | A | 355.0 | -5.0 | 7.0 | 84.0N | 55.0E | -- | -- | 8 |
| early Tertiary | 6 | 58 | AR | 315.2 | -1.3 | 20.0 | 45.0N | 36.0E | -- | -- | 8 |
| Cretaceous | 3 | 29 | A | 346.0 | -32.0 | 21.7 | 71.0N | 347.0E | 13.7 | 24.4 | 8 |
| Cretaceous/ upper Jurassic | -- | 8 | T | 325.0 | -5.0 | 6.1 | 55.0N | 32.0E | 3.6 | 7.2 | 7 |
| Southeast Arm | | | | | | | | | | | |
| lower Cretaceous/ upper Jurassic | -- | 4 | T | 230.0 | -62.0 | 8.8 | -- | -- | -- | -- | 7 |
| **MALAY** | | | | | | | | | | | |
| Cretaceous | -- | 16 | A | 136.0 | -31.0 | 9.4 | 44.0N | 35.0E | -- | -- | 9 |
| **PHILIPPINES** | | | | | | | | | | | |
| Plio-Pleistocene(a) | 23 | 162 | ATR | 358.9 | 16.2 | 4.2 | 85.2N | 314.7E | 1.8 | 3.0 | 10, 11, 12 |
| late Miocene(a) | 7 | 47 | ATR | 357.7 | 17.2 | 6.0 | 80.8N | 218.1E | 3.6 | 5.5 | 10 |
| late Miocene(b) | 10 | 56 | ATR | 16.7 | 28.3 | 7.2 | 73.8N | 210.9E | 3.9 | 5.1 | 10 |
| early-mid Miocene(a) | 15 | 126 | ATRF | 23.9 | 17.6 | 7.6 | 65.0N | 215.4E | 3.5 | 5.8 | 10, 13 |
| early-mid Miocene(b) | 8 | 21 | ATR | 302.4 | 11.3 | 16.9 | 29.5S | 196.3E | 8.9 | 14.5 | 10, 12 |
| Oligocene*(a) | 6 | 43 | AT | 57.6 | 16.6 | 6.3 | 33.4N | 209.6E | 3.4 | 4.8 | 10 |
| Eocene(a) | 10 | 78 | ATR | 115.4 | 8.2 | 11.2 | 23.0S | 198.9E | | | 12 |
| Cretaceous | 3 | 17 | ATF | 297.1 | 15.8 | 14.6 | 25.6S | 199.8E | 7.7 | 8.1 | 10 |
| Jurassic | 5 | 23 | ARF | 75.7 | 27.7 | 19.4 | 17.2N | 198.6E | 13.6 | 15.2 | 10 |

Stability-A: demagnetized by alternating field method, T: demagnetized by thermal method, P: presence of reversal test, F: presence of fold test, α95: radius of 95% confidence circle, dp and dm semiaxis of oval of 95% confidence on pole position.
a) Data used for construction of Fig. 3C.
b) Data excluded from construction of Fig. 3C. This data was excluded because of possibility of being related to collisions
*) Oligocene data from Zambales region of central Luzon


1) Chamalaun and Sunata, 1982
2) Chamalaun, 1977a
3) Chamalaun, 1977b
4) Schmidtke et al., 1985
5) Haile, 1979
6) Haile, 1977
7) Haile, 1978
8) Sasajima et al., 1980
9) McElhinney et al., 1977
10) McCabe et al., (J. G. R., in press)
11) McCabe et al., 1982
12) Fuller et al., 1983
13) McCabe et al., 1983

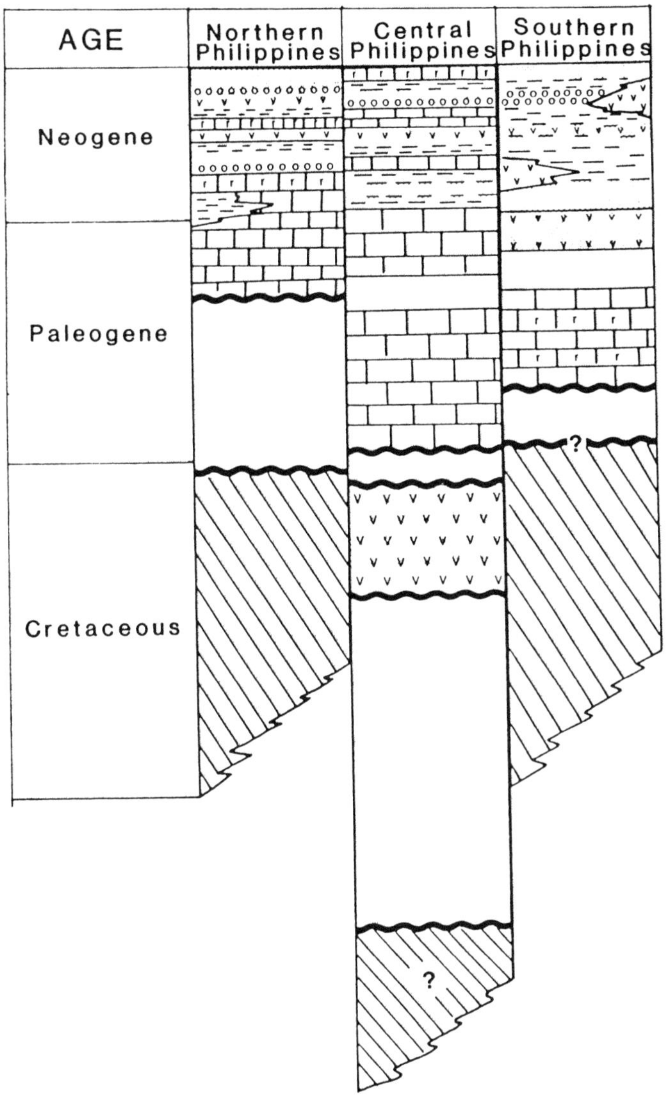


Figure 6—Generalized stratigraphic columns for northern, central, and southern Philippines. These stratigraphic columns suggest that the various portions of the Philippine Archipelago (except the North Palawan Continental Terrane, discussed below) show Cretaceous to Recent island arc/convergent margin tectonic and sedimentological affinities without continentally influenced sedimentation. The stratigraphic legend is on Figure 2.

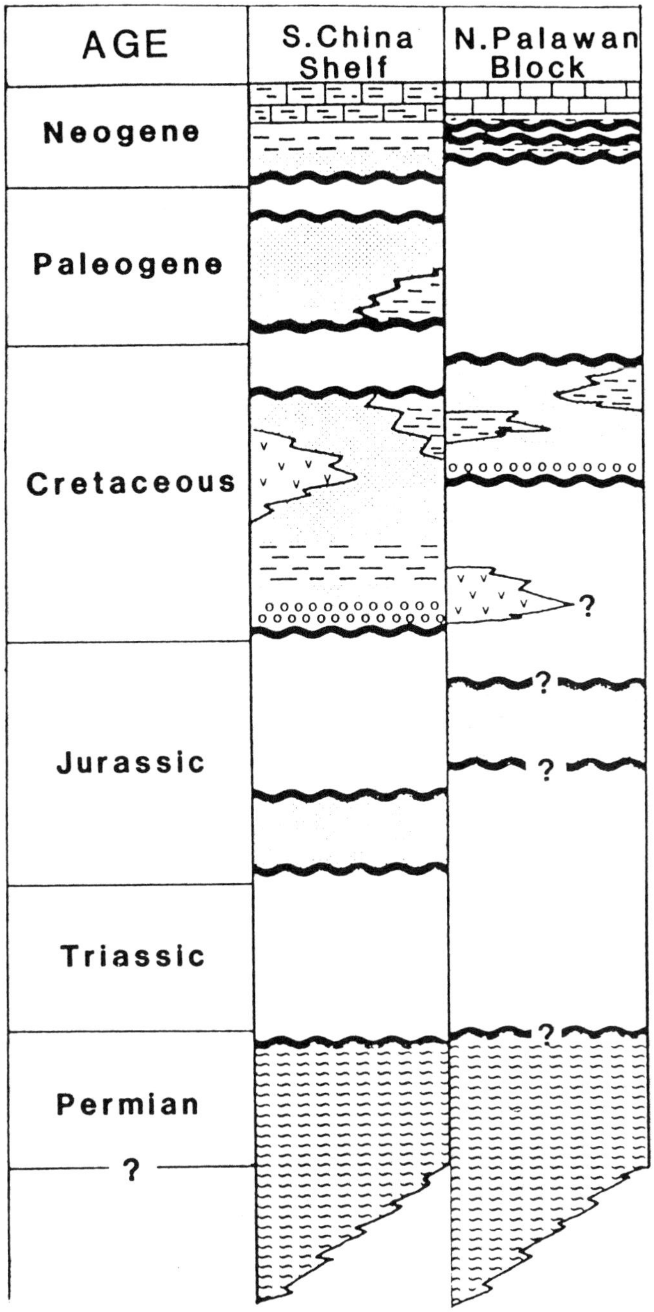


Figure 7—Generalized stratigraphic columns for the South China Sea shelf and the North Palawan Continental Terrane (NPCT) (after Taylor and Hayes, 1980; Holloway, 1982). The mid-Tertiary shift toward marine sedimentation on North Palawan and the presence of the uppermost mid-Miocene regional unconformity in the Palawan area that is not correlatable to the South China shelf suggest that the NPCT rifted off of South China during the late Tertiary (Hamilton, 1978, 1979; Taylor and Hayes, 1980; Holloway, 1982). The stratigraphic legend is on Figure 2.

The Neogene rifting of North Palawan from South China is stratigraphically recorded as: (1) the mid-Tertiary shift toward marine sedimentation on North Palawan with continued continentally influenced sedimentation on the South China shelf; and (2) the presence of an uppermost mid-Miocene regional unconformity in the Palawan area that is not correlatable with the South China shelf (Holloway, 1982). Stratigraphic data further suggest the North Palawan Continental Terrane collided with the western Philippines by late Miocene, as suggested by the presence of coarse clastic sands and boulder conglomerates apparently derived from North Palawan in portions of the central Philippines (Mindoro Island) (Andal, 1966; Hamilton, 1979; unpublished data from Phillips Petroleum, 1981; Karig, 1983; McCabe, Almasco, and Yumul, 1985; Sarewitz and Karig, in press).

Numerous paleomagnetic studies have been reported on the Philippine Arc (Hsu, 1972; DeBoer et al., 1980; McCabe et al., 1982a, b; Fuller et al., 1983; Almasco and Fuller, 1983; McCabe et al., 1987). For the most part, each of these studies was carried out on different sections of the Philippine Arc. Since the Philippines represent a composite terrane composed of an amalgamated collection of island arcs and ophiolite fragments (Hamilton, 1978, 1979; Balce et al., 1981; Moore and Silver, 1983; Karig, 1983; McCabe, Almasco, and Yumul, 1985; Hawkins et al., 1985), these studies are not internally consistent. Even with this overall lack of consistency, some general trends are emerging in all studies. The most noted of these trends is that the Philippines have translated northward since the Eocene (Hsu, 1972; DeBoer et al., 1980; Uyeda and McCabe, 1983; Fuller et al., 1983; McCabe et al., in press). Tertiary paleomagnetic data from the Philippine Sea Plate (Louden, 1977) suggest that it has moved northward and rotated clockwise into its present location. Uyeda and McCabe (1982a) and McCabe et al. (in press) note the close correspondence between the paleoinclinations of the Philippine Arc and the Philippine Sea Plate (Figure 8), which suggests that these plates may have been coupled during a large part of the Tertiary.

The major inconsistencies in Philippine paleomagnetic data obtained by various workers is in the declination values. Most recently, McCabe et al (1982b) and Fuller et al (1983) argue for a counterclockwise rotation of the northern Philippines. These arguments were originally based on early to middle Miocene data from northern Luzon (Bagiuo region). Recent studies report that much of the pre-late Miocene data from northern Luzon are remagnetized (McCabe, 1984b; Fuller, 1985). Until future work is completed on pre-late Miocene data from this region, evidence for the counterclockwise rotation of northern and central Luzon must thus be treated as suspect. Data from central Luzon (Zambales region) show large deflections in declination values (Fuller et al., 1983) which suggests that this region has undergone a large post-Eocene rotation. In addition to the Eocene data, Fuller et al. (1983) also report the results from an Oligocene site from Zambales. The data from this unit (Dec. 81 °, Inc. 9 °) show an intermediate direction from the Eocene data and suggest a clockwise rotation for the region. Since this original paper, we have sampled numerous dikes from the Zambales Ophiolite. Although the age of these dikes is uncertain, reported K-Ar age data (DeBoer et al., 1980) suggest they are middle to late Oligocene. These dikes show stable, single-component magnetizations whose mean directions (Table 1) suggest a clockwise rotation for the region. The fact that these dikes and the Oligocene sediments from Zambales are both intermediate between the present field direction and the Eocene direction from the Zambales Ophiolite suggests that the region has undergone a post-Eocene clockwise rotation. A Tertiary apparent polar wander path for the Philippines (excluding Palawan) is shown in Figure 3c.

Because there is a close correspondence between the paleolatitudinal data from the Philippine Archepelago and the Philippine Sea Plate, and the fact that both the Philippines and the Philippine Sea Plate appear to have undergone large post-Eocene clockwise rotations, it is suggested that the Philippine Arc was part of or coupled to the Philippine Sea Plate during most of the Tertiary. Presently, the Philippine Sea Plate and the Philippine Arc are separated by an active convergent boundary, the Philippine Trench (Cardwell, Isacks, and Karig, 1980). Although this trench presently decouples the central and southern parts of the Philippines from the Philippine Sea Plate, the lack of active volcanoes along the southern and central parts of this trench (Philippines Bureau of Mines, 1964; Lewis, Hayes, and Mrozowski, 1982) and the fact that the Benioff zone associated with the trench extends to only 100 km depth (Cardwell, Isacks, and Karig, 1980) suggest that subduction along this trench began sometime in the late Neogene.

The late Jurassic poles of the North Palawan Continental Terrane (McCabe et al., in press) and South China are shown in Figure 3d. An examination of this figure shows that the VGP

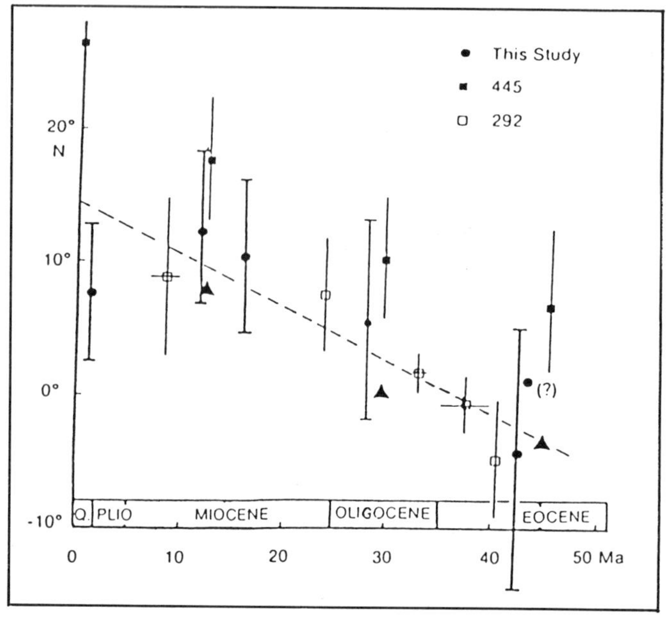


Figure 8—Comparison of the Cenozoic paleolatitudes from the Philippine Arc (Uyeda and McCabe, 1982; Fuller et al., 1983; McCabe et al., 1987) and the results from DSDP site 292 (Louden, 1977) and DSDP site 445. Because of the error limits on the Eocene data, we cannot at present constrain if the Eocene paleolatitude is north or south of the equator. In the figure, an error bar is placed on south value for Eocene paleolatitude. The Eocene paleolatitude for the north of the equator solution is shown by a point to the right of the Eocene error bar marked with a question mark. Until data are collected from more of the Tertiary section, this problem will remain unresolved. Closed triangles are latitudes of site 445 shifted southward by 10 °. Figure is from McCabe et al. (1987).

of Mindoro is displaced southward from the late Jurassic VGP of South China. The inclination data suggest that during the late Jurassic both these regions were at the same latitude (around 20 °). The data are consistent with a model that the North Palawan Continental Terrane was part of South China during the Jurassic and has rifted southward after the Jurassic.

## DISCUSSION

We have attempted to use stratigraphic and paleomagnetic data to relate Timor and the Sula Microcontinental fragment to the northern Australian continental shelf, western Sulawesi to southeastern Borneo, and the North Palawan Continental Terrane to the southeast China shelf. We have also suggested that the Philippines are a complex amalgamation of Cretaceous and younger arc-related terranes that has moved into its present position from the southeast. On the other hand, southern Palawan, the Sulu archipelago, the North Arm and the eastern portion of Sulawesi, and the inner Banda Arc represent Eocene and younger arcs and uplifted subduction melanges which were built in-place and have overprinted the region. We are therefore implying that the Banda, Celebes, and Sula Basins represent a once-continuous Cretaceous to Eocene ocean basin that has been subsequently dissected into these present-day separate entities by middle to late Tertiary tectonic processes.

Numerous workers in the past have tried to reconstruct the tectonic evolution of this region (Uyeda and Miyashiro, 1974; Uyeda and Ben-Avraham, 1972; Ben Avraham and Uyeda,

1973; Ben-Avraham, 1978). These earlier syntheses usually place a series of N-S-oriented transform fault boundaries between the Pacific and Indian Basins and place oceanic crust between these transform boundaries. By this type of reconstruction, smaller plates, such as the Philippine Sea Plate, are explained by initiation of subduction caused by changing plate motions. These models, however, fail to address other problems such as:

(1) Evidence for a Cretaceous and early Paleocene volcanic arc on Luzon Island (Balce et al., 1981; Karig, 1983), and on the islands of Cebu and Marinduque (Balce et al., 1981; McCabe, Almasco, and Yumul, 1985) all suggest that some type of subduction boundary existed in this region since the Cretaceous or even earlier.

(2) Stratigraphic and structural studies from eastern Mindanao strongly suggest that an arc-arc collision took place in the middle Tertiary (Moore et al., 1981a; Wright et al., 1981; Hawkins et al., 1985; Lewis, Hayes, and Mrozowski, 1982). Along-strike continuation of these rocks north to Samar and the Bicol (Balce et al., 1981; Karig, 1975, 1983; McCabe, Almasco, and Yumul, 1985) and south into the Molucca Sea (Silver and Moore, 1978; Hamilton, 1979; Moore et al., 1981b; Moore and Silver, 1983) suggests that this collision may be represented along the entire eastern Philippine Arc. Such a collision infers that a missing plate existed in this region during part of the Tertiary.

(3) Numerous Tertiary microplates such as the Caroline Plate (Bracey, 1975, 1983), plate boundaries such as the Ayu Trough (Weissel and Anderson, 1978), and the southern subbasin of the West Philippine Sea (Mrozowski, Lewis, and Hayes, 1982) all complicate any easy connection between the Pacific Ocean and the Indian Ocean.

Here, for simplicity, we will not attempt to address the complex problem of the relationship of the Pacific Ocean to the Indian Ocean. Instead, we will only model the region between northern Australia and southeast Asia including the Sunda Shelf. In Figures 9-13, we discuss the evolution of the southeast Asian margin through time. Simple plate-tectonic theory predicts that plate boundaries should terminate at other plate boundaries. Because of the existence of limited pre-Neogene data for portions of southeast Asia and the superposition of late Cenozoic tectonic complexities in this region, we are not able to connect all of our proposed plate boundaries. Because of this lack of understanding of some of these early plate boundaries, we show some of our plate margins dying out without the necessary connection to other margins. These terminations are marked with question marks on Figures 9-13. It is hoped that future work in the region will help to further delineate some of these boundaries and add new ones to the figure. We hope that this simple model will offer a new way to look at this complex region. This schematic representation, which is greatly simplified, does not propose to explain the effects of smaller scale tectonic events which occurred in the region.

## Cretaceous

Figure 9 is a schematic sketch of a possible arrangement of plate boundaries in the late Jurassic and Cretaceous. In the figure, Australia is located off the figure, farther to the south in its pre-rift Gondwana position. Paleomagnetic data and stratigraphic studies (discussed above) suggest that the pre-late Jurassic positions of the islands of Sula, Timor, Sumba, Ceram, and Buru were part of the northern Australian-New Guinea margin (Visser and Hermes, 1962; Audley-Charles, Carter, and Milson, 1972; Hamilton, 1973, 1978, 1979; Katili, 1975, 1978; Carter, Audley-Charles, and Barber, 1976; Barber, Audley-Charles, and Carter, 1977; Norvick, 1979). Paleomagnetic data from Mesozoic rocks from Timor (Chamalaun and Sunata, 1982) and Sumba (Otofuji et al., 1980) yield steep inclinations, which suggest that both Timor and Sumba may have formed farther to the south. A more southern origin is consistent with these islands forming a part of the northern continental margin of Australia.

Detailed analysis of magnetic anomalies off northwestern Australia detected ENE-trending anomalies of late Jurassic age (M25-M22) which become older to the south (Falvey, 1972; Sclater and Fisher, 1974; Larson, 1975; Heirtzler et al., 1978). DSDP leg 27 results (Veevers et al., 1974) report late Jurassic sediments overlying basaltic crust in the Warton Basin. These studies confirm the ages indicated by the magnetic anomalies and are consistent with the interpretation that a Mesozoic spreading system existed off northern Australia during the late Mesozoic.

The Cretaceous granitiod rocks in southern China and Vietnam (Jahn, Chen, and Yen, 1976; Hamilton, 1979) infer that a north-dipping subduction zone existed under this region during the Cretaceous. The direction and timing of this subduction suggest that an oceanic basin was being subducted under southern China. We suggest that this basin was the northern limb of the ocean basin that existed north of Australia. Paleoinclination data from Jurassic rocks collected from the eastern part of the North Palawan Continental Terrane (McCabe et al., 1987) suggest that Palawan and southern China shared a common paleolatitude. In addition, oil company proprietary data show that the Cretaceous basement of the North Palawan Continental Terrane is subduction related, which is also consistent with this terrane being located along the South China margin during this period.

Mesozoic paleomagnetic data from Malaya, Borneo, and western Sulawesi (McElhinny, Haile, and Crawford, 1974; Haile, 1978; Haile, McElhinny, and McDougall, 1977) suggest that these regions were a single tectonic unit which was far to the north of Australia and hence not a portion of southern Gondwana. Cretaceous plutonism in both southeastern Borneo and Sulawesi (Hamilton, 1978, 1979) suggest that an east- to southeast-facing subduction system existed under southeastern Sunda during the Cretaceous. This subduction is shown in Figure 9, where we rotated Borneo and Sulawesi back into their precounterclockwise rotation position. In this model, we are faced by the same problem that confronted other workers—how to relate the subduction systems between Borneo and southern China. In this figure, we follow other workers (e.g., Holloway, 1982) and link the two north-dipping subduction zones by placing a 1000-km transform fault between the two systems.

The Cretaceous age mafic arc volcanism, marine sediments, and dioritic plutonism on Cebu indicate that the Philippine Arc (or fragments of the present Philippine Arc) was likely an oceanic island arc system during the late Mesozoic. Although a limited amount of Cretaceous paleomagnetic data from the Philippines suggest a more equatorial latitude for the Cretaceous arc than its present position (McCabe et al., 1987), these data are still very limited and hence do not allow us to infer the Cretaceous position of the Philippine Arc. At least

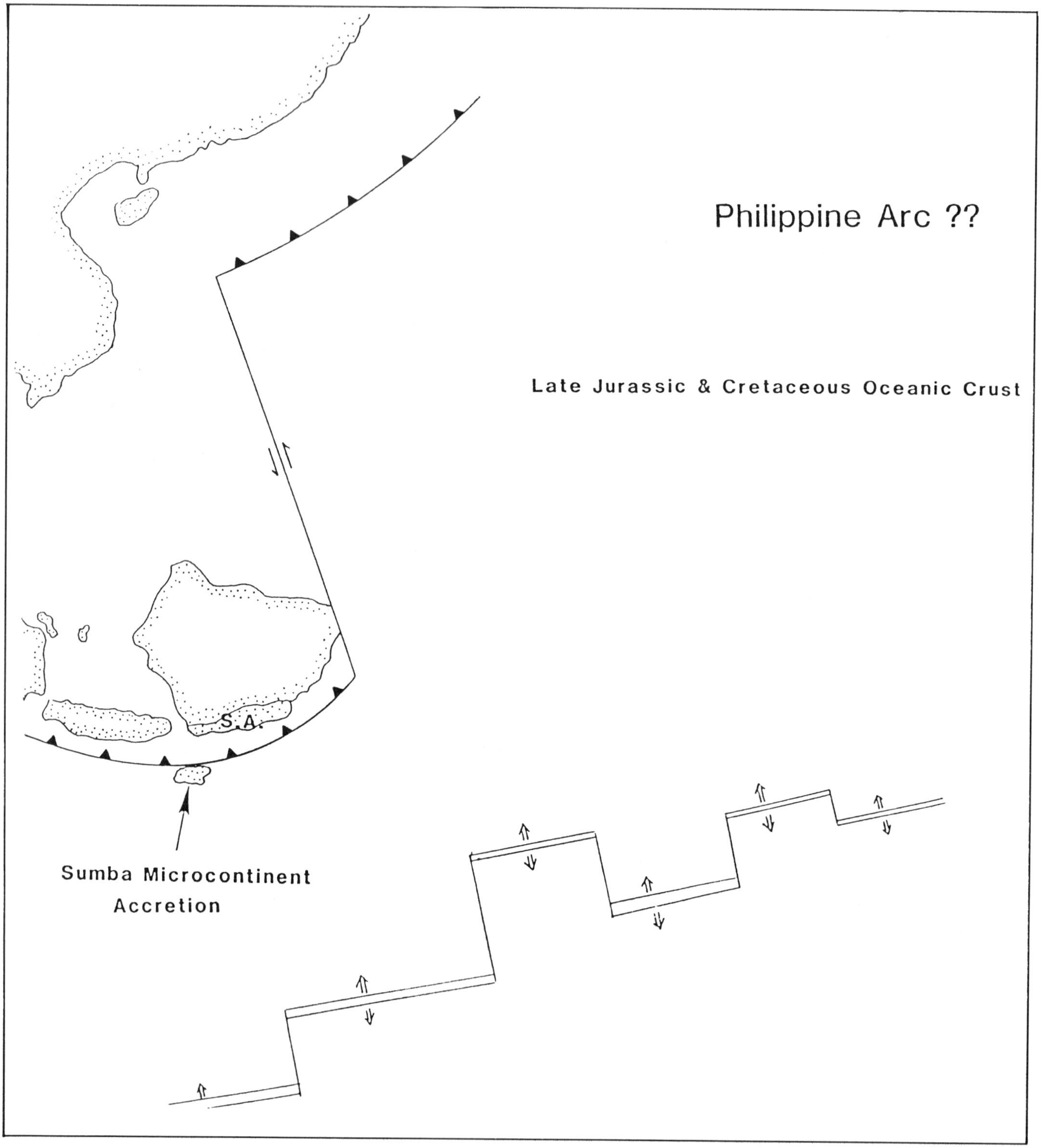


Figure 9—Cretaceous reconstruction of the southeast Asian margin.

two scenarios are presently possible for the Philippine Arc. Scenario one has been suggested by Holloway (1982) and recently by Stephan et al., (1986) and Mitchell, Hernandez, and Dela Cruz (1986). These workers continue the subduction zone under southern Borneo eastward to include the Cretaceous proto-Philippine Arc. This scenario requires that the Philippine Arc rotated counterclockwise during the Tertiary and also requires that the continental arc (Borneo) continues eastward and becomes an oceanic arc (Philippine Arc) to the east. The second possible origin is that the Philippine Arc existed as an arc somewhere east of the Mesozoic ocean basin. As we have no constraints on the polarity of the Cretaceous subduction system in the Philippines (McCabe, Almasco, and Ymul, 1985) and do not have enough paleomagnetic data to better constrain the Mesozoic translational history of the archipelago, we are unable to judge between these two possible origins. However, since northeastern Borneo and South Palawan are characterized by Tertiary continental related sediments (Hamilton, 1979), while the Philippine arc lacks early to middle Tertiary continental sediments, we favor the second model suggesting that the Philippine arc and Borneo were not a part of the same plate boundary.

## Eocene

In the early Tertiary (Figure 10), Australia begins to separate from Antarctic (Weissel and Hayes, 1972; Weissel, Hayes, and Herron, 1977). Subduction begins under northern Borneo and southern Palawan. As a result of this subduction, Borneo begins its counterclockwise rotation (Haile, McElhinny, and McDougall, 1977) and consumes Paleocene oceanic crust in the proto-South China Sea Basin. The Banda, Celebes, Sulu Basin spreading center has migrated northward from its Cretaceous position, and is at equatorial latitudes. Eocene paleomagnetic data from the Philippines (Uyeda and McCabe, 1982; Fuller et al., 1983; McCabe et al., 1987) suggest that the western part of the present-day Philippine Arc is near equatorial latitudes and has undergone a large clockwise rotation of greater than 90°. Paleoinclination data from the Philippines and the Philippine Sea Plate show a similar northward drift for these two regions from Eocene to present. This similar northward drift between the Philippines and Philippine Sea Plate (Uyeda and McCabe, 1982), combined with the fact that both areas share a common clockwise rotation, suggests that the Philippines formed the western margin of the Philippine Sea Plate during the mid-Tertiary. Rotating the Philippine Arc back to this Eocene position, as suggested by the paleomagnetic data, shows that the arc lies to the southeast of its present position. We suggest that a north-dipping subduction zone developed along the southern boundary of this old arc (Figure 10). Uyeda and McCabe (1982) suggested a similar scenario in which they proposed that this subduction developed in response to a late Eocene collision in the Bicol portion of Luzon or was caused by the late Eocene reorganization of the Pacific Plate. This model is consistent with models proposed by other authors, who suggest a middle Tertiary collision along the present eastern boundary of the Philippines (Karig, 1975; Balce et al., 1981; Moore and Silver, 1983; McCabe, Almasco, and Yumul, 1985; Hawkins et al., 1985).

## Oligocene

Australia continues its northward drift during this period (Figure 11). The Makassar Strait, between Borneo and Sulawesi, begins opening as a back-arc basin behind the eastward-advancing South Arm of Sulawesi during the middle Tertiary (Hamilton, 1979). This rifting is evidenced by: (1) the occurrence of transgressive marine clastic sedimentary rocks in the Kutai Basin of eastern Borneo (van Bemmelen, 1949; Hamilton, 1979); (2) the occurrence of early to middle Tertiary volcanic rocks on the South Arm of Sulawesi (van Bemmelen, 1949; van Leeuwen, 1975; Katili, 1978; Sukamto, 1975a; Hamilton, 1979); (3) a Tertiary subduction melange in eastern Sulawesi (Hamilton, 1979; Katili, 1978); and (4) middle Tertiary north-trending extensional faulting in the eastern margin of the Java Sea (Bishop, 1980). Paleomagnetic studies from Borneo (discussed above) suggest that Borneo was continuing its counterclockwise rotation until some time late in the Tertiary. The timing of the rotation is further supported by stratigraphic data. Hamilton (1979) shows that Cretaceous and Tertiary granitoid and melange belts in northern Borneo are deflected in a counterclockwise sense from the main northwest-southeast Sunda trend.

In the Philippines, granitoid plutonism of Oligocene and Miocene age is restricted along the western margin of Luzon and the central Philippines (McCabe et al., 1985) during this period. This suggests that an eastward-dipping subducting slab occurring along the western margin of the archipelago during this time. Magnetic anomalies in the South China Sea (Taylor and Hayes, 1980, 1983) and published stratigraphic studies from the Palawan shelf (Holloway, 1982) indicate that the North Palawan Continental Terrane began its southward migration during this period. This southward migration is also supported by paleomagnetic data from Jurassic rocks in southern Mindoro (McCabe et al., 1987).

## Early to Middle Miocene

Australia continues its northward drift (Figure 12). The Banda Sea is segmented by the development of a large E-W shear along the northern edge of the Australia-India Plate (Silver and Smith, 1983). Buton, the Banda Ridges (Silver et al., 1985), and the Sula Microcontinent (Hamilton, 1979; Silver and Smith, 1983) translate westward toward eastern Sulawesi. The westward translation of these blocks isolates the South Banda Basin from the North Banda Basin and begins to separate the Celebes and Banda Seas. Seismic-reflection data, lack of seismic activity in Southern Palawan and northwest Borneo, possible Miocene thrusting of ophiolites in South Palawan (Hamilton, 1979), and the occurrence of highly deformed pre-middle Miocene rocks overlain by undeformed late Neogene sedimentary and limestone sequences on South Palawan (Balce et al., 1981) suggest that the Palawan Trench became inactive during this period. Although the cause of inhibition of this subduction system is unknown, recent tectonic models of the region infer that a Miocene collision event occurred between this trench and submerged continental blocks in the South China Sea (Hamilton, 1979; Ludwig et al., 1979; Taylor and Hayes, 1980, 1983). As a result of the cessation of subduction in the Palawan Trench, the counterclockwise rotation of Borneo was inhibited, as suggested by the Miocene paleomagnetic data (Schmidtke et al., 1985).

Magnetic anomaly data in the South China Sea show that spreading was inhibited during this period (Taylor and Hayes, 1980, 1983). The timing of the inhibition of spreading, the age of suggested collisions along the margin of South Palawan, and the collision of the North Palawan Continental Terrane with the eastward-dipping subduction system located along the western margin of the Philippine Arc suggest that these collisions may have inhibited South China Sea spreading (Uyeda and McCabe, 1983; McCabe et al., 1982a; Fuller et al., 1983). As a result of this collision, the northward drift of the Philippines was inhibited (Uyeda and McCabe, 1982a), as suggested by paleoinclination data.

## Late Miocene to Recent

Australia begins to collide with the eastern edge of the Banda Arc (Figure 13). Timor is formed along the northern edge of this collision zone, either as an uplifted "rootless" accretionary prism (Hamilton, 1978, 1979); Von der Borch, 1979) or as a series of nappes which are made from parts of the downgoing plate, including the Australian margin and

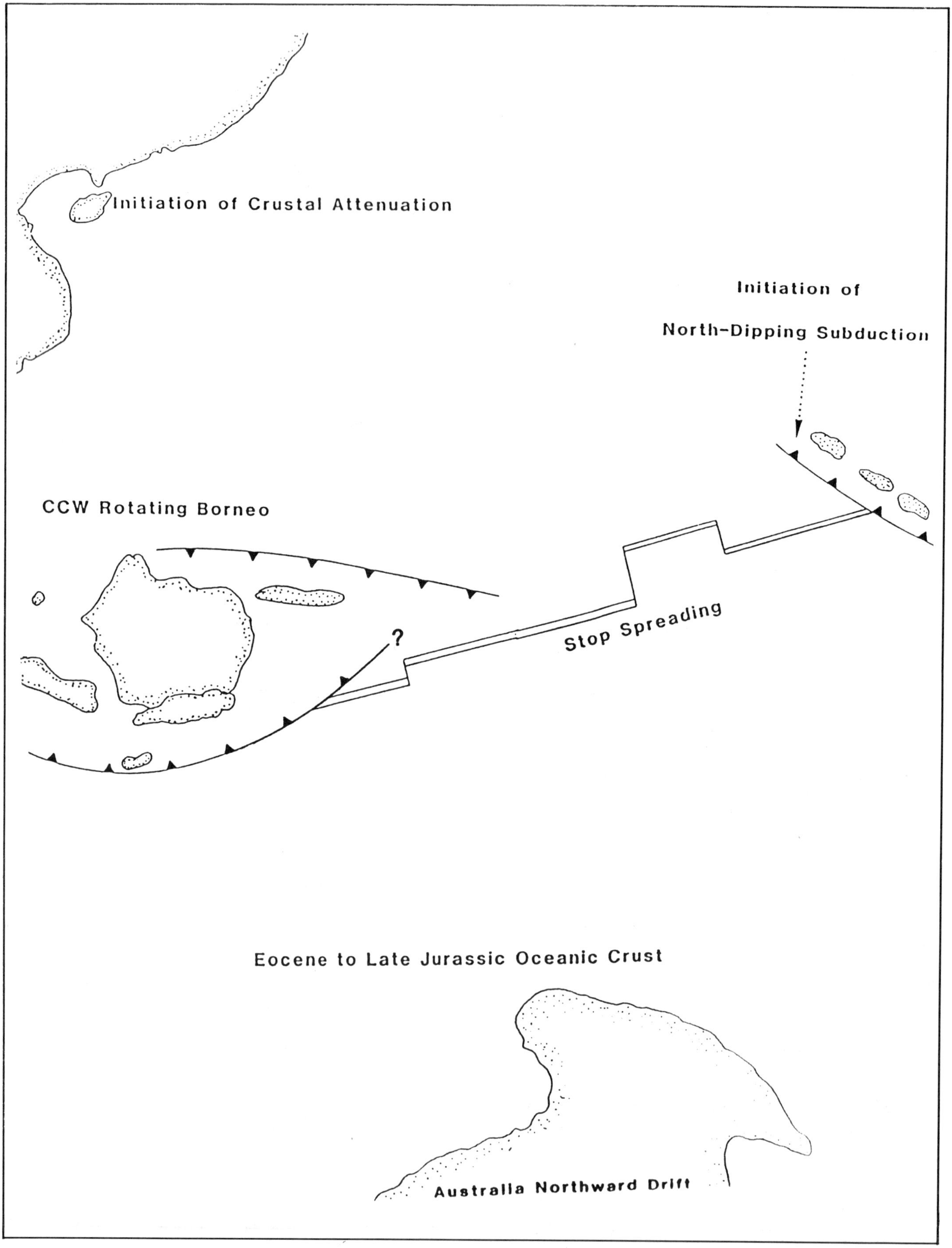


Figure 10—Eocene reconstruction of the southeast Asian margin.

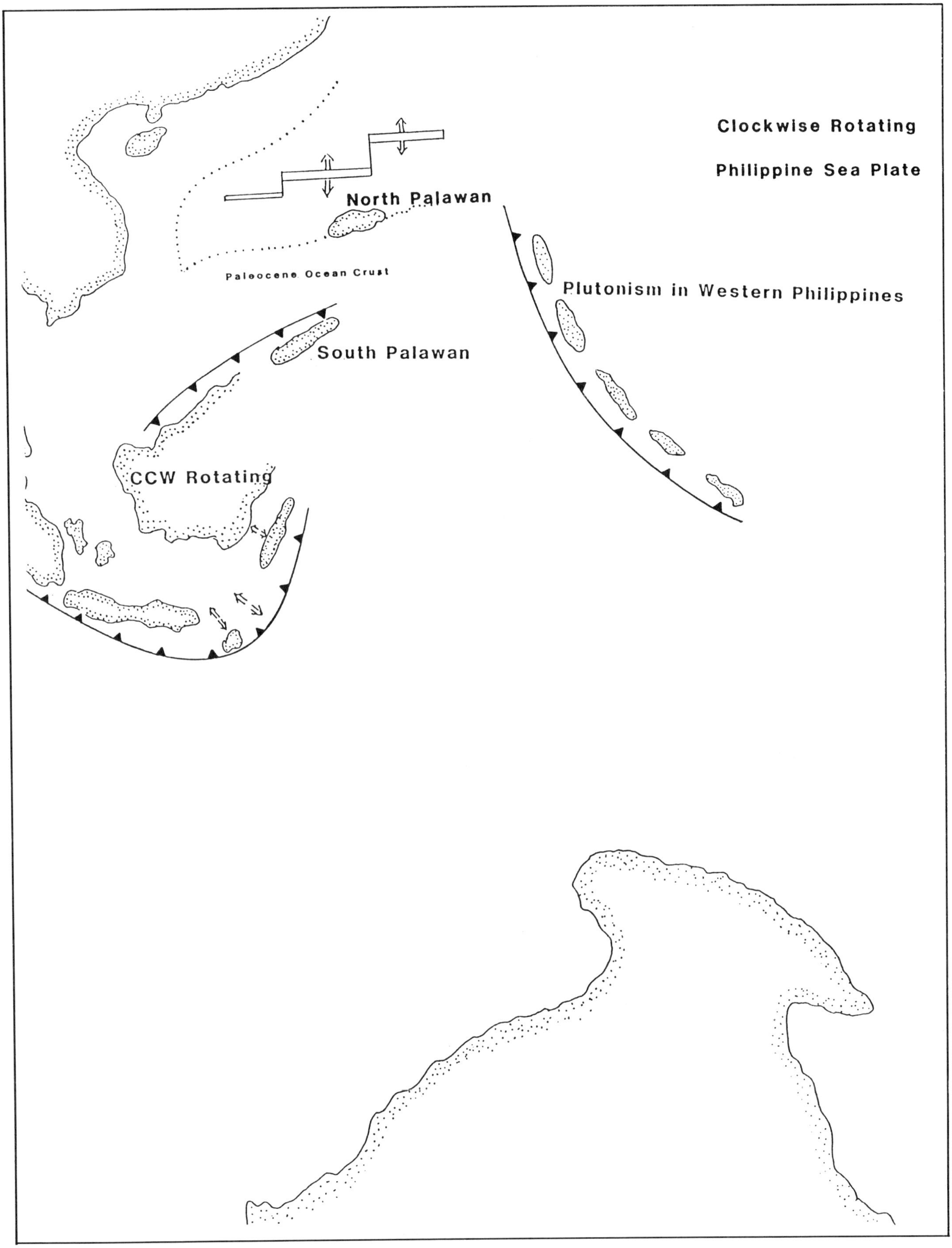


Figure 11—Oligocene reconstruction of the southeast Asian margin.

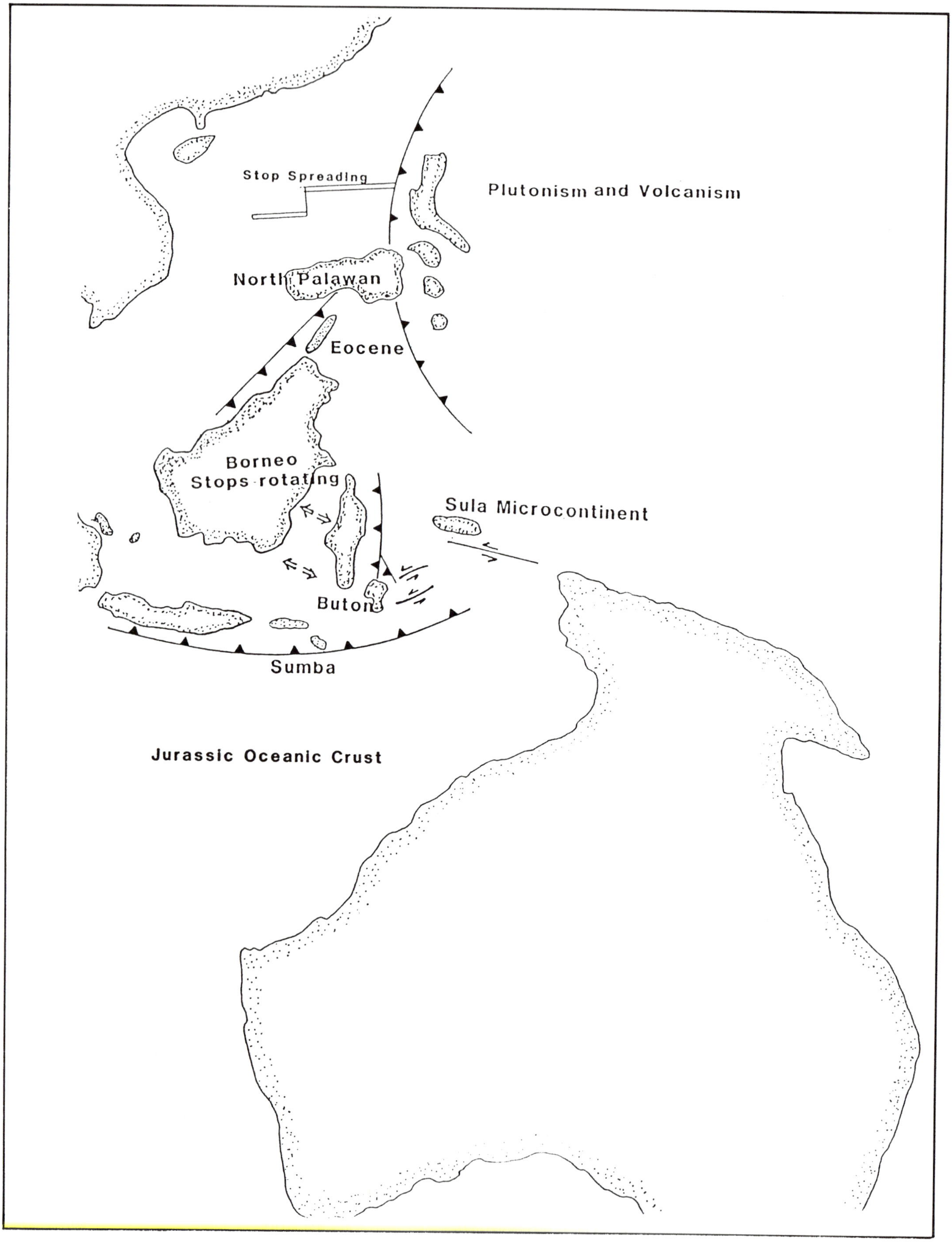


Figure 12—Early to middle Miocene reconstruction of the southeast Asian margin.

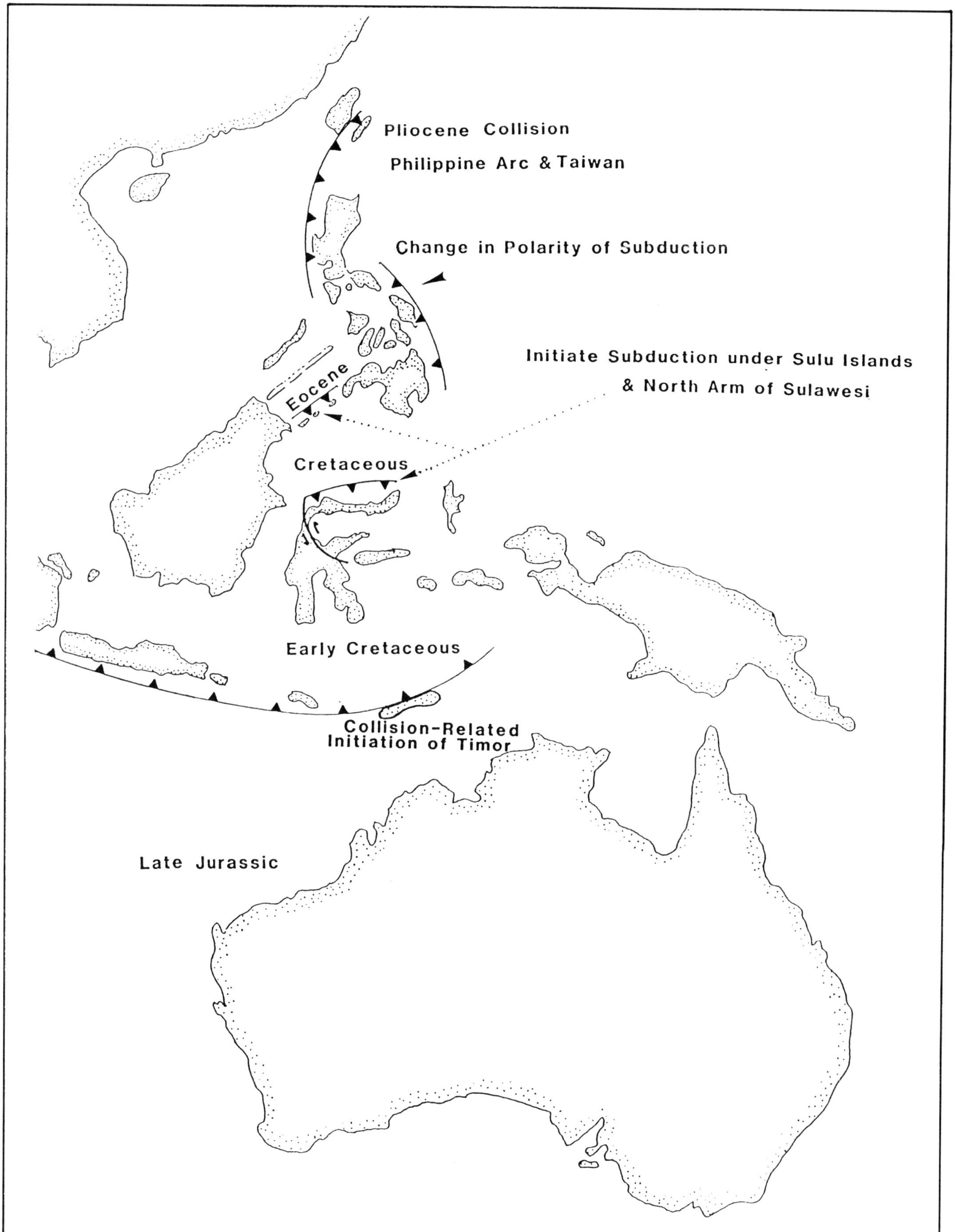


Figure 13—Late Miocene to Recent configuration of the southeast Asian margin.

parts of the accretionary prism of the upper plate (Audley-Charles, 1983, 1986). In the northwestern portion of the Banda Sea, the collision between the Sula Microcontinent and eastern Sulawesi is in its later stages (Visser and Hermes, 1962; Hamilton, 1979; Silver and Smith, 1983). As a result of this collision, ophiolitic bodies are thrust eastward toward the microcontinent, forming the Southeast and East Arms of Sulawesi (Hamilton, 1979). As a further result of this collision, the polarity of the subduction zone flips, and southward-dipping subduction begins under the North Arm (Silver et al., 1983). These two events—the translation and collision of the Sula Microcontinent with Sulawesi and the resulting change from an east-dipping subducting slab to a north-dipping slab and south-dipping subduction slab—isolated the Banda and Celebes Seas. To the north, the collision of the North Palawan Continent Terrane with the western Philippine Arc results in a reversal of polarity in the Philippines (Uyeda and McCabe, 1982). This reversal history is recorded in the age of volcanic rocks throughout the archipelago. Plutonic and volcanic rocks of middle to late Tertiary ages are generally restricted to the western side of the archipelago. The eastern edge of the Philippines is characterized by pre-middle Tertiary igneous basement, which is unconformably overlain by late Miocene to Recent volcanic rocks (Bales et al., 1981; McCabe, Almasco, and Yumul, 1985). The reversal in the polarity of the subduction trenches, which is still in progress, is evidenced by a gradual slowing of subduction along the western margin of the Philippines (along the Manila-Negros Trench systems) and initiates subduction along the eastern margin (Philippine Trench) of the archipelago. The occurrence of late Neogene (Philippines Bureau of Mines, 1964) volcanic rocks in the Sulu Archipelago that are built upon an early Tertiary ophiolitic basement suggest that subduction began in late Neogene in this region. As a result of this subduction, the Sulu Sea became isolated from the Celebes Sea.

## REFERENCES CITED


Allen, G.A., L.G.G. Pearce, and W.E. Gardner, 1978, A regional interpretation of the Browse Basin: APEA Journal, v. 18, pt. 1, p. 23-33.

Almasco, J., and M.D. Fuller, 1984, Paleomagnetic studies in Luzon, N. Philippines: Eos, American Geophysical Union Transactions, v. 65. p. 862.

Andal, D.R., 1966, A. report on the discovery of fusulinids in the Philippines: Philippine Geologist, v. 20, p. 14-22.

Audley-Charles, M.G., 1965, Permian paleogeography of the northern Australia-Timor region: Paleogeography, Paleoclimatology, Paleoecology, v. 1, p. 297-305.

Audley-Charles, M.G., 1968, The geology of Portuguese Timor: Geological Society of London Memoir 4, 76 p.

Audley-Charles, M.G., 1983, Reconstruction of eastern Gondwanaland: Nature, v. 306, p. 48-50.

Audley-Charles, M.G., 1985, The Sumba enigma: Is Sumba a diapiric fore-arc nappe in process of formation?: Tectonophysics, v. 119, p. 435-449.

Audley-Charles, M.G., 1986, Rates of Neogene and Quaternary tectonic movements in the Southern Banda Arc based on micropaleontology: Journal of the Geological Society of London, v. 143, p. 161-175.

Audley-Charles, M.G., D.J. Carter, and J.S. Milson, 1972, Tectonic development of eastern Indonesia in relation to Gondwana dispersal: Nature, v. 239, p. 35-39.

Balce, G.R., et al., 1981, Metallogenesis in the Philippines: Geological Survey of Japan Report 261, p. 125-148.

Barber, A.J., M.G. Audley-Charles, and D.J. Carter, 1977, Thrust tectonics of Timor: Geological Society of Australia Journal, v. 24, p. 51-62.

Ben-Avraham, Z., 1978, The evolution of marginal basins and adjacent shelves in east and southeast Asia: Tectonophysics, v. 45, p. 269-288.

Ben-Avraham, Z., and S. Uyeda, 1973, The evolution of the China Basin and the Mesozoic paleogeography of Borneo: Earth and Planetary Science Letters, v. 18, p. 365-376.

Bishop, W.F., 1980, Structure, stratigraphy, and hydrocarbons offshore southern Kalimantan, Indonesia: AAPG Bulletin, v. 64, p. 37-58.

Bracey, D.R., 1975, Reconnaissance geophysical survey of the Caroline Basin: GSA Bulletin, v. 86, p. 775-784.

Bracey, D.R., 1983, Geophysics and tectonic development of the Caroline Basin, U.S. Naval Oceanographic Office Technical Report TR-283, 75 p.

Cardwell, R.K., B.L. Isacks, and D.E. Karig, 1980, The spatial distribution of earthquakes, focal mechanism solutions, and subducted lithosphere in the Philippines and northeastern Indonesian islands, *in* D.E. Hayes, ed., The tectonic and geologic evolution of southeast Asian seas and islands: American Geophysical Union Geophysical Monograph 23, p. 1-35.

Carter, D.J., M.G. Audley-Charles, and A.J. Barber, 1976, Stratigraphical analysis of island arc-continental margin collision in eastern Indonesia: Geological Society of London Journal, v. 132, p. 179-198.

Chamalaun, F.H., 1977a, Paleomagnetic evidence for the relative positions of Timor and Australia in the Permian: Earth and Planetary Science Letters, v. 34, p. 107-112.

Chamalaun, F.H., 1977b, Paleomagnetic reconnaissance result from the Maubisse Formation, East Timor, and its tectonic implication: Tectonophysics, v. 42, p. T17-T26.

Chamalaun, F.H., and W. Sunata, 1982, The paleomagnetism of the western Bondon Arc system, *in* J. Briden, ed., Paleomagnetic research in southeast and east Asia: Bangkok, Thailand, CCOP Project Office, p. 162-194.

Crostella, A.A., 1977, Geosynclines and plate tectonics in Banda Arcs, eastern Indonesia: AAPG Bulletin 61, p. 2063-2081.

DeBoer, J., et al., 1980, The Bataan orogene: Eastward subduction, tectonic rotations, and volcanism in the western Pacific (Philippines): Tectonophysics, v. 67, p. 251-282.

Embleton, B.T., 1981, A review of paleomagnetism of Australia and Antarctica, *in* M.W. McElhinny and D.A. Valencio, eds., Paleoreconstruction of the Continents: American Geophysical Union, 194 p.

Falvey, D.A., 1972, Sea-floor spreading in the Wharton Basin (northeast Indian Ocean) and the breakup of eastern Gondwanaland: Australian Petroleum Exploration Association Journal, v. 12, pt. 2, p. 86-88.

Fuller, M.D., 1985, Paleomagnetism in an accretionary margin Luzon, northern Philippines: Journal of Geodynamics, v. 2, p. 141-158.

Fuller, M.D., R. McCabe, I. Williams, A. Zanoria, J. Almasco, and R. Encina, 1983, Paleomagnetism of Luzon, *in* D.E. Hayes, ed., the tectonic and geologic evolution of southeast Asian seas and islands: American Geophysical Monograph Series 27, p. 79-94.

Gollner, E.R.D., 1924, Versiag over de vitkomsten van mijnbouwkundig - geologische ondersoekingan op Poeloe Laoet: Jaarbock Mijnwezen Nederlandsch Oost-Indie, 50, pt. 1, p. 4-52.

Gribi, E.A., Jr., 1973, Tectonics and oil prospects of the Moluccas, eastern Indonesia: Geological Society of Malaysia Bulletin, v. 6, p. 11-16.

Haile, N.S., 1979, Rotation of Borneo microplate completed by Miocene: Paleomagnetic evidence: Warta Geologi 5, p. 19-22.

Haile, N.S., 1978, Reconnaissance paleomagnetic results from Sulawesi, Indonesia, and their bearing on paleogeographic reconstructions: Tectonophysics, v. 46, p. 77-85.

Haile, N.S., M.W. McElhinny, and I. McDougall, 1977, Paleomagnetic data and radiometric ages from the Cretaceous of West Kalimantan (Borneo), and their significance in interpreting regional structure: Geological Society of London Quarterly Journal, v. 133, p. 133-144.

Hamilton, W., 1973, Tectonics of the Indonesian region: Geological Society of Malaysia Bulletin, v. 6, p. 3-10.

Hamilton, W., 1978, Tectonic map of the Indonesia region: U.S. Geological Survey Miscellaneous Investigations Series Map 1-875-D, 1:5,000,000.

Hamilton, W., 1979, Tectonics of the Indonesia region: U.S. Geological Survey Professional Paper 1078, 345 p.

Hashimoto, W., 1981, Geologic development of the Philippines, *in* T. Kobayashi et al., eds., Geology and paleontology of southeast Asia (vol. 22): University of Tokyo Press, p. 83-170.

Hashimoto, W., and T. Sato, 1973, Geologic structure of North Palawan and its bearing on the geological history of the Philippines, *in* T. Kobayashi, et al., eds., Geology and paleontology of southeast Asia (vol. 13): University of Tokyo Press, p. 145-161.

Hawkins, J.W., G.F. Moore, R. Villamor, C. Evans, and E. Wright, 1985, Geology of the composite terranes of east and central Mindanao, *in* D. G. Howell, Circum-Pacific Council for Energy and Mineral Resources Earth Science Series Vol. 1, p. 437-463.

Heirtzler, J.R., et al., 1978, The Argo abyssal plain: Earth and Planetary Science Letters, v. 41, p. 21-31.

Hermes, J.J., 1968, The Papuan geosynclines and the concept of geosynclines: Geologie en Mijnbouw, v. 47, p. 81-97.

Hermes, J.J., 1974, West Irian, *in* A.M. Spencer, ed., Mesozoic-Cenozoic orogenic belts: Geological Society of London Special Paper 4, p. 475-490.

Holloway, N.H., 1982, North Palawan block, Philippines—Its relation to Asian mainland and role in evolution of South China Sea: AAPG Bulletin 56.

Hsu, I., 1972, Magnetic properties of igneous rocks in the northern Philippines: St. Louis, Mo., Washington University, unpublished Ph.D. dissertation.

Jahn, B.M., P.Y. Chen, and T.P. Yen, 1976, Rb-Sr ages on granitic rocks in southeastern China and their tectonic significance: GSA Bulletin 86, p. 763-776.

Jarrard, R., and S. Sasajima, 1980, Paleomagnetic synthesis for southeast Asia: Constraints on plate motions *in* D. E. Hayes, ed., The tectonic and geologic evolution of southeast Asian seas and islands: American Geophysical Union, Geophysical Monograph Series 23, p. 293-316.

Karig, D.E., 1975, Basin genesis in the Philippine Sea: Initial Reports of the Deep Sea Drilling Project, v. 31, p. 857-880.

Karig, D.E.., 1983, Accreted terranes in the northern part of the Philippine Archipelago: Tectonics, v. 2, p. 211-236.

Katili, J.A., 1975, Volcanism and plate tectonics in the Indonesian island arcs: Tectonophysics, v. 26, p. 165-188.

Katili, J.A., 1978, Past and present geotectonic position of Sulawesi, Indonesia: Tectonophysics, v. 45, p. 289-322.

Kaye, P., 1972, Untitled discussion: APEA Journal, v. 12, pt. 2, p. 53-54.

Krause, D.C., 1965, Submarine geology north of New Guinea: GSA Bulletin, v. 76, p. 27-42.

Larson, R.L., 1975, Late Jurassic sea-floor spreading in the eastern Indian Ocean: Geology, v. 3, p. 69-71.

Lee, C.S., and R.J. McCabe, 1986, The Banda-Celebes-Sulu Basin: A trapped piece of Cretaceous-Eocene oceanic crust?: Nature, v. 322, p. 51-54.

Lewis, S.D., D.E. Hayes, and C.L. Mrozowski, 1982, The origin of the West Philippine Sea Basin by inter-arc spreading, *in* G.R. Balce and A.S. Zanoria, eds., Geology and tectonics of the Luzon-Marianas region: Philippine Seatar Committee Special Publication, v. 1, p. 31-51.

Louden, K., 1977, Paleomagnetism of DSDP sediments, phase shifting of the magnetic anomalies and rotation of the west Philippine Basin: Journal of Geophysical Research, v. 82, p. 2989-3002.

Ludwig, W.J., et al., 1979, Profiler-Sonobuoy measurements in the South China Sea Basin: Journal of Geophysica Research, v. 84, p. 3505-3518.

McCabe, R., 1984a, Implications of paleomagnetic data on the collision related bending of island arcs: Tectonics, v. 3, p. 409-428.

McCabe, R., 1984b, Paleomagnetism of the Central Philippines and of Late Neogene rocks from Luzon, University of Tokyo, unpublished Ph.D. dissertation, 225 p.

McCabe, R., and S. Uyeda, 1983, Hypothetical model for the bending of the Mariana arc, *in* D.E. Hayes, ed.: American Geophysical Union Geophysical Monograph 27, p. 281-293.

McCabe, R., J. Almasco, and W. Diegor, 1982a, Geologic and paleomagnetic evidence for a possible Miocene collision in western Panay, central Philippines, Geology, v. 10, p. 32-329.

McCabe, R., J. Almasco, and G. Yumul, 1985, Terranes of the Central Philippines, *in* D.H. Howell, ed., Tectonostratigraphic terranes of Circum-Pacific Region: Circum-Pacific Energy and Mineral Resources Earth Science Series Vol. 1, p. 421-436.

McCabe, R., J. Almasco, H. Shibuya, W. Zanoria, H. Pagauitan, and M. Torii, 1982b, Evidence against late Neogene rotations of the island of Luzon, *in* J. Briden ed., Paleomagnetic research in southeast and east Asia: Bangkok, Thailand, CCOP Project Office, p. 117-129.

McCabe, R., E. Kikawa, J. Cole, A. Malicse, P. Baldauf, J. Yumul, and J. Almasco, 1987, Paleomagnetic results from Luzon and the central Philippines, Journal of Geophysical Research, v. 92, p. 555-580.

McCaffrey, R., E.A. Silver, and R.W. Raitt, 1981, Seismic refraction studies in the east arm, Sulawesi-Banggai Islands region of eastern Indonesia, *in* A.J. Barber, and S. Wiryosuyono, eds., The geology and tectonics of eastern Indonesia: Indonesian Geological Research and Development Centre Special Publication 2, p. 321-325.

McElhinny, M.W., N.S. Haile, and A.R. Crawford, 1974, Paleomagnetic evidence shows Malay Peninsula was not a part of Gondwanaland: Nature, v. 252, p. 641-645.

Mitchell, A.H.G., F. Hernandez, and A.P. Dela Cruz, 1986, Cenozoic evolution of the Philippine Archipelago: Journal of Southeast Asian Earth Science, v. 1, p. 3-22.

Moore, G.F., and E.A. Silver, 1983, Collision processes in the northern Malucca Sea, *in* D.E. Hayes, ed., The tectonic and geologic evolution of southeast Asian seas and islands: American Geophysical Monograph Series 27, p. 360-372.

Moore, G.F., et al., 1981a, Geology of southern Mindanao, Philippines: EOS (American Geophysical Union Transactions), v. 62, p. 1086.

Moore, G.F., et al., 1981b, Geology of the Talaud Islands, Molucca Sea collision zone, northeast Indonesia: Journal of Structural Geology, v. 3, p. 467-475.

Mrozowski, C.L., S.D. Lewis, and D.E. Hayes, 1982, Complexities in the tectonic evolution of the West Philippine Basin: Tectonophysics, v. 82, p. 1-24.

Norvick, M.S., 1979, The tectonic history of the Banda Arcs, eastern Indonesia, A review: Geological Society of London Journal, v. 136, p. 519-527.

Otofuji, Y., S. Sasajima, S. Nishimura, A. Dharma, and F. Hehuwat, 1981, Paleomagnetic evidence for clockwise rotation of the northern arm of Sulawesi, Indonesia: Earth and Planetary Science Letters, v. 54, p. 272-280.

Philippine Bureau of Mines, 1964, Geological map of the Philippines: Scale 1:1,000,000, 9 sheets.

Pigram, C.J., Surono, J.B. Supandjono, 1985, Origin of the Sula Platform, eastern Indonesia: Geology, v. 13, p. 246-248.

Powell, D.E., 1976, The geological evolution of the continental margin off northwest Australia: APEA Journal, v. 16, pt. 1, p. 13-23.

Ranneft, T.S.M., R.M. Hopkins, A.J. Froelich, and J.W. Gwinn, 1960, Reconnaissance geology and oil possibilities of Mindanao: AAPG Bulletin, v. 44, p. 529-568.

Sarewitz, D.R., and D.E. Karig, (in press), Processes of allocthonous terrane evolution, Mindoro Island, Philippines: Tectonics.

Sasajima, S., S. Nishimura, K. Hirooka, Y. Otofuji, T. van Leeuwen, and F. Hehuwat, 1980, Paleomagnetic studies combined with fission-track datings on the western arc of Sulawesi, East Indonesia: Tectonophysics, v. 64, p. 163-172.

Schmidtke, E., J.R. Dunn, M.D. Fuller, and R. Haston, 1985, Preliminary paleomagnetic results from Sabah, East Malaysia: Eros (American Geophysical Union Transactions, v. 66, p. 864.

Schmidtke, E., J. Almasco, J.R. Dunn, M.D. Fuller, R. Haston, and R. Landgraf, 1986, Paleomagnetic results from Sabah and Sarawak, East Malaysia: AAPG Bulletin, v. 70, p. 934.

Sclater, J.G., and R.L. Fisher, 1974, Evolution of the east central Indian Ocean, with emphasis on the tectonic setting of the Ninetyeast Ridge: GSA Bulletin 85, p. 683-702.

Silver, E.A., and J.C. Moore, 1978, The Molucca Sea collision zone, Indonesia: Journal of Geophysical Research, v. 83, p. 1681-1691.

Silver, E.A., and R.B. Smith, 1983, A comparison of terrane accretion in modern southeast Asia and the Mesozoic North American cordillera: Geology, v. 11, p. 198-202.

Silver, E.A., R. McCaffrey, and R.B. Smith, 1983, Collision, rotation, and the initiation of subduction in the evolution of Sulawesi, Indonesia: Journal of Geophysical Research, v. 88, p. 9407-9418.

Silver, E.A., J.B. Gill, D. Schwartz, H. Prasetyo, and R.A. Duncan, 1985, Evidence for a submerged and displaced continental borderland, North Banda Sea, Indonesia: Geology, v. 13, p. 687-691.

Stephan, J.F., R. Blanchet, C. Rangin, J. Pelletier, J. Letouzey, and C. Muller, 1986, Geodynamic evolution of the Taiwan-Luzon-Mindoro belt since late Eocene: Tectonophysics, v. 125, p. 245-268.

Sukamto, R., 1975a, Geologic map of Indonesia, Ujung Pandang: Indonesia Geological Survey, scale 1:1,000,000. Sheet 8.

Sukamto, R., 1975b, The structure of Sulawesi in the light of plate tectonics: Association of Indonesian Geologists, Regional Conference on Geology and Mineral Resources of Southeast Asia, Jakarta, 15 p.

Taylor, B., and D.E. Hayes, 1980, The tectonic evolution of the South China Sea Basin, *in* D.E. Hayes, ed., The tectonic and geologic evolution of southeast Asian seas and islands: American Geophysical Union Geophysical Monograph 23, p. 89-104.

Taylor, B., and D.E. Hayes, 1983, Origin and history of the South China Sea Basin, *in* D.E. Hayes, ed., The tectonic and geologic evolution of southeast Asian seas and islands: American Geophysical Union, Geophysical Monograph Series 27, p. 23-56.

Teichert, C., 1970, Marine fossil invertebrate faunas of the Gondwana region: Second Gondwana Symposium, Pretoria 1970, Proc. and Papers, South Africa Council for Scientific and Industrial Research, p. 125-138.

Uyeda, S., and Z. Ben-Avraham, 1972, Origin and development of the Philippine Sea: Nature, Phys. Sci., 240, p. 176-178.

Uyeda, S., and A. Miyashiro, 1974, Plate tectonics and the Japan Islands: A synthesis: GSA Bulletin 85, p. 1159-1170.

Uyeda, S., and R. McCabe, 1982, A possible mechanism of episodic spreading of the Philippine Sea, *in* M. Hashimoto and S. Yueda, eds., Accretion tectonics in Circum-Pacific regions: Tokyo, Terra Scientific Publishing Company, p. 291-306.

van Bemmelen, R.W., 1949, The geology of Indonesia: The Hague, Netherlands, Govt. Printing Office, 732 p.

van Leeuwen, T.M., 1975, Geological investigation of KP 190 (Biru) south Sulawesi: P.T. Rio Tinto Indonesia (unpublished report), 34 p.

Veevers, J.J., 1969, Sedimentology of the Upper Devonian and Carboniferous platform sequence in the Bonaparte Gulf Basin: Australia Bureau of Mineral Resources, Geology and Geophysics Bulletin 109, 86 p.

Veevers, J.J., D.A. Falvey, L.V. Hawkins, and W.J. Ludwig, 1974, Seismic reflection measurements of northwest Australian margin and adjacent deeps: AAPG Bulletin 58, p. 1731-1750.

Visser, W.A., and J.J. Hermes, 1962, Geological results of the exploration for oil in Netherlands New Guinea: Konikl, Nederlands Geol. Mijnbouw Genoot., Verh., Geol. Ser., 20, 265 p.

Von der Borch, C.C., 1979, Continent-island arc collision in the Banda Arc: Tectonophysics, v. 54, p. 169-193.

Waterhouse, J.B., 1972, The evolution, correlation, and paleogeographic significance of the Permian ammonoid family Cyclolobidae: Lethaia, v. 5, p. 251-270.

Weissel, J.K., and D.E. Hayes, 1972, Magnetic anomalies in the southeast Indian Ocean, *in* D.E. Hayes, ed., Antarctic oceanology II: Australia-New Zealand Sector: American Geophysical Union Antarctica Research Series 19, p. 234-249.

Weissel, J.K., and R.N. Anderson, 1978, Is there a Caroline Plate?: Earth and Planetary Science Letters, v. 41, p. 143-158.

Weissel, J.K., D.E. Hayes, and E.M. Herron, 1977, Plate tectonic synthesis: The displacements between Australia, New Zealand, and Antarctica since late Cretaceous: Marine Geology, v. 25, p. 231-277.

Wright, E., et al., 1978, East Mindanao ophiolite belt: Petrology of volcanic series rocks: EOS (American Geophysical Union Transactions), v. 62, p. 1086.

CHAPTER 31

# SOUTHEAST ASIA: A TECTONIC TRIPTYCH

**Richard W. Murphy**
*Walton-On-Thames, United Kingdom*

ABSTRACT

The geology of Southeast Asia is so complicated in detail that the conceptual imperative is to simplify. As a first order approximation, I sketch some critical elements of three moments in the geotectonic history of Southeast Asia—Upper Paleozoic, Middle Cretaceous, and Holocene.

Upper Paleozoic tilloids extending roughly north-south from at least 22° N in Burma to south of the Equator in Sumatra were originally deposited at high southern latitudes in a general east-west orientation in eastern Gondwanaland along the north rim of present-day Australia. Subsequent northward drift of several tens of degrees and clockwise rotation of about 90° impose a geometric constraint on all possible reconstructions of Southeast Asia. The Tethyan elements of Southeast Asia have an overall northward-drifting, clockwise-rotating history.

Paleozoic-Mesozoic magmatic arcs with genetically related subduction complexes, ophiolites, microcontinents, and suture zones can be traced from the Assam Syntaxis southeastward through Southeast Asia to a position in medial Borneo, where they are truncated. A north-south Central Borneo Transform, slicing off the ends of these arcs and transporting them left-laterally into the northern Pacific region, was terminated by the Meratus Suture, an arc-ophiolite system which was obducted in late Cenomanian time onto the region between present-day west Java and eastern Borneo. Subsequent arcs project east of Borneo into eastern Indonesia, signifying the cessation of major left-lateral transform movement in central Borneo.

Holocene Southeast Asia is an exceedingly complex appendage to the southeastern margin of Eurasia. With the northern attachment of Eurasia as a relatively fixed boundary, the other three boundaries to the continental Sunda core of Cretaceous age are spongy, soft-margin compressional zones with large amounts of strike-slip faulting. A gross oversimplification of the current tectonics characterizes the western margin as a right-lateral transform margin, the eastern margin as left lateral, and the southern margin, facing the northward-drifting Australian continental landmass, as compressional. While rejecting any rigid plate solution to Southeast Asia, I have included a "cartoon" as a device to gain insight into current regional tectonics.

## INTRODUCTION

This is more an essay than a scientific publication. I was fortunate enough to be asked to fill a suddenly vacant slot at the Circum-Pacific Conference in Singapore in August 1986, and the ensuing cerebral activity coupled with the stimulus of the papers being presented at the conference generated this paper. After 25 years of wrestling with the geology of Southeast Asia I have come to recognize that the critical element in understanding Southeast Asia is not how much data you accept, but what kinds of information you discount and what kinds you emphasize. I am focusing on three distinct Phanerozoic time periods because I believe that the insight obtained by understanding these three moments sheds much light on the overall development of the region and places positive and necessary constraints on speculation about other alternative scenarios.

Because of the origin of this paper and its formulation as an essay, customary documentation of much of the material introduced—for example, in Figures 2 and 6—is omitted.

## UPPER CARBONIFEROUS

Stauffer (1983) reviewed the Paleozoic history of Southeast Asia and pointed out two salient elements:

(1) the existence of a Lower Paleozoic core of shelfal to oceanic sedimentary suites extending from west to east across the Malay Peninsula (Jones, 1968). The shelfal facies contain quartzose sediments which could have only been derived from an adjacent Precambrian continental landmass. No such landmass is now adjacent nor apparent.
(2) the significance of a prominent zone of pebbly mudstones, extending from Burma to Sumatra, and newly reinterpreted to be of glacial origin.

The conclusion is that Southeast Asia is a detached continental fragment which developed during the Early to Middle Paleozoic as a margin to a Precambrian continental craton, underwent an episode of glaciation in Carboniferous time,

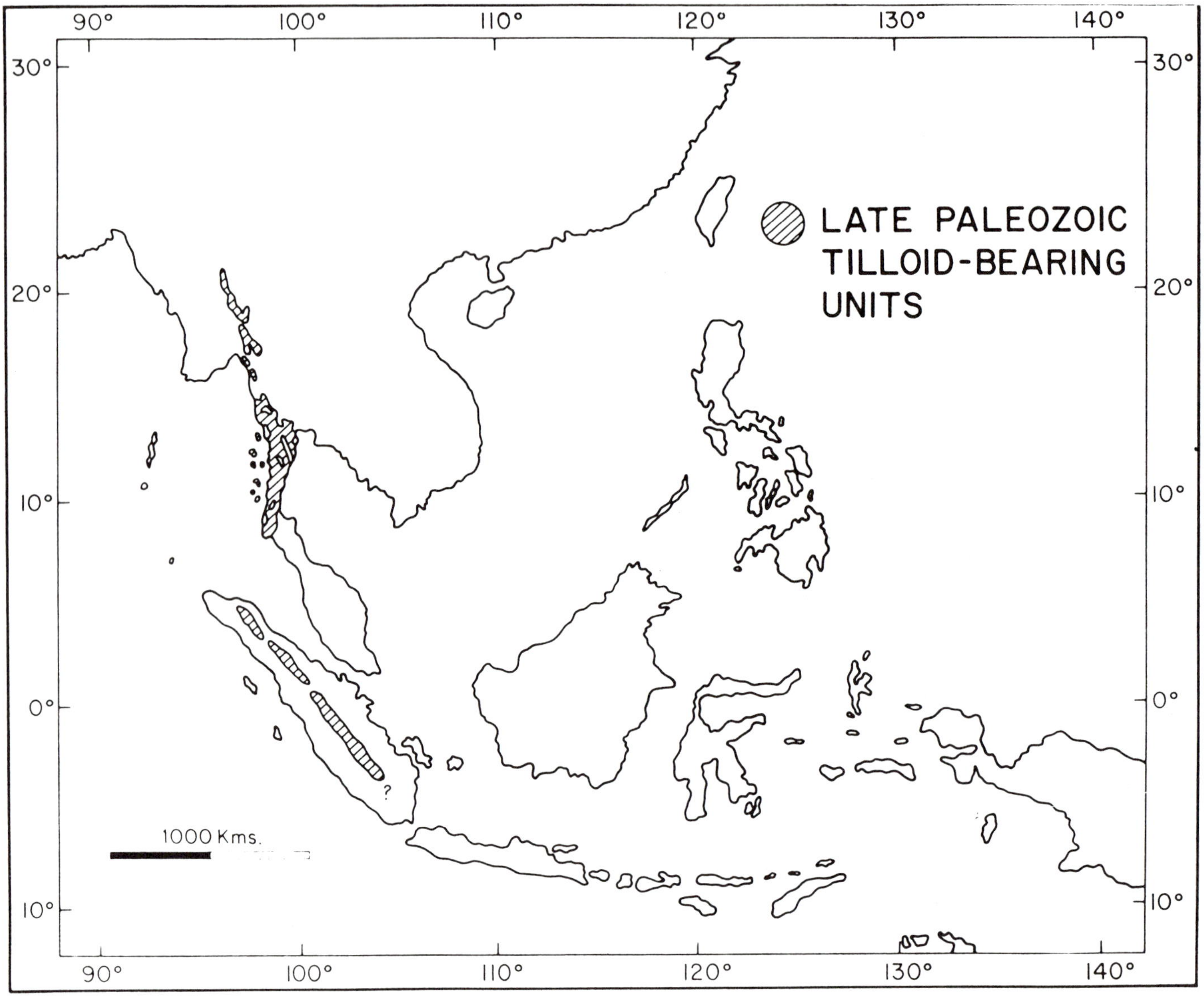


Figure 1. Late Paleozoic tilloid-bearing units in Southeast Asia.

separated from the parent continent, and has since migrated to its present position as an allochthonous mass. Furthermore, since the present outcrop of tilloid-bearing units stretches over 26° of latitude from Burma to Sumatra (Figure 1), and since a glacial deposit could not have been formed in a north-south orientation of such magnitude, a large degree of rotation—from an original latitudinal orientation to its present longitudinal orientation—is required.

The dual requirements of a Precambrian parent and an east-west orientation, as well as the history of the Tethys oceans (Senghör, 1984), leads to the inescapable conclusion that the core of Southeast Asia was originally attached to eastern Gondwanaland (Audley-Charles, 1983) and has migrated subsequently through the various Tethys oceans to its present position in Southeast Asia.

These observations on the Carboniferous status of the core of Southeast Asia, if correct, invoke an argument of great simplicity and elegance about the subsequent history of Southeast Asia. In essence, it fingerprints the overall development of the entire Tethyan wing of the Southeast Asia complex. No amount of internal rotations, small plate activity, or local plate tectonic solutions can negate the gross historical story—northward movement of about 50-60° and clockwise rotation of about 90°.

## CENOMANIAN

The history of Southeast Asia subsequent to the deposition of the Carboniferous tilloids is one of an episodic accretion of magmatic arcs with their attendant ophiolites, flysch and molasse sequences, and derivative pyroclastic sedimentary rocks. These arcs are numbered and depicted on Figure 2 and listed in Table 1.

In a larger context, the visible Southeast Asian arcs are but extensions of similar arc elements which can be traced along the southern margin of Eurasia from the Crimea in the west to Borneo in the east (Stöcklin, 1977, Senghör, 1984). Although individual correlations are very difficult, the overall continuity is quite clear. The southern margin of Eurasia has been a compressional margin since the Late Paleozoic, and dozens or scores of arcs and microcontinents have been ripped off from northeastern Gondwanaland and accreted to Eurasia after passing through various versions of the Tethys Ocean (Senghör, 1984).

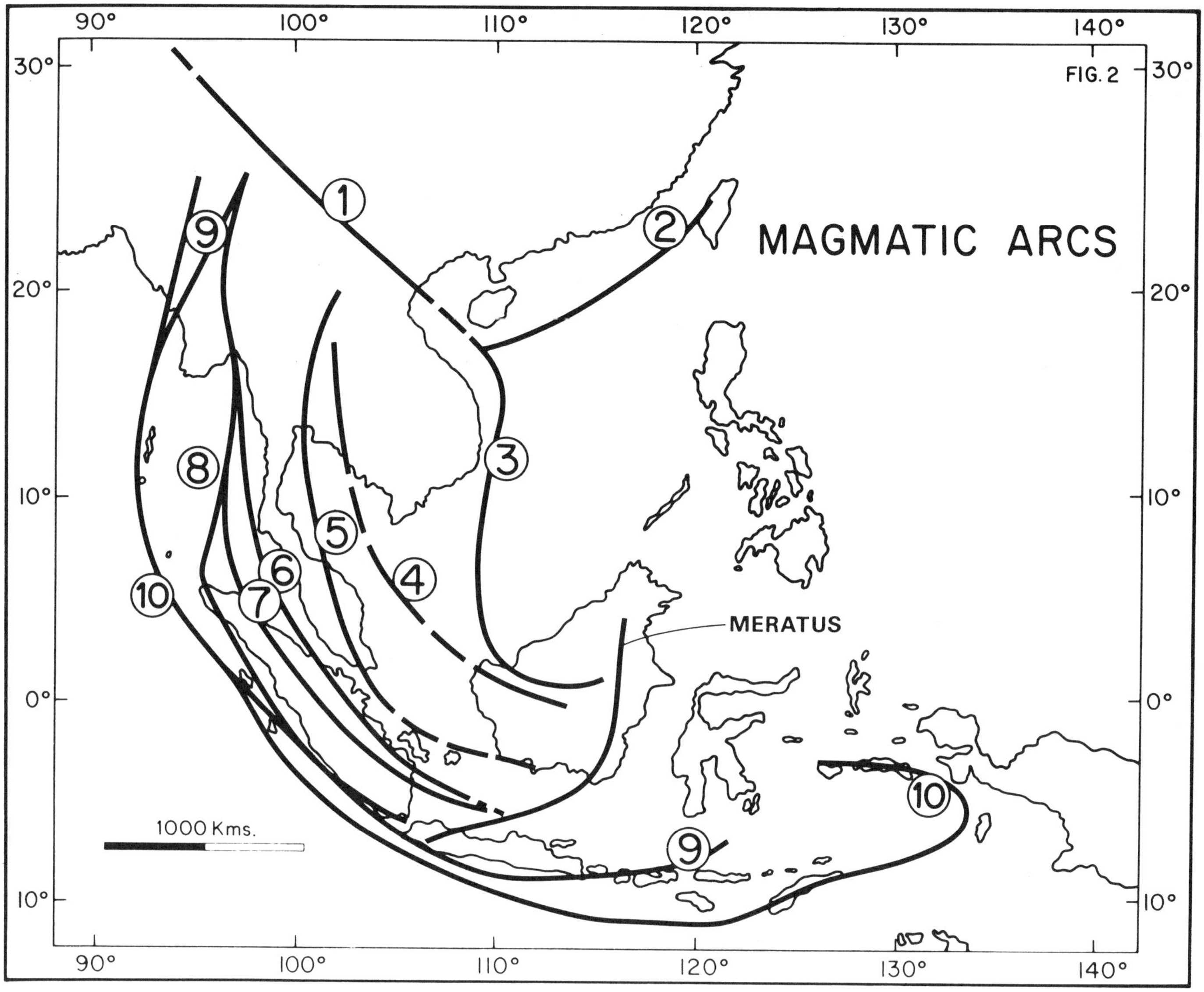


Figure 2. Magmatic arcs in Southeast Asia (see Table 1).

The Southeast Asian arcs are the easternmost remaining extensions of these great arc systems. They appear to terminate in central Borneo in a collage of north-trending Cretaceous to Tertiary melange terranes.

One of the fundamental questions in the geology of Southeast Asia is the relationship between the older, east-trending arcs and the younger, north-trending melanges. I think the key lies in the Meratus Mountain system, an arc which collided with Boreno in the late Cenomanian or early Turonian. Documentation for this has been provided by the Doctoral dissertation of Nafrizal Sikumbang (1986), whose field mapping has established that the Meratus ophiolite complex was tectonically emplaced over volcanic turbidites which carry Cenomanian fossils. The ophiolite is overlain unconformably by Turonian sandstone with fragments of the ophiolite.

The Meratus Mountains plunge northward under the very thick Cenozoic sedimentary rocks of the Kutei Basin, and their northward continuation cannot be traced with certainty. Toward the southwest, however, melange outcrops on the Karamunjava Islands in the Java Sea, on the island of Java west of Jogjakarta, and also in the southwestern corner of Java appear to belong to the same arc system (Hamilton, 1979).

If these scattered outcrops are parts of a single arc system—an assumption supported by basement trends in the Java Sea (Hamilton, 1979, Figure 46)—they define a Cenomanian arc system 1500 km long, which trends northeast-southwest obliquely across the east-trending older arcs.

An interpretation of these relationships is shown in Figures 3 and 4. The arcs shown in Figure 3 are arcs 1 to 7 of Figure 2, but Southeast Asia has been rotated back and reinserted into a more internal position with respect to South China and Eurasia. This device enables us to visualize more accurately the probable south-facing orientation of Southeast Asia during the Cenomanian, prior to the collision of India and Eurasia (Tapponier et al., 1982).

A hypothetical Central Borneo Transform fault truncates all the older arcs at an angle of about 90°. From regional considerations the most likely sense of movement along the fault was left lateral, slicing off the eastern ends of the older arcs and transporting them as rafted remnants northward, where they are presumably lodged in the accumulation of terranes in

**Table 1** **SOUTHEAST ASIAN MAGMATIC ARCS**

| Arc Number | Name | Age |
|---|---|---|
| 1. | Red River | Upper Triassic Suture<br>Oligo-Miocene strike slip zone |
| 2. | Southeast China | Mesozoic |
| 3. | East Vietnam - Lupar | Cretaceous-Eocene |
| 4. | East Thai - West Borneo | Carboniferous - Late Triassic |
| 5. | West Thai - East Malaya | Ordovician - Triassic |
| 6. | East Burma - Mergui | Mesozoic |
| 7. | West Burma - Sumatra | Early Cretaceous |
| 8. | Mergui - West Sumatra | Late Creatceous |
| 9. | Burma - Andaman - Barisan | Late Creatceous - Tertiary |
| 10. | Andaman - Mentawai - Banda Sea | Miocene - Holocene |

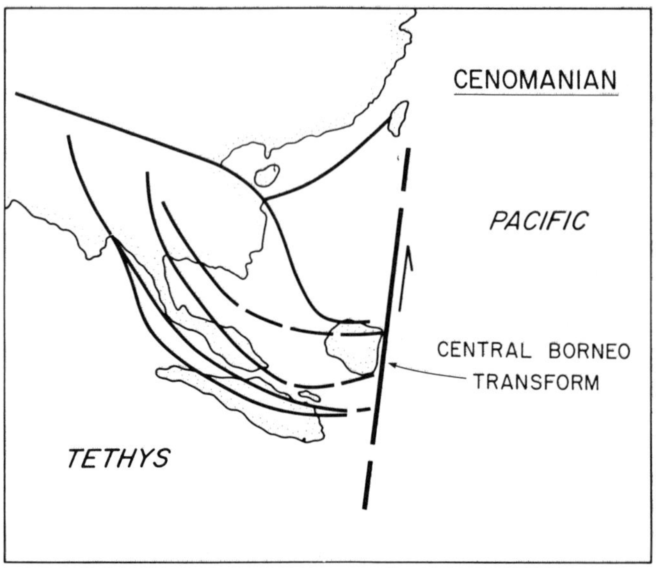

Figure 3. Cenomanian tectonics, Southeast Asia.

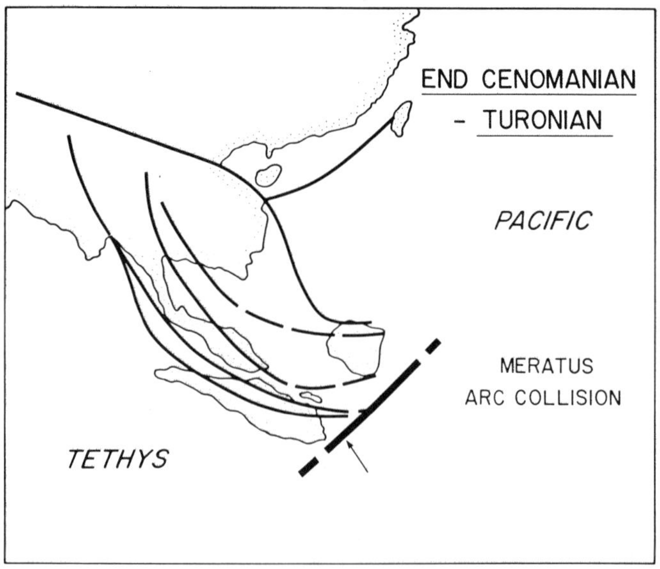

Figure 4. End of Cenomanian and Turonian tectonics, Southeast Asia.

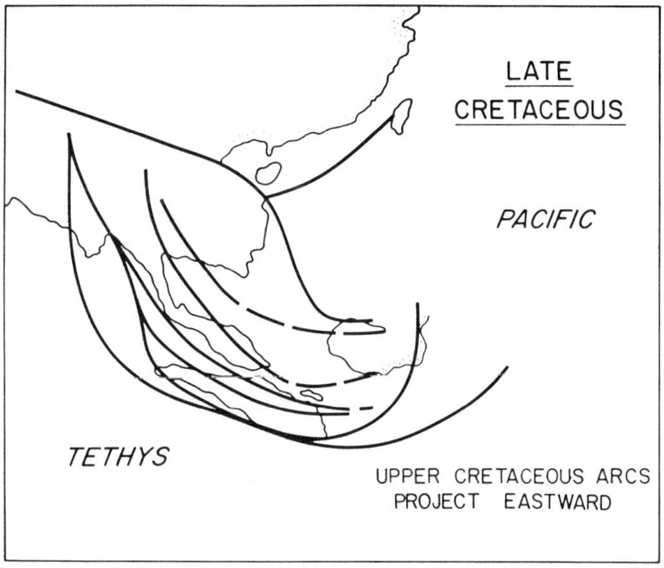


Figure 5. Late Cretaceous tectonics, Southeast Asia.

the north Pacific margin. Since the arcs were the products of activity within the Tethyan realm, and the left-lateral movement along the Central Borneo Transform fault is essentially a Pacific phenomenon, the fault itself marks the boundary between Tethys and the Pacific.

The arrival of the Meratus arc in the late Cenomanian or early Turonian (Figure 4) is interpreted as the event which sealed off the Central Borneo Transform. All arcs younger than Turonian (Figure 5) continue around Borneo into eastern Indonesia.

Implied in the reconstruction portrayed here is a plate motion reorganization on a regional, if not worldwide, scale. Unpublished (but public domain) work at Exxon Production Research Company in the late 1970s postulated such a worldwide plate reorganization at about 93 Ma, in the Cenomanian. Such an event would fit neatly with the thesis put forth here, assuming that plate motion vectors as well as the timing are suitable.

## HOLOCENE

We complete this essay by a selective examination of some salient active features of regional tectonics.

Figure 6 is a compendium of some of the major compres-

Figure 6. Major transcurrent faults in Southeast Asia.

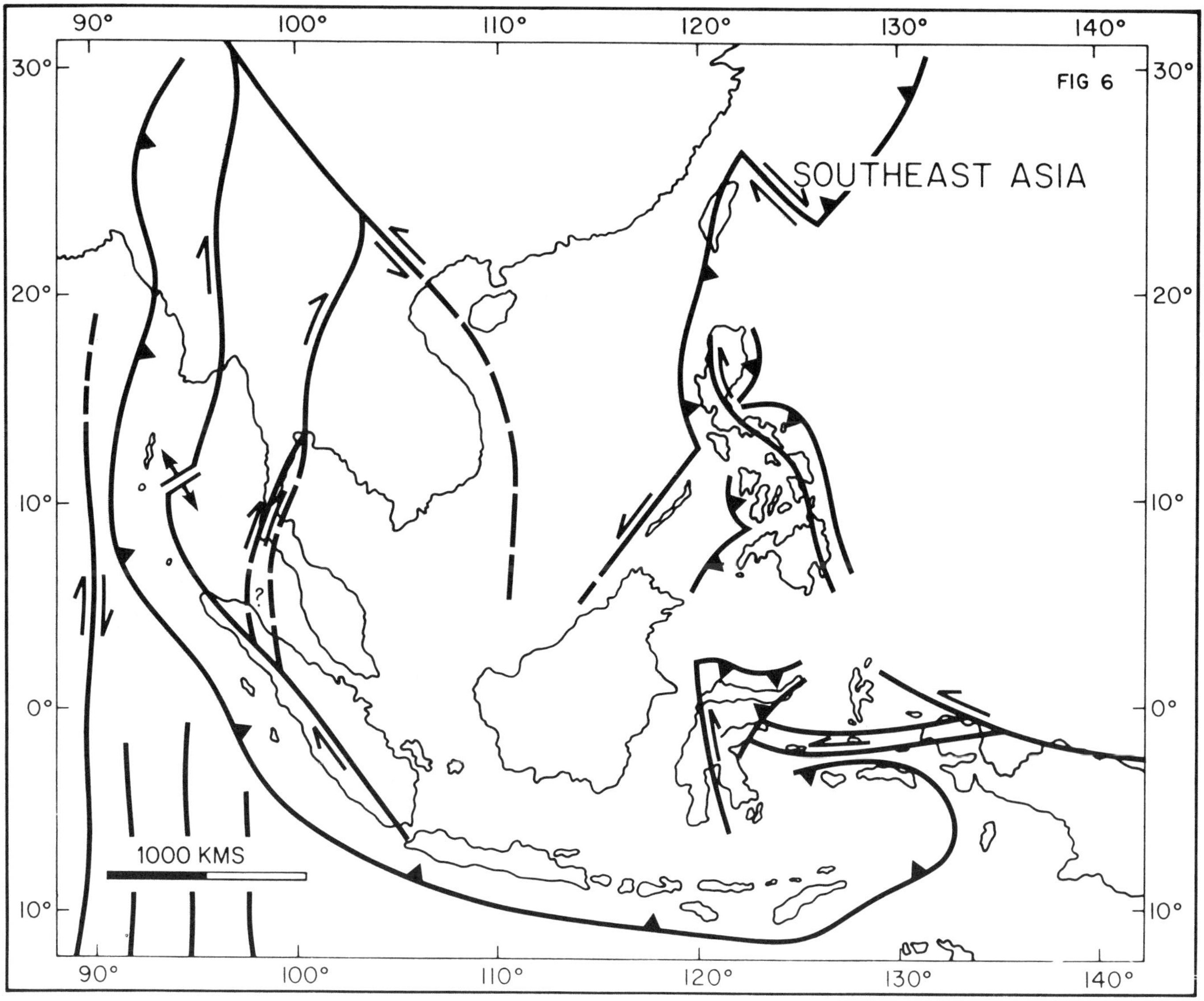

sional margins and throughgoing wrench faults in Southeast Asia. Particular emphasis is placed on the requirement that the wrench faults be "throughgoing," that is, that they are long, continuous features which serve as platelet or block boundaries, or which merge with each other to form tectonic boundaries of regional significance.

If this compilation is reasonably correct, one important observation emerges—all of the throughgoing faults west of the Sunda Strait (between Sumatra and Java) are right lateral; all those east of it are left lateral. The entire western margin of Southeast Asia—a zone as much as 500 km wide—is riddled with long, subparallel right-lateral faults. The right-lateral sense of offset is synthetic with the relative plate motion between the Indian plate and Southeast Asia during the northward progression of India from Gondwanaland to its collision with Eurasia. They are genetically related.

The collision of India with Eurasia, the subsequent crustal thickening, and the indentation and extrusion (after the model of Tapponier et al, 1982) have resulted in profound and deeply penetrating effects upon western Southeast Asia. Continuing compression between India and Eurasia causes right-lateral strike-slip faulting and probable clockwise rotation of internal segments.

These marginal effects have been accentuated by the extrusion of Southeast Asia. As the core of Sunda is extruded, it rotates clockwise, making the angle of convergence with respect to an essentially northward-moving India/Australian plate progressively more acute. As the angle of convergence decreases, right-lateral strike-slip faulting is emphasized over sheer compression. The north-striking wrench faults bite deeper into the Sunda core. As crustal blocks are detached and shoved toward the north, extensional opening of the Andaman Sea results.

The opposite margin, facing the Pacific Ocean, is a much more complex assemblage of island arcs and marginal seas together with active plate boundaries, which are both compressional and transform. Between Taiwan and New Guinea, however, the throughgoing wrench faults are all left lateral in offset, a mirror image to the right-lateral faults west of the Sunda Strait. Since there is no active spreading along this margin today, the driving forces must be the west-northwestward Pacific plate motion and the northward-converging Indo-Australian plate approach.

Figure 7 is a "cartoon" suggested by the relationships briefly sketched above. No one, certainly not myself, would maintain that Southeast Asia is in any sense a rigid plate, but if Southeast Asia *were* a rigid plate attached as a projecting appendage to Eurasia, and if the adjacent margins of Eurasia were compressional—as they are (India-Tibet collision on the West and Ryukyu-Japan Trench on the East)—then the only stable plate boundary configuration would be fault-trench-fault, as shown in Figure 7.

If there is any validity in this concept, it suggests a surprising degree of coherence in a region which is arguably the foremost example in the world of currently active nonrigid plate tectonics.

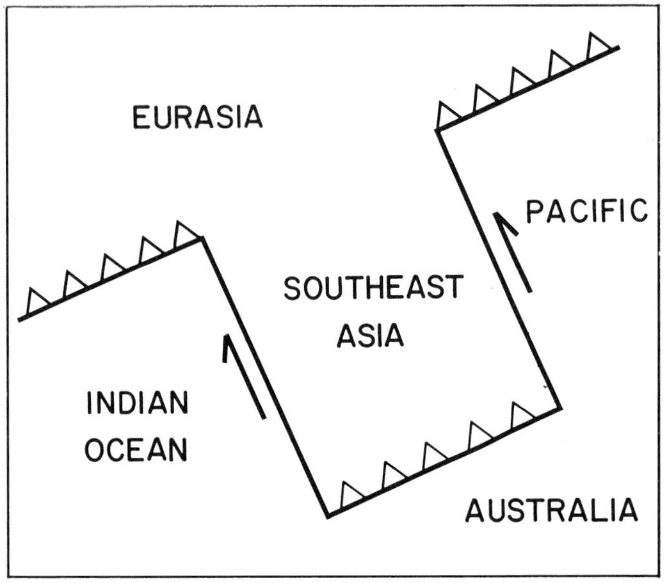

Figure 7. Cartoon of rigid-plate solution for Southeast Asia.

The three panels of the triptych are now open. I hope that the suggestions made here concerning the origin, internal construction, and current tectonic activity of Southeast Asia will help provide a framework for further research in this geologically fascinating corner of the world.

## REFERENCES CITED

Audley-Charles, M.G., 1983, Reconstruction of eastern Gondwanaland: Nature, v. 306, p. 48-50.

Hamilton, W., 1979, Tectonics of the Indonesian region: U.S. Geological Survey Professional Paper 1078, 345 p.

Jones, C. R., 1968, Lower Paleozoic rocks of Malay Peninsula: AAPG Bulletin, v. 52, p. 1259-1278.

Senghör, A.M. Celâl, 1984, The Cimmeridge orogenic system and the tectonics of Eurasia: GSA Special Paper 195,82p.

Sikumbang, N., 1986, Geology and tectonics of the Pre-Tertiary rocks in the Meratus Mountains, Southeast Kalimantan, Indonesia: Egham, Surrey, UK, Royal Holloway and Bedford New College, unpublished Ph.D. dissertation.

Stauffer, P.H., 1983, Unraveling the mosaic of Paleozoic crustal blocks in Southeast Asia: Geologische Rundschau, v. 72, p. 1061-1080.

Stöcklin, J., 1977, Structural correlation of the Alpine ranges between Iran and Central Asia: Memo. h. ser. Soc. geol. France, n. 8, p. 333-353.

Tapponier, R., et al., 1982, Propagating extrusian tectonics in Asia: new insights from simple experiments with plasticine: Geology, v. 10, p 611-616.

CHAPTER 32

# MARGINAL DEFORMATION OF CRUSTAL PLATES AS A KEY TO CRUSTAL MOTION, CRUSTAL SPIRALS, AND THE DRIVING FORCE

**Barry G.M. Wood**[1]
*Marathon International Petroleum (G.B.) Ltd.*
*London, United Kingdom*

## ABSTRACT

Plate boundaries are of three types—compressional, extensional, and shear—and can be recognized respectively by tectonic assemblages associated with subduction, rifting, and buckle-folding. Boundary type is related to plate motion.

The boundaries of five plates are examined and relative and absolute motions are considered. Each plate is found to have a component of northwest absolute motion.

The earth's plates line up end to end to form continuous northwesterly spiralling segmented bands encircling the globe, the lateral margins of which form the well-documented crustal geosutures. Segments (plates) of these crustal ribbons are of different sizes and composition, and a direct relationship between size and extent of deformation is recognized. The Pacific and Eurasian Plates dominate in size, marginal deformation, and mobility. The Pacific Plate is the most mobile, the Eurasian Plate the least.

The southwest margins of these two plates form the transpressional Tethyan system across which segments of more southerly tails of the crustal spirals—specifically the African, Indian, and Australian Plates—have been drawn north and spun counterclockwise in response to the northward buckle-folding of the crustal strip containing the Pacific and Eurasian Plates. The crest of the fold lies approximately north of the Indian Plate. Northwest drift of the Pacific Plate is driving the system.

The recognition of an organized integrated global system driven by the westward drift of the dominant plate suggests that the real driving force is a differential lag of the earth's crust to the earth's rotation. In addition, the recognition of the mechanical pattern of plate deformation provides a powerful graphical method for determining plate paleopositions, certainly back through the Mesozoic and perhaps into the Paleozoic.

## INTRODUCTION

Plate boundary recognition has been critical to the concept of drifting crustal plates since it was first recognized that "mirror image" continental margins faced each other across vast oceans. It was not, however, until seismic-activity maps were compiled in the 1960s that geologists began to outline with confidence the world's major crustal plates and the notion of plate tectonics gained acceptance. Recent years have seen a sustained effort to reconstruct the early positions of these plates using first geological and biological evidence with limited success; then, with increasing promise, paleomagnetics. This chapter offers a new insight to plate scale mechanical deformation and the driving force behind it, leading, it is hoped, to even better reconstructions.

[1]Presently with Marathon Oil Company, Houston, Texas.

Continental drift dealt with the wandering continental fragments. Plate tectonics developed to take into consideration the interaction of the mosaic of interdependent crustal plates including the intervening oceans. Following a similar sequence to develop the ideas of this paper, deformational changes at the leading, lateral, and trailing edges of single plates will be considered before examining the mosaic of crustal plates, their interdependency, and their physical reactions to, what turns out to be, a common stress.

## MARGINAL DEFORMATION

Two simple assumptions set the stage for the examination of marginal deformation—first, that the plate in question is moving; second, that it is moving faster than any adjacent plate. The margins of this plate can then be defined in terms of leading, lateral, and trailing edges (Figure 1A). It is self evident that

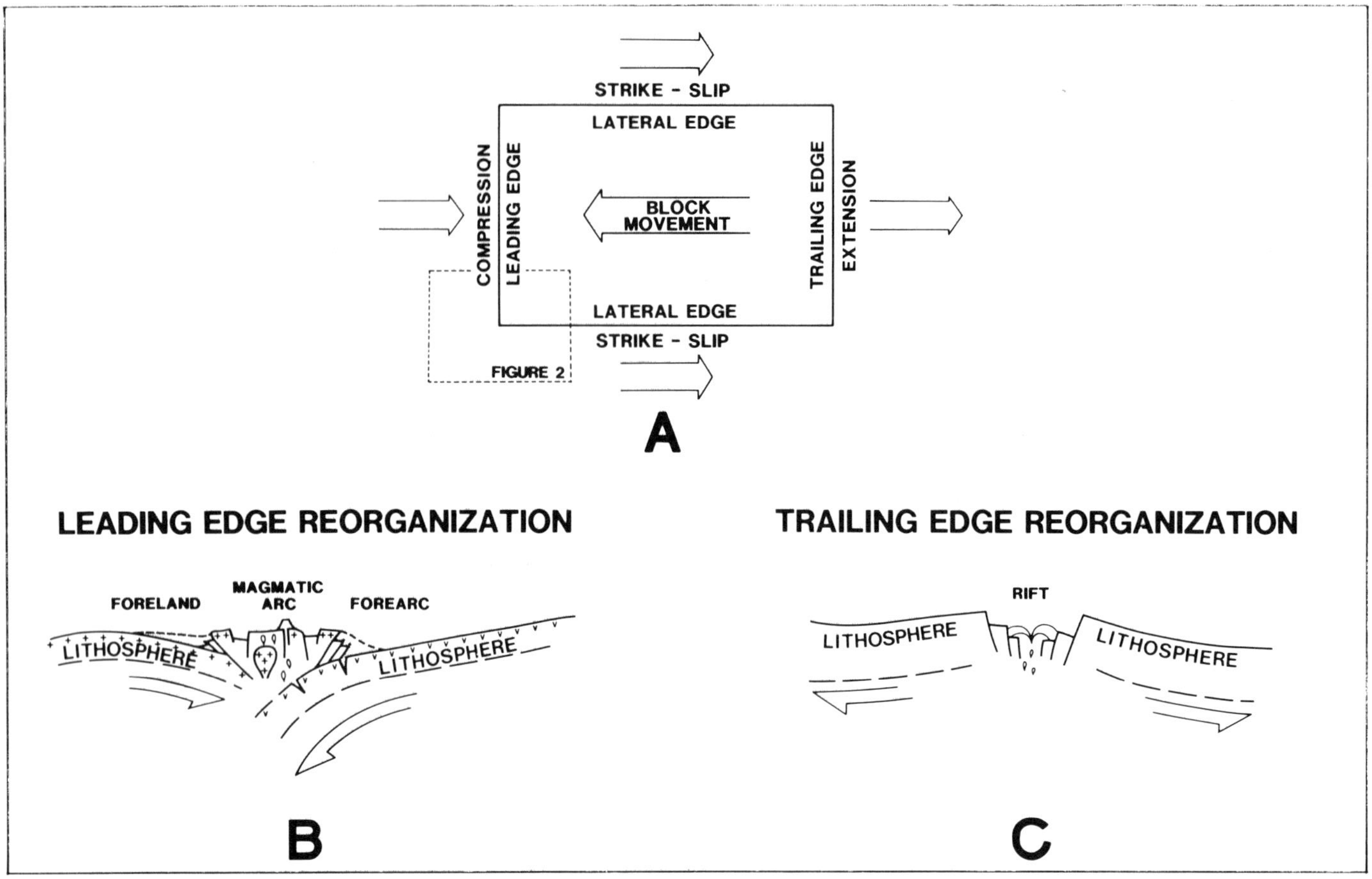


Figure 1. A. Schematic to demonstrate the relationship of compression, extension, and strike-slip at the margins of a moving plate. B. Schematic to demonstrate deformation due to direct compression along an oceanic-continental "Pacific Type" margin. C. Schematic to demonstrate zone of extension between diverging plates. Here adjacent plate composition is not a critical factor.

the leading edge will be under compression, that the lateral edges will be dominated by shearing, and that the trailing edge will be under tension.

Leading-edge deformation has been well documented (Dewey and Bird, 1970; Toksoz and Bird, 1977; Bally, 1983). The composition of the colliding plates ultimately determines underthrusting or overthrusting relationships, and in the oceanic-continental collision example of Figure 1 this margin will be the site of a subduction trench and a major magmatic arc created during the shock of block shortening (Figure 1B).

Trailing-edge deformation has been equally well documented (Moore, 1973; Mann et al., 1983; Lister, Etheridge, and Symonds, 1986). As this margin pulls away from the adjacent plate, the intervening rift becomes the site of emplacement of new crust from below (Figure 1C). Although the mechanism of emplacement may be poorly understood, the magnetic signature of this area leaves a detailed history of the time and direction of movement. This form of record at spreading centers exists only back into the middle Mesozoic, but it has significantly improved recent plate reconstructions. Other promising paleomagnetic techniques are now being developed for older and more varied sequences of rock (Fuller, 1984) and should soon add a wealth of information about older plate positions.

Lateral-margin deformation, being far more complex, is less well understood. Simple strike-slip faulting and associated structuring has been amply documented (Moody, 1973; Ron, Freund, and Garfunkel, 1984; Harding, 1985); however, this is but one stage in the sequential and progressive development of a lateral plate margin involving folding, thrusting, shearing, and buckle-folding (Wood, 1985). The concept of arc development along the length of a plate's lateral margin by buckle-folding (Figure 2) is central to the following argument. Movement toward the leading edge of a moving plate is increasingly restricted as leading-edge resistance becomes more prevalent. The increase in compression manifests itself in a compensatory decrease in plate movement and an increase in the zone of deformation (Figure 2A). Shear systems propagate within the lateral margins of adjacent plates, forming long, slender strips of crust which shear off the plate margin, are end-loaded during continued movement, and buckle away from the plate margin (Figures 2B, 2C). These buckle-folded strips at first become single mobile arcs (Figure 2C), which sweep laterally away from the plate margins, increasing the width of the zone of marginal deformation. Multiple-arc complexes develop as single complexes and are continually carried forward piggyback on the moving plate toward that plate's leading edge and, because of undercutting and passive secondary rotation, continue to migrate and rotate out over the underriding plate (Figure 2B). From opposite lateral margins these complexes merge toward the central areas of the leading edge of the plate, creating confusion between leading-edge subduction and migrating arcs as both actions occur simultaneously.

Wood (1985) has documented this sequential development and behavior of arcs in Southeast Asia (Figure 3), a region at

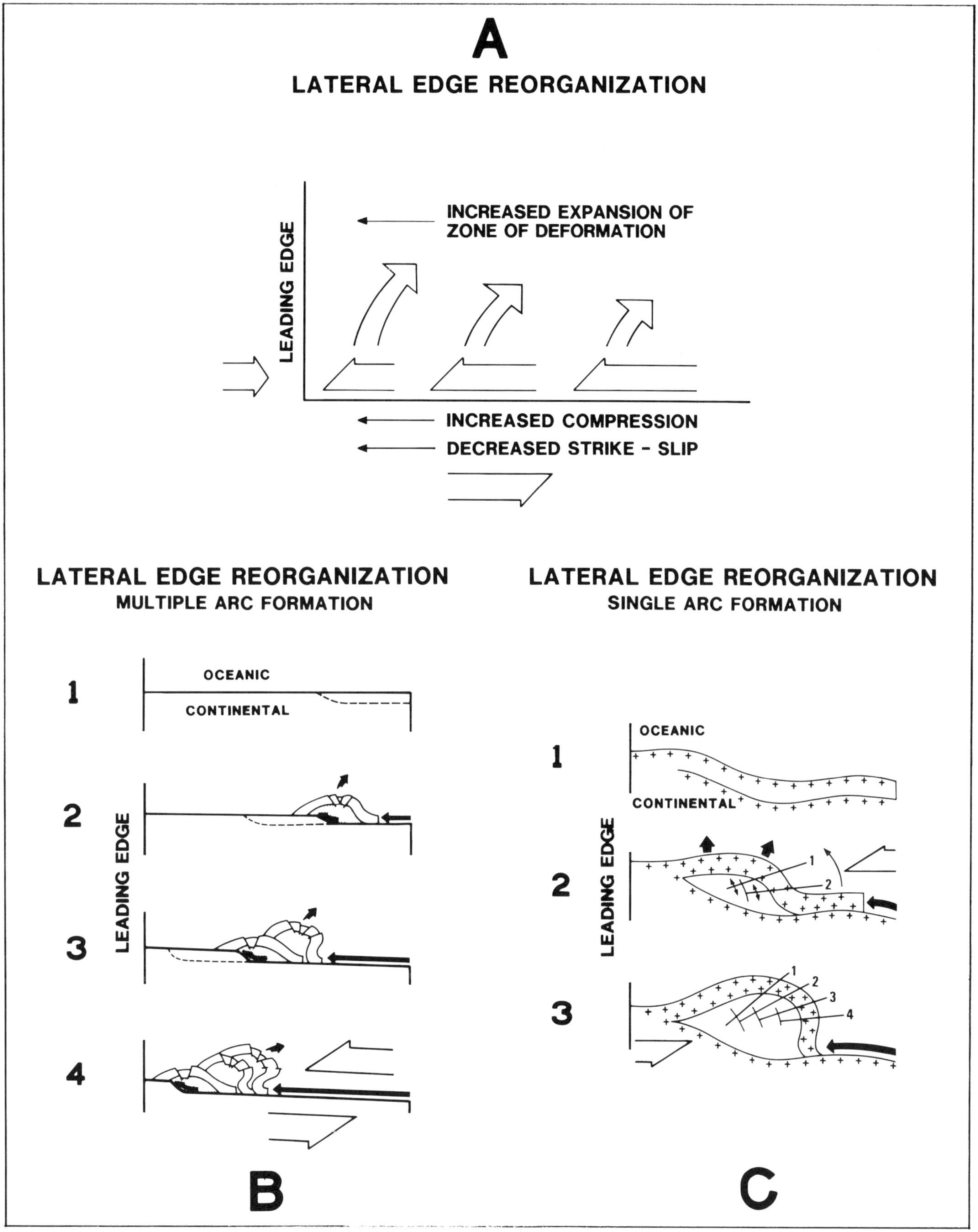


Figure 2. A. Relative movement of material during the reorganization along a plate's lateral margin near its leading edge. B. Schematic showing multiple-arc formation by undercutting shears and passive secondary rotation of early formed arcs. C. Formation of an arc-backarc basin complex showing sequential formation and rotation of spreading centers and radial expansion of the compressed arc.

the southwest forward lateral margin of the Pacific Plate which has been overridden by the reversely migrating arc-backarc complexes. Paleomagnetic evidence demonstrates the clockwise rotation and sequential opening from oldest to youngest, of the Philippines, South China, and Indo-China backarc basins, respectively, at a time of tremendous left-lateral strike-slip faulting along the northern Australian margin. These complexes have, as do others around the world, favored movement out over the Oceanic Plate in this oceanic-continental boundary situation.

The South Alaskan-Aleutian area lies at the northwest limit of the northeast lateral margin of the Pacific Plate (Figure 4). Southward buckling of the Aleutian Arc (Figure 4B) was initiated in the late Jurassic and continued into the lower Cretaceous (magnetic anomalies M1-M13), (Pitman, Larson, and Herron, 1974; Marlow and Cooper, 1980; McLean, 1979) as the Asian and North American Plates collided (Richards, 1974). Deep rifts of the South Alaskan Plate, filled during the late Cretaceous and Tertiary, formed at a time of undercutting and secondary passive counterclockwise rotation of the Aleutian Plate southward out over the Pacific Plate (Figure 4C). Late Tertiary (late Miocene to Recent) single-arc formation concurrently developed to the southeast, resulting in the formation of the Juan de Fuca Plate.

The concept of buckle-folding of sheared slivers of crust is not unique to oceanic-continental margin contacts. Continental-continental collision and subsequent arc formation by shearing is exemplified in Western Europe (Figure 5). Major single-arc complexes, the Carpathian facing north and the Atlas/Apenninic and Aegean Arcs facing south, have formed by the buckle-folding of three sheared slivers of continental crust which were end-loaded during the right-lateral transpressional movement along the European-African boundary. These arc-backarc complexes have developed on alternate sides of this boundary since the late Cretaceous (Horvath, Berckhemer, and Stogena, 1981; Schwan, 1980; Burollet, 1978; De Bremaecker, Huchon, and Le Pinchon, 1982; Pescatore and Slaczka, 1984). Of note is a similarity in their size, time of development, and apparent radially expanding nature. The hypothesis of buckle-folding of elongate lithospheric platelets aptly explains the coexistence and areal distribution of compression and expansion.

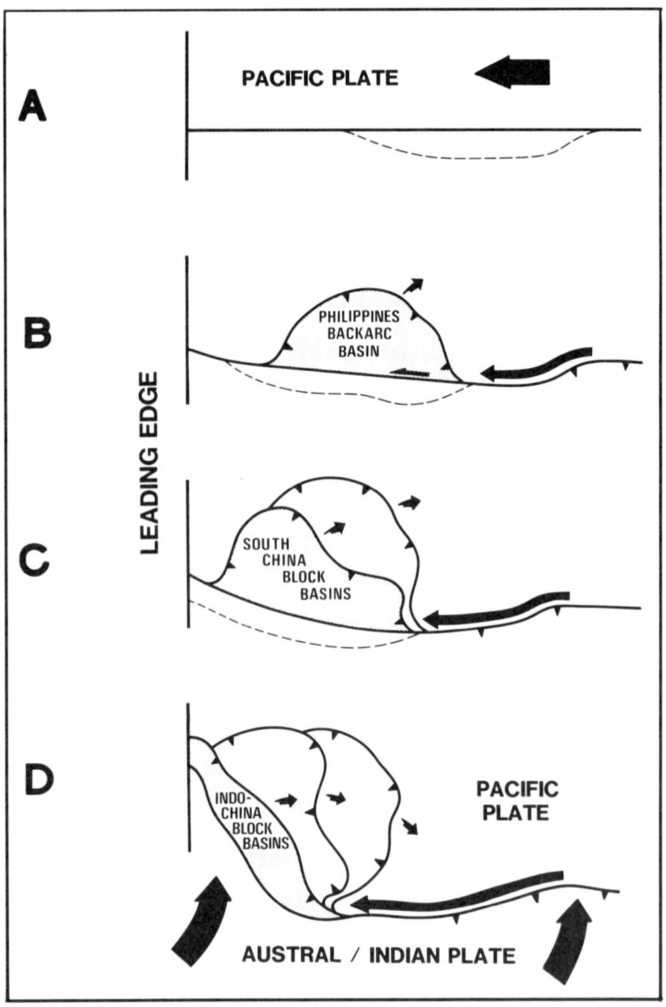


Figure 3. Sequential development of Southeast Asian backarc basins by repeated undercutting and buckle-folding of marginal strips. Note how the movement of the Austral/Indian Plate has further modified the final orientations of the basin.

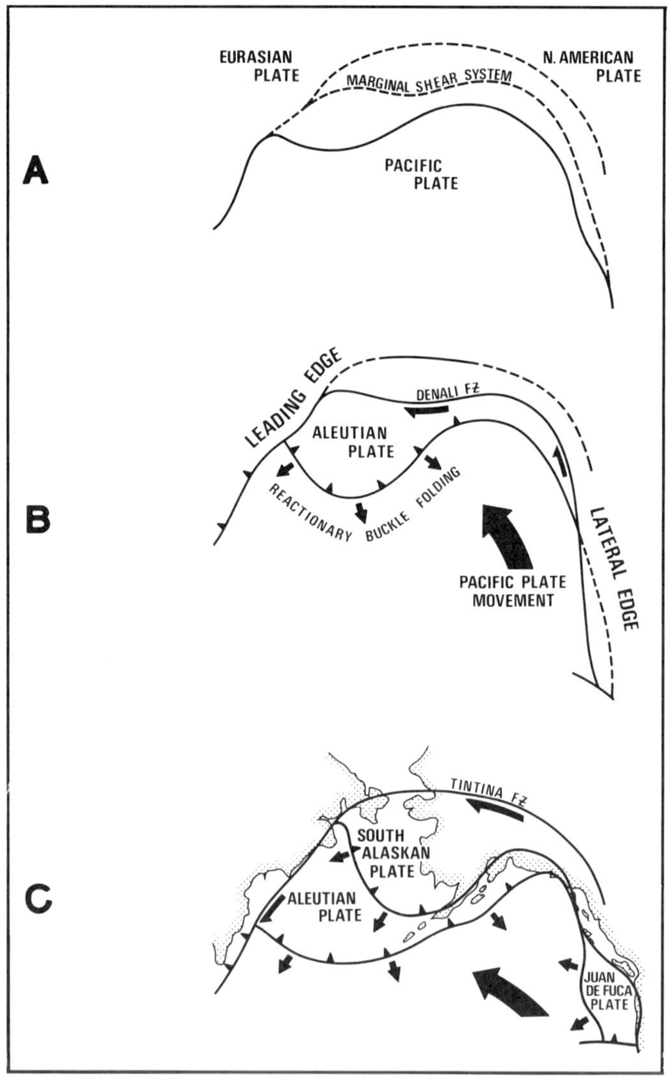


Figure 4. Sequential development of the Aleutian and South Alaskan backarc basins.

## APPARENT MOTIONS OF CRUSTAL PLATES

Turning to an examination of crustal plates and their margins, the apparent motions of several of the earth's major plates will be demonstrated. The Pacific and Eurasian plates dominate in surface area and marginal deformation and will be reviewed first. The African, Indian and Australian Plates will then be shown to be alike in their paleomotions, motions which are reactionary to the movements of the Pacific and Eurasian Plates.

### Pacific Plate

The boundaries and internal features of the Pacific Plate are shown in Figure 6. The west-northwest margin of this plate is dominated by deep subduction trenches and compressional mobile belts, major units of a collage of tectonic features which have come to be representative of compressional subduction boundaries (Bally, 1983) and, in this chapter, indicators of a compressional leading plate margin. Along the southwest margin of the plate, left-lateral displacement is evident in Indonesia at the margin of the Australian Plate (Hamilton, 1979). In this same area, where the Pacific Plate meets the Eurasian and Austral/Indian Plate, forward lateral-edge deformation has resulted in multiple buckle-folding of arc complexes and a clockwise rotation to the northeast of these complexes out over the oceanic Pacific Plate away from the continental Australian Plate. Lithospheric material, in this region of the lateral margin near the leading plate edge, is escaping from the forward corner of the Pacific Plate rather than subducting beneath the Eurasian Plate.

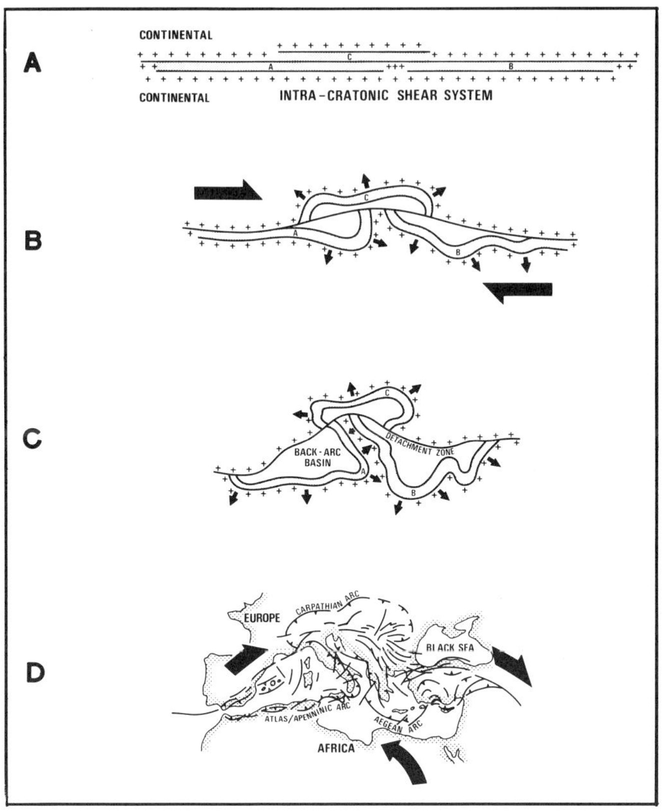


Figure 5. Single-arc development along a continental-continental lateral strike-slip margin. In this example a north component of motion of the African Plate results in a transpressive plate contact enhancing the development of the arcs.

To the east-southeast along this southern margin, the Magellan Sea Mount Chain (M.S.M.C.) and Marshall Islands (M.I.F.Z.) of Melanesia mark the positions of transform faults which, by transform concepts, were originally nearly perpendicular to their spreading ridge. Sea-floor magnetic anomalies (Pitman, Larson, and Herron, 1974) indicate progressively eastward younging of the sea floor toward the presently active spreading center at the East Pacific Rise connecting the transforms of the southwest Pacific to an origin at the East Pacific Rise.

The traces of these and other parallel transforms presently arc northwestward away from the Pacific Plate margin and terminate in the region of the east-southeast-trending Mid-Pacific Mountains. Upper Jurassic and Cretaceous sea floor (ages defined by magnetic finger-printing) appears to have rotated clockwise about a pole near the midpoint of the northern half of the East Pacific Rise and swept away from the Australian/Pacific boundary.

Along the North American west coast, lateral-edge reorganization is comparable with that found in Southeast Asia, although a mirror image due to right-lateral motion between the North American and Pacific Plates. Single- and multiple-arc formation, already described (Figure 4), is indicative of a progression from trailing to leading edge from the southeast to the northwest.

Of special interest in the north and central Pacific regions are the traces of the tremendous transform faults. These parallel features trend almost due east-west, where they emerge from beneath the North American Plate, but immediately curve smoothly southward and terminate in the region of the east-southeast-trending Mid-Pacific Mountains, the Line Island Chain (L.I.), or the Tuamotu Archipelago (T.A.). Upper Jurassic and Cretaceous sea floor appears to have this time rotated counterclockwise while sweeping away from the plate margin about a pivot located near lat. 45 °S, long. 140 °W, a point where a sharp bend occurs in the magnetic Anomalies 31 and 32 (Pitman, Larson, and Herron, 1974). The whole North and Central Pacific sea floor appears to be collapsing west and south over the West Pacific. Another point of rotation appears to be in the Gulf of Alaska, where again magnetic Anomalies 25 through 32 have been sharply bent westward (Pitman, Larson, and Herron, 1974) in response to buckling of the Aleutian Plate (A.P.). There is a definite resistance to subduction.

The East Pacific Rise, from the Gulf of California south to the El Tanin Fracture Zone (E.T.F.Z.), marks the trailing edge of the Pacific Plate. Across the El Tanin Fracture the spreading center is sharply offset to the west, indicating a major boundary between adjacent plates. This boundary fracture zone disappears to the west at the Tonga Trench but apparently reappears to the northwest, possibly as the fracture zone controlling the Magellan Sea Mount Chain. The northeast corner of the Australian Plate appears to have overridden the west-moving Pacific Plate, which accounts for the counterclockwise whirlpool structure of the Fiji Plateau region (F.P.).

In summary, the Pacific Plate has its leading edge to the west-northwest and its trailing edge to the east-southeast, and is slipping west-northwest between the Australian and North American Plates. Mobilized lateral material is moving piggyback with the plate and collapsing toward the center of the plate as it is compressed against the Eurasian Plate; this

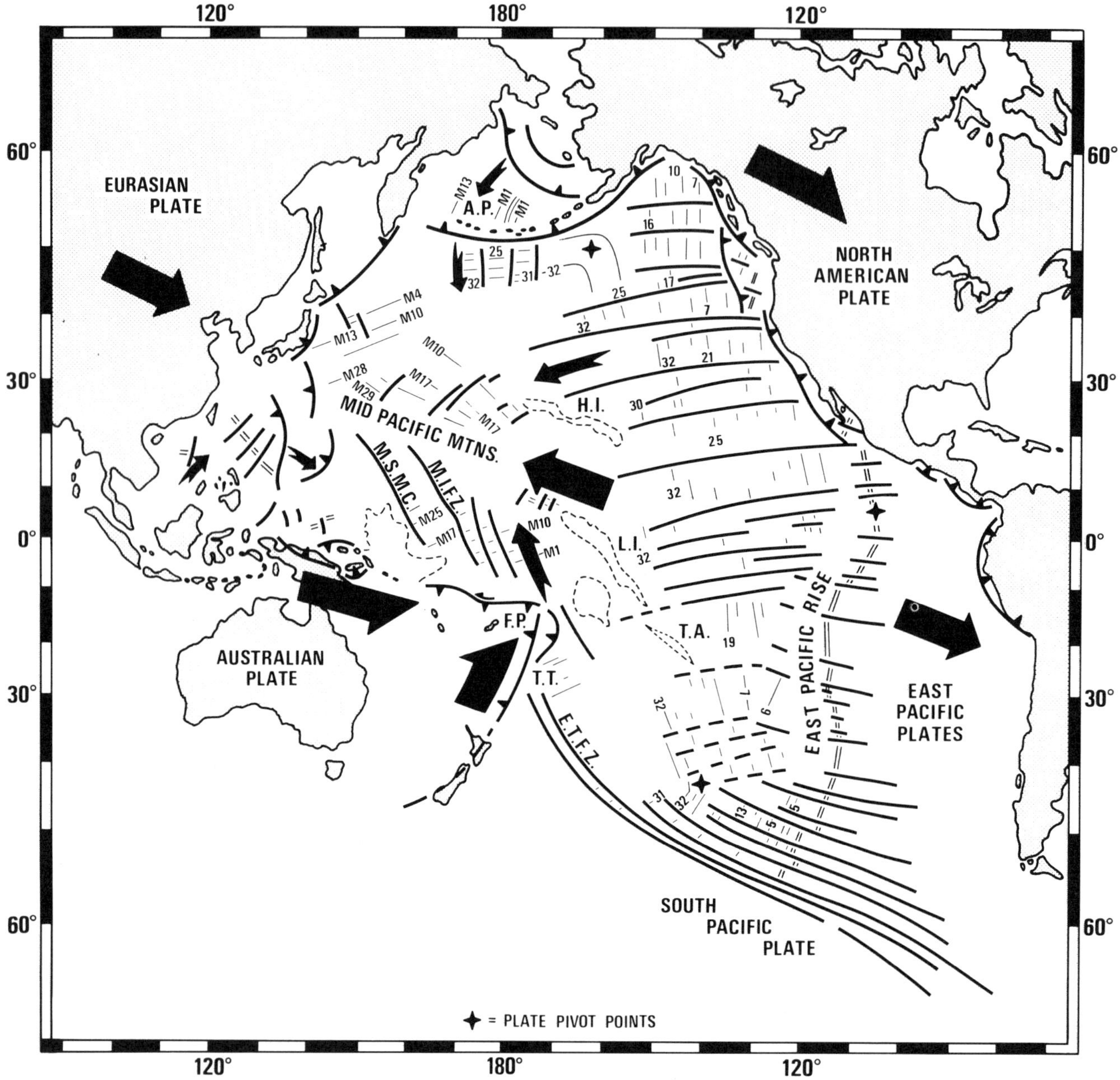


Figure 6. West-northwest movement of the Pacific Plate and resultant tectonic reoganization of the plate. M.S.M.C. - Magellan Sea Mount Chain; M.I.F.Z. - Marshall Islands Fracture Zone; E.T.F.Z. - El Tanin Fracture Zone; F.P. - Fiji Plateau; T.T. - Tonga Trench; A.P. - Aleutian Plate; H.I. - Hawaiian Islands; L.I. - Line Islands; T.A. - Tuamotu Archipeligo. Stars indicate intraplate pivot points, and numbers refer to magnetic anomalies. Large, blocky arrows indicate apparent plate motion; smaller arrows indicate intraplate structural reorganization.

explains the concentration of the oldest remaining plate material near the center of the western margin of the Pacific Plate and major mountain ranges down the center line of the plate.

## Eurasian Plate

The Eurasian Plate presents a potential engima (Figure 7). Its apparent and absolute motions appear at odds until the premises that a plate must be in motion and that it must be moving faster than the adjacent plates are more closely considered. Intuitively, all crustal plates cannot satisfy conditions if plate motions are interdependent. One plate must inherently become a focal point which has no apparent motion or, as in the case of the Eurasian Plate, apparently reverse motion relative to the others.

The southeastern and southwestern margins of the Eurasian Plate have already been described (Figures 3, 5), the southeastern margin being strongly compressional and the southwest margin an active right-lateral transpressional edge. The Atlantic spreading center marks a well-defined trailing edge. The northeast margin is less well defined but appears to be represented by the shear system extending from the Spitzbergen Fracture of the North Atlantic, southeastward across the Barents Sea and the north end of the Ural Mountains and along the southern margin of the Siberian Platform. Apparent present movement is to the southeast relative to adjacent plates.

Internal structuring of the Eurasian Plate suggests a different picture. Toward the end of the Paleozoic, Central and Western Europe was under a regime of counterclockwise

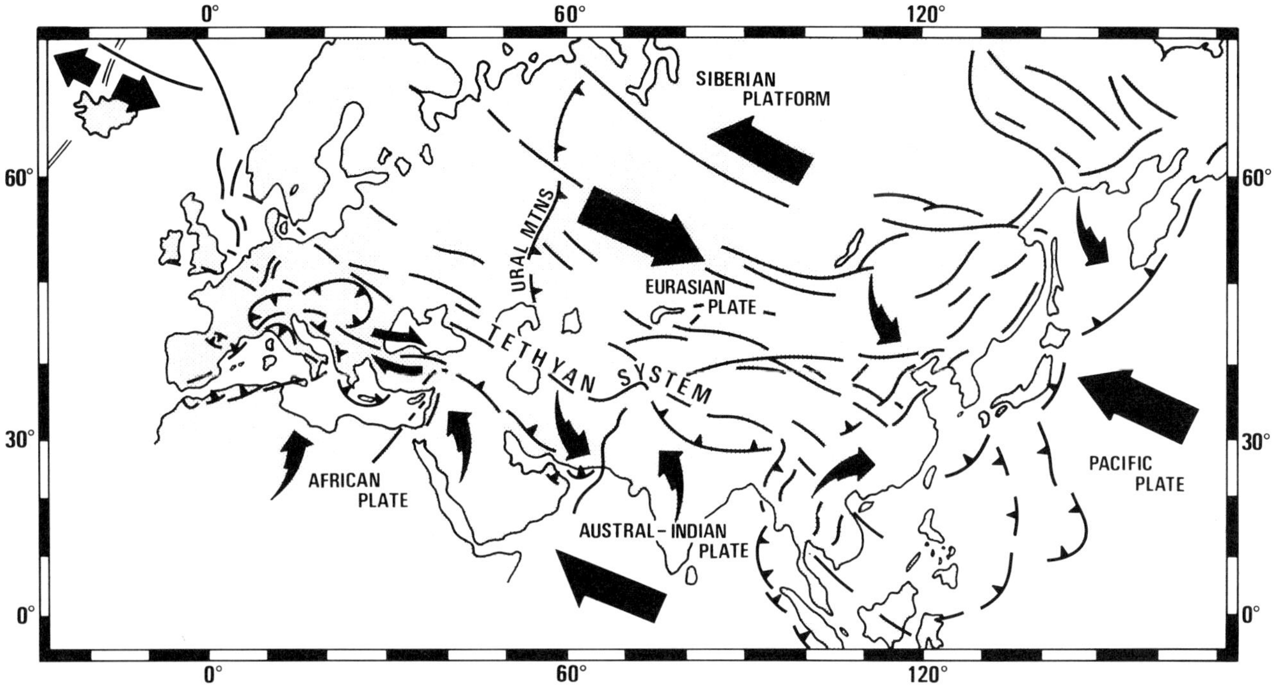


Figure 7. Apparent east-southeast motion of the Eurasian Plate and resultant tectonic reorganization along the Tethyan System and the Pacific/Eurasian Plate interface. Large, blocky arrows represent apparent relative motions; smaller arrows indicate secondary reorganization.

rotational strain resulting in large-scale left-lateral transpressional shearing (Bachtadse, Heller, and Kroner, 1983); and in Central Europe the Urals attest to plate-scale northwest- by southeast-oriented late Paleozoic compression (Peterson and Clarke, 1983). Meanwhile, in eastern Asia, extensional fault-block tectonics dominated Paleozoic structuring (Chang et al., 1980) suggesting that, prior to the destruction of the Tethys and the collision of Laurasia and Gondwana in the Mesozoic, the Eurasian Plate had an apparent net northwesterly movement.

Mesozoic and Cenozoic absolute plate motion is more difficult to document, yet recent articles on structuring in the North Sea area (Gowers and Saebøe, 1985) indicate that wrench tectonics has played more than a passing role. East Asian extensional tectonics is well documented (Change et al., 1980; Zhu and Chen, 1980).

Although diminished, the same processes that shaped Paleozoic Europe may still be active and the Eurasian Plate may presently continue to have, albeit small, a net westward drift. Meanwhile, adjacent plates have outpaced it in the Mesozoic and Tertiary, resulting in the Pacific Plate pressing from behind and the southern plates squeezing past to the south in an overall procession of westward-driving plates.

It is recognized that the argument for net west movement of Eurasia is presently weak, but certainly its resistant nature versus the excessive mobility of the equally large oceanic Pacific Plate is illuminating and raises the strong possibility that mantle rooting and plate size influences ultimate plate movement. That is to say, the larger the continental plate the deeper its rooting and the more immobile that plate may become. This suggests that the Eurasian Plate has become, by virtue of its size and composition, in relative terms the most immobile of all plates. In contrast, a large oceanic plate such as the Pacific, because of its shallower rooting and more established upper mantle low-velocity layer on which to glide (Anderson and Bass, 1986; Lowman, 1986a, 1986b), becomes increasingly more mobile. Ultimately, a plate the size of the Pacific must become the most active and dominate crustal plate.

These considerations appear at first to support a fixist concept of continental drift (Lowman, 1986a, 1986b); however, an alternative hypothesis will be offered later in this chapter. First, several plates under the influence of the Pacific and Eurasian Plates will be examined.

## Indian Ocean Plates

The Indian Oceanic area is best described by considering it to be composed of five subplates (Figure 8), which are differentiated by age and orientation as determined by magnetic fingerprinting (Johnson, Powell, and Veevers, 1976; McKenzie and Sclater, 1971; Falvey, 1972; Coffin, 1985). The oldest plates (upper Jurassic to lower Cretaceous and lower Cretaceous to Paleocene) are paired across the present spreading ridge. The youngest plate (Paleocene to Recent) is the elongate strip of oceanic crust extending from the South Pacific to the Gulf of Aden and is itself split by the presently active northwest-southeast-trending spreading ridge.

Considering that magnetic lineations run roughly parallel to the spreading ridge of their origin and looking only at the subplates north of the presently active spreading ridge, then it can be inferred that the oldest spreading ridge (upper Jurassic to lower Cretaceous—magnetic anomalies M25-M1) is now oriented approximately northeast, whereas the spreading ridge for the lower Cretaceous to Paleocene (magnetic anomalies 32-21) now lies almost east-west. The presently active ridge trends approximately southeast.

Restoration of these subplates through time involves a clockwise rotation of the older plates, sequentially bringing the related spreading ridges into a position identical to the

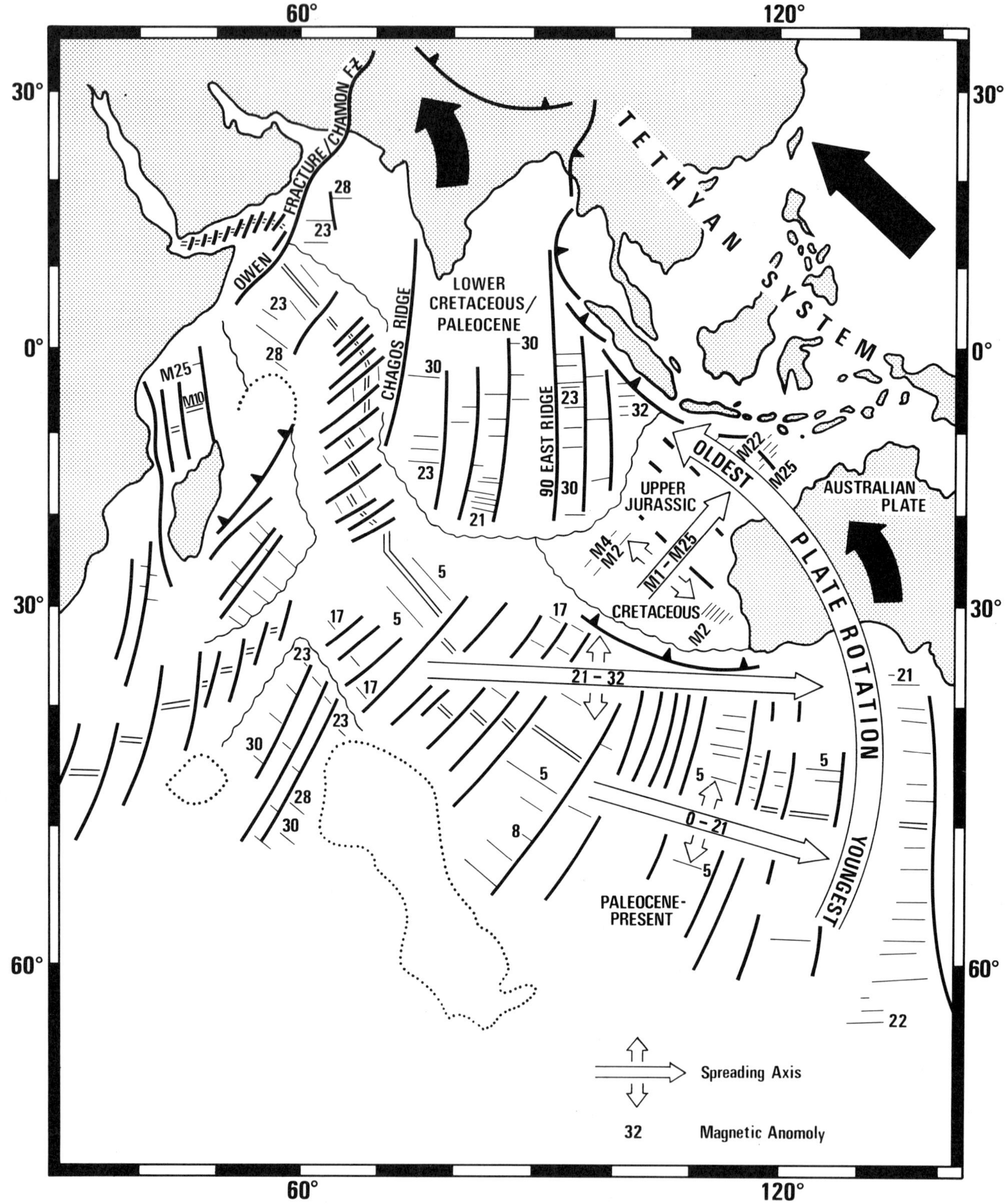


Figure 8. Sequential rotation of Indian Plates. The continuous wavy lines outline the assemblage of subplates of the Indian Ocean region. These subplates are paired across two ancient spreading centers and the presently active ridge. Subplates to the north show a sequential counterclockwise rotation of progressively older plates due to transpression along the Tethyan system and compression from the South Pacific.

present southeast-oriented ridge if the Antarctic Plate is held fixed. Conversely, each active spreading position was successively initiated in an identical southerly location, migrated north with a counterclockwise rotation, and died while a new spreading center formed to the south.

The two jumps in spreading location can be dated, the first at approximately Anomaly 32 time (75 Ma) and a second at about Anomaly 21 time (53 Ma). This final spreading began between Australia and Antarctica and propagated with time to the northwest, unzipping to about the Chagros Ridge by about 35 Ma, where a 45° shift in spreading directions signaled the event (McKenzie and Sclater, 1971). These rifting

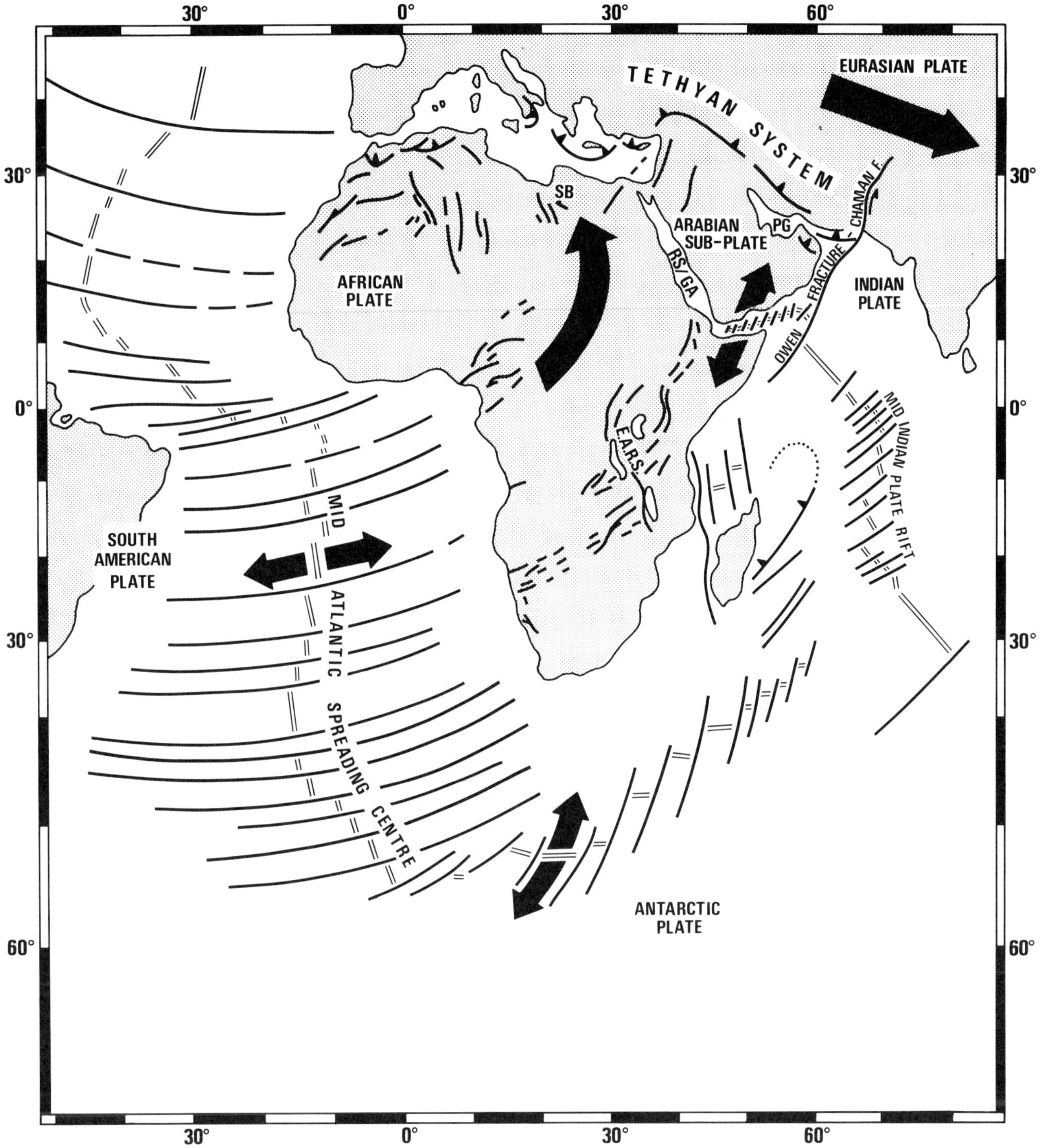


Figure 9. Counterclockwise rotation of the African Plate and resultant extension and compression. S.B. - Sirte Basin; R.S./G.A. - Red Sea/Gulf of Aden Rift; E.A.R.S. - East Africa Rift System.

events coincide with increased Pacific Plate activity (Schwan, 1980).

The oldest block to the north of the present Indian Ocean spreading center has been rotated as much as 90° counterclockwise since the latest Jurassic. The thesis advanced here is that the developing plates, caught by a left-lateral transpressional Tethyan Shear System to the north, were dragged northwest and spun counterclockwise. The resultant rotational torque placed the southeastern margin of these plates under radial tension coincident with the South Pacific Plate impinging on it from the southeast. The spreading ridges have therefore opened as tension fractures in response to this northwest-oriented stress.

## African Plate

The African Plate (Figure 9) is triangular, its margins being the transpressional Tethyan System in the north, the Atlantic extensional system in the west and southwest, and the more complex shear and extensional boundary with the Indian oceanic plates to the southeast.

The key to understanding the movement of this plate lies in interpreting the apparent anomalous extensions within the eastern and southern areas. If the Africa plate were moving north to northwest under its own power or under the influence of a push from the south, then these areas should be under compression. Instead, a pulling action from the north is more consistent with the observed extensional patterns. However, this fails to account for the observed counterclockwise rotation of blocks such as the Arabian subplate (Irving and Tarling, 1961; Tarling, 1970; Beydoun, 1982) and indeed the African Plate as a whole (Smith, Hurley, and Briden, 1981), resulting in the radial tensioning of the southeastern margin. An action consistent with all motions and observed fault patterns is one whereby the African Plate is being jammed against the Tethyan Boundary by a northwest-oriented stress, resulting in the present right-lateral transpressional shearing. This action lifts the plate northwest, spins it counterclockwise to the west with the northwest corner acting as a pivot, and places the entire southeastern margin under tension. The rifts of Mid-Africa (from the East African Rift System to the Sirte Basin) may be precursors of a developing spreading center as the Red Sea/Gulf of Aden Rift rotates to the north to its death in the Tethyan System. The Indian Ocean spreading ridge and its extension into the Red Sea is in reality the center of a Modern Tethys!

### Australian Plate

Northward counterclockwise spiraling action is again clearly visible for the Australian Plate (Figure 10). The Tethyan Shear System extends across the northern compressional leading edge of the plate, and the spreading center between Australia and Antarctica marks the position of the trailing margin. To the east the major right-lateral Alpine fault system of New Zealand delimites the eastern margin of the block. An active western margin is not present as the Australian and Indian Plates, as already demonstrated, have been locked together as a unit since about Anomaly 21 time (53 Ma).

The fact that the entire crustal strip, from the Owen fracture/Chaman Fault trend west of the Indian Plate, east, certainly as far as the continental Australian Plate, must be considered as a unit since the late Paleocene/early Eocene, sheds doubt on the validity of singling out the movement of the Indian Plate in the Alpine tectonic development of much of Asia (Tapponier and Molnar, 1976; Tapponier et al., 1982; McKenzie and Sclater, 1971). This massive elongate crustal strip has been moving north and rotating counterclockwise at least since the late Paleocene and possibly as early as the Cretaceous (Austin and Williams, 1977).

Coincident with this rotation, the entire stripe from India to Australia appears to be buckling south. The Diamantina Trench south of the Broken Ridge off the southwest corner of Australia indicates southward compression. This lies directly south of the south-vergent Sumatra/Java Trench and approximately at the midpoint along the northwest-southeast length of the Austral/Indian Plate. To the southeast, at the leading edge near Tonga, compression has a north vergence demonstrated by the eastern corner of the Austral-Indian Plate overriding the Pacific Plate. To the northwest the plate is again converging northeastward in the region of the Himalayas. The entire plate is buckling in the horizontal plane into an arc convex to the south because of end-loading from the southeast. This action is consistent with the recognition of a northwest-directed stress within the plate from earthquake data (Weissel, Anderson, and Geller, 1980) and absolute plate motion studies (Minster and Jordan, 1978).

Having downplayed the primary role of slip-line field theory (Tapponier and Molnar, 1976), the concept does become exceedingly useful in the consideration of large- and small-scale secondary effects of plate motions. The Tasman spreading center between Australia and New Zealand was active between Anomaly 32 (upper Cretaceous) and Anomaly 25 (Paleocene), a period coincident with the development of the lower Cretaceous-Paleocene northern Indian Plate (Figure 8) and activity on the Alpine Fault of New Zealand. Applying Tapponier and Molnar's concept by considering the Australian Continent as a north-moving rigid indenter then a reconstruction to, say, Anomaly 25 time, would necessitate rotating the Tasman Sea area counterclockwise while at the same time rotating the Indian Continental Plate clockwise. This would, at Anomaly 25 time (Paleocene), bring the spreading centers of the Tasman Sea and the Indian Ocean into alignment and directed southeast-northwest. The Coral Sea spreading ridge, also active during this upper Cretaceous to Paleocene time, would restore southward with little rotation, thus completing a northwest-oriented spreading ridge extending across the top of the southerly-restored Australian Continent. This is consistent with the known northward-facing Atlantic Margin type tectonic setting at that time for the region from Irian Jaya and Papua-New Guinea to New Zealand.

Therefore, as the Australian Continent lifted north away from Antarctica and rotated counterclockwise, blocks to the east were swept away from its path to the north and east and spun clockwise. Blocks to the west, locked to the Australian Plate, lifted north and rotated counterclockwise with it. Coincident with, although possibly predominantly in the later stages of this internal plate reorganization, the entire Indo-Australian Plate buckled south because of end-loading from the South Pacific.

## GLOBAL INTERACTIVE SYSTEM AND CRUSTAL RIBBONS

Plate-edge deformation has resulted in a set of related structures associated with progressive reorganization. Three systems dominate the earth's surface—(1) lateral shear (2) compressional convergent and (3) tensional divergent—and are plotted on Figure 11. Dominating the map are the lateral shear systems exemplified by the Tethyan System, extending from Europe through Asia into Southeast Asia. The west-northwest extension of this system crosses the North Atlantic and extends along the northeast margin of the North American Plate at the dislocation of the North American Craton and the Canadian Arctic Islands. To the southeast the Tethyan System is continuous with the El Tanin Fracture Zone and carries eastward, passing between South America and Antarctica before terminating at the Atlantic-Indian Ridge, a divergent system. Another dominant shear system is the Pelusium System (Neev and Hall, 1982), which extends from the Eastern Mediterranean southwest across Africa to the Atlantic equatorial fracture zones, where it merges with the shear system

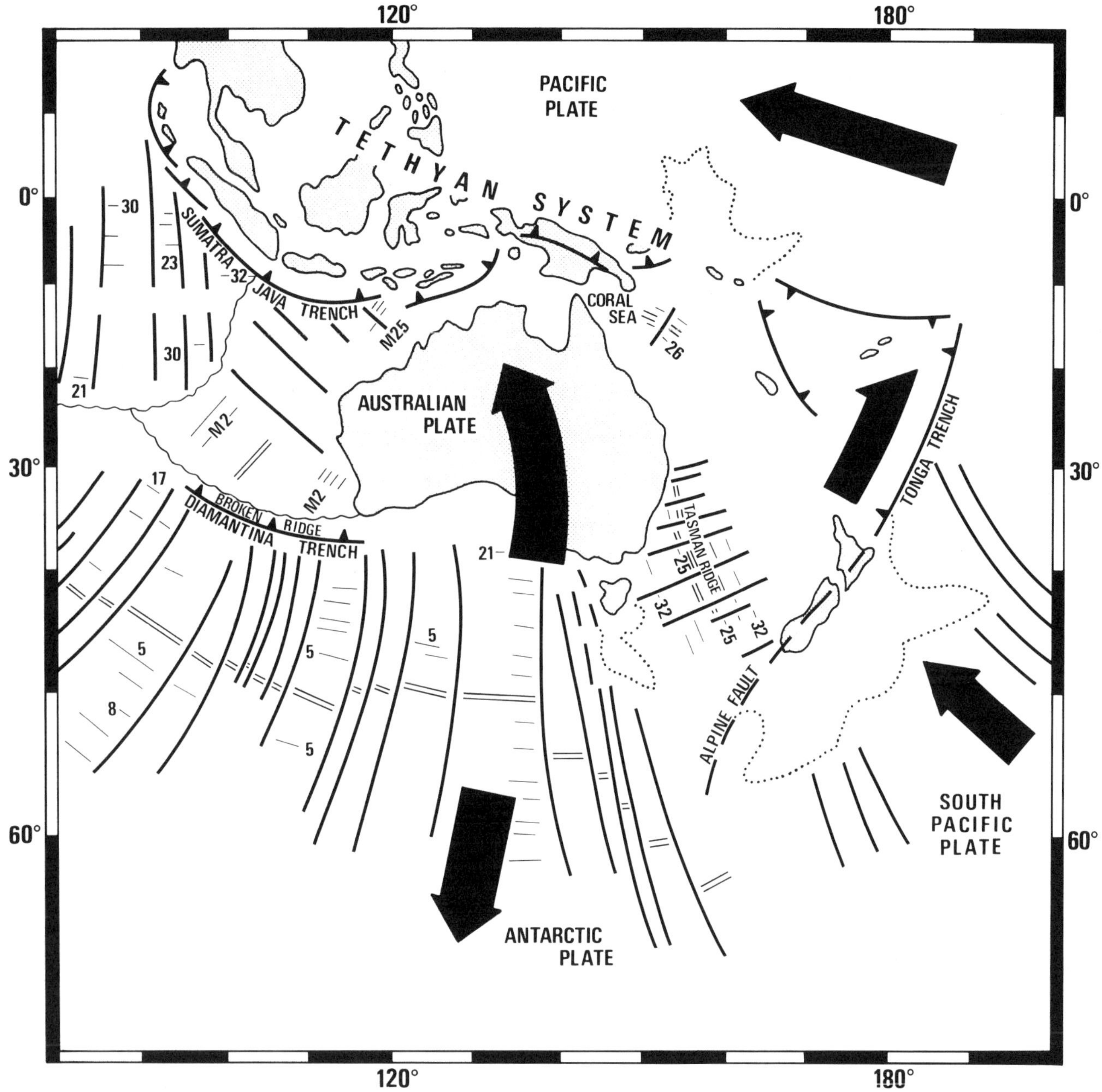


Figure 10. Counterclockwise rotation of the Australian Plate due to the left-lateral transpression along the Tethyan System. Secondary indenter tectonics is also demonstrated by the opposite rotations of the Tasman Sea area and Indian Ocean subplates away from a north-moving Australian Plate.

along the southwest margin of the North American Plate. This latter system extends from the Caribbean to Alaska.

Along the lengths of these global shear systems there is a change in dominance of strike-slip versus transpression. The section of the Tethyan System between Southeast Asia and Europe is extremely complex and its width is substantial. To the southeast, from eastern Indonesia to Fiji, the zone narrows until along the El Tanin Fracture it is not only extremely narrow but also is seismically very quiet. To the northwest from Europe the same phenomenon occurs. A dominance of strike-slip, a lessening of seismic activity, and a narrowing of the zone of deformation is typical of the shear system as it crosses the North Atlantic and passes along the northeast margin of the North American Plate. This modification of complexity, width, and seismic activity is coincident with a distinct disparity of motions of adjacent plates. To the north of the Tethyan System, the Pacific and Asian Plates are both moving generally northwest. However, the motions of the southern plates differ dramatically, from nearly perpendicular along the margins of the African, Indian, and Australian Plates to near parallel along the lateral edges of the North American and South Pacific Plates.

Convergent systems are exemplified by the Western Pacific and Western South American complexes, and the Tethyan System along the north to northeast margins of the African, Indian, and Australian Plates is also convergent when referenced to the southern plates. These compressional convergent margins, marking the leading edge of individual plates, all lie perpendicular to the lateral shear systems. Tensional divergent systems, exemplified by the mid-ocean spreading

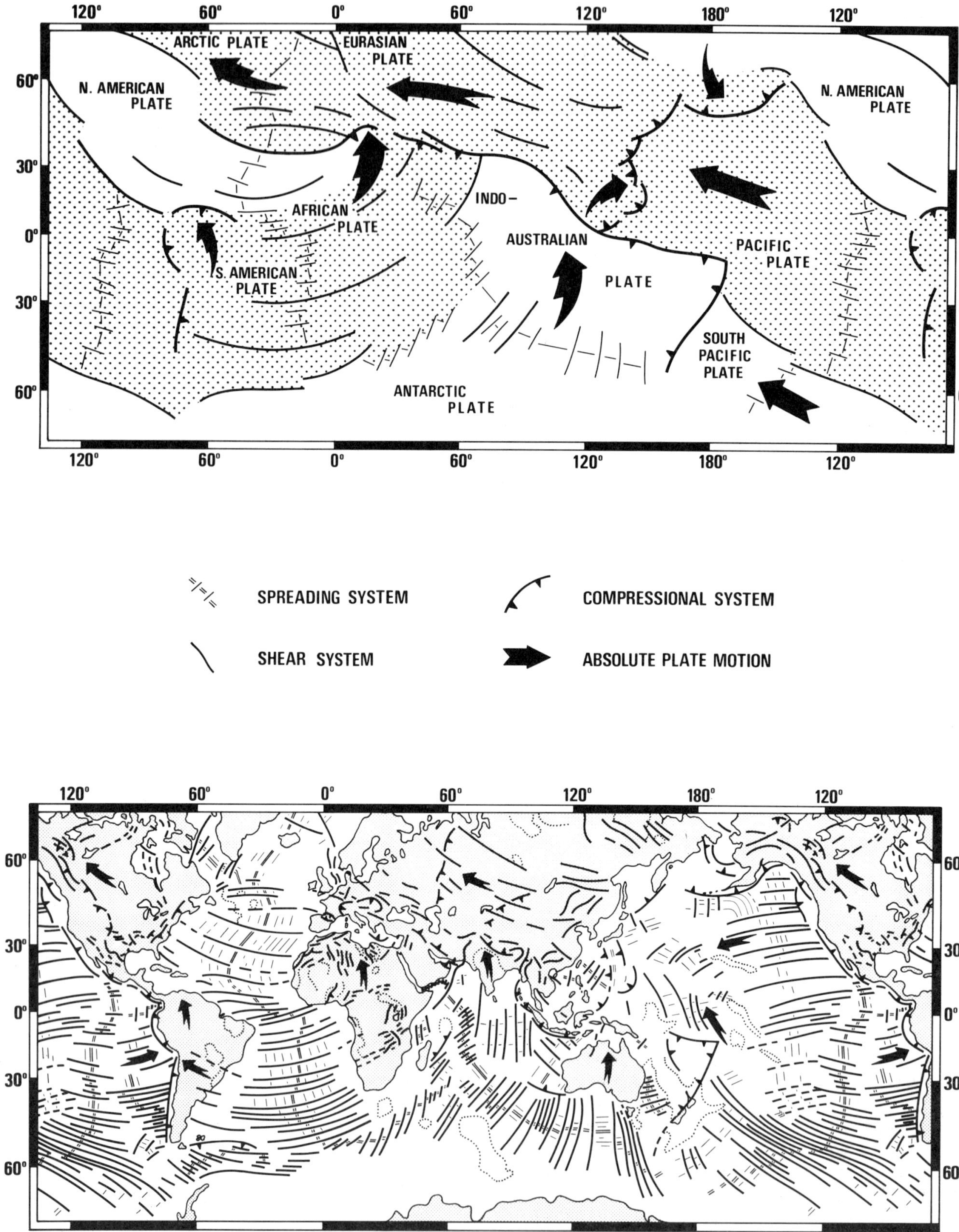


Figure 11. The northward buckle-folding of the Pacific/Eurasian/Arctic crustal strip by end-loading from the Pacific and the incorporation of the African, Indian, and Australian Plates into the heart of the fold.

centers, and marking the trailing edges of moving crustal plates also lie perpendicular to the lateral shear systems.

Figure 11 demonstrates that plate margin lateral shear systems are contiguous and break the earth's crust into long, continuous ribbons which spiral northwesterly about the globe (Tethyan Spiral of O'Driscoll, 1980). These ribbons consist of, along their length, alternating oceanic and continental plates of different sizes. By far the largest and most dominant are the Eurasian and Pacific Plates. If intensity of plate margin deformation is an indicator of absolute plate motion, then the Pacific Plate dominates as the most active, whereas the Eurasian Continental Plate appears to be the most resistant to motion. Beginning in the north and working southeastward, the most continuous ribbon consists of the Canadian Arctic Plate, the Eurasian Plate, the Pacific Plate, the South American Plate, and the Africa Plate. All but the African Plate have a simple demonstrable west-northwest motion. The African Plate at the southern tail of this ribbon has moved strongly northwest, while spinning counterclockwise like the adjacent Indian and Australian Plates. The lateral shears of this tail plate swing eastward and then northeastward through as much as 90° before they terminate against the Tethyan System south of the Eurasian Plate. This almost perpendicular relationship of the shear systems along the Tethyan System prompted O'Driscoll (1980) to envisage a second system of geosutures, whereas the same convergent pattern was recognized by Neev and Hall (1982) as a converging spiral with an "eye" stretching the length of the Alpine-Himalayan orogenic belt. Neither ideas, however, could be linked to a common origin.

These global "geosutures," if viewed as contiguous lateral margins of individual building blocks placed end to end, then form crustal ribbons which can be themselves considered units of higher order. It is proposed that these ribbons will react to end loading in a similar way as the much smaller strips of sheared crust along the lateral margins of crustal plates (Figure 2C).

The Pacific and Eurasian Plates are a good example of this phenomenon. They lie end to end running from the equatorial Pacific more than 23,000 km (14,300 mi) to the North Atlantic, and cover more than 40% of the earth's surface. They form a portion of crustal ribbon which has buckled northward, drawing and spinning the African, Indian, and Australian Plates into the heart of the fold. The Indian Oceanic Plates form the core of this tremendous buckle-fold. The Pacific Plate is end-loading the ribbon and driving the system.

## SUMMARY

1) Direction and intensity of movement of individual crustal plates can be determined by examining deformational patterns at each plate's margins. The leading edge of a plate is dominated by compressional features, whereas lateral margins are dominated by shear systems and buckle-folded complexes. The trailing edge is the site of extension and rifting.

2) Intensity of plate deformation is related to plate size, relative movement, and composition of adjacent plates.

3) Marginal systems, compressional convergent, lateral shear and tensional divergent, are not haphazardly distributed. Lateral shear systems form continuous spiraling geosutures which break the earth's crust into continuous ribbons. The compressional and tensional systems separate the ribbons of crust perpendicularly into segments which form the recognized crustal plates. Ribbon segments (plates) are not of equal size or composition and do not move at equal rates.

4) All plates have a persistent west-directed component of motion.

5) The Pacific and Asian Plates presently dominate crustal deformation. The Pacific Plate has end-loaded and distorted the ribbon of crust stretching from the Central Pacific to NW Europe by buckling it into a convex north-facing fold

6) The plates to the south of this fold have been incorporated into the core of the fold and distorted by counterclockwise rotation. The southeastern margins of these plates have been radially tensioned.

7) During rotation, these core plates have rifted along their southeastern tensioned margins. The sites of this rifting have not been random, and have periodically jumped south in order to maintain a west-northwest-directed spreading center, inferring an active west-northwest-directed stress.

8) Indenter tectonics (Tapponier and Molnar, 1976) is an active but secondary mechanism of deformation. The primary stress is the northwestward push of the Pacific Plates.

## CONCLUSIONS

The rhegmatic patterns of the earth's crust can best be characterized as the outline of a northwesterly spiraling segmented ribbon composed of plates of different sizes and composition. The lateral margins of this band form the well-known Tethyan Shear System. There is a direct relationship between the extent of deformation along a plate margin and plate size. The Pacific Plate dominates in size and marginal deformation. The tail of this north-spiraling ribbon is deformed in the Southern Hemisphere, most notably in the area of the Africa, Indian, and Australian Plates. Each southern plate has a strong north component of motion as well as a counterclockwise spin. The plates appear to have rotated in response to the northward buckling of the Pacific and Eurasian Plates. Buckling has been due to end-loading by the Pacific Plate.

The crest of the buckle-fold lies approximately north of the Indian oceanic region. Continental plates have responded in brittle fashion and the intervening areas filled with new ocean floors. The history of this ultimate fold is recorded in the magnetic lineations of the Indian Ocean floor. The fold has developed since the early Cretaceous in three distinct phases, each phase related to a west-northwest-to northwest-directed compressional pulse. These phases relate to recorded movements of the Pacific Plate.

The direct relationship between marginal deformation and plate size, the pervasiveness of a westward component of motion for all plates, and the apparent relationship of the dominant westward-moving Pacific to the folding of the crustal ribbons indicate the driving force to be differential lag of the crustal plates to the earth's spin. Other independent evidence gathered over the years has also suggested a connection of the earth's spin with lithospheric plate motion (Nelson and Temple, 1972; Winterer, 1973; Solomon, Sleep, and Richardson, 1975; Austin and Williams, 1977).

This chapter offers a hypothesis to explain and describe the interconnection of plate motions and the earth's spin. This hypothesis provides a powerful tool for sound mechanical reconstructions of paleoplate positions and perhaps a test of other methods. The challenge now lies in verification.

## ACKNOWLEDGMENTS

I wish to thank Marathon for permission to publish this paper. I am indebted to Colin Cockle who single handedly drafted all the figures, to Colette O'Connor for typing the manuscript, and to the Marathon staff for their valuable suggestions.

## REFERENCES CITED

Anderson, D.L., and J.D. Bass, 1986, Transition region of the Earth's upper mantle: Nature, v. 320, p. 321-328.

Austin, P.M., and G.E. Williams, 1977, Tectonic development of late Precambrian to Mesozoic Australia through plate motions possibly influenced by the earth's rotation: Journal of the Geological Society of Australia, v. 25, pt. I, p. 1-21.

Bachtadse, V., F. Heller, and A. Kroner, 1983, Palaeomagnetic investigations in the Hercynian mountain belt of central Europe: Tectonophysics, v. 91, p. 285-299.

Bally, A.W., 1983, Structural styles and the evolution of sedimentary basins: AAPG Short Course, Society of Exploration Geophysicists Annual Convention.

Beydoun, Z.R., 1982, The Gulf of Aden and NW Arabian Sea, *in* A.E.M. Nairn and F.G. Stehli, eds., The ocean basins and margins, vol. 6, The Indian Ocean: New York, Plenum Press, p. 253-313.

Burollet, P.F., 1978, From Atlantic to Pacific, main hypothesis of the Alpine system: SEAPEX Proceedings, v. IV, p. 50-59.

Chang, W., D. Zhong, F. Ma, H. Sun, and J. Zhang, 1980, Fault block tectonics and oil resources in China, *in* J.F. Mason, ed., Petroleum geology in China: Tulsa, Oklahoma, PennWell Publishing Company, PennWell Books, p. 116-131.

Coffin, M.F., III, 1985, Evolution of the conjugate east African-Madagascan margins and the western Somali Basin: New York, Columbia University unpublished Ph.D. dissertation.

De Bremaecker, J., P. Huchon, and X. Le Pichon, 1982, The deformation of Aegea: A finite element study: Tectonophysics, v. 86, p. 197-211.

Dewey, J.F., and J.M. Bird, 1970, Plate tectonics and geosynclines: Tectonophysics, v. 10, p. 625-638.

Falvey, D.A., 1972, Sea-floor spreading in the Wharton Basin (northeast Indian Ocean) and the breakup of eastern Gondwanaland: Australian Petroleum Exploration Association Journal, v. 12, p. 86-88.

Fuller, M., 1984, Paleomagnetism, Short course on methods and application: Santa Barbara, University of California unpublished seminar manual, 126 p.

Gowers, M.B., and A. Saebøe, 1985, On the structural evolution of the Central Trough in the Norwegian and Danish sectors of the North Sea: Marine and Petroleum Geology, v. 2, p. 298-319.

Hamilton, W., 1979, Tectonics of the Indonesian region: U.S. Geological Survey Professional Paper 1978, 354 p.

Harding, T.P., 1985, Seismic characteristics and identification of negative flower structures, positive flower structuring, and positive structural inversions: AAPG Bulletin, v. 69, p. 582-600.

Horváth, F., H. Berckhemer, and L. Stogena, 1981, Models of Mediterranean back-arc basin formation: Philosophical Transactions of the Royal Society of London, A. 300, p. 383-402.

Irving, E., and D.H. Tarling, 1961, The paleomagnetism of the Adam volcanics: Journal of Geophysical Research, v. 6, p. 549-556.

Johnson, B.D., C. McA. Powell, and J.J. Veevers, 1976, Spreading history of the eastern Indian Ocean and Greater India's northward flight from Antarctica and Australia: GSA Bulletin, v. 87, p. 1560-1566.

Lister, G.S., M.A. Etheridge, and P.A Symonds, 1986, Detachment faulting and the evolution of passive continental margins: Geology, v. 14, p. 246-250.

Lowman, P.D., Jr., 1986a, Plate tectonics with fixed continents: a testable hypothesis-I: Journal of Petroleum Geology, 8, 4, p. 373-388.

Lowman, P.D., Jr., 1986b, Plate tectonics with fixed continents: a testable hypothesis-II: Journal of Petroleum Geology, 9, 1, p. 71-88.

Mann, P., D.C. Bradley, D.C. Burke, and M.R. Hempton, 1983, Development of pull-apart basins: Journal of Geology, v. 91, n. 5, p. 529-554.

McLean, H., 1979, Review of petroleum geology of Anadyr and Khatyrka basins, USSR: AAPG Bulletin, v. 63, n. 9, p. 1467-1477.

McKenzie, D.P., and J.G. Sclater, 1971, The evolution of the Indian Ocean since the late Cretaceous: Geophysical Journal of the Royal Astronomical Society, 25, p. 437-528.

Marlow, M.S., and A.K. Cooper, 1980, Mesozoic and Cenozoic structural trends under the southern Bering Sea shelf: AAPG Bulletin, v. 64, no. 12, p. 2139-2155.

Minster, J.B., and T.H. Jordan, 1978, Present day plate motions: Journal of Geophysical Research, v. 83, n. B11, p. 5331-5337.

Moody, J.D., 1973, Petroleum exploration aspects of wrench fault tectonics: AAPG Bulletin, v. 57, n. 3, p. 449-476.

Moore, D.G., 1973, Plate-edge deformation and crustal growth, Gulf of California structural province: GSA Bulletin, v. 84, p. 1183-1906.

Neev, D., and J.K. Hall, 1982, A global system of spiraling geosutures: Journal of Geophysical Research, v. 87, n. B13, p. 10689-10708.

Nelson, T.H., and P.G. Temple, 1972, Mainstream mantle convection: A geologic analysis of plate motion: AAPG Bulletin, v. 56, p. 226-246.

O'Driscoll, E.S.T., 1980, The double helix in global tectonics: Tectonophysics, v. 63, p. 397-417.

Pescatore, T., and A. Slaczka, 1984, Evolution models of two flysch basins: The northern Carpathians and the southern Apennines: Tectonophysics, v. 106, p. 49-70.

Peterson, J.A., and J.W. Clarke, 1983, Geology of the Volga-Ural petroleum province and detailed description of the Romashkina and Arlan oil fields: U.S. Geological Survey Open-File Report 83-711, 90 p.

Pitman, W.C., III, R.L. Larson, and E.M. Herron, 1974, The age of the ocean basins: GSA (two charts and summary statement).

Richards, H.G., 1974, Tectonic evolution of Alaska: AAPG Bulletin, v. 58, p. 79-105.

Ron, H., R. Freund, and Z. Garfunkel, 1984, Block rotation by strike-slip faulting: Structural and paleomagnetic evidence: Journal of Geophysical Research, v. 89, n. 87, p. 6256-6270.

Schwan, W., 1980, Geodynamic peaks in Alpino type orogenies and changes in ocean-floor spreading during the late Jurassic-late Tertiary time: AAPG Bulletin, v. 64, p. 359-373.

Smith, A.G., A.M. Hurley, and J.C. Briden, 1981, Phanerozoic paleocontinental world maps: Cambridge, Cambridge University Press, 102 p.

Solomon, S.C., N.H. Sleep, and R.M. Richardson, 1975, On the forces driving plate tectonics: Inference from absolute plate velocities and intraplate stress: Geophysical Journal of the Royal Astronomical Society, v. 42, p. 769-801.

Tapponier, P., and P. Molnar, 1976, Slip-line field theory and large scale continental tectonics: Nature, v. 264, p. 319-324.

Tapponier, P., G. Pelitzer, A.Y. Le Dain, and R. Armija, 1982, Propagating extrusion tectonics in Asia: New insights from simple experiments in plasticine: Geology, v. 10, p. 611-616.

Tarling, D.H., 1970, Palaeomagnetism and the origin of the Red Sea and Gulf of Aden: Philosophical Transactions of the Royal Society of London, A. 267, p. 219-226.

Toksoz, M.N., and P. Bird, 1977, Formation and evolution of marginal basins and continental plateaus, *in* M. Talwani and W.C. Pitman III, eds., Island arcs, deep sea trenches and back-arc basins: American Geophysical Union, Maurice Ewing Ser. I, p. 379-393.

Weissel, J.K., R.N. Anderson, and C.A. Geller, 1980, Deformation of the Indo-Australian plate: Nature, v. 287, p. 284-287.

Winterer, E.L., 1973, Sedimentary facies and plate tectonics of equatorial Pacific: AAPG Bulletin, v. 57, p. 265-282.

Wood, B.G., 1985, The mechanics of progressive deformation in crustal plates—a working model for Southeast Asia: Geological Society of Malaysia Bulletin 18, p. 55-99.

Zhu, X., and H.J. Chen, 1980, Tectonic evolution of Chinese petroleum basins: Revue de l'institut Francais du Petrole, March-April, v. XXXV, n. 2.

# SEDIMENTARY BASIN DEVELOPMENT ABSTRACTS

DEWEY, JOHN F., University of Oxford, Oxford, England, United Kingdom, and GARRY D. KARNER and WALTER C. PITMAN, Lamont-Doherty Geological Observatory, Palisades, NY

Relationship Between Small and Large Sedimentary Basins: The Scale Spectrum

Realizing that detachments influence lithospheric extension helps us define the fundamental elements responsible for basin formation. In addition to strain rate, crustal thickness, and lithospheric thickness, these factors are (1) stretching distribution within the upper plate of the detachment, (2) stretching distribution within the lower plate, and (3) depth distribution of the detachment(s), where lateral strain must be balanced within the lithosphere during extension. Intracrustal duplex systems may reflect previous compressional events, or they may be produced by upper-crustal flaking in strike-slip regimes. Basin initiation represents a mechanical process (e.g., hanging-wall collapse), whereas subsequent basin development is primarily an isostatic process controlled by the thermal and mechanical properties of the lithosphere. However, subcrustal lithospheric involvement and, hence, heating are minimal until flexure becomes unimportant, which depends on basin wavelength, lithospheric temperature structure, and hence, rigidity.

This wavelength or scale dependency may help explain the difference between the abrupt but short-lived subsidence of small basins, such as the Ridge basin of southern California, and the general negative exponential subsidence characteristic of passive continental margins. Further, the rheological implications of depth-dependent stretching, with thermal modification of the lithosphere during extension, ultimately controls lithospheric strength and, hence, the actual rift/breakup configuration. Together, factors 1 and 2 generate rift and thermal subsidences, the final form of which depends on the flexural properties of the lithosphere. The exact relationship between factors 1 and 2 determines the spatial configuration between rift and sag basins and temporary uplifts. We illustrate these concepts and their implications with examples from California, Nevada, Brazil, Turkey, and Australia.

HAYES, DENNIS E., Lamont-Doherty Geological Observatory, Palisades, NY

Structure and Origin of South China Basin

The southern margin of China and the northern margin of Reed Bank and associated blocks represent the boundary between normal oceanic crust and subsided transitional crust that has been rifted, thinned, and perhaps intruded by igneous rocks. The intervening South China Sea basin is composed of two subbasins. The eastern basin was demonstrably formed by approximately north-south sea-floor spreading in the middle Tertiary.

The western basin of the South China Sea is characterized by northeast-southwest tectonic trends and is much narrower than the eastern basin. Any plausible reconstruction of the eastern subbasin to its predrift configuration creates a large overlap of the inferred zones of transitional crust bordering the western subbasin. Because there is no obvious structural discontinuity in the deep-water parts of the two subbasins, one can hypothesize that the eastern basin opening was initially accommodated in the west by anomalously large continental crustal stretching. The fault-bounded microcontinental blocks that occupy much of the margins of the western basin are believed to be underlain by transitional crust of intermediate thickness and composition. The geometry of the deep basin as defined by the seaward limit of the continent-ocean boundary (COB) zone provides constraints on the east-to-west variations in crustal stretching of both the northern and southern rifted margins of the South China Sea. Although the southwest subbasin was also formed by sea-floor spreading processes, the processes were active only during the late phase of spreading in the east. Either the southwest subbasin spread obliquely, or the late spreading phase changed from a north-south trend to northwest-southeast.

HOWELL, DAVID G., U.S. Geological Survey, Menlo Park, CA

Terranes and Resources

Continents increase in mass because new material is added (primarily fragments of volcanic islands, seamounts, and hydrogenous and biologic sediments), formed in oceanic settings. Admixed with this material are continental fragments and piles of terrigenous sediment; both of these affect continental recycling but not net growth. Terranes are fault-bounded fragments found in orogenic belts. The rate of accretion calculated for the Circum-Pacific region since the breakup of Pangea during the early Mesozoic is 2-2.5 $km^3$/year, or as much as 80 $m^3$/sec. An average global growth rate needed to account for the entire mass of continents ($7.6 \times 10^9$ $km^3$) since the early Archean ($3.8 \times 10^9$ years) is 2.0 $km^3$/year; however, the flux of continental growth remains a topic of great controversy.

Although ocean crust contributes little to the mass of continents, the generative cycle creates the "conveyor belt" that leads to the formation of sialic material, which leads to the accretion and dispersive kinematics resulting in the growth and shaping of continents. The isochrons deduced from magnetic stripes in the modern oceans (about $300 \times 10^6$ $km^2$ in area) indicate an average age for ocean crust of only 70 Ma. Oceanic crustal rocks form at a rate of 800 $m^3$/sec, assuming a 6-km thick crust. If plate tectonic processes have been operative since 3.8 Ga, even at modern rates, as many as 27 world-ocean crusts would have been born at spreading centers and lost back to the mantle in subduction zones.

The processes linking tectonics, sedimentation, and resources are not clearly understood, but as we learn more about the geologic and physical properties of terranes, as well as their kinematics and dynamic histories, our prospecting perspicacity should improve. Some explanations are already available. Campeche, the Persian Gulf, and Prudhoe Bay are all passive-margin sequences tectonically overrun by foreland thrust and fold environs. Southern California and the western Australia offshore represent dispersive settings—transcurrent and orthogonal rifting, respectively. Sumatra and Java display compressional features, resulting from the subduction of the India-Australia plate; these features are superposed over expansion rifts formed when India collided with southern Asia. Other untested conditions may also have petroleum accumulations.

# MARINE FRONTIERS

CHAPTER 33

# REPORT ON GLOBAL GEOSCIENCE TRANSECTS PROJECT

J.W.H. Monger

*Canada Geological Survey, and Coordinator GGT*

*Vancouver, Canada*

The term "transect" herein refers to a cross section showing the composition and structure of the Earth's crust and, if possible, lower lithosphere, that incorporates and integrates all available geological and geophysical data. Transects lie along corridors up to a few thousand kilometers in length that are selected by regional experts so as to cross major crustal features. Their primary purpose is to elucidate the nature and evolution of the crust.

The Global Geoscience Transects (GGT) project was conceived by the Inter-Union Commission on the Lithosphere (ICL) in August 1985 in Tokyo, and is modelled after the Decade of North American Geology continent:ocean transects program. Subsequently, letters announcing GGT were sent out by ICL to heads of International Lithosphere Program (ILP) National Committees, and drew a widespread and favorable response. In addition, Muawia Barazangi (United States) was asked by ICL to prepare a global map of possible transect lines; theslocations were determined by reference to major geological structures and to previously implemented and proposed transects, with emphasis on continental regions, although including some marginal seas and island arcs. This map served as a focus for discussion at a workshop held in December 1985 in San Francisco, which was attended by representatives of several developing countries as well as scientists who had been involved in the North American transects program. At this meeting, the writer was asked by ICL President Fuchs to coordinate GGT.

The year 1986 has been devoted to getting GGT "off the ground." The most critical task at this time is to encourage scientists to start drawing transects. At least initially, guidelines developed for the North American transects are to be followed, although these may well become modified as the project progresses. Ideally and initially, the following items are to be included: (1) a geological strip map of the 100-km-wide transect corridor, scale 1:500,000 or 1:1,000,000; (2) a "data base display," using same scale(s) as map where appropriate, (vertical = horizontal), and comprising a geological cross section, gravity, and magnetic profiles and seismic-reflection and seismic-refraction records and other ancillary geophysical and geochemical (isotopic) data; (3) a "time-space" diagram showing the evolution of the crust of the transect as determined largely from surface geological relationships; (4) the transect proper, which attempts to integrate all of these features in a crustal model; and (5) an accompanying explanatory pamphlet, with references. To encourage the use of this format, the writer attended a European Geotraverse meeting in West Germany in March, discussed the project, and suggested preparation of transect lines in addition to the previously adopted north-south EGT line.

In August, ICL met in conjunction with the Fourth Circum-Pacific Energy and Mineral Resources Conference in Singapore. Representatives of several ILP National Committees brought proposed transect lines to this meeting. These lines, and others proposed by interested individuals and organizations (e.g., EGT in western Europe and SEATAR in southeast Asia) which have agreed to coordinate their efforts with GGT, are shown on the accompanying map (Figure 1). The map draws attention to regions where few transects have been proposed, partly because they cross international boundaries, and efforts are being made at present to encourage scientists to propose transect lines in these areas and to link up with lines in adjacent regions.

Also at the Singapore meeting, ICL established a new coordinating committee (CC-7), which is concerned solely with GGT. The committee will be made up of about 10 regional coordinators in addition to the writer as chairman. The functions of CC-7 are to encourage compilation of transects within and between interested countries, ensure that data and interpretation are presented uniformly so as to permit direct comparison of crust in different parts of the world, and ultimately to edit the transects for publication. Subsequent to the Singapore meeting, a form requesting numeric transect location data and brief transect descriptions has been sent out to ILP National Committees, various international groups, and interested individuals. This material will be compiled to form a GGT catalogue, to be used to locate transects accurately on maps of various projects, and to guide the project as a whole.

A key part of GGT is the exchange of information and ideas and discussion of problems by the scientists doing transect compilations. In 1982-84, the writer was responsible for compiling a transect that was part of the North American transect program. An important, rewarding, and creative part of this program was the annual workshop at which transect compilers were able to discuss their ideas for transect presentation and mutually explore problems of transect compilation. The outcome of these workshops became the guidelines noted

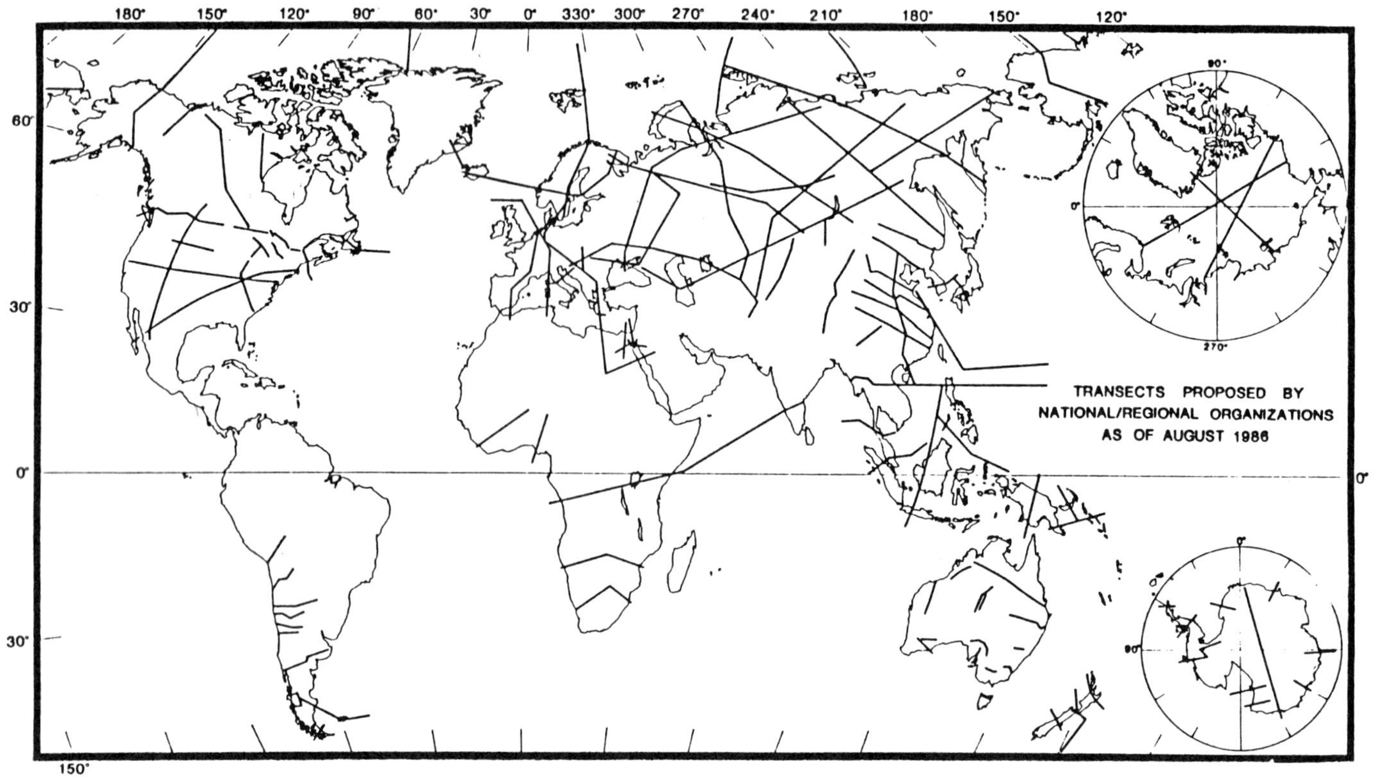


Figure 1. Transects proposed by national/regional organizations as of August, 1986. This map is incomplete, tentative and transect locations are approximate. Transect location data have been supplied by the following sources that are listed alphabetically by regions/countries: *Africa/Middle East*: S. Riad, Egypt, B. Rumvegeri, Zaire; *Antarctic*: I.W.D. Dalziel/Scientific Committee for Antarctic Research; *Arctic*: Arctic subcommittee of ICL; *Australia*: R. Rutland/National Committee for ILP; *China*: National Committee for ILP; *India*: H.K. Gupta; *Japan*: K. Tamaki/National Committee for ILP; *New Zealand*: F. Davey; *North America*: Lithoprobe Steering Committee, Trans-Alaska Lithosphere Investigation, W.R. Van Schmus, U.S.A., D.B. Stewart, U.S.A., P.F. Hoffman, Canada; *South America*: Argentinian National Committee, U. Cordiani, Brazil; *Southeast Asia*: J.R. Curray/Studies in East Asian Tectonics and Resources; *USSR*: National Committee for ILP, E. Kozlovsky; *Western Europe*: European Geotraverse, P. Giese *F.R.G.*, M. Von Knorring, *Finland*.

above that form the starting rules for GGT. It is essential that this process of consultation and discussion be part of GGT. New problems not addressed in the North American program (which was concerned with the Phanerozoic margins of the continent) can be anticipated, such as the treatment in GGT of the extensive Precambrian rocks in platform areas with a Phanerozoic cover. These problems can only be resolved after scientists have wrestled with them and discussed their findings with others working on similar problems. For GGT to be a viable and truly "global" project, as many scientists as possible who are compiling transects must attend such workshops and contribute to the project directly. To this end, a GGT workshop is planned as part of the ILP contribution to the XIX General Assembly of the International Union of Geodesy and Geophysics in August 1987 in Vancouver. Regional workshops are planned for the Latin American Geological Congress in Brazil and, tentatively, for African and Middle Eastern countries in 1988. The writer would welcome suggestions for others. A major deadline for GGT will be the 28th International Geological Congress in 1989, in Washington, D.C., U.S.A., where there will be GGT symposia and poster sessions with initial drafts of transects.

# MARINE FRONTIERS ABSTRACTS

AUZENDE, J. M., Ifremer Centre de Brest, Brest, France, J. DANIEL and J. P. EISSEN, Orstom Centre de Nouméa, Nouméa, New Caledonia, J. P. FOUCHER,* Ifremer Centre de Brest, Brest, France, and B. PONTOISE and J. RECY, Orstom Centre de Nouméa, Nouméa, New Caledonia

SEAPSO Project in New Hebrides-Fiji-Tonga Area

The SEAPSO project, organized jointly by Ifremer and Orstom, was conducted on the R/V *Jean-Charcot* from October 1985 to January 1986. The new data concern multinarrow-beam bathymetry, gravity, magnetics, 3.5-kHz echo sounding, single-channel seismic reflection, sediment coring, rock dredging, hydrocasts, and sea-floor photographic observations. The purpose of the cruise was to conduct regional bathymetric and geologic surveys of selected targets within three major themes.

First, we considered tectonics of the interaction between aseismic ridges and subduction zones. Tectonic effects were studied at: (1) the intersection of the Loyalty Islands Ridge and the New Hebrides volcanic arc, which represents an early stage of the phenomenon; (2) the frontal collision of the d'Entrecasteaux Ridge and the New Hebrides subduction zone; and (3) the oblique collision of the Louisville Ridge and the Tonga volcanic arc.

Second, we considered tectonics and magmatism of back-arc basins. The back-arc troughs of the New Hebrides Ridge appear as simple or double grabens that may represent an incipient stage of back-arc basin development. The Lau and North Fiji basins are examples of young and mature back-arc basins, respectively.

Third, we considered oceanic accretion and hydrothermalism in back-arc basins. First-order objectives in the North Fiji and Lau basins were to identify crustal accretion centers and to search for hydrothermal activity on a regional scale.

AUZENDE, J. M., Ifremer Centre de Brest, Brest, France, J. P. EISSEN, Orstom Centre de Nouméa, Nouméa, New Caledonia, and SEAPSO LEG III Shipboard Scientific Team

Accretion and Hydrothermalism in North Fiji Basin, Southwest Pacific

The North Fiji basin is a marginal basin surrounded by: (1) the New Hebrides island arc in the west, which ends at the Matthew-Hunter Ridge in the south; (2) the Hunter fracture zone in the southeast; (3) the Fiji Islands in the east; and (4) the complex Vitiaz Trench system in the north. The investigations in this area suggest a complex opening has been active for approximately the past 10 m.y.

The detailed bathymetry and the structure of the central part of this basin are still poorly known. The major element consists of an axial ridge. The trend of this ridge is imprecise, but it is probably oriented north-south near 173.5°E, at least between 21° and 15°S. The other remarkable feature consists of a complex system of ridges and faulted blocks running along the western edge of the Fiji Archipelago. This system has been interpreted by some authors as an accretion ridge.

The purposes of the third leg of SEAPSO cruise onboard the R/V *Jean-Charcot* (December 2-24, 1985) are as follows: (1) to explore the typical structures of the North Fiji basin between 21° and 17°S (axial ridge, seamounts, fracture zones, and the complex border west of Fiji), using multibeam echosounder (Seabeam), seismic reflection, magnetic, and gravimetric surveys; and (2) to complete Seabeam coverage and samplings (such as dredging, coring, water sampling, and bottom photographs) of two to three small box-shaped areas on the axial ridge, in order to localize and eventually characterize hydrothermal vents.

The various morphotectonic, petrologic, chemical, and geodynamic results of this survey will be discussed.

BEIERSDORF, HELMUT, Bundesanstalt für Geowissenschaften und Rohstoffe, Hannover, Federal Republic of Germany, and JÖRG ERZINGER, Justus-Liebig-Universität, Giessen, Federal Republic of Germany

Bathymetry, Tectonics, and Stratigraphy of Northeastern Slope of Manihiki Plateau, Pacific Ocean

An area of 1,800 km$^2$, about 60 km northwest of Rakahanga (Cook Islands) at the northeastern slope of the Manihiki Plateau, was surveyed with a multibeam echo sounder and a 3.5-kHz subbottom profiler. One of two 300-m deep canyons marks the trace of a northeast-dipping listric fault. The fault results from a rotation of the crustal block between the plateau margin and the canyon. The block tilting has caused the exposure of basaltic basement and a 200-m thick layer of volcaniclastic rocks lying above it. This same layer was cored at DSDP Site 317, 200 km away at the center of the high plateau part of the Manihiki Plateau. The thickness and distribution of this layer indicate a significant explosive volcanic event occurred during the Late Cretaceous in this region.

BLANCHET, R., and J. F. STEPHAN, GIS Océanologie et Géodynamique, Brest, France, C. RANGIN, CNRS, Université de Paul et Marie Curie, Paris, France, D. BALADAD, Bureau of Energy Development, Manila, Philippines, PH. BOUYSSE, Bureau de Recherches Géologiques et Minières, Orléans, France, M. P. CHEN, National Taiwan University, Taipei, Taiwan, P. CHOTIN, Université de Paul et Marie Curie, Paris, France, J. Y. COLLOT and J. DANIEL, Orstom Centre de Nouméa, Nouméa, New Caledonia, J. M. DROUHOT, Ifremer, Jakarta, Indonesia, B. MARSSET, GIS Océanologie et Géodynamique, Brest, France, B. PELLETIER, Orstom Centre de Nouméa, Nouméa, New Caledonia, M. RICHARD, GIS Océanologie et Géodynamique, Brest, France, and M. TARDY, Université de Savoie, Chambéry, France

From Subduction to Collision: Results of French POP2 Program on Taiwan-Philippine Festoon

A sea-beam, seismic, magnetic, and gravimetric survey was conducted with the R/V *Jean-Charcot* in three key regions off the Taiwan-Philippine festoon in the western Pacific: (1) Ryukyu active margin and its junction with Taiwan; (2) northern part of the Manila Trench and its junction with the Taiwan tectonic prism; and (3) southern termination of Manila Trench in front of Mindoro Island. Transitions between active subduction along the Manila Trench and collision of Taiwan and Mindoro, and relations between active subduction and extension in the Okinawa-Ryukyu and the northeastern Taiwan systems are particularly studied.

BOGDANOV, N. A., Institute of the Lithosphere, USSR Academy of Sciences, Moscow, USSR

Geologic Evolution of the Bering Sea Komandorksy Deep Basin

The deep-water Komandorsky basin is located in the southwestern part of the Bering Sea. On the east, it is separated from the Aleutian basin by the submerged Shirshov Ridge; on the west, it is bordered by structures of the north Kamchatka accretionary prism. The Komandorsky basin is characterized by strongly dissected relief of it acoustic basement, which is overlain by a 1.5 to 2.0-km thick sedimentary cover.

The western part of the basin is occupied by a rift zone, which is characterized by modern seismicity and high heat flow. It is considered to be the axial zone of Miocene-Pleistocene spreading. On the north terrace of the Komandorsky island arc, traced active volcanos provide evidence that subduction is occurring under the arc from the north. The spreading rift zone is reflected on the continent in Miocene-Pleistocene volcanic rocks, characterized by typical oceanic tholeiitic composition.

The Komandorsky basin formed as a result of spreading during the Maestrichtian. Spreading within the basin occurred during the early and middle Oligocene and the late Miocene. East and west of the spreading axis, accretionary prisms formed. The latter are observed along the western flank of the Shirshov Ridge and on the eastern sides of the Kamchatka Peninsula and Koraginsky Island.

BOGDANOV, NIKITA, Institute of the Lithosphere, USSR Academy of Sciences, Moscow, USSR

Soviet Deep Drilling Program. (*No abstract available*)

BOIRAT, JEAN-MICHEL, Bureau de Recherches Géologiques et Minières and GIS Océanologie et Géodynamique, Brest, France, PHILIPPE MALETERRE and JEAN-FRANÇOIS STEPHAN, GIS Océanologie et Géodynamique, Université de Bretagne Occidentale, Brest, France, GUILLERMO R. BALCE, BMG, Manila, Philippines, and LOUIS GUILLOUX, Bureau de Recherches Géologiques et Minières, Orléans, France

Miocene to Holocene Mineralization Associated with Manila Trench Subduction in Central Cordillera, Luzon, Philippines

The Central Cordillera of Luzon is part of the plutonic-volcanic arc associated with the eastward subduction of the South China Sea basin along the Manila Trench. Successive episodes of intrusion and volcanism occurred from the early Miocene to the Holocene. Hydrothermal activity related to these plutonic-volcanic complexes is responsible for various mineralizations, among which five types can be distinguished: (1) porphyry copper with Cu-Au-(Ag)-(Mo); (2) replacement deposit with Zn-Cu-Au-(Ag)-(Pb)-(Cd); (3) vein-type copper-arsenic sulfide deposit with Cu-As-Au-(Ag)-(Te); (4) disseminated mineralization with Au-Ag-(Cu)-(Te); and (5) vein-type deposit with Au-Ag-(Cu)-(Te).

Types 1 and 2 correspond to the oldest (7.3-3.6 Ma) and the deepest mineralization into the plutonic-volcanic complexes. Types 4 and 5 correspond to the youngest (< 1 Ma) and the most surficial complexes. Age versus depth of emplacement distribution of the mineralization is due to the intense Pliocene-Quaternary uplift and correlative erosion of the Central Cordillera. This uplift is associated with the strike-slip motion of the Philippine fault and is contemporaneous with the collision that occurred during the same period, between the southern tip of the arc and the North Palawan block.

BORNHOLD, B. D., and M. T. BRANDON, Pacific Geoscience Centre, Sidney, British Columbia, Canada, R. M. CLOWES, University of British Columbia, Vancouver, British Columbia, Canada, R. G. CURRIE and E. E. DAVIS, Pacific Geoscience Centre, Sidney, British Columbia, Canada, D. M. HUSSONG, Hawaii Institute of Geophysics, Honolulu, HI, R. D. HYNDMAN, Pacific Geoscience Centre, Sidney, British Columbia, Canada, R. P. RIDDIHOUGH, Earth Physics Branch, Ottawa, Ontario, Canada, and G. C. ROGERS and C. J. YORATH, Pacific Geoscience Centre, Sidney, British Columbia, Canada

Recent Acoustic Studies of Western Canadian Continental Margin

A regional survey of the western Canadian continental margin from the central Queen Charlotte Island, 52°40′N, to the Strait of Juan de Fuca, 47°40′N, has been completed with the acoustic imaging system SeaMARC II. These data, combined with single-channel and multichannel seismic reflection data, reveal many new insights concerning the deep structure of the subduction margin off Vancouver Island. Clearly evident in the imagery are the deformation of sediments at the base of the slope, the surface expression of seismically active faults, the mass wasting of sediment frequently observed at the base of the slope, and the erosional canyons and sediment transport channels on the slope and adjacent abyssal plain. The variability in the surficial and deep structures along the length of the margin is great and corresponds well with the postulated variations in the local ocean/continent motion vectors: motion along the southern Queen Charlotte Islands margin is primarily transform (about 55 mm/year) with a small component of convergence (about 10 mm/year); motion south of the triple junction at the Wilson Knolls is convergent but at a very slow rate (about 10 mm/year); and motion along the central and southern Vancouver Island margin is nearly orthogonal to the coast and more rapid (about 40 mm/year).

BRUNS, T. R., R. VON HUENE, and R. D. CULOTTA, U.S. Geological Survey, Menlo Park, CA, and S. D. LEWIS and J. W. LADD, Lamont-Doherty Geological Observatory, Palisades, NY

Geology and Petroleum Potential of Shumagin Continental Margin, Western Gulf of Alaska

Interpretations of multichannel seismic reflection data indicate that the Shumagin continental margin seaward of the Border Ranges fault is underlain by two major seismic sequences, separated by an erosional unconformity beneath the shelf and by the time-correlative conformity seaward. Rocks above the unconformity are late Miocene and younger. Rocks below the unconformity can be as young as middle Miocene beneath the outer shelf and slope, seaward of a paleoshelf break. However, beneath the shelf they are primarily Late Cretaceous turbidites of the Shumagin Formation and Paleocene granodiorite.

Late Miocene and younger structures of the Shumagin margin include Shumagin, Sanak, and Unimak basins and Unimak Ridge, a midslope structural high. Strata in Sanak and Unimak basins were deposited on a subsiding outer shelf and slope, and trapped behind Unimak Ridge and its now-buried structural continuation. Sanak and Unimak basins are in part bounded by northwest-trending extensional faults that parallel both the early Tertiary Beringian margin and a transverse tectonic boundary that segments the forearc. These faults may have developed during collapse and extension along the southeastward continuation of the old Beringian margin, analogous to the processes that created the Bering Shelf basins.

The most promising areas of the Shumagin margin for petroleum potential are Sanak and Unimak basins, which contain strata 8 and 4.5 km thick, respectively, and beneath the outer shelf and slope. Paleogene source rocks like those on the adjacent Alaska Peninsula may be preserved offshore, seaward of the inferred paleoshelf break. Reservoir rocks might have formed from granitic-rich erosional products derived during Oligocene and Miocene erosion of the shelf plutons.

CHASE, T. E., B. A. SEEKINS, J. D. YOUNG, and J. A. WAHLER, U.S. Geological Survey, Menlo Park, CA

SOPAC Marine Geology Atlases

The U.S. Geological Survey conducted a series of marine geologic and geophysical cruises in the southwest Pacific Ocean in 1982 and 1984 as part of a program with participation by Australia and New Zealand. These two SOPAC expeditions obtained various data, which have been compiled into a series of charts and thematic products for the offshore areas of Tonga, Fiji, Vanuatu, the Solomon Islands, and Papua New Guinea. The maps and charts presently being compiled or revised combine previously collected data with information from the SOPAC expeditions. Regional charts at a scale of approximately 1:3 million are included, and more detailed coverage is available at 1:1 million.

Additional geologic information—such as gravity, magnetics, and possibly sediment isopachs—is provided on overlays to the topographic base charts. Reproductions of the seismic reflection data are also included, and tracklines with both time marks and shotpoints will permit correlation with the analog and digital seismic records.

CHUN YEUNG YAN and LOREN W. KROENKE,* Hawaii Institute of Geophysics, University of Hawaii, Honolulu, HI

Cenozoic Reconstruction of Southwest Pacific

Poles of opening and spreading rates for some of the well-studied marginal basins in the southwest Pacific have been redetermined. Times of opening range from Late Cretaceous-Paleocene in the Tasman basin to middle Pliocene in the Bismarck Sea. The observed magnetic lineations in most of these basins show a relatively short duration of opening and relatively small area of total opening. Most of the smaller basins are bounded by troughs and arcuate island chains, some of which are inferred to be trenches and volcanic arcs situated along paleoconvergent boundaries. At least four successive paleoconvergent boundaries are believed to have formed between the Pacific and the Indian-Australian plates during the Cenozoic. Combining the newly determined poles of opening, spreading rates, and paleoplate boundary locations, a series of palinspastic maps of the southwest Pacific have been constructed for these times, relative to a fixed hot-spot frame of reference for both the Pacific and Indian-Australian plates.

COOPER, ALAN K., U.S. Geological Survey, Menlo Park, CA, and FREDERICK J. DAVEY, Department of Scientific and Industrial Research, Wellington, New Zealand

Sedimentary Basins in Ross Sea, Antarctica

The Ross Sea lies in the Pacific sector of the Antarctic continental margin. Three major sedimentary basins (from east to west, the Eastern, Central, and Victoria Land basins) lie beneath the broad, deep (400-1,000 m) continental shelf of the Ross Sea. These north-south-trending basins occur in the extensionally deformed region between East and West Antarctica. Multichannel seismic reflection (MCS) surveys have been conducted over these basins since 1980 by West German, French, Japanese, and U.S. expeditions.

The MCS and previous geophysical surveys have shown that the three basins contain 5-6 km of sedimentary rock, possibly Late Cretaceous and younger. An additional 6-8 km of sedimentary and volcanic rock lies within the deeper parts of the Victoria Land basin. The basins are separated by uplifted and eroded basement ridges covered by thin sedimentary sections. Each basin has distinct characteristics, commonly related to its extensional origin. The Eastern basin, a 300-km wide basement trough, is deepest along its west flank and structurally opens into the Southern Ocean. This basin is filled, in part, by an upper sequence of seaward-prograding Miocene(?) and older(?) layered sedimentary rock. The Central basin, a 100-km wide depression, is bounded by high-angle basement faults and, like the Eastern basin, is filled with a nearly flat sedimentary section. A positive gravity anomaly, possibly rift related, coincides with the axis of the basin.

The Victoria Land basin (VLb) is structurally the most complex. MCS data, collected by the U.S. Geological Survey in 1984, show extensive deformation (faulting, uplift, intrusion) of the layered sedimentary section within the 150-km wide asymmetric basin. An axial rift zone, possibly Paleogene to Holocene in age, comprises a sediment-filled graben and an adjacent magmatically intruded arch. The rift is superimposed upon older basin structures and deeply buried rocks of possible late Mesozoic and older age. The VLb and rift zone parallel the Transantarctic Mountains for nearly 300 km from northern Victoria Land to Ross Island, an active volcanic center, and possibly farther southward.

Petroleum hydrocarbons are unknown from the Ross Sea region, with the possible exception of ethane gas recovered by the Deep Sea Drilling Project. Previous model studies, based on estimated sediment thickness, assumed temperature gradients, and postulated seismostratigraphy, indicate that hydrocarbons could be generated at depths of 3.5-6 km within the sedimentary section. However, this hypothesis cannot be verified without further geologic and geophysical data from the Ross Sea region.

COOPER, ALAN K., MICHAEL S. MARLOW, and DAVID W. SCHOLL, U.S. Geological Survey, Menlo Park, CA

Frontier Basins of Bering Sea

The Bering Sea contains three hydrocarbon provinces: shelf, slope, and abyssal basin. The shelf province, which is underlain by three exceptionally large sedimentary basins (Bristol, St. George, and Navarin) has been extensively covered by geophysical surveys. Exploratory wells in this province have not discovered economic hydrocarbon deposits.

Frontier basins, with future economic prospects, lie beneath the mostly unexplored slope and abyssal provinces of the Aleutian and Bowers basins of the Bering Sea. These expansive deep-water basins cover an area nearly two-thirds the size of Alaska. Sedimentary rocks of late Mesozoic(?) and Tertiary age make up the 3 to 11-km thick section that covers the entire area. These pelagic and fine-grained terrigenous rocks overlie slowly subsiding oceanic crust, which was trapped in the Bering Sea after the Aleutian Ridge formed during the Eocene.

Thick (6-11 km) sedimentary sections and high basement relief (0.5-1.0 km) underlie parts of the 1,700-km long Beringian-Koryak slope and rise. At the base of the margin, nearly flat strata of the Aleutian basin drape and onlap faulted acoustic-basement rock. Sparse information from the Deep Sea Drilling Project and U.S. Geological Survey dredging indicates that source and reservoir rocks may exist within the upper and, possibly, lower parts of the sedimentary section. Beneath the central part of the abyssal basin, the sedimentary sequence is only 2-4 km thick, and buried basement relief is greater than 1.0 km. Near-surface temperature gradients are high (average = 58°C/km), and acoustic anomalies that may be indicative of hydrocarbons (velocity-amplitude anomalies, VAMPs) are present.

The slope and abyssal basins of the Beringian-Koryak margin have received minimal exploratory attention because water depths that cover the area are typically 2,000-3,700 m. Advances in deep-water production technology will eventually make these frontier basins attractive for hydrocarbon exploration.

DALZIEL, I. W. D., Institute for Geophysics, University of Texas, Austin, TX, S. W. GARRETT, British Antarctic Survey, Cambridge, England, United Kingdom, A. M. GRUNOW, Lamont-Doherty Geological Observatory, Palisades, NY, R. J. PANKHURST and B. C. STOREY, British Antarctic Survey, Cambridge, England, United Kingdom, and W. R. VENNUM, Sonoma State University, Rohnert Park, CA

Southern Rim of Pacific Ocean Basin: Southern Andes to Southern Alps

Between the southern Andes of Tierra del Fuego and the Southern Alps of New Zealand lies the least accessible and geologically least explored part of the Pacific Ocean basin. A joint United Kingdom-United States project was initiated in 1983 to elucidate the geologic history and structure of the Pacific margin of Antarctica from the Antarctic Peninsula to Pine Island Bay at approximately long. 105° W. The first season (1983-1984) of this West Antarctic Tectonics Project was spent in the Ellsworth-Whitmore crustal block, and the second (1984-1985) in the Thurston Island crustal block. The project involves structural and general field geology, petrology, geochemistry, paleomagnetism, and airborne geophysics (magnetics and radar ice echo sounding). A final geologic season will be spent in the Pensacola Mountains of the Transantarctic Range in 1987-1988. Currently, airborne geophysical work will continue near the Byrd Subglacial basin and the Bentley Subglacial Trough.

This display will feature new paleomagnetic data indicating that little relative motion has occurred between the Antarctic Peninsula and the Ellsworth-Whitmore crustal blocks, but movement of both blocks has occurred relative to the East Antarctic craton. The favored reconstruction restores them with a 15°-20° counterclockwise rotation and a northward translation of up to approximately 10° of latitude, to a position on the Pacific side of the Falkland Plateau-Cape Fold Belt-Coats Land junction.

Geochemical data indicate that orthogneisses exposed at Haag Nunataks represent a Proterozoic cratonization event, and aeromagnetic data demonstrate that related rocks occur beneath the ice from the Ellsworth Mountains to the Antarctic Peninsula. Thus, the above reconstruction is geologically acceptable in that it places these rocks close to crystalline Precambrian rocks of similar age in the Falkland Islands and along the coast of Coats Land.

The Ellsworth-Whitmore block consists mainly of folded Paleozoic strata of the Gondwana cover sequence cut by middle Jurassic peraluminous "S-type" granites. These are crustal anatectic (or at least highly contaminated) melts signifying an intracontinental thermal event of substantial proportions that seems to herald the breakup of the Gondwanaland supercontinent.

The subsequent late Mesozoic-Cenozoic history of the Ellsworth-Whitmore block is recorded in geophysical evidence for sub-ice fault-bounded crustal rifts, some of which are probably young features related to Cenozoic alkali magmatism and the recent marked uplift of the Ellsworth Mountains, the highest on the continent (5,000 m).

Finally, plans are to continue the project westward into Marie Byrd Land (long. 105°-180°) as a cooperative United Kingdom-New Zealand-United States venture, to complete a study of the entire southern rim of the Pacific Ocean basin.

DAVEY, F. J., Department of Scientific and Industrial Research, Wellington, New Zealand, M. HAMPTON, U.S. Geological Survey, Menlo Park, CA, and K. LEWIS, New Zealand Oceanographic Institute, Department of Scientific and Industrial Research, Wellington, New Zealand

Convergent Plate Margin East of North Island, New Zealand

The Indian-Pacific plate boundary passes along the eastern margin of North Island, New Zealand, with the Pacific plate being thrust under the Indian plate to the west. The continental slope forming the Indian plate margin is broad with a well-formed series of trench slope basins and intervening ridges along the continental slope and shelf, subparallel to the margin, and continuing onto land. Multichannel seismic reflection data recorded across this margin show a thick (2.5-km) sedimentary section overlying oceanic basement in the deep-water part of the profile, and part of this sedimentary section is apparently being subducted under the accretionary prism. At the toe of the continental slope, nascent thrusts, often showing little apparent offset but a change in reflection amplitude, occur over a broad region. Well-defined trench slope basins show several episodes of basin formation and thrusting and are similar to structural interpretations for adjacent onshore basins. A bottom simulating reflector, which may delineate a gas-hydrate layer, can be traced over the midslope part of the profile. A major reflector, interpreted as the base of the accretionary prism, can be traced discontinuously to the coast where it coincides with the top of a zone of high seismicity, considered to mark the top of the subducted Pacific plate.

DAVIS, E. E., and R. G. CURRIE, Pacific Geoscience Centre, Sidney, British Columbia, Canada, H. VILLINGER, Institute for Polar Research, Bremerhaven, Federal Republic of Germany, W. D. GOODFELLOW, Canada Geological Survey, Ottawa, Ontario, Canada, D. M. HUSSONG, Hawaii Institute of Geophysics, Honolulu, HI, W. B. F. RYAN, Lamont-Doherty Geological Observatory, Palisades, NY, and S. R. HAMMOND, National Oceanic and Atmospheric Administration, Newport, OR

Regional Acoustic Imagery and Detailed Geophysical Studies of Northern Juan de Fuca Ridge

Acoustic image and high-resolution swath mapping studies have been conducted over the northern Juan de Fuca Ridge system from the central Juan de Fuca Ridge at lat. 46°N to the triple junction of the ridge with the Queen Charlotte Islands continental margin at lat. 52°N. Results of the surveys have been published as 1:250,000 compilations and 1:50,000 detailed sheets of acoustic image mosaics and 10-m contour bathymetric maps. Of particular interest is the information that influences processes of hydrothermal mineralization. In general, little direct information is present in the imagery; only the exact location and nature of the currently active rift axis are seen. However, this information is important, since the along-strike variations in crestal morphology, the recency and level of volcanic activity, and the degree of post-volcanic extension must be known in order to understand any hydrothermal system observed along the axis.

An exceptional case has been studied in detail at the northern Juan de Fuca Ridge, lat. 49°30′ N, where Pleistocene turbidite sediments bury the axial valley, and structures suspected to be of hydrothermal origin were observed in the acoustic imagery over faults that bound the valley. Subsequent higher resolution acoustic imagery, seismic reflection profiling, and heat-flow studies indicated that the features were of hydrothermal origin. A 2.5-m piston core sample of massive sulfides from one of the structures confirmed this speculation. The setting in which these relatively large (500-m diameter, 60-m surficial relief, 350-m relief above basement) structures occur is remarkably similar to that where large sediment-hosted sulfide deposits are found on land, although we do not know what the bulk composition of the structures is, and thus whether the analogy is complete.

EITTREIM, STEPHEN L., and MONTY A. HAMPTON, U.S. Geological Survey, Menlo Park, CA

Prebreakup Sedimentary and Volcanic Sequences on Rifted Continental Margin of Wilkes Land, Antarctica

The transition zone from continental crust to oceanic crust on the Wilkes Land, Antarctica, margin was surveyed with 24-fold multichannel seismic reflection along 1,800 line-km in 1984. The northward transition from extended continental crust to normal oceanic crust is characterized by a progression through three distinct volcanic sequences. The first, deposited on extremely thin continental crust, is interpreted to be a flood basalt, presumably deposited during the late rifting stage. The second, which overlaps the first at its northern end, is a seaward-dipping reflector sequence that forms anomalously thick oceanic crust. The third, which in turn onlaps the second, is normal oceanic crust with a rough upper surface and no internal reflectors.

A deep marginal basin, containing synrift deposits and topped by a breakup unconformity, lies landward of these volcanic sequences in the region of attenuated continental crust. These synrift deposits have a seismic velocity of 3.6 ± 0.5 km/sec, and are as thick as 4 km in the deepest part of the basin adjacent to the edge of oceanic crust. Because the breakup that culminated the rifting phase between Antarctica and Australia occurred at about 96 Ma, the synrift deposits are Early Cretaceous and perhaps Late Jurassic in age. The deepest sequence on attenuated continental crust is a highly stratified, block-faulted unit of 4.2 ± 0.4 km/sec seismic velocity as much as 4 km thick that is bounded below by shallow-dipping faults or decollements. The fault blocks are back- and fore-rotated, and the offset of marker horizons indicates extension not exceeding 25%. Whether this deepest stratified unit is prerift or synrift is debatable. Its relatively high seismic velocity, erosional upper surface, and lack of syndepositional faulting indicate prerift. However, the relatively mild extension to which it was subjected, with block rotations rarely exceeding 10°, makes it less likely to be a prerift depositional unit, as studies in known extensional regions normally show block rotations that produce steep dips and extension commonly exceeding 25%.

Moho reflectors can be followed from beneath oceanic crust landward to a depth of 20 km beneath attenuated continental crust. Steps in the Moho along its descent beneath the continent suggest subcropping crustal shear zones or faults that accommodated the thinning of the overlying attenuated continental crust.

EXON, N. F., Bureau of Mineral Resources, Canberra, Australia, and M. S. MARLOW, U.S. Geological Survey, Menlo Park, CA

The Undrilled New Ireland Basin in Papua New Guinea

The arcuate, west-northwest-trending, mostly offshore New Ireland basin is 900 km long and about 160 km wide, and extends northeastward from Manus Island, New Hanover, and New Ireland. The prospective eastern part of the basin is structurally simple and contains up to 7 km of sedimentary strata. Our study used 4,000 km of multichannel seismic data, including 2,000 km of 1984 R/V *S. P. Lee* data acquired under a tripartite aid program, other geophysical data, outcrop geology, and seabed samples.

The basin formed in a forearc between a southerly Eocene to early Miocene volcanic arc, and a northerly outer-arc high bounding the Manus Trench. Its southern margin drops down to the back-arc Manus basin, which commenced spreading in the Pliocene. North of Manus Island, the New Ireland basin contains areas of deformed strata that have apparently been accreted to the Manus arc by south-dipping thrust faults. In places these strata are overlain by shallowly buried lava flows, which may represent attempted spreading.

The sedimentary sequence in the eastern part of the basin is interpreted to contain thick Oligocene to early Miocene volcaniclastic sediments, overlain by 1,000-2,000 m of Miocene shelf carbonates, overlain by 2,000 m of overburden. The presumed shelf carbonates could contain both source and reservoir rocks. The *Lee* line 401 revealed a flat, high-amplitude reflector or bright spot in an anticlinal core 1,700 m beneath the seabed in water 2,500 m deep off New Ireland, suggesting that hydrocarbons have been generated in New Ireland basin.

FALVEY, D. A., Bureau of Mineral Resources, Canberra, Australia, K. HINZ, Bundesanstalt für Geowissenschaften und Rohstoffe, Hannover, Federal Republic of Germany, J. B. WILLCOX, N. F. EXON, P. A. SYMONDS, and P. E. WILLIAMSON, Bureau of Mineral Resources, Canberra, Australia

Prospective Frontier Basins off Eastern Australia

Eleven thousand kilometers of high-quality multichannel seismic reflection data have been gathered in four poorly known, but prospective areas off eastern Australia. The data were collected by the Bureau of Mineral Resources (BMR) using R/V *Rig Seismic* and by the Bundesanstalt für Geowissenschaften und Rohstoffe (BGR) using F. S. *Sonne*. The four areas are the Otway basin off Victoria and South Australia, and the west Tasmanian margin, both partly explored, and the unexplored Lord Howe Rise and Queensland Plateau.

The Otway basin has an area of 100,000 $km^2$, more than half of which is in depths exceeding 500 m. Its Cretaceous and Cenozoic sedimentary sequence is up to 10,000 m thick and is cut by large coast-parallel normal faults. The oldest marine strata are Cenomanian. The basinal area off west Tasmania covers 40,000 $km^2$, two-thirds of it in offshelf depths. It contains up to 6,000 m of Cretaceous and Cenozoic sequences similar to those of the Otway basin. Both Otway and west Tasmanian areas formed during the breakup of Australia and Antarctica, and exhibit complex structuring due to wrenching imposed on the simpler extensional features.

The Lord Howe Rise is a ribbon of continent off eastern Australia, about 2,000 km long and 400 km wide. Much of its crest lies in water depths of 750-1,200 m. Up to 4,500 m of Mesozoic and Cenozoic sequences has been identified, and extensive faulting, related to the formation of the Tasman Sea, has formed rift basins and horst and graben areas. Simple extension was apparently dominant in the south, and oblique extension in the north.

The Queensland Plateau covers 200,000 $km^2$, half of which is in water shallower than 1,000 m. The plateau behaved as a stable block during and after the Paleogene spreading episode, which formed the Coral Sea to the northeast. About 1,000 m of latest Cretaceous and Cenozoic sequences are present above a planated surface on the plateau, and up to 5,000 m in the flanking Queensland and Townsville Troughs to the southwest. The sequence beneath the planated surface is believed to contain Cretaceous rift-fill sediments in places.

FISHER, MICHAEL A., U.S. Geological Survey, Menlo Park, CA, and JEAN-YVES COLLOT, JACQUES RECY, and JACQUES DANIEL, Orstom Centre de Nouméa, Nouméa New Caledonia

Structure Imposed on New Hebrides Island Arc by Subduction of d'Entrecasteaux Zone, a Major Submarine Mountain Chain

The northwest-trending New Hebrides island arc marks the juncture between the subducting Australia-India plate on the west and the overriding Pacific plate and North Fiji basin on the east. The d'Entrecasteaux zone (DEZ), which consists of two subparallel, highstanding (1-2 km) ridges, trends eastward across the India-Australia plate and collides with the central part of the arc. Recently, considerable new geologic and geophysical information—including dredged rock samples, seismic refraction and multichannel seismic reflection data, Seabeam bathymetry, and locally detailed magnetic and gravity data—has been obtained over this collision zone. Our studies use these data to investigate the structure and evolution of this collision zone. Migrated multichannel seismic data show highly varied rock structure beneath the arc's trench slope within the collision zone. The north ridge of the DEZ is apparently being subducted without much disruption of the Bougainville Spur, an enigmatic feature that may contain obducted oceanic rocks. This spur juts above the lower slope of the arc, but a recognizable subduction zone does not extend beneath it. Highly detailed Seabeam bathymetric data have recently been acquired and are being integrated into the study of lower-slope structure.

Directly east of where the DEZ and the New Hebrides arc collide, the Aoba basin lies encircled by a relatively shallow-water (0 to 2,000-m depth), arc-summit platform. This basin is overlain by water 2-3 km deep and contains 5-6 km of sedimentary rocks. Theoretical studies and interpretation of multichannel seismic data suggest that the basin may have deepened substantially during collision of the DEZ and arc. Hence, understanding the tectonic processes that attend arc-ridge collisions will not only contribute to our general understanding of the development of intra-arc basins, but also may help determine whether hydrocarbons have been generated within the Aoba basin.

GARDNER, J. V., D. A. CACCHIONE, D. E. DRAKE, B. D. EDWARDS, M. E. FIELD, M. A. HAMPTON,* H. A. KARL, and D. S. MCCULLOCH, U.S. Geological Survey, Menlo Park, CA, and N. H. KENYON and D. G. MASSON, Institute of Oceanographic Sciences, Wormley, England, United Kingdom

Sonograph Mosaic of U.S. West Coast Exclusive Economic Zone

The Geological Long-Range Inclined Asdic (GLORIA) side-scanning sonar system was used to obtain data that were compiled as an image-enhanced acoustic mosaic, similar to an aerial photograph, of the sea floor from the edge of the continental shelf to 200 nmi offshore within the U.S. Exclusive Economic Zone off California, Oregon, and Washington. The mosaic clearly displays the large-scale geomorphic and sedimentologic features of the sea floor, including spreading centers, seamounts, fracture zones, sediment fans, continental-slope canyons, and abyssal-plain channels. Hundreds of seamounts (some previously uncharted) dot the deep sea floor, and many have large summit craters and attendant volcanic flows. The major Nitinat, Astoria, Delgada, and Monterey sediment fans are traversed by lengthy channel-levee complexes that extend from morphologically diverse canyons on the adjacent continental slope. Areally extensive sediment-wave fields occur adjacent to the complexes. Some channels on the abyssal plain are straight whereas others are highly sinuous, suggesting that various channel-forming processes occur on the sea floor. The contrast between the transform tectonic regime south of Cape Mendocino and the convergent tectonic regime to the north is particularly apparent on the mosaic. The linear basement ridges that were generated at the Gorda and Juan de Fuca spreading centers and were later moved apart by sea-floor spreading are abruptly truncated by the

Mendocino and Blanco fracture zones that accommodate horizontal slip between adjacent lithospheric plates.

GERRARD, CARLA S., GEORGE H. GELB, and ALLEN LOWRIE, U.S. Navy, China Lake, CA

GEOTEC—A Promising Energy Alternative

The Pacific basin borders are characterized by active plate boundaries in which geothermal energy, associated with volcanic arcs, is prevalent. High-quality (270°-306°C) geothermal resources have been developed in areas such as New Zealand and Japan, but lower quality (less than 200°C) resources have not been tapped for electrical power generation.

If these lower quality geothermal heat sources were used in conjunction with cold ocean-water heat sinks, the resulting increased temperature differences would improve energy conversion efficiencies so as to be economically attractive. The technology of solar-driven ocean thermal energy conversion (OTEC) may be combined with geothermal power system technology to produce a geothermal/ocean thermal energy conversion (GEOTEC) system.

Geothermal reconnaissance has shown a promising resource at the base of three clustered volcanoes on Adak Island, located along the Aleutian arc of Alaska. Geophysical surveys revealed areas of abrupt and steep gravitational gradients and high electrical conductivity over the volcanoes, indicating possible subsurface magma. Geochemical analyses indicated a potential reservoir temperature of 180°C. The nearby Bering Sea has a surface temperature ranging between 3° and 8°C year-round and would provide for high heat reinjection.

The combination of geothermal and ocean sink resources could result in a GEOTEC plant having an overall thermal efficiency approaching one-third that of the present diesel-electric generating system at the Naval Air Station on Adak. The more than 5,000 permanently stationed personnel at the station consume over 8 million gal of JP-5 fuel for space heating and electrical power. Presently, electric energy on Adak costs 250-350 mils/kilowatt-hour. Preliminary estimates of the cost of electric power from a GEOTEC plant are about 200 mils/kilowatt-hour. A GEOTEC plant would also be a secure alternative energy source for the U.S. Navy.

GINSBURG, R. N., H. E. CLIFTON, and R. J. WEIMER, SEPM Steering Committee, Global Sedimentary Geology, Tulsa, OK

Global Sedimentary Geology Program

The Society of Economic Paleontologists and Mineralogists, in collaboration with the International Association of Sedimentologists and the International Union of Geological Sciences Committee on Sedimentology, is developing a new international study under the provisional title of Global Sedimentary Geology Program (GSGP).

Initially, three research themes are being considered: (1) event stratigraphy—the documentation of examples of mass extinctions, eustatic fluctuations in sea level, major episodes of volcanisms, and changes in ocean composition; (2) facies models in time and space—an expansion of the existing data base of examples of facies models (e.g., deltas, fluvial deposits, and submarine fans) and global-scale study of the persistence of facies at various times in geologic history; and (3) sedimentary indices of paleogeography and tectonics—the use of depositional facies and faunas in paleogeography and in assessing the timing, locus, and characteristics of tectonism. Plans are being developed to organize pilot projects in each of these themes.

Research in offshore areas will contribute to all three themes, especially to the second and third. For example, far too few of the world's deltas and submarine fans have been studied in detail. Existing sedimentary indices of paleolatitude and circulation must be refined, and new indices found. Also, the sedimentary indices of tectonism, especially provenance, need increased calibration.

With a global-scale focus on sediments, sedimentary rocks, and their contained floras and faunas, the GSGP will encourage sedimentologists, stratigraphers, paleontologists, and other geoscientists to expand research into new areas, to widen international collaboration, and to promote the exchange of information and ideas.

GONZALEZ, EDUARDO, Empresa Nacional del Petróleo, Santiago, Chile

Petroleum Possibilities in Continental Margin off Central Chile

The continental margin off central Chile, from Valparaiso to Valdivia, encompassing an area of 100,000 km$^2$, has been the target of exploratory activity by Empresa Nacional del Petróleo since 1970. Exploratory drilling began in 1972. By August 1984, total exploratory efforts had resulted in drilling 14 offshore wells and acquiring 12,130 km of seismic reflection lines. A biogenic gas accumulation was discovered in the "F" well. Because these attempts to find oil were unsuccessful and because drilling costs have escalated, exploratory activities have been curtailed.

Forearc basins off central Chile are characterized by low geothermal gradient and a sedimentary filling of Cretaceous and Tertiary strata. Tertiary sequences are characterized by low organic carbon content, immature humic-type organic matter, and a biogenic gas potential. Cretaceous sequences are characterized by higher organic carbon content, good reservoir rocks, and fair to good source rocks. The organic matter is sapropelic, with vitrinite and liptinites, and is favorable for oil and gas generation. Seismic and well data suggest that Mesozoic and Cenozoic sedimentary rock sequences filling the basins (more than 4,000 m thick at the shelf edge) extend 40-70 km beyond the present shelf edge. Mesozoic rocks deposited on the slope may generate petroleum and gas that could migrate upslope and accumulate in traps associated with the faulted basement highs and graben-type depressions existing at the shelf edge.

This geologic setting favors the development of large petroleum accumulations along the shelf edge and graben on the sedimentary basins off central Chile.

GREENE, H. G., F. L. WONG, D. A. FALVEY, A. MACFARLANE, and SCIENTIFIC STAFF of R/V of *S.P. Lee*, U.S. Geological Survey, Menlo Park, CA

Basin Development and Petroleum Prospects of Central New Hebrides Arc

Tectonic processes resulting from active subduction and collision of the d'Entrecasteaux Ridge with the central New Hebrides arc have created thick sedimentary basins that could be promising targets for hydrocarbon concentration. Two extensive intra-arc basins, the Central basin of Vanuatu and the Vanikolo basin, have recently been mapped in detail and appear to have developed in the last 3 m.y. primarily from subduction and arc-polarity reversal. The Central basin of Vanuatu is being modified by the ridge-arc collision and is still accumulating sediment, whereas the Vanikolo basin has no physiographic expression and is completely filled with sediment.

Quaternary volcanism has separated the Central basin of Vanuatu into two physiographic basins (North Aoba and South Aoba basins). The Central basin is flanked on the west by five smaller insular basins associated with the islands of Malakula and Espiritu Santo. Although North and South Aoba basins lie in water depths as great as 3 km, migration paths into the shallower water basins appear to be present, and structures that could trap hydrocarbons occur. More than 5 km of sediment of Miocene and younger age were deposited in the deeper basins, whereas 2-km thick sediments of late Miocene and younger age occur near the islands. The most promising hydrocarbon basin is Malakula basin, which lies on the eastern shelf of Malakula Island, where onshore source rock analyses indicated potential gas concentrations.

North of the Central basin of Vanuatu lies the largest basin in the region, the Vanikolo basin, covering more than 14,000 km$^2$ and containing more than 6 km of probable Miocene and younger sediment. Water depths are more than 2 km in the deepest part of this basin, but much of the margins lie in water less than 1 km deep. Structures along the margins could trap hydrocarbons, especially along the western side, and are accessible to drilling. More work must be done in Vanikolo basin to assess hydrocarbon source rocks, migration, and heat flow before a comprehensive hydrocarbon evaluation can be made. However, the extent of the basin and the sediment thickness are unusual for an active island-arc setting; further study is needed before it can be discounted as a potential petroleum target.

GREENE, H. GARY, U.S. Geological Survey, Menlo Park, CA, and MICHAEL P. KENNEDY, California State Division of Mines and Geology, Menlo Park, CA

California Continental Margin Map Series

A newly published map series depicting geologic data in the offshore regions of California will help evaluate marine energy and mineral resources. This series represents the first standardized marine geologic maps to be produced for the West Coast of North America. It is compiled and published cooperatively by the California Division of Mines and Geology and the U.S. Geological Survey with assistance from the California Coastal Zone Commission. The series consists of 28 sheets at a scale of 1:250,000, covering seven offshore areas along the entire coast from the Mexican border to Oregon. It includes a bulletin illustrating symbols used, idealized stratigraphic columns, a master legend, and a comprehensive bibliography.

For each of the seven areas mapped, four separate sheets printed in color depict: (1) sea-floor geology and geohazards, (2) gravity and magnetic contours overlaid on geology, (3) earthquake epicenters and fault-plane solutions, and (4) data sources including tracklines of geophysical data used in the map compilation and the location of offshore well holes and actively producing petroleum rigs. These sheets include most of the up-to-date data published for the offshore region of California as well as newly interpreted geophysical data evaluated by the U.S. Geological Survey.

HAYES, DENNIS E., Lamont-Doherty Geological Observatory, Palisades, NY

Regional Synthesis of Marine Geologic and Geophysical Data from Offshore Circum-Antarctic

Preliminary maps of bathymetry, free-air gravity, magnetic anomalies, sediment thickness, and crustal structure have been compiled from all available data sources for the Circum-Antarctic region. The maps cover the region from the Antarctic continent to lat. 30°S, and have been produced on a Polar Stereographic Projection in common grid sectors and at a nominal scale of 1:8 million.

Because certain sectors of the offshore Antarctic region are sparsely surveyed, SEASAT data have been used to provide more complete gravity coverage. The 5-min × 5-min DBDB bathymetric data file compiled by NAVOCEANO has been used to create a computer-generated bathymetric map for the region. This bathymetric map has been updated using data collected from recent cruises. The maps will be presented at reduced scale with focus given to those sectors that incorporate the Pacific regions.

The overall compilation is the result of data input and collaboration of various scientists from several countries, most notably, Federal Republic of Germany, France, United Kingdom, New Zealand, Australia, Norway, and Japan.

HEIN, JAMES R., U.S. Geological Survey, Menlo Park, CA, FRANK T. MANHEIM and WILLIAM C. SCHWAB, Woods Hole Oceanographic Services, Woods Hole, MA, and DAVID A. CLAGUE, U.S. Geological Survey, Menlo Park, CA

Cobalt-Rich Ferromanganese Crusts from Central Pacific

Ferromanganese crusts coat most hard-rock substrates on seamounts, ridges, and plateaus in the Central Pacific basin. Crusts from shallow water (< 2,500 m) are rich in the strategic and economically important metals manganese, cobalt, nickel, and platinum. We collected samples from the U.S. Exclusive Economic Zone surrounding the Hawaiian, Johnston, Palmyra, and Marshall Islands. Crust thicknesses are variable, Necker Ridge having the greatest average thickness of 4 cm. The dominant mineral in all crusts is $\delta MnO_2$ (vernadite); quartz, plagioclase, and apatite are minor phases. Substrate rocks are dominantly volcanic breccia, with lesser amounts of hyaloclastite, basalt, phosphorite, limestone, and volcaniclastic sandstone. Ferromanganese-encrusted volcanic breccias commonly show two or three stages of reworking. Sediment reworking either by gravity flow or bottom-current activity on the flanks of the submarine edifices partly controls the crust thicknesses by burial and dissolution, or by erosion, whereupon growth of the crusts begins again. Talus dredged from seamount slopes is coated with crusts half as thick as crusts on rocks recovered from nearby outcrops. Regional variations in crust chemistry and mineralogy can be related to various oceanographic and geologic conditions. For example, quartz content in the crusts varies with the zonal wind patterns and is eolian, whereas apatite is most common in crusts formed near zones of high biologic productivity. In the Line Islands, cobalt, nickel, and manganese increase and iron decreases with proximity to the equator in the outer layers of the crusts. Of the areas we studied, the Exclusive Economic Zones of Johnston Island and the Marshall Islands have the greatest economic potential.

HELSLEY, C. E., B. KEATING, E. DECARLO, G. MCMURTRY, F. CAMPBELL, L. KROENKE, P. JARVIS, and M. PRINGLE, Hawaii Institute of Geophysics, Honolulu, HI

Resource Assessment of Cobalt-Rich Ferromanganese Crusts in Central and Western Pacific

Studies of the distribution, variability, and geochemistry of ferromanganese crusts within the Exclusive Economic Zone (EEZ) of Hawaii and southwestern Pacific nations indicate that ferromanganese crusts are widely distributed within the EEZs and that the crusts provide an important potential economic resource. Ferromanganese crust thicknesses vary widely, but often show maximum thicknesses of 12 cm on older seamounts. Elemental abundances and depth relationships observed are similar to results from other central and western Pacific seamount studies. The dominant crystalline manganese mineral phase is two-peak $\delta MnO_2$. The substrates on which the manganese crusts occur include basalts, hyaloclastites, and sediments, with hyaloclastites and basalts the most common substrates. The seismic reflection studies and SeaMARC II data analyses indicate rock type varies considerably on a local scale on EEZ seamounts.

HERZER, RICHARD H., New Zealand Geological Survey, Lower Hutt, New Zealand, PETER F. BALLANCE, University of Auckland, Auckland, New Zealand, JAMES W. COLE, Victoria University, Wellington, New Zealand, NEVILLE F. EXON, Bureau of Mineral Resources, Canberra, Australia, ANDREW J. STEVENSON, U.S. Geological Survey, Menlo Park, CA, and DAVID TAPPIN, Ministry of Lands Survey and Natural Resources, Nuku'alofa, Tonga

Seismic Stratigraphy, Sediments, and Basin History of Tonga Forearc Basin, Late Eocene to Pleistocene

Four seismic reflectors (A, B, C, V) define primarily unconformity-bounded sedimentary sequences. Basement (V) is a block-faulted surface, apparently of Eocene volcanics. Above this, an upper Eocene sequence (CV) mainly buries the fault topography, pinching out locally on fault-block and volcanic highs along the eastern side of the basin. This sequence includes volcaniclastics and, on paleohighs, shallow-water limestones. Overlying this is a widespread upper Oligocene-lower Miocene sequence (BC), which also thins and pinches out locally against the high eastern side of the basin. Volcaniclastics are common, but limestones may occur locally. Seismic interpretations indicate little faulting during deposition of this sequence; prominent lenticular bodies could be either sills or reefs. Sequence AB, of middle and late Miocene to early Pliocene age, is composed of volcanopelagics deposited when the Lau arc was active, adjacent to the Tonga platform. No volcanic centers are seen in this sequence in the forearc, but shallow intrusions are common. Major tensional faulting developed toward the end of this depositional cycle. The uppermost sequence (SBA), of late Pliocene to Pleistocene age, also comprises volcanopelagic sediments. The volcanics are derived from the nearby Tofua arc, which developed with the opening of the Lau back-arc basin. Doming and tensional faulting in the late Pliocene-Pleistocene raised parts of the Tonga forearc basin, allowing wide reef platforms to develop.

Dredge samples were used to supplement the age control from wells, on Tongatapu and outcrops on 'Eua. All dredged rocks consist of two components in varying proportions: biogenic ooze and volcanic ash. Many samples indicate downslope reworking and mixing as gravity flows. Gravity flows have carried carbonate well below the present carbonate compensation depth. Much of the platform is mantled with recent pumice. Diagenetic effects include common alteration of glass to smectites and zeolites, precipitation of zeolites and clays in cavities, minor growth of secondary carbonate, and rarely, prehnite alteration and carbonate veining. Cretaceous rocks dredged from the lower trench slope—calcareous and siliceous oozes and silica-rich tuffs—are exotic. They are inferred to have been transferred from the Pacific plate, with volcanic and mafic igneous rocks, by offscraping from the downgoing Louisville Ridge.

HINZ, KARL, ENRICO H. K. KEMPTER, and HANS-ULRICH SCHLÜTER, Bundesanstalt für Geowissenschaften und Rohstoffe, Hannover, Federal Republic of Germany

Geology of Southern Palawan-Balabac Area: Results of Marine Geophysical Studies in Northwest Sulu Sea and in South China Sea

According to our interpretation, two major terranes are superimposed in the Balabac-south Palawan area. An autochthonous Palawan continental terrane extends from the Dangerous Grounds and Reed Bank to beneath the islands of Palawan and Balabac, and parts of the Sulu Sea. An allochthonous imbricated terrane covers an area of approximately 170 × 350 km between the southwestern prolongation of the Cagayan Ridge in the Sulu Sea and the Palawan Trough in the South China Sea, including southern and central Palawan and the Balabac Island. This originally accreted terrane formed outside the Balabac-southern Palawan area, but has been thrust onto the Palawan continental terrane since the early Miocene. If the allochthonous terrane is underlain by the continuation of the Palawan continental terrane, as we believe, new possibilities for hydrocarbon exploration will develop.

HOLMES, MARK L., JANET L. MORTON, DAVID A. CLAGUE, U.S. Geological Survey, Menlo Park, CA, ROBERT W. EMBLEY, National Oceanic and Atmospheric Administration, Newport, OR, DAVID A. CACCHIONE, U.S. Geological Survey, Menlo Park, CA, and DALLAS H. ABBOTT, Oregon State University, Corvallis, OR

Morphology and Mineral Resources of Gorda Ridge and Blanco Fracture Zone, Northeast Pacific Ocean

The Gorda Ridge is a 300-km long segment of the Pacific mid-ocean ridge system located 250 km west of the northern California and southern Oregon coastline. It is bounded on the north by the Blanco fracture zone and on the south by the Mendocino Ridge. This slow to medium-rate spreading axis exhibits numerous morphologic similarities to the slow-spreading Mid-Atlantic Ridge and is unique in this respect among Pacific ridge crests.

The axis of Gorda Ridge is characterized by a deep (3,200-3,800 m) axial valley flanked by steep ridges rising to depths of 2,000-2,500 m. The ridge is divided into three parts by right-lateral offsets of the axis at lat. 41°37′N and 42°26′N. SeaMARC II and GLORIA acoustic imagery and Seabeam bathymetry show the structural and volcanic features associated with active ridge-crest processes, including: (1) high reflectivity of the valley floor, indicating recent volcanism; (2) numerous small volcanic cones, discs, and pit craters; and (3) long continuous fault scarps on the ridge flanks that are parallel or subparallel to the ridge axis. Structures at the northern end of Gorda Ridge bend westward where they abut the Blanco fracture zone. Escanaba Trough, the southernmost ridge segment, contains more than 500 m of Pleistocene turbidites and hemipelagic sediment. Several discrete volcanic centers (3-6 km in diameter and spaced along the trough axis at intervals of 15-20 km) pierce the sediment fill of Escanaba Trough. Heat-flow values exceeding 1,200 mW/m$^2$ were measured over two of these intrusion zones.

Dredge hauls from the northeastern flank of a large extrusive body at lat. 40°45′N recovered mudstone slabs, basaltic glass, and about 4.4 kg of massive sulfides consisting primarily of unoxidized pyrrhotite and minor galena, sphalerite, and chalcopyrite. Bottom photographs of the axial-valley floor in the northern Gorda Ridge segment at lat. 42°45′N also indicate the existence of hydrothermal deposits, although no samples were recovered from that site. Four deep (4,500 m) extensional basins in the Blanco fracture zone are structurally analogous to pull-apart basins found along large strike-slip faults on continents and in the Gulf of California. Subsurface intrusive bodies, and domed and tilted turbidites in one Blanco basin (Cascadia Deep) indicate active volcanism and the possibility of associated hydrothermal systems.

HONZA, EIICHI, Japan Geological Survey, Ibaraki, Japan, JOCK KEENE, Sydney University, Sydney, Australia, and DONALD L. TIFFIN, CCOP/SOPAC, Suva, Fiji

### Plate Boundaries in Solomon Sea Region and Subduction at Trenches

The Solomon Sea region is part of an interactive buffer zone of small plates between the Australia and Pacific plates. The Solomon Sea plate is subducting both in the New Britain Trench on the north and in the Trobriand Trough on the south. Subduction in the Trobriand Trough is at a rate of less than 2 cm/year. A low rate of oblique convergence would explain the present lack of recorded seismicity under the Trobriand platform. The Trobriand Trough is filled with thick sediments. The depth to oceanic basement is approximately the same as in the New Britain Trench. The difference in the surficial depth to the trenches results from the greater thickness of the trench fill in the Trobriand Trough.

Accretion is observed throughout the New Britain Trench, gradually increasing toward the west, and is finally extended to the Huon Ridge forming the Finsch Deep behind the ridge. These distinct subduction-related processes document the structural development of: (1) sediment subduction accompanied by minor accretion in the east; (2) subduction-accretion farther west; and (3) collision in the west.

The age of the Solomon Sea plate is refined using new magnetic, heat-flow, and water-depth data; in particular, Oligocene magnetic anomalies are identified within the age range of 28-39 Ma.

HUSSONG, DONALD M., and BRIAN TAYLOR, Hawaii Institute of Geophysics, University of Hawaii, Honolulu, HI, LAVERNE D. KULM, Oregon State University, Corvallis, OR, and THOMAS W. C. HILDE, Geodynamics Research Program, Texas A&M University, College Station, TX

### Detailed Surveys of Offshore Peru Margin

The complex and highly variable structure of the submarine continental margin of central Peru is revealed by geophysical surveys and geologic sampling completed in 1985. The surveys were conducted in preparation for deep scientific drilling to be undertaken by the Ocean Drilling Program in November-December 1986. More than 11,000 km$^2$ of sea floor were mapped using the SeaMARC II side-scan sonar and bathymetry system; 1,500 km of multichannel seismic reflection profiles and 4,000 km of single-channel seismic data were shot; and many coring, dredging, and heat-flow stations were obtained. The data permit construction of detailed three-dimensional geologic maps of the region. These maps show that the ancient metamorphic rocks of South America extend close to the trench axis and apparently have undergone a history of truncation and subsidence related to the subduction of the Nazca oceanic plate. Adjacent segments of the Peru forearc have dramatically different structure and appear to have had differing tectonic histories. The margin is disrupted by extensive (primarily tensional) faulting; the larger faults extend perpendicular to the strike of the trench and often serve as conduits for diapirs and mud volcanoes. Living chemosynthetic clams were dredged from 3,800-m depth along one of these fault trends, suggesting that active fluid venting occurs at depth on the continental wall of the Peru Trench.

KATILI, JOHN A., CCOP, Jakarta, Indonesia

### SEATAR Transect Project

The main objectives of the SEATAR program originally were defined as: (1) to determine the location characteristics and significance of the principal tectonic features of the continental margins and associated structural elements of east and southeast Asia; (2) to relate metalliferous ore deposits to the major tectonic features and plate boundaries, particularly convergent and shear boundaries; and (3) to analyze the characteristics of various types of sedimentary basins and their hydrocarbon habitat in terms of their position relative to plate margins, and to study the geologic, geochemical, and geothermal history factors governing the transformation of organic matter into hydrocarbons in small basins.

The investigation was concentrated along several so-called transects, sections of the earth's crust where the processes operating in the earth's interior manifest themselves in several geologic and geophysical features. Six transects were originally defined in the region of east and southeast Asia, followed by other transects crossing the western and eastern parts of Indonesia.

These transect study projects covered the major areas of plate interaction, each cutting across an arc-trench system and adjacent basins and continental areas. The results from individual studies could be coordinated to give a comprehensive multidisciplinary picture of this critical part of the earth's crust.

At a special meeting held in Bangkok in 1985, the coordinators agreed to present the final maps and sections of the transects at the Circum-Pacific Conference, and the final report may be published in 1987.

Each transect will consist of a separate volume that includes a folio and maps and sections in standard format and printed in color. Furthermore, the separate volumes in the series will be published by the same publisher and released at the same time. If this schedule is to be met, then funding must be requested, in-kind inputs must be provided, and an overall coordinator must be employed.

LETOUZEY, J., and L. SAGE,* Institut Français du Pétrole, Rueil-Malmaison, France

### Geologic and Structural Map of Eastern Asia

A synthesis of the onshore and offshore geologic data of eastern Asia, prepared by the Institut Français du Pétrole (IFP), has allowed the construction of geologic and structural maps for this region. These maps include three color sheets (scale = 1:2.5 million) and three plates of geologic and structural cross sections. Located between lat. 4° and 35°N, and long. 106° and 132°E, the maps cover the following geographic areas: East and South China Sea, Sulu Sea, West Philippine basin and onshore neighboring terrains, Kyushu and Ryukyu Islands, the China margin, Taiwan Island, Vietnam, North West Borneo, and the Philippines. The maps synthesize seismic interpretations, oil well data, geologic work in south Japan, Taiwan, Borneo, and the Philippines, and recent data published between 1976 and 1985.

Besides exhibiting detailed bathymetry, they display large structural units (oceanic crust, active margins, large tectonic trends, faults, subduction and collision trenches, and troughs), folding and thickness of total sediment in meters (above economic basement) in onshore and offshore sedimentary basins, and onshore geology revised with field data.

Twenty-four geologic cross sections (scale = 1:1.25 million, vertical exaggeration × 6) intersect ocean margins, important basins, and the different structural domains. They are based on seismic profiles, well data, and available onshore and offshore geologic data. These cross sections show basement composition and structures, different tectonic and sedimentary domains, and the structure and thickness of different sedimentary deposits (such as age, unconformities, and geologic structures). Maps and cross sections will be published in early 1987.

MORGAN, HUGH M., Western Mining Corporation, Ltd., Melbourne, Australia

### Geology, Structure, and Volcanic Processes of Southern Juan de Fuca Ridge

The southern Juan de Fuca Ridge is a medium-rate spreading center (total opening rate = 6 cm/year) in the northeast Pacific Ocean, 500 km west of central Oregon. A 25-km segment of the ridge has been the focus of a 6-year study by the U.S. Geological Survey. The ridge crest displays well-developed cross-axis symmetry with 80 to 100-m high ridges formed by predominantly inward-facing fault scarps and a relatively smooth 1-km wide axial valley floor. Multichannel seismic reflection profiles across the axis indicate the presence of a crustal magma chamber roof, 1-2 km wide and 2.3 km beneath the valley floor. The seismic reflection data are insufficient to determine the along-axis continuity of the magma chamber. SeaMARC II side-scan imagery, extensive photographic surveys, and observations from the *Alvin* submersible reveal a continuous linear depression 30-50 m wide and 10-30 m deep that bisects the axial valley floor through the study area. The depression is offset in places and may locally have short overlapping strands.

The depression is characterized by sharp rims, steep walls, and a floor of mostly rough-surfaced lava and lineated sheet flows. It is flanked by a region of subsided, fresh lava flows. These flows, in turn, are surrounded by sheet flows containing numerous small collapse pits. Unpitted sheet flows occupy the marginal part of the axial valley floor. The depression appears to be the site of voluminous fissure eruptions that fed the lavas on the surrounding valley floor. Active hydrothermal vents are present within the axial depression, but have not been observed elsewhere on the axial valley floor. This axial depression appears to be a unique feature among medium-rate spreading centers that have been studied to date.

MORTON, JANET L., and TRACY L. VALLIER,* U.S. Geological Survey, Menlo Park, CA

Geophysical and Geologic Studies at Valu Fa Ridge, Lau Basin, Southwest Pacific Ocean

The Lau basin is an actively spreading back-arc basin located west of the Tonga subduction zone and Tofua volcanic arc in the southwest Pacific Ocean. In the southern part of the basin is Valu Fa Ridge, a north-northeast-trending ridge located approximately 40 km west of the Tofua volcanic arc axis. There, spreading is occurring at a rate of 7 cm/year. Seven multichannel seismic reflection profiles over Valu Fa Ridge show a strong reflection 3.5 km beneath the sea floor. The inverted polarity of the reflection indicates that it is from a low-velocity zone, which is interpreted as the top of a flat, 2 to 3-km wide, axial magma chamber. The reflection coefficient indicates a large acoustic impedance contrast between the material above and below the reflector, possibly because a thin gas-rich layer is present at the top of the chamber.

Two dredge hauls from the crest of Valu Fa Ridge recovered highly vesicular andesites that are relatively homogeneous, both mineralogically and chemically. Dredge-1 rocks average 57.5% $SiO_2$, 11.8% FeO (total Fe as FeO), 3.3% MgO, and 0.52% $K_2O$; dredge-2 rocks have average values of 55.5% $SiO_2$, 12.9% FeO (total Fe as FeO), 3.8% MgO, and 0.43% $K_2O$. Average isotopic values are $^{87}Sr/^{86}Sr$ 0.70330 ± 2; $^{143}Nd/^{144}Nd$ 0.51303 ± 2; $^{206}Pb/^{204}Pb$ 18.65 ± 2; $^{207}Pb/^{204}Pb$ 15.55 ± 1; and $^{208}Pb/^{204}Pb$ 38.34 ± 4. Trace-element averages (ppm) are Sr 167, Rb 7, Cs 0.18, La 4.2, Eu 1.3, Lu 0.56, Y 37, and Zr 78. Isotopic data are similar to data from rocks of the Tofua arc and overlap the relatively enriched MORB fields. High Cs, Rb, Th, U, and K contents are characteristic of an arc component in the melt. These chemical data indicate a small but significant subduction-zone effect on the Valu Fa andesites.

The presence of hydrothermal manganese crusts as much as 10 mm thick and nontronite in the dredge hauls, elevated concentrations of particulate and total dissolvable manganese in the water column above Valu Fa Ridge, and small bottom-water temperature anomalies indicate that hydrothermal circulation is occurring at the ridge.

NESS, GORDON E., Oregon State University, Corvallis, OR, FRANCISCO SUAREZ V., CICESE, Baja California Norte, Mexico, and GUSTAVO CALDERON R., Direccion General de Oceanografia/IOM, Colima, Mexico

Geophysical Structures and Late Neogene History of Gulf and Peninsular Province of the Californias

Researchers from the Continental Margins Study Group at Oregon State University, the Direccion General de Oceanografia, and CICESE have completed a 10-year reconnaissance-scale geophysical survey of the Pacific Exclusive Economic Zone of Mexico. Six cruises, aboard research vessels of the Armada de Mexico, have been completed, and approximately 50,000 nmi of survey track lines have been run. From these data and from other published sources, we have compiled a series of onshore-offshore maps of the region extending 1,500 nmi from Point Conception in southern Alta California to Lazaro Cardenas in Michoacan. These maps include: *Physiography,* with 200-m contours of bathymetry and topography; *Free-Air Gravity Anomalies,* with 10-mgal contours of both onshore and offshore gravity: *Marine Magnetic Anomalies and Oceanic Crustal Isochrons,* with 50-nT contours over continental crust and 1-m.y. isochrons over oceanic crust; *Seismo-Tectonics,* with 1963-1982 WWSSN seismicity superimposed on interpreted onshore-offshore faults.

Slow gulf rifting, at least in the south, began as early as about 12-14 Ma. Oceanic crust was first exposed there by about 9-10 Ma. Pacific-North American relative motion was distributed between faults in the gulf and faults on the Pacific side of Baja California until about 4-5 Ma. This rifting history is consistent with the history of changes in volcanic chemistry along the gulf side of Baja. Initial rifting in the southern gulf was diffuse, involving many small, independently active crustal blocks. Present-day faulting, onshore in northern Baja, in the northern and central gulf, and onshore in Jalisco and Colima, also involves many small, independently active fault blocks and slices that are similar to those occurring onshore in southern Alta California.

SCHMIDTKE, E., J. ALMASCO, J. R. DUNN, M. FULLER,* R. HASTON, and R. LANDGRAF, University of California, Santa Barbara, CA

Paleomagnetic Results from Sabah and Sarawak, East Malaysia

Borneo lies in a central position in the southeast Asian seas and islands, so its tectonic history is an important aspect of the geologic evolution of the region. We obtained paleomagnetic results from Sabah, north of Borneo, and Sarawak, southwest of Borneo. The results indicate a strong counterclockwise rotation between the late Oligocene and Pliocene, for the region near Kinabalu. Further north, in the Kudat region, a small clockwise rotation was found. Thus, the paleomagnetic results differ in these two regions as does the tectonic strike. Paleomagnetic data from post-tectonic intrusions in Sarawak indicate that no late Cenozoic rotations have occurred. When considered in conjunction with earlier data from Borneo, the results suggest that the island has been deformed heterogeneously, so regions with different tectonic rotations are separated by shear zones, such as the Lupar line and the Kinabalu fault. The tectonic implications of the results will be discussed in the light of other paleomagnetic results from around the South China Sea.

SIBUET, J. C., Ifremer Centre de Brest, Brest, France, J. LETOUZEY, Institut Français du Pétrole, Rueil-Malmaison, France, B. MARSSET, GIS Océanologie et Géodynamique, Brest, France, M. DAVAGNIER (affiliation not available), J. P. FOUCHER and H. BOUGAULT, Ifremer Centre de Brest, Brest, France, and L. DOSSO, R. MAURY, and J. L. JORON (affiliation not available)

Tectonic Evolution and Volcanism of Okinawa Trough

The Okinawa Trough is a back-arc basin formed by extension of the east China continental lithosphere behind the Ryukyu Trench system. The age of marine deposits drilled in the northern Okinawa Trough indicates a Miocene age for the splitting of the volcanic arc and the first tensional movements. The POP 1 cruise of the R/V *Jean-Charcot* (September-October 1984) provided new evidence concerning the two main periods of extension as recognized by Kimura (*Marine and Petroleum Geology,* 1985). Tilted fault blocks in the northern Okinawa Trough trend north 40°-60° and belong to the early Pleistocene phase (2-0.5 Ma). The present-day phase is characterized over the entire basin by normal faults oriented 80°N in the north and 90°N in the south. In the southern Okinawa Trough, most of the deformation occurs along linear, subparallel, en echelon depressions intruded by volcanic ridges associated with positive magnetic anomalies. The system of volcanic ridges ends northeast of Okinawa Island in a series of parallel volcanic ridges named the VAMP (volcanic arc-rift migration processes) area, which merges into an active volcanic chain extending north to Japan. Chemical analyses of the vesicular basalts dredged on the back-arc basin display flat to enriched rare-earth patterns. The niobium-tantalum negative anomalies reflect a subduction signature. A good positive correlation between strontium isotopic compositions and concentrations suggests a contamination effect.

SILVER, ELI A., DONALD L. REED, DARLA H. HEIL, and JILL TAGUDIN, University of California, Santa Cruz, CA

Dynamics of an End-On Collision, Interpreted from SeaMARC II Swath Mapping and Migrated Seismic Reflection Data on North Margin of Panama

We have imaged nearly the entire North Panama thrust belt (NPTB) to determine the dynamics of the Panama arc colliding end-on with South America. Observations support a simple flexural beam model in which eastward collision of the arc against the continent is transformed into northward relative motion of the NPTB. Fold axes and thrust faults within the NPTB show remarkable uniformity in strike (075°), changing only on the east margin across a major fault and on the west margin across a sharp flexure. The belt increases in width (and presumably convergence) eastward, as predicted by the beam model. Displacement along the Uraba fault zone separated a thin sliver of the NPTB, and two earthquake focal mechanisms on the fault show right-oblique slip. Vergence of thrust sheets within the accretionary wedge is dominantly landward in much of the central part of the thrust belt, indicating low basal shear stresses and high fluid pressures along the decollement. Mud volcanoes within the accretionary wedge attest to elevated fluid pressures. Because of an exceptionally well-developed and widespread gas hydrate layer, we suggest that pore fluids are a mixture of water and methane. However, near the central part of the thrust belt, a concentrated zone of mud volcanoes has breached the crests of seaward-verging anticlines. We propose that the zone of anomalous seaward-verging thrust sheets results from widespread release of pore fluids from the decollement, leading to increased basal friction.

SILVER, ELI A., University of California, Santa Cruz, CA, and BRIAN TAYLOR,* Hawaii Institute of Geophysics, Honolulu, HI

## Proposed Program for Scientific Drilling in Western Pacific

The Western Pacific Panel of the Ocean Drilling Program recommends that the following scientific problems take highest priority for the next round of scientific drilling in the western Pacific region: (1) origin and development of marginal basins, and their implications for the tectonic history of western Pacific arc systems; (2) origin and tectonic history of oceanic forearcs, with implications for tectonic erosion and ophiolite emplacement; and (3) structure and mechanics of accretionary tectonics. High-priority marginal basins include Japan, South China, Banda, and Sulu Seas. Oceanic forearcs include the Bonin, Vanuatu, Okinawa-Ryukyu, and Lau forearcs. High-priority accretionary wedges include the Nankai and Sunda regions. Other problems have a high scientific merit, such as serpentine diapirism in forearcs and arc reversal in the Solomons, but as yet, the potential for solution by drilling is unclear. The panel encourages input on these and other objectives, especially from citizens of western Pacific countries.

SNAVELY, PARKE D., JR., U.S. Geological Survey, Menlo Park, CA

## Tertiary Evolution and Petroleum Potential of Oregon-Washington Continental Margin

The Oregon-Washington continental margin was the site of a deep marginal basin in which more than 7,000 m of Tertiary sedimentary and volcanic rocks accumulated. Oceanic basalts of Paleocene to early Eocene age form the basin floor and are interpreted to represent eruptions in an elongate trough formed by rifting of the continental margin. Middle Eocene turbidite sandstone overlapped both the oceanic crust and the pre-Tertiary rocks of the Klamath Mountains, thus indicating that suturing of the Coast Range–Olympic terrane to North America was about 50 Ma.

Oblique convergence between the Farallon and North American plates occurred during most of the middle Eocene to middle Miocene. Sedimentation, punctuated by episodes of volcanism, was essentially continuous in a forearc basin whose axis lay along the present inner continental shelf. The oblique interaction between the plates was interrupted by two periods of more head-on convergence during the middle late Eocene and late middle Miocene. Thick accretionary melange wedges of Eocene and of late Oligocene to late middle Miocene ages were formed during these strongly compressive episodes. Geochemical analyses indicate that the accretionary melange wedges, which crop out along the west side of the Olympic Peninsula and beneath the adjacent shelf, have the highest potential for oil and gas generation. They are the source rocks for numerous gas seeps and oil and gas shows in exploratory wells, and for the 12,000 bbl of 38.9° API paraffin-based oil produced from a well drilled on the southwest Washington coast. Potential exploration targets exist where the Eocene and Oligocene-Miocene melanges are underplated to a position beneath the lower Eocene oceanic basalt. Hydrocarbons generated in the melanges could migrate upward into structures in strata that overlie the basalt in the upper plate.

STEVENSON, ANDREW J., and TRÁCY L. VALLIER, U.S. Geological Survey, Menlo Park, CA

## Evolution, Structure, and Petroleum Potential of Aleutian Subduction Complex

Along the southern margin of the Aleutian arc is a large accretionary wedge or subduction complex of offscraped terrigenous and pelagic sediments. We believe the wedge has been accumulating since the late Miocene onset of increased sediment supply to the Aleutian Trench, initiated by glaciation in Gulf of Alaska drainages. Sediment incorporated into the accretionary body was originally deposited in abyssal plain and trench axis settings, and contains both pelagic ooze and thick turbidite facies. Adjacent to the central Aleutian Islands, (long. 172°–177°W), a 300-km long mass of offscraped sediment (36,000 $km^3$) forms a tectonic wedge 6-8 km thick that contains many potential structural and stratigraphic traps for oil and gas accumulations. Seismic reflection profiles reveal antiforms, normal and reverse fault traps, and stratigraphic pinch-outs. The source potential of these deposits is low to moderate with an estimated total organic content of 0.3-0.6%. Both potential reservoir turbidite sands and clay seals exist in the accretionary complex. Heat-flow values inferred from indirect measurements place the oil-generation window at 3-6.5 km. An active water drive to sweep petroleum generated at depth into trapping structures is a likely consequence of the compaction and tectonic dewatering of the accreted trench sediment. Although prerequisite conditions exist for the formation and entrapment of oil and gas deposits in the Aleutian accretionary wedge, the actual existence of large deposits of petroleum in an active tectonic zone is impossible to predict.

TAIRA, ASAHIKO, and HIDEKAZU TOKUYAMA, Ocean Research Institute, University of Tokyo, Tokyo, Japan

## Geotectonic Framework of Japanese Convergent Margin

A synthesis of recent marine geophysical and geologic investigations around Japan uses data obtained by the Ocean Research Institute of the University of Tokyo (including the Japan-France KAIKO Project), the Hydrographic Office and Geological Survey of Japan, and the Japan National Oil Company. This work reveals the contrasting characteristics of forearc and back-arc morphotectonics around Japanese subduction zones. From the north, the following features can be summarized.

The Kuril Trench off Hokkaido shows a pile of trench deposits as much as 2-sec two-way traveltime in thickness. Despite this thickness, there is little evidence of deformation related to the accretionary processes. We propose that subduction of the thick trench fill is occurring at this boundary.

The Japan Trench is characterized by massive slumping processes and debris accumulation at the toe of the overriding plate. The Dai-ichi Kashima seamount at the trench has been broken in half by normal faulting related to bending of the oceanic lithosphere into the subduction zone. Forearc subsidence is also evident on various profiles. We believe that the inner wall of the Japan Trench has been eroded tectonically at least from the Miocene to the Pliocene.

Three plates interacting at the trench triple junction (Japan, Sagami, and Ogasawara Trenches) created a complex framework of convergent margin tectonics, including highly oblique convergence, forearc sliver collision, and pull-apart basin.

The conspicuous features observed in the Izu-Bonin arc are forearc ophiolite (serpentinite diapir) and back-arc rifting close to the volcanic arc. As a result of intraplate shortening related to the Izu-Bonin arc collision against Honshu, the Shikoku basin oceanic crust was ruptured and uplifted, forming a high called the Zenisu Ridge. The Nankai Trough has been filled by turbidites transported through the Suruga Trough area from the Izu collision zone. The trench fill is being offscraped and accreted to the landward side. A series of forearc depressions developed along the Nankai Trough is the consequence of east-west contraction resulting from northeast and southwest Honshu collision at the Fossa Magna.

The Japan Sea is composed of three sedimentary basins that are underlain by oceanic crust and many topographic highs of continental fragments. A rapid rotation of the Japanese arc at 15 Ma created this back-arc basin, which probably is now being consumed at the northeast Japan margin with development of a new subduction boundary.

The Ryukyu Trench is morphologically similar to the Japan Trench. The forearc basin along the Ryukyu arc shows a contrasting style of development: normal compression to the north versus oblique convergence to the south divided by the Miyako depression. The back-arc basin of the Ryukyu arc, Okinawa Trough, is the rare example of ongoing back-arc continental rifting. The north-south extension has started since the late Miocene.

TAYLOR, B., Hawaii Institute of Geophysics, Honolulu, HI, K. CROOK, Australia National University, Canberra, Australia, J. SINTON, Hawaii Institute of Geophysics, Honolulu, HI, and the SHIPBOARD PARTY of *Mauna Wave*, 8517 and 18

## Fast Spreading and Sulfide Deposition in Manus Back-Arc Basin

The detailed location of the active plate boundaries in the Manus basin, north of the New Britain arc-trench system, is revealed by geophysical surveys and bottom sampling/photography completed in January 1986. More than 40,000 $km^2$ of sea floor was mapped using the SeaMARC II side-scan sonar and bathymetry system. The Weitin fault on New Ireland extends northwest, offshore to the eastern end of three overlapping, en echelon, sigmoidal spreading centers north of the Gazelle Peninsula. These spreading centers link westward with a rhomb-shaped microplate bounded on the northeast and southwest by northwest-trending transform faults, on the south by two en echelon rift grabens with axial basalt flows, and on the northwest by a 100-km long, oblique spreading, curving, propagating spreading center. The fabric of the transferred or possibly rotated crust of the 12,000 $km^2$ microplate trends north-south. The southern boundaries of the microplate are not teleseismic.

A new type of plate boundary, herein termed an "extensional transform zone" (ETZ), was discovered. It links the southwest end of the oblique spreading center (trend 050°) to the left-lateral transform fault (trend 115°), which bounds the northeastern edge of the Willaumez Rise. The 90-km long ETZ is characterized by overlapping, right-stepping en echelon spreading ridges (trend 075°) cut by a Reidel shear (trend 105°) synthetic to the western transform. These features may be explained in terms of the second-order strain ellipse associated with the transform fault to the west.

Dredges and bottom-camera surveys of the main spreading center (opening rate > 100 mm/year) revealed fresh glassy pillow basalts and sheet flows, with occasional fissures and a light dusting of sediment. The first photographs of metalliferous sulfide chimneys in the western Pacific were taken on January 5, 1986. A new species of gastropods dominated the otherwise typical vent fauna.

VEDDER, J. G., U.S. Geological Survey, Menlo Park, CA, J. B. COLWELL, Bureau of Mineral Resources, Canberra, Australia, and T. R. BRUNS and A. K. COOPER, U.S. Geological Survey, Menlo Park, CA

Regional Offshore Geology of Central and Western Solomon Islands and Bougainville, Papua New Guinea

The central and western Solomon Islands and the Bougainville regions are parts of a complex island-arc system that includes an intra-arc basin and remnants of both forearc and back-arc depositional wedges. These features formed in response to episodic Cenozoic tectonism along the convergent boundary between the Pacific and Australia-India plates. Presumed early Tertiary southwest-directed subduction of the Pacific plate and associated arc magmatism were curtailed by impingement of the leading edge of the Ontong Java Plateau. Aprons of back-arc and forearc sediment were derived from highstanding parts of the arc during the late Oligocene and early Miocene. Late Tertiary arc-polarity reversal and northeastward-directed subduction of the Woodlark spreading system caused a renewal of island-arc magmatism that completed the construction of the Central Solomons Trough as an enclosed intra-arc basin.

Interpretations of multichannel profiles from 1982 and 1984 CCOP/SOPAC Tripartite Cruises of the research vessel R/V *S. P. Lee* indicate that the Central Solomons Trough is a composite intra-arc basin containing as much as 5.5 km of late Oligocene(?) and younger sedimentary rocks. As many as five lenticular seismic-stratigraphic units can be identified on the basis of unconformities and abrupt velocity changes. Late Miocene and younger folds and faults deform the northeast and southwest flanks of the basin. Profiles across the Kilinailau Trench show Ontong Java Plateau rocks covered by 2-4 km of trench sediment. The inner trench wall consists of folded, upfaulted, and rotated blocks of trench and forearc strata. The deep-water basin northwest of Bougainville is a southeastward extension of the New Ireland forearc basin, the southern margin of which is formed by a subsided part of the early Cenozoic arc. There, Oligocene(?) and younger basin strata, as much as 7 km thick, are deformed by pre-Pliocene faults and folds. The Buka–Green Island Ridge east of the basin is part of the late Cenozoic arc that cuts across the older arc beneath Bougainville.

Most of the conditions essential for hydrocarbon accumulation and entrapment are present in the Central Solomons Trough. However, the occurrence of suitable source and reservoir rocks has not been demonstrated. Water depths of more than 800 m pose a further deterrent to exploratory drilling.

VON HUENE, ROLAND, U.S. Geological Survey, Menlo Park, CA, and JOHN MILLER, U.S. Geological Survey, Denver, CO

Seismic Images of Modern Convergent Margin Tectonic Structure

An atlas of 14 seismic sections shows major tectonic features across various convergent margins. All sections are at the same scale and have been processed to a research level. Such processing produces clear images of sediment accretion at the front of some margins as well as images of sediment accretion from below. Most margins show sediment subduction. The results of tectonic erosion are seen in midslope areas as buttresses against which the accretion complex is stacked. The images of structures indicating tectonic erosion are about as common as the images of structure indicating accretion.

VON STACKELBERG, ULRICH, Bundesanstalt für Geowissenschaften und Rohstoffe, Hannover, Federal Republic of Germany, and SHIPBOARD SCIENTIFIC PARTY of *Sonne*

Hydrothermal Sulfide Deposits in Back-Arc Spreading Centers in Southwest Pacific

During *Sonne* cruise SO-35 (December 6, 1984, to February 6, 1985), two areas in the Lau basin and two in the North Fiji basin were investigated, to identify oceanic spreading centers and indications of hydrothermal activity within a convergent plate boundary setting. For bathymetric and geophysical mapping of the survey areas, a multibeam echo sounder, a 3.5-kHz subbottom profiler, and a proton magnetometer were used. Ridges and seamounts were sampled with dredges and observed with a TV/photo sledge. Gravity, free-fall, and piston cores were taken within the sediment-covered areas. Areas with fresh lava flows were found in the Lau and North Fiji basins. The fresh volcanic rocks in the Lau basin occur at narrow ridges of a back-arc spreading center. Their occurrence in the North Fiji basin is possibly better explained by a "leaky transform fault." Hydrothermal sulfides, the first seen in that region, impregnate these fresh volcanic rocks in both basins. Sulfides were also recovered from off-axis seamounts in the Lau basin. Precipitates of manganese oxides and iron silicates of hydrothermal origin are common at active and fossil spreading zones in both basins, as well as on seamounts.

ZHAI GUANGMING, Scientific Research Institute of the Ministry of the Petroleum Industry, Beijing, People's Republic of China

Maps and Sections of Petroleum Geology of Continental Shelf of China

The distribution of the sedimentary basins of the continental shelf of China and the geology of adjacent regions will be presented in map displays. Other maps will exhibit the major structures and formations of pre-Tertiary basement rocks. Accompanying these displays are three sedimentary facies maps for the Bohai Bay, East China Sea, and South China Sea basins. Examples of offshore oil and gas traps, of which there are more than 10 types, will be exhibited. Some of these traps are not known in the east mainland of China. Seismic sections revealing the subsurface stratigraphy and structures of Bohai Bay, South Yellow Sea, East China Sea, Pearl River Mouth, Ying Ge Hai, Beibu Bay, and other regions will augment the map displays.

# CIRCUM-PACIFIC MAP PROJECT

CHAPTER 34

# FACTS AND PERSPECTIVES OF THE CIRCUM-PACIFIC MINERAL RESOURCES MAP SERIES

**José Corvalán**
*Department of Geology and Geophysics*
*University of Chile*
*Santiago, Chile*

ABSTRACT

The Circum-Pacific Mineral Resources Map Series is intended to present a simplified but significant geologic-tectonic framework with which different mineral deposits are associated. These are represented according to their type, size, and shape; for some deposits, the age of mineralization is given. All the important deposits are shown, and in several regions small deposits or occurrences, although unexploitable at present, are incorporated. In this sense, the maps do not always represent the present resource situation. However, they serve to identify favorable areas for the occurrence of certain minerals and thus help in exploration planning and resource assessment. The maps incorporate sea-floor mineral occurrences, which include ferromanganese nodules, hydrothermal sulfides, phosphorites, and heavy-mineral sands.

The maps are intended to be simple and objective in order to provide solid data which, complemented with information given through other Circum-Pacific maps, should constitute an important document to stimulate the study of the distribution characteristics and genesis of ore deposits. Examples to show these perspectives are included.

## GENERAL BACKGROUND

As planned since the Circum-Pacific Map Project started in 1974, the Mineral Resources Map Series was conceived to depict the distribution of mineral deposits within the Circum-Pacific Region, according to their type, size, shape, and age, and the geologic environment with which they are associated.

Because the type of information to be compiled varied widely from one region to another, and uniformity of cartographic representation was desirable for all quadrants of the map, the first steps taken were to decide on the legend and symbology to be used. Guidelines were adopted that could generally be modified according to the needs of each region. No attempt was made to differentiate metallogenic provinces or regions, and only extensive districts are outlined.

The maps are intended to be as objective as possible, showing the factual situation both on land and on the sea floor. On land, commodities are shown by shape and color of symbol, the deposit type by ticks, and the size of the deposit by the size of the symbol. Mineralization ages for some deposits are indicated by double ticks, in clockwise order from older to younger. Where only an older or younger age limit can be determined, this is indicated by a hook facing clockwise or counterclockwise, respectively. Large, medium, and small deposits are separated based on limits in metric tons of metal or mineral contained, totaling past production and/or reserves (Tables 1A, 1B).

In several regions, small deposits or occurrences, although unexploitable at present, were also incorporated because they were considered as sources of valuable information. In this sense, the maps do not always represent the present resource situation but the additional information may help in exploration planning.

The deposit type classification adopted initially, rather than being based on morphology, became a mixture of form and genesis of the deposit. This did not fit well for some Australian deposits as noted by Chan (1985) who thinks, for instance, that because vein and stockwork deposits commonly overlap, these should be grouped together, or that magmatic cumulates as such should be included in the magmatic category and not with the stratabound. It is true that a certain classification may be misleading in one region although appropriate for another, and some differences are expected in the cartographic representation in different quadrants.

Selection of the background to be used for these maps has also been controversial—whether strictly geologic, geologic with selected structural elements, or tectonic. The Tectonic Map Series is still being compiled and is not yet available; the geologic background with selected structural elements has been taken mostly from the Geologic and Plate-Tectonic Map Series. However, a slightly different approach is also expected since some quadrants, such as the Northwest, are incorporating geologic units mostly based on age rather than on lithology. The geologic background being worked out for the

**Table 1a. Land Resources.**

**Commodities** Shown by shape and color of symbol.

| | ● | ■ | ◆ | ♦ | ⬬ |
|---|---|---|---|---|---|
| *red* | W | Sn | Sn W | Be | Nb(Ta)U |
| *orange* | Cu(AuAg) | Mo | CuMo | CuZn(PbAgAu) | CuNi |
| *yellow* | Au(Ag) | Ag(PbZnCuAu) | AuAg, AgAu | Precious gems | Semi-precious gems |
| *yellow green* | S | FeS | Nitrate | Li | B |
| *green* | Cr | Ni(CoCuPt) | Co(Ni) | Pt group | Asbestos |
| *blue green* | Na | K | | Gypsum-anhydrite | Magnesite |
| *blue* | Pb | Zn | PbZn, ZnPb | PbZn(AgCuAu) | |
| *heliotrope* | Hg | Sb | F | Ba | Sr |
| *light brown* | U | V | UV, VU | RETh | Al |
| *dark brown* | Fe | Mn | Ti | FeTi(V) | P |

⬭ Outline of extensive district

**Deposit size** By size of symbol. Limits in metric tons of metal or mineral contained unless otherwise specified. Past production and(or) reserves totaled.

| Size | ○ Large > | ○ Medium > | ∘ Small |
|---|---|---|---|
| Aluminum (bauxite) | 100,000,000 | 1,000,000 | |
| Antimony | 50,000 | 5,000 | |
| Asbestos | 10,000,000 | 100,000 | |
| Barite ($BaSO_4$) | 5,000,000 | 50,000 | |
| Beryllium (BeO) | 1,000 | 10 | |
| Boron ($B_2O_3$) | 10,000,000 | 100,000 | |
| Chromium ($Cr_2O_3$) | 1,000,000 | 10,000 | |
| Cobalt | 20,000 | 1,000 | |
| Copper | 1,000,000 | 50,000 | |
| Fluorite ($CaF_2$) | 5,000,000 | 50,000 | |
| Gold | 500 | 25 | |
| Graphite | 1,000,000 | 10,000 | |
| Gypsum-anhydrite | 100,000,000 | 5,000,000 | |
| Iron (ore) | 100,000,000 | 5,000,000 | |
| Lead | 1,000,000 | 50,000 | |
| Lithium ($Li_2O$) | 100,000 | 10,000 | |
| Magnesium ($MgCO_3$) | 10,000,000 | 100,000 | |
| Manganese (tons of 40% Mn) | 10,000,000 | 100,000 | |
| Mercury (flasks) | 500,000 | 10,000 | |
| Molybdenum | 500,000 | 5,000 | |
| Nickel | 500,000 | 25,000 | |
| Niobium-Tantalum ($R_2O_5$) | 100,000 | 1,000 | |
| Phosphate ($P_2O_5$) | 200,000,000 | 200,000 | |
| Platinum group | 500 | 25 | |
| Potassium ($K_2O$) | 10,000,000 | 1,000,000 | |
| Precious gems | 10 | 1 | |
| Pyrite ($FeS_2$) | 20,000,000 | 200,000 | |
| Rare earths ($RE_2O_3$) | 1,000,000 | 1,000 | |
| Semi-precious gems | 100 | 10 | |
| Silver | 10,000 | 500 | |
| Sodium (salts) | 10,000,000 | 1,000,000 | |
| Strontium (salts) | 1,000,000 | 10,000 | |
| Sulfur | 10,000,000 | 100,000 | |
| Thorium | 10,000 | 1,000 | |
| Tin | 100,000 | 5,000 | |
| Titanium ($TiO_2$) | 10,000,000 | 1,000,000 | |
| Tungsten | 10,000 | 500 | |
| Uranium | 10,000 | 100 | |
| Vanadium | 10,000 | 500 | |
| Zinc | 1,000,000 | 50,000 | |

**Table 1b. Deposit type and age of mineralization.**

**Deposit type** Shown by tick(s) on basic symbol.

- Veins and shear-zone fillings
- Stratabound, including magmatic cumulates
- Stockworks, including "porphyry" deposits
- Magmatic and irregular massive deposits
- Skarn or greisen deposits
- Sandstone (red bed) deposits
- Laterite deposits (surficial chemical concentrations)
- Placer deposits (surficial mechanical concentrations)

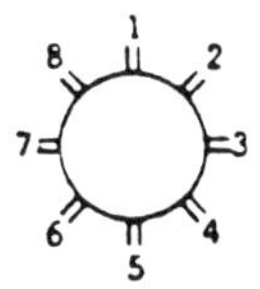


**Age of mineralization** Geologic age of mineralization shown by double ticks on some deposits. Age known only as "not younger than" or "not older than" indicated by a hook on the age tick.

1. Precambrian
2. Cambrian-Middle Devonian
3. Late Devonian-Early Triassic
4. Middle Triassic-Jurassic
5. Cretaceous (except latest)
6. Latest Cretaceous-Eocene (Laramide)
7. Oligocene-Pliocene
8. Post-Tertiary

Southwest Quadrant combines three basic age units—Phanerozoic, Proterozoic, and Archean—with a variety of rock environments, where the emphasis is on depicting the rock type giving the geologic age by letter symbol, as done on the Northeast and Southeast Quadrants. Of course, the number of rock units will vary from one region to another, but the general scheme is supposed to remain as simple as possible (Table 2 illustrates the units adopted for the NE Quadrant already published). The Geologic and Tectonic Map Series will serve to complement the minerals maps if more detailed information is needed.

Sea floor resources on these maps include occurrence, abundance, and chemical data on manganese nodules; occurrences of polymetallic sulfide on the sea floor and subsea floor; submarine hot springs; and occurrences of phosphorite and heavy-mineral sands.

Manganese nodule abundance varies widely; some areas contain none, whereas in other areas the nodules cover more than 50% of the sea floor. Their copper, nickel, cobalt, and manganese content is also shown; chemical data are from dredge or core samples of nodules or crusts (Table 3 shows the cartographic representation of sea-floor resources).

Also shown on these maps are the sea-floor sediment distribution, grouped into four main types selected from the compilations made by F. McCoy, (Lamont-Doherty Geological Observatory) for all the quadrants.

## PROGRESS IN THE MINERAL RESOURCES MAP SERIES

The 1:10,000,000-scale maps include, as in the other series, the four quadrants and the Antarctic Region. The Northeast Quadrant map was published in 1984 and the Southeast Quadrant map is in cartographic preparation.

Considerable time and effort have been spent by many scientists in obtaining, compiling, reviewing, and transferring new information on mineral deposits to the adopted base map scale. The scale of source maps, either published or especially made available to the project, varied widely. The information was first transferred to a 1:1,000,000 or to a 1:5,000,000 scale. Selection of the data to be portrayed on the 1:10,000,000 scale has been a difficult task, especially in regions such as the Central Andes where there is a high concentration of mineral deposits in a very narrow belt. In some places, this resulted in grouping of several small deposits into one medium size deposit, or several occurrences into one small deposit, in order not to lose information which might be significant in the future or that might be useful for evaluating prospective regions.

Upon completion of the five regional maps of the series, expected in 1987, a summary 1:17,000,000-scale map of the entire Pacific Basin will also be compiled.

## PERSPECTIVES OF THE MINERALS MAP SERIES

Besides bringing up-to-date information on mineral deposits, their distribution and geological environment, the fact that all the significant known deposits will be shown on these 1:10,000,000-scale maps, makes them of considerable use in regional resources research and assessment. The synoptic views of the basin-wide occurrence of selected deposit types that they can provide may be extremely helpful in evaluating possible reasons for their distribution and for identifying areas where more study and exploration may be advantageous.

On this integrated version, the polymetallic sulfide deposits along the East Pacific Rise and the Galápagos Rift zone, the most exciting mineral discoveries of recent years, appear to form in areas of medium to fast spreading (Figure 1). The spreading zone of the Eastern Pacific seems to be the most favorable area for additional discoveries, where the potential could be expected to be very large. These deposits consist of various amounts of iron, zinc, copper, and other sulfides, with appreciable quantities of silver and some gold.

Something similar can be said for the post-Paleozoic porphyry copper deposits. The Circum-Pacific Region contains the world's largest resources of copper in the large porphyry deposits distributed along or near converging plate margins or, as it happens in North and South America, near margins that were converging when the mineralization took place in the Mesozoic and Early Tertiary. However, different geologic-tectonic evolution in certain segments along these margins may be responsible for the apparent nonexistence of these types of deposits in regions such as the southern part of the Central Andes and the Southern Andes (Figure 2).

The distribution of tin and tungsten deposits does not follow the same plate-tectonic pattern as the porphyry copper deposits. In some regions tin and tungsten occur together in the

**Table 2. Rock Units selected for the NE Quadrant Minerals map.**

Land geology. Colors and patterns suggest the essential nature of the rocks. Geologic ages are indicated, where feasible, by letters keyed to those in the Correlation Diagram (boundaries from Geologic Map Series).

| Color | Symbol | Unit |
|---|---|---|
| (yellow) | | Surficial deposits |
| (orange) | a b c | Essentially undeformed, principally subaerial volcanic rocks: a) felsic, b) intermediate to mafic, c) mafic |
| | a b | Platform cover rocks of Phanerozoic ages: a) marine, b) continental |
| | | Basinal and marginal deposits, slightly to strongly deformed |
| (a, light green) (b, dark green) | a b | Deformed sedimentary and volcanic rocks of phanerozic age, unmetamorphosed or little metamorphosed: a) "miogeosynclinal", b) "eugeosynclinal" |
| (blue) | | Metamorphic complexes: most are of Phanerozoic age |
| (purple) | | Proterozoic sedimentary and volcanic rocks, in general not highly metamorphosed |
| (light brown) | | Precambrian crystalline rocks |
| (red) | a b c | Intrusive igneous rocks of Phanerozoic age: a) felsic, b) intermediate to mafic, c) mafic |
| | | Ultramafic rocks |

same deposit (Bolivia, China, Japan, Alaska), but in others there is little or no correlation between them. Ideas to explain the distribution and genesis of deposits like these are needed (Reinemund, Guild, and Addicott, 1982). In the meantime the Mineral Resources Map Series tries to assemble as many facts as possible in order to contribute to generation of new ideas.

Other important facts that become evident on these maps are the rather abrupt changes in the abundance and type of mineral deposits when passing from one region to another, as between the Central and Northern Andes. No doubt the occurrence and distribution of mineral deposits along the Andean Belt is controlled by regional longitudinal structures and by magmatic and tectonic processes operating through time. These are better explained by plate tectonics, considering that subduction of oceanic crust under different velocities, positions, and inclinations would have conditioned the parallel longitudinal mineralized bands. These constitute the metallogenic subprovinces distinguished along the belt (Ericksen,

**Table 3. Legend to seafloor resources.**

Manganese Nodules

Abundance data. Photographic station showing percent of sea floor covered by nodules (in black).

- 0%
- 0-1%
- 1-10%
- 10-25%
- 25-50%
- >50%

10 — Abundance contour in percent

Occurrence data. Sediment cores used in conjunction with bottom photographs to estimate sea-floor coverage.

O Nodules absent

x Nodules present

Chemical data. Dredge or core samples of nodules or crusts.

+ >1.8% nickel plus copper (red)

+ 1.0-1.8% nickel plus copper (blue)

+ <1.0% nickel plus copper (green)

$+_{Co}$ >1.0% cobalt (purple)

$+_{Mn}$ >35% manganese (brown)

These data points were not used in estimating sea-floor coverage.

Outline of area for which nodule analyses average 1.8 percent or more nickel plus copper (red)

Other Sea-floor Deposits

Polymetallic sulfide occurrence

On sea floor

471 Sub-sea floor (in DSDP core)

Submarine hot spring

Phosphorite (brown)

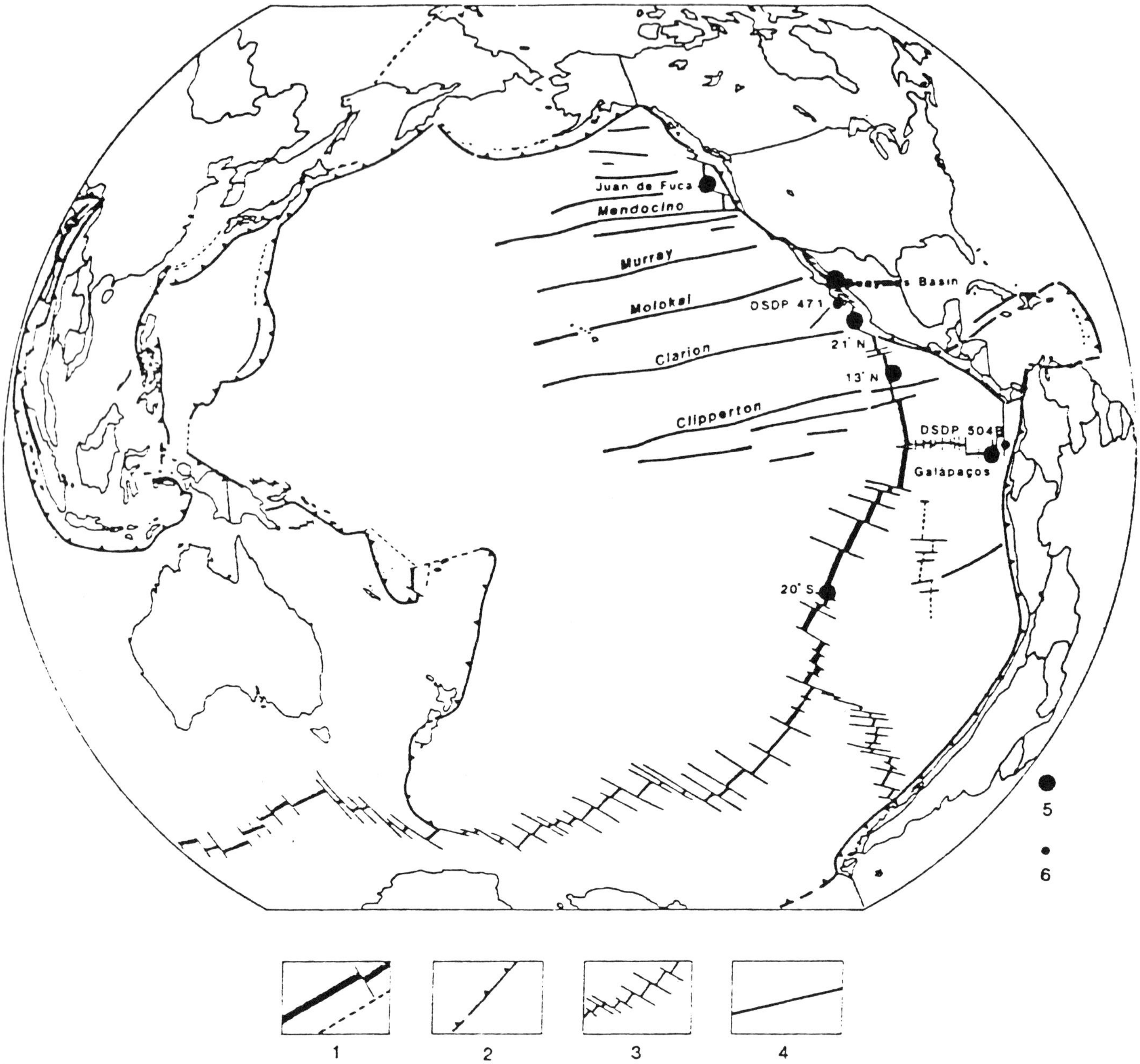


Figure 1. Polymetallic sulfide deposits within the Pacific. 1: spreading ridge; 2: converging plate margin; 3: transform faults; 4: major fracture zone; 5: deposits on sea floor; 6: sulfides found in DSDP core (after Reinemund, Guild, and Addicott, 1982).

1976; Sillitoe, 1972). Ericksen (1976) recognizes five major subprovinces: iron, copper, polymetallic, tin, and gold, from west to east. However, in the Northern Andes this differentiation is more difficult to determine or is less evident (Sillitoe, 1972), and the number of deposits is contrastingly less and of a different nature; platinum, chromium and nickel placer deposits are the most important.

Some iron and copper deposits do exist in Ecuador and Colombia. These could suggest the continuation of these subprovinces into that region but under a quite different geologic environment. The geologic framework of this segment of the Andes differs greatly from that of the Andes south of the Huancabamba deflexion (Gansser, 1973). There, the Coastal and Western Cordilleras are dominated by extensive basic to ultrabasic ophiolitic complexes with deep-water marine sequences, constituting what has even been considered a tectonostratigraphic terrane (Howell, Jones, and Schermer, 1985). This geologic-tectonic environment does not continue south into Perú.

The close relationship between the occurrence of mineral deposits and their geologic environment, although simplified, stands out clearly on these maps. For instance, the band of the most important iron deposits in Chile and Perú is related to Jurassic-Lower Cretaceous units, and the main belt of porphyry copper deposits follows the distribution of Cretaceous-Lower Tertiary formations and intrusive rocks. Explorations

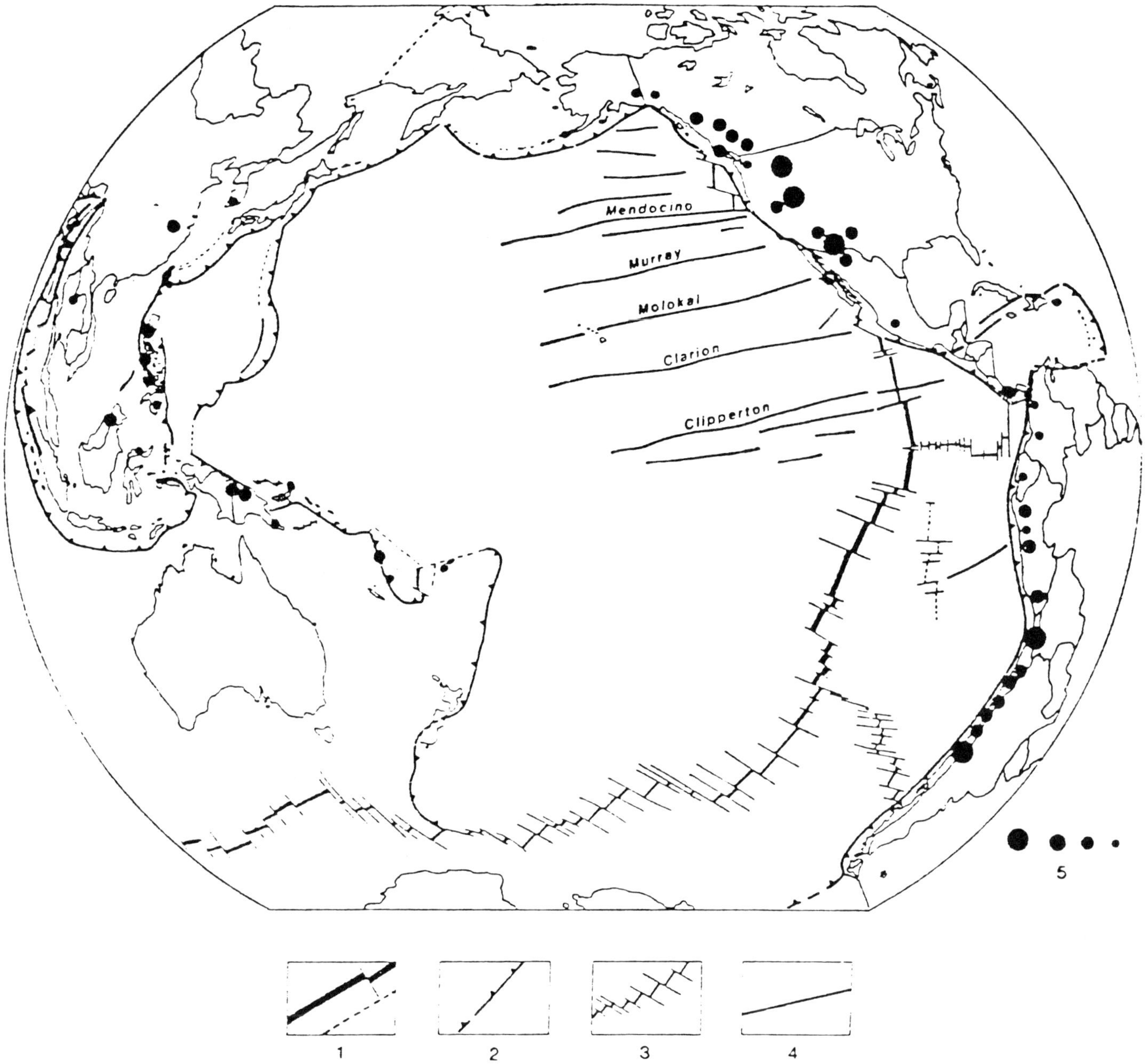


Figure 2. Post-Paleozoic porphyry copper deposits and districts. 1: spreading ridge; 2: converging plate margin; 3: transform faults; 4: major fracture zone; 5: deposits and districts; symbol size indicates relative size of deposits (after Reinemund, Guild, and Addicott, 1982).

conducted in this latter environment during the last 5 years have led to several important discoveries such as the Escondida porphyry copper in Northern Chile (Ortiz et al., 1985) and the detection of a chain of very important copper anomalies in the Western Cordillera of northernmost Perú (Cobbing et al., 1981) north of Michiquillay; this may continue into south-central Ecuador (Figure 3).

The Circum-Pacific Map Committee expects that the updated synthesis that the Mineral Resources Map Series will provide will be useful not only for evaluating prospective economic environments and research and exploration planning but also for educational purposes, which is an aim of all the map series that are being prepared.

## REFERENCES CITED

Chan, R., 1985, Modification of compilation guidelines for the mineral resources map of the Southwest Quadrant, *in*: Addicott, W.O., Scope and status of the Circum-Pacific Map Project: U.S. Geological Survey, Open-File Report 85-267.

Cobbing, E.J., W.S. Pitcher, J.J. Wilson, J.W. Baldock, W.P. Taylor, W. McCourt, and N.J. Snelling, 1981, The geology of the Western Cordillera of Northern Perú: London, Institute of Geological Sciences Overseas Memoir 5.

Ericksen, G.E., 1976, Metallogenic provinces of southeastern Pacific Region, Circum-Pacific Energy and Mineral Resources Conference: AAPG Memoir 25, p. 527-538.

Gansser, A., 1973, Facts and theories on the Andes: Journal of the

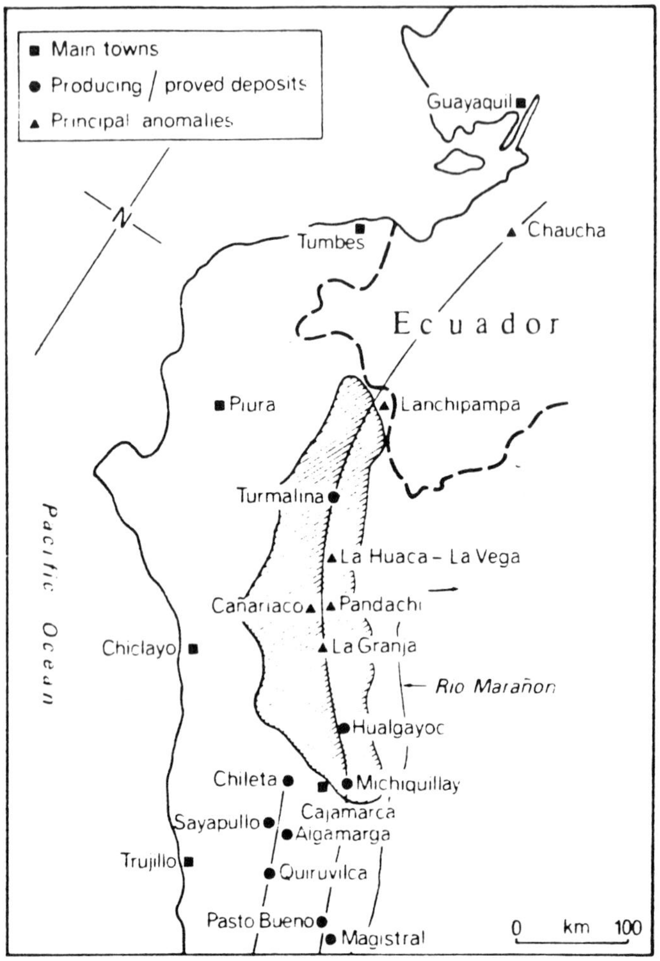


Figure 3. Regional trends of mineralization in northern Perú (after Cobbing et al., 1981).


Geological Society of London, v. 29, p. 92-131.

Howell, D.G., D.L. Jones, and E.R. Schermer, 1985, Tectonostratigraphic terranes of the Circum-Pacific Region: Tulsa, Circum-Pacific Council for Energy and Mineral Resources, 28 p.

Ortiz, F.J., J.D. Lowell, N.D. Rojas, and P.J. Burns, 1985, Reseña técnica del descubrimiento del yacimiento de pórfido cuprífero Escondida, II Region, Chile, *in* Actas IV Congreso Geológico Chileno, t.IV, p. 3-703–3-729.

Reinemund, J.A., P.W. Guild, and W.O. Addicott, 1982, Circum-Pacific Map Project: Framework for international assessment, *in* Transactions, Third Circum-Pacific Energy and Mineral Resources Conference: Tulsa, AAPG, p. 677-694.

Sillitoe, R., 1972, Relation of metal provinces in western South America to subduction of oceanic lithosphere: GSA Bulletin, v. 83, p. 813-818.

CHAPTER 35

# THE CIRCUM-PACIFIC MAP PROJECT GEOLOGIC MAP

**H.F. Doutch**
*Bureau of Mineral Resources*
*Canberra, Australia*

Work by regional panels on the five sheets of the 1:10,000,000-scale Geologic Map of the Circum-Pacific Region is virtually complete. The Geologic Map Series covers all of the Australian and Antarctic continents, the eastern part of Asia, the western halves of North and South America, and all of the Pacific Ocean and its islands. The sheets covering the Americas and the eastern Pacific (Drummond, 1984; Corvalán, 1985) are already published, and the rest (Craddock, 1987; Doutch, 1987; Nishiwaki, 1987) are in proof stage. Each sheet is accompanied by a set of Explanatory Notes. A 1:17,000,000-scale geologic map covering the whole of the Pacific basin is in a very early stage of preparation.

The project area includes six of the major and seven of the minor crustal plates that encrust this planet and also a large number of its terranes, and this is reflected in the land geology correlation chart of the map's explanation. The chart is fundamental to an understanding of the map. It began as a simple summary, lumping the standard world geological systems within the area into 21 time-rock units. During the 8 years or so of trying to live with the problems that this generated, the map and correlation chart, beginning with North America, have been developed into a strato-tectonic synthesis of 52 units.

The strato-tectonic approach has not turned the Circum-Pacific Geologic Map into another version of its Tectonic Map, although the two map series now have a certain degree of commonality. In essence, the land geology of the Geologic Map is depicted by unconformity- or fault-bounded stratigraphic units that are, or resemble, the supergroups of the International Stratigraphic Guide or the megaunits that have been suggested from time to time; each unit is in overall contrast with all of its neighbors.

The Circum-Pacific Geologic Map is not meant to stand in isolation from the other project maps. The five regional panels have always been in agreement that it should consist of units that could be used to derive the background for the project's Mineral Resources and Energy Resources maps. To a great extent this has fallen into place painlessly, as the Geologic Map's strato-tectonic units are commonly mineral or hydrocarbon specific. Notwithstanding this happy coincidence, the geologic bases for the resources maps do differ somewhat between sheets in order to deal with continental peculiarities.

So much for concepts and generalities. In detail, panels have been at pains to ensure that the characteristic geology of their regions is clear to map users. On the eastern margin of the Pacific they have been concerned to show the well-known Cenozoic and Mesozoic stratigraphic developments in the Americas associated with plate collision and subduction, and also the more enigmatic and for the most part older allochthonous terranes (for which internal stratigraphic sequences are all important). Therefore, Cenozoic volcanics and Mesozoic batholiths feature prominently; so do Cenozoic marine basins, also noteworthy because many of them are oil- and gas-bearing. Mesozoic and Paleozoic terranes are perhaps not so easy to pick out—the arcuate faults of northwestern North America and unit color discontinuities provide a first indication of their presence. Precambrian shield areas have been made obvious, and in South America the contrasts between them, flanking Cenozoic sediments, and the Andean chain are quite striking.

The complexities of the geology around the western margin of the Pacific, including its island arcs, are more difficult to deal with. The simple model developed for the Americas cannot be used without creating undesirable distortions or even false constructs.

In the southwest part of the Pacific basin, Australia and Oceania geology is treated as two sets of stratigraphies. One is appropriate to the continent's Precambrian shields, the Paleozoic "suspect" terranes accreted thereto, and the Mesozoic-Cenozoic epicratonic and passive-margin deposits. The other covers the mainly Cenozoic active-margin volcanics and marine sediments of the island arcs. The two stratigraphies overlap in the islands of New Guinea, New Caledonia, and New Zealand.

As Australia was so intimately involved in the breakup of Gondwana, ante- and post-breakup stratigraphic sequences within the continent are shown as two special passive-margin-associated strato-tectonic units. One is the sum of all so-called "rift"/"pre-drift" sequences, the other the sum of all "drift" sequences (which are more obvious on the map because they commonly prograde over "rift" sequences). How far they should be extended inland must be a matter of definition, and many will no doubt consider that the author has been overly generous. Some drift sequences are shown also in New Zealand, but no attempt has been made to extend the treatment to that other major Gondwana fragment, Antarctica.

The greater bulk of passive-margin-associated sequences occurs in the present continental-shelf deposits of Australia and Antarctica, and major oceanic features such as the Lord Howe Rise east of Australia appear to contain considerable volumes of "rift" sequences. Similar sequences are found in the vast shelf areas of Southeast Asia, for example the Sunda Shelf. However, despite the important oil and gas fields currently being exploited in many western Pacific shelf locations, and the potential still to be explored, the regional panels felt that too little is known yet to permit systematic depiction of shelf geology.

Although Gondwana links are more than tenuous, a somewhat similar passive- and active-margin picture is shown by the map in Asia and its peripheral island arcs. Overlaps between older continental crust and newer collision-margin products, so well displayed in Japan, extend into the continent in places; example, the Mesozoic batholiths of eastern China, where they are succeeded by passive-margin deposits along China's coast as sea-floor spreading opened the South China Sea.

South of China, on the other side of the strato-tectonic fault discontinuity along the Red River in Vietnam, a rough correlation with eastern Australia is afforded by the Paleozoic fold belt of western Malaysia, Thailand, and Burma; the adjacent Mesozoic epicratonic sequence of Thailand, Laos, Kampuchea, and Vietnam; and the fringing and overlapping Cenozoic passive-margin deposits.

The older parts of the continent can readily be interpreted from the map to be crustal agglomeration around Precambrian nuclei resulting from plate (and terrane) collisions.

The regional panel for Asia has had a much bigger task coordinating input from the various countries on their map sheet area than have other panels, and major new contributions were being received until very recently. The panel has cooperated much more closely than the others with the Commission for the Geologic Map of the World (CGMW), and has developed a certain relationship with the United Nations Economic and Social Commission for Asia and the Pacific (ESCAP), particularly its Committee for the Coordination of Offshore Prospecting in East Asia (CCOP).

Antarctic geology has to be something of a frustrating contrast to that of the rest of the Pacific basin. Not only are there far fewer people working on it, but strato-tectonic treatment of stratigraphy at 1:10,000,000 scale must have less meaning in this continent than in any other, and terrane analysis in western Antarctica is necessarily obscure. Nevertheless, the project map clearly distinguishes between the Precambrian igneous and metamorphic rocks of the East Antarctic Shield, and the largely Phanerozoic sedimentary sequences of the Transantarctic Mountains fold belt, West Antarctica, and the Antarctic Peninsula.

Further, in the spirit of the strato-tectonic and terranes concepts, it is probably safe to roughly correlate—but hardly equate—Phanerozoic units across the split between Gondwanan Antarctica and Australia: the closest relationship so far postulated is for the (rifting related?) Jurassic dolerites of Victoria Land and Tasmania. In the opposite direction along the strike of Phanerozoic outcrop in Antarctica, Australia-like geology gives way, not all that obviously on the map, to the microcontinents/?terranes of Marie Byrd and Ellsworth Lands, which in turn adjoin the Cenozoic-Mesozoic subduction products of the Antarctic Peninsula.

For the Geologic Map as a whole the Phanerozoic age range colors used are reasonably familiar; those for the Precambrian are perhaps less familiar. However, igneous rocks are not allotted special colors but are shown by patterns in black overprinted on an age color. The same approach is used for metamorphic rocks. Metamorphic rocks have been reduced to three broad classifications—low temperature and pressure, high temperature, and high pressure. The map explanation lists rock types and age ranges; the Explanatory Notes contain supplementary information and documentation that could not be included on the map.

Although it is very natural that land geology should be the first thing to catch the attention on the Circum-Pacific Geologic Map—especially as it is emphasized by the bright colors used for the purpose—perhaps a more important aim of the project is to summarize the latest information about the geology of the ocean floor.

To put together everything known about oceanic deposits and bedrock would produce a virtually unreadable map. The regional panels agreed, at an early stage, that age of the basaltic oceanic crust distribution, as deduced from oceanic magnetic anomaly patterns, was as appropriate to the Tectonic Map as to the Geologic Map. It was also evident that there was enough information, although somewhat patchy, to make a 1:10,000,000-scale sea-floor-sediment map of the Pacific. So oceanic crustal ages will appear on the Tectonic Map sheets—albeit in event packages—and the Geologic Map shows "unconsolidated sediments recovered primarily by coring and presumably exposed on the ocean floor at the sediment-water interface....They are not necessarily of Holocene age, nor are the deposits necessarily the result of Holocene sedimentary processes" (statement in Explanatory Notes that accompany each map sheet). This statement is by Floyd McCoy of Lamont-Doherty Geological Observatory, who, with the help of a Lamont team and Fran Mills, map project cartographer, was responsible for preparing the sea-floor-sediment component for all five map-sheet areas. The 13 units in the map explanation seem clear cut. However, the map user should read the Explanatory Notes closely, as they are essential for understanding the map properly.

Sea-floor-sediment units are based on a primary data base of sediment descriptions derived from smear-slide analyses and a secondary data base developed from published and unpublished sources. The classification developed for project purposes was modified from the Murray and Renard (1981) scheme of classification.

The smear-slide is said to offer a quick standardized method of microscopically determining the relative abundances of minerals and of biogenic skeletal components, and to be an excellent way of assessing proportions of silts and clays. For this project the method was supplemented by calcium carbonate measurements. The technique was primarily applied to the uppermost sediment of more than 4500 Lamont-Doherty cores from all over the Pacific basin.

Some smear-slide descriptions from other sources were included in the primary data base, but others that were uncertain were included in the heterogenous information of the secondary data base. Although this has a coherent framework supplied by the two-thirds of it derived from World Data Bank compilations, it also accommodates information from published literature, unpublished information, map compilations, and historical records such as oceanographic expedition reports. Much of this information is contradictory, inconsistent, or incomplete.

The sediment classification scheme starts with a major boundary at one part nonbiogenic rock to two parts biogenic. Nonbiogenic rocks are divided into gravel, sand, silt, and clay (including pelagic clay) following the Wentworth scale, but without intermediate classes such as silty clay. Biogenics are either calcareous or biosiliceous skeletal debris. The simplest mixtures of biogenic and nonbiogenic rocks covered by the scheme are calcareous and biosilicic clays; add some silt and you have biosilicic mud or marl (i.e., marl ooze—ooze being at least 60% biogenic debris plus clay, which may also consist in part of biogenic material). In addition, calcareous and volcanic gravel, sand, and silt are recognizable around atolls and volcanic islands. Diatomaceous silt dominates in some localities. On the map calcareous deposits are shown in blue; siliceous in yellow; volcanics and clays in brown; and gravel, sand, and silt also in yellow.

Primary controls in establishing trends of, and boundaries between, sediment types were bathymetry and regional depth variations of the calcite compensation depth (CCD). Calcareous ooze is presumed present, and is mapped, on seamounts and other sea-floor prominences, wherever their tops are at least 2000 m above the regional calcium carbonate compensation level.

The project maps show clearly the dominance of carbonates along the Equator and siliceous oozes and muds in high latitudes. More generally, first impressions are that elsewhere pelagic clays are the most common deposit in the northern hemisphere, but in the southern hemisphere they are confined to the east of the boundary between the Pacific and Australia-India plates, with extensive areas of calcareous sediments separating the clays derived from the Australian and South American continents. Although a 1:17,000,000-scale Pacific Basin Geologic Map has not yet been prepared, a project map at that scale has been published that shows sea-floor sediment and manganese nodule distribution. Consequently, sea-floor sediment may be omitted from the 1:17,000,000-scale Geologic Map in favor of oceanic crustal ages.

Although sea-floor sediment constitutes the greater part of ocean geology on this map, two other kinds of seabed information are also depicted.

Deep Sea Drilling Project (DSDP) sites are shown on the maps together with specially prepared columnar sections for many, if not most, of the sites. In the Explanatory Notes there is a summary of how the columnar sections from DSDP sites shown on the map were selected and compiled. Choice was on the basis of total depth of penetration, age of the oldest sediment, and whether or not basement, either basaltic or continental, was reached. In compiling the columns, DSDP's sediment classification scheme was retained, and the lithological patterns used illustrate it; geological colors are the same as for land units—absence of color indicates uncored zones of unknown age or sections of core of undetermined age.

Other aspects of offshore geology are also included in the Explanatory Notes. On the map, selected "bedrock" samples are shown by triangles labeled by geologic age and lithology. Samples come not only from outcrop, much of it basic igneous, but also from dredge hauls and gravity and piston cores, some bottoming in Tertiary or Cretaceous deposits. Map-element author George W. Moore reminds us of the age distribution of oceanic crust, noting the lack of cover on the newest parts of it or where it is exposed by faulting, and that fault scarps and trench slopes give access to gabbro, greenschist, and serpentinite outcrops. He briefly discusses the origin of oceanic plateaus, such as the Ontong Java plateau which seems to be of similar age to crust underlying it, and seamounts, including those of presumed hotspot origin such as the Hawaiian Ridge, which is younger than its underlying crust. He also points out that in general sediment cover is thickest near the continents, adjacent to which there are commonly thick Quaternary deposits; however, in the central Pacific basin early Tertiary strata crop out or lie within 1 m of the sea floor, in part because of dissolution of calcium carbonate being faster than biologic productivity and in part because of erosion by bottom currents. Apart from sample localities, much of this information can be deduced from the map only with some effort, and Moore's notes are in the nature of a supplement to it.

The mention of such individual contributions reflects a situation made possible largely through the good will of contributors' parent organizations. Special acknowledgment for their participation and encouragement should go to the Circum-Pacific Council, United States Geological Survey, and the American Association of Petroleum Geologists. The project would not have progressed very far without the personal dedication of headquarters staff, in particular Director John A. Reinemund, Maurice J. Terman, and former General Chairman Warren O. Addicott. Further, since 1975, regional panel chairmen have provided an essential link with geologists all over the Pacific basin, and perhaps it is mainly due to their efforts that this Geologic Map illustrates a reasonable and up-to-date consensus about the nature of the area.

## REFERENCES CITED

Corvalán, J., 1985, Geologic map of the Circum-Pacific Region, Southeast Quadrant: Circum-Pacific Council for Energy and Mineral Resources, scale 1:10,000,000.

Craddock, C., 1987, Geologic map of the Circum-Pacific Region, Antarctica Sheet: Circum-Pacific Council for Energy and Mineral Resources, scale 1:10,000,000.

Doutch, H.F., 1987, Geologic map of the Circum-Pacific Region, Southwest Quadrant: Circum-Pacific Council for Energy and Mineral Resources, scale 1:10,000,000.

Drummond, K.J., 1984, Geologic map of the Circum-Pacific Region, Northeast Quadrant: Circum-Pacific Council for Energy and Mineral Resources, scale 1:10,000,000.

Murray, J., and A.F. Renard, 1891, Report on deep-sea deposits based on the specimens collected during the voyage of H.M.S. *Challenger* in the years 1872 to 1876, *in* C.W. Thomson and J. Murray, eds., Report on the scientific results of the voyage of H.M.S. *Challenger* during the years 1873-1876: New York, Johnson Reprint Corporation, p. 30-47.

Nishiwaki, C., 1987, Geologic map of the Circum-Pacific Region, Northwest Quadrant: Circum-Pacific Council for Energy and Mineral Resources, scale 1:10,000,000.

CHAPTER 36

# ENERGY RESOURCES MAP OF THE NORTHEAST QUADRANT: ENERGY RESOURCES MAP SERIES, CIRCUM-PACIFIC MAP PROJECT

**Kenneth J. Drummond**
*Mobil Oil Canada, Ltd.*
*Calgary, Canada*

ABSTRACT

Maps portraying the energy resources of the Circum-Pacific Region are being compiled for each of the quadrants and Antarctic region at a scale of 1:10,000,000 and a Pacific Basin Map sheet at a scale of 1:17,000,000. The maps will show the occurrence of petroleum, coal, and geothermal resources overprinted on a simplified geologic/tectonic background. Color-coded sediment isopachs for sedimentary basins and oceanic areas indicate the age and thickness of sediments. Generalized stratigraphic sections show the sedimentary section for selected basins. Explanatory notes booklets including supplementary data are being prepared to accompany the maps.

The first of the Energy Map Series to be published is the Energy Resources Map of the Northeast Quadrant. This will be followed by the Southeast Quadrant and the other map sheets. The series of maps will show the interrelationship between geologic and tectonic features as they relate to the occurrence of energy resources for a region covering nearly half the globe.

## INTRODUCTION

The Energy Resources Map Series is being compiled to show the relationship of the major energy resources to the tectonic framework of the Circum-Pacific Region. The maps show oil and gas fields, oil shale, oil sands, coal deposits according to rank, and geothermal convection systems.

The Energy Resources Map of the Northeast Quadrant of the Circum-Pacific Region (Drummond, 1986) is the first in a series of five overlapping 1:10,000,000-scale energy resources map sheets. The map is accompanied by an "Explanatory Notes Booklet," with additional data, explanations, and references that could not be depicted on the face of the map. Other maps in the series will be the Northwest, Southwest, and Southeast Quadrants, and the Antarctica Region. Upon completion of these, the Energy Resources Map of the Pacific Basin will be prepared at a scale of 1:17,000,000. The Energy Resources Map Series is designed to be as factual as possible, with a minimum of interpretation. The small scale of the maps necessitates an enormous simplification of both the background information and the energy resource data; hence they can give only a general impression of the distribution, character, and geologic environment of these resources. Nevertheless, the maps will provide a unified overview of the energy resources of the Circum-Pacific Region.

The Energy Resources Map of the Northeast Quadrant is intended to show the relationship of geology, tectonics, and crustal dynamics to the major energy resources of oil, gas, oil sands, oil shale, coal, and geothermal. The geologic background depicts the relevancy of the "economic basement" to the sedimentary basins. Depicted in a generalized format are Precambrian basement, igneous intrusives, volcanic cover, and deformed sedimentary fold belts. Sediment isopachs color coded according to age are shown in both continental and oceanic areas. Representative stratigraphic columns are keyed to some of the major petroleum producing basins of the northeast Pacific margin.

## REGIONAL FRAMEWORK, NORTHEAST PACIFIC

Figure 1 is a very simplified tectonic features map to show the relationship between the major terranes; the craton with bordering mobile belt, the accretionary terrane, and bordering oceanic basins. Shown are the major plate boundaries, significant fracture zones, and representative Holocene volcanoes.

The regional elements of the continental blocks include shield, platform, mobile belt, and continental shelf. The plat-

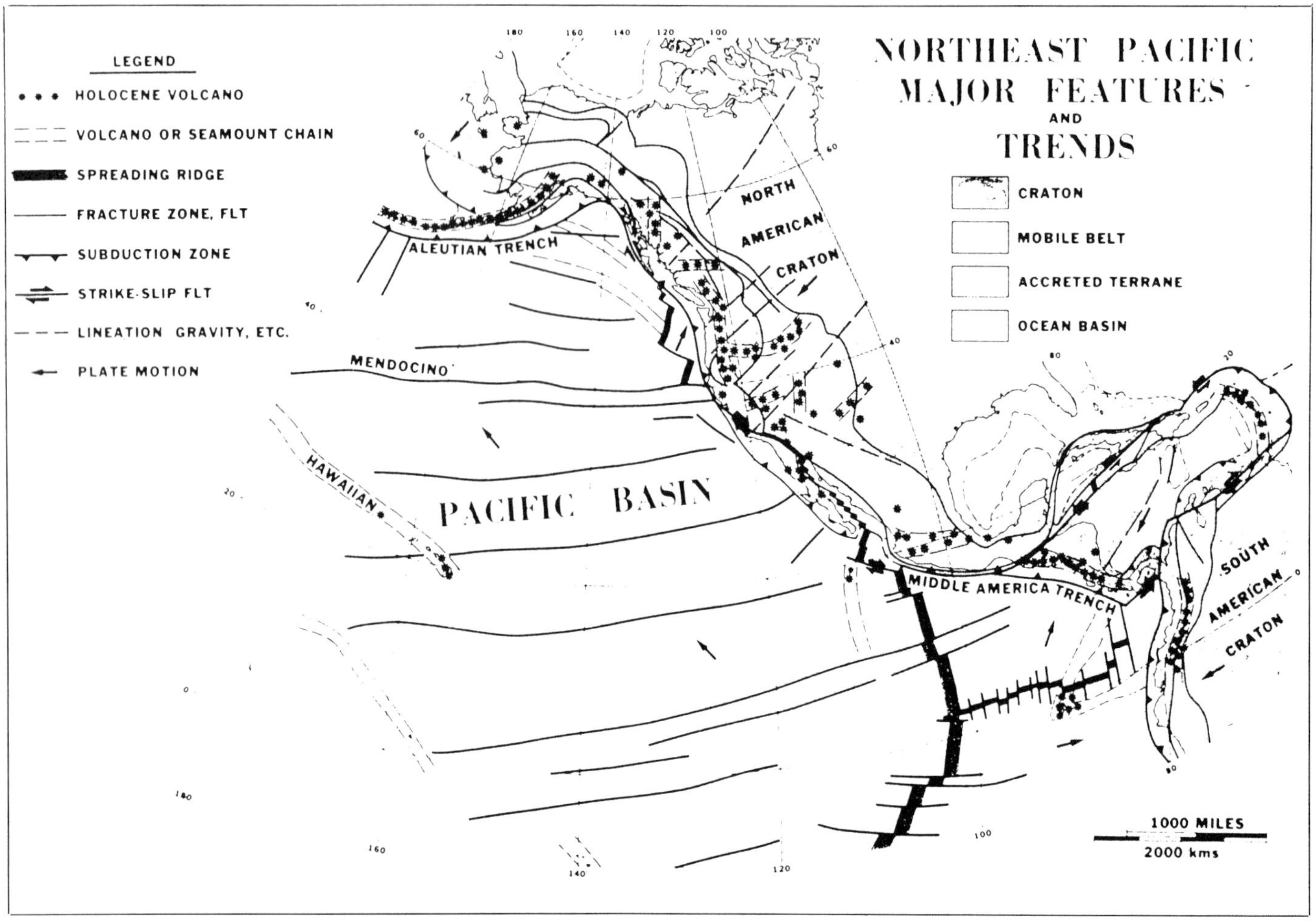


Figure 1. Tectonic features map.

form areas are flanked by the mobile belts, which pass westward into deeper water slope and oceanic deposits with associated volcanics. The western regions of North and South America are comprised of several large blocks of accreted terrane. Superimposed on this are syntectonic deposits of Jurassic-Cretaceous age, and extensive Cenozoic plateau volcanics. Late Cenozoic volcanic chains are associated with the major trenches, such as Aleutian, Lesser Antilles, and Central America. Transverse linear late Cenozoic volcanic belts include Yellowstone and Trans-Mexico. Coastal plain deposits which flank the continental blocks are extensive along passive regions such as the Gulf of Mexico and the Arctic and are generally narrow and limited along the active Pacific margin.

Oceanic tectonics are dominated by the long linear fracture zones, spreading centers, major trenches or subduction zones, and large strike-slip faults. Superimposed on the overall tectonic grain are chains of volcanic islands, including the Hawaiian, Welker-Kodiak, Marquises, and Line Islands.

## NORTHEAST PACIFIC VOLCANISM

Major volcanism, flood basalts, and volcanic arcs show definite trends for both continental and oceanic regions. Cenozoic plateau volcanics are commonly co-extensive with major transverse trends of the Cordillera. Volcanoes active, or inferred to have been active, in the Holocene occur in definite trends and clearly show the relationship to present-day plate interactions. The trend of Tertiary volcanic chains, such as the Aleutian, Central America, and Lesser Antilles, parallel the associated trenches and are related to subduction of the oceanic crust. The northwesterly trending volcanoes and seamounts, as Hawaiian and Kodiak, showing an age progression were formed as the Pacific plate moved over fixed hot spots.

## SEDIMENTARY BASINS OF THE NORTHEAST QUADRANT

Thick accumulations of sedimentary rocks, mostly of Mesozoic age, occur all along the continental margin of the Northeast Pacific Quadrant. The sedimentary basins (Figure 2) can be subdivided into three major classifications: interior basins (cratonal and foreland), intramontane or fold belt basins, and marginal basins either directly or indirectly related to oceanic areas. The interior platforms comprise a wedge of sediments thickening westward to 7 to 11 km along the front of the Cordillera. The basin areas are formed by a series of depressions between the Cordillera and the shields, commonly separated by basement arches.

The intramontane basins are sedimentary basins of mid-Jurassic to late Tertiary age that occur within the major Cordilleran uplifts. Sediments were deposited in these basins that

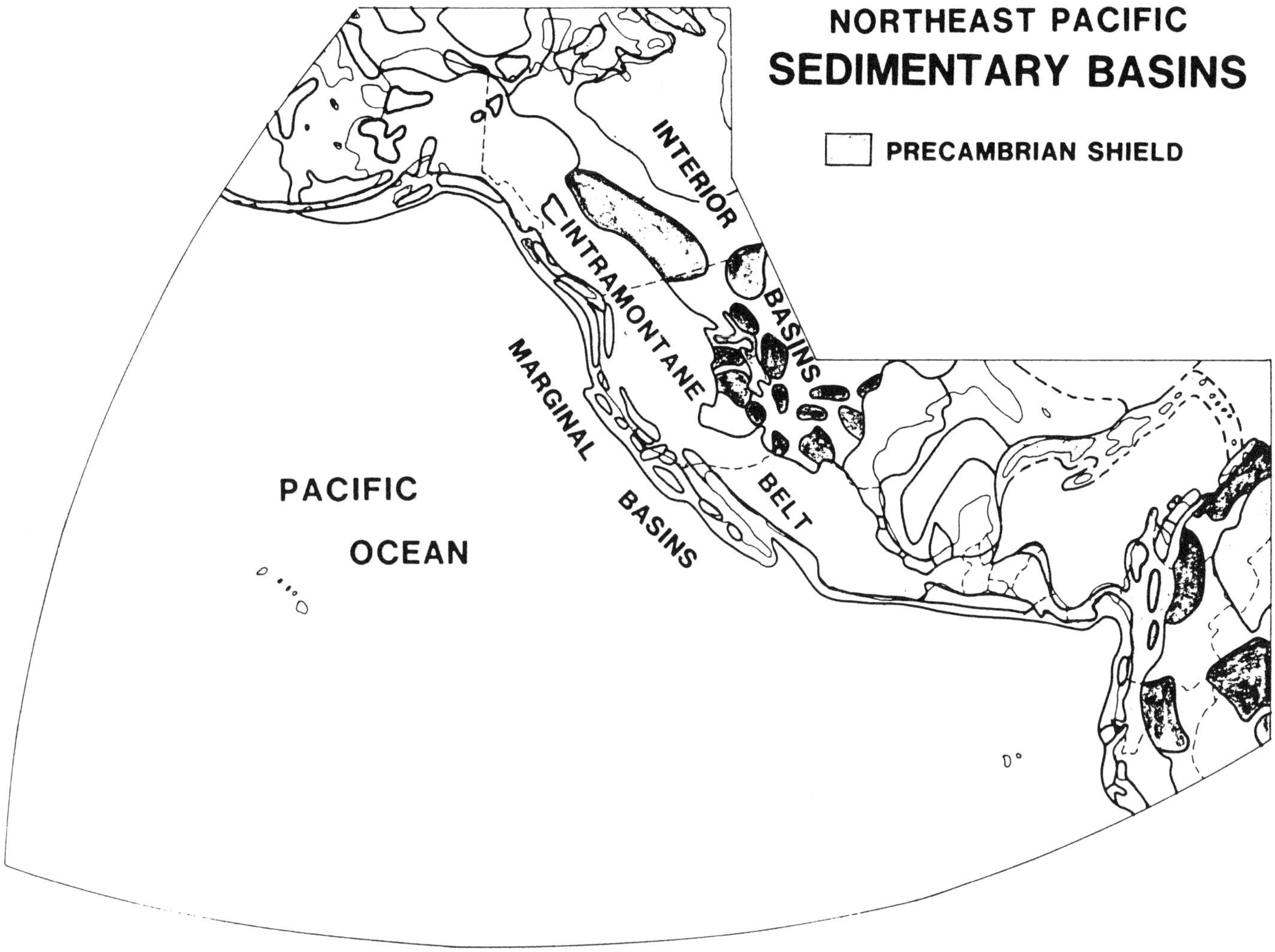


Figure 2. Sedimentary basins of the Northeast Quadrant.

developed by localized subsidence, both during and after the major deformation.

Marginal basins include all basin areas bordering the North and South American continents. These include Circum-Pacific, Bering Sea, Gulf of Mexico, and Circum-Caribbean basins. In this paper I will only discuss the Northeast Pacific marginal basins.

Thick postorogenic continental and shallow-marine sequences are deposited in basins along the northeast Pacific continental margin. Age of sediments ranges from Middle Mesozoic to late Tertiary and Recent. The marginal basins of the Northeast Quadrant extend some 12,800 km (8000 mi) from northeast Peru to the Alaska Peninsula.

## ENERGY RESOURCES

### Oil and Natural Gas

Figure 3 shows those basins which contain giant fields with estimated ultimate potential of more than 500 million barrels of oil or 3 trillion cubic feet of natural gas. Indicated are 25 basins containing 87 giant oil fields with a total cumulative production of 90.8 billion barrels, which represents 43.4% of the cumulative production for countries of the Northeast Quadrant (Table 1). The giant fields contain approximately 42.6% of the total initial established oil reserves of the Northeast Quadrant.

There are 29 giant gas fields with a total estimated ultimate potential of 244.6 trillion cubic feet. This represents 20.4% of the total of 1197.4 trillion cubic feet initial established gas reserves of the Northeast Quadrant.

Index maps and the tabulation of oil and gas production and reserves for these fields are included in the Explanatory Notes Booklet (Drummond, 1986) accompanying the Northeast Energy Resources Map.

A total of 10 productive oil and gas basins (Figure 4) are associated with the Pacific Margin of the Northeast Quadrant. These basins, which are mainly onshore, include: (1) Cook Inlet, (2) Sacramento, (3) San Joaquin, (4) Santa Maria, (5) Salinas, (6) Ventura, (7) Los Angeles, (8) Daule, (9) Progreso, and (10) Talara.

Basins of the Northeast Pacific margin have an estimated cumulative production to the end of 1984 of 22.04 billion barrels of oil and 30.32 trillion cubic feet of natural gas. The largest resources are in the basins of Southern California; other significant occurrences are in the Cook Inlet basin of Alaska and in the coastal basins of Southern Ecuador and Northern Peru (Figures 5, 6, 7).

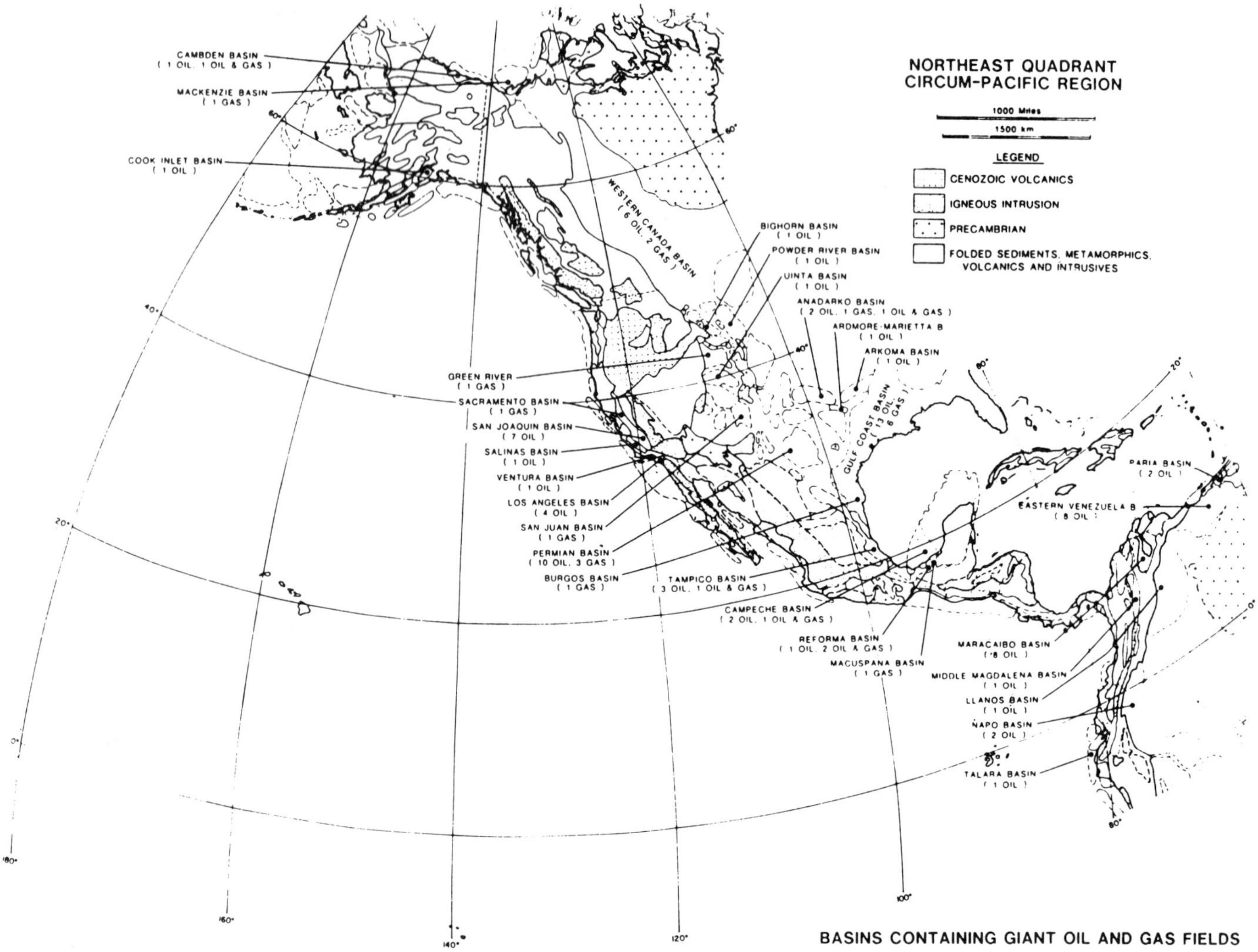


Figure 3. Basins containing giant oil and gas fields, Northeast Quadrant.

TABLE 1

ESTIMATED INITIAL AND REMAINING RESERVES
BY COUNTRY
NORTHEAST QUADRANT, CIRCUM-PACIFIC REGION

| | INITIAL ESTABLISHED | | CUM.PROD.(DEC.31,1984) | | RESERVES(DEC.31,1984) | |
|---|---|---|---|---|---|---|
| COUNTRY | OIL(MMB) | GAS(BCF) | OIL(MMB) | GAS(BCF) | OIL(MMB) | GAS(BCF) |
| CANADA | 17498.7 | 148869.0 | 11522.3 | 49167.6 | 5976.4 | 99701.4 |
| U.S.A. | 168701.9 | 827389.6 | 140255.9 | 629926.6 | 28446.0 | 197463.0 |
| MEXICO | 61187.4 | 98527.0 | 11927.4 | 21825.0 | 49260.0 | 76702.0 |
| GUATEMALA | 20.0 | | 10.6 | | 9.4 | |
| TOTAL N.AMERICA | 247408.0 | 1074785.6 | 163716.2 | 700919.2 | 83691.8 | 373866.4 |
| COLOMBIA | 3603.6 | 8290.0 | 2403.6 | 4156.0 | 1200.0 | 4134.0 |
| ECUADOR | 1857.4 | 3674.0 | 1025.9 | 748.0 | 831.5 | 2926.0 |
| PERU | 2119.0 | 3246.0 | 1448.5 | 2254.0 | 670.5 | 992.0 |
| VENEZUELA | 66877.0 | 94118.0 | 38843.0 | 35318.0 | 28034.0 | 58800.0 |
| TRINIDAD | 2739.9 | 13319.0 | 2201.9 | 3019.0 | 538.0 | 10300.0 |
| BARBADOS | 3.4 | | 2.8 | | 0.6 | |
| TOTAL N.W. S.AMERICA | 77200.3 | 122647.0 | 45925.7 | 45495.0 | 31274.6 | 77152.0 |
| TOTAL NORTHEAST QUADRANT | 324608.3 | 1197432.6 | 209641.9 | 746414.2 | 114966.4 | 451018.4 |

SOURCES: DEGOLYER&MCNAUGHTON, 1985, WORLD OIL, AMER.ASS.PETR.GEOL, U.S.DEPT. OF ENERGY, U.S.G.S.PROF.PAPER 885, CAN.PETR.ASSOC.

## Cook Inlet Basin

The Cook Inlet basin is a fault-bounded basin located in an arc-trench gap setting between the volcanic arc, represented by the Alaska Range, and deformed Upper Cretaceous turbidites to the southeast. Basin fill comprises Middle and Upper Jurassic sediments, unconformably overlain by Upper Cretaceous and Tertiary sediments. The major oil and gas fields shown are located on north-south-trending anticlines. The largest oil field is McArthur River with cumulative production of 499 million barrels and total estimated ultimate reserves of 570 million barrels. The total cumulative production from the Cook Inlet basin to December 31, 1984 is 1.36 billion barrels of oil. The largest gas field is Kenai, with an estimated ultimate reserve of 2.5 trillion cubic feet.

## Southern California Basins

Onshore basins of Southern California contain large accumulations of oil and gas in thick turbidite sequences primarily of Miocene age. Offshore in the Southern California borderland, equivalent rocks are much thinner in all but the nearshore basins. The Ventura basin with a very thick section of Tertiary sediments extends offshore into the Santa Barbara Channel. The offshore Santa Maria basin has received considerable recent activity, and significant discoveries have been made in the Miocene Monterey fractured shale. The Los

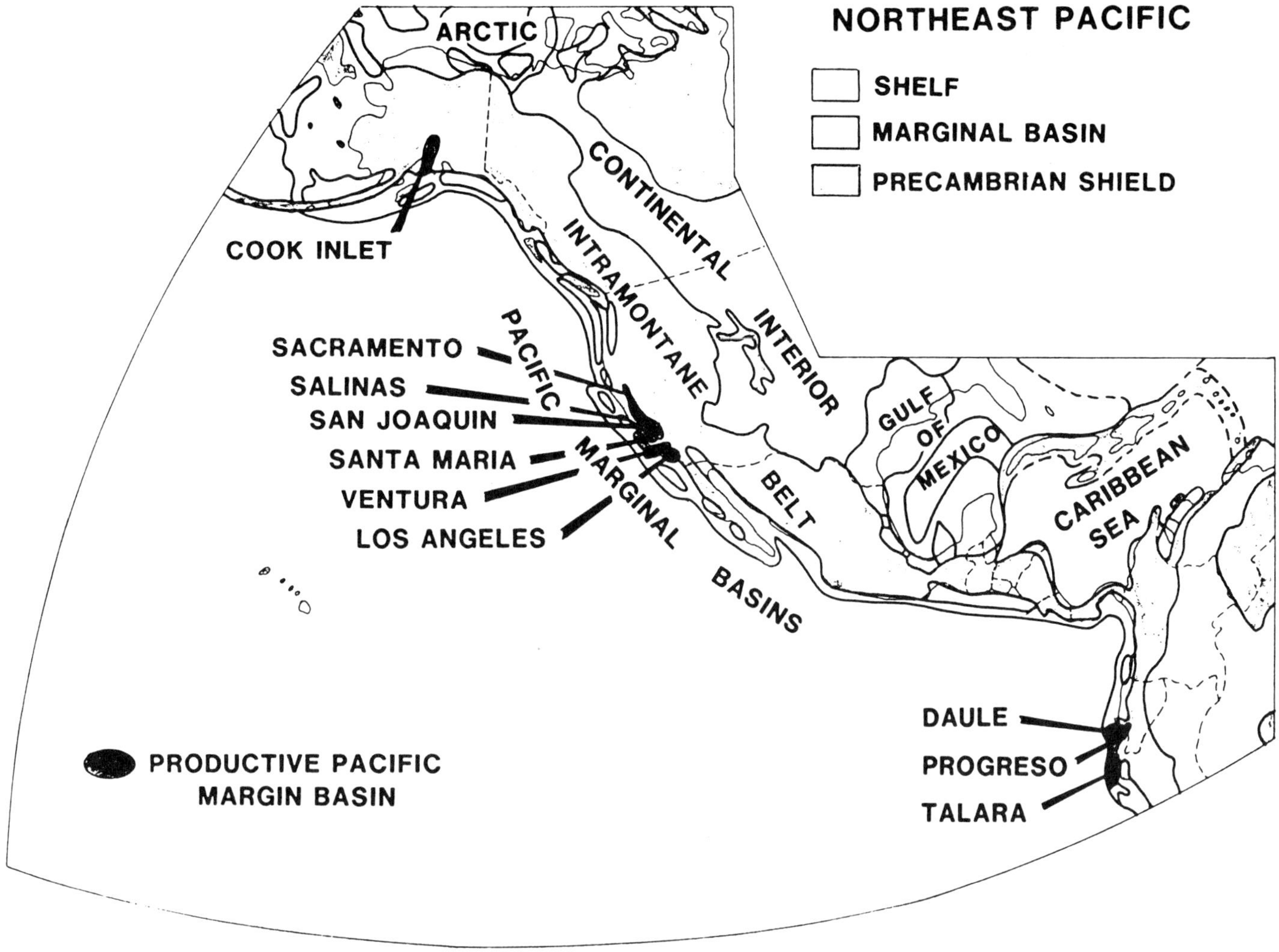


Figure 4. Productive Pacific marginal basins.

Angeles basin is one of the most prolific oil production areas per square mile in the world. Oil was discovered in 1880 and as of December 31, 1984, total production had reached 4.2 billion barrels from about 1400 square miles.

Cumulative production to the end of 1984 from the Southern California basins is 19.37 billion barrels of oil and 27.3 trillion cubic feet of marketed natural gas. There are 15 giant oil fields with ultimate recoverable reserves of more than 500 million barrels.

### Coastal Ecuador and Peru

The productive coastal basins of Southern Ecuador and Northern Peru include the Daule, Progreso, and Talara. The Daule basin covers the western part of the Santa Elena Peninsula and extends into the offshore. The largest oil field, Ancon, had a cumulative production of 115.8 million barrels to the end of 1984.

The Progreso basin, containing as much as 6000 m of post-Oligocene sediments, extends from the Gulf of Guayaquil, Ecuador, into northwest Peru. A potentially giant gas field, Amistad, occurs in this basin, lying offshore in the Gulf of Guayaquil.

The Talara basin, Peru, is bounded on the east and southeast by the Coastal Range, and extends into the offshore. The basin, containing more than 8000 m of Campanian to late Eocene sediments, produces mainly from Eocene deltaic, fluvial, and turbidite sandstones. The Talara basin has produced about 85% of Peru's oil production to date, mainly from the giant La Brea-Parinas field, discovered in 1869. Total oil production from La Brea-Parinas to December 31, 1984 is 910.7 million barrels.

The cumulative oil production to December 31, 1984 from coastal Ecuador and Peru is about 1.24 billion barrels.

## Oil Sands and Oil Shale

Oil sands are oil-impregnated sands whose oil is not recoverable by conventional well-bore methods. The major oil sands, shown in Figure 8, are in Canada and Venezuela, and smaller deposits are found in the western United States. The oil sand deposits of Alberta contain an estimated 1080 billion barrels of oil in place, of which 35 billion are considered economically recoverable through current methods. To the end of 1984 cumulative production of synthetic crude oil was 459 million barrels. The Orinoco tar belt in Venezuela is estimated to contain approximately 1.2 trillion barrels of heavy oil in place.

The principal oil shale deposits of the western United States are in the Piceance basin in Colorado and the Uinta basin in

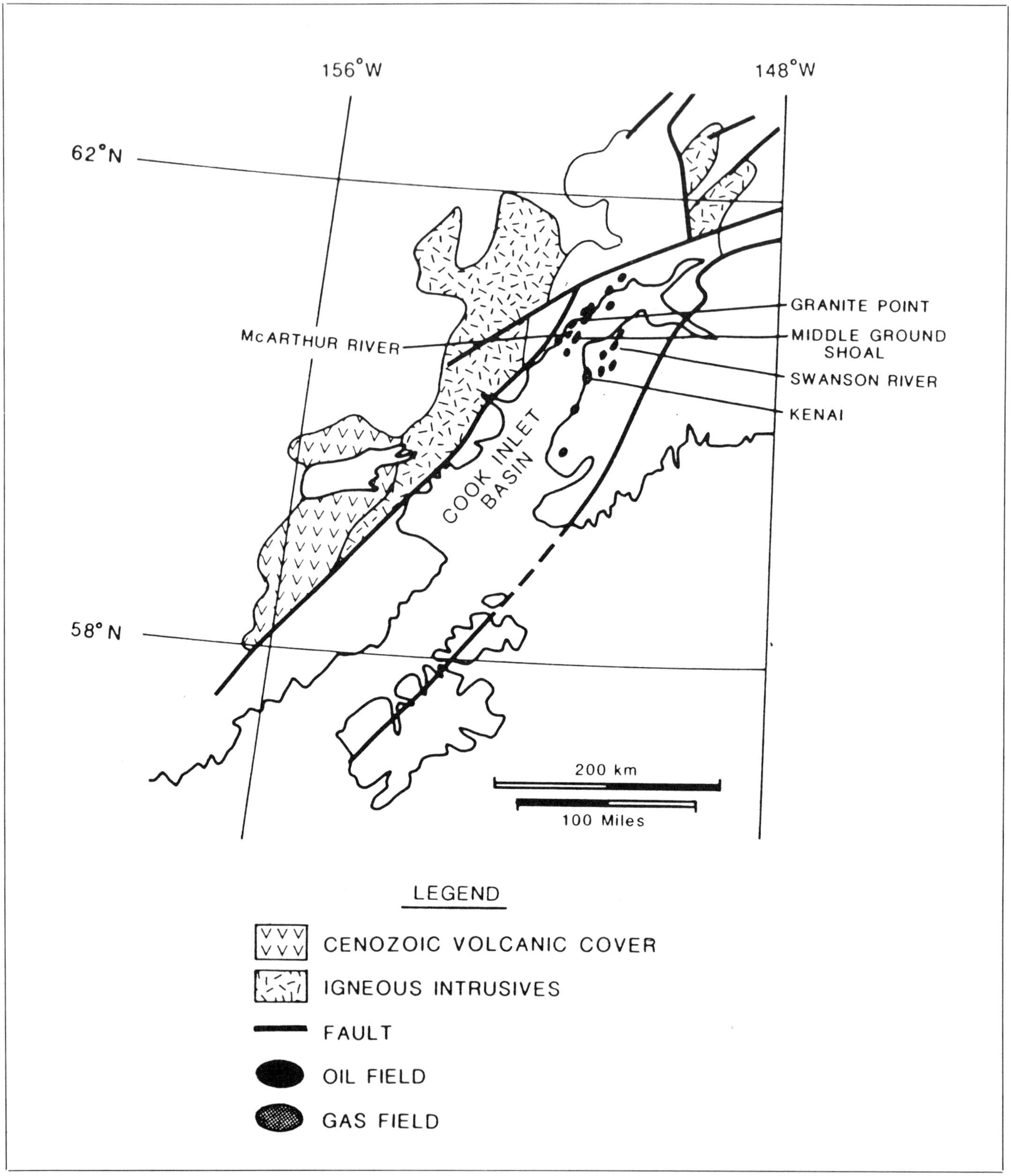


Figure 5. Oil and gas fields, Cook Inlet basin, Alaska.

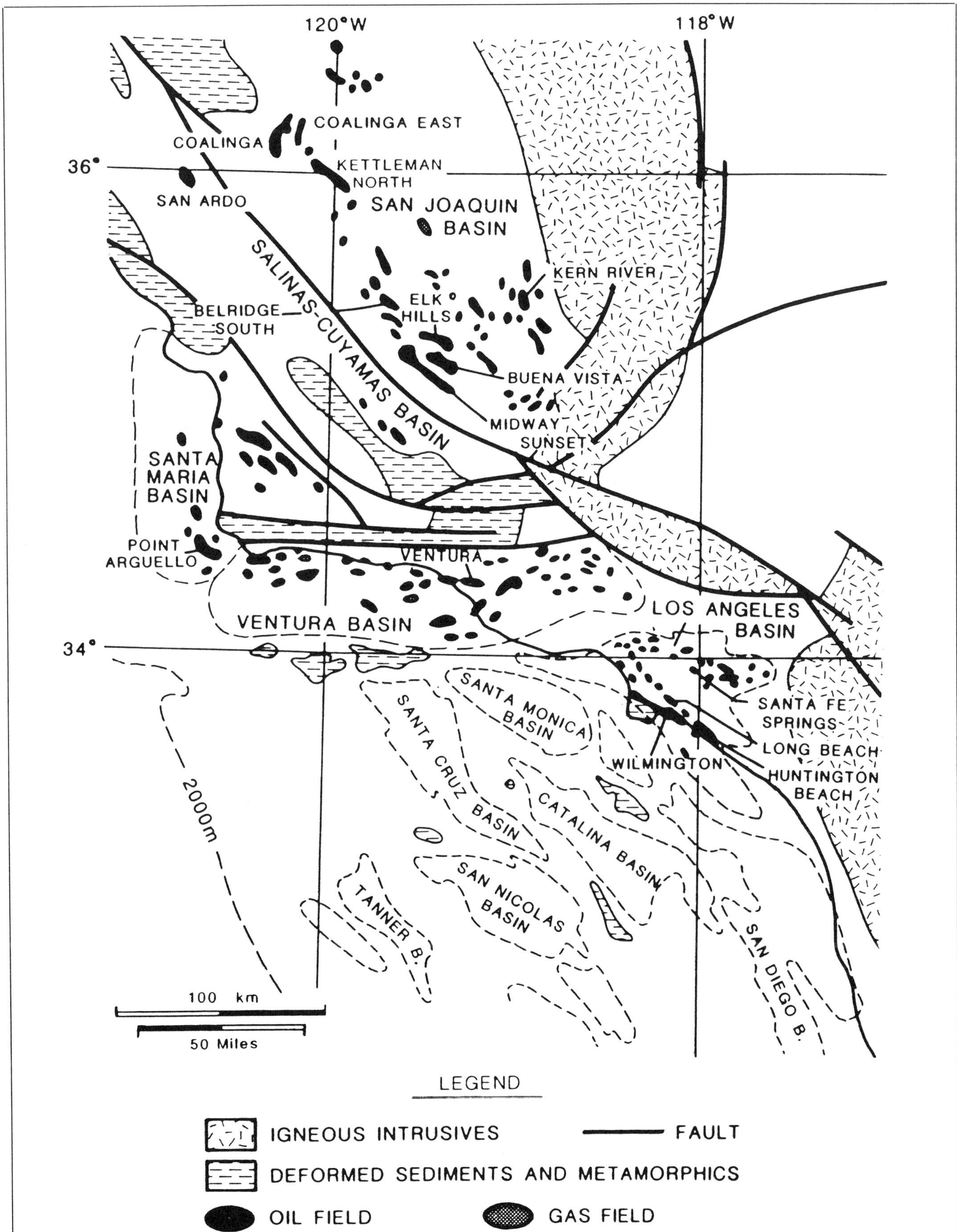


Figure 6. Oil and gas fields, Southern California.

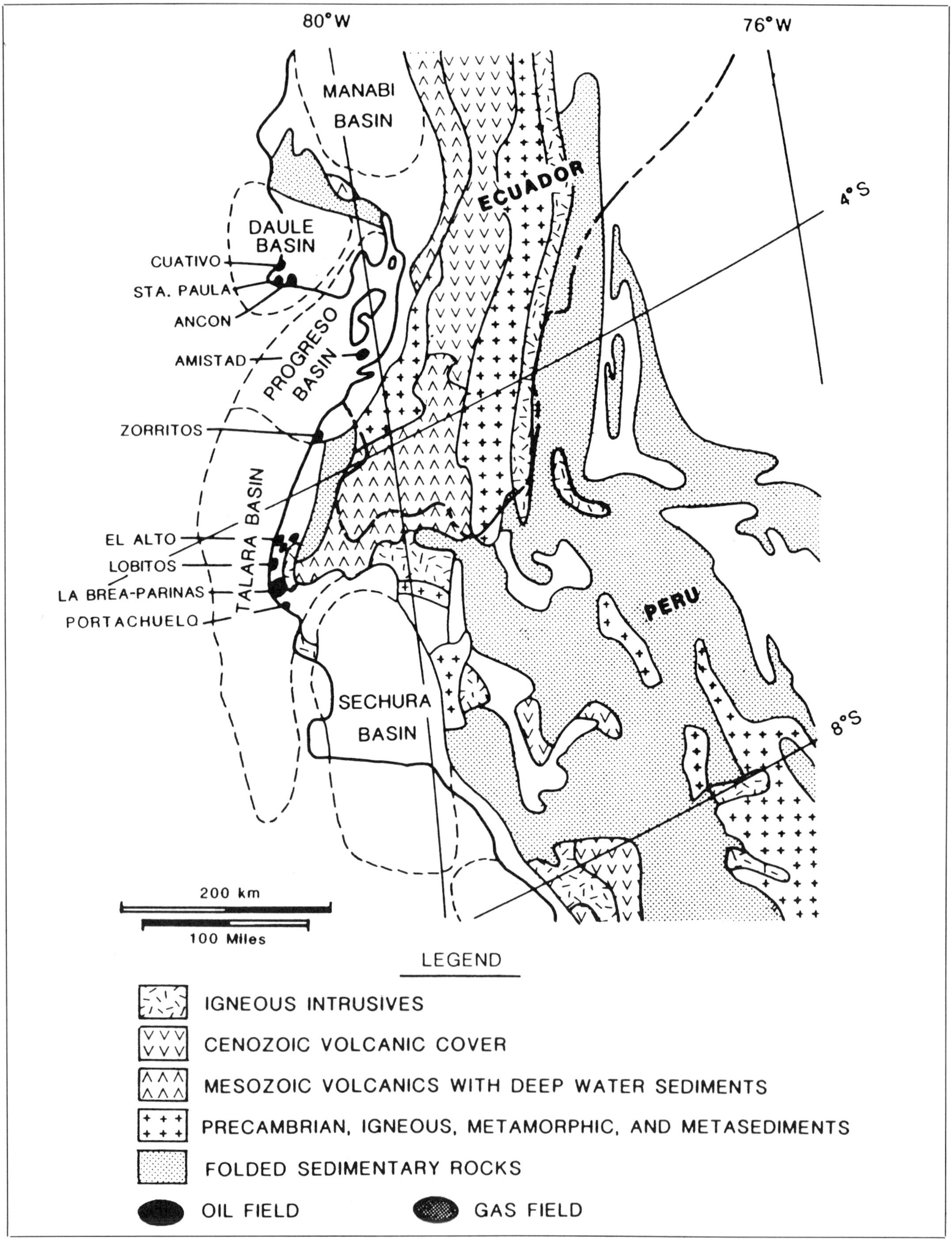


Figure 7. Oil and gas fields, coastal Ecuador and Peru.

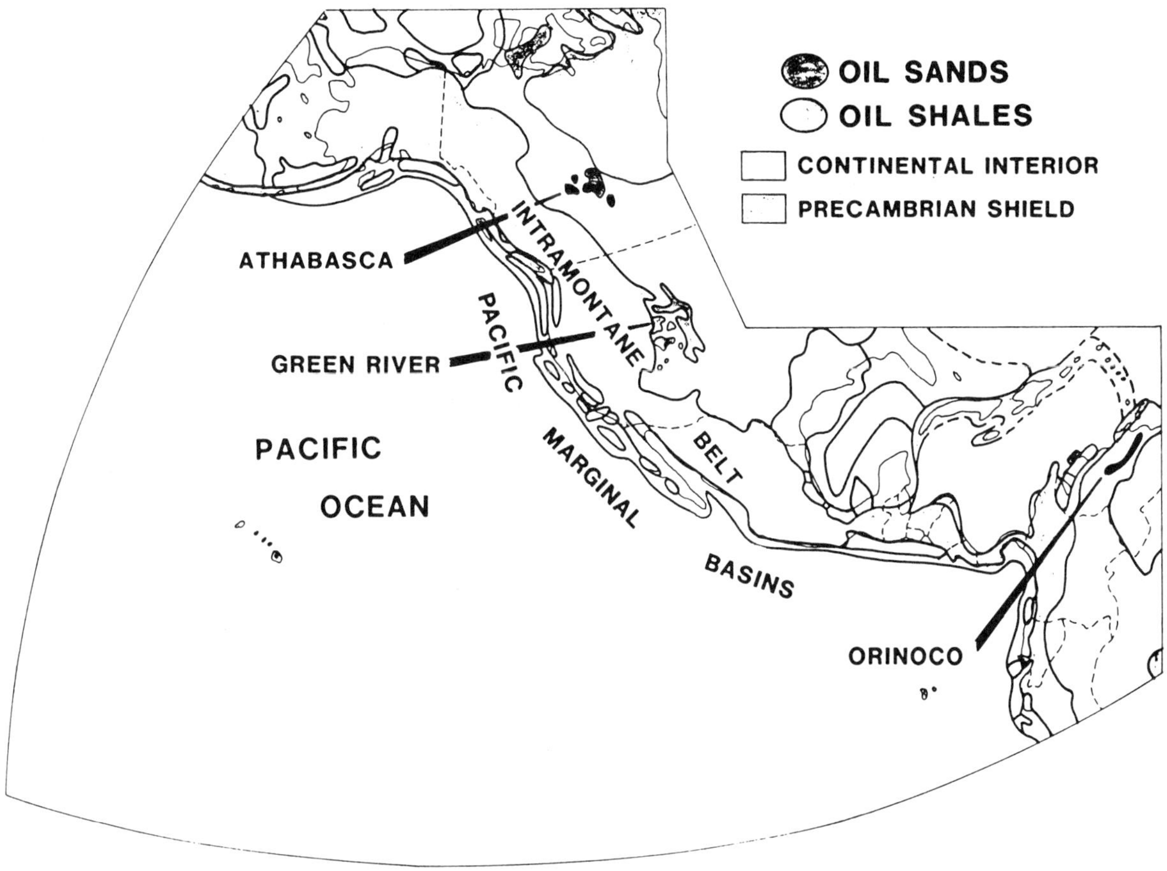


Figure 8. Oil sands and oil shale of the northeast Pacific.

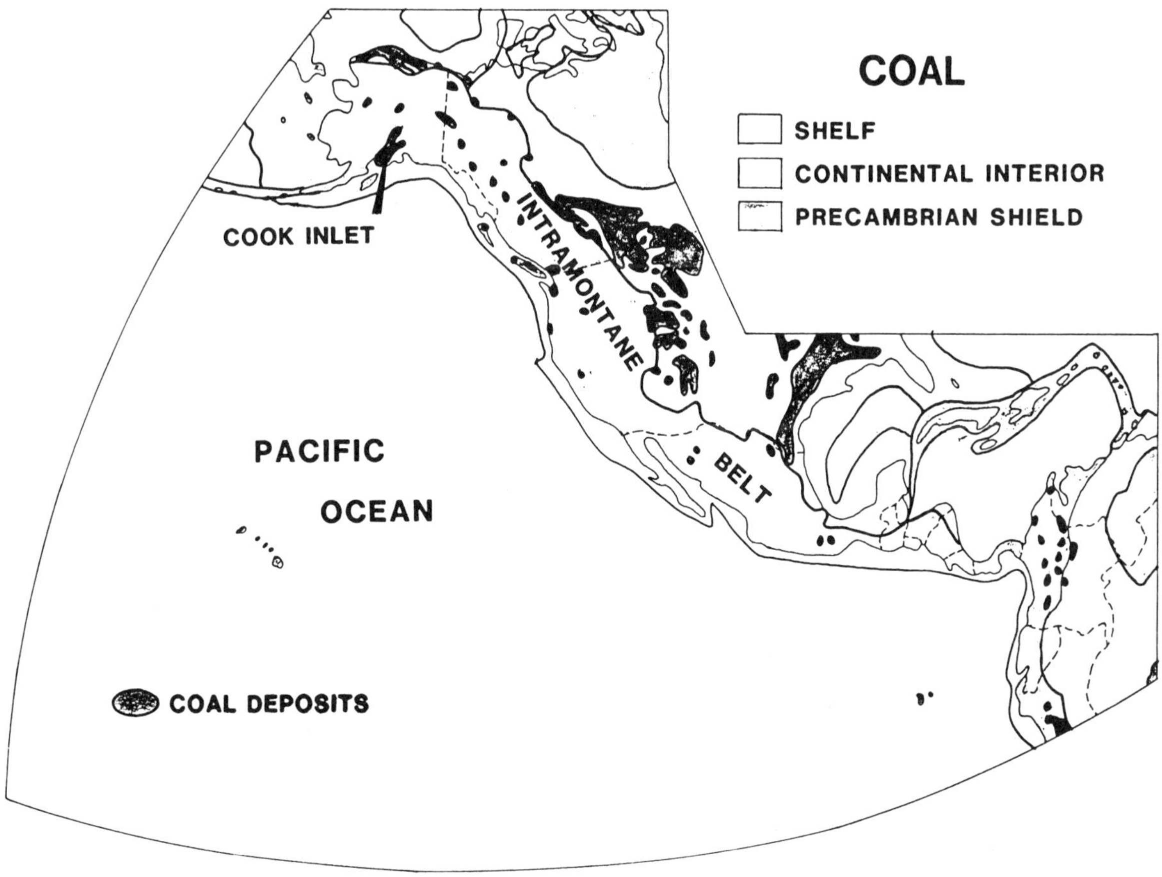


Figure 9. Coal deposits of the northeast Pacific.

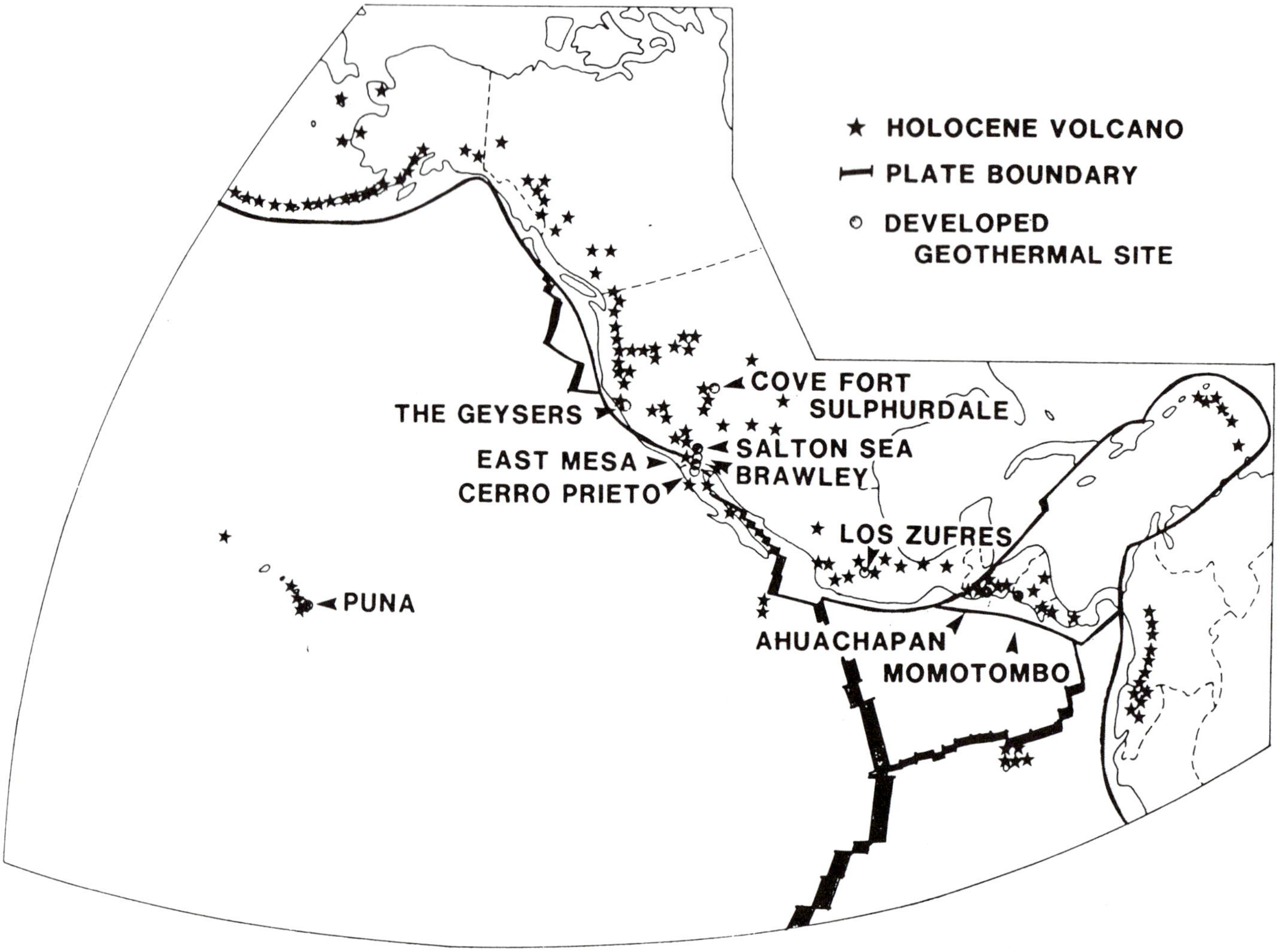


Figure 10. Geothermal energy of the northeast Pacific.

Utah. A number of research projects are under way in these areas. The Green River Formation in these areas contains an estimated 1.8 trillion barrels of identified shale oil resource (Culbertson and Pitman, 1973). There are no known deposits of any significance of oil shale associated with the Pacific margin of the Northeast Quadrant.

## Coal

Major coal deposits (Figure 9) occur throughout much of the Northeast Quadrant. Most of the larger deposits and production are associated with the interior basins and bordering foreland thrust belts. The more significant coal deposits of the Pacific margin, with only minor production at present, occur in the Cook Inlet-Susitna basin area of Alaska, the Queen Charlotte Islands, and Vancouver Island of western Canada, as well as in Washington, Oregon, California, Mexico, Costa Rica, Columbia, Ecuador, and Peru. Coal production in 1984 was 253.6 million short tons from the western United States and 43.9 tonnes from Western Canada.

## Geothermal

The Pacific margin of the Northeast Quadrant is generally a zone of subduction, with associated volcanism. Figure 10 shows the relationship of the major plate boundaries and Holocene volcanoes to the main geothermal energy sites. Developed geothermal sites of the Northeast Quadrant include: the Geysers, Cove Fort Sulphurdale, Salton Sea, Brawley, and East Mesa of the western USA; Cerro Prieto and Los Zufres, Mexico; Ahuachapan, El Salvador; Momotombo, Nicaragua; and Puna, Hawaii. Several others are currently under development.

The Energy Resources Map of the Northeast Quadrant is the first to be published. The series of maps will show the interrelationship between geologic, tectonic, and geodynamic features as they relate to the occurrence of energy resources for a region covering nearly half the globe. This should be a very valuable series to help evaluate the energy resources of the future.

## REFERENCES CITED

Culbertson, W.B., and J.K. Pitman, 1973, Oil shale, *in* United States Mineral Resources: U.S. Geological Survey Professional Paper 820, p. 495-503.

Drummond, K.J., 1986, Energy resources map of the Circum-Pacific Region, Northeast Quadrant: AAPG, scale 1:10,000,000.

CHAPTER 37

# MODE OF SUBDUCTION, COLLISION, AND TYPES OF METALLOGENESIS

**Chikao Nishiwaki**
*Northwest Quadrant Panel of the Circum-Pacific Map Project*
*Tokyo, Japan*

**Seiya Uyeda**
*Earthquake Research Institute*
*University of Tokyo*
*Tokyo, Japan*

## INTRODUCTION

Since the publication of the Circum-Pacific Plate Tectonic Maps in 1982, the theory of plate tectonics has enjoyed much refinement and reinforcement in many fields of theoretical studies and applications in resources exploration. At the Circum-Pacific theater its characteristic features of the consuming end of the plate motion, namely subduction, collision, terrane accretion, and amalgamation, have been continuously investigated. We would like to summarize some of the recent research in this field for a few regions of Asia and try to deduce some prospects for application.

One of the writers (Uyeda) distinguished two types of subduction, namely the Chilean-type and the Mariana-type, by their contrasting styles, stress fields, accompanying seismicity, and other geophysical features, and urged the study of the comparative subductology as one of the basic fields of investigation (Uyeda and Kanamori, 1979; Uyeda, 1982). The authors also noted that the regions with these two types of subduction are commonly associated with contrasting types of base metal mineralization, namely, the high-stress Chilean-type subduction with its porphyry copper deposits and the Mariana-type subduction in an extensional stress environment with Kuroko and related massive sulphide deposits. They occur complementary, as their genesis are basically favored by the contrasting stress field of these modes of subduction at the time and place of emplacement (Uyeda and Nishiwaki, 1980).

In the meantime, others recognized that the collision and accretion of exotic tectonostratigraphic terranes have made up western North America (Jones et al., 1983). This recognition and their methods of investigation spread rapidly all through the Circum-Pacific area. It also stimulated study of many other large continental collision areas, like India-Tibet and China.

The first seminar on Accretion Tectonics in the Circum-Pacific region was held in Hokkaido, Japan, in August 1981. Other meetings were held in Stanford, California, in 1983 and in Sydney, Australia, in 1985, and the Preliminary Tectonostratigraphic Terrane Map of the Circum-Pacific Region with an explanatory text was published by the Circum-Pacific Council in 1982. A careful look at this map will reveal that most of the continents at the consuming end of the spreading oceanic plate of the Pacific have been and are growing oceanward by receiving and accreting continental fragments, island arcs, and seamounts. These buoyant features, commonly accompanied by miogeoclinal sediments on their shores when they arrive at the old shores of the continents, collide and are partly thrusted under the continents as if they are pulled down by the leading subducting heavy oceanic lithosphere.

Himalaya-Tibet, Indochina-Sundaland, Southeast China, and Central Japan were selected in this chapter for the purpose of illustrating the polycyclic collision/accretion process, which includes buoyant continental crust subduction, magmatism and mineralization.

## COLLISION/ACCRETION IN THE HIMALAYA-TIBET REGION

As a result of 3 years of investigations by the French-Chinese Joint expedition (Allegre et al., 1984), which includes 400 km of north-south deep seismic profiling, much new tectonic information has been added. Some of their results pertinent to the purpose of this paper are discussed below.

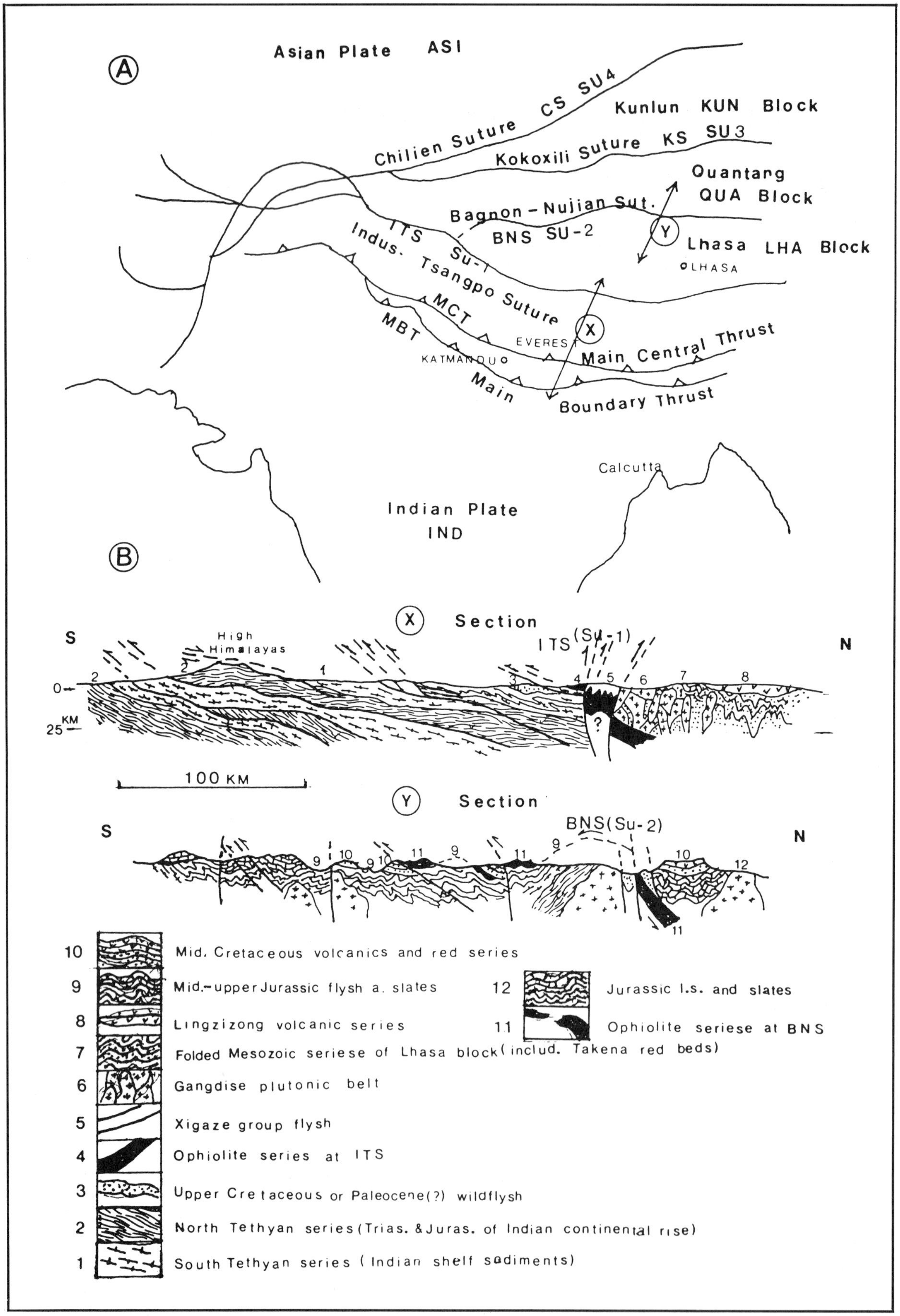


Figure 1. Himalaya and Southern Tibet. A, Main tectonic blocks and boundaries. B, N-S sections. (After Allegre et al., 1984).

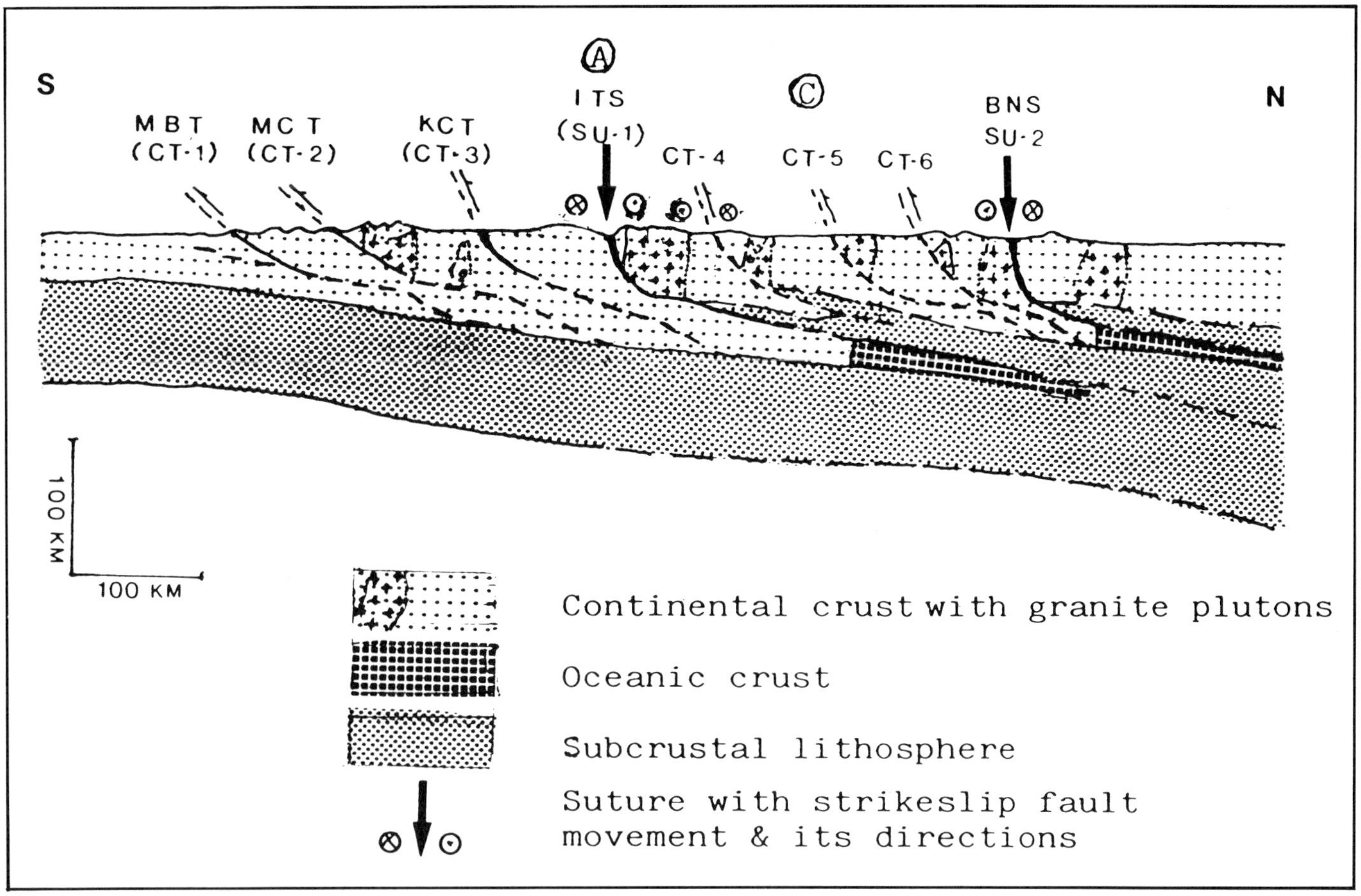


Figure 2. Schematic model for the structure of the lithosphere in the Himalaya-South Tibet region (After Allegre et al., 1984). A 400-km-long traverse of deep seismic-reflection profiling revealed the two superimposed Mohos at point C (Yangbajain) and north of BNS.

(1) Southward from the southern periphery of the Tarim Basin to the Main Boundary Thrust at the northern end of the Indian-plate, there are five accretionary terranes as shown in Figure 1A, which shows successive accretion events. Figures 1A and 1B show their mutual positions. The Lhasa block has a Precambrian basement and is covered by Mesozoic and lower Tertiary formations. It was situated in the tropics in the late Triassic, and its Mesozoic plant paleontology shows more affinity with Europe, North Africa, or even Sechuan, China, than with Gondwana. Lhasa had a paleolatitude of 12 °N at 80 MYBP and 15 °N at about 50 MYBP, and has moved northward nearly 20 ° since the touch of India to Asia.

(2) As shown in Figure 1B, the geologic profiles from the main boundary thrust (MBT) through the Lhasa terrane (LHA) to the Bagnon Nugiang suture (BNS) show that north-dipping subduction as well as most of the general structures are low angle, except the high-angle reactivated strike-slip fault at Indus Tsangpo suture and Bagnon Nujiang suture (BNS). These two sutures originally must have been low-angle thrusts, but they terminated their thrust activity when the accretion was completed. When the next block collided again, the former suture was reactivated as high-angle strike-slip faults. The conversion to a high-angle strike-slip fault is a common feature in accretionary terrane zones of western North America and China. Also remarkable are the crustal shortening and thickening caused by various types of folding and thrusting, including large-scale nappe.

(3) A 400-km-long north-south deep seismic-reflection profile complemented by a fan profile was successfully made. The crust is shown to be as thick as 50 to 70 km. Remarkable features are (1) 20-km steps of Moho exactly under the strike-slip active faults at two sutures, ITS and BNS, on the south and north boundaries of the Lhasa block (ITS is at point A on Figure 2) and (2) crustal repetition, which is seen as two superimposed Mohos observed at the north end of the Lhasa block (point C on Figure 2). The thickness of the crust between the two observed Mohos is 20 km. Thus the Tibetan Moho has very complex topography, as shown by the seismic-reflection profile in Figure 2.

(4) There are two kinds of major plutonic activity. One is that of Gangdise (Transhimalayan) plutonic belts, which comprise numerous bodies of gabbro, diorite, and granite intruded between 95 and 40 m.y. ago. Their initial Nd, Sr, and Pb isotope compositions are complex, suggesting a mixture of mantle and crustal derivatives. This may be subduction-related complex on the continental crust, similar to the Sierra Nevada batholith. Granitoids of this belt are mostly magnetite-type or I-type with sporadic S-type or ilmenite-type (Ishihara, 1981, 1984). The other type of plutonic activity is represented by granite plutons of lesser size, that are frequently aligned parallel to the suture line. Many of them are leucocratic and contain two micas. Their geochemistry including isotope study of Nd, Sr, and Pb indicate that they may be mostly the S-type, anatectic granite. The inherited Pb-component in the zircons in the

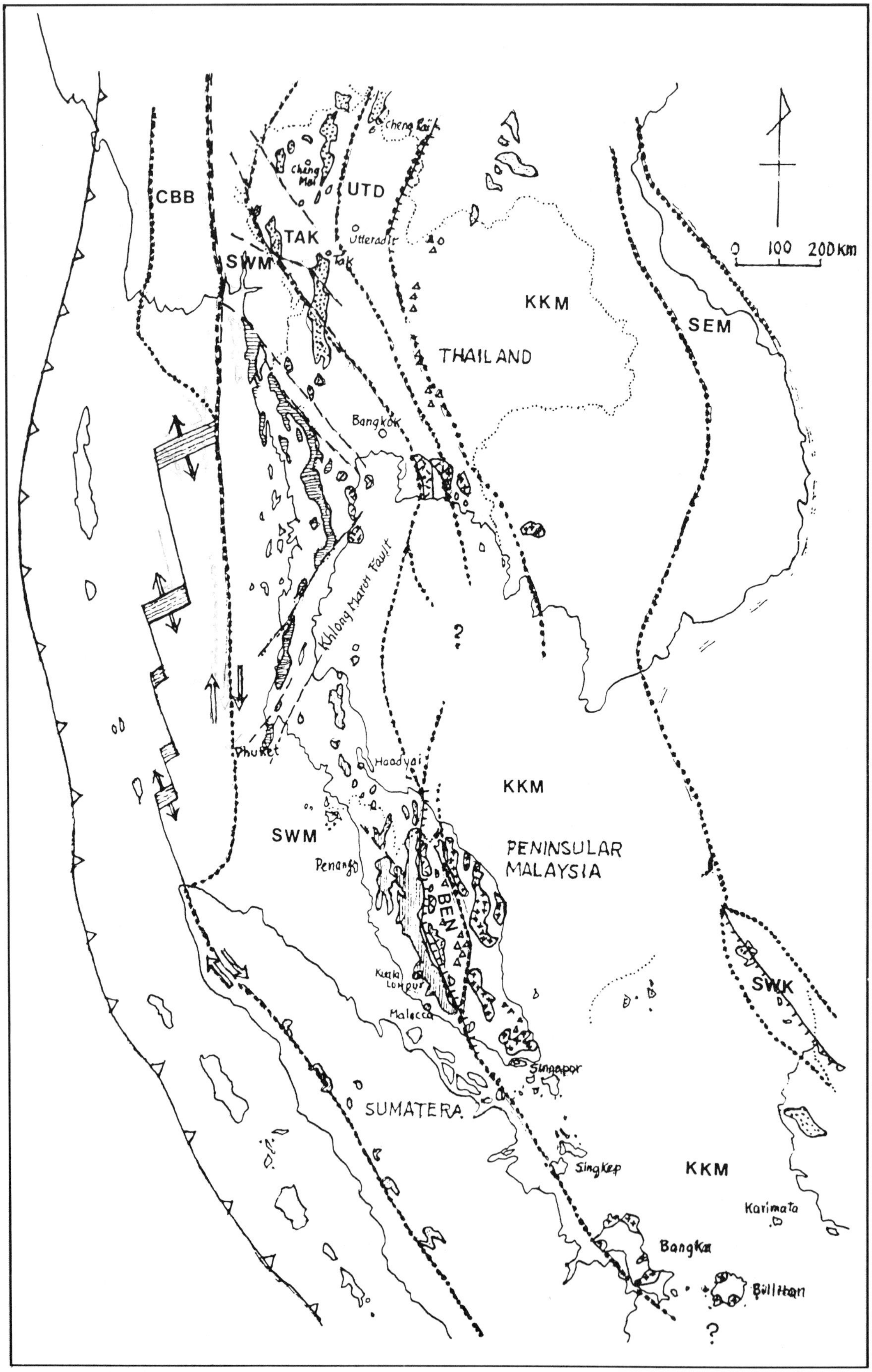


Figure 3. Distribution of granites and tectonostratigraphic terranes in Burma-Thai-Malaysia-Sumatra of northern Sundaland. Distribution of granites from Hutchison, 1983.

granite yielded a minimum age of 2200 my near Everest and also in the northern part of the Lhasa block. Ages of leucogranites of the High Himalayas and those intruded in Tethyan sediments farther north are between 15-30 m.y. They are of anatectic origin, belonging to the ilmenite-type.

## TECTONIC DEVELOPMENT OF INDOCHINA TO SUNDALAND

The tectonic development of Southeast Asia is another example of accretion tectonic history. This was the subject of a recent paper (Gatinsky, Hutchison, and Minh, 1984) which provides much interesting information. The great Sundaland between the Pacific Ocean and the Tethys Sea is postulated to have been formed by a collage of lands from Gondwana and Pacifica.

Peninsular Malaysia plays an important role in the early tectonic development of Southeast Asia. The Eastern Belt of the peninsula was affected by island-arc-type calc-alkaline igneous activity in the Precambrian basement during the early Permian to the late Triassic (280-200 m.y. ago). (See Figure 3 and Hutchison, 1983.) The terrane extends northward to Cambodia, Laos, and beyond, and southward to the Banka-Billiton islands in the Sunda Sea. Granite constitutes about 35% of the calc-alkali plutons, and is mostly Triassic in age and S-type anatectic in origin. The Banka and Billiton area is rich in tin. To the west of the Eastern Belt is a graben, and beyond it toward the west are metamorphic mountains and an ophiolite zone (BEN, Figure 3) which border the main range farther west. According to Hutchison (1983), after the development of back-arc basins (BEN), tectonic forces from the west pushed the continental fragment in the Tethys Sea that extended farther west to collide with the Eastern Belt, the above-mentioned island arc, during the end of the Permian and into the Triassic period (Figure 3). This Hercyno-type collision event produced a large amount of anatectic, S-type "tin granites" of the typical ilmenite-type. They intruded the folded epicontinental crust of thick miogeoclinal Paleozoic sediments resting on top of the hidden Proterozoic basement. This is the Main Range, which occupies the western half of the peninsula. The northern extension of the Main Range of the peninsula extends into Thailand but is divided by the Khlong Marui fault group and its northwestward extension. To the west of this transcurrent fault group is an extensive anatectic S-type epizonal granite again intruded in the similar Paleozoic-Mesozoic crust that had come from the Indian Ocean to collide with and become accreted to the Main Range. The transcurrent fault group may roughly represent the old suture.

These granitoids are younger, mostly Cretaceous (123-93 m.y. by Rb:Sr isochron), belonging to the ilmenite-type, and are rich in tin. On the map (Figure 3) they are designated as "Western belt Cretaceous epizonal granites."

East and north of this terrane of Cretaceous granites, there are a number of granite batholiths in folded Paleozoic phyllites and schists of northern Thailand. These granite batholiths are dated as late Triassic. All of them are S-type anatectic granites which originated from well-matured, upper crustal, probably Phanerozoic sediments (initial $^{87}Sr/^{86}Sr$ ratio = 0.723-0.734). In spite of their purely S-type nature, they yield very little tin, most probably due to the shallow anatectism. These granitoids are shown as north Thailand granite on Figure 3.

The major characteristics of the present partly submerged great Sundaland are: (1) Its complex history of building up a mosaic work of allochtonous blocks and slices of various tectonostratigraphic terranes attached to the Asian continent, Indochina, and Borneo. It contains island arcs, continental fragments, thick piles of "miogeoclinal" sediments at passive continental margins, and fold belts commonly sutured with ophiolites. Its present borders are mostly steep fault lines, which commonly are active at present as strike-slip faults.

(2) Prevalence of a large amount of granitoid intrusions, commonly as large batholiths, mainly anatectic S-type (Figure 3).

(3) World's heaviest concentration of tin, typically associated with tungsten, in several granitoid zones as mentioned above; this is geochemically and economically remarkable.

Recent isotope research seems to provide useful information on the origin and process of the tin concentration. For geochronology of granite intrusion and crystallization, Rb-Sr and K-Ar methods was extensively used.

Legend

Eastern Belt, epizonal calc-alk. series gabbro, granodiorite to granite with tin. Late Permian to late Triassic.

Central Belt, Na- rich granitoid, practically no tin, and metamorphic rocks. Ophiolite at western borber. Permian to late Triassic.

Main Range Belt granite. Early Permian to predominantly late Triassic, S-type and rich in tin. Deep seated and sheared on east, becoming Epizonal towards west to Penang Island.

North Thailand granite of continental crust deviation. S-type but less tin. Predominantly lower to upper Triassic.

Western belt Cretaceous epizonal granite. S-type and rich in tin between Phuket and Ranong and offshore.

Other granitoids.

Main area of late Paleozoic to early Mesozoic volcanism.

Ophiolite line (Suture zone).

Main major fault lines.

Active trench.

SWM KKM Name of accretion terrane after Howell, Jones et al. and Terrane boundary.

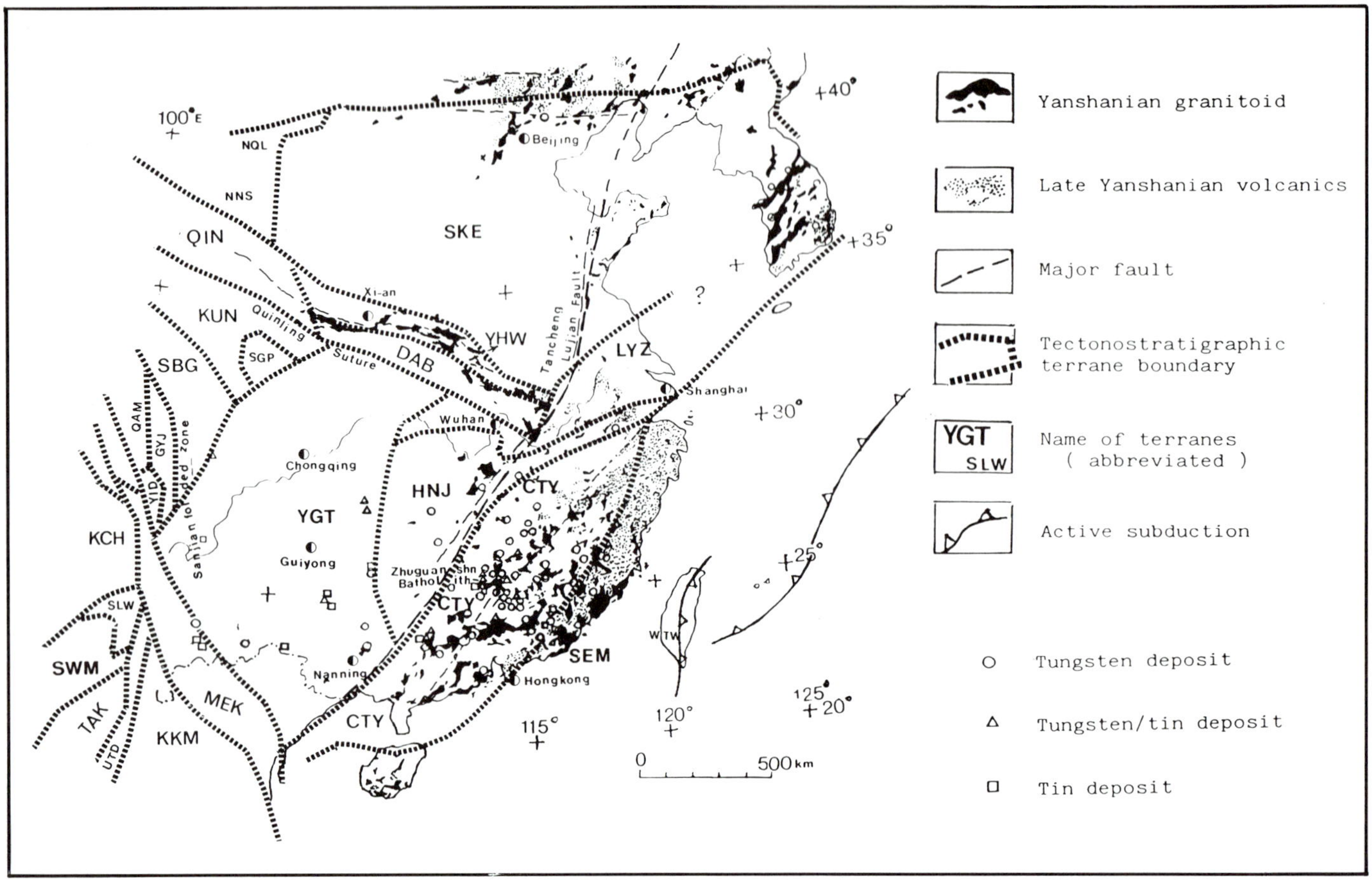


Figure 4. Distribution of the Yanshanian volcanic and granitic rocks, tungsten/tin deposits, and tectonostratigraphic terranes in Southeast China and Korea.

The initial $^{87}Sr/^{86}Sr$ as summarized below were used to judge the nature of the original rocks, protoliths, of the anatectic granites.

| | | |
|---|---|---|
| Malay Peninsula | | |
| Eastern Belt | Cal-alkali granite | 0.706-0.708 |
| | S-type granite | 0.7102 + |
| | Billiton, S-type | 0.7152 |
| Main Range | S- & ilmenite-type | 0.710-0.712 |
| Phuket Cretaceous granite | S-type | 0.707-0.7086 |
| North Thai Zone | S-type | 0.723-0.737 |

Higher values of the initial $^{87}Sr/^{86}Sr$ means anatectic granite magma of highly evolved or matured sedimentary crust origin. Hutchison postulated that if the granite is magnetite-type and its initial $^{87}Sr/^{86}Sr$ ratio is less than 0.706, this granite is of mantle origin and is not a likely source of tin deposits. Too high a value, such as 0.72 or more, is also not very favorable.

The U-Pb zircon method was also extensively tried for determining chronology of granite intrusions and also for the age of inherited zircon from the original rock before anataxis (Liew and Page, 1985). These results seem to be successful for the S-type granites of the Eastern Belt and the Main Range of Peninsular Malaysia. The values obtained were reconfirmed by another method using Nd (Neodemium) model age calculated based on a depleted mantle evolution model ($T^{Nd}_{DM}$) (Liew and McCulloch, 1985). The ages (in million years) of the protoliths thus obtained by different methods are:

| | Main Range | Eastern Belt |
|---|---|---|
| U-Pb zircon | 1500-177$^{Myr}$ | 800-1350$^{Myr}$ |
| T(Nd/DM) | 1410-1770 | 900-1400 |
| | (Lower Proterozoic) | (Upper Proterozoic) |

Main Range granite has average higher values of tin. Hence, this taken together with the metamorphic grades of the host rock and other geologic features, leads us to suppose that they were a better source of tin than the older Proterozoic crust at the basement (Gatinsky, Hutchison, and Minh, 1984; Hutchison, 1983).

## SOUTHEAST CHINA

Another tectonically complex accretional region with remarkable granitoid development is Southeast China (Hunan, Guangxi, Guangdon to Fujian Provinces of Figure 4) (Zhang, Liou, and Cokman, 1984).

It is supposed that at the middle of the Paleozoic era several continental fragments (YGT, HNJ, CTY terranes) were amalgamated together, and in the Permian the present south Yangtze block represented by YGT was at latitude 2° N, according to paleomagnetic data (Howell et al., 1985).

In the late Triassic when a continental fragment probably of Gondwana origin and the present Main Range of the Malayan

Peninsula collided and accreted to the old Sundaland, here in China, the amalgamated continental fragment, Yangtze block, supposedly began to collide and accrete to the Sino-Korea block of the present Northern China. At about the same time, another subduction-collision complex, the Indochina (which is the northern extension of an accreted mass of the Eastern Belt and the Main Range of the Malayan Peninsula), began to accrete to the Yangtze block from the west. By some time in the Jurassic, the framework of Southeast Asia as well as that of Southeast China had been built. The age of this tectonism is called the Yanshanian period.

Figure 4 is a map of Southeast China showing terranes, igneous activity of Yansharnian age, and the distribution of tungsten and tin deposits.

**Figure 5(B) illustrates the amalgamation of the YGT (original Yangtze terrane), the LYZ (Lower Yangtze), the HNG (Hunan-Jiangzi), and the CTY (Cathaysian fold system). These are mostly Precambrian-based continental fragments except for CTY, which is supposed to be a pre-Devonian arc built upon the Precambrian basement. After the amalgamation of these terranes, probably in the late Paleozoic, they made a quick northward drift and accreted to the Sino-Korea block, represented by the SKE terrane on Figure 4, with YHW (a broad, folded zone with much Yanshanian pluton suturing) and DAB (a probable microcontinent caught between YGT and SKE, and bounded by Ophiolite) as a very broad folded belt.**

The basic northward plate motion probably continued through this period. This resulted in the accretion of the SEM (Southeast Maritime) terrane in the latest Triassic(?), which was probably followed by subduction of the Pacific plate. The subduction might have reactivated the deep tectonic and igneous activities of the old amalgamated heterogeneous blocks. In particular, the compressional force associated with the later subduction/collision process would have reactivated the suture between HNJ and CTY as an active strike-slip Tancheng-Lyjian fault, similar to the strike-slip faults of the ITS or BNS sutures at the Tibet-Himalaya accretion boundary; or the fault group near Phuket, which was reactivated in the Cretaceous on the western coast of the Thai-Malaysia area.

At the same time, the extensive plutonism of the granitoids might have also been caused by polycyclic reactivation of crustal tectonism during the Yanshanian period. About 20% of South China is occupied by granitoid exposures of various ages, about 60% of which belong to the Yansharian age. Almost all of them are anatectic in origin, in contrast with the prevalent older granites of metasomatic or granitization origin, which occur more abundantly toward the northern Yangtze block. The general age trend is older to the north, even among the Yansharian granites. In the Nanling area, which is famous for its many large tungsten producers, earlier Yanshanian granite is a predominant rock type; in Fujian province toward the sea coast younger granites, associated with more extrusives, are prevalent. More than half of these anatectic granites belong to the ilmenite-type as in Malaysia, and the link between sediment-gene anatectic mica granite and the tungsten/tin mineralization seems to be very firm. In contrast with the tin-dominated deposits in the Thai-Malaysia-Indonesia zone (except for Burma, where we have a range of tin and tungsten to pure tungsten deposits toward the north), Chinese tungsten deposits have more tin, mostly cassiterite, as a very common coproduct. China even has a number of large tin mines.

Upper Proterozoic to lower Paleozoic basement sediments, especially littoral sandstone and some shale, commonly show high anomalous values—from more than 10 to 100 times the average—of tungsten content in areas some 10 to 30 km away from the mine in Guangxi. Similar high anomalous contents are also reported in Jiangxi and other mine areas. One Paleozoic strictly stratiform tungsten deposit in Guangxi is thought to be a fossil placer deposit (Ishihara, 1984).

It is very probable that these basement Caledonide sediments underwent anataxis at the low-angle subduction of the collided continental fragments or arcs, and thus generated large amounts of later Mesozoic anatectic granites (Figure 4). The anatectic granites themselves in this area generally have a high content of tungsten, typically 5 to 80 times the average. The more highly differentiated granites commonly show the higher values.

Polycyclic-events such as metamorphism, anataxis, and gravitational sedimentary concentration are the main forces that enhance crustal tin enrichment. Concentration of tin/tungsten high enough for economical mining cannot be realized by the direct and simple-stage differentiation from deep mantle source magma. The ultimate enrichment to economic ore-grade deposits depends mainly on the crystal-liquid partition coefficient, which indicates how far the metals can remain concentrated in fluid at the time of magma crystallization. Field observations and recent research revealed that less oxygen instability, owing to the existence of more carbon compounds in organic sediments or otherwise, and higher content of volatiles, such as halogen compounds, in the solidifying magma makes favorable conditions for metal concentration. Tectonic environments such as subduction of continental lithosphere in the region of repeated collision/accretion seems to be very favorable to cause polycyclic reactivation of high-mountain building, severe weathering and erosion, deep burial and metamorphism of sediments, and remelting of the differentiated crust, such as granite anataxis. The area treated in this chapter and its extension toward the north along the Circum-Pacific coast, namely to South Korea, Sikhote-Allin to most of the Pacific far east of the USSR, and the Seward Peninsula, Alaska, are in such a tectonic environment. If this cycle occurs in sedimentary formations that have a slightly high tin/tungsten content, there is a good chance for the formation of economic-grade concentrations. Metal is concentrated by geochemical processes, as mentioned above, but tectonism will create favorable conditions for enhancing the necessary chemical processes.

## CENTRAL JAPAN

Japan has an example of buoyant subduction which preceded the collision of the volcanic arc of the Philippine Sea plate with the Japanese arcs (Figure 6). From Miocene time (30-15 MYBP), the northern end of the Philippine Sea plate, probably a submarine andesitic volcanic pile with marine sediments intercalated, began to subduct northwestward under the "Fossa Magna." To the east of the zone is the Kanto plain, under which the Pacific plate is subducting from the Japan trench. Two subducting plates meet where the Philippine Sea plate terminates on top of the Pacific plate (note in Figure 6 the depth of intersection in km). To the west, it subducts under the Eurasian plate.

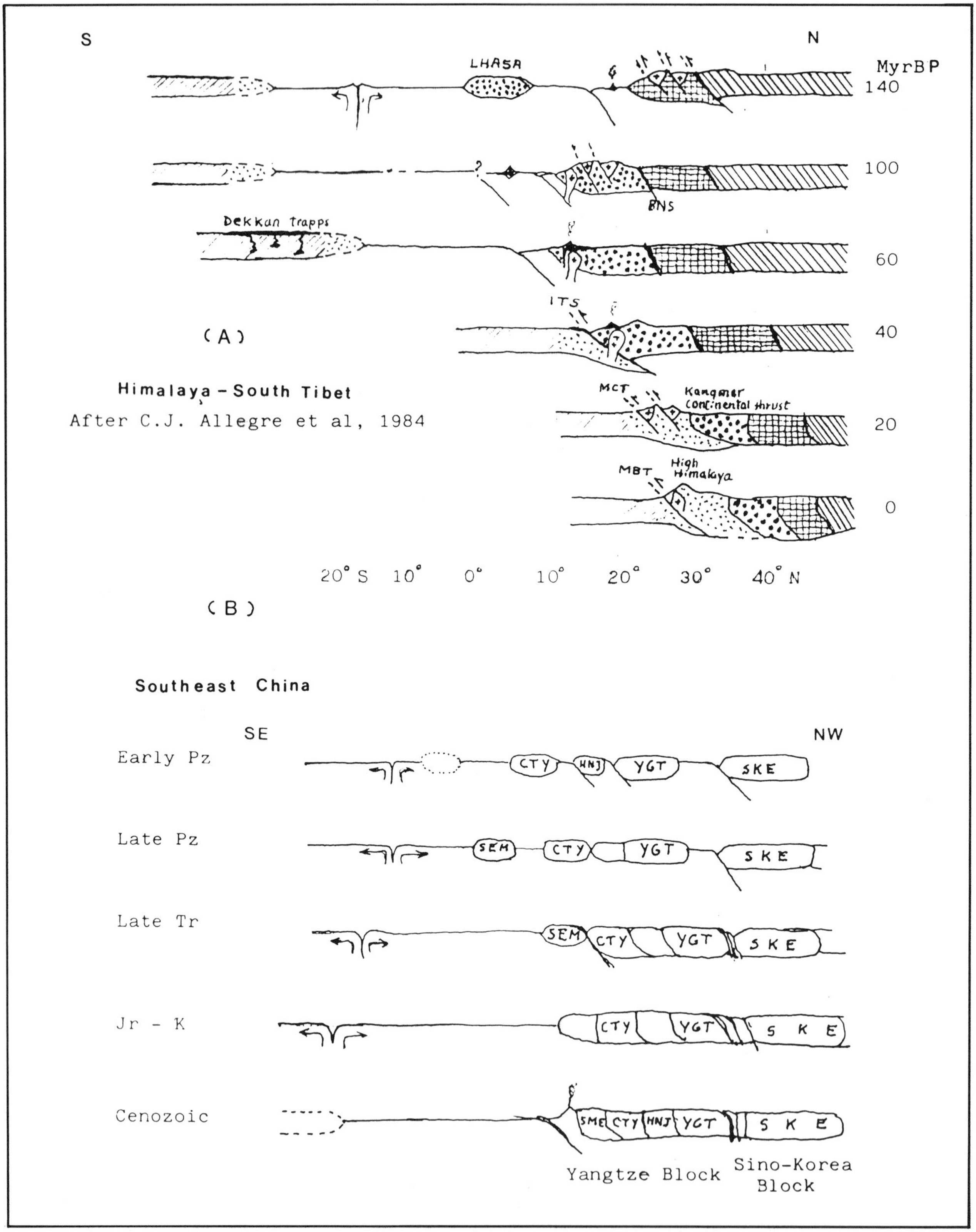


Figure 5. Cartoon to represent tectonic evolution of (A) Himalaya-South Tibet and (B) Southeast China.

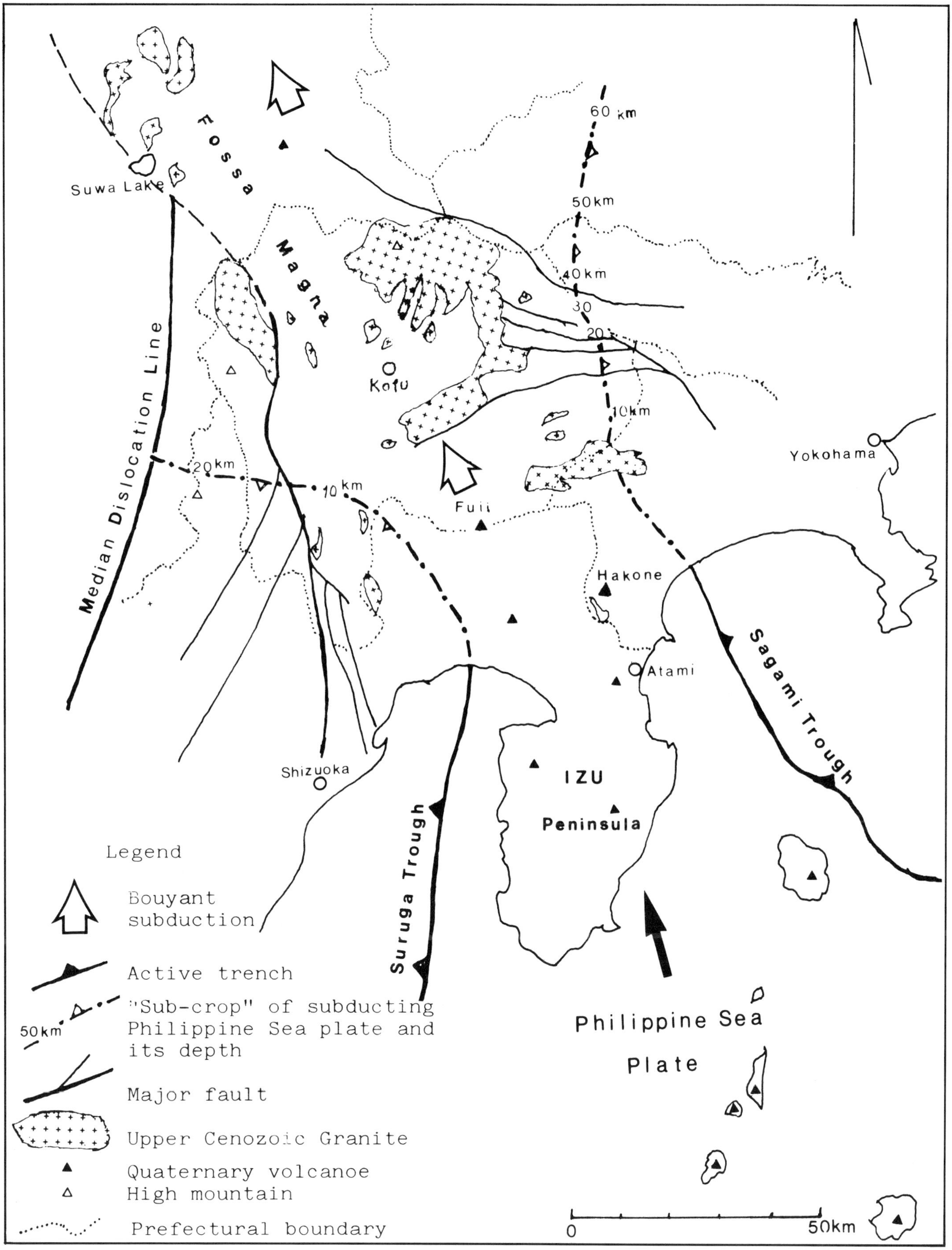


Figure 6. Buoyant subduction and collision of Philippine Sea plate with the central Japan Arc.

The "Fossa Magna" itself is a sort of narrow graben with parallel faults. On the western sides of this low are roughly en echelon high folded mountain ranges which form the highest mountain ranges of Japan; these mountains are still upheaving at about 1mm/year. The dip angle of subduction in front of the collision zone appears to be about 40°-45°; and the zone is 200 km long at the most (M. Ishida, private communication). Many young anatectic granites with metamorphic aureoles are present here. From their chronology (5-10 MYBP) and petrology, it is inferred that they were buried to a depth of several kilometers at the time of intrusion. It is very probable that during 4-5 m.y. time in the latest Cenozoic, an upheaval of 1000 m per million years and rapid erosion took place (Kaizuka, 1984).

There are no high mountains like the Himalayas and no metal concentration, but this small buoyant subduction made tens of kilometers of thick deep crust and resulted in a large amount of anataxis of crustal material at depth. The upheaval and rapid erosion in this area is similar to many other examples of large continental-scale buoyant subduction described in the foregoing sections.

## CONCLUSIONS

(1) One of the important discoveries of recent plate tectonic investigations on the active consuming end of the plate activity is that of the polycyclic repetition of collision/accretion of large allochthonous terranes from the south extending all through the southern shore of Eurasia.

(2) This tectonic evolution involves oceanic plate subduction; collision of varied exotic buoyant features such as continental fragments, island arcs, and seamounts; accretion of these terranes to the continent with the help of buoyant subduction; high-mountain building with doubling and thickening of the crustal materials; metamorphism and magma genesis by large-scale anataxis of crustal materials; and rapid and extensive erosion and epicontinental sediment pile formation.

(3) In the regions of Tibet-India, Malaysia-Indochina-Sundaland, and Southeast China, at least three or four times of accretion tectonism have been repeated since the late Paleozoic, resulting in an extensive growth of the Eurasian continent to the south.

(4) Among numerous geologic events during this tectonic evolution, granitoid anataxis is selected in this report, because anatectic granitoids seem to occur extensively and characteristically at the buoyant subduction areas, and these anatectic granites show close genetical relationship with tin/tungsten metallogenesis. In particular, the region of Burma-Thailand-Malaysia-Indonesia and Southeast China, together with South Korea and the Far-East Pacific part of the USSR, which have similar polycyclic accretional tectonic history along the Pacific coast, contains about 70% of the tin and 50% of the tungsten resources and production of the world.

(5) The favorable source rock of these anatectic tin/tungsten granites seems to be clastic shallow sea sediments which have slightly high anomalous contents of these metals. The likely candidates are Proterozoic sediments in Malaysia, and Proterozoic to Lower Paleozoic sediments in China. Proterozoic basement, with Paleozoic miogeoclinal cover and Mesozoic orogenism with anataxis, is the most common tectonic sequence.

## REFERENCES CITED


Allegre, C.J., V. Coutillot, P. Tapponier, A. Hirn, M. Mattauer, C. Coulon, L. Tindon, and Xiao Xuchange, 1984, Structure and evolution of the Himalaya-Tibet orogenic belt: Nature, v. 307, no. 5946, p. 7-22.

Gatinsky, Yu. G., C.S. Hutchison, and N.N. Minh, 1984, Tectonic evolution of southeast Asia: Moscow, 27th IGC, Tectonics of Asia Reports, v. 5.

Howell, D.G., E.R. Schermer, D.L. Jones, Z.-Ben Avraham, and E. Scheibner, 1985, Preliminary tectonostratigraphic terrane map of the Circum-Pacific region (scale 1/17,000,000) and explanatory notes to accompany the map: Circum-Pacific Council for 1985, AAPG.

Hudson, T.L., 1984, Tin granites of Seward Peninsula, Alaska: Proc. 6th Quadrennial IAGOD Symposium, p. 113.

Hutchison, C.S., 1983, Multiple Mesozoic Sn-W-Sb granitoid of southeast Asia: GSA Memoir 159, p. 35-59.

Ishihara, S., 1981, The granitoid series and mineralization: Economic Geology, 75th Anniversary vol., p. 458-484.

Ishihara, S., 1984, Granitoid series and Mo/W-Sn mineralization in East Asia: Geological Survey of Japan Report No. 263, p. 173-208.

Jones, D.L., D.G. Howell, P.C. Coney, and J.W.H. Monger, 1983, Recognition, character and analysis of tectonostratigraphic terranes in western North America, *in* M. Hashimoto and S. Uyeda, eds., Accretion tectonics in the Circum-Pacific regions, p. 21-35.

Kaizuka, S., 1984, Landforms in and around the south Fossa Magna and their tectonic processes of growth: Quaternary Research, v. 23, no. 2, p. 55-70.

Kuang, C.T., Wang Zhang-gang, and Yu Xue-Yuan, 1980, Genesis of granitic rocks in South China and related mineralization: Mining Geology Special Issue No. 8, p. 189-196.

Liew, T.C., and M.T. McCulloch, 1985, Genesis of granitoid batholith of Peninsular Malaysia and implication for models of crustal evolution: Evidence from a Nd-Sir isotope and U-Pb Zircon study: Geochimica et Cosmochimica Acta, v. 49, p. 587-600.

Liew, T.C., and R.W. Page, 1985, U-Pb zircon dating of granitoid plutons from the west coast province of Peninsular Malaysia: Journal of the Geological Society of London, v. 142, p. 515-526.

Uyeda, S., 1982, Subduction zones: An introduction to comparative subductology: Tectonophysics, v. 81, p. 133-159.

Uyeda, S., and H. Kanamori, 1979, Back-arc opening and the mode of subduction: Journal of Geophysical Research, v. 84, p. 1049-1061.

Uyeda, S., and C. Nishiwaki, 1980, Stress field, metallogenesis and mode of subduction, *in* D.W. Strangeway, ed., The continental crust and its mineral deposits: Geological Association of Canada Special Paper 20.

Zhang Zh, M., J.G. Liou, and R.G. Cokman, 1984, An outline of the plate tectonics of China: GSA Bulletin, v. 95, p. 293-312.

CHAPTER 38

# CIRCUM-PACIFIC TECTONIC MAPS AS AN AID FOR ENERGY AND MINERAL RESOURCES EXPLORATION

**Erwin Scheibner**
*Geological Survey of New South Wales*
*Sydney, Australia*

ABSTRACT

The Circum-Pacific Tectonic Maps are part of an integrated series of maps showing the relationship of known hydrocarbon and mineral resources to the geology, tectonics, and crustal dynamics of the Pacific region. The tectonic maps distinguish areas of oceanic and continental crust and active plate margins. With the search for hydrocarbons in mind, further subdivision is into basement complexes, deemed to be too deformed to contain hydrocarbons, and cover sequences. Symbols in red mark active plate boundaries and colored patterns show tectonic units at active plate margins. The tectonic development of oceanic crust is shown by episodes of sea-floor spreading which correlate with the rift and drift sequences at passive continental margins and episodes of tectonic activity at active margins. Color codings show the age of deformation of basement complexes, transitional tectonic (molassic) and reactivation sequences, and inception of platform basins. Colored patterns show the age of inception of continental-margin (rift and drift) sequences.

As the tectonic setting is one of the determining factors in the formation of deposits of nonrenewable resources, the Circum-Pacific Tectonic Maps should be important in energy and mineral exploration.

## INTRODUCTION

The aim of the Circum-Pacific Map Project is to produce an integrated series of maps (geographic, plate-tectonic, geologic, tectonic, geodynamic, mineral-resources, and energy-resources) showing the relationship of known energy and mineral resources to the geology, tectonics, and crustal dynamics of the Pacific region.

The need to publish the tectonic maps in the Circum-Pacific Map Project was perceived by the Project's scientific coordinator, Maurice Terman. In 1978 he produced an open-file report which became the basis for discussion on how to approach the tectonic-map compilation (Terman, 1978). The principles of the Circum-Pacific Tectonic Map compilation were developed during lengthy constructive discussions with the panel chairmen and technical advisers to the Circum-Pacific Map Project. The final guidelines were adopted in December 1982 (see U.S. Geological Survey Open-File Report 83-64).

The Circum-Pacific Tectonic Map Series has been designed to complement the other map series with the intention of aiding the exploration for energy and mineral resources. The aim is to produce tectonic maps which are as objective as possible. Each of the quadrant maps gives a complete tectonic synopsis of one-eighth of the Earth's surface. At the same time, enough detail has been retained to illustrate the specific differences between individual regions. The data contained in these maps, combined with the data in other thematic map series of the Circum-Pacific Map Project, can be used to test existing conceptual models that try to explain the origin and distribution of energy and mineral resources. Such studies are expected to lead to improved conceptual models and to help in practical exploration for nonrenewable resources.

## PRINCIPLES OF THE CIRCUM-PACIFIC TECTONIC MAP SERIES

The principal subdivisions on the tectonic maps show the contrasting *oceanic* and *continental crustal domains* along with the *active plate margins*, which represent the transition between the two domains and mark the region where the continental crust is formed. With the search for hydrocarbons in mind, the above three subdivisions have been further subdivided into: (1) *basement complexes*—which are generally deemed to be too deformed to contain hydrocarbons, and (2) *cover sequences*—those which include transitional tectonic (molassic) and reactivation sequences, basins at active plate margins, passive continental-margin sequences, and platform basins.

## Oceanic Crustal Domain

The basement complexes of this domain, i.e., the *oceanic crustal rocks*, have been divided according to their age of formation. Successive episodes of sea-floor spreading are color coded. These episodes of sea-floor spreading appear to correlate not only between distant oceanic regions but also with the tectonic development of passive and active plate margins. The spreading episodes are thought to reflect changes in plate kinetics and plate interactions. Undifferentiated, or more precisely, oceanic crust of unknown age, is shown in gray.

On the Southwest Quadrant of the Circum-Pacific Tectonic Map, 10 successive episodes of sea-floor spreading are shown, and a brief description (Moore and Scheibner, in press) of some of the tectonic events with which they are associated is given in Table 1.

The magnetic lineations shown are mainly those compiled by Xenia Golovchenko, Roger Larsen, Walter Pitman, and Nabuhir Isezaki, as printed on the Plate Tectonic Map of the Circum-Pacific Region, Pacific Basin Sheet, 3rd printing, 1985. On the Southwest Quadrant Tectonic Map these lineations have been amended in accordance with published and unpublished data. In order to provide a complete interpretation for the oceanic crust, the magnetic lineations have been interpolated into areas without data. For the region south and west of Australia the computer-predicted ages of Johnson and Veevers (in Veevers, 1984) have been used. The boundary within the Cretaceous magnetic quiet-zone (KMQZ) is the most contentious and will remain so until sea-floor imaging by systems such as Gloria has been completed.

*Oceanic plateaus*, shown by an overprint pattern, are characterized by anomalously thick crust in contrast with the surrounding normal oceanic crust. Some oceanic plateaus represent anomalously thick oceanic crust formed at spreading axes, whereas others are epiliths formed by intraplate igneous activity. (Submarine plateaus which are microcontinents are included in the continental crustal domain.)

*Oceanic volcanic islands and seamounts* include traces of hotspots and other intraplate igneous activity. The rock type, which is mostly basic volcanic, is shown by an overprint symbol; the age is shown by color, with gray for unknown ages. These complexes represent igneous cover of the oceanic crust.

*Sedimentary cover on the oceanic crust* is indicated by isopachs.

## Active Plate Margins

At active plate margins the basement/cover classification is only locally applied. Allochthonous microcontinents are classified as older basement complexes. Tectonically emplaced oceanic crustal basement is shown wherever the scale allows. Three main subdivisions of active plate margins have been depicted.

*Magmatic arcs or chains* include volcanic-island arcs, continental-margin arcs, volcanic rifts, and other forms of magmatic chains related to plate interaction at plate margins. The composition of the volcanic and plutonic rocks is shown by overprint symbols (patterns), and the color of these symbols indicates the age.

*Fore-arc basin sediments* are shown by a pattern, and the color of the pattern indicates the time of onset of sedimentation. Isopachs indicate the thickness of the strata.

*Accretionary-prism rocks, including melanges*, are shown by a pattern, and the color indicates the time of onset of subduction.

## Continental Crustal Domain

The following subdivisions of the continental crustal domain are depicted on the map: basement rocks, rocks of transitional-tectonic (molassic) and reactivation sequences (basins), and cover sequences.

*Basement rocks* comprise deformed sedimentary and volcanic rocks (those generally deemed to be too deformed to contain hydrocarbons), and metamorphic and igneous rocks of orogenic belts, mobile belts, metamorphic belts, and igneous belts. Overprints indicate lithologies of igneous rocks as in the Geologic Map Series. Ultramafics are shown in black, and ophiolites in purple. Grade of metamorphism is shown by black overprints, identical to those on the geologic maps. Solid colors indicate age of deformation, metamorphism, or igneous emplacement.

*Transitional tectonic sequences* are defined as rocks formed immediately following major deformation in orogenic regions and immediately preceeding platform-strata deposition (Doutch, 1972; Scheibner, 1976). These complexes are commonly referred to as "late orogenic" or molassic. Sedimentary rocks include molassic conglomerates, redbeds, and other continental and shelf sediments which commonly accumulated in foreland basins, fore-, back- or intra-deeps, and grabens. The igneous rocks include mostly felsic volcanics, bimodal volcanics in rifts, and postkinematic granites and associated volcanics.

The age of these sequences is indicated by a colored dot and dash band that follows the boundaries, and, if the complexes are deformed, by horizontal colored lines that indicate the age of deformation.

*Reactivation sequences* comprise deposits which are unrelated to the immediately preceeding tectonism, except for inherited structural control, i.e., localization along old dislocations. These sequences appear to be related to plate collisions, as exemplified by the China-type basins (Bally and Snelson, 1980), which are related to collision of the Indian plate with the Asian plate. Rocks include some fore-deep, graben, and molasselike deposits and rift-related volcanics (Molnar and Tapponnier, 1978).

*Cover sequences* comprise platformal strata, continental-margin deposits, and intraplate igneous rocks. Continental and shelf platformal strata include remains of sedimentary basins and also blanket deposits not related to tectonic downwarps and basins. Boundaries in black indicate the limits of basin preservation, and parallel color bands indicate the age range of deposits, usually the age of basin inception. Within boundaries, pastel tints show the age of the basement where this is known. Thickness of strata is shown by isopachs.

*Continental-margin deposits* include rift and drift or breakup sequences. Rift sequences are thick deposits which occur either in rift valleys or grabens at continental margins preceding continental breakup and sea-floor spreading. Colored patterns indicate the age of onset of sedimentation which

Table 1. EPISODES OF SEAFLOOR SPREADING SHOWN ON THE CIRCUM-PACIFIC TECTONIC MAP SERIES (Moore & Scheibner, in press)

| Letter symbol | Magnetic anomalies (as on Plate-Tectonic Map Series) | Age in mil. years(Ma) | | Some of the more notable events characterizing the spreading episode Plate-Tectonic Map Series |
|---|---|---|---|---|
| QT | 0-3 | 0-4 Ma | 3 | (4Ma) Early Pliocene: Gulf of California started opening; Lau Basin started opening; Drake Passage stopped opening. |
| Tm | 3-6 | 4-20 Ma | 4 | (8Ma) Late Miocene: Scotia Sea stopped opening. |
| | | | 5C | (17Ma) Middle Miocene: South China Sea stopped opening; West Mariana Basin stopped opening; Japan Sea stopped opening at this time approximately. |
| To | 6-13 | 20-37 Ma | 6 | (20Ma) close to Miocene-Oligocene boundary: South Fiji Basin stopped opening; East Pacific Rise at Nazca Plate jumped west; Galapagos Rift was established; Red Sea started opening. |
| | | | 9 | (29Ma) Middle Oligocene: Caroline Basin stopped opening. |
| ToTe | 13-18 | 37-43 Ma | 13 | (37Ma) Early Oligocene: South Fiji Basin started opening; South China Sea started opening; Japan Sea probably started opening; Baffin Bay and Labrador Sea stopped opening. |
| | | | 16 | (40Ma) Late Eocene: Philippine Sea stopped opening. |
| Te | 18-24 | 43-54 Ma | 18 | (43Ma) Late Eocene: Emperor bend indicates that the local motion of the Pacific Plate changed from north to northwest; Laramide Orogeny ended; Mendocino Fracture Zone changed trend; subduction began at New Caledonia and Tonga. |
| Tpa | 24-27 | 54-62 Ma | 24 | (54Ma) earliest Eocene: Tasman Sea stopped opening; major reorientation in Gulf of Alaska; Kula Plate believed to have expired; Aleutian subduction believed to have begun; Norwegean Sea started opening. |
| TK | 27-KMQZ | 62-95 Ma | 27 | (62Ma) Early Paleocene: Coral Sea began to open. |
| | | | 29 | (65Ma) Tertiary-Cretaceous boundary: Chile Rift established. |
| | | | 33 | (80Ma) Cenomanian: Intrusion of the Sierra Nevada batholith ceased. |
| Ku | KMQZ-M 0 | 95-113 Ma | KMQZ | (95Ma) Cenomanian: Australia began to separate from Antarctica; Tasman Sea began to open; seafloor spreading realigned east of Ninetyeast Ridge; Canada Basin, Alaska stopped opening approximately at this time; Greenland began to separate from North America. |
| K | M 0-M10N | 113-124 Ma | M 0 | (118Ma) Aptian: Reorganization of seafloor spreading in the region of Manihiki and Ontong-Java Plateaus; emplacement of the Sierra Nevada batholith began. (124Ma) Hauterivian, change in velocity of seafloor spreading in western Pacific (K. Tamaki, pers. comm.). |
| KJ | M10N-JMQZ | 124-170 Ma | M11 | |
| | | | | (136Ma) Valanginian: Cuvier Basin began to open; Canada Basin began to open; South Atlantic began to open. |
| | | | M26 | (161Ma) Oxfordian: seafloor spreading north of Australia began. |
| | | | JMQZ | (170Ma) Bathonian: North-central Atlantic began to open; spreading may have begun on the north side of the Chatham Rise; |

is closely related to the age of rifting. Drift or breakup sequences accumulated over the rift sequences and the rest of the contemporaneous shelf, which developed after the breakup. Color patterns indicate the age of the onset of sedimentation. Isopachs indicate thickness of strata.

*Intraplate igneous rocks* include plateau basalts, anorogenic extrusives and intrusives related to the rifting of continental lithosphere (rift valleys), and hotspot activity. The rock type is indicated on the maps by patterns identical to those on the geologic maps, and the color of the patterns indicates the age of extrusion or intrusion.

The names of structural units (morphotectonic units) have also been added (in grey) because of their tectonic importance.

## Graphic Symbols

Apart from some additional symbols, the graphic symbols used in the Tectonic Map Series are identical to those in the Plate-Tectonic Map Series. For subduction zones, transform faults, spreading centers, and faults, red indicates active structures and black inactive. Magnetic anomalies are dashed where approximate, and short-dashed where interpolated and computer-predicted.

The boundaries of tectonic units are dashed if they are approximate or are in oceanic areas. The continent ocean boundary (COB) is generally shown by the limit of oceanic crust, but in special cases by a short-dashed boundary line. The inner edge of the continental crust affected by rifting and breakup is shown by a dot-dash boundary where such data are available. Symbols for faults and folds follow international standards. Dikes and dike swarms are shown wherever the scale of the map allows. A special symbol has been devised to depict selected palaeomagnetic data. It consists of a small circle with a bar showing the orientation of the declination; the sector between true north and the declination orientation is colored to indicate the time (age) of rotation. A dashed line normal to the declination bar is labeled with the value of the palaeolatitude.

## Time-Space Plot

The time-space plot provides an index to the tectonic units shown on the maps. Each tectonic unit has been assigned an alpha-numeric symbol which further helps in identification of individual tectonic units. The standard color coding of the tectonic map subdivision enables the reader to comprehend the tectonic evolution of regions for which time-space plots are shown (see Figure 1).

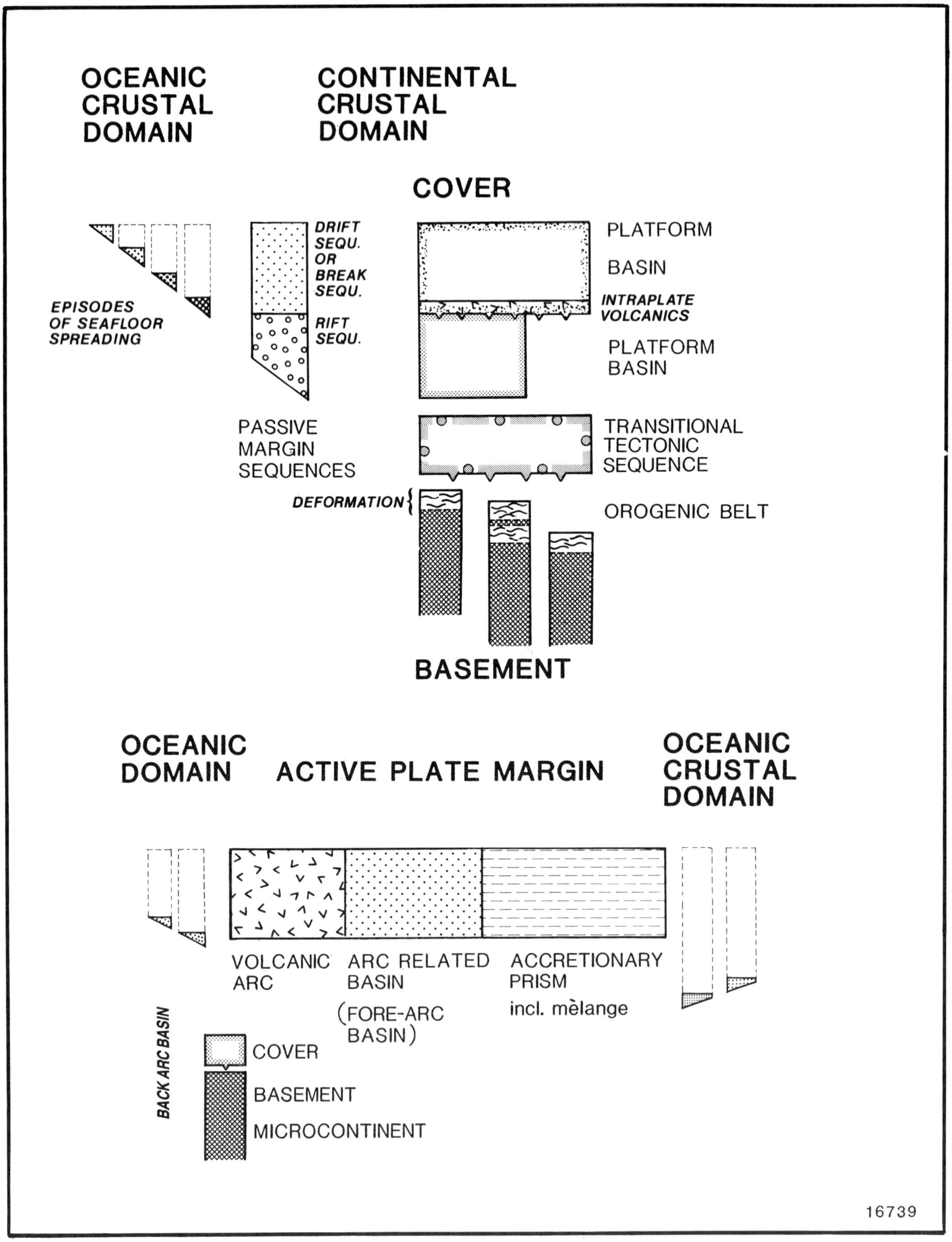


Figure 1.—Method used to construct time-space plots for various tectonic settings.

Table 2. A sample page of the Tectonic Map Explanatory Notes

| Color | Letter Symbol | Structural Name & Age Span | Tectonic Setting (interpretation & brief description) | Source of data-References |
|---|---|---|---|---|
| (934) | $Pzm^9$ | Georgina, Ngalia, Amadeus & Officer Basins 410-300 Ma | Cratonic cover - platformal basin, molasse-like sediments | Geol. Soc. Aust. 1971; Wilford et al. 1981 |
| | $Pzm^{10}$ | Tuhua Orogen late part 410-395 Ma | Orogenic domain, cf. $Pz^5$ terminal deformation & metamorphism Middle Devonian; granites Devonian - Carboniferous | Cooper 1979; Copper & Grindley 1982; Suggate 1978 |
| | $Pzm^{11}$ | Irian-Jaya, Birds Head basement rocks | Orogenic domain, probably originally part of the Tasman Fold Belt System | Dow & Sukamto 1984; Pigram & Panggabean 1983 |
| (920) | RP & $RPz_2$ | TRIASSIC TO PERMIAN, TRIASSIC TO LATE PALEOZOIC AND LATE PALEOZOIC UNIT | | |
| (919) | $Pz_2$ $RPz^2_2$ & $RP^{12}$ | Canning, Bonaparte Gulf and Officer Basins 360-185 Ma | Continental margin infra-rift sequences | Gee et al. 1979; Geol. Soc. Aust. 1971; Wilford et al. 1981 |
| | $Pz^2_2$ & $RPz^2_2$ | New England Fold Belt (New England & Yarrol Provinces) | Orogenic domain, active plate margin; Late Carboniferous to Early Permian magmatic arc & accretionary prism; deformation & metamorphosed in Mid Permian (also affected the earlier part of the fold belt $Pzm^6$). Post-kinematic granite & felsic volcanics | Day et al. 1978; Day et al. 1983; Leitch, 1974 Scheibner 1974; in press |
| | $Pz^3_3$ & $RP^2_2$ | Hodgkinson-Broken River Fold Belt, Georgetown & Coen Inliers | Post-kinematic orogenic granites & assoc. volcanics in the fold belt and adjacent basement inliers | Day et al. 1978; Day et al. 1983; Henderson & Stephenson 1980 |

## Explanatory Notes

The Explanatory Notes describe all the tectonic units shown on the map, as well as the unique alpha-numeric symbols, the structural name and age span, tectonic setting (i.e., interpretation and brief description of the tectonics), and finally sources of data (references). In some cases the brief description of tectonics suggests multiple hypotheses, and this, of course, is the most interpretive and contentious part of the Tectonic Map Series. An example is provided by a sample page from the Explanatory Notes (Table 2).

## TECTONIC MAP SERIES AS AN AID FOR ENERGY AND MINERAL EXPLORATION

The basic rationale is straightforward. It is generally recognized that the tectonic setting is one of the determining factors in the formation of deposits of nonrenewable resources. From this point of view the tectonic maps should be important in exploration. There is, however, a problem in that the tectonic maps have been produced by academically oriented researchers biased toward structures rather than the origin and distribution of deposits. For this reason, especially designed tectonic maps as a base for metallogenic maps or mineral-occurrence maps are really useful (Warren, 1972 Barnes et al., 1986). In our case, there is an advantage that the tectonic maps do not stand alone but are complementary to and are complemented by the other Circum-Pacific Map Series. The fact that the major subdivision, the "basement rocks," is defined as being too deformed to contain hydrocarbons, should be helpful in the search for energy resources. There is, however, a possibility of migration of fluid hydrocarbons into basement complexes if they occur as upper plates along thrust faults over productive sequences.

Experience has shown that worldwide major hydrocarbon deposits commonly occur in sequences similar to those depicted as transitional-tectonic and remobilization sequences in the Circum-Pacific Tectonic Map Series. The potential of these sequences in the Circum-Pacific region remains to be realized.

This tectonic map, of course, highlights the subject of the latest gold rush—the young volcanic island arcs with their epithermal gold deposits—as well as other magmatic chains which localize a variety of metal deposits. Suture zones between tectonostratigraphic terranes are another specific setting for the localization of gold deposits, the importance of which has only recently been appreciated (Albers, 1983).

The scale of the Circum-Pacific Tectonic Maps is a drawback to their usefulness in practical exploration. The real usefulness of these maps is in their synoptic nature. They should be useful in the designation of long-term strategies in exploration and especially in the testing of conceptual models (Mitchell and Garson, 1981; Sawkins, 1984) that attempt to explain the genesis of deposits of nonrenewable resources. It is hoped that ultimately new models might emerge from comparative studies combining all the data available in the maps of the Circum-Pacific Map Project. Discovery of even one major deposit would more than pay for all the monetary expenditure connected with this international project.

## ACKNOWLEDGMENTS

This manuscript benefited from critical reviews by Helena Basden and Dennis J. Pogson. Permission to publish was given by the Secretary, New South Wales Department of Mineral Resources.

## REFERENCES

Albers, J., 1983, Distribution of mineral deposits in tectonostratigraphic terranes and cratonal rocks in western United States: Canadian Journal of Earth Sciences, v. 20, p. 1019-1029.

Bally, A.W., and S. Snelson, 1980, Facts and principles of world petroleum occurrences: Realms of subsidence: Canadian Society of Petroleum Geologists Memoir, v. 6, p. 9-94.

Barnes, D.P., K.F. Gaukroge, E.A. Smith, P.R. Johnson, and E. Scheibner, 1986, Selected mineral occurrences of the Arabian Shield, Riofinex Geological Mission: Jiddah, Ministry of Petroleum and Mineral Resources Technical Record RF-TR-06-1, scale 1:1,000,000.

Doutch, H.F., 1972, The concept of transitional tectonism in Australia: Canberra, Geological Society of Australia Joint Specialist Groups Meeting, Abstracts, p. E10-E12.

Mitchell, A.H.G., and M.S. Garson, 1981, Mineral deposits and global tectonic setting: London, Academic Press, 405 p.

Molnar, P., and P. Tapponnier, 1978, Active tectonics of Tibet: Journal of Geophysical Research, v. 83, p. 5361-5375.

Moore, G.W., and E. Scheibner, in press, Oceanic tectonic units, Tectonic Map of the Circum-Pacific Region, *in* E. Scheibner and T. Sato, eds., Explanatory notes for the Circum-Pacific Tectonic Map, Southwest Quadrant: Tulsa, AAPG, scale 1:10,000,000.

Sawkins, F.J., 1984, Metal deposits in relation to plate tectonics: Berlin, Springer-Verlag, 325 p.

Scheibner, E., 1976, Explanatory notes on the Tectonic Map of New South Wales: Sydney, Department of Mines, Geological Survey of New South Wales, 283 p.

Terman, M.J., 1978, Tectonic maps of the Circum-Pacific Map Project: U.S. Geological Survey Open-File Report, 10 p.

U.S. Geological Survey, 1983, Circum-Pacific Map Project: Status Report: Open-File Report 83-64, 60 p., appendices.

Veevers, J.J., (ed.), 1984, Phanerozoic earth history of Australia: Oxford, Claredon Press, 418 p.

Warren, R.G., 1972, A commentary on the Metallogenic Map of Australia and Papua New Guinea and Metallogenic Map, scale 1:5,000,000: Australia Bureau of Mineral Resources, Geology and Geophysics Bulletin 145, 85 p.

# CIRCUM-PACIFIC MAP PROJECT ABSTRACTS

MOORE, GEORGE W., U.S. Geological Survey, Menlo Park, CA

## Tectonic Processes Inferred from New Circum-Pacific Geodynamic Map

The geodynamic maps prepared by the Circum-Pacific Map Project show crustal thickness, active plate boundaries, volcanic centers, historical faulting, and earthquake focal mechanisms, all of which record past and present lithospheric deformation. The sheets also show free-air gravity anomalies and the state of lithospheric stress, both indicative of structural imbalances that may cause future deformation. Lithospheric-stress directions from focal mechanisms within oceanic plates indicate that the oceanic crust is under horizontal compression everywhere except along plate boundaries. This compression limits possible plate-driving mechanisms and seems to relegate the process of pull by sinking heavy lithospheric slabs into subduction zones to that of a local factor only.

Two wavelengths of gravity and topographic anomalies are conspicuous on the map sheets: anomalies with wavelengths of about 3,000 km that seem to be perpendicular to spreading axes, and knuckle-shaped anomalies, which I previously named condyles, with wavelengths of about 300 km that are prominent along the outer swells of oceanic trenches and seamount ridges. The longer wavelength anomalies probably have one of the following causes: convection within the mantle, irregularities on the core of the earth, systematic lithospheric density variations created at spreading axes, or fossil subduction zones overridden by other plates. The condyles probably reflect the natural wavelength under compression of the flexurally rigid outer layers of the earth.

Because the geodynamic maps combine continental and oceanic data, they are particularly effective in resolving the dynamics of continental margins. For example, the merging of terrestrial gravity with oceanic gravity from SEASAT radar altimetry has revealed many prospective oil-bearing sedimentary basins that were previously unknown, such as those around the perimeter of Antarctica.

# GEOTHERMAL ENERGY

CHAPTER 39

# RECENT TRENDS IN GEOTHERMAL SCIENCE AND DEVELOPMENT: NEW ZEALAND

**R. W. Henley**
*Bureau of Mineral Resources*
*Canberra, Australia*

**J. W. Hedenquist**
*D.S.I.R., Wairakei, Taupo,*
*New Zealand*

ABSTRACT

Recent deep exploration of the Mokai and Rotokawa geothermal fields in New Zealand, the latter previously investigated to only 1000 m, has yielded spectacular success based initially on an integrated scientific exploration program. Development of these fields will depend on resolution of problems related to high silica contents, and at Rotokawa, on high gas contents. Potential geothermal resources farther west in the Taupo Volcanic Zone have also received initial exploration.

At Broadlands, the 100 $MW_e$ Ohaaki Power Station is due for completion in 1988 and will be the first of the New Zealand fields to involve full reinjection. At Wairakei, thermal degradation by widespread dilution after 30 years exploitation has prompted investigation of future reinjection and deep exploration options.

Recent study has also focused on the effects of domestic exploitation at Rotorua, resolution of casing corrosion problems at Broadlands, and the correlation of active geothermal systems with fossil hydrothermal ore-forming systems.

The effects of changing funding policy in geothermal science in New Zealand may have a detrimental effect on long-term prospects for geothermal exploration and technology development, both in New Zealand and worldwide.

## INTRODUCTION

Recent geothermal exploration in the Taupo Volcanic Zone, North Island, New Zealand, has focused on the re-evaluation of undeveloped fields through deep drilling (>2000 m). The extension of primary exploration into the western part of the Taupo Volcanic Zone, where geothermal features are few, is largely being guided by geophysical surveys. This has resulted in spectacular successes in the proving of New Zealand's hottest field to date (>330 °C) at Rotokawa, previously explored only to about 1000 m, and the discovery of a major new resource at Mokai, 20 km northwest of Wairakei. The locations of these and other major New Zealand fields are shown in Figure 1. Chemical data from this new phase of exploration already provide interesting challenges to modeling the origin of geothermal systems, as well as establish guidelines for future exploration in New Zealand and elsewhere. Significant advances have also been made in the correlation of these active systems with those in similar volcanic terrane in the geologic past, and have provided important guidelines for current precious metal exploration.

Other foci for research have been provided by the need to monitor chemical and physical changes in the production zones of developed fields resulting from their large-scale exploitation; and by the need to evaluate a series of development criteria for the 100 $MW_e$ Ohaaki power project, due for completion at Broadlands in 1988.

In this paper recent results from these investigations are reviewed, with particular emphasis on geochemistry and inferences useful in the exploration and development of geothermal resources outside New Zealand. Hedenquist (1986) has reviewed the geological, geochemical, and geophysical research on geothermal systems in the Taupo Volcanic Zone, and recent progress in reservoir engineering in New Zealand is discussed by Grant et al. (1983).

## EXPLORATION

In addition to budgetary restraint, a moratorium extending to 11 undeveloped geothermal fields (Figure 1) has restricted exploration drilling in recent years. This restriction, presently under review, is designed to preserve a wide range of natural thermal features for New Zealand's heritage. Most of the exploration drilling has therefore been in unrestricted areas such as Mokai, Rotokawa, and Ngawha and most recently at Ngatamariki, Magakino, and Horohoro. These are areas with few or no surface manifestations (Figure 1).

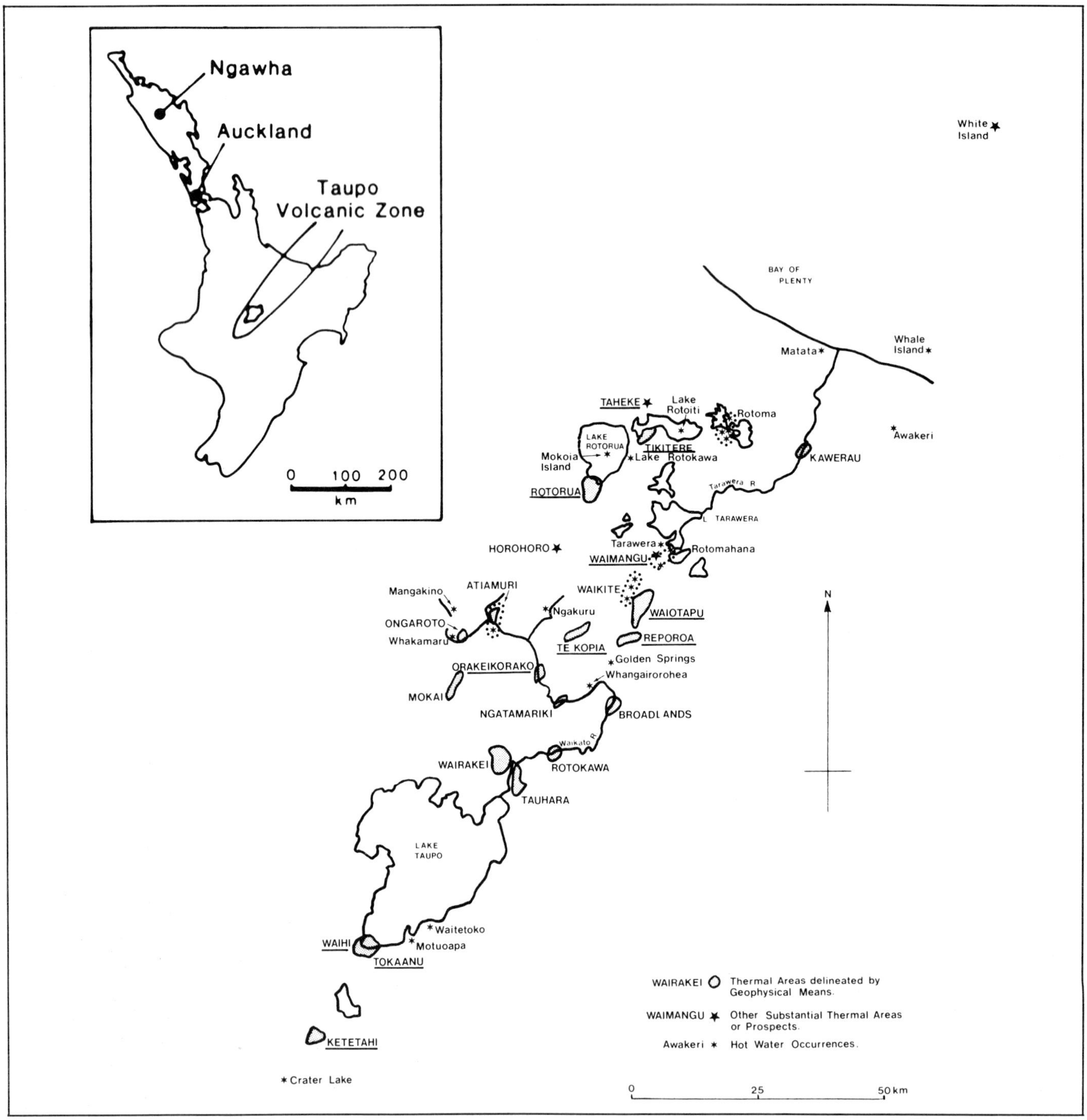


Figure 1. Location of known geothermal fields in the North Island, New Zealand. Underlining indicates fields currently closed to exploration.

Drilling at Mokai in 1982 located aquifer temperatures of 290-315° C at about 1650 m; MK5, drilled in 1983, proved to be New Zealand's most productive well, with a discharge of 700 tonnes/hour (about 25 $MW_e$ equivalent output). Figure 2 shows the distribution of exploration wells and natural features at Mokai. The combination of exploration geophysics (dominated by resistivity traversing) and geochemistry is particularly instructive in this field. Although fumarolic activity was known at Mokai and recorded in early geologic reconnaissance programs, the lack of substantial natural features and ad hoc assumption that fields west of Wairakei were in general older and of lower resource potential in comparison with those in the east prevented further evaluation in this region. In 1980, as part of a regional resistivity project by the Geophysics Division of the Department of Scientific and Industrial Research (D.S.I.R.), a large resistivity anomaly was recognized circumscribing the known thermal area (Bibby et al., 1981). Natural features were few but sufficient to provide, through standard solute geothermometry (Fournier, 1981), evidence of subsurface temperatures of approximately 175 °C immediately south of the northern group of hot dilute chloride springs. Gas geothermometers indicated higher feed tempera-

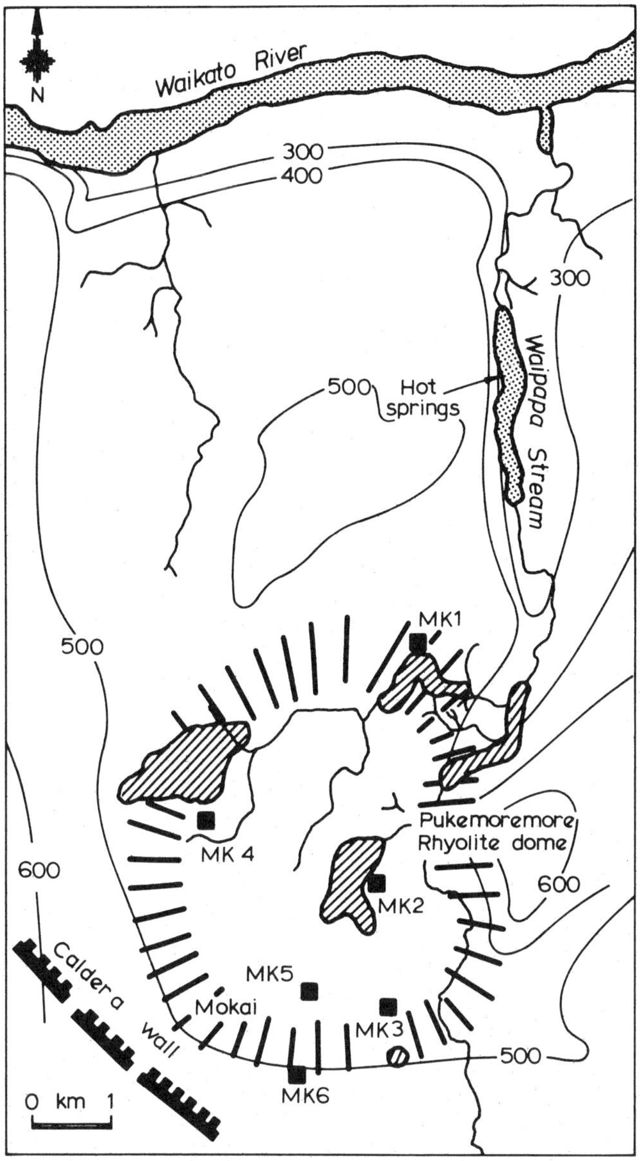


Figure 2. Distribution of geothermal features (hot springs, fumaroles, etc.), exploration wells, and the resistivity gradient boundary at Mokai. Low resistivities extend well to the north of the gradient boundary indicating the extent of the intermediate temperature reservoir (see text). Cross hatching indicates steam-heated thermal areas; solid squares are geothermal wells. Modified from Bibby et al., 1981, and Hulston et al., 1981.

tures (> 200 ° C) for the few fumaroles to the south (Hulston et al., 1981). The distribution of these features thus suggests the possibility of an economic resource at greater than 220 ° C flowing northward with progressive dilution by ground water in the manner suggested for some other geothermal fields in the region (Healy and Hochstein, 1973).

The first exploration well, MK1, confirmed the large intermediate-temperature reservoir of ~ 170 ° C. Subsequent deeper wells to the south located the high-temperature resource (Henley and Plum, 1985). Temperature inversions in MK1 and MK4 and pressure measurements in all wells then corroborated the lateral flow concept deduced during exploration.

The extensive dispersion of the high-temperature anomaly by mixing with ground water, as demonstrated at Mokai, is important for exploration elsewhere, particularly in high relief terrane which will develop large hydraulic gradients. In many exploration strategies over-emphasis is placed on the use of solute geothermometers as reliable indicators of maximum resource temperature, when the basis of these geothermometers is such that only the last zone of fluid-rock equilibration is recorded in the fluid chemistry. Different geothermometers respond at different rates to mineral-fluid reactions such that, for example, the silica geothermometer generally records shallow interactions while the "slower" NaKCa geothermometer, with its corrections for magnesium content (Fournier, 1981), may record somewhat deeper reaction equilibria, as in the intermediate-temperature reservoir at Mokai. Gas geothermometers tend to respond even more slowly and, provided due care is taken in their interpretation, any record deeper fluid-rock interaction. Fortunately at Mokai, the presence of fumaroles and the opportunity for some gas geothermometry, in conjunction with geophysical data and the preliminary interpretation of local hydrology, provided the confidence required to proceed with exploration drilling.

In other regions such as the Cascades, U.S.A. (Muffler, Bacon, and Duffield, 1982), as well as in the many other areas of volcanic terrane in the Circum-Pacific, higher relief and greater rainfall may combine to prevent such surface manifestations. The use of the geothermometers in "distal" warm to hot springs may not, in isolation, provide such a basis for further exploration, just as their geothermometry may not accurately reflect temperatures in their laterally displaced (and hotter) reservoirs.

Drilling in other fields along the west flank of the Taupo Volcanic Zone has not been as successful to date as at Mokai. At Atiamuri, adjacent to an extensive fossil system characterized by widespread alteration (Henneberger, 1986), drilling has failed to locate temperatures greater than 170 ° C in an area with some hot springs and a broad resistivity anomaly. These features may be derived from an extensive lateral flow of heated ground water extending south from Horohoro.

At Ngatamariki, another area characterized by a broad resistivity anomaly (Bennie, 1983) and minor hot springs, four deep wells have recently been drilled, though environmental constraints are postponing output testing until 1987. NM2 is the deepest geothermal well (2749 m) in New Zealand and of some interest is the intersection at 2649 m of a quartz microdiorite, the first hyperbyssal intrusive yet encountered in the Taupo Volcanic Zone.

Prior to 1984, drilling at Rotokawa, located in the eastern part of the Taupo Volcanic Zone, had been confined to a depth of ~ 1000 m. Three wells indicated a rhyolite-ignimbrite-ashflow sequence similar to that at Wairakei, 10 km to the west (Steiner, 1977); temperatures up to 315 ° C were measured, but the wells were not found to be permeable. Although natural features in this field are abundant, they provide few geochemical data useful for exploration. Most are steam-heated waters, steam condensates, or acidic sulfate-chloride hot springs whose waters have undergone extensive shallow interaction with rock and are characterized by a relatively high magnesium content. The presence of an estimated 5 to 10 million tonnes of native sulfur in the surficial acid alteration zone and underlying lacustrine sediments does, however, testify to a substantial heat flux as vapor (containing $H_2S$) from the deep system, and suggests a relatively high gas content in the deep fluid. The measured natural heat flow is in

TABLE 1

| Geothermal Field | Upflow Temp $^{o}C$ | Cl mg/kg | Gold solubility $CO_2$ mg/kg | $H_2S$ mg/kg | Minimum Heat flux from surface values MW |
|---|---|---|---|---|---|
| Wairakei & Tauhara | 250 | 1600 | 366 | 29 | 420 |
| Mokai | 320 | 1800 | 220 | 68(max) | 100 |
| Rotokawa | 320 | 500 | 13370 | 230 | 200 |
| Broadlands | 300 | 1000 | 12030 | >68 | 100 |
| Waiotapu | 300 | 1000 | 1858 | 102 | 600 |
| Kawerau | 280 | 900 | 9777 | 102 | 50 |
| Ngawha* | 233 | 1257 | 17400 | 121 | 55 |

excess of 200 $MW_{(t)}$ (Allis, 1980).

Recent drilling of four deep wells in the northern half of the Rotokawa system has identified permeable zones with temperatures in excess of 330 °C (Henley and Middendorf, 1985). These wells are presently undergoing long-term testing for a preliminary evaluation of the energy potential of the system.

The Ngawha field in Northland (Figure 1) remains enigmatic despite recent drilling but similarly has been shown to harbor high temperatures at depth (Mongillo, 1983). Previously the field had been considered to be gas-rich (possibly "vapor-dominated" sensu lato), of exotic chemistry (high boron, mercury, and ammonia content) and of low permeability relative to systems in the Taupo Volcanic Zone. However, much of this enigma must originate from its remote location and the consequent absence of any sustained exploration activity.

As at Rotokawa, recent exploration at Ngawha has concentrated on a more accurate definition of the field boundary and reservoir structure. The recent drilling of nine wells emphasized the absence of a uniform reservoir within the Mesozoic graywackes and argillites. The last and deepest well, NG13, although recording low permeability, did encounter temperatures in excess of 320 °C at 2255 m. As in the Taupo Volcanic Zone systems, a strong component of lateral outflow is considered to exist at Ngawha, in this case as a consequence of the presence of a "cap" 445 to 730 m thick of Cretaceous sediments. Little deep-system fluid penetrates this sequence, and natural features are dominated by steam and gas discharges which through time are responsible for the disseminated cinnabar deposits which have been mined from time to time. In contrast to the Taupo Volcanic Zone systems convective flow from depth is very low (Table 1).

## REGIONAL STRUCTURE

The deep investigation drilling at Mokai and Rotokawa has provided important data for the understanding of the geologic and geochemical setting of geothermal systems in the Taupo Volcanic Zone. Table 1 shows data for minimum heat flux, gas content, salinity, and temperature for the deep upflow of the Taupo Volcanic Zone fields and for Ngawha in Northland. These data suggest that systems within the eastern Kawerau and Taupo-Reporoa grabens are significantly higher in gas content (largely $CO_2$) than those to the west (Henley, 1986a). Surface observations for the undrilled fields to the west do not indicate any divergence from this pattern, and this suggests that the geochemical pattern is related to the tectonic structure rather than the age of the systems. The gas compositions of the eastern fields are also distinctive (e.g., in the abundance of their higher hydrocarbons; Giggenbach, 1986) and suggest equilibration temperatures for these components of up to 800 °C. Such a distribution of compositions may well occur in other volcanic belts, and this raises questions about the coupling of crustal processes in the Taupo Volcanic Zone; their resolution requires new investigations of the deep thermal structure, extensional tectonics, and related magmatism. Similar problems arise in the understanding of the distant Ngawha system, which is also distinguished by its high gas content. Important to the resolution of these questions is the recent publication of deep seismic and regional resistivity data for the Taupo Volcanic Zone (Stern and Davey, 1985; Geophysics Division, 1985); new gravity and geodetic data are also providing fresh insights to the rates of extensional tectonics (Allis, personal communication) and the location of possible earlier volcanic centers or calderas in the Taupo Volcanic Zone (Rogan, 1982).

At shallower levels evidence of alteration style, fluid chemistry, and the hydrology of geothermal systems (Giggenbach, 1980, 1981, 1984, 1986) provides an important basis for the understanding of fossil hydrothermal systems in ancient volcanic terrane. Of greatest significance has been the recognition by Brown (1986) of gold-rich (60 kg/tonne) chalcopyrite scales in wells at Broadlands and Kawerau. Together with field experiments, these metal-rich scales demonstrate that aquifer fluids are close to saturation with respect to gold, silver, and copper. This in turn resolves paradoxes concerning the length of time for epithermal ore deposition and depositional mechanisms which had arisen from earlier data (Brown, 1986; Henley, 1986b). It is now clear that at the low flow rates of typical hot springs discharging chloride waters (10-20 kg/s), significant gold-silver deposits of several million ounces may form in response to shallow boiling in zones of focused outflow. At Waiotapu, for example, Hedenquist and Henley (1985) argue that 0.1 million ounces of gold may have deposited below the active Champagne Pool since its origin by hydrothermal eruption 900 years ago. Recognition of these active ore-forming systems then provides much needed quantitative confidence for the exploration for precious metals in young volcanic terrane.

## GEOTHERMAL DEVELOPMENT

A low economic growth rate during the last decade, along with a surplus of electrical generating capacity (largely from hydroelectric development in the South Island), combined to reduce demand for geothermal development in New Zealand during the early 1980s. Future planning for the use of natural gas from the offshore Maui field may also affect the future use of geothermal reserves, although some generation will be required to make up a temporary shortfall in generating capacity. For this reason a staged development has been proposed at Mokai; together with proposals for the commercial sector development of Tauhara and Rotokawa, these may total around 100 $MW_e$ by 1990. Despite encouraging temperatures at depth at Ngawha in the latest drilling, an earlier decision to develop a 50 to 100 $MW_e$ power station has now been reversed. This deferral for at least 15 years, without continuation of a baseline drilling program to further define the deep system, reduces the long-term prospect for public sector development of this field. However, the New Zealand Government has recently restructured the Ministry of Energy, putting the Gas and Geothermal Trading Group and the Electricity Department on a corporate basis. This has opened the exploration and development of New Zealand's geothermal resources to the private sector, which should result in many economically attractive ventures proceeding.

The completion of the Ohaaki geothermal power project (100 $MW_e$) at Broadlands is scheduled for 1988. Unique to this New Zealand field is the use of a natural draft cooling tower, which will also be used to control gas dispersal to the atmosphere. Because of environmental constraints on discharge to the Waikato River, reinjection of all waste water will be practiced.

Ohaaki flash plant design has been influenced by the necessity to minimize silica scaling in pipelines prior to reinjection (Henley, 1984). Only minor supersaturations are reached by two-step flashing from reservoir temperatures of ~270 °C to the separation temperature of 155 °C (4.4 b.g.), since the alkalinity of the flashed liquid increases its silica solubility. However, in the hotter fields now explored in the Taupo Volcanic Zone, development depends upon resolution of scaling problems which can be anticipated for the higher silica contents involved. In these fields aquifer temperatures in excess of 300 °C result in silica saturations of about 1.6 (relative to amorphous silica) in waste waters resulting from flashing to 150 °C. Higher pressure steam separation to avoid reaching silica supersaturation would involve unacceptable waste of energy. The very high saturations reached by stepwise flashing to atmospheric pressure may obviate this problem if future research can prove that the resulting fast silica nucleation and growth will allow the controlled removal of the precipitate, perhaps using dissolved air flotation (Shannon, Owers, and Rothbaum, 1982). Coupled with chemical methods now available for removal of arsenic, current disposal problems (e.g. Rothbaum and Anderton, 1975; Shannon, Owers, and Rothbaum, 1982) may then be resolved while allowing a maximum energy recovery. Silica scaling and waste treatment are clearly major areas for research against which the future development of geothermal power may be balanced, both in New Zealand and elsewhere.

Resolution of the waste water disposal problem by reinjection is itself of greater importance than the possibility of maintaining aquifer pressures to minimize a decline in productivity. Recent tracer injection studies at Wairakei (McCabe, Barry, and Manning, 1983), along with chemical and physical trends monitored during 25 years of exploitation, suggest that reinjection near the production field would certainly result in cool water returns and consequent thermal damage in the production aquifer. Similar detrimental effects have been observed elsewhere (Ahuachapan, El Salvador; Otake, Japan; Puhagen and Tongonan, Philippines). At Broadlands, data from a long-term draw down test in 1971 suggested that the exsolution of $CO_2$ from aquifer fluid would buffer the rate of pressure drop (Grant, 1977); in contrast, the absence of significant dissolved gas in Wairakei fluids resulted in a rapid and widespread pressure drop. At Broadlands it is hoped (and partially confirmed by tracer and interference tests) that injection into the high-level Ohaaki rhyolite southwest and above the main production zones will minimize such short-term cool-water returns.

The relatively high $CO_2$ content of the aquifer fluid at Broadlands (1–4 wt % at 300 °C; Sutton and McNabb, 1977) also leads to other specific problems for development. In general, high $CO_2$ contents imply relatively high $H_2S$ contents, so that greater control of gas emission is required than in a low gas field such as Wairakei. Following an extensive experimental program (Hart, Dutton, and Glover, 1980), direct contact condensers will be employed in association with a 105 m natural draft cooling tower; the latter was also designed to promote the dispersal of injected waste gas into a plume penetrating the atmospheric inversion layer normally present in the area. An additional consequence of the high $CO_2$ content of the fluid is that shallow (100 to 500 m) ground waters heated by gas-rich steam rising from the deep system becomes enriched in $CO_2$. These slightly acidic (pH ~5.5) waters are responsible for extensive clay-rich alteration zones and for a recently recognized problem with casing and cement corrosion in Broadlands wells at shallow levels (Hedenquist and Stewart, 1985).

At Wairakei (Figure 3), the 180 $MW_e$ power development

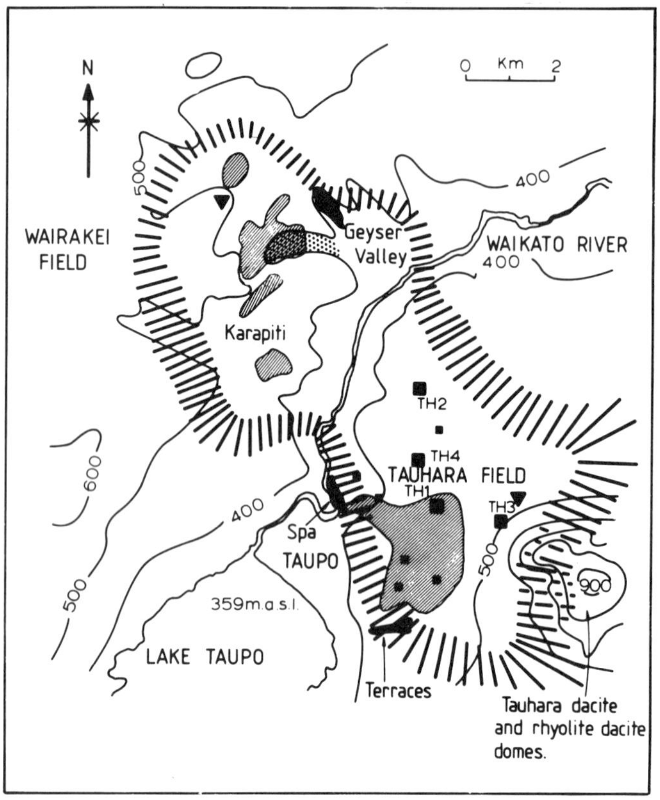


Figure 3. Distribution of thermal features and the resistivity gradient boundary at Wairakei and Tauhara. Note that the pressure loss disturbance following initial exploitation at Wairakei affected both fields and was expressed by growth of areas of steaming ground and some shallow hydrothermal eruptions. Subsequently shallow cold-water infiltration occurred from the northeast (Geyser Valley) and south, with consequent production enthalpy decline; hot-water recharge was from the west and south, the latter encompassing the Tauhara aquifer. The upflow zone is indicated by the triangle symbol. Cross-hatching indicates steam-heated thermal areas; solid squares are geothermal wells. From Henley, 1986b.

commissioned in 1959 resulted in widespread aquifer boiling (Allis, 1981) and subsequent infiltration by cool surficial waters. This resulted in loss of production enthalpy and well-head pressure. As well as the question of reinjection to avoid further contamination of the Waikato River by waste disposal, current field management is concerned with the prediction of the future run-down over the next 15 years and the economics of further deep exploration in the western portion of the field. The latter is a prospect encouraged by recent deep exploration successes elsewhere in the region. Short-term exploitation of high-level steam zones which have resulted from the aquifer pressure loss is now under way from some recently drilled wells.

The Tauhara field may soon be developed jointly by public and private sector interests, the first such venture in New Zealand. In this field (Figure 3) it has already been shown that a substantial aquifer pressure drop occurred because of exploitation at Wairakei (Allis, 1983) and that this was accompanied by major changes in surface hydrology, such as local hydrothermal eruptions, temperature increases in hot springs and in shallow aquifers used for domestic heating, etc. The increased mass withdrawal related to small-scale development (10-20 $MW_e$) at Tauhara is not likely to cause further substantial changes at the surface in the vicinity of Taupo. However, it may affect the recharge to the nearby Wairakei production field; this potential conflict is an issue not fully considered at this time in New Zealand.

Nonelectric developments in New Zealand have been slow to proceed. Horticultural projects are already established at Wairakei, Kawerau, and Rotorua, and increased steam usage is anticipated by the Tasman Pulp and Paper Mill at Kawerau. Since use for electrical generation alone wastes some 95% of the total available energy, a more efficient use of geothermal energy in the future should be encouraged through direct use. This would require changes in the Government pricing of geothermal energy, which currently is tied to alternative "equivalent" energy sources. The most widespread use of low-grade heat has been at Rotorua, where many homes and hotels and a research institute have operated individual bores for many years. The total drawoff, estimated at 32,000 tonnes/day in winter, has been recently shown to have affected the field hydrology (by up to −0.5 bars); this loss of hydraulic head has likely been the predominant cause of the demise of many geysers and hot springs in the Whakarewarewa area (Mahon, 1985). More closely controlled exploitation is anticipated in the future once the necessary regulatory authority is established, and this may focus on the use of downhole heat exchangers (Allis and James, 1980), communal bore systems, and greater efficiency of energy use and bore operation (Drew et al., 1985; Brown and Gould, 1985).

## FUTURE PERSPECTIVES

New Zealand is fortunate to possess an abundant hydro and fossil fuel energy base in addition to its many geothermal systems. This situation, in conjunction with the low economic growth of recent years, is unlikely in the immediate term to lead to substantial demand for electricity or industrial energy from geothermal resources. Certainly there will be very little demand that specifically needs geothermal-based energy. An exception may be in the recent growth of the New Zealand timber industry, which has been accommodated by expansion of the Kinleith (near Mangakino) and Kawerau Mills. These will be sufficient to handle most of the pulp production from the large timber resource reaching maturity in the East Taupo forests in the 1990s, although additional sawmill capacity will be required to meet the export demand for sawn timber. The location of geothermal resources in the western part of the Taupo Volcanic Zone provides an option for direct or electric energy supply to the Kinleith plant.

There is presently a moratorium on exploration drilling in many of the undeveloped Taupo Volcanic Zone fields, which will remain in place indefinitely or until a balanced program of development and conservation is established. This program will also include provision for waste-water disposal and public submission prior to commencement of exploration drilling. There is ample justification for this, inasmuch as a substantial proportion of New Zealand's geysers and other geothermal features have already been destroyed by geothermal and hydroelectric development since 1950. In our opinion, early declaration of National Park status for all fields in the Rotorua-Waimangu-Waiotapu-Orakeikorako quadrangle would both preserve the remaining important geothermal features and open and encourage future exploration outside this region.

In addition, present Ministry of Energy budgetary constraints limit deep exploration drilling to about three wells per year, with signs of further cuts in the very near future. All of these restraints on exploration in New Zealand must ultimately have a long-term flow-on effect for development as well as for geothermal science at all levels. This last point, as discussed below, may also have wider implications for worldwide geothermal exploration and development.

Geothermal systems are an integral component in the evolution of the earth's crust. As such, they are responsible for the formation of the majority of the world's base and precious metal ore deposits from the Archean to the present. Therefore, it is of concern that outside New Zealand so much geothermal drilling data together with fluid-rock geochemical studies should be restricted for proprietary reasons. In many cases few scientific studies have accompanied (or in some cases not even preceded!) exploration and development. These practices clearly waste invaluable opportunities for advancing knowledge of these important crustal processes and, from a more commercial standpoint, advancing efficiency in exploration and development.

On the basis of public sector development and freedom of information in New Zealand throughout the last 40 years, New Zealand science and technology has provided a focus for worldwide geothermal energy development and evaluation. The decline of the government-funded exploration and development program in New Zealand, particularly in terms of advancements in science and new technology, may therefore have a wider effect on the future of geothermal energy usage than within New Zealand alone. By the same token, the trend toward cost recovery in public sector science, both in New Zealand and elsewhere, has a detrimental effect on information transfer and long-term initiatives in both science and technology. Cost recovery research is by definition allied to only short-term gains able to be realized by the private sector. We see no hopeful signs for the resolution of this problem other than the hope that the strength of a nation's social and economic success may in the long term be measured against its support of not only the arts but of the sciences, both of which are essentially trivial in cost but immeasurable in benefit.

## ACKNOWLEDGMENTS

The convenors of the 4th Circum-Pacific Energy and Mineral Resources Conference are thanked for their invitation to present this paper in Singapore. R.W. Henley wishes to thank Drs. S.W. Kieffer and K.L. Brown for helpful comments during the preparation of this review paper.

## REFERENCES CITED

Allis, R.G., 1980, Heat flow, *in* M.P. Hochstein and T.M. Hunt, eds., Guide to geophysics of the volcanic and geothermal areas of the North Island, New Zealand: The Royal Society of New Zealand Miscellaneous Series 3, p. 47-48.

Allis, R.G., 1981, Changes in heat flow associated with exploitation of Wairakei geothermal field, New Zealand: New Zealand Journal of Geology and Geophysics, v. 24, p. 1-19.

Allis, R.G., 1983, Hydrologic changes at Tauhara geothermal field, New Zealand: Department of Scientific and Industrial Research, Geophysics Division Report 193, 46 p.

Allis, R.G., and R. James, 1980, A natural convection promoter for geothermal wells: Geothermal Energy, v. 8, p. 41-45.

Bennie, S.L., 1983, Geophysical investigation of the Ngatamariki geothermal area: Department of Scientific and Industrial Research, Geophysics Division Report 192, 35 p.

Bibby, H.M., G.B. Dawson, H.H. Rayner, V.M. Stagpoole, and D.J. Graham, 1981, Geophysical investigations of Mokai geothermal field: Department of Scientific and Industrial, Geophysics Division Report 184, 41 p.

Brown, K.L., and T.A. Gould, 1985, The use of calcite antiscalants in the Rotorua geothermal field: Proceedings, 7th New Zealand Geothermal Workshop, University of Auckland, November, p. 129-132.

Brown, K.L., R.W. Henley, and H. Plum, 1985, Aquifer boiling and dilution in the Wairakei geothermal field due to exploitation 1950-1981: Department of Scientific and Industrial Research, Chemistry Division Report, in press.

Drew, S., B. Simpson, R. Robinson, and D. Paul, 1985, The Rotorua geothermal field; A report of the geothermal monitoring programme: Department of Scientific and Industrial Research, Ministry of Energy, and Ministry of Works and Development, New Zealand, 48 p.

Fournier, R.O., 1981, Application of water geochemistry to geothermal exploration and reservoir engineering, *in* L. Rybach and L.P.J. Muffler, eds., Geothermal systems: Principles and case histories: New York, John Wiley and Sons, p. 109-143.

Geophysics Division, 1985, Sheet U17 - Wairakei. Electrical resistivity map of New Zealand, 1:50000, Nominal Schlumberger array spacing 1000 m: Department of Scientific and Industrial Research, New Zealand.

Giggenbach, W.F., 1980, Geothermal gas equilibria: Geochimica et Cosmochimica Acta, v. 44, p. 2021-2032.

Giggenbach, W.F., 1981, Geothermal mineral equilibria: Geochimica et Cosmochimica Acta, v. 45, p. 393-410.

Giggenbach, W.F., 1984, Mass transfer in hydrothermal alteration systems: Geochimica et Cosmochimica Acta, v. 48, p. 2693-2711.

Giggenbach, W.F., 1986, The use of gas chemistry in delineating the origin of fluids discharged over the Taupo Volcanic Zone: Proceedings of Symposium 5, International Volcanological Congress, Hamilton, p. 47-50.

Grant, M.A., 1977, Broadlands - A gas-dominated geothermal field: Geothermics, v. 6, p. 9-29.

Grant, M.A., P.F. Bixley, M.J. O'Sullivan, and J.D. Leaver, 1983, Recent developments in reservoir engineering in New Zealand: Proceedings, 9th Workshop on Geothermal Reservoir Engineering, Stanford University, p. 107-114.

Hart, W., L.R. Dutton, and R.B. Glover, 1980, Pilot plant studies for design of the Ohaaki geothermal power station condensers: New Zealand Institute of Engineers Transactions, v. 7, p. 95-99.

Healy, J., and M.P. Hochstein, 1973, Horizontal flow in hydrothermal systems: New Zealand Journal of Hydrology, v. 12, p. 71-82.

Hedenquist, J.W., 1986, Geothermal systems in the Taupo Volcanic Zone: Their characteristics and relation to volcanism and mineralisation, *in* I.E.M. Smith, ed., Late Cenozoic volcanism in New Zealand: The Royal Society of New Zealand Bulletin 23, p. 134-168.

Hedenquist, J.W., and R.W. Henley, 1985, Hydrothermal eruptions in the Waiotapu geothermal system, New Zealand: Their origin, associated breccias and relation to precious metal mineralization: Economic Geology, v. 80, p. 1640-1669.

Hedenquist, J.W., and M.K. Stewart, 1985, Natural $CO_2$-rich steam-heated waters at Broadlands, New Zealand: Their chemistry, distribution and corrosive nature: Transactions of the Geothermal Resources Council Annual Meeting, Hawaii, v. 9, p. 245-250.

Henley, R.W., 1984, Chemistry and silica scaling potential of multiple flash geothermal waters: Geothermics, v. 12, p. 307-312.

Henley, R.W., 1986a, Ore transport and deposition in epithermal environments, *in* H. Hervert, ed., Stable isotopes and fluid processes in mineralisation: Geological Society of Australia Special Publication, in press.

Henley, R.W., 1986b, The geothermal frameworks of epithermal deposits: Society of Economic Geologists Reviews in Economic Geology, v. 2, in press.

Henley, R.W., and K.I. Middendorf, 1985, Geothermometry in the recent exploration of Mokai and Rotokawa geothermal fields, New Zealand: Transactions of the Geothermal Resources Council Annual Meeting, Honolulu, Hawaii, v. 9, p. 317-324.

Henley, R.W., and H. Plum, 1985, Chemistry of geothermal fluids discharged from exploration wells at Mokai, New Zealand: Zeitschrift der Deutschen Geologischen Gesellschaft, v. 136, p. 235-251.

Henneberger, R.C., 1986, Ohakuri fossil epithermal system, New Zealand, *in* R.W. Henley, J.W. Hedenquist, and P.J. Roberts, eds., Guide to the active epithermal (geothermal) systems and precious metal deposits of New Zealand: Monograph Series on Mineral Deposits, No. 26, Gebrüder Bornträeger, Berlin, in press.

Hulston, J.R., R.W. Henley, R.B. Glover, and M.A. Cox, 1981, Stable isotope and geochemical reconnaissance of the Mokai geothermal system, Taupo Volcanic Zone: Proceedings of the 3rd Geothermal Workshop, University of Auckland, p. 81-86.

McCabe, W.J., B.J. Barry, and M.R. Manning, 1983, Radioactive tracers in geothermal underground water flow studies: Geothermics, v. 12, p. 83-110.

Mahon, W.A.J. (ed.), 1985, The Rotorua geothermal field: Technical Report of the Geothermal Monitoring Programme, 1982-1985: Department of Scientific and Industrial Research, Ministry of Energy, Ministry of Works and Development, 522 p.

Mongillo, M.A., 1983, The Ngawha field: A review of the scientific investigations: Department of Scientific and Industrial Research, Geothermal Report 8, 241 p.

Muffler, L.J.P., C.R. Bacon, and W.A. Duffield, 1982, Geothermal systems of the Cascade Range: Proceedings of the Pacific Geothermal Conference, University of Auckland, New Zealand, p. 337-344.

Rogan, M., 1982, A geophysical study of the Taupo Volcanic Zone, New Zealand: Journal of Geophysical Research, v. 87, p. 4073-4088.

Rothbaum, H.P., and B.H. Anderton, 1975, Removal of silica and arsenic from geothermal discharge waters by precipitation of useful calcium silicates: 2nd U.N. Symposium on the Development and Use of Geothermal Resources, San Francisco, v. 2, p. 1417-1425.

Shannon, W.T., W.R. Owers, and H.P. Rothbaum, 1982, Pilot scale solids/liquid separation in hot geothermal waters using dissolved air flotation: Geothermics, v. 11, p. 43-58.

Steiner, A., 1977, The Wairakei geothermal area; North Island, New Zealand: New Zealand Geological Survey Bulletin 90, 139 p.

Stern, T.A. and F.J. Davey, 1985, Crustal structure studies within the central North Island: The Central Volcanic Region: Department of Scientific and Industrial Research, Geophysics Division Report 207, 47 p.

CHAPTER 40

# GEOTHERMAL RESOURCES IN THE ISLANDS OF THE SOUTHWEST PACIFIC

**W.A.J. Mahon**
*Geothermal Energy New Zealand, Ltd.*
*Auckland, New Zealand*

## ABSTRACT

Many islands in the southwest Pacific are areas of present or historical volcanism and contain zones of geothermal activity. The volcanism and associated geothermal activity are related to areas of crustal consumption and subduction zones. Only in a few cases have the geothermal activity and the geothermal systems producing the activity been investigated in any detail. As a result, the size of the geothermal energy reserves of the region is unknown.

The development of the geothermal resources of the islands is dependent, both practically and economically, on the uses to which the energy can be directed and on the location of the resources relative to the power load centers on the various islands. Since population centers are likely to be located in the safest volcanic-risk zones, it is frequently impracticable and very expensive to use local geothermal energy.

Joint UNDP/World Bank Energy Sector Assessment Programmes have been completed for a large number of the southwest Pacific Islands. Whereas geothermal energy is mentioned in some of these reports as a possible local energy resource, the lack of knowledge of the sizes of these resources makes it difficult to assess the role of geothermal energy in the development of the region. This could only be done if a full scientific reconnaissance and investigation programme of the resources was carried out.

This paper describes some of the geothermal resources of the region and possible strategies for developing and using these resources. Likely development and utilization problems are investigated and discussed.

## INTRODUCTION

The "Pacific Ring of Fire" is a term coined by volcanologists over the last decades to describe the extensive volcanic activity that occurs on the tectonic plate boundaries surrounding the Pacific Ocean. These areas of active volcanism are also zones of active geothermal systems, many of which have been developed for electric power generation over the last 20 years. Of the 4764 MW of electricity generated by geothermal energy at the end of 1985, about 4100 MW, or 86%, was generated from the Pacific area.

The southwest and west Pacific areas are noted for the line of Island Arcs which stretches from New Zealand in the southwest to Japan in the west (Figure 1). These volcanic islands, which occur over the intersections of the west Pacific plate with the Australia-India plate in the southwest and the Eurasian plate in the northwest, have very high reserves of geothermal energy capable of being converted into electrical energy. To date 1333 MW of electricity, or 28% of the total world geothermal electricity, comes from the southwest and west Pacific areas.

Major geothermal developments, for both electrical and nonelectrical uses, are taking place in New Zealand, Indonesia, the Philippines, and Japan and certainly the overall contribution of geothermal electricity to the first three countries is, and will be in the future, very significant. It has been estimated that if New Zealand developed its available geothermal resources over the next 20 years, the contribution to the total generating capacity of the country could be as high as 14 to 16%.

There are many other islands in the southwest Pacific where volcanic activity and/or geothermal systems occur. They occur in a region of diverse tectonic features that include crustal subduction and associated island arc formation, back arc basins, and elevated areas of sea floor in which new crust is being formed as well as continental type land masses. Excluding New Zealand, the best known of these islands are the Kermadec Islands, the islands of Tonga, the Samoan Islands, the Fiji Island group, New Caledonia and the New Hebrides-Santa Cruz Islands, the Solomon Islands, Bougainville Island, and Papua New Guinea. The Hawaiian Islands, although in the central Pacific, are included in this region.

An excellent summary of the volcanic and tectonic geology of these areas and the occurrence of geothermal activity on the islands was given by Cox (1980). He presents an impressive list of references in his paper illustrating the large amount of scientific work that has been done in the general area.

From his study and review of the local geothermal resources, Cox concluded that:

a) the geothermal systems with the highest reservoir temperatures are within active volcanic provinces,

b) low-temperature systems occupy recently extinct volcanic terrains and nonvolcanic environments that have undergone tectonism,

c) the type of surface activity and chemistry of the thermal fluids characterize the type of system, and

d) many of the geothermal systems in the region are at present (1980) unsuitable for power production because of isolation, unfavorable fluid chemistry, or the possibility of eruption; however, several undeveloped areas show promise as geothermal fields.

## SIZE OF GEOTHERMAL RESOURCES

Geothermal Energy New Zealand, Limited, and Kingston, Reynold, Thom and Allardice, two private New Zealand companies, have collectively been engaged in investigating the geothermal resources of the island of EFATE in Vanuatu. Their field experience has indicated the difficulty of traversing the local terrain and carrying out a comprehensive inventory of the local thermal manifestations. It seems very likely, certainly after reading accounts of the geothermal areas of the Solomons and Papua New Guinea, that considerable reconnaissance and exploration work is still required before a definite assessment of the geothermal energy potential of the areas as a whole can be made. It is likely that the values of natural heat flow that are available from the various resources are extremely tentative and almost certainly low.

It has taken New Zealand and Japan many years to assess their own countries' geothermal potential. This has been accomplished with technical expertise built up over considerable periods of time and within environments which are reasonably accessible. Since the Pacific Islands are not particularly accessible and since very few have the necessary inhouse geothermal expertise, any assessment of the local and regional geothermal resources will have to be made by an external agency. This could well be financed through some form of regional aid programme. Needless to say, there is strong evidence to indicate that the regional geothermal energy resource is significantly large.

The present value of gold and the popularity that gold and silver exploration and mining are undergoing in the 1980s may aid in the longer term assessment of the region's geothermal resources. The association of epithermal gold deposits and geothermal and hydrothermal activity is well known, and already a large number of mining prospects have been identified on islands in the southwest Pacific. As mining prospects are investigated, geological and geochemical details of the local geothermal resources should become available. Shallow prospecting wells will allow estimates of heat flow and geothermal gradients to be assessed. Hopefully, in the longer term, a much better assessment of the region's geothermal reserves and potential will become available.

## NATURE OF THE GEOTHERMAL RESOURCES

On the basis of the chemical compositions of the surface hot waters discharged from springs and seepages and of the steam discharged from fumaroles, Cox (1980) recognized five different types of geothermal systems. These can be enumerated as follows (see Figure 2):

1. Active volcanic systems—emissions are wholly or partly magmatic fluids. High to very high temperatures are present (100 ° to 800 ° C). Condensed fluids are normally highly acidic.
2. Recently active volcanic system—emissions are neutral to slightly alkaline sodium chloride waters which have a mainly meteoric water origin. Steam emissions, if they exist, originate form the hot chloride water and contain mainly $CO_2$ and $H_2S$. High temperatures are present (100 ° to 374 ° C).
3. Steam systems—systems produce mainly steam with minor amounts of water. Waters at depth could be sodium, sulphate, bicarbonate waters underlain by neutral sodium chloride waters.
4. Systems present in old volcanic rocks which have retained some residual heat and/or have a higher than normal geothermal gradient through them. Mainly low temperatures (25 ° to 180 ° C).
5. Waters which discharge from types 2 and 4 systems near the coast can be contaminated with sea water. Alternatively, systems which develop near the coast are likely to contain hot chemically modified sea water as the primary water present.

The higher temperature systems tend to occur on the western side of the Island Arcs.

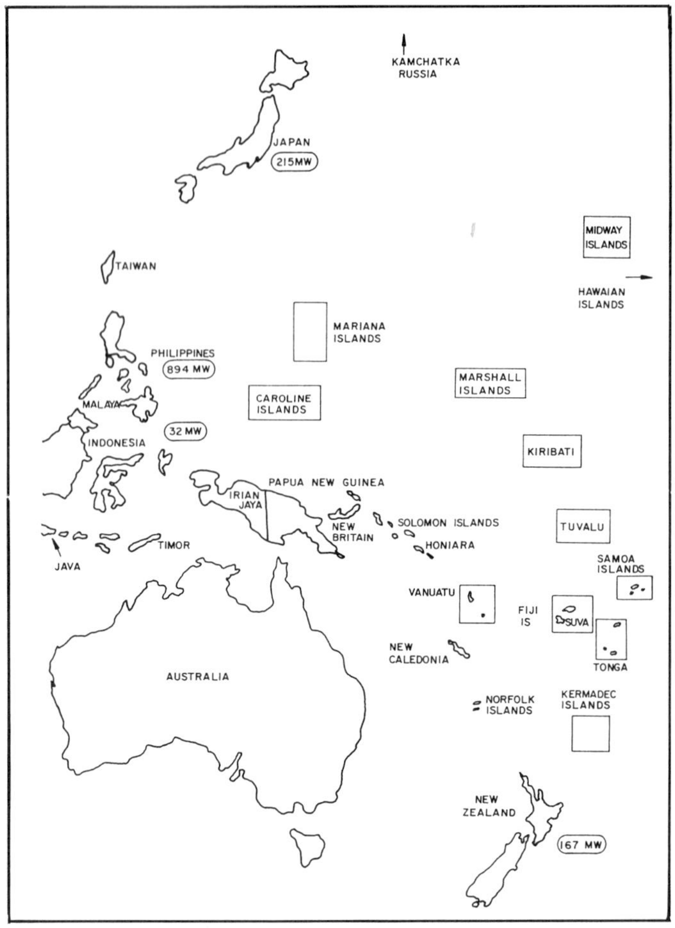


Figure 1. Islands of the west and southwest Pacific.

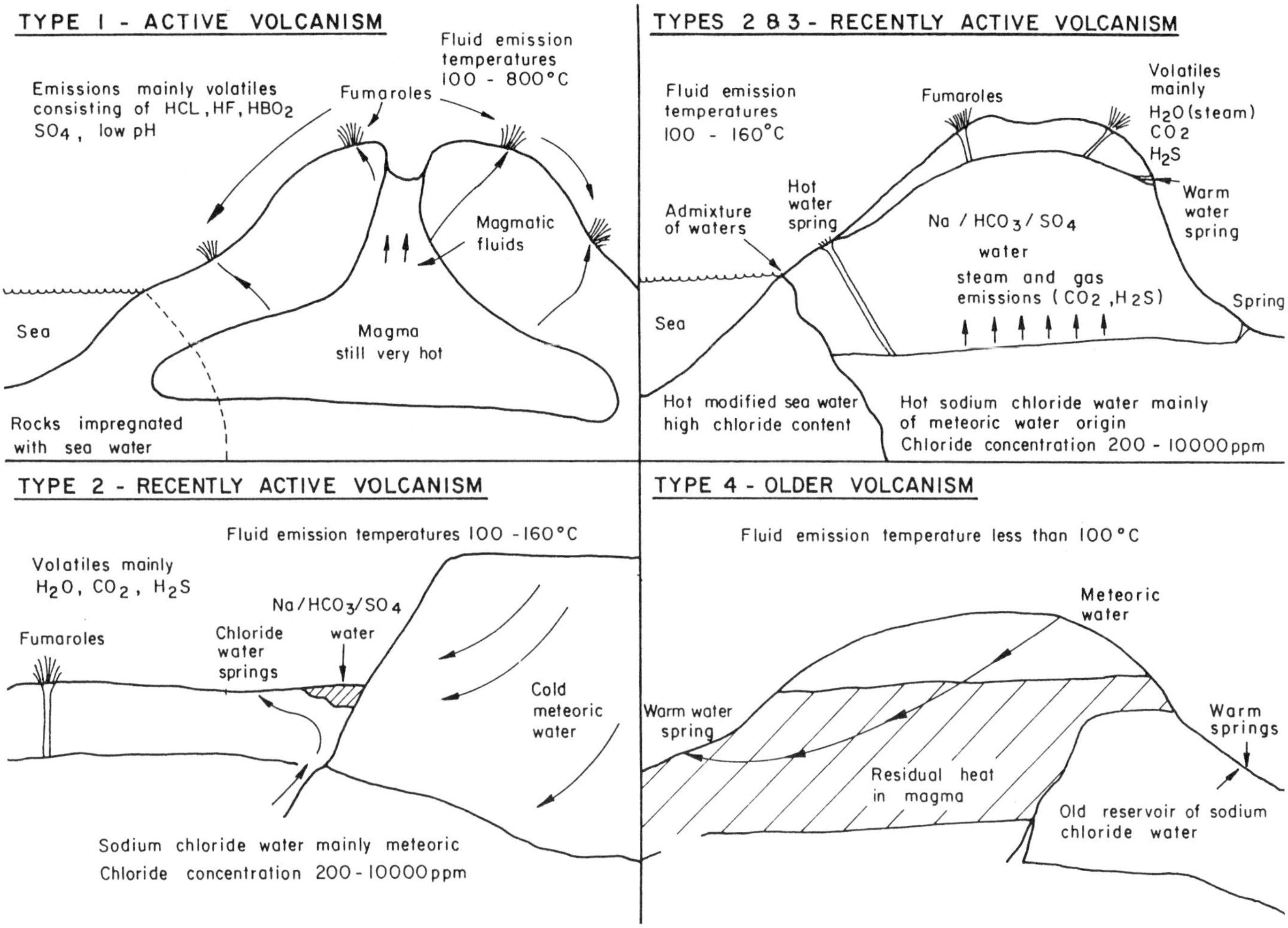


Figure 2. Different types of geothermal systems in the southwest Pacific.

## REASONS FOR DEVELOPMENT

At the present time there are no real reasons or incentives to assess or develop the geothermal resources of many of the Island areas. In fact, this incentive has decreased over the last period with the combination of falls in oil price and the large emphasis on petroleum-produced energy on most of the islands. There are also other energy resources on many of the islands, including hydro, natural gas, and wood that may, in the longer or even shorter term, be more sensible to develop than geothermal energy.

Owing to the lack of essential infrastructures such as transportation and communication on many of the islands, electric energy demand has been concentrated mainly in the urban areas. These areas are commonly situated away from the geothermal resources, making the costs of transmission of electricity and development of the resources prohibitive.

The main reasons for assessing the geothermal potential of the islands is to aid in long-term energy planning, to assess industrial growth possibilities, and to move away from expensive petroleum-based electricity generation. There are no reasons to believe that the present low price of oil will continue and that in the future high oil prices would not have an adverse effect on local and regional development. The existing electrical capacities of the different islands and the predicted load increases over the next 30 years are also commensurable and favorable to small-scale geothermal developments.

## DEVELOPMENT STRATEGIES

Considerable worldwide experience is available for developing geothermal systems (such as types 2, 3 and 4, listed earlier) for electricity generation. Experience for developing sea water systems is presently being obtained from the work at Fushime in Japan, Milos in Greece, and in Iceland. It is unlikely that active volcanic areas will be capable of development for some time, since the hot fluids present in these areas are normally highly acidic and corrosive.

The widely spread populations of the islands and the inaccessibility of many of the island hinterlands eliminates any possibility of fully integrated electrical power systems on most islands. Small generating units (coal, diesel, or geothermal) are probably the only viable possibility except in areas of concentrated population (capital cities) or where large captive power requirements are needed.

The small mobile geothermal turbine would be ideal to

meet the predicted electricity load forecasts for many of the Pacific Islands in the next 20 or so years. These turbines are readily available, sometimes off the shelf, from a number of manufacturers. They range in generating capacity from 250 KW to 5000 KW. They can be either condensing or noncondensing turbines, are simple to install, and are easy to run and require little maintenance. Minimum skills are needed to operate the machines.

Geothermal development costs have increased considerably over the last 10 years. A substantial proportion of the increase is due to the stricter environmental regulations controlling the discharge of geothermal waste (hot water, gas, etc.) to the atmosphere. In many developments local regulations require that the waste hot water and steam condensate be injected back into the ground. This necessitates the drilling of expensive reinjection wells, which typically have a limited life. However, the location of many of the geothermal systems in the Islands, close to the sea, would allow discharge of waste directly into the ocean, where this is environmentally acceptable, and thus would reduce development costs and make geothermal-generated electricity very competitive.

It is impossible to predict geothermal energy costs, in terms of cents/kw, in the islands, since costs will be very strongly site-related. For a reasonably accessible development site close to the sea and no major infrastructure requirements (roading, etc.), a figure of US$1000 per installed Kilowatt is possible. It is difficult to compare this figure with costs of other forms of electricity generation or, for that matter, the cost of importing oil. However, during the period of high oil prices, it was assessed that generating 7.5 MW of electricity in the Solomons using wood-fired boilers would save the country US$45-53M in the period 1987-2000, or $3.4 to 4 million per year in oil import costs.

## SUMMARY

A considerable amount of further exploration work is required before the geothermal potential of the southwest Pacific can be properly assessed. This will have to be completed before the significance of the geothermal resources of the region can be judged against other local energy sources. The predicted electrical loads for many of the islands over the next 20 years would be admirably met by small geothermal generating units. Nonelectrical uses of the energy, such as refrigeration and industrial processing, could well prove important for the region in the longer term.

## REFERENCE CITED

Cox, M.E., 1980, Geothermal occurrences in the southwest Pacific: UN ESCAP, CCOP/SOPAC Technical Bulletin 3, p. 197-219.

CHAPTER 41

# A GEOCHEMICAL OVERVIEW OF THE GEYSERS (CALIFORNIA) GEOTHERMAL RESERVOIR

**A.H. Truesdell**
*U.S. Geological Survey*
*Menlo Park, CA*

**W.T. Box, Jr.**
*Geysers Geothermal Company*
*Santa Rosa, CA*

**J.R. Haizlip**
*GEO Operator Company*
*Santa Rosa, CA*

ABSTRACT

Examination of fieldwide data for The Geysers geothermal reservoir shows relatively small variations in geology, geophysics, and steam temperature and pressure. Steam composition, however, shows large gradients of gas content (from 100 to <65,000 ppm gas by weight) and $\delta^{18}O$ (from −7 to +3 permill SMOW) from the southeast to the northwest of the field. The areal extent (about 3 to 7 km wide by 18 km long) and the depth and thickness (from 0.5 km to 2-6 km deep) are defined by steam productivity, induced microseismicity, and (less definitely) geologic boundaries. Fluid circulation at The Geysers is dominated by vertical upward flow of steam and downward flow of condensate (the "heat pipe" mechanism) over short distances, by lateral ("Rayleigh") steam flow and condensation over intermediate distances, and by flushing of connate brines with meteoric water over most of the field. This flushing produced the major gas and isotope gradients.

## INTRODUCTION

The Geysers in northern California (Figure 1) is one of two known great vapor-dominated geothermal systems. (The other is Larderello in Tuscany, Italy; smaller vapor-dominated fields occur in Japan, Indonesia, and elsewhere). Unlike the more common hot-water systems, these systems produce steam with little or no liquid water, although most of this steam results from vaporization of liquid in the reservoir. Fieldwide studies have been made of the geophysics and surface and subsurface geology of The Geysers, but very little has been published on the chemical and isotopic compositions of fluids and reservoir rocks. Large-scale variations in the isotopic and chemical compositions of steam from Larderello (Celati et al., 1973; D'Amore et al., 1977) have led to models for fluid circulation and steam origin (D'Amore and Truesdell, 1979; Calore et al., 1982) that have been useful in guiding exploration and estimating reserves. Although variations in steam composition at The Geysers have been known informally and were suggested by study of chemical variations in altered reservoir rocks (Lambert, 1976; Sternfeld, 1981), only the fieldwide variation of steam isotope compositions has been published (Haizlip, 1985). It is our purpose to present a general description of The Geysers reservoir, to show (with new data) that large fieldwide gradients in isotope and gas compositions exist in Geysers steam, and to discuss the origin of these gradients.

## THE NATURE OF THE GEYSERS RESERVOIR FLUID

Wells at The Geysers generally produce nearly saturated steam. This steam may be slightly wet (contain entrained droplets of liquid), especially during very early production, or it may be slightly superheated. The widely accepted model of "vapor-dominated" systems presented by White et al. (1971)

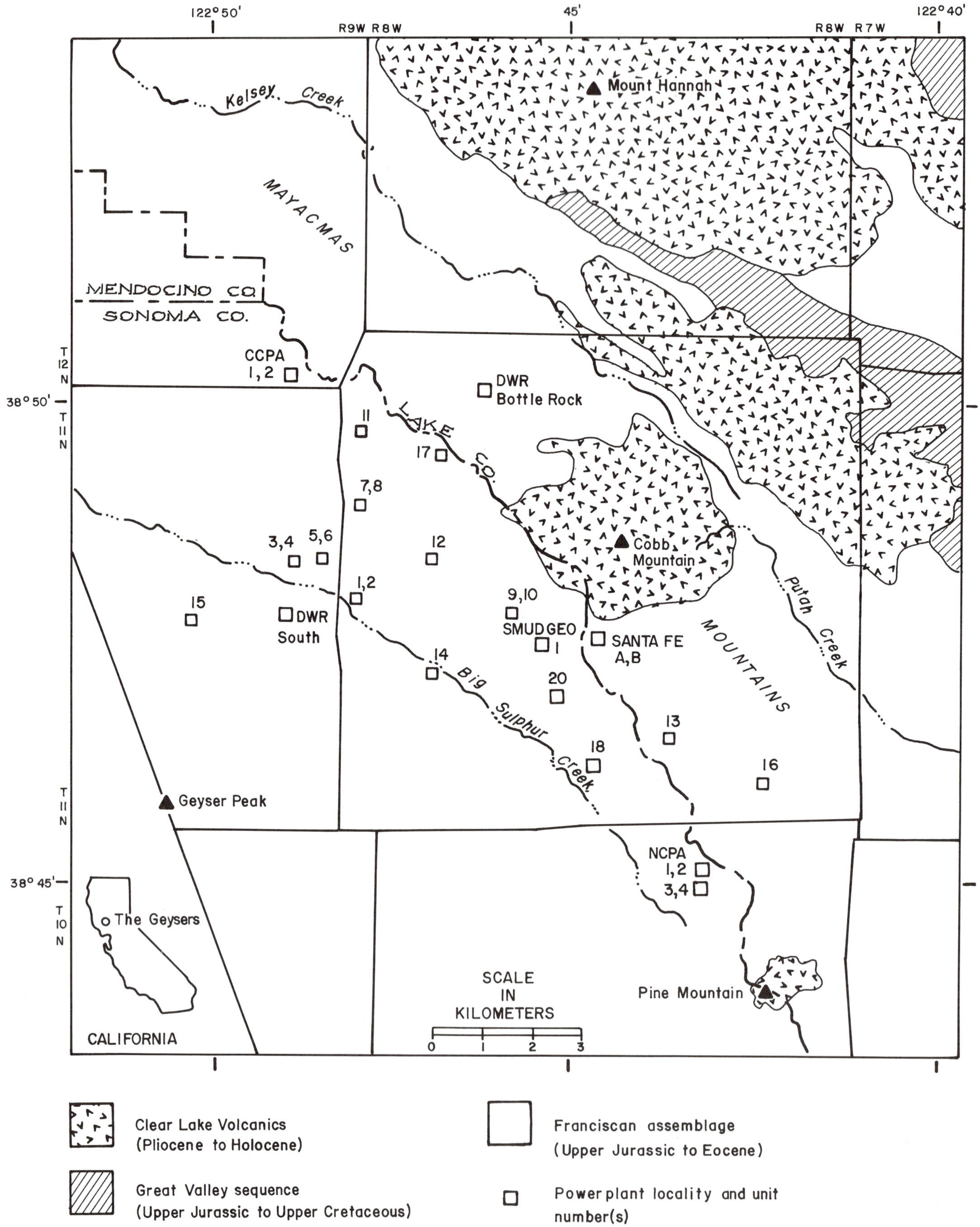


Figure 1. Location of The Geysers field and power plants with generalized regional geology (modified from Brook, 1981).

and Truesdall and White (1973) suggests that the reservoir contains highly mobile steam in large fractures and essentially immobilized liquid water adsorbed on surfaces and held in small fractures (akin to a "cracked sponge," Weres et al., 1977). In zones affected by exploitation, pressure decreases assisted by heat transferred from the rock cause most of this water to boil. As exploitation proceeds, pressures continue to drop and boiling extends further from the well. Boiling of liquid continues until the rock is dry or until the only liquid remaining is sealed from connection with the fracture network (as for example in fluid inclusions), is held in very small cracks by capillarity, or has a high salt content. Water in very small cracks or containing large concentrations of salt has a lower vapor pressure than dilute "bulk" water in larger cracks and may remain after bulk water is boiled off. In a continuous, permeable reservoir, pressure drops are transmitted long distances from wells and the immobilized water in a large volume of rock boils, possibly without producing a totally dry zone near the well. Thus steam remains close to saturation. This is probably the situation for most of The Geysers. In much of Larderello there is less water in the reservoir and lower permeability, so for most wells the near-well volume is dried out completely. Steam originating further away undergoes nearly isothermal decompression near the well to produce superheated steam.

In areas of The Geysers unaffected by previous exploitation (except northwest), most wells produce a water and steam mixture for days or months after the start of production and then produce saturated or slightly superheated steam. This behavior could be the result of initial production from a liquid-rich condensate zone near the top of the reservoir or of steam production from a two-phase liquid-steam reservoir with some mobile liquid and an initial pressure drop insufficient to cause total vaporization.

Natural fluid circulation in vapor-dominated reservoirs has been described (White et al., 1971), and simulated experimentally (Sondergeld and Turcotte, 1977), and mathematically (Preuss, 1985) as resembling that in a "heat pipe." In a heat pipe, heat is rapidly transmitted by a fluid that vaporizes at one end and condenses at the other, with counter flows of vapor and liquid along the pipe. This heat pipe effect produces high heat flow (to $9 \times 10^{-6}$ cal•cm$^{-2}$•sec$^{-1}$) observed over The Geysers; Urban et al., 1976) without net mass flux. A similar mechanism (essentially a horizontal heat pipe) produces the lateral steam flow and condensation responsible for the areal variations in steam composition observed at Larderello and parts of The Geysers (D'Amore and Truesdell, 1979).

The heat pipe effect is important for the fluid reserves of The Geysers because most steam produced originates in the reservoir from boiling of liquid that is either fixed (adsorbed on mineral surfaces, held in small pores, and fractures closed downward) or draining downward from zones of condensation.

## GEOLOGY OF THE GEYSERS RESERVOIR

The surface geology of The Geysers was mapped by McLaughlin (1978), and the subsurface geology was discussed by McLaughlin (1981) and Thomas (1981). The Geysers reservoir is generally developed within a thick, areally extensive body of graywacke of the Jurassic Franciscan assemblage that is overlain by a tectonic "thrust assemblage" (also Franciscan) and intruded and underlain to an unknown depth by a Pliocene or younger siliceous igneous rock locally known as "felsite." The overlying thrust assemblage consists of bodies of metagraywacke, melange, greenstone, chert, blueschist, and serpentine. The felsite locally contains steam (as do parts of the thrust assemblage) but to a lesser extent than the main graywacke. Massive felsite is encountered in wells in the southeast and south-central Geysers and probably exists at greater depths farther north. Felsite dikes have been encountered in wells throughout The Geysers field.

The main graywacke is intensely faulted and fractured, and productivity of the reservoir is related to the number, size, and continuity of these fracture networks. This is because the intrinsic porosity and permeability of this graywacke are very small (<3% and <1 md, respectively), and fluids must be contained and move in fractures rather than in the rock matrix. Jurassic to Tertiary thrust faulting has imparted a horizontal or subhorizontal fracture permeability to The Geysers reservoir and overlying rocks. Later San Andreas style wrench faulting (Tertiary to present) has produced locally intense, vertical fracturing (i.e., along Big Sulphur Creek and Squaw Creek), which are often used as drilling targets in highly deviated wells. Stress solutions of current seismic activity indicate that the directions of maximum and minimum stress in The Geysers are consistent with those along the San Andreas fault to the west (Oppenheimer, 1986). Fracture density in the main graywacke appears to decrease at the reservoir margin, and the remaining fractures are often sealed with alteration minerals in this area. Temperatures in some wells drilled outside The Geysers reservoir, but near the margin of the field, are higher than the reservoir temperature, indicating conductive heat flow in this area consistent with the model presented by White et al. (1971).

The petrologic and isotopic studies of Lambert (1976) and Sternfeld (1981) document a complex thermal history for the field, with three separate hydrothermal systems differing in temperature, reservoir fluid, and type of system. The earliest hydrothermal system was developed in the Jurassic and Cretaceous soon after deposition of the Franciscan. Temperatures were 170 °C to 200 °C and the reservoir fluid was isotope-shifted connate sea water ($\delta^{18}O = +3$ to $+7$) heated by deep burial in a normal geothermal gradient. Much later, Pliocene to recent igneous activity related to the Sonoma (2.9 to 5.3 Ma) and Clear Lake (0.01 to 2.1 Ma) volcanics again heated The Geysers reservoir to form a liquid-dominated geothermal system with temperatures of 220 °C to 320 °C and fluids of meteoric origin with $\delta^{18}O$ values of $-2.5$ to $-0.5$ affected by oxygen isotope shift. The fluid appears to have been isotopically lighter in the south-central Lakoma Fame area (Lambert's data) than in the north-central Sulphur Bank area (Sternfeld's data). This difference is interesting in light of the present variations in steam compositions discussed later. Alteration and vein minerals now found in The Geysers reservoir were mainly formed by these earlier liquid-dominated systems.

The last documented hydrothermal system corresponds to the modern Geysers vapor-dominated reservoir. Fluids changed little in $\delta^{18}O$ but temperatures (~250 °C) generally decreased. Although few hydrothermal minerals are deposited by steam and condensate, these fluids can dissolve and alter existing minerals. Calcite, common in upper zones, is rare within the vapor reservoir because $CO_2$-saturated condensate is acidic and dissolves carbonate as $HCO_3$, which is

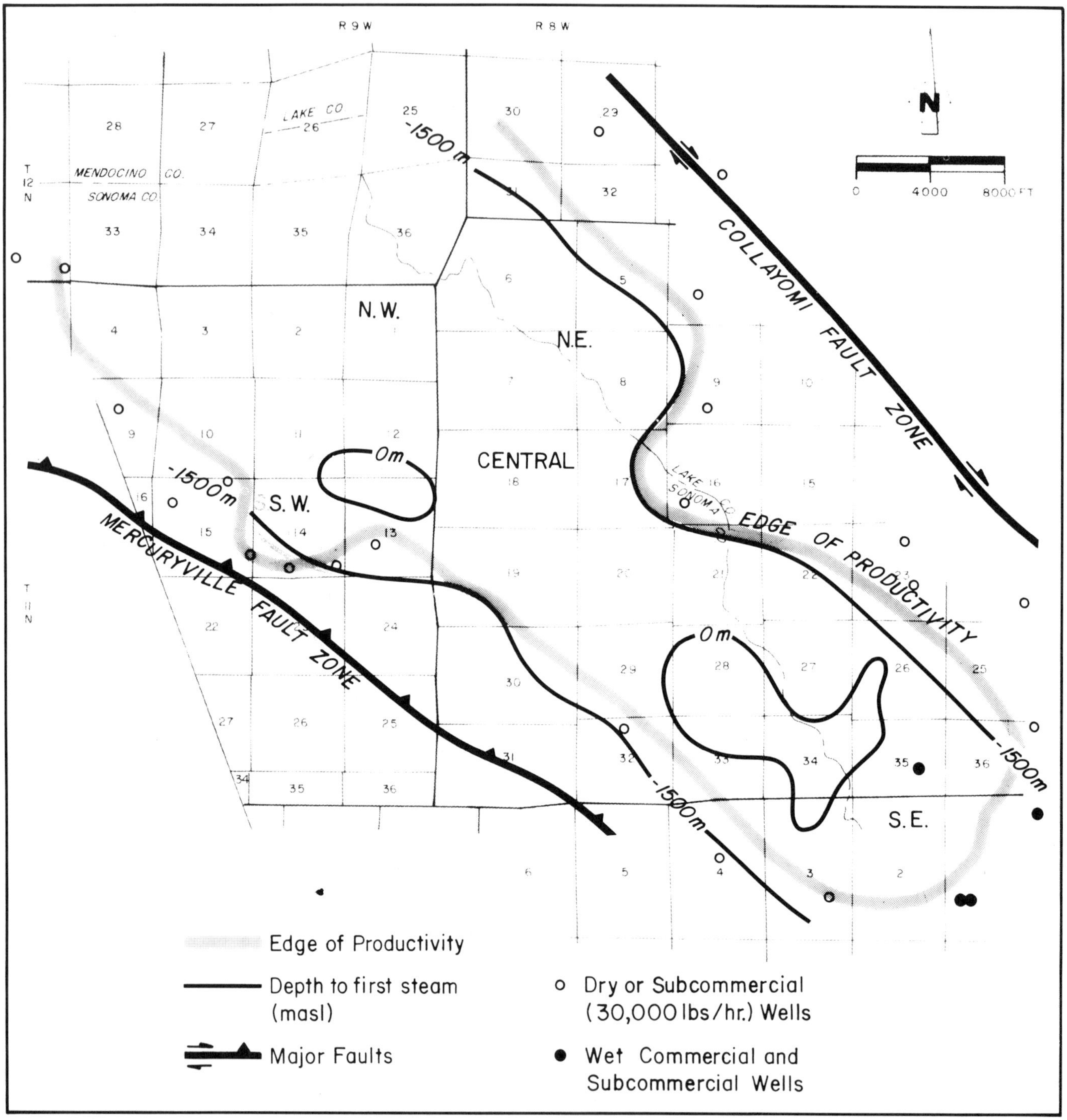


Figure 2. The areal extent of The Geysers reservoir as indicated by the maximum extent of productive (> 15 tons/hr) wells and the – 1500-m asl depth to first steam. Also shown are nonproductive wells and major faults located close to the field boundaries.

carried down to boiling zones where calcite is precipitated and $CO_2$ is regenerated. Similar leaching processes probably affect quartz and other silicates where freshly formed $CO_2$-charged condensate aggressively dissolves rock in zones of condensation. Since these zones are near the top and sides of the reservoir at the distal ends of steam-flow paths, mineral solution may be an important means of enlarging the reservoir. This process would also cause deposition of dissolved minerals in the deep boiling zones, which would rapidly decrease porosity and fluid flow if fluid conduits were not kept open by continuing tectonic movement.

## EXTENT OF THE GEYSERS RESERVOIR

Successful wells at The Geysers are contained in an irregular southeast–northwest-trending area about 18 km long and 3 to 7 km wide (Figure 2). The depth to steam varies from less than 200 m to more than 3000 m. The depth to the first occurrence of steam is in the public record and has been used by Thomas (1981) to divide the field into "'anomalies" (or highs) that appear to coincide in part with "pressure sinks" reported by Lipman et al. (1977). The productive steam field was suggested by McLaughlin (1981) to be limited to the "main graywacke" and bounded by the Mercuryville fault on the southwest and the Collayami fault on the northeast. In Figure 2 we indicate locations of unproductive wells (either low permeability or flooded with liquid water) along with depth to first steam (generalized from Thomas, 1981) and the Mercuryville and Collayami faults. The boundary of The Geysers reservoir is not yet determined to the northwest and other boundaries are not everywhere well located.

The close correspondence of productive (> 15 tons/hour) wells and first steam entries at depths less than − 1500 m asl (Figure 2) suggests that reservoir productivity decreases rapidly across near-vertical boundaries. This is also shown by the number of dry or subcommercial wells drilled just outside these boundaries to depths exceeding 3000 m (or greater than 2000 m below sea level) and by induced microearthquakes in the west-central production areas that extend below 3 km depth and have a sharp cutoff to the southwest (Eberhardt-Phillips and Oppenheimer, 1984).

The thickness of the producing reservoir is much less well known. The upper limit of the reservoir is known from the elevations of first steam entries as compiled by Thomas (1981), who also compiled the elevation of the top of the "main graywacke" reservoir rock (see discussion of geology). The lower limit is not known from drilling. No well has definitely penetrated below the reservoir and no deep water table has been detected. Most models of vapor-dominated systems suggest a boiling water-saturated zone under the vapor-dominated reservoir; such a zone has been detected chemically and by drilling at Larderello. There is no direct evidence of a boiling water table at The Geysers, but we know indirectly that boiling liquid zones must exist from the high heat flow produced by condensing steam and the low mass flow out of the system—the "heat pipe" effect. The best direct evidence of the nature of the deeper parts of The Geysers reservoir (below drilled depths) comes from induced microseismicity.

Induced seismicity is characteristic of the central and northeast Geysers (Bufe et al., 1981; Eberhart-Phillips and Oppenheimer, 1984; Oppenheimer, 1986). Detailed seismic study of The Geysers started in 1975, and the latest published data are for 1982; most of the following discussion is based on Eberhart-Phillips and Oppenheimer (1984).

In the central Geysers area (Figure 1), frequent (greater than 10 a day) microearthquakes occur in the reservoir at depths of 0 to 6 km (Figure 3). Microseismic activity also occurs in the southern part of The Geysers but with much lower frequency. The microearthquakes occur both continuously and in swarms and are related to production and not to injection of condensate. The area with the most microearthquakes was developed earliest (units 1–8 in the central Geysers) and includes the area of major natural thermal activity. In the central and southwestern Geysers the opening of new power plants was followed within a year by microseismicity that increased rapidly in frequency.

Although the clearly induced nature of the microseismicity is not encouraging for use as an exploration tool, it is very useful as an indication of the dimensions of the reservoir. Whatever the mechanism, the production of steam triggers the microseismicity so that the locations of the earthquakes indicate the locations of steam production. Using this as a guide, we see that steam is produced rather evenly from depths of 1 to 3 km beneath areas feeding units 9, 10, 14, and 15 and from depths of 0.2 to 6 km beneath the area feeding units 1–8 (Figure 3). Between and outside these zones of greatest seismicity are relatively aseismic zones that may correspond to areas of less production, as between units 1–8 and units 9 and 10, or may be less seismic for unknown reasons, as at unit 13. (Both very low and very high permeability may be associated with reduced microseismicity.) Although most seismicity is connected to exploited areas, increasing seismicity was observed after 1978 in previously aseismic zones to the north of exploited areas, suggesting lateral flow of steam from a distance. Pressure drops observed in areas drilled but not yet produced also indicate lateral drainage of steam to producing areas (sometimes a matter of economic concern at The Geysers).

The seismic zones are more extensive vertically than horizontally and show no concordant lower boundary or concentration at a particular depth that might indicate boiling from a condensate layer or a deep water table. Except in the units 1–8 area, the depth of the seismicity is comparable to the depth of the wells, with the shallowest seismicity corresponding to the depth of the first steam entries. In the units 1–8 area (the "Big" Geysers) seismicity is much deeper than drilled depths, which may be the result of much greater total production over a longer time period. The existence of a shallow reservoir in this area (Lipman et al., 1977) is also reflected in the microseismicity. The depth of seismicity appears to increase with time and production, although this is not well documented. Recent shallow microseismicity in the southwest Geysers may indicate production of steam from an overlying water-saturated condensate zone.

The Collayami and Mercuryville fault zones were suggested by McLaughlin (1981) to bound the reservoir on the northeast and southwest. This does not seem to be exactly true, although the reservoir boundaries are parallel to these faults (Figure 2) and may be controlled in part by associated structures. Near unit 15 the limit of productivity and the southwest edge of microseismicity coincide with the Mercuryville fault zone, but in other areas this fault zone is not a reservoir boundary. To the northeast of The Geysers field the induced seismicity appears to end southwest of the Collayami fault zone (FIgures 2, 3). This is consistent with the known limits of commercial production in this area and indicates that the Collayami fault is not the reservoir boundary, although other parallel faults may be.

## GASES IN GEYSERS STEAM

A small number of gas analyses have been reported for steam from The Geysers. Allen and Day (1927) reported that steam from six of the shallow steam wells drilled from 1921 to 1925 contained notably less $CO_2$ and more $H_2$ and $CH_2$ than steam samples from Lassen Park. The difference was probably due to the presence of far more organic matter in The Geysers sedimentary reservoir rock than in the all-volcanic Lassen system. Other than providing this observation, the gas analyses were not very useful to Allen and Day. This situation continued until the early 1980s with small numbers of analyses reported by Barnes et al. (1973), Weres et al. (1977), Nehring (1981), and Truesdell et al. (1981) without well locations and without significant interpretation.

The use of gas chemistry in geothermometry had been successfully applied to hot-water systems for many years (e.g., Hulston and McCabe, 1962), but gases from vapor-dominated systems did not appear to be in equilibrium. Some progress

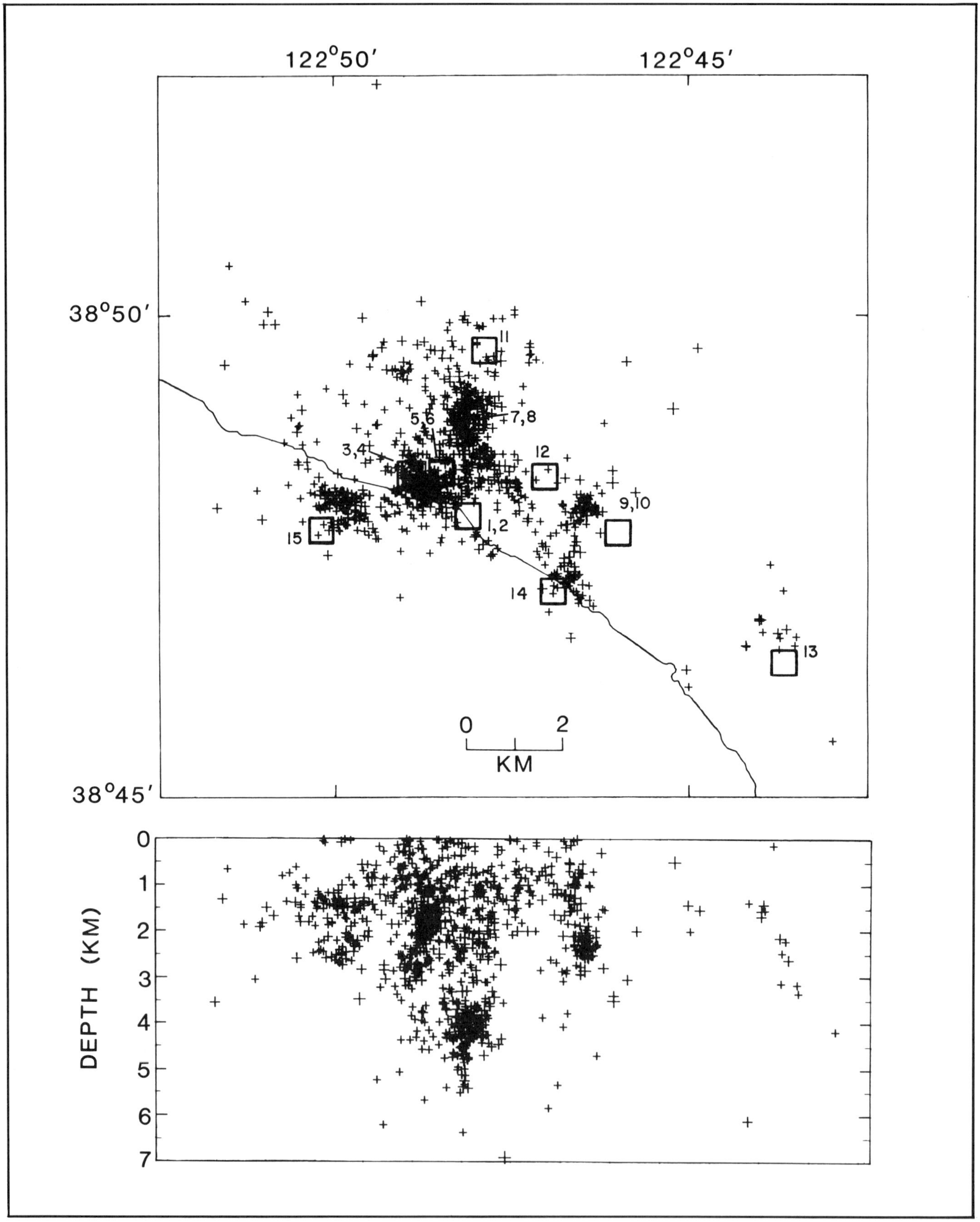


Figure 3. Locations of epicenters of microseisms measured from January through April 1981 in plan view and projected on an east-west section. Power plants operating in 1982 are also shown. (Modified from Eberhart-Phillips and Oppenheimer, 1984).

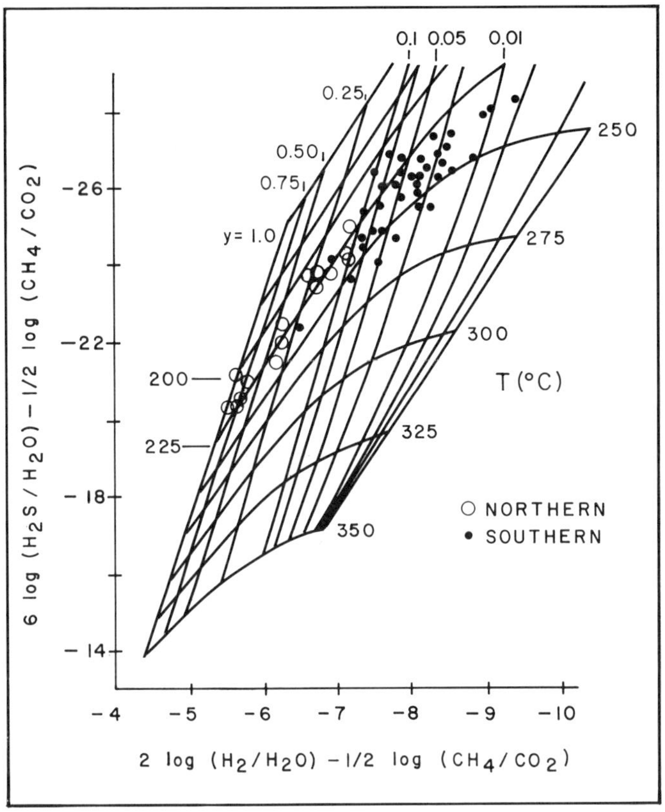


Figure 4. Reservoir temperatures and steam saturation for two areas of The Geysers calculated from gas geothermometers. Adapted from D'Amore and Truesdell (1985).

(more at Larderello than at The Geysers) was made by D'Amore and Truesdell (1980) in applying semi-empirical gas geothermometers, but understanding of The Geysers gas chemistry awaited recognition that wellhead steam was a mixture of reservoir vapor and vaporized reservoir liquid.

Gases at equilibrium in reservoir liquid and vapor will have quite different concentrations (according to solubilities) in each phase. Mixtures of vapor and vaporized liquid will, therefore, have gas concentrations that cannot be related to equilibrium in either phase unless the reservoir temperature or the fraction of vapor (or liquid) is known. Using a method that Giggenbach (1980) developed for hot-water systems, D'Amore et al. (1982) and D'Amore and Celati (1983) showed that by combining two gas equilibria with gas solubilities, both the reservoir temperature and the vapor fraction or effective reservoir steam saturation (called "y") could be calculated. D'Amore and Truesdell (1985) used this method to show that different parts of The Geysers differed greatly in y but relatively little in temperature (Figure 4). Southeast Geysers samples showed y values from 0.005 to 0.1, indicating that produced steam originated almost entirely from vaporized liquid, but samples from the central and southwest Geysers showed y values of 0.1 to 1.0, indicating a much larger contribution of reservoir vapor to produced steam.

The variation in y is directly related to the variation in gas/steam ratio. Representative gas analyses are given in Table 1 for low, medium, and high total gas/steam ratios from three areas of The Geysers.

## FIELDWIDE PATTERNS IN GASES AND ISOTOPES

Studies of the fieldwide gas and isotope chemistry of Larderello steam (Celati et al., 1973; D'Amore et al., 1977) show that, in general, gas concentrations (gas/steam, $NH_3$) increase from the center of the field toward the edge while concentrations of substances more soluble in liquid ($^{18}O$, $H_3BO_3$, Cl) decrease in this direction. These patterns were interpreted by D'Amore and Truesdell (1979) to result from lateral steam movement and condensation that could be modeled as a Rayleigh condensation process. In this process the concentration of each constituent (C) is related to its original concentration Co and the fraction of remaining steam M/Mo by the equation

$$(C/Co) = (M/Mo)^{(1/B-1)}$$

in which B is the distribution coefficient of the constituent between vapor and liquid ($C_v/C_l$). Limited data from The Geysers were presented that support this process. This model of convective circulation of steam and condensate was used to explain observed gradients in the depth to the liquid saturated zone at the margins of the Larderello field (Calore et al., 1982) and to indicate an extension of the field to the east (R. Celati, oral communication, 1981).

From the isotope data of Haizlip (1985), fragmentary published gas data, and many unpublished analyses, it is known that large concentration gradients of gas and isotope constituents exist at The Geysers. Understanding the natural processes which caused these gradients will advance our knowledge of the field and possibly assist exploitation, as it did at Larderello.

With the cooperation of some of the major steam producers (GEO Operator Co., Geysers Geothermal Co., Unocal, and Sante Fe Geothermal), maps have been drawn of gas/steam ratios (in ppm gas by weight) and $\delta^{18}O$ in steam from the entire drilled Geysers field (Figures 5, 6). These maps are highly generalized and represent different densities of data in various parts of the field. Early production data representative of original conditions were sought. Steam compositions indicating influence of injected water have been rejected as have analyses from unusually shallow (gas rich) and subcommercial (low flow rate) wells. The maps show variations only in gas/steam ratios and $\delta^{18}O$); other variations in isotopes are shown in Figure 7 and in gases in Table 1.

### Non-Rayleigh Gradients

Geysers steam shows ranges in gas and isotope compositions much larger than those observed at Larderello. The range in gas/steam is from about 150 ppmw (parts per million by weight) to more than 65,000 ppmw (Table 1); the published range at Larderello is from 10 to 120 liters STP/kg (approximately 17,000 to 200,000 ppmw). The range in $\delta^{18}O$ is also larger at The Geysers, −7 to +3 permil, compared with −6 to 0 at Larderello. The magnitude and direction of these composition gradients at The Geysers require a different interpretation.

The Rayleigh condensation model is consistent with composition gradients in the central and Serrazzano zones at Larderello (about two-thirds of the entire field). At The Geysers

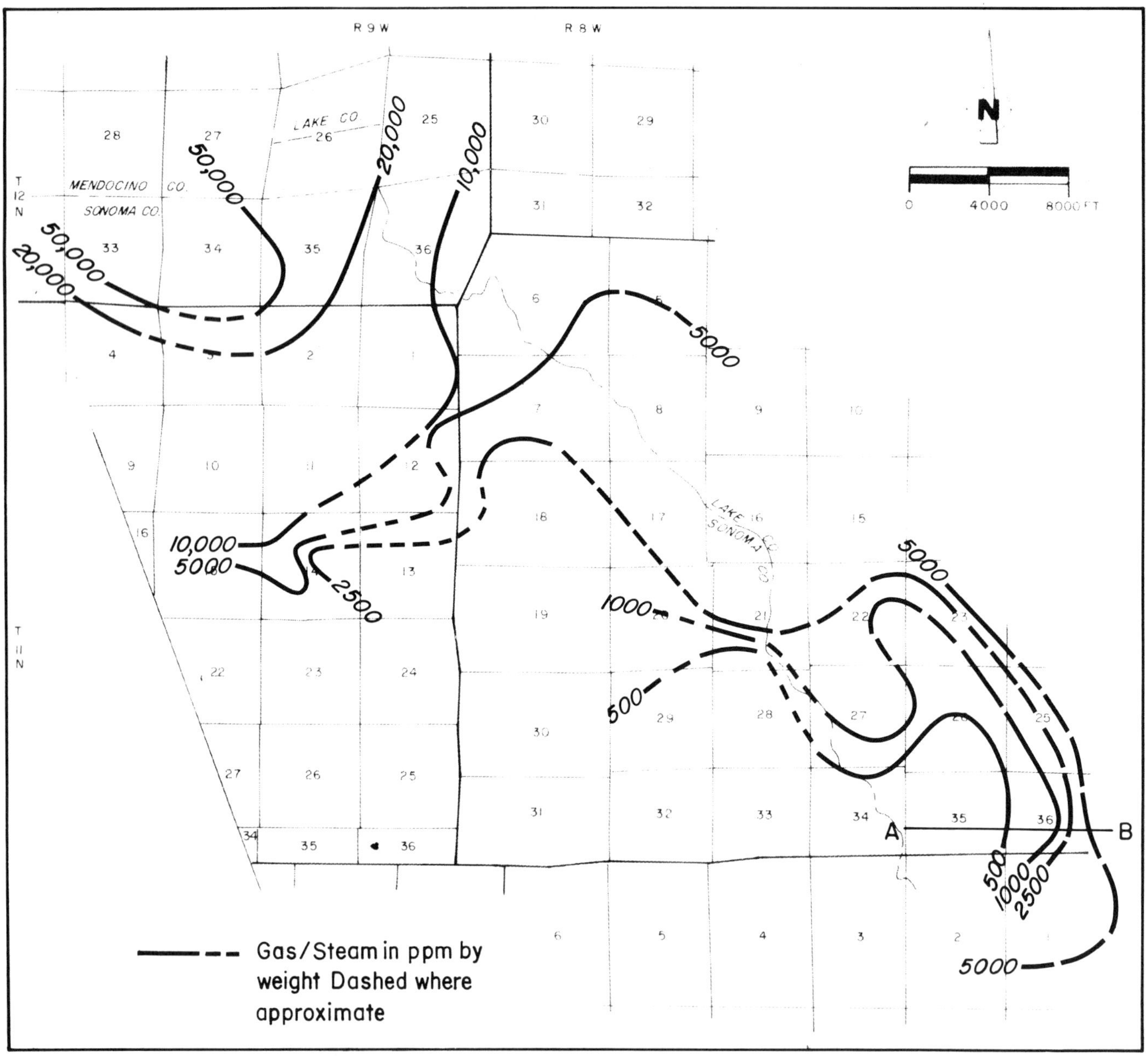


Figure 5. Gas/steam ratios for early production from wells at The Geysers, expressed as parts per million by weight. Data from the U.S. Geological Survey and steam-producing companies including GEO Operator, Geysers Geothermal, Unocal Geothermal, and Sante Fe Geothermal.

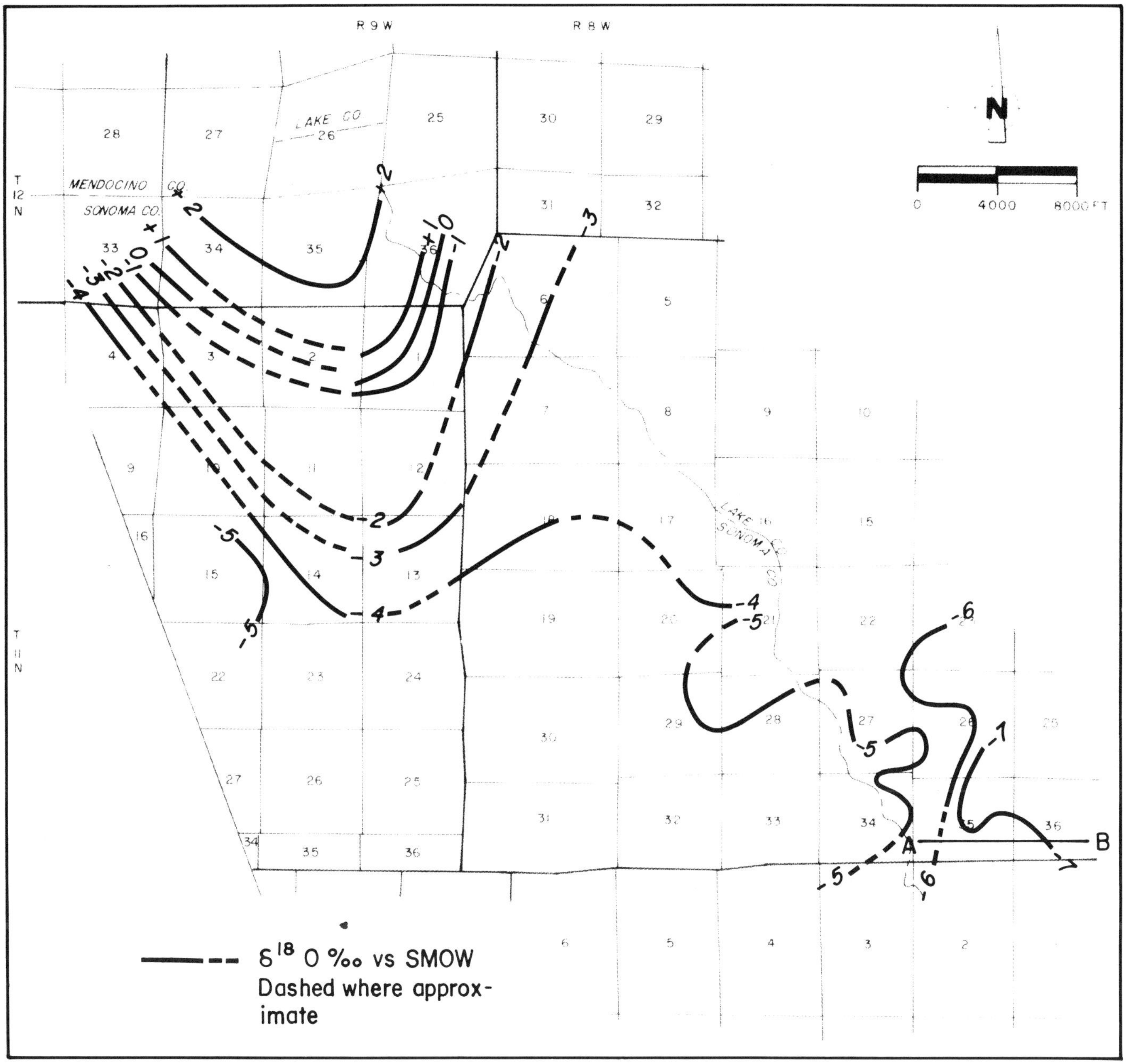


Figure 6. Oxygen-18 compositions of early production steam from The Geysers. Data sources as in Figure 5.

**Table 1. Representative gas analyses.**

| | High | | Medium | | Low | |
|---|---|---|---|---|---|---|
| | ppm (weight) | mole% w/o $H_2O$ | ppm (weight) | mole% w/o $H_2O$ | ppm (weight) | mole% w/o $H_2O$ |
| **Northwest Geysers** | | | | | | |
| Total gas | 65,223 | | 29,422 | | 8,410 | |
| $CO_2$ | 55,500 | 74.3 | 26,600 | 71.8 | 7,450 | 69.8 |
| $H_2S$ | 1,710 | 2.96 | 958 | 3.34 | 356 | 4.31 |
| $NH_3$ | 576 | 1.99 | 414 | 2.89 | 251 | 6.08 |
| Ar | -- | -- | .86 | .0026 | .11 | .0012 |
| $N_2$ | 560 | 1.18 | 270 | 1.15 | 48.8 | .717 |
| $CH_4$ | 2,580 | 9.49 | 1,020 | 7.56 | 161 | 4.13 |
| $H_2$ | 347 | 10.1 | 225 | 13.3 | 73.1 | 14.9 |
| S/G (mole) | | 30.7 | | 66.1 | | 227 |
| **Central and Southwest Geysers** | | | | | | |
| Total gas | 13,524 | | 5,422 | | 2,616 | |
| $CO_2$ | 11,500 | 62.5 | 4,460 | 58.5 | 2,080 | 53.6 |
| $H_2S$ | 662 | 4.65 | 310 | 5.26 | 184 | 6.12 |
| $NH_3$ | 223 | 3.13 | 101 | 3.41 | 151 | 10.0 |
| Ar | <2 | <0.01 | 2.0 | 0.03 | .7 | .02 |
| $N_2$ | 153 | 1.3 | 78.3 | 1.61 | 35.5 | 1.44 |
| $CH_4$ | 851 | 12.7 | 412 | 14.8 | 131 | 9.24 |
| $H_2$ | 133 | 15.8 | 57.3 | 16.4 | 34.8 | 19.5 |
| S/G (mole) | | 133 | | 185 | | 608 |
| **Southeast Geysers** | | | | | | |
| Total gas | 982 | | 443 | | 143 | |
| $CO_2$ | 734 | 51.9 | 322 | 51.1 | 94.7 | 43.2 |
| $H_2S$ | 116 | 10.6 | 75 | 15.3 | 36.4 | 21.4 |
| $NH_3$ | 30 | 5.46 | 34 | 13.9 | .3 | .37 |
| Ar | 1.0 | .08 | .03 | .006 | .01 | .0062 |
| $N_2$ | 46 | 5.17 | 3.3 | .82 | 2.5 | 1.79 |
| $CH_4$ | 43 | 8.33 | 3.4 | 1.54 | 6.4 | 8.06 |
| $H_2$ | 12 | 18.4 | 5.3 | 18.3 | 2.4 | 23.8 |
| S/G (mole) | | 1,724 | | 3,870 | | 11,100 |

this model appears to apply to the southern half of the field. Data are most complete for the southernmost end, where steam flow appears to be from west to east, as indicated by gradients in gases and isotopes. These gradients show increases to the east in total gas and $NH_3$ and decreases in $\delta^{18}O$ and B (Figure 8). Data supplied by Unocal for part of the south-central Geysers show the expected change in $CO_2$, although this could not be compared with changes in other constituents (D'Amore and Truesdell, 1979). This model does not work, however, for the central and northwest Geysers. The general increase of *both* gas and oxygen-18 in steam from the south-central Geysers toward the northwest is inconsistent with the condensation model, which requires changes in oxygen-18 and gas concentrations to be opposite. In addition, the magnitude of the observed changes, if caused by condensation, would require more than 99.8% condensation (based on gas/steam) in the northern part of the reservoir, which seems highly unlikely. Isotope and gas gradients are less definite for the south-central Geysers, and data toward the field boundaries are missing or fragmentary. Faced with explaining the oxygen-18 data, Haizlip (1985) rejected processes involving the fluid alone (boiling, condensation, etc.) and concluded that only a south-to-north increase in the amount of remnant connate water or in the rock/water ratio could explain the isotope gradient.

The isotopic similarity of steam from the northwest Geysers with connate or metamorphic waters from the Sulphur Bank mine and Wilbur Spring, 25 and 40 km northwest of The Geysers, was noted by Haizlip (1985). These mineralized thermal waters were described by White et al. (1972) as originating from the metamorphism of connate water (original $\delta^{18}O$ and $\delta D = 0$) contained in Jurassic marine sediments, with $\delta^{18}O$ values increased to $+4$ to $+5$ by exchange with silicates and carbonates and $\delta D$ reduced to $-20$ to $-30$ by exchange with low-deuterium marine clays. The Sulphur Bank and Wilbur Spring waters are similar in $\delta^{18}O$ to the connate(?) fluids described by Lambert (1976) and Sternfeld (1981), as once present in the central Geysers, and fall approximately along the $\delta^{18}O$ trend of northwest Geysers steam (Figure 7).

Gas compositions are less distinctive. No analyses of Wilbur Spring gases are available, but Sulphur Bank mine gas with mainly $CO_2$, $CH_4$, $N_2$, and $H_2S$ (Nehring, 1981). This gas, collected at the surface, may have lost $H_2S$ and $H_2$, but the high $CH_4$ contents and light $\delta^{13}C$ in $CO_2$ are similar to Geysers samples ($-14$ for Sulphur Bank; about $-11$ to $-16$ for The Geysers; Huebner, 1981, and unpublished data).

## Flushing by Meteoric Recharge

The major gradients in gas and isotope composition of steam are almost certainly due to meteoric recharge entering the reservoir from the southeast. Wells drilled at the southeast

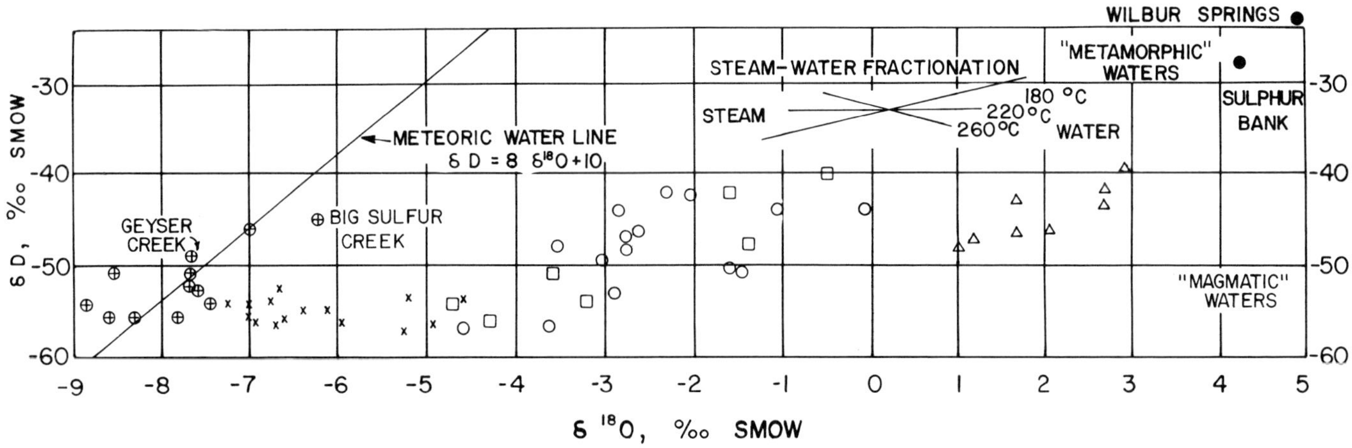


Figure 7. Oxygen-18 and deuterium compositions of early production steam from The Geysers (Data sources as in Figure 5). Symbols: X southeast; □ central; o west; Δ northwest; o meteoric waters; • connate or metamorphic waters.

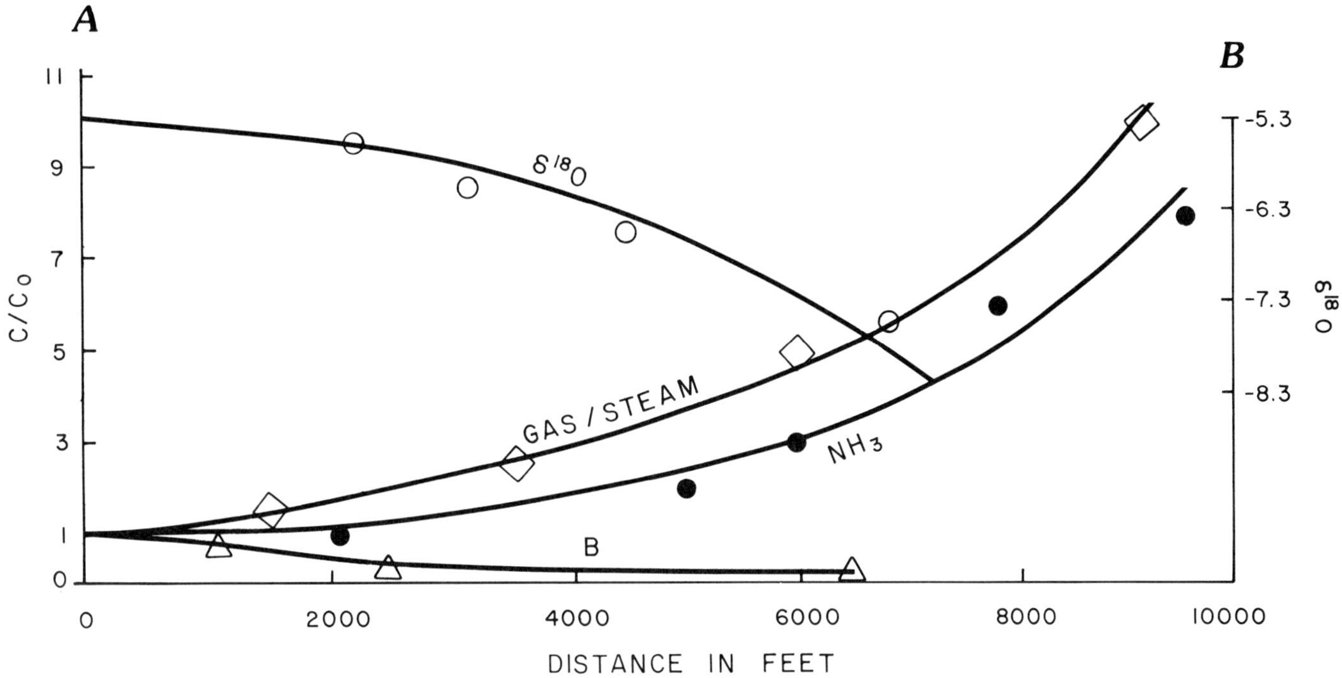


Figure 8. Gradients in steam composition across part of the southern Geysers. Locations of cross section line A-B shown in Figures 5 and 6.

and east edges of The Geysers field are known to produce steam-water mixtures or water alone from the same depth as the steam reservoir a few hundred meters away. The similarity between isotopic compositions of local meteoric water and steam from the southeast Geysers (Figure 7) together with the low gas contents of steam in this area suggest that this part of the field was the site of massive recharge to the system. Recharge adjacent to outcrops of the reservoir rock has been observed at Larderello, with increases in tritium and decreases in gas and $\delta^{18}O$ in steam observed over periods of a few years (Calore et al., 1982; D'Amore et al., 1983). The lack of tritium in the steam at the southeast Geysers suggests that rechard is not a rapid process related to present exploitation (as at Larderello), but has occurred slowly over a longer period or from older waters (> 60 years).

Low gas content, $\delta^{18}O$ and y values, and high steam flows in the southern Geysers all suggest that a large proportion of the reservoir fluid consists of liquid water. This liquid water must be held in small fractures and pores of the reservoir rock coexisting with vapor in larger openings, because neither drilling nor seismic observations favor the existence of a large, well-defined water-saturated zone within the productive volume of the reservoir. In the northern Geysers there is also no evidence of a well-defined water-saturated zone even to the 5–6-km maximum depths of induced seismicity in the oldest producing zone. In this part of The Geysers, high gas, $\delta^{18}O$ and y, and lower productivity suggest much lower liquid water contents in the productive reservoir.

## Boiloff, Then Flushing

The apparent lack of a deep water-saturated zone in any part of The Geysers suggests that the system has undergone a nearly complete boiloff of its original liquid. This is in agreement with geologic, petrologic, and simulation studies, which all suggest that The Geysers started as a hot-water saturated reservoir and became vapor-dominated through a deficit of fluid recharge to discharge (see for example, White et al., 1971; Lambert, 1976; Preuss, 1985). This process does not necessarily tend to a stable steady state; the vapor-dominated zone can extend downward as long as their is liquid to displace. It is possible that a seismic event or other disturbance could allow liquid to enter the underpressured vapor-dominated reservoir from the side where adequate permeability exists. If liquid water entered a nearly dry vapor-dominated reservoir, its effect would depend on its quantity and the depth of entry. Moderate quantities and shallow entry would favor recharge of liquid suspended in the reservoir, while large quantities and deep entry could produce a liquid saturated zone that might be large enough to reverse the process of vapor domination. In either case, with time the liquid would move throughout the reservoir. The movement of suspended liquid would depend on vaporization and condensation and would be relatively slow unless hastened by removal (production) of steam as observed with recent recharge at Larderello. Venting of steam and resulting lateral pressure gradients must be important in causing lateral movement of liquid away from areas of recharge. The relative locations of recharge areas and vents have apparently influenced the flow of recharge at The Geysers.

## Gradients Related to Vents

Although surface alteration is found throughout The Geysers, the greatest concentration is in the area of the Big Geysers (units 1–8), along with the largest natural vents. The distribution of vent areas may explain the smaller gradients in $^{18}O$ and gas between the southeast and central Geysers and the steepening gradients to the northwest. Driven by meteoric recharge from the southeast, gases and high $\delta^{18}O$ steam would have migrated gradually to these vents and left the system. Strong venting in the central Geysers (the Big and Little Geysers areas) would have allowed extensive flushing and homogenization of fluid compositions between this area and the southeast Geysers, while less venting to the northwest would have allowed less flushing and homogenization.

If recharge waters entered the system from the southeast, reservoir rocks in this area would have undergone the greatest alteration of their oxygen isotopes to reach equilibrium with fresh (unaltered) meteoric water. Reservoir rocks further to the northwest would have been progressively less altered as oxygen-18 in recharge waters increased through exchange with rock or mixture with connate formation waters. The influx of meteoric recharge would also be expected to leach minerals with which it was not in equilibrium. Sulfide minerals and organic matter would be oxidized by air-saturated meteoric recharge; calcite would be dissolved through the refluxing action of steam condensation and downward flow of steam condensate. The existence of Rayleigh condensation gradients at the southern margin of The Geysers (Figures 5, 6, 8) that are counter to gradients resulting from flushing suggests that flushing occurred and was complete before the Rayleigh gradients were established. Exploration of this chronology may be possible through petrologic studies.

With continued exploitation of The Geysers, reservoir pressure differentials between the reservoir and surrounding ground water will increase and, as at Larderello, renewed recharge may occur and tritium may appear. For this reason the use of tritium as an artificial tracer should be avoided.

# CONCLUSION

The Geysers geothermal field is large and geologically and geochemically complex, with a history that is only starting to be understood. Natural circulation at The Geysers is locally dominated by the "heat pipe" mechanism, in which upward-flowing steam condenses and produces downward-flowing condensate driven by volcanic heat supplied to yet undiscovered boiling zones. Larger scale circulation patterns based on lateral steam flow away from boiling zones (and condensate back-flow), indicated by "Rayleigh" composition patterns, appear to dominate at Larderello and in parts of The Geysers. Large-scale circulation at The Geysers appears to have been dominated by flushing with meteoric water probably initiated by catastrophic breakthrough of liquid at the southeastern margin of the system. In this process recharge has progressively removed oxygen-18 and gases (including gas-forming minerals) from the reservoir, with effects diminishing from the southeast, where steam is almost gas free and similar in $\delta^{18}O$ to local meteoric water, to the northwest, where steam is gas- and oxygen-18 rich. The exact roles of residual connate waters and high rock/water ratios in the original fluid compositions flushed by meteoric recharge are not yet known. The association of high permeability with highly flushed fluids suggests that the flushing has had physical as well as chemical effects on the reservoir or that flushing was facilitated by existing high permeability.

# ACKNOWLEDGMENTS

We wish to thank the GEO Operator Co., Geysers Geothermal Co., Unocal Geothermal Co., and Sante Fe Geothermal Co. for the use of data on steam compositions. We wish to thank our colleagues, especially Don White, Dave Oppenheimer, and Franco D'Amore for stimulating discussions, and Cathy Janik and Bob Fournier for careful reviews. Steve Mason, Barbara Bernegger and Mary Ann Mikus were indispensable to the production of this paper.

# REFERENCES

Allen, E.T., and A.L. Day, 1927, Steam wells and other thermal activity at The Geysers, California: Carnegie Institution of Washington Publication 378, p. 106.

Barnes, Ivan, M.E. Hinkle, J.B. Rapp,, C. Heropoulos, and W.W. Vaughn, 1973, Chemical composition of naturally occurring fluids in relation to mercury deposits in part of north-central California: U.S. Geological Survey Bulletin 1382-A, 19 p.

Brook, C.A., 1981, Variability and sources of hydrocarbon sulfide and other gases in steam at The Geysers, *in* R.J. McLaughlin, and J.M. Donnelly-Nolan, eds., Research in the Geysers-Clear Lake geothermal area, northern California: U.S. Geological Survey Professional Paper 1141, p. 193-203.

Bufe, C.G., S.M. Marks, F.W. Lester, R.S. Ludwin, and M.C. Stickney, 1981, Seismicity of The Geysers-Clear Lake region, *in* R.J. McLaughlin and J.M. Donnelly-Nolan, eds., Research in the Geysers-Clear Lake geothermal area, northern California; U.S. Geological Survey Professional Paper 1141, p. 129-137.

Calore, C., R. Celati, F. D'Amore, and P. Noto, 1982, Geochemical evidence of natural recharge in the Larderello and Castelnuovo areas: Proceedings, 8th Workshop on Geothermal Reservoir Engineering, Stanford, p. 323-328.

Celati, R., P. Noto, C. Panichi, P. Squarci, and L. Taffi, 1973, Interactions between the steam reservoir and surrounding aquifers in the Larderello geothermal field: Geothermics, v. 2, n.3-4, p. 174-185.

D'Amore, F., and C. Celati, 1983, Methodology for calculating steam quality in geothermal reservoirs: Geothermics, v. 12, n. 2-3, p. 129-140.

D'Amore, F., and A.H. Truesdell, 1979, Models for steam chemistry at Larderello and The Geysers: Proceedings, 5th Workshop on Geothermal Reservoir Engineering, p. 283-297.

D'Amore, F., and A.H. Truesdell, 1980, Gas thermometry for drillhole fluids from vapor dominated and hot-water geothermal fields: Proceedings, 6th Workshop on Geothermal Reservoir Engineering, Stanford, p. 351-360.

D'Amore, F., and A.H. Truesdell, 1985, Calculation of geothermal reservoir temperatures and steam fraction from gas compositions: 1985 International Symposium on Geotherm Energy, Geothermal Resources Council Transactions, v. 9, pt. 1, p. 305-310.

D'Amore, F., R. Celati, and C. Calore, 1982, Fluid geochemistry applications in reservoir engineering (vapor-dominated systems): Proceedings, 8th Workshop on Geothermal Reservoir Engineering, Stanford, p. 295-308.

D'Amore, F., R. Celati, (?) Ferrara, and C. Panichi, 1977, Secondary changes in the chemical and isotopic composition of the geothermal fluids in Larderello field: Geothermics, v. 5, p. 153-163.

D'Amore, F., R. Celati, C. Calore, and R. Bertrami, 1983, Effects of natural recharge on gas composition in the Larderello-Castelnuovo area: Proceedings, 9th Workshop on Geothermal Reservoir Engineering, Stanford, p. 221-227.

Eberhart-Phillips, D., and D.H. Oppenheimer, 1984, Induced seismicity in The Geysers geothermal area, California: Journal of Geophysical Research, v. 89, n. B-2, p. 1191-1207.

Giggenbach, W.F., 1980, Geothermal gas equilibrium: Geochimica et Cosmochimica Acta, v. 44, p. 2021-2032.

Haizlip, J.R., 1985, Stable isotope composition of steam from wells in the northwest Geysers, The Geysers, Sonoma County, California: 1985 International Symposium on Geothermal Energy, Geothermal Resources Council Transactions, v. 9, pt. 1, p. 311-317.

Huebner, M., 1981, Carbon-13 isotope values for carbon dioxide gas and dissolved carbon species in springs and wells in The Geysers-Clear Lake region, *in* R.J. McLaughlin and J.M. Donnelly-Nolan, eds., Research in the Geysers-Clear Lake geothermal area, northern California: U.S. Geological Survey Professional Paper 1141, p. 211-213.

Hulston, J.R., and W.J. McCabe, 1962, Mass spectrometer measurements in the thermal areas of New Zealand, Part 1, Carbonate dioxide and residual gas analyses: Geochimica et Cosmochimica Acta, v. 26, p. 383-397.

Lambert, S.J., 1976, Stable isotope studies of some active geothermal systems: Pasadena, California Institute of Technology unpublished Ph.D. dissertation, p. 387.

Lipman, S.C., C.J. Strobel, and M.S. Gulati, 1977, Reservoir performance of The Geysers field, *in* Proceedings of the Larderello workshop on geothermal resource assessment and reservoir engineering: Geothermics, v. 7, p. 209-219.

McLaughlin, R.J., 1978, Preliminary geologic map and structural sections of the central Mayacmas Mountains and the Geysers steam field, Sonoma, Lake, and Mendocino Counties, California: U.S. Geological Survey Open-File Map 78-389, scale 1:24,000.

McLaughlin, R.J., 1981, Tectonic setting of pre-Tertiary rocks and its relation to geothermal resources in The Geysers-Clear Lake area, Northern California: U.S. Geological Survey Professional Paper 1141, p. 3-23.

Nehring, N.L., 1981, Gases from springs and wells in The Geysers-Clear Lake area, *in* R.J. McLaughlin, and J.M. Donnelly-Nolan, eds., Research in the Geysers-Clear Lake geothermal area, northern California: U.S. Geological Survey Professional Paper 1141, p. 205-209.

Oppenheimer, D.H., 1986, Extensional tectonics at The Geysers geothermal area, California: Journal of Geophysical Research (in press).

Preuss, K., 1985, A quantitative model of vapor-dominated geothermal reservoirs as heat pipes in fractured porous rock: Geothermal Resources Council Transactions, v. 9, pt. II, p. 353-359.

Sondergeld, C.H., and D.L. Turcotte, 1977, An experimental study of two-phase convection with applications to geological problems: Journal of Geophysical Research, v. 82, p. 2045-2053.

Sternfeld, J.N., 1981, The hydrothermal petrology and stable isotope geochemistry of two wells in The Geysers geothermal field, Sonoma County, California: University of California, Riverside Report UCR/IGPP 81/7, M.S. thesis, 202 p.

Thomas, R.P., 1981, Subsurface geology, *in* A reservoir assessment of The Geysers: California Department of Conservation, Division of Oil and Gas Report TR-27, p. 9-20.

Truesdell, A.H., and D.E. White, 1973, Production of superheated steam from vapor-dominated geothermal systems: Geothermics 2, v. 3-4, p. 154-173.

Truesdell, A.H., M. Nathenson, and G.A. Frye, 1981, Downhole measurements and fluid chemistry of a Castle Rock steam well, The Geysers, Lake County, California: Geothermics, v. 10, n. 2, p. 103-114.

Urban, T.C., W.H. Diment, J.H. Sass, and I.M. Jamieson, 1976, Heat flow at The Geysers, California: Proceedings, Second United Nations Symposium on the Development and Use of Geothermal Resources, San Francisco, 1975, v. 2, p. 1241-1245.

Weres, O., K. Tsao, and B. Wood, 1977, Resource technology and environment at The Geysers: Lawrence Berkeley Laboratory Report LBL-5231.

White, D.E., L.J.P. Muffler, and A.H. Truesdell, 1971, Vapor-dominated hydrothermal systems compared with hot-water systems; Economic Geology, v. 66, n. 1, p. 75-97.

White, D.E., I. Barnes, and J.R. O'Neil, 1972, Thermal and mineral waters of nonmeteoric origin, California Coast Ranges: GSA Bulletin, v. 84, p. 547-560.

CHAPTER 42

# JAPAN'S GEOTHERMAL RESOURCES

**Yuzaburo Yoshimura**
*Japan Petroleum Exploration Co., Ltd.*
*Tokyo, Japan*

ABSTRACT

For Japan, which has a scarcity of energy resources, its geothermal areas are deemed to be an important indigenous energy source. Japan, with an abundance of younger volcanoes, has a high potential for geothermal energy. However, though 20 years have elapsed since the first geothermal power plant was constructed at Matsukawa, the development of geothermal energy resources has been slow, with the present total output at only 213.7MW. At present, there are nine geothermal plants in Japan. The locations are shown in Figure 1, and the details of each are described in Table 1. The outputs of the Suginoi and Kirishima Kokusai Hotels are small because the two hotels constructed the facilities for private use only. Since the construction of the Mori geothermal plant in 1982, there have been no major geothermal power plants constructed.

## PRESENT STATUS OF GEOTHERMAL DEVELOPMENT

Following the oil crises, many private enterprises took up projects to develop geothermal energy. Several areas have already been confirmed to produce some steam, and feasibility studies for commercial production are being carried out. They are located mainly in the Tohoku district (northern part of mainland) and Kyushu Island areas, as shown on Figure 2 and described below.

### Tohoku District

**Sumikawa Area**

The Sumikawa area, about 1.5 km west of Ohnuma, a geothermal power plant, being explored by Mitsubishi Metal Company, Ltd. Four exploratory wells ranging from 450 to 1550 m deep were drilled in this area, and all were successful. The maximum temperature is 300 °C, and the potential of this area is assessed to be more than 100 MW. Two to three exploratory wells are scheduled to be drilled during this fiscal year, and a power plant of 50MW is planned for the first stage of development.

**Kakkonda No. 2 Unit**

Kakkonda No. 2 Unit is in an area adjacent to the Kakkonda No. 1 geothermal facilities and is also being developed by Japan Metals and Chemical Company, Ltd. Plans for a 50MW plant were drawn in 1985 and approvals have subsequently been granted. Thus far, two production wells have been drilled. The average depth of the wells is 1500 m and the average temperatures is 240 °C.

**Uenotai**

Exploratory wells 600 to 1500 m deep have been drilled by the Dowa Mining Company. The maximum temperature is 300 °C, and a 25MW power plant is planned for the first stage of development.

**Nishiyama**

Following the successful result of a promotion survey (Figure 3) by the New Energy Development Organization (NEDO), Okuaizu Geothermal Company is exploring the area. Exploration wells 700- to 2000-m-deep have been drilled, and the maximum temperature is 310 °C. The potential of the area is assessed to be 50MW.

### Kyushu Island

**Takigami**

Idemitsu Geothermal Company is exploring the area and has drilled four exploratory wells ranging from 2000 to 3000 m deep. Several exploratory wells are scheduled to be drilled during this fiscal year. The potential of this area is expected to be 25MW to 50MW.

**Hachobaru No. 2**

In the adjacent area to the Hachobaru No. 1 power plant, Kyushu Electric Power Company, Ltd. is developing its No. 2 plant. Several exploratory wells have been drilled and have already confirmed half the amount of steam required to operate a 55 MW power plant.

**Kirishima**

Nippon Steel Corporation and Nittetsu Mining Company are jointly exploring the area. Seven exploratory wells ranging

Table 1. Status of geothermal power stations in Japan.

| Name of Power Station | Capacity MW | Conversion Type | Start-Up Date | National Park |
|---|---|---|---|---|
| MATSUKAWA | 22.0 | DS | 1966 OCT | In Side |
| ŌTAKE | 12.5 | SF | 1967 AUG | In Side |
| ŌNUMA | 8.6 | SF | 1974 JUN | In Side |
| ONIKOBE | 12.5 | SF | 1975 MAR | In Side |
| HATSUCHŌBARU | 55.0 | DF | 1977 JUN | In Side |
| KAKKONDA | 50.0 | SF | 1978 MAY | In Side |
| SUGINOI-HOTEL | 3.0 | SF | 1981 AUG | In Side |
| MORI | 50.0 | DF | 1982 NOV | Out Side |
| KIRISHIMA KOKUSAI-HOTEL | 0.1 | SF | 1984 FEB | In Side |
| TOTAL | 213.7 MW | | | |
| | Note: DS=Dry Steam. SF=Single Flush. DF=Doubl Flush. | | | |

from 1000 to 2000 m deep have been drilled. The potential of the area is assessed to be 50MW.

**Fushime**

This area is located in the southernmost part of Kyushu Island. Japan Petroleum Exploration Company, Ltd. is exploring the area. Thirteen exploration wells have been drilled in the area, and the maximum temperature was more than 370 °C. The potential of the northern part of the area is assessed to be 30MW.

### Summary

Totaling the above, there is a possibility that plants capable of generating 300MW to 350MW can be constructed within several years.

One of the obstructions facing geothermal development is National Parks. Eight percent of Japanese land is either national parks or quasinational parks, and most of the prospective geothermal areas are located in these parks. All the existing geothermal power plants, except one, are constructed and are operating within such parks. The Environmental Agency approved six locations which had been surveyed 15 years ago, but they have not approved any areas within the parks for development since then.

## GOVERNMENTAL PROJECTS FOR GEOTHERMAL DEVELOPMENT

The Japanese government established NEDO in 1980. This agency is carrying out a number of projects to promote the development of geothermal energy.

### Nationwide Survey of Geothermal Resources

To promote the future development of geothermal energy, a nationwide survey was carried out in order to delineate pro-

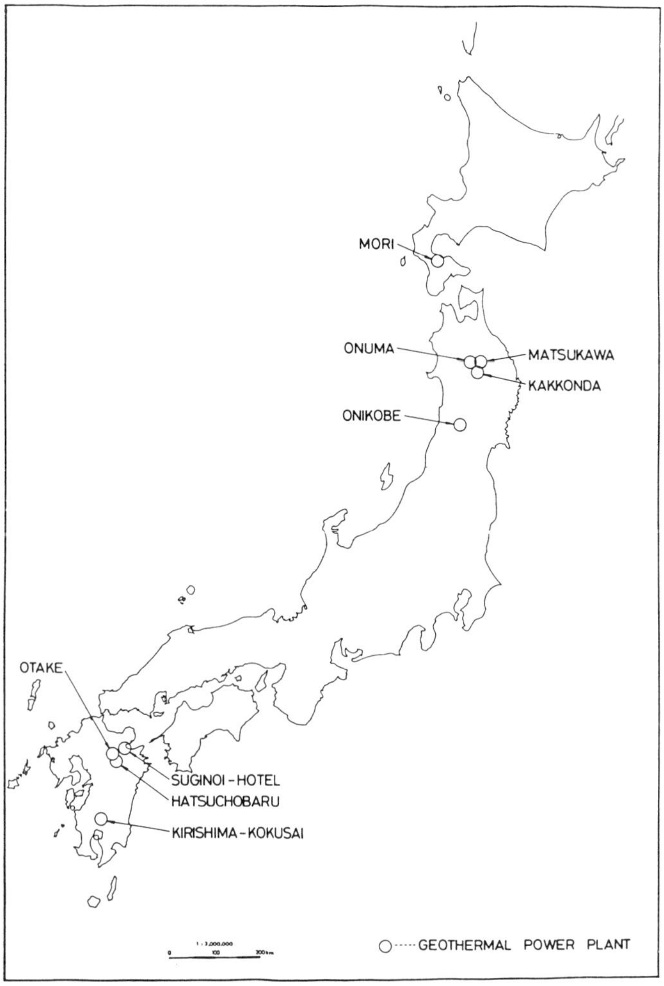


Figure 1. Geothermal power plants in Japan.

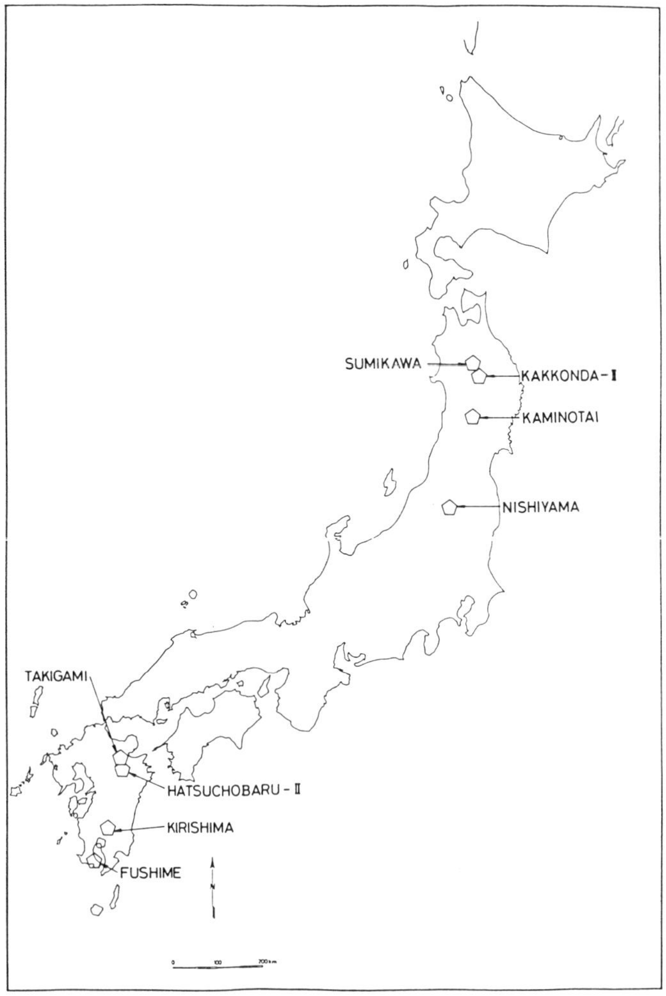


Figure 2. Geothermal areas being studied for feasibility of commercial production.

spective areas. The first phase of the survey was performed for 4 years during the period from the latter half of 1980 through to the middle of 1984. To implement the nationwide survey in this limited period, the following three surveys were performed.

## Airborne Radar Surveys

By using Synthetic Aperture Radar, 36,300 km of lines covering the whole country were surveyed.

## Airborne Magnetic Surveys

About 150,000 km of lines covering the whole country were surveyed.

## Gravity Surveys

The data obtained from these surveys were compiled with existing data that were acquired during exploration for petroleum and metals. This resulted in a coverage of approximately 50% of the area of Japan.

## Results of Surveys

"The Map of Prospective Geothermal Areas" was made by a comprehensive interpretation of the lineament distribution map (from airborne radar surveys), Curie point depth map (from airborne magnetic surveys), the gravity basement map (from gravity surveys), and existing data on geology, geophysics, wells, and hot springs. The map is drawn on a scale of 1:200,000 shows the areas broken down into six classifications—areas of three types of hydrothermal convection systems (related to young igneous intrusion), areas of deep hot-water sources, and those of two types showing possible high temperatures. The map also shows Curie point depth contours, locations of hot springs and their temperature, geological information on the basement, Paleogene and Neogene sedimentary basins, and distribution of Quaternary volcanic rocks.

The survey is now in its second phase, which commenced in 1984 as a 3-year project. The first phase had mainly to do with techniques of remote sensing, whereas the second phase is applying detailed surface surveys including geology, geophysical surveys, and heat-flow surveys. The subject survey areas had been screened down to four areas out of the prospective areas selected during the first phase. The results to be

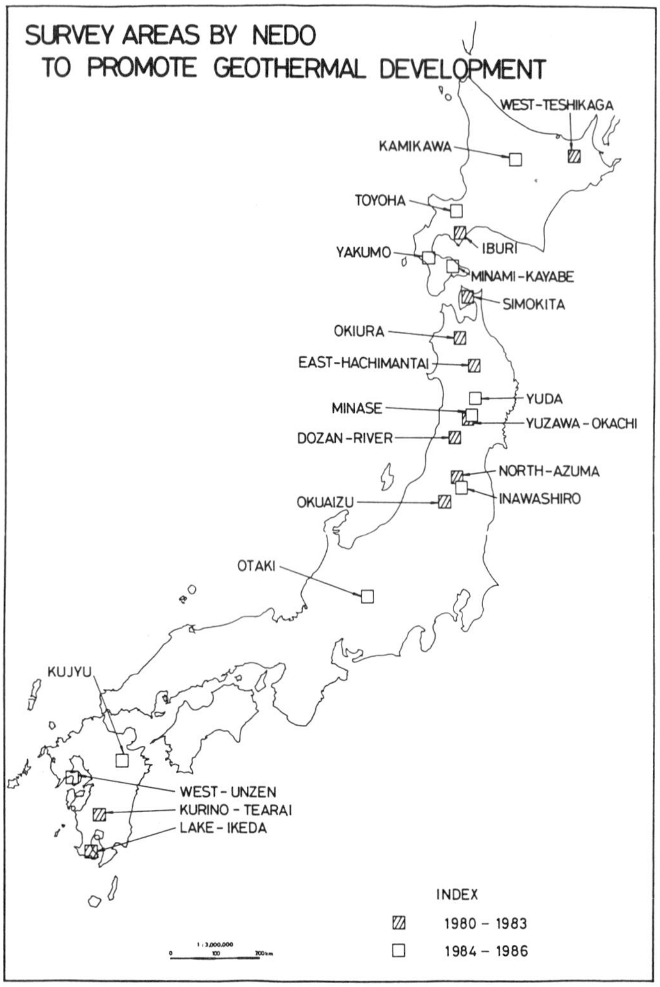


Figure 3. Areas surveyed by NEDO to promote geothermal development.

obtained from the surveys during the second phase hopefully will provide us with guidelines to make the proper use of the map of prospective geothermal areas of development of geothermal resources in the future.

## SURVEYS TO PROMOTE GEOTHERMAL DEVELOPMENT

There are still prospective geothermal areas that lack data and are of high risk, where private enterprises have not undertaken surveys. NEDO is taking a leading role to implement the surveys of these remaining areas. The aim is to confirm the prospectiveness of these areas through the surveys, followed by guiding private enterprises to promote the development of geothermal resources.

This survey was started in 1980, and already 12 areas have been surveyed. At present, nine areas are being surveyed. Three prospective areas are selected every year, and each area is studied more than 3 years. The studies include geology, geochemistry, geophysical surveys, heat-flow surveys using 400- to 500-m wells, and seven exploratory wells with a depth of 1000 to 1500 m. Evaluation of each area is made by interpreting the above results comprehensively. Under the present situation where the economics of geothermal projects are critical owing to the drastic drop of oil prices, these surveys are very important to encourage private enterprise.

Besides implementing these surveys, NEDO is promoting studies for the development of technologies including geothermal hot-water supply systems, extraction of geothermal energy from hot dry rock (which includes cooperation in projects being conducted in Fenton Hill in United States), surveys to confirm geothermal exploration technology, and development of geothermal reservoir evaluation technology. Furthermore, studies for the utilization of geothermal resources not only for power generation but also for multipurpose applications are being actively undertaken, especially in the northern part of the country.

# GEOTHERMAL ENERGY ABSTRACTS

SUMITRAMIHARDJA, ARIFIN, Pertamina, Jakarta, Indonesia, and D. ROBERT and KASAP IBRAHIM, Beicip, Rueil-Malmaison, France

## Geothermal Resource Development for Electric Power Generation in Indonesia: Results and Future Promises

Indonesia is one of the largest developing countries in southeast Asia; therefore, energy demand tends to increase continuously. Fortunately, large amounts of energy resource potentials are available, among which is energy from geothermal resources. Some of these energy resources comprise exportable commodities such as oil, natural gas, and coal; others are for domestic consumption such as hydrothermal and geothermal energy.

During the next several years the Indonesian government intends to acceler ate development of nonexportable energies used to generate electrical power in order to save exportable energies that can bring foreign currencies. Therefore, geothermal has become a priority goal. Moreover, this type of energy is of particular interest because Indonesia has a large geothermal energy potential related to the Circum-Pacific volcanic belts. These geothermal manifestations are spread throughout almost the entire archipelago, except the island of Kalimantan.

Geothermal exploration in Indonesia began in 1929 when some shallow wells were drilled in Kamojang, West Java. Actual exploration for geothermal energy to generate electricity commenced in 1972. Preliminary reconnaissance surveys were made by the Volcanological Survey of Indonesia.

In 1982, the state oil company, Pertamina, was placed in charge of exploration and development activities for geothermal energy in different fields, either by its own activities or in the form of joint-operation contracts with foreign companies. In addition, the state electrical company, PLN, is responsible for installing a power plant to generate and distribute electricity.

Presently, several projects are at different stages of maturity. Some fields are in an exploration stage, and others are already developed. Geothermal fields have mainly been developed in Java, e.g., Kamojang has a 30-MW power plant, in use since February 1982, that is being upgraded to 140-MW capacity, scheduled for completion in 1987. Dieng is being developed for 55-110 MW, and Gunung Salak for 110 MW.

Fields in the exploration stages are Darajat, Banten, Cisolok in West Java, and another in Lahendong, Menado, North Sulawesi. Exploration studies are in progress in 15 fields among the 77 promising geothermal areas found throughout Indonesia.

In this paper, important data about exploration studies, brief summaries about development histories, and actual status of developing fields will be presented.

VASQUEZ, N. C., and Z. F. SARMIENTO, PNOC-Energy Development Corporation, Manila, Philippines

## Reservoir Engineering Applications for Development and Exploitation of Geothermal Fields in the Philippines

After a geothermal well is completed, several tests and downhole measurements are conducted to help evaluate the subsurface fluid and reservoir properties intersected. From these tests, a conceptual model of the well can be developed by integrating data from the various parts of the field. This paper presents the completion techniques applied in geothermal wells, as well as the role of reservoir engineering science in delineating a field for development. Monitoring techniques and other reservoir engineering aspects of a field under exploitation are also discussed, with examples from the Philippines.

# VOLCANOLOGY

CHAPTER 43

# COST-TO-BENEFIT RATIO OF VOLCANIC FORECASTING IN MEXICO: LOOKING FOR RELIABLE BUT INEXPENSIVE MONITORING TECHNIQUES

**Servando De la Cruz-Reyna**
*Instituto de Geofisica*
UNAM
*Ciudad Universitaria, Mexico*

ABSTRACT

Mexico is a country with a significant level of volcanic activity. Nearly 100 eruptions have been produced from 14 volcanoes in historical times. The types of eruptions have ranged from discrete lava emissions to destructive explosions, including the birth of four volcanoes. This activity represents a sample of the Quaternary volcanism which has produced a few thousand cones across the country. A simple analysis of these data yields an average of 19 eruptions per century, from which about one-third are of explosive character (VEI $\geq$ 3). The most destructive activity of the current century is that of El Chichón in 1982. The Plinian eruptions of March 28 and April 3 and 4 caused ashfall, pyroclastic flows, and surges. These eruptions left 1750 dead and 26,000 homeless and devastated 400 $km^2$ of land. Banana and cocoa crops were damaged over an area of 1000 $km^2$ and several thousand cattle were lost. A prior investment of about 0.1% of the material losses in an efficient volcano warning system could have reduced damage to a minimum. This experience must be used in the development of monitoring systems which can work within the country realistically. Two main problems have to be solved: the geographical scatter of volcanic sites (for instance, El Chichón and Colima are 1200 km apart, and Tacaná and Tres Vírgenes are 2500 km apart) and the low proportion of geoscience centers (only two in the country). The lack of local centers with trained scientists makes it necessary to look for ad-hoc solutions such as long-distance telemetry to a central observatory and back-up by developing local groups.

## INTRODUCTION

The active volcanoes of Mexico are widespread (Figure 1). Volcanic activity in the form of geothermal manifestations can be found as far northwest as Cerro Prie to near the Mexico–U.S. border at Mexicali, and as far southeast as the recent volcanic earthquakes and new fumarole in Tacaná Volcano at the Mexico-Guatemala border. Roughly halfway between these sites is the Mexican Volcanic Belt (MVB), which contains thousands of Tertiary and Quaternary cones. Some are as small as Chapultepec, rising 45 m over the very center of downtown Mexico City; and some are as large as Citlaltépetl (also known as Pico de Orizaba), a huge stratovolcano 5700 m above sea level, the highest elevation in Mexico. Outside the volcanic belt are other important volcanoes like those in the Revillagigedo Islands, one of which, Bárcena, was born in 1952. Tres Vírgenes is located near the middle of Baja California Peninsula, and of course El Chichón, the formerly inconspicuous volcano, is located in the middle of the "volcanic gap" between the southeastern end of the MVB and the northwestern end of the Central American volcanic chain. Several of these volcanoes have shown a variety of types of activity in historical times, ranging from fumarolic or gentle effusions to destructive, explosive eruptions. During the last two millenia more than 100 eruptions have occurred, at least 30 of which were strong enough to damage human settlements and land. At least 13 others have been reported which produced different levels of damage.

Although the eruption rate is not as frequent as in other parts of the world, the volcanic risk is scattered over a large area and there is no responsible official body specifically budgeted for volcano monitoring and risk evaluation. This has forced University research programs to try to cover such tasks. Under their limited capabilities in budget, manpower,

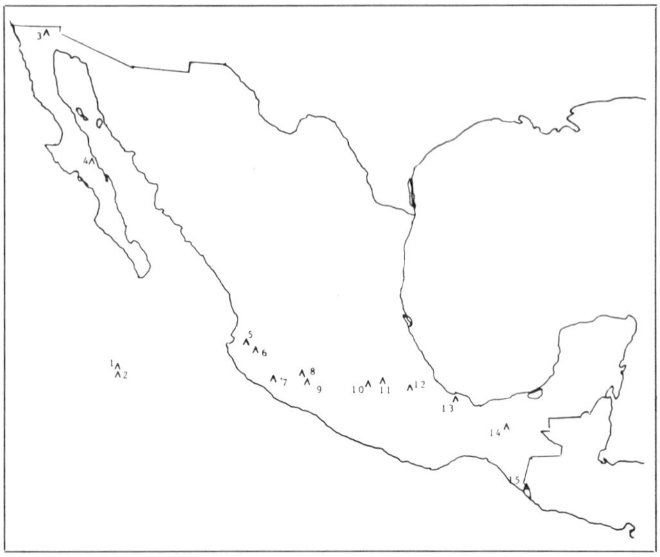

Figure 1. Index map showing some of the recently active volcanoes and geothermal fields of México. (1) Bárcena (2) Socorro (3) Cerro Prieto (4) Tres Vírgenes (5) Sanganguey (6) Seboruco (7) Colima (8) Paricutín (9) Jorullo (10) Xitle (11) Popocatépetl (12) Citlaltépetl (13) San Martín Tuxtle (14) El Chichón (15) Tacaná.

and logistics, University programs must look for affordable monitoring and risk evaluation techniques that can provide at least a minimum of coverage for preparedness and disaster prevention.

## EL CHICHÓN AND TACANÁ: TWO DIFFERENT CASES

On March 23, 1982, a phone call from a (now non-existent) Mexican federal government office dealing with urban emergencies informed the Institute of Geophysics (UNAM) that rumbles and earthquakes were reported by the people living near El Chichón volcano, and that a team of scientists was needed to make an investigation. Could such a group be ready to travel to Chiapas on March 25, when a helicopter would be available there, to visit the rather inaccessible areas around the volcano? A field crew of five persons with three portable drum recording seismographs was prepared to go. On March 25 another call asked the Institute to change the field trip to March 29 because of helicopter problems. On March 28 at about 11:30 p.m. a phreato-magmatic explosion initiated the strongest episode of eruptive activity in more than 6 centuries, and all attempts to reach the volcano area some 700 km southeast of Mexico City were frustrated by the closing of all airports in southeastern Mexico. Only with the help of PEMEX was it possible to arrive at the distressed region, by combining short flights with road travel. In the evening of March 30 a headquarters was set up in Villahermosa, capital city of the state of Tabasco, 80 km due north of the volcano, which was covered by a fine (2 to 5 mm) layer of clear gray ash. On March 31 the seismic stations started recording within a radius of 20 km around the north side of the volcano. During the following days more portable seismographs from the Institutes of Geophysics and Engineering (UNAM) were set up around and closer to the volcano. Data from these instruments were supplemented with the a posteriori analysis of information recorded at the Chicoasen Network, a telemetric seismic network operating since 1980 to study induced seismicity caused by impounding dams in a complex 50 km south of El Chichón. The combined data from all these sources provided information about the seismic evolution of the volcano up to 3 months before the eruptions. These results are reported in Havskov et al. (1983).

The most destructive eruptions occurred one week after the first, on April 3 and 4, 1982. Witnessing them and learning later of the damage they caused was most depressing. During the field work, we were in close contact with the evacuation efforts and were aware that some towns and villages on the south side of volcano had no means of communication. El Chichón volcano and the characteristics of the 1982 eruptions are described in Luhr, Carmichael, and Varekamp (1984), Varekamp, Luhr, and Pretegaard (1984), Rose et al. (1984), Sigurdsson, Carey, and Espíndola (1984), Tilling et al. (1984), and De la Cruz-Reyna et al. (1985). The social and psychological damage done on the Zoque, a Mayan-related ethnic group which has inhabited the area for centuries, has not yet been evaluated. It seems quite likely that the Zoque underwent the effects of at least one of the previous eruptions, as suggested by the semantic content of the Zoque name of El Chichón, "PIOWACHUWE," "old women who burns."

The experience of this disaster is rooted deep in the minds of citizens and authorities in the states of Chiapas and Tabasco. After the emergency was over, there were very frequent calls from the governments of these states asking for evaluations of the mildest seismic activity reported anywhere in their territories. This was a hard job because Chiapas has the highest level of seismicity in the country. But recently, in December 1985 and January 1986, significant rumbling and trembling were reported from some towns around Tacaná volcano, a 4050-m-high stratovolcano located on the border between Mexico and Guatemala. The analysis of seismic records obtained with portable seismographs showed that the activity had two sources: one originated in a fault system about 25 km northeast of Tacaná, and the other from directly below the volcano itself. When the state authorities were informed of this situation, a civil protection plan was developed in a short time, and now officials at all levels and citizens have instructions on what to do if an emergency arises, and the relocation of some remote settlements is under way. As a matter of fact, on May 8, 1986, a phreatic explosion opened a small crater in the northwestern flank of the volcano, and a change from yellow to orange alert was issued. However, the government of Chiapas, as with many other states, makes decisions about population mobilization on the recommendations of the scientific group. And the scientific group has to base its recommendations on the readings and recordings it obtains from available monitoring instruments. And so the main problem is how a small research group with little resources can handle the task of monitoring volcanoes spread over 3000 km.

## MONITORING: ARE THERE CHOICES?

Theoretically, the proper solution to such a problem would be to develop strong groups in local universities or research centers or government offices, so they could be in charge of volcanic observatories to monitor the status of volcanoes. In practice, however, and in a country the size of Mexico such a task may take 20 years to implement, assuming the funding was available. And presently there are 14 high-risk volcanoes, and an average population of tens of thousands is within the hazard range of each volcano. In such situations not many choices are left. For some volcanoes, like Tacaná and Colima, limited deformation studies have been started. Radon detectors have been implanted at Ceboruco, Colima, Tacaná, and San Andrés. To prevent using all of the budget just for travel expenses, the only short-term solution to this problem seems to be long-distance telemetry to a central observatory. Local groups as, for example, the University of Colima's electronics and solid state physics group, is helping by providing monitoring services. The University of Colima has no Earth Science division.

Summarizing, a feasible first step for the monitoring of at least nine of the higher risk volcanoes in Mexico seems to be the development of long-distance telemetry lines transmitting a minimum of seismic and deformation information to a processing center in Mexico City, with the back-up of local groups (presently Colima and Tapachula). Actually, a cooperative UNAM-CICESE (Ensenada) program is under way to produce the equipment (except transducers and data processing) required for this project. Since one of the main objectives of this plan is the prevention of large eruptions, special emphasis is being placed on radon monitoring. This inexpensive technique has showed good results in detecting crustal stress changes, provided that external factors are properly filtered out (e.g., see De la Cruz-Reyna et al., 1985; Segovia et al., 1986).

## REFERENCES CITED

De la Cruz-Reyna, S., M. Mena, N. Segovia, J.F. Chalot, J.L. Seidel, and M. Monnin., 1985, Radon emanometry in soil gases and activity in ashes from El Chichón volcano: PAGEOPH, v. 123: 407-421.

Havskov, J., S. De la Cruz-Reyna, S.K. Singh, F. Medina, and C. Gutierrez, 1983, Seismic activity related to the March-April 1982 eruptions of El Chichón volcano, Chiapas, Mexico: Geophysical Research Letters, v. 10: 293-296.

Luhr, J.F., I. S.E. Carmichael, and J.C. Varekamp, 1984, The 1982 eruptions of El Chichón volcano, Chiapas, Mexico: Mineralogy and petrology of the anhydrite-bearing pumices: J. Volcanol. Geotherm. Res., 23, 69-108.

Rose, W.I., T.J. Bomhorst, S.P. Halsor, W.A. Capul, C. Plumley, S. De la Cruz-Reyna, M. Mena, and R. Mota, 1984, Volcán El Chichón, Mexico: Pre-1982 S-rich eruptive activity: J. Volcanol. Geotherm. Res., v. 23: 147-167.

Segovia, N., S. De la Cruz-Reyna, M. Mena, M. Romero, J.L. Seidel, M. Monnin, E. Malavassi, E. Barquero, J. Fernández, G. Avila, R. Vander Laat, L. Ponce, and G. Juárez, 1986, Radon variations in active volcanoes and in regions with high seismicity: Internal and External Factors: Nucl. Tracks and Rad. Meas. (in press).

Sigurdsson, H., S.N. Carey, and J.M. Espíndola, 1984, The 1982 eruptions of El Chichón volcano, Mexico: Stratigraphy of pyroclastic deposits. J. Volcanol. Geotherm. Res., v. 23: 11-37.

Tilling, R.I., M. Rubin, H. Sigurdsson, S. Carey, W.A. Duffield, W.I. Rose, 1984, Holocene eruptive activity of El Chichón volcano. Chiapas, Mexico: Science, v. 247, 747-749.

Varekamp, J.C., J.F. Luhr, and K.L. Pretegaard, 1984, The 1982 eruptions of El Chichón volcano (Chiapas, Mexico): Character of the eruptions, ash-fall deposits, and gas phase. J. Volcanol. Geotherm. Res., v. 23, 39-68.

CHAPTER 44

# MINERALIZATION ASSOCIATED WITH VOLCANIC-RELATED HYDROTHERMAL SYSTEMS IN THE CIRCUM-PACIFIC BASIN

**Jeffrey W. Hedenquist**
*Geothermal Research Center, D.S.I.R.,*
*Wairakei, Taupo, New Zealand*

## ABSTRACT

Epithermal ore deposits hosted by volcanic rocks and formed by volcanic-related hydrothermal systems are becoming increasingly important gold producers in Circum-Pacific countries, with many recent discoveries. One such type of epithermal deposit, termed low sulfidation, is characterized by reduced sulfur. This type of deposit formed in an environment analogous to active geothermal systems, and it is now recognized that in the low sulfidation system, gold precipitates from boiling fluids focused into zones of high permeability. Exploring for this type of deposit can be greatly aided by detailed mapping of alteration. The other group of gold deposits reviewed here is characterized by relatively oxidized sulfur, termed high sulfidation. These types of deposits remain somewhat enigmatic, because they are poorly studied. However, indications are that their formation is closely related to an acid magmatic fluid, with mineralization therefore more intimately associated with volcanism than is the case for the low sulfidation type. Further exploration for this style of mineralization will be greatly assisted through integrated geologic and geochemical (particularly paragenetic) studies of this type of deposit and of active volcanic systems.

## INTRODUCTION

This paper proposes a framework which will hopefully lead to a better understanding of hydrothermal gold mineralization hosted by volcanic rocks and intimately associated with volcanism. This approach needs to be based on a rigorous characterization of the chemistry of the hydrothermal ore fluids. One group of deposits reviewed here has a deduced paleochemistry similar to geothermal systems, and as such is relatively well understood. The sulfur present in the mineralizing fluids is in the reduced state, and these are termed "low sulfidation" systems; sulfide is the principal ligand for gold transport. Another group is more enigmatic, though there are strong indications that it is more intimately related to volcanic rather than geothermal fluids. Sulfur is relatively oxidized, the fluids are acid, and they are referred to as "high sulfidation" systems; under these conditions, chloride may also be an important ligand for gold transport.

There is much potential for further discovery of both types of deposits in the Circum-Pacific; however, an inadequate understanding of both the controls on mineralization and the geologic environment of emplacement of the high sulfidation systems hampers exploration. In this chapter, I will briefly discuss the distinctive characteristics of both low and high sulfidation gold deposits. I will then attempt to identify the problems hindering the same high confidence of understanding of the high sulfidation systems that we have with the low sulfidation systems. Along the line of the analogous active systems, used to such advantage in the study of low sulfidation deposits, I will review the chemistry of volcanic fluids, largely from Japan and New Zealand, and their interaction with the rock matrix. This approach will be invaluable to eventually understanding the controls on the high sulfidation style of mineralization, and to the development of exploration criteria.

### Background

A dramatic increase in the exploration for volcanic-hosted ore deposits, particularly of gold, has led to a wealth of discoveries since the early 1970s in the Circum-Pacific (Figure 1). A majority of these deposits formed in the near-surface, or epithermal, portion of hydrothermal systems; essential components of these convective systems are magmatic heat sources intimately associated with subduction-related volcanism. Most discovered epithermal deposits formed in the Tertiary or Recent; however, new Permo-Devonian discoveries in Australia indicate that preservation of the volcanic host is the only necessary aspect in retaining epithermal mineralization. Examples of these recent epithermal discoveries, some with spectacular gold grades or tonnages, include Hishikari (Japan), Lihir Island and Porgera (Papua New Guinea), Kelian (Indonesia), Babelthuap (Palau), Temora (Australia), Golden Cross and Martha Hill (New Zealand), El Indio (Chile), Round

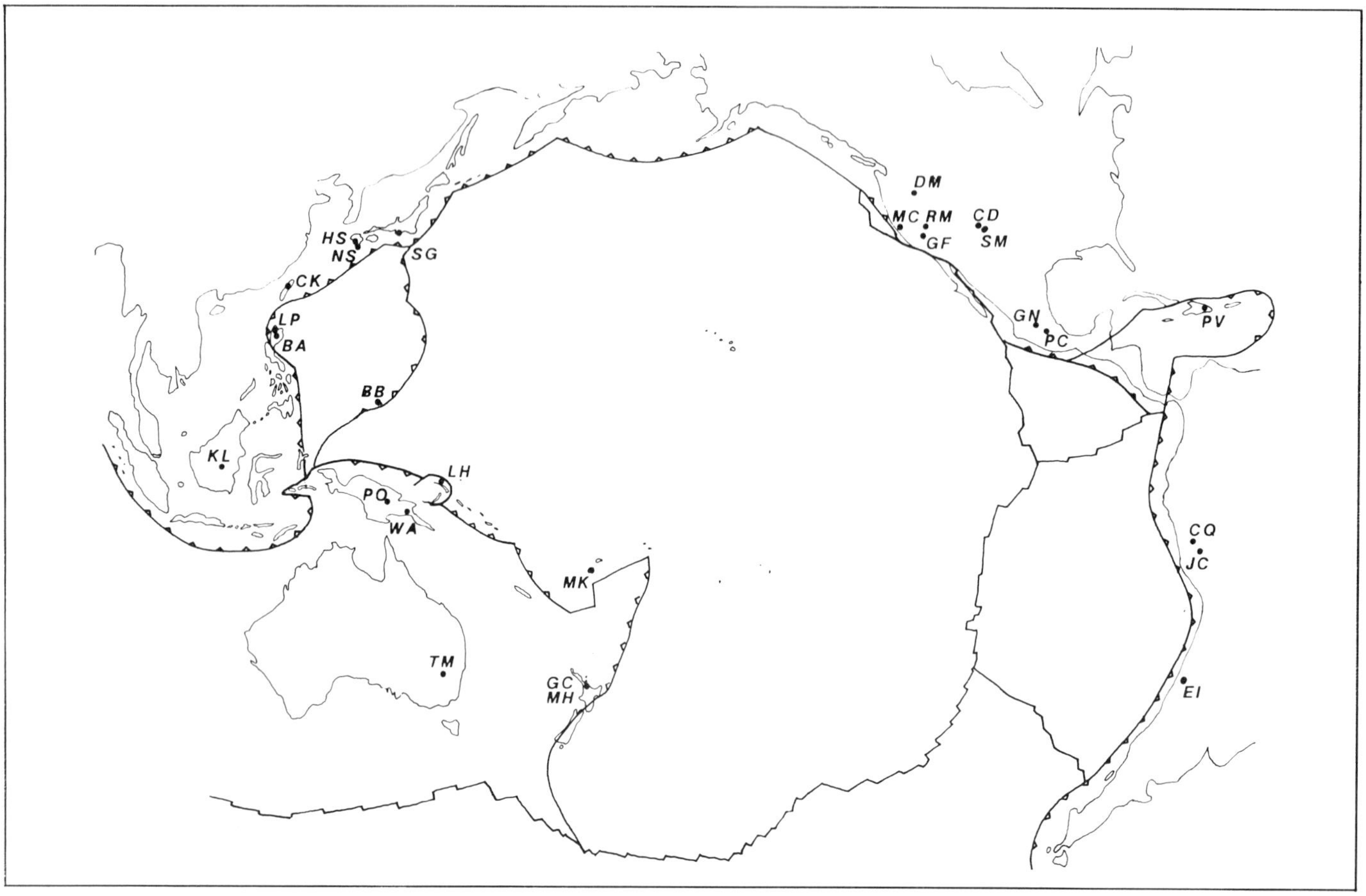


Figure 1. Tectonic map of the Circum-Pacific basin, showing ocean ridges, transforms and lines of subduction (teeth on upper plate). Also shown are the locations of some of the epithermal deposits discussed in the text; Round Mountain (RM), McLaughlin (MC), Goldfield (GF), Creede (CD), Delamar (DM), Summitville (SM), Pueblo Viejo (PV), Pachuca (PC), Guanajuato (GN), Julcani (JC), Colqui (CQ), El Indio (EI), Golden Cross (GC), Martha Hill (MH), Mt. Kasi (MK), Temora (TM), Lihir (LH), Wau (WA), Porgera (PO), Kelian (KL), Babelthuap (BB), Baguio (BA), Lepanto (LP), Chinkuashih (CK), Nansatsu-type (NS), Hishikari (HS), Seigoshi (SG).

Mountain (Nevada), and McLaughlin (California). Some of these deposits (e.g., Round Mountain and Martha Hill) are developments on historic workings, with previous subeconomic mineralization becoming ore with the rise in gold prices. Discoveries such as Lihir, Kelian (Ferguson, 1986) and Babelthuap (Miller and Rytuba, 1986) reflect the high level of exploration in the southwest Pacific.

Many detailed studies on the characteristics of epithermal deposits have already been completed, and these characteristics have been compiled by several reviewers (Sillitoe, 1977, 1981; Buchanan, 1981; Berger and Eimon, 1983; Giles and Nelson, 1984; Hedenquist and Reid, 1985; Hayba et al., 1986). There are great differences in the characteristics of these deposits; this largely reflects the variety of physical controls on fluid flow near the surface (e.g., topography, rock permeability, fracture properties, etc.).

Research on geothermal systems has led to a recognition of the equivalence of the low sulfidation epithermal environment to that which can be observed in active systems (White, 1955, 1981; Henley and Ellis, 1983). Application of this analogy has provided a quantitative understanding of the processes related to gold transport and deposition in epithermal systems (Hedenquist and Henley, 1985a; Henley, 1986), thus improving our ability to search for such deposits. Therefore, the thread which is now available to tie together the apparent wide variety of low sulfidation deposits is based on the identification of the chemistry of the ore fluid through alteration mineralogy and fluid inclusion studies. The fluid chemistry controls the type and quantity of metals which are transported, determines the processes which will cause mineralization, and also to a large extent controls the location of the mineralization; characteristics of the host rocks, except as they affect fluid flow, are secondary to mineralization.

In contrast, the alteration mineralogy of another group of volcanic-hosted gold (and copper) deposits, formed in the epithermal environment, indicates that the fluids responsible for mineralization may have been intimately related to volcanism; they were very acid, variably saline, and had a significant portion of the sulfur present as sulfate. Where the host volcanics and mineralization have been dated, they are of the same age within analytical uncertainty. The type example often used (Bethke, 1984) is Goldfield (Nevada); other similar deposits include Summitville (Colorado), Chinkuashih (Taiwan), Lepanto (Philippines), Frieda River (Papua New Guinea), and Mt. Kasi (Fiji). Their possible relationship to porphyry copper mineralization (particularly of the high gold type) and associated alteration has been noted (Sillitoe, 1983; Sillitoe and Gappe, 1984). El Indio and Lihir, mentioned earlier, have some of the mineralogical characteristics of this group, though evidence (see below) suggests that these were overprinted by the fluids mainly responsible for gold mineralization. The Nansatsu-type deposits of Kyushu, Japan (e.g.,

TABLE 1—Categories and examples of typical epithermal precious metal deposits hosted by volcanics in the Circum-Pacific.

**Gold-silver deposits**

Veins, stockworks, and disseminations in intermediate to silicic volcanic rocks

| Low sulfidation | High sulfidation | Intermediate characteristics and/or overprinted |
|---|---|---|
| Golden Cross and Martha Hill, New Zealand | Goldfield, Nevada | El Indio, Chile |
| Round Mountain, Nevada | Summitville, Colorado | Nansatsu-type (Iwato, Akeshi, Kasuga), Japan |
| McLaughlin, California | Chinkuashih, Taiwan | Pueblo Viejo, Dominican Republic |
| Oatman, Arizona | Lepanto, Philippines | Temora, Australia |
| Baguio District, Philippines | Mt. Kasi, Fiji | Lihir and Porgera, Papua New Guinea |
| Golden Plateau, Australia | Frieda River, Papua New Guinea | |
| Kushikino, Hishikari, and Seigoshi, Japan | | |
| Babelthuap, Palau | | |
| Kelian, Indonesia | | |
| Wau, Papua New Guinea | | |
| Vanua Levu, Fiji | | |

**Silver-gold deposits (commonly base-metal-rich)**

Veins, stockworks, and disseminations in intermediate to silicic volcanic rocks

| Low sulfidation | High sulfidation |
|---|---|
| Virginia City, Rochester, and Tonopah, Nevada | Julcani, Peru |
| Delamar, Idaho | Red Mountain, Colorado |
| Creede and Sunnyside, Colorado | |
| Bodie, California | |
| Pachuca, Guanajuato, and Tayoltita, Mexico | |
| Colqui, Peru | |

Akeshi, Iwato, Kasuga), Pueblo Viejo (Dominican Republic), as well as Temora, all have some of the characteristics of this group.

## Classification

This paper distinguishes the two groups of deposits (as well as the active systems) by the terms "low sulfidation" and "high sulfidation" (also termed "adularia-sericite" and "acid-sulfate," respectively, by Hayba et al., 1986), with low and high referring to the relative oxidation state of sulfur in the mineralization fluid. There is no inference in these terms as to the amount of total sulfur in the fluid, though the high sulfidation systems commonly have abundant sulfide and sulfate minerals present (Sillitoe, 1983, uses the term "massive sulfide"). In contrast, low sulfidation systems generally have a restricted amount of total sulfide deposited in the ore zone (Buchanan, 1981). Thus, the low sulfidation system has fluid with sulfur present at an oxidation state of $-2$ (i.e., as $H_2S$), whereas the sulfur present in the high sulfidation systems has an oxidation state up to $+4$ (i.e., essentially all $SO_2$).

As will be argued below, the chemistry of the fluid, and particularly the sulfur chemistry, is integral to the formation of epithermal deposits, and is used in distinguishing the deposits. The gold-silver ratio, widely variable in epithermal deposits (Graybeal, 1981), also reflects the chemistry of the fluids, specifically the chloride-sulfide ratio (Henley, 1986) and is also used in classification (Table 1). The host rock and the form of the deposit (i.e., veins, stockworks, disseminations) can be highly variable but are relatively unimportant (Hedenquist and Reid, 1985), except on a local scale and as secondary factors in determining whether or not an ore deposit will form; they are not considered in the classification scheme. An exception to this statement may be the carbonate-hosted replacement deposits (Tooker, 1985). They remain enigmatic because of the relatively few rigorous geochemical studies existing, and are not considered further. This classification is used only for the purpose of assisting discussion; in nature there will be a range of geochemical characteristics between the end member systems.

Finally, the definition of "epithermal" is used rather loosely here simply to recognize the existence of an environment near the earth's surface (usually $<1$ km deep) where many physical and chemical changes take place in a hydrothermal system which will be conducive to mineral deposition, the precious metals in particular. Because of boiling point for depth constraints in hydrostatic systems, the temperatures are $<300\,°C$ where these epithermal changes are taking place.

## LOW SULFIDATION SYSTEMS

A large proportion of the discovered epithermal deposits, as well as a majority of geothermal systems, are of the low sulfidation type; $H_2S$ is the dominant sulfur species entering the epithermal portion of the convecting hydrothermal system. Mineralization occurred at temperatures of 170 to 270 °C, at 50 m to as much as 1000 m depth.

The principal fluid in these well-studied systems, and that responsible for metal transport and deposition, is a dilute "chloride" fluid; chloride is the dominant anion and is accompanied by $Na^+$, $K^+$, $Ca^{++}$ and silica, variable gas concentrations (mainly $CO_2$, which can dominate the chloride in amount, and $H_2S$), plus minor concentrations of other elements including metals. The water, which is principally meteoric (Craig, 1963; O'Neil and Silberman, 1974), circulates to depths as great as 5 to 10 km, becomes heated by a magmatic body, and convects to the surface (Figure 2). As it circulates through the host rock, it acquires many of its constituents through fluid-rock interaction, leaving behind at shallower depths a characteristic fingerprint of propylitic alteration. Any contribution of magmatic gases to the hydrothermal system is neutralized beneath the epithermal environment (Giggenbach, 1981); however, some components such as $CO_2$, helium,

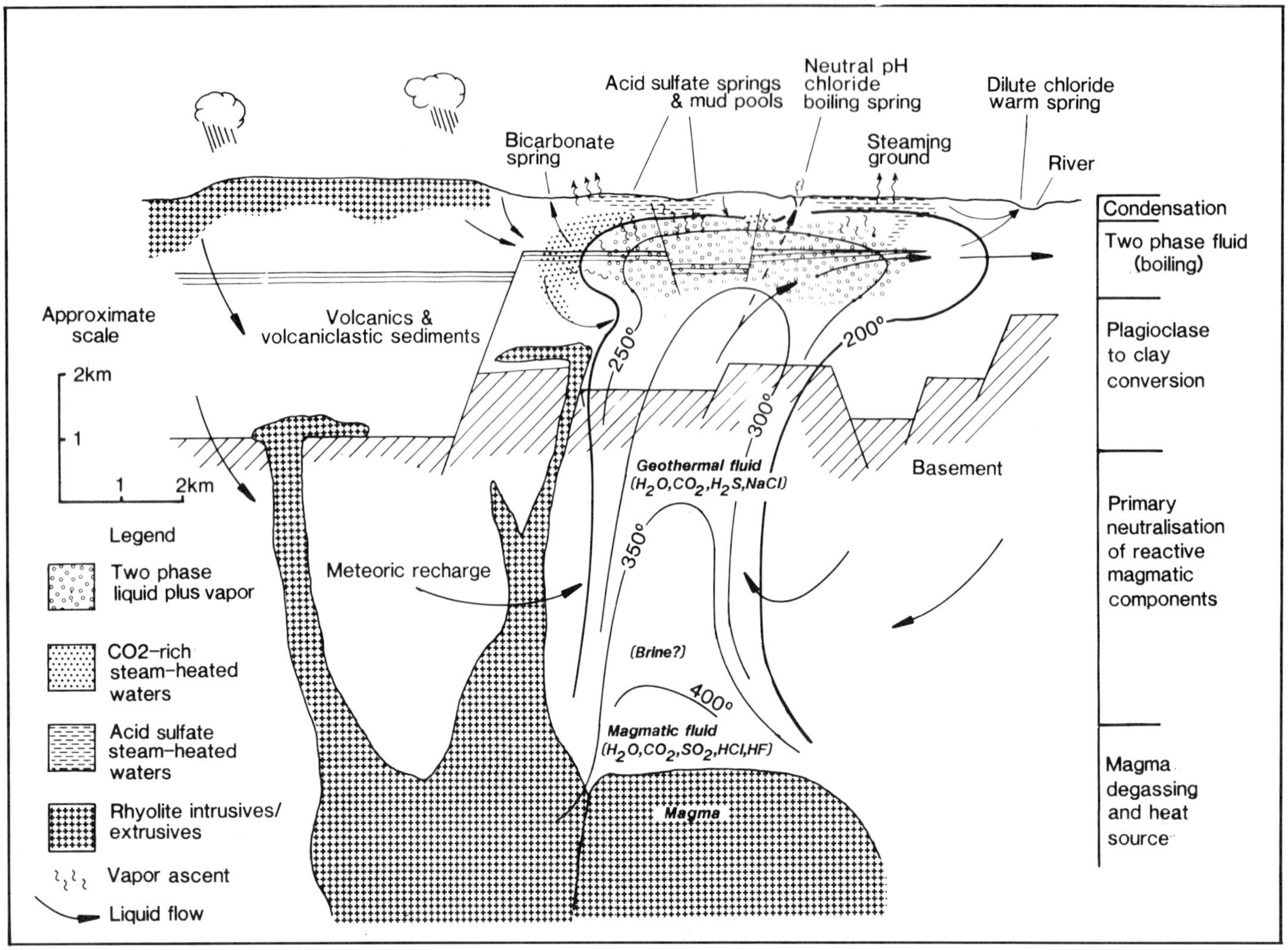


Figure 2. Schematic cross section of a meteoric-dominated hydrothermal system, the type associated with the formation of a low sulfidation epithermal deposit. Note the different fluids formed near the surface. A gold deposit, if formed, will occupy only a very small portion of this system, near one of the principal channels controlling upflow of chloride fluid. Modified from Henley and Ellis (1983), Giggenbach (1981), and Hedenquist (1986).

and the sulfur can be traced in part to a magmatic source through their isotopic signatures (reviewed by Hedenquist, 1986). The source of metals remains a question, though it is not a constraint on understanding the controls on mineralization, as fluid chemistry and not source rock ultimately determines the mineralization potential of a fluid.

The alteration assemblage resulting from the chloride fluid commonly includes quartz, adularia, K-mica, chlorite, calcite, epidote, pyrite, albite, zeolites, and base metal sulfides along with precious metals. The chemistry of the fluid deduced from this mineral assemblage is plotted on an $f_{H_2}$ ($f_{O_2}$)-pH diagram (Figure 3a); these are some of the critical factors in identifying the stability of the various metal complexes.

The two most important physical processes affecting fluid chemistry in geothermal systems, and by analogy having occurred in the low sulfidation epithermal deposits, is boiling and dilution as the liquid approaches the surface. As the liquid ascends into the epithermal environment, it will eventually intersect the boiling point curve. For pure water, this occurs at about 1000 m for a 300 °C liquid, whereas with 4 wt% $CO_2$ in solution (the maximum observed in these systems; Hedenquist and Henley, 1985b), boiling begins at 2200 m. As the fluid boils, the gases fractionate strongly into the vapor; this causes severe changes to the residual liquid. Loss of $CO_2$ causes the pH to increase (Figure 3a), shifting the mineral-fluid stability toward that of adularia (a mineral commonly noted to be associated with gold in veins); the increase in pH may also cause calcite to precipitate. Loss of $H_2S$, and to a lesser extent the shift in pH, will also result in a drastic decrease in gold bisulfide solubility, with concomitant gold deposition (Hedenquist and Henley, 1985a). In these low-salinity, reduced, and neutral pH fluids, gold chloride complexing is insignificant (Figure 3a).

Besides causing mineral deposition, boiling will also release steam to migrate independently from the liquid to the surface. If the steam encounters ground water, the steam will condense and the $H_2S$ will oxidize, forming steam-heated acid sulfate waters at 100 to 150 °C (Figure 2). This acid sulfate fluid will result in a near-surface zone of advanced argillic alteration, characterized by such minerals as alunite, kaolinite, native sulfur and cristobalite. While the system is active the buoyant upflow of neutral pH liquid will keep the relatively cool acid waters on its margin. If the acid water does penetrate the system, it will be unlikely to heat to temperatures above about 200 °C because of heat and mass balance constraints. Furthermore, and very importantly, oxidation of $H_2S$ can only occur in the near-surface vadose zone.

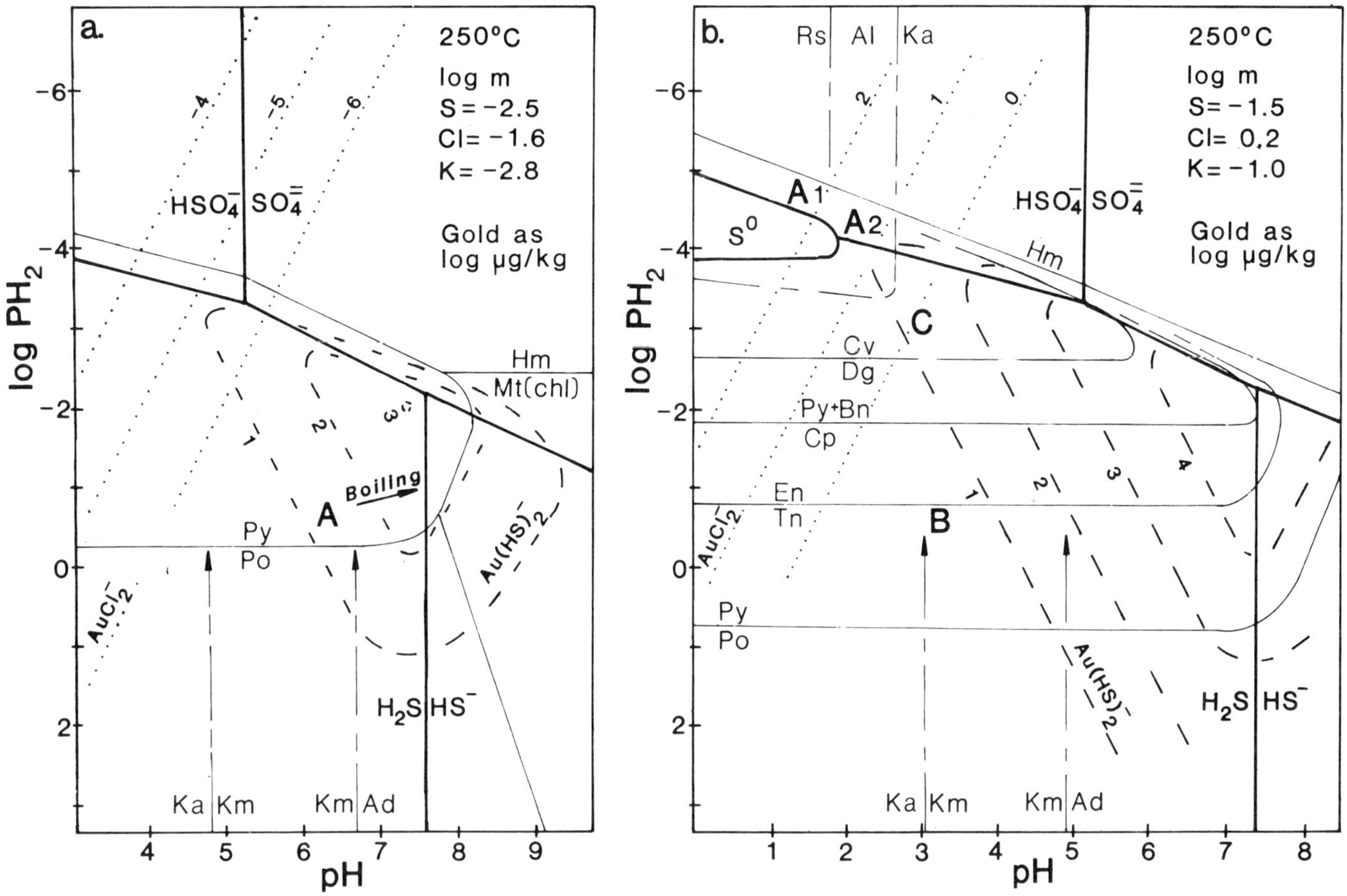


Figure 3a. $P_{H_2}$ versus pH diagram showing the relative stabilities of the minerals commonly occurring in low sulfidation systems, and the stable sulfur species. The chemistry of the mineralizing fluid, observed from active systems and deduced from deposit mineralogy, is shown as "A." Under these conditions, gold bisulfide solubility (shown as log ug/kg-ppb) greatly exceeds gold chloride solubility, with the former being responsible for gold transport. Boiling will cause an increase in pH, because of $CO_2$ loss, and a great decrease in gold solubility (below that shown) because of $H_2S$ loss. The diagram has been modified from Hedenquist and Henley (1985a) for the conditions shown. Abbreviations include pyrite (Py), Pyrrhotite (Po), hematite (Hm), magnetite (Mt), chlorite (chl), kaolinite (Ka), k-mica (Km), and adularia (Ad). b.

b. Similar to a., but constructed for the higher sulfur, chloride, and potassium concentrations existing in the high sulfidation environment; the total sulfur concentration is constrained by the size of the native sulfur (S°) field (Stoffregen, 1985). Stoffregen (1985) deduced the conditions related to alteration as follows: extreme base leaching, or residual silica (Rs) - $A_1$; alunite (Al) - kaolinite halos $A_2$; deep mineralizing fluid where chalcopyrite (Cp) and tennantite (Tn) are stable - B; and the environment of main stage gold deposition, where enargite (En), covellite (Cv), diagenite (Dg), and bornite (Bn) are stable - C. Under the main stage conditions (C), gold chloride solubility is becoming significant with respect to gold bisulfide, limiting the processes which can cause gold deposition.

If the steam condenses into ground water below the zone of oxidation, the $H_2S$ will not be appreciably oxidized, but $CO_2$ will be absorbed, forming a marginal steam-heated $CO_2$-rich fluid (Figure 2) with a pH slightly less than that of the deep liquid. This fluid has been observed in active systems (Hedenquist and Stewart, 1985), and its presence has been deduced from fluid inclusion and mineralogical evidence in epithermal deposits (Hayba, 1984). It will have nil salinity, temperatures of 140 to 180 °C, and will be stable with the moderate temperature clays (interlayed illite-montmorillonite), which will form a "halo" around the deep upflow. This relatively cool fluid, along with cool ground waters, commonly serves as a dilutant of the deep fluid. If the mineralizing fluid has a relatively high salinity, such as in the silver-base metal deposits (Henley, 1986), this dilution may lead to ore deposition.

Differences between deposits can generally be attributed to differences in fluid flow, with focusing mechanisms such as hydrothermal eruption vents or faulting being important in the development of ore, as specific flow channels will experience greater pressure drops, with accentuated boiling and gas loss. This is the factor which causes a gold ore deposit to be restricted to 1/100 to 1/1000 of the area of the system as a whole, the latter often encompassing many $km^2$ (Figure 2). Extensive lateral flow can result when there is a strong hydraulic gradient due to topography (Henley and Ellis, 1983), and this may result in disseminated mineralization within an aquifer.

The multitude of variation within the epithermal environment precludes the development of any single model, or even a series of models, which can be used as successfully in exploration as the models for porphyry copper deposits. Rather, exploration must be based on a sound conceptual understanding of all the factors which can combine to form a deposit in the epithermal environment. This will allow each prospect to be assessed on its own merits, and not on the degree of fit to the currently favored model. Specific aspects to consider in the exploration of these deposits (discussed by Hedenquist and Henley, 1985a) include a detailed determination of alteration mineralogy and fluid inclusion studies to identify paleo flow patterns and zones of boiling, the latter conducive to gold deposition.

## HIGH SULFIDATION SYSTEMS

Gold mineralization in high sulfidation ore deposits is generally associated with enargite (or its lower temperature dimorph luzonite); in places it is associated with pyrite, tennantite-tetrahedrite, covellite and/or alunite; and commonly it is associated with a vuggy silica which is the product of extreme hydrolytic base leaching (Meyer and Hemley, 1967). The ore zones, which commonly are strongly localized by structural features (Sillitoe, 1983) and are associated with hydrothermal breccias, have a ubiquitous though narrow halo (generally less than tens of meters across) which grades from the leached residual silica (if it is present) outward to quartz-alunite, quartz-kaolinite, clays (interstratified illite-montmorillonite), and into a district scale zone of propylitic alteration (Ransome, 1909; Huang, 1955; Steven and Ratte, 1960; Harvey and Vitaliano, 1964; Urashima, Saito, and Sato, 1981; Sillitoe, 1983; Stoffregen, 1985). Locally important as constituents of ore in the wallrock are kaolinite, dickite, pyrophyllite, diaspore, K-mica, native sulfur, barite and anhydrite.

### Mineralogical and Fluid Inclusion Data

The mineralogy will eventually allow specific conclusions to be made about the chemistry and temperature of the fluids responsible for formation of this type of deposit. However, with the exception of the recent detailed study by Stoffregen (1985) on the Summitville deposit, there are no data available on the exact spatial and temporal relationship between the gold mineralization and the gangue mineralogy in high sulfidation deposits of the this type. This means that it is not possible to fully and confidently determine the chemistry of the ore fluids in a variety of related deposits, their source(s) and the process(es) related to mineralization.

However, the chemistry of the ore fluid can be limited by consideration of the phase relationships between the minerals in question (Hemley and Jones, 1964; Hemley et al., 1969, 1980; Knight, 1977). Figure 4 is slightly modified from Hemley et al. (1969) and demonstrates the relative stabilities of alunite, kaolinite, and K-mica. At low pH values, caused by sulfuric acid, kaolinite becomes unstable with respect to alunite; if the $K^+$ concentration in the fluid is high enough, the kaolinite stability between alunite and K-mica is eliminated.

At high temperatures (~280 °C; Hemley et al., 1980) in equilibrium with quartz, pyrophyllite replaces kaolinite. However, under acid conditions the existing silica polymorph is commonly cristobalite or even amorphous silica, allowing kaolinite-pyrophyllite coexistence to temperatures as low as 160 °C. If diaspore is present with pyrophyllite, the experimental data indicate minimum coexistence temperatures of 280 °C (Hemley et al., 1980). Although these minerals are present together at temperatures as low as 200 °C in acid zones in the Bacon-Manito geothermal system, Luzon (Reyes, 1985), fluid inclusion data indicate they were probably formed at somewhat higher temperatures, supporting in general the experimental data.

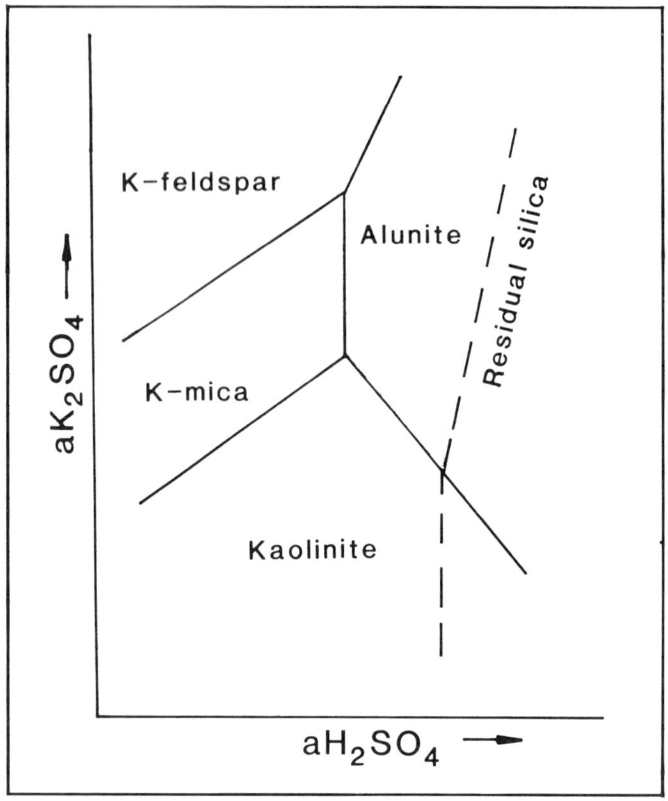


Figure 4. Representation of the relative stabilities of minerals under conditions of increasing potassium and decreasing pH. The field of residual silica, where kaolinite and alunite are soluble, depends on the aluminum concentration (Stoffregen, 1985). Modified from Hemley et al. (1969).

At present, the constraint on salinities in these systems is poor, because of the lack of suitable material for fluid inclusion analyses. Available data are from secondary inclusions in quartz phenocrysts at Summitville and other deposits (Bruha and Noble, 1983); the results indicate temperatures of 200 to 300 °C and salinities of 7 to 21 wt% NaCl, commonly averaging ~10 wt%, with daughter minerals common. These temperatures were confirmed by Stoffregen (1985) for Summitville on fine quartz crystals intergrown with sulfide mineralization in the vuggy silica zone; however, Stoffregen found 4 to 6 wt% NaCl in inclusions. Although fluid inclusions, in conjunction with mineralogy, indicate temperatures similar to those in the low sulfidation systems, the salinities (at least at some stage during alteration) range widely, and indicate the presence of a chloride brine. Fluid inclusion data on quartz associated with the overprinting stage of El Indio (Jannas and Araneda, 1985), on post-mineralization quartz at Mt. Kasi (Turner, 1986), and in the Nansatsu-type deposits (Izawa and Urashima, 1983) all indicate temperatures in the range of 200 to 300 °C, but apparent salinities are all less than 2 wt% NaCl, similar to the low sulfidation systems. If all these data are for mineralization which is post high sulfidation, they suggest greatly decreasing salinities (possibly due to the intrusion of a meteoric convective cell?) with time.

### Chemistry of the Fluids

Stoffregen (1985) has provided an excellent study on the paragenetic relationships between gold and alteration at Summitville, and has taken the first step in placing the geochemistry of the high sulfidation systems into a rigorous understanding. Stoffregen has suggested, based on his assessment of aluminum mobility, that a pH as low as 1.7 to 2.0 may be necessary to produce the vuggy silica alteration type. The lack of hematite and the late occurrence of native sulfur also restrict the chemistry of the leaching fluid, as does the presence of pyrite and alunite and lack of kaolinite. This information, plotted on another $f_{H_2}$-pH diagram under the salinity and total sulfur conditions of interest (Figure 3b) helps to restrict the chemistry of this alteration type. During reaction with the wallrock, pH will increase, resulting in the observed zonation of mineralogy away from the acid fluids, out through kaolinite and illite stability to the propylitic host alteration.

Stoffregen (1985) has recognized that the vuggy silica alteration at Summitville decreases and pinches out at depth. Furthermore, through a detailed textural and grade versus alteration study, Stoffregen has shown that gold mineralization is not restricted to the vuggy silica zone and is unlikely to be related to this alteration; rather, gold is more closely related to high-level covellite, enargite, and luzonite. This high-sulfur sulfide assemblage, requiring a relatively low pH in places where native sulfur is present, overlies and may be slightly later than a more reduced sulfide assemblage of chalcopyrite and tennatite, which is devoid of gold. Stoffregen has proposed that the early acid leaching at Summitville assisted in focusing subsequent fluid flow by generating permeability (Brimhall and Ghiorso, 1983); this is not, however, a prerequisite at Summitville, where ore is also in breccia zones, or in other similar deposits lacking the silica residue (Sillitoe, 1983; Stoffregen, 1985).

This detailed study of Summitville, which should serve as an example for further studies, along with the patchy information from similar deposits, suggests gold transport by relatively oxidizing and acid fluids, though not necessarily the most acid fluid in the system. Stoffregen (1985) has concluded that gold was transported as a bisulfide complex; however, under higher salinities and possibly lower pH values required for other deposits, chloride complexing will compete with bisulfide complexing for gold transport (Figure 3b), thus limiting any discussion of specific depositional mechanisms at this time. Stoffregen (1985) has eliminated the possibility of gold deposition through reaction with country rocks and a pH increase by noting the lack of reaction with wallrocks in some of the highest grade ore shoots. Some of the possible causes of deposition include temperature and/or pressure drop, phase separation, or mixing with another fluid.

### Origin of the Ore Fluids

The origin of the sulfur in the Goldfield deposit has been determined from isotopic studies to be magmatic, probably contributed as $SO_2$ (Field, 1966; Jensen, Ashley, and Albers, 1971; Bethke, 1984); a similar conclusion may be made for Chinkuashih (Folinsbee et al., 1972), as well as for some vein alunite deposits in Utah (Cunningham et al., 1984). The possible relationship of these deposits with intrusives also related to porphyry copper deposits (Gustafson and Hunt, 1975; Brimhall, 1980) has been noted and commented on by several workers (e.g., Sillitoe, 1973, 1975, 1983; Wallace, 1979; Brimhall and Ghiorso, 1983). The characteristics of one group of porphyry deposits, those which are gold-rich (Sillitoe, 1979; Sillitoe and Gappe, 1984; Sillitoe and Ciceron, 1985), may have a particular bearing on and relationship to these high-level gold-copper deposits. In the case of the Butte porphyry deposit (Brimhall, 1980; Brimhall and Ghiorso, 1983), the collapse of a meteoric convective cell on the magmatic system resulted in a concentration of copper protore into veins; this collapse was preceded by the development of an advanced argillic alteration related to acid, magmatic fluids. Hence, the origin of fluids in the high sulfidation system has often been attributed to at least a partially magmatic source (for the $SO_2$ and other acid volatiles, and probably for a large portion of the water and other dissolved components such as NaCl). Examination of the range in chemistry of volcanic fluids presently discharging, and the way in which they interact with their host rocks, may provide the critical constraints to realistically model the chemistry of the high sulfidation ore fluids. This will assist our understanding of the variation in the chemistry of the magmatic ore fluids, e.g., how they interact with the enveloping meteoric waters, and may suggest controls on ore deposition and evolution of the system.

## CHEMISTRY OF VOLCANIC DISCHARGES

The availability of high-quality and complete analyses of volcanic discharges is surprisingly limited, though the difficulty and danger associated with sampling up to 1000 °C roaring fumarole vents on a volcano which may be in an eruption phase can be appreciated. However, there is a high standard of data available for discharges from Showashinzan for the past 30 years (Mizutani and Sugiura, 1982), and for other andesite to dacite-related volcanic fumaroles in Japan (Sakai and Matsubaya, 1977; Ueda, 1983; Kiyosu, 1983, 1985; Kiyosu and Kurahashi, 1983, 1984) and their associated hot springs. In addition, Giggenbach (1975, 1982, 1986) has maintained a detailed surveillance of fumarolic compositions from White Island (New Zealand) since 1970. The variability of these discharges has provided much insight into the processes occurring at depth in the hydrothermal system related to this volcano (Giggenbach, in prep.), and serves as the basis for many of the following ideas.

At Showashinzan, following an eruption of a dacite cone in 1944, fumarolic compositions have changed markedly, owing to a gradual decrease in the primary volcanic gas content and an increase in the meteoric water component (Mizutani and Sugiura, 1982); temperatures also decreased from a maximum of 1000 °C to 550 °C over a period of 33 years.

Japan has many acid sulfate-chloride-dominated geothermal systems related to active volcanism. Sakai and Matsubaya (1977) have shown that one source of sulfur in some of these systems is sulfate in marine sediments or ocean water, and that assimilation of ocean water by the Satsuma-Iwojima magma chamber is possible (Ueda, 1983). However, in many systems sulfur and other gases are contributed by a magma, regardless of initial source (Sakai and Matsubaya, 1977; Kiyosu and Kurahashi, 1984). There is commonly evidence for much meteoric water contamination in the acid fluids which eventually discharge as hot springs or low-temperature (~ 100 °C) fumaroles (Kiyosu, 1983, 1985).

The $\delta^{34}S$ values of sulfate (or sulfate minerals) in these systems commonly are enriched relative to the sulfide, and along with $\delta^{18}O$ data suggest derivation from the disproportion of

$SO_2$ (Sakai and Matsubaya, 1977; Kiyosu and Kurahashi, 1984).

$$4SO_2 + 4H_2O \rightarrow 3H_2SO_4 + H_2S \quad (1)$$

This reaction, taking place at high temperature, results in isotopically heavy sulfate and concomitantly light sulfide. This reaction is also likely responsible for large accumulations of sulfate during the Archean, some of it associated with gold mineralization (Hattori and Cameron, 1986). In contrast, sulfate produced by the surficial oxidation of $H_2S$ in the low sulfidation system

$$H_2S + 2O_2 \rightarrow H_2SO_4 \quad (2)$$

is distinct in that it will have an isotopic value similar to the sulfide, since isotopic fractionation does not occur at the low temperatures (~ 100 °C) of oxidation due to kinetic factors. These two types of sulfate are termed here primary and secondary hypogene, respectively. Their production can be distinguished from that of supergene sulfate originating from the weathering oxidation of sulfide

$$2FeS_2 + 7H_2O + 9O_2 \rightarrow 2Fe_2O_3{\cdot}3H_2O + 4H_2SO_4 \quad (3)$$

from their $\delta^{34}S$ and $\delta^{18}O$ values (Bethke, 1984).

## Cyclic Chemistry of Volcanic Discharges

As discussed previously, there was a dramatic evolution of the gas chemistry of the volcanic discharges from Showashinzan after eruption. In the case of White Island, Giggenbach (1975, 1982) noted the cyclic behavior of gas chemistry (Table 2) through several periods of heating and cooling (ranging from ~ 800 °C down to ~ 100 °C). During such a period of cooling, the gas chemistry changes from that typical of a magmatic gas in which the average oxidation state of the sulfur is ~ +3 (i. e. mainly $SO_2$), and HC1, HF, and $H_2$ are significant, to one which resembles much more closely that of a low sulfidation geothermal discharge, in which the reactive gases $SO_2$, HC1, and HF have been absorbed or neutralized within the vent system (Table 2; Figure 2); $H_2S$ and $CH_4$ greatly increase as the oxidized magmatic fluids are reduced through their interaction with the wallrock hosting the hydrothermal system. During periods of heating, HC1 and HF, as well as native sulfur, are revolatilized and once the vent system has been cleared of these products, the end member— + magmatic gas, containing high $CO_2$, $SO_2$, $H_2$, and CO (but lower HC1 than during the heating period) —reaches the surface.

During a high-temperature discharge, the system bears little relationship to that envisaged for the high sulfidation hydrothermal system. However, during waning stages, or lateral to the principal discharge vent as the self-adjusting envelope of saline waters surrounding the vent (Giggenbach, 1986) begins to interact with the magmatic gases (Figure 5), a situation such as that proposed by Stoffregen (1985) for the Summitville deposit may develop.

During the high-temperature discharge of White Island, the total sulfur $\delta^{34}S$ is relatively constant, and is dominated by $SO_2$. At lower temperatures, subsurface deposition of isotopically light elemental sulfur increases the $\delta^{34}S$ of the surface discharge; the opposite occurs when the sulfur is remobilized (Giggenbach, 1982). The $\delta^{34}S$ of sulfate and $H_2S$ from Japanese volcanic-related hot springs closely reflects the initial $SO_2/H_2S$ of the magmatic gas (Kiyosu and Kurahashi, 1987); the light $H_2S$ and heavy sulfate provide further evidence for their origin from the disproportionation of $SO_2$.

Giggenbach (in prep.) has determined that the chemical composition of the volcanic gases from White Island and several other volcanoes, once they leave their magmatic parent, is controlled by the tendency of the high-temperature oxidized gases to remain in equilibrium with a cooler, more reducing host rock. During rapid, high-temperature venting of these gases, they are largely insulated from the wallrock and the $SO_2/H_2S$ gas buffer and total pressure tends to maintain control of the redox conditions, with the gases largely reflecting the oxidation state of iron in the magma. With decreasing pressure $SO_2$ converts to $H_2S$, though the fluid cannot fully reduce the $SO_2$, resulting in it "borrowing" against the rock if it can (Giggenbach, personal communication, 1986). Therefore, once the volcanic gases begin to mix with a condensed phase such as a chloride brine, their ascent slows and they interact more with the rock matrix. This has the effect of reducing the gases but in turn oxidizing the rock; disequilibrium in the reduction of $SO_2$ to $H_2S$ can lead to $SO_2$ disproportionating and gives the brine a highly acid nature. With sufficient fluid-rock interaction, the system may approach a lower temperature (~ 300 °C) equilibrium and the fluids enter the reduced, low sulfidation environment. In this situation much of the sulfur has been deposited in the rock (greatly increasing the C/S ratio),the fluids are completely reduced, and they have been greatly diluted by waters in a meteoric convection cell; the magmatic fluids in turn contribute their heat to the system (Figure 5).

# DISCUSSION

According to Giggenbach (personal communication, 1986), the ascent of volcanic gases, and their transition from an oxidized (high sulfidation) to reduced (low sulfidation) state is a "battle of the buffers, in which each achieves a partial victory" (though at the cold end the rock always wins). The system never comes to equilibrium, but depending on the degree and duration of fluid-rock interaction and the length of the reaction path, the whole hydrothermal continuum from volcanic to geothermal (high to low sulfidation) conditions may exist in one system. The high sulfidation systems are fluid dominated, resulting in the fluid remaining acid, oxidizing, and hot. In contrast, the fluids in the low sulfidation systems are dominated by the rock, resulting in their chemistry becoming neturalized and reducing as they cool (compare Figures 2, 5).

It is the position in the volcanic hydrothermal system where the fluids have condensed and are reacting greatly with the rock (resulting in hydrolytic leaching) which interests us in terms of an analogy to the high sulfidation gold deposit. If the degassing magma is relatively close to the surface, and/or the level of principal fluid-rock interaction is taking place where temperatures are ~ 300 °C, the acid brine may result in the alteration associated with high sulfidation gold deposits. With continued evolution of this system (increased neutralization at depth, waning of the magmatic source, an increase in the meteoric water component), the more reduced conditions envisaged by Stoffregen (1985) to be associated with gold transport may develop. However, the exact process leading to

TABLE 2—Gas composition of fluids discharged from geothermal wells and from volcanic fumaroles (mmol/mol gas) in New Zealand (Data from Henley and Ellis, 1983, and Giggenbach, 1982).

| Source | Temperature °C | Gas fraction in total discharge | $CO_2$ | $S_t$ | $\bar{n}$[1] | HCl | HF | $NH_3$ | He | $H_2$ | Ar | $O_2$ | $N_2$ | $CH_4$ | CO |
|---|---|---|---|---|---|---|---|---|---|---|---|---|---|---|---|
| Broadlands Well 25 | 290 | 0.005 | 960 | 14 | −2.0 | <0.1 | <0.01 | 2.20 | 0.002 | 1.5 | 0.022 | <0.01 | 7.8 | 14.20 | <0.0 |
| Wairakei Average well | 260 | $2.0 \times 10^{-4}$ | 917 | 44 | −2.0 | na | na | 0.60 | na | 0.8 | na | na | 1.5 | 0.90 | na |
| Waiotapu Well 7 | 225 | $6.1 \times ^{-4}$ | 900 | 56 | −2.0 | na | na | 0.40 | na | 0.2 | na | na | 2.1 | 0.60 | na |
| Ketetahi[2] fumarole | 138 | 0.080 | 901 | 31 | −2.0 | <0.1 | <0.01 | 2.28 | 0.002 | 27.9 | 0.010 | <0.01 | 11.6 | 26.70 | <0.0 |
| Ngauruhoe fumarole | 520 | 0.101 | 696 | 179 | +1.6 | 31.0 | 2.6 | 0.31 | 0.004 | 28.4 | 0.150 | <0.01 | 42.4 | 0.01 | 0.4 |
| White Island fumarole | 620 | 0.165 | 617 | 297 | +3.2 | 59.0 | 0.66 | 0.13 | 0.006 | 24.1 | 0.004 | <0.01 | 3.3 | 0.04 | 0.5 |
| White Island fumarole | 115 | 0.025 | 820 | 156 | −1.8 | 3.2 | 0.21 | 0.06 | 0.070 | 0.6 | 0.070 | <0.01 | 10.2 | 8.90 | <0.0 |

[1] $\bar{n}$ = average oxidation state of sulfur.
[2] The Ketetahi fumarole is on the flank of Mt. Ngauruhoe, about 5 km from the high-temperature volcanic discharge. This relatively low temperature discharge is similar in environment to that of the low sulfidation fumaroles shown schematically in Figure 5.

gold deposition, given the apparent shift to more oxidizing conditions where gold bisulfide and chloride solubilities differ in roughly opposite manners, is still not apparent. Furthermore, these ideas remain poorly constrained when applied to other deposits where there is no apparent postsilica leach episode to associate with the gold (e.g., Nansatsu-type), the silica leach is absent (e.g., Lepanto; Gonzales, 1956), the minerals associated with gold and their parageneses are unknown (most deposits), or there appears to be a significant interlude between the high sulfidation system and gold mineralization (e.g., at El Indio, as discussed by Walthier, Sirvas, and Araneda, 1984; and on Lihir, as reported by Williamson, 1983), with mineralization apparently not coeval with enargite.

## SUGGESTIONS FOR FURTHER WORK

There have been many suggestions as to the origin of high sulfidation deposits, beginning with Ransome (1909). Some of these ideas have been incorporated here, and those elucidated by the exemplary study of Stoffregen (1985) are the best constrained by hard evidence and geochemical calculation. However, it will be some time before the full range of variation exhibited by the high sulfidation gold deposits is understood to the same degree as the low sulfidation systems. This is due in part to the lack of detailed paragenetic studies of these deposits, but it also reflects the fact that their active analog, the volcanic hydrothermal system, has not been studied in the detail of geothermal systems. As such, this paper is also a less-than-satisfactory discussion of the topic. Following are several recommendations for lines of research to close this gap of understanding the high sulfidation volcanic systems, and in so doing, to improve our ability to search for volcanic and subvolcanic-hosted hydrothermal ore deposits.

1. The most glaring shortcoming is the lack of information on detailed mineral parageneses from the whole range of deposits which appear to "fit" the high sulfidation classification. This includes study to determine the minerals directly associated with gold mineralization, separating out the multiple events that are commonly present; this may be best approached from a combined lithologic-grade evaluation, as initially developed by Brimhall for studying porphyry copper remobilization (outlined in Stoffregen, 1985).
2. In conjunction with mineralogical study, much information can be gained from fluid-inclusion studies, particularly on physical changes in the nature of the ore fluid. However, suitable material for fluid-inclusion study is generally lacking in these deposits, particularly that associated with gold. Bruha and Noble (1983) partially circumvented this problem by using secondary inclusions in quartz phenocrysts in the wallrock. Stable-isotope studies, particularly of sulfur and oxygen (Bethke, 1984) may be suitable for distinguishing distinct events in the evolution of a system.
3. It goes without saying that mine to district and regional scale geologic mapping is absolutely essential in reconstructing the setting of mineralization. Too often mapping (if done at all) stops at the outer limits of mineralization. Only by extending this mapping will a correlation be able to be made with active volcanoes.
4. The next step is to map active volcanoes, specifically focusing on types and extent of alteration. Although there is much work done on volcanoes, classic volcanologists quite often overlook alteration as not integral to their primary purpose of understanding the magmatic system and its eruptive activity. Furthermore, geologists involved in mapping altered and ancient volcanic terrane should be given experience in active volcanic areas (and vice versa).
5. Despite the valiant efforts by a few workers on the characterization and monitoring of volcanic gases, more work is needed, particularly on temporal and spatial variations. Furthermore, efforts must be made following the pioneering work by Giggenbach to extend our understanding of how these gases relate to the overall volcanic hydrothermal system (Moore, Jakobsson, and Norrman, 1986), and how the volcanic and geothermal environment is intimately related. This must be done hand in

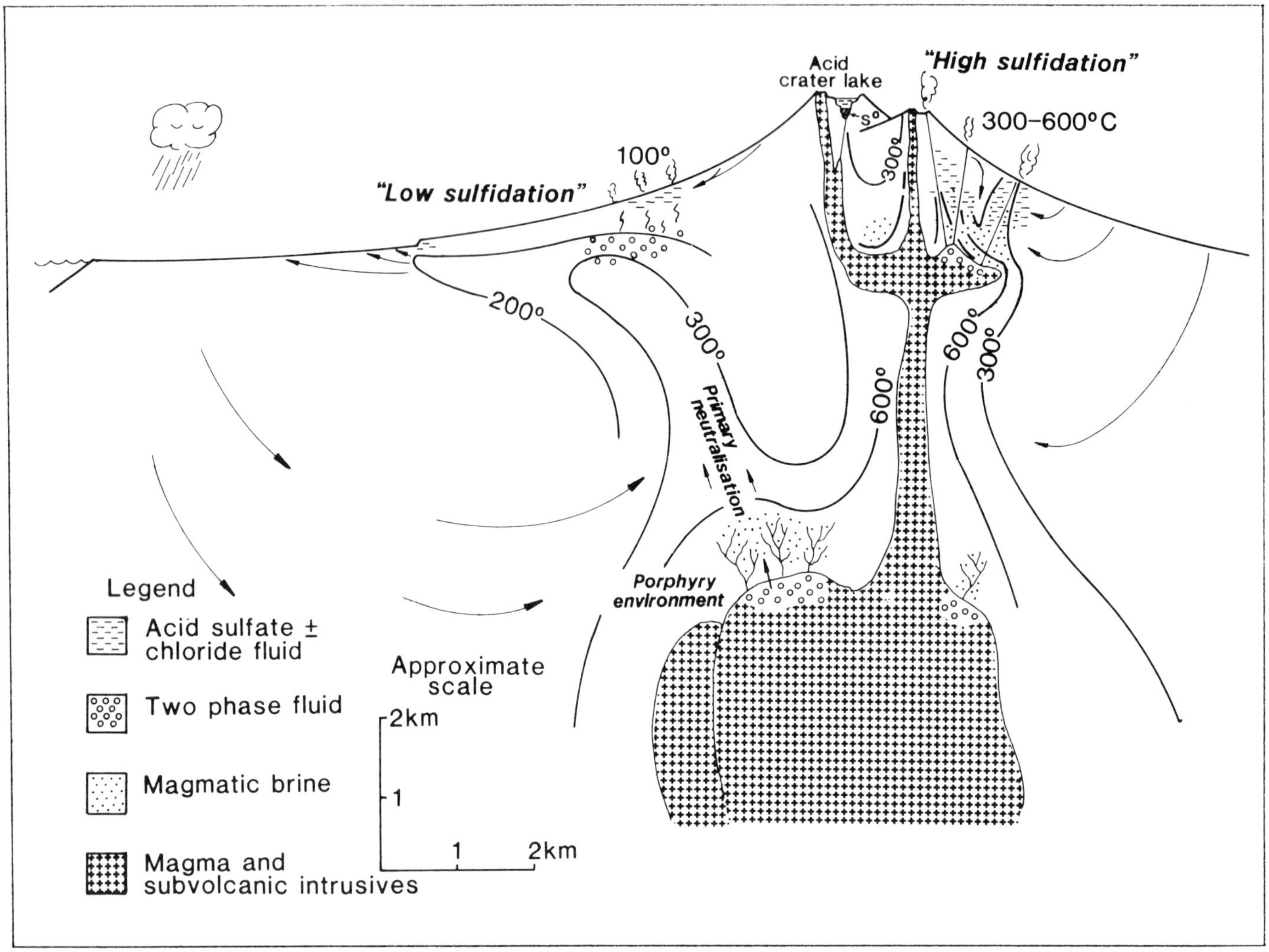


Figure 5. Schematic representation of the fluid flow paths and processes present within a high sulfidation system related to active andesite-dacite volcanism. The region of potential gold mineralization related to this system is relatively high in the volcanic sequence. Also shown is the environment deduced for porphyry mineralization. The region of low sulfidation hydrothermal activity overlies a region where magmatic fluids have been neutralized through interaction with the host rock and greatly diluted by meteoric fluids. There can be much lateral flow in this high topographic setting, causing the neutral pH chloride discharge to be greatly displaced from the region of upflow and first boiling. The potential for gold mineralization in this low sulfidation environment is closer to (and beneath) the ~100 °C advanced argillic alteration, rather than related to the silicification in the outflow path. Adapted from several sources, including Sillitoe (1975) and Henley and Ellis (1983).

hand with an understanding of the chemistry of fluids related to both the low and high sulfidation deposits to approach the solution from both directions—the initial and ongoing system and the final product.

6. Once an understanding of the last point is well under way, we can more fully examine the importance of specific magma types, as proposed by Burnham and Ohmoto (1980), on the characteristics of the hydrothermal fluid and thus the ore deposit. For example, relatively oxidized ("I" type) magmas were proposed to be more intimately related to the formation of gold-rich copper porphyries, and this type of discriminatory tool could have great potential in exploration. However, it is possible that the low sulfidation systems can be related to "S" type magmas as well, without an intermediate oxidized, high sulfidation hydrothermal system as suggested here for "I" type magmas. Examination of magmatic and other controls on the total gas concentrations in low sulfidation systems will also provide guidelines for exploration (Hedenquist and Henley, 1985b).
7. Further examination of the relationship between differences in the epithermal environment with that of the porphyry systems, begun geologically by Sillitoe and geochemically by Brimhall, is also warranted. It is interesting to consider the importance of a magmatic-related protore to the formation of epithermal deposits of all types; at present this must remain necessarily speculative (Brimhall and Ghiorso, 1983).
8. Although much is known about stabilities of minerals related to the high sulfidation systems through the pioneering work by Hemley and others, there still remain several problems, one of which concerns the mobility of aluminum and iron species in the acid fluids (Stoffregen, 1985). Only through a quantitative understanding of the chemistry of the fluids related to the transport of metals under the appropriate conditions will we be able to determine the processes responsible for mineralization, and thus understand the controls on the formation of

these ore deposits and how to successfully look for them.

## ACKNOWLEDGMENTS

I would like to thank Dick Henley and Werner Giggenbach for much useful discussion on many aspects of this topic and Sharon Thorne for typing the manuscript. In particular, Werner Giggenbach is acknowledged for carving a geochemical path through the forest of hydrothermal systems, placing candles at particularly dark corners to light the way.

## REFERENCES CITED

Berger, B. R., and P. Eimon, 1983, Conceptual models of epithermal precious metal deposits, *in* W. C. Shanks, III, ed., Cameron Volume on Unconventional Mineral Deposits: AIME Society of Mining Engineers, p. 191-205.

Bethke, P. M., 1984, Controls on base and precious metal mineralization in deeper epithermal environments: U.S. Geological Survey Open-File Report 84-890, 14 p.

Brimhall, G. H., Jr., 1980, Deep hypogene oxidation of porphyry copper potassium-silicate protore at Butte, Montana: A theoretical evaluation of the copper remobilization hypothesis: Economic Geology, v. 75, p. 384-409.

Brimhall, G. H., Jr., and M. S. Ghiorso, 1983, Origin and ore-forming consequences of the advanced argillic alteration processes in hypogene environments by magmatic gas contamination of meteoric fluids: Economic Geology, v. 78, p. 73-90.

Bruha, D. J., and D. C. Noble, 1983, Hypogene quartz-alunite ± pyrite alteration formed by moderately saline, ascendant hydrothermal solutions (abs.): GSA, Rocky Mountain Section, Abstracts with Programs, v. 15, p. 325.

Buchanan, L. J., 1981, Precious metal deposits associated with volcanic environments in the Southwest, *in* W. R. Dickson and W. D. Payne, eds., Relations of tectonics to ore deposits in the southern Cordillera: Arizona Geological Society Digest, v. 14, p. 237-262.

Burnham, C. W., and H. Ohmoto, 1980, Late-stage processes of felsic magmatism: Mining Geology Special Issue 8, p. 1-11.

Craig, H., 1963, The isotopic geochemistry of water and carbon in geothermal areas, *in* E. Tongiorgi, ed., Nuclear geology in geothermal areas: Consiglio Nazionale Delle Ricerche, Pisa, p. 17-53.

Cunningham, C. G., R. O. Rye, T. A. Steven, and H. H Mehnert, 1984, Origins and exploration significance of replacement and vein-type alunite deposits in the Marysvale volcanic field, west central Utah: Economic Geology, v. 79, p. 50-71.

Ferguson, K. J., 1986, The Kelian gold prospect, Kalimantan, Indonesia, *in* Volcanism, hydrothermal systems and related mineralization: Hamilton, New Zealand, International Volcanological Congress, Proceedings of Symposium 5, p. 41-46.

Field, C. W., 1966, Sulfur isotopic method for discriminating between sulfates of hypogene and supergene origin: Economic Geology, v. 61, p. 1428-1435.

Folinsbee, R. E., K. Kirkland, A. Nekolaichuk, V. Smejkal, 1972, Chinkuashih—A gold-pyrite enargite-barite hydrothermal deposit in Taiwan: GSA Memoir 135, p. 323-335.

Giggenbach, W. F., 1975, Variations in the carbon, sulfur and chlorine contents of volcanic gas discharges from White Island, New Zealand: Bulletin Volcanologique, v. 39, p. 15-27.

Giggenbach, W. F., 1981, Geothermal mineral equilibria: Geochimica et Cosmochimica Acta, v. 45, p. 393-410.

Giggenbach, W. F., 1982, The chemical and isotopic composition of gas discharges from New Zealand andesitic volcanoes: Bulletin Volcanologique, v. 45, p. 253-255.

Giggenbach, W. F., 1986, The chemistry of carbon gases including hydrocarbons in high temperature fumarole discharges from White Island, New Zealand: Journal of Geophysical Research, in press.

Giles, D. L., and C. E. Nelson, 1984, Principal features of epithermal lode gold deposits of the Circum-Pacific rim: Honolulu, Hawaii, Transactions of the Third Circum-Pacific Energy and Minerals Resource Conference, August 22-28, 1982, AAPG, p. 273-278.

Gonzales, A., 1956, Geology of the Lepanto copper mine, Mankayon, Mountain Province, *in* A. R. Kinkel, Jr., et al., eds., Copper deposits of the Philippines: Philippines Bureau of Mines Special Projects Series Publication 16, p. 17-50.

Graybeal, F. T., 1981, Characteristics of disseminated silver deposits in the Western United States, *in* W. R. Dickinson and W. D. Payne, eds., Relations of tectonics to ore deposits in the southern Cordillera: Arizona Geological Society Digest, v. 14, p. 271-281.

Gustafson, L. B., and J. P. Hunt, 1975, The porphyry copper deposit at El Salvador, Chile: Economic Geology, v. 70, p. 857-912.

Harvey, R. D., and C. J. Vitaliano, 1964, Wallrock alteration in the Goldfield district, Nevada: Journal of Geology, v. 72, p. 564-579.

Hattori, K., and E. M. Cameron, 1986, Archaen magmatic sulphate: Nature, v. 319, p. 45-47.

Hayba, D. O., 1984, Documentation of thermal and salinity gradients and interpretation of the hydrologic conditions in the OH vein, Creede, Colorado (abs.): GSA Abstracts with Programs, v. 16, p. 534.

Hayba, D. O., P. M. Bethke, P. Heald, and N. K. Foley, 1986, The geological, mineralogical and geochemical characteristics of volcanic-hosted epithermal deposits, *in* B. R. Berger and P. M. Bethke, eds., Geology and geochemistry of epithermal systems: Society of Economic Geologists, Reviews in Economic Geology, v. 2, in press.

Hedenquist, J. W., 1986, Geothermal systems in the Taupo Volcanic Zone of New Zealand: Their characteristics, and relation to volcanism and mineralization, *in* I. E. M. Smith, ed., Late Cenozoic volcanism in New Zealand: Royal Society of New Zealand Bulletin 23, p. 134-168.

Hedenquist, J. W., and R. W. Henley, 1985a, Hydrothermal eruptions in the Waiotapu geothermal system, New Zealand: Their origin, associated breccias and relation to precious metal mineralization: Economic Geology, v. 80, p. 1640-1668.

Hedenquist, J. W., and R. W. Henley, 1985b, The importance of $CO_2$ on freezing point measurements of fluid inclusions: Evidence from active systems and implications for epithermal ore deposition: Economic Geology, v. 80, p. 1379-1406.

Hedenquist, J. W., and F. W. Reid, 1985, Epithermal gold: Short Course Notes, Earth Resources Foundation, The University of Sydney, 316 p.

Hedenquist, J. W., and M. K. Stewart, 1985, Natural $CO_2$-rich steam-heated waters at Broadlands, New Zealand: Their chemistry, distribution and corrosive nature: Honolulu, Hawaii, Transactions of the Geothermal Resources Council Annual Meeting, v. 9(2), p. 245-250.

Hemley, J. J., and W. R. Jones, 1964, Chemical aspects of hydrothermal alteration with emphasis on hydrogen metasomatism: Economic Geology, v. 59, p. 538-569.

Hemley, J. J., P. B. Hostetler, A. J. Gude, and W. T. Mountjoy, 1969, Some stability relations of alunite: Economic Geology, v. 64, p. 599-612.

Hemley, J. J., J. W. Montoya, J. W. Marinenko, and R. W. Luce, 1980, Equilibria in the system $A1_2O_3$-$SiO_2$-$H_2O$ and some general implications for alteration/mineralization processes: Economic Geology, v. 75, p. 210-228.

Henley, R. W., 1986, The geothermal frameworks of epithermal deposits, *in* B. R. Berger and P. M. Bethke, eds., Geology and geochemistry of epithermal systems: Society of Economic Geologists Reviews in Economic Geology, v. 2, in press.

Henley, R. W., and A. J. Ellis, 1983, Geothermal systems, ancient and modern: Earth Science Reviews, v. 19, p. 1-50.

Huang, C. K., 1955, Gold-copper deposits of the Chinkuashih mine, Taiwan, with special reference to the mineralogy: Acta Geologic Taiwanica, no. 7, p. 1-20.

Izawa, E., and Y. Urashima, 1983, Gold-silver deposits in southern Kyushu, Japan: Proceedings of the Sendai, Metal Mining Institute of Japan/Australian Institute of Mining and Metallurgy Joint

Symposium, p. 97-111.

Jannas, R. R., and G. R. Araneda, 1985, Geologia de la Veta Indio Sur 3500: Una estructura tipo bonanza del yacimiento El Indio, Chile: Preprint presented at the Fourth Chilean Geological Congress, Antofagasta, August 19-24, 23 p.

Jensen, M. L., R. P. Ashley, and J. P. Albers, 1971, Primary and secondary sulfates at Goldfield, Nevada: Economic Geology, v. 66, p. 618-626.

Kiyosu, Y., 1983, Hydrogen isotopic compositions of hydrogen and methane from some volcanic areas in northeastern Japan: Earth and Planetary Science Letters, v. 62, p. 41-52.

Kiyosu, Y., 1985, Isotopic composition of acid sulfate-chloride waters and volcanic steam from some volcanoes in northeastern Japan: Journal of Volcanology and Geothermal Research, v. 26, p. 25-36.

Kiyosu, Y., and M. Kurahashi, 1983, Origin of sulfur species in acid sulfate-chloride thermal waters, northeastern Japan: Geochimica et Cosmochimica Acta, v. 47, p. 1237-1245.

Kiyosu, Y., and M. Kurahashi, 1984, Isotopic geochemistry of acid thermal waters and volcanic gases from Zaō Volcano in Japan: Journal of Volcanology and Geothermal Research, v. 21, p. 313-331.

Knight, J. E., 1977, A thermochemical study of alunite, enargite, luzonite and tennantite deposits: Economic Geology, v. 71, p. 1321-1336.

Meyer, C., and J. J. Hemley, 1967, Wall rock alteration, *in* H. L. Barnes, ed., Geochemistry of hydrothermal ore deposits: New York, Holt, Rinehart and Winston, p. 166-235.

Miller, W. R., and J. J. Rytuba, 1986, An occurrence of an epithermal gold system, Republic of Palau, Western Pacific (abs.), *in* Volcanism, hydrothermal systems and related mineralization: Hamilton, New Zealand, International Volcanological Congress Proceedings of Symposium 5, p. 106.

Mizutani, Y., and T. Sugiura, 1982, Variations in chemical and isotopic compositions of fumarolic gases from Showashinzan volcano, Hokkaido, Japan: Geochemical Journal, v. 16, p. 63-71.

Moore, J. G., S. P. Jakobsson, and J. O. Norrman, 1986, Investigation of Surtsey Volcano: EOS (American Geophysical Union Transactions), v. 67, p. 74.

O'Neil, J. R., and M. L. Silberman, 1974, Stable isotope relations in epithermal Au-Ag deposits: Economic Geology, v. 69, p. 902-909.

Ransome, F. L., 1909, The geology and ore deposits of Goldfield, Nevada: U.S. Geological Survey Professional Paper 66, 258 p.

Reyes, A. G., 1985, A comparative study of acid and neutral pH hydrothermal alteration in the Bacon-Manio geothermal area, Philippines: The University of Auckland, unpublished M. Sc. thesis, 258 p.

Sakai, H., and O. Matsubaya, 1977, Stable isotopic studies of Japanese geothermal systems: Geothermics, v. 5, p. 97-124.

Sillitoe, R. H., 1973, The tops and bottoms of porphyry copper deposits: Economic Geology, v. 68, p. 799-815.

Sillitoe, R. H., 1975, Lead-silver, manganese, and native sulfur mineralization within a strato volcano, El Queva, northwest Argentina: Economic Geology, v. 70, p. 1190-1201.

Sillitoe, R. H., 1977, Metallic mineralization affiliated to subaerial volcanism: A review, *in* Volcanic processes in ore genesis: Geological Society of London, Special Publication 7, p. 99-116.

Sillitoe, R. H., 1979, Some thoughts on gold-rich porphyry copper deposits: Mineralium Deposita, v. 14, p. 161-174.

Sillitoe, R. H., 1981, Ore deposits in Cordilleran and island-arc settings, *in* W. R. Dickinson and W. D. Payne, eds., Relations of tectonics to ore deposits of the southern Cordillera: Arizona Geological Society Digest, v. 14, p. 49-70.

Sillitoe, R. H., 1983, Enargite-bearing massive sulfide deposits high in porphyry copper systems: Economic Geology, v. 78, p. 348-352.

Sillitoe, R. H., and C. A. Ciceron, Jr., 1985, Geological characteristics and evolution of a gold-rich porphyry copper deposit at Guinaoang, Luzon, Philippines: Asia Mining, p. 15-26.

Sillitoe, R. H., and I. M. Gappe, Jr., 1984, Philippine porphyry copper deposits: Geologic setting and characteristics: CCOP Technical Publication 14, 89 p.

Steven, T. A., and J. C. Ratte, 1960, Geology and ore deposits of the Summitville district, San Juan Mountains, Colorado: U.S. Geological Survey Professional Paper 343, 70 p.

Stoffregen, R. E., 1985, Genesis of acid-sulfate alteration and Au-Cu-Ag mineralization at Summitville, Colorado (including sections on supergene alteration and clay mineralogy of the deposit): Berkeley, University of California, unpublished Ph. D. thesis, 205 p.

Tooker, E. W., ed., 1985, Geologic characteristics of sediment and volcanic-hosted disseminated gold deposits—Search for an occurrence model: U.S. Geological Survey Bulletin 1646, 150 p.

Turner, S., 1986, Fluid inclusion, alteration and ore mineral studies of an epithermal vein system: Mount Kasi, Vanua Levu, Fiji, *in* Volcanism, hydrothermal systems and related mineralization: Hamilton, New Zealand, International Volcanological Congress Proceedings of Symposium 5, p. 87-96.

Ueda, A., 1983, Chemical and isotopic analyses of sulfur in volcanic rocks and its application to volcanic systems of Satsuma-Iwojima: Okayama University, unpublished Ph. D. thesis, 185 p.

Urashima, Y., M. Saito, and E. Sato, 1981, The Iwato gold ore deposits, Kagoshima Prefecture, Japan: Mining Geology Special Issue 10, p. 1-14.

Wallace, A. B., 1979, Possible signatures of buried porphyry-copper deposits in middle to late tertiary volcanic rocks of western Nevada: Nevada Bureau Mines and Geology Report 33, p. 69-76.

Walthier, T. N., E. Sirvas, and G. R. Araneda, 1984, The El Indio gold, silver, copper deposit, Region of Coquimbo, Chile: Preprint presented at the Mining Congress Convention, Phoenix, 24 p.

White, D. E., 1955, Thermal springs and epithermal ore deposits: Economic Geology 50th Anniversary Volume, p. 99-154.

White, D. E., 1981, Active geothermal systems and hydrothermal ore deposits: Economic Geology, 75th Anniversary Volume, p. 392-4223.

Williamson, A., 1983, Thermal activity on Lihir Island, New Ireland Province: Geological Survey of Papua New Guinea Report 83/15, 24 p.

CHAPTER 45

# MONITORING AND FORECASTING OF VOLCANIC ERUPTIONS WITH REFERENCE TO ACTIVE VOLCANOES IN INDONESIA

**Adjat Sudradjat**
*The Volcanological Survey of Indonesia*
*Department of Mines and Energy*
*Jakarta, Indonesia*

ABSTRACT

The state of monitoring and prediction of volcanic activities is summarized in this chapter. Examples are given from experiences of the Volcanological Survey of Indonesia, and include the Galunggung volcano, Merapi, Kelut, Colo, and Sangeang Api.

Seven methods of volcano monitoring are presently applied for active volcanoes in Indonesia—visual observation, seismicity, ground deformation, gravity and magnetic studies, self-potential studies, petrochemistry, and geochemical monitoring.

At some Indonesian volcanoes including Galunggung, Merapi, and Colo, seismic monitoring augmented by other methods has been proven to be effective, particularly for predicting individual eruptions after the initial eruptive phase.

The main constraints on the development of volcano monitoring in Indonesia include the fact that there is a large number of active volcanoes and skill and expertise are limited. Baseline data for the high-risk and active volcanoes are therefore still lacking. The preparation of volcanic hazard maps and the continuous awareness of potential volcanic eruptions by both public authorities and the local population are important to mitigate volcanic danger.

## INTRODUCTION

Monitoring and forecasting of volcano eruptions are closely related, for by carrying out monitoring, one may be able to forecast a volcanic eruption. In turn, the forecasting will lead people to the possibility of escaping from the volcanic danger caused by eruption or, in other words, people may be able to minimize the volcanic risk.

Similar to other natural events, volcanic activity is a very complex physical and chemical process involving a large number of parameters. Therefore, the prediction of a volcanic eruption is very difficult. To date, volcanic monitoring has revealed only a small portion of that natural process, and the remainder remains poorly understood.

Several authors have written papers on forecasting of volcanic eruptions in the last 15 years, including Decker (1978; 1986), Gorshkov (1971), Minakami (1971), Civetta et al. (1974), Walker (1974), Fournier d'Albe et al. (1979), and Martin and David (1982). These papers comprehensively discuss volcanic behavior that may lead to the forecasting of a volcanic eruption. The present chapter highlights the methods of volcano monitoring and cases of volcanic eruption forecasting. Reference is made to Indonesian volcanoes where such methods are applied. Two recent eruptions, namely those of Galunggung in 1982 and Colo volcano in 1983, are referred to most because of the monitoring results that may lead to the forecasting of the volcanic eruption.

## VOLCANIC ERUPTION FORECASTING

Volcanic eruption forecasting is the main objective of almost every institution dealing with volcanoes. However, eruption forecasting is difficult, and success stories are mixed with failures to predict eruptions, or the embarrassment of evacuations after which an expected eruption did not occur ("false alarm").

Volcanic eruption forecasting may be approached by several methods. Fundamental research may be the basis for long-range as well as short-range volcanic eruption forecasting. By studying a volcano's deposits one may reconstruct and understand the behavior of the volcano in the past, and may extrapolate that behavior into the future. Age dating techniques available at present, such as C-14, K/Ar, dendrochronology, varve count, lichenometry, magnetism, tephrochronology, hydration rind and fission track, provide the detailed chronological succession of tephra deposits; hence, the eruption history covering thousands of years may be constructed. One successful long-term forecast based mainly on geological investigations coupled with historical records concluded in 1978 that Mount St. Helens, located in the state of Washington (U.S.A.), was likely to erupt within 100 years, perhaps even before the end of this century. And as history has proven, Mount St. Helens erupted on May 18, 1980 and devastated some 600 $km^2$ north of the volcano.

Another technique of long-range volcanic eruption forecasting is statistical analysis of the historical record. The problem of this approach is the availability and reliability of the records, which differ from one volcano to another. Few historical records of active volcanoes in Indonesia, for example, describe events before the 16th century. The longest historical record of volcanic eruption in the world may be that of Santorin, which extends back to centuries B.C.

The reliability of the record might be questioned. The trend of volcanic eruption frequency has been very much influenced by the advance of communication technology as noted by Simkin et al. (1981). Furthermore, some recorded eruptions did not actually occur, and a few reported volcanoes do not exist at all. Tjia (1977) examined the active volcano called Umsini, located in Irian Jaya, Indonesia, which is registered as an active volcano in Neuman van Padang's Catalog of Active Volcanoes (1951). Apparently sedimentary rocks of Early Tertiary age were found in the area. The year 1006 A.D. has been questioned as the occurrence of the catastrophic eruption of Merapi, which destroyed the Mataram Kingdom and forced King Airlangga to flee. Recent study on tephrochronology and C-14 age dating of Merapi deposits and 8th to 10th century temples around Merapi, reveals somewhat older age eruptions. Therefore, the statistical method might only be successful if based on the most reliable part of the historical record. Klein (1982) and Decker (1986) recently analyzed the eruption patterns of Kilauea and Mauna Loa volcanoes in Hawaii, where historical records are relatively complete and reliable. The average repose time at Kilauea is 501 days, whereas that of Mauna Loa is 1412 days.

Short-range forecasting of volcanic eruptions relies on volcano monitoring. The monitoring methods which are currently known are outlined in this chapter.

## VOLCANO MONITORING METHODS

Volcano monitoring is aimed at detecting changes in the behavior of a volcano. By studying the trend of the changes, compared with baseline data, a volcanic eruption may be forecast. To establish baseline data, observations should be made for several years, or even tens of years. A volcano with long repose periods needs a longer period of baseline observation than does a frequently erupting volcano.

Modern instrumental observation only began at the beginning of this century and is still developing rapidly. Space technology has also provided the possibility of using a satellite platform to monitor volcanoes. The plumes of some volcanoes located in remote areas in Indonesia have been observed by Sawada (written communication, 1983) by means of the GMS Japanese weather satellite. The plume of the erupting Galunggung volcano was monitored through a NOAA satellite (Katili and Sudradjat, 1984). Malingreau and Kaswanda (1986) monitored and analyzed the plume emitted by Colo volcano in 1983.

The advance in electromagnetic technology has only developed in the second half of this century. Modern baseline data, therefore, may only cover 20 to 30 years, which is not always adequate for a sound interpretation of any changes leading to forecasting. The electromagnetic seismograph installed in Kelut volcano in Indonesia, for example, has only been monitoring the volcano since the 1970s. The most recent eruption of 1966 was recorded by a mechanical seismograph instead, from which seismic records are different. Thus seismic data obtained by present instruments might be difficult to interpret based only on the seismic baseline data from the Kelut volcano.

The important monitoring methods are categorized as follows:

A. Visual Observations
B. Geophysical Methods
  1. Seismic monitoring
  2. Thermal radiation measurement
  3. Ground deformation
  4. Geomagnetic study
  5. Gravity observation
  6. Geoelectric measurement
C. Geochemical methods
  1. Gas emission observation
  2. Water chemistry observation
  3. Petrochemical analysis

Some of the methods have been established, and others are being developed. Seismic monitoring belongs to the first group. The pioneering works by Omori (1911) in Usu volcano, and subsequently continued by Minakami et al. (1951), have made seismic monitoring the most reliable method of volcano monitoring at present. Minakami (1967) recognized several types of volcanic earthquakes, known at present as A, B, and C types of Minakami.

Geomagnetic and geoelectric measurements were only begun in the late 1960s or early 1970s. Unusual magnetic noise prior to the 1971 Keanakakoi crater eruption of Kilauea have been observed.

Under geomagnetic surveillance, the magnetotelluric method is now being developed. Geoelectric survey at present includes self-potential measurement, resistivity measurement using dipole-dipole method, and conductivity measurement. These methods are based on a relationship between the electrical properties of rocks and the temperature.

Geochemical monitoring is becoming increasingly important, particularly after the successful $SO_2$ monitoring at Mount St. Helens during the 1980 eruption (Casadevall et al., 1981). Petrochemical analysis of rocks is important for long-term forecasting.

### Visual Observation

This is the most common and very basic observation which has been carried out as far back as centuries B.C. In this century the technique is still being used. The phenomena being observed are plume emission, including color, pressure, height etc., vegetation, and surface condition, lake water character, volume, gas bubbling, etc. Those phenomena might also be monitored by applying modern techniques, for example, using a short circuit TV camera telemetered for a limited distance, or using a satellite (nonimaging system) for a longer distance. Observable changes have in some cases led to forecasts of eruptions, but in most cases the eruptions or the initial phases of the eruptions have already occurred when the phenomena are observed. Visual observations can be subjective and misleading; the well-trained and experienced observer will increase the objectivity of the findings.

## Geophysical Method

### Seismic Monitoring

The most applied volcano monitoring technique at present is the seismic method. The development of seismometers for short-period quakes, since they were first invented by Omori in 1911, has been very rapid. Mechanical seismographs using a huge pendulum were replaced by sensitive electromagnetic seismographs. Later, the wire connecting transducer and recorder was also replaced by radio waves. Thus the problem of locating the seismometer close to the dangerous area is alleviated. In isolated small volcanic islands, where there is no safe space to monitor the volcano, a telemetric system is a device to monitor the volcano from a neighboring island. The Colo eruption, Central Sulawesi, Indonesia is an example of monitoring such volcanoes (Katili and Sudradjat, 1984) which are numerous in the Indonesian archipelago. Krakatau volcano is another example where a telemetric seismograph is needed.

The seismic monitoring technique detects ground fracturing and vibration caused by the movement of magma. The host rock may amplify or otherwise attenuate the wave, or even produce the new source of waves. The characteristics of the propagating wave are largely determined by physical properties of the rocks along the path from the activity center (hypocenter) to the monitoring device.

In this paper, the writer will not treat the theoretical aspect of seismic monitoring. Interested readers may consult papers dealing with this matter, among others by Kubotera (1965), Minakami (1967, 1974), Dibble (1965), and others.

Usu and Asama volcanoes, Japan, have been very intensively studied in terms of their seismicity. From his observation experience gained over almost 30 years, Minakami (1967) identified several groups of wave signatures recorded at the seismometer. Relatively deep focus earthquakes originating from a volcanic activity are called *A type* earthquakes, shallower ones are *B type*; the earthquakes accompanying a volcanic eruption are called *eruption quake* or *continuous tremor*. Explosive eruptions produce *air-shock waves* that can be recorded on seismometers.

In the Usu volcano the frequency of an A type quake dominated the earlier stage of the 1944-1945 eruption. The B type quake became dominant at the later stage. The concentration of hypocenters moved from a depth of around 2 to 7 km to several hundred meters below the surface. The same phenomenon was also observed in Arenal volcano, Costa Rica, during the 1968 eruption, at Merapi volcano during its 1969 eruption, and most recently at Galunggung volcano during the 1982 eruption (Katili and Sudradjat, 1984; Suparto, 1986).

The first stage of the activity at Galunggung (April 1982) was characterized by dominantly A type earthquakes with foci distributed in the area at depths of 3 to 8 km below the volcano. The hypocenter distribution formed an elliptical shape with a long axis of 10 km and a short axis of 6 km. The long axis was inclined toward the southeast at an angle of around 30 °.

The activity of May to August was accompanied by continued occurrence of A type earthquakes but with a predominant number of B type events, which indicated that the magma front had become shallower, perhaps at depths of 3 km or less. The latest stage of the activity, from August 1982 to January 1983, was characterized by a B type volcanic earthquake. Volcanic tremors saturated the first stage of activity until May and continued to June. In April and May, amplitudes were high and later decreased in June. Several days before an eruption took place, a tremor usually disappeared, then recurred several hours prior to the eruption. The tremors consistently decreased in amplitude at each individual eruption during those months. During the activities in June and July, tremors again disappeared and then recurred in the latest stage of the activity, beginning in August. The amplitudes were lower compared with those of April and May but showed regular precursory patterns in relation to the eruptions.

In the Colo volcano, Central Sulawesi, the volcano island was jolted by earthquakes almost continuously for several days. The largest earthquake, on July 14, 1983, measured 6.4 on the Richter scale. Nine days later an eruption took place which devastated the entire island. No casualties occurred because the 7000 inhabitants had been evacuated a few days after the first shocks were felt.

In Mount St. Helens intense local seismicity was detected 7 days before a first small phreatic eruption took place on March 27, 1980. The seismic swarm continued with high and early constant energy release. Bursts of volcanic tremors began on March 31 and continued intermittently through April 5, April 12, and May 8, and ended with the disastrous eruption of May 18, 1980. At the June 12 eruption a short-term forecast was issued, based on the appearance of tremors about 9 hours before the eruption took place.

In Hawaiian volcanoes (Koyanagi, Endo, and Ebisu, 1975; Decker et al., 1983) seismicity has been a very important precursor. The 1975 summit eruption of Mauna Loa was preceded by increasing seismicity beneath the summit region since 1974. The seismicity recurred and increased rapidly in 1983, and was followed by the eruption 2 years later, on March 25, 1984.

Continuing seismicity sometimes occurs without ending in an eruption. The most recent examples are at Long Valley, California, which had been shaken for more that 4 years starting from 1979 and gradually decreased by 1983, but which remained one of the most seismically active localities in California until recently. The same is true of Pozzuoli, Campi Flegrei, Italy (1983-1984), Rabaul (1983-1984), and examples from Indonesia such as Lawu (1981), Tangkuban Parahu (1984), and most recently Lamongan (1985). Tangkuban Parahu seismicity increased in mid-1984 and continued for months. The temperature at the solfataric fields of Kawah Ratu rose from a normal level of 110 °C to about 160 °C. At the same time a slight inflation was also detected. The volcanic swarms started to decline by the end of the year and were back to background value at the beginning of 1985 (VSI annual report, 1985).

### Thermal Radiation Measurement

Thermal radiation measurement might lead to the forecasting of a volcanic eruption. For example, Minakami (1974) noted a gradual increase of geothermal temperatures at the floor of Aso volcano, Japan, prior to the eruption. However, not all volcanic eruptions are preceded by increasing thermal radiation. Conversely, increasing temperature of the solfataric field, lake water of geothermal spots around the volcanic centers is not always followed by an eruption.

A temperature increase in the crater lake of Taal volcano led volcanologists of the Philippine Commission on Volcanology to predict the eruption of Taal volcano in 1967, based on the experience obtained a year earlier.

Taal lake water had never shown a temperature exceeding 33 °C from 1961 to 1964. On the first week of July 1965, however, the temperature increased steadily and reached 45° within 26 days. This temperature was stable until the first week of August and later dropped to 43 °C on September 25. Three days later, September 28, a catastrophic eruption took place and killed 190 persons living near the newly formed crater.

Another rise of temperature began in March 1966. The alarm was subsequently issued. After reaching 40 °C in the last week of June, Taal volcano erupted on July 5, 1966. No casualties were claimed, thanks to the timely warning issued by the Commission.

A false alarm happened at Kelut volcano, East Java, Indonesia. Here a crater lake containing 3 million cubic meters of water is monitored. The temperature had been stable at 32 °C since the last eruption of 1966 until 1972, when the temperature began to increase. The temperature exceeded 40 °C by the end of 1973, rose to 45 °C by mid-1974, and remained stable at this level for almost 6 months. Warning was issued when the temperature exceeded 40 °C. The temperature began to decline after 1975, and no eruption occurred. The temperature was back to normal at the level of 32 °C in mid-1976. The average repose period has already been exceeded, so intensive monitoring and societal preparation is being carried out at Kelut volcano.

Thermal radiation measurement can be done directly as well as indirectly. Infra-red thermometers are used to measure the temperature from a distance. The infra-red image scanner is able to collect data on the distribution of heat at the surface. Satellite platforms and telemetric techniques are now used to transmit temperature data. The conventional thermometer and thermocouple are still important.

**Ground Deformation Measurement**

Movement of magma, particularly in large amounts at shallow depths, is manifested in topographical changes at the surface. The rising magma deforms the earth's surface; the increasing pressure produces the upheaval of the volcanic body or "inflation." On the other hand, decreasing pressure caused by the retreating magma or shrinkage due to cooling will produce deflation or subsidence.

Displacement typically has both vertical and horizontal components. Such deformation produces tilting of the ground, which is measurable with tiltmeters.

The magnitude of the earth deformation depends upon the amount of the accumulated pressure and also upon the stress-response of host rocks or the volcanic body. Magma characteristics play an important role as to the nature of the pressure. Basaltic magma, such as that from Hawaiian volcanoes, will produce a gradual accumulation of pressure, so the deformation might be detected a reasonable time prior to the eruption. Moreover, the host rock where the stress is accumulated consists of basaltic lava which is relatively homogeneous, forming a shield and easily reacting to the pressure.

Ground deformation studies, therefore, have been successful at Kilauea and Mauna Loa volcanoes in Hawaii. As early as 1926, attempts were made to correlate temperature and earthquakes.

Several ground deformation devices have been developed, as follows:

a. Tilt Measurement
- watertube system
- mercury-pool tiltmeter (3-dimensional electric leveling)
- spirit level
- lake level tiltmeter

b. Vertical Displacement
- role tape measurement
- leveling instrument
- satellite geodetic positioning
- tide gauges

c. Horizontal Displacement
- triangulation survey
- roll tape measurement
- electro distance measurement (EDM)
- satellite geodetic positioning

Using Hawaiian volcano ground deformation monitoring, the Halemaumau eruption of 1967 was detected one month before the eruption by shortbase tiltmeters. Similarly, measurements of horizontal strain at the caldera had shown the continuous growth (dilatation) almost 2 years before the October 1967 summit eruption, and before flank eruptions at the end of 1968 and the beginning of 1969.

In intermediate and silicic volcanic rocks, rising magma might deform the ground surface as a cryptodome, as during the 1943-1945 eruption of the Usu volcano. A large upheaval up to 50 m in the months from early 1944 to June 1944 preceded an eruption that began from a fissure on top of the upwarp. The uplift continued until late 1945, when a final height of 150 m was reached, producing Syowa-Sinzan (roof mountain) dome. This activity was followed by the extrusion of a viscous lava dome 100 to 200 m high.

In Indonesia, ground deformation measurements were begun in the late 1970s. Benchmarks for ground deformation were installed in several volcanoes to be monitored. The "Dry tilting" or spirit leveling technique has been used at several volcanoes. Tangkuban Parahu volcano in West Java has shown continuous inflation totaling 600 microradians during the last 5 years (Dvorak and Said, 1986). When the seismicity increased in 1984, the first stage alarm was issued; fortunately, the activity ended with a small phreatic eruption at Kawah Baru.

Deformation changes without eruptions have been noted in some other places, such as in Rabaul caldera, Papua New Guinea, during a 1983-1984 crisis (McKee et al., 1985), and in Pozzuoli, Campi Flegrei, Italy from 1982 to 1984 (Bianchi et al., 1984). Here the earth surface uplifted up to 160 cm within that period. This might be related to the subsurface movement of the magma.

Some phases of the Galunggung eruption, West Java, Indonesia, were preceded by deformation (Katili and Sudradjat, 1984; Matahelumual, 1986). An electronic distance-measuring instrument was used here to detect the dilatation or the shortening of lines to several points on the slope of the volcano. The measuring distance covers some 10 km, and dilatation of around 4 to 16 ppm was detected prior to the eruptions. Although the method was used only to supplement seismic monitoring, it was significant contribution to the forecasting of individual volcanic eruptions during the crisis.

**Miscellaneous Geophysical Methods**

Other geophysical methods that will be briefly discussed in this section consist of geomagnetic study, gravity observation and geoelectric, e.g., self-potential and conductivity measurement. Some of the techniques have been carried out for long time; for example, geomagnetic study has been done since 1951 in some Japanese volcanoes by Rikitake and Yokoyama (1956) and others, and also in Hawaiian volcanoes. However, the method has not been operationally used for volcano monitoring for short-term forecasting. The experiment carried out in Galunggung during the 1982 eruption, and also in Merapi since 1981, has not produced any promising results. Some geomagnetic variations have been detected in the individual Galunggung eruptions; however, the preceding phenomena have not been detected.

Geomagnetic study is also applied to reveal the subsurface structure of volcanic areas. The survey may be done onland, airborne, or shipborne. For deeper structures, magnetotelluric surveys are usually carried out.

Gravity observation may be used to detect subsurface structure and to monitor changes in gravity associated with volcanic activity. Gravity surveys can also be used to detect elevation changes and may be correlated with the deformation measurement.

Geoelectric measurement is used to measure the electric resistivity of the rocks, which in turn will reflect the temperature. Resistivity will decrease as temperature increases or magma approaches. Resistivity measurements using dipole-dipole have been experimentally carried out in Kilaeua and Oshima. Another geoelectric method is the self-potential survey to measure the fluctuations of self-potential values, which theoretically correspond to the temperature of the rock or the hydrothermal system. In the Merapi volcano, Central Java, the experiment has been done since 1981. During the eruption of Api Siauw in 1985 self-potential measurement was done across the south flank down from the active vent. A significant change was detected in the slope of the volcano suspected to be the weak zone, where the flank eruption may take place.

Conductivity measurement was just recently started in the Kelut volcano, East Java, Indonesia. Here the conductivity of the crater lake water is monitored and the result is transmitted to the observatory 8 km from the crater lake. By measuring the conductivity, which reflects the concentration of dissolved ions, it is hoped that changes of salinity can be monitored (McGee, Sutton and Supriatman, 1986).

## Chemistry and Geochemistry

**Gas Emission Observation**

Volcanic gas analysis has been important since the successful monitoring of Mount St. Helens in 1980. Earlier work by Tazieff and associates, which was concentrated on the volcanic gas monitoring in many volcanoes such as Nyiragonggo, Africa; Mount Etna; Italy and Merapi, Indonesia has also contributed considerably to the state of the art of volcanic eruption forecasting. At Kilauea, volcanic gases currently are monitored at 25 sites analyzing more than 10 different species (Greenland, 1984; Tilling and Bailey, 1985).

During the 1980 eruption of Mount St. Helens, a correlation spectrometer (COSPEC) was used to monitor the sulfur dioxide emission in airborne and ground-base modes. Emission of sulfur dioxide and carbon dioxide helped in the prediction of an explosive and nonexplosive eruption of this volcano (Casadevall et al., 1981).

In Japan, emission studies of $SO_2$, HCl, and $H_2$ at the Sakurajima volcano were carried out. It was concluded that the HC1/ $SO_2$ ratio might be correlated with the occurrence of volcanic seismicity.

In the Merapi volcano, Central Java, Indonesia, gas monitoring includes the measurement of amounts of $CO_2$, CO, $SO_2$, $H_2$ and $N_2$ emitted from fixed points at the solfataric field at the Merapi summit. The measurement is done at an interval of twice a month. Additionally, radioactive gas concentration of radon ($Rn^{222}$) is also measured.

$SO_2$ measurement using COSPEC is done on a weekly interval or shorter, depending upon the Merapi activity and the weather conditions. It may be concluded that the significant parameters in gas emission detection at Merapi are the $SO_2$ flux and the ratio of CO over $CO_2$, which reflect the increasing temperature of the solfataric field.

In the 1982 Galunggung eruption the COSPEC was used. The airborne mode was attempted during the first 2 weeks of August 1982. It soon became apparent that it was almost impossible for the aircraft to fly underneath the eruption plumes, which usually rose no higher than 400 to 1000 m above the vent. The method was then abandoned and replaced by a fixed ground measurement station.

The results have been encouraging, as evidenced by sharp increases in $SO_2$ before or during eruptions. In general the average rate of $SO_2$ emission increased from August (200 to 400 tons/day) until September (2000 tons/day) 1982 and declined by the end of September. However, a sharp increase occurred in the first week of October, which was followed by an eruption. Later, the rate of $SO_2$ emission again gradually increased from 200 tons/day to a peak of around 1000 tons/day by early November. This phenomenon might be interpreted as a manifestation of rapid solidification and degassing of the molten magma reaching the surface.

**Water Chemical Composition**

Water chemical composition monitoring can be done from springs around the volcano, rivers, and also crater lakes. The most significant results on water chemical composition monitoring are perhaps obtained from water in the crater lake. The lake water absorbs gas emitted form the crater's bottom. A monitoring device to detect temperature was placed in the lake. The convection currents of the lake water were also studied.

In the Kelut volcano, East Java, the Volcanological Survey of Indonesia installed telemetric equipment to measure chemical as well as physical properties of the lake water. Water volume fluctuation, pH, and conductivity are automatically measured. Previously pH, temperature and water volume were measured by means of conventional and telemetric devices since 1974 (Moegiono, 1980).

Temperature was one of the significant changes preceding 1951 eruption, but as pointed out earlier in this chapter, temperature gave a false alarm in 1974. Volume, air, temperature, and chemical composition have to be taken into account in the temperature interpretation of lake water.

**VOLCANIC ERUPTION FORECASTING**

A. FUNDAMENTAL RESEARCH APPROACH
- GEOLOGICAL RECORD

B. ERUPTIVE RECORD
- HISTORICAL RECORD
- STATISTICAL APPROACH

C. VOLCANO MONITORING
- PHYSICAL METHOD
- GEOPHYSICAL METHOD
- CHEMISTRY & GEOCHEMISTRY

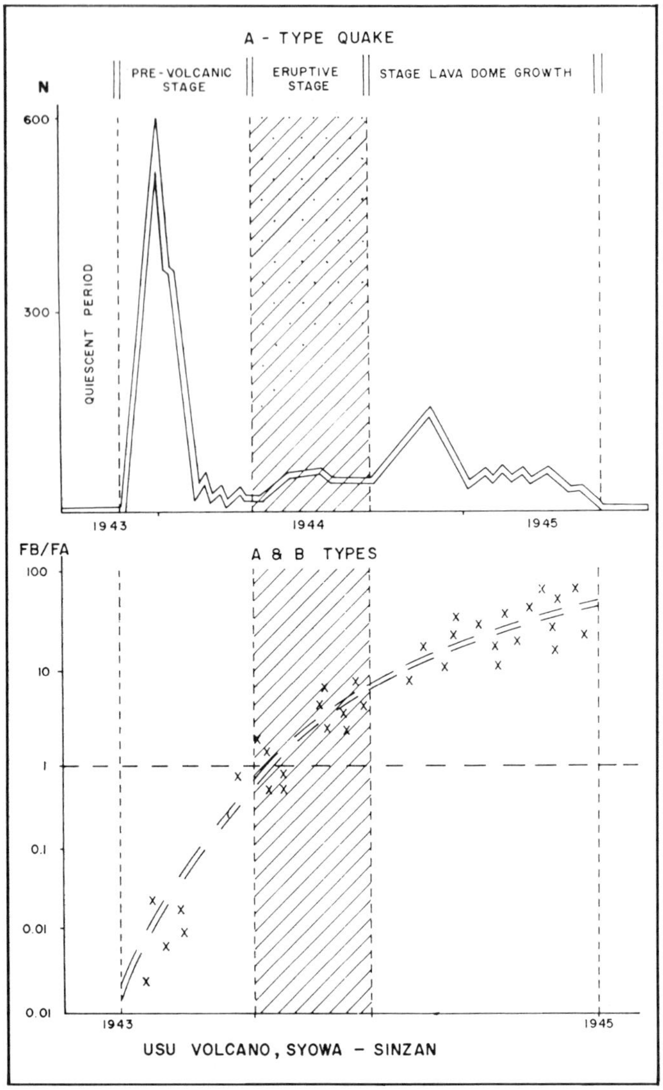


The chemical composition of the lake water contains $Mg^{2+}$ and $Cl^-$ where the ratio between $Mg^{++}/Cl^-$ perhaps will be an indicative parameter.

The hydrophone to monitor the rate of gas bubble raising up from the lake's floor was also tried (Bercy, Sabroux, and Suparwoto, 1983). The result was encouraging, since the rate of gas bubble indicates the increasing activity. The experiment is now continuing.

### Petrochemical Analysis

Petrochemistry is applied to the monitoring of volcanoes for long-term forecasting. Petrochemical analysis of the volcanic products deposited on the volcanic slope shows the evolution of magma composition in the past. The petrochemical interpretation of the Krakatau's volcanic rocks is an example of this type of monitoring. It is believed that the silica content of the rocks produced by Krakatau will progressively increase before a caldera-forming eruption occurs. On the basis of the petrochemical analysis of the rock sequence in the Krakatau complex, van Bemmelen came to the conclusion that the repeated cataclysmic eruptions of Krakatau were depicted in the $SiO_2$ content found in the layer. $SiO_2$ content of about 67% is believed to be critical for an occurrence of a destructive eruption. It was interpolated that an eruption similar to that of 1883 will likely occur in more than 400 years to come. At present the rock composition produced by Krakatau eruptions since the birth of Anak Krakatau in 1927 has ranged from 50% at the earliest development to 53% in the last 10 years.

A similar technique was used in monitoring the individual eruptions of the Galunggung volcano in 1982. A very significant decrease was found, from 54% of $SiO_2$ content at the beginning of the crisis (April) to about 50% in July 1982. New composition was perhaps injected by August 1982, as reflected in the sharp increase of $SiO_2$ content during the eruption of this month. In the following months the $SiO_2$ content decreased gradually from 54% back to 48% by the end of the activity which produced lava flows in January 1983.

## CONCLUDING REMARKS

Volcanic eruption forecasting is one of the most crucial parts of the volcanological science. The result of this scientific work is directly evaluated and will help save human lives. There are many success stories about volcanic eruption forecasting; unfortunately, there are also many unsuccessful ones which ended in embarrassment, chaos, or even more tragic conclusions.

The Mount St. Helens eruption in 1980 has brought a new horizon in the volcano monitoring technique and its application. Although no distinct short-term warnings were noted preceding the main eruption of May 18, 1980, the subsequent individual eruptions later have been successfully forecast. The experiences gained in volcano monitoring in Japan, particularly in the field of seismicity, in the USSR during the Bezymianny 1956 eruption, and in various fields of volcano monitoring in Hawaiian volcanoes that frequently erupt, have been the solid foundation for the development of the art of forecasting of volcanic eruption. In addition scientists from Japan, Italy, and France who have been working in Asama,

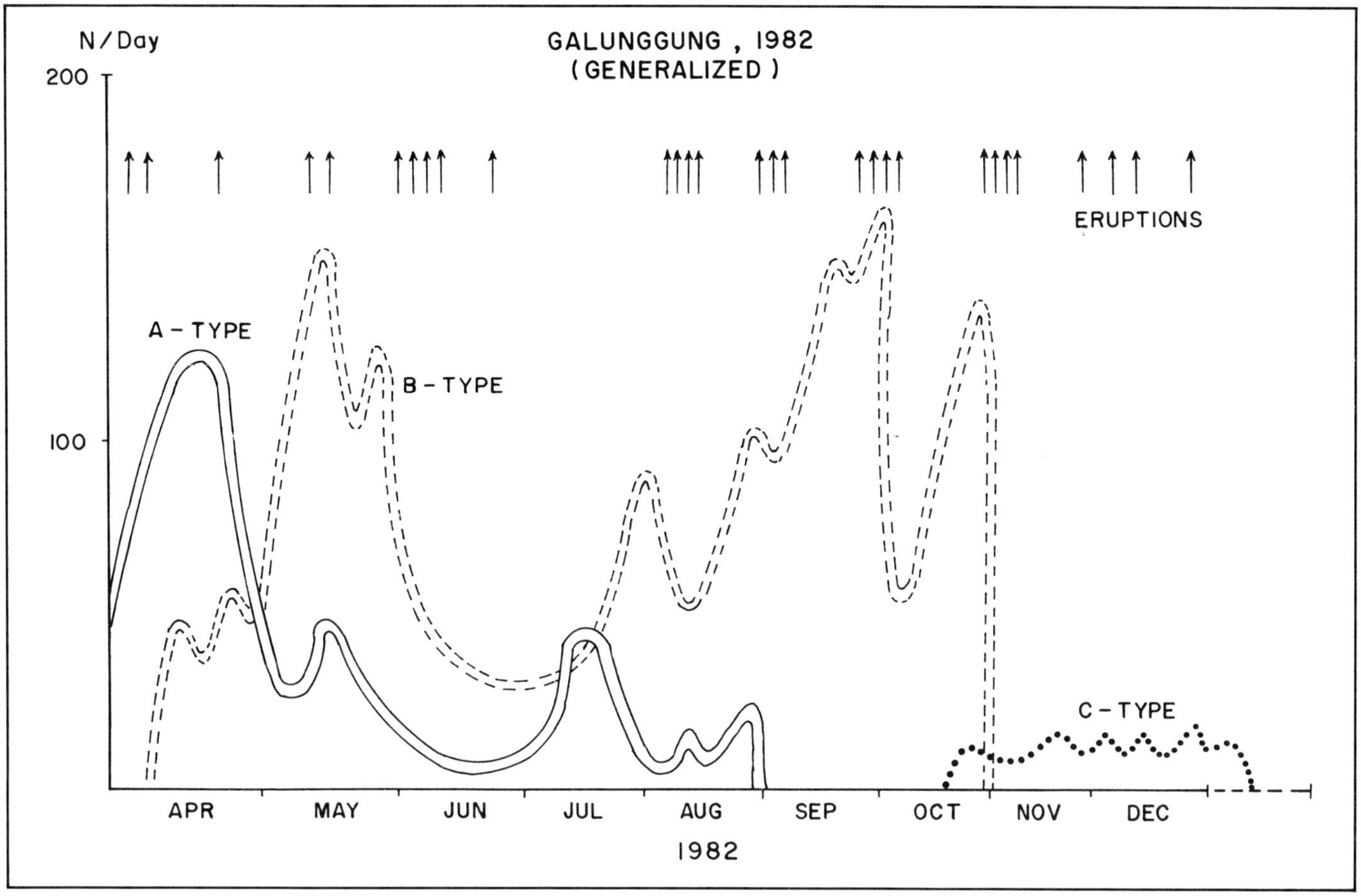
N/Day
GALUNGGUNG, 1982
(GENERALIZED)
200
100
ERUPTIONS
A-TYPE
B-TYPE
C-TYPE
APR
MAY
JUN
JUL
AUG
SEP
OCT
NOV
DEC
1982

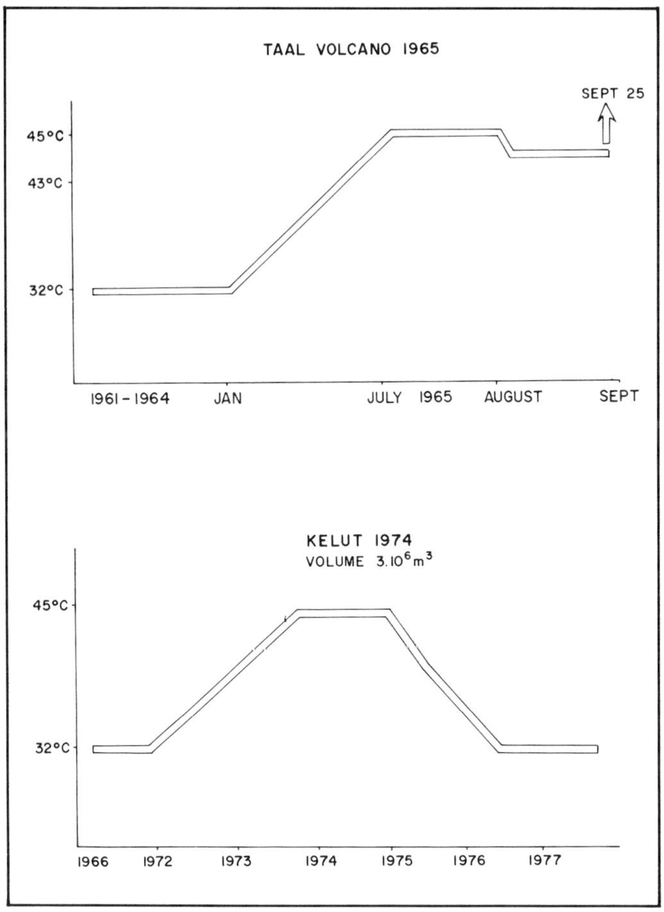
TAAL VOLCANO 1965
SEPT 25
45°C
43°C
32°C
1961-1964
JAN
JULY 1965
AUGUST
SEPT
KELUT 1974
VOLUME 3.10⁶m³
45°C
32°C
1966
1972
1973
1974
1975
1976
1977

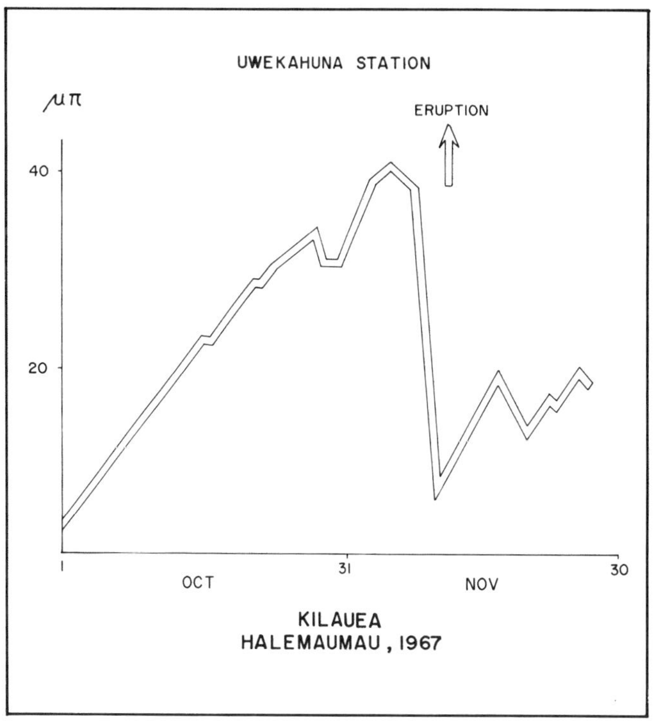
UWEKAHUNA STATION
μπ
ERUPTION
40
20
1
31
30
OCT
NOV
KILAUEA
HALEMAUMAU, 1967

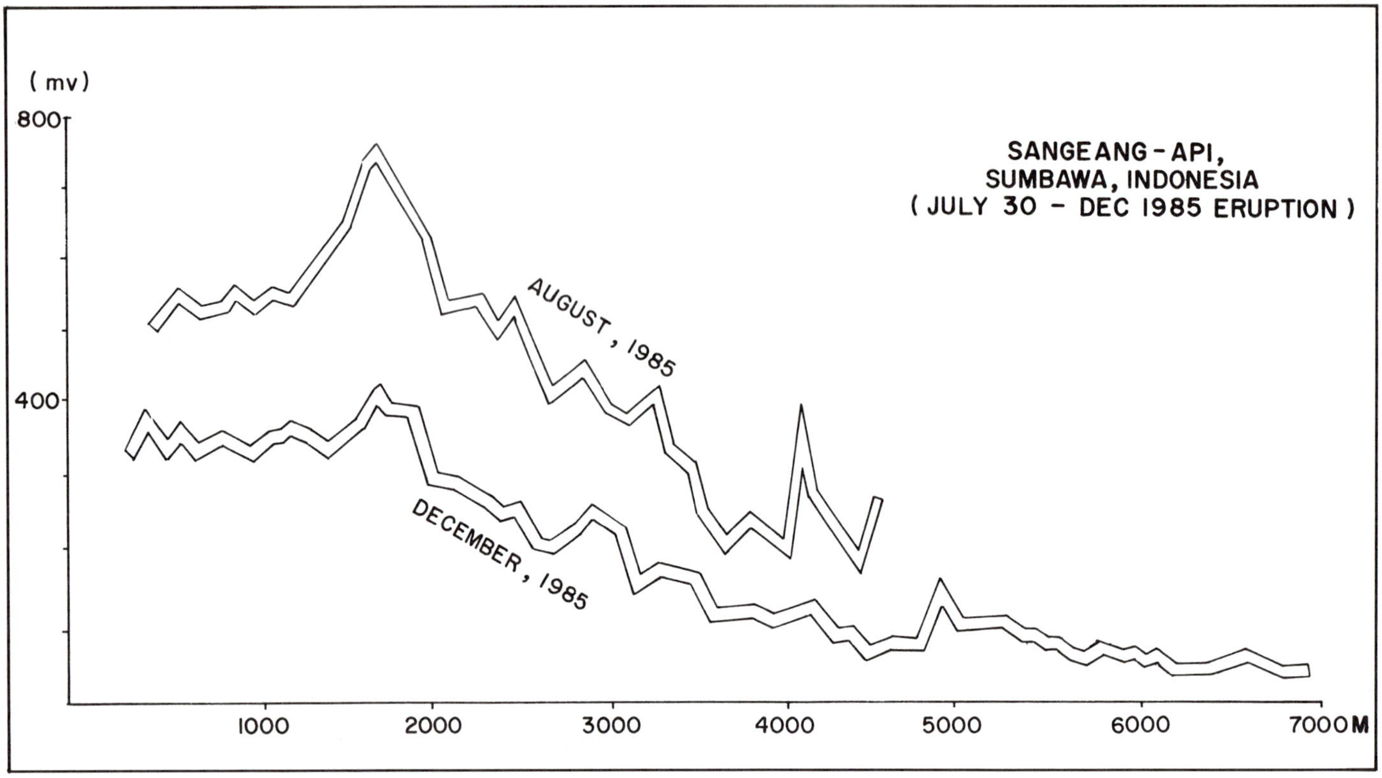

(mv)
800
400
SANGEANG-API,
SUMBAWA, INDONESIA
(JULY 30 - DEC 1985 ERUPTION)
AUGUST, 1985
DECEMBER, 1985
1000
2000
3000
4000
5000
6000
7000M


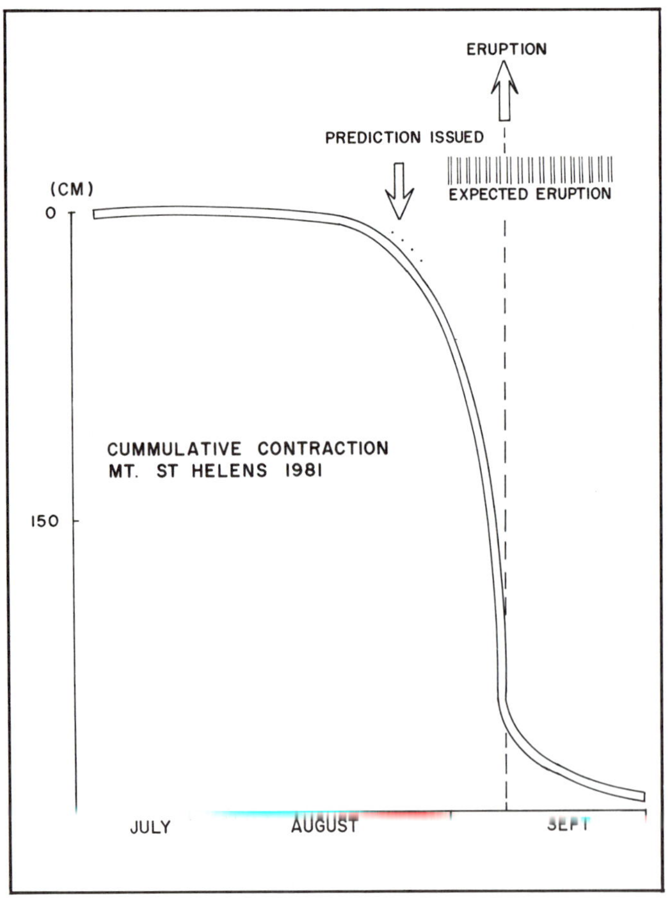

ERUPTION
PREDICTION ISSUED
EXPECTED ERUPTION
(CM)
0
150
CUMMULATIVE CONTRACTION
MT. ST HELENS 1981
JULY
AUGUST
SEPT


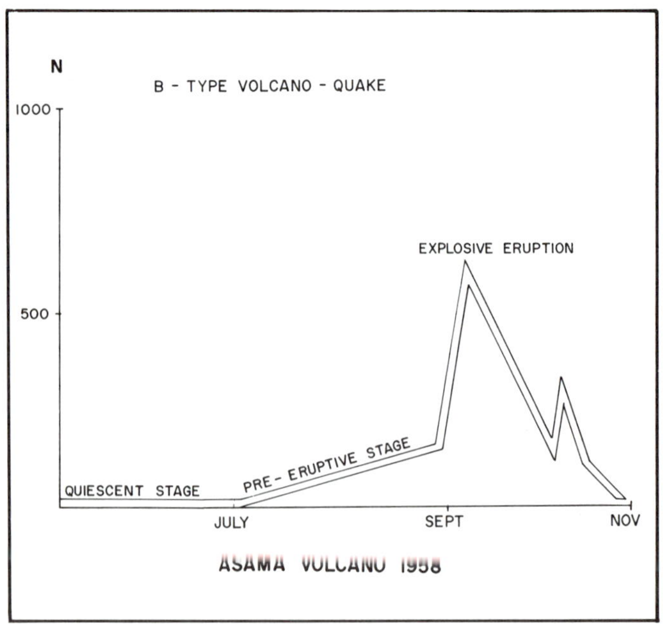

N
B - TYPE VOLCANO - QUAKE
1000
500
EXPLOSIVE ERUPTION
QUIESCENT STAGE
PRE-ERUPTIVE STAGE
JULY
SEPT
NOV
ASAMA VOLCANO 1958

Usu, Mount Etna, and Mount Vesuvius contribute significantly to the development of monitoring techniques and volcano eruption forecasting, notably in the fields of seismicity and geochemistry.

Since volcanic eruption forecasting involves human lives at stake, it is important to incorporate societal measures in the volcano hazard mitigation program. A national or international program of volcano monitoring and forecasting is therefore highly recommended. The provisions for baseline data and skillful and professional volcanologists and scientists are imperative. With the increasing awareness among people in the world, it is hoped that the volcanic disaster, such as that of Armero, Colombia recently, might be minimized in the future.

## ACKNOWLEDGMENTS

The author wishes to acknowledge the VSI staffs for their contributions of data collected in the field, and also the U.S. Geological Survey-Volcanological Survey of Indonesia cooperative program. Special thanks go to Dr. Christ Newhall, who reviewed the manuscript.

## REFERENCES

Barberi, F., D.P. Hill, F. Innocenti, G. Luongo, and M. Treuil, 1984, The 1982-1984 Bradyseismic crisis at Phlegraean Fields (Italy), Bulletin Volcanologique, v. 47, n. 2 (special issue), 411 p.

Bercy, C., J.C. Sabroux, and G. Suparwoto, 1983, Underwater noise survey in the crater lake of Kelut volcano (Indonesia), *in* Forecasting volcanic events: Elsevier, p. 529-543.

Bianchi, R., C. Federico, G. Giberti, G. Sartoris, and R. Scandone, eds., 1984, Modelling of subsurface ground deformations in the Phlegraean Fields volcanic area, Italy, *in* The 1982-1983 Bradyseismic Crisis at Phlegraean Fields (Italy), Bulletin Volcanologique, v. 43, n. 2, p. 321-330.

Casadevall, T.J., D.A. Johnston, D.M. Harris, W.I. Rose, Jr., L.L. Malinconico, R.E. Stoiber, T.J. Bornhorst, S.N. Williams, Laurel Woodruff, and J.M. Thompson, 1981, $SO_2$ emission rates at Mount St. Helens from March 29 through December 1980, The 1980 eruptions of Mount St. Helens, p. 143-207.

Civetta, L., P. Gasparini, G. Luongo, and A. Rapolla (eds.), 1974, Development in solid earth geophysics *in* Physical volcanology, New York, Elsevier, 333 p.

Decker, R.W., 1978, State of the art in volcano forecasting: Geophysical Prediction, Stud. Geophys., p. 47-57.

Decker, R.W., 1986, Forecasting volcanic eruptions: Ann. Rev. Earth Sci., v. 14, p. 267-291.

Decker, R.W., R.Y. Koyanagi, J.J. Dvorak, J.P. Lockwood, A.T. Okamura, et al., 1983, Seismicity and surface deformation of Mauna Loa volcano. Hawaii: Eos (American Geophysical Union Transactions), v. 64, p. 545-547.

Dibble, R.R., 1965, Volcanic seismology, *in* New Zealand volcanology, central volcanic region: New Zealand Geol. Handb. Inf. Ser., 50, p. 93-98.

Dvorak, J., and H. Said, 1986, Review of the deformation measurements in the last 5 years: USGS-VSI Cooperative Program Internal. Rep., unpubl.

Fournier d'Albe, E.M., H. Tazieff, B. Booth, P.E. Baker, J.E. Griest, et al., 1979, Volcano prediction: Journal of the Geological Society of London, v. 136, p. 321-359.

Gorshkov, G.S., 1971, Prediction of volcanic eruptions and seismic methods of location of magma chambers—A review: Bull. Volcanol., v. 35, n. 1, p. 198-211.

Greenland, L.P., 1984, Gas composition of the January 1983 eruption of Kilauea volcano, Hawaii: Geochimica et Cosmochimca Acta, v. 48, p. 193-198.

Katili, J.A., and A. Sudradjat, 1984, The devastating 1983 eruption of Colo volcano, Una-una Island, Central Sulawesi, Indonesia, Geol. Jahrb. Reiche. A 75, p. 27-47.

Klein, G.W., 1982, Pattern of historical eruptions at Hawaiian volcanoes: Journal Volcanol. Geotherm. Res., v. 12, p. 1-35.

Koyanagi, R.Y., E.T. Endo, and J.S. Ebisu, 1975, Re-awakening of Mauna Loa volcano, Hawaii, Geophys. Res. Lett., 2, p. 1-35.

Kubotera, S., 1965, Volcanic earthquake: Bull. Volcanol. Soc. Japan, v. 10, p. 91-99.

MacDonald, G.A., 1960, Prediction of eruption of Hawaiian volcanoes: Bull. Volcanol, Ser. II, v. 23, 210 p.

Malingreau, J.P., and I. Kaswanda, 1986, Monitoring volcanic eruptions in Indonesia using weather satellite data: The Colo eruption of July 28, 1983: J. of Volcanol. and Geoth. Res., v. 27, p. 179-194.

Martin, R.C., and J.F. Davis (eds.), 1982, Status of volcanic prediction and emergency response capabilities in volcanic hazard zones of California: California Division of Mines and Geology Special Paper 63, 275 p.

Matahelumual, J.,, 1986, Pengukuran deformasi Galunggung: Penyelidikan Letusan Galunggung 1982.

McGee, Kenneth A., Andrew J. Sutton, and I. Supriatman, 1986, Establishment of a telemetred lake monitoring system at G. Kelut, USGS-VSI Cooperative Program Internal Rep., unpubl.

McKee, C.O., P.L. Lowenstein, P. de Saint Ours, B. Talai, J.J. Mori, and I. Itikarai, 1985, Seismic and ground deformation crisis at Rabaul caldera, September 1983 to October 1984: Volcano News, 1985 (19-20), p. 3.

Minakami, T., 1967, Earthquake originating in volcanoes: Journal of the Seismological Society of Japan, v. 20, p. 177-181.

Minakami, T., (ed.), 1971, The surveillance and prediction of volcanic activity: Paris, Unesco, Earth Sci. Ser., n. 8, 116 p.

Minakami, T., 1974, Seismology of volcanoes in Japan, *in* L. Civetta et al., eds., Physical volcanology: New York, Elsevier, p. 198-211.

Minakami, T., T. Ishikawa, and K. Yagi, 1951, The 1944 eruption of volcano Usu in Hokkaido, Japan: Bull. Volcanol. Ser. II, 11, p. 45-157.

Moegiono, R., 1980, Sistem Telemetri G. Kelut, Laporan Kerjasama Gadjah Mada - Direktorat Vulkanologi, unpubl, 12 p.

Okada, Hm., H. Watanabe, H. Yamashita, and I. Yokoyama, 1981, Seismological significance of the 1977-1978 eruptions and the magma intrusion process of Usu volcano, Hokkaido: Journal of Volcanology and Geothermal Research, v. 9, p. 311-334.

Omori, F., 1911, The Usu-san eruption and earthquakes and elevation phenomena: Bull. Imp. Earth Invest. Comm., v. 5, p. 1-38.

Simkin, T., L. Siebert, L. McClelland, D. Bridge, C. Newhall, and J.H. Latter, 1981, Volcanoes of the world: Stroudsburg, Pennsylvania, Hutchinson Press, 232 p.

Suparto, S., 1986, Pemonitoran Seismik, Penyelidikan Letusan Galunggung 1982.

Tilling, R.I., and R.A. Bailey, 1985, Volcano hazards program in the Unites States: Journal of Geodynamics, p. 425-446.

Tjia, H.D., 1977, Hasil Penyelidikan G. Umsini, Irian Jaya 1976, Arsip Direktorat Vulkanologi, unpubl, 30 p.

Volcanological Survey of Indonesia, 1986, Annual report: Bulletin of the Volcanological Survey of Indonesia, for the year 1985-1986.

Walker, G.P.L., 1974, Volcanic hazards and the prediction of volcanic eruptions: Geological Society of London Miscellaneous Publication 3, p. 23-41.

# VOLCANOLOGY ABSTRACTS

BYKOVSKAYA, E. V., A. O. SOBOLEV, A. A. TARKNAEV, and A. P. KARPINSKY, All-Union Geological Research Institute, Leningrad, USSR

## Peculiarities of Volcanism of Marginal-Continental Volcanic Belts of East Asia

The eastern Sikhote-Alin and Okhotsk-Chukhotskiy volcanic belts represent a linear structure extending for thousands of kilometers. They developed at the margin of the Asian continent during the middle Cretaceous and formed over a heterogenous basement of older, folded rocks. Both volcanic belts developed in multiple stages, and the magmatic evolution of both was the opposite of normal magmatic trends. Thus, a rhyolite-granite association characterizes the oldest stage of development; andesite-granodiorite, the mature stage; and basaltic associations, the youngest stage. Consequently, their structural style also changes—from the development of ring volcanic-tectonic depressions, calderas, and domes during the oldest and mature stages, to the development of grabens, linear depressions, and shield volcanoes during the youngest stage.

Significant metallogenic properties of these belts are related to the development of gold-silver, silver-tin, tin and associated metals, mercury-antimony, porphyry molybdenite-copper, and fluorite mineralizations.

Geochemical zonation parallels the trend of the volcanic belts, with the expected increase in potassium and aluminum oceanward, and an increase in sodium continentward. The copper and molybdenum ores are concentrated along the axis of the volcanic belts, whereas tin and its related metals association are concentrated along the western (external) parts of the volcanic belts. Transverse zonation is seen in the variations in $SiO_2$, $K_2O$ + $Na_2O$, MgO, and other types of metallic associations. For example, gold and silver mineralization is associated with paleouplifts.

The close relationship between magmatism and the formation of tin, and their distribution in the contexts of subsurface geology, fault-block tectonics, and the host country rocks, are demonstrated by the results of the geological and geophysical studies presented.

MUFFLER, L. J. PATRICK, U.S. Geological Survey, Menlo Park, CA

## Volcanic Processes Influencing Development of Geothermal Resources

High-temperature geothermal resources occur primarily in zones of volcanism along spreading ridges, above subduction zones, and at intraplate melting anomalies. In these regions of crustal magmatism, thermal energy is transferred upward by conduction and by the coupled processes of magma movement and hydrothermal circulation of meteoric and ocean water. Favorable geothermal targets occur where basaltic magma injected repeatedly into the crust creates long-lived magmatic systems producing intermediate and silicic magmas through a complex set of processes dominated by fractionation and crustal contamination. These derivative magmas of low density and high viscosity are more likely than basalt to accumulate in the upper crust as stable heat sources driving overlying hydrothermal convection systems.

Geothermal reservoir geometry in volcanic terranes results from the complex interaction of active volcano-tectonic processes and older volcanic stratigraphy and structure. Hot springs tend to occur at relatively low elevations at the distal ends of obliquely upflowing tongues of hot water, commonly along structurally favorable zones. Frequently, the principal geothermal reservoir is a deeper zone of primary hydrothermal upflow that is offset toward the youngest volcanic rocks. Therefore, in developing a geothermal field, the ages, volumes, and spatial relations of young volcanic rocks must be considered.

Volcanic hazards must be considered in siting and constructing surface geothermal facilities. Both the potential for various types of eruptions and the impact of magma-induced ground deformation on sensitive structures must be evaluated.

SIMKIN, TOM, and LINDSAY MCCLELLAND,* Smithsonian Institution, Washington, D.C.

## Eruptions in Space and Time: Durations, Intervals, and Comparison of World's Active Volcanic Belts

A computerized data bank, compiled over the last 12 years at the Smithsonian Institution, allows summaries to be made of Holocene volcanism. The Scientific Event Alert Network tracks current volcanic activity. However, the record of most volcanoes is poor before the last 100 years, and some eruptions still pass unreported.

The time interval since the previous eruption can be calculated for 4,835 of the 5,564 compiled eruptions. The median interval is 5.0 years, but much longer intervals commonly precede unusually violent eruptions. For the 25 most violent eruptions in the file (with known preceding interval), the median interval is 865 years. Of the historic eruptions in this group, 50% resulted in fatalities.

The interval between an eruption's start and its most violent paroxysm may be measured in months or years, but it is usually short. Of the 205 larger eruptions for which data are available, 92 had the paroxysmal event within the first day of the eruption, allowing little time for emergency preparations after the eruption's opening phase.

To compare the recent vigor of different volcanic belts, we calculated the number of years in which each volcano was active in the last 100 years, summed these for each belt, and divided by belt length. Another index of recent vigor is the number of recognized Holocene volcanoes divided by belt length. A third index is the number of large explosive eruptions (volcanic explosive index $\geq$ 3) of the last 100 years, again normalized by belt length. These three measures correlate reasonably well, serving to contrast vigorous belts such as Kamchatka, Central America, and Java with relatively quiet belts such as the Cascades, South Sandwich Islands, Greece, and southern Chile.

# MINERALS

CHAPTER 46

# NEOGENE-QUATERNARY VOLCANISM AND MINERALIZATION IN THE CENTRAL ANDES

**George E. Ericksen**
*U.S. Geological Survey,*
*Reston, VA*

**V. Raul Eyzaguirre**
*Empresa Minera del Centro del Peru*
*Lima, Peru*

**Fernando Urquidi B.**
*American Embassy*
*La Paz, Bolivia*

**Raul Salas O.**
*Servicio Nacional de Geología y Minería*
*Santiago, Chile*

ABSTRACT

Eruptive centers, which include calderas, stratovolcanos, and flow-dome complexes in the Neogene-Quaternary volcanic complex of the central Andean region of northwestern Argentina, western Bolivia, northern Chile, and southern Peru, are potential targets for hydrothermal mineral deposits. The volcanic complex, consisting chiefly of rhyolitic ash-flow tuffs and andesitic lavas, extends over an area of about 300,000 $km^2$ within a 1,000,000 $km^2$ area that is one of the world's great mineral provinces. Because many deposits of Cu, Pb, Zn, Sn, W, Sb, Bi, Ag, and Au in this area are chiefly of pre-Pliocene age, it had been believed that most of the Neogene-Quaternary volcanic rocks represented post-mineralization cover. However, recent studies have shown that metalliferous deposits as young as Pleistocene are associated with the eruptive centers. Mineral deposits in these centers were emplaced during the waning phases of volcanic activity, and available radiometric ages show they were emplaced chiefly within an interval of not more than 1 m.y. after the major eruptive phase of the center.

## INTRODUCTION

Polymetallic Sn deposits, epithermal Ag-Au deposits, and polymetallic Ag-base-metal deposits, some of which have been exploited since the 16th century, are by far the most economically important types of deposits associated with the Neogene-Quaternary volcanic rocks in the central Andes. Among these are the polymetallic Sn and Sn-Ag deposits in southern Bolivia, which are associated with porphyry stocks and vent breccias of former stratovolcanos or dome complexes, and Ag-Au and base-metal veins associated with dome complexes, calderas, and fossil thermal-spring systems. Other types of deposits are (1) porphyry-type Cu, Ag-Cu, and Au-Cu deposits associated with porphyry stocks in deeply eroded stratovolcanos; (2) native S, Ag- and Sn-bearing veins, covellite impregnations, and presumed magnetite-hematite flows in the upper parts of stratovolcanos; and (3) manganese oxide, Sb, wood tin, and supergene uranium deposits associated with thermal-spring systems, some of which are still actively depositing metals. In addition, volcanic rocks, chiefly the rhyolitic ash-flow tuffs, and associated thermal springs contain abundant water-soluble salts, and are the chief sources of the widespread saline deposits of this region, which include world-class resources of Li and B.

This is a preliminary phase of a proposed investigation of the relationship of mineral deposits and Neogene-Quaternary volcanism in the central Andean region of southern Peru, northern Chile, western Bolivia, and northwestern Argentina. The information presented has been abstracted from many published reports, chiefly of the last 10 years, and from unpublished information from the authors of this report. The volcanic rocks under consideration include rhyolite ash-flow tuffs or ignimbrites that have originated at many caldera centers and andesitic lavas from hundreds of stratovolcanos. These late Tertiary and Quaternary rocks cover a 300,000 $km^2$ area in the western Andean Highlands (Figure 1). They obscure about a third of the central Andean region, one of the most heavily mineralized areas on earth, where there are many metalliferous deposits of hydrothermal origin ranging in age from Triassic to Miocene. These deposits contain about a third of the world's resources of copper, chiefly in some 20–30 known porphyry copper deposits, and world-class resources of lead, zinc, silver, gold, antimony, bismuth, tin, and tungsten. Obviously, many deposits remain to be found beneath the late Tertiary-Quaternary volcanic cover.

Investigations during the past 10–15 years have shown that the Neogene-Quaternary volcanic rocks are not largely post-mineralization, as had been previously thought, but that many eruptive centers in them have metalliferous mineral deposits. We believe that metalliferous deposits are far more numerous and widespread in eruptive centers in the central Andes than is now known, and that they can be found by geophysical, geochemical, and geological investigations.

Although many papers about the young volcanic rocks of the central Andes have been published since the early 1970s, relatively few are concerned with the mineral deposits in them. One of the milestone papers is that of Sillitoe (1973) showing the relation of porphyry copper deposits to stratovolcanos, which is largely based on Sillitoe's work in Chile and Argentina. Francis et al. (1983) discussed the relationship between mineralization and silicic volcanism in the central Andes. Sillitoe and Bonham (1984) compiled information about the relation of mineral deposits and volcanic landforms worldwide, which includes several examples from the central Andes. Noble et al. (in press) discuss silver and gold deposits associated with calderas and stratovolcanos in southern Peru. Several reports (Sillitoe et al., 1975; Angus et al., 1977; Francis et al., 1981; Grant et al., 1977, 1980) discuss the relationship of polymetallic tin deposits in southern Bolivia to Miocene volcanism. Reports describing metalliferous deposits in specific eruptive centers include those of Brodtkorb and Coira (1986), Caelles et al. (1971), Chase (1948), Clark (1970), Dávila (1981), McKee et al. (1975), Noble and McKee (1982), Sillitoe (1975), Noble and Silberman (1984), Petersen et al. (1977), and Peterson et al. (1983).

Geochemical studies of ash-flow tuffs offer clues to the nature and history of silicic magmas involved in caldera-type eruptions, which may provide clues to the existence of buried mineral deposits in intracaldera areas. For example, unusual tin-rich two-mica tuffs at Macusani, Peru (Noble et al., 1984), and in the Morococala volcanic field, southern Bolivia (Ericksen et al., 1985), are believed to be the extrusive equivalents of Sn-Li pegmatites of Sn-granites. We believe that subvolcanic intrusives in the eruptive centers of these tuffs are favorable exploration targets for Sn deposits.

# VOLCANIC ROCKS

## Character and Distribution of Eruptive Centers

The Neogene-Quaternary eruptive centers, which are the loci of most of the known mineral deposits associated with the young volcanic rocks of the central Andes, include (1) composite stratovolcanos, (2) calderas, (3) dome complexes, and (4) subvolcanic intrusions, breccia pipes, and hydrothermal alteration zones not related to known volcanic landforms. Several hundred stratovolcanos are present in the region. They range from conical peaks showing little or no erosion, to deeply eroded complexes having only remnants of the original cone. The volcanic cones are the most prominent landforms of the region, generally rising to altitudes of 5000–6000 m and having summits that are 1000–2000 m above the surrounding terrain. Several volcanic cones are more than 6000 m high, and the highest, Ojos del Salado, reaches 6800 m.

Widespread ash-flow tuffs, which many specialists in Andean geology formerly believed to be related to fissure-type eruptions, are now recognized as having resulted chiefly from explosive eruptions from many calderas (Baker, 1981; Gardeweg and Ramírez, 1984; Kussmaul et al., 1977). Calderas were first recognized in the central Andes in 1972 (Noble et al., 1974). During the following decade, a total of more than 30 known and inferred[1] calderas were recognized Figure 2). Many of these calderas have resurgent domes clearly outlined by ring-fracture zones. Others are more obscure because of superimposed stratovolcanos and flow-dome complexes, and still others are broad domal ash-flow tuff fields without an obvious collapse structure, which Baker (1981) called ignimbrite shields.

Dome complexes have been recognized in and near several eruptive centers, where they may or may not be genetically related to metallic mineralization. Perhaps the best examples of domes in a mining district are those at Julcani, Peru, where 30 interpenetrating domes have been recognized (Noble and Silberman, 1984). Metallic mineralization took place during a late phase of dome emplacement. Domes also are found along the northern edge of the Arcata mining district, southern Peru (Noble et al., in press). The Sn-Ag deposits at Oruro, Bolivia, are associated with a dome complex (Chase, 1948; Sillitoe and Bonham, 1984).

Several Neogene-Quaternary eruptive centers are characterized by subvolcanic intrusions and hydrothermal alteration zones. These include subvolcanic intrusions and hydrothermal breccias that either did not reach the surface, or, if they did, formed small volcanic cones or domes that were subsequently destroyed by erosion. The polymetallic tin deposits of southern Bolivia are associated with subvolcanic intrusions of this type. Francis et al. (1983) proposed that many of these tin deposits are related to former dome complexes. Other deposits were formed in a near-surface geothermal system not associated with a recognized volcanic landform.

[1]About a fourth of the calderas shown in Figure 2 are inferred calderas, which include calderas identified in LANDSAT images but not visited, and the ignimbrite shields of Baker (1981).

## Ages

The volcanic rocks and associated mineral deposits considered in this report range in age from earliest Miocene (23 Ma) to Holocene (< 10,000 Ma). Stratovolcanos and calderas were active throughout most of this time, but rhyolitic ash-flow tuff eruptions from calderas were dominant during the Miocene and early Pliocene, whereas andesitic lavas and pyroclastic eruptions from stratovolcano centers were dominant after the early Pliocene (Gardeweg and Ramírez, 1984). The youngest known ash-flow tuff sheet in the central Andes has been dated at 0.77 m.y. (Baker and Francis, 1978). Many of the stratovolcanos have fumarolic native sulfur deposits of probable late Pleistocene and Holocene age. Several of the stratovolcanos have had historic eruptions (Casertano, 1963), and some volcanos now show fumarolic activity.

In general, the level of erosion of the stratovolcanos is indicative of their relative ages. Gardeweg and Ramírez (1984) suggested that the most deeply eroded volcanos, which are those not having recognizable cones, are chiefly of Miocene age, whereas progressively less-eroded cones are Pliocene and Pleistocene in age. Surprisingly, only a few of the volcanos show evidence of intense Pleistocene glaciation, indicating that an arid climate persisted in this region during the Pleistocene.

Mineralization generally occurred less than a million years after the major eruptive phase of a given volcanic center (Noble et al., in press; Francis et al., 1983). In most cases, it accompanied late-phase intrusive activity characterized by subvolcanic porphyry intrusions, hydrothermal breccias, fumarolic activity, and thermal-spring systems.

## Major and Trace Element Chemistry

The volcanic rocks of the central Andes are predominantly calc-alkaline. The ash-flow tuffs consist chiefly of quartz latites and rhyodacites, but range from dacites to high-silica rhyolites. On the other hand, the lavas and other ejecta from stratovolcanos are chiefly andesites and dacites, but locally range from basalts to rhyolites (Baker and Francis, 1978; Gardeweg and Ramírez, 1984; Lahsen, 1982; Thorp and Francis, 1979). Some of the stratovolcanos in southern Bolivia and northwestern Argentina are enriched in K, and have been classified as shoshonites (Deruelle, 1978). For convenience of discussion, we will refer to ash-flow tuffs as rhyolites and lavas as andesites. Also, rhyolitic pyroclastics—ash flows, ash falls, and ignimbrites—are lumped together as ash-flow tuffs.

Our studies of ash-flow tuffs in the Bolivian tin belt, which extends from southern Peru to northern Argentina, show them to differ from tuffs elsewhere in the central Andes in being highly peraluminous ($Al_2O_3/Na_2O + K_2O + CaO > 1$) and being rich in Sn, Li, and other trace elements such as Rb, Cs, Be, B, and U, relative to other ash-flow tuffs of the region. Peraluminous two-mica tuffs in the Macusani volcanic field, southern Peru (Figure 1), which have been dated at 4.2 Ma, contain 30–60 ppm Sn and as much as 260–300 ppm Li (Noble et al., 1984). Similar, late Miocene (8.5–6.4 Ma; Koeppen et al., in press) two-mica tuffs in the Morococala volcanic field, southern Bolivia (Figure 1), are anomalously rich in Sn and Li, some containing more than 25 ppm Sn and 400 ppm Li. A younger cordierite-bearing tuff in the southern part of the Morococala field is similarly enriched in Sn and Li. Other ash-flow tuffs in the Morococala field, from a still younger eruptive center to the north, and similar tuffs in the Los Frailes field to the south, also are peraluminous. These tuffs are, however, less so than the two-mica and cordierite-bearing tuffs, and have slightly elevated values for Sn (as much as 7 ppm) and Li (as much as 100 ppm), relative to the normal calc-alkaline tuffs (< 2 ppm Sn) elsewhere in the central Andes.

We believe that the above ash-flow tuffs reflect the unique composition of magmas in the Bolivian tin belt, from which the tin deposits were derived. The Andean tin deposits are confined to this belt, and with few exceptions, tin is absent in mineral deposits elsewhere in the central Andes. Furthermore, known tin mineralization took place over a period of nearly 200 m.y., from late Triassic to Miocene, and the restriction of tin mineralization to a relatively narrow belt (at no place is it more than 150 km wide) over a long span of time indicates a tin-enriched source of the magmas, probably in the lower crust, or perhaps the upper mantle. Noble et al. (1984) pointed out that the Macusani tuff is similar in composition to S-type granites, which would imply a lower crustal source.

## Deformation

The Neogene-Quaternary volcanic rocks of the central Andes show widespread block faulting and tilting associated with Andean uplift, and some of the Miocene rocks have been folded locally (Noble et al., 1974). Some of the numerous closed basins within the volcanic terrain of this region are fault basins (Stoertz and Ericksen, 1974). Extensive ash-flow tuffs sheets of Pliocene age along the western Andean front in Chile have been tilted westward as much as 5° from their original near-horizontal position.

Compressional deformation of the volcanic rocks is most evident in southern Peru, where open, upright folds are present in volcanic rocks of Miocene age. The best documented history of Miocene deformation in this region is that of McKee and Noble (1982) for the Julcani-Huachacolpa-Castrovirreyna area. The volcanic rocks in this region were deformed into a series of folds during a pulse of compressional deformation beginning during the early Miocene interval of 19.5 and 17 m.y. and ending in middle Miocene (12.5–12 m.y.). Younger volcanic rocks in this area are undeformed, but those in the nearby area of Ayacucho were deformed during the intervals of 11.5-9.5 m.y. and 7–5.5 m.y.

## Mineral Deposits

The mineral deposits genetically related to Neogene-Quaternary volcanism include the following types (Table 1): (1) Ag-Au veins, (2) polymetallic Ag-base-metal veins, (3) polymetallic Sn veins, (4) porphyry-type Cu, Ag-Cu, and Sn deposits, (5) manganese-oxide deposits, (6) magnetite-hematite flows, (7) native S, wood-tin, and covellite deposits, and (8) uranium deposits. In addition, the volcanic rocks and associated thermal springs have been the sources of most of the saline constituents of the widespread salars and saline lakes of the central Andes. These salars contain major resources of lithium and boron. Thermal springs in the volcanic terrain of northwestern Argentina are actively depositing the borate mineral ulexite, along with manganiferous travertine (Sillitoe, 1975). We have observed a thermal spring in Bolivia that is actively depositing stibnite, and another in northern Chile that is depositing pyrite or marcasite.

The major metal production from deposits in Neogene-Quaternary eruptive centers has been of Ag, Au, and base metals from several deposits in southern Peru; Sn and Ag from

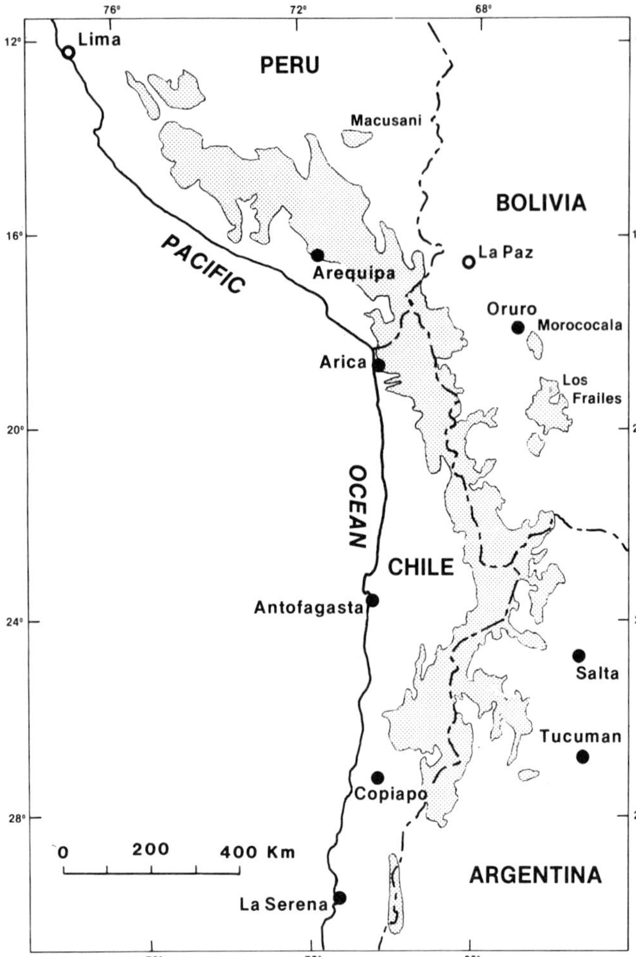


Figure 1. Distribution of Neogene-Quaternary volcanic rocks (stipple pattern) in the central Andes. From Gardeweg and Ramírez (1984) and the Instituto de Geología y Minería del Perú (1975).

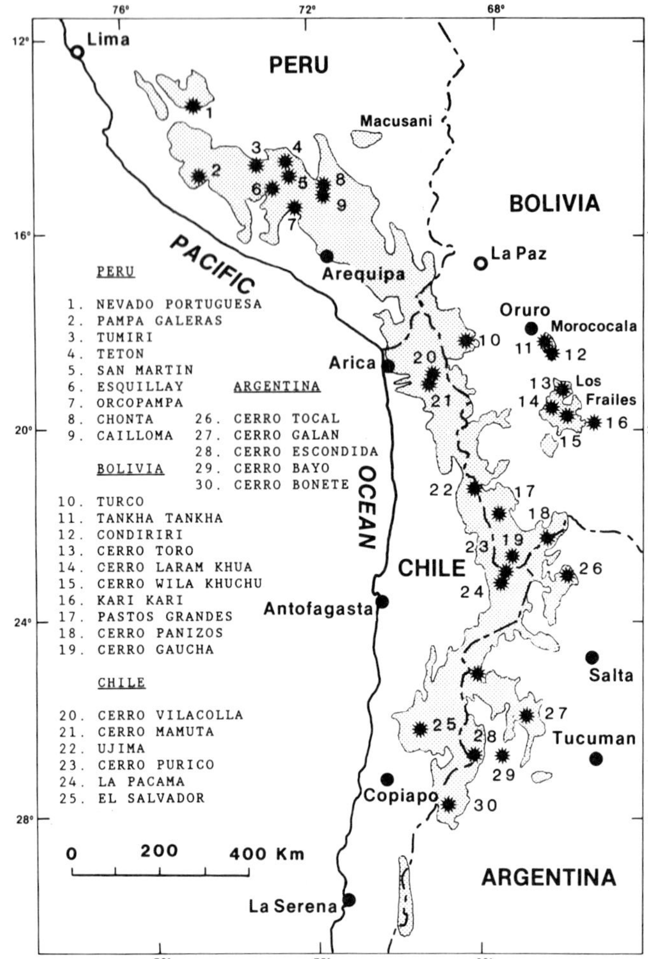


Figure 2. Known and inferred calderas (numbered symbols) in the central Andes. Locations of calderas from Baker (1981), Francis et al. (1983), Noble et al. (in press), and Ericksen et al. (unpublished data, 1986).

many deposits in southern Bolivia; and Ag and Au from the El Indio and Choquelimpie deposits in Chile. In southern Peru, the mining districts of Julcani and Huachocolpa (Figure 5) have produced large amounts of polymetallic, Ag-rich base-metal ores, and mines at Cailloma, Orcopampa, and Arcata (Figures 4 and 5) have been major producers of Ag-Au ores. Of the Sn-Ag deposits in southern Bolivia, the most famous is Cerro Rico de Potosí (Table 1), which ranks as one of the largest, if not the largest, Ag producer in the world. Llallagua, Bolivia (Table 1), a porphyry tin deposit, is the world's largest lode-tin deposit. All these deposits are currently being exploited, and several have been exploited intermittently since Colonial times. El Indio is the most recent discovery, having been discovered in the late 1970s.

## Alteration and Ore Mineralogy

The mineral deposits associated with the Neogene-Quaternary volcanic centers in the central Andes show various types and intensities of hydrothermal alteration and various characteristic suites of primary ore minerals. The most intense alteration is that in the tin-bearing porphyry stocks of southern Bolivia, which characteristically consist chiefly of a quartz-sericite-tourmaline rock. In contrast, the epithermal Ag-Au deposits of southern Peru (Table 1) commonly show weak to moderate propylization, and wall rocks of veins may show locally intense argillic alteration. For example, the dacitic lavas that are the host of the veins at Arcata (Table 1) are propylitized, and veins have halos as much as 25 m wide in which the lavas have been transformed to a white rock consisting almost entirely of dickite. The altered porphyry stock at the Choquelimpie deposit (Table 1) contains the clay mineral nacrite in addition to quartz and tourmaline.

Epithermal Ag-Au and base-metal deposits in southern Peru consist of sulfides and sulfosalts of Ag, Cu, Pb, Sb, As, and Bi. Zn occurs as sphalerite, and Au, with rare exception, occurs as native gold or is contained in pyrite. Ag occurs chiefly in the ruby silver minerals and argentiferous tetrahedrite/tennantite and galena. In a typical Ag-Au deposit, Ag is on the order of 10 to 100 times as abundant as Au, which characteristically is present in amounts of only a few grams per tonne. However, a few exceptionally rich Ag-Au veins contain several ounces of gold and several hundred ounces of silver per tonne. The richest gold vein now being mined, Veta Sur at the El Indio deposit, Chile (Table 1), averages 150–200 g Au, 50–150 g Ag, and 1–5% Cu; assay values of more than a kilogram of Au per tonne are common. In the base-metal deposits, galena and sphalerite are the principal Pb and Zn minerals,

Table 1. Mineral deposits associated with Neogene-Quaternary volcanic centers in the central Andes. (1) indicates deposits examined by authors of this chapter.

STRATOVOLCANOS

| Deposit | Type | Age of mineralization | Volcanic environment | Sources of information |
|---|---|---|---|---|
| 1. Ccarhuaraso (Peru) | Au-Ag veins | 1.2 Ma | Dacitic lavas in eroded volcano | Noble et al. (in press) |
| 2. Choquelimpie (Chile) | Porphyry Ag-Cu | Pliocene (?) | Brecciated porphyry stock in roots of deeply eroded andesitic stratovolcano | Salas et al. (1966); (1) |
| 3. Aucanquilcha (Chile) | Fumarolic covellite and native sulfur | Pleistocene-Holocene | Andesite breccia in vent zone at top of stratovolcano | (1) |
| 4. El Laco (Chile) | Magnatite-hematite flows | Pleistocene | Magnetite-hematite lavas at 5 localities on flanks of andesitic stratovolcano | Ruiz et al. (1965); (1) |
| 5. Salle (Argentina) | Sb-Ag-Au-Sn impregnations | 8.9–8.5 Ma | Hydrothermal breccia in eroded andesitic stratovolcano | Brodtkorb and Coira (1986) |
| 6. El Quevo (Argentina) | Pb-Ag veins | Pleistocene | Dacite flows in central part of glaciated stratovolcano | Sillitoe (1975) |
| 7. Farallon Negro (Argentina) | Porphyry Cu-Au; Mn-Au veins | 7.9–7.1 Ma | Dacite porphyry stocks in roots of deeply eroded andesitic stratovolcano | Angelelli et al. (1970); Caelles et al. (1971); Guilbert et al. (1986) |
| 8. Many deposits | Fumarolic native S | Pleistocene | Andesite flows and breccias; Holocene at tops and flanks of volcanic cones | Ruiz et al. (1965); (1) |

CALDERAS

| Deposit | Type | Age of mineralization | Volcanic environment | Sources of information |
|---|---|---|---|---|
| 9. Nevado Portuguesa (Peru) | Ag-Pb-Zn-Cu veins | 1.9 Ma | Resurgent dome of caldera centered on large composite andesitic volcano (3.7 Ma) | Noble and McKee (1982) |
| 10. Tumiri (Peru) | Ag-Au veins | Late Miocene | Altered andesite flows (?) associated with Tumiri caldera | Noble et al. (in press); (1) |
| 11. Teton-Santo Domingo (Peru) | Ag-Au veins | Late Miocene | Rhyodacite tuffs and lavas associated with Teton caldera | Noble et al. (in press); (1) |
| 12. San Martin-Farallon (Peru) | Ag-Au veins | Late Miocene | Rhyolitic ash flow tuffs associated with San Martin caldera | Noble et al. (in press); (1) |
| 13. Esquillay (Peru) | Ag-Au veins | Late Miocene | Rim (?) of Esquillay caldera | Noble et al. (in press); (1) |
| 14. Sucuitambo (Peru) | Ag-Au veins | 11.4–10.5 Ma | Resurgent dome of Chonta caldera (11.4 Ma) | Peterson et al. (1983); (1) |
| 15. San Miguel (Peru) | Ag-base metal veins | 11.4–10.5 Ma | Ring-fracture zone of Chonta caldera | Peterson et al. (1983); (1) |
| 16. Paco Paco (Peru) | Ag-Au veins | 11.4–10.5 Ma | Ring-fracture zone of Chonta caldera | Peterson et al. (1983); (1) |
| 17. Cailloma (Peru) | Ag-Au veins | 16.5 Ma | Rim of Chonta caldera near intersection with younger Cailloma caldera | Dávila (1981); Noble et al. (in press); (1) |

Table 1. Continued.

| | | | | |
|---|---|---|---|---|
| 18. Orcopampa (Peru) | Ag-Au veins | 17 Ma | Ash flow tuffs (19.5 Ma) in resurgent dome of Orcopampa caldera | Silberman et al. (1985); Noble et al. (in press); (1) |
| 19. Coyuma (Bolivia) | Pb-Zn-Ag veins | ~8.5 ma | Quartz latite dome on west side Tankha Tankha caldera | United Nations (unpublished report, 1982); (1) |
| 20. Pumpuri (Bolivia) | Pb-Zn-Ag-Sn veins | Middle (?) Miocoene | Dacite flows in or near ring-fracture zone of Pumpuri caldera | United Nations (unpublished report, 1982); (1) |
| 21. Cerro Rico de Potosí (Bolivia) | Sn-Ag veins | 13.8 Ma | Quartz latite porphyry stock in ring fracture zone of Kari Kari caldera (20.8 Ma) | Francis et al. (1971); Sillitoe et al. (1975); Grant et al. (1979) |
| 22. Illimani (Bolivia) | Pb-Zn-Ag veins | Early (?) Miocene | Ash flow tuffs in resurgent dome of Kari Kari caldera | Ahlfeld and Schneider-Scherbina (1964); (1) |
| 23. Cord. Andacaba (Bolivia) | Sn-Ag and Pb-Zn | Early (?) Miocene | Ash flow tuffs in resurgent dome of Kari Kari caldera | Ahlfeld and Schneider-Scherbina (1964); (1) |
| 24. Cunurana (Bolivia) | Sn-Ag veins | Miocene | Granodiorite stock in ash flow tuffs of Kari Kari caldera | Ahlfeld and Schneider-Scherbina (1964); (1) |
| 25. El Salvador (Chile) | Porphyry Cu | 41 Ma | Porphyry stocks in ring-fracture zone of a Paleocene caldera | Gustafson and Hunt (1975); Francis et al. (1983); Clark et al. (1985) |
| | | DOMES | | |
| 26. Julcani (Peru) | Ag-Cu-Si-Pb-Zn-W-Au veins | 9.0–8.4 Ma | Ash flow tuffs (10.3 Ma) cut by quartz latite domes | Petersen et al. (1977); Noble and Silberman (1984); Shelnutt and Noble (1985) |
| 27. Huachocolpa (Peru) | Pb-Zn-Ag veins | Between 8.2 and 4.0 Ma | Dacitic flows, breccias, and domes (10.4–8.2 Ma) | McKee et al. (1975); Birnie and Petersen (1977) |
| 28. Arcata (Peru) | Ag-Au veins | 5.0–4.5 Ma | Dacite flows and rhyolitic domes | Noble et al. (in press); (1) |
| 29. Oruro (Bolivia) | Sn-Ag veins | Miocene | Brecciated porphyry stocks and coeval quartz latite flows | Chase (1948); Grant et al. (1979); Sillitoe and Bonham (1984) |
| 30. San Cristobal de Lipez (Bolivia) | Pb-Zn-Ag veins and disseminations | Miocene | Dacite flows and domes | Jacobson et al. (1969); Sillitoe and Bonham (1984) |
| 31. Pan de Azucar (Argentina) | Pb-Zn-Ag vein | 12 Ma | Dacite and quartz latite pyroclastics and domes | Coira (1979); Sillitoe and Bonham (1984) |
| | | BRECCIAS AND ALTERATION ZONES OF INFERRED STRATOVOLCANOS OR DOMES | | |
| 32. Berguenguela (Bolivia) | Ag-Cu-Pb-Zn veins | Miocene | Rhyolite breccia pipe in andesite flows (26–10.5 Ma) | Evernden et al. (1977); McNamee (oral communication, 1986) |
| 33. Llallagua (Bolivia) | Porphyry Sn | 21.1–20.6 Ma | Brecciated porphyry stock in Paleozoic sedimentary rocks | Sillitoe (1975); Grant et al. (1979); Francis et al. (1983) |
| 34. Colquechaca (Bolivia) | Ag-Pb-Zn-Sn veins | 22–21.5 Ma | Vent (?) in quartz latite flows of former volcano | United Nations (unpublished report, 1982); Grant et al. (1980) |

Table 1. Continued.

| | | | | |
|---|---|---|---|---|
| 35. Porco (Bolivia) | Sn-Ag-Zn veins | Miocene | Rhyodacite tuffs and porphyry stocks of flow dome complex | United Nations (unpublished report, 1982) |
| 36. Chocaya (Bolivia) | Zn-Pb-Ag-Sn veins | 12.5 Ma | Rhyodacite pyroclastics and flows in vent zone (?) of eroded volcano | Sillitoe et al. (1975); Grant et al. (1980); Francis et al. (1983) |
| 37. Chorolque (Bolivia) | Sn veins | 16.2 Ma | Breccia pipe in coeval flows and pyroclastics | Sillitoe et al. (1975); Grant et al. (1979) |
| 38. La Joya (Bolivia) | Ag-Au-Cu stockwork | Miocene | Quartz porphyry stock in pre-Miocene sedimentary rocks | Ahlfeld (1967) (1) |
| THERMAL SPRING SYSTEMS IN VOLCANIC TERRANE | | | | |
| 39. Several deposits (Bolivia) | Wood tin | Late Miocene | Quartz latite ash flow tuffs in Los Frailes volcanic field | Servicio Geologica de Bol. (unpublished report, 1982) |
| 40. (Many) (Bolivia) | Supergene uranium | Late Miocene-Pliocene (?) | Quartz latite ash flow tuffs in Los Frailes volcanic field | A. A. Farfan (unpublished report, 1981); (1) |
| 41. Milluri (Bolivia) | Sb veins | ~8.5 Ma | Quartz latite ash flow tuffs in Los Frailes volcanic field | United Nations (unpublished report) |
| 42. Machicado (Bolivia) | Sb veinlets | Active thermal springs | Paleozoic sandstone, Tihua window in Los Frailes field | United Nations (unpublished report, 1982); (1) |
| 43. Many deposits (Chile, Argentina, Bolivia) | MnO veins and impregnations | Quaternary | Thermal springs associated with stratovolcanos; some lacustrine accumulations | Salas et al. (1966); Cruzat (1970); Sillitoe (1975); (1) |
| 44. El Indio (Chile) | Au-Ag-Cu veins | 17.2–15.9 Ma | Dacitic tuffs; capping of siliceous sinter and fumarolic native S; mineralization in open geothermal system | Araneda (1984); Jannas and Araneda (1985) |

whereas chalcopyrite and tetrahedrite/tennantite, and, in some, enargite, are dominant Cu minerals. Some deposits also contain significant amounts of Pb sulfosalts, bismuthinite, and stibnite. The principal gangue minerals of both the Ag-Au and base-metal deposits are quartz, pyrite, rhodochrosite, rhodonite, and barite.

The tin deposits of southern Bolivia are characteristically polymetallic, containing cassiterite as the dominant Sn mineral, and a variety of sulfides, sulfosalts, and sulfostannates of Ag, Cu, Pb, Zn, Bi, Sb, and As. Depending upon the relative abundances of the principal metals, the Sn deposits are classed as either Sn-Ag, Sn-Zn, and Sn-Bi types.

The porphyry copper deposits discussed in this report—El Salvador, Chile, and Farallon Negro, Argentina (Table 1)—show propylitic, argillic, and potassic alteration typical of porphyry coppers in general. Chalcopyrite is the dominant primary Cu mineral, but significant amounts of bornite, enargite, and tetrahedrite/tennantite may be present. At least one of the porphyry bodies at Farallon Negro has exceptional Au values of 0.7 g/metric ton.

Many of the deposits show characteristic mineral zoning, which may be district-wide lateral zoning of many veins or both lateral and vertical zoning within single veins. For example, many of the polymetallic Sn deposits of southern Bolivia have cassiterite-rich central zones that grade laterally into Ag-rich base-metal zones. Cerro Rico de Potosí (Table 1) also shows vertical zoning wherein bonanza-type primary Ag ores were encountered in upper mine levels and high-grade Sn ores in deeper levels. Petersen et al. (1977) reported district-wide zoning at Julcani, Peru (Table 1), to be as follows (from the district center outwards): (1) wolframite and Au-bearing pyrite; (2) enargite, pyrite, and tetrahedrite/tennantite; (3) Ag- and Bi-sulfosalts and bismuthinite; (4) galena; and (5) Pb sulfosalts, orpiment, and realgar. Veins in the Arcata district, Peru (Table 1), show a characteristic vertical zoning, having rich Ag ores confined to a mid-depth range of 200–300 m, which grades upwards into a low-Ag pyrite zone and downwards into a low-Ag galena-sphalerite zone.

## DEPOSITS ASSOCIATED WITH STRATOVOLCANOS

Mineral deposits associated with stratovolcanos are shown in Figure 3 and briefly described in Table 1. The precious- and base-metal deposits in the upper parts of volcanic cones show a variety of structures and mineral associations. For example, the Aucanquilcha covellite occurrence, Chile (Table 1), consists of impregnations and masses of covellite and pyrite, cut by veinlets of native sulfur, in andesite breccia at the margin of the vent zone on top of Volcan Aucanquilcha. In contrast, El Quevo, Argentina, described by Sillitoe (1975), consists chiefly of lenses and masses of Ag-bearing galena within a fault zone high on the southwestern flank of Volcan Nevados Pastos Grandes. The deposit is exposed in a glacial cirque, and Sillitoe (1975) estimated that it was emplaced at a depth of 400–500 m below the original volcano surface. The Salle deposit consists of Sn-Ag- base-metal veins in an eroded volcanic cone (Brodtkorb and Coira, 1986).

The sulfide deposits in deeply eroded stratovolcanos—Choquelimpie and Farallon Negro (Table 1)—are associated with subvolcanic intrusions and related intrusive or hydrothermal breccias. The Choquelimpie deposit consists of disseminations and veins of Ag- and Au-bearing base-metal sulfides in and near a porphyry stock exposed in the base of a deeply eroded andesitic stratovolcano 8–10 km in diameter. The stock shows widespread hydrothermal breccia, and is pervasively argillized and silicified. A semiquantitative spectrographic analysis of a typical ore specimen, which we collected from an active open pit at Choquelimpie in 1984, shows 70 ppm Ag, 200 ppm Cu, 200 ppm Pb, and 150 ppm Sb. Cabello (1986) cited ore grades of 150–300 ppm Ag and 0.5–3 ppm Au, which were reported by A. Thomas (unpublished report, 1973). Cabello also suggested that Choquelimpie is associated with a caldera, which we believe to be unlikely.

The Farallon Negro district, Argentina, has several small subvolcanic intrusions with porphyry-type mineralization (Caelles et al., 1971), which evidently were emplaced in the root zone of a complex stratovolcano that has been almost completely destroyed by erosion. Caelles et al. (1971) cite a radiometric age for this volcanic complex of 10.7 m.y., and an age of one of the porphyry intrusions of 7.9 m.y. They further estimated the age of mineralization to be between 7.9 and 7.1 m.y. Guilbert et al. (1986) show that one of these bodies, the Bajo La Alumbrera deposit, is a porphyry Cu-Au deposit that contains an estimated 300 million tonnes of ore averaging 0.49% Cu and 0.7 g Au/ton. At 1987 prices, the gold in this deposit is worth more than the copper.

In addition to the porphyry-type deposits, a steeply dipping Au-bearing manganiferous zone is present in the Farallon Negro district. This zone has an outcrop length of about 2 km and ranges from 1 to 20 m in thickness. It forms a prominent black ledge, for which the Farallon Negro (black cliff) district is named. Angelelli et al. (1970) report the primary vein material in the manganiferous zone to consist of Au-bearing pyrite, sparse base-metal sulfides, and manganiferous carbonates and oxides. They state that considerable Au had been produced from the zone, and estimate that ore-grade material being recovered in 1965 contained 6 g Au and 114 g Ag/tonne as well as small amounts of Cu, Pb, and Zn.

The unique El Laco magnetite-hematite deposits, which are considered to have been emplaced as lava flows (see Ruiz et al., 1965), occur at five localities on the flanks of Volcan El Laco, a complex andesitic stratovolcano in northern Chile (Figure 3, Table 1). The deposits contain large amounts of high-grade iron ore, perhaps totaling more than 100 million tonnes. They consist of vuggy, finely to coarsely crystalline magnetite and hematite. The magnetite-hematite ore contains abundant, near-vertical, open tubes, which appear to be gas-escape tubes. These tubes are lined with octahedral crystals of magnetite, as much as 5 cm in diameter. These crystals and others in vuggy ore are partly to wholly altered to hematite, which suggests that magnetite was the dominant primary mineral of the flows, and that alteration to hematite took place during the subsequent fumarolic phase. However, the deposits also contain primary hematite in the form of tabular hexagonal crystals as much as 5 cm in diameter.

Published reports on the El Laco deposits all conclude that they were emplaced as flows of iron-oxide magmas. Park (1961) was the first to report on these deposits, and he suggested that these magmas were segregations highly charged with gas and that they were intruded at shallow depth and locally reached the surface to form flows. They have been considered to be dome flows of iron-oxide magma. Frutos and Oyarzún (1975) suggested the iron-oxide magma originated by fusion of iron formation that they presumed existed at depth in this area.

Small manganese oxide deposits of Quaternary age and of probable thermal-spring origin are widespread in northern Chile and have been reported to occur in Bolivia (Ahlfeld and Schneider-Scherbina, 1964) and Argentina (Angelelli et al., 1970; Sillitoe, 1975). These deposits consist of surficial layers and veins and impregnations in bedrock at the sites of former thermal springs, and as thin layers interbedded with clastic sediments and diatomite at the sites of former lakes and ponds (Salas et al., 1966; Cruzat, 1970). The manganese oxides in the lacustrine deposits probably were discharged by nearby thermal springs. Such deposits are found on the flanks of stratovolcanos as well as in adjacent volcanic terranes. Some thermal springs are actively depositing manganiferous travertines in northern Chile (Cruzat, 1970) and northwestern Argentina (Sillitoe, 1975). The principal manganese ores consist of pyrolusite, psilomelane, and wad (Salas et al., 1966; Cruzat, 1970). The manganese oxide deposits are small and low grade; only a few have been mined and small tonnages of hand-selected manganese ore recovered (Salas et al., 1966; Sillitoe, 1975).

Many stratovolcanos in the central Andes have fumarolic sulfur deposits in vent areas on their tops and flanks. The largest number of such deposits are in Chile and Bolivia (Figure 4). Argentina has fewer deposits, and Peru has still fewer. The sulfur occurs as cement, veins, and irregular masses in argillically altered, andesitic to dacitic pyroclastics and brecciated or fractured lavas. Active fumaroles are present at some deposits where they are now depositing native sulfur. Several sulfur deposits were being worked during the early 1980s, and in 1984 Chile and Bolivia produced 54,000 tons and 2300 tons of native sulfur, respectively, whereas Peru produced 100 tons (U.S. Bureau of Mines, 1986). Argentina had no recorded production of native sulfur in that year, but had produced small amounts in previous years.

## DEPOSITS ASSOCIATED WITH CALDERAS

Comparison of Figures 2 and 3 shows that only a few of the many calderas in the central Andes are known to have associated metalliferous mineral deposits. However, among these deposits are several that rank among the most productive mineral deposits of the region. These include the Cailloma and Orcopampa, Peru, Ag-Au deposits; the Cerro Rico de Potosí, Bolivia, Sn-Ag deposit; and the Paleocene El Salvador

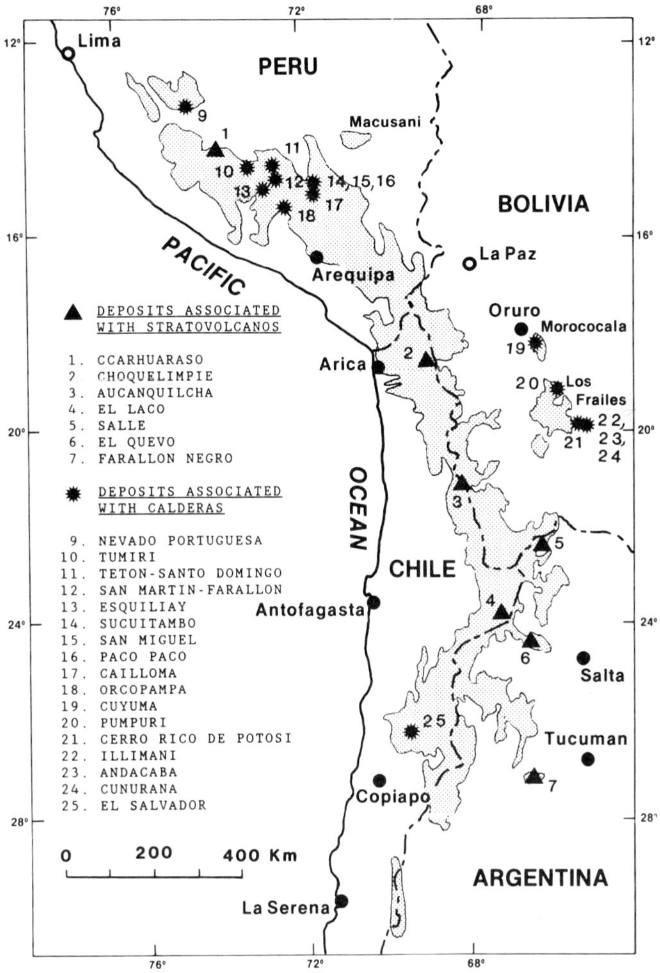


Figure 3. Distribution of metalliferous deposits associated with stratovolcanos and calderas in the central Andes.

porphyry copper deposit, Chile (Table 1). Significant amounts of Ag and Au also have been produced from Sucuitambo and other districts in the Chonta caldera (Table 1) and small amounts from Nevado Portuguesa and Esquillay, Peru. Moderately large amounts of Ag, Pb, and Zn have been recovered from the Illimani mine, Sn and Ag from Cunarana, and Pb and Ag from other mines in the Cordillera Andacaba, all of which are within the resurgent dome of the Kari Kari caldera, Bolivia (Table 1). Veins in the Pumpuri district (Table 1), which apparently are older than the associated caldera in the Los Frailes volcanic field, have had a moderately large production of Sn, Ag, Pb, and Zn. The Coyuma deposit (Table 1), on the western side of the resurgent dome of the Tankha Tankha caldera in the northern part of the Morococala volcanic field, has had a small production of Ag and Pb.

The mineral deposits associated with calderas in southern Peru consist chiefly of Ag-Au veins have varying amounts of Cu, Pb, and Zn sulfide and sulfosalt minerals. Ag is more abundant in these deposits than is Au, and the dominant Ag-bearing minerals are pyrargyrite, proustite, and argentiferous tetrahedrite-tennantite. Of these deposits, by far the most important economically are those of the Cailloma district, which has produced about 100 million ounces of Ag, and Orcopampa, which has produced about 50 million ounces (Silberman et al., 1985).

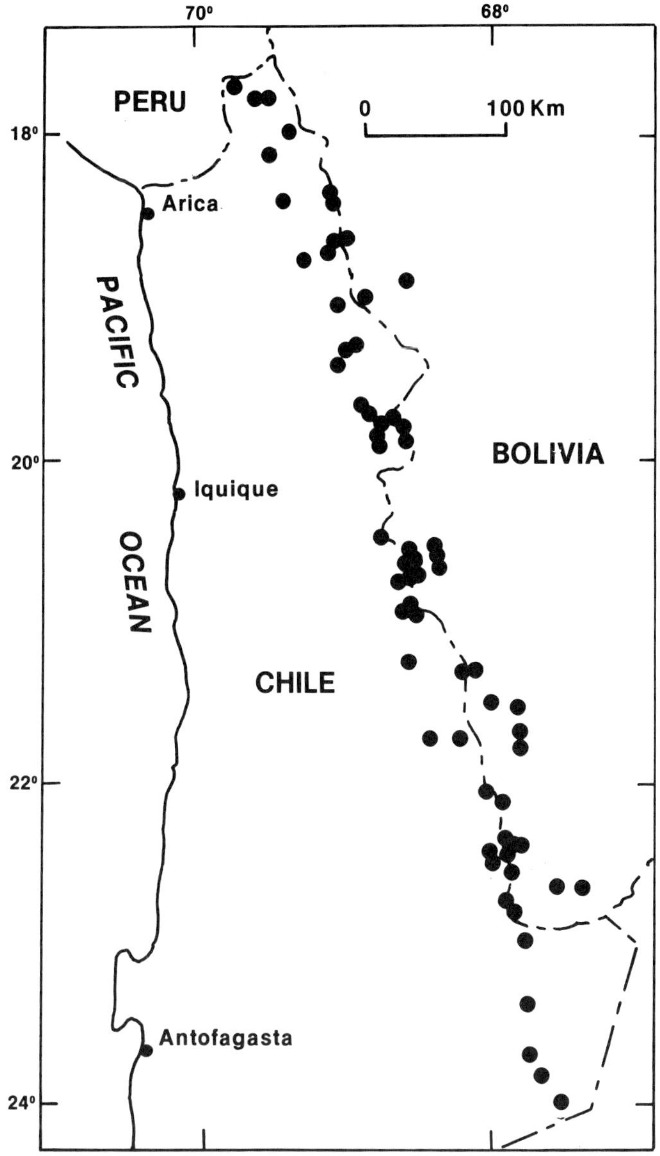


Figure 4. Distribution of native sulfur deposits associated with stratovolcanos in Bolivia and Chile. Each symbol indicates one or more deposits associated with a single volcano. Modified from Montes de Oca (1982) and Ruiz et al. (1965).

The Kari Kari caldera (Figure 2) was considered to be a batholith until Francis et al. (1981) recognized that it consisted predominantly of ~20 Ma ignimbrites within a resurgent caldera. Several mineral deposits occur within the resurgent dome of this caldera, and Francis et al. (1981) suggest that the porphyry stock with which the famous Cerro Rico de Potosí Sn-Ag deposit is associated was emplaced in the ring-fracture zone of the Kari Kari caldera. The Cerro Rico mineralization has been dated at 13.8 Ma, and it is estimated that the stock was emplaced not more than a million years earlier. The Cerro Rico deposit is among the giant silver producers of the world. Total production is estimated to have been between 20,000 and 50,000 tons of Ag since discovery in 1545 (United Nations, unpublished report, 1982). Currently, Cerro Rico is one of the major Sn mines of Bolivia. Cerro Rico consists chiefly of polymetallic Sn-Ag-base-metal veins, having a complex suite of minerals dominated by cassiterite, silver sulfosalts, and many base-metal and iron sulfides and sulfosalts.

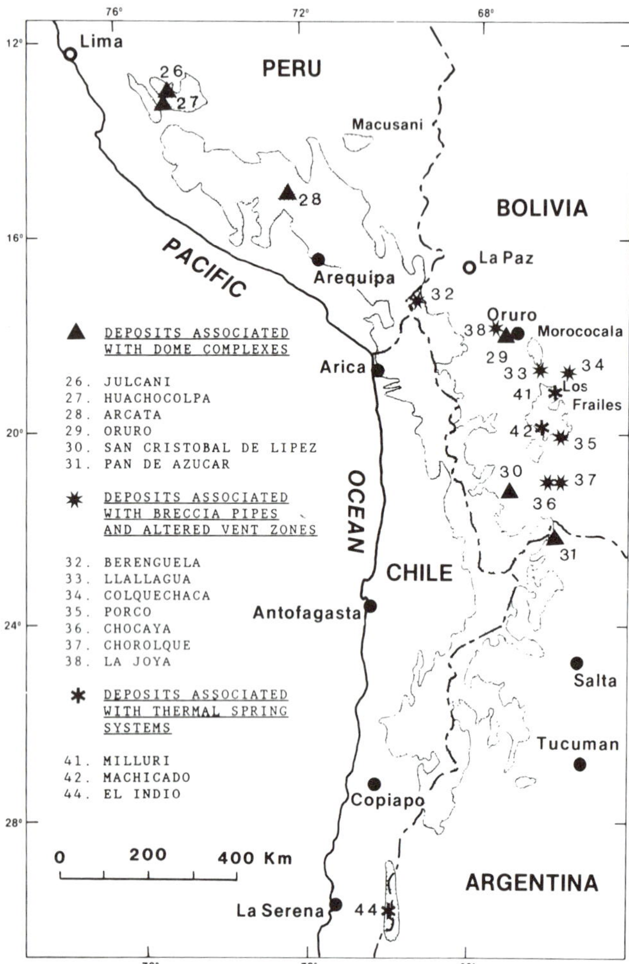

Figure 5. Mineral deposits associated with dome complexes and with altered vent zones, breccia pipes, and thermal-spring systems not related to known volcanic landforms.

Several small mining districts in the Cordillera Andacaba, in the southern part of the Kari Kari complex, have polymetallic veins in which either Sn or Pb-Zn are the dominant metals. Veins in the Illimani district to the north have Ag-bearing Pb-Zn sulfide ores.

The Pumpuri mining district in the Los Frailes volcanic field, Bolivia, and the El Salvador porphyry copper deposit, Chile (Figure 2 and Table 1) may not be genetically related to the calderas with which they are associated. The Pumpuri district is in volcanic rocks of presumed Miocene age exposed in a window eroded in the late Miocene Los Frailes ash-flow tuffs. It is not known whether mineralization took place during an early phase of caldera development or whether it is unrelated to caldera magmatism. The El Salvador deposit consists of a mineralized porphyry stock complex in the ring-fracture zone of a rather obscure caldera that was identified by Francis et al. (1983) in LANDSAT images. The age of mineralization is about 41 Ma, about 20–25 m.y. younger than the caldera, and Clark et al. (1985) believe that mineralization at El Salvador is not genetically related to the caldera.

Most of the other calderas in the central Andes either lack surface expression of hydrothermal alteration and mineralization, or have only small alteration zones and sparse associated metallic minerals. We believe that hydrothermal mineral deposits are associated with late-stage intrusions of many of these calderas, but have not been found because erosion has not been sufficiently deep to expose either the deposits themselves or major associated alteration halos. Geochemical and geophysical studies should prove useful in locating such buried deposits.

Mineral deposits in calderas would not be expected to accumulate during the explosive eruptive phase of caldera development but, rather, during the waning phases of resurgence and intracaldera intrusion. Nevertheless, the geochemical signatures of ash-flow tuffs may give clues to the chemical nature of mineralizing fluids associated with the source magmas. Such is the case for the Sn-rich tuffs at the Macusani and Morococala volcanic fields (see Introduction), which we believe are indicative of tin-rich magmas similar to those that formed the porphyry intrusions with which the Sn deposits of the Central Bolivian Tin Belt are genetically related.

## DEPOSITS ASSOCIATED WITH DOMES

Several mining districts in the central Andes are associated with intrusive-extrusive complexes whose only surface expressions are dome-flow complexes (Figure 5 and Table 1). The intrusions tend to contain hydrothermal breccias, and breccia-filled volcanic vents may be present. Mineral deposits associated with domes occur in the underlying intrusive bodies in the surrounding host rocks, and, rarely, in the domes themselves. In addition, some of the deposits associated with subvolcanic intrusions in eruptive centers not having known volcanic landforms, discussed in the next section, may have been associated with domes that subsequently have been destroyed by erosion.

The Julcani and Arcata Ag districts, Peru, and the Oruro Sn-Ag district, Bolivia, are by far the most productive deposits associated with the known dome complexes listed in Table 3. These districts have had long histories of production and are currently among the major active mining districts of the region. The Oruro deposit, which has been active since the 16th century, produced more than 271 million oz Ag up to 1948 (Chase, 1948) and an additional one million oz from 1948 to 1985 (COMIBOL, unpublished data, 1986). The Oruro district also has been one of the major producers of Sn in Bolivia during the 20th century, and mining currently is chiefly of low-grade Sn-Ag ore. Production of Ag, Pb, and Zn from San Cristobal, Bolivia, and Pan de Azucar, Argentina, both of which are associated with domes, has been relatively small.

The Julcani district, Peru, shows the best documented sequence of geologic events in the development of a mineralized eruptive center characterized by domes. This sequence of volcanic and hydrothermal events began at about 10.1 Ma and continued for about 0.7 Ma (Shelnutt and Noble, 1985). The main volcanic event, at about 10.1 Ma, consisted of eruption of 15–20 $km^3$ of dacite and rhyodacite pyroclastics from a central vent now buried beneath a complex of younger volcanic rocks (Noble and Silberman, 1984; Shelnutt and Noble, 1985). This stage of explosive volcanic activity was followed by emplacement of a complex of 30 or more interpenetrating dacite and rhyodacite domes and associated pyroclastic eruptive material (Shelnutt and Noble, 1985). Following this phase of dome emplacement, many dikes of "fluidized" breccia were emplaced in the central part of the district and subsequently were altered to dense rock consisting almost entirely of fine-grained quartz, tourmaline, and pyrite (Noble and Silberman, 1984; Shelnutt and Noble, 1985). These dikes cut some domes but pre-date the main stage of mineralization. Fluidized dike emplacement and alteration were followed by a series of volcanic, hydrothermal, and tectonic events during which the principal Ag-base-metal ores were formed (Noble and Silberman, 1984). This was followed by emplacement of post-mineral dikes and domes (Noble and Silberman, 1984).

The Oruro mining district is centered on an elongate hill consisting of a late Tertiary intrusive-extrusive complex surrounded by lower Paleozoic sedimentary rock. The most comprehensive study of this district is that of Chase (1948), who first recognized that the complex intrusive body in this district was capped by coeval extrusive rocks, probably a dome complex. According to Chase, the complex consists of several intensely altered quartz latite stocks, dikes, irregular intrusive masses, and lavas. Hydrothermal breccias are widespread. The ores at Oruro occur as well-defined veins, stockworks, and mineralized breccias in the porphyry intrusive complex. They consist chiefly of cassiterite, stannite, Ag-sulfides, and Ag-rich base-metal sulfides and sulfosalts.

Of the other mining districts in which domes occur, the Ag-Au veins at Arcata were emplaced in a sequence of dacitic lavas during a phase of mineralization genetically related to intrusive activity. This intrusive activity resulted in emplacement of domes along the northern side of the district (Noble et al., in press). Mineralization is similar to that of the Cailloma district. The Huachacolpa district, southern Peru (Figure 5; Table 1), has been a major producer of Pb and Zn from fissure-filling veins in late Tertiary volcanic rocks. McKee et al. (1975) reported the presence of domes in the Huachacolpa district. Ag-bearing Pb-Zn sulfide veins at San Cristobal de Lipiz, Bolivia, are in a complex of dacite intrusions, hydrothermal or intrusive breccias, lavas, and ash-flow tuffs of late Pliocene and Pleistocene age (Jacobson et al., 1969). Sillitoe and Bonham (1984) reported domes at San Cristobal. They also identified the Pb-Zn-Ag vein at Pan de Azucar, Argentina, as being related to a dome complex.

## DEPOSITS IN OTHER TYPES OF VOLCANIC CENTERS

A number of metalliferous deposits in the central Andes are associated with altered and brecciated porphyry intrusions, vent breccias, and hydrothermal alteration zones that are presumed to represent eruptive centers where former volcanic landforms have been destroyed by erosion. By far the most productive mineral deposits in such eruptive centers are the polymetallic Sn deposits of southern Bolivia and the El Indio Au-Ag deposit, Chile (Figure 5, Table 1). Sillitoe et al. (1975) and Francis et al. (1983) have suggested that the porphyry stocks with which the polymetallic Sn deposits are associated represent the roots of former stratovolcanos or dome complexes. Araneda (1982) suggested that El Indio is associated with a caldera, but the available evidence for such an association is not conclusive.

Most of the polymetallic Sn deposits of southern Bolivia are associated with intensely altered—silicified, sericitized, and tourmalinized—porphyry stocks, many of which have abundant hydrothermal breccias. These stocks occur as isolated bodies in lower Paleozoic (Ordovician-Devonian) and Mesozoic sedimentary rocks. As the result of intense silicification and tourmalinization, they are resistant to erosion and tend to form conical hills that rise above the surrounding terrane of less resistant sedimentary rocks.

The tin deposits of southern Bolivia have complex mineralogy, but Sn, Ag, and Bi minerals are most important economically. The minerals occur chiefly as fissure-filling veins, generally not more than a meter or two thick, and local, closely spaced veinlets. Some stocks show pervasive mineralization similar to that of the porphyry-copper deposits, and Sillitoe et al. (1975) have suggested that these bodies are the tin-bearing analogs of porphyry copper deposits. The Llallagua deposit (Figure 5, Table 1) is the best example of such a deposit. In the early part of the present century, veins at Llallagua were exploited for relatively high-grade Sn ore, but in the 1940s the mining system was changed over to block caving of low-grade Sn ore, a method used widely in porphyry-copper mines. In addition to Llallagua, Sillitoe et al. (1975) proposed that Potosí, Oruro, and Chorolque are examples of porphyry-tin deposits. Chorolque differs from other Sn deposits in southern Bolivia in that the tin mineralization is largely confined to a cylindrical breccia pipe about 1 km in diameter that is surrounded by coeval pyroclastics. Sillitoe et al. (1975) suggested that the Chorolque complex is the erosion remnant of a former large volcano.

The El Indio Au-Ag deposit, which was not discovered until the late 1970s, ranks as one of the major recent mineral discoveries of the central Andes. A brochure by the St. Joe Minerals Corp. (1984), the principal owner of El Indio, gives the following information about mining and grade-reserve statistics. Direct-shipping high-grade Au ore was first extracted in 1979. A processing plant for treatment of low-grade Au-Ag ore began operating in 1981. By the end of 1982, the company had blocked out 4.3 million metric tons (MT) of ore averaging 14.0 g Au/MT, 122 g Ag/MT, and 5.03% Cu. Some of this ore is high-grade direct-shipping ore, and the above reserves include 62,284 MT of such ore averaging 248.5 g Au/MT, 113 g Ag/MT, and 2.72% Cu.

El Indio has been described in several published reports, of which the most comprehensive are those of Araneda (1982, 1984), Jannas and Araneda (1985), and Walthier et al. (1986). The following data about El Indio was extracted from these reports. El Indio consists of Au-Ag-Cu-bearing veins within a zone of hydrothermally altered (argillic, sericitic, and silicic) andesitic to rhyolitic volcanic rocks and small granodiorite intrusions of late Oligocene to Miocene (27–11 Ma) age. The principal host rock is ash flow tuff. Some 20–30 similar alteration zones are in a 150–200-km belt extending north and south of El Indio. The volcanic rocks are folded and faulted, and are cut off in the west by a major longitudinal reverse fault. Although the relationship of El Indio to a caldera, as Araneda (1982) suggested, remains uncertain, it is evident that the veins were emplaced in a geothermal system marked by thermal-spring and fumarolic activity at the surface, as indicated by the presence of siliceous sinter and fumarolic sulfur cappings (Araneda, 1984). Two types of veins are recognized at El Indio, one of which consists predominantly of Au-Ag-bearing quartz and the other of Au-Ag-bearing massive enargite (Areneda, 1984). The veins extend over a vertical range of about 800 m, from the surficial thermal-spring sinters to the deepest mine levels.

The La Joya Ag-Au deposit in southern Bolivia (Figure 5, Table 1) consists of an altered and mineralized dacite porphyry stock, which resembles the porphyry stocks with which the tin deposits of the nearby Central Bolivian Tin Belt are associated. However, La Joya does not contain Sn minerals. La Joya is in a longitudinal range of hills that extends northward for more than 100 km within the Altiplano. The stock is exposed as a conical hill about 100 m high and 500 m in diameter, rising above the flat alluvial plain of the Altiplano. The primary ore, averaging about 24 g Ag and 1.4 g Au, consists of closely spaced veinlets and impregnations of Ag- and Au-bearing pyrite and chalcopyrite. A 60-m thick oxide capping, having Ag and Au values similar to those of the primary ore, was being mined in the early 1980s and treated by a heap-leaching process.

Ag- and Au-bearing porphyry intrusions similar to La Joya are known in the vicinity of La Joya and elsewhere in the Altiplano to the north. Although some of these deposits have been exploited for Au and Ag in the past, little is known about them. We believe that these deposits and other porphyry intrusions in this region have a potential for significant amounts of low-grade Ag-Au ore.

Ash-flow tuffs of Los Frailes volcanic field (Figure 1) contain small wood-tin deposits and uranium deposits. Wood-tin occurs as fracture-filling veinlets of fibrous to dense, locally nodular cassiterite, generally not more than a centimeter or two in thickness. The veinlets tend to be concentrated along linear fracture zones and may occur over areas of many thousands of square meters. Streams draining such areas have small placer deposits. Both lode and placer deposits of wood-tin have been mined, but total production has been insignificant. Several small uranium occurrences are in the southwestern part of the Los Frailes volcanic field and one of these, the Cotaje deposit, was being mined in the early 1980s (A.P. Farfan, unpublished report, 1981; Leroy et al., 1985). Leroy et al. (1985) distinguished four types of uranium occurrences in the Los Frailes area, as follows: (1) Magmatic—U-rich horizon (5–10 cm thick) in unaltered dacitic tuff; (2) Hydrothermal—U accumulations related to faults in highly altered (kaolinized and silicified) tuff; Cotaje is of this type; (3) U occurrences related to hot and cold springs; and (4) U occurrences in unconsolidated detritus. Farfan (unpublished report, 1981) reported the following uranium minerals to be present: uraninite, autunite, torbernite, boltwoodite, and uranophane-beta. These minerals are associated with Fe and Mn oxides. Farfan further reported the highest uranium values at Cotaje to be 1.2–2.5% $U_3O_8$, and more typical values at Cotaje and elsewhere of 0.05–0.9% $U_3O_8$.

## CONCLUSIONS

The principal mineral deposits associated with eruptive centers of Neogene-Quaternary volcanic rocks of the central Andes are hydrothermal precious- and base-metal deposits, polymetallic tin deposits, magnetite-hematite flows, and fumarolic sulfur deposits. The relationship of many of these deposits to eruptive centers was not generally recognized prior to the 1970s. Before that time, geologists believed that much of the late Tertiary and Quaternary volcanic rocks represented post-mineralization cover, and that the major potential for mineral deposits was beneath these rocks. To a certain extent, this thinking continues to guide mineral exploration in this region. However, accumulated information about mineral deposits, nature and ages of eruptive centers, and geochemistry of these young volcanic rocks indicates them to be prime exploration targets. Such exploration should be concerned primarily with the search for buried mineral deposits in eruptive centers, particularly collapse calderas, deeply eroded stratovolcanos, flow-dome complexes, and fossil geothermal-spring systems. In addition, many mineral deposits of Miocene and Pre-Miocene age can be expected to occur beneath the volcanic cover, and some of these, particularly those beneath thin volcanic cover, can be found by modern geologic mapping and application of geophysical methods.

## REFERENCES CITED


Ahlfeld, F., 1967, Metallogenic epochs and provinces of Bolivia, the tin province (part 1), the metallogenic provinces of the Altiplano (part II): Mineralum Deposita, v. 2, p. 291-311.

Ahlfeld, F., and A. Schneider-Scherbina, 1964, Los yacimientos minerales y de hidrocarburos de Bolivia: Departamento Nacional de Geología Bol 5, 388 p.

Angelelli, V., J.C. Fernandez Lima, A. Herrera, and L. Aristarain, 1970, Descripción del mapa metalogenético de la República Argentina, minerales metaliferous: Dirección Nacional de Geología y Minería Anales XV, 172 p.

Araneda G., R., 1982, El Indio, yacimiento de oro, plata y cobre, Coquimbo, Chile: Minerales, v. 37, p. 5-13.

Araneda, G., R., 1984, Nuevos aportes al conocimiento de la geología de El Indio, yacimiento de oro, plata y cobre, Coquimbo, Chile: Minerales, v. 39, p. 27-39.

Arenas F., M.J., 1975, Geología de la Mina Orcopompa y alrededores, Arequipa: Sociedad Geológica del Perú, Bull., v. 46, p. 9-24.

Baker, M.C.W., 1977, Geochronology of Upper Tertiary volcanic activity in the Andes of north Chile: Geologishe Rundschau, v. 66, p. 455-464.

Baker, M.C.W., 1981, Nature and distribution of Upper Cenozoic ignimbrite centres in the central Andes: Journal of Volcanology and Geothermal Research, v. 11, p. 293-315.

Baker, M.C.W., and P.W. Francis, 1978, Upper Cenozoic volcanism in the central Andes—ages and volumes: Earth and Planetary Science Letters, v. 41, p. 175-187.

Birnie, R.W., and U. Petersen, 1977, The paragenetic association and compositional zoning of lead sulfosalts at Huachocolpa, Peru: Economic Geology, v. 72, p. 983-992.

Brodtkorb, M.K., and B. Coira, 1986, The Salle epithermal deposit and its paragenesis, Province of Jujuy, Argentina *in* Mineral paragenesis: Theophrastus Publications, p. 251-260.

Cabello, J., 1986, Precious metals and Cenozoic volcanism in the Chilean Andes: Journal of Geochemical Exploration, v. 25, p. 1-19.

Caelles, J.C., A.H. Clark, E. Farrar, S.L. McBride, and S. Quirt, 1971, Potassium-argon ages of porphyry copper deposits and associated rocks in the Farallon Negro-Capillitas district, Catamarca, Argentina: Economic Geology, v. 66, p. 961-964.

Casertano, L., 1963, General characteristics of active Andean volcanoes and summary of their activities during recent centuries: Seismological Society of America Bull., v. 53, 1415-1433.

Chase, F.W., 1948, Tin-silver veins of Oruro, Bolivia: Economic Geology, v. 43, p. 333-383 and 435-470.

Clark, A.H., 1970, An occurrence of the assemblage native sulphur-covellite "$Cu_{5.5x}Fe_xS_{6.5x}$," Aucanquilcha, Chile: American Mineralogist, v. 55, p. 913-918.

Clark, A.H., E. Ferrar, and M. Zentilli, 1985, Discussion of "Relationships between mineralization and silic volcanism in the central Andes" by P.W. Francis, C. Halls, and M.C.W. Baker: Journal of Volcanology and Geothermal Energy, v. 24, p. 353-358.

Coira, B.L., 1979, Descripción geológica de la Hoja 3c, Abra Pampa, Provincia de Jujuy: Servicio Geológico Nacional, Argentina, Bol. 170, 90 p.

Coira, B., J. Davidson, F. Mpodozis, and V. Ramos, 1982, Tectonic and magmatic evolution of the Andes in northern Argentina and Chile: Earth-Science Reviews, v. 18, p. 303-332.

Cruzat O., A., 1970, Genesis of manganese deposits in northern Chile: Economic Geology, v. 65, p. 681-689.

Dávila M., D., 1981, Cailloma: Caldera tipo Valle: Sociedad Geológica del Perú, Bol. No. 68, p. 25-31.

Deruelle, B., 1978, Calc-alkaline and shoshonitic lavas from five Andean volcanoes (between latitudes 21°45′ and 24°30′ S) and the distribution of the Plio-Quaternary volcanism of the south-central and southern Andes: Journal of Volcanology and Geothermal Research, v. 3, p. 281-298.

Deruelle, B., 1982, Petrology of the Plio-Quaternary volcanism of the south-central and meridional Andes: Journal of Volcanology and Geothermal Research, v. 14, p. 77-124.

Ericksen, G.E., R.L. Smith, R.G. Luedke, M. Flores, A. Espinosa, F. Urquidi, and F. Saravia, 1985, Preliminary geochemical study of ash-flow tuffs in the Morococala and Los Frailes volcanic fields, central Bolivian tin belt: U.S. Geological Survey Open-File Report No. 85-258, 12 p.

Evernden, J.F., S.J. Kriz, and C. Cherroni, 1977, Potassium-Argon ages of some Bolivian rocks: Economic Geology, v. 72, p. 1042-1061.

Fernández, A., P.K. Hormann, S. Kussmaul, J. Meave, H. Pichler, and T. Subieta, 1973, First petrologic data on young volcanic rocks of SW Bolivia: Tschermaks Mineralogische und Petrographische Mitteilungen, v. 19, p. 149-172.

Francis, P.W., and M.C.W. Baker, 1978, Sources of two large ignimbrites in the central Andes: Some Landsat evidence: Journal of Volcanology and Geothermal Research, v. 4, p. 81-87.

Francis, P.W., M.C.W. Baker, and C. Halls, 1981, The Kari Kari caldera, Bolivia, and the Cerro Rico stock: Journal of Volcanology and Geothermal Research, v. 10, p. 113-124.

Francis, P.W., C. Halls, and M.C.W. Baker, 1983, Relationships between mineralization and silicic volcanism in the central Andes: Journal of Volcanology and Geothermal Research, v. 18, p. 165-190.

Frutos J., J., and J. Oyarzún M., 1975, Tectonic and geothermal evidence concerning the genesis of El Laco magnetite lava flow deposits: Economic Geology, v. 70, p. 988-990.

Gardeweg, M., and C.F. Ramírez, 1984, Volcanismo cenozoico superior del Altiplano chileno: Servicio Nacional de Geologia y Mineria (Chile), Miscelanea, No. 4, p. E1-E31.

Gardeweg, M., S. Ishihara, Y. Matsuhisa, K. Shibata, and S. Terashima, 1984, Geochemical studies of upper Cenozoic igneous rocks from the Altiplano of Antofagasta, Chile: Geological Society of Japan Bull., v. 35, p. 547-563.

Grant, J.N., C. Halls, E. Avila, and G. Avila, 1977, Igneous geology and the evolution of hydrothermal systems in the sub-volcanic tin deposits of Bolivia: Geologic Society of London Special Publication, v. 7, p. 117-127.

Grant, J.N., C. Halls, W. Avila, and W.J. Snelling, 1979, K-Ar ages of igneous rocks and mineralization in part of the Bolivian tin belt: Economic Geology, v. 74, p. 838-851.

Grant, J.N., C. Halls, S.M.F. Sheppard, and W. Avila, 1980, Evolution of the porphyry tin deposits of Bolivia: Tokio, Mining Geology Special Issue, No. 8, p. 151-173.

Guilbert, J.M., A. Allison, A. Stults, and B.F. Suchomel, 1986, The geology of Bajo La Alumbera and related porphyry copper-gold deposits, Catamarca Province, Argentina: New Orleans, 115th AIME Annual Meeting, Program, p. 43.

Gustafson, L.B., and J.P. Hunt, 1975, The porphyry copper deposit at El Salvador, Chile: Econ. Geol., v. 70, p. 857-912.

Instituto de Geología y Minería del Perú, 1975, Mapa geológica del Perú: Lima, Instituto de Geología y Minería del Perú, Map, scale 1:1,000,000.

Jacobson, H.S., C. Murillo, L. Ruiz, O. Tapia, H. Zapata, H. Alarcon, E. Delgadillo, and C. Velasco, 1969, Geology and mineral deposits of the San Cristobal district, Villa Martín province, Potosí, Bolivia: U.S. Geological Survey Bull. 1273, 22 p.

Jannas R., R., and R. Araneda G., 1985, Geología de la veta Indio Sur 3,500; una estructura tipo bonanza del yacimiento El Indio: Revista Geológica de Chile, no. 24, p. 49-62.

Klerkx, J., S. Deutsh, H. Pichler, and W. Zeil, 1977, Strontium-isotope composition and trace-element data bearing on the origin of the Cenozoic volcanics of the central and southern Andes: Journal of Volcanology and Geothermal Research, v. 2, p. 49-71.

Koeppen, R.P., R.L. Smith, M.J. Kunk, M. Flores, R.G. Luedke, and (?) Sutter, in press, The Morococala volcanics: Highly peraluminous rhyolite ash flow magmatism in the Cordillera Oriental, Bolivia: Geological Society of America Abstracts with Programs, v. 19.

Kussmaul, S., P.K. Hormann, E. Ploskonka, and T. Subieta, 1977, Volcanism and structure of southwestern Bolivia: Journal of Volcanology and Geothermal Research, v. 2, p. 73-111.

Lahsen, A., 1982, Upper Cenozoic volcanism and tectonism in the Andes of northern Chile: Earth-Science Reviews, v. 18, p. 285-302.

Leroy, J., B. George-Aniel, and E. Pardo-Leyton, 1985, Deposits and radioactive anomalies in the Sevaruyo region (Bolivia) *in* Uranium deposits in volcanic rocks: International Atomic Energy Agency, Proceedings of a Technical Committee Meeting, El Paso, Texas, April 2-5, 1984, p. 289-300.

McKee, E.H., and D.C. Noble, 1982, Miocene volcanism and deformation in the Western Cordillera and high plateaus of south-central Peru: Geological Society of American Bull., v. 93, p. 657-662.

McKee, E.H., D.C. Noble, U. Petersen, M. Arenas F., and A. Benavides Q., 1975, Chronology of late Tertiary volcanism and mineralization, Huachocolpa district, central Peru: Economic Geology, v. 70, p. 388-390.

Montes de Oca, I., 1982, Geografía y recursos naturales de Bolivia: La Paz, Bolivia, published by the author, 628 p.

Noble, D.C., and E.H. McKee, 1982, Nevado Portuguesa volcanic center, central Peru: A Pliocene central volcano-collapse caldera complex with associated silver mineralization: Economic Geology, v. 77, p. 1893-1900.

Noble, D.C., and M.L. Silberman, 1984, Evolucíon volcánica y hidrothermal y cronología de K/Ar del Distrito Minero de Julcani: Sociedad Geológica del Perú, 60th Aniv. vol., part 5, 35 p.

Noble, D.C., M. Arenas F., and D.W. Charlton, 1974, Nevado Portuguesa volcanic center, central Peru: A late Pliocene collapse caldera complex with associated silver mineralization: Geological Society of America Abstracts with Programs, v. 6, p. 230.

Noble, D.C., V.R. Eyzaguirre, and E.H. McKee, in press, Precious-metal mineralization of Cenozoic age in the Andes of Peru *in* Geology of the Andes and its relation to hydrocarbon and mineral resources, G.E. Ericksen and J. A. Reinemund, eds.: American Association of Petroleum Geologists.

Noble, D.C., E.H. McKee, E. Farrar, and U. Petersen, 1974, Episodic Cenozoic volcanism and tectonism in the Andes of Peru: Earth and Planetary Science Letters, v. 21, p. 213-220.

Noble, D.C., U. Petersen, E.H. McKee, M. Arenas, A. Benavides Q., 1972, Cenozoic volcano-tectonic evolution of the Julcani-Huachocolpa-Castrovirreyna area, central Peru (abs.): The Geological Society of America Abstracts with Programs, v. 4, no. 7, p. 613.

Noble, D.C., T.A. Vogel, P.S. Peterson, G.P. Landis, N.K. Grant, P.A. Jezek, and E.H. McKee, 1984, Rare-element enriched, S-type ash-flow tuffs containing phenocrysts of muscovite, andalusite, and sillimanite, southeastern Peru: Geology, v. 12, p. 35-39.

Park, C.F., 1961, A magnetite "flow" in northern Chile: Economic Geology, v. 56, p. 431-436.

Petersen, U., D.C. Noble, M.J. Arenas, and P.C. Goodell, 1977, Geology of the Julcani mining district, Peru: Economic Geology, v. 72, p. 931-949.

Peterson, P.S., D.C. Noble, E.H. McKee, and V.R. Eyzaguirre, 1983, A resurgent, mineralized caldera in southern Peru: Preliminary report: American Geophysical Union, Transactions v. 64, p. 884.

Ruiz F., C., L. Aguirre, J. Corvalán, C. Klohn, E. Klohn, and B. Levi, 1965, Geología y yacimientos metalíferos de Chile: Instituto de Investigaciones Geológicas (Chile), 305 p.

Salas O., R., R.E. Kast, F. Montecinos P., I. Salas Y., 1966, Geología y recursos minerales del departamento de Arica: Instituto de Investigaciones Geológicas (Chile), Bol. no. 21, 114 p.

Shelnutt, J.P., and D.C. Noble, 1985, Premineralization radial dikes of tourmalinized fluidization breccia, Julcani district, Peru: Economic Geology, v. 80, p. 1622-1632.

Silberman, M.L., E.H. McKee, and D.C. Noble, 1985, Age of mineralization at the Cailloma and Orcopampa silver districts, southern Peru: Isochron/West, no. 43, p. 17-18.

Sillitoe, R.H., 1973, The tops and bottoms of porphyry copper deposits: Economic Geology, v. 68, p. 799-815.

Sillitoe, R.H., 1975, Lead-silver, manganese, and native sulfur mineralization within a stratovolcano, El Queva, northwest Argentina: Economic Geology, v. 70, p. 1190-1201.

Sillitoe, R.H., and H.F. Bonham, Jr., 1984, Volcanic landforms and ore deposits: Economic Geology, v. 79, p. 1286-1298.

Sillitoe, R.H., C. Halls, and J.N. Grant, 1975, Porphyry tin deposit in Bolivia: Economic Geology, v. 70, p. 913-927.

St. Joe Minerals Corporation, 1984, San José and El Indio: St. Joe Minerals Corporation, 21 p.

Stoertz, G.E., and G.E. Ericksen, 1974, Geology of salars in northern Chile: U.S. Geological Survey Prof. Paper 811, 65 p.

Thorp, R.S., and P.W. Francis, 1979, Variations in Andean andesite compositions and their petrogenetic significance: Tectonophysics, v. 57, p. 53-70.

U.S. Bureau of Mines, 1986, Minerals Yearbook, 1984, v. 3, Area report: International: U.S. Bur. Mines, 1111 p.

CHAPTER 47

# BULK TONNAGE GOLD MINERALIZATION IN EASTERN QUEENSLAND, AUSTRALIA

**David J. Horton**
*D.J. Horton & Associates*
*Chapel Hill, Queensland, Australia*

## ABSTRACT

Since 1979, heightened exploration for gold in eastern Queensland has resulted in the re-evaluation and new discovery of several primary (nonalluvial) bulk tonnage gold deposits. These include: (a) Kidston - 44.4 m.t. @ 1.79 g/t Au, 2.22 g/t Ag (b) Red Dome - 13.8 m.t. @ 2.0 g/t Au, 0.46% Cu, 5.25 g/t Ag (c) Mt. Leyshon - 16.5 m.t. @ 1.5 g/t Au.

The above deposits are currently in various stages of production or development. Together with a substantial number of subeconomic prospects, these deposits are of the same tectonodepositional style. They are postorogenic in relation to the orogen in which they occur and are associated with subaerial acid volcanic or subvolcanic rocks. The deposits are found in Devonian to Early Cretaceous continental and volcanic arch volcanics. Deposits of Permo-Carboniferous age are most common. The deposits occur as breccia pipes, hydrothermal volcanic vents, proximal extruded deposits, and as related skarn deposits.

The Mount Morgan Mine, formerly the largest bulk tonnage gold mine in eastern Queensland, recently ceased production. Until the end of 1979, it had produced more than 238,000 kg gold and more than 360,000 t copper from 51 m.t. ore. This deposit is an example of precratonic (or orogenic) mineralization associated with submarine acid volcanism. Deposits of this category are related to Kuroko-type massive sulfide mineralization forming as gold-rich footwall stringer zones or pipes. Auriferous deposits of this type are found along dominantly submarine volcanic arcs (protoisland arcs) and in volcanic rifts. They range in age from Cambro-Ordovician to Early Permian.

Although numerous examples of nonvolcanic-related gold mineralization occur in eastern Queensland, none have realized economic potential as bulk tonnage deposits (i.e., those greater than 5 m.t.).

## INTRODUCTION

The bulk of Queensland's past gold production has been derived from deposits in Paleozoic and Mesozoic host rocks, and only relatively minor amounts have been produced from those in Proterozoic and Cenozoic units (Ishaq, 1985). Almost all gold produced from Cenozoic deposits has come from alluvial or paleoalluvial sources.

The Paleozoic and Mesozoic Tasman Orogenic Zone (or Tasman Fold Belt) of eastern Australia contains a wide variety of different styles of primary (nonalluvial) gold mineralization. However, only a few of these varieties have formed economic or near-economic bulk tonnage deposits—i.e., those with reserves greater than 5 m.t.

All significant bulk tonnage gold deposits found to date in eastern Queensland are hosted by Paleozoic and Mesozoic acid volcanics, subvolcanics, and related breccias. These include the Kidston deposit (Figure 1, no. 2), currently Australia's single largest gold producer, and the Red Dome and Mount Leyshon deposits, which are due to begin production this year. The Mount Morgan Mine, which recently ceased production from primary ore, can also be classified as a bulk tonnage gold deposit hosted by acid volcanic rocks.

On the world scene, volcanic-hosted gold deposits are most abundant in Archean greenstone terranes of many Precambrian cratons and in the Cenozoic volcanic-plutonic arcs of the Circum-Pacific belt. Paleozoic and, to a lesser extent, Mesozoic volcanic-hosted deposits are less common and are generally poorly documented.

The mineral deposits of eastern Australia most closely resemble those of the Circum-Pacific belt. Many styles of mineralization are common to both regions (e.g., porphyry copper and Kuroko-style massive sulfide deposits). However, in many cases, significant differences can be found between comparable deposit types, and this is particularly true of volcanogenic gold deposits.

## TECTONIC AND METALLOGENIC SETTING

The Tasman Orogenic Zone of eastern Australia developed marginal to, and on the eastern edge of, a Precambrian craton during the Paleozoic and Mesozoic Eras. In Queensland, it consists of three separate orogens (or fold belts). These are shown on Figure 1. The complex history of each of these orogens can be subdivided into three basic tectonic regimes (Rickard and Scheibner, 1975) which represent stages of progressive cratonization:

(a) a precratonic (or orogenic) regime characterized by a Pacific-type mobile zone composed of volcanic arcs, volcanic arches, rifts, and marginal interarc and backarc basins;
(b) a transitional (or mountain-building) regime associated with the emplacement of late-orogenic to postorogenic volcanic and plutonic rocks; and
(c) a cratonic (or continental platform) regime.

These tectonic regimes may be correlated with metallogenic regimes of precratonic, transitional, and cratonic character (Scheibner, 1975). However, for the present purpose the latter two regimes can be combined and termed postorogenic in contrast with precratonic regimes, which are orogenic.

A simplified summary of the tectonic and metallogenic framework of each of the three Queensland orogens is given in Table 2. All three orogens had differing tectonic histories up until the Middle Devonian, although connecting links can be recognized between them after this time (Murray, 1984). In particular, a major Permo-Carboniferous metallogenic event accompanying igneous activity introduced precious- and base-metal mineralization into most of northeast Queensland as well as parts of the Precambrian craton to the west. The principal metallogenic event to affect southeast Queensland occurred during the Permo-Triassic.

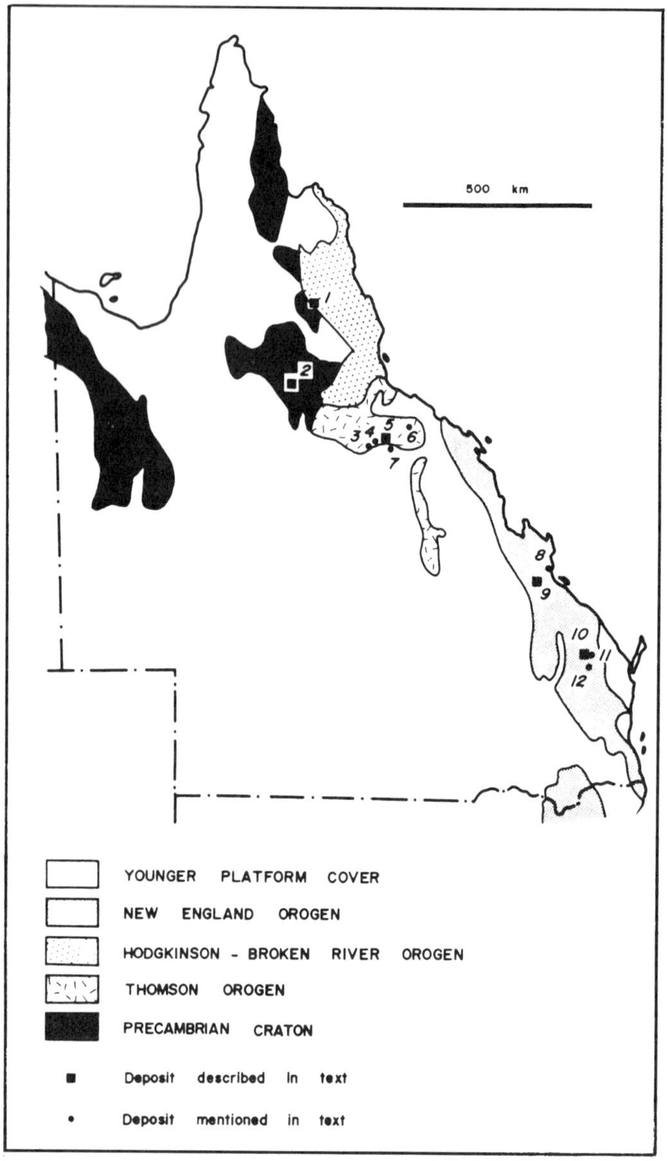


Figure 1. Structural framework of eastern Queensland showing locations of deposits mentioned in the text. Deposit details are shown in Table 1.

## ACID VOLCANIC BULK TONNAGE GOLD MINERALIZATION

Two principal categories of acid volcanic bulk tonnage gold mineralization are recognized in eastern Queensland. One category is deposits associated with postorogenic, dominantly subaerial continental and volcanic arch volcanics and subvolcanic intrusions. These deposits exhibit a close spatial and temporal association with porphyry-type copper deposits (Horton, 1982). The second category is deposits associated with orogenic submarine volcanics. These deposits occur as stringer zones or pipes to Kuroko-type massive sulfide mineralization. Horton (1984) recognized a third category of acid volcanic-hosted bulk tonnage gold mineralization associated with postorogenic subaerial and shallow-marine volcanic rift and marginal-sea volcanic rocks. Deposits of this type have been recognized previously only in New South Wales.

### Postorogenic Acid Volcanic Mineralization

Horton (1984) has recognized several styles of bulk tonnage gold mineralization associated with postorogenic acid volcanic rocks (Figure 2). Economic deposits have yet to be found in all categories, but they include high-level gold-rich porphyry-type copper deposits (e.g., Coalstoun, Figure 1), breccia pipes (e.g., Kidston, Mount Leyshon, Mount Wright), hydrothermal/volcanic vents (e.g., Mount Ophir), proximal extruded (or disseminated stratabound) mineralization (e.g., Mount Rawdon), and skarn mineralization (e.g., Red Dome). The deposits can be found throughout eastern Queensland in Devonian to Early Cretaceous continental and volcanic arch volcanic rocks.

There is a great deal of overlap in mineralization/alteration style between deposits of each category. Ignoring skarn and porphyry-type copper deposits, alteration products are typically kaolin and sericite (argillic alteration) with lesser chlorite and epidote (propylitic alteration). Silicification is common but potassic alteration and alunite development are either rare or have not been recognized. Pyrite and possibly arsenopyrite are ubiquitous. Base and precious metals include copper, lead, zinc, gold, and silver. Other metals such as bismuth and molybdenum are present in some deposits.

Brief descriptions of several of these deposits follow.

*Kidston (18°53′S; 144°08′E)*

At Kidston, gold mineralization occurs around the margins of an oval-shaped breccia pipe with lateral dimensions

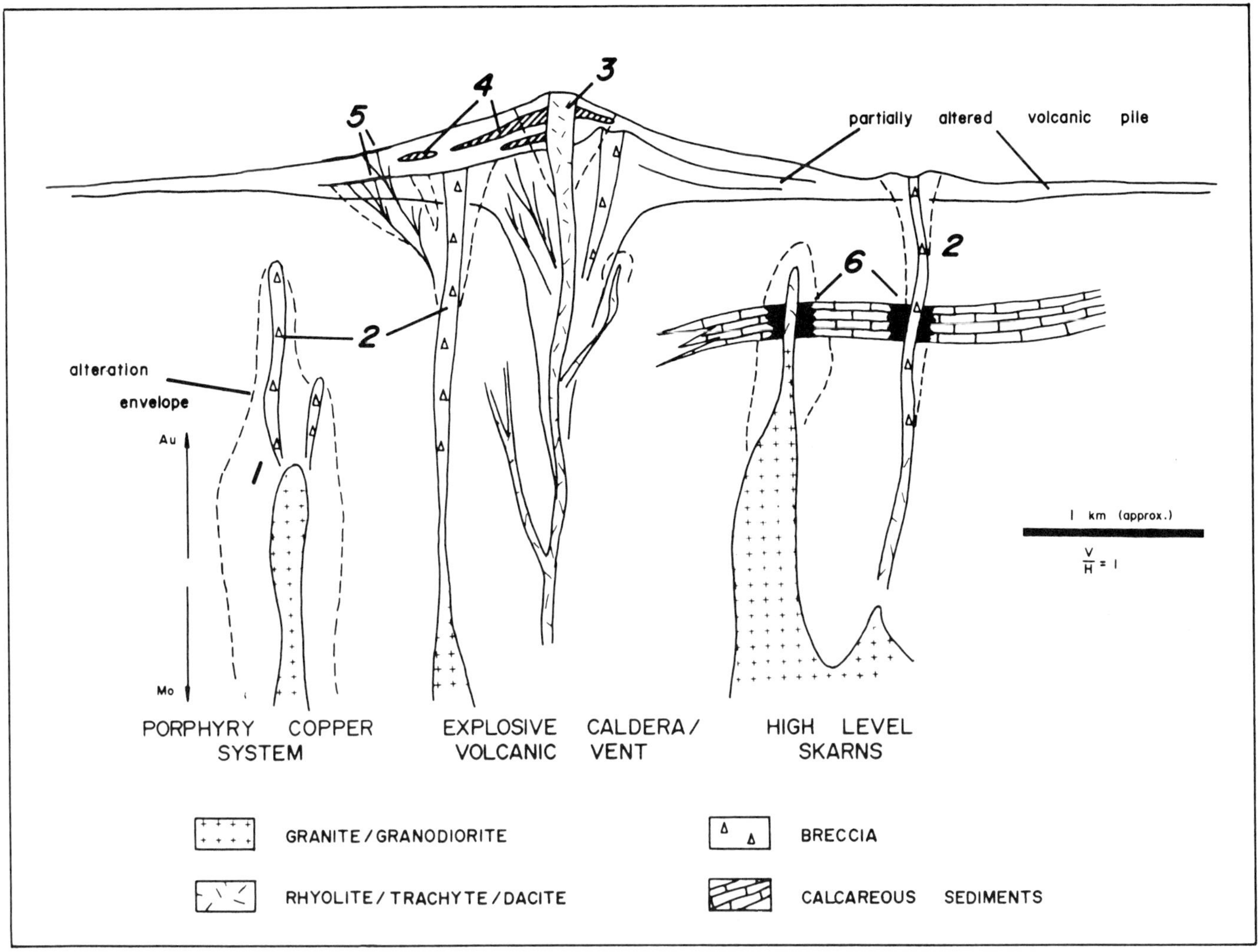


Figure 2. Generalized model of postorogenic subaerial acid volcanic and subvolcanic hosted bulk tonnage gold deposits in eastern Queensland. (1) auriferous high-level porphyry copper; (2) mineralized breccia pipe; (3) mineralized hydrothermal/volcanic vent; (4) proximal extruded mineralization; (5) ?hot springs/epithermal vein systems (yet to be documented in eastern Queensland); (6) skarn mineralization).

of 1300 by 900 m. The breccia pipe is associated with Middle Carboniferous rhyolitic intrusions and is hosted by Proterozoic granodiorite and metamorphic rocks (Figure 3). It is funnel-shaped in cross section and is surrounded by a 10 to 20 m zone of crackle breccia overlapped by sets of parallel-sheeted quartz veins. Breccia fragments within the pipe can be grouped into three principal lithological domains having gradational boundaries. These comprise a thin outer monomictic zone reflecting the lithology of the adjacent country rock, an inner bimictic zone where both rhyolite and adjacent country rock fragments are evident, and a core of polymictic breccia where mixing is greatest and rhyolite, granodiorite, and metamorphic fragments are evident. Breccia fragment size increases toward the margins of the pipe.

The breccia pipe is considered to have had an explosive origin with prebreccia events including the formation of stockwork quartz veins and localized breccias cemented by quartz and tourmaline. Crenulate bands of interlayered quartz crystals and micrograpite have also been found in breccia fragments. Postbreccia events involved cavity infilling between fragments with quartz, carbonate, pyrite, minor pyrrhotite and sphalerite, and gold. The sheeted quartz veins are also postbreccia and are auriferous.

The principal orebody is the Wises Hill deposit (Figure 3), which is as much as 200 m wide and has a known vertical depth of 300 m. Most of the gold occurs in the free state in a deposit which has a relatively low sulfide content. Total ore reserves stand at 44.4 m.t. grading 1.79 g/t gold and 2.22 g/t silver (Mustard, 1984; Wilson, 1984; Bain and Withnall, 1980.

*Red Dome (17°05′S; 144°22′E)*

Gold-copper mineralization at Red Dome is related to the intrusion of Carboniferous quartz feldspar porphyry (rhyolite) dikes and stocks into Siluro-Devonian reef limestones, interbedded volcanic rocks, and sedimentary rocks.

Hypogene gold mineralization occurs in quartz vein stockworks and silicified zones within quartz feldspar porphyry and sandstone. It also occurs within calc-silicate skarn zones hosted by porphyry, limestone, andesite, and clastic sedimentary rocks. Deep weathering of the deposit to depths of 150 m has produced an overlying ferruginous

Table 1: List of deposits mentioned or described in text (see Figure 1 for locations)

| No. | Deposit | Age | Mineralization Style | References |
|---|---|---|---|---|
| 1 | Red Dome | L. Carboniferous | skarn, stockwork | Joyce et al (1984) |
| 2 | Kidston | M. Carboniferous | breccia pipe | Mustard (1984); Wilson (1984) |
| 3 | Liontown | Cambro-Ordovician | v.m.s.* | Levingston (1974) |
| 4 | Highway | Cambro-Ordovician | stringer zone to v.m.s.* | Kay (1984) |
| 5 | Mt Leyshon | Permo-Carboniferous | breccia pipe | this paper |
| 6 | Mt Wright | L. Carboniferous | breccia pipe | Clarke (1971); Levingston (1974) |
| 7 | Pajingo | ?L. Carboniferous | quartz veins | Louthean et al (1986) |
| 8 | Mt Chalmers | E. Permian | v.m.s.* | Taube and van der Helder (1983) |
| 9 | Mt Morgan | M. Devonian | pipe to v.m.s.* | Taube (1980) |
| 10 | Mt Rawdon | Triassic | proximal extruded deposit | Cayzer and Leckie (1984); Horton et al (1980) |
| 11 | Mt Ophir | Triassic | volcanic vent | Ellis (1968) |
| 12 | Coalstoun | E. Triassic | porphyry copper | Ashley et al (1978); Horton (1982) |

*volcanogenic massive sulfide

Table 2: Simplified portrayal of the tectonic and metallogenic framework of eastern Queensland. The age ranges depicted are approximate as different parts of each orogen were affected by various tectonic and metallogenic events at slightly different times.

| OROGEN | AGE RANGE OF OROGENIC REGIME | PRINCIPAL METALLOGENIC EVENTS: Orogenic (or pre-cratonic) | PRINCIPAL METALLOGENIC EVENTS: Post-Orogenic (transitional and cratonic) |
|---|---|---|---|
| THOMSON | Cambrian to Silurian | Cambro-Ordovician (#3,4*) | Silurian to Middle Devonian |
| | | | Middle to Late Carboniferous** (#5,6,7*) |
| HODGKINSON-BROKEN RIVER | Ordovician to Middle Carboniferous | Late Silurian to Devonian | Middle Carboniferous to Early Permian** (#1,2*) |
| NEW ENGLAND | Silurian to mid-Permian | mid-Devonian (#9*) | Late Permian to mid-Triassic (#10,11,12*) |
| | | Early Permian (#8*) | Early Cretaceous |

* Associated deposits. Numbers correspond to those on Figure 1 and Table 1.

**This metallogenic event also affected parts of the Precambrian Georgetown Inlier.

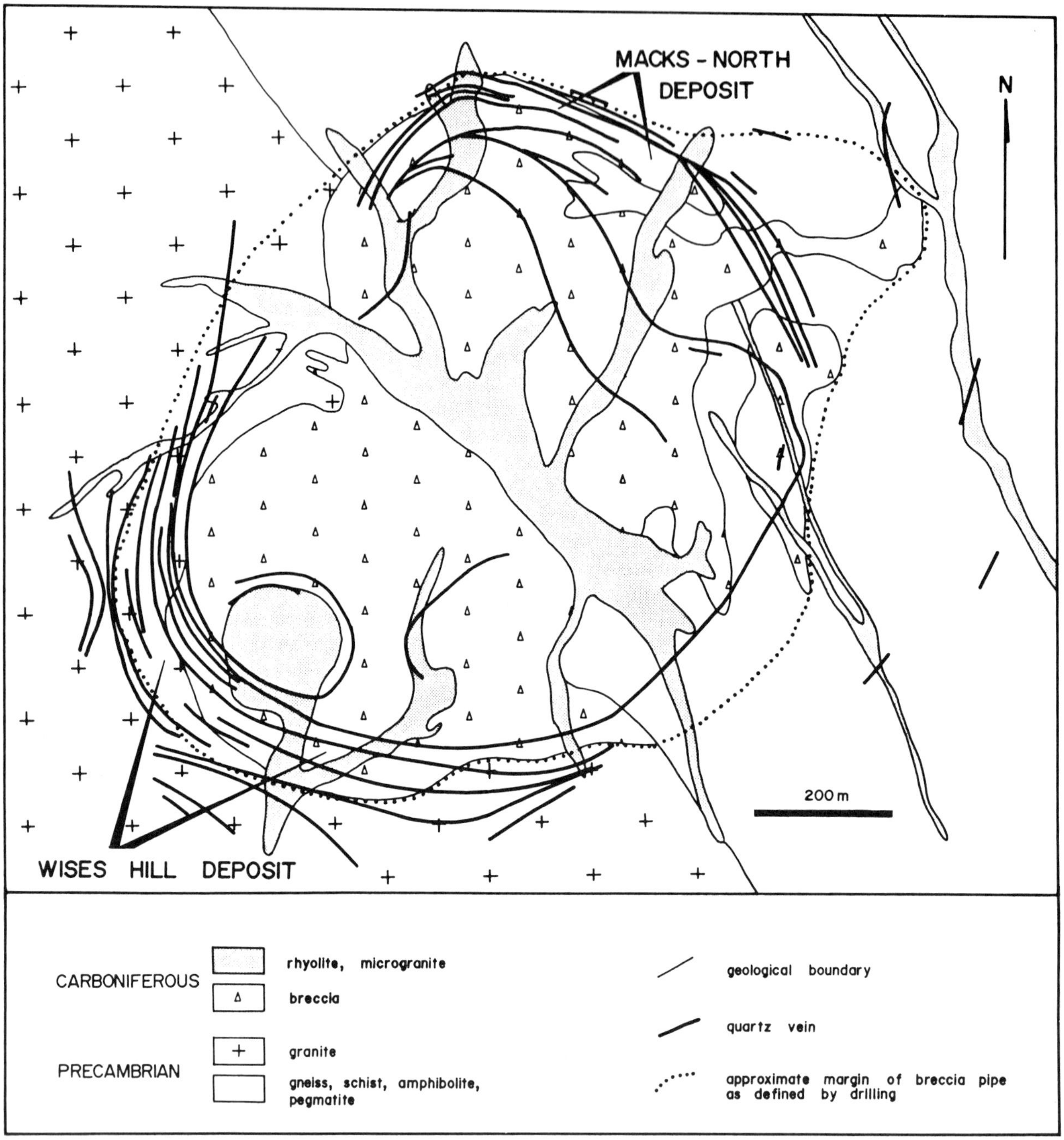


Figure 3. Geology of the Kidston deposit (after Bain and Withnall, 1980).

breccia containing significant gold-copper mineralization (Joyce, Karjalainen, and Johnson, 1984).

Alteration assemblages related to the gold mineralization show no easily recognizable zonation because of the wide lithological variety of host rocks. However, alteration assemblages related to the various rock types are as follows (Amoco Minerals Australia Co., written communication, 1984): (a) *Pure limestone*: garnet-magnetite-wollastonite-calcite skarn. (b) *Calcareous sandstones, conglomerate, and porphyry*: quartz, garnet, diopside, calcite, and clay alteration of feldspars. (c) *Andesite*: quartz, garnet, calcite, and chlorite. Based on a cut-off grade of 0.7 g/t Au, the deposit has reserves of 13.8 m.t. at 2.0 g/t Au, 0.46% Cu, and 5.25 g/t Ag.

*Mount Leyshon (20°18′S; 146°16′E)*

The Mount Leyshon deposit is located on the contact between Cambro-Ordovician metasedimentary rocks and Ordovician granites. It is situated within a Permo-Carboniferous volcanic breccia pipe, roughly oval in shape, with lateral dimensions of 1400 by 800 m (Figure 4).

The pipe comprises a wide variety of breccia types intruded by contemporaneous rhyolitic and trachytic dikes and plugs. The earliest breccia phase, a layered chlorite matrix breccia characterized by either granite or metasediment clasts, is presumed to predate the mineralization and is largely distributed outside the pipe. The later breccia phases are essentially polymictic in character with gradational contacts between recognized types.

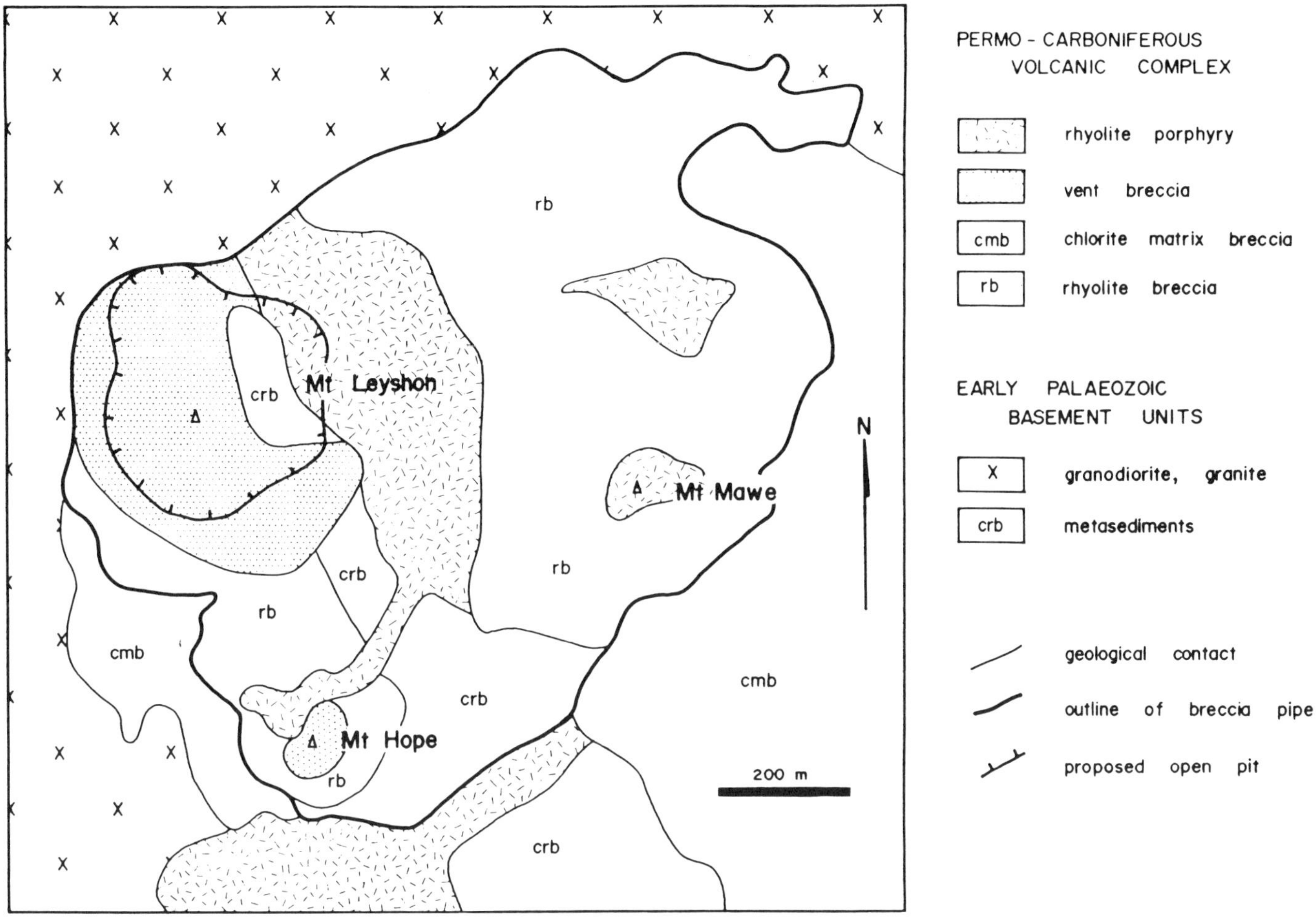


Figure 4. Geology of the Mount Leyshon deposit (after Noranda Pacific Ltd. Prospectus, 1984).

Disseminated sulfide mineralization is present throughout the whole breccia pipe but is more intense where stronger brecciation and alteration suggest proximity to a volcanic vent. Several volcanic vents containing more rounded "milled" fragments have been recognized within the volcanic pipe, and some of these are significantly auriferous.

The main ore zone is situated on Mount Leyshon itself. Oxide ore is distributed mainly in the lower section of the weathered profile along a broad interface between underlying breccias and trachytic "cap" rock. The primary ore zone lies predominantly within sulfide-rich polymictic breccia and milled breccia. Host rock mineralogy is largely quartz, sericite, and chlorite. Alunite has been recognized in surface outcrop. Pyrite with lesser sphalerite, galena, and chalcopyrite is also present in the primary zone.

The Mount Leyshon deposit contains combined probable and possible ore reserves of 16.5 m.t. grading an average of 1.5 g/t Au (Noranda Pacific Ltd and Pan Australian Mining Ltd prospectuses, 1984 and 1983, respectively, and internal company reports; published with the permission of the companies).

*Mount Rawdon (25°16'S; 151°46'E)*

The Mountain Rawdon deposit occurs in hydrothermally altered acid pyroclastic rocks of Triassic age in close proximity to a probable volcanic vent. The pyroclastics are composed largely of dacitic to rhyolitic agglomerate and lapilli tuff which have been intruded by numerous comagmatic but postmineralization dikes and sills. The probable vent site for these volcanic rocks is now occupied by a rhyolite (felsite) plug. Minor brecciation is locally evident.

Alteration is extensive and consists largely of pervasive surface kaolinization, sericitization, and pyritization, with lesser development of chlorite and epidote.

Pyrite with attendant gold and silver mineralization is distributed throughout the pyroclastics as disseminations and weakly developed stockworks (Cayzer and Leckie, 1984; Horton, Krosch, and Dixon, 1980). Two phases of mineralization have been recognized—the first (predominantly pyrite) was probably syngenetic with the pyroclastics, and the second (pyrite- gold-silver-weak Cu, Pb, Zn) appears to be largely veinlet and fissure controlled. The distribution of higher gold values within the pyroclastic rocks as narrow, vertically layered zones dipping away from the postulated vent site points to the bulk of the gold mineralization being originally syngenetic. There is strong evidence to suggest local remobilization of precious metals occurred as a consequence of the emplacement of the rhyolite dikes and sills.

Resources at Mount Rawdon have been established at 30 to 40 m.t. of material grading 1.3 g/t Au (Louthean et al., 1986) although a smaller resource of 7.1 m.t. of 1.75 g/t Au has also been defined (Brooks, 1986).

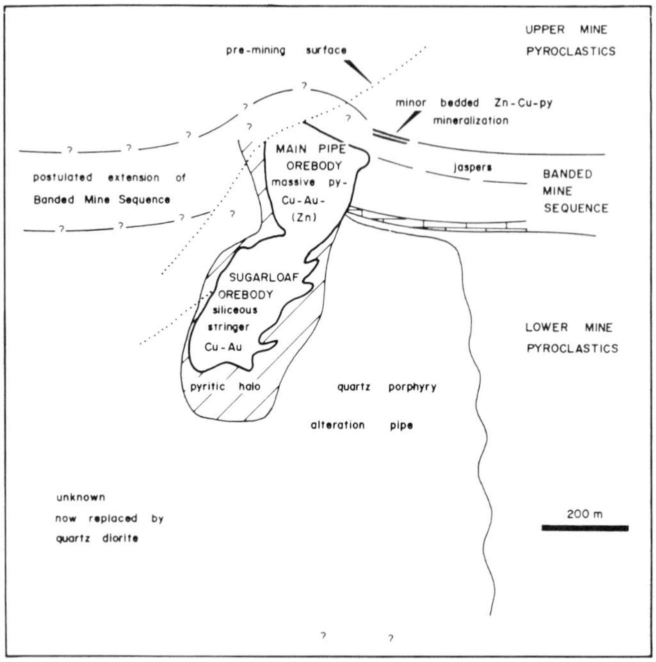


Figure 5. Reconstruction of the Mount Morgan deposit with respect to regional dip and the Slide Fault zone which separates the two orebodies. This generalized longitudinal section looks north-northwest (after Taube, 1980).

## Orogenic Acid Volcanic Mineralization

Several authors (e.g., Scheibner and Markham, 1976; Taube and van der Helder, 1983) consider that the numerous acid volcanic-hosted massive sulfide deposits in eastern Australia are similar to the Kuroko-type massive sulfide deposits of Japan. In eastern Queensland, these deposits range in age from Cambro-Ordovician in north Queensland to Early Permian in central and southern Queensland.

Most acid volcanic massive sulfide deposits in eastern Queensland contain minor amounts of gold (generally less than 2 g/t). In deposits containing significant amounts of gold, the gold is usually found within stringer zones or pipes underlying proximal stratiform ore lenses. Alteration products associated with footwall mineralization of this type are typically pyrite, quartz, chlorite, and sericite. Some gold deposits hosted by orogenic acid volcanic rocks may represent feeder zones to now-eroded or poorly formed massive sulfide lenses. The Highway deposit (Figure 1, no. 4) is considered such an example. Almost all orogenic ore deposits have undergone some degree of regional metamorphism.

Deposits of this type in eastern Queensland which contain appreciable quantities of gold mineralization are Mount Morgan, Mount Chalmers, Liontown, and Highway. Only Mount Morgan is considered to be a bulk tonnage deposit.

*Mount Morgan (23°38′S; 150°23′E)*

The Mount Morgan Mine closed in 1981 after almost 100 years of continuous production. It yielded more than 238,000 kg gold and more than 350,000 t of copper from 51 m.t. of ore. It has been Australia's single biggest gold producer. The deposit consists of two orebodies and lies within Middle Devonian volcanic and pyroclastic rocks deposited during the orogenic regime of the New England Orogen. It comprises the "Main Pipe" orebody consisting of stringer mineralization invading pyritic massive sulfides, and the adjacent "Sugarloaf" orebody which contains disseminated and siliceous stringer mineralization. Ore zone minerals include pyrite, pyrrhotite, chalcopyrite, magnetite, and sphalerite in that order of abundance associated with gold, gold tellurides, and silver minerals.

The host rocks consist of quartz feldspar porphyry, crystal lithic tuffs, cherty ash tuffs, and some true chert and jasper. Alteration around the Sugarloaf orebody consists of a zone of replacement-style silicification. Within this zone, a strong pyritic halo surrounds the orebody.

Reconstruction of the original stratigraphy (Figure 5) indicates that the two orebodies form part of a vertical pipe-like body overlain in part by minor stratiform pyritic beds. These beds contain minor amounts of copper-lead-zinc mineralization and grade laterally away from the pipe into a jasperoidal sequence (Taube, 1980). It is now generally accepted that this deposit is a volcanogenic massive sulfide deposit. This conclusion is supported by isotopic evidence (Golding and Wilson, 1981).

## DISCUSSION

Several workers (e.g., Guild, 1972; Hutchinson and Hodder, 1972) have attributed a close association between porphyry copper and massive sulfide mineralization to evolutionary developments within particular orogenic belts. Although both types of deposits are generated in dominantly calc-alkaline volcanic-plutonic arcs, they are not found together in space and time. In eastern Australia, Kuroko-type massive sulfide deposits and porphyry-type copper-molybdenum deposits are confined to orogenic and postorogenic regimes, respectively. Sillitoe (1980) contends that porphyry copper and massive sulfide deposits are incompatible, partly because they are associated with andesitic and rhyolitic volcanics, respectively. In eastern Australia, this argument does not hold because the volcanics associated with both types of mineralization are rhyolitic to dacitic. Hutchinson and Hodder (1972) also note that massive sulfide deposits contain a metal suite similar to that of porphyry copper deposits. On this basis, it could be argued that both types of mineralization have a common basic origin, and that mineralization styles are dependent on such factors as depositional environment or crustal thickness.

Comparisons between the acid volcanic bulk tonnage gold deposits of eastern Queensland and those outside eastern Australia are fraught with difficulties. In the case of the Mount Morgan deposit, Kuroko-type affinities have been noted (Taube, 1980), although some comparisons have been made with porphyry copper-type breccia pipes (Cornelius, 1969). The large auriferous pipe at Mount Morgan is atypical of volcanogenic massive sulfide deposits. Porphyry copper deposits are known in present-day island arcs but are absent in the much thinner volcanic arcs (protoisland arcs) of the precratonic regimes of the Tasman Orogenic Zone. Mount Morgan is considered to have been emplaced along a Devonian island arc (Day, Murray, and Whitaker, 1978) and this may account for its "hybrid" characteristics.

On the basis of mineralization styles (i.e., breccia pipes, skarns, etc.), many similarities can be observed between the postorogenic gold deposits of eastern Queensland and certain gold deposits of the Circum-Pacific belt. For example, the Golden Sunlight deposit in Montana (Sillitoe, 1983) occurs in a breccia pipe and contains similar tonnages and grades to the Kidston deposit. However, other styles of gold mineralization found within the Circum-Pacific belt have yet to be recognized in eastern Queensland. There is a distinct lack of disseminated replacement mineralization in eastern Queensland. Morrison (1984) noted that there is also a noticeable lack of gold occurrences in the major rhyolite calderas of north Queensland. These calderas are of the collapse type.

Volcanic-hosted epithermal vein-type and hot springs deposits (White, 1981; Berger and Eimon, 1982; Bonham, 1986) have not been documented in eastern Australia. However, several vein systems, particularly in central-east Queensland, could possibly represent the lower portions of such systems. The recently discovered Pajingo deposit (Figure 1, no. 7) may be of this type. Little information is available on this deposit, although preliminary reserve figures are estimated at 1.43 m.t. grading 12.6 g/t Au and 40 g/t Ag (Brooks, 1986).

Although the only documented gold deposits in eastern Queensland with bulk tonnage characteristics are associated with acid volcanic rocks, there is some potential for discovering other types of bulk tonnage gold mineralization. Chert-associated gold mineralization has been found at two localities in the Hodgkinson Basin in north Queensland (Ishaq, 1985; Keevers, 1984). Values in the chert are generally low but range up to 1 g/t Au. Other sedimentary rocks which may be host to bulk tonnage gold deposits include ferruginous varieties such as those in Cape York Peninsula (Ishaq, 1985).

## CONCLUSIONS

All major bulk tonnage gold deposits in eastern Queensland are associated with rhyolitic volcanic and subvolcanic rocks and range in age from Cambro-Ordovician to Early Cretaceous. The deposits are either postorogenic and contemporaneous with porphyry-type copper mineralization or orogenic representing the underlying stringer zone or pipe to Kuroko-type massive sulfide mineralization.

The most common styles of postorogenic gold mineralization are breccia pipes, although disseminated stratabound mineralization, gold-rich porphyry copper deposits, and high-level skarns are recognized.

Although similarities exist between the deposits of eastern Queensland and certain deposits of the Circum-Pacific belt, overall the deposits of the two regions are dissimilar. This feature may be, in part, a product of different volcanic host lithologies (rhyolite in eastern Queensland as opposed to andesite to rhyolite in the Circum-Pacific belt). Many types of epithermal gold deposits found in the Circum-Pacific belt (e.g., vein-type and disseminated replacement deposits) have yet to be documented in eastern Queensland.

## ACKNOWLEDGMENTS

I wish to express my appreciation to the following individuals and companies for the assistance given in providing information and material for both the presentation and this paper: Amoco Minerals Australia Co., Max Baker, Dick Cayzer, Harry Mustard, Noranda Pacific Ltd., Pan Australian Mining Ltd., Alex Taube, and Chris Torrey.

The manuscript benefitted greatly from discussions with, and reviews by, Shamim Ishaq and Lloyd Hamilton.

## REFERENCES CITED

Ashley, P.M., W.G. Billington, R.L. Graham, and R.C. Neale, 1978, Geology of the Coalstoun porphyry copper prospect, southeast Queensland, Australia: Economic Geology, v. 73, p. 945-965.

Bain, J.H.C., and I.W. Withnall, 1980, Mineral deposits of the Georgetown region, northeast Queensland, *in* R.A. Henderson and P.J. Stephenson, eds., The geology and geophysics of northeastern Australia: Geological Society of Australia, Queensland Division, p. 129-148.

Berger, B.R., and P. Eimon, 1982, Comparative genetic model characteristics of epithermal gold-silver deposits: American Institute of Mining Engineers Preprint 82-13, 20 p.

Bonham, H.F., 1986, Models for volcanic-hosted epithermal precious metal deposits; a review, *in* Volcanism, hydrothermal systems, and related mineralizations: International Volcanological Congress, Auckland, New Zealand, Proceedings of Symposium 5, p. 13-17.

Brooks, J.H., 1986, Mineral exploration and mining developments in Queensland in 1985 and outlook for 1986: Queensland Government Mining Journal, v. 87, p. 89-93.

Cayzer, R., and J.F. Leckie, 1984, Mount Rawdon: Exploration of a bulk low grade gold deposit (abs.): Gold in Queensland Symposium, Abstracts, University of Queensland, Brisbane, p. 14.

Clarke, D.E., 1971, Geology of the Ravenswood 1 mile Sheet area, Queensland: Geological Survey of Queensland Report 53, 49 p.

Cornelius, K.D., 1969, The Mount Morgan Mine, Queensland—A massive gold-copper pyritic replacement deposit: Economic Geology, v. 64, p. 885-902.

Day, R.W., C.G. Murray, and W.G. Whitaker, 1978, The eastern part of the Tasman Orogenic Zone: Tectonophysics, v. 48, p. 327-360.

Ellis, P.L., 1968, Geology of the Maryborough 1:250,000 Sheet area: Geological Survey of Queensland Report 26, 101 p.

Golding, S.D., and A.F. Wilson, 1981, An oxygen and carbon isotopic study of some gold deposits of eastern Australia: Australasian Institute of Mining and Metallurgy, n. 278, p. 13-21.

Guild, P.W., 1972, Massive sulfide vs. porphyry deposits in their global setting: Mining and Metallurgical Institute of Japan - AIME Preprint G13.

Horton, D.J., 1982, Porphyry-type copper and molybdenum mineralization in eastern Queensland: Geological Survey of Queensland Publication 378, 59 p.

Horton, D.J., 1984, Framework of acid volcanic hosted bulk tonnage gold mineralization in eastern Australia (abs.): Gold in Queensland Symposium, Abstracts, University of Queensland, Brisbane, p. 3-4.

Horton, D.J., N.J. Krosch, and O. Dixon, 1980, Departmental investigations of the Swindon gold prospect, Mount Perry: Geological Survey of Queensland, Record 1980/8, 26 p.

Hutchinson, R.W., and R.W. Hodder, 1972, Possible tectonic and metallogenic relationships between porphyry copper and massive sulfide deposits: Canadian Mining and Metallurgical Bulletin, v. 1972, p. 34-40.

Ishaq, S., 1985, Gold in Queensland—A review of production, exploration, and potential: Queensland Government Mining Journal, v. 86, p. 72-77.

Joyce, P.J., H. Karjalainen, and L. Johnson, 1984, The geology of the Red Dome deposit (abs.): Gold in Queensland Symposium, Abstracts, University of Queensland, Brisbane, p. 15.

Kay, J.R., 1984, The Highway gold mine, Charters Towers—Submarine volcanogenic gold-barite stringer mineralization modified by lateritic weathering (abs.): Gold in Queensland Symposium, Abstracts, University of Queensland, Brisbane, p. 18.

Keevers, R.E., 1984, Some occurrences of gold-bearing exhalites associated with submarine mafic volcanics in Paleozoic rocks of northeastern Queensland and northeastern New South Wales: Gold in Queensland Symposium, Abstracts, University of Queensland, Brisbane, p. 16-17.

Levingston, K.R., 1974, Ore deposits and mines of the Charters Towers 1:250,000 Sheet area, north Queensland: Geological Survey of Queensland Report 57, 103 p.

Louthean, R., S. Bell, P. Maher, and D. Wilkinson, 1986, Gold, *in* R. Louthean, ed., Register of Australian mining 85-86: Resources Information Unit, Perth, p. 195-300.

Morrison, G.W., 1984, Styles of bulk tonnage gold mineralization in north Queensland: Gold in Queensland Symposium, Abstracts, University of Queensland, Brisbane, p. 5-6.

Murray, C.G., 1984, Metallogeny and tectonic development of the Tasman Fold Belt system in Queensland: Geological Society of Australia, Abstracts, n. 12, p. 391-394.

Mustard, H., 1984, Brecciation and mineralization at the Kidston gold deposit, north Queensland: Geological Society of Australia, Abstracts, n. 12, 394-395.

Rickard, M.J., and E. Scheibner, 1975, The philosophical basis and terminology for tectonic nomenclature: Geological Society of Australia, Tectonics and Structural Geology Newsletter, n. 4, p. 16-29.

Scheibner, E., 1975, An outline of the tectonic development of New South Wales with special reference to mineralization, *in* N.L. Markham and H. Basden, eds., The mineral deposits of New South Wales: Geological Survey of New South Wales, p. 1-39.

Scheibner, E., and N.L. Markham, 1976, Tectonic setting of some stratabound massive sulfide deposits in New South Wales, *in* K.H. Wolf, ed., Handbook of stratabound and stratiform ore deposits, Volume 6: Amsterdam, Elsevier, p. 55-77.

Sillitoe, R.H., 1980, Are porphyry copper and Kuroko-type massive sulfide deposits incompatible?: Geology, v. 8, p. 11-14.

Sillitoe, R.H., 1983, Styles of low-grade gold mineralization in volcano-plutonic arcs: Nevada Bureau of Mines and Geology Report 36, p. 52-68.

Taube, A., 1980, The Mount Morgan copper-gold mine, *in* G.W. Hofmann, ed., 1980 Field Conference, Mount Morgan, Rundle Range, Yeppoon area: Geological Society of Australia, Queensland Division, p. 40-50.

Taube, A., and P. van der Helder, 1983, The Mount Chalmers mine and environment, *in* Permian Geology of Queensland: Geological Society of Australia, Queensland Division, p. 387-399.

White, D.E., 1981, Active geothermal systems and hydrothermal ore deposits, *in* B.J. Skinner, ed., Seventy-Fifth Anniversary Volume, 1905-1980: Economic Geology, p. 392-423.

Wilson, G.I., 1984, The Kidston gold mine (abs.): Gold in Queensland Symposium, Abstracts, University of Queensland, Brisbane, p. 13.

CHAPTER 48

# THE OLYMPIC DAM COPPER-URANIUM-GOLD DEPOSIT, SOUTH AUSTRALIA

**J.H. Lalor**
*Western Mining Corporation Ltd.*
*Preston, Australia*

ABSTRACT

The Olympic Dam copper-uranium-gold deposit is located 650 km north-northwest of Adelaide in South Australia and was discovered in 1975, when RD1, Western Mining Corporation's initial stratigraphic hole on the Stuart Shelf, intersected 38 m @ 1.05 % Cu. The deposit is in the basement beneath 350 m of unmineralized, flat-lying late Proterozoic to Cambrian sediments. It is coincident with gravity and magnetic anomalies and the intersection of tectonic lineaments. The early Proterozoic sediments hosting the mineralization within the basement are predominantly sedimentary breccias. Two types of copper sulphide mineralization have been defined—stratabound and younger transgressive. Both types contain uranium, rare earths, gold and silver. The deposit has an areal extent exceeding 20 km$^2$, and the maximum vertical thickness is 350 m. The estimated resource is 2000 million tonnes of 1.6% copper, 0.06% uranium oxide, and 0.6 grams per tonne gold.

## INTRODUCTION

The Olympic Dam copper-uranium-gold deposit appears to be a new type of hydrothermal stratabound sediment-hosted ore deposit. It is located 650 km north-northwest of Adelaide in South Australia and was discovered in 1975, when RD1, the initial stratigraphic hole on the Stuart Shelf, intersected 38 m @ 1.05% Cu. Since that time, a programme of evaluation has been in progress. Following completion of the Final Feasibility Study, construction commenced in March 1986. The study was based on an annual production rate of approximately 2000 tonnes of $U_3O_8$, 55,000 tonnes of copper, and 90,000 ounces of gold, and the total capital cost was estimated in the range of $550-600 million.

The deposit has an areal extent exceeding 20 km$^2$ and the maximum vertical thickness of mineralization 350 m. The orebody consists of a probable ore reserve of 450 million tonnes of 2.5% copper, 0.08% uranium oxide, 0.6 grams per tonne gold, and 6 grams per tonne silver contained within a resource of 2000 million tonnes of 1.6% copper, 0.06% uranium oxide, and 0.6 grams per tonne gold. The Olympic Dam orebody can also be considered as a hematitic iron deposit containing a large amount of intimately associated copper, uranium, gold and light rare earths. The rare earths and hematite are not considered as being potentially economic in the initial exploitation of the deposit but constitute a significant resource.

This paper summarizes earlier papers on the discovery and geology of the Olympic Dam deposit and incorporates new data generated from a major program of underground development.

## REGIONAL GEOLOGY

The Olympic Dam deposit occurs within the Stuart Shelf region, which is separated from the Adelaide Geosyncline to the east by a major zone of north-south faulting termed the Torrens Hinge Zone (Figure 1). The Stuart Shelf consists of a relatively thin, flat-lying sequence of Late Proterozoic (Adelaidean) and Cambrian sediments overlying middle Proterozoic and possibly older basement which hosts the Olympic Dam mineralization. These sediments have not been affected by the deformation event which complexly deformed the Adelaide Geosyncline 500 m.y. ago. To the west and southwest, the basement is interpreted to consist of Gawler Domain rocks, but no reliable correlation between the Olympic Dam succession and specific units of the exposed Gawler Domain can be made.

A generalized stratigraphic table is as follows:

| Age | Unit | Lithology |
|---|---|---|
| Cambrian | Andamooka Limestone | Massive, locally bedded, limestone |
| | Unconformity | |
| Upper Proterozoic | Arcoona Quartzite | White and red, massive, cross-bedded sandstones |

| Age | Unit | Lithology |
|---|---|---|
| Upper Proterozoic | Corraberra Sandstone | Red, massive, cross-bedded sandstone |
| Upper Proterozoic | Tregolana Shale | Thinly laminated red and green shales |
| Upper Proterozoic | Whyalla Sandstone | Cream cross-bedded sandstone |
| | Unconformity | |
| Upper Proterozoic | Pandurra Formation | Brown and white, massive and bedded sandstone |
| | Unconformity | |
| Middle Proterozoic | Basement | — |

In the Olympic Dam area the cover sediments are about 350 m thick. The Tregolana Shale lies unconformably on basement, as the Whyalla Sandstone/Pandurra Formation are absent.

## DISCOVERY

The discovery history is based on the written documentation of events leading to the drilling of drill hole RD1, and on subsequent events up to and including the drilling of hole RD10. Hearsay, opinions, and recollections have been, in fairness to all, ignored and individuals are not mentioned. The discovery of Olympic Dam was very much a team effort.

Preliminary studies began in mid-1972 with the commencement of a geological literature study of the Proterozoic and early Paleozoic rocks of South Australia. The study was aimed at defining targets for sediment-hosted copper deposits, and it employed conceptual modelling of the origin, transport, and precipitation of copper in sediment-hosted copper deposits. The key part of the model was that copper in continental tholeiitic basalts became readily mobile and available to circulating hydrothermal waters during oxidative (that is hematite-producing) burial-metamorphic alteration. The basalts, as proposed copper source rocks for the sediment-hosted copper deposits, were considered one of the principal targeting parameters for defining areas with potential to contain sediment-hosted copper deposits.

In 1974 the Stuart Shelf was regarded as a prime target, because it was considered to be underlain by appropriately altered basalts, to contain appropriate host rocks, and, importantly, because it had not been extensively explored for copper occurrences. In addition, it was considered to have certain broad structural and lithological similarities to the setting of the Zambian Copperbelt, although the conceptual model being used in the South Australian study was not thought to be directly applicable to the Zambian Copperbelt.

Figure 2 shows the region selected for the initial study. The geological literature assessment pinpointed a number of discrete areas interpreted to contain appropriately altered basalts. The study was followed by an assessment of all publicly available exploration data on the smaller areas. This was completed in early 1974 with the definition of several areas as potential exploration targets.

The conceptual model clearly could not define targets in regions of flat-lying or nearly flat-lying rocks, and additional target definition criteria were as follows: (1) Ore would be localized in basins adjacent to major faults intersecting the sedimentary succession above altered basalt piles; (2) Regional magnetic and gravity interpretations would define areas which might contain concealed basalt piles: (3) Lineament analysis would define major tectonic zones and associated major faults and, more particularly, targets associated with the intersection of such zones.

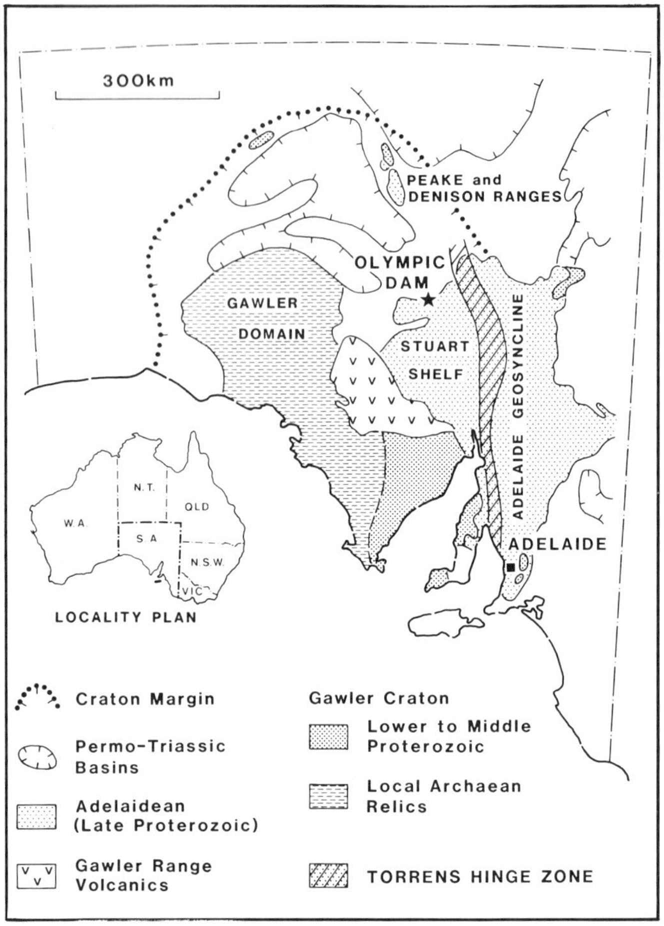


Figure 1. Geologic plan of the southern part of South Australia showing the main lithologic and tectonic features.

Field inspections of outcrops of altered basaltic lavas and inspection of diamond drill cores of altered lavas from the Stuart Shelf stratigraphy were carried out in early 1974. Emphasis was placed upon documentation of hematite-bearing alteration assemblages, as this was essential to the conceptual model. Lavas were sampled wherever possible to check for copper depletion.

Initially, work was planned to focus on the southern part of the Stuart Shelf, where the occurrence of appropriately altered lavas had been documented. However, the northern part of the Stuart Shelf was also considered to be of interest because of the occurrence of pronounced positive regional gravity anomalies with associated positive magnetic anomalies interpreted from Bureau of Mineral Resources data. These anomalies were interpreted from Bureau of Mineral Resources data. These anomalies were interpreted to be caused by concealed piles of basalt equivalent to those around the southern margin of the Stuart Shelf and were thus indicative of possible source rocks for concealed copper deposits in the overlying sediments.

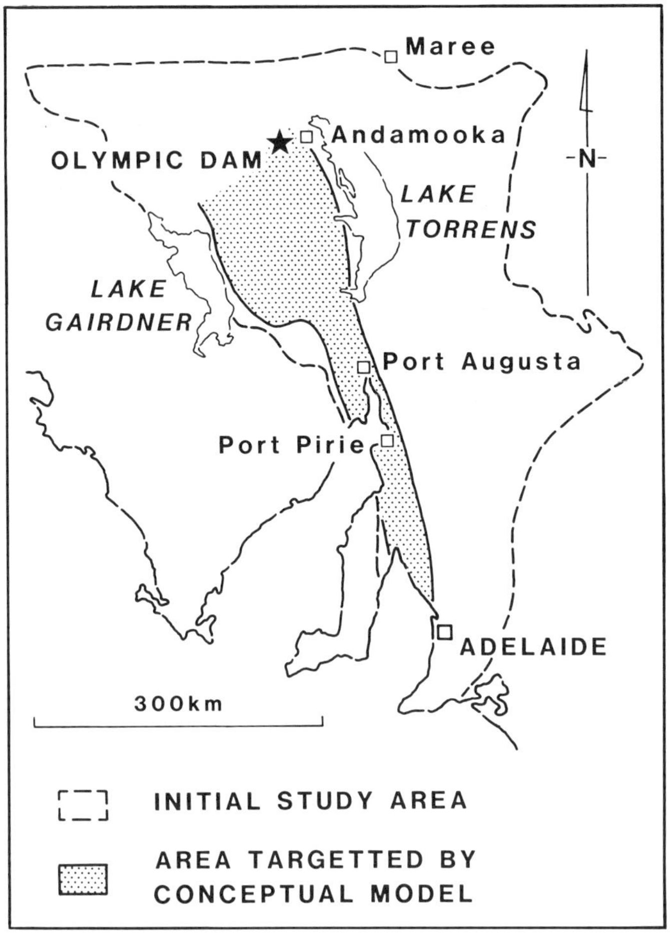


Figure 2. Locality plan showing the Olympic Dam deposit, and the initial study area.

Ground acquisition on the Stuart Shelf was being seriously considered in mid-1974 based on: (1) The Whyalla Sandstone/ Pandurra Formation unconformity could have been one control on mineralization; (2) Basement highs could have been a control; (3) Areas selected had to be underlain by suitably altered basalts; (4) Lineament studies would be used to define areas of interest.

Toward the end of 1974 most of the data assessment and interpretation on the Stuart Shelf area had been completed, and it was clear that testing of the Stuart Shelf targets would require stratigraphic diamond drilling. The reasons for the proposed stratigraphic drilling of these targets were documented as follows:

(1) The Whyalla Sandstone/Pandurra Formation unconformity, on which copper mineralization in the Mt. Gunson area was located, probably existed in the area (Figure 3).

(2) The geophysical interpretation suggested that the targets included anomalies similar to, but larger than, those that occurred in the Mt. Gunson area;

(3) Sediment-hosted copper deposit source rocks probably underlay the region, and a local thickening was probable near the northern projection of the Torrens Hinge Zone;

(4) Favorable tectonic targets had been defined, four of which coincided with the geophysical anomalies.

An Exploration Licence was applied for in January 1975 and granted in May 1975. Ground work followed immediately to define sites for stratigraphic diamond drill holes on two of the targets, at Olympic Dam and Acropolis. Although these were ranked as being the most promising of all the targets, a detailed written listing of the reasons for such a choice was never made.

Ground exploration work consisted of magnetometer traversing by Western Mining Corporation and refraction and reflection seismic surveys by the South Australian Department of Mines (Figure 4). The ground magnetometer traverse extended from the Olympic Dam target to Coronation Dam. Because of time constraints, ground magnetometer traversing was not carried out as Acropolis prior to the drilling of the first stratigraphic hole on the anomaly.

The experimental seismic survey was carried out around Olympic Dam to establish whether seismic methods normally applied in younger sedimentary basins could be used to map horizons within Proterozoic sediments. The majority of the survey used refraction techniques, although approximately 1 km of reflection seismic was shot.

The interpretation of the seismic survey data revealed a single flat interface at 300 m depth, an unexpected and unexplainable result at that time. The interface has subsequently been shown to be the contact between the Pandurra Formation and the overlying Tregolana Shale.

The two stratigraphic drill holes were programmed to test the sedimentary sequence above the inferred source rocks, as well as to test for the presence of the source rocks themselves. It was decided that at least a 50-m thickness of the source rocks should be intersected, to allow for differences in what was predicted to be a variable alteration pattern. The holes were targeted to intersect near the edge of the body producing the magnetic anomaly and within the tectonic target and corridor (Figure 3).

Drill hole RD1 was begun on the 14th June, 1975, and was completed on the 31st July, at a depth of 411 m. The first lithological log of this hole is as follows:

| | |
|---|---|
| 0.00 - 37.91 m | Andamooka Limestone |
| 37.91 - 197.45 m | Arcoona (Simmens) Quartzite |
| 197.45 - 209.95 m | Corraberra Sandstone |
| 209.95 - 334.85 m | Woomera (Tregolana) Shale |
| 334.85 - 335.73 m | Red, medium- to coarse-grained hematitic sandstone with heavy mineralization cross-stratigraphic, trace chalcopyrite |
| 335.73 - 411.00 m | Coarse-grained altered hematitic amygdaloidal basic volcanics (Roopena Volcanics?) |

The rocks described as hematite-altered basalts were subsequently identified as hematite-bearing granite breccias. At the time of completion of RD1, however, there was a desire to see the appropriate source rocks, and they bore sufficient resemblance to hematite-altered fragmental amygdaloidal lavas to be identified as such. These rocks were chip sampled at 15-cm intervals for chemical analysis, which is standard Western Mining Corporation procedure to check for anomalous response, but in this case was also to check for evidence of copper depletion. On the 22nd September, 1975, the analytical data were received, and they revealed that the interpreted "depleted" lavas contained an 189-m interval from 335 m to 353 m assaying 930 ppm copper and a thickness of 38 m from 353 m to 391 m assaying 1.05% copper. The copper was in the

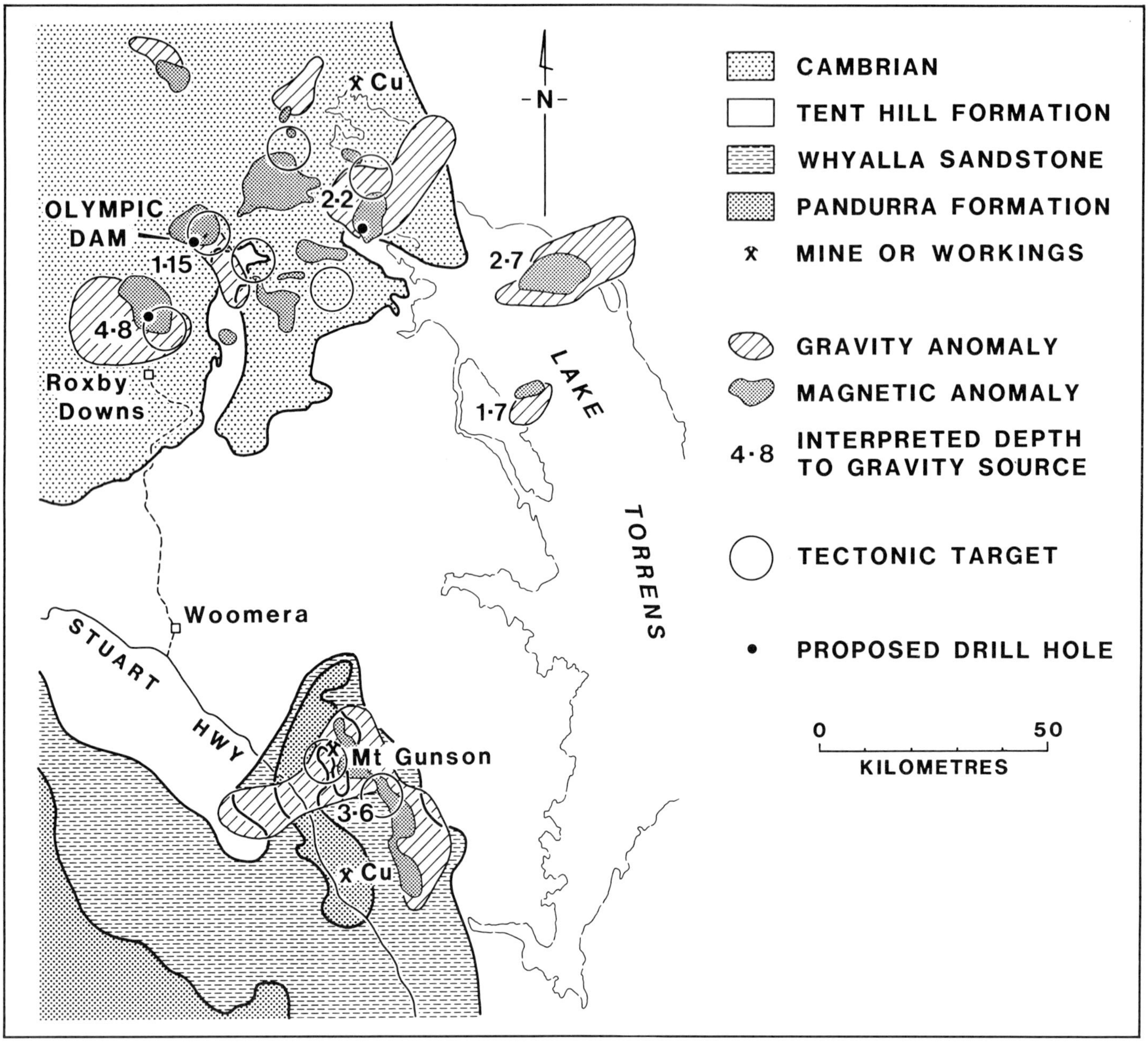


Figure 3. Summary plan showing ingredients for proposed drill hole localities (Drawn September 1974).

presumed source horizon, not the host horizon, and it is interesting to speculate whether exploration would have continued if the hole had only been drilled a few metres into "basement" below the sedimentary sequence.

A preliminary examination of sawn faces of drill core resulted in recognition of their silica-rich nature, but no positive identification of the nature of copper mineralization. However, the rocks were petrographically examined in the Western Mining Corporation Geological Research Laboratory at Kalgoorlie, Western Australia, and the report read as follows:

"The origin of these rocks is still highly problematical, and although having perhaps most features in common with a poorly sorted clastic sediment, the possibility that the rocks were derived from an acid igneous rock should certainly not be discounted. The mineralized rocks in hole RD1 are quartz and hematite-rich, apparently fragmental, and hydrothermally altered. The nature of the parent rock remains obscure, but appears to have been derived from acid to intermediate plutonic rocks. Hydrothermal alteration is extreme, and the mineral assemblage produced has some similarities to that seen in high level breccia pipes such as those in Chile. Copper mineralization occurs as bornite-chalcocite-digenite disseminated throughout the sericite and hematite-rich ground mass of the rocks. In some instances the sulphides are obviously open space fillings that have subsequently been infilled by fluorite."

The unusual nature of the mineralization was apparent at this early stage. Drill hole RD2 was sited on the Acropolis target, and began immediately after hole RD1 was completed. It

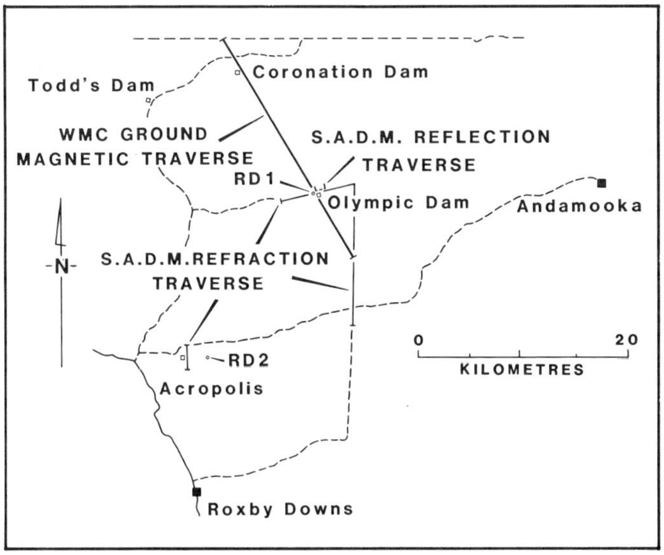


Figure 4. Location plan showing ground geophysical traverses.

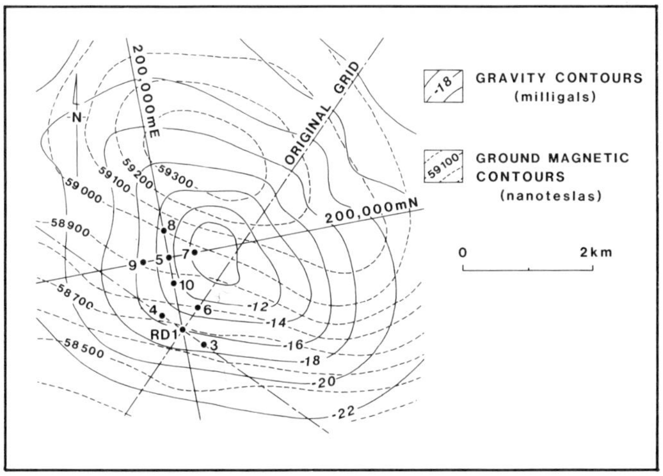


Figure 5. Plan showing gravity and magnetic contours with diamond drill hole locations.

was stopped at a depth of 513 m, within Tregolana Shale. It is not known whether, the hole would have intersected basement at 560 m if drilling had been continued. The unusual nature of the Olympic Dam mineralization and the large size of the geophysical anomaly indicated that there was potential for a large body of mineralization. The copper content of the mineralization intersected in RD1 was of relatively low tenor and probably uneconomic, but the large size of the untested area warranted a continuation of drill testing.

Ground gravity and magnetic traversing were then carried out along a surveyed base line and two cross lines (Figure 5) in the Olympic Dam-Acropolis area. The survey broadly defined the gravity and magnetic anomalies on which the early drilling in the Olympic Dam area was based and was generally in agreement with later detailed surveys (Figure 5).

Drilling of step-out diamond drill holes resulted in the discovery of mineralization grading up to 1% copper. However, it was not until RD10, sited 730 m north of RD1 (Figure 5), intersected 170 m of 2.12% copper from a depth of 348 m, that the potential for a large body of economic mineralization was indicated. The drill core was visually spectacular, containing thick zones of disseminated bornite and chalcopyrite in hematite-rich polymict breccias.

RD10 was the first hole to contain significant uranium. Analytical scans of preceding holes had indicated relatively minor amounts of gold, silver, and uranium, but RD10 contained 580 ppm (0.58kg/t) $U_3O_8$ coincident with the copper intersection.

The success of RD10 enabled an enlarged programme of drill testing. This inevitably led to the discovery, in RD17, of the large elongate zone of potential ore-grade mineralization in the northern and eastern parts of the gravity anomaly on which the current mining project is based.

The discovery was the result of:

(1) The integration of all available techniques—conceptual models, geology, geophysics, tectonics;

(2) A confidence, a boldness, and an enthusiasm in the team; and above all, good communications;

(3) A considered and intelligent persistence.

These characteristics were apparent at all stages of the discovery history. For example, boldness was obvious in the decision to drill not only inferred host stratigraphy, but to drill 50 m into interpreted source rocks to see if they were appropriately altered. Boldness was obvious in the decision to drill in this isolated area, the decision being helped by the knowledge that here was a large area in which no subsurface testing had been carried out. Enthusiasm smoothed the path for these decisions.

The interplay between the involved geoscientists and management was important. It was made easier by the confidence of the geoscientists and management in each other—a confidence engendered by good communications. This confidence grew with time, especially during the discussions on data generated by the early field reconnaissance studies, and its facilitated decision leading to drilling.

That persistence was an essential ingredient is obvious. The drilling of a successful discovery hole after nine diamond drill holes which were either bereft of mineralization, or contained clearly subeconomic mineralization, required persistence. The persistence was helped by genuine scientific curiosity. In summary, the discovery history is characterized by a basic confidence in scientific staff, a boldness, an enthusiasm, a genuine sense of scientific curiosity, and finally persistence. These were the ingredients of success.

## OLYMPIC DAM DEPOSIT

### Stratigraphy

The Proterozoic sediments comprising the basement sequence in the Olympic Dam area are predominantly unmetamorphosed sedimentary breccias. The breccias have been classified into two broad groups, monomict and polymict, on the basis of clast type. The polymict group contains most of the significant mineralization. Many different clast compositions occur in the polymict breccias, including altered basement granite; felsic, intermediate, and mafic volcanics; various types of hematite-rich rocks including banded iron-formation and hematitic siltstone; arenite, carbonate, fluorite, barite, and sulfides. Reworked clasts of breccia are common, particularly toward the top of the sequence. Clast sizes range up to 12 m but most are in the range of 1 to 3 cm. The main matrix types are crystalline black and gray hematite, granular brown hematite, and earthy red hematite.

The sediments have been divided into formal lithostratigraphic units—the Olympic Dam Formation and the Greenfield Formation. Both formations have several members, as illustrated in the following table. Contacts between the formations and members within formations are generally conformable, but disconformable contacts have been observed in some areas. These relate to reworked breccias, igneous dikes, massive iron-rich bodies, discordant polymict breccias, and two informal lithostratigraphic units—Recycled Breccias and Discordant Units (D.E. Roberts, personal communication, 1986).

The Recycled Breccias are hematite-rich or polymict sedimentary breccias which overlie the formal stratigraphic units. Discordant Units consist of igneous dykes and sills ranging from intensely altered felsic dykes to weakly altered dolerites and hematite-rich and polymict breccia bodies, which crosscut older sedimentary units.

**DISCORDANT AND REWORKED UNITS**

Hematite-rich and polymict breccias; dykes and sills.

| UNIT | MAXIMUM THICKNESS | LITHOLOGY |
|---|---|---|
| GREENFIELD FORMATION | | |
| Volcanic Member | 250 m | Altered felsic volcanics, volcanic breccias, conglomerates. Hematite-rich breccias, bvanded iron-formation. In part veined by silican and barite and strongly hematized and silicified. |
| Breccia Member | 500 m | |
| OLYMPIC DAM FORMATION | | |
| Upper Granite Breccia Member | 1000 m | Granite and polymict breccias with some black hematite-dominated brerccias; bedded and graded units common in parts; chalcocite-bornite, chalcopyrite-pyrite mineralized zones. |
| Brooks Member | 300 m | |
| Whenan Member | 350 m | |
| Black Hematite Member | 150 m | |
| Lower Granite Breccia | 800 m+ | |

Figure 6 is a generalized geological plan showing structural and lithological relationships at the $-420$ m level.

## BASEMENT GRANITE

The basement granite is present to the west, south, and north of the deposit and is presumed to underlie the Olympic Dam Formation. It is an anorogenic A-type, medium-coarse grained alkali feldspar granite with minor fine-grained phases. The granite consists almost entirely of potassium feldspar and quartz with less than 10% albite and ferromagnesian minerals. The feldspars are pink to brick red in color owing to hematite alteration, and in some areas they are totally replaced by hematite.

## STRUCTURE

The Olympic Dam mineralization is confined to a sedimentary sequence in a fault-bounded trough informally called the Olympic Dam graben (Figure 6), which has no surface expression in the flat-lying cover sediments. The graben has a northwest trend with a length exceeding 7 km; the width is about 4 km in the central part and narrows to about 0.5 km at the ends. The base of the sedimentary sequence in the graben has not been intersected in drilling, giving the sequence a minimum thickness of 1 km. The underground development has exposed zones of faulting and strong shearing (D.E. Roberts, personal communication, 1986).

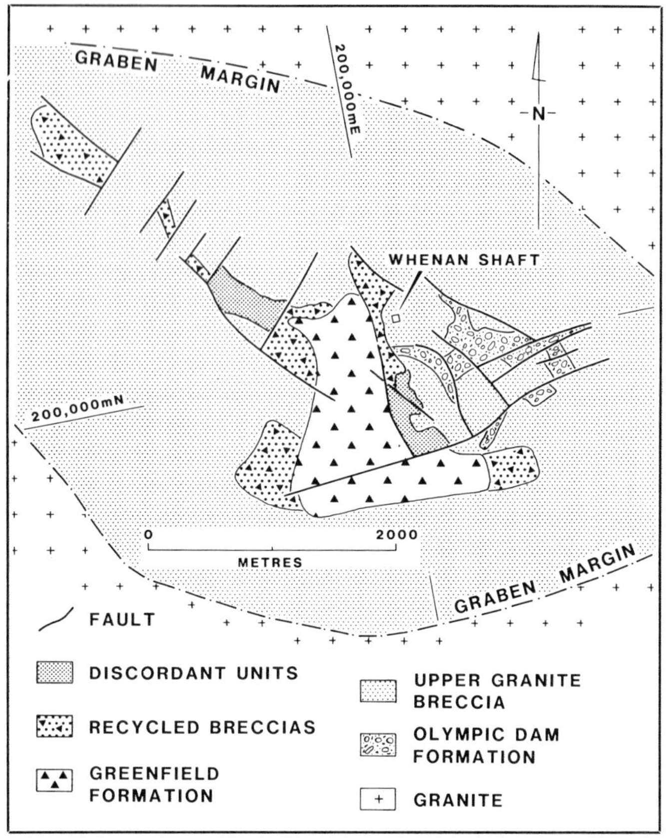


Figure 6. Generalized geologic plan of the Olympic Dam deposit at the main underground development level.

## ALTERATION

All rocks in the Olympic Dam graben are altered to different degrees. The alteration pattern is complex, particularly in the upper parts of the stratigraphy, and the dominant types are hematite, sericite, chlorite, and silica. Minor carbonate and potassium feldspar alteration are present. Black hematite alteration is the most common, followed by red-brown hematite, chlorite, and silica. A large number of samples have been semi-quantitatively analyzed for alteration mineralogy, and the results indicate that the intensity of all types of alteration increases towards major structural zones, particularly those trending northwest, and around the margins of the Greenfield Formation.

## MINERALIZATION

The complex association of copper, uranium, rare earths, ,gold, silver, and iron mineralization in the Olympic Dam deposit is one of its unique features. Mineralized zones range from steeply dipping discontinuous veins less than 1 m thick, through moderately dipping lenses 30 to 50 m thick with strike and dip dimensions in the order of hundreds of metres, to flat-lying zones more than 200 m thick with areal extents of hundreds of thousands of square metres.

## COPPER MINERALIZATION

Two distinct types of copper mineralization are defined on the basis of sulfide assemblage, rock association, and stratigraphic position: (1) stratabound bornite-chalcopyrite-pyrite mineralization confined to the Olympic Dam Formation and (2) chalcocite-bornite mineralization in lenses and crosscutting veins in both the Recycled Breccias and Discordant Units, with lesser amounts in the Olympic Dam and the Greenfield Formations. Significant uranium, rare-earth and gold mineralization is associated with both types. The two types can occur individually or together (D.E. Roberts, personal communication, 1986).

There is a well–developed vertical zoning of sulfides, grading from a relatively sulfur-rich copper-poor pyrite (chalcopyrite) assemblage at the base of the Black Hematite Member, to a relatively copper-rich sulfur-poor bornite-chalcopyrite assemblage at the top of the Whenan Member, and in the Brooks Member of the Olympic Dam Formation.

There is no clear evidence of supergene enrichment. However, zones displaying evidence of oxidation and leaching of copper sulfides occur throughout the entire mineralized sequence.

## URANIUM MINERALIZATION

Three main uranium-bearing mineral species have been identified. Uraninite is the most abundant; coffinite is subordinate to uraninite in abundance; and brannerite accounts for only a minor proportion of the uranium mineralization. On a deposit scale uranium mineralization is very closely associated with copper zones. However, the distribution and form of the uranium mineralization varies with the type of copper mineralization. In stratabound copper zones, uranium is evenly distributed; whereas in transgressive copper zones, uranium-rich veins and segregations are common and complex textural associations with copper-sulphides are observed.

## GOLD MINERALIZATION

Free gold is associated with both types of copper sulfide mineralization, but the concentration is relatively low in the stratabound types. Exploration on the eastern side of the Greenfield Formation has defined a separate gold-rich zone containing a series of irregular ore lenses. The zone occurs within an area of increased fracturing, shearing, and felsic dyke occurrence. The lithologies hosting gold mineralization range from matrix-poor breccias to hematite matrix-rich polymict breccias. Overlapping zones of intense sericite, hematite, and quartz alteration are associated with zones of tectonic brecciation, shearing, and foliation.

## DISCUSSION

The delineation of the Olympic Dam deposit is still incomplete, and attempts to explain the genesis of the deposit therefore continue to be speculative. However, the initial event leading to the formation of the Olympic Dam deposit appears to have been the development of a thick pile of sedimentary breccias (the Olympic Dam Formation) in a fault-controlled, northwest-trending trough. Volcanism accompanied the development of the sedimentary pile and was possibly bimodal, as both felsic and mafic clasts occur in the breccias. Following the formation of the complex sedimentary pile, the main phase of stratabound mineralization occurred in association with further volcanic and hydrothermal activity, the distribution of mineralization being controlled by permeability-reactivity factors. The major renewal of tectonic and igneous activity indicated by the influx of volcanic material and the reworking of the pre-existing sediments was accompanied by a second phase of mineralization resulting in transgressive high-grade copper-uranium zones and some local stratabound zones of mineralization in the recycled breccia units.

## REFERENCES

Lalor, J.H., 1984, Olympic Dam, South Australia—The discovery history (abs.): Geological Society of Australia Abstracts 12, p. 321.

Roberts, D.E., and G.R.T. Hudson, 1983, The Olympic Dam copper-uranium-gold deposit, Roxby Downs, South Australia: Economic Geology, v. 78, p. 799-822.

Roberts, D.E., and G.R.T. Hudson, 1984, The Olympic Dam copper-uranium-gold deposit, Roxby Downs, South Australia: A reply: Economic Geology, v. 79, p. 1944-45.

CHAPTER 49

# MINERALS OVERVIEW

**H.M. Morgan**
*Western Mining Corp., Ltd.*
*Melbourne, Australia*

It is a particular privilege to be invited to give this introductory paper at this important conference. My topic is not an easy one, and because of the difficult times we face, it brings to mind the comment made by the distinguished political philosopher Michael Oakeshott who, when speaking of politics, said:

> In political activity, then, men sail a boundless and bottomless sea: there is neither harbour for shelter nor floor for anchorage, neither starting-place nor appointed destination. The enterprise is to keep afloat on an even kneel; the sea is both friend and enemy; and the seamanship consists in using the resources of a traditional manner of behavior in order to make a friend of every hostile occasion.
>
> There is not a great deal of difference, I think, between political activity, and finding, producing, and selling minerals. There is more rationality, to be sure, in the minerals industry, than in politics, but ultimately both politics, and mining, and minerals exploration depend upon the ideas which people hold and the beliefs which motivate them—what Oakeshott calls "traditional manners of behaviours."

For 5 years now the hard-rock miners have been trying to understand why base metal prices will not, at a time when the U.S. economy has been, according to the accepted indicators, growing strongly, move up from levels equivalent to the depression prices of the 30s. Steel prices have likewise been depressed since 1981, and within the last month LTV Corp., the number two U.S. steel company, filed for Chapter 11. A number of other steel companies are expected to follow suit.

The world oil markets are dominated by discussions of Middle East politics, and characterized by a volatility, indeed an instability, which makes the life of those who work in the oil exploration industry difficult indeed.

Since the price of oil is the benchmark which determines, at least in the long term, the prices for other fuels, including ultimately uranium, we can see that the instability of the oil market has implications which spread out far beyond the oil exploration industry (Figure 1).

It is the conventional wisdom of the day that commodity prices generally, and metal prices particularly, will remain depressed for the remainder of this decade, and pick up sometime in the 1990s. According to this view, the correct strategy for the minerals industry today is to improve its efficiency and gear itself for an expansion in the 90s.

Let me put a much more pessimistic view. It is certainly possible that metal prices will rise dramatically this decade, and that they will do so because the drums of war will be heard more persistently, more ominously, more convincingly.

The increasing indifference shown by political leaders in every quarter of the globe, to the spreading political consequences of protectionism, of subsidy of over production, and of economic sanctions against unfashionable countries, can only result, if they are unchecked, in the rekindling of martial spirit, in a reborn determination to compete on the battlefield, if peaceful competition in trade and commerce is to be outlawed. Economic warfare—sanctions, protection, subsidies, trade barriers, and blockades—has very often in the past led to military adventures, and it would be foolish to think it will not happen again.

Despite the possibility of a military-led recovery in metal prices, let me remind you that it is the hallmark of good generals and admirals that they avoid a battle if at all possible. Battles and wars are unpredictable things, and in these days of thermonuclear possibilities, such prudence is to be doubly commended. Whatever influence the minerals industry around the world can bring to bear against those forces which, if unleashed, will carry us into war, should be firmly applied.

These comments may seem to some unduly pessimistic, even alarmist. I see no point in averting our gaze from clear and apparent dangers. Our industry has been, for centuries, internationalist in its thinking and action, and this conference is a fine example of this tradition. It is an appropriate place to say these things.

Let me put forward this morning, at least as a basis for discussion and debate, the principal ideas which in my view have brought us to our present difficulties. If we can understand something of the metaphysics which underlie the movements in prices and demands for the commodities we produce, and the political forces which accompany them, then at the least we can take some evasive action if the metaphysics go further wrong. We can even take up some of these fundamental issues ourselves, and attempt to influence ideas and beliefs.

So let us take a long view. Let us go back to the First World War and consider first the way in which those crucial institutions, governments, have increased enormously in power and activity.

The easiest way of understanding the growth of government since 1910 is to consider the growth of the share of

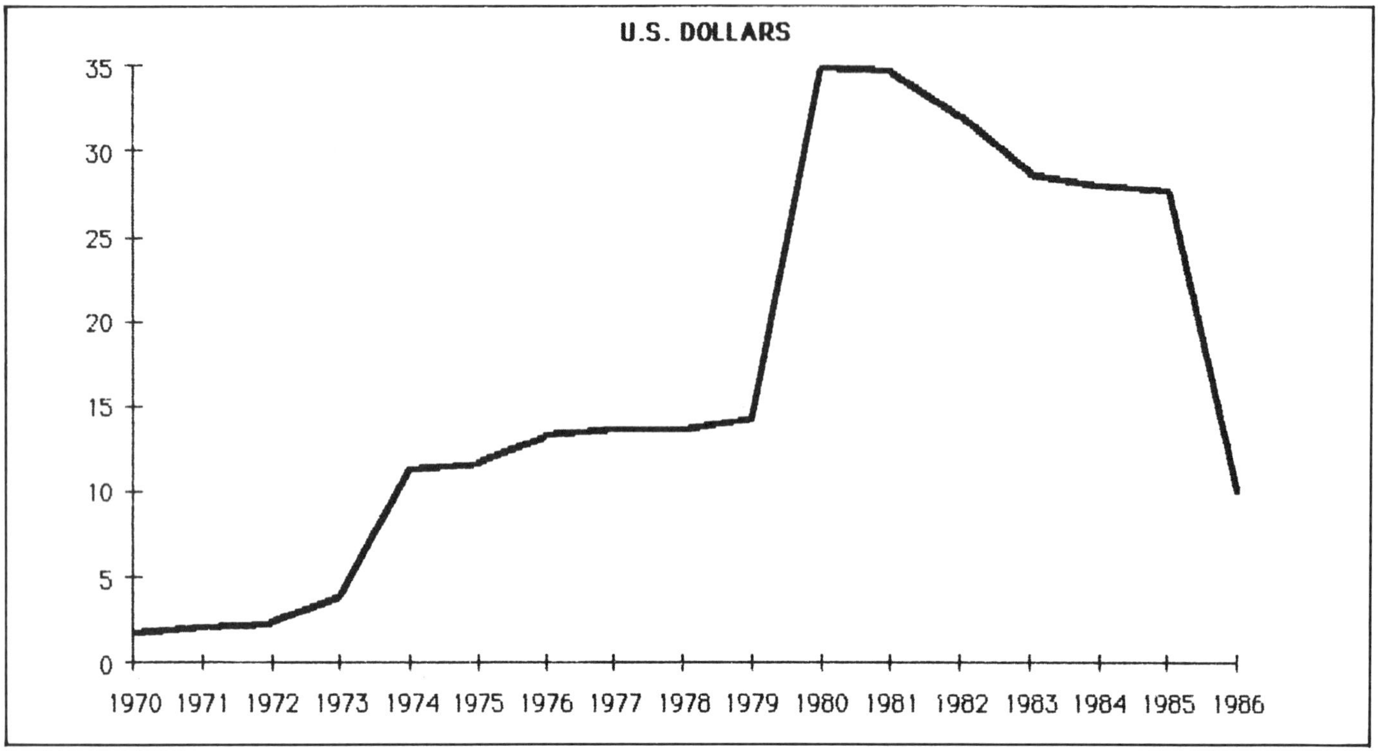


Figure 1. Oil prices, 1970-1985.

G.D.P. under government control, and I take as examples Australia and the U.K.

In Australia, before the Great War of 1914-18, governments consumed approximately 15% of G.D.P., and approximately half of this expenditure went into railway construction and other infrastructure development.

Today in Australia, nearly 45% of G.D.P. is in government hands. Expenditure on infrastructure development is still of the order of 8% of G.D.P. as noninvestment expenditure in government hands has increased from approximately 7% of G.D.P. to 35% of G.D.P. in 75 years.

In Britain, between 1840 and 1900, government expenditure never exceeded 11% of G.D.P. In the immediate pre-First World War period, which was characterized by a substantial naval build-up, government expenditure increased to 12 and 14%. By 1918, the last year of the war, it was 52%. During 20s and 30s it was 25 to 30%. But in 1975 it was 59%, more than in 1918.

Britain is an outstanding example of a Western democracy addicted to overwhelming government expenditure, but the fate of the Soviet Union, and subsequently Eastern Europe, is also relevant to my theme. Prior to the Great War, Russia was a major grain exporter, a rapidly industrializing country, with an economy characterized by the growth of property rights and the development of capitalist enterprise. Nineteenth-century Russia was also a society which produced great writers, poets, musicians, and performing artists. This came to an abrupt conclusion after Lenin's success in 1917. Today, the contrast between the life of the ordinary citizen in the Soviet bloc and the life of the ordinary American, is a contrast which the Soviet authorities do their best to keep suppressed and unconsidered.

The First World War had a profound effect on the relationship between the citizen and the state in Western societies. The efficacy of government authority in planning, organizing, directing, coordinating, virtually the whole of society with the nations which contested that awful struggle, led not merely to an acceptance of greatly increased state power, but amongst many intellectual elites led to the idealization of it.

Nationalism was the great ideological enthusiasm of 19th century Europe, and the First World War, destroying as it did all of the old constraints containing that enthusiasm, left government power, the symbol and expression of nationalist sentiment, largely unconstrained.

One of the most influential apologists in the West for the beneficence, and efficacy, of state power and government authority was John Maynard Keynes, the pre-eminent British economist of the 1930s and 1940s.

Keynes was a complex figure, and whether you are a committed monetarist, or a supply sider, whether you are an under-consumptionist (and therefore a pump primer), or an anti-inflationist, you can still quote Keynes to support your position.

Despite the potential eclecticism of Keynesian economics, the important consequences of Lord Keynes, as we consider the minerals industry around the world today, are twofold.

First, Keynes made chronic government deficits not only popular but respectable. Up until the Great War, governments, at least in Britain, took it as an article, if not of faith, then of cast iron convention, that their budgets could not run into deficit.

Keynes made government deficits an intellectually acceptable method of stimulating economic activity allegedly in order to increase employment. In Keynes' original formulation, deficits were supposed to be incurred on the down side of the business cycle, and counter-balanced by surpluses on the up side. But for a politician primarily concerned with re-election, the up side never came—it was always after the next election.

In Britain the continuing use of deficits led to the prospects of national bankruptcy in the mid-1970s and recourse to the IMF.

The British Prime Minister, James Callaghan, summarized what had happened in his address to the British Labour Conference on 28 September 1976:

> We used to think that you could spend your way out of a recession, and increase employment by cutting taxes and boosting government spending. I tell you, in all candour, that that option no longer exists, and that, insofar as it ever did exist, it only worked by...injecting bigger doses of inflation into the economy, followed by higher levels of unemployment as the next step. ... That is the history of the past 20 years.

Today there are few nations not weighed down with intractable problems of government deficits and growing debt. Singapore is, notably, one of the few. Japan, the greatest saving nation in the world, is now supporting much of the Western world's debt.

The other significant Keynesian legacy was the abandonment of gold as a measure and store of value.

In Keynes' General Theory we find:

> It is interesting to notice that the characteristic which has been traditionally supposed to render gold especially suitable for use as the standard of value, namely, its inelasticity of supply, turns out to be precisely the characteristic which is at the bottom of the trouble.

and also his celebrated, contemptuous remarks on gold mining:

> If the Treasury were to fill old bottles with banknotes, bury them at suitable depths in disused coalmines which are then filled up to the surface with town rubbish, and leave it to private enterprise on well-tried principles of laissez-faire to dig the notes up again (the right to do so being obtained, of course, by tendering for leases of the note-bearing territory), there need be no more unemployment and, with the help of the repercussions, the real income of the community, and its capital wealth also, would probably become a good deal greater than it actually is. It would, indeed, be more sensible to build houses and the like; but if there are political and practical difficulties in the way of this, the above would be better than nothing. The analogy between this expedient and the gold-mines of the real world is complete.

This is absurd, indeed dangerous, nonsense, but it is nonsense which influenced governments and their advisers all around the world for 40 years or more.

Keynes' legacy, however unintended it might have been, is the chart showing the inverted prices of gold, and silver, since the 2nd War for four currencies (Figures 2, 3).

Thus the combination of the First World War and the Keynesian legacy in economics has produced for us very large governments, unconstrained by the normal household restrictions of living within one's means, and at the same time, free to debauch the currencies for which they are responsible.

Since governments are, inevitably, run by ordinary human beings, almost all governments have yielded to the temptations resulting from this new freedom from constraint.

Since the last war, up until the early 1980s, most of the Western world had known nothing but chronic inflation. The degree of inflation has varied from time to time and from country to country. In the 18-year period of 1950-1968, it was suppressed, at least in terms of gold price, as U.S. gold reserves were slowly run down by apprehensive allied governments. (Private citizens were not allowed to buy and sell gold.) The general conclusion, however, is that few people under 45, outside Japan and Singapore, can personally recall sound money.

Inflation has had two important and very negative consequences.

The first consequence is that of distorting and misinforming investment decisions. If money is always declining in value, and declining in an unpredictable way, it becomes important to invest early rather than late, to build up inventories quickly, and to borrow in order to do these things.

This attitude reached a logical conclusion when stories about oil tankers anchored indefinitely beyond the horizon of the U.S. coast, if not actually steaming backwards, became the accepted conventional wisdom in 1973 and 1974.

Not only was money something to be out of, but debt became something to be in. The greater your debt, the more acute was your financial perception. Such an attitude, of course, is quite destructive of social and economical development.

The next stand in the tapestry of my argument, is the 1972 Club of Rome Study "The Limits to Growth."

It is only 14 years ago since this 20th-century update of 18th-century Malthusian doctrine was published, and it is difficult now for us to comprehend how such manifest nonsense could have, so quickly, become so widely and passionately believed. The Club of Rome cult took on a religious respect, with prophets such as Erlich and Commoner telling us that Western Society would run out of everything we needed, for 20th-century living, within the lifetime of virtually everyone who could listen to them. The very air, in some extreme prophecies, would be depleted of oxygen, and the population explosion would leave us with standing room only.

Let me quote two, quite representative, passages from this work of only 14 years ago:

> Given present resource consumption rates the great majority of the currently important non-renewable resources will be extremely costly 100 years from now. ...
>
> Many people are concluding on the basis of mounting and reasonably objective evidence that the length of life of the biosphere as an inhabitable region for organisms is to be measured in decades rather than hundreds of millions of years. This is entirely the fault of our own species.

The theme of the Club of Rome report is that of a sinful human race exploiting, beyond all rationality, a finite resource base, and facing imminent exhaustion of vital materials as a penalty for transgression.

It is important to reiterate what everyone in this audience knows extremely well. The Club of Rome authors were as fundamentally wrong as it is possible to be.

Let me quote the figures.

In the decade from the early 1970s to 1982 bauxite reserves increased by 35%, copper reserves by 25%, and nickel reserves by 25%. These are conservative, U.S. Bureau of

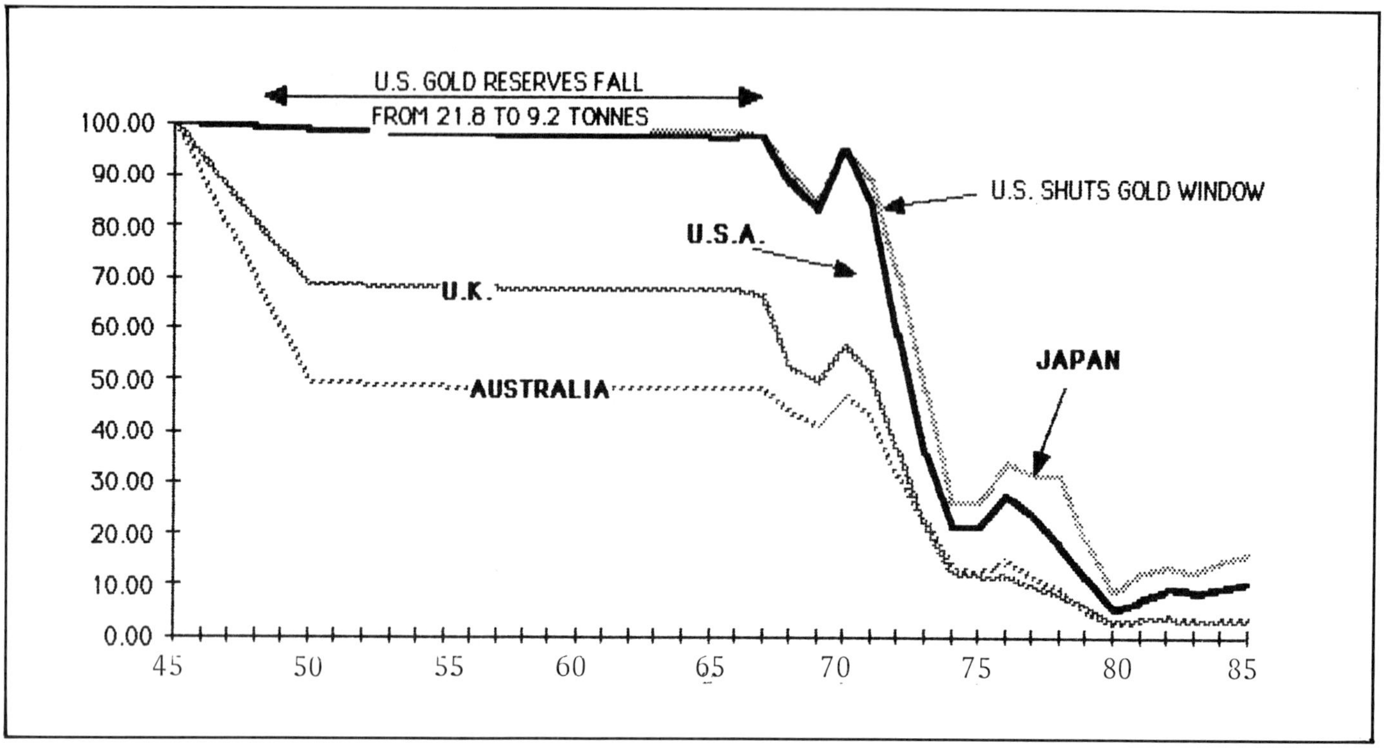


Figure 2. Currency values based on gold, 1945-1985.

Mines figures.

For uranium, Western World Reasonably Assured Reserves have increased from 0.85 MT to 1.7 MT, a figure which would need to be vastly increased to reflect the new discoveries in Canada and Australia.

For coal, in 1972 the British Petroleum estimate was 269 BT of oil equivalent; in 1982 it was 441 BT of oil equivalent, nearly a two-fold increase which provides a handsome 220 years of reserves at today's consumption rate. Gas increased by nearly 70%, and oil, with a small increase of some 6%, was the only mineral to show a reduction in the number of years consumption covered by reserves. It dropped from 32 to 31 years. Since 1982, as oil consumption has continued to decline, this time constant has doubtless increased.

While you and I are aware of these figures, and understand the implications of them, the important point is that they are not known, still less understood, by what is called "informed, educated opinion."

The Club of Rome Study with all its computer output graphs and tables, with all its footnotes and bibliographic detail, with its carefully worded caveats and qualifications emerges, with hindsight, as the leitmotif of a pied piper's repertoire leading us into world-wide, centralized energy and minerals planning, regulation and control. The Law of the Sea Convention with its "common heritage of mankind" rhetoric, with its proposals for regulated markets and international production quotas in nickel, cobalt and other base metals, was the lineal offspring of the Club of Rome report.

It is difficult, even now, for us to understand how extraordinarily influential Club of Rome thinking was. The first oil shock of 1973 was at least to some significant degree, a consequence of Club of Rome advocacy.

Even today illusions of ultimate scarcity and exhaustion, perhaps not across the whole range of minerals but for some particular commodities, can be found in some circles, particularly government circles. Because of the inertia of public opinion, despite the fact we now find ourselves in a period of general glut of almost everything, there is still an underlying, widely held view, which believes that resources are scarce. Politicians reflect, in a subliminal way, this public opinion.

We in the mining industry also have to note that as we were persuaded, in the mid 1970s, that we were running out of resources, we were also encouraged, as a corollary, in the belief that minerals and energy resources had a very special status. Paradoxically, this high status encouraged those people most deeply influenced by Club of Rome doctrines, the "greenies" and the conservationists, into a frenzy of political and protest activity aimed at restricting the development of major projects.

A good example of this frenzy has been evident at Roxby Downs, near the center of South Australia, where my company, Western Mining Corporation, discovered a very large, medium-grade, copper/uranium/gold orebody in 1975. Because of the uranium content of the orebody, we have been under constant attack from anti-nuclear power lobby groups who demand, in religious terms, that we "leave uranium in the ground." The pressure groups do not understand that black-banning the uranium reserves at Roxby Downs cannot influence the arms race, cannot affect the nuclear power programme, and can only give Australia a poor international reputation in world forums and earn international rebuke as an unreliable supplier of important commodities.

Public opinion surveys tell us that the overwhelming majority of Australians are in favor of mining and exporting uranium. The greatest segment of opposition comes from people with tertiary education. That is an important and influential group which we have failed to win over, and it is interesting to speculate why it is that the most highly educated section of the

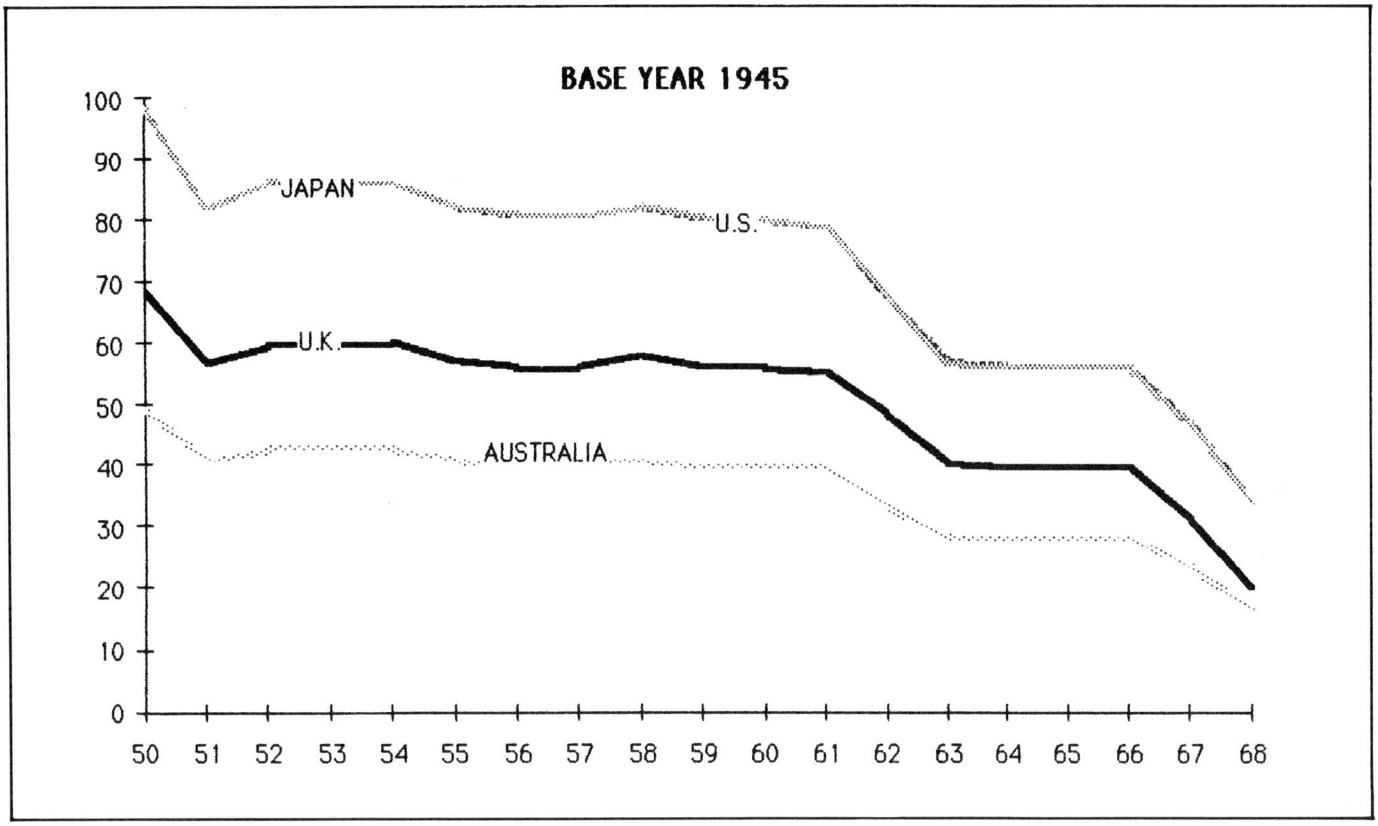


Figure 3. Currency values based on silver, 1950-1968.

community should be most deeply committed to irrational phobias and superstitions.

There is a common element in the thinking of the protesters at Roxby Downs and in the thinking of the Multilateral Development Banks and the IMF, for example, which have continued to subsidize base metal production in the LDCs, despite protests from other producers who are squeezed by these subsidies. That common element is the Club of Rome doctrine that a mine is enormously valuable, and that control of the mine—be it financial, managerial, or political—invests the controller with great power.

Thus mine head frames become, I regret to say, not just steel structures with stresses and strains, but talismans, religious symbols, with magical properties.

This point, seemingly obvious, has to be emphasized. If it is widely believed that copper mines, for example, are sources of great wealth and power, arising from the development of supposedly scarce, perhaps "sacred" sources, then governments, despite all logic, will feel tempted or even obliged to nationalize and expropriate them; banks will find it fashionable to lend money on the security of them; and those who manage them will pursue production policies that are often perverse. Such perverse behavior will, nonetheless, be fully encouraged by those whose financial security depends upon the efficacy of the Club of Rome myth.

Thus two important influences in the last 50 years, Keynes's General Theory of 1936, and the Club of Rome "Limits to Growth" of 1972, have helped to produce for us, in 1986, a world economy that has been without a stable measure of value for the best part of 25 years; that has built up unprecedented levels of debt which are reflected not only in obligations from nation to nation, but more significantly, I think, in obligations from the present generation to succeeding generations. Every child born into citizenship of the U.S. acquires, I understand, along with citizenship, a debt obligation of between 50 and 60 thousand U.S. dollars.

Thus from a period in the sixties and seventies, when escalating investment programs, inventory build up and buoyant commodity prices became the everyday experience of business, we are now in a situation of over capacity, very lean stockpiles, no sign of improvement in commodity prices, and debt obligations which hem us in on every side.

After the binge comes the hang-over.

What of the future?

When we fully understand that economic activity has no existence apart from the hopes, the fears, the understanding, the traditional manners of behavior, of the people who work, save, invest, and consume, we can see some encouraging, as well as alarming, portents in changes in belief and language around the world. Let me quote some examples:

One hopeful omen is the discrediting of the Club of Rome. Because of general commodity glut persisting now for at least 5 years, Malthusian doctrines describing a special status for minerals, requiring international control and regulation, will be discredited for the rest of this century, and hopefully, well into the next. We should not forget how much harm they have perpetrated, and we should not allow others, notably the IMF and the international development banks, to forget.

Another hopeful portent is the beginning of a move to the restoration of sound money. President Reagan, in his State of the Union message early this year, directed his Treasury Secretary, Jim Baker, to determine if a new international meeting should take place to discuss "the role and relation of our currencies," by which, I take it, he meant the restoration of internationally sound, internationally accepted money.

The President's statement is the outward sign of intense debate and concern within the U.S. about the debilitating and impoverishing effect of the absence of a reliable currency.

The minerals industry, because it forms a significant component of international trade, would benefit very greatly from sound money. We ought to be doing all we can to promote that cause.

In my own country, we are moving, within a few months, to an historic decision to reduce protection on our most outrageously protected industries, the Footware Clothing and Textile (FCT) industries. Australia would benefit, economically, if every person employed in those industries were to be retired on a life pension set at Average Weekly Earnings. For many people in those industries, this would mean an increase in income. It is an extraordinary thing that a country which has had very many people with outstanding intellectual capacity, should have impoverished its citizens to this degree, for so many years. But it is now beyond question that this has been the case. It will still take all the resources of intellect and will which our responsible Minister, Senator Button, can bring to bear to move Australia from the debilitating road of self impoverishment, to the morale-boosting road of international competitiveness, which is also, of course, the highway of international good neighborliness. This decision will be an historic watershed in the economic history of Australia and of the West Pacific.

The question which dominates all discussions of the economic future of the West Pacific, indeed, perhaps, for the world has a whole, is the immediate prospects for China. Will China become a free economy, emulating Hong Kong, and Singapore or, contrawise, will Hong Kong be swallowed up into economic decline and decay?

If the former proves to be true, if we are about to enter into an extraordinary period of economic growth in China, the whole world will benefit, and the countries of the West Pacific will benefit particularly. The Chinese themselves of course, will benefit most of all. Whether this much-to-be-desired good fortune becomes reality depends primarily on the ideas, hopes, and ambitions of the influential sectors of Chinese opinion.

Inextricably linked to events and opinion in China is the great debate going on in the U.S., in Japan, and one would hope even in Western Europe, over national policies regarding trading relations between sovereign states. The tide of rising protectionism is our most serious problem.

In the U.S., at the moment, the protagonists for mercantilism, for protection, for trade barriers against foreigners, for subsidies to domestic exporters, seem to be winning every congressional battle. Sometimes they miss out, by just a handful of votes, in obtaining the two-thirds majority which would override a presidential veto. Maybe these votes are prearranged, an outsider cannot tell. Nonetheless, it is a melancholy thing to read of the events of the 1920s in the U.S., of the Hawley-Smoot Protection Act of 1930, of the subsequent dramatic contraction in world trade, of the chain reaction of impoverishment that spread around the world, of the drift into World War, and in reading of these things to see, in so many crucial ways, a repeat performance going on 60 years later. The spirit of Hawley-Smoot is not merely alive and well in Washington, and other major capitals. It is winning battle after battle.

The participants at this conference will be discussing the latest in earth science technology, minerals exploration, and development. This conference programme is an outstanding example of the extraordinary success of 20th-century science and technology, not only in intellectual terms, and that is impressive enough, but in carrying the economics of the world up over the impediments and against the forces that now loom so large.

I have been realistic, I think, in my comments today. In Australia the leaders of the mining industry have been accused by some journalists of preaching doom and gloom, and some may think that realism is inappropriate at an international forum such as this.

Let me conclude, therefore, by reminding you, as Edmund Burke remarked, that all government is based on opinion. Whether we have war or peace, trade or protection, impoverishment or prosperity, will depend on opinion, and that this great industry with its roots deep in antiquity, and its contemporary techniques at the forefront of modern science, is at least as well placed to influence and even change opinion, as any other industry. The mineral industry needs us to meet this challenge more urgently than ever before. I commend it to your consideration.

CHAPTER 50

# GRANITE PROVINCES IN THE SOUTHEAST ASIAN TIN BELT

**P.E.J. Pitfield**
**E.J. Cobbing**
*British Geological Survey*
*Nottingham, United Kingdom*

**M.C.G. Clarke**
*U.K. Geothermal Project*
*Nairobi, Kenya*

**D.I.J. Mallick**
*British Geological Survey*
*Nottingham, United Kingdom*

**L.H. Teoh**
*Pejabat Penyiasatan Kajibumi*
*Wisma Persekutuan*
*Kelantan, Malaysia*

## ABSTRACT

Four granite provinces have been delineated, each with its own distinctive pattern of mineralization.

| | | |
|---|---|---|
| Eastern Province | I-type | Magnetite-cassiterite skarns ± base metal sulfides with antimony in Thailand. |
| Main Range Province | S-type | Endogenous greisen-bordered vein swarms of cassiterite and wolframite. |
| North Thailand Migmatitic Province | S-type | Endogenous vein and skarn replacement scheelite and fluorite deposits with some tin and local antimony. |
| Western (Peninsular Thailand-Burma) Province | S- & I-type | Endogenous greisen-bordered vein swarms and pegmatites of cassiterite and wolframite. |

The granite provinces are clearly linked with different geological terranes which form three separate and approximately parallel strips bounded by structural discontinuities at the western margin of the Sundaland craton. Geochemical polarity for the granites of all provinces has been established, which is chiefly characterized by a northward trend toward less differentiated granitoids. This is accompanied by a northerly decline in the importance of tin and a rise in the importance of tungsten, antimony, and fluorite.

## INTRODUCTION

The area popularly known as the Southeast Asian Tin Belt is a region of complex geology with epochs of plutonism and mineralization ranging from Permian to Cretaceous or Tertiary in age. Although tin is widely distributed, it is localized within certain parts of the belt and is absent in others. Thus, the term Tin Belt accurately denotes the importance of the region for the production of that metal, but it does not provide an adequate perspective for the region as a whole. However, the term likely will endure for reasons of historic usage and familiarity.

The objective of the study reported here is to describe systematically the granite geology of the region, and to relate the various patterns of mineralization to differences in the geology and geochemistry of the granites.

Regional granite provinces within the Tin Belt have been recognized by Hutchinson and Taylor (1978) and Beckinsale (1979), and these form the basis for a first order subdivision of the granitoids of the region. In what follows, the regional classification is expanded and the provinces are more precisely defined on the basis of data gathered during the present study (Figure 1 and Table 1).

1. *Eastern Province*—Small batholiths of zoned and unzoned plutons of a compositionally expanded suite ranging from gabbro to monzogranite but dominated by monzogranite which is of I-type affinity and Permo-Triassic age (Bignell and Snelling, 1977; Hutchinson, 1977). These granites are associated with base metal mineralization and, in Peninsular Malaysia, with tin deposits. Small unmineralized I-type plutons of Cretaceous age are present in Peninsular Malaysia (Bignell and Snelling, 1977).

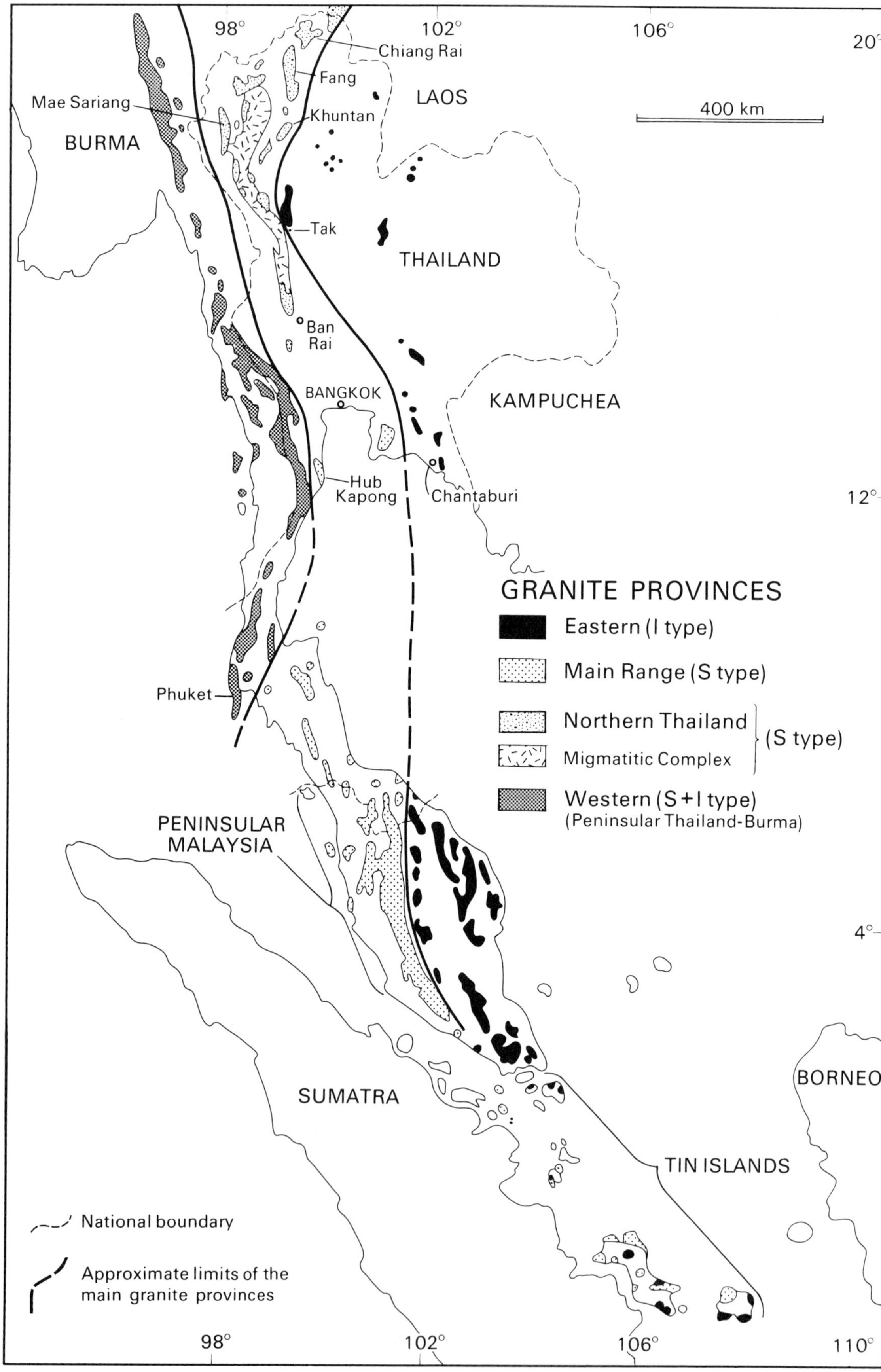


Figure 1. Distribution of granite provinces within the Southeast Asian Tin Belt.

TABLE 1 Characteristics of the granite provinces in the Southeast Asian Tin Belt

| | Western Province (Peninsular Thailand-Burma) | North Thailand Migmatite Province | Main Range Province | Eastern Province |
|---|---|---|---|---|
| Age of envelope | Carboniferous-Permian | Cambrian-Permian | Cambrian-Devonian | Carboniferous-Triassic |
| Lithology of envelope | Clastic sedimentary rocks | Clastic sedimentary rocks with shelf carbonates | Clastic sedimentary rocks with shelf carbonates | Clastic sedimentary rocks with shelf carbonates and calc-alkaline volcanics |
| Granite Type | S and I | S | S | I |
| Granite Age (Ma) | 82-98 Phuket<br>130 Mae Lama | 200-244 | 200-230 | 200-280 and 70-80 |
| Initial $^{87}Sr/^{86}Sr$ ratio | 0.714-0.740 Phuket<br>0.709 Mae Lama | 0.722-0.733 | 0.711-0.719 | 0.708-0.712 |
| | (Putthapiban & Gray 1983 Beckinsale 1979) | (Beckinsale 1979) | (Hutchison 1977) | (Hutchison 1977) |

2. *Main Range Province*—Major batholiths and large complex plutons of restricted compositional range comprise a suite of S-type granites of mainly Triassic age which are mostly tin bearing (Bignell and Snelling, 1977; Liew and Page, 1985).

3. *North Thailand Migmatite Province*—Major S-type granite batholiths and plutons of Triassic age (Teggin, 1975; Braun et al., 1976; Hutchinson, 1983) are associated with a migmatite terrain of regional extent which has been variously designated as basement complex (Teggin, 1975), as an anatexitic aureole of Paleozoic granite (Braun, Hahn, and Maronde, 1981), and as Precambrian (Veeraburus, 1982). Recent fieldwork indicates that the migmatites are probably of Permo-Triassic age. Small Cretaceous I-type plutons are sporadically distributed in this province.

4. *Western Province (Peninsular Thailand-Burma)*—Large batholiths of S-type tin-bearing granites and smaller unmineralized I-type plutons are present, all of which are of Cretaceous age (Garson et al., 1975; Beckinsale, 1979).

The subdivision of the granites of Peninsular Malaysia into three belts by Hutchinson (1977) has been reduced to two by recognition that the granites of the Eastern and Central belts are similar. These have now been unified into the Eastern Province, although they do have distinctly different patterns of mineralization (Table 2).

Granite plutons and granite units within complex plutons were distinguished in the field by differences in their primary textures (Cobbing et al., 1986). Granites with secondary textures, designated as two-phase granitoids, were recognized within plutons in all provinces, though in different proportions, and appear to represent an early stage in the process leading to mineralization (Cobbing et al., 1986). They are most abundant in the Main Range Province.

Each granite province is made up of many individual plutons (Cobbing et al., 1986). However, in addition to the regional differences in patterns of their granite geology, there are also differences in the regional geology of the crustal segments which host each granite province (Table 1). The constituent plutons of the provinces, although geologically and geochemically distinctive on an individual basis, provide collective geochemical and mineralization patterns which are distinctive for each province. It is these patterns in particular which will be discussed in the following pages.

## GEOCHEMISTRY

The geochemistry of the four regional granite provinces is based on the results of x-ray diffraction (XRF) analyses for 38 elements on 390 rock samples including 173 from Peninsular Malaysia, 83 from the Indonesian Tin Islands, and 134 from Thailand. The lower limit of determination is 0.01% for the major oxides and 1 or 2 ppm for all trace elements except Ba (10 ppm), Ce (5 ppm) La (4 ppm), and Zr (3 ppm). Concentrations which were found to be below the lower limit of determination (LLD) have been put to half the value of the determination limit. To complete the coverage within Thailand and present a balanced picture of all granite divisions, this data set has been augmented by selected geochemical data from Garson et al. (1977), Ishihara et al. (1980), Mahawat (1982), and Putthapiban and Gray (1983). Rocks with less than 10% modal quartz have been excluded from the discussion in order to maintain a valid level of comparability.

I- and S-type granites can be successfully distinguished by their field characteristics (Chappell and White, 1974; Pitcher, 1983; Cobbing et al., 1986). The major element geochemical discriminations advocated by Chappell and White (1974) and White and Chappell (1977, 1983) have been used by various authors for discussion of the relative I and S affinities of granitoids of the Thai-Malay peninsula (Ishihara et al., 1980; Liew, 1983; Putthapiban and Gray, 1983; Teoh, 19856; Cobbing et al., 1986). I-type granitoids can generally be distinguished from S-types by their relatively high CaO and $Na_2O$ and lower $K_2O$ over a large range of $SiO_2$ content. In the Southeast Asia

Table 2. Metallogenic Subprovinces in the Southeast Asian Tin Belt.

| GRANITE PROVINCE | METALLOGENIC SUB-PROVINCE | GEOLOGICAL ENVIRONMENT | COMPOSITE METAL ASSOCIATIONS | TYPE MINERALOGICAL ASSOCIATIONS | PRIMARY ORE TYPE | TYPE LOCALITIES |
|---|---|---|---|---|---|---|
| **WESTERN THAI-BURMA PROVINCE** | PHUKET (Khlong Marui-Ranong Fault Block) | Permo Carboniferous continental margin pebbly mudstones with slaty cleavage | **Sn±W**<br>**Sn±Li±Ta**±Nb±Y±REE's | Cassiterite-Wolframite<br>Cassiterite-Lepidolite | Greisen/Stockworks<br>Pegmatites | Khao Khanim, Thosung<br>Phangnga fault zone |
| | TENASSERIM (Ranong-Three Pagodas) Fault Block | | **Sn-W**±Cu±Mo±Bi±Pb±Zn±Fe±F | Cassiterite-Wolframite-(Chalcopyrite-Molybdenite-Bismuth) | Vein systems<br>Stockworks±Greisen | Mergui, Tavoy — Burma<br>Ta Ko Pithong, Thailand |
| | SHAN (Three Pagodas-Moei Fault Block) | Shan Dolomite Group Mainly Ordovician to Triassic shelf limestones with subordinate pebbly mudstones | **W &/or Sn**±Sb±Cu±Pb±Zn±Mo±Bi±As±Fe | Wolframite±Scheelite-Cassiterite-(Arsenopyrite) | Vein systems/Greisen ±Skarn | Mawchi, Burma<br>Mae Lama, Thailand |
| **NORTHERN THAILAND MIGMATITIC PROVINCE** | CENTRAL MIGMATITIC BELT AND EASTERN GRANITIC RANGE | Cambrian to Carboniferous shelf clastics and limestones of greenschist facies domed and cored by the migmatite complex. Upright folded Eastern range bounded by major faults | **W &/or Sn**±Sb±F±As±Fe±Zn±Cu | Scheelite±Arsenopyrite<br>Cassiterite-Scheelite±Wolframite | Greisen-bordered vein swarms/Stockworks<br>Skarn | Samoeng, Chiang Mai<br>Thung Laung, Lampang<br>Doi Mok, Wiang Pa Pao |
| | FANG AND CHIANG MAI BASINS (PING RIVER VALLEY) | Cambrian to Carboniferous shelf clastics and limestones with Mesozoic to Cenozoic cover in fault bounded synclinoria and basins | **F &/or Sb**±Pb±Zn±As (±Mn) (±Ba) | Fluorite±chalcedony<br>Stibnite-quartz<br>Sphalerite-Galena-Arsenopyrite-Pb/Sb/Cu Sulphosalts | Vein systems<br>Fault Breccias<br>Replacement±Bedded deposits | Nam Mae Pai<br>Li-Ban Hong<br>Mae Wa |
| **MAIN RANGE PROVINCE** | PENINSULAR THAILAND (Khlong Marui-Bentong-Raub Fault Block) | Cambrian to Carboniferous shelf clastics in lowest greenschist facies with cover of Permian Shelf limestones and Triassic sandstones | **Sn &/or W**±Sb±Cu±Pb±F<br>**Sn-As-Fe** | Wolframite (ferberite)-Stibnite<br>Cassiterite±Wolframite±Chalcopyrite<br>Cassiterite-Arsenopyrite-Pyrite<br>Cassiterite-Malayaite | Breccia pipes<br>Vein systems±Greisen<br>Stockworks<br>Skarn | Khao Soon<br>Yod Nam<br>Labu, Pinyok — Border with Malaysia |
| | NORTH MAIN RANGE PENINSULAR MALAYSIA | Cambrian to Devonian shelf clastics and limestones flanked eastwards by calc-alkaline volcanics passing to deep water pelites deformed in greenschist facies and overlain by Carboniferous to Triassic shelf carbonates and clastics | **Sn-Nb (-Ta) &/or Fe**<br>**Sn±W**±Cu±Pb±Sb±Zn ±Bi±Mo±Hg±F(±Au)<br>**Sn-As-Fe** | Cassiterite-Columbite<br>Cassiterite±Scheelite±Wolframite<br>Cassiterite±Scheelite±Stannite±Complex Sulphides<br>Cassiterite±Malayaite±Sn Andradite | Pegmatites<br>Massive Greisens<br>Greisen-bordered vein swarms/Stockworks<br>Xenothermal deposits ±Skarns (pipes) | Gunong Jerai, Kedah<br>Keladang Range, Perak<br>Kramat Pulai, Lahat-Menglembu, Chenderiang — Kinta Valley |
| | SOUTH MAIN RANGE PENINSULAR MALAYSIA | Similar to N. Main Range sub-province but with no volcanics | **Sn-Nb-Ta**±Be±U±REE's<br>**Sn±W**±Cu±Zn±Pb±Sb ±Mo±Fe±Be (±Au) | Cassiterite-Columbite±Monazite<br>Cassiterite±Wolframite±Scheelite±Stannite±Base metal Sulphides and Sulphosalts±Beryllium silicates<br>Cassiterite±Malayaite±Scheelite | Pegmatites<br>Massive Greisens<br>Greisen-bordered vein swarms/Stockworks<br>Xenothermal deposits±Skarns (Stratabound) | Bakri area, W. Johore<br>Sungei Besi, Kampung Pandan — Kuala Lumpur, Selangor |
| | Main Range Type TIN ISLANDS INDONESIA | Amphibolite facies pelites and hornblende schists overlain by Permo-Triassic flysch locally attaining greenschist facies±Triassic rhyodacitic volcaniclastics. | **Sn±W** | Cassiterite±Wolframite±Base Metal Sulphides | Greisen-bordered vein swarms/Stockworks and pipes | Dabo, Singkep<br>Tikus, Belitung |
| **EASTERN PROVINCE** | NORTHERN AND CENTRAL THAILAND | Carboniferous to Permian clastics, carbonates and calc-alkaline volcanics with cover of Triassic red beds and Triassic-Jurassic calc-alkaline volcanics. | **Cu±Pb±Zn**±Fe±Au±Mo (±Mn)<br>**Sb**±Pb±Zn±W±Cu±F (±Ba) | Chalcopyrite-Malachite<br>Magnetite-Hematite<br>Ferberite-Stibnite-Fluorite | Porphyries<br>Bedded/Skarn<br>Vein systems<br>Breccia Pipes | Phu Hin Lek Fai, Loei<br>Pha Khan, Doi Ngom — Phrae |
| | TERENGGANU AND E PAHANG PENINSULAR MALAYSIA (E of Lebir Fault) | Carboniferous to Permian shelf clastics with subordinate calc-alkaline volcanics, moderately deformed and covered by Jurassic molasse. | **W±Sn±Cu**±Fe±Mn±As±Zn±Pb±Bi±Mo±Au<br>**Fe-Sn**±Mn±Cu±Zn<br>**Cu &/or Pb±Zn**±Fe ±Bi±Au | Wolframite (ferberite) ±Cassiterite±Base Metal Sulphide±Molybdenite<br>Cassiterite-Magnetite/Pyrrhotite | Greisen-bordered and composite vein systems/Stockworks<br>Magnetite Skarn<br>Sulphide Skarn | Chendrong<br>Sungei Lembing<br>Bukit Besi<br>Chye Heng Loong |
| | CENTRAL BELT PENINSULA MALAYSIA (Bentong-Raub-Lebir Fault Block) | Permo-Triassic clastics and volcaniclastics with important calc-alkaline volcanicity. Greenschist facies and local amphibolite facies metamorphism. Jurassic molasse. | **Au±Sb**±W±As<br>**Au-Cu-Pb-Zn**±Sn±Fe±Ag ±Mn±Hg | Gold±Stibnite±Scheelite<br>Cassiterite-Stannite-Ag<br>Galena-Chalcopyrite±Cinnabar | Vein systems<br>Xenothermal replacement deposit<br>Porphyries? | Raub<br>Manson, Kelantan |
| | E. JOHORE PENINSULAR MALAYSIA | Permo-Triassic clastics strongly deformed in upper greenschist facies. Undeformed Triassic rhyodacitic volcaniclastics. | **Fe±Sn**±W±As±Cu±Bi ±Mn±F (±Au)<br>**Cu±Pb**±Zn±Sn (±Fe±U) | Magnetite-Cassiterite±Fluorite ±Scheelite<br>Chalcopyrite-Bornite-Galena | Vein systems<br>Magnetite Skarn<br>Sulphide Skarn | Pelepah Kanan<br>Bukit Payong |
| | Eastern Province Type TIN ISLANDS INDONESIA | Permo-Triassic flyschoid sediments and Permo-Triassic rhyodacitic volcaniclastics. | **Sn-Fe**<br>**Cu±As &/or Mo** | Cassiterite-Pyrrhotite-Magnetite | Bedded±Skarn<br>Vein Systems | Batu Besi, Kelapa Kampit — Belitung |

tin belt the $Na_2O$ versus $K_2O$ diagram of White and Chappell (1983) has proved quite satisfactory for distinguishing I- and S-type granitoids, except where there has been late-stage independent alkali mobility associated with redistribution of residual fluids, volatile fluxing, and feldspathisation. This is particularly evident in secondary textured S-types, which tend to plot in the relatively $Na_2O$-rich I-type field on this diagram (Cobbing et al., 1986).

A special feature of S-type granitoids is their peraluminous character, whereas I-types are dominantly metaluminous (Chappell and White, 1974). In Southeast Asia, however, alumina-based criteria are less effective; even after adjustment of the boundary conditions relating to the excess alumina content of S-types, no entirely satisfactory discriminant can be found (Teoh, 1985; Cobbing et al., 1986).

The usefulness of this mineralogical and major element classification is therefore restricted in that it depends on too few variables and is overly simplistic in its extended interpretation as an indicator of particular tectonic settings (Pitcher, 1983). In these circumstances it is appropriate to consider the use of trace elements, which has been done by Pearce, Harris, and Tindle (1984) and Harris, Pearce, and Tindle (1986) with the objective of discriminating between granites from different tectonic settings. Accordingly, an assemblage of incompatible elements of petrogenetic significance has been plotted against normalized, hypothetical Ocean Ridge Granite (Figure 2) and yields patterns which may be directly compared with Andean granites of similar $SiO_2$ content (Atherton and Sanderson, 1985).

All four regional granite provinces yield profiles which demonstrate selective enrichment in the LIL elements and depletion in heavy rare-earth elements (Figure 2), and which are consistent with the geochemical signature of subduction-related (I-type) and collision-related (S-type) magmatism (Pearce et al., 1984). Among these trace elements, Rb is the best discriminant element to differentiate between volcanic arc (or I-type) and syncollisional (or S-type) granitoids (Pearce, Harris, and Tindle, 1984). Comparing granite types of similar $SiO_2$ content (Figure 2), the S-types are notably enriched in Rb, and are therefore characterized by high ratios with the depleted incompatible elements such as Zr, Ti, R, and Sr. Discrimination using these Rb-based parameters is improved by plotting them against some measure of the differentiation which largely removes the effects of fractionation and therefore relates more closely to the composition of the original melt. In Southeast Asia the abundances of the high-field-strength elements in the more differentiated granites is rather erratic, and consequently Rb ratios with these elements such as Rb/Zr are less useful discriminants (Harris, Pearce, and Tindle, 1986). It has been found, however, that Rb ratios with certain LIL elements, and especially with Sr, are very effective when plotted against the Thornton and Tuttle differentiation index. Such plots (Figure 3) achieve good discrimination between I- and S-type granitoids and are particularly useful in distinguishing Eastern Province granitoids from the Main Range and other S-type provinces of the Southeast Asian Tin Belt. They have been found to be effective at high levels of DI and with granites of both Permo-Triassic and Cretaceous age. If $SiO_2$ or K/Rb are used as indices of fractionation (Shaw, 1968, 1970), plots of Rb/Sr do not achieve the same degree of separation of these contrasted granite provinces (Hutchinson, 1983).

In an attempt to discriminate between granites from different tectonic settings, Pearce, Harris and Tindle (1984) have used Rb-(Nb + Y) and Rb-(Yb + Ta) projections, which demarcate fields designated as ocean ridge granite (ORG), volcanic arc granite (VAG), within-plate granite (WPG) and syncollisional granite (syn-COLG).

In this discussion, only the RB-(Nb + Y) plot of Pearce, Harris, and Tindle (1984) has been used (Figure 4); although there is a sound petrogenetic rationale behind this discrimination, one should be aware of the limitations. Whereas Y has been shown to be an effective discriminator for granite classification (Ramsay, Odell, and Drysdall, 1986), the high values for some highly evolved granites can lead to their being plotted in the WPG field of the diagram, although they are clearly members of a volcanic arc or syncollisional granite population.

## PETROGENETIC DISCRIMINATION OF THE GRANITOID PROVINCES

Geochemically, the Eastern Province granitoids comprise an extended series of calc-alkaline affinity as demonstrated on AFM plots (Cobbing et al., 1986), the large range of $SiO_2$ contents, Thornton-Tuttle differentiation indices, and the relatively systematic major element variations on Harker diagrams. This is especially true when individual batholiths are considered.

The Main Range, North Thai, and Western Province S-type granites, in contrast with those of the Eastern Province, are compositionally restricted, with high $SiO_2$ contents (Figure 5); also, with the exception of $Al_2O_3$ and $TiO_2$, variation diagrams generally exhibit a larger scatter and less well-defined trends. Comparison of the different segments of the Main Range, North Thai and Eastern provinces, respectively (Figure 5) indicates a progressive stepwise northward reduction in $SiO_2$ content. The North Thai Migmatite Province in particular has the lowest mean $SiO_2$ content of all S-type provinces, and is generally much less differentiated (Figure 5).

On the ORG normalized plots (Figure 2) the Eastern Province granitoids exhibit overall enrichment in all the incompatible trace elements except Ba relative to the Andean volcanic arc granitoids of similar $SiO_2$ content. Moreover, selective enrichment in Rb and Th relative to Nb and Ta, and the light rare earth elements (LREE) relative to yttrium (Y) and heavy rare earth elements (HREE), together with the large negative Ba anomaly, suggest an appreciable crustal contribution for the rocks of this province. Some crustal anatexis in the generation of these granitoids is consistent with the initial $^{87}Sr/^{86}Sr$ isotope evidence (Liew, 1983; Priem et al., 1975; Teggin, 1975) and is further suggested by the Eastern Province granitoids overlapping into the "within-plate" granite field on Figure 4. By comparison with the Eastern Province, the S-type granitoid provinces are generally enriched in Rb and to a lesser extent in Ta, Nb, Pb, Sm, Sn, U, and W, whereas Ba, Sr, and to some extent Zr are depleted. The ORG normalized profiles on Figure 2 are similar, but steeper and more exaggerated, especially for the LIL elements.

Increasing specialization associated with the development of secondary textures, which are commonly but not always accompanied by tin mineralization, leads to still further enrichment in Nb, Rb, Sn, U, W and impoverishment of Ba, Sr, and Zr in southeast Asia (Cobbing et al., 1986) in common with other tin-mineralized granite regimes (Tischendorf, 1977; Beus and Grigorian, 1977). However, Y and HREE show some

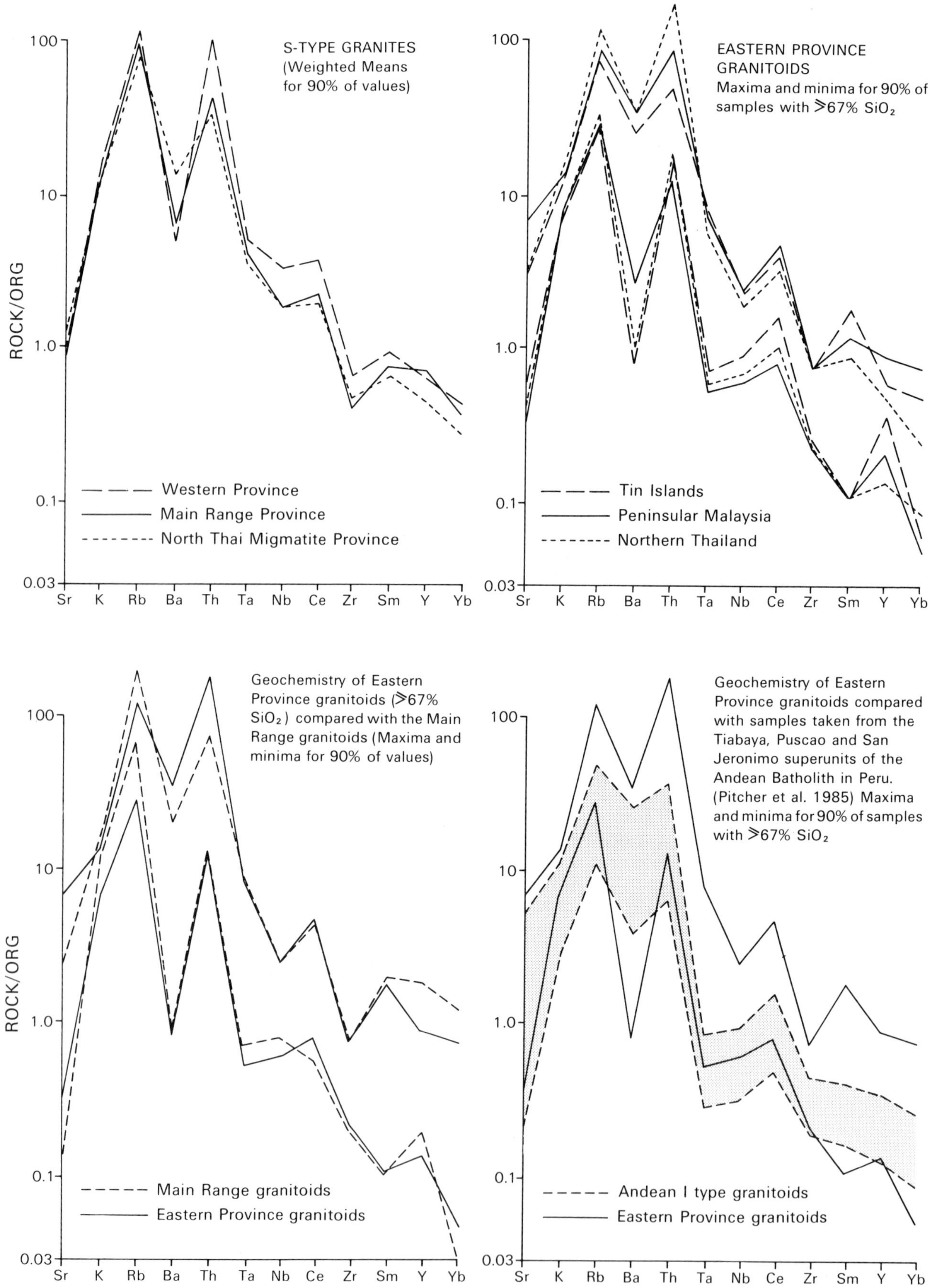


Figure 2. Hypothetical Ocean Ridge Granite (ORG) normalized geochemical patterns for the granite provinces and subdivisions of Southeast Asia compared with the Andean Batholith in Peru. (Normalizing factors from Pearce, Harris, and Tindle, 1984).

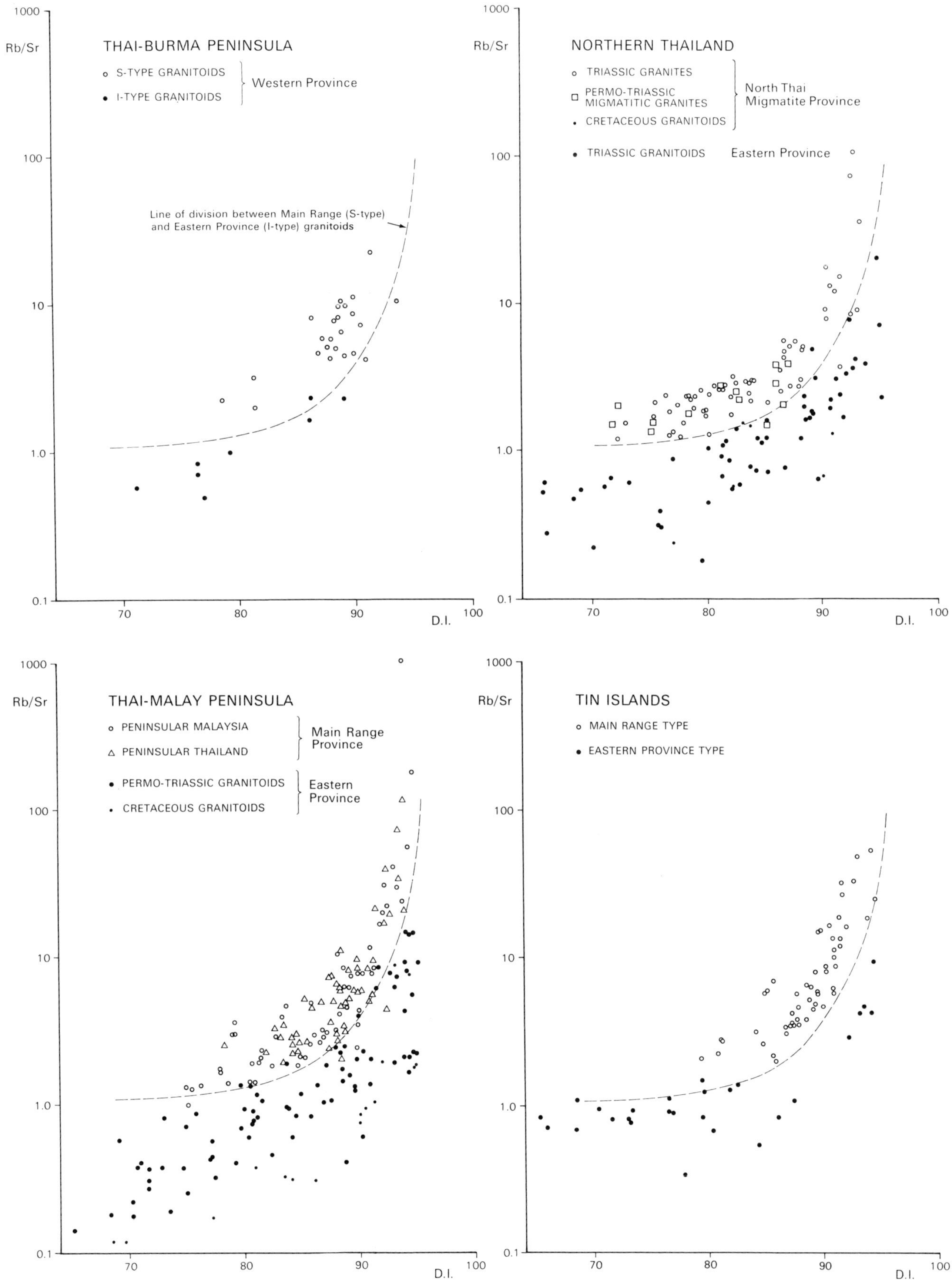


Figure 3. Rb/Sr vs DI discriminant plots for the granite provinces of Southeast Asia with discriminant line for I- and S-type granites empirically determined according to the data from Peninsular Malaysia and the Tin Islands Indonesia.

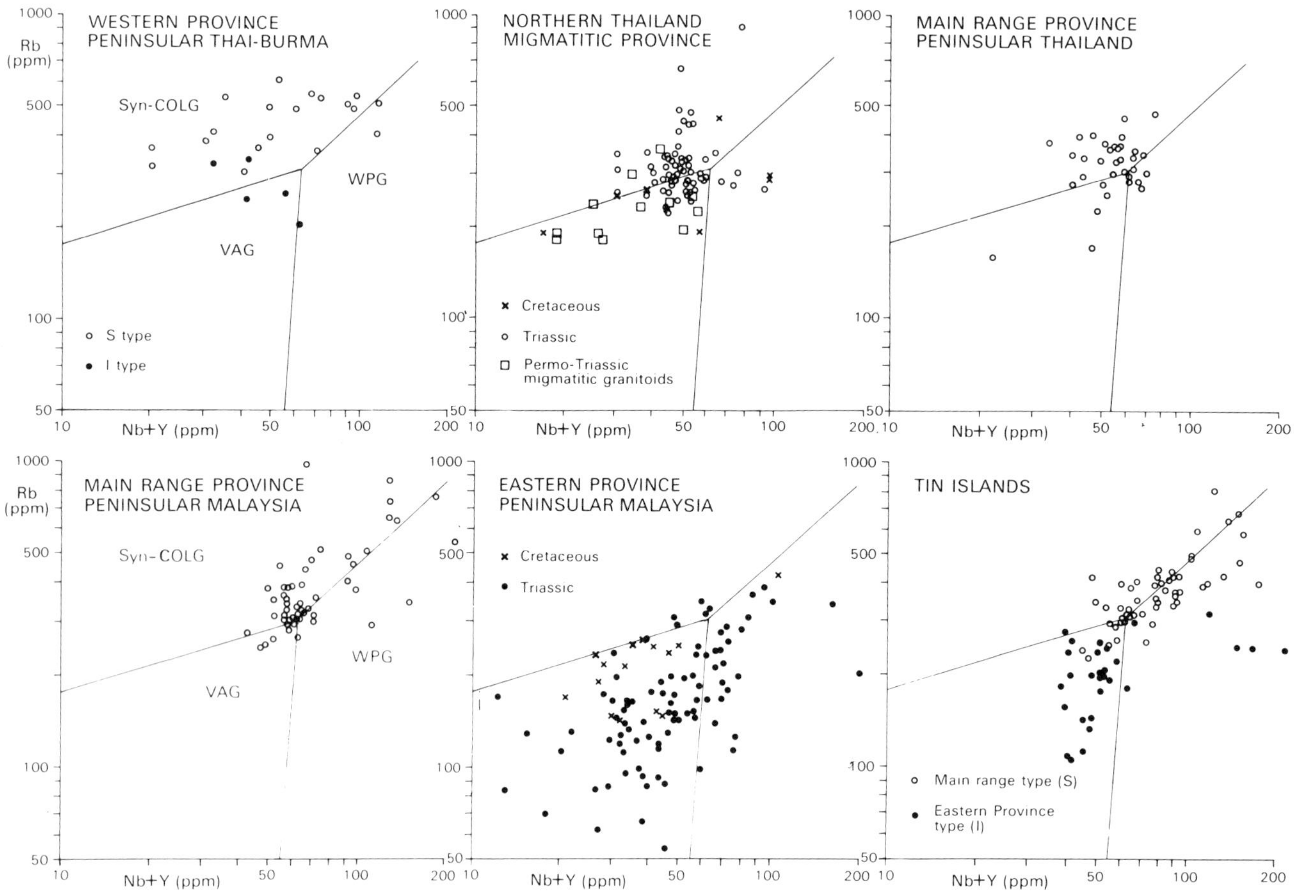


Figure 4. Rb vs (Nb + Y) discriminant plots for the granite provinces of Southeast Asia showing the tectonic classification suggested by Pearce, Harris, and Tindle (1984).

anomalous enrichment relative to LREE and the normalizing composition.

The more restricted range of concentration of the S-type granitoids allows mean values to be computed to a relatively high level of significance for each province. Comparison of the ORG normalized means for each S-type granite province shows distinctive differences in their chemistry. Relative to the Main Range profile, the North Thai Migmatite Province exhibits generally lower Rb, Th, Ta and rare earth elements (REE) and higher Ba, Sr, and Zr which is consistent with the less differentiated character. The Western Province, by comparison, exhibits the highest K, Rb, Th, Ta, Nb, Zr, and LREE and the lowest Ba. In this case, enrichment in Th and Zr appears to be somewhat anomalous relative to the normal differentiation trend. These diagrams demonstrate that the most highly evolved S-type granite magmas are those from the Western Province, that those from the North Thai Migmatite Province are the least evolved, and that those from the Main Range Province occupy an intermediate position. These differences also correspond to different degrees of mineralization within the provinces.

## Rb/Sr vs DI

On the Rb/Sr vs DI discriminant plots (Figure 3), I-type granitoids have consistently lower Rb/Sr ratios than S-type granites over the range of differentiation indices utilized. An empirically determined discriminant line derived from the Peninsular Malaysia data set is effective in separating I-types from S-types in all provinces including those regions with mixed I-S populations, such as the Tin Islands and Western Province, and of Triassic and Cretaceous ages.

In comparing the distribution of points for the three main geographical segments of the Eastern Province, it is notable that Peninsular Malaysia exhibits a more linear trend toward lower Rb/Sr ratios in the least differentiated granites, which suggests an approach to a mantle-dominated source region for the more basic units. In the Tin Island and North Thailand plots, however, there is some indication of a different evolutionary path; below DI 80, a flatter profile comprising a higher, more restricted range of Rb/Sr ratios predominates, which may indicate a greater contribution from a crustal source region.

The overall more highly evolved character of the S-type provinces compared with the Eastern Province is apparent in that all points lie in the compositional field above DI 72 and an Rb/Sr ratio of unity. The Peninsular Thai and Tin Island segments in particular are highly differentiated (> DI 78), with Rb/Sr ratios generally above 2.0. Peninsular Malaysia exhibits a greater compositional range, reflecting a mixed population in which the least differentiated members are representatives of the Taiping and Bt Tinggi plutons (Cobbing et al., 1986). In contrast with the Main Range distribution, S-type granitoids of the North Thai Migmatite Province mostly plot

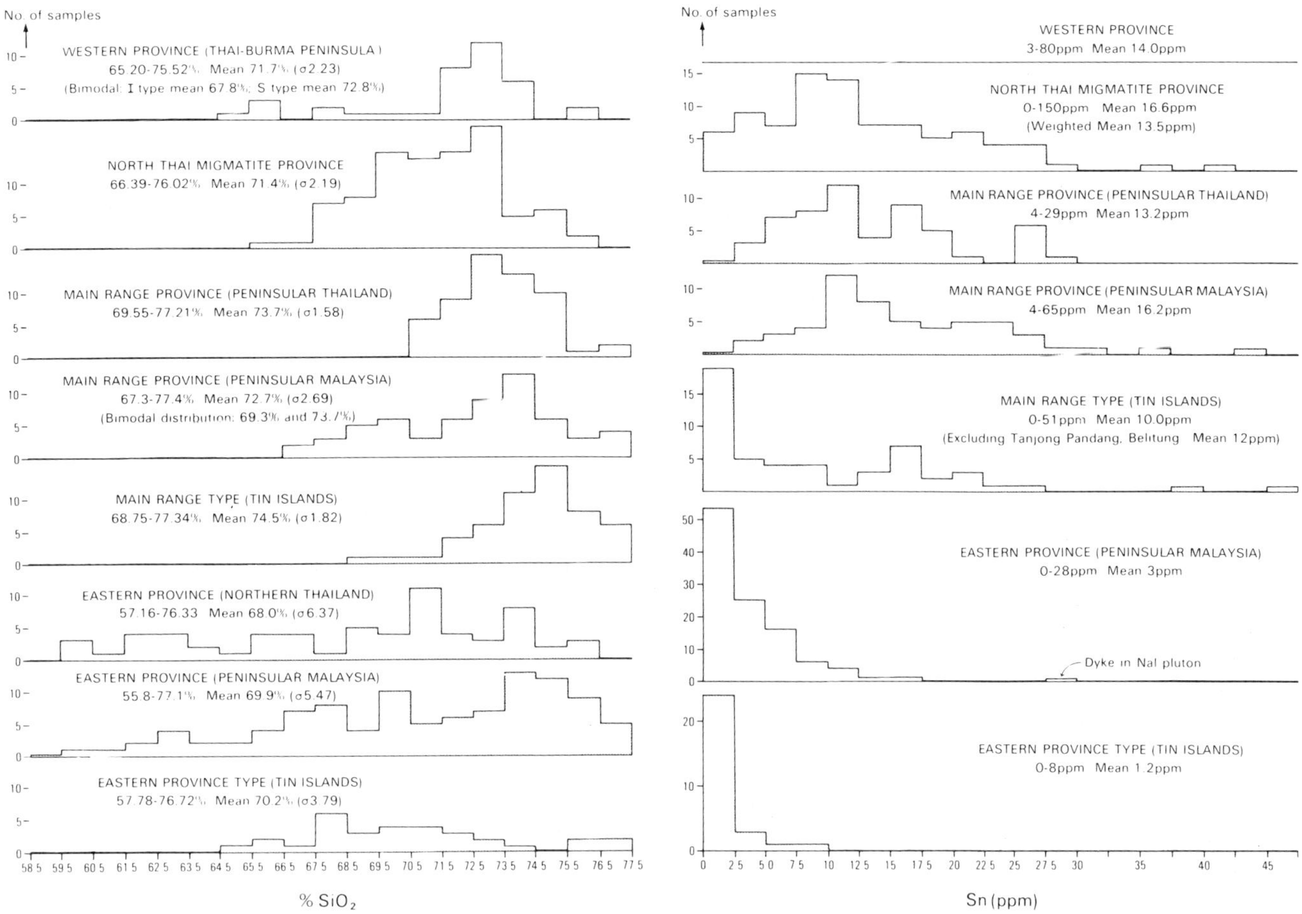


Figure 5. Histograms for $SiO_2$ and SN for the granite provinces and their subdivisions. Bimodality in the distribution of $SiO_2$ within the Main Range Province of Peninsular Malaysia is due to the contribution from the Taiping and Bt Tinggi plutons (Cobbing et al., 1986) which yield a mean of 69.3%.

below DI 85, where they yield a narrow range of Rb/Sr ratios and give a fairly flat profile. This striking uniformity is evident in other aspects of their geochemical and isotopic composition (Beckinsale et al., 1979). It is also notable that most of the early migmatitic granitoids and primary phases of the main batholiths plot in this field without any compositional separation, suggesting that they are derived from a similar source region. Above DI 85 the distribution of the North Thai late differentiates is comparable with that of the Main Range, indicating a similar evolutionary path toward the mineralized end members.

In the Western Province this empirical discriminant is effective in separating I- and S-type granitoids. Rb/Sr ratios in the I-types are below 2.5, whereas in the more primitive S-types they range from 2 to 4, and in the more highly evolved and generally mineralized S-types from 4.0 upward. In comparison with the Eastern Province, the I-types from the Western Province do not reach such high levels of differentiation and are not mineralized. Relative to the other S-type provinces, the Western Province S-types which are highly mineralized do not appear to attain such extreme levels of differentiation or Rb/Sr fractionation.

## Rb vs (Nb + Y)

The Rb vs (Nb + Y) discriminant diagram of Pearce, Harris, and Tindle (1984) is generally effective in distinguishing I- and S-type granitoids, which is achieved mainly as a function of Rb content. Thus the Eastern Province granitoids are mainly localized in the volcanic arc field, whereas the Main Range S-types lie chiefly within the field of syncollisional granite (Figure 4). The limited spread of plots from both provinces into the WPG field corresponds to two-phase granites and microgranite-mesogranite derivatives spatially associated with mineralization. In some I-types the high (Nb + Y) values are paralleled by distinctly higher REE contents and can be directly related to abnormally high concentrations of accessory minerals such as allanite. By contrast, the Triassic granites of the North Thai Province which give anomalously high (Nb + Y) values cannot be related specifically to secondary features.

The three geographical segments of the Main Range Province exhibit distinct populations which show a northward diminution in (Nb + Y) content from the Tin Islands to Peninsular Thailand. This northward trend parallels the fall in modal (and to some extent mean) $SiO_2$ content (Figure 5) and, although partly fractionation dependent, may suggest some kind of variation in the source region or level of partial melting within the Main Range Province. The trend is accentuated in North Thai Province, where the main concentration of points straddle the syn-COLG-VAG axis. The early migmatitic granitoid phase, which may well be syncollisional, plots preferentially in the VAG field. Clearly some caution needs to be exercised in the interpretation of this feature in terms of a tectonic-geochemical signature.

The Western Province S-types, which plot almost entirely in the syncollisional field in response to the uniformly high Rb content (> 300 ppm), are further characterized by a wide spectrum of (Nb + Y) values (Figure 4). This, in combination with their profiles on the ORG diagram (Figure 2), emphasize the rather unusual character of the S-type granites of this province. The I-types are marginally more fractionated than those in the Eastern Province and have higher Rb and HREE and lower Sr, Ba, and LREE. They span the VAG-syn COLG fields on Figure 4 and in fact plot in almost exactly the same region as the North Thai S-type granites.

## CRETACEOUS POSTOROGENIC GRANITES

Cretaceous granites are widely scattered throughout the region beyond the confines of the Western Province. They appear to be fairly abundant in the North Thai Migmatite Province and in the Peninsular Malaysia sector of the Eastern Province but have not been identified in the Main Range Province and the Indonesian Tin Islands. With respect to their host rocks they are clearly post orogenic but, with the exception of the Mae Yan complex of North Thailand, do not exhibit any marked alkaline affinity. The majority of these granites resemble orogenic I-types in their petrographic and geochemical affinities. For the Cretaceous granitoids from the North Thai Province this is clearly indicated on the Rb/Sr vs DI diagram (Figure 3), where they are separated from the S-type granitoids by the empirically-derived discriminant line. On the Rb vs (Nb + Y) diagram (Figure 4) they occupy virtually the same field as the migmatitic S-types and plot mainly in the VAG field, with only the alkaline rocks of the Mae Yan complex falling well within the WPG field. Cretaceous granites from the Eastern Province in Peninsular Malaysia also plot within the VAG field. On the Rb/Sr vs DI diagram (Figure 3) they form a distinct band within the I-type field and are characterized by some of the lowest Rb/Sr ratios.

## MINERALIZATION

The distribution of different types of ore deposits and different mineral and metal associations within the region are well documented (Figure 6; Hosking, 1973, 1977; Hutchinson and Taylor, 1978) and there is abundant evidence that the primary mineralization is spatially associated with specific granites. In particular, the distinctive associations and styles of mineralization of the four regional granitoid provinces have been outlined by Hutchinson and Taylor (1978) and Cobbing et al. (1986) and are summarized in Table 2.

The Eastern Province is characterized by stanniferous and base metal vein deposits of Cornish type, magnetite-cassiterite skarn, replacement or bedded deposits, and both stratabound and porphyry-type base metal sulphide ± gold deposits. The other provinces are characterized by endogenous greisen-bordered stanniferous vein swarms, pegmatites, and massive greisens of disseminated ore localized in the apical or marginal zones of plutons which have yielded extensive alluvial tin fields.

Regional trends along the length of the tin belt indicate an overall northward increase in the relative importance of copper and antimony in the Eastern Province and in tungsten, antimony, and fluorite in the S-type granite provinces, which is accompanied by a decline in the importance of tin deposits (Figure 6; Hutchinson and Taylor, 1978).

The North Thai Migmatite Province in particular does not have major tin fields like the Main Range and Western provinces, and the North Thai segment of the Eastern Province is entirely devoid of tin deposits. In the Southern Shan states, in northwest Thailand, and in certain parts of the Thai-Burma Peninsula, deposits associated with the granitoids have yielded more tungsten than tin, or are exclusively tungsten or antimony producers. This contrasts with the Indonesian parts of the tin belt, which are relatively deficient in tungsten with only two production centres known.

The northward change in metallogenesis for both the Eastern and S-type granite provinces broadly parallels the stepwise change in geochemical signature and overall decrease in $SiO_2$ content of the component granitoids (Figures 2, 5) along the belt. These overall trends are illustrated by subdivision of the provinces into metallogenic subprovinces (Table 2), which in some cases seem to reflect more detailed geochemical differences within the provinces.

### Eastern Province

Despite the wide compositional differences within the Eastern Province (Figure 2), particular segments display distinctive geochemical and metallogenic characteristics, suggesting that the source regions for the parent magmas and their evolutionary paths were also different. This is most clearly demonstrated in that the Eastern belt of the Peninsular Malaysia segment is the locus for tin deposits, whereas the neighboring Central Belt is distinctly auriferous (Figure 6; Table 2). Gold-antimony-mercury mineralization within the Central belt mostly occurs in lodes occupying faults and high-level fracture zones within a variety of rock types, but it is genetically related to the underlying granites (Tantisukrit, 1975). In comparing their whole-rock geochemistry, it is notable that the granitoids of the Central belt have marginally higher K, Rb, Pb, and Nb, and lower Cr contents than other Eastern Province suites. Th and Sr contents occupy a large range but are higher in mean absolute abundance. Initial $^{87}Sr/^{86}Sr$ ratios of below 0.708 for granitoids of the Central belt (Bignell and Snelling, 1977) are significantly lower than for the remainder of the Eastern Province, suggesting a predominant mantle-lower crust derivation for this subprovince.

The most widespread and characteristic type of tin deposit within the Eastern belt is the cassiterite-magnetite skarn. This is in marked contrast with the Main Range Province, where the primary iron ore deposits are devoid of tin, and the stanniferous skarns contain tin-bearing andradite or malayaite but are nowhere associated with magnetite or any other primary iron minerals (Hosking, 1977).

The cassiterite-magnetite skarns have been traditionally regarded as granite-related replacement deposits, though some stratabound lodes may be classed as volcanogenic (Hutchinson and Taylor, 1978). The other main types of tin deposits within the Eastern belt, which have no counterparts in the Main Range Province, are the stanniferous lodes of the Cornish type, which are mineralogically and structurally complex with considerable strike and downdip extension. All the major vein systems and the largest of the sheeted vein/stockwork

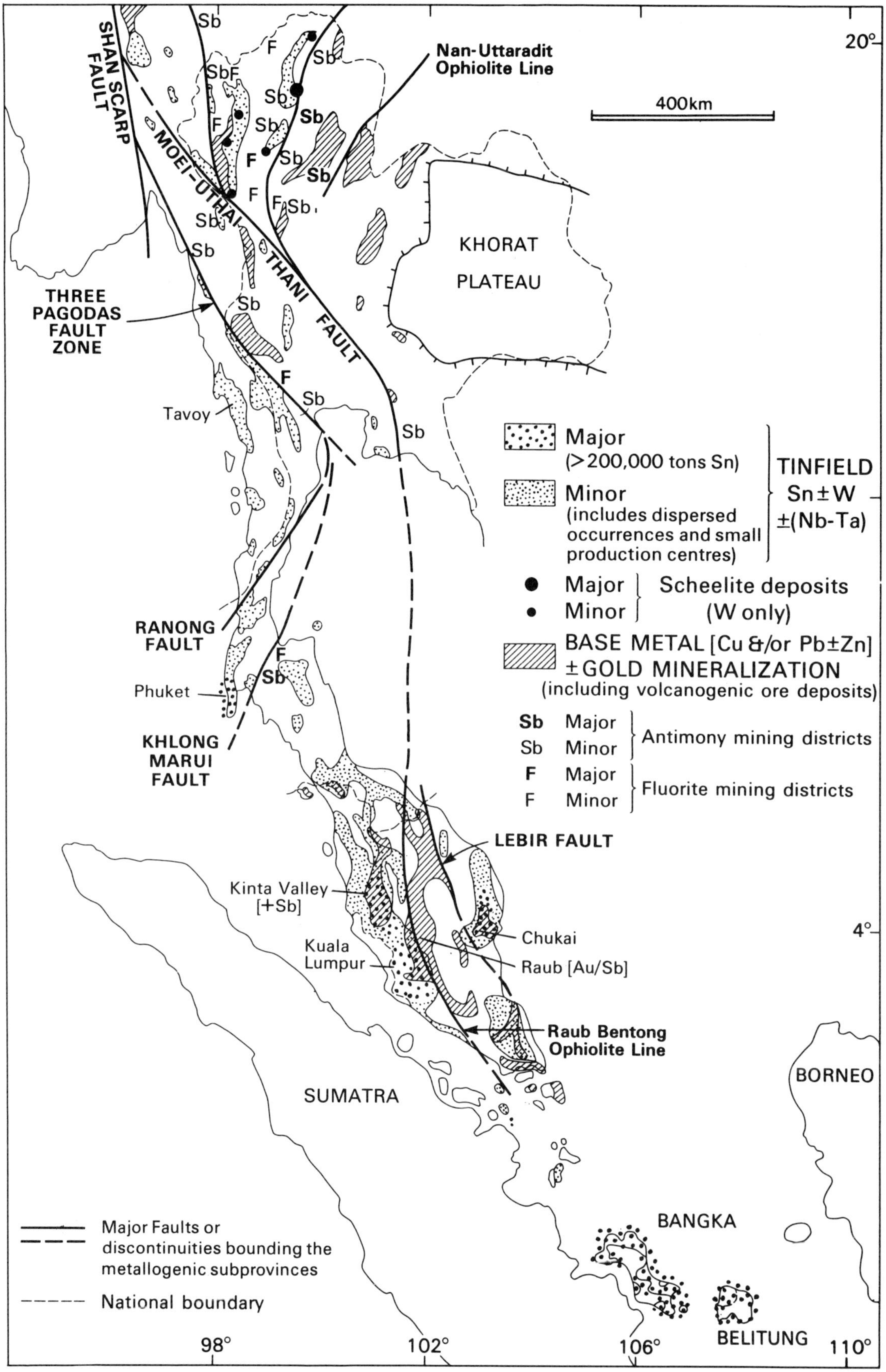


Figure 6. Distribution of dominantly granitoid related mineral deposits within the Southeast Asian Tin Belt. Data based on Hutchinson and Taylor (1978), Bender (1983), and the mineral resource maps of Burma (1977), Peninsular Malaysia (1976), and Thailand (1975).

systems are confined to the Chukai tin field in the Terengganu-Pahang subprovince (Figure 6; Table 2).

In this subprovince the mineralized granitoids east of the Lebir fault are generally enriched in $SiO_2$ (66.7–77.1%), $K_2O$ and Rb, and are depleted in Ba, Sr, Co, Cu, Sn, V, and all other major elements except $Na_2O$ in comparison with unmineralized granites, indicating their higher degree of differentiation.

Mineralized and seemingly unmineralized granitoids of the Johore subprovince also generally have a highly evolved I-type geochemistry and a close geochemical affinity with granitoids of the Chukai tin field, especially with regard to the low Ba, Sr, Cr, and V and the relatively high HREE values (Cobbing et al., 1986). Furthermore, irrespective of any demonstrable local metallogenic association, the Rb/Sr ratios for these granitoids plot on or very close to the line delimiting the compositional field of Main Range S-types (Figure 3). The Eastern Province I-type granitoids in the Indonesian Tin Islands are not directly associated with tin mineralization. Compared with the Peninsular Malaysian segment, they have lower overall LIL element contents and generally higher HFS element abundances (Figure 2) and are comparable with unmineralized granites in the Terengganu-Pahang subprovince.

The Eastern Province granitoids of Northern Thailand, which are not tin-related, can be divided geotectonically by the Nan-Uttaradit ophiolite line into two metallogenic associations (Figure 6; Table 2). The Loei-Chiengkarn sector to the east of the line consists of a granodiorite-tonalite suite with a typical I-type porphyry copper/base metal and gold association (Jacobson et al., 1969). West of the ophiolite line, granitoids of the Phrae-Lampang area are associated with antimony-dominated base metal mineralization which mostly takes the form of vein and replacement deposits localized along faults and breccia zones in a variety of rock types. Compared with the Malaysian and Indonesian segments, granitoids within this metallogenic subprovince have some of the highest K, Rb, Th, and U values, whereas Ba and Sr exhibit a similar range of values to the Tin Island I-types (Figure 2). The generally lower degree of differentiation and higher Rb/Sr values at $DI > 75$ (Figure 3) and marginally higher initial $^{87}Sr/^{86}Sr$ ratios indicate that relative enhancement in the radioelement contents is not so much due to fractionation during their magmatic evolution but is ascribed to a greater crustal component, such as LIL element-enriched subcontinental lithosphere, in the magma genesis.

## Main Range Province

In contrast with the Eastern Province, the Main Range granitoids exhibit much less variation in the type and style of mineralization. The principal primary ore occurrence is in the form of endogenous greisen-bordered vein systems of limited dip and strike length associated with the cusps of the late two-phase granites. The most distinctive types of tin deposits within this terrane, however, are the massive greisen systems with pervasive disseminated mineralization of the granite roof zone and the much less common stanniferous pegmatites in which Nb/Ta species may be present. In Peninsular Malaysia strongly telescoped xenothermal stanniferous veins and pipes are also developed, many of which carry an impressive assemblage of ore minerals including several varieties of stannite. These xenothermal deposits, which are best developed in the Kinta Valley and Kuala Lumpur tin fields (Figure 6), may not be genetically related to the Triassic granites, but rather are associated with Upper Cretaceous to Tertiary igneous activity (Hosking, 1977).

Breccia pipe deposits at Khao Soon (Peninsular Thailand) and Tikus (Belitung) are rare but are important tungsten producers within the province. They are characterized by wolframite, mainly as ferberite, and cassiterite mineralizations overlapped by low-temperature ore-mineral assemblages. Like the xenothermal deposits, they appear to have formed in near-surface environments but are distinguished in that they show no reverse zoning and the paragenetic sequence is normal. Of particular note is the striking similarity of the tungsten-antimony breccia pipe of Khao Soon to the Doi Ngom deposit in the Eastern Province of Northern Thailand, which shows that this type of tungsten mineralization is nonspecific to the Main Range Province.

Although they are of heterogeneous distribution, Nb, Ta, Th, U, Y, Zr, and REE's are clear petrogenetic associates of tin, which are best developed as separate mineral species in the pegmatites and some massive greisens. Tungsten and the base metals, however, appear to have a less strong paragenetic consanguinity with the tin deposits, and on a local scale tin and tungsten ore types may exist separately. In Peninsular Thailand the northwestern part of the Main Range Province constitutes a separate mineral district in which wolfram, antimony, other base metals and fluorite are locally important with or without tin.

The primary granitoid geochemistry of the Main Range segments which host the major tin fields is not significantly different from one segment to another except for the whole-rock tin content, which appears to be markedly lower and in many cases is undetectable in some of the S-type granites of Bangka and Belitung in the Indonesian Tin Islands (Figure 5). The Tanjon Pandang pluton, Belitung, is particularly anomalous in having low tin content, but it has a pattern of mineralization which is quite common in other parts of the Main Range in that extensive placer deposits are proximally derived from bodies of specialized two-phase granitoids within the main pluton. This pattern is also present in Peninsular Malaysia, where relatively primitive S-type plutons such as Taiping and Bukit Tinggi host small bodies of highly specialized two-phase derivatives which are rich tin producers and contribute to the place deposits (Cobbing et al., 1986).

## North Thailand Migmatite Province

This province is not a major tin producer like the other S-type granite provinces, but is characterized by a number of relatively small tin and/or tungsten deposits which are similar in type to the primary greisen systems of the Main Range Province. The most distinctive mineralization is, however, the skarn replacement scheelite and the endogenous and exogenous vein deposits with fluorite and antimony assemblages (Figure 6). Furthermore, the major granitoid divisions and some individual morphostructural segments of the province are characterized by specific mineral associations such that tin-, tungsten-, antimony-, lead/zinc-, and fluorite-dominated deposits are arranged in zones parallel to the long axes of the granitic ranges. Tin predominates in the axial zones of the granite-cored major anticlinoria; tungsten and antimony occur mainly in the limb or contact zones and are associated with major fault lines; and the complementary sedimentary basins and graben are marked by the incoming of lead-zinc and fluorite deposits. A similar symmetrical regional zonation occurs within the Main Range Province in that part of Perak, Malaysia which embraces the Kinta Valley (Hosking, 1973).

Tungsten mineralization is particularly well developed in the marginal zones of the North Thai Province adjacent to the Eastern and Western Provinces.

Some occurrences of tungsten and fluorite which do not conform to the zonal pattern of ore deposits within the Central Migmatite Belt appear to be associated with Late Cretaceous intrusions or major faults that have been initiated or reactivated during the Cretaceous.

Because the North Thai granitoids are more compositionally restricted than Main Range Province granitoids, it is difficult to establish geochemical patterns at subprovince level with the data available at present. However, the higher initial Sr ratios and Rb/Sr ratios at DI values below 85 indicate a crustal source region.

### Western Thai-Burma Province

The Cretaceous S- and I-type granitoids readily subdivide into three discrete metallogenic subprovinces, all of which are bounded or transected by major faults. Throughout this province the tin-tungsten mineralization is exclusively associated with the S-type granitoids and particularly with the latest, most fractionated and most peraluminous S-type granitoid suite in the area (Putthapiban and Gray, 1983). The southernmost subprovince (Phuket-Phangnga-Ranong), which is the most distinctive and productive tin field of this province, has also been the most thoroughly investigated. In addition to endogenous massive greisen or disseminated deposits, such as the Haad Som Pan deposit near Ranong (Aranyakanon, 1961), the Phuket area is distinctive for its major development of large stanniferous lepidolite pegmatites and mica-tourmaline pegmatites (Garson et al., 1977). These deposits locally carry significant amounts of columbite/tantalite, wolframite, monazite, and other REE) and Y-bearing minerals in close association with cassiterite, and the metallogenic association is reflected in the host granite geochemistry by the generally higher K, Rb, Th, Ta, Nb, Zr, and REE contents (Figure 2).

The Tenasserim and Shan subprovinces of Burma and immediately adjacent parts of Thailand are characterized by a particularly close association of tin and tungsten and a general northward increase in the relative importance of tungsten. The primary mineralization is mostly in the form of hydrothermal greisen-bordered vein systems and stockworks which are confined to granite country rock contact zones with little continuation at depth. SniW ratios for different deposits typically range from 2:1 to 1:3 (for example, at Mawchi, Shan, Hermingyi, and Tavoy) (Bender, 1983); wolframite is commonly localized in the veins and cassiterite is disseminated in the greisen. The closest association is at Mawchi, where both wolframite and cassiterite occur in the same quartz veins. Mineralization within the Tenasserim and to a lesser extent the Shan subprovinces is further attended by a distinctive base metal assemblage that includes molybdenum and bismuth (Table 2). Important differences in the metallogenic nature of these two subprovinces is the additional occurrence of vein deposits with only wolframite or stibnite in the E Shan segment. The predominance of tungsten and antimony mineralizations indicates a certain metallogenic affinity with the North Thai Province.

Geochemistry of the Cretaceous granitoids within the NW Thailand sector of the Shan subprovince shows them to be highly differentiated S-types (i.e., DI values 85-92) which plot in the same Rb/Sr vs DI compositional field as the highly evolved S-type granite suites of Phuket, but have significantly lower contents of Th, Ta, Nb, Zr, Y, and rare earth (RE) elements.

Although the Shan subprovince exhibits certain metallogenic affinities with the neighboring North Thai Province, there are notable differences between the associated granitoids. This is particularly evident with respect to the Mae Lama stock, which yields an initial $^{87}Sr/^{86}Sr$ ratio of c 0.708 (Beckinsale et al., 1979) suggestive of a crustally contaminated mantle source or genesis from extensively altered oceanic lithosphere.

## DISCUSSION

### Tectonic Setting and Granite Classification

The granite provinces are clearly linked with different geological terranes which form three separate, and approximately parallel, strips, each bounded by structural discontinuities or sutures at the westward margin of the Sundaland craton. This clearly favors a collage tectonic interpretation for the region (Ridd, 1980). In broad terms, the tectonic setting is evidently collisional; however, a volcanic arc setting is appropriate for the Eastern Province, which may have undergone some precollisional volcanic arc plutonism.

The granitoids from the provinces of Permo-Triassic age are adequately classified by the geochemical scheme of Pearce, Harris, and Tindle (1984), which argues that the compositions of granites are determined by the tectonic settings in which they form. Thus, the granites of the Main Range and North Thai provinces mainly fall in the syncollisional field on Figure 4, whereas the Eastern Province granites lie in the volcanic arc field. The system, however, is less successful in classifying the mixed S- and I-type populations of the Tin Islands and the Western province; here, two tectonic settings are identified in each province (Figure 4), whereas in reality only one exists. In areas such as these, alternative explanations are required for the differences in granite composition, such as differences in the composition of the source region (Chappell and White, 1974).

Systematic geochemical polarity, which has been established from all provinces, is characterized chiefly by a northward trend toward less differentiated granitoids. It is accompanied by equally systematic differences in mineral assemblages, characterized by a northerly decline in the importance of tin and a rise in the importance of tungsten, antimony, and fluorite. In particular, there seems to be a regional concordance between tungsten, antimony, and fluorite.

Th northerly variation in both granite geochemistry and mineralization assemblages strongly suggests that the mineralization is closely linked to the granite geology rather than to some alternative control. This may be particularly illustrated by differences between the Main Range and the North Thai provinces. From the similarities in the spatial arrangement and style of ore deposition, it would appear that differences in type and extent of metallogenesis between these provinces is related to the composition of the granitoids and, by inference, that of their source region. The restricted compositional range of primary phases and the very high initial Sr ratios of the granitoids of the North Thai Province suggest their derivation from highly evolved, upper continental crust (Hutchinson, 1983). This possibility, in conjunction with the hypothesis that the lower crust is the main source region for tin (Hutchinson, 1983), may be supported by the U/Pb studies of Liew and Page (1985) and Liew and McCulloch (1985), which concluded that the Main Range granitoids were derived by recycling of mid-Proterozoic crystalline basement from the lower crust.

The northward variation affects all provinces equally, even though the granite provinces are linked to specific terrains for which the present best explanation is in terms of collage tectonics. This implies that the cause of variation lies either in a change in the composition of the source region in each terrain, or in the process of granite generation in each terrain. Moreover, the metallogenic differences are not restricted to one epoch, suggesting that the crustal infrastructure and the underlying reasons for variation remained constant throughout the Mesozoic.

These underlying similarities in all terranes and in all provinces are perhaps another expression of the fact that a tectonic/geochemical classification for the granites of the region is not completely successful, particularly with respect to the Western Province and the Indonesian Tin Islands, where mixed S- and I-type granite populations prevail; this is further illustrated by the almost complete spectrum of composition between I- and S-type granitoids for the region as a whole (Figures 2 to 4). Moreover, whereas the granites of the Eastern Province are not typically Andean (Figure 2), all the S-type granites and those from the Main Range Province in particular are highly comparable in their composition with other S-type regimes, such as the European Hercynides, as in SW England (Darbyshire and Shepherd, 1985).

Similarities between provinces are particularly well illustrated on the multi-element plots (Figure 2) where the volcanic arc granites of the Eastern Province have profiles which, although different from those for the S-type granites, resemble them more than they do those for Andean volcanic arc granites. This suggests that the crustal source region for each province is similar, even though the terranes may have been tectonically distinct. It may also suggest that all the terranes originally formed part of the same continental block which became fragmented, with each terrane having a subsequently different geological history. Cretaceous plutonism, although particularly concentrated in the mixed S- and I-type Western Province, is distributed over the whole of the remainder of the region, with the possible exception of the Main Range Province and the Indonesian Tin Islands. It seems to produce plutons of I-type affinity but with no particularly marked contrasts with the earlier Triassic plutonism. In northern Thailand small plutons appear to be associated with fluorite mineralization, whereas in the Main Range Province the distinctive xenothermal mineralization in the Kinta Valley and Kuala Lumpur tin fields has been attributed by Hosking (1977) to Cretaceous or Tertiary igneous activity. Granite plutonism of this age has not, however, been recognized in the Main Range Province, though there is a pattern of discordant K/Ar ages relative to Rb/Sr ages (Bignell and Snelling, 1977). Cretaceous plutons recognized in the Eastern Province do not appear to be associated with mineralization.

## CONCLUSION

Geochemical N-S polarity, which has been established for the granitic components of all the provinces recognized, is accompanied by differences in the mineral assemblages of the mineral deposits present in all provinces. This suggests an underlying unity for the source regions and for the process of granite generation in all provinces of both Permo-Triassic and Cretaceous age. This conclusion may lead to new insights or constraints upon the currently favored collage tectonic interpretation for orogenic processes in the region.

## ACKNOWLEDGMENTS

The data presented here are the result of a 5-year research programme initiated by the Overseas Development Administration of the UK in support of the joint CCOP-IOC project Studies in East Asian Tectonics and Resources (SEATAR). The project was jointly conducted by the Geological Survey of Malaysia, the Geological Research and Development Centre of Indonesia, the Department of Mineral Resources of Thailand, and the British Geological Survey. The UK input was funded by the Overseas Development Administration of the UK. We particularly thank Mr. Santokh Singh and Mr. Gwyn Hughes for their positive support at the outset. This paper is published by permission of the Director of the British Geological Survey (N.E.R.C.).

## REFERENCES CITED

Aranyakanon, P., 1961, The cassiterite deposit of Haad Som Pan, Ranong province, Thailand: Royal Department of Mines, Thailand, Report of Investigation No. 4, p. 1-182.

Atherton, M.P., and (?) Sanderson, 1985, The chemical variation and evolution of the super-units of the segmented Coastal Batholith, *in* W.S. Pitcher, M.P. Atherton, E.J. Cobbing, and R.D. Beckinsale, eds., Magmatism at a plate edge—The Peruvian Andes: Blackie & Son, Glasgow, UK, John Wiley & Sons, New York, p. 208-227.

Beckinsale, R.D., 1979, Granite magmatism in the tin belt of Southeast Asia, *in* M.P. Atherton and J. Tarney, eds., Origin of granite batholiths: Orpington, Kent, Shiva Publishing, p. 34-44.

Beckinsale, R.D., S. Suensilpong, S. Nakapadungrat, and J.N. Walsh, 1979, Geochronology and geochemistry of granite magmatism in Thailand in relation to a plate tectonic model: Journal of the Geological Society of London, v. 136, p. 529-540.

Bender, F., 1983, Geology of Burma: Berlin-Stuttgart, Gebruder Borntraeger, 293 p.

Beus, A.A., and S.V. Grigorian, 1977, Geochemical exploration for mineral deposits: Applied Publishing Ltd.

Bignell, J.D., and N.J. Snelling, 1977, Geochronology of Malayan granites: Overseas Geological and Mineral Resources No. 47, 72 p.

Braun, E.V., L. Hahn, and H.D. Maronde, 1981, Geological map of Northern Thailand: Hannover, Federal Institute for Geosciences and Natural Resources, scale 1:250,000.

Braun, E.V., C. Besang, W. Eberle, W. Harre, H. Kreuzer, J. Lenz, P. Muller, and I. Wendt, 1976, Radiometric age determinations of granites in Northern Thailand: Geol. Jahr., Reihe B, Heft 21, p. 171-204.

Chappell, B.W., and A.J.R. White, 1974, Two contrasting granite types: Pacific Geology, v. 8, p. 173-174.

Cobbing, E.J., D.I.J. Mallick, P.E.J. Pitfield, and L.H. Teoh, 1986, The granites of the Southeast Asian Tin Belt: Journal of the Geological Society of London, v. 143, p. 537-550.

Darbyshire, D.P.F., and T.J. Shepherd, 1985, Chronology of granite magmatism and associated mineralization, SW England: Journal of the Geological Society of London, v. 142, p. 1159-1177.

Garson, M.S., B. Young, A.W.G. Mitchell, and R.A.R. Tait, 1977, The geology of the tin belt in Peninsular Thailand around Phuket, Phangnga and Takua Pa: Institute of Geological Sciences, Overseas Memoir No. 1, 112 p.

Harris, N.B.W., J.A. Pearce, and A.G. Tindle, 1986, Geochemical characteristics of collision-zone magmatism, *in* M.P. Coward and A.C. Ries, eds., Collision tectonics: Geological Society Special Publication No. 19, p. 67-81.

Hosking, K.F.G., 1973, The primary tin mineralization patterns of West Malaysia: Bulletin of the Geological Society of Malaysia, v. 6, p. 297-308.

Hosking, K.F.G., 1977, Known relationships between the "hard-rock" tin deposits and the granites of Southeast Asia: Bulletin of the Geological Society of Malaysia, v. 9, p. 141-157.

Hutchinson, C.S., 1977, Granite emplacement and tectonic subdivision of Peninsular Malaysia: Bulletin of the Geological Society of Malaysia, v. 9, Special Circum-Pacific Volume, p. 187-207.

Hutchinson, C.S., 1983, Multiple Mesozoic Sn-W-Sb granitoids of southeast Asia: GSA Memoir 159, p. 35-60.

Hutchinson, C.S., and D. Taylor, 1978, Metallogenesis in S.E. Asia: Journal of the Geological Society of London, v. 135, p. 407-428.

Ishihara, S., H. Sawata, K. Shibata, S. Terashima, S. Arrykul, and K. Sato, 1980, Granites and Sn-W deposits of Peninsular Thailand, *in* S. Ishihara and S. Takenouchi, eds., Granite magmatism and related mineralization: Mining Geology Special Issue 8, p. 223-241.

Jacobson, H.S., C.T. Pierson, T. Danusawad, T. Japakasetr, B. Inthuputi, C. Siriratanamongkol, S. Prapassornkul, and N. Pholphan, 1969, Mineral investigations in northeastern Thailand: U.S. Geological Survey Professional Paper 618, 96 p.

Liew, T.C., 1983, Petrogenesis of the Peninsular Malaysian granitoid batholiths: Canberra, Australian National University, unpublished Ph.D. thesis.

Liew, T.C., and M.T. McCulloch, 1985, Genesis of granitoid batholiths of Peninsular Malaysia and implications for models of coastal evolution: Evidence from a Nd-Sr isotopic and U-Pb zircon study: Geochimica et Cosmochimica Acta, v. 49, p. 587-600.

Liew, T.C., and R.W. Page, 1985, U-Pb zircon dating of granitoid plutons from the West Coast Province of Peninsular Malaysia: Journal of the Geological Society of London, v. 142, p. 515-526.

Mahawat, C., 1982, The petrology and geochemistry of the granitic rocks of the Tak batholith, Thailand: Liverpool, Liverpool University, unpublished Ph.D. thesis, 210 p.

Pearce, J.A., N.G.W. Harris, and A.G. Tindle, 1984, Trace element discrimination diagrams for the tectonic interpretation of granitic rocks: Journal of Petrology, v. 25, p. 956-983.

Pitcher, W.S., et al., 1985, Magmatism at a plate edge—The Peruvian Andes: Blackie & Sons, Glasgow, UK, John Wiley & Sons, New York, 328 p.

Priem, H.N.A., N.A.I.M. Boelrijk, E.H. Bon, E.H. Habeda, E.A.Th. Verdurman, and R.H. Vershure, 1975, Isotope geochronology in the Indonesian Tin Belt: Geologie en Mijnbouw, v. 54, p. 61-67.

Putthapiban, P., and C.M. Gray, 1983, Age and tin-tungsten mineralization of the Phuket granites, Thailand: Proceedings of the Conference on Geology and Mineral Resources of Thailand (DMR Bangkok), p. 1-10.

Ramsay, C.R., J. Odell, and A.R. Drysdall, 1986, Felsic plutonic rocks of the Midyan region, Kingdom of Saudi Arabia—II. Pilot study in chemical classification of Arabian granitoids: Journal of African Earth Sciences, v. 4, p. 79-85.

Ridd, M.F., 1980, Possible Palaeozoic drift of S.E. Asia and Triassic collision with China: Journal of the Geological Society of London, v. 137, p. 635-640.

Shaw, D.M., 1968, A review of the K-Rb fractionation trends by covariance analysis: Geochimica et Cosmochimica Acta, v. 32, p. 573-601.

Shaw, D.M., 1970, Trace element fractionation during anatexis: Geochimica et Cosmochimica Acta, v. 34, p. 237-243.

Teggin, D.E., 1975, The granites of Northern Thailand: Manchester, University of Manchester, unpublished Ph.D. thesis.

Teoh, L.H., 1985, Geochemistry of the Peninsular Malaysian granites: Proceedings of the 1st Symposium of IGCP Project 220 on Correlation and Resource Evaluation of Tin/Tungsten Granites in Southeast Asia and the Western Pacific Region.

Tischendorf, G., 1977, Geochemical and petrographic characteristics of silicic magmatic rocks associated with rare-element mineralization, in M. Stenprok, L. Burnot, and G. Tischendorf, eds., Metallization associated with acid magmatism: v. 2, Prague, p. 41-96.

Veeraburus, M., 1982, Geological map of Thailand: Bangkok, Department of Mineral Resources, scale 1:1,000,000.

White, A.J.R., and B.W. Chappell, 1977, Ultrametamorphism and granitoid genesis: Tectonophysics, v. 43, p. 7-22.

White, A.J.R., and B.W. Chappell, 1983, Granitoid types and their distribution in the Lachlan Fold Belt, southeastern Australia: GSA Memoir 159, p. 21-34.

CHAPTER 51

# ECONOMIC ASSESSMENT AND REVIEW OF THE U.S. EXCLUSIVE ECONOMIC ZONE (EEZ)

**Reid T. Stone**
**Timothy J. MacGillvray**
*U.S. Department of the Interior*
*Washington, D.C.*

ABSTRACT

The Department of the Interior, through its Minerals Management Service, is presently developing overall plans for a leasing program of nonenergy minerals in the EEZ including phosphorites, manganese, cobalt, titanium, PGM, heavy-mineral sands, and sand and gravel. This chapter describes (1) nonenergy mineral resources found in the U.S. EEZ; (2) currently estimated economic potential of those resources, by type; (3) future potential of nonenergy resources, given the studies conducted to date; and (4) actions the U.S. government is taking to encourage exploration and development of those mineral resources.

The chapter specifically discusses reasons for establishing Federal-state-industry task forces to assess the economic and environmental criteria used in formulating specific leasing proposals. The need to provide an economic climate acceptable for investment in this new frontier area within our EEZ is outlined in the context of the Outer Continental Shelf Lands Act.

## INTRODUCTION

The Circum-Pacific Conference has become a significant meeting not only to those of us involved in the development of our mineral resources in the Pacific area, but to the rest of the world. The strategic importance of those resources can be evidenced not only by the attendance at this meeting, but by the international activities taking place throughout the Pacific area. My discussion will focus on the United States Exclusive Economic Zone/Outer Continental Shelf (EEZ/OCS), and I will review the Department's regulations and leasing proposals now under consideration.

Development worldwide of our marine mineral resources (not only in the U.S. EEZ/OCS but the rest of the world) will depend largely on their economic viability and accessability. These resources must be made available at a competitive cost by the nations which govern that access. Realistically, we believe that most of these resources will be developed near shore in relatively shallow water depths in the immediate future. The technology is available, and the markets are waiting for such commodities as sand and gravel or the placer minerals. In the more distant future, the deeper and more unique deposits, such as the cobalt-rich manganese crusts found throughout the central Pacific basin, will be developed. To a major extent, industrial need and price will determine the rate of technology development required to make these types of commodities available and attractive.

## BACKGROUND

The Outer Continental Shelf Lands Act (OCSLA) of 1953, amended in 1978, charges the Secretary of the Department of the Interior (DOI) with administering mineral exploration and development on the OCS. Under this Act, the Secretary is directed to encourage exploration and development of the OCS in order to achieve national economic and energy policy goals, assure national security, reduce dependence on foreign sources, and maintain a favorable balance-of-payments in world trade. Section 8(k) of the OCSLA authorizes leasing of oil, gas, sulfur and other minerals on the OCS on the basis of competitive bidding, and under such terms and conditions as may be prescribed at the time of offering the area for lease. With respect to EEZ/OCS minerals leasing, the following elements must be considered: (1) balancing an orderly mineral resource development with protection of the human, marine, and coastal environments, (2) preservation and maintenance of competition, and (3) receipt of fair market value.

The creation of the U.S. Exclusive Economic Zone occurred in 1983 when President Reagan issued his Proclamation confirming our sovereign jurisdiction over all submerged lands to a line 200 miles off our coasts and including the island State of Hawaii. The U.S. EEZ covers an area larger than the entire land mass of the United States, and it contains a vast mineral storehouse.

The President's EEZ Proclamation provides the foundation

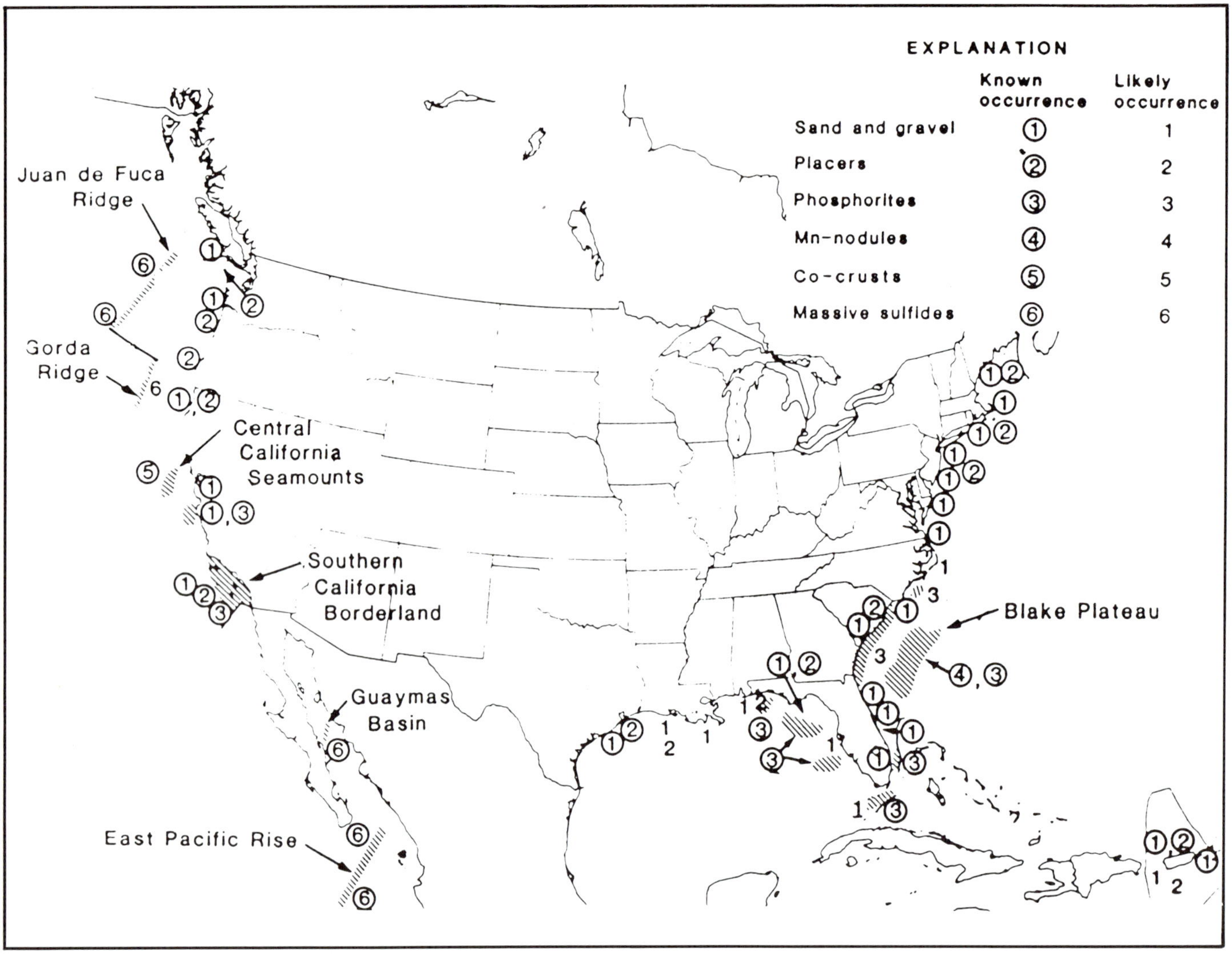


Figure 1. Nonenergy mineral occurrences located offshore U.S. mainland (Modified from Ballard and Bischoff, 1984, USGS Circular 929).

for the Minerals Management Service's (MMS's) nonenergy mineral leasing program and regulatory initiative. This Proclamation confirmed U.S. jurisdiction over 3 billion acres (12 million km$^2$) contiguous to U.S. coastal states, territories, and possessions. Within the EEZ, the U.S. has sovereign rights for the purposes of exploring, developing, conserving, and managing the mineral and other natural resources of the seabed. Both the U.S. Bureau of Mines and the U.S. Geological Survey are supporting our efforts to evaluate and define these resource potentials.

In 1983, MMS established the Office of Strategic and International Minerals (OSIM) in Long Beach, California, to develop policy and implement the program for the exploration, development, and recovery of nonenergy minerals from the EEZ/OCS. The program being developed by OSIM provides for a case-by-case examination of lease proposals, based on industry interest and national needs, and involves extensive consultation with State governments. Before any decision regarding lease offerings is made, the specific marine environments and mining technologies involved in each case will undergo detailed study. The goal of the nonenergy minerals leasing program is to encourage exploration and development of a variety of mineral resources within the EEZ/OCS, while at the same time ensuring environmental protection and full public and State participation.

During the past 40 years the United States has become increasingly dependent on mineral imports. The U.S. is more than 50% dependent for 21 of 32 essential minerals used in our economy. Among the critical minerals listed are cobalt, manganese, platinum, and chromium. These minerals are generally considered our most important strategic and critical minerals, since they play a vital role in our economy and national security and are largely obtained from other countries. In addition to cobalt, chromium, platinum, and manganese, the U.S. is also heavily dependent on imports for a wide variety of other minerals such as tin, nickel, tungsten, and zinc, and it is becoming increasingly more dependent on imports of copper, iron, and lead, minerals for which the U.S. has been historically self-sufficient.

Growing dependence on mineral imports contributes greatly to our U.S. balance-of-payments problem. Between 1981 and 1984, the U.S. trade deficit in nonenergy minerals more than doubled from $9 billion to $19 billion. Moreover, the deficit represents a 10-fold increase between 1973 and

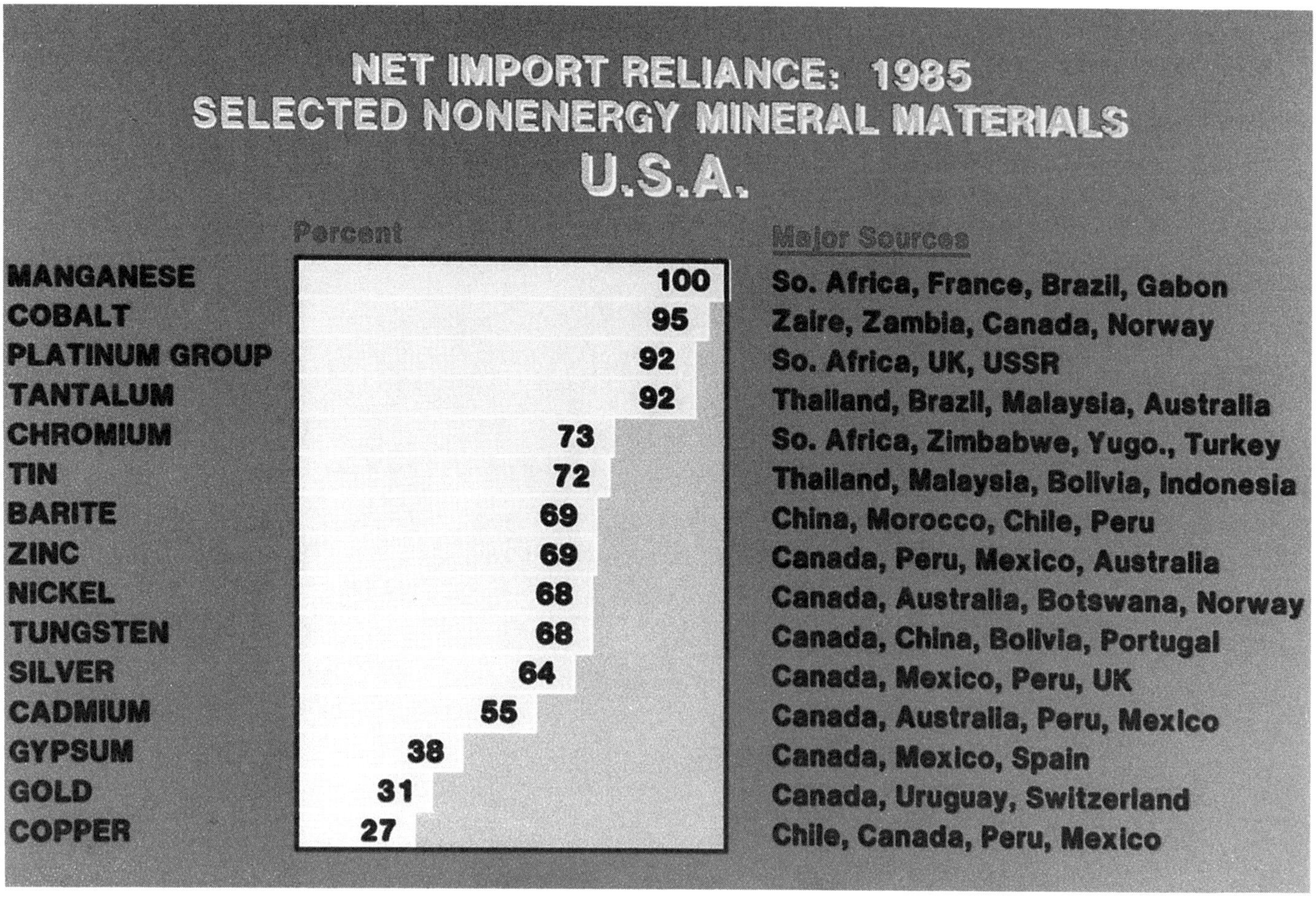


Figure 2. 1985 U.S. import dependency on certain nonenergy minerals.

1984. This represents a tremendous dollar drain on the U.S. economy, an outflow of investment dollars, and lost employment opportunities. By the year 2000 the Bureau of Mines has estimated that several of these resources may increase in their demand and further place our dependence on them beyond our control (see Figure 2).

Many nations have recognized the mineral potential and are now obtaining large quantities of minerals from the seabed. Companies in Great Britain and Japan, for example, have been mining large quantities of sand and gravel from the seabed for decades. Other large-scale mining operations for tin are also being conducted off the coast of Indonesia and Thailand. These nations are leading the pioneering efforts to develop new technologies for marine mining. If the U.S. is going to stimulate the development of technology to explore and develop the marine mineral resources, it must proceed now to develop a regulatory framework that encourages exploration and development of minerals within the EEZ/OCS.

## GENERAL REGULATORY FRAMEWORK

The MMS is currently in the process of developing a three-tier regulatory framework for administering exploration and development of nonenergy minerals in the EEZ/OCS. Under this framework separate regulations are being promulgated for prospecting, leasing, and operations. The rules being developed by MMS will address requirements mandated by Congress and the President. These requirements include assuring receipt of fair market value, balancing mineral development with protection of the environment, as is mandated by OCSLA, and encouraging greater participation by the States and the general public.

Currently, joint Federal-State task forces have been established to evaluate any potential lease offering, as we are now considering for the Hawaii-Johnston Island EEZ for the cobalt-rich manganese crusts, and to recommend to the Secretary a program of geological, biological, and socioeconomic studies to investigate environmental impacts and prepare an environmental impact statement (EIS) for any lease offering. These feasibility studies will be supported by funding and/or services provided by MMS and other Federal and State agencies. Industry will be invited to assist the task force, collect resource data, and participate in the refinement of leasing proposals to be presented to the public for review and comment.

Development of any regulations for offshore mining will be keyed to a case-by-case examination of leasing proposals. As noted earlier, prior to any decision regarding a lease offering, the specific marine environments and mining technologies involved in each case will undergo detailed study. As this process unfolds, there will be an opportunity to adopt additional requirements in the form of stipulations/conditions on an individual lease sale basis during the exploration, development, and production phases. In this case-by-case approach, items

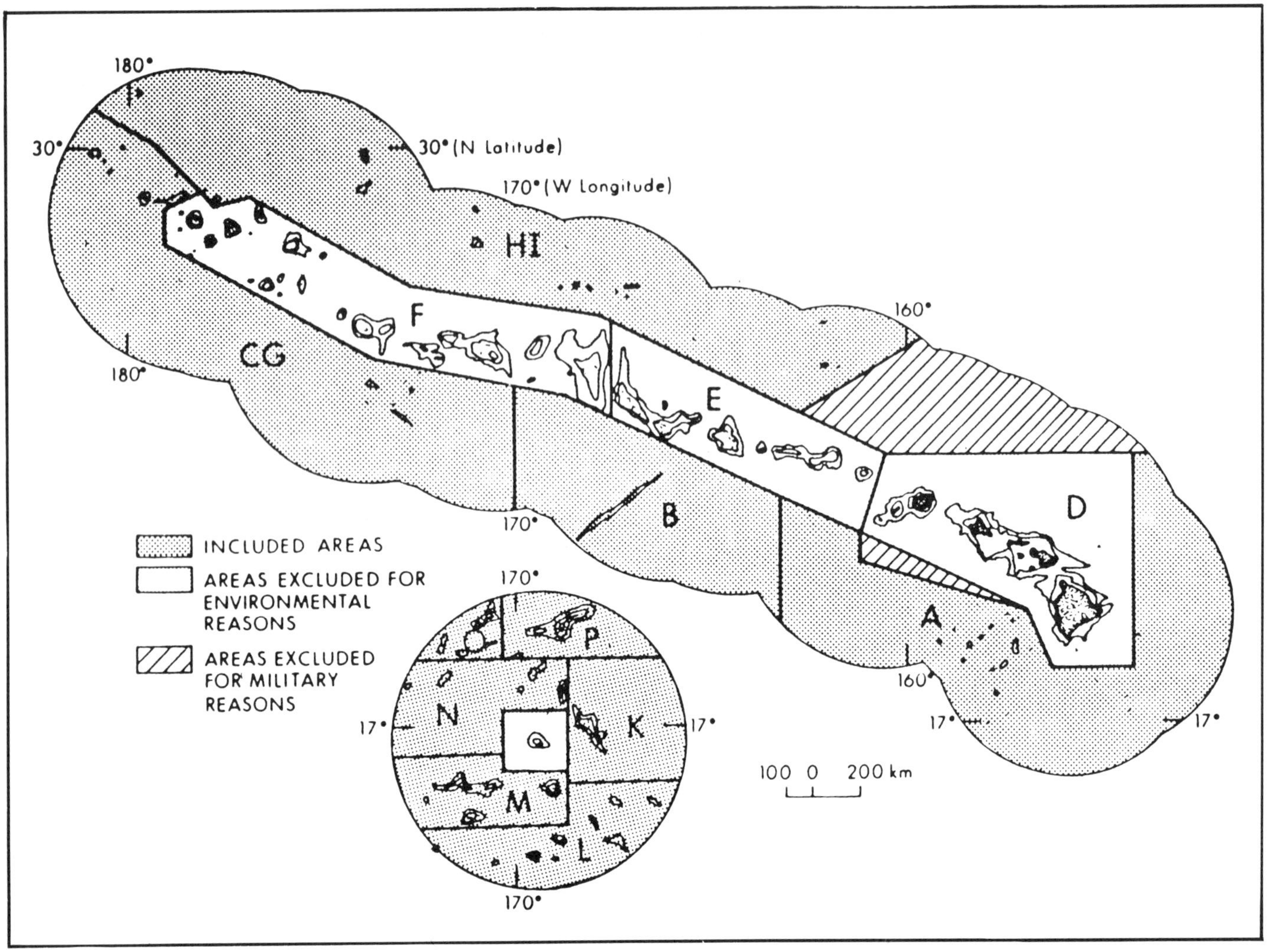


Figure 3. Hawaii and Johnston Island EEZ potential lease tracts.

common to all forms of mining and all commodities will be covered by prospecting, leasing, and/or operating regulations. Specific commodities will be covered by lease stipulation, and site-specific issues will be addressed as stipulations included in the exploration, development, and production plans approved by MMS. This regulatory process will provide maximum flexibility by allowing lease terms and conditions to have uniquely tailored leasing provisions for each offering. This flexibility is needed because there are more than 80 different mineral commodities to which regulations may apply.

Following the publication of a draft EIS, such as the Hawaii/Johnston Island DEIS, a public comment and hearing period occurs. Affected State(s), industry, the general public, and all other interested parties may submit comments on the Draft EIS. Comments will be addressed in the Final EIS for the lease offering. A "Proposed Notice of Sale" will be published in the *Federal Register*, which specifies the Secretary's proposed decisions regarding terms and conditions of the lease offering, including the size and primary term of the leases and applicable rents, royalties, stipulations, and conditions. After the proposed notice is issued, the Governors of the affected State(s) and local government will have 60 days to submit comments to the Secretary regarding the proposed lease offering. The Secretary shall accept the recommendations of the Governor and may accept the recommendations of the local government if he determines that they provide for a reasonable balance between the national interest and the well-being of the citizens of the affected State. The Secretary's decisions on the lease offering would then be specified in the "Final Notice of Sale" published in the *Federal Register* at least 30 days prior to the scheduled date of the offering at public lease sale.

Proposed regulations would call for mineral leases to be issued for primary terms of no less than 10 years and no more than 30 years. However, a mineral lease may remain in force as long as it is in production. Size of lease tracts will not be limited to the standard offshore oil and gas tract (5760 acres, or 23.3 $km^2$). Nonenergy mineral lease tracts may be considerably larger or smaller depending on the physical nature of the resource and market conditions. For example, mineral deposits which occur in thin layers, such as cobalt-rich manganese crusts in the Hawaiian EEZ, typically require considerably more acreage than a thick sand and gravel deposit located along the Atlantic coast. It is the intent of the proposed regulations to provide lease tracts of sufficient size to support economic rates of production for a standard mine life of 20 to 30+ years.

Proposed regulations will also provide for multiple leases of different minerals covering the same areas, as long as the

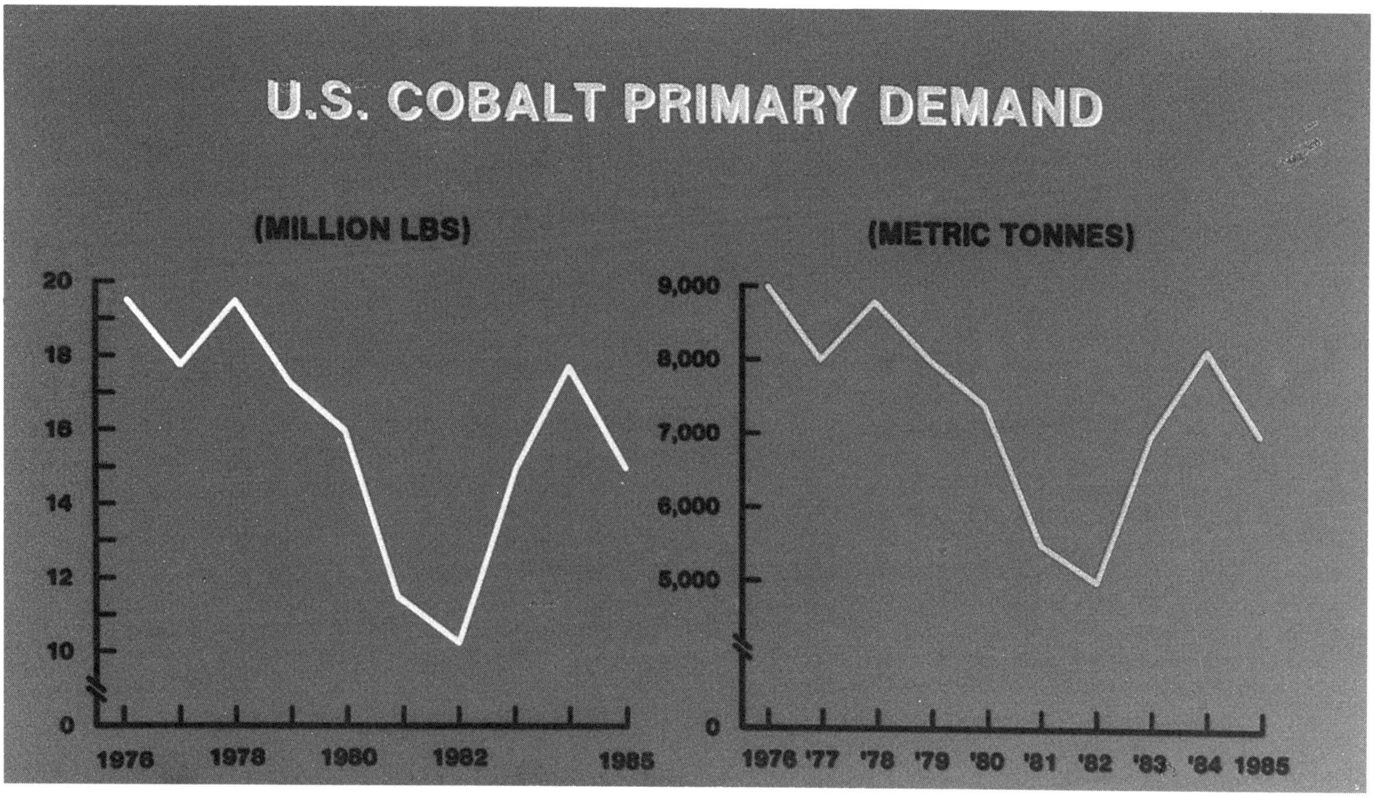


Figure 4. U.S. cobalt primary demand. (Source: MMS/OSIM and Dept. of Commerce).

operations of one lessee do not unreasonably interfere with those of another.

As shown in Figure 3, the Hawaii-Johnston Island EEZ has been tentatively separated into nine potential lease tracts—each believed to contain sufficient resources for a 20-year + economic mine life.

## AN EXAMPLE

Demand for cobalt in the U.S. has been cyclical, as reflected in Figure 4, with 1982 being the lowest in several decades. Substitutability, supply changes, and instability overseas have resulted in fluctuating U.S. demand and stockpile requirements. Nonetheless, cobalt consumption has been averaging about 16 million lbs/yr, and its use is expected to increase and diversify between now and the year 2000. In order for mining companies to warrant the tremendous capital expenditure needed for new mines (now estimated at more than $200 million U.S.) as well as facilities required, certain financial incentives will be needed:

- Minimum front-end acquisition costs
- Credit for prelease exploratory work
- Credit for postlease costs

A bonus or acquisition payment to permit exploration prior to investment is a standard practice in onshore and offshore lease sales. Several options might be included as incentives to ease the burden of those expenses prior to knowing more about the cobalt and receiving a return on investment. Crediting could be used to reduce acquisition costs by the amount owed in direct relation to investments made during the exploration phase. Consider a potential lease area of 3000 $km^2$ for one of the tracts, a manganese crust thickness of 2.5 cm, a 66% coverage, and an estimated recovery of crusts of 70%. With these potential leasing area characteristics, if a firm were to submit verifiable costs for a specific lease tract, and they totaled $1 million over a defined period prior to a lease sale, and the acquisition payment at the time of the sale was $2 million, Figure 5 shows that the remaining payment over the allowable 10-year period (because of deferred action) totals approximately $1 million since the credit figure per year has offset the other half due in those years. Thus, crediting could offset half the total acquisition costs (in our example) if DOI were to permit a 50% level of crediting.

Prelease crediting could be used to permit the investor to conduct exploration and resource evaluation studies prior to commitment of development and other major property acquisition costs. Exploration expenses incurred prior to property acquisition costs might be credited in some manner against either the acquisition costs or later rights involved in holding the property title through postlease credits. Use of this crediting concept has been shown as an incentive for mineral resource development.

Both crediting and deferred acquisition payments are currently under study in DOI, and they may be allowed prior to the initial public lease sale of nonenergy minerals, now anticipated for CY 1987. Notices prior to any lease sale, including the allowable crediting percentage(s) for the various minerals to be developed in the U.S. EEZ/OCS, will be published in the *Federal Register*.

Cobalt-rich manganese crust samples were obtained during exploration activities in the Hawaii-Johnston Island EEZ. A typical vessel for conducting such activities in the EEZ is the German ship Sonne. Japan and Russia are also active in the

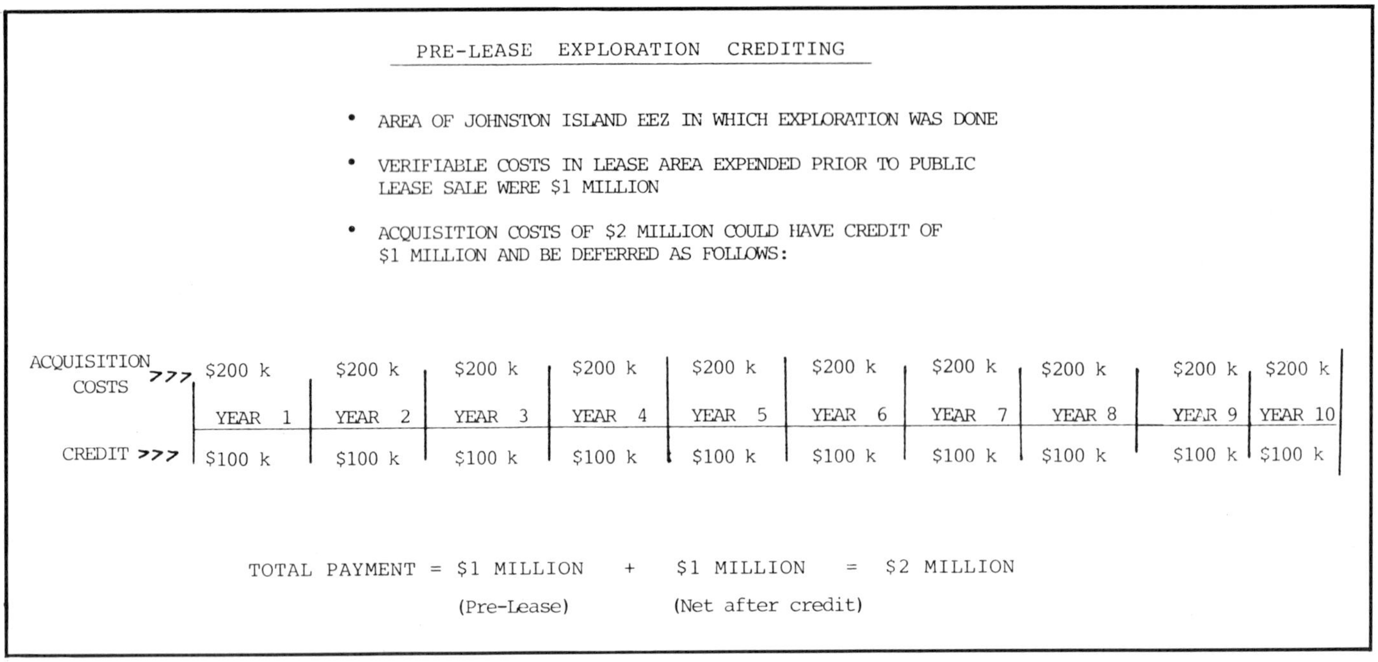


Figure 5. Example of prelease exploration crediting (Source: MMS/OSIM).

Pacific area evaluating these resources. Private firms form Germany, Great Britain and the U.S. are also evaluating various mineral resources in our EEZ.

## SUMMARY

We recognize that the demand for minerals in our economy today is essential to meet the nation's future requirements. These requirements not only include our present needs, but those new innovative demands on strategic material supplies brought on by new-found uses of outer space and the oceans. We need to highlight the following:

- Minerals are vital to our economy.
- United States is dependent on imports.
- Demand is increasing for minerals.
- Onshore mineral supply is decreasing.
- Trade deficit is growing.

The Department of the Interior is committed to making these new-found resources in our EEZ/OCS available to aid in solving these problems.

## REFERENCES


Bettenberg, W.D., 1985, The DOI and MMS—Three decades of leasing experience in the EEZ: Exclusive Economic Zone Symposium, Washington, D.C.

Fischer, David W., 1986, Returns to society from deepsea hard mineral resource development, unpublished draft.

MacGillvray, T.J., 1986, Economic and financial implications of crediting for minerals in the EEZ, MMS unpublished staff report.

Nyhart, J.D., L. Antrim, A. Capstaff, A.D. Kohler, and D. Leshaw, 1978, A cost model of deep ocean mining and associated regulatory issues: MIT Sea Grant Program, Cambridge, Massachusetts, March, 1978.

Smith, John B., 1985, Managing nonenergy marine mineral development—Genesis of a program: Ocean 85 Conference, November, 1985.

Smith, John B., and T.J. MacGillvray, 1987, Developing a regulatory framework for leasing minerals on submerged federal lands: Marine Mining Journal, Special Issue (in press).

Smith, John B., B.R. Holt, and R.G. Paul, 1985, Current status of leasing proposals for the U.S. Exclusive Economic Zone: Offshore Technology Conference, Houston, Texas, May, 1985.

U.S. Bureau of Mines, 1986, Mineral facts and problems, 1985 Edition: U.S. Department of the Interior, Washington, D.C.

U.S. Department of the Interior, 1985, Solicitor finds Interior Department has authority to issue mineral leases within EEZ Off U.S. coast: Department of the Interior News Release, Washington, D.C.

Weed, E.G.A., A. Kopec, F.G. Adinolfi, and T.J. Rowland, 1985, Nonenergy minerals of the Atlantic and Caribbean Exclusive Economic Zone: MMS unpublished staff report.

CHAPTER 52

# AUSTRALIA'S NEW DIAMOND SEARCH

**E.W.J. Tyler**
*Ashton Mining Limited*
*Melbourne, Victoria, Australia*

ABSTRACT

Diamonds have been recorded in Australia since the 19th century, but only since the 1960s has the systematic search for economic deposits of diamonds been carried out. The writer initiated his search for diamonds in 1969 and formed the exploration venture that led to the discovery of the Argyle Diamond Mine in the Kimberley district in Western Australia. This is the world's most concentrated source of naturally occurring diamonds. The mine came into full scale production December 1985.

This chapter traces the history of the discovery from the first find of a diamond in 1973, through the locating of the West Kimberley diamond pipes in 1976, to the discovery of Argyle in 1979. It discusses the changes in search techniques from the classical technique developed in South Africa, to the statistical approach using microdiamonds, which in 1985 found a further series of pipes some 900 km east of Argyle in the Northern Territory.

## INTRODUCTION

Diamonds have been recorded from mining in Australia since the 19th century and, reputedly, the first diamond was found by E.H. Hargraves in 1851 from near Bathurst in New South Wales. Diamonds were found from time to time in gold, tin, and sapphire wash, and were recovered as a by-product from these 19th century alluvial mining fields. By the end of the century, diamonds had been reported from many parts of New South Wales, and up to 1973 a total production of 207,836 carats had been reported. This represented more than 99% of Australia's total production to that time.

Elsewhere throughout Australia, diamonds had been found, but up to 1970 the total recorded in Western Australia was 70 stones—less than 20 carats. The logic of mineral exploration would have led those looking for primary diamonds to concentrate on New South Wales, but this has not been successful. It is the success of the search for diamond-bearing pipes in Northern Australia which is the subject of this chapter.

### Work of Professor R.T. Prider

During the 1930s, Professor Prider worked at Cambridge on a suite of unusual minerals which occurred in a group of rocks—leucite lamproites—found in the West Kimberley district of Western Australia.

These rocks were considered to represent Jurassic volcanism in the Fitzroy Trough (Figure 1), a graben developed along the southwestern margin of the Precambrian Kimberley Block. The presence of many volcanic vents had been reported by Fitzgerald in 1907. They rise above the surrounding plains and are conspicuous topographic features.

Prider (1960) described a number of rock types in the complex series, and was of the opinion that these leucite lamproites indicated the presence of depleted mantle peridotite below the Fitzroy Graben. He considered that such rocks might be consanguineous with kimberlites, the well-described source rock for primary diamonds.

## WORK OF TANGANYIKA HOLDINGS LIMITED

Tanganyika Holdings Limited ("Tanganyika"), a London-based subsidiary of Tanganyika Concessions Limited, a mining finance company with a Central African tradition, became involved in Australia through its association with Union Miniere du Haut Katanga, the Belgium mining and minerals processing company. Union Miniere, following the loss of its Congolese (now Zaire) assets, sought to diversify worldwide and embarked upon international programmes of exploration and acquisition. In 1969 a branch of Tanganyika was established in Australia in the wake of Union Miniere, with a list of nonconflicting activities. Essentially, Union Miniere would look for base metals, and Tanganyika would look for precious metals and diamonds.

Tanganyika carried out a number of studies on the occurrence of diamonds in Australia. I.M. Paltridge, the Western Australia Manager, located C.B. Smith and commissioned him to carry out a study of the Kimberley district in Western Australia and report on its potential for diamonds.

Smith, who had worked with De Beers in South Africa, commented favorably on the region, and the present author—a

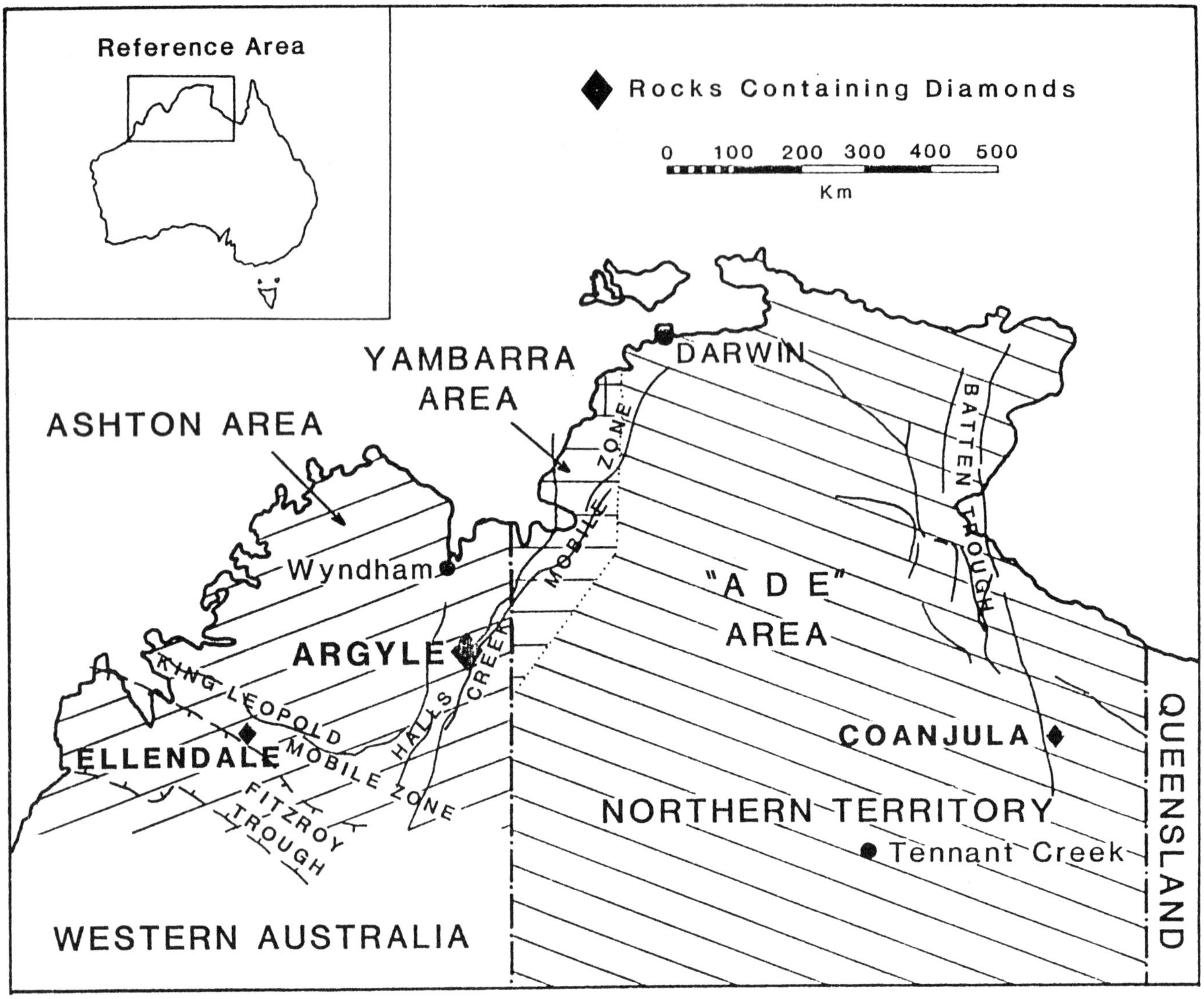


Figure 1. Map showing principal areas explored for diamonds, and the main structural elements.

one-time student of Professor Prider—requested Smith to design and cost an initial search programme.

Various parties had carried out diamond exploration in the Kimberleys during the 1960s, but their results were inconclusive. Nevertheless, during 1971, Tanganyika sought to interest other companies to join its search in the region. Sibeka, a Belgium diamond producer and a corporate affiliate of Tanganyika, expressed a willingness to join. A.O. Australia Pty. Ltd. (a subsidiary of London Tin Corporation), Jennings Mining Limited, and Northern Mining Corporation N.L., also requested participation. Each party secured a 20% interest in what became the Kalumburu Joint Venture, for a commitment to spend $20,000.

## SYSTEMATIC DIAMOND SEARCH

Because of the exceedingly low crustal abundance of diamond, the search for primary kimberlite diamond deposits has been by the location of minerals which act as pathfinders to kimberlite. Kimberlitic rocks, the host rocks for primary diamonds, are complex hybrid rocks with both mantle and crustal xenoliths. The matrix consists of partially melted upper mantle material. The rocks contain minerals originating in those high temperature and pressure regimes that occur deep in the earth. In classical practice, a trail of indicator minerals ("indicators") is followed up-drainage to the kimberlite source. The concentration of diamonds, even in a kimberlite which is payable, may be no more than 5 grams in 100 tonnes of rock. However, the classical kimberlite contains high-pressure indicators in far greater abundance, and it is the trails of these minerals which are the targets in the typical diamond search.

A kimberlite is a soft, easily decomposed rock, typically rich in olivine and containing the indicators phlogopite, garnet, pyroxene, ilmenite, and spinel. Because of its ready decomposition, it tends not to outcrop, but where intersected by drainages it becomes a ready source of added material to the fluvial load.

The indicators all have a specific gravity above 2.9 and where stream velocity is checked, these minerals will concentrate into the stream bed. The soil profiles above a kimberlite will also contain the same minerals, and where the surface is exposed to wind action, a concentration of heavy minerals containing such indicators may be located. A process called bioturbation, whereby termites and small creatures transport indicators through tens of meters of surface cover from an underlying kimberlite, will also delivery minerals to the surface to be concentrated by wind action.

The location of sites where indicators may be concentrated rather than dispersed becomes an integral part of the diamond search technique. In areas with active drainage, "trap sites" (where well-packed gravel has developed in the bed of streams), will provide a rich harvest of heavy minerals, and sampling programmes for diamonds are designed to locate and sample such sites. The objective is to produce a sample highly biased in heavy minerals, such that one grain of a telltale indicator may be present in the concentrate.

Samples taken from trap sites are usually screened in the field to remove the oversize, and may be variously jigged and panned down at the sample site or taken back to a concentrate preparation unit elsewhere in the field. The cleaned dry concentrate is then sent to a laboratory where it is separated in a heavy liquid (SG 2.9) and the heavy minerals are prepared for examination.

The heavy minerals are sized by screening and then examined on a grain-by-grain basis under a low-power binocular microscope, seeking to identify any one of the known indicators. The ease with which such an examination can be carried out is a function of the size of the particles being viewed. Grains of + 1 mm size are relatively easy to examine, but the examination of material smaller than 0.4 mm is much more arduous. Much of the concentrate from African and Australian drainages contains flood proportions of limonite from the laterite profiles, and this presents special problems for the mineral observer.

When a particular concentrate is identified as positive, that drainage is resampled on a much closer spacing, seeking to identify which particular tributary introduces the indicator, and this is then traced to its source. It may take several sorties to a positive drainage to finally locate the tributary of interest.

At the head of the drainage, a depression could mark the location of the kimberlite but, invariably, geophysics or loaming of the surface will be required to narrow the search to finality. It should be noted that, throughout the whole of this process, diamonds may not necessarily have been seen. In many instances diamond will be observed for the first time when kimberlite is processed for the extraction of the stones.

The mining of a payable diamond deposit may rest on the discovery of a kimberlite containing small quantity of very high value diamonds, or possibly one with a large quantity of low-value diamonds. At the beginning of the search there is no way of knowing if kimberlite will be discovered, far less whether it will contain diamonds, and, if it is diamondiferous, whether it will contain sufficient quantity and quality to permit mining. Diamond is, indeed, the most elusive of all minerals.

## KALUMBURU JOINT VENTURE 1972-1976

### Selected Area

Globally, it can be shown that diamond-bearing kimberlites have been emplaced in cratonic areas which have been stable for 1800 million years.

The Kimberley Craton is such an area, and the selected portion covered some 200,000 km$^2$ in the center of the block. The active regions—the Halls Creek Mobile Zone and Fitzroy Trough, bordering the Kimberley Craton on the eastern and southern margins—were outside the selected area.

### Objective

The objective was to locate a diamondiferous kimberlite through a carefully planned search programme. The primary tasks were twofold. First, to place a broad-spaced sampling network over the active drainages of the block and second, to attempt to identify one grain of a recognized indicator mineral from the samples collected. The tas was to be accomplished within a budgeted expenditure of $100,000.

The initial programme required the collection of some 1600, 8-kg gravel and loam samples from drainage and photo features described by Smith in his photogeological survey of the search area.

### Procedures

The Kimberley Craton covers an area the size of Great Britain at the north of Western Australia. It has virtually no infrastructure, and although the area is broken up into a series of cattle stations and Aboriginal Reserves, the human habitation is minimal. The stations are not highly productive, and as a result the investment in roads and tracks is minimal. Road construction is difficult because of outcropping Proterozoic quartzites, and in many instances it is only possible following stratigraphic horizons of shales and volcanics. Effectively, helicopter transport is the only available means of travel, and any sampling programme must be helicopter borne.

The use of helicopters places a high collection cost on each sample obtained. It becomes imperative, therefore, to ensure that each sample is properly taken and that maximum bias is achieved to optimise the chance of the recovery of an indicator. Sample collection becomes a specialized task and, under such a high-cost regime, sample treatment requires careful study.

It was determined that the processing of such high-cost samples would be best carried out under standard procedures capable of rigorous laboratory control. Whereas under African conditions processing to a concentrate was mainly carried out in the field, the Kalumburu Joint Venture from the outset bore the extra freight cost of moving the whole sample to the laboratory to ensure that its processing was carried out under standard conditions.

Ideally, the rate of sample examination in the laboratory had to be such that "interesting" results could be followed up before field parties moved on to other locations. This was not always possible, and in the early stages, while the means of dealing with flowsheet problems were being resolved, delays of up to a year were experienced in processing difficult samples.

The object in the diamond search laboratory is to prepare a clean, sized, heavy concentrate that can be examined on a grain-by-grain basis by keen-eyed mineral observers trained to spot the telltale minerals using a binocular microscope. The removal of extraneous material without diminishing the value of the remaining part (bearing in mind that one mineral grain

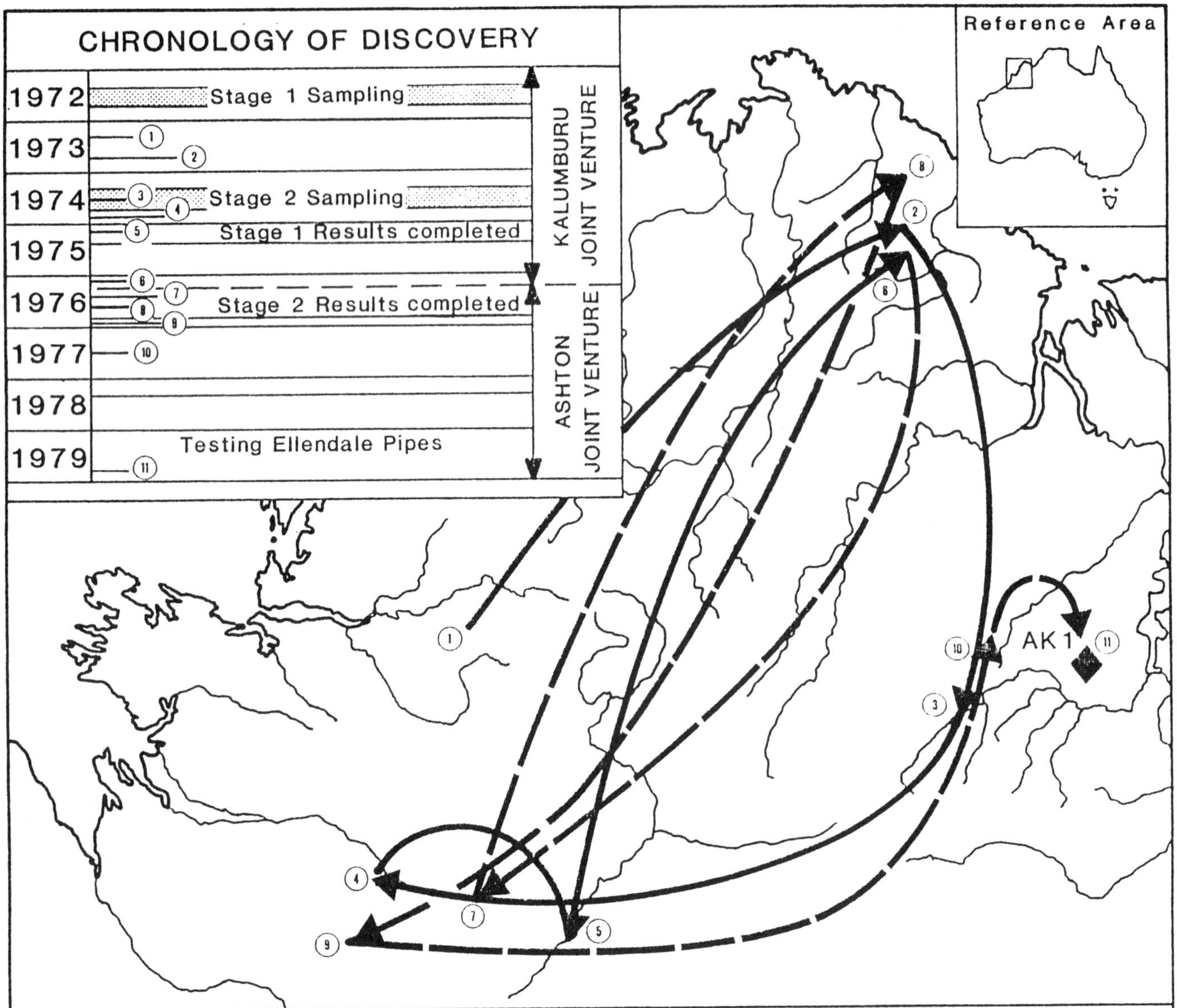


MILESTONES IN DISCOVERY

(1) APRIL 1973– Chrome diopside on Charnley Sheet
(2) AUGUST 1973– Diamond and Indicator assemblege–Pteropus Creek
(3) JUNE 1974– Picroilmenite, Pyrope, Chrome Diopside–Wilson River
(4) OCTOBER 1974– Pyrope and Chromite–Mt. Percy and Mt. Abbot
(5) JANUARY 1975– Chromite and Chrome diopside–Prairie Hill and Leopold Downs
(6) JANUARY 1976– Breccia on Pteropus Kimberlite
(7) MAY 1976– Big Spring No.1 KImberlite
(8) SEPTEMBER 1976– Skerring Kimberlite
(9) NOVEMBER 1976– Ellendale 4 Kimberlite
(10) JUNE 1977– Wilson River and Maude Creek Kimberlites
(11) OCTOBER 1979– Argyle Kimberlite (AUSTRALIA'S FIRST DIAMOND MINE)

Figure 2. Milestones in discovery of diamonds.

may be the key that unlocks the door of discovery), is a challenge for such a laboratory.

## Discoveries

The Kalumburu Joint Venture's first field season began in May 1972, and the collection of the (1600) samples was completed by August that year. The wide spacing of the samples, one sample per 125 $km^2$, and the small sample size (in the first instance only 8 kg), set special standards on sample processing to meet the objective of the recovery of one grian of an indicator mineral.

It was decided that the whole sample, after desliming and screening, would be separated in the heavy liquid tetrabromoethane ("TBE"), which has a specific gravity of 2.9. The heavy concentrate was washed, dried, and subjected to microscopic examination.

It was soon apparent that flood proportions of limonite, particularly from samples collected in areas where laterites had formed over the Carson Volcanics, gave an almost impossible task to the mineral observer. Methods for removing this limonite taxed the ingenuity of the laboratory.

## Charnley Chrome Diopside

Figure 2 shows the chronology of discovery. Sample examination began late in 1972, and in April 1973 on the Charnley 1:250,000 sheet, the first grain of chrome diopside was discovered. The chemistry of the rock-forming minerals of kimberlites is well documented, and matching diopspide analyses with those from other provinces was an essential part of the construction of the exploration data base. Table 1 shows the $Cr_2O_3$ and CaO determinations for the first chrome diopsides investigated, and compares them with two African chrome diopsides from kimberlite. They were deemed similar.

## Pteropus Creek

In August 1973, a most significant discovery was made in the laboratory. In drainage sample M109, collected by field geologist Maureen Muggeridge, a small diamond weighing 0.008 carats was discovered, and accompanying it were pyrope garnet, picroilmenite, chromite, and kimberlitic zircon. The indicators were relatively fresh, suggesting no great distance of transport, and it was concluded that a kimberlite must be in the vicinity.

Table 2 shows the analyses of the mineral grains as first calculated.

The garnets were lilac colored, and in terms of garnet endmember molecules were high in pyrope (66.0% - 70.7%); the high chrome in grains 2 and 3 shows 2.6% and 4.5% knorringite in the molecule. By comparison with kimberlitic garnets, aluminous magnesiochromites, and picroilmenite from elsewhere, the mineral suite clearly came from a diamondiferous kimberlite. The presence of chromite was to prove most important.

The sample had been collected inside a small dome in folded Proterozoic sandstones, drained by a small creek, subsequently named Pteropus Creek (after the fruit bats, its most regular inhabitants).

## Wilson River

It was not until June 1974 that the next significant site was identified. This was in the Wilson River on the southeastern margin of the craton. Here gravel sampling showed pyrope garnet, chrome diopside, and picroilmenite. Table 3 shows typical analyses of the mineral assemblage from the Wilson River sites.

The garnets were orange to orange-red, chromium-bearing titanium-rich pyrope. The freshness of the grains indicated a kimberlite source nearby.

**Table 1. Chemical composition of chrome diopside.**

| Location | $Cr_2O_3$% | CaO% |
|---|---|---|
| Kalumburu-Charnley 1 | 0.65 | 20.7 |
| Kalumburu-Charnley 2 | 0.76 | 21.2 |
| South Africa-Lys Mine | 0.40 | 20.7 |
| Zaire-Katsha Kimberlite | 0.30 | 20.7 |

**Table 2. Chemical composition of indicators from Pteropus Creek (in percent).**

| | Garnet | | | Chromite | | | Picroilmenite | |
|---|---|---|---|---|---|---|---|---|
| Grain No. | 1 | 2 | 3 | 1 | 2 | 3 | 1 | 1 |
| $SiO_2$ | 41.7 | 41.4 | 41.4 | - | - | - | nd | nd |
| $TiO_2$ | 0.2 | 0.2 | 0.1 | 0.28 | 0.17 | 0.27 | nd | nd |
| $Al_2O_3$ | 20.7 | 19.7 | 18.7 | 21.7 | 14.0 | 21.9 | nd | nd |
| $Cr_2O_3$ | 3.4 | 4.7 | 6.2 | 44.7 | 54.6 | 45.8 | nd | nd |
| $Fe_2O_3$ | 1.0 | 1.0 | 1.0 | 21.4 | 19.9 | 21.4 | nd | nd |
| FeO | 7.5 | 7.5 | 6.5 | - | - | - | nd | nd |
| MgO | 19.8 | 19.3 | 19.6 | 11.9 | 10.0 | 12.2 | 6.65 | 13.3 |
| CaO | 5.4 | 5.9 | 6.2 | - | - | - | nd | nd |
| MnO | 0.3 | 0.3 | 0.3 | - | - | - | nd | nd |
| | 100.0 | 100.0 | 100.0 | 99.98 | 98.67 | 101.57 | - | - |

Analyses by Amdel; calculations by C. B. Smith (1973).

**Table 3. Chemical composition of indicators from Wilson River (in percent).**

| | Garnet | | Picroilmenites | | Chrome Diopside |
|---|---|---|---|---|---|
| Grain No. | 1 | 2 | 1 | 2 | 1 |
| $SiO_2$ | 40.6 | 40.7 | - | - | 54.0 |
| $TiO_2$ | 0.77 | 0.94 | 53.0 | 53.1 | 0.15 |
| $Al_2O_3$ | 20.6 | 21.0 | - | - | 0.8 |
| $Cr_2O_3$ | 1.25 | 0.17 | 1.11 | 0.69 | 1.65 |
| FeO | 9.94 | 11.6 | 34.6 | 35.2 | 2.73 |
| MgO | 22.3 | 22.3 | 10.9 | 10.6 | 16.7 |
| CaO | 4.35 | 4.37 | - | - | 21.8 |
| $Na_2O$ | - | - | - | - | 1.58 |
| | 99.81 | 101.08 | 99.61 | 99.59 | 99.41 |

Analyses by Amdel.

**Table 4. Chemical composition of indicators from the South and Central Kimberleys (in percent).**

| | Mt. Abbot | | | Prairie Hill | | Central Plateau | |
|---|---|---|---|---|---|---|---|
| | Garnets | | Chrome Diopside | Chrome Diopside | Chromite | J12 Chromite | |
| Grain No. | 1 | 2 | 1 | 1 | 1 | 1 | 2 |
| $SiO_2$ | 39.8 | 41.3 | 53.4 | 51.4 | - | - | - |
| $TiO_2$ | 0.58 | 0.06 | 0.09 | 1.05 | 0.99 | 0.66 | 0.75 |
| $Al_2O_3$ | 21.4 | 19.4 | 2.37 | 0.23 | 5.93 | 20.1 | 19.8 |
| $Cr_2O_3$ | 0.59 | 5.68 | 3.06 | 0.75 | 59.3 | 44.6 | 44.6 |
| FeO | 14.2 | 7.52 | 2.20 | 2.44 | 19.8 | 21.3 | 21.6 |
| MgO | 20.6 | 19.9 | 15.2 | 17.5 | 13.7 | 12.6 | 12.1 |
| CaO | 3.40 | 4.88 | 19.0 | 23.6 | - | - | - |
| $Na_2O$ | - | - | 2.84 | 0.47 | - | - | - |
| $K_2O$ | - | - | 0.07 | nd | - | - | - |
| ZnO | - | - | - | - | 0.08 | 0.11 | 0.09 |
| | 100.57 | 98.74 | 98.23 | 97.44 | 99.80 | 99.37 | 98.94 |

Analyses by Amdel.

## Lamproites and the Southern Kimberleys

In May 1974, a more detailed sampling programme was begun in the Northeast Kimberley in the vicinity of Pteropus Creek to locate the kimberlite, but the results of the 1972 sampling programme were only slowly emerging from the laboratory.

Professor Prider's conclusions concerning the leucite lampriotes made sampling necessary both on and around these volcanic rocks, to study both the mineralogy of the rocks in situ and their outwash. Mt. Abbot had been described as a cryptovolcanic breccia, a dome of shattered sediments, stained green as a consequence of the decomposition products of some thin veining. It was presumed that at Mt. Abbot a leucite lamproite had failed to breach the surface; the sediments had shattered, but the main emplacement of lamproite was concealed and only narrow veins of lamproite had surfaced. Mt. Abbot had been described by Carter (1973) as something of an enigma, where indicator minerals had first been observed and then failed to be reported in further work.

Progress on the laboratory examination of the Stage 1 samples reached the Lennard River 1:250,000 sheet by October 1974, and this nearly completed the circular path of results which began on the Charnley sheet some 19 months earlier. In samples taken at Mt. Percy (a lamproite plug) and Mt. Abbot, pyrope garnet and spinel were reported, whereas on the Central Plateau chromites similar to those at Pteropus Creek were discovered.

By January 1975, abundant chromite and chrome diopside had been reported from the vicinity of lamproites at Prairie Hill and Old Leopold Downs. It was clear that a province with kimberlite indicator minerals was developing along the southern margin of the Kimberley Craton, but at this point no diamond had been found. Analyses of mineral grains from this southern area and from the Central Plateau are shown in Table 4.

Garnet 1 was a bright-orange garnet, a chrome-bearing titanium-rich pyrope; garnet 2 was a violet garnet, a chrome-rich pyrope. Similar mauve pyrope was found in a creek draining Mt. Percy. The chrome diopside from Mt. Abbot was considered typically kimberlitic, with high $Al_2O_3$ and $Na_2O$, whereas those from the Prairie Hill lamproites were perceived as being highly calcic and low in $Al_2O_3$ and $Na_2O$, probably a lower pressure form.

The leucite lamproites, which had been crushed and tested for indicators, had not shown chromite. However, the presence of aluminous magnesiochromite could be traced from Pteropus Creek in the Northeast Kimberley through the Central Plateau to the Southern Kimberleys and across the Fitzroy Valley to Mt. Abbot. If it did not occur in lamproites, was it pointing to widespread distribution of kimberlitets? The importance of this chromite, found with diamonds at Pteropus

Creek and other indicators elsewhere, was a closely guarded joint venture secret. It was believed that chromite would ultimately lead to a diamond pipe. It did so.

### Pteropus Breccia

The year 1975 was one in which the laboratory facilities were expanded, and earth-moving and bulk-testing equipment (heavy-media-separation plant, jigs, etc.) were purchased and made operational. This put a considerable financial burden on the Kalumburu joint venturers. In August 1975, Dr. Morelli of Sibeka was panning in the King George River when he recovered a diamond. He expressed the opinion that "Now we have the fox by the tail." The site is known as "Morelli's Fox."

In January 1976, Muggeridge, Paltridge and Smith (1976) reported that "An outcrop of a curious breccia has been discovered near Pteropus Creek at the southern end of the dome structure. ... From its field appearance, this breccia may well be vent material from a kimberlite volcano." It seemed likely that the source of the diamond and indicators in sample M109 had been found.

There was now considerable pressure from some of the joint venturers to develop an alluvial mining operation to treat the gravels in the areas where diamonds had been found. The Stage II sampling carried out in the Northeast Kimberley in 1974 had discovered a number of diamonds, and as a way to alleviate the increasing costs of exploration, the notion of cash flow appealed. The alternative was to invite a sixth party to join the exploration programme and to charge a premium for admission.

An exploration subsidiary of CRA Limited ("CRA") became a party to the Joint Venture in February 1976 and agreed to provide the next $1.6 million of exploration funding to secure a 35% interest. Tanganyika would remain Manager of the now named Ashton Joint Venture until CRA had made its full contribution, when all of Tanganyika's staff would transfer to CRA and it would then assume management.

## ASHTON JOINT VENTURE 1976–

### Objective

The primary objective of the Ashton Joint Venture was to discover a payable kimberlite.

The Kalumburu Joint Venture had discovered small diamonds in several parts of the Northeast Kimberley, yet at the time of the formation of the Ashton Joint Venture, not one diamond of commercial size had been discovered. The reason for this was a function of the sample size and the laboratory flow sheet. The laboratory studied closely material from 0.4 mm to 1.0 mm, where statistics demonstrated the optimal chance of single-grain discovery, but such diamond sizes are well below the commercial sizes required to make a mine.

A payable kimberlite must contain a series of diamond sizes much larger than anything found by the Kalumburu Joint Venture. CRA's consultant, R. Baxter-Brown, specified the discovery of a 0.25-carat stone to ensure that Kimberley diamonds were not simply confined to small sizes.

### Discoveries

A small bulk sampling plant with a trommel and Pleitz jig was established on the King George River near Morelli's Fox. This stretch of the river was judged to have the best chance of locating a 0.25-carat stone. In July 1976, only shortly after the onset of processing, a 0.4-carat stone was discovered.

### Big Spring Kimberlites

Early in 1975, attention had been focused toward the Southern Kimberleys, where sampling in Pigeon Creek showed chromite and phlogopite. At the head of this drainage (at Big Spring), near the Leopold Downs Homestead, a lampriote had been described by Derrick and Gellatly (1972). In May 1976 the first kimberlite-Big Spring No. 1 Pipe—was discovered nearby. Loaming in the oil revealed diamond and, subsequently five additional pipes were discovered. Bulk testing of the pipes in 1977 was disappointing, as all the diamonds were small.

### Skerring Kimberlite

The 1974 Stage II sampling in the lower King George River had demonstrated an abundance of picroilmenite. These results became available early in 1976, and at the site of sample IP41, an updrainage program using field recovery techniques was tried. In this instance the trail of indicators was followed up by panning gravel samples in the field. At IP41 the concentration of picroilmenite was quite easy to follow, and the trail was traced up Scumble Creek and lost in Red Herring Creek. Loaming at the headwaters revealed large pieces of ilmenite in places, and on the 17th of September 1976, Chris Smith located the kimberlite, appropriately named Skerring (Plate I). The body itself was a blow on a kimberlite dyke, about 400 m long and 75 m wide at its widest point. A large zircon was discovered which was named "The Star of the Skerring" before it was realized it was not a diamond. Bulk testing of Skerring in 1977 showed that it was not diamondiferous.

### Ellendale Kimberlites (Lamproites)

Follow-up sampling of the chromite and pyrope anomaly found near Mt. Percy led to a tributary at the headwaters of Mt. North Creek. Here, in Chris Smith's sample CS712, two diamonds were reported. Frank Hughes, CRA's geological consultant, went back to the area and, in November 1976, the first of 45 kimberlitic bodies at Ellendale was discovered.

Although this creek was draining the Ellendale No. 4 pipe (a large, albeit currently subeconomic diamond deposit), it is important to note that CS712 was the first sample in the drainage to report diamond, and (it was subsequently realized) the site was on the pipe itself. The diamond explorer must guard against the anticipation of discovery of + 0.4-mm diamonds in his search. Drainage samples right up to a significant diamond deposit can give no advance assurance of success.

Subsequent discoveries of the Ellendale, Fitzroy, and Calwynyardah pipes were substantially by aeromagnetic surveys, and the interpretation of the aeromagnetic data was well developed by Jenke (1983). The distribution of the bodies at

Ellendale showed them to be developed along basement faults in what could be called tension gashes in the Lennard Shelf. The diamondiferous bodies were far from the center of the Kimberley Craton, the classical location, and indeed their location was in the mobile zones bounding the craton.

Many of the kimberlites, which on detailed examination by the petrographers were renamed olivine lamproites, were found to be diamondiferous. However, on bulk sampling, only two—Ellendale No. 4 and Ellendale No. 9—had diamond concentrations approaching being payable. The diamonds are good quality, but the deposits have yet to be brought to production.

### Wilson River and Maude Creek Kimberlites

In 1977-1978 other companies, encouraged by the success of discovery, entered the diamond search, and exploration title covered much of the Southern Kimberleys.

The diamond exploration team's main work in this period was to secure under title all the kimberlites in the Ellendale and Fitzroy provinces. The anomalies located by aeromagnetic work had to be mapped and drilled prior to evaluation by bulk sampling, which began on August 1978.

The Halls Creek Mobile Zone was outside the search area, although the 1974 Wilson River discoveries were on its fringe. During 1977, follow-up work on the Wilson River located kimberlites at Duck Creek near the original sites, and prospecting to the north located the Maude Creek kimberlite. The existence of these bodies and the Ellendale and other Southern Kimberley discoveries in the mobile zone led to a decision to explore systematically the Halls Creek Mobile Zone.

### Argyle AK1

During the time evaluation work was taking place in the Southern Kimberleys, exploration in the Halls Creek Mobile Zone had begun, and from a drainage sample collected by Maureen Muggeridge and her team in Smoke Creek, the laboratory reported two diamonds in August 1979. This was the key sample which led to the Argyle Diamond Mine.

Follow-up sampling toward the headwaters of Smoke Creek showed a trail of diamonds leading to a Devonian conglomerate, and initially it was believed that the diamonds were being reworked from these sediments—there were no classical indicators. However, the diamond finds, in ever increasing abundance, were seen to extend above these Devonian rocks. On the 2nd of October 1979, Frank Hughes with Warren Atkinson (managing the Joint Venture) walked onto and identified AK1, the kimerlitic body which was destined to become the world's largest diamond mine. The Argyle Diamond Mine, in its geological setting, is located almost on the eastern boundary of the Halls Creek Mobile Zone. The diamonds occur in the crater of an ancient volcano, which was emplaced some 1200 million years ago.

Detailed descriptions of the kimberlitic rocks discovered in the period 1976-1984 can be found in Atkinson, Hughes, and Smith (1984), "A Review of the Kimberlitic Rocks of Western Australia."

## YAMBARRA JOINT VENTURE - (NORTHERN TERRITORY)

### Objectives

The search area of both the Kalumburu and Ashton Joint Ventures was confined to Western Australia, and when the significance of the mobile zones as search areas was defined in 1977, AO proposed to CRA that together they should explore the northeastern extension of the Halls Creek Mobile Zone into the Northern Territory (Figure 1). AO was now part of Malaysia Mining Corporation BHD ("MMC"), which had acquired the whole of the share capital of the London Tin Corporation. In addition, at the instigation of AO, a new Australian company, Ashton Mining Limited ("Ashton") was in the throes of creation, in which MMC would have a majority interest.

Ashton was to become a significant company in Australia. Not only is it one of the world's largest diamond producers, but it is also a prominent explorer for diamonds and precious metals.

However, in 1977, AO and CRA decided against joint exploration in the Northern Territory. CRA went elsewhere and discovered kimberlites at Wandagee in northwest Western Australia; and it became Ashton, along with Aberfoyle Limited, who created the Yambarra Joint Venture for the defined objective to explore the northeastern extension of the mobile zone for diamonds.

### Work Programme

During September to November 1978, a field programme of gravel sampling was undertaken over a series of exploration licences held in the Joint Venture area (Figure 1). Although chromite and diamonds were discovered, difficulty was experienced in securing accesss into some Aboriginal Reserves, so that much of this area is yet to be assessed.

## AUSTRALIAN DIAMOND EXPLORATION JOINT VENTURE (Australia-wide)

### Participants

One of the consequences of the failure of AO and CRA to agree on exploration in the Northern Territory was that Ashton had to build an additional diamond laboratory to carry out determinative work on both the Yambarra Joint Venture samples, and on samples collected by Ashton in Western Australia outside the Ashton Joint Venture area. The provision of a new facility permitted Ashton to seek other participants to join in the search for diamonds elsewhere in Australia. Aberfoyle Limited and A.O.G. Minerals Limited joined the search and a work programme began late in 1979.

## Objectives

The Joint Venture under the direction of Herman Zeissink sought to investigate sites in Australia where diamond had been reported, but where few details were available. Also, it was proposed to examine areas where regional geology indicated a possible crustal setting for kimberlites.

The presence of chromites leading away from the Yambarra Joint Venture area prompted further investigation into the Northern Territory to the east of the Argy le Diamond Mine. This search in the Northern Territory became the main thrust of the Australian Diamond Exploration Joint Venture ("ADE").

## Discoveries

Whereas the exploration in the Kimberleys in Western Australia had occurred on land forms which were well dissected, the Northern Territory is much more sparsely drained. Drainages that exist are the low-energy type, with limited coarse material in transport. This presented new problems, for the gravel sampling technique relies on well-packed gravel being available.

These drainages move mostly sand-size particles (less than 1 mm), and the gullies are fed by sheet wash from the interfluves. It quickly became apparent that an exploration technique to locate indicators that relied on examination of particles $-1$ $+0.5$ mm would be unsatisfactory in this instance. Most of the material recovered was $-0.5$ mm, with relatively small amounts of coarser material, and a method had to be developed to deal with this fine material in the laboratory.

## Western Finds

By the middle of 1980, diamond had been recovered in the western part of the Northern Territory, along with some chromite. Attempts to track the diamonds to source were fruitless; it appeared that the diamonds were moving along the top of old land surfaces, and that they could be recovered at places where there was a change of slope.

During 1981, the activity moved farther east, where diamonds with pyrope garnets were located. This occurrence seemed to suggest a nearby source, but again the trail was lost because of difficult terrain with no drainages.

In 1982 the search again moved westward toward the Western Australia border, where diamonds were located in what were finally recognized as palaeochannels. Trends in microdiamond ($-0.4$ mm) abundance, which the laboratory was now recovering from the samples, started to be recognized. In November 1982, Zeissink reported: "There is no doubt in my mind that a major kimberlite field comprising several pipes will be discovered."

## Coanjula Pipes

By March 1983, modifications to the distribution of diamonds by wind action was advocated, and extensive sampling occurred to examine the effects of such a transport mechanism. The notion that diamonds were being carried by rivers into internal lakes, and subsequently deflated into dunes, and finally degraded by wadis, was tested. Again, Zeissink predicted: "By June a kimberlite source will be within reach or microdiamonds will be shown to be ubiquitous like chromite."

The collation of 1980-1983 sampling results into a comprehensive study of the frequency of microdiamond discovery in the Northern Territory is shown in Figure 3. A distinct pattern is seen, with some parts of the sampled area showing a great abundance of microdiamonds. This statistical approach showed that the place of highest microdiamond abundance lies in a 170-km-long zone extending from Coanjula Creek to Brunette Downs. Diamond must be the best indicator for diamond, and this anomaly seemed so striking that in September of 1983, an airborne magnetic survey was carried out to cover its full extent.

When the results became available toward the end of that year, 40 magnetic anomalies with characteristics suggestive of kimberlites were shown to underlie black soil at the northeastern end of the diamond anomaly. It appeared as though diamonds whose source was in these magnetic bodies had been carried by inland drainage southwestward toward Brunette Downs and into Lake Sylvester. They had then been redistributed along drainage both to the northwest (old course of the Victoria River?) and southeast and, perhaps ultimately, widely distributed by wind action related to a changing lake which covered the area to the north and east of Tennant Creek.

By April 1984, the available sample results showed that diamond abundance in Queensland, although not as high as at Coanjula, were sufficiently encouraging as to require further investigation. The Queensland diamond search was placed in a separate investigation—the Barkly Joint Venture.

With the discovery of the magnetic anomalies in an area of high microdiamond abundance, the three parties to ADE sought to introduce another party to assist with financing further exploration. In November 1984, Australian Diamond Exploration N.L. ("ADEX") made a public offering of shares to enable it to participate in the acquisition of a 25% interest in ADE.

## Rock Types

Early in 1985, a programme of drilling these anomalies began. A series of 17 diatremes has been discovered. The rocks are unusual; they are not classical kimberlites but are best described as kimberlitic, following Smith (1984). They show evidence of extensive metasomatism with carbon dioxide, water and potash, and pose problems for the petrographer and mineragrapher because of the extent of this alteration. The rocks are highly contaminated with crustal xenoliths and xenocrysts, and whole-rock geochemistry is highly variable. These varied rocks make a basic to ultrabasic potassic, highly explosive, volcaniclastic series. Key constituents are two generations of olivine, phlogopite, pyroxene, magnetite, and an alkali feldspar set in a much altered groundmass. It has been suggested by P.H. Nixon (1985) that the rocks are lamproitic and appear to have differentiated into basic and ultrabasic types before being invaded and mixed with trachytic magmas—the product of metasomatism (fennitisation) associated with a carbonatite magma. Plugs of unusual chrome-bearing gabbroic rocks and a metapyroxenite (komatiite) have also been intersected in drilling.

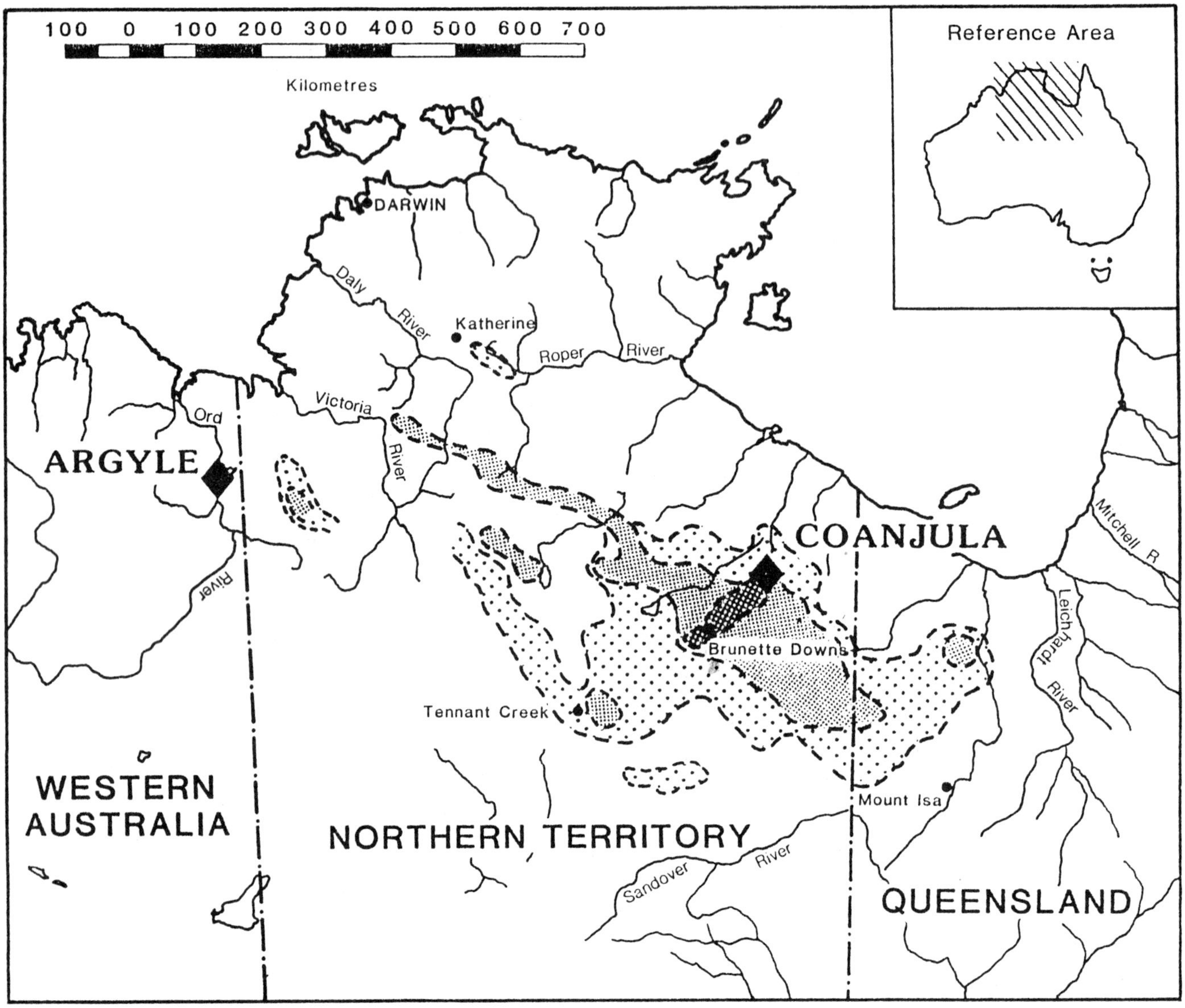


Figure 3. Map showing regional distribution of microdiamonds in the Northern Territory and part of Queensland.

Table 5 shows the range of compositions of the rocks at Coanjula, and Table 6 the contents of selected trace elements.

The trace element assembly shown in the first four rock types presented is similar to the kimberlitic rocks described elsewhere in Australia. Note the lower trace element content in rocks 224 and 348.

## Mineral Assemblage

The preparation of crushed rock concentrate has produced small numbers of mineral grains suitable for analysis under the microprobe. Generally the metasomatic alteration has destroyed the integrity of the mineral grains, and this has made determinative work difficult.

Olivine, $Fo_{87}$-$Fo_{93}$, and phlogopite Mg/(Mg + Fe) 0.47-0.91 are found in many of the rocks. The phlogopite contains 3-8 wt% $TiO_2$, 14-18 wt% $Al_2O_3$, and as much as 1.25 wt% $Cr_2O_3$. Picroilmenite with 2-9 wt% MgO and rutile and anatase are the main titanium minerals. Garnets, which are rare, represent Groups 5 and 9 of the Dawson and Stevens (1975) classification. Pyroxenes include clinopyroxene, 0.3-1.8 wt% $Cr_2O_3$, O-1.8 wt% $Na_2O$; and orthopyroxene, 2.1-14.5 wt% $Al_2O_3$ and as much as 0.9 wt% $Cr_2O_3$. The spinels are represented by aluminous magnesiochromite, typically 24 wt% $Al_2O_3$, 14 wt% MgO, and 42 wt% $Cr_2O_3$ with 0-1.6% $TiO_2$, with magnetite and titanomagnetite. Megacrysts of a titaniferous amphibole, kaersutite, were reported in some of the rocks. Rarely moissanite is found.

Some typical analyses of mineral grains recovered from Coanjula rock concentrates are shown in Table 7.

## Diamond Source

Although the quantities of material treated from the diatremes has been small at the time of this writing, microdiamonds have not been recovered from these rocks. The orthopyroxenes and clinopyroxenes, if they were in equilibrium, plot in the diamond stability field, and diamonds are abundant in the soils covering the area.

Recent drilling has disclosed abundant microdiamonds near

**Table 5. Chemical analysis of whole rocks from Coanjula (in percent).**

| Drill Hole<br>Rock No. | CJ16<br>119 | CJ8<br>43 | CJ12<br>92 | CJ54<br>345 | CJ54<br>348 | MD9<br>224 |
|---|---|---|---|---|---|---|
| $SiO_2$ | 30.00 | 40.00 | 44.20 | 53.20 | 47.7 | 44.70 |
| $TiO_2$ | 2.60 | 3.00 | 2.95 | 3.05 | 0.30 | 0.26 |
| $Al_2O_3$ | 10.30 | 13.00 | 10.50 | 8.30 | 8.65 | 6.55 |
| $Fe_2O_3$ | 12.50 | 3.40 | 5.35 | 18.50 | 2.80 | 2.95 |
| FeO | 6.85 | 8.80 | 8.20 | 0.25 | 7.05 | 8.15 |
| MnO | 0.18 | 0.09 | 0.10 | 0.06 | 0.04 | 0.04 |
| MgO | 19.90 | 13.40 | 12.40 | 2.50 | 19.70 | 26.40 |
| CaO | 5.25 | 3.00 | 7.05 | 0.59 | 2.00 | 4.10 |
| $Na_2O$ | 0.23 | 1.19 | 1.69 | 0.09 | 0.26 | 0.10 |
| $K_2O$ | 1.25 | 0.77 | 2.25 | 0.58 | 1.82 | 0.01 |
| $P_2O_5$ | 0.44 | 0.53 | 0.54 | 0.22 | 0.10 | 0.002 |
| $H_2O+$ | 4.20 | 6.50 | 3.60 | 6.50 | 5.65 | 6.90 |
| $H_2O-$ | 4.00 | 3.30 | 1.30 | 6.00 | 2.10 | 0.10 |
| $CO_2$ | 6.00 | 1.80 | 1.50 | 0.09 | 0.07 | 0.25 |
| Total | 103.70 | 98.78 | 101.63 | 99.93 | 98.24 | 100.51 |
| S | 0.008 | 0.002 | 0.088 | 0.005 | 0.005 | 0.028 |
| F | 0.140 | 0.130 | 0.100 | 0.110 | 0.040 | 0.070 |

Analyses by Amdel.

Rock No.

119 Highly altered (olivine-phyric) volcanic.
43 Lamproite-related "kimberlite."
92 Altered megacrystic volcanic with mantle xenoliths.
345 Chrome-rich, very-fine-grained porphyritic volcanic.
348 Altered syenogabbro with chromite, amphibole, and phlogopite.
224 Metapyroxenite (komatiite).

**Table 6. Contents of selected trace elements of rocks from Coanjula.**

| Sample No. | Ba | Li | Rb | Sr | Th | U | Zr | Nb | Y | La | Ce | Sc | V | Cr | Co | Ni | Mn |
|---|---|---|---|---|---|---|---|---|---|---|---|---|---|---|---|---|---|
| 119 | 495 | 140 | 18 | 150 | 4 | 4 | 250 | 75 | 40 | 60 | 160 | 20 | 290 | 315 | 50 | 205 | 1400 |
| 43 | 1360 | 130 | 16 | 315 | 6 | 4 | 250 | 75 | 35 | 60 | 100 | 20 | 330 | 745 | 65 | 295 | 695 |
| 92 | 715 | 170 | 78 | 480 | 8 | 8 | 285 | 70 | 35 | 40 | 120 | 20 | 210 | 475 | 60 | 210 | 805 |
| 345 | 550 | 7 | 150 | 78 | 4 | 15 | 280 | 100 | 380 | 1340 | 1020 | 30 | 5 | 1280 | 70 | 500 | 495 |
| 348 | 290 | 30 | 64 | 30 | 4 | 4 | 75 | 4 | 8 | 20 | 40 | 20 | 85 | 1860 | 45 | 365 | 315 |
| 224 | 10 | 4 | 2 | 30 | 4 | 4 | 45 | 4 | 6 | 20 | 20 | 20 | 65 | 865 | 75 | 1000 | 305 |

Analyses by Amdel.

**Table 7. Contents of selected mineral grains (in percent).**

| Grain No. | Phlogopite<br>908(3) | Ilmenite<br>908(1) | Garnet<br>908(1) | Cpx<br>1310(1) | Opx<br>1310(2) | Chromite<br>857 |
|---|---|---|---|---|---|---|
| $SiO_2$ | 38.78 | 0.22 | 41.65 | 52.00 | 56.93 | - |
| $TiO_2$ | 6.03 | 49.05 | - | 0.41 | - | 0.81 |
| $Al_2O_3$ | 16.39 | 1.02 | 23.32 | 5.21 | 2.08 | 23.15 |
| $Cr_2O_3$ | 0.26 | 0.30 | 0.24 | 1.14 | 0.68 | 42.80 |
| FeO | 9.76 | 42.37 | 11.B5 | 5.58 | 5.11 | 18.36 |
| MgO | 17.55 | 6.26 | 17.34 | 18.50 | 34.13 | 13.99 |
| CaO | 0.53 | 0.17 | 5.23 | 16.15 | 0.91 | 0.13 |
| $K_2O$ | 9.99 | - | - | 0.15 | - | - |
| $Na_2O$ | 0.73 | - | - | 0.68 | - | 0.59 |
| MnO | - | 0.34 | 3.37 | - | 0.15 | - |
| | 100.02 | 99.73 | 100.00 | 99.82 | 99.99 | 99.83 |

the diatremes in Proterozoic sediments in the Batten Trough (Figure 1). Very high microdiamond concentrations are developed across a zone 200 m thick. The diamond morphologies do not suggest that the diamonds are detrital in the sediments and a primary origin is inferred, but the mode of introduction of the diamonds has yet to be identified. The fact of the setting in a zone of crustal disturbance similar to Ellendale and Argyle is perhaps significant.

## CONCLUSION

Even after 17 years successfully searching for diamonds in Australia, it must be said that the search is in its infancy. There are many environments in Australia where existing search tools are inadequate and new techniques must be developed. The rush to explore in the north has by now abated, leaving only those with a stake in the diamond world to carry on

Plate I. On Skerring Pipe.

Plate II. Argyle Diamond Mine—March 1986.

developing their exploration methods. Many have fallen by the wayside.

It is interesting to speculate whether the Kalumburu Joint Venture would have survived if, instead of processing the samples in a clockwise manner around the Kimberleys, the direction had been counterclockwise; and if at the outset, the Southern and Eastern Kimberleys had been shown to contain only indicators, but no diamonds. It was of paramount importance that sample M109, showing a diamond and chromite—in addition to indicators—was found at the start of the laboratory processing, or the participants could easily have lost interest.

Although all exploration carries a strong element of luck, it is said that you can create your own luck by diligence. The standardization of laboratory procedures to ensure that mineral recovery is optimal on the one hand, and a view on the significance of results that is not too rigid, combined to make exploration successful. Wilfred Jones, who has controlled the laboratories for most of the period, has always insisted "we must be flexible" and this flexibility permitted the changes in method from the classical to the microdiamond technique.

Persistence brings its own reward, and the progress back and forth across the Kimberleys, when each time a little more was added to the data bank, was the persistence which finally led to Argyle. This $A465 million development (Plate II), which began operations in December 1985, is the crown of success thus far, but its production is substantially low-value diamonds.

The task remains to find a diamond mine to rival those of Southern Africa in terms of quality. Perhaps it will be found in the Northern Territory at Coanjula, or in Queensland where in recent weeks the Barkly Joint Venture has found small but striking gemstones. It is certain that the search will go on to find more of this rare and beautiful stone.

## ACKNOWLEDGMENTS

The writer wishes to thank all those geoscientists who have been involved with him in the search for this most elusive of minerals over the past 17 years. This paper represents an account of their efforts.

## REFERENCES CITED

Atkinson, W.J., F.E. Hughes, and C.B. Smith, 1984, A review of the kimberlitic rocks of Western Australia, *in* J. Kornprobst, Kimberlites. 1: Kimberlites and related rocks: Proceedings of the Third International Kimberlite Conference, Elsevier, Amsterdam, p. 195-224.

Carter, J.D., 1973, Diamond exploration in Western Australia: Western Australia Geological Survey Annual Report, p. 73-79.

Dawson, J.B., and W.E. Stephens, 1975, Statistical classification of garnets from kimberlite and associated xenoliths: Journal of the Geological Society of Australia, v. 83, p. 589-607.

Derrick, G.M., and D.C. Gellatly, 1972, New leucite lamproites from the West Kimberley, Western Australia: B.M.R. Australian Bulletin, v. 125, p. 103-119.

Jenke, G., 1983, The role of geophysics in the discovery of the Ellendale and Fitzroy Kimberlites: Third Biennial Conference, Australian Society of Exploration Geophysicists.

Muggeridge, M.T., I.M. Paltridge, and C.B. Smith, 1976, Progress Report No. 13, Kalumburu Joint Venture: Private Report, Tanganyika Holdings Limited.

Nixon, P.H., 1985, Coanjula diamonds—An overview: Private Report, Ashton Mining Limited.

Prider, R.T., 1960, The leucite lamproites of the Fitzroy Basin, Western Australia: Journal of the Geological Society of Australia, v. 6, p. 71-118.

Smith, C.B., 1973, Progress Report No. 5, Kalumburu Joint Venture: Private Report, Tanganyika Holdings Limited.

Smith, C.B., 1984, What is a kimberlite?, *in* Glover and Harris, eds., Kimberlite occurrence and origin: A basis for conceptual models in exploration: Seminar organized by the Department of Geology, University of Western Australia, in cooperation with University Extension, University of Western Australia, p. 1-18.

# MINERALS ABSTRACTS

COBBING, E. J., D. I. J. MALLICK, P. E. J. PITFIELD, D. P. F. DARBYSHIRE, and L. H. TEOH, British Geological Survey, Keyworth, England, United Kingdom, and G. S. M. IPOH, Pepak, Malaysia

## Granites of Southeast Asian Tin Belt

The objective of this study is to establish the essential granite geology of the southeast Asian tin belt, distinguishing plutons within batholiths and characterizing them by their component units, field relationships, lithology, texture, petrology, geochemistry, geochronology, and mineral potential. To date, approximately 180 plutons have been recognized and partly or entirely described in the above terms. In addition, four granite provinces have been delineated, each with its own distinctive mineralization pattern, as follows: (1) the Main Range province, endogenous greisen-bordered vein swarms of cassiterite and wolframite; (2) Eastern province, magnetite-cassiterite skarns ± base metal sulfides with antimony in Thailand; (3) Western (peninsular Thailand–Burma) province, endogenous greisen-bordered vein swarms and pegmatites of cassiterite and wolframite; and (4) North Thailand Migmatitic province, endogenous vein and skarn replacement scheelite and fluorite deposits with some tin and local antimony.

In all provinces, but particularly in the Main Range, granitoids designated as two-phase variants have been recognized where xenocrysts and xenoliths of coarse, primary texture granite are enclosed in and corroded by an invasive, equigranular quartzo-feldspathic matrix. These rocks form an essential part of the granite sequence in all provinces and have probably resulted from the infiltration and disruption of the host granite by late-stage magmatic fluids.

Whole-rock geochemistry from peninsular Malaysia shows that the granites from the Main Range and Eastern provinces comprise two contrasted suites, which correspond approximately to the I- and S-type categories advocated by Chappell and White. In addition, individual plutons within batholiths in the two provinces have distinct geochemical parameters. Variation diagrams of plutons having the intrusive sequence primary texture granite–two-phase granite–microgranite show linear trends with increasing $SiO_2$, $Na_2O$, Rb, W, Sn, and U, and decreasing Sr, Ba, Th, and all other major elements.

CRONAN, D. S., Marine Mineral Resources Programme, Imperial College, London, England, United Kingdom

## Potentially Economic Manganese Nodule Deposits in Western Equatorial Pacific

Manganese nodules of principal mining company interest are located in the so-called Clarion-Clipperton zone in the northeastern equatorial Pacific where the following environmental conditions prevail: (1) biological productivity in the surface waters elevated to 50-100 grams of carbon per square meter per year ($g/m^2$/year); (2) sea-floor depth near or below the calcium carbonate compensation depth (CCD); and (3) an absence of turbidite sedimentation.

Relevant data for the western equatorial Pacific indicate that these conditions are fulfilled in parts of this region also. First, biological productivity is sufficiently elevated to promote high-grade nodule formation more or less symmetrically across the equator in a wedge widening from the equator near the Gilbert Islands to more than lat. 10° N at the Line Islands. Second, much of this high productivity area, especially in the west, is below the CCD. Third, away from the island groups, turbidite sedimentation should be minimal. These observations indicate that the central Pacific basin should contain areas where conditions are conducive to the formation of potentially economic nodules. Parts of these areas fall within the Exclusive Economic Zones of island nations of the region.

Analysis of the limited available nodule data for the western equatorial Pacific confirms that high-grade and occasionally abundant nodule deposits are present in areas predicted by the above criteria. Therefore, areas of potential prospectivity can be defined for the region, which should serve as a basis for future prospecting there for manganese nodules.

ERICKSEN, GEORGE E., and ROBERT L. SMITH, U.S. Geological Survey, Reston, VA, V. RAUL EYZAGUIRRE, Empresa Minera del Centro del Peru, Lima, Peru, FERNANDO URQUIDI B., American Embassy, La Paz, Bolivia, and RAUL SALAS O., Servicio Nacional de Geología y Minería, Santiago, Chile

## Neogene-Quaternary Volcanism and Mineralization in Central Andes

The Neogene-Quaternary volcanic province in the central Andean region of southern Peru, western Bolivia, northern Chile, and northwestern Argentina contains one of the world's greatest accumulations of silicic ash-flow tuffs and andesitic stratovolcanoes, both of which are attractive targets for mineral deposits. The region is a major mineral province in which many deposits of copper, lead, zinc, antimony, bismuth, tin, tungsten, silver, and gold are older than the widespread volcanic rocks. Although the volcanic rocks generally represent postmineralization cover, they contain additional mineral deposits, including: (1) magnetite flows, lead-zinc-silver, copper, manganese, and sulfur on the flanks and tops of eroded volcanoes; (2) epithermal silver and gold associated with subvolcanic intrusions and thermal springs; (3) wood-tin in ash-flow tuffs and lava flows; and (4) supergene uranium in ash-flow tuffs. The volcanic rocks contain abundant water-soluble salts and are the chief sources of the widespread saline deposits of this region, which include major resources of lithium and boron.

Our ongoing studies show fundamental differences in the volcanic rocks of the central Andes, and indicate that recently discovered calderas and deeply eroded stratovolcanoes may contain buried mineral deposits. For example, the igneous rocks in the Bolivian tin province are distinctly different geochemically from those in the Chilean copper province. The Bolivian volcanic suites are dominantly quartz latites and low-silica rhyolites, whereas the Chilean suites are dominantly calc-alkaline dacites. The Bolivian rocks are higher in virtually all lithophilic elements and show a moderate to strong peraluminous trend culminating in a rare tin-rich two-mica tuff that may be the extrusive equivalent of a tin-bearing two-mica granite.

FOUQUET, YVES, Ifremer Centre de Brest, Brest, France

## Polymetallic Sulfide on East Pacific Rise Between Latitude 21° N and 21° S

Sulfide deposits and hydrothermal activity have been studied by Ifremer along the East Pacific Rise during numerous surface and diving cruises between 1978 and 1985 in several areas: lat. 21° N, 13° N, 17° 30′ S, 18° 30′ S, 20° S, and 21° 30′ S. Although the sulfide mineralogy and geochemistry are similar in these zones, the proportion of the different paragenetic associations is variable. Deposits occur in two main locations: on the ridge or on seamounts within a few kilometers of the ridge axis. The largest deposits are localized on seamounts where hydrothermal activity is likely to be focused in space and time. In contrast, the small deposits on the axis are scattered along several kilometers. The most extensive deposit on the East Pacific Rise occurs at lat. 13° N, on the flank of a seamount 6 km off the axis. This single sulfide lense is estimated to be 800 m long, 400 m wide, 10 m thick in places, and to contain between 2 and 5 million MT of sulfide. However, the total mass of more than 100 small deposits on the axis at lat. 13° N does not exceed 100,000 MT. The size of the largest sea-floor deposits is similar to that of single lenses that are mined in fossil volcanogenic massive orebodies on land, such as Cyprus type and Kuroko type. An important difference between fossil deposits on land and present-day sea-floor deposits is that the latter are still growing. Thus, sulfides deposits on the East Pacific Rise are at different stages of maturity depending on their position in the volcano-tectonic cycle, and maturity, paragenesis, and geochemistry of deposits are related.

FRANKLIN, JAMES M., Geological Survey of Canada, Ottawa, Ontario, Canada

## Modern Sea-Floor Hydrothermal Deposits: Their Relationship to Preserved Deposits

Four types of massive sulfide (VMS) deposits are in active spreading ridges. (1) The smallest (a few tonnes in zinc-rich isolated chimneys and small mounds), in moderately volcanically active ridge crests, are confined to the centers of axial valley(s). (2) Larger deposits (1 million MT of cupriferous pyrite) occur on marginal faults to axial valleys, in volcanically active rifts. (3) Large deposits (2 million MT of cupriferous pyrite) occur on faulted off-axis seamounts. (4) Very large deposits in sediment-filled failed-rift valleys form mounds up to 1 km in diameter, beneath a thin pelite cap.

Studies of these deposits clarify numerous characteristics of VMS deposits. (1) Deposits "grow" and are compositionally refined by highly efficient precipitation within the mound through mixing of cold seawater with hot metalliferous fluid. (2) The amount of associated alteration is partly related to the volume of cold seawater refluxing through the immediate footwall. (3) Alteration composition is controlled by the amount of refluxing cold seawater and the degree of phase separation in the hydrothermal fluid. At low pressure, the footwall is carbonated by a carbon dioxide-rich phase that separates and moves ahead of the metalliferous residual fluid. (4) Generation of a 400°C fluid within the substratum induces a broad semiconformable alteration.

A subvolcanic intrusion, sufficiently large to sustain a hydrothermal system, is a necessary heat source; the size and abundance of deposits are functions of the size and shape of this intrusion. Also, faults that penetrate to near the magma zone, and substrata that are relatively impermeable are necessary.

HALLBERG, J. A., Division of Mineralogy, CSIRO, Wembley, Australia, and B. FEHLBERG, Valient Consolidated Ltd., Belmont, Australia

### Tectonic Control of Archean Epigenetic Gold Mineralization in Kalgoorlie-Menzies-Leonora Gold Corridor North of Kalgoorlie, Western Australia

Major gold deposits in the Kalgoorlie-Menzies-Leonora "gold corridor" lie within or marginal to an early rift that forms part of a complex, linear zone of crustal extension separating two volcano-sedimentary basins in the Eastern Goldfields province of the Archean Yilgarn block, Western Australia. Rifting was accompanied by ultramafic, mafic, and felsic igneous activity, and by the deposition of a distinctive suite of sedimentary rocks derived primarily from felsic volcanic source areas flanking the rift. Metamorphosed acid hydrothermal alteration assemblages rich in sulfate are found in ultramafic extrusive rocks at the rift margin near several of the largest mining centers; the alteration is similar to that associated with modern geothermal systems.

The early rift zone is a favored site for epigenetic gold mineralization due to a combination of factors including: the establishment of geothermal systems at rift margins; site preparation during penetrative, ductile deformation and dynamic-style metamorphism; accessibility to metasomatizing fluids during periodic reactivation; and the presence of multiple magma sources during prolonged tectonism.

The recognition of an extensional tectonic setting for gold deposits in the Kalgoorlie-Menzies-Leonora gold corridor, and the documentation of the lithotectonic features that characterize this environment lead to some understanding of the localization of gold from a regional standpoint. This study offers a distinct focus for exploration devoted to the discovery of blind gold deposits in covered areas of the Western Australian shield.

KOUDA, RYOICHI, and YOSHINORI MIYAZAKI, Japan Geological Survey, Ibaraki, Japan

### Polygonal Linear Patterns from Satellite Images and Epithermal Gold Deposits in Southern Kyushu, Japan

Remote-sensing image data from Landsat series satellites show some advantages over land surveys, as follows: (1) synchronous (at the local time during 9:30-10:30) and wide-scale area (180 × 180 km); (2) repeatability (every 16-18 days); and (3) multispectral bands (4 bands of visible and near-infrared for Multispectral Scanner, MSS, and 7 bands of visible to infrared for Thematic Mapper, TM).

One disadvantage is the relatively low resolution (79 × 79 m for MSS, and 30 × 30 m for TM). Several MSS images of southern Kyushu have been studied to extract linear patterns through digital processings. Comparing the enhanced results, many linear patterns are visible and confirmed by ground surveys, and some patterns are polygonal in shape. We assumed that these linear patterns are associated with deeper structures in the earth's crust. For example, the inner and outer shape of the Quaternary Kakuto caldera depicts a hexagonal pattern in land surveys. Recent Ebino earthquakes are considered a result of deep activity of the Kakuto cauldron. Consequently, some linear patterns may be polygonal in shape, and may reflect calderas on the surface or unexposed cauldrons in the deep.

Many epithermal deposits in the world are associated with cauldrons and intrusive igneous rocks. Therefore, we assume the following: (1) linear patterns from satellite images can reflect the earth's crust fracture system; (2) the linear patterns that reflect cauldrons and intrusives can show some polygonal shapes; and (3) ore-forming fluids that form epithermal ore deposits are transported under the control of fracture systems and, hence, force the sites of the epithermal ore deposits to be near the fracture systems.

If these assumptions are correct, then sites of epithermal ore deposits should be strongly associated with polygonal linear patterns as shown by satellite images. When distributions of known epithermal gold deposits in southern Kyushu are overlain on the map of linear patterns from satellite images, many polygonal linear patterns are observed near known deposits, suggesting that these three assumptions are correct. Moreover, a few polygonal linear patterns that are not associated with known deposits should be prospected for mineral economics.

PRELAT, ALFREDO, E., Unocal Science and Technology Division, Brea, CA

### Landsat Digital Data as Tool for Mining Exploration and Geologic Mapping in Coastal Plain of Malaysian Peninsula

Landsat Multispectral Scanner (MSS) data were used for systematic mapping of the coastal plain of the Malaysian Peninsula. The study was conducted to evaluate the use of Landsat digital data to assist the geologist in the exploration and mapping of the coastal plain of the Malaysian Peninsula. The brightness of the four Landsat MSS channels and six ratios were recorded for each pixel, and were subsequently used as input to the supervised classification technique. Several lithologic units were selected to define training groups, and the remaining study area was classified into the preselected categories. The results indicated that using Landsat digital data combined with geologic and topographic information can improve the interpretation and quality of the geologic maps. The Landsat information was an effective tool to delineate the structural features and recognize different geologic patterns that can be defined as exploration targets. The most useful Landsat parameters to study the geology in the area were channels 5 and 7 and their ratio 7/5. Vegetation areas were enhanced by the ratio 6/4. Further fieldwork in the area will confirm the results of the present study.

# ILP SYMPOSIUM ON RESEARCH ACTIVITIES AND RESULTS OF THE ILP

# CHAPTER 53

# THE INTERNATIONAL LITHOSPHERE PROGRAM—ACHIEVEMENTS AND CHALLENGES

**Karl Fuchs**
*Geophysical Institute*
*Karlsruhe University*
*Karlsruhe, Federal Republic of Germany*

## INTRODUCTION

The International Lithosphere Program (ILP) is a joint venture of the two Geoscience Unions (IUGG and IUGS) within the International Council of Scientific Unions (ICSU). It is within the tradition of cooperation in the solid-earth sciences as highlighted in the Upper Mantle Program and the Geodynamics Project. The ILP is directed by an Inter-Union Commission on the Lithosphere (ICL). It is one of the largest programs in the solid-earth sciences, directly involving more than 400 scientists within the ICL and many more within the National Committees of the member countries. The complete title of the ILP is: Dynamics and Evolution of the Lithosphere—The Framework for Earth Resources and the Reduction of Hazards. The subtitle emphasizes the involvement of the ILP with resources and with hazards. ICL wants to strengthen the ties to the exploration industries. It has started a number of projects which require this kind of cooperation for mutual benefit. This was a prime motivation for ILP to accept the invitation to cosponsor the Circum-Pacific Mineral and Energy Conference 1986 in Singapore.

The main objectives of the ILP are:

- Difference between continental and oceanic lithosphere
- Present plate motions; dynamic models of driving mechanisms
- Intraplate phenomena
- Physics and chemistry of plate interaction
- Lithosphere evolution
- Strengthened ties to applications: Resources, Hazards
- Strengthened geosciences in developing countries.

ICL is operating in six Working Groups and seven Coordinating Committees. Table 1 is the new structure of the Inter-Union Commission on the Lithosphere as established after the first 5 years of ICL under the leadership of its former President Raymond A. Price and its former Secretary-General Edward A. Flinn. The results of these first 5 years will be published in the Geodynamics Series of the American Geophysical Union and the Geological Society of America. ICL was reviewed by the two Unions and by ICSU and restructured. In addition, at the joint request of the presidents of IUGG and IUGS, Lal and Hutchinson, the ICSU General Assembly at Berne, Switzerland in September 1986 resolved to extend the mandate of the Inter-Union Commission on the Lithosphere beyond 1990 subject to five-year reviews.

This decision marks a milestone in the cooperation of the two Unions in the field of solid-earth sciences. Earth scientists are now spared the frustrating effort of rebuilding the international form of cooperation every 10 years from the beginning, at the closing of the previous program. Instead, they can now use the well-established links of communication. Furthermore, ICL can now look farther into the future and start or sponsor long-range programs which extend beyond the original lifetime of ICL. So much for the organization of ICL.

## HOW DOES ILP WORK?

ICL is best described by how it works. Joint symposia on the occasion of the General Assemblies or the Congresses of the two Unions or their Associations form an essential element of ICL's interdisciplinary communication. In the years between, ICL is seeking opportunities to join other major international conferences in the solid-earth sciences, e. g. , with the mineral industry or within developing countries. The large meetings are necessary to communicate with a large number of scientists from the general field of solid earth. However, the in-depth discussion among specialists focusing on selected topics takes place at small interdisciplinary meetings or workshops very much in the style of the Penrose, Chapman or Alfred Wegener Conferences. Here are the titles of a few such meetings recently supported by ICL: Deep Reflection Survey of the Continental Crust; Lower Crust; Deep Drilling; Sedimentary Basins; Continental Rifting; Oceanic Lithosphere.

Table 1: INTER-UNION COMMISSION ON THE LITHOSPHERE

**Bureau:**

| | |
|---|---|
| President: | K. Fuchs (Federal Republic of Germany) |
| Secretary General: | H.J. Zwart (Netherlands) |
| Members: | A. Beus (U.S.S.R.)<br>W. Fyfe (Canada)<br>H. Gupta (India)<br>C. Lomnitz (Mexico)<br>R. Oxburgh (United Kingdom) |

**Representatives:**

L. Ernster, ICSU (Sweden)
V.I. Keilis-Borok, IUGG (U.S.S.R.)
V. Sibrava, UNESCO (France)
R. Sinding-Larsen, IUGS (Norway)

**Past President:**

R.A. Price (Canada)

**Working Group (WG) and Task Group (TG) Chairmen and Vice Chairmen**

| | | |
|---|---|---|
| WG-1 | Recent Plate Movements and Deformation | K. Nakamura – C. (Japan)<br>E.A. Flinn – V.C. (USA) |
| WG-2 | The Nature and Evolution of the Continental Lithosphere | A.L. Yanshin – C. (USSR)<br>R.W.R. Rutland – V.C. (Australia)<br>A. Beus – Secr. (USSR) |
| | TG-A Variations in the Nature and Evolution of Mobile Belts | P. Hoffman – C. (Canada)<br>P. Tapponier – V.C. (France) |
| | TG-B Plate Motions and Orogeny Through Time | C. Scotese – C. (USA)<br>L. Zonenshain – V.C. (USSR) |
| | TG-C Thermal, Mechanical, Chemical and Magmatic Evolution of the Continental Lithosphere | N. Arndt- C. (FRG)<br>G. Schubert – V.C. (USA) |
| WG-3 | Intraplate Phenomena | P. Ziegler – C. (Netherlands)<br>N. Logatchev – V.C. (USSR) |
| WG-4 | The Nature and Evolution of the Oceanic Lithosphere | T.W.C. Hilde – C. (USA)<br>S.M. Zverev – V.C: (USSR) |
| WG-5 | Palaeo-environmental Evolution of the Oceans and Atmosphere | J. Thiede – C. (FRG)<br>L.A. Frakes – V.C. (Australia) |
| WG-6 | Structure, Physical Properties, Composition, and Dynamics of the Lithosphere-Asthenosphere System | J. Oliver – C. (USA)<br>F. Stacey – V.C., Physical Properties (Australia)<br>N. Sobolev – V.C., Composition (USSR)<br>R. Peltier – V.C., Dynamics (Canada) |

**Coordinating Committtee (CC) and Subcommittees (SC) Chairmen and Vice Chairmen**

| | | |
|---|---|---|
| CC-1 | Environmental Geology and Geophysics | G. Kiersch – C. (USA)<br>I. Gubin– V.C. (USSR) |
| CC-2 | Mineral and Energy Resources | P. Stokke – C. (Norway)<br>L. Montadert – V.C. (France) |
| CC-3 | Geosciences Within Developing Countries | H. Gupta – C. (India)<br>B. Katz – V.C. (Australia) |
| CC-4 | Continental Drilling | F. Stehli – C. (USA)<br>O.L. Kuznetsov – V.C. (USSR)<br>R. Andrews – Secr. (USA) |
| CC-5 | Data Centres and Data Exchange | A. Gvishiani – C. (USSR)<br>O. Dottin – V.C. (France) |
| CC-6 | National Representatives | M. Berry – C. (Canada)<br>V.C. to be elected |
| | SC-1 The Himalayan Region | H. Gupta – C. (India)<br>Ma Xingyuan – V.C. (China) |
| | SC-2 The Artic Region | G.L. Johnson – C. (USA)<br>C.E. Keen – V.C. (Canada) |
| | SC-3 Andean Tectonics | U.G. Cordani – C. (Brazil) |
| CC-7 | Global Geoscience Transects | J.W.H. Monger – C. (Canada)<br>V.C. to be elected |

In addition to these meetings, ICL tries to emphasize major research initiatives, which can be regarded as key projects. A key project is defined as a critical test of an important hypothesis or as an experiment closing crucial gaps in our knowledge. These are the presently defined key projects—Reflection survey of the continental lithosphere the world over; Long-range seismic profiling; Continental drilling, European geotraverse (EGT); Xenoliths and metasomatism; ELAS; Global seismic networks; Lower crust; Global geoscience transects; World stress map.

Some will be discussed in detail later.

At the beginning of a global tectonic cycle, the problem of the breakup of continents requires a solution which does not directly emerge from the framework of plate tectonics. It requires knowledge not only of the rheology of the continental lithosphere and asthenosphere but also of the interaction with dynamic processes in the mantle below and with neighboring plates. Rifts seem to start preferentially on continents; Vink, Morgan, and Zhao (1984) attribute this to the reduced strength of the continental lithosphere, which is caused by its weak rheological properties induced by the special properties of the lower crust on continents.

This continental lower crust is another focal point of the ILP. The widespread investigation of the crystalline continental basement by seismic-reflection surveys (another key project of ILP) unexpectedly revealed a highly reflective lower crust beneath an almost transparent upper crust (Barazangi and Brown, 1986). The laminated, strongly reflecting lower crust appears especially in regions of recent or past extensional regimes. Although the laminae are excellent markers of the lower crust, their origin is still controversial. However, they seem to be connected, directly or indirectly, to the reduced viscosity of the lower crust during their time of formation. The lower crust is related to many tectonic events on the continents such as formation of sedimentary basins, plateau uplift, orogenies, intraplate seismicity, and graben formation. Its composition is studied from xenoliths and outcrops. Mechanisms of transport of volatiles through the lower crust are still under intensive debate. Diffusion along grain boundaries is certainly to be considered, but more effective mechanisms like crack propagation in viscous media with nonuniform distribution of strain would be more effective. Earthquakes observed in the laminated lower crust indicate that opening of cracks may be an effective mechanism even under conditions of high pressure and reduced viscosity in the lower crust.

The primary cause driving the breakup of a continent is found in the mantle beneath the lithosphere. The new technique of seismic tomography (Anderson and Dziewonski, 1984) revealed for the first time gross lateral heterogeneities from a systematic analysis of travel time residuals and variations of surface-wave dispersion. These heterogeneities are most likely imaged regions of warm and cold material in the convecting mantle; this is most clearly seen at the mid-ocean ridges and at subduction zones. Apart from these dominating regions, there are other heterogeneities which have no immediate relationship to surface tectonic features. However, in the mantle beneath the Red Sea, one of the largest heterogeneities with warm material is located. The relationship of such a heterogeneity to the rifting process in the Red Sea is still obscure. Inasmuch as the estimate of the lifetime of such tomographic anomalies ranges over several orders of magnitude (10 to 1000 Ma), one cannot even be sure whether the mantle anomalies existed beneath the Red Sea when the rifting was initiated about 30 Ma, or whether their positions coincided when rifting started. It could become a future challenge to interdisciplinary research in the solid-earth sciences to date geochemical mantle reservoirs which could provide constraints to convection modelling of the mantle and an improved estimate of how the pattern of seismic tomography evolves with time. A new door is opening to a multidimensional model for a better understanding of how the earth works.

A Federation of Global Seismic Networks has been formed recently and is affiliated with ICL. Its purpose is to install a modernized digital network of broadband seismic stations which will allow an improved high-resolution tomography of the earth interior.

The mobility of the earth established in the past two decades within the framework of plate tectonics can be sensed today directly by intercontinental distance measurements. Very Long Base-line Interferometry (VLBI) of cosmic radio signals by stations on various continents and Satellite Laser Ranging (SLR) leads to rates of continental distance change (Christodoulidis et al. , 1985) very close to those determined from other geophysical methods (Minster and Jordan, 1978). This is certainly only a first step, but it indicates that geodesists will provide in the near future data on present-day deformation of the earth's surface. Regional networks, like the WEGENER network in the Eastern Mediterranean, will be set up in parts of the globe which have large deformation rates. With a large number of Global Positioning Stations (GPS), it will become possible to monitor "tectonics in realtime."

Another global project of ILP is the "World Stress Map," which is prepared by a task group under the leadership of Mary-Lou Zoback within Working Group 4 "Intraplate Phenomena. " It is the intention to compile all available, reliable observed data on stress distribution in the lithosphere from various sources—earthquake focal mechanisms, analysis of bore-hole elongation data, and hydrofrac measurements. Such a stress map of the world will lead to a recognition of stress provinces; it may provide constraints for convection modeling and could become of practical importance for engineers and the mineral industry. The data compilation will heavily rely upon the analysis of 4-arm caliper data from bore holes of the mineral industry. An appeal has been made by the Inter-Union Commission on the Lithosphere to the mineral industry for cooperation in this project for mutual benefit.

Especially close cooperation with the industry and with the geological surveys is also required in another worldwide project of ICL, i. e. , the project of Global Geoscience Transects (GGT). In this project all available geophysical and geological data along about 200 corridors through crucial tectonic features will be compiled worldwide on the same scale and with the same color code down to the base of the crust. This will form the basis for a global comparison of similar tectonic features (Ref. J. Monger, this volume).

The most ideal transects are those which, in addition to deep seismic sounding data, are connected with deep drill holes. The plan for deep exploration of the lithosphere as proposed for the USSR by Koslovsky (1984) forms an example for such an ideal situation.

Deep continental drilling is a key project of ICL with a special Coordinating Committee 4 under chairman Frank Stehli. ICL offers the only international platform for deep drilling on continents. It may be a new task of CC-4 to define criteria for internationally accepted drilling sites which are necessarily located within the countries of the funding agencies.

## CONCLUSION

These were but a few examples of global programs within the Inter-Union Commission on the Lithosphere. All these programs require international and interdisciplinary cooperation on a long-term basis. Such a cooperation in the solid-earth sciences has been placed on a firm footing by the ICSU decision to extend the ICL mandate without a termination clause.

We are living in a fascinating time for the earth sciences. The rate of development is very fast. New tools of data acquisition and analysis are becoming rapidly available. In such a world ICL's main goals are: firstly, communication within ICL and with all participating countries; secondly, contacts with the industries; and last but not least, keeping the earth sciences together for an improved understanding of how the earth works.

## REFERENCES CITED


Anderson, D.L., and A.M. Dziewonski, 1984, Seismic tomography: Scientific American, v. 251, no. 4, p. 60-68.

Barazangi, M., and L. Brown, 1986, Reflection seismology: The continental crust: American Geophysical Union, Geodynamics Series, v. 14, 339 p.

Christodoulidis, D.C., D.E. Smith, R. Kolenkiewicz, S.M. Klosko, M.H. Torrence, and P.J. Dunn, 1985, Observing tectonic plate motions and deformations from satellite laser ranging: Journal of Geophysical Research, v. 90, p. 9249-9263.

Koslovsky, Y.A., 1982, Kola super-deep: Interim results and prospects: Episodes, v. 5, no. 4, p. 9-11.

Minster, J.B., and T.H. Jordan, 1978, Present-day plate motions: Journal of Geophysical Research, v. 83, p. 5331-5354.

Vink, G.E., W.J. Morgan, and W.L. Zhao, 1984, Preferential rifting of continents: A source of displaced terraines: Journal of Geophysical Research, v. 89, p. 10072-10076.

CHAPTER 54

# GLOBAL TECTONICS AND RESOURCES

**W.S. Fyfe**
*The University of Western Ontario*
*London, Canada*

ABSTRACT

As our knowledge of plate tectonic convective processes has improved, so has our ability to understand and find resources. Plate tectonic processes control virtually all resources, from metals to fuels to soils. The new emphasis on the environment, as the ultimate resource, demands improved knowledge of the mass-energy fluxes from the deep earth which influence the biomass. In particular, we need to understand the variations in the deep fluxes, on a range of time scales. On this water-cooled planet, almost half the internal energy production is transferred to fluids in the crust, which produce major chemical fluxes.

## INTRODUCTION

With few exceptions, the upper few kilometers of the earth's surface are water saturated. Voids in the near-surface crust, caused by pore spaces and fractures, are filled with aqueous fluids of highly variable salinity. Although the mass of water in the hydrosphere-cryosphere systems is rather well known, the fluid inventory for the lithosphere is not well quantified. In the lithosphere in general, volatiles are contained in liquid or gas phases, but some components, like $CO_2$ and $H_2O$, and even halogen and nitrogen compounds, are fixed in common minerals. From what we know of the distribution of such phases in the crust, the mass of most volatiles in the solid earth is similar to, or even greater than, the mass in the atmosphere-hydrosphere system. (Fyfe, 1983).

One of the great challenges of modern earth science is to understand and quantify all mass-energy fluxes of our planetary system, and to understand the fluctuations in such fluxes on all relevant time scales. These fluxes influence the chemistry, and changes in the chemistry, of all major geospheres. It is here that the Lithosphere project is linked to the new Global Change or Geosphere-Biosphere Project, in that the nature of the Biosphere (and hence $O_2$, $CO_2$, $CH_4$, etc) responds to changes in the chemistry of the environment, the fluxes of elements essential to, or toxic for, living species.

The energy flux through the surface is transported by thermal conductivity and mass flow. Our present models of continental heat flow are based almost exclusively on conductive models. However, studies of the sea-floor environment, over the past decades, have clearly shown that energy transport by fluid motion is of great significant in all oceanic crust, both young and old. On continents, we know well that, whenever the gradient exceeds about $50\,°C/km^{-1}$, hot spring activity is common and observable. It is difficult to imagine that fluid convection is not possible for the more normal gradients of $20–30\,°C/km^{-1}$, for Straus and Schubert (1977) have shown that the adiabatic gradient is greatly exceeded for all normal gradients. At a recent nuclear waste meeting in Canada, Durrance (1986) reported crustal convection cells in southwest England. These multi-km cells were demonstrated by study of helium and radon fluxes and heat flow. In a continental situation, deep fluxes must be diluted or concealed by near-surface flow of rainwater in the outer permeable layers. Only very deep measurements of heat flow are likely to be valid.

We were all educated to believe that cracks-pores close at high pressures. This is equivalent to saying that all deep fluids are totally expelled from deep lithosphere levels which, in turn, implies finite permeability at deep levels. Most deep materials contain fluid inclusions, commonly becoming increasingly rich in $CO_2$ with depth. We known that, in many areas of the deep crust, the electrical conductivity is much higher than would be expected for assemblages of dry silicate minerals (Gough, 1986), and seismic velocities commonly are too low. We also know that hydrocarbons can be stored at substantial depths for times exceeding $10^8$ years, and recently Jackson et al. (1986) report hydrocarbons in rocks as old as 1. 7 Gyr. We are confronted with observations which do not fit any simple pattern of depth-permeability-fluid content variation. We do not yet understand mass-energy fluxes in the deep crust.

With the development of the present ideas of deep convection (plate tectonics), we have begun to appreciate some aspects of the mass flows associated with such convection.

We have begun to appreciate how such mass flows produce resources, from things like copper and gold, even to soil. We have become much better at the task of finding resources, perhaps too good, as reflected in the recent report on World Resources (World Resource Institute, 1986), where metals and energy resources receive only slight attention.

We are at a most interesting point in the description of earth processes. Over the next decades, via satellites and long base line arrays, we will directly observe surface plate motions and the local variability of such motions. With modern tools for observation of chemical and isotopic fluxes (e.g., $^{10}$Be, $^{3}$He, and the like), we will be able to associate chemical change with the plate motions. Chemical and geophysical observations, coupled with better deep sampling, via drilling, should produce vastly improved models of mass-energy transfer processes, the key to the global resource base. In what follows, I would like to examine some aspects of the plate-tectonic cycle, with respect to chemical transport processes.

## OCEAN RIDGES - NEW LITHOSPHERE

Ocean-ridge magmatism produces new oceanic crust and lithosphere, with the near-surface manifestation being basaltic volcanism, basaltic dyke injection, and deeper intrusion—the ophiolite succession (Emiliani, 1981). Cooling, crystallization, and contraction of this new crust, from 1200° C to a few hundred degrees, causes a large permeability by cracking processes. Studies of the fluid dynamics of such new ridge materials clearly show that convective sea-water cooling must occur (Lister, 1972; Wolery and Sleep, 1976) and is required to explain the heat flow patterns which involve large-scale (multi-km) convection cells. Metamorphic and alteration patterns in rocks influenced by the water cooling (see Spooner and Fyfe, 1973; Rona et al., 1983) indicate thermal gradients exceeding 100° C/km$^{-1}$. The ultimate proof for such flow systems has come from the numerous observations of the spectacular hot springs (black smokers) on the sea floor, with discharge temperatures at times exceeding 300° C. Such hot fluids, commonly near seawater in salinity, carry enhanced levels of a number of metals (Fe, Mn, Cu, Ag, Zn, etc. ) which can precipitate to form ores, or be dispersed into ocean water and eventually precipitated in Mn-Fe oxides.

The classic literature of ocean chemistry involves a model where most input to the oceans occurs via the river flux off the continents, with the chemical and biological precipitation processes in the oceans maintaining the final chemical balance. But it is now becoming clear that the hydrothermal fluxes through the ocean crust are commonly important, or even dominant, for many important species ($SiO_2$, K, Cu, Mn, Fe, etc. ). Some of these species (K, $SiO_2$) may be bioessential, whereas others (Cu, Mn, V, Zn, etc. ) may provide the source for the enzyme catalysts of certain biomass species. Perhaps one of the first clues to the significance of this mass flux was provided by studies of the $^{87}$Sr/$^{86}$Sr systematics of ocean waters and younger sediments. Present world rivers feed strontium isotopes with a ratio of 0. 711, whereas submarine hot springs feed water with a mantle ratio of about 0. 702. Present ocean water, with a ratio 0. 709, shows the balance between the two fluxes. However, what is of immense significance is that the sedimentary record shows that this ratio has changed substantially over time (Burke et al., 1982). With respect to $^{87}$Sr, the present ocean is more enriched than it has been for a very long period of time, indicating enhanced erosion off evolved continental crust. It increasingly appears that continental tectonics (integrated elevations) control such variation (e.g., the Himalayan orogeny). But what is true for strontium systematics must be true for other species; and variation in the chemical fluxes to oceans will be reflected in biomass, which in turn will be reflected in $O_2$-$CO_2$ systematics, etc. ; all parts of the global system are connected!

## OFF-RIDGE SYSTEMS

Although convective cooling with free venting of thermal waters into the oceans is now well established, much less is known about circulation in older ocean floor, where a relatively impermeable sediment blanket covers the warm crust. The thickness and nature of the sediment cover varies with time and the proximity to sediment sources of the spreading center. One of the first cases of such a system, the Juan de Fuca ridge, was described by Davis and Lister (1977). On the basis of heat-flow measurements and seismic data on sediment thickness, they showed that closed convection must be occurring in the relatively permeable basaltic crust beneath the sediments. Anderson, Hobart, and Langseth (1979) described heat-flow patterns in the Indian Ocean, showing similar evidence for fluid circulation in crust exceeding 50 Ma in age. There is now increasing evidence for such convection in many older parts of the ocean floors, a process which may alter the magnetic signatures (Levi and Riddihough, 1986). We have recently found evidence for magnesium addition to pelagic clays off the Hawaiian Islands, a process which could be attributed to slow influx of sea water (Tazaki, Fyfe, and Heath, 1986).

Such observations are of great significance (Figure 1). If convection continues in old crust, where the heat flow approaches normal continental values and the permeability should be less than in the young ridge environment (because of fracture filling, etc. ), then a similar flux should occur in continental crust, where the permeability and bed thickness are similar. And, in a layered crust of varied permeability, not necessarily a simple function of depth, there is the possibility of very complex convective patterns.

In terms of chemical transport of metals, a very interesting situation arises when the lower ultramafic layers of the ocean crust cool below 500° C and react with water to form serpentine. Large-scale formation of serpentinite is observed in all the great ophiolite complexes. A particularly interesting case is now available from the deep drilling of the Cyprus ophiolite.

MacDonald and Fyfe (1985) studied diffusion of $H_2O$ through serpentine, and showed that the diffusion coefficients are so large that it is very difficult not to hydrate peridotite, if water is available in a deep marine environment. The chemical potential gradients for $H_2O$ across the boundaries, $H_2O$-serpentinite-periodotite, are so large that water is literally sucked in by the chemical reaction. Further, as the hydration is accompanied by a 40% volume expansion, the chemical reaction produces a volume strain which may lead to intense fracturing and continuous permeability creation. The evidence for such intense fracturing processes is seen in most serpentinites.

As shown by MacDonald and Fyfe (1985), when sea water reacts with peridotite to form serpentine, the end product is a highly saline brine produced by preferential extraction of water. Vibetti, Kerrich, and Fyfe (1985) have now found the evidence for such brines in fluid inclusions in the deeper zones of the Cyprus ophiolite. Finally, it should be noted that the hydration process is highly exothermic, so that a significant contribution to crustal energy production is possible (Fyfe and Lonsdale, 1981).

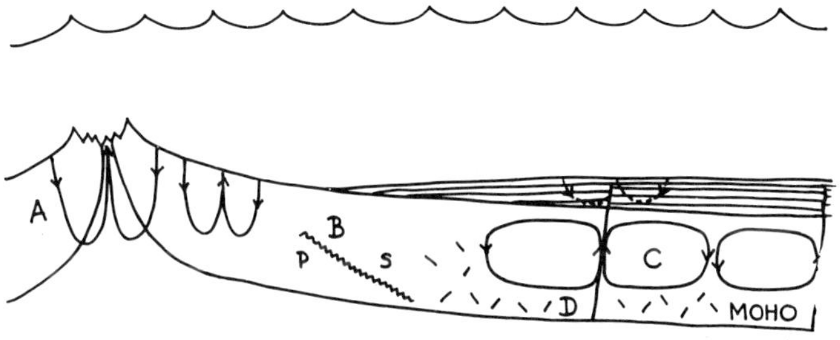


Figure 1. Convective cooling systems in the sea-floor environment. A, cells with free discharge to the surface; C, closed cells under a sediment cover. At B, the peridotite (P) to serpentine (S) reaction occurs, producing a highly fractured and faulted system (D).

When these parts are all put together, quite new possibilities for formation of mineral deposits appear. In older ocean crust, if convection continues beneath relatively impermeable cover, rising fluids may deposit minerals near the sediment interface, a site for chemical mixing, and one where the thermal gradient may steepen. Further, it is well known that the solubility of most metals—like Zn, Cu, Ag, Mn, Fe—rises rapidly with salinity, so that the saline byproducts of serpentinization will have greatly enhanced capacity for metal transport. Further, the rising metal rich fluids are less subjected to dispersion which may occur in the near-ridge free-venting situation. That such processes may be important for ore deposition has been stressed by Edmond and von Damm (1983), and Barriga (1983) has shown that, for some of the greatest sulphide ore deposits of the planet (e.g., Aljustrel, Portugal, Cu-Fe-Zn-Au), the ores appear to have formed in a submarine environment, beneath a thin sediment cover.

We have much to learn about the chemical fluxes over the vast surface areas of the planet formed by the sediment-ophiolite structure, the most common type of crust of the planet. But, as we have begun to appreciate that ridge fluxes influence ocean chemistry, it seems possible that the quiet, cooler saline fluxes off ridge may also be significant.

It must be stressed that the alteration processes that occur in the ocean crust must eventually have a profound effect on upper mantle processes. The ocean crust that is eventually subducted is not the same as that which forms the new crust at ridges. It is volatile loaded ($H_2O$, $CO_2$, S, C1) and enriched in elements such as Na - K - Mg - K - U (Fyfe and Lonsdale, 1981). We are slowly learning that deep recycling is a much larger scale process than considered in geochemical models of a decade ago (e.g., Allègre and Turcotte, 1986).

## SUBDUCTION

The creation of new oceanic lithosphere at ridges is more or less balanced by subduction of old ($\pm$ 100 Ma) and modified crust and lithosphere at trenches. Geochemists and isotope specialists now increasingly agree that the upper mantle is modified by such subduction (see Allègre and Turcotte, 1986). New observations on trench structure have shown how ocean sediments may also be subducted in substantial quantities (Uyeda, 1983). A recent spectacular confirmation of deep sediment subduction has been provided by Chopin (1984) in his work on very high pressure silica-rich rocks found in the Italian Alps. These rocks, in many ways reminiscent of marine sediments in composition, contain the high-pressure polymorph of $SiO_2$, coesite, implying burial to at least 100 km. Thus, during the Alpine orogeny, surface rocks such as the pillow lava eclogites of Zermatt (Bearth, 1959) have been dragged down 100 km or so, and quite rapidly returned to the surface (see Fyfe, 1986a). We know little of these processes which must profoundly influence mantle rheology and chemistry. We are increasingly aware that there is significant mixing of all components—atmosphere-hydrosphere-crust and mantle (Fyfe, 1983). Of enormous significance is the recycling of volatiles ($H_2O$, C-S-N species, etc. ) and heat-producing elements (U, K) into the interior. Just one example may suffice to show the scale. If subducted ocean crust contains about 5% $H_2O$ (the problem is the mass of subducted sediment and serpentine), then, with a crustal subduction rate of 10–15 $km^3\ a^{-1}$, a water mass 1.5–2.2 $\times 10^{15}$ g $a^{-1}$ is subducted and the ocean mass is recycled every billion years or less. Similar considerations apply to surface $CO_2$ and S species. Given faster recycling rates for a younger, more energetic planet, it seems that our volatile elements may have been recycled at least something like 10 times during earth history.

The present convective mixing dynamics associated with plate tectonics has interesting implications. Increasingly, we consider an early very hot outer surface with magma oceans (Matsui and Abe, 1986). Common thermodynamic wisdom implies that a hot body emits gases, whereas a cooling body absorbs volatiles. If our planet is mixing while it cools, there must be a tendency for volatile fixation to increasingly occur. The ocean mass should be reduced, along with $CO_2$. Similar arguments apply to species like $SiO_2$ concentrated in continental crust. The masses of the volatile, fusible components of the outer geospheres can hardly be constant in a partially mixed cooling body.

We now know a little of how the return flow of volatiles modifies the crust, and produces resources. Mantle above subduction zones is softened by volatile injection, eventually producing andesite magmas. The enhanced heat flow associated with such magma rise causes fusion of continental crust, with production of granite magmas. The entire array of processes can produce final fractions of metal-enriched products (such as Li, Be, Sn, W, Ta, and U pegmatites) and water cooling of plutons can produce the giant copper deposits of the world, and sources of geothermal energy.

Another process of profound significance which is receiving increasing attention involves the magma underplating of continental crust (Figure 2). We were all educated to believe that

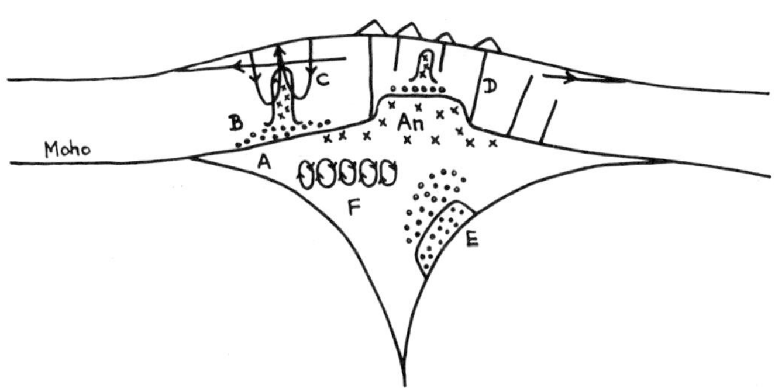


Figure 2. Dense mantle magma underplating of continental crust. The magma chambers (A) convect and differentiate, producing anorthosites (An). Dense crustal components (E) founder and degas or melt. The crust is metamorphosed and melted near the Moho (B) and produces plutons, which are water-cooled at high levels (C). Dykes and vein systems occur above the underplate (D).

the thermal gradient in continental crust is near 20–30 ° C/ $km^{-1}$. This implies temperatures at the continental Moho near 600 ° C, a temperature where slightly damp crust will begin to soften and melt. Yet, time after time, metamorphic petrologists find evidence for continental rocks metamorphosed at temperatures near 1000 ° C (see Sandiford and Powell, 1986, who report granulites produced a temperatures of at least 1000 ° C). How are such temperatures achieved?

Work on the compressibility of magmas (see Herzberg, Fyfe, and Carr, 1983) has shown that, when ultramafic and basaltic magmas are compressed, their Moho (10 kbar) density can be higher than typical continental crust. Thus, dense mantle magmas rising beneath continental crust may penetrate this light crust only if it is coupled to its upper mantle. But, once the continental crust softens at the Moho, because of enhanced heat flow, such magmas should float off the crust, forming giant intrusions near the Moho. After such intrusions, the Moho region will be subjected to a massive prograde metamorphic event with partial melting. Such a process is required to explain many types of observed continental magmatic and metamorphic processes.

During an event where dense basic mantle magmas near 1200 ° C penetrate the continental Moho, processes must occur which lead to resources. Mafic parts of the continental crust may founder into the underplate region. Their metamorphism and degassing at very high temperatures will produce very high temperature gases such as $CO_2$, Co, COS, HC1, and HF—with unique metal-carrying capacities. The same gases, on cooling (e.g., $2CO \rightarrow C + CO_2$), may account for the graphite-sulphide veins found in crustal granulites (see Fyfe, 1986a). Such fluids may provide a transport mechanism, to introduce metals from the mantle into the crust, metals like gold and platinum.

Further, above the Moho, with the greatly enhanced thermal gradients, metamorphic dehydration events may lead to the mobilization of metals like gold. Many consider that most of the world's great primary gold deposits, associated with quartz veining and $CO_2$ transport, are formed from such hot metamorphic or even igneous fluids (Fyfe and Kerrich, 1985; Burrows, Wood, and Spooner, 1986). Similar processes must also occur when random mantle hot spots arise beneath continental crust.

## CONTINENTAL COLLISIONS

The situations discussed above, cooling at ridges and volatile-fluxed subduction magmatism, are becoming well known, as are the ore deposits associated with such phenomena (Cu-Zn, massive sulphides, gold, tungsten, etc. ). However, there are other fluid transport processes associated with plate tectonic phenomena which are still little understood. One of the most significant and common processes that modifies continental crust and involves vast mobilization of fluids is that of continental collision. It is now certain that continental collision, overthrust tectonics, has been a major process in crustal modification. We are at present living in the middle of one of the greatest of these events, the Himalayan orogeny.

The ongoing (5 cm $a^{-1}$) collision of India and Asia has produced a region 3000 × 1500 km, where the continental crust is 60-70 km thick (Allègre et al., 1984). The thickening is created by a series of major thrust structures, which are reflected in a jagged, saw-blade Moho (Hirn et al., 1984). Although details may be complex, the net effect of the process, in terms of fluids, is like that of taking most of continental Australia (30 km thick) and heating it by 300 ° C and compressing it by 10 kbars. In such a process, the basal half of the overthickened crust will be substantially dessicated by release of metamorphic $H_2O$-$CO_2$ and, finally, by melting at the base. The fluids generated by the thrust processes lubricate the thrusts and catalyze rock deformation (Hubbert and Rubey, 1959; Fyfe and Kerrich, 1985; Oliver, 1986).

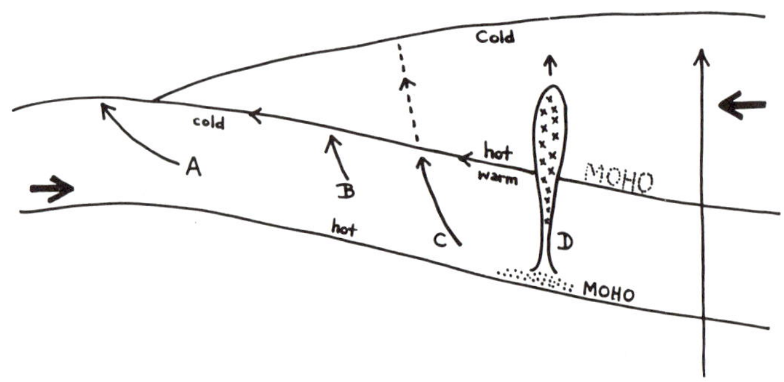


Figure 3. Processes during a large continental collision. At (A) fluids, including water and hydrocarbons, are expelled in front of the thrusts. As the underplate becomes hotter, metamorphic fluids ($H_2O$-$CO_2$) are evolved (B and C). As the Moho becomes hotter, melting may occur, producing granitic plutons (D). Finally, warming in the sub-Moho region may produce melts representing mantle contaminated by earlier subduction prior to collision.

The masses of fluids which may be involved are very large indeed. Imagine that a crustal block 3000 × 1500 × 30 $km^3$, containing an average of 4% $H_2O$ and 1% $CO_2$ (in hydrates and carbonates) loses its volatiles. The $H_2O$ released is about $1.6 \times 10^{22}$ g, a mass similar to the ice caps. The $CO_2$ released by reactions like $CaCO_3 + SiO_2 \rightarrow CaSiO_3 + CO_2$, is more than 4000 times atmospheric $CO_2$ and 100 times the mass of $CO_2$ in oceanic bicarbonate. Such processes may perturb the outer environmental systems (Figure 3).

Collisions also create high topography and cause accelerated erosion. These influences are probably reflected in the ocean sediment $^{87}Sr/^{86}Sr$ record, which so beautifully integrates the record of runoff and hydrothermal events in the ocean crust (Burke et al., 1982).

The flow of fluids in thrusts must be a major chemical transport process. Given that a metal is transported in solution at 1 ppm, a million-tonne metal deposit requires $10^{18}$ g of fluid, 1000 $km^3$ (the annual flow of the Yangtze River) for formation. The potential for ore formation in the Himalayan collision is thus related to 10,000 Yangtzes of fluid! We have recently described a very major uranium deposit formed by this simple process (Lobato et al., 1983). Oliver (1986) has suggested that the same types of processes may be involved in the concentration of oil and gas and certain types of lead-zinc-silver deposits.

It is of interest that similar processes may occur with large strike-slip faults where local override occurs. Such structures have been proposed for the Alpine fault of New Zealand (Allis, 1981). Mineralization must be possible on major strike-slip and thrust faults (Fyfe, 1986a).

## EROSION

The major elevations of the continental surface are caused by subduction, collision, and rifting processes. Removal of surface materials, by solution and particle transport, would rapidly lead to a great global swamp, were it not for the convective mass-energy fluxes from the interior, which periodically form new areas of high elevation.

Erosion and weathering processes create the fertile soils for our agriculture. In general, areas of high, sustained productivity are related to volcanic phenomena or recent erosional events. Old, stable, low-elevation regions are seldom of high fertility (Fyfe et al., 1983). Biosphere processes produce the great deposits of carbonaceous materials trapped in sediments, and are commonly associated with metal concentration processes (Fyfe, 1986b). Low-grade metal concentrations from igneous or metamorphic events are commonly reconcentrated during weathering (gold in the Carajas district of the Amazon, phosphates and rare earths in the tropical weathering of carbonatites, alluvial tin or gold). In a sense, all our resources are related to the flow of fluids, either meteoric near-surface fluids or fluids from deep sources. And the flow regimes, driven thermally or gravitationally, are ultimately related to the interior convective processes.

## CONCLUDING STATEMENT

The mass-energy fluxes of the earth influence the environment of the biomass. Although some of such surficial flows are moderately quantified, the deep fluxes are not. Ocean-ridge cooling processes are now recognized to be of great significance in this respect, but much less is known about fluxes associated with subduction and deep continental flow systems. Further, we know little about time variation of the deep flow systems. Although plate tectonic processes can be modeled as steady state, they surely are not on time scales relevant to life processes. One of the great future challenges for those who study the deep convective processes is to quantify the time-variation of chemical transport into the surface environment. Here, the Lithosphere project must complement the new Geosphere-Biosphere project.

## REFERENCES


Allègre, C.J., and D.L. Turcotte, 1986, Implications of a two-component marble-cake mantle: Nature, v. 323, p. 123-127.

Allègre, C.J., et al., 1984, Structure and evolution of the Himalayan-Tibet orogenic belt: Nature, v. 307, p. 17-22.

Allis, R.G., 1981, Continental underthrusting beneath the Southern Alps of the New Zealand: Geology, v. 9, p. 303-307.

Anderson, R.N., M.A. Hobart, and M.C. Langseth, 1979, Geothermal convection through oceanic crust and sediments in the Indian Ocean: Science, v. 204, p. 828-832.

Barriga, F.J.A.S., 1983, Hydrothermal metamorphism and ore genesis at Aljustrel, Portugal: University of Western Ontario, Toronto, Ph.D. thesis (unpublished), 368 p.

Bearth, P., 1959, Uber eklogite, glaukophanschiefer und metamorphe pillowlaven: Schweiz. Mineral. Petrogr. Mitt., 39, 267-286.

Burke, W.H., R.E. Denison, E.A. Hetherington, R.B. Koepnick, H.F. Nelson, and J.B. Otto, 1982, Variation of seawater $^{87}Sr/^{86}Sr$ throughout Phanerozoic time: Geology v. 10, p. 516-519.

Burrows, D.R., P.C. Wood, E.T.C. Spooner, 1986, Carbon isotope evidence for a magnetic origin for Archaean gold-quartz vein ore deposits: Nature, v. 321, p. 849-854.

Chopin, C., 1984, Coesite and pure pyrope in high-grade blueschists of the Western Alps: A first record and some consequences: Contrib. Mineral. Petrol., v. 86, p. 107-118.

Davis, E.E., and C.R.B. Lister, 1977, Heat flow measured over the Juan de Fuca ridge: Evidence for widespread hydrothermal circulation in a highly heat transportive crust: Journal of Geophysical Resource, v. 82, p. 4845-4860.

Durrance, E., 1986, Helium in soil gas: An alternative method of mapping groundwater circulation in fractural plutonic rock (abs.): Atomic Energy Canada Conference, Mont Ste. Marie, Quebec.

Edmond, J.M., and K. von Damm, 1983, Hot springs on the ocean floor: Scientific American, v. 248, no. 4, p. 78-93.

Emiliani, C., 1981, The oceanic lithosphere: New York, Wiley-Interscience, 1738 p.

Fyfe, W.S., 1983, Subduction and the geochemical cycle: Tectonophysics, v. 99, p. 271-277.

Fyfe, W.S., 1986a, Fluids in deep continental crust: American Geophysical Union, Geodynamics Series 14, p. 33-39.

Fyfe, W.S., 1986b, Organisms, minerals and ore deposits: Proceedings Denver Region Exploration Geological Society symposium, Organics and ore deposits, p. 1-6.

Fyfe, W.S., and R. Kerrich, 1984, Gold: Natural concentration processes, in Rotterdam, The geology, geochemistry and genesis of gold deposits: R.P. Foster, ed., A.A. Balkema, p. 99-128.

Fyfe, W.S., and R. Kerrich, 1985, Fluids and thrusting: Chemical Geology, v. 49, p. 353-362.

Fyfe, W.S., and P. Lonsdale, 1981, Ocean floor hydrothermal activity, in C. Emiliani, ed., The oceanic lithosphere: New York, Wiley-Interscience, p. 589-638.

Fyfe, W.S., B.I. Kronberg, O.H. Leonardos, and N. Olorunfemi, 1983, Global tectonics and agriculture: A geochemical prospective: Agriculture, Ecosystems, and Environment, v. 9, p. 383-399.

Gough, D.I., 1986, Seismic reflections, conductivity and stress in the continental crust: Nature, v. 323, p. 143-144.

Herzberg, C.T., W.S. Fyfe, and M.J. Carr, 1983, Density constraints on the formation of the continental Moho and crust: Contributions to Mineralogy and Petrology, v. 84, p. 1-5.

Hirn, A., et al., 1984, Crustal structure and variability of the Himalayan border of Tibet: Nature, v. 307, p. 23-25.

Hubbert, M.K., and W.W. Rubey, 1959, Role of fluid pressure in mechanics of over-thrust faulting: GSA Bulletin 70, p. 115-166.

Jackson, M.J., T.G. Powell, R.E. Summons, and I.P. Sweet, 1986, Hydrocarbon and petroleum source rocks in sediments as old as $1.7 \times 10^9$ years: Nature, v. 322, p. 727-729.

Levi, S., and R. Riddihough, 1986, Why are marine magnetic anomalies suppressed over sedimented spreading centres?: Geology, v. 14, p. 651-654.

Lister, C.R.B., 1972, On the thermal balance of a mid-ocean ridge: Geophysical Journal of the Royal Astronomical Society, v. 26, p. 515-535.

Lobato, L., J.M.A. Forman, W.S. Fyfe, R. Kerrich, and R.L. Barnett, 1983, Uranium enrichment in Archaean basement associated with overthrusting: Nature, v. 303, p. 235-237.

Matsui, T., and Y. Abe, 1986, Evolution of an impact-induced atmosphere and magma ocean on the accreting earth: Nature, v. 319, p. 303-306.

MacDonald, A.H., and W.S. Fyfe, 1985, Rate of serpentinization in seafloor environments: Tectonophysics, v. 116, p. 123-135.

Oliver, J., 1986, Fluids expelled tectonically from orogenic belts: Their role in hydrocarbon migration and other geologic phenomena: Geology, v. 14, p. 99-103.

Rona, P.A., K. Boström, L. Laubier, and K.L. Smith, 1983, Hydrothermal processes at seafloor spreading centres: New York, Plenum Press, 796 p.

Sandiford, M., and R. Powell, 1986, Pyroxene exsolution in granulites from Fyfe Hills, Enderby Land, Antarctica: Evidence for 1000° C metamorphic temperatures in Archaean continental crust: American Mineralogist, v. 71, p. 946-954.

Spooner, E.T.C., and W.S. Fyfe, 1973, Sub-seafloor metamorphism, heat and mass transfer: Contributions to Mineralogy and Petrology, v. 42, p. 287-304.

Straus, J.M., and G. Schubert, 1977, Thermal convection of water in a porous medium: Effects of temperature and pressure-dependent thermodynamic and transport properties: Journal of Geophysical Research, v. 82, p. 325-333.

Tazaki, K., W.S. Fyfe, and G.R. Heath, in press, Palygorskite formed on montmorillonite in North Pacific deep-sea sediments.

Uyeda, S., 1983, Comparative subductology: Episodes, v. 1983, p. 19-24.

Vibetti, N.J., R. Kerrich, and W.S. Fyfe, 1985, Presence of Ca-Na dominated hypersaline brines in quartz calc-silicate veins of Troodos ophiolite (Cyprus) (abs.): Program abstract, December meeting, A.G.U., San Francisco.

World Resources Institute, 1986, World resources 1986: New York, Basic Books, Inc., 349 p.

Wolery, T.J., and N.H. Sleep, 1976, Hydrothermal circulation and geochemical flux at mid-ocean ridges: Journal of Geology, v. 84, p. 249-275.

# APPENDIX 1: HIGH-RISK VOLCANOES

# Mitigating Volcanic Hazards in Developing Nations through a Program of Integrated Communications, Training, and Field Studies

Ad-Hoc Working Group for the Mitigation of Volcanic Hazards in the Circum-Pacific

## SUMMARY

Hazards at some Circum-Pacific volcanoes are effectively mitigated by existing programs, but hazards at many more are partly or wholly unmitigated. The tragedy at Nevado del Ruiz in Colombia in 1985 is symptomatic of a great need for further mitigation efforts. Effective mitigation of volcanic hazards is best done by national programs, supported by international efforts.

We propose a project of integrated communication measures, training, and field studies on high-risk volcanoes of the Circum-Pacific region. The project aims are (1) to increase training opportunities in volcanology, particularly at a postgraduate, practical level; (2) as an integral part of that training, to gather baseline data needed to mitigate hazards at the highest-risk volcanoes of participating countries; and (3) to prepare video and other source materials that can more effectively communicate information about volcanic hazards and their mitigation to public officials and to the general public. All of the above will increase national self-sufficiency in coping with volcanic hazards.

Three levels of project activity are proposed. The lowest level is a simple, inexpensive ($US $10^4$/yr) program involving annual "roundtable" discussions on volcanic hazards, establishment of a clearinghouse for information on training opportunities and simple designs of monitoring equipment, and promotion of conferences using satellite communications (such as PEACESAT).

The second level of project activity (funded on the scale of $US $10^5$/yr) consists of volcanological short courses and field workshops, the production of educational videos on topics such as volcanic hazards and monitoring techniques, the promotion of exchange visits for observatory personnel for in-service training and experience, and increasing the opportunities for positions in current degree programs at existing institutions.

The third level of activity is creation of a Circum-Pacific Institute of Volcanology ($US $10^6$/yr) that would be established within an existing university earth science program in the region, to offer classroom, laboratory, and group field training in volcanology. As the final part of their training, participants would conduct thesis field studies on high-risk volcanoes in their home countries, or by invitation in other countries. The Institute would have up to 25 trainees at any given time, and about 100 high-risk volcanoes would be studied in the first 5 years of the Institute. Results from these studies would be communicated fully by trainees to local authorities, through national volcano or geological agencies. Some contributions would be required from the host country; the remaining cost would be supported by coordinated bilateral aid agreements, supplemented where possible by UN funding.

All three levels of work are badly needed, and we are ready to work with the Circum-Pacific Council to locate funding and begin work on these projects, to prevent the next volcanic crisis from becoming a volcanic disaster.

## INTRODUCTION

About 50 volcanoes are active each year, threatening the lives and property of millions of people. People living on the slopes and at the foot of volcanoes are the most severely threatened, and some must be evacuated to areas of lower risk. Those living farther away are in less immediate danger, but their cities and towns, crops, industrial plants, transportation network, and electrical grids may be damaged by ash and by such processes as increased sedimentation in rivers. The highest risk volcanoes are those which are likely to erupt within decades or less, those which are not presently well studied or monitored, and those which are surrounded by a large population. Volcanic disasters since 1700 A.D. involving a thousand or more fatalities are given in Table 1. Twenty-three of the 26 volcanic disasters were from Circum-Pacific volcanoes, and two were in the adjacent (and geologically closely related) Caribbean region.

The potential severity of volcanic hazards was tragically demonstrated in November 1985 by Nevado del Ruiz, in Colombia. After a year of premonitory unrest, a relatively small explosive eruption melted snow and ice at the summit of the volcano. Meltwater joined water from rain-saturated ground downslope to form volcanic mudflows (lahars) that swept through several towns and killed approximately 30,000 people. Monitoring of Nevado del Ruiz was too little and too late, and even more importantly, education of local officials was insufficient, resulting in tragic inaction.

Volcanic hazards can be substantially mitigated if:

(1) national and local governments are committed to mitigating volcanic hazards;

(2) scientists responsible for such mitigation know how to collect and interpret geologic, geophysical, and geochemical data about volcanoes;

(3) sufficient equipment is available to these scientists;

(4) field studies are begun long before eruptions occur, and

(5) coordination with civil authorities is established before a crisis arises.

Geologic studies are needed to define likely types and magnitude of future eruptions. Baseline monitoring is needed to define the normal, nonworrisome fluctuations in activity of high-risk volcanoes. A continuation of that monitoring is required to ensure timely warning of when an eruption is likely to take place and, if possible, the likely nature of that eruption. All of this information must then be conveyed to civil authorities, who can use it to plan prudent land use, emergency evacuations, and other mitigation measures.

The level of funding required to mitigate volcanic hazards of a country depends on the level of volcanic activity, the size of the population, and the value of land and property at risk. The level of funding actually available depends on the government's (and local population's) awareness of volcanic hazards, and their willingness to accept the risk of an unanticipated eruption. Governments usually balance costs of mitigation against likely costs of failure to mitigate. Some countries are keenly aware of volcanic hazards and invest heavily in their mitigation; this is especially true where investment in volcanic hazards work is cost-effective even before considering the value of human lives. Other countries are equally aware of the hazards but decide to bear the risk rather than spend on adequate mitigation, or are unaware of volcanic hazards and so do little or nothing to mitigate them. We know in general principle how to mitigate volcanic hazards, but details of specific volcanoes remain poorly understood, and much more work remains to be done.

Worldwide, funding for mitigation efforts has been intermittent and usually inadequate. Numerous volcano observatories or agencies have been established only after volcanic disasters. Too often, concerned scientists cannot find adequate funding until a volcano temporarily increases public awareness of volcanic hazards, and brings a clamor for protection against those hazards.

The program of volcanic hazards mitigation proposed here is designed specifically to encourage commitments of appropriate government bodies:

— To provide substantially more opportunities for training than are presently available.

— To make timely baseline and crisis studies of high-risk volcanoes, and

— To disseminate those results thoroughly to civil authorities and thence to the public.

The program will rely primarily on trainees working under close supervision from their advisors, with emphasis on "hands-on" training and data gathering.

This proposed program of volcanological training and field studies is only a part—albeit a critical part—of the overall effort needed for volcano-hazards mitigation in the Circum-Pacific region. The program is intended to supplement national, bilateral, and other international efforts, including those of the International Association of Volcanology and Chemistry of the Earth's Interior (IAVCEI). Specifically not covered by this volcanological training and field studies program are:

— Quick responses to volcanological crises (e.g., Mayon in 1984; Nevado del Ruiz in 1985).

— Creation and maintenance of national volcanological surveys or observatories.

— Acquisition of monitoring equipment for other than instructional purposes.

— Bilateral aid programs that supply equipment and on-the-job training.

— Existing degree programs in earth sciences.

## EXISTING EFFORTS TO MITIGATE VOLCANIC HAZARDS IN THE CIRCUM-PACIFIC

Present efforts to mitigate volcanic hazards in the Circum-Pacific are in five main categories:

(1) Studies of past eruptions—to aid in preparation of volcanic hazards zonation maps;

(2) Surveillance of active volcanoes—to warn of increased unrest or imminent eruptions;

(3) Research on volcanic processes—to aid in (1) and (2);

(4) Land use planning—to discourage age development in areas of high volcanic risk;

(5) Development of contingency plans—to plan for evacuations, emergency communications, and so forth, should volcanic crises arise.

The first three of these activities can be done by volcanologists alone; the last two are done largely by civil authorities, with advice from volcanologists.

Some countries of the Circum-Pacific have large national programs for volcanic hazards mitigation, and some have nationally funded programs of international scope. Details of these are given in Appendix 1B. Even the largest programs are constantly faced with problems in maintaining budgets for this work. In other countries, budgetary limitations permit only rudimentary efforts, and in still other countries volcanic hazards mitigation is given such a low priority that virtually nothing at all is done. Volcanic disasters in any single country are infrequent, and politicians are rarely under pressure to fund mitigation measures until after a disaster has occurred.

Indeed, some nations of the region prefer to deal with the consequences of a volcanic disaster rather than commit resources toward prevention of disasters.

International programs can be a valuable supplement to local and national programs. UNESCO has a regular program for natural hazards, including activities entitled "Assessment and Prediction of Natural Hazards" and "Risk Mitigation." Under these activities, UNESCO has sponsored preparation and publication of sourcebooks on Surveillance and Prediction of Volcanic Activity (1971), Volcanic Hazards Zonation (1984), and Volcanic Emergency Management (cosponsored with UNDRO, 1985). In addition, UNESCO has funded short courses in volcano observation and monitoring, principally for participants from the Philippines, Indonesia, and Papua New Guinea, but including participants from Costa Rica and Colombia. The first of these short courses was held in the Philippines in late 1984; the second was held in Indonesia in early 1986.

One recent UNESCO proposal is for training and research in volcanology in developing countries of the Western Pacific. As early as the 1975 meeting of IUGG in Grenoble, resolutions were passed favoring establishment of a volcanological institute in the western Pacific, at which young volcanologists from developing nations of the western Pacific could pursue courses on special topics in volcanology. A team of volcanologists from Australia, Japan, Indonesia, and the Philippines, for UNESCO's Regional Office of Science and Technology for Southeast Asia (ROSTSEA), proposed a "summer school" in volcanology. As of August 1986, funding prospects from UNESCO/UNDP are poor.

A related but philosophically different proposal, also to UNESCO, calls for a program for quick assistance by international teams of volcanologists, during volcanic crises that cannot be handled by a particular country's own resources. The proposal for an "International Mobile Early Warning System for Volcanic Eruptions and (Related) Earthquakes (IMEWS)" was formulated by three regional working groups (Southeast Asia and the Pacific, Latin America and the Caribbean, and Europe/Africa) that met specifically to assess the need and feasibility of quick international assistance during volcanic crises. The proposal was submitted by UNESCO (Paris) to the United Nations Environmental Programme (UNEP), but as of August 1986 if has not been funded, and prospects for imminent funding are not good. On the occasion of the volcanic crisis at Nevado del Ruiz, Colombia, 1985, UNESCO dispatched an inspection team in cooperation with the UN Disaster Relief Coordinator (UNDRO) and the World Organization of Volcano Observatories (WOVO).

Some working groups in the International Association of Volcanology and Chemistry of the Earth's Interior (IAVCEI) are involved in efforts to mitigate volcanic hazards. In particular, the IAVCEI Working Group on the Mitigation of Volcanic Disasters is compiling information about precursors to historical eruptions of the world, and promoting several lines of communication between volcanologists and public officials. WOVO maintains a directory of observatory resources and acts as a referral service for observatories in need of assistance during routine operation and during volcanic crises. The IAVCEI Working Group on Explosive Volcanism has investigated how it might mitigate the hazards of ash to aircraft, but no working program has yet evolved. All efforts of IAVCEI Working Groups are on the initiative and funding of individual members.

From 1977 to 1984, Pacific-region volcanologists from Australia on the west to the United States on the east held satellite conferences about once every month on the PEACESAT network. Subjects ranged from investigations of specific volcanoes to reviews of current activity around the entire network, and from nuts and bolts questions about how to fix a particular piece of equipment to discussions of optimum collaboration between scientists and civil defense officials. The satellite (ATS-1 of NASA) has drifted from its original orbit and now serves all of the Americas, but no longer reaches the western Pacific. PEACESAT (University of Hawaii) is seeking access to one or more satellites that could serve the original clientele, their new Latin American clientele, and more potential clientele in the far western Pacific (Philippines, Indonesia, China). Valuable information for volcanic hazards mitigation was exchanged several years ago, and a broadened network would be even more valuable for future mitigation efforts.

The Circum-Pacific Council for Energy and Mineral Resources, a group that promotes international geoscience projects that aid in the development of the Pacific region, has inquired as to how it might help volcanic hazards mitigation efforts in the Circum-Pacific. A meeting in Hilo, Hawaii, in June 1984 reviewed hazards mitigation efforts as of that time (Souther et al., Episodes, 1984), and the present proposal is in follow-up to the meeting in Hilo and a working group meeting in Singapore in August 1986.

Numerous Circum-Pacific universities offer training that is useful for volcanic hazards mitigation. However, very few of them offer the full spectrum of specializations used in volcanic hazards mitigation. For example, many universities have programs in volcanic geology, but few have programs in volcano geophysics, and those with the latter may lack the former. To our knowledge, no university program in the Circum-Pacific region has as its specific goal the mitigation of volcanic hazards.

A few training programs include the teaching of techniques of hazards mitigation, but these are not, in general, integrated with programs of field work to actually mitigate those hazards. Notable exceptions include graduate study at Victoria University of Wellington, and on-the-job training in Indonesia, Papua New Guinea, and the Philippines.

## NEED FOR FURTHER HAZARDS MITIGATION EFFORTS IN THE CIRCUM-PACIFIC

The outstanding fact is that despite the above-mentioned efforts, volcanic hazards are effectively mitigated in only a few parts of the Circum-Pacific. For many high-risk volcanoes, we still lack the geologic, geophysical, and geochemical data, and coordination with public officials that are needed to effectively mitigate hazards. There are still many volcanoes like Nevado del Ruiz in Colombia and El Chichon in Mexico that are poorly known and ignored until it is too late.

Particular needs are:

*Training*: Two levels of training are needed. The first level should be designed for volcano observers who are mostly high-school graduates, to improve their understanding of volcanic processes and monitoring techniques. The second level is for college degree holders who need specialized courses in volcanology and related fields, preferably leading to postgraduate degrees. One special need is training that will demonstrate the symbiotic importance of integrating geologic, geophysical, and geochemical data to interpret and forecast activity of a volcano; too often, these various lines of data are collected by separate individuals, in separate institutions or agencies, and are not fully integrated.

Exchange of technicians and scientists is also needed, especially before and during eruptions, to share experience and techniques on handling volcanic emergencies. This is an unstructured but excellent training mechanism for understanding volcanoes in eruption and for upgrading responses to volcanic crises.

*Equipment*: No program for the mitigation of volcanic hazards can succeed without at least a minimum of monitoring and other observational equipment. Even a strong group of well-trained scientists or volcano watchers can do little if they do not have the instruments required to measure the level of volcanic activity. Many developing countries of the Circum-Pacific area are rich in volcanoes but poor in monitoring equipment, and any training program must be supported by appropriate equipment.

To acquire monitoring devices for all countries of the Circum-Pacific is an enormous and expensive task. Most equipment will need to be obtained from national budgets and bilateral cooperation or assistance programs. However, an inexpensive and realistic approach is for developing countries to build as much of their own monitoring equipment as possible. With hardware designs and clear instructions, some equipment can be fabricated rather than bought. For example, the lack of good telemetry has frequently hampered monitoring attempts. If designs for modulators, transceivers, and discriminators could be made available, many countries could fabricate this equipment for themselves and greatly increase the use of radio-telemetry in their monitoring. A program that is directed toward the production of simple and reliable designs of monitoring components, and the dissemination of existing designs, would be an efficient way to satisfy needs for some key items of monitoring equipment.

*Baseline Field Studies*: A third essential component of hazards mitigation is baseline geologic, geophysical, and geochemical information about high-risk volcanoes. Such information about the past and present state of a volcano is needed to interpret its present activity and to anticipate its future activity. When a volcano becomes restless, it is too late to start volcanostratigraphic studies, map the effects of previous eruptions, gather baseline geophysical and geochemical data with which to interpret the geometry and contents of subvolcanic magma and hydrothermal plumbing systems, or attempt comparisons of present unrest and routine, baseline activity. Studies such as these take months to years, and the time available during volcanic crises is usually days to weeks, or rarely, months.

*Public Education*: The best monitoring and hazard assessment is of little use unless the results are communicated in a timely manner to the responsible civil authorities and unless these authorities are prepared to act promptly. Greatly increased efforts are needed to educate public officials, the news media, and thus the general public, about volcanic hazards.

## PROGRAM PROPOSAL: GOALS AND GENERAL APPROACH

The needs for volcano hazards mitigation outlined in the previous section cannot be met by any one program, and indeed are being addressed in part by a variety of national, bilateral, and international efforts. There remain, however, key areas that are not being covered by present or planned efforts and which are addressed here. We propose a program of integrated communication measures, training, and field studies. This program is designed to reduce loss of life and property during and after volcanic eruptions, to minimize false alarms about volcanic hazards, and thus to minimize unnecessary economic and social disruption.

Program activities, listed in the next section, are designed to be:

1. *Communications-intensive*, with efforts to improve communications among volcanologists, and between volcanologists and the general public.
2. *Problem-oriented*. Participants should define problems, plan how to solve them, implement the plans, and write up the results. Study toward solving specific problems is of more interest than training in methodology alone.
3. *Field-oriented, with emphasis on hands-on, how-to-do-it training*. Lectures should be limited to the necessary minimum, and should be integrated with the field study.
4. *Multifaceted, multidisciplinary*. The program will require a team examination of relationships between several types of data—for example, seismic, ground deformation, petrologic, geochemical, stratigraphic, and hydrologic. Far too often, geoscientists do not know what their colleagues are doing, or why, and how colleagues can help.
5. *Scientifically critical*. There should be a strong emphasis on skepticism, questioning one's own data and the results of others. Education in many countries is still by rote memorization, and a training course will have much greater long-term returns if it can get trainees to be critical about what they see, read, and hear.
6. *Strongly oriented toward self-help and self-sufficiency*. The task of mitigating volcanic hazards requires years of geologic studies, continuous surveillance of active volcanoes, and frequent interaction with local public safety officials. These activities are much better handled by national scientists than by foreign scientists.

We have divided the proposed activities into three levels: (1) those that can be done within existing programs and facilities, at relatively low cost (\$US $10^4$/yr); (2) those that are fundamentally new projects, that could be done at a moderate level of funding (\$US $10^5$/yr); and (3) the cornerstone, but also the largest item in our proposal (\$US $10^6$/yr), a Circum-Pacific Institute of Volcanology that would offer training and coordinate thesis projects on high-risk volcanoes of the region.

### Level A, Low-Cost Projects

Items on the smallest, simplest scale are for improving communications and the exchange of ideas and information among Circum-Pacific volcanologists. We propose that the following be established.

1. *A clearinghouse for information on both organizational matters (training opportunities, potential funding programs, etc.) and scientific/technical information (new hardware/software, recent publications, current status of volcanic activity, etc.)*. Some of these information needs are currently handled through newsletters of IAVCEI, WOVO, and the IAVCEI Working Group for the Mitigation of Volcanic Disasters. Looking toward the future, an international computer bulletin

board, accessible by local telephone call from major cities around the Pacific, would provide effective, rapid communication without requiring any one person or institution to take sole responsibility for collection and dissemination of information. Costs: computer connect time, annual subscription fees, within-country long-distance telephone charges.

2. *A regular program of satellite conferences, perhaps monthly, for Circum-Pacific volcanologists.* Since NASA's ATS-1 satellite drifted into the eastern Pacific, trans-Pacific PEACESAT conferences have been impossible. There are other satellites which could serve the volcanology and other PEACESAT conferences, but NASA has not as yet made channels and time available for these purposes. Such conferences provide the opportunity for frequent discussions without incurring the high costs in time and money of travel to international meetings. Costs: nominal long-distance telephone charges to ground stations, and new (but relatively low-cost) radio links for those volcanological organizations presently without access to PEACESAT.

3. *An annual roundtable for discussion of Circum-Pacific volcanism.* Activities could range from discussion of common problems and exchange of information on latest advances in volcanic hazards mitigation to coordinated several-day field studies by a carefully constituted group of volcanologists. Costs: travel and per diem for some participants.

## Level B, Intermediate-Cost Projects

We envision several activities that are relatively modest in cost and great in potential benefit. These would require an annual budget in the range of $100,000-$200,000 per year and would operate in conjunction with the activities in Level A. Activities of Level B include: (1) short courses and field workshops, (2) exchange of personnel for in-service training and eruption observation, (3) promotion of volcanological research and education in existing degree programs, and (4) videocassette education.

At this intermediate level, staff services will be required: minimally only part-time clerical help, but optimally a full-time activity coordinator. This role is important not only in coordinating workshops, field courses, and exchanges of scientific and technical personnel, but also in coordinating this training with various bilateral aid efforts in volcano-hazard mitigation.

*Short courses and field workshops:* As an interim solution to the need for specialized training of volcano observers in Indonesia, the Philippines, and Papua New Guinea, a series of short-term (2-4 weeks duration), nondegree technical training in the form of short courses and field workshops have been successfully carried out in the Philippines (1984) and in Indonesia (1986). These courses have been funded through grants from UNESCO, with partial support from the host countries and (for two participants from Colombia) from USAID. The focus of these courses has been in-country training with attention to high-risk volcanoes. The principal purpose of these courses and workshops has been to provide hands-on training in the use of modern equipment of volcano monitoring, interpretation of the data collected using that equipment, and the recognition and interpretation of recent volcanic deposits to infer what future eruptions might include, at selected high-risk volcanoes in the host countries (Mayon volcano in the Philippines and Merapi volcano in Indonesia). Participants in these courses have included volcanological observers and geologists from the Philippines, Indonesia, Papua New Guinea, Costa Rica, and Colombia.

To date these workshops have focused their attention on the western Pacific region. Each workshop has depended on local initiative for organizing and running the workshop on an informal basis. Notice of the courses was sent to volcanological institutions in the region, and no attempt was made to advertise these courses widely.

In 1987, a workshop on lahars will be held in the Philippines and a short course on volcano monitoring will be held in Papua New Guinea. UNESCO's Regional Office for Science and Technology in Southeast Asia (ROSTSEA) will pay costs for participants from the Philippines, Indonesia, and Papua New Guinea. We seek travel support for additional participants from countries of the eastern Pacific who are not eligible for ROSTSEA support. For subsequent years, we seek CPCEMR's help to sustain funding from UNESCO and other sources to continue and expand these short courses and workshops.

Costs for a 2- to 4-week-long workshop, with 10 host country participants and 10 regional participants, is between $(US)20,000 and $30,000, excluding salary for lecturers.

2. *Exchange of personnel for in-service training and eruption observation*: International exchanges of personnel are relatively inexpensive and mutually beneficial. Each visitor brings experiences to share and learns from his/her hosts. Some exchanges trade experience with many volcanoes and eruptions for training in new techniques for volcanic study. Some training topics are better handled on an individual, one-on-one basis than in a short-course or workshop setting. Other exchanges bring together workers with similar skills during volcanic crises, e.g., sharing of trained eruption observers during eruptions. Public officials who must deal with volcanic crises could both learn and contribute during volcanic crises in countries other than their own. Eruptions provide a unique, superb setting for some types of training, and it is rarely feasible to schedule a short-course or workshop to correspond to actual eruptions.

Accordingly, we recommend that funds be designated for the exchange of personnel for in-service training and eruption observation. Costs for this type of exchange could vary widely, depending on the number of personnel seeking training and the frequency and nature of eruptions in any one year. We estimate an annual budget of $(US)50,000 for 10, month-long visits at $(US)5000/visit.

3. *Promotion of volcanological research and education in existing degree programs*: Many universities in the Circum-Pacific region have departments of earth science, with graduate programs related to volcanology. It would be possible with limited funding to support additional graduate students to undertake research at high-risk volcanoes. It is estimated that an average cost of support for graduate students is $(US)10,000/year; an average M.Sc. takes 2 years and Ph.D. 4 years. Most universities could accept one or two graduate students without additional staffing, and it should therefore be possible to place 20 graduate students in universities around the Pacific rim, each studying a high-risk volcano during the period of 1987-1991. If 50% were Ph.D. students, the total cost would be $(US)600,000 over the 4-year period (i.e., $(US)150,000/year). Some of the funding could come directly from bilateral aid programs. The benefit of this approach is that it could be implemented immediately using available training facilities.

4. *Videocassette education*: Although public officials and the general public in volcanic areas may recognize the existence of volcanic hazards in a general way, only rarely will they have a sharp awareness of specific volcanic hazards. The city of Armero in Colombia was a tragic example of this. A volcanic hazard map may be a graphic and lucid document to a volcanologist, but it is an abstract piece of paper to most laymen. Videocassettes showing the general nature of volcanism, illustrating dramatic examples of specific volcanic hazards and outlining response measures (including evacuation), could be invaluable in public education and in effecting timely reactions to increased volcanic activity.

Videocassettes can be prepared with the narrative in one language (e.g., English) and subsequently can be translated into any appropriate language. The advantage over other media lies primarily in the ability to communicate information through the television set. In addition, videocassettes have great advantages over film in ease of editing. In short, videocassettes provide a powerful and flexible means of public education.

Dr. Maurice Krafft, an eminent volcanologist-filmmaker, is prepared to produce two 20-minute video tapes—one on volcanic hazards in general and one with source materials on specific hazards—for the cost of production only (using his own and other existing footage at no extra charge). Production costs are $(US)43,000 for each 20-minute video tape.

Videocassettes could also be a powerful tool in training for volcanologists, especially for showing examples of eruptions and practical, "how-to-do-it" matters of volcano monitoring. A student trained in a particular technology (e.g., electronic distance measurement) would find a videocassette module to be a valuable reinforcement and reference after his/her return to the home country. In addition, the videocassette would also be a powerful teaching aid in disseminating the technology to persons unable to attend a formal training session or to those who did not have the appropriate language capability to complete a training course in English. Training videocassettes could be particularly valuable in training technical people on their home ground. A picture, particularly a moving one on television, is worth a thousand words; a technique is learned much more easily and thoroughly by watching than by reading or listening, particularly when the medium of video allows repetitive observation and freezing frames of particularly important steps. The USGS's Volcano Early-Warning Disaster Assistance Program (VDAP) plans to prepare some volcano training videocassettes.

## Level C, Circum-Pacific Institute of Volcanology

A major goal of any Circum-Pacific program in volcano hazards must be the development of national self-sufficiency; each country must become able to carry out and maintain its own program of volcano hazards mitigation. To achieve this goal of national self-sufficiency, it is essential that a critical number of indigenous personnel of developing countries be brought to a level of volcanological expertise sufficient to (a) make effective use of bilateral aid programs, and (b) maintain national programs without indefinite external support.

Training is needed at all levels, from the volcano watcher level through the technician level to the engineering, scientific, and managerial level. Much of the needs at the lower levels can be met by on-the-job training, bilateral aid programs, and field workshops, as outlined in previous sections. Similarly, the highest levels can partly be met by existing or evolving advanced-degree programs. But it is at the intermediate level where *there is a critical need for broad training in practical volcanology*. The needed training is beyond the general geological training of a bachelor's degree, but not at the specialized academic level represented by a masters or doctoral program. The demand for this broad training in practical volcanology is large and exists in all countries of the Circum-Pacific region.

The need for this broad, practical volcanological training cannot be met on a country-by-country basis, because the educational and volcanological resources of developing countries are too limited. Instead, this need must be met on a regional basis, essentially by the pooling of educational and volcanological resources of Circum-Pacific countries.

The best mechanism for pooling these resources would be to establish a Circum-Pacific Institute (or School) of Volcanology. Details are given in Appendix 1C. The curriculum of this school would focus on a 1-year course covering the full spectrum of volcanology (fundamentals, monitoring, and hazard assessment) and leading to the award of a "Diploma in Volcanology" certifying the achievement of a standard level of competency. This diploma would qualify the graduate to return to his home country as a knowledgeable volcanological practitioner, and in addition would represent the basic qualification for further volcanological training at the masters or doctoral level.

It would be advantageous to house the Circum-Pacific Volcanological Institute in an existing university or universities. This is important to obviate creation of new infrastructure, to minimize the need for new faculty, and to enable trainees to receive diplomas and degrees from a recognized university. To the extent possible, training should be closely coordinated with a nearby volcano observatory.

Operation would be similar to that established at the Geothermal Institute, University of Auckland, which has proved to be highly successful (see Appendix 1-D).

Trainees for the Volcanological Institute would be welcome from all Circum-Pacific countries, but special preferences would be given to participants from developing countries. Instruction would come from the host country or countries, with supporting staff drawn from other countries.

The different requirements of the western Pacific countries and the Latin American countries suggest that it may be advantageous to house the Volcanological Institute at two locations within the Circum-Pacific region. Reasons for this include the following:

- Language: English is the lingua franca"—the language of commerce, science, and international communication—of the western Pacific. In contrast, Spanish is the common language of volcanic areas of Latin America.
- Distance: The difficulties and expense of travel from Latin America to the western Pacific and vice versa will mitigate against students and scientists from one region fully participating in the program of an Institute located in the other region.
- Environment: Most of the volcanoes in low latitudes of the western Pacific are in tropical environments with heavy vegetation, deep soils, and free of snow, ice, and glaciers. Some in low latitudes of the western Pacific are the same, but many others are in desert climates, are lit-

tle eroded, and are at such high elevations that snow, ice, and glaciers are major logistical considerations. Effective training has to be adapted to the regional needs of these two contrasting environments.

If it is decided to have two campuses of the Institute, one should be in the eastern Pacific (Latin America) and the other in the western Pacific, with some exchange of Institute staff between the two locations.

## Conclusions

The proposal sets out to:
— Achieve national self-sufficiency in coping with volcanic hazards.
— Gather data to mitigate hazards at high-risk volcanoes.

This would be achieved by means of a three-stage program with three increasing levels of funding:

*Level A* (cost \$US $10^4$ per year):
— Roundtable discussions on volcanic hazards
— Clearinghouse for information on training opportunities
— Exchange of simple designs for monitoring equipment
— Promotion of satellite communication (e.g., PEACESAT)

*Level B* (cost \$US $10^5$ per year):
— Organization of short courses in developing countries
— Operating field workshops
— Production of educational videotapes on volcanic hazards and monitoring techniques
— Promotion of exchange visits for observatory personnel
— Increase of the number of positions in existing degree programs

*Level C* (cost \$US $10^6$ per year):
— Establishment of a Circum-Pacific Institute of Volcanology

# APPENDIX 1A: HIGH-RISK VOLCANOES: A PROPOSAL FROM THE INTERNATIONAL ASSOCIATION OF VOLCANOLOGY AND CHEMISTRY OF THE EARTH'S INTERIOR

The eruption of Nevado del Ruiz on November 13, 1985, caused the death of more than 25,000 people. The potential for tragedies of similar or greater dimensions exists at many other volcanoes, especially in developing countries. The International Association of Volcanology and Chemistry of the Earth's Interior (IAVCEI), on the occasion of the International Volcanological Congress in New Zealand, strongly urged that programs to reduce the risk of major volcanic catastrophes be initiated and given lasting support.

An inventory of especially dangerous volcanoes was recently prepared by IAVCEI and UNESCO. About 100 high-risk volcanoes were identified; most of these lie around the Pacific Ocean Basin in developing Third World countries. IAVCEI has also recently commissioned a task force that will develop a broadly defined strategy to mitigate the hazards of high-risk volcanoes.

Effective programs that would lay a foundation for long-term mitigation of volcanic hazards must include:

(1) Training of scientists from developing countries having such volcanoes. This is paramount in order to bring about long-term self-sufficiency in volcano monitoring within such countries;

(2) Studies of previous eruptive behavior, preparation of hazard-zonation maps, hazard information and training for people living near high-risk volcanoes, and advance interactions with responsible civil authorities;

(3) Intensified monitoring of dangerous volcanoes showing current signs of unrest; and long-term baseline monitoring studies of other high-risk volcanoes.

We ask the research and aid agencies of the Developed Countries to pool their financial resources to initiate and maintain the proposed program on HIGH-RISK VOLCANOES. The amount of money needed is small compared with many scientific and aid programs, but long-term stable funding is needed. This program is needed now; we cannot wait for another disaster like that at Nevado del Ruiz.

IAVCEI Executive Committee—February 1986

# APPENDIX 1B: EXISTING EFFORTS AT HAZARDS MITIGATION IN THE CIRCUM-PACIFIC

## National Programs

**New Zealand/Kermadecs**

New Zealand has 10 potentially active volcanoes, plus Raoul Island in the Kermadecs. Much excellent work has been done by New Zealand volcanologists, especially in the field of tephrochronology and volcanic stratigraphy. However, these studies found a largely scientific audience until October 1980, after the eruption of Mount St. Helens, when the New Zealand Civil Defense agency formed a Planning Committee on Volcanic Hazards. That committee reported on the likelihood of various types of eruptions and hazards at 8 volcanic centers, including several large silicic centers (Dibble et al., 1985; Latter, 1986). Hazards zonation maps were prepared for some centers.

Nearly half of New Zealand's potentially active volcanoes are presently monitored, though not as intensively as wished by New Zealand scientists. White Island and the Tongariro volcanoes (Ruapehu, Ngauruhoe, and Tongariro) are monitored by DSIR, Victoria University, and the Tongariro National Park. The Taupo center, the Okataina/Haroharo center (including Tarawera), and Mt. Taranaki (Mt. Egmont) are monitored less intensively by DSIR, with assistance at Taranaki from the Egmont National Park. Auckland University has recently installed a seismic network in the Auckland volcanic field.

An immediate warning system for lahars has been installed on the ski slopes of Mt. Ruapehu, to warn skiers when the crater lake overflows. A warning would also be conveyed to the National Railways, which experienced a tragic accident in 1953 when a lahar swept away a railway bridge just before a passenger train passed by. Another serious eruption occurred at Mt. Tarawera in 1886, and New Zealand volcanologists organized a very simulating and successful international volcanological congress on the occasion of the Tarawera centennial. One item of particular concern to scientists and authorities in New Zealand is the infrequent but extremely dangerous occurrence of large eruptions from silicic centers, like that which occurred about 1800 years ago from the Taupo center. Latter (1986) recently estimated that early warnings of eruptions in New Zealand could save approximately 16% of what would otherwise be lost, or about $(US)390,000 per year out of a projected loss of $(US)2.3 million per year.

New Zealand universities, together with DSIR, offer excellent facilities for training in volcanology. Auckland and Victoria Universities each have several specialists in volcanology, and smaller numbers of specialists teach at Waikato, Massey, Canterbury, and Otago Universities. All of these universities offer graduate degrees in the earth sciences. In addition, New Zealand has a 1-year diploma course in geothermal studies (Geothermal Institute at the University of Auckland) and plans to start one in volcanology (Victoria University) next year. The latter course is well suited to students from developing nations who need a 1-year course in lieu of or before graduate study in volcanology. Diploma training in geothermics at Auckland has been jointly funded by UNDP, the New Zealand Government's Ministry of Foreign Affairs, and the University of Auckland, and is specifically designed for students from developing countries as well as from New Zealand.

### Tonga

There are 11 recognized volcanoes in the Tongan group—most of which are submarine, and with a relatively high population for the land area (average 134 people/km$^2$), there is a danger. Particular risk occurs at Tofua and Niuafou'ou. There is no established volcanological group, although one geologist (David Tappin) is supplied from UK. Information on the state of Tonga volcanoes could be sent via satellite to New Zealand.

### Samoa

Four volcanoes have been active in historic times.

### Australia

Australia has no active volcanoes, but does have Quaternary volcanoes along its east coast from Queensland to Victoria, some of which have erupted within the last 15,000 years. An active research community is documenting that activity, but there is no immediate need for further action by civil authorities.

Australia has a historically close relationship with Papua New Guinea, and continues to provide the Rabaul Volcano Observatory with assistance when requested. Australia also has a special concern about the safety of international flights to and from Australia that pass through volcanic areas. Two Australia-bound jumbo jets were severely damaged after flying into ash clouds from Galunggung volcano, Indonesia, in 1982, and Qantas continues to make expensive re-routings of its aircraft when volcanic activity is likely along or upwind of their flight paths. An advisory group known as VULCAN/AUS advises Australian aviation authorities on hazards from volcanic eruptions.

### Papua New Guinea

Papua New Guinea has approximately 14 active volcanoes, and others that have erupted within the past several hundred years but not in the short period of historical reporting. The Volcanological Observatory at Rabaul (Department of Minerals and Energy) is the focus for all volcanological work in Papua New Guinea. The Observatory was established and originally staffed by Australia; from 1975 to 1981 Papua New Guinea gradually assumed the operation of the observatory, but still retains a strong, largely informal link with the Australian BMR. The Observatory now has two national geologists, and more will be added when candidates are available. The Observatory is responsible for monitoring all of Papua New Guinea's very active volcanoes, so the small staff of the Observatory is sometimes spread thinly around the country.

One particularly challenging crisis occurred from 1983-1984 in the Rebaul caldera; intense seismicity, updoming of the caldera floor, and other changes foreshadowed a possible eruption. An alert was called and some residents and businesses left Rabaul for a period of some months; however, unrest declined in 1984, the alert was downgraded, and as of the time of this writing (August 1986) no eruption has taken place. At large silicic centers such as this, potential danger is high but prediction is especially difficult, and the chance of a false alarm is also high.

The University of Papua New Guinea, in Port Moresby, offers a geology major but does not offer a specialization in volcanology.

### Solomon Islands and Vanuatu

The Solomon Islands have 5 active volcanoes. A government geologist in Honiara makes visual inspections of volcanoes as required. Savo is monitored by a 3-station network, and could be a serious risk.

Vanuatu has 12 active volcanoes, and has from time to time had modest monitoring efforts. The French Overseas Technical Program, ORSTOM, monitors Matthew Island by radiotelemetry.

### Indonesia

There are 128 active volcanoes in Indonesia, of which 70 are situated in densely populated areas. Thirty-eight of these are monitored continuously by the Volcanological Survey of Indonesia (VSI), most by an observatory on or near each volcano, equipped with one seismograph. Other monitoring methods are visual observations, studies of ground deformation (spirit-level tilt measurement and precise distance measurement), geochemical studies (of gas and condensates from fumaroles, water from crater lakes and hot springs, and eruption products), and geophysical methods (self-potential and geomagnetic variation). After large explosive eruptions satellite imagery has been used to monitor the movement of eruption plumes and to provide warning of tephra fall.

In addition to direct monitoring, the VSI prepares volcanogeologic and volcanic-hazard maps; the former are used to better understand the behavior of the volcano, and the latter are used by local authorities to evaluate potential dangers and to direct evacuations in case of eruption.

From January 1980 through September 1985 there were 35 volcanic eruptions in Indonesia. Nine of these were relatively large eruptions that threatened life and property, and required evacuation of local residents. No lives were lost during this period, but 190,000 people were evacuated and approximately 14,000 homes were damaged, primarily by mudflows and pyroclastic flows. The largest eruption was from Galunggung, West Java, which affected more than 900,000 km$^2$, required evacuation of 75,000 persons and caused more than $(US)130 million in property damage.

VSI participates in international assistance projects with several countries and the Asian Development Bank. Assistance has been provided by the U.S. Geological Survey (USGS) and the U.S. Agency for International Development (USAID), the French Centre National de Recherches Scientifiques (CNRS), the Japan International Co-operation Agency (JICA), the University of Pisa and the Consiglio Nazionale della Richerche (CNR) of Italy, and Geothermal Institutes of Italy and New Zealand.

Geology degrees are offered by the Institut Teknologi Bandung (ITB), Padjajaran, Gadjak Mada, Hassanudin, Trisakti, and Pembangunan Universities; volcanology courses are offered at each, and one can pursue an M.S. or Ph.D. with a specialization in volcanology at ITB. In 1986, VSI hosted the second UNESCO-funded regional volcanology training course in Yopyakarta.

### Philippines

The Philippines has 20 active volcanoes and more than 50 others that are geologically young and have thermal manifestations. The Philippine Institute of Volcanology and Seismology (PHIVOLCS) is responsible for most volcanological work, and is joined by the Philippine National Oil Company (PNOC) and the Bureau of Energy Development (BED) in connection with geothermal exploration, and research and development activities. PHIVOLCS' volcano-related work consists largely of geologic mapping of young volcanoes, maintaining observatories at the five most active volcanoes (Taal, Mayon, Bulusan, Canlaon, and Hibok-Hibok), conducting studies on economic minerals and rocks associated with younger volcanoes and older volcanic terranes, and undertaking topical research projects. Geologic maps are made to serve two objectives: eruption prediction and hazards assessment. At the observatories, seismic and visual monitoring are the principal methods, but these are supplemented by temperature measurements at Taal, Canlaon, and Bulusan, and recently by electronic distance measurements at all five volcanoes. These monitoring data are collated and processed, and together with historical data are compiled into "Pre-eruption volcano profiles."

Both PNOC and BED undertake exploration and development of geothermal areas for large-scale power generation. PHIVOLCS conducts limited exploration activities in subeconomic areas, and develops technologies for nonelectrical use of geothermal steam. In geothermal energy development, the Philippines has an installed generating capacity of 894 MW and a total generating capacity of 1478 MW from five major geothermal fields (Tiwi, Makiling-Banahaw, Tongonan, Palimpinon, and Bacon-Manito).

The Philippines experiences only 1 eruption every 2 or 3 years, on the average, but these active volcanoes are in densely populated areas. To minimize economic and social dislocation of a vast number of people, the PHIVOLCS staff must give timely warnings, and also allow residents to return to their homes as soon as it is safe. PHIVOLCS and its predecessor, the Commission on Volcanology (COMVOL), have, in practice, made decisions about when and where evacuations must be made; civil authorities have followed their recommendations. Such an arrangement is gratifying, but also places volcanologists in a de facto position of weighing volcanic hazard against social and economic factors, and of deciding what level of risk the public should be allowed to take.

In September 1984 an eruption of Mayon volcano threatened more than 500,000 people. As many as 73,000 people were evacuated during the peak of activity, and although most were allowed to return home within a few weeks, those who lived in areas subject to lahars were kept evacuated for several months.

Several universities offer degrees in geology, and the University of Philippines (UP) offers graduate degrees. Volcanology and geothermal energy are taught as subjects in the graduate school at UP. In 1984, PHIVOLCS hosted a Regional Training Course for Volcano Observers, sponsored by UNESCO and attended by trainees from Indonesia, Papua New Guinea, Costa Rica, and the Philippines. The Philippines has also offered to host a UNESCO-funded summer school in volcanology, and the Philippines and Indonesia have a Volcanological Exchange Programme under the auspices of UNESCO (see International efforts at volcanic hazards mitigation in the Circum-Pacific).

### Marianas

There are at least 8 active volcanoes in the Marianas, including several submarine centers.

### Japan

In Japan, 77 volcanoes are known to have erupted in historical time or have the potential for activity in the near future. These volcanoes lie principally on Kyushu, Honshu, Hokkaido, and in the Bonin and Volcano Islands.

The country maintains a large program of volcanic hazards mitigation, under the umbrella of the Coordinating Committee for the Prediction of Volcanic Eruptions, which was established according to a resolution of the Geodetic Council of Japan. Principal participants are universities, the Japan Meteorological Agency, the Geological Survey of Japan, the Maritime Safety Agency, the National Research Center for Disaster Prevention, the Geographical Survey Institute, and administrative officials from the Science and Technology Agency, National Land Agency, and Ministry of Education, Science and Culture. This committee meets about three times per year, and more frequently in the case of volcanic eruptions, to coordinate the activities of various member observatories and to make on-the-spot recommendations, such as evacuation of residents, to administrative agencies. Research and monitoring are funded by a National Project for Prediction of Volcanic Eruptions, which has operated since 1974.

Geologic mapping and studies of previous eruptions are done by universities and by the Geological Survey of Japan. Research-oriented volcano observatories are operated by the National Research Center for Disaster Prevention, and by universities—e.g., at Usu by Hokkaido University; at the volcanoes of the Tohoku district by Tohoku University; at Asama, Oshima, and Kirishima by the Earthquake Research Institute, University of Tokyo; at Kusatushirane by Tokyo Institute of Technology; at Aso and Sakurajima by Kyoto University; and at Unzen by Kyushu University. Surveillance-oriented, largely seismic observatories are maintained by the Japan Meteorological Agency. The Maritime Safety Agency is concerned with submarine volcanism such as that which destroyed the research ship Kaiyo-maru No. 5 in 1952, and the Geographical Survey Institute maintains an extensive geodetic network that includes many active volcanoes.

Historically, Japanese civil authorities have not encouraged preparation of volcanic hazards maps, as such maps raise political controversy. However, this reluctance may be changing, and the Kagoshima Prefectural Government, the Sakurajima Volcano Observatory, and other scientists are organizing a meeting in 1988 that will specifically aim to improve communications between volcanologists and civil authorities.

Many universities in Japan offer specialized training in volcanology, through the Ph.D. level. In addition, Japan maintains the International Institute for Seismology and Earthquake Engineering (IISEE), which offers training in volcanic physics within their program for general seismology studies.

**Union of Soviet Socialist Republics**

The USSR has approximately 55 active volcanoes, 35 in the Kuriles and 20 in Kamchatka. The government maintains a large volcano research program at the Institute of Volcanology in Petropavlovsk (Kamchatka), and smaller programs in Sakhalin, Moscow, and other cities. Programs include volcano mapping, geophysical and geochemical surveillance of selected volcanoes, basic research on volcanic and tectonic processes of the region, and research on relationships between volcanic and hydrothermal activity and mineral deposits. Most Kamchatkan and Kurile volcanoes are in sparsely populated areas, so hazards mitigation has been a by-product of volcano research in the USSR.

Universities in Moscow and other major cities offer graduate training for students from the USSR and abroad. A number of Latin American scientists have studied in the USSR, and some USSR volcanologists have conducted cooperative research in Mexico, Nicaragua, and other countries of Central and South America.

**Canada**

Four young stratovolcanoes and several tens of smaller basaltic cones occur in British Columbia. One or more large silicic centers lie in the northern Yukon, near the border with Alaska. The population around these volcanoes is sparse, and eruptions have occurred in Canada within historical time. Scientists from the Geological Survey of Canada and the University of British Columbia are studying the geology and petrology of Canada's young volcanoes, but the only monitoring of these volcanoes is a by-product of the regional seismic network.

Because of Canada's proximity to volcanoes in the State of Washington, especially Mount Baker, Canadian civil defense and aviation authorities have participated actively in contingency planning and briefings about unrest at nearby U.S. volcanoes.

Most universities in Canada offer degrees in geology, and some (e.g., the University of British Columbia and the University of Western Ontario) offer courses in volcanology. The Pacific Science Center in British Columbia has prepared educational displays about volcanic activity of the Cascades.

The Canadian International Development Agency (CIDA) has provided assistance during several recent volcanologic crises, including Galunggung (Indonesia, 1982) and Nevado del Ruiz (Colombia, 1985).

**United States of America**

The U.S. has 27 active volcanoes in the Aleutians, 47 on the Alaska Peninsula, around the Cook Inlet, and in SE Alaska, 6 in the Cascades, and 5 in Hawaii. Many others have erupted within the past 1000 years. Most volcanic hazards mitigation work is conducted by the U.S. Geological Survey, which operates 2 observatories—the Hawaiian and Cascades Volcano Observatories—and maintains a program of geologic, geochemical, and geophysical research on volcanoes of Hawaii, the Cascades, and Alaska.

Another important effort is by the Smithsonian Institution, which maintains a large data bank on volcanoes of the world. The geography and eruption histories of more than 1000 volcanoes active within the past 10,000 years have been entered into a computerized Volcano Reference File and published in Volcanoes of the World, and maps and photographs of active volcanoes are being gathered in the Smithsonian Volcanological Archives. The Smithsonian also maintains a directory of scientists who are familiar with each volcano or with specialized study techniques, and the Scientific Event Alert Network (SEAN) gathers and disseminates information on current volcanic activity around the world.

Several universities also make major contributions toward volcanic hazards mitigation, including the University of Washington (seismic monitoring of volcanoes in Washington and northern Oregon) and the University of Alaska (seismic and other monitoring of Augustine volcano). Geologic and petrologic studies are conducted by researchers from many universities. Several state geological agencies study the geology of volcanoes in their state, but only California and Alaska operate their own monitoring stations.

Geologic and hazards maps have been prepared for Hawaiian volcanoes and for most large Cascades volcanoes, and in some instances the geologic maps have been revised several times over by successive workers.

By far the greatest volcanic crisis in U.S. history occurred in 1980 at Mount St. Helens. Two months of dramatic precursory activity were climaxed by a massive landslide and laterally-directed blast. Six hundred km$^2$ of forest were devastated, but much of that had been evacuated, and the eruption occurred on a Sunday rather than a workday, so only 57 people died. Dome growth and small explosions continue as of May 1986. Another potential volcanic crisis occurred at Long Valley caldera, eastern California, beginning in about 1978, peaking in 1983, and declining gradually to the present time. The unrest is partly tectonic, but potentially serious eruptions have occurred in this large volcanic center as recently as 500 years ago.

The U.S. Geological Survey, Los Alamos Scientific Laboratories, several universities, and USAID are involved in a number of bilateral cooperative programs with foreign volcanological surveys. Current programs are with Indonesia, the Philippines, Italy, Colombia, Ecuador, Guatemala, El Salvador, and Costa Rica. Researchers from U.S. universities work in these and many other volcanic areas. Most of the U.S. international efforts have been of a long-term, institution-building nature, but some have involved assistance during volcanic crises (e.g., Ruiz). A recent agreement between the USGS and the U.S. Office of Foreign Disaster Assistance (OFDA/USAID) provides support for precrisis through postcrisis activities in Latin America, including maintenance by the USGS of a cache of monitoring equipment, development of a data bank on potentially active volcanoes, limited baseline monitoring, and development of liaisons with volcanological centers in Colombia, Ecuador, and Costa Rica.

Many universities in the U.S. offer graduate degrees in geology, and several (University of California at Santa Barbara and Santa Cruz, University of Washington, University of Hawaii, University of Alaska, Dartmouth College, Michigan Technological University, University of Rhode Island, Arizona State University, University of Texas at Arlington, Louisiana State University, and others) offer courses and specializations in volcanology. The U.S. Geological Survey offers, from time to time, short courses on geologic hazards mitigation, and opportunities for individual training at the Cascades and Hawaiian Volcano Observatories.

**Mexico**

The recent eruption of El Chichon volcano in Mexico focused attention on the volcanoes of Mexico, and a modest program of volcano monitoring has been carried out in that country since 1982, mostly through the effort of Dr. Servando de la Cruz and his colleagues of the Nacional University of Mexico City. The present economic difficulties of Mexico threaten the existence of the national monitoring efforts. Locally, the national hydroelectric agency operates seismic networks near their hydroelectric reservoirs, and it was this network that detected the first sign that El Chichon was coming back to life. Most other studies of Mexican volcanoes are geologic or petrologic, conducted by university researchers from Mexico, the U.S., France, the USSR, and other countries. The capital city, Mexico, Distrito Federal, lies within a field of geologically young cinder cones.

The seismic and fumarolic activity that started in 1986 in Tacana volcano, on the Mexico-Guatemala border, has triggered a new wave of concern, particularly among authorities of the state of Chiapas, and of the Guatemalan government. Mexico and Guatemala have designed compatible prevention and preparedness plans, and the Chiapas plan is to be used as a model for a national plan.

**Central America**

Central America, from Panama to Guatemala, contains nearly 70 volcanoes, of which about 35 have had historical activity (since the mid-1500s). Active volcanoes are found in all countries except Belize and Honduras. In the past decade, the unstable political climate in Guatemala, El Salvador, and Nicaragua has affected the efforts of government and university agencies to monitor volcanic activity in those countries. Only Costa Rica has a bonafide national plan of volcano monitoring. Many Central American volcanoes have been prospected for their geothermal resource potential. Such evaluations typically involve geological mapping and geophysical studies including seismic studies.

Guatemala has about 23 volcanoes, of which 6 have had activity during the past 500 years. Traditionally, the Instituto Geografico Militar has been responsible for observing and reporting on the activity of Guatemala volcanoes with reports supplemented by geologists from U.S. universities, especially Dartmouth College. Recently, the Instituto de Sismologia, Vulcanologia, Meteorologia e Hidrologia (INSIVUMEH, Guatemala City) has borne the responsibility of monitoring volcanic activity in Guatemala. One permanent seismic station was recently installed near Santiaquito volcano, another station was installed at Tacana volcano under a bilateral Guatemala-Mexico disaster prevention program, and other seismic swarms have been monitored at the Amatitlan caldera, Fuego volcano, and near Sta. Rosa.

El Salvador contains about 19 volcanoes, 6 of which have had historical activity. Once again, political instability in El Salvador during the past decade has seriously reduced the frequency and quality of volcano monitoring efforts. The government seismological agency, Centro de Investigaciones Geotecnicas, is responsible for monitoring El Salvador's active volcanoes but encounters logistical problems. The government hydroelectric/geothermal power agency employs at least one geologist with special training in volcanology.

Nicaragua contains about 22 volcanoes, of which about 10 have had historical activity. The recent political instability of Nicaragua has made it difficult to continue monitoring of these volcanoes on a regular and systematic basis. A national seismic network, originally installed by the U.S. Geological Survey following the 1972 Managua earthquake, remains in operation despite withdrawal of USGS support for the network. The Instituo Nicaraguense de Estsudios Territoriales (INETER) is responsible for geologic studies and monitoring of Nicaraguan volcanoes.

The 11 volcanoes of Costa Rica receive the best coordinated monitoring surveillance in Latin America. The Costa Rican Observatory of Volcanology and Seismology at the Nacional University at Heredia maintains a modern seismic network with many seismic stations on or near the active volcanoes. This network is operated in cooperation with the University of California-Santa Cruz. In addition, geologists of the ISV maintain and routinely reoccupy deformation monitoring efforts at 4 to 6 of the volcanoes. Periodic visual observations are made at about 8 of the volcanoes. In the event of increased activity at a volcano, ISV geologists are prepared to respond to the activity with increased seismic and visual observations. The government hydroelectric power agency, Instituo Costarricense de Electricidad (ICE), employs one or more geologists to assess volcanic hazards to its operations. Collaborative Costa Rican-U.S. studies in tephrochronology are revealing new information about explosive eruptions of Arenal and other Costa Rican volcanoes.

Panama contains a single volcano, Baru, with a record of historical activity.

**South America**

In South America, more than 140 volcanoes occur along the Andean chain, spread unevenly through six countries including Bolivia, Chile, Argentina, Peru, Ecuador, and Colombia. Only Chile, Ecuador, and recently Colombia have organizations which regularly report on the activity of volcanoes in their countries. These reporting efforts rely largely on the personal initiative of several individual scientists, and they do not represent systematic monitoring efforts of any national organizations. Except for the newly formed observatory at Nevado del Ruiz in Colombia, there are no volcano observatories or observation outposts located in South America.

In the wake of the Ruiz disaster, Colombia has begun a program to evaluate the activity of its dozen active volcanoes. Presently all activity of the newly formed Observatorio Vulcanologico de Colombia of INGEOMINAS is focused on monitoring the continuing activity at Ruiz (see article by D. Herd in *Eos* of May 13, 1986). The new observatory was established with assistance from Spain, USAID and the U.S. Geological Survey, and Japan. Situated high in a skyscraper in the city of Manizales, it has an unmatched combination of unobstructed view, line-of-sight radio reception, and proximity to the communications, transportation, etc. of a large city.

Volcanology has not traditionally been offered in Colombian universities, but that may change as a result of the Ruiz eruption. At least one seismologist from the National University in Bogota will include volcanic seismology in his courses.

Ecuador contains 11 active volcanoes along its Andean chain and an additional 16 volcanoes in the Galapagos archipelago located about 1000 km from the continent. Several historical eruptions of Cotopexi volcano account for nearly all of the several thousand casualties related to volcanic activity in Ecuador. Volcano monitoring in Ecuador is actively carried out by the Department of Geology in the Escuela Politecnica Nacional in Quito, under the direction of Dr. Minard Hall. Currently three of Ecuador's Andean volcanoes are monitored using seismic instrumentation. None of the Galapagos volcanoes are monitored instrumentally, but the Charles Darwin research station on Santa Cruz usually sends personnel to verify and observe eruptive activity once reports of activity have been received. A spirit-level tilt station near Fernandina has been remeasured several time in the past decade.

Peru contains about 10 volcanoes, 5 of which have had recorded historical activity. Of particular note is the volcano El Misti located near Arequipa, the second largest city in Peru. A seismic station is operated by the University in Arequipa, and visual observations have been made regularly by a retired geologist. Heightened activity of El Misti was noted in 1984-85.

The young volcanoes of Bolivia (numbering at least 14) are poorly known. Petrological studies have been carried out at some Bolivian volcanoes, and geothermal prospecting has also demonstrated that a number of Bolivian volcanoes contain active geothermal systems.

Chile contains more than 80 volcanoes, about 30 of which have had historical eruptions. There is no national program to monitor Chilean volcanoes, although the Department of Geology and Geophysics of the University of Chile in Santiago does have a Volcanological Service which takes the lead in reporting on the activity of Chilean volcanoes. Universities in other Chilean cities operate seismic stations near several volcanoes. Recent casualties from Chilean volcanic eruptions have come mostly from mudflows, caused by melting of snow or glacial ice.

Two of the best examples of a link between large tectonic earthquakes and volcanic eruptions are found in southern Chile; Villarrica (?) began erupting in 1906, just before a great earthquake near Valparaiso, and Carran (near Puyehue) began to erupt several hours after the great 1960 Chilean earthquake. Thus efforts to anticipate volcanic eruptions should recognize the chance that eruptions will be triggered by unanticipated tectonic earthquakes, and conversely, unrests at volcanoes might be a precursor to a great earthquake.

Argentina contains only two volcanoes in the Andes with reported historical eruptive activity. Several other volcanoes in Argentina have been prospected for geothermal exploitation. One individual, Dr. Jorge Viramonte of the University of Salta, has been interested in the activity of Argentina volcanoes.

### European Countries

Many European countries have cooperative bilateral geology programs in the Circum-Pacific region, and some of these programs include a volcanological component. In addition, many researchers from European universities have volcano-related projects in the Circum-Pacific region.

European universities continue a long tradition of offering graduate training to students from overseas, including many students from Circum-Pacific nations.

# APPENDIX 1C: DETAILS OF THE PROPOSED CIRCUM-PACIFIC INSTITUTE OF VOLCANOLOGY

## Types of Training

Training within the Institute of Volcanology would be at two levels:

1.1 Diploma in Volcanology (1 year): Instruction for the Diploma would include classroom and laboratory training, analysis of volcanic products, and the use and maintenance of volcano monitoring instruments. Training would include group field study of one or more volcanoes, to introduce trainees to field techniques and, at the same time, gather data needed to mitigate hazards of the particular target volcano(es). Study topics would include the eruptive history of a target volcano, its geophysical and geochemical activity (baseline and crisis levels), and study of any recent eruption of that volcano. Group field study would emphasize practical techniques that will bridge any potential gap between what trainees learn in lectures and what they will need to apply during their independent project, described next. Group study would be in the country or countries hosting the Institute, at sites chosen to illustrate the importance of multidisciplinary approaches to the study of volcanoes. This would constitute at least several weeks during the academic year.

The independent project would involve preliminary field work, preferably in the trainee's home country prior to beginning the Diploma course, and subsequent analyses and interpretation during the course of the Diploma. Independent projects would attack specific, limited but important, hazards-related problems on high-risk volcanoes.

1.2 Masters and Doctoral Programs (2-4 years): The Institute would encourage graduate studies at M.Sc. or Ph.D. level for students with appropriate background qualifications. Some of the students could be selected from those graduating with a Diploma from the Institute.

The thesis topics would be selected after consultation between the Institute and appropriate agency in the student's home country. Field theses should be on high-risk volcanoes. Thesis projects would be designed specifically to gather the information that one needs to mitigate hazards during a volcanic crisis. Such information includes the eruptive histories of these volcanoes, baseline and crisis-level geophysical and geochemical activity of these volcanoes, recent eruptions, and other topics which will help to mitigate hazards of these and similar volcanoes.

Instructors would visit field thesis areas to advise on unresolved problems and on possible improvements in the study, and to verify the quality of the thesis work. Communication between the student, supervisor, and agency responsible for monitoring the volcano under study should be maintained throughout the duration of the project.

The Institute should encourage publication of theses or parts thereof, either in international journals or through a rapid but widely distributed internal information series of the Institute.

Graduate students should be encouraged to discuss their

work with appropriate authorities in the country of the thesis area.

## Academic Standards and Selection Procedure

It is essential that a high academic standard is set and maintained by the Institute. For practical reasons, this will ensure that countries who send students will benefit as much as possible when a qualified graduate returns to carry out field studies as well as assist in local training. Furthermore, these people may eventually be called upon to make extremely important decisions, and their capability as well as their credentials will determine how effective they are in hazard mitigation (i.e., whether or not they are listened to by public officials).

The initial selection procedure is critical in identifying students with the basic qualifications, experience, enthusiasm, and ability to handle a year of intensive overseas training. Selection in initial stages of the Institute will be based on recommendations by the managers from participating countries, in conjunction with a degree of personal knowledge by the Institute staff and/or other recommendations. Assisting in this selection will be curriculum vitae details and the student's ability in or potential to learn the language of the Institute.

During the course of study, continuous assessment is necessary to monitor the program of the student, but in the case of the Diploma a final examination will be necessary. A minimum standard will have to set and those who do not meet this standard will receive only a certificate of attendance.

## Potential Sites for Location of Institute(s)

Because the main emphasis of the Institute is on academic training, it would be most appropriate for the Institute to be located within a University or a closely affiliated University Observatory, preferably within a Geology or Geophysics department.

The following criteria should be considered:

a. There should be an existing infrastructure into which the Institute can be integrated.

b. A significant proportion of the Institute staff should be available locally, to reduce cost.

c. Active volcanoes should be relatively accessible for group field training (hopefully within some existing monitoring programs).

d. Some field and laboratory equipment, particularly that used for monitoring, would already exist in or be available to the University.

e. Opportunity for continued postgraduate study by qualified students should be available.

f. There should be a potential source of bilateral aid to help expand and/or continue the Institute.

g. Living and working environment for students and staff should be conducive to study, with as few external complications as possible.

h. Location should be easily accessible (time and cost), particularly for potential travel to and from home-country field areas (for postgraduate research projects).

i. Specialized language-training programs should be available at or near the Institute.

j. The possibility should exist for studying a volcano during group training that is similar to volcanoes in the students' home countries.

Universities that satisfy many of the above criteria include:

*Eastern Pacific*:

Costa Rica (UNA Heredia)

Mexico (UNAM)

United States (University of California, Santa Barbara and Santa Cruz) (University of Hawaii, Hilo and Honolulu)

*Western Pacific*:

New Zealand (Victoria University)

Philippines (University of the Philippines)

Indonesia (ITB; Gadjah Mada)

Japan (University of Tokyo, Kyoto University, Hokkaido University)

## Number of Trainees and Number of Volcanoes Studied

It is anticipated that at least 100 students would be trained to diploma level in a 5-year period and at least 25 students would be enrolled for M.Sc. and Ph.D. courses. This would enable at least 25 high-risk volcanoes to be studied together with limited research into other volcanoes through Diploma projects. It is likely that if an Institute were established at more than one location, more students could be accommodated and more volcanoes could be studied than if it were sited at a single location.

## Staffing

A director would need to be appointed, and secretarial and/or administrative facilities would have to be provided. Additional scientific staff are also necessary, but only one or two of these need be permanent to provide continuity and coordination. Others could be drawn from local university and government staff and visiting teaching fellows. Because of the likely diversity of trainees, a number of tutors will need to be appointed on a limited term basis. It may be possible that the M.Sc. and Ph.D. students registered in the Institute can fill this role. An advisory panel should be established to coordinate details of course structure, criteria for student selection, etc.

## Budget

Costs listed in the budget are on the basis of a single site with a building provided but with no staff or equipment. We would recommend that the Institute be established at two existing universities. Some equipment should be available and some staff may be shared with existing facilities. This would reduce the cost. In addition, it is anticipated that bilateral aid may provide support for some graduate students, thereby further reducing the overall cost.

# APPENDIX 1D: GEOTHERMAL INSTITUTE, UNIVERSITY OF AUCKLAND

The Geothermal Institute, University of Auckland, New Zealand, can serve as an example of how successful a Volcanological Institute could be in training scientists from countries with high-risk volcanoes, and establishing a significant research program at the same time. The Geothermal Institute was established in 1979 and has graduated an average of 25 scientists and engineers each year with a Diploma in Geothermal Energy Technology. The 160 alumni have come from 20 countries with geothermal potential, and their training has been funded from a variety of sources. The UNDP set up the Institute and provided funded from a variety of sources. The UNDP set up the Institute and provided 90% of the funding, with 10% from the New Zealand Department of Foreign Affairs. UNDP funding is now approximately 50%, with the balance being taken up by New Zealand Foreign Affairs (bilateral aid) and private industry. Some overseas companies fully sponsor their staff to attend.

The Diploma course covers a 10-month period of intensive course work, first general and then increasing in specialization, and culminates in a 6-week student project and workshop presentation. Final examinations constitute about half of the grade. Because of the high academic standards pursued by the Director and staff, the Diploma is recognized worldwide as a significant credential.

Due to the heavy workload, and common language problems (alleviated to some extent by a prior 3-month language course in Wellington), the 5 fulltime staff are assisted by 6 to 8 tutors, many of whom are Institute graduates who have continued research studies toward an M.Sc. or Ph.D. degree. In addition, nearly 50 geothermal experts from government and industry in New Zealand contribute to the lecture and fieldwork program. The Institute has helped to fill the extreme shortage of trained geothermal scientists and engineers in several developing countries with exploration and development programs, leading to higher quality and a greater self-sufficiency in areas of data gathering and interpretation as well as in planning and management. The Institute was originally established to offer a 1-year course, and evolved after the first 2 years to also offer research opportunities. At present, there are 25 Diploma students in the Institute and 15 M.Sc. and Ph.D. students with research thesis work usually based on home-country geothermal systems and related problems. The Institute has evolved with time, modifying its program as necessary to best meet the needs of its students, and as funds became available.

## APPENDIX 1E:

## MEMBERS OF AD-HOC WORKING GROUP FOR MITIGATION OF VOLCANIC HAZARDS IN THE CIRCUM-PACIFIC, CPEMR, SINGAPORE, AUGUST 17-22, 1986

Tom Casadevall
U.S. Geological Survey/Volcanological Survey of Indonesia, J1. Diponegoro 57, Bandung, Jawa Barat, Indonesia

Jim Cole
Geology Department, Victoria University of Wellington, Private Bag, Wellington, New Zealand

Servando de la Cruz Reyna
Instituto de Geofisica, Universidad Autonoma de Mexico, 04510 Mexico, D.F., Mexico

Jeffrey Hedenquist
Department of Scientific and Industrial Research, Geothermal Research Centre, Private Bag, Taupo, New Zealand

R.W. (Wally) Johnson
Division of Petrology and Geochemistry, Bureau of Mineral Resources, Geology, and Geophysics, P.O. Box 378, Canberra, A.C.T. 2601, Australia

Kosuke Kamo
Sakurajima Volcano Observatory, Kyoto University, Sakurajima, Kagoshima, Japan

Lindsay McClelland
Museum of Natural History, NHB-MRC 129, Smithsonian Institution, Washington, D.C., 20560 USA

L.J. Patrick Muffler
U.S. Geological Survey, MS 910, 345 Middlefield Rd., Menlo Park, California 94025 USA

Chris Newhall, David A. Johnston
Cascades Volcano Observatory, U.S. Geological Survey, 5400 MacArthur Blvd., Vancouver, Washington 98661 USA

Raymundo Punongbayan
Philippine Institute of Volcanology and Seismology, 6th Floor, Hizon Building, 29 Quezon Ave., Quezon City, Philippines

Adjat Sudradjat
Volcanological Survey of Indonesia, J1. Diponegoro 57, Bandung, Jawa Barat, Indonesia